PALEOTECTONICS AND SEDIMENTATION
in the Rocky Mountain Region, United States

AAPG Memoir 41

AAPG Memoir 41

PALEOTECTONICS AND SEDIMENTATION
in the Rocky Mountain Region, United States

Edited by
James A. Peterson

Published by

The American Association of Petroleum Geologists

Tulsa, Oklahoma 74101, U.S.A.

Library of Congress Cataloging-in-Publication Data

Paleotectonics and sedimentation in the Rocky
   Mountain Region, United States.

   (AAPG memoir ; 41)
   Bibliography: p.
   Include index.
   1. Geology--Rocky Mountains--Congresses.
2. Geology--West (U.S.)--Congresses. 3. Paleogeography
--Rocky Mountains--Congresses. 4. Paleogeography--
West (U.S.)--Congresses. I. Peterson, James A.
II. American Association of Petroleum Geologists.
III. Series.
QE79.P35 1986   557.8   86-17467
ISBN O-89181-319-5

Association Editor: James Helwig
Science Director: Ronald L. Hart
Project Editors: Victor V. Van Beuren and Anne Thomas
Production and Design: Custom Editorial Productions, Inc.
                    Cincinnati, Ohio

PREFACE

The papers presented in this volume are the outgrowth of a symposium program on Paleotectonics and Sedimentation at the Rocky Mountain Section AAPG–SEPM Annual Meeting in Billings, Montana, September, 1983. Widespread interest in this subject was shown not only by the enthusiastic response received during planning of the program, but also by the fact that attendance at the Billings sessions overflowed the meeting room, necessitating a move to larger facilities. Most of the papers solicited for the 1983 symposium are included in the volume, but for completeness of coverage several other invited papers are also included.

The Rocky Mountains are a natural laboratory for studying the relationship between paleotectonics and sedimentary processes, with a wide spectrum of regional and local paleotectonic provinces involving a variety of sedimentary facies and depositional environments. Important studies addressing this subject have been conducted with increased frequency over the past decade or two, many of them focusing on the value of this work in exploration for economic resources, particularly petroleum, coal, phosphate, and other mineral deposits. The addition of a great volume of subsurface data in the postwar years had added greatly to the regional and local knowledge of the paleostructural history of the region. Likewise, a century or more of surface mapping and geologic study, beginning with the early mineral and natural resource explorations of the west, has accumulated an enormous volume of geologic information on structure, stratigraphy, paleontology, economic deposits, and other data. Much of this work has been published in field conference guidebooks, government publications, and other journals, and much has been presented at professional meetings through the years. Thus, it seems timely and appropriate to gather together a significant coverage of this knowledge into a special volume on the subject.

The papers included in this volume include: (1) overviews of tectonic terranes and sedimentary facies, lineaments and their tectonic implications, and Rocky Mountain paleogeography through geologic time; (2) geographic coverage of the northern, middle, and southern Rocky Mountains of the United States; (3) a relatively complete geologic coverage, ranging from Proterozoic to Cenozoic; (4) phases of sedimentary facies development and their relationship to paleostructural events, including syntectonic conglomerates, fluvial, deltaic, and coastal sedimentation, blanket sandstones, fine-grained clastic sediments, carbonate–evaporite associations, and coal and petroleum deposits. The range of authorship includes industry, state and federal government, university, and independent researchers. Some papers are resource oriented, others are more theoretical or principle oriented.

The Billings symposium and resulting publication were planned by the editor in collaboration with Donald L. Smith, whose tragic death prior to the symposium delayed completion of the work. Because of Dr. Smith's great interest and involvement in the subject matter of this volume and his significant contribution of ideas and support during the planning stages of the project, this volume is respectfully dedicated to his memory.

James A. Peterson
U.S. Geological Survey and
University of Montana
Missoula, MT 59812

CONTENTS

Part I
REGIONAL OVERVIEW

Rocky Mountain Paleogeography Through Geologic Time

J. A. Peterson
University of Montana
Missoula, Montana

D. L. Smith[1]
Montana State University
Bozeman, Montana

The main basins and uplifts of the Rocky Mountain region are outlined by a generalized thickness map of the total Phanerozoic sedimentary cover. Twelve maps showing thickness, generalized sedimentary facies, and probable emergent terrigenous clastic source areas are presented, covering the late Precambrian, Cambrian, Ordovician, Silurian, Devonian, Mississippian, Pennsylvanian, Permian, Triassic, Jurassic, Cretaceous, and Cenozoic.

In this paper we present thickness and facies maps that we have compiled from many sources over a period of several years. These maps include one that shows the present-day positions of basins and uplifts in the Rocky Mountains (Figure 1), followed by the twelve maps showing the thickness, sedimentary facies, and paleotectonic elements of the Rocky Mountains from late Precambrian to Cenozoic time (Figures 2–13). A lengthy discussion of each map is not included; each is presented at face value for interpretation by the interested observer. Detailed discussions, presentation of data, and analyses of sedimentary and tectonic processes can be found in the references cited below for each map (listed in Bibliography). Additional papers not cited here but that are relevant to topics relating to Rocky Mountain paleogeography are also given in the Bibliography.

In addition to personal data files, the main sources of information for the maps are as follows:

Rocky Mountain Basins and Uplifts (Figure 1): Haun and Kent (1965), Peterson (1965), and Jensen (1972).

Late Precambrian (Figure 2): Williams (1953), Hansen (1957), Wallace and Crittenden (1969), Crittenden et al. (1971), Stewart (1972), Beus et al. (1974), Ruppel (1975), Ruppel et al. (1975), Stewart and Suczek (1977), Reynolds and Lindsey (1979), Sears et al. (1982), Dutch (1983), and Winston (this volume).

Cambrian (Figure 3): Kottlowski (1965), Adler (1971), Crittenden et al. (1971), Balk (1972), Hintze (1973), Greenwood et al. (1977), Stewart and Suczek (1977), Peterson (1977a, 1981, 1984b), and Ross and Ross (this volume).

Ordovician (Figure 4): Kottlowski (1965), Adler (1971), Foster (1972), Hintze (1973), Greenwood et al. (1977), Ross (1977), Peterson (1977a, 1981, 1984b), Witzke (1980), and Ross and Ross (this volume).

Silurian (Figure 5): Kottlowski (1965), Adler (1971), Gibbs (1972), Hintze (1973), Greenwood et al. (1977), Poole and Sandberg (1977), Peterson (1977a, 1981, 1984b), and Ross and Ross (this volume).

Devonian (Figure 6): Kottlowski (1965), Adler (1971), Baars (1972), Hintze (1973), Greenwood et al. (1977), Loucks (1977), Poole and Sandberg (1977), Peterson (1977a, 1981, 1984b), Beus (1980), and Ross and Ross (this volume).

Mississippian (Figure 7): Kottlowski (1965), Adler (1971), Craig (1972), Hintze (1973), Rose (1976), Sando (1976), Poole and Sandberg (1977), Greenwood et al. (1977), Peterson (1977a, 1981, 1984b), Armstrong and Mamet (1978), Craig and Connor (1979), Armstrong et al. (1980), Skipp and Hall (1980), and Ross and Ross (this volume).

Pennsylvanian (Figure 8): Kottlowski (1965), Roberts et al. (1965), Adler (1971), Mallory (1972), Hintze (1973), McKee et al. (1975), Greenwood et al. (1977), Rich (1977), Peterson (1977a, 1981, 1984b), Skipp and Hall (1980), De Voto (1980), and Ross and Ross (this volume).

Permian (Figure 9): Kottlowski (1965), Roberts et al. (1965), McKee et al. (1967), Adler (1971), Rascoe and Baars (1972), Hintze (1973), Greenwood et al. (1977), Peterson (1977b), Skipp and Hall (1980), Peterson (1980, 1981, 1984a, b), Wardlaw (1980), and Ross and Ross (this volume).

Triassic (Figure 10): McKee et al. (1959), MacLachlan (1972), Hintze (1973), and Peterson (1981, 1984b).

Jurassic (Figure 11): McKee et al. (1956), Peterson (1972, 1981), Hintze (1973), and Imlay (1980).

---

[1]Deceased, July 1983.

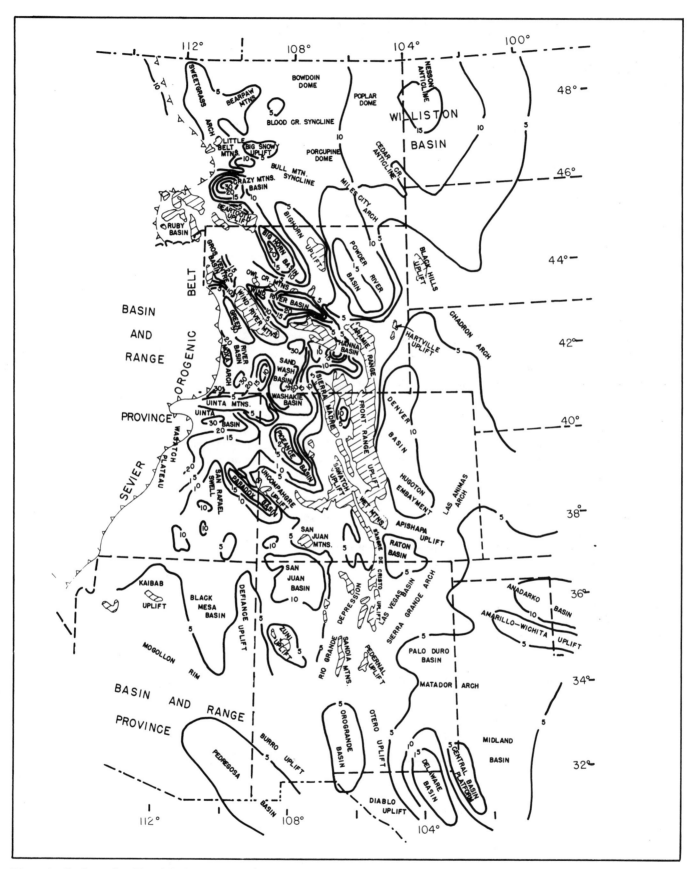

Figure 1—Basins and uplifts of the Rocky Mountains, showing approximate thickness of total Phanerozoic sedimentary cover in thousands of feet. Areas of exposed Archaean rocks east of the thrust belt are shown by cross-hatching; eastern edge of thrust belt is shown by barbed line.

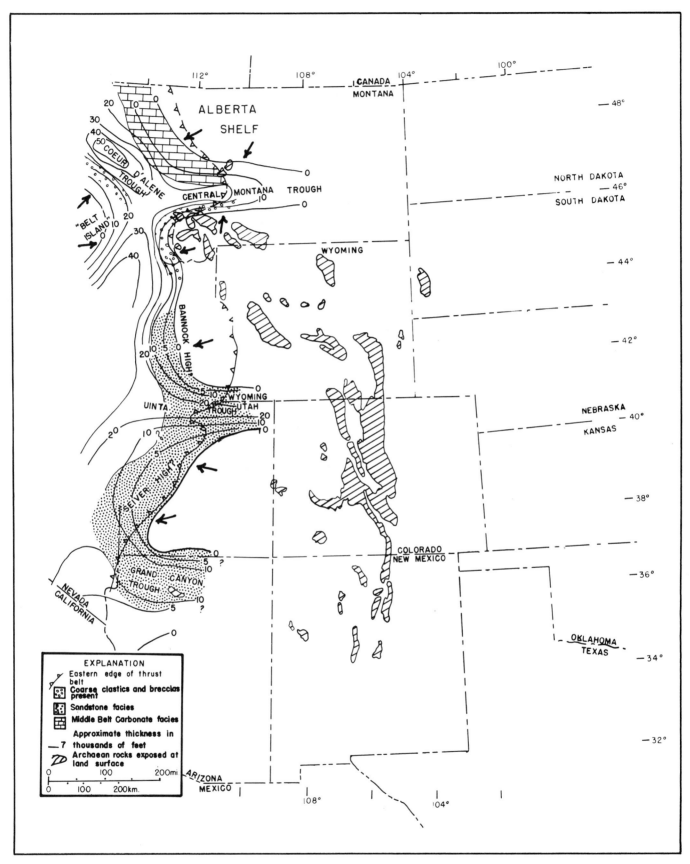

Figure 2—Late Precambrian, showing approximate thickness, general sedimentary facies, and main paleotectonic elements. Data are palinspastically restored in the western thrust belt; arrows indicate probable transport directions of terrigenous clastic sediments.

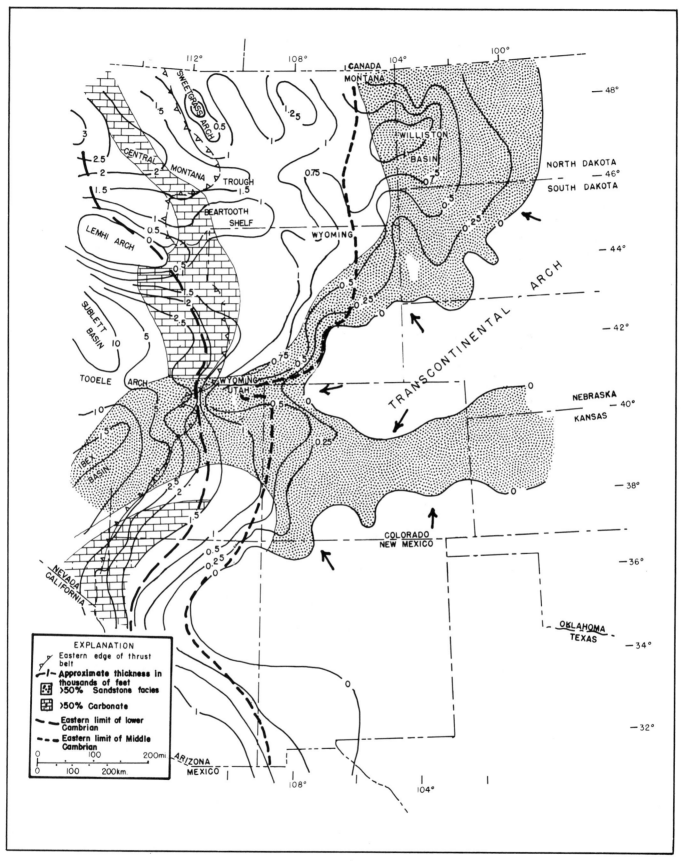

Figure 3—Cambrian System, showing approximate thickness, general sedimentary facies, and main paleotectonic elements. Data are palinspastically restored in the western thrust belt; arrows indicate probable transport directions of terrigenous clastic sediments. Approximate eastern limits of Lower and Upper Cambrian sedimentary rocks are shown by dashed lines.

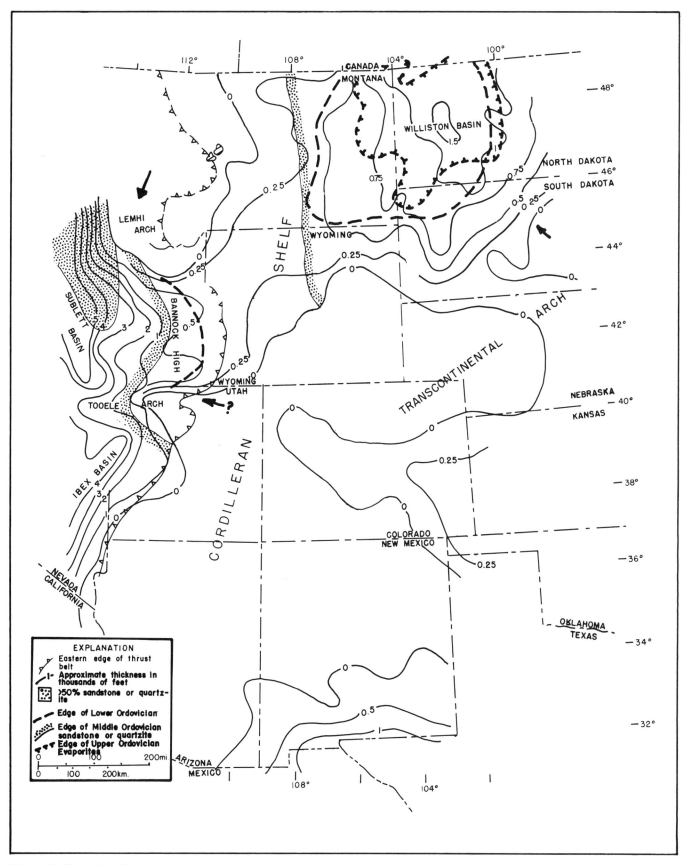

Figure 4—Ordovician System, showing approximate thickness, general sedimentary facies, and main paleotectonic elements. Data are palinspastically restored in the western thrust belt; arrows indicate probable transport directions of terrigenous clastic sediments.

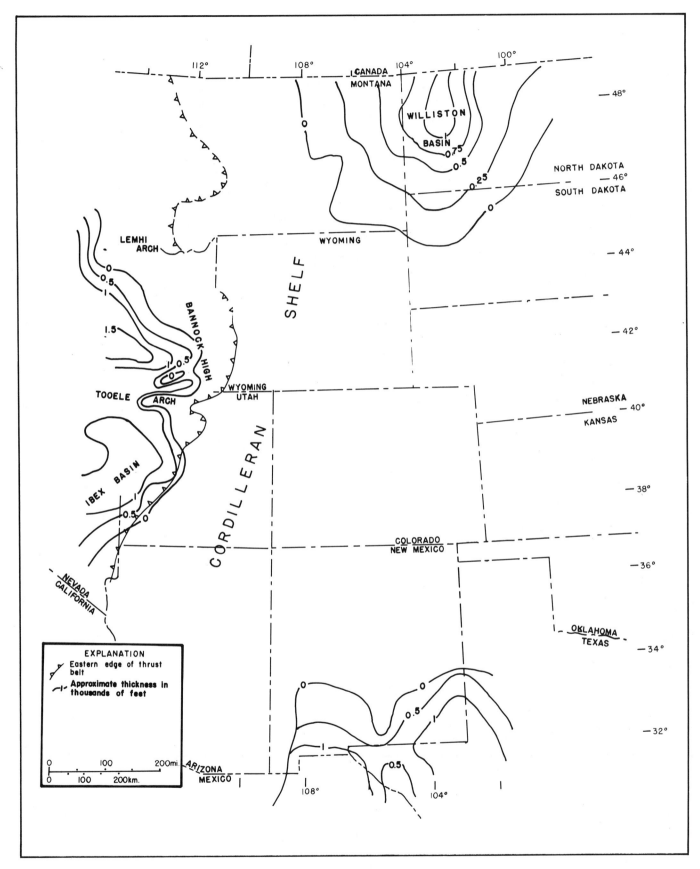

Figure 5—Silurian System, showing approximate thickness and main paleotectonic elements. Rocks are primarily marine dolomite in most areas. Data are palinspastically restored in the western thrust belt.

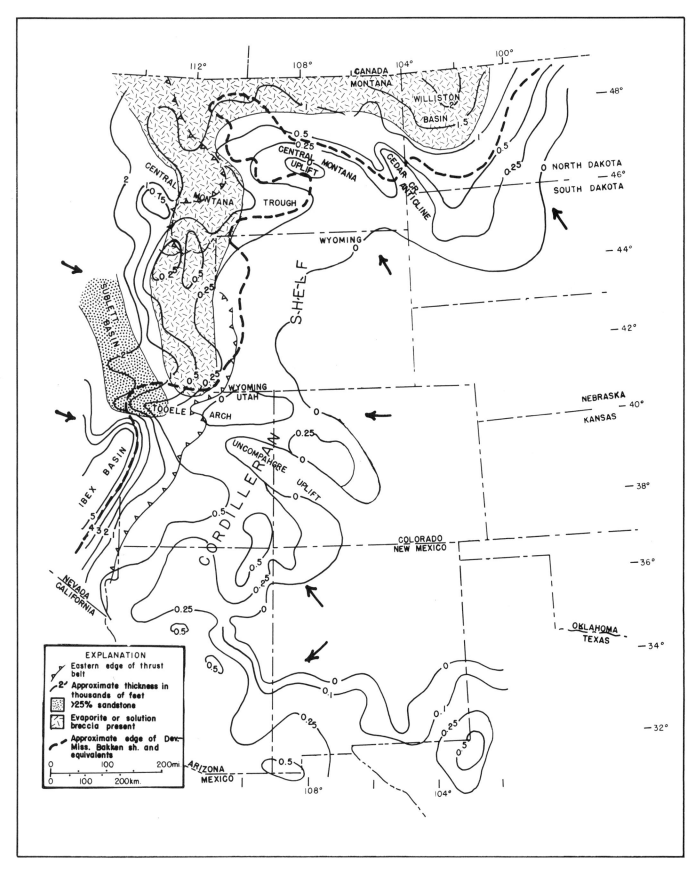

Figure 6—Devonian System, showing approximate thickness, general sedimentary facies, and main paleotectonic elements. Data are palinspastically restored in the western thrust belt; arrows indicate probable transport directions of terrigenous clastic sediments.

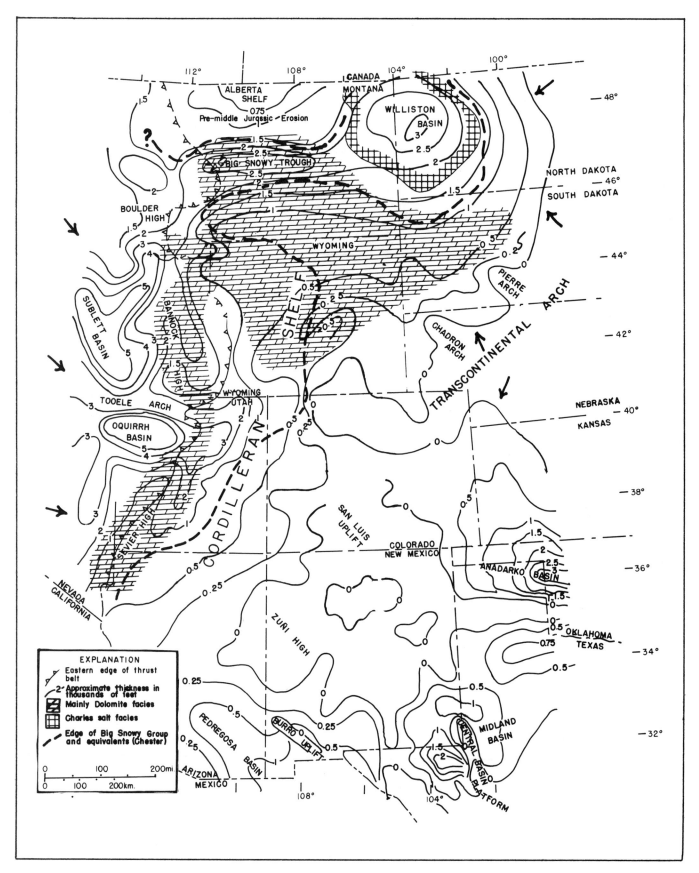

Figure 7—Mississippian System, showing approximate thickness, general sedimentary facies, and main paleotectonic elements. Data are palinspastically restored in the western thrust belt; arrows indicate probable transport directions of terrigenous clastic sediments.

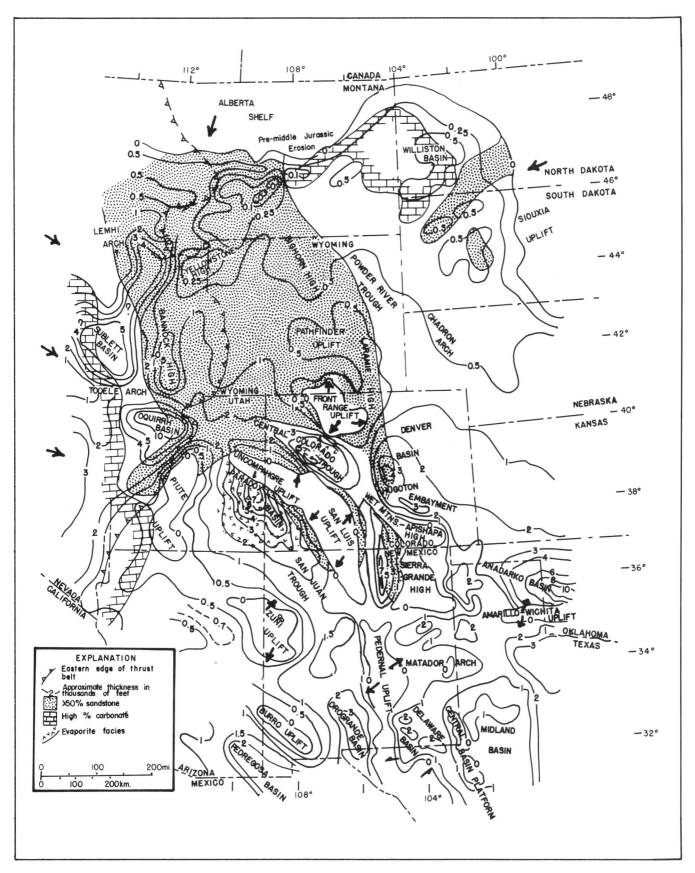

Figure 8—Pennsylvanian System, showing approximate thickness, general sedimentary facies, and main paleotectonic elements. Data are palinspastically restored in the western thrust belt; arrows indicate probable transport directions of terrigenous clastic sediments.

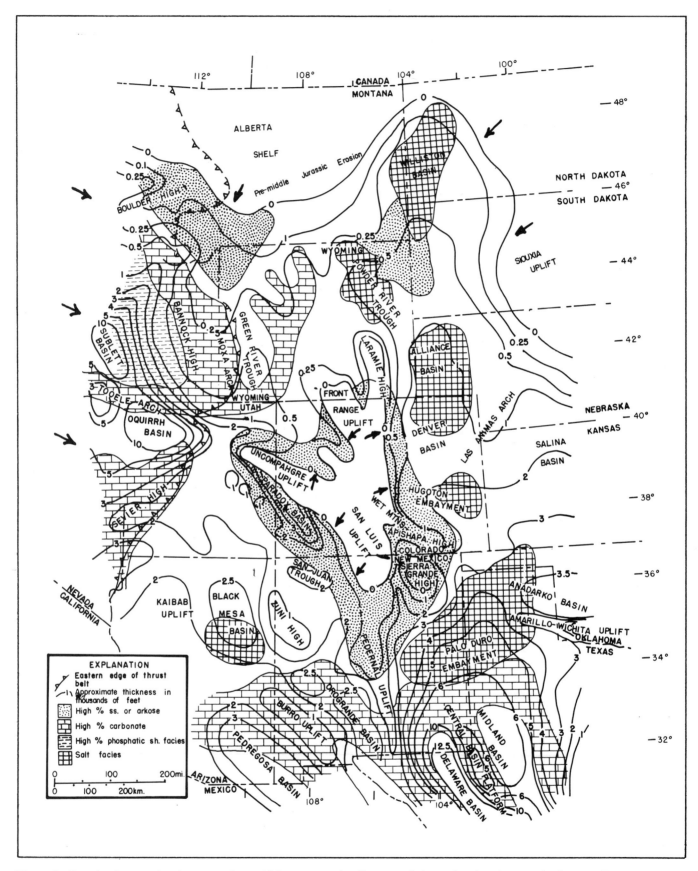

Figure 9—Permian System, showing approximate thickness, general sedimentary facies, and main paleotectonic elements. Data are palinspastically restored in the western thrust belt; arrows indicate probable transport directions of terrigenous clastic sediments.

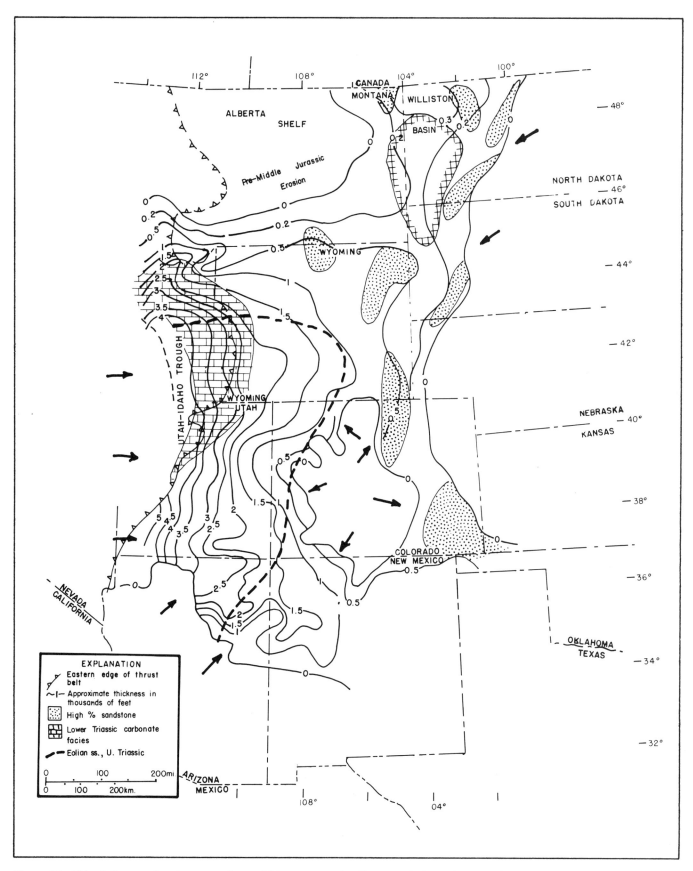

Figure 10—Triassic System, showing approximate thickness, general sedimentary facies, and main paleotectonic elements. Data are palinspastically restored in the western thrust belt; arrows indicate probable transport directions of terrigenous clastic sediments. Approximate western limit of known Triassic sedimentary rocks in Idaho is shown by dashed line.

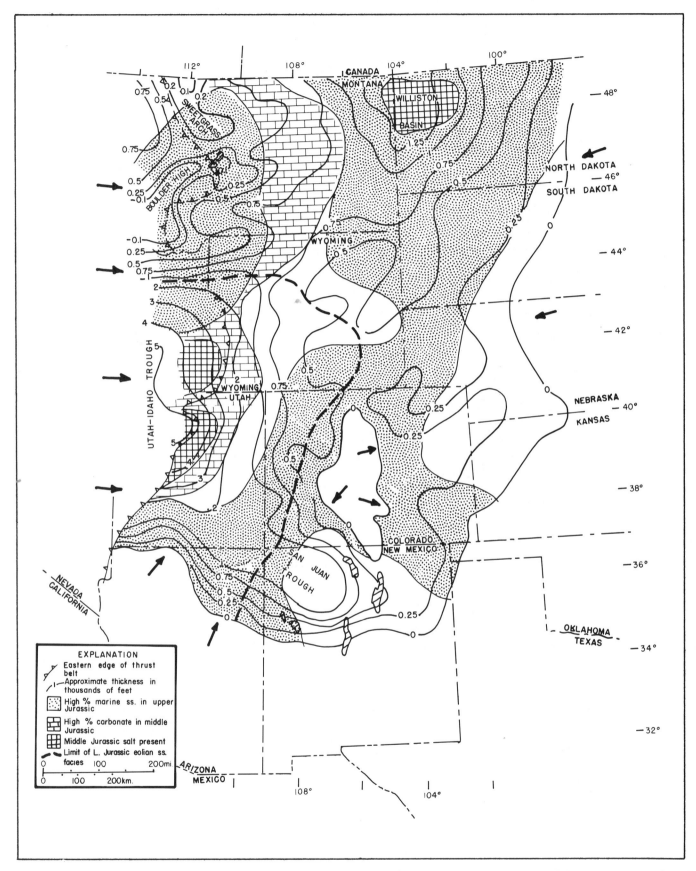

Figure 11—Jurassic System, showing approximate thickness, general sedimentary facies, and main paleotectonic elements. Data are palinspastically restored in the western thrust belt; arrows indicate probable transport directions of terrigenous clastic sediments.

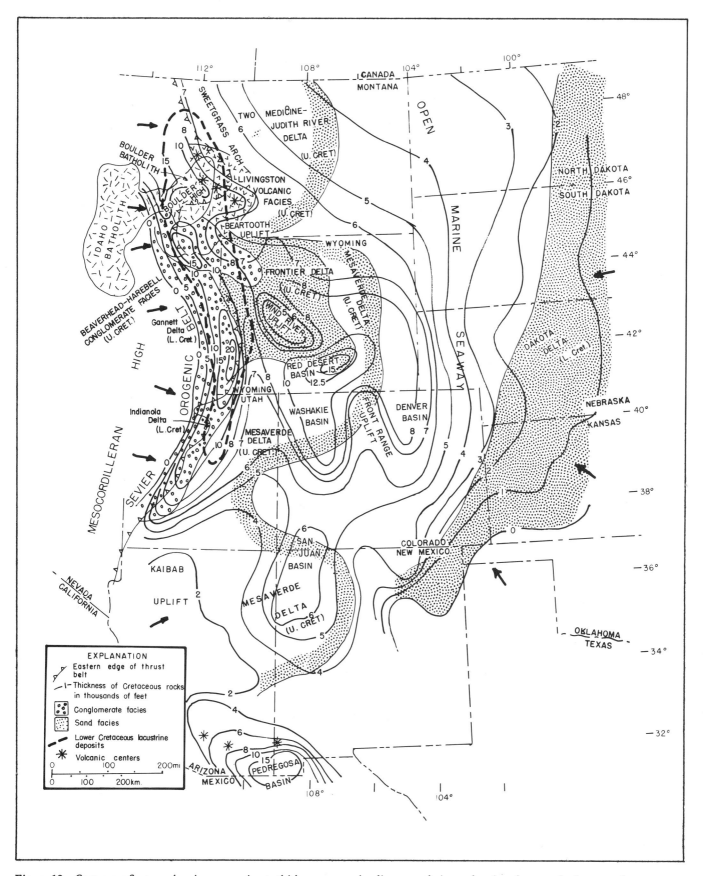

Figure 12—Cretaceous System, showing approximate thickness, general sedimentary facies, and main paleotectonic elements. Arrows indicate probable transport directions of terrigenous clastic sediments.

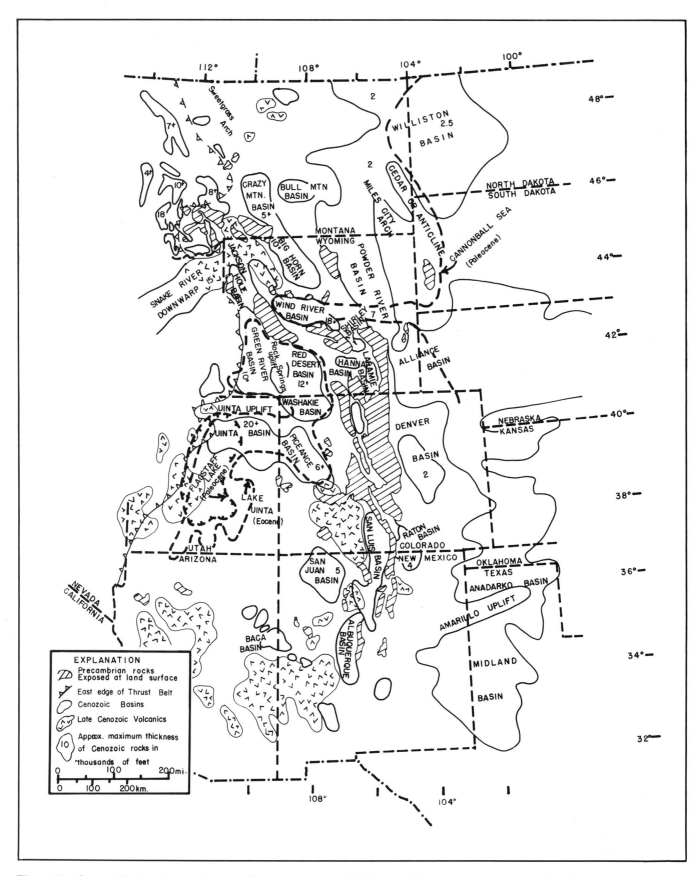

Figure 13—Cenozoic Basins, showing location of basins, approximate thickness of Cenozoic rocks, and areas of late Cenozoic volcanism.

Cretaceous (Figure 12): Kottlowski (1965), Armstrong and Oriel (1965), Adler (1971), McGookey et al. (1972), Greenwood et al. (1977), and Peterson (1981).

Cenozoic (Figure 13): Kuenzi and Fields (1971), McDonald (1972), and Chapin and Cather (1981).

BIBLIOGRAPHY

Adler, F. J., 1971, Future petroleum provinces of the mid-continent, Region 7, *in* I. H. Cram, ed., Future petroleum provinces of the United States—their geology and potential: AAPG Memoir 15, v. 2, p. 985–1042.

Armstrong, A. C., and B. L. Mamet, 1978, The Mississippian System of southwestern New Mexico and southeastern Arizona, *in* Guidebook, 29th Field Conference: New Mexico Geological Society, p. 183–192.

Armstrong, A. K., B. L. Mamet, and J. E. Repetski, 1980, The Mississippian System of New Mexico and southern Arizona, *in* T. D. Fouch, and E. R. Magathan, eds., Paleozoic paleogeography of west-central United States: Rocky Mountain Section, Society of Economic Paleontologists and Mineralogists, p. 82–99.

Armstrong, F. C., and S. S. Oriel, 1965, Tectonic development of Idaho–Wyoming thrust belt: AAPG Bulletin, v. 49, p. 1847–1866.

Baars, D. L., 1972, Devonian System, *in* Geologic Atlas of the Rocky Mountain Region: Denver, Colorado, Rocky Mountain Association of Geologists, p. 90–99.

Balk, C. L., 1972, Cambrian System, *in* Geologic Atlas of the Rocky Mountain Region: Denver, Colorado, Rocky Mountain Association of Geologists, p. 60–75.

Beus, S. S., 1980, Late Devonian (Frasnian) paleogeography and paleoenvironments in northern Arizona, *in* T. D. Fouch and E. R. Magathan, eds., Paleozoic paleogeography of west-central United States: Rocky Mountain Section, Society of Economic Paleontologists and Mineralogists, p. 55–69.

Beus, S. S., R. R. Rawson, R. O. Dalton, G. M. Stevenson, S. Reed, and T. M. Daneker, 1974, Preliminary report on the Unkar Group (Precambrian) in Grand Canyon, Arizona, *in* T. N. V. Karlstrom, G. A. Swann, and R. L. Eastwood, eds., Geology of northern Arizona, Part I—regional studies: Geological Society of America, Rocky Mountain Section 27th Annual Meeting, Guidebook, p. 34–53.

Chapin, C. E., and S. M. Cather, 1981, Eocene tectonics and sedimentation in the Colorado Plateau–Rocky Mountain area, *in* W. R. Dickinson and W. D. Payne, eds., Relations of tectonics to ore deposits in the Southern Cordillera: Arizona Geological Society Digest, v. xiv, p. 173–198.

Craig, L. C., 1972, Mississippian System, *in* Geologic Atlas of the Rocky Mountain Region: Denver, Colorado, Rocky Mountain Association of Geologists, p. 100–110.

Craig, L. C., and C. W. Connor, 1979, Paleotectonic investigations of the Mississippian System in the United States: U.S. Geological Survey Professional Paper 1010, 559 p.

Crittenden, M. D., F. E. Schaeffer, D. E. Trimble, and L. A. Woodward, 1971, Nomenclature and correlation of some Upper Precambrian and basal Cambrian sequences in western Utah and southeastern Idaho: Geological Society of America Bulletin, v. 82, p. 581–602.

De Voto, R. H., 1980, Pennsylvanian stratigraphy and history of Colorado, *in* H. C. Kent and K. W. Porter, eds., Colorado Geology: Rocky Mountain Association of Geologists, p. 71–101.

Dutch, S. I., 1983, Proterozoic structural provinces in the north-central United States: Geology, v. 11, p. 392–393.

Foster, N. H., 1972, Ordovician System, *in* Geologic Atlas of the Rocky Mountain Region: Denver, Colorado, Rocky Mountain Association of Geologists, p. 76–85.

Gibbs, F. A., 1972, Silurian System, *in* Geologic Atlas of the Rocky Mountain Region: Denver, Colorado, Rocky Mountain Association of Geologists, p. 86–90.

Greenwood, E., F. E. Kottlowski, and S. Thompson III, 1977, Petroleum potential and stratigraphy of Pedregosa basin: comparison with Permian and Orogrande basins: AAPG Bulletin, v. 61, p. 1448–1469.

Hansen, W. R., 1957, Precambrian rocks of the Uinta Mountains (Utah): Intermountain Association of Petroleum Geologists 8th Annual Field Conference Guidebook, p. 48–52.

Haun, J. D., and H. C. Kent, 1965, Geologic history of Rocky Mountain region: AAPG Bulletin, v. 49, p. 1781–1800.

Hintze, L. H., 1973, Geologic history of Utah: Brigham Young University Geology Studies, v. 20, pt. 3, 181 p.

Imlay, R. W., 1980, Jurassic paleobiogeography of the conterminous United States in its continental setting: U.S. Geological Survey Professional Paper 1062, 134 p.

Jensen, F. S., 1972, Thickness of Phanerozoic rocks, *in* Geologic Atlas of the Rocky Mountain Region: Denver, Colorado, Rocky Mountain Association of Geologists, p. 56.

Kottlowski, F. E., 1965, Sedimentary basins of south-central and southwestern New Mexico: AAPG Bulletin, v. 49, p. 2120–2139.

Kuenzi, W. D., and R. W. Fields, 1971, Tertiary stratigraphy, structure and geologic history Jefferson basin, Montana: Geological Society of America Bulletin, v. 82, p. 3374–3394.

Loucks, G. G., 1977, Geologic history of the Devonian, northern Alberta to southwest Arizona, *in* Rocky Mountain Thrust Belt geology and resources: Wyoming Geological Association Guidebook, p. 119–134.

McDonald, R. E., 1972, Eocene and Paleocene rocks of the southern and central basins, *in* Geologic Atlas of the Rocky Mountain Region: Denver, Colorado, Rocky Mountain Association of Geologists, p. 243–256.

McGookey, D. P., et al., 1972, Cretaceous System, *in* Geologic Atlas of the Rocky Mountain Region: Denver, Colorado, Rocky Mountain Association of Geologists, p. 190–228.

McKee, E. D., et al., 1956, Paleotectonic maps, Jurassic System: U.S. Geological Survey Miscellaneous Geological Investigations Map I-175, 6 p.

McKee, E. D., et al., 1959, Paleotectonic maps of the Triassic System: U.S. Geological Survey Miscellaneous

Geological Investigations Map I-300, 33 p.

McKee, E. D., S. S. Oriel, et al., 1967, Paleotectonic maps of the Permian System: U.S. Geological Survey Miscellaneous Geological Investigations Map I-450.

McKee, E. D., E. J. Crosby, et al., 1975, Paleotectonic investigations of the Pennsylvanian System in the United States: U.S. Geological Survey Professional Paper 853, 349 p.

MacLachlan, M. E., 1972, Triassic System, *in* Geologic Atlas of the Rocky Mountain Region: Denver, Colorado, Rocky Mountain Association of Geologists, p. 166–176.

Mallory, W. W., 1972, Regional synthesis of the Pennsylvanian System, *in* Geologic Atlas of the Rocky Mountain Region: Denver, Colorado, Rocky Mountain Association of Geologists, p. 111–127.

Peterson, J. A., 1965, Introduction, Rocky Mountain sedimentary basins: AAPG Bulletin, v. 49, p. 1779–1780.

———, 1972, Jurassic System, *in* Geologic Atlas of the Rocky Mountain Region: Denver, Colorado, Rocky Mountain Association of Geologists, p. 177–189.

———, 1977a, Paleozoic shelf-margins and marginal basins, western Rocky Mountains—Great Basin, United States, *in* Rocky Mountain Thrust Belt, geology and resources: Wyoming Geological Association 29th Annual Field Conference, p. 135–153.

———, 1977b, Permian paleogeography and sedimentary provinces, west central United States, *in* T. D. Fouch and E. R. Magathan, eds., Paleozoic paleogeography of the west-central United States: Rocky Mountain Section, Society of Economic Paleontologists and Mineralogists, p. 271–292.

———, 1980, Permian paleogeography and sedimentary provinces, west-central United States, *in* T. D. Fouch and E. R. Magathan, eds., Paleozoic paleogeography of west-central United States: Rocky Mountain Section, Society for Economic Paleontologists and Mineralogists, p. 271–292.

———, 1981, General stratigraphy and regional paleostructure of the western Montana overthrust belt, *in* T. E. Tucker, ed., Southwest Montana Field Conference and Symposium: Montana Geological Society, p. 5–35.

———, 1984a, Permian stratigraphy, sedimentary facies, and petroleum geology, Wyoming and adjacent areas, *in* J. Goolsby and D. Morton, eds., The Permian and Pennsylvanian geology of Wyoming, 35th Annual Field Conference Guidebook: Wyoming Geological Association, p. 26–64.

———, 1984b, Stratigraphy and sedimentary facies of the Madison Limestone and associated rocks in parts of Montana, Nebraska, North Dakota, South Dakota, and Wyoming: U.S. Geological Survey Professional Paper 1273-A, p. A1–A34.

Poole, F. G., and C. A. Sandberg, 1977, Mississippian paleogeography and tectonics of the western United States, *in* J. H. Stewart, C. H. Stevens, and A. E. Fritsche, eds., Paleozoic paleogeography of the western United States: Pacific Section, Society of Economic Paleontologists and Mineralogists, p. 67–85.

Rascoe, B., Jr., and D. L. Baars, 1972, Permian System, *in* Geologic Atlas of the Rocky Mountain Region: Denver, Colorado, Rocky Mountain Association of Geologists, p. 143–165.

Reynolds, M. W., and D. A. Lindsey, 1979, Eastern extent of the Proterozoic Y Belt Basin, Montana: U.S. Geological Survey Professional Paper 1150, 68 p.

Rich, M., 1977, Pennsylvanian paleogeographic patterns in the western United States, *in* J. H. Stewart, C. H. Stevens, and A. E. Fritsche, eds., Paleozoic paleogeography of the western United States: Pacific Section, Society of Economic Paleontologists and Mineralogists, p. 87–112.

Roberts, R. J., M. D. Crittenden, Jr., E. W. Tooker, H. T. Morris, R. K. Hose, and T. M. Cheney, 1965, Pennsylvanian and Permian basins in northwestern Utah, northeastern Nevada, and south-central Idaho: AAPG Bulletin, v. 49, p. 1926–1956.

Rose, P. R., 1976, Mississippian carbonate shelf margins, western United States: U.S. Geological Survey Journal of Research, v. 4, p. 449–466.

Ross, R. J., Jr., 1977, Ordovician paleogeography of the western United States, *in* J. H. Stewart, C. H. Stevens, and A. E. Fritsche, eds., Paleozoic paleogeography of the western United States: Pacific Section, Society of Economic Paleontologists and Mineralogists, p. 19–38.

Ruppel, E. T., 1975, Precambrian Y sedimentary rocks in east-central Idaho: U.S. Geological Survey Professional Paper 889-A, 23 p.

Ruppel, E. T., R. J. Ross, Jr., and D. Schleicher, 1975, Precambrian Z and Lower Ordovician rocks in east-central Idaho: U.S. Geological Survey Professional Paper 889-B, p. 25–34.

Sandberg, C. A., and W. J. Mapel, 1967, Devonian of the northern Rocky Mountains and plains, *in* International Symposium on the Devonian System: Alberta Society of Petroleum Geologists, p. 843–877.

Sando, W. J., 1976, Mississippian history of the northern Rocky Mountain Region: U.S. Geological Survey Journal of Research, v. 4, p. 317–338.

Sears, J. W., P. J. Graff, and G. S. Holden, 1982, Tectonic evolution of lower Proterozoic rocks, Uinta Mountains, Utah and Colorado: Geological Society of America Bulletin, v. 93, p. 990–997.

Skipp, B., and W. E. Hall, 1980, Upper Paleozoic paleotectonics and paleogeography of Idaho, *in* T. D. Fouch and E. R. Magathan, eds., Paleozoic Paleogeography of the west-central United States: Rocky Mountain Section, Society of Economic Paleontologists and Mineralogists, p. 387–422.

Stewart, J. H., 1972, Initial deposits in the Cordilleran geosyncline: evidence of a late Precambrian (850 m.y.) continental separation: Geological Society of America Bulletin, v. 83, p. 1345–1360.

Stewart, J. H., and C. A. Suczek, 1977, Cambrian and latest Precambrian paleogeography and tectonics in the western United States, *in* J. H. Stewart, C. H. Stevens, and A. E. Fritsche, eds., Paleozoic paleogeography of the western United States: Pacific Section, Society of Economic Paleontologists and Mineralogists, p. 1–18.

Wallace, C. A., and M. D. Crittenden, Jr., 1969, The stratigraphy, depositional environment and correlation of the Precambrian Uinta Mountain Group, western Uinta Mountains, Utah, *in* J. B. Lindsay, ed., Geologic Guidebook of the Uinta Mountains: Intermountain Association of Geologists 16th Annual Field Conference, p. 127–141.

Wardlaw, B. R., 1980, Middle–Late Permian paleogeography of Idaho, Montana, Nevada, and Wyoming, *in* T. D. Fouche, and E. R. Magathan, eds., Paleozoic paleogeography of the west-central United States, Rocky Mountain Section, Society of Economic Paleontologists and Mineralogists, p. 353–362.

Weimer, R. J., 1984, Relation of unconformities, tectonics, and sea-level changes, Cretaceous of western interior, U.S.A., *in* J. S. Schlee, ed., Interregional unconformities and hydrocarbon accumulations: AAPG Memoir 36, p. 7–35.

Williams, N. C., 1953, Late Precambrian and early Paleozoic geology of western Uinta Mountains: AAPG Bulletin, v. 37, p. 2734–2742.

Witzke, B. J., 1980, Middle and Upper Ordovician paleogeography of the region bordering the Transcontinental arch, *in* T. D. Fouch and E. R. Magathan, eds., Paleozoic paleogeography of the west-central United States: Rocky Mountain Section, Society of Economic Paleontologists and Mineralogists, p. 1–18.

Influence of Tectonic Terranes Adjacent to the Precambrian Wyoming Province on Phanerozoic Stratigraphy in the Rocky Mountain Region[1]

J. J. Tonnsen
General Hydrocarbons, Inc.,
Billings, Montana

The perimeter of the Archean Wyoming province can be generally defined. A Proterozoic metamorphic belt surrounds the province and separates it from the Archean Superior province to the east. The western margin lies under the western overthrust belt but the province extends at least as far west as southwestern Montana and southeastern Idaho. The province is bounded on the north by a regionally extensive terrane composed of Archean rocks that were apparently remobilized by Proterozoic tectonic events. The southern mobile belt does not appear to contain rocks as old as Archean. The tectonic response of these Precambrian lithostructural terranes to vertical and horizontal tectonic stress fields has influenced Phanerozoic stratigraphic facies distributions. An analysis of the major unconformities in the stratigraphic record in light of the Precambrian lithostructural history of the region discloses new evidence concerning the stratigraphy of the Rocky Mountain region. The sedimentation patterns appear to have been influenced by differential and recurrent uplift of the terranes making up the ancient substructure. A correlation between the tectonic terranes and the localization of regional hydrocarbon accumulations has been observed. A model is discussed that has been useful in basin analyses and trend mapping for petroleum exploration.

INTRODUCTION

Petroleum resources are not uniformly distributed in the Rocky Mountain region. Statistical information for oil reserves indicates that the Wyoming area has a total of 76 billion bbl of oil in the cumulative production, discovered reserves, and future reserves categories (Figure 1). In contrast, the states immediately surrounding Wyoming contain oil reserves of 37 billion bbl. These adjacent areas include the Denver, Uinta, Paradox, and Williston basins. The statistics do not include the western overthrust belt potential. One objective of this work was to learn whether there might be any regional structural or stratigraphic reason for the apparent unequal distribution of reserves. The future petroleum potential of the United States was last reviewed by industry geologists over a decade ago (Cram, 1971).

During the past several years many advances have been made in understanding the Precambrian history of the earth (Harrison and Peterman, 1984). There also is better understanding of the causes and nature of the cyclical changes in global sea level and how the changes affected the distribution of the Phanerozoic sedimentary veneer on the craton. The Rocky Mountain regional stratigraphic studies to date have been primarily concerned with the east–west transgressive and regressive movements of the ocean on and off the craton and continental shelf from ocean basin to cratonic interior. The cycling and the sequences of unconformities have been observed in a north–south direction on the western shelf and are discussed in this paper.

This work is a regional synthesis combining outcrop and subsurface data with numerous concepts. The information for this overview consists of geologic reports, maps, aerial and satellite photos, geophysical data, subsurface maps and cross sections, and field notes that I have gathered while working in Montana, Wyoming, and Colorado since 1958. The data were brought together in conjunction with petroleum exploration work performed for the Montana Power Company in the northern Rocky Mountain region

---

[1]Preliminary versions of the figures in this paper were used in the poster session of the same title, that won the Poster Session Award at the 1983 AAPG Rocky Mountain Section meeting in Billings, Montana.

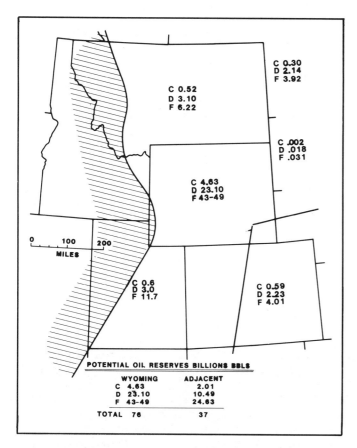

Figure 1—Distribution of potential oil reserves (billion bbl) in the Rocky Mountain region exclusive of the western overthrust belt. The reported total cumulative production (C), discovered reserves (D), and future anticipated reserves (F) for the Wyoming area is over twice the total for the adjacent areas. Data were assembled from Cram (1971) and Dolton (1981) and updated by the AAPG North American Developments and World Energy Developments annual issues for 1968 to 1984. The data do not include the western overthrust belt, which is west of the line shown.

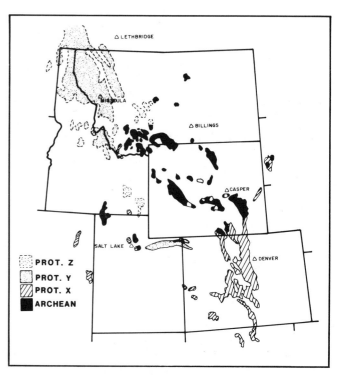

Figure 2—Archean rocks outcrop mainly in Wyoming and southwestern Montana. Early Proterozoic (X) rocks outcrop in Colorado and South Dakota, middle Proterozoic (Y) sediments outcrop in northwestern Montana, and late Proterozoic (Z) sediments outcrop in southeastern Idaho. Six outcrops of Archean greenstone rocks (small unshaded areas) are grouped close to each other in central Wyoming and are also shown on Figure 4A.

during 1976–1981. Reports such as those relating to the age of the global changes of sea level (Vail et al., 1977) and of the coincidence of tectonic events with transgressive episodes of the sea (Johnson, 1971) have been used. The regional stratigraphic observations were tied to observations of regional anisotropy in the crystalline basement. Several geologic atlases were used, including Haun and Kent (1965), Mallory (1972b), and Cook and Bally (1975).

It has been discovered that the distribution of the strata and unconformities in the north–south direction coincides closely with the regional distribution of Precambrian terranes. The Phanerozoic sedimentary environments and stratigraphic relationships were apparently profoundly influenced by a lithostructural framework that relates back to the late Archean and early and middle Proterozoic.

PRECAMBRIAN BACKGROUND

Only about 10% of the crystalline basement in the region is exposed (Figure 2), but studies of the outcrops and basement drill hole samples have advanced the under-

standing of the Precambrian greatly in recent years (Muehlberger, 1980). The Precambrian history embraces about 4 b.y., which is almost eight times longer than Phanerozoic history. The divisions of Precambrian time (Figure 3) are adapted from Harrison and Peterman (1982, 1984). Slightly different ages mark the end of the divisions of Precambrian time from place to place on the continent. The main events occurring at about 2.5 b.y. ago (thermal events marking the end of the Archean), about 1.8 b.y. ago (metamorphic and thermal events), and after about 1.6 b.y. ago (extensional events) are emphasized (Catanzaro, 1967). There are events much older than 2.5 b.y. old, and there are numerous other events that occurred over a span of several tens or hundreds of millions of years before and after the reference dates, but the reference dates are used here for convenience of discussion.

The Rocky Mountain region is divided here into five inferred lithostructural terranes, which are depicted schematically in Figure 4. Figure 4D is a composite base map summarizing the significant trends, and this map was used in the stratigraphic study. The terranes discussed are, from north to south, Montana block (M), Central Montana block (CM), Wyoming block (W), Southern Wyoming block (SW), and Colorado block (C). Key references used were the basement map of North America (Kinney and Flawn, 1967), the basement map of the United States (Bayley and Muehlberger, 1968), the magnetic map of the United States (Zietz, 1982), and the gravity map of the United States

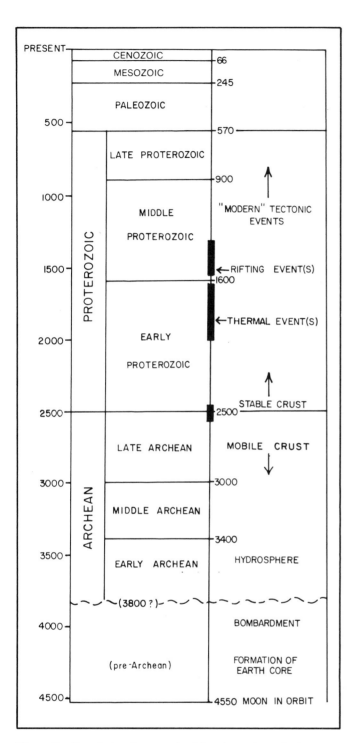

Figure 3—Correlation chart emphasizing the main geologic events: transition from mobile to stable crust (about 2.5 b.y. ago), early Proterozoic orogenic era (about 1.8 b.y. ago), and middle Proterozoic extensional tectonic era. The transition from a more mobile crust to a more stable crust might have occurred earlier in the Archean. Ages on scale are in millions of years. Modified from Harrison and Peterman (1982).

(Lyons and O'Hara, 1982). Some of the many sources of data used in the interpretation and construction of the maps are shown as numbers in Figure 4 and given in the legend.

Archean Structures

The basement rocks in Wyoming are composed of gneiss, granite, and greenstone that have uniform old (2.5–2.8 b.y. or greater) radiometric age dates (Figure 2). This basement has been termed the Precambrian Wyoming province by numerous authors (Condie, 1976). The basement to the east, north, and west is composed of gneiss, granite, meta-sedimentary rocks, and metavolcanic rocks that contain a wider range of radiometric age dates (about 2.8–1.8 b.y.). These latter rocks are Archean rocks that have been metamorphosed or otherwise altered more intensely by younger events than were the Archean rocks in Wyoming. Crust of Archean age has not been found south of Wyoming in western Nebraska (Lidiak, 1972) and Colorado (Tweto, 1980). The Archean structural relationships are illustrated in Figure 4A.

The region contains a series of Archean volcanic arc complexes, and the structural systems trend predominantly northeastward. An Archean greenstone facies is found in central Wyoming among the arcs (Houston, 1971; Condie, 1972). A sedimentary sequence of complex origin was deposited in southwestern Montana (D. W. Mogk, personal communication, 1984). Archean supracrustal sedimentary, volcanic, and intrusive rocks were added. The inferred locations of early continental margins that probably became suture zones during the process of amalgamation are shown by the northeast-trending lines in western Montana and Idaho in Figure 4.

The Archean margin on the north was somewhere northwest of the present-day Precambrian crystalline outcrops in southwestern Montana (Erslev, 1981). This suture might have been in the general area of Missoula, possibly coinciding with the Great Falls tectonic zone (O'Neill and Lopez, 1985). I suggest, however, that the suture might have been even farther northwest along the Moyie–Dibble Creek trend in Canada.

An oceanic plate (Figure 4A) probably existed south of Wyoming, although there might have been an Archean terrane here that was tectonically destroyed or removed before early Proterozoic time.

The northeast trends possibly correlate to the Archean structural trends of the Superior province in eastern North Dakota, Minnesota, and Canada (Peterman and Goldich, 1982). A suggested reconstruction showing the great "web" of Archean and earliest Proterozoic supracrustal rock that accumulated in the wide region located among the Superior, Wyoming, and Slave (northern Canada) Archean core provinces was illustrated by Sears and Price (1978). Volcanic arc complexes, such as those presumed to have formed in Wyoming, were common in late Archean time, and these linear masses were pushed together into long, lenticular features forming microcontinents of sialic crust that were preserved.

Early Proterozoic Structures

The Archean ended about 2.5 b.y. ago with major global thermal events (Windley, 1977). The change from a more

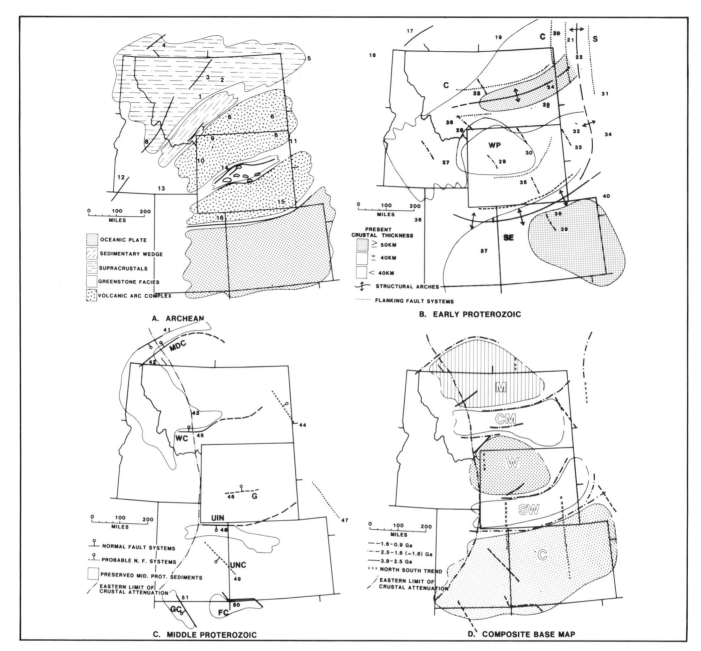

Figure 4—A. Archean (3.8–2.5 b.y.) geologic relationships. B. Early Proterozoic (2.5–1.6 b.y.) structural features, which coincide closely with present crustal thickness anomalies, are the Wyoming province (WP), the Superior province (S), the Churchill Province (C), and the Southeastern U.S. terrane (SE). The stable Archean core in Wyoming (oval near center of map) was not affected by early Proterozoic metamorphic events. C. Middle Proterozoic fault systems, which include Moyie-Dibble Creek (MDC), Willow Creek (WC), Granite (G), Uinta (UIN), Uncompahgre (UNC), Four Corners (FC), and Grand Canyon (GC). D. Composite base map, which synthesizes the Precambrian structural features. A north–south trend (x's and dots) is possibly of early Proterozoic age and is discussed in the text. The five Precambrian paleotectonic blocks are Montana (M), Central Montana (CM), Wyoming (W), Southern Wyoming (SW), and Colorado (C). Differential and recurrent tectonic uplift among these blocks influenced Phanerozoic sedimentation. Ga is billion years old. Sources (shown as numbers on maps): General—King, 1976; Windley, 1976; Muehlberger, 1980. Archean (3.8–2.5 b.y.)—1. Catanzaro, 1967; 2. Peterman, 1981a; 3. O'Neill and Lopez, in press; 4. Monger et al., 1982; 5. Peterman and Goldich, 1982; 6. Peterman, 1981b; 7. Erslev, 1981; 8. Wilson, 1981; 9. Casella, 1964; 10. Reed and Zartman, 1973; 11. Lidiak, 1971; 12. Sears and Price, 1978; 13. Armstrong, 1975; 14. Condie, 1972; 15. Houston, 1978; 16. Sears et al., 1982; Bryant, in press. Early Proterozoic (2.5–1.6 b.y.)—17. Cavanaugh and Seyfert, 1977; 18. Duncan, 1978; 19. Burwash et al., 1962; 20. Camfield and Gough, 1977; 21. Hajnal and Fowler, 1982; 22. Peterman and Hedge, 1964; 23. Woodward, 1970; 24. Smith, 1978; 25. Dutch, 1983; 26. James and Hedge, 1980; 27. Kanasewich, 1966; 28. Schmidt and Hendrix, 1981; 29. Condie, 1972; 30. Houston, 1971; 31. Baragar and Scoates, 1981; 32. Zartman et al., 1964; 33. Goldich et al., 1966; 34. Van Schmus and Bickford, 1981; 35. Hills and Houston, 1979; 36. Stewart, 1972; 37. Condie and Martell, 1983; 38. Warner, 1978; 39. Tweto, 1980; 40. Lidiak, 1972. Middle Proterozoic (1.6–0.9 b.y.)—41. McMechan, 1981; 42. Kanasewich et al., 1969; 43. Stewart, 1976; 44. Clement, 1955; 45. McMannis, 1963; 46. Peterman and Hildreth, 1978; 47. Lidiak et al., 1966; 48. Wallace, 1972; 49. Baars and Stevenson, 1981; 50. Case and Joesting, 1962; 51. Walcott, 1889; Van Gundy, 1946; Anderson, 1967.

mobile crust to a more stable crust had a profound effect on the nature of the styles of global tectonism that occurred after about 2.5 b.y. Many questions remain regarding the nature of the plate tectonic processes that might have operated during Archean time; however, by early Proterozoic time, the crust was less mobile and thus more brittle tectonic deformation began to occur.

During the period from about 2.5 to 1.9 b.y. ago, there were few major tectonic, thermal, or metamorphic events and supracrustal sedimentation continued to increase the sedimentary mass on the continental crustal platforms. After about 1.9 b.y. ago and extending to about 1.7 b.y., renewed global tectonic activity and major thermal and metamorphic events occurred. Terms such as Penokean, Hudsonian, and Churchill have been used to describe the events occurring at about 1.8 b.y. The Archean Slave province in northwestern Canada was brought closer to the Wyoming and Superior cores (Cavanaugh and Seyfert, 1977), and the closing of the web region among these three Archean terranes was taken up by a series of northeast-trending suture zones (Lewry and Sibbald, 1980).

These events caused tectonism to affect the Rocky Mountain region from the north and northwest, thus repeating the northeast trend of the Archean structures. However, the events were complex and tectonism also occurred from other directions (probably from the western and southwestern margins) to contribute other trend axes. The interpretation of this geologic history is illustrated in Figure 4B. The Wyoming province (WP) was isolated as a geologic entity by the early Proterozoic events that created the Churchill province (C), added the southeastern United States terrane (SE), and adhered the western region to the Superior province (S). Northeast to east–west trends are dominant, but a northwest trend (indicated by dash-dot line on Figure 4B) also developed. Flanking fault systems are inferred to have developed on the sides of the orogenic belts. Thus, a metamorphic or orogenic "rind" engulfed the stable, lenticular pod (shaded areas on Figure 4B).

The early Proterozoic metamorphic rind or aureole has been documented around the periphery of the Superior province by Baragar and Scoates (1981), who also give an excellent discussion describing how the early Proterozoic events isolated the stable Archean core terrane. The boundary of the Wyoming province is less well known, but has been studied at several places: Black Hills (Goldich et al., 1966), southern Wyoming (Hills and Houston, 1979), Uinta Range (Sears et al., 1982; Bryant, 1985), and Teton Range (Reed and Zartman, 1973). The western boundary of the Wyoming province was affected by younger crustal attenuation and thrusting and is not as well delineated as the other sides are. A description of the Precambrian geologic history of the Wyoming province is contained in Condie (1976).

The southern Wyoming border was the site of a postulated obduction margin (Hills and Houston, 1979). The nature of the marginal events on the north side of the stable Wyoming core is not known, but an obduction margin is not likely (Wilson, 1981; Fountain and Desmarais, 1980).

I have interpreted the central Montana system along the northern boundary of the stable Archean core on the basis of the crustal thickness (Smith, 1978), gravity, and magnetic data (Dutch, 1983). This system trends toward central Idaho. The crustal thickness anomaly in central Idaho (Kanasewich, 1966) is a northwest-trending middle Proterozoic feature (Armstrong, 1975; Ruppel and Lopez, 1984). Further evidence for the interpretation shown is the diorite intrusive in the Little Belt Mountains and the apparent absence of such bodies elsewhere across southern Montana. This diorite intrusive (the Pinto Diorite) is related to early Proterozoic events (Woodward, 1970), and such intrusive rocks are common around the periphery of the Wyoming province (Reed and Zartman, 1973).

Why certain parts of the Archean terranes were spared the early Proterozoic remobilization events and remained stable is not yet clear.

Middle Proterozoic Structures

The principal middle Proterozoic structures were extensional (Figure 4C). The major normal fault systems and some probable fault systems are shown in Figure 4C. Middle Proterozoic strata are preserved adjacent to many of these systems. Note the eastern limit of crustal attenuation in Figure 4C. Early Proterozoic events ended after 1.7 b.y. ago, and by 1.5 b.y. ago, the North American continent was affected by a continent-wide rift episode (Burke et al., 1978). In Idaho, western Utah, and Nevada the continent was rifted sufficiently to form a stagnant intracontinental basin. This basin was filled with the middle Proterozoic Belt Supergroup sediments (Obradovich et al., 1984). The continent then rifted along this weakness (about 800 m.y. ago), and part of the continent separated and drifted away (Stewart, 1976; Sears and Price, 1978).

The fault systems (Figure 4C) mark the general location of the boundaries of the major Precambrian lithostructural terranes that influenced later sedimentary history on the western shelf. The Moyie–Dibble Creek system is at the north end of the study area in southern Alberta. The Willow Creek system also was active and has been mapped (McMannis, 1963). The Granite Mountains fault system might also have been active during middle Proterozoic time based on a study by Peterman and Hildreth (1978). The fault system in the Uinta Mountains has been reported along the Colorado–Utah state line in the Uinta basin by Wallace (1972) and Bryant (1985). The systems at the southern end of the study area include the Uncompahgre, reviewed by Baars and Stevenson (1981); the Four Corners, noted by Case and Joesting (1962); and the systems in the Grand Canyon, mapped by Walcott (1899) and reviewed by Van Gundy (1946) and Beus et al. (1974).

During Precambrian time, these systems divided the cratonic margin into subparallel blocks that were high in central Wyoming and successively lower to the north and south. The western terminus of each block closely coincided with the eastern end of the attenuated continental crust (Stewart, 1976), and these structures are well inland from the western post-separation margin of the continent (Sears and Price, 1978). The blocks are schematically depicted in block diagram form in Figure 5.

The fault systems do not appear to be randomly developed structures. The Willow Creek and Uinta systems flank the Wyoming province. The Granite system is located along the northern margin of the part of the Wyoming

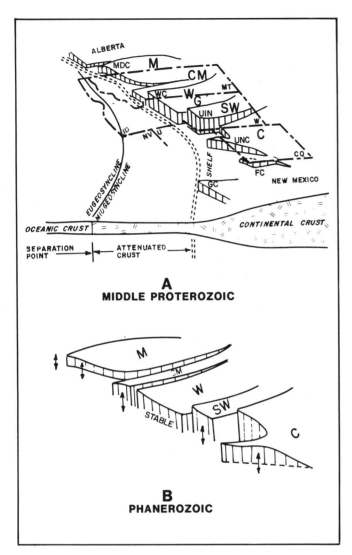

Figure 5—Diagram depicting general structural relationships of paleotectonic blocks during middle Proterozoic time (A) and the reversal that occurred during the Phanerozoic (B). Differential and recurrent tectonic adjustments among the structural blocks created more complicated structural relationships than shown here. Fault systems are shown on the map in Figure 4C, and the block name abbreviations are defined in Figure 4D. The double dotted line is the eastern limit of the attenuated crust as shown in Figures 4C and D, and this feature is modified from Stewart (1976).

province that was influenced by marginal events in early Proterozoic time. This system is also the approximate southern boundary of the Archean greenstone facies in central Wyoming. A special Precambrian geologic explanation for the Moyie–Dibble Creek, Uncompahgre, Four Corners, and Grand Canyon systems is not known, but a coincidence with early or middle Proterozoic events is suspected.

A minimum number of paleostructural systems and trends are shown in Figure 4. These structural systems are not single features. They are complex and poorly exposed, thereby obscuring their precise character. It is interesting to note, however, how many structural systems were developed

by the end of middle Proterozoic time, as well as the curved or jagged nature of most of the terrane boundaries.

Structural trends or systems derived from the Precambrian events reviewed here are synthesized in a composite base map (Figure 4D), which was used to review the stratigraphy. The late Archean, early Proterozoic, and middle Proterozoic trends are emphasized because most of the structural systems that delineate the Phanerozoic paleotectonic features were formed during these eras.

The five lithostructural terranes or paleotectonic blocks delineated in Figure 4D are the Montana, Central Montana, Wyoming, Southern Wyoming, and Colorado blocks. The Wyoming block is the stable Archean core, and the Central Montana and Southern Wyoming blocks make up the Archean rocks that were remobilized to form the metamorphic rind in early Proterozoic time. The Montana block is the Archean–early Proterozoic terrane on the north, and the Colorado block is the Proterozoic terrane on the south.

The north–south trend included on the composite map (Figure 4D) might have developed in conjunction with the early and middle Proterozoic structures (Figures 4B and 4C). The Churchill collision with the Superior and Wyoming provinces was such that the Churchill tectonic front spread into the embayments and other indentations around the periphery of the stable Archean core. The tectonic front in Manitoba apparently pushed against the Superior province and spread the plates on the west side of the province southward (Gibb, 1983). This process might have affected the basement farther south, thus causing the several north–south trends observed in the Rocky Mountain region. The Nesson anticline in North Dakota, the Rock Springs uplift in Wyoming, and the Colorado Front Range fault are a few examples of Laramide north–south structures that might be descendant from these enigmatic early Proterozoic features.

Also shown in Figure 4D are trends that project from one terrane to another (dotted lines), which create approximately linear trends that appear to extend long distances within the continental basement.

Differential and Recurrent Uplift

Evidence suggests that the paleotectonic blocks (Figure 5) were differentially and recurrently uplifted with respect to one another during Phanerozoic time. When the oceans were at lowstand (Figure 5A), the tectonic activity was also at a minimum and extensional tectonic adjustment tended to occur on the continents. As the oceans moved to highstand, the coincident increased epeirogenesis affected the more mobile blocks in Montana and Colorado (Figure 5B). The result was to concentrate sediments with fewer unconformities over the Wyoming block. As discussed later, more unconformities occur over the Montana, Central Montana, Southern Wyoming, and Colorado blocks than over the Wyoming block. The youngest Precambrian block (Colorado) contains the most severe unconformities, suggesting possibly greater tectonic mobility here than in the other blocks. The two rind blocks (Central Montana and Southern Wyoming) reacted differently, perhaps reflecting the different nature of the early Proterozoic events. The Central Montana block was

often the place of sag or extensional tectonic events during
the Phanerozoic, and a trough structure developed here even
though at other times this block was a tectonic high. The
Southern Wyoming block, however, was persistently a
tectonic highland.

PHANEROZOIC UNCONFORMITIES

I reviewed many detailed paleotectonic, paleogeologic, and
stratigraphic maps showing Phanerozoic stratigraphy for
this study and prepared numerous maps embracing the
geologic record from late Precambrian to late Eocene time.
Selected stratigraphic intervals are presented to illustrate
that the concepts involved occurred often throughout the
Phanerozoic. Much of the information is highly inter-
pretative, and there are numerous controversial or as yet
unknown aspects to the maps presented. The principal
authors from which the maps were modified are noted, but
many additional sources were referenced.

The anisotropic character of the basement and the
changes through time in the nature and position of the
continental margin orogenic events around the periphery of
the North American continent (Gilluly, 1967) affected the
distribution of the strata and the unconformities.

The observations of the distribution of major unconformities in the record is important in this study, not the
cause of the sea level changes. Numerous explanations have
been advanced for the cyclical changes in sea level during the
Phanerozoic, including the global transfer of water, basinal
drying or filling by water, and sea floor spreading. The
studies continue to generate interesting new data and
opinions, and it is recognized here that the episodic global
changes in sea level are not always due to the same
processes. The model used here emphasizes the
tectonomechanical crustal flexures.

As the sea periodically flooded the continent (probably
resulting from acceleration of sea floor spreading activity at
the oceanic ridges), the lower topographic areas in Montana
and Colorado were flooded first (Figure 6A). As the flooding
progressed, marginal tectonism caused structural growth of
the craton, which affected the Montana and Colorado areas
more than the Wyoming area. The spreading and flooding
processes act in concert (Vail et al., 1977). Accelerated
impingement of oceanic crust against the continent and
possible movements in the mantle forced elevation of the
paleotectonic blocks, and this occurred contemporaneously
with the eustatic rise of sea level. The maximum stand of
sea level was thus approximately coincident with the
maximum uplift of mobile blocks along the continental
margin and in the interior as well.

In the Rocky Mountain region, the less stable Montana
and Colorado blocks and the Southern Wyoming and Central
Montana blocks became elevated concurrently with the
periodic transgressions of the ocean. Consequently, the
stable Wyoming block apparently became the favored
depositional center (Figure 6B), while the more mobile
Montana and Colorado areas were often uplands and only
shallowly inundated by the epeiric seas. More unconformities
are in the record in these two regions than are in the record
over the Wyoming block (Figure 6C). Also, the Southern

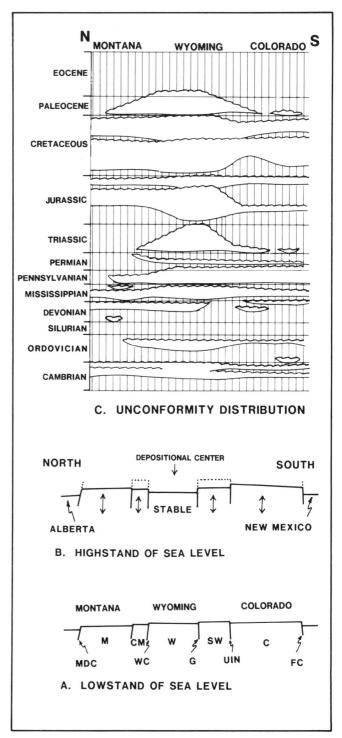

Figure 6—Tectonic model for Precambrian lithostructural terranes.
Fault systems and paleotectonic blocks are defined in Figure 4. As
the oceans periodically transgressed from lowstand (A) to highstand (B) and the shelf was flooded, there usually was coincident
structural growth in the more mobile blocks north and south of the
more stable Wyoming block. Differential adjustments in the
Central Montana and Southern Wyoming blocks (dotted lines)
complicated the stratigraphic patterns. The general distribution of
unconformities in the north–south direction is also shown (C).
Vertical ruling indicates a hiatus. Stratigraphic correlations for
Montana, Wyoming, and Colorado are from Balster (1971), Love
and Christiansen (1980), and Kent and Porter (1980), respectively.

Wyoming and Central Montana blocks were often tectonically adjusted lower or higher than the Wyoming block, creating either a trough at times or an upland.

A stratigraphic chart summarizing the stratigraphy in the north–south direction in the Rocky Mountain region is shown in Figure 7. Selected examples of the stratigraphic effect of the paleotectonic model and the recurrent nature of it throughout Phanerozoic time are discussed for the Late Cambrian, Middle–Late Ordovician, Late Devonian–Early Mississippian, Late Mississippian, Pennsylvanian, Permian, Middle Triassic, Middle Jurassic, early Late Cretaceous, Late Cretaceous, and early Tertiary age strata in the Rocky Mountain region (Figures 8, 9, and 10).

Cambrian and Early Ordovician (Sauk) Sequence

A uniform sandstone to shale to carbonate sedimentary cycle is evidenced in the Cambrian units over the Wyoming block (Figure 7A), and there are more lithologic varieties and unconformities distributed over the other blocks (Figure 8A). The paleotectonic base map (Figure 4D) was used for construction of Figure 8, and it is evident that the Precambrian structural trends influenced the distribution of the strata. Stratigraphic change in the Cambrian formations from northwest to southeast across the approximate location of the Willow Creek fault system has been described by Graham and Suttner (1974). There is generally less lithologic variety, greater thickness, and a greater concentration of limestone facies in the Late Cambrian strata over the Wyoming block.

Another observation can be made from Figure 7A. A major unconformity in the northern Cordillera resulted from the Cambrian orogeny in Canada (McCrossan and Porter, 1973). The double cycle of shale and carbonate sedimentation during the Cambrian over the Montana and Central Montana blocks suggests that the orogenic pulses in the Canadian Cordillera affected the terranes north of Wyoming and that the stable nature of the Wyoming province might have insulated the southern terranes from the orogeny to the north.

Middle and Late Ordovician and Silurian (Tippecanoe) Sequence

The Big Horn Formation (and equivalents) comprise the Middle and Late Ordovician strata on the shelf (Figure 7A). Structural growth in the Montana, Central Montana, Southern Wyoming, and Colorado blocks resulted in removal (Figure 8B) of the Ordovician and Silurian deposits from most of the Rocky Mountain region, with the exception of the Wyoming block and the Williston basin (Fuller, 1961). Most of the Silurian rocks are missing, but the strata that are present suggest that a depositional pattern occurred that was similar to the Ordovician erosion.

Devonian and Mississippian (Kaskaskia) Sequence

A paleotectonic model commonly described for the Kaskaskia sequence in the Rocky Mountain region is termed the "Wyoming shelf." This shelf was open to the north and northwest and had an upland area to the south in Colorado (Figure 7B). The Antler orogeny (Ketner, 1983) dominated the western continental margin. The shelf developed during Devonian and Mississippian time because of the coincident

development of a structural upland in the Colorado area and a structural sag in the Montana area. The shelf is observed in the isopach patterns of these strata (Baars, 1972; Craig, 1972). The tectonic model used here suggests that there also should have been structural growth in the Montana area, but there apparently was none. However, the important marginal orogenic events of the Kaskaskia sequence were not only compressive marginal tectonic events, but also marginal basin rifting events. Major marginal rift discontinuities occurred beginning in Late Devonian time with the onset of rifting occurring along the east coast of North America and across the northern hemisphere from eastern Greenland to the Sverdrup basin in the arctic (McCrossan and Porter, 1973). A possible rift-related structural sag feature developed across central Montana on the north side of the stable Wyoming province during Late Devonian and Mississippian time. This side is more proximate to the arctic rifting events than the Wyoming and Colorado areas. Did the arctic rifting events affect only the north side of the Wyoming province? Furthermore, the onset of continental collision along the eastern and southeastern margins of the continent would have tended to maintain structural growth in the Colorado area. Apparently the tectonic stability of the Wyoming block, dating from early Proterozoic time, effectively insulated the adjacent terranes from tectonism at opposite ends of the continent.

Some structural growth did occur in the Montana area as evidenced by the distribution of lithofacies on the Bakken Formation and equivalents (Figure 8C). This black shale facies was deposited widely across the Montana area, but was partially eroded as structures developed in Montana, and during the later stages of deposition a sandy facies developed. The sea flooded onto the shelf progressively during Devonian and Early Mississippian time, and the black shale facies was deposited as this transgression advanced to the continental interior. Coincidently, however, the continued sea floor spreading and marginal tectonic events began to affect the mobile Montana (and Colorado) area, and structures were locally developed that caused a change in facies to a clastic sequence. The Montana and Central Montana blocks were not as prominent as the Colorado and Southern Wyoming blocks during this cycle (as previously discussed), and notice (Figure 8C) that a wide beach was apparently established over the stable Wyoming block. The structural trends outlined on the paleotectonic base map (Figure 4D) effectively controlled the distribution of these strata.

During Mississippian time, a greater thickness of strata was deposited in the Montana area (Figure 8D). The Big Snowy uplift was an important positive area that modified the distribution of Mississippian strata notwithstanding the general development of a broad basin in central Montana (Smith, 1982). The presence of the positive block in the broad basin is an important observation, because the model suggests that there was some positive structural growth in the Montana area during this part of the Kaskaskia sequence and Smith's (1982) study indicates that there was. Sporadic uplift in central Montana continued throughout deposition of the Chester Big Snowy Group (Jensen and Carlson, 1972).

During Chester time, the coincidence of maximum oceanic flooding and a culmination of tectonism occurred. The sea

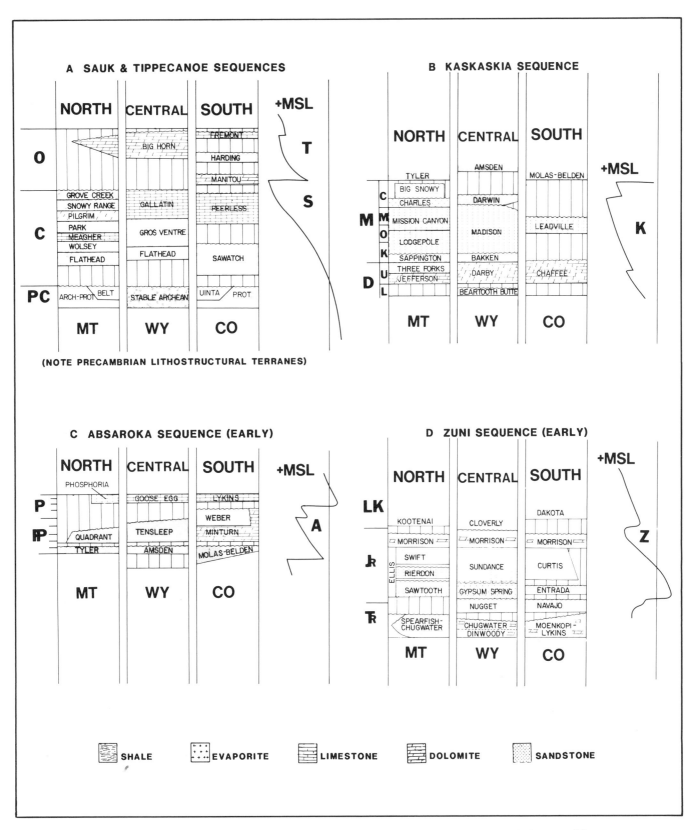

Figure 7—Stratigraphic correlation chart showing distribution of selected unconformities in a north–south direction. The most unconformities and the greatest variation in lithology are in the formations over the Montana and Colorado areas, and the most complete stratigraphic record was deposited over the stable Wyoming block. The Precambrian Archean and Proterozoic substructure is noted in part A (Belt and Uinta indicate middle Proterozoic sediments). Curve on right of each diagram is variation in mean sea level (MSL) adapted from Vail et al. (1977). The sequences (Sloss, 1963) are noted next to these curves (S, Sauk; T, Tippecanoe; K, Kaskaskia; A, Absaroka; Z, Zuni). Vertical ruling indicates a hiatus.

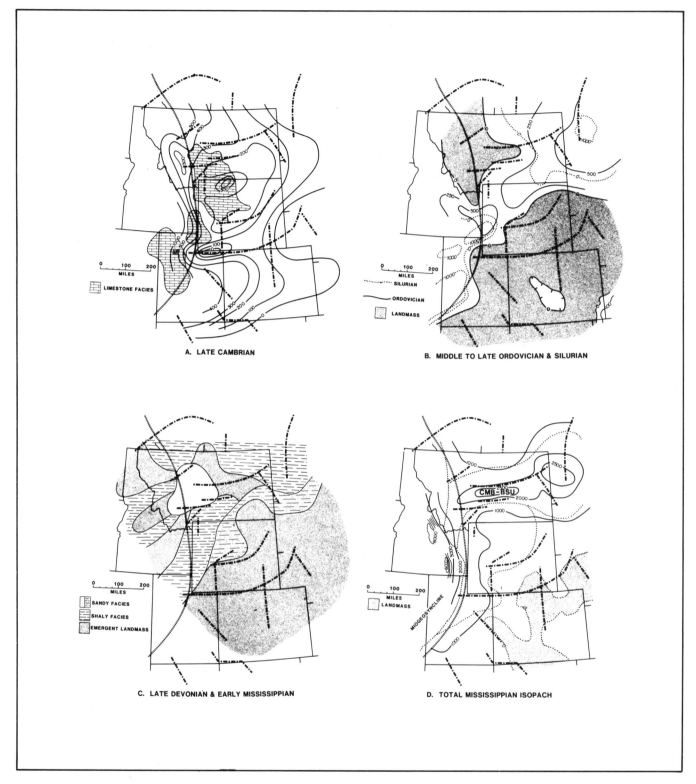

Figure 8—Four examples from early Paleozoic stratigraphy illustrating how the Precambrian terranes influenced the distribution of sedimentary rocks. A. Isopach map of Late Cambrian strata. Note the greater concentration of limestone facies over the Wyoming block. Modified from Lochman-Balk (1972). B. Isopach map of Middle to Late Ordovician and Silurian strata. Modified from Foster (1972) and Gibbs (1972). C. Paleogeologic map of Late Devonian and Early Mississippian strata (Bakken Formation and equivalents). Modified from Sandberg (1963), Baars (1972), and Loucks (1977). D. Total Mississippian isopach map (contour interval in feet) showing the greater thickness of strata in the Central Montana basin (CMB). The Big Snowy uplift (BSU) also developed in this broad trough. Modified from Craig (1972) and Roberts (1979).

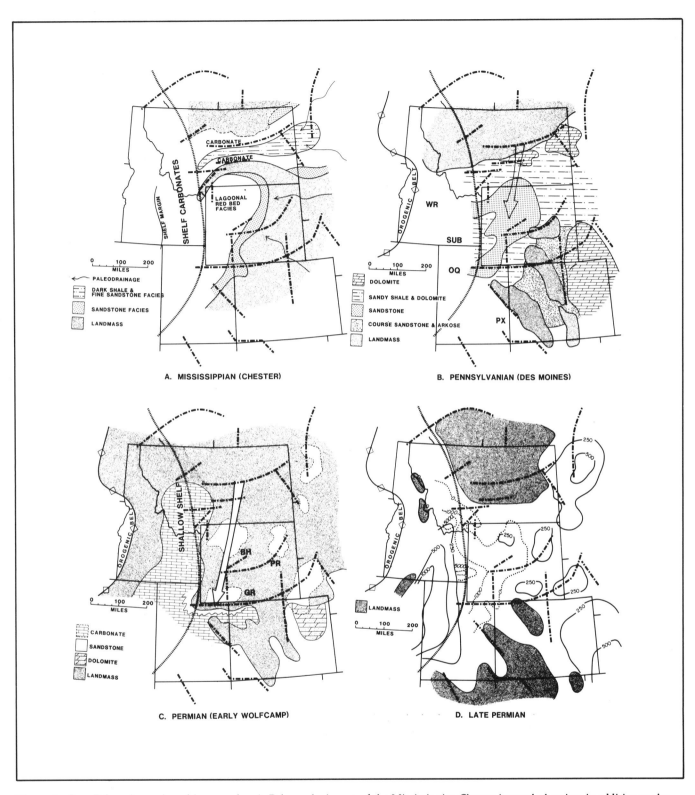

Figure 9—Late Paleozoic stratigraphic examples. A. Paleogeologic map of the Mississippian Chester interval, showing, in addition to the Central Montana basin facies, a concentration of lagoonal red bed facies over the Wyoming block. Modified from Craig and Connor (1979) and Sando (1976). B. Paleogeographic map showing the depositional setting for the Desmoinesian (Pennsylvanian) sandstones. Arrow shows main direction of sediment transport. The Wyoming shelf was open to the south-southeast. The Paradox (PX), Oquirrh (OQ), Sublette (SUB), and Wood River (WR) basins are depositional troughs along the western margin of the shelf. Modified from Mallory (1972a). C. Map showing early Wolfcampian (Permian) sandstone deposited in Colorado and Utah from sources in Montana and Wyoming (arrow). Proto-Greater Green River (GR), Big Horn (BH), and Powder River (PR) basins were developed during Permian Guadalupian time by extensional events. Modified from McKee and Oriel (1967). D. The Late Permian isopach map (contour interval in feet) shows a concentration of black shales and clastic strata over the Wyoming block. Modified from McKee and Oriel (1967). Organic carbon isograds (dotted lines; contour interval in kg/m^2) are from Claypool et al. (1978).

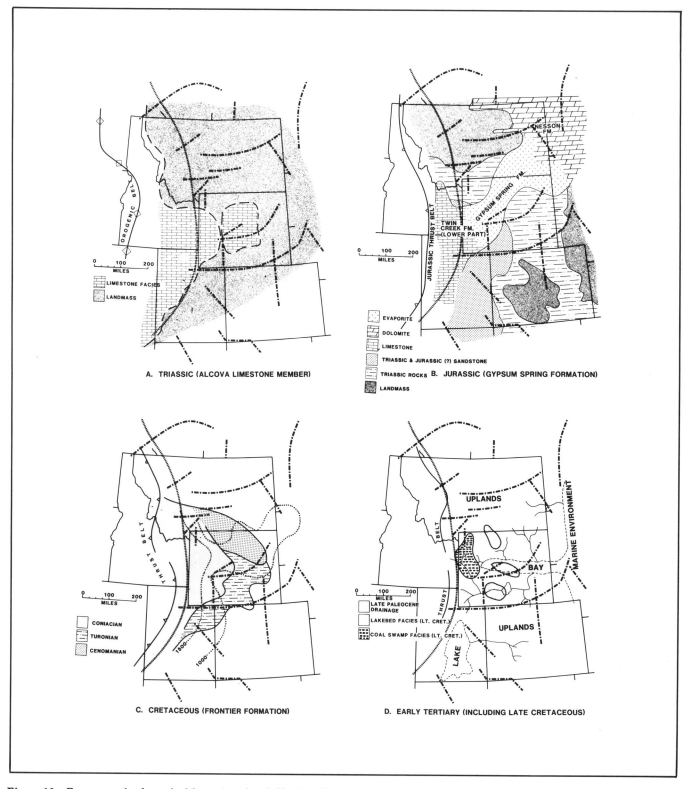

Figure 10—Four examples from the Mesozoic and early Tertiary illustrating the apparent persistence of the model throughout Phanerozoic time. A. Paleogeologic map showing the Alcova limestone member of the Chugwater Formation (Triassic) below the Tr-2 unconformity. Modified from Pipiringos and O'Sullivan (1978). B. Map of the Jurassic Gypsum Spring Formation (and equivalents) below the Jr-2 unconformity evidencing more uniform deposition over the Wyoming area and the continuing appearance of uplands in the Montana and Colorado areas causing numerous unconformities to develop there. Modified from Pipiringos and O'Sullivan (1978). C. The tectonically derived sediment of the Frontier Formation (Cretaceous) was concentrated in greater thickness in Wyoming than in the Montana and Colorado areas (isopachs in feet). Modified from Cobban and Reeside (1952) and McGookey (1972). D. Map of early Tertiary (late Paleocene) paleodrainage, and the general location of the large Late Cretaceous lakes and coal swamps. Modified from McGookey (1972) and McDonald (1972).

persisted on the shelf but structural features also developed (Figure 9A). The paleogeologic map depicts Sando's (1976) late Chester cycle III, phase 5, a period of maximum shelf carbonate deposition in the geosyncline. The nature of these strata in northern Montana is not known due to pre-Jurassic erosion (Maughan and Roberts, 1967). A broad lagoon developed over the stable Wyoming block as anticipated by the model.

The end of Chester time marked the end of the Kaskaskia sequence, and as tectonism subsided the sea retreated to a lowstand (Vail et al., 1977). Drainages developed in central Montana; and as the sea again advanced during the early part of the Absaroka sequence, the basal Pennsylvanian Tyler Formation channel sandstones were deposited (Maughan, 1984). In the Colorado area the Molas Formation (DeVoto, 1980) formed, like the Amsden Formation in Wyoming and Montana, as a regolitic terra rosa on upland surfaces. The overlying Belden Formation (Brill, 1958) was deposited, like the Tyler Formation in Montana, in a structural sag or trough feature that developed in the Colorado block.

Pennsylvanian to Triassic (Absaroka) Sequence

During the Absaroka cycle the earlier Wyoming shelf structure reversed from open to the north-northwest (Kaskaskia sequence) to open to the south-southeast by Late Pennsylvanian time (Figure 7C). The Quadrant, Tensleep, and Weber formations are composed of clean, fine-grained quartz sandstone, and this lithology is distributed across the shelf in a pattern (Figure 9B) predictable from the model presented.

Structural growth was great in the Montana area. Notice in Figure 9B how the early Pennsylvanian basin in Montana was eroded along the central Montana structural trends shown on the paleotectonic base map (Figure 4D). Depositional basins developed in Colorado, Utah, and adjacent areas, and one explanation for the change in conditions there might be the onset of the Ouachita orogeny along the southern margin of the continent (Kluth and Coney, 1983). As foreland basins developed in Oklahoma, Arkansas, and Texas in response to the Ouachita events, adjacent parts of the craton were depressed toward the interior of the continent including the Colorado area in the Rocky Mountain region. The Montana area appears to have been insulated from this tectonism by the stable Wyoming block. Tectonism occurring along the continental margin in western Idaho might have contributed to uplift on the northern side (Poole and Sandberg, 1977; Dickinson, 1977). A pattern of structural growth in the Montana area, more passive stability in the Wyoming area, and differential uplift, basin development, and faulting in the Colorado area, all altered the shelf from its earlier paleotectonic setting.

The Weber Formation is partly contemporaneous with the Quadrant and Tensleep formations, and sands derived from the older Quadrant Formation in Montana were deposited in younger parts of the Weber Formation in Colorado and Utah (Figure 9C). The main body of this sandstone formation was deposited over the stable Wyoming block as illustrated in Figure 9B (Saperstone and Ethridge, 1984). The Tensleep Formation in Wyoming apparently was also a source for the Permian upper Weber and Lyons formations in Utah and Colorado (Maughan, 1980).

The change in paleotectonic setting from open to the north-northwest to open to the south-southeast and the foreland deformation caused by the Ouachita orogeny added great complexity to the paleotectonic structures in the Rocky Mountain region. A brief extensional era following termination of middle Permian tectonism caused the development of discrete basins in Wyoming. The general outline of these basins is shown in Figure 9C. Permian Guadalupian dolomite and evaporite strata are concentrated in the proto-Greater Green River, Big Horn, and Powder River basin structures in Wyoming.

The final Permian depositional episode initially involved differentiation of basins and uplifts and coincident transgressive deposition. Structural growth interrupted deposition of a black shale sequence in a manner similar to that which occurred during deposition of the Bakken Formation. Figure 9D shows the paleogeologic distribution of the shales and the interpretation of their organic carbon content (Claypool et al., 1978). The distribution of these strata mimics the distribution of the Ordovician Big Horn Formation (Figure 8B) across the Wyoming block between the two tectonically active regions of Montana and Colorado.

Depositional patterns are difficult to discern in the Triassic terrestrial sedimentary rocks (MacLachlan, 1972). However, a pattern consistent with the model exists in the areal distribution of the beds below the Tr-2 unconformity (terminology of Pipiringos and O'Sullivan, 1978). The Alcova limestone member of the Chugwater Formation is correlative, or closely so, with the Portneuf limestone member of the Thaynes Formation (Koch, 1976). This sandy limestone formation is shown in Figure 10A, and the formation has been described by Picard et al. (1969). The minor but very widespread Tr-2 unconformity eroded the upper part of the Alcova limestone member (Pipiringos and O'Sullivan, 1978). The sea was apparently low during Late Permian and Early Triassic time, but beginning with late Early and Middle Triassic time the sea again began to flood the continent (Vail et al., 1977); the Alcova limestone member represents the initial transgression. The presumed tectonic pulses responsible for the flooding kept the Montana and Colorado areas elevated relative to the Wyoming block. In general, the Triassic sedimentary rocks are very uniform over the Wyoming block and are more variable or absent over the Colorado and Montana areas.

The paleotectonic blocks did not rebound structurally from the foreland and interior continental deformation caused by the Ouachita orogeny, and the Permian western shelf remained open to the south-southeast throughout Mesozoic time. This might explain the later transgression of the Cretaceous seas across the Colorado Paleozoic paleotectonic highlands.

By Mesozoic time, more detail was introduced into the sedimentary patterns by the development of new paleotectonic trends that were parallel or subparallel to the Precambrian trends shown in Figure 4D. By early Mesozoic time the shelf had been tectonically affected from many directions horizontally and also by several episodes of vertical tectonism. Many additional trends would have to be added to the composite base map to explain all of these sedimentary patterns. What is important, however, is the apparent persistence of paleotectonic influence from the original Precambrian structures.

Jurassic and Cretaceous (Zuni) Sequence

The great lithologic variation in the Jurassic strata often obscures analysis of regional facies patterns. Numerous subtle paleotectonic features developed in the Jurassic embayment in the Rocky Mountain region (Imlay, 1980).

The Jurassic Gypsum Spring Formation and equivalents is an evaporitic unit situated mainly over the Wyoming block (Figure 7D). This formation is below the Jr-2 unconformity (Pipiringos and O'Sullivan, 1978), and the pattern of distribution (Figure 10B) indicates that the Precambrian trends illustrated on the paleotectonic base map (Figure 4D) appear to have influenced the distribution of these strata.

The Jurassic Sundance Formation (Figure 7D) was deposited in the same general place as the Gypsum Spring Formation. Sand was broadly deposited in the Curtis Formation in the Colorado area, indicating an upland there relative to the Wyoming block during this time. The more variable lithology of the Ellis Group (Figure 7D) in Montana illustrates the tectonic instability of the Montana area relative to Wyoming at this same time (Peterson, 1981). The basin and block combination structure of Mississippian time (Figure 8D) is repeated, and the Central Montana block was elevated in spite of the existence of a general basinal form in the Montana area. The Ellis Group of marine rocks, therefore, was deposited in a basin across central Montana that had a structural block along its axis that modified the depositional patterns in the strata. The similarity in depositional setting between the Ellis Group in Jurassic time and the Big Snowy Group of Late Mississippian time (Figure 7B) was recognized and described by Walton (1946).

A coal basin developed across central Montana as a distinct facies of the Jurassic Morrison Formation (Peterson, 1981). This apparently indicates a persistence of the sag structure during Late Jurassic time. The basal Cretaceous terrestrial sediments also contain distinctive stacked channel sandstones in this same area across central Montana (McGookey, 1972). A major drop in sea level occurred at the end of Jurassic time (Vail et al., 1977), and it appears that structurally controlled coal (Jurassic) and channel sandstone (Cretaceous) facies accumulated here much like the channel sandstones of the Tyler Formation accumulated during lowstand in Early Pennsylvanian time (Figure 7C).

The Frontier Formation is given here as an example of the possible applicability of the paleotectonic model to an analysis of the Cretaceous strata. The main body of the Frontier Formation (Cobban and Reeside, 1952) appears to be thicker over the Wyoming block, having possibly been directed there preferentially by structural growth in the Montana and Colorado areas (Figure 10C). Currents in the Cretaceous seaway were important to the regional distribution of this sediment, but the apparent concentration of the strata over the Wyoming block is consistent with the tectonic model presented.

Tertiary (Tejas) Sequence

The early Tertiary (late Paleocene) map, which also shows the Upper Cretaceous Campanian and Maestrichtian paleogeology, provides evidence for the apparent persistence of the model through the end of the Zuni sequence and even into the Tejas sequence (Figure 10D). Even today, the greatest thicknesses of Tertiary rocks anywhere in the Rocky Mountain region are in the Wyoming basins (Love et al., 1963).

CONCLUSION

A comparison of the distribution of Phanerozoic sediments in the Rocky Mountain region discloses that there are lithologic variations and numerous unconformities in the strata over the Montana and Central Montana blocks and great variation but more pronounced unconformities in the record over the Southern Wyoming and Colorado blocks. It also shows that the sedimentary record is most complete over the Wyoming block (the stable Archean core). The regional stratigraphic patterns indicate that the fault systems, paleotectonic blocks, and the tectonic model described here provide a lithostructural framework for detailed area and local studies.

The distribution of oil reserves on the shelf (Figure 1) shows a concentration of the discovered oil to be in Wyoming. The unconformity patterns suggest that different exploration approaches are necessary for different places in the Rocky Mountain region because of the regional variation and anisotropy in the crystalline basement substructure. For example, the Permian phosphatic black shales of the western Wyoming area (Figure 9D) are very important to a consideration of the origin of petroleum in the Rocky Mountain region (Stone, 1967). If a large quantity of petroleum was generated in early or middle Mesozoic time from these shales in western Wyoming and southeastern Idaho (Sheldon, 1967), it appears from the model presented here that, generally speaking, the oil could have migrated to eastern Wyoming but that structural barriers might have existed in southern Montana and northern Colorado to prevent efficient migration of the oil into central Montana and central Colorado. Furthermore, oil that did migrate into Montana and Colorado is probably trapped by different trapping mechanisms than in Wyoming. More petroleum will perhaps be found in the Montana and Colorado areas by pursuing the stratigraphic opportunities that arise from the more numerous unconformities in those areas.

The relationship between Archean cores or nuclei and the adjacent tectonic terranes has been noted for other continents (Hurley and Rand, 1969; Piper, 1982). Suggested general types of observed cores and examples of them are summarized in Figure 11, but further detail about these is not presented here. The Archean and Proterozoic age distinctions are not completely known for many of these old and stable rock areas on earth, and the cores are of different sizes. The listing is offered only as a preliminary observation and to demonstrate that the model described here might have some possible application to structural and stratigraphic mapping for petroleum exploration on a global basis.

A consideration of the Precambrian ancestry of observed trends may help to develop a geologically sensible explanation for the Phanerozoic paleotectonic features and offers a theme to pursue for more detailed exploration leads. This regional study provides a possible model for structural and stratigraphic interpretations of gravity, magnetic, seismic, air photo, satellite imagery, and geochemical data

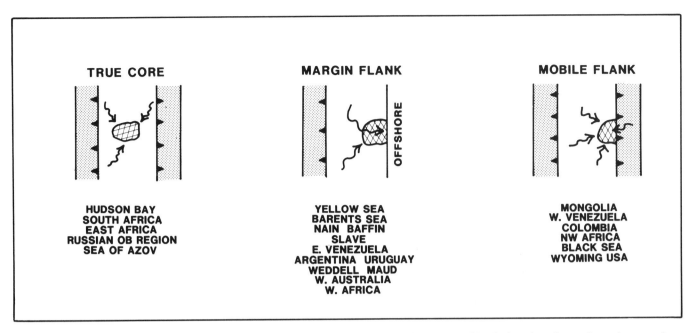

Figure 11—World occurrences of stable Archean core terranes. The cores are not always located in the interior of a continent (true core) but can be on an inactive continental margin (margin flank) or adjacent to a younger active terrane or margin (mobile flank). The figure is compiled from numerous sources, and the radiometric age dates of many of these rocks are not yet conclusive. The shaded areas with sawteeth depict the younger tectonic fronts, and the arrows show the general paleodrainage patterns.

used in petroleum exploration in the Rocky Mountain region.

ACKNOWLEDGMENTS

I wish to acknowledge the help of Edwin K. Maughan, U.S. Geological Survey (Denver); Robert A. Chadwick, Montana State University (Bozeman); Robert J. Weimer, Colorado School of Mines (Golden); and Edward T. Ruppel, U.S. Geological Survey (Denver) regarding the manuscript and figures. Zell E. Peterman, U.S. Geological Survey (Denver); Kent C. Condie, New Mexico Institute of Mining and Technology (Socorro); David W. Mogk, Montana State University (Bozeman); and David A. Lopez, North American Resources Company (Billings), provided suggestions concerning the Precambrian interpretation. Others who contributed valued comments are Robert F. Grabb and Michael B. Bryant, Cenex Exploration and Production (Billings); David Love, U.S. Geological Survey (Laramie); William J. Perry, Jr., U.S. Geological Survey (Denver); and Laurence L. Sloss, Northwestern University (Evanston). While I remain responsible for the interpretations and speculations presented herein, I am grateful for their thoughtful and constructive criticism. Thanks, too, go to Dick Lohrenz for the geological drafting and to Sue Besel for typing the manuscript. One key to the Precambrian is the continued accumulation of samples for radiometric age dating. More samples are needed from boreholes into the basement. Cores or large cuttings are very helpful and many Precambrian geologists are available to come to the well site and supervise the collection of a sample.

REFERENCES CITED

AAPG North American Developments and World Energy Developments, 1968–1984, Annual Issues.

Anderson, C. A., 1967, Precambrian wrench fault in central Arizona: U.S. Geological Survey Professional Paper 575-C, p. C60–C65.

Armstrong, R. L., 1975, Precambrian (1500 m.y. old) rocks of central Idaho—the Salmon River arch and its role in Cordilleran sedimentation and tectonics: American Journal of Science, v. 275-A, p. 437–467.

Baars, D. L., 1972, Devonian system, *in* Geologic atlas of the Rocky Mountain region: Rocky Mountain Association of Geologists, p. 90–99.

Baars, D. L., and G. M. Stevenson, 1981, Tectonic evolution of the Paradox basin, Utah and Colorado, *in* D. L. Wiegand, ed., Geology of the Paradox Basin: Rocky Mountain Association of Geologists, p. 23–31.

Balster, C. A., ed., 1971, Catalog of stratigraphic names for Montana: Montana Geological Society and Montana Bureau of Mines and Geology Special Publication 54, 448 p.

Baragar, W. R. A., and R. F. J. Scoates, 1981, The Circum-Superior Belt: a Proterozoic plate margin?, *in* A. Kroner, ed., Precambrian plate tectonics: Elsevier, Amsterdam/New York, p. 297–325.

Bayley, R. W., and W. R. Muehlberger, 1968, Basement rock map of the United States, exclusive of Alaska and Hawaii: U.S. Geological Survey Map., scale 1:2,500,000.

Beus, S. S., R. R. Rawson, R. O. Dalton, G. M. Stevenson, S. Reed, and T. M. Daneker, 1974, Preliminary report on the Unkar Group (Precambrian) in Grand Canyon, Arizona, *in* T. N. V. Karlstrom, G. A. Swann, and R. L.

Eastwood, eds., Geology of northern Arizona, Part I. Regional studies: Geological Society of America Rocky Mountain Section, 27th Annual Meeting Guidebook, p. 34–53.

Brill, K. G., Jr., 1958, The Belden Formation, *in* B. F. Curtis, ed., Symposium on Pennsylvanian rocks of Colorado and adjacent areas: Rocky Mountain Association of Geologists, p. 102–105.

Bryant, B., 1985, Structural ancestry of the Uinta Mountains, Utah, *in* M. D. Picard, ed., Uinta Basin Geologic Resources: Utah Geological Association Publication 11.

Burke, K. L., J. F. Delano, A. Dewey, W. S. F. Edelstein, K. D. Kidd, A. M. C. Nelson, J. Sengor, and J. Stroup, 1978, Rifts and sutures of the world: unpublished report to Geophysics Branch, ESA Division of National Aeronautics and Space Administration and the GSFC, Greenbelt, Maryland.

Burwash, R. A., H. Baadsgaard, and Z. E. Peterman, 1962, Precambrian K-Ar dates from the western Canada sedimentary basin: Journal of Geophysical Research, v. 67, p. 1617–1626.

Camfield, P. A., and D. I. Gough, 1977, A possible Proterozoic plate boundary in North America: Canadian Journal of Earth Sciences, v. 14, p. 1229–1238.

Case, J. E., and H. R. Joesting, 1962, Precambrian structures in the Blanding basin and Monument upwarp, southeast Utah: U.S. Geological Survey Professional Paper 424-D, p. D287–D291.

Casella, C. J., 1964, Geologic evolution of the Beartooth Mountains, Montana and Wyoming, Part 4, Relationship between Precambrian and Laramide structures in the Line Creek area: Geological Society of America Bulletin, v. 75, p. 969–985.

Catanzaro, E. J., 1967, Correlation of some Precambrian rocks and metamorphic events in parts of Wyoming and Montana: Mountain Geologist, v. 4, p. 9–21.

Cavanaugh, M. D., and C. F. Seyfert, 1977, Apparent polar wander paths and the joining of the Superior and Slave provinces during early Proterozoic time: Geology, v. 5, p. 207–211.

Claypool, G. E., A. H. Love, and E. K. Maughan, 1978, Organic geochemistry, incipient metamorphism, and oil generation in black shale members of Phosphoria Formation, western interior United States: AAPG Bulletin, v. 62, p. 98–120.

Clement, J. H., 1955, The Pine Field, Dawson, Fallon, Prairie, and Wibaux counties, Montana: AAPG Geological Record, Rocky Mountain Section, p. 165–171.

Cobban, W. A., and J. B. Reeside, Jr., 1952, Frontier Formation, Wyoming and adjacent areas: AAPG Bulletin, v. 36, p. 1913–1961.

Condie, K. C., 1972, A plate tectonics evolutionary model of the South Pass Archean greenstone belt, southwestern Wyoming: Proceedings of 24th International Geological Congress, Montreal, p. 104–112.

————, 1976, The Wyoming province in the western United States, *in* B. F. Windley, ed., The early history of the earth: John Wiley, New York, p. 499–510.

Condie, K. C., and C. Martell, 1983, Early Proterozoic metasediments from north-central Colorado:

metamorphism, provenance, and tectonic setting: Geological Society of America Bulletin, v. 94, p. 1215–1224.

Cook, T. D., and A. W. Bally, 1975, Stratigraphic atlas of North and Central America: Princeton University Press, Princeton, New Jersey, 272 p.

Craig, L. C., 1972, Mississippian system, *in* Geologic atlas of the Rocky Mountain region: Rocky Mountain Association of Geologists, p. 100–110.

Craig, L. C., and C. W. Connor, 1979, Paleotectonic investigations of the Mississippian System in the United States: U.S. Geological Survey Professional Paper 1010, 559 p.

Craiglow, C., 1983, Tectonic significance of Ross Pass Fault Zone, Central Bridger Range, Montana (abs.): AAPG Bulletin, v. 67, p. 1333.

Cram, I. H., ed., 1971, Future petroleum provinces of the United States—their geology and potential: AAPG Memoir 15, 1496 p.

Davidson, A., 1972, The Churchill Province, *in* R. A. Price, and R. J. W. Douglas, eds., Variations in tectonic styles in Canada: Geological Association of Canada Special Paper 11, p. 381–434.

DeVoto, R. H., 1980, Pennsylvanian stratigraphy and history of Colorado, *in* H. C. Kent, and K. W. Porter, eds., Colorado geology: Rocky Mountain Association of Geologists, p. 71–101.

Dickinson, W. R., 1977, Paleozoic plate tectonics and the evolution of the Cordilleran continental margin, *in* Paleozoic paleogeography of the western United States: Society of Economic Paleontologists and Mineralogists Pacific Section, Pacific Coast Paleogeography Symposium, p. 137–155.

Dolton, G. L., ed., 1981, Estimates of undiscovered recoverable conventional resources of oil and gas in the United States: U.S. Geological Survey Circular 860, 87 p.

Duncan, I. J., 1978, Rb/Sr whole rock evidence for three Precambrian events in the Shuswap complex, southeast British Columbia: Geological Society of America Abstracts with Programs, v. 10, p. 392–393.

Dutch, S. I., 1983, Proterozoic structural provinces in the north-central United States: Geology, v. 11, p. 478–481.

Erslev, E. A., 1981, Petrology and structure of the Precambrian metamorphic rocks of the southern Madison Range, southwestern Montana: Ph.D. Thesis, Harvard University, Cambridge, Massachusetts, 124 p.

Foster, N. H., 1972, Ordovician system, *in* Geologic atlas of the Rocky Mountain region: Rocky Mountain Association of Geologists, p. 76–85.

Fountain, D. M., and N. R. Desmarais, 1980, The Wabowden terrane of Manitoba and the pre-Belt basement of southwestern Montana: a comparison, *in* Guidebook of the Drummond-Elkhorn areas, west-central Montana: Montana Bureau of Mines and Geology Special Publication 82, p. 35–46.

Fuller, J. G. C. M., 1961, Ordovician and contiguous formations in North Dakota, South Dakota, Montana and adjoining areas in Canada and the United States: AAPG Bulletin, v. 45, p. 1334–1363.

Gibb, R. A., 1983, Model for suturing of Superior and Churchill plates: an example of double indentation tectonics: Geology, v. 11, p. 413–417.

Gibbs, F. A., 1972, Silurian system, *in* Geologic atlas of the Rocky Mountain region: Rocky Mountain Association of Geologists, p. 86–90.

Gilluly, J., 1967, Chronology of tectonic movements in the western United States: American Journal of Science, v. 265, p. 306–331.

Goldich, S. S., E. G. Lidiak, C. E. Hedge, and F. G. Walthall, 1966, Geochronology of the midcontinent region, United States, 2, northern area: Journal of Geophysical Research, v. 71, p. 5389–5408.

Graham, S. A., and L. J. Suttner, 1974, Occurrence of Cambrian islands in southwest Montana: Mountain Geologist, v. 11, p. 71–84.

Gries, R., 1981, Oil and gas prospecting beneath the Precambrian of foreland thrust plates in the Rocky Mountains: The Mountain Geologist, v. 18, p. 1–18.

Hajnal, A., and C. M. R. Fowler, 1982, The earth's crust under the Williston Basin in eastern Saskatchewan and western Manitoba, *in* J. E. Christopher, and J. Kaldi, eds., Fourth International Williston Basin Symposium: Saskatchewan Geological Society, p. 13–18.

Ham, W. E., and J. L. Wilson, 1967, Paleozoic epeirogeny and orogeny in the central United States: American Journal of Science, v. 265, p. 332–407.

Harrison, J. E., and Z. E. Peterman, 1982, Geochronometric units for divisions of Precambrian time: AAPG Bulletin, v. 66, p. 801–804.

———, 1984, Introduction to correlation of Precambrian rock sequences: U.S. Geological Survey Professional Paper 1241-A, 7 p.

Harrison, J. E., A. B. Griggs, and J. D. Wells, 1974, Tectonic features of the Precambrian Belt basin and their influence on post-Belt Structures: U.S. Geological Survey Professional Paper 866, 15 p.

Haun, J. D., and H. C. Kent, 1965, Geologic history of Rocky Mountains: AAPG Bulletin, v. 49, p. 1781–1800.

Hills, F. A., and R. S. Houston, 1979, Early Proterozoic tectonics of the central Rocky Mountains North America, *in* Contributions to Geology: University of Wyoming, Laramie, v. 17, p. 89–109.

Houston, R. S., 1971, Regional tectonics of the Precambrian rocks of the Wyoming Province and its relationship to Laramide structure, *in* A. R. Renfro, ed., Wyoming tectonics and their economic significance: Wyoming Geological Association, p. 19–27.

———, ed., 1978, A regional study of rocks of Precambrian age in that part of the Medicine Bow Mountains lying in southeastern Wyoming—with a chapter on the relationship between Precambrian and Laramide structure: Geological Survey of Wyoming Memoir 1, 167 p.

Huebschman, R. P., 1973, Correlation of fine carbonaceous bands across a Precambrian stagnant basin: Journal of Sedimentary Petrology, v. 43, p. 688–699.

Hurley, P. M., and J. R. Rand, 1969, Predrift continental nuclei: Science, v. 164, p. 1229–1242.

Imlay, R. W., 1980, Jurassic paleobiogeography of the conterminous United States in its continental setting: U.S. Geological Survey Professional Paper 1062, 134 p.

James, H. L., and C. E. Hedge, 1980, Age of the basement rocks of southwest Montana: Geological Society of America Bulletin, v. 91, p. 11–15.

Jensen, F. S., and K. P. Carlson, 1972, The Tyler Formation in the subsurface, central Montana, *in* Geologic atlas of the Rocky Mountain region: Rocky Mountain Association of Geologists, p. 128–130.

Johnson, J. G., 1971, Timing and coordination of orogenic, epeirogenic, and eustatic events: Geological Society of America Bulletin, v. 82, p. 3263–3298.

Kanasewich, E. R., 1966, Deep crustal structure under the plains and Rocky Mountains: Canadian Journal of Earth Sciences, v. 3, p. 937–945.

Kanasewich, E. R., R. M. Clowes, and C. H. McCloughan, 1969, A buried Precambrian rift in western Canada: Tectonophysics, v. 8, p. 513–527.

Kent, H. C., and K. W. Porter, eds., 1980, Colorado geology: Rocky Mountain Association of Geologists, 258 p.

Ketner, K. B., 1983, Reply to mid-Paleozoic age of the Roberts thrust unsettled by new data from northern Nevada: Geology, v. 11, p. 618.

King, P. B., 1976, Precambrian geology of the United States: an explanatory text to accompany the geologic map of the United States: U.S. Geological Survey Professional Paper 902, 85 p.

Kinney, D. M., and P. T. Flawn, 1967, Basement map of North America: AAPG and U.S. Geological Survey, scale 1:5,000,000.

Kluth, C. F., and P. J. Coney, 1983, Reply to plate tectonics of the Ancestral Rocky Mountains: Geology, v. 11, p. 121-122.

Koch, W. J., 1976, Lower Triassic facies in the vicinity of the Cordilleran hingelines: western Wyoming, southeastern Idaho, and Utah, *in* J. G. Hill, ed., Geology of the cordilleran hingeline: Denver, Rocky Mountain Association of Geologists Symposium, p. 203–218.

Lewry, J. F., and T. I. I. Sibbald, 1980, Thermotectonic evolution of the Churchill Province in northern Saskatchewan: Tectonophysics, v. 68, p. 45–82.

Lidiak, E. G., 1971, Buried Precambrian rocks of South Dakota: Geological Society of America Bulletin, v. 82, p. 1411–1420.

———, 1972, Precambrian rocks in the subsurface of Nebraska: Nebraska Geological Survey Bulletin, v. 26, 41 p.

Lidiak, E. G., R. F. Marvin, H. H. Thomas, and M. N. Bass, 1966, Geochronology of the midcontinent region, United States, 4, eastern area: Journal of Geophysical Research, v. 71, p. 5427–5438.

Lochman-Balk, C., 1972, Cambrian system, *in* Geologic atlas of the Rocky Mountain Region: Rocky Mountain Association of Geologists, p. 60–75.

Loucks, G. G., 1977, Geologic history of the Devonian, northern Alberta to southwest Arizona, *in* Rocky Mountain thrust belt geology and resources: Wyoming Geological Association Guidebook, p. 119–134.

Love, J. D., and A. C. Christiansen, 1980, Preliminary correlation of stratigraphic units used on 1° x 2° geologic quadrangle maps of Wyoming, *in* Stratigraphy of Wyoming: Wyoming Geological Association, p. 279–282.

Love, J. D., P. O. McGrew, and H. D. Thomas, 1963,

Relationship of latest Cretaceous and Tertiary deposition and deformation to oil and gas in Wyoming: AAPG Memoir 2, p. 196–208.

Lyons, P. L., and N. W. O'Hara, 1982, Gravity anomaly map of the United States, exclusive of Alaska and Hawaii: Society of Exploration Geophysicists, scale 1:2,500,000.

MacLachlan, M. E., 1972, Triassic system, *in* Geologic atlas of the Rocky Mountain Region: Rocky Mountain Association of Geologists, p. 166–176.

Mallory, W. W., 1972a, Regional synthesis of the Pennsylvanian System, *in* Geologic atlas of the Rocky Mountain Region: Rocky Mountain Association of Geologists, p. 111–127.

———, ed., 1972b, Geologic atlas of the Rocky Mountain region: Rocky Mountain Association of Geologists, 331 p.

Maughan, E. K., 1980, Permian and lower Triassic geology of Colorado, *in* H. C. Kent, and K. W. Porter, eds., Colorado Geology: Rocky Mountain Association of Geologists Guidebook, p. 103–110.

———, 1984, Paleogeographic setting of Pennsylvanian Tyler Formation and relation to underlying Mississippian rocks in Montana and North Dakota: AAPG Bulletin, v. 68, p. 178–195.

Maughan, E. K., and A. E. Roberts, 1967, Big Snowy and Amsden groups and the Mississippian–Pennsylvanian boundary in Montana: U.S. Geological Survey Professional Paper 554-B, 27 p.

McCrossan, R. G., and J. W. Porter, 1973, The geology and petroleum potential of the Canadian sedimentary basins—a synthesis, *in* R. G. McCrossan, ed., The future petroleum provinces of Canada: Canadian Society of Petroleum Geologists Memoir 1, p. 589–720.

McDonald, R. E., 1972, Eocene and Paleocene rocks of the southern and central basins, *in* Geologic atlas of the Rocky Mountain region: Rocky Mountain Association of Geologists, p. 243–256.

McGookey, D. P., 1972, Cretaceous system, *in* Geologic atlas of the Rocky Mountain Region: Rocky Mountain Association of Geologists, p. 190–228.

McKee, E. D., and S. S. Oriel, 1967, Paleotectonic maps of the Permian System: U.S. Geological Survey Professional Paper 515, 271 p.

McMannis, W. J., 1963, LaHood Formation—a coarse facies of the Belt Series in southwestern Montana: Geological Society of America Bulletin, v. 74, p. 407–436.

McMechan, M. E., 1981, The middle Proterozoic Purcell Supergroup in the southeastern Rocky and southeastern Purcell Mountains, British Columbia and the initiation of the cordilleran miogeocline, southern Canada and adjacent United States: Bulletin of Canadian Petroleum Geology, v. 29, p. 583–621.

Monger, J. W. H., R. A. Price, and D. J. Tempelman-Kluit, 1982, Tectonic accretion and the origin of the two major metamorphic and platonic welts in the Canadian Cordillera: Geology, v. 10, p. 70.

Muehlberger, W. R., 1980, The shape of North America during the Precambrian, *in* Continental Tectonics: National Academy of Sciences, Washington, D.C., p. 175–183.

Obradovich, J. D., R. E. Zartman, and Z. E. Peterman, 1984, Update of the geochronology of the Belt Supergroup, *in*

S. W. Hobbs, ed., Belt Symposium II: Montana Bureau of Mines and Geology Special Publication 90, p. 82–84.

O'Neill, J. M., and D. A. Lopez, 1985, Character and regional significance of Great Falls Tectonic Zone, east-central Idaho and west-central Montana: AAPG Bulletin, v. 69, p. 437–447.

Peterman, Z. E., 1979, Geochronology and the Archean of the United States: Economic Geology, v. 74, p. 1544–1562.

———, 1981a, Archean gneisses in the Little Rocky Mountains, Montana, *in* Shorter contributions to isotope research in the Western United States: U.S. Geological Survey Professional Paper 1199-A, p. 1–6.

———, 1981b, Dating of Archean basement in northeastern Wyoming and southern Montana: Geological Society of America Bulletin, v. 92, p. 139–146.

Peterman, Z. E., and C. E. Hedge, 1964, Age of basement rocks from the Williston Basin of North Dakota and adjacent areas: U.S. Geological Survey Professional Paper 475-D, p. 100–104.

Peterman, Z. E., and R. A. Hildreth, 1978, Reconnaissance geology and geochronology of the Precambrian of the Granite Mountains, Wyoming: U.S. Geological Survey Professional Paper 1055, 22 p.

Peterman, Z. E., and S. S. Goldich, 1982, Archean rocks of the Churchill basement, Williston Basin, North Dakota, *in* Fourth International Symposium on Williston Basin: Saskatchewan Geological Society Special Publication 6, p. 11–12.

Peterson, J. A., 1981, General stratigraphy and regional paleostructure of the western Montana overthrust belt, *in* T. E. Tucker, ed., Southwest Montana Field Conference and Symposium, Montana Geological Society, p. 5–35.

Picard, M. D., A. Aadland, and L. R. High, 1969, Correlation and stratigraphy of the Red Peak and Thaynes formations, western Wyoming and adjacent Idaho: AAPG Bulletin, v. 53, p. 2274–2289.

Piper, J. D. A., 1982, The Precambrian paleomagnetic record—The case for the Proterozoic super-continent: Earth and Planetary Science Letters, v. 59, p. 61–89.

Pipiringos, G. N., and R. B. O'Sullivan, 1978, Principal unconformities in Triassic and Jurassic rocks, western interior United States—a preliminary survey: U.S. Geological Survey Professional Paper 1035-A, 29 p.

Poole, F. G., and C. A. Sandberg, 1977, Mississippian paleogeography and tectonics of the western United States, *in* Paleozoic paleogeography of the western United States: Society of Economic Paleontologists and Mineralogists Pacific Section, Pacific Coast Paleogeography Symposium, p. 67–85.

Reed, J. C., Jr., and R. E. Zartman, 1973, Geochronology of Precambrian rocks of the Teton Range, Wyoming: Geological Society of America Bulletin, v. 84, p. 561–582.

Reynolds, M. W., and D. A. Lindsey, 1979, Eastern extent of the Proterozoic Y Belt Basin, Montana: U.S. Geological Survey Professional Paper 1150, p. 68.

Richardson, G. B., 1913, The Paleozoic section in northern Utah: American Journal of Science, v. 36, p. 406–416.

Roberts, A. E., 1979, Northern Rocky Mountains and adjacent Plains region, Chapter N, *in* Paleotectonic investigations of the Mississippian System in the United States: U.S. Geological Survey Professional Paper 1010, pt. 1, p. 249–271.

Ruppel, E. T., and D. A. Lopez, 1984, The thrust belt in southwest Montana and east-central Idaho: U.S. Geological Survey Professional Paper 1278, 41 p.

Sandberg, C. A., 1963, Dark shale unit of Devonian and Mississippian age in northern Wyoming and southern Montana: U.S. Geological Survey Professional Paper 475-C, p. 17–20.

Sando, W. J., 1976, Mississippian history of the northern Rocky Mountains region: U.S. Geological Survey Journal of Research, v. 4, p. 317–338.

Saperstone, H. I., and F. G. Ethridge, 1984, Origin and paleotectonic setting of the Pennsylvanian Quadrant sandstone, southwestern Montana, *in* J. Goolsby, and D. Morton, eds., The Permian and Pennsylvanian Geology of Wyoming: Wyoming Geological Association Guidebook, p. 309–331.

Schmidt, C. J., and T. E. Hendrix, 1981, Tectonic controls for thrust belt and Rocky Mountain foreland structures in the northern Tobacco Root Mountains–Jefferson Canyon area, southwestern Montana, *in* T. E. Tucker, ed., Southwest Montana Field Conference and Symposium: Montana Geological Society, p. 167–180.

Sears, J. W., and R. A. Price, 1978, The Siberian connection: a case for Precambrian separation of the North American and Siberian cratons: Geology, v. 6, p. 267–270.

Sears, J. W., P. J. Graff, and G. S. Holden, 1982, Tectonic evolution of lower Proterozoic rocks, Uinta Mountains, Utah and Colorado: Geological Society of America Bulletin, v. 93, p. 990–997.

Sheldon, R. P., 1967, Long-distance migration of oil in Wyoming: The Mountain Geologist, v. 4, p. 53–65.

Slack, P. B., 1981, Paleotectonics and hydrocarbon accumulation, Powder River Basin, Wyoming: AAPG Bulletin, v. 65, p. 730–743.

Sleep, N. H., 1976, Platform subsidence mechanisms and "eustatic" sea-level changes: Tectonophysics, v. 36, p. 45–56.

Sloss, L. L., 1963, Sequences in the cratonic interior of North America: Geological Society of America Bulletin, v. 74, p. 93–114.

———, 1984, Comparative anatomy of cratonic unconformities, *in* J. S. Schlee, ed., Interregional unconformities and hydrocarbon accumulation: AAPG Memoir 36, p. 1–6.

Smith, D. L., 1982, Controls on carbonate accumulation in the shelf to basin transition, Lodgepole Formation, central and south-central Montana, *in* J. E. Christopher, and J. Kaldi, eds., Fourth International Williston Basin Symposium, Saskatchewan Geological Society, p. 245–246.

Smith, R. B., 1978, Seismicity, crustal structure, and intraplate tectonics of the interior of the western Cordillera, *in* R. B. Smith, and G. P. Eaton, eds., Cenozoic tectonics and regional geophysical of the western Cordillera: Geological Society of America Memoir 152, p. 111–144.

Stewart, J. H., 1972, Initial deposits in the Cordilleran geosyncline: evidence of a late Precambrian (850 m.y.) continental separation: Geological Society of America Bulletin, v. 83, p. 1345.

———, 1976, Late Precambrian evolution of North America: Plate tectonics implication: Geology, v. 4, p. 11–15.

Stone, D. S., 1967, Accumulation theory, Big Horn Basin: AAPG Bulletin, v. 51, p. 2056–2114.

Tweto, O., 1980, Precambrian geology of Colorado, *in* H. C. Kent, and K. W. Porter, eds., Colorado Geology: Denver, Rocky Mountain Association of Geologists, p. 37–46.

Vail, P. R., R. M. Mitchum, and S. Thompson, 1977, Seismic stratigraphy and global changes of sea level, Part 4: Global cycles of relative changes of sea level, *in* C. E. Payton, ed., Seismic stratigraphy—applications to hydrocarbon exploration: AAPG Memoir 26, p. 83–97.

Van Gundy, C. E., 1946, Faulting in east part of Grand Canyon of Arizona: AAPG Bulletin, v. 30, p. 1899–1909.

Van Schmus, W. R., and M. E. Bickford, 1981, Proterozoic chronology and evolution of the mid-continent region; North America, *in* A. Kroner, ed., Precambrian plate tectonics: Elsevier, Amsterdam/New York, p. 261–296.

Walcott, C. D., 1889, Study of a line of displacement in the Grand Canyon in northern Arizona: Geological Society of America Bulletin, v. 1, n. 1, p. 49–64.

Wallace, C. A., 1972, A basin analysis of the upper Precambrian Uinta Mountain Group, Utah: Ph.D. thesis, University of California, Santa Barbara, 412 p.

Walton, P. T., 1946, Ellis, Amsden, and Big Snowy Group, Judith Basin, Montana: AAPG Bulletin, v. 30, p. 1294–1305.

Warner, L. A., 1978, The Colorado Lineament: a middle Precambrian wrench fault system: Geological Society of America Bulletin, v. 89, p. 161–171.

Weimer, R. J., 1984, Relation of unconformities, tectonics, and sea-level changes, Cretaceous of western interior, U.S.A., *in* J. S. Schlee, ed., Interregional unconformities and hydrocarbon accumulation: AAPG Memoir 36, p. 7–35.

Wilson, M. L., 1981, Origin of Archean lithologies in the southern Tobacco Root and northern Ruby Ranges of southwestern Montana, *in* T. E. Tucker, ed., Southwest Montana Field Conference and Symposium: Montana Geological Society, p. 37–44.

Windley, B. F., ed., 1976, The early history of the earth: John Wiley, New York, 619 p.

———, ed., 1977, The evolving continents: John Wiley, New York.

Woodward, L. A., 1970, Time of emplacement of Pinto Diorite, Little Belt Mountains, Montana: Earth Science Bulletin, v. 3, p. 15–26.

Zartman, R. E., J. J. Norton, and T. W. Stern, 1964, Ancient granite gneiss in the Black Hills, South Dakota: Science, v. 145, p. 479–481.

Zietz, I., 1982, Composite magnetic anomaly map of the United States, Part A: conterminous United States: U.S. Geological Survey Map GP-954-A, scale 1:2,500,000.

Lineaments and Their Tectonic Implications in the Rocky Mountains and Adjacent Plains Region

E. K. Maughan
W. J. Perry, Jr.
U. S. Geological Survey
Denver, Colorado

Two orthogonal systems of lineaments reflect recurrent structural movement in the basement rocks of the Rocky Mountains and adjacent plains—the area of the middle Phanerozoic Cordilleran continental shelf. The shelf lay between the Transcontinental arch and the Cordilleran continental margin (or miogeocline) from Arizona and New Mexico to Montana and North Dakota. Major tectonic features that affected middle Phanerozoic sedimentation on this shelf were the Big Snowy trough and the Williston basin, the Ancestral Rocky Mountains, and the Zuni uplift. Some of the boundaries of these and other structural elements in the shelf area seem to be related to an orthogonal system of two perpendicular sets of lineaments, one trending northeast and the other northwest. The first parallels the northeast-trending Cordilleran miogeocline in Nevada, and the other parallels the northwesterly continuation of the miogeocline north of Idaho. The two sets are roughly parallel to the Transcontinental arch and the margin of the Canadian shield, respectively. A second, less important orthogonal system is oriented north–south and east–west.

Differential vertical movements of rectangular basement blocks bounded by the fracture systems were apparently propagated upward through the strata, forming a variety of structures. Movements of the blocks at different times in the Phanerozoic influenced erosion, deposition, and lithofacies of the sediments and thus influenced the distribution of sedimentary minerals and petroleum deposits and later igneous intrusions, volcanism, and hydrothermal ores. The lineaments appear to have been avenues for passage of hydrothermal solutions, so their identification should help guide exploration for mineral deposits. Tectonic movements in the Phanerozoic seem to have been mainly along the northeast and northwest system of lineaments, but the north–south and east–west system also influenced the formation of Laramide structures and the present landscape in the Rocky Mountains. Deformation associated with the two systems is probably related to events at the North American plate margin or to incipient continental fragmentation of the plate.

INTRODUCTION

A relationship between rectilinear structural, depositional, and topographic features in the Rocky Mountains and adjacent plains has been noted by many geologists. Most previous workers have shown evidence for an orthogonal system of lineaments that trends approximately east–west and north–south (N 75–85° W and N 5–15° E) and is related to Laramide structural features. A few others have indicated an orthogonal system trending northwest and northeast (N 30–40° W and N 50–60° E) locally within the region. Most of these workers either have shown or inferred a relationship of the lineaments to recurrent movements of Precambrian basement structures.

The Phanerozoic sedimentary rocks of this region comprise deposits that formed in response to epeirogenic movements commonly associated with a tectonic fabric inherited from older structures. The tectonic fabric of this part of the North American craton in the northern Rocky Mountain region and adjacent areas apparently consists of subrectangular crustal blocks bounded by approximately orthogonal fracture systems that follow the same trends as the lineament systems just described. Flexing of the cratonic crust during Phanerozoic time seems to have been accommodated by recurrent vertical and lateral movement along these ancient fracture systems in the Precambrian basement. These tectonic responses and their effects on the Phanerozoic strata shifted through time and varied in intensity. Comprehensive documentation of these ancient fractures and their genesis is beyond the scope of this paper. Our purpose here is to identify the major lineaments as a step toward understanding the distribution of ancient crustal fracture systems and related geologic structures that influenced the deposition of sediments and preservation of rocks in the Rocky Mountains and adjacent plains region.

Linear structural trends that affected rocks throughout the region are indicated by the alignment of a variety of paleotopographic and geologic features that seem to have been tectonically active at various times during the middle Phanerozoic (Late Devonian–Triassic) (Maughan, 1983). Primarily, these lineaments are indicated in middle Phanerozoic rocks by linear depositional limits, aligned changes in thickness and lithofacies, and erosional truncation or depositional thinning along linear trends. Conclusive evidence that these trends relate to fracture zones in basement rock is elusive, however, and the existence of some of these features remains uncertain. Indeed, the identification of some of the lineaments may satisfy several of the 32 Rabelaisian rules for Linesmanship (linear "glo-art") elucidated by Wise (1982). Nevertheless, deposition, facies, and preservation of older Cambrian–Middle Devonian) and younger (post-Triassic) Phanerozoic rocks also seem to have been affected along many of the same lineaments as those identified in the middle Phanerozoic rocks. Furthermore, many Laramide structures, sedimentary deposits, volcanic vents, and present-day drainage courses are coincidentally aligned.

The term *lineament* was first used by Hobbs in 1904 to describe linear geographic features, such as streams, dry valleys, ravines, straight coast lines, visible lines of fractures, zones of fault breccias, and boundaries between formations that aligned with each other. In 1912 (p. 227), Hobbs wrote:

> These significant lines of landscapes, which reveal the hidden architecture of the rock basement, are described as *lineaments*.... It is important to emphasize the essentially composite expression of the lineament...; but in every case it is some surface expression of a buried fracture.

Hobbs' characteristics of lineaments have been reiterated more recently in the definition by O'Leary et al. (1976). We use this definition except that we extend the concept to include paleotopographic and paleogeologic features wherever they can be discerned in the subsurface as well as at the surface.

The definition of lineament just given contrasts with a common corruption of the term, which applies it to broad bands of structures. The Montana lineament (or Lewis and Clark line), Texas lineament, and Colorado lineament are broad linear bands of structural and related topographic features similar to those of the shear zone in western Nevada known as the Walker lane. We therefore define the term *lane* as it was applied by Locke et al. (1940), to mean a broad, linear structural zone or geomorphic zone. The term does not necessarily imply wrench fault tectonics, although strike-slip faulting is commonly associated with these zones. Because this term seems appropriate for wide and composite linear features, we show these broader lines and lineaments on Figure 1 as the Walker lane, Lewis and Clark lane, Texas lane, and Colorado lane. These structural lanes seem to be bounded by major lineaments (in the strictest sense), which are narrow, rectilinear to slightly curvilinear bands that incorporate the features outlined by Hobbs (1904) and by O'Leary et al. (1976) as described above.

PREVIOUS WORK

Many authors have related structures in the sedimentary cover to the reactivation of lines of weakness in the Precambrian basement. However, most of these studies, such as the one by Thomas (1974), have been concerned chiefly with the interpretation of Laramide features. Another example is Chamberlin's (1945) interpretation of Laramide structures in northern Wyoming and southern Montana; he noted rectangular patterns outlined by structural contour lines and suggested that basic control was by intersecting sets of steeply dipping shear planes in the basement. Evidence for the association of Laramide drape folding and basement block faulting in the Wyoming structural province was presented in a volume devoted to this topic (Matthews, 1978). Some examples of geologic reports dealing with the reactivation of Precambrian basement faults in pre-Laramide Phanerozoic time include the following. Thom (1952) recognized reactivation of Precambrian basement faults by thickness variations in Paleozoic through Lower Cretaceous rocks, but did not identify specific faults or other structural features. Other examples of a relationship between middle Phanerozoic rocks and reactivated basement faults or inferred faults have been locally described for central Montana by Cooper (1956), for western South Dakota and adjacent areas by Maughan (1966), for northeastern Wyoming and adjacent areas by Brown (1978), for western Wyoming and adjacent areas by Peterson (1981, 1985), for western Nebraska by Moore and Nelson (1974), and for southwestern Colorado by Weimer (1980). Baars and Stevenson (1981, and Stevenson and Baars, this volume) have shown a relationship between Precambrian structures and lineaments in Paleozoic and Mesozoic rocks in southwestern Colorado and adjacent areas in Utah. Sonnenberg and Weimer (1981) note the control on sedimentation and preservation of Paleozoic and Mesozoic rocks by northwest and northeast linear trends on the east flank of the Colorado Front Range and in the adjacent Denver basin. Schmidt and Hendrix (1981) recognize northwest-trending Laramide faults in southwestern Montana that have a Precambrian ancestry. Erslev (1982) has identified a northeast-trending mylonite zone in Precambrian rocks of southwestern Montana that is nearly parallel with the Snowcrest trough (Figure 1). This zone is a major, linear structural discontinuity that is coincident with a northeast-trending lineament and parallel with the trough and other lineaments described below.

MAJOR TECTONIC ELEMENTS

The major tectonic elements in the northern Rocky Mountain region are illustrated in Figure 1. West of the Rocky Mountain region the axis of the Cordilleran miogeosynclinal trough, shown by Roberts and Thomasson (1964), approximately parallels the Paleozoic continental margin (miogeocline) as interpreted from strontium isotopes by Kistler and Peterman (1973). In Nevada, the geosynclinal axial trend is about N 40° E, and it makes an approximately right angle bend in southern Idaho to about N 30° W in central and northern Idaho. East of the Rocky Mountain region the northeast-trending Transcontinental arch and the northwest-trending curvilinear edge of the Precambrian exposures of the Canadian shield north of the Williston basin are approximately parallel with the

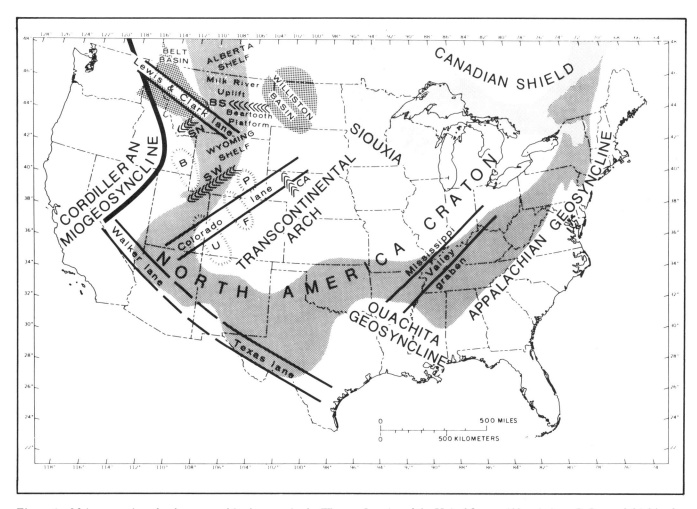

Figure 1—Major tectonic and paleogeographic elements in the Western Interior of the United States. Abbreviations: B, Bannock highland; BS, Big Snowy trough; CA, Chadron arch; F, ancestral Front Range uplift; P, Pathfinder uplift; SN, Snowcrest trough; SW, Sweetwater trough; U, ancestral Uncompahgre uplift; and Z, Zuni uplift. The light shaded area is the margin of Precambrian rocks in the Canadian shield; the dark shaded area is the outer margin of the North American craton covered by sedimentary rocks. Modifed from Mallory (1972a). Used with permission of the Rocky Mountain Association of Geologists.

northeast and northwest legs, respectively, of the miogeosynclinal axis. The western continental shelf of North America (Cordilleran shelf) lay between the miogeocline and the Canadian shield and Transcontinental arch.

Harrison et al. (1974) showed the Precambrian Belt basin to project into the craton eastward from Idaho, through western Montana, and toward the intracratonic Williston basin. The Central Montana trough developed along this trend in early Paleozoic time (Peterson, 1981), and the Central Montana uplift was a Middle Devonian feature in the same area that had roughly the same trend. The Big Snowy trough, which developed along this trend during Mississippian–Pennsylvanian time, divided the Alberta shelf on the north from the Wyoming shelf on the south (Maughan, 1984). The Snowcrest trough in southwestern Montana (Maughan and Perry, 1982; Saperstone and Ethridge, 1984) and the Sweetwater trough (Mallory,

1972b), of somewhat lesser magnitude in northeastern Utah and south-central Wyoming, projected northeastward into the craton. Carboniferous rocks along the Sweetwater trough are thicker than those on adjacent areas of the shelf.

The Ancestral Rocky Mountain system is made up of a few northwest-trending uplifts that projected from the northwest flank of the Transcontinental arch to the vicinity of the Sweetgrass trough. The ancestral Uncompahgre and the ancestral Front Range are the principal uplifts of the Ancestral Rocky Mountain system (Figure 1). The Pathfinder uplift and the Chadron arch are included as part of this system because they similarly influenced the deposition and the extent of preservation of the middle Phanerozoic rocks adjacent to the ancestral Uncompahgre and Front Range.

Farther north, the Bannock highland, Milk River uplift, and Beartooth platform were other tectonically positive areas at times during late Paleozoic and early Mesozoic time.

LINEAMENTS

Many structures of the Rocky Mountains and adjacent parts of the Great Plains region seem to be related to two systems, each composed of approximately perpendicular sets of lineaments. One system trends approximately N 30–40°W and N 50–60° E, and the other, which appears to be a secondary system, trends approximately N 5–15° E and N 75–85° W. The major lineaments, shown in Figure 2, have been identified in this report by the characteristics proposed by Hobbs (1904, 1912) and reiterated by O'Leary et al. (1976). For the most part, these lineaments probably result from recurrent movement along ancient faults or major fracture systems in the crystalline basement rocks that lie beneath the Paleozoic rocks of the shelf (Prucha et al., 1965; Maughan, 1966).

Northwest and Northeast System

The northeast-trending set of lineaments is approximately parallel with the northeast trend of the miogeocline in Nevada and with the Transcontinental arch in Nebraska, Colorado, and New Mexico. Similarly, the northwest-trending lineaments are approximately parallel with the miogeocline in Idaho and with the edge of the Canadian shield.

A major northeast structural trend was originally noted as the Wyoming lineament by Ransome (1915, p. 294–295). He applied the name to:

(a) great transverse line . . . defined by (1) the break in the continuity of the Laramide System in central Wyoming, (2) the south end of the Wasatch Range, (3) the northwest edge of the Colorado Plateaus, (4) the south end of the Sierra Nevada, and (5) the notable change of trend of the coastal ranges between the Mohave Desert and Point Conception. It may be worthy of note that the Black Hills of South Dakota lie on the same line. This great line or zone, which may conveniently be referred to as the *Wyoming lineament*, runs nearly northeast and southwest.

Blackstone (1956) depicted a "Wyoming lineament" across the Laramie Range about 60 km southeast of the lineament indicated by Ransome (1915). The continuation of Blackstone's lineament to the northeast of the Laramie Range passes along the northwest flank of the Hartville uplift, and it seems to continue to the southern flank of the Black Hills. To the southwest, Blackstone's Wyoming lineament passes through the Hanna basin and along the north edge of the Medicine Bow Mountains and the Sierra Madre.

Data collected in southeastern Wyoming subsequent to the observations of Blackstone have indicated the continental importance of a narrow linear zone yet farther southeast and about 140 km southeast of Ransome's Wyoming lineament. Named the Sybille lineament (Maughan, 1983), it is especially notable because it is the boundary between Archean rocks about 2.4 b.y. old north of Sybille Creek and Proterozoic rocks that were emplaced about 1.4 b.y. ago south of Sybille Creek in the Laramie Range (Condie, 1969). This Archean–Proterozoic boundary aligns southeastward with the Mullen Creek–Nash Fork shear zone in the Medicine Bow Mountains and in the Sierra Madre (Houston et al., 1968). Northeastward it aligns with

northeast-trending faults along the southeast margin of the Hartville uplift and with the southeast margin of the Black Hills. The magnitude of this lineament is indicated by its probable northeast continuation through South Dakota (Shurr, 1979) to the Lake Superior region as indicated by Warner (1978).

The Wyoming lineament indicated by Maughan (1967) is a broad, diffuse zone between Ransome's line and the Sybille Creek trend that influenced the areal extent and facies of Permian rocks in southeastern Wyoming. This zone was also an important structural platform that subtly controlled the topography that defined one edge of the shallow Alliance basin and halite deposition in that basin during Permian time (Maughan, 1966).

A northeastward extension of the Colorado mineral belt (Tweto and Sims, 1963) from the Front Range, approximately parallel to the trend of the various northeast-trending Wyoming lineaments, was also important in the control of Permian salt deposition. Designated the Front Range mineral belt trend by Maughan (1966), minor tectonic movement along this lineament seems to have determined topography that controlled the southeast margin of the Alliance basin along the northeast extension of this lineament. The Front Range mineral belt lineament includes the structural zone that Warner (1978) indicated to have localized the Idaho Springs–Ralston shear zone. Sonnenberg and Weimer (1981) noted the relationship of the mineral belt lineament to the accumulation of oil in certain fields of the Denver basin.

A Precambrian wrench fault system was proposed by Warner (1978) for the Colorado lane; Precambrian strike-slip movements along this structural zone were compared by him to Phanerozoic (primarily Mesozoic and Cenozoic) longitudinal wrench fault systems. All of the northeast-trending lineaments identified in this paper (Figure 2) probably have a common Precambrian origin. The contrast in age between Archean and Proterozoic terrains across the Mullen Creek–Nash Fork shear zone strongly suggests that the Sybille lineament had been a crustal boundary and that it had the greatest amount of strike-slip movement in the region, if the origin of the lineaments is correctly related to a wrench fault system. The Front Range mineral belt lineament and other northeast-trending lineaments identified here also may have originated by strike-slip faulting similar to that along the Sybille lineament, but of somewhat lesser magnitude. In contrast to the presumed strike-slip Precambrian movements that occurred along some of the basement-involved structural lineaments, the effects on middle Phanerozoic rocks seem to have resulted primarily from vertical tectonics.

The Mullen Creek–Nash Fork shear zone and the Idaho Springs–Ralston shear zone bound a northeast-trending complex of Precambrian structures that Warner (1978) designated as the Colorado lineament; we call it the Colorado lane. This approximately 160-km-wide zone is not a lineament in the sense of Hobbs (1904), but is a structurally complex belt bounded by the Sybille and the Front Range mineral belt lineaments. Shear zones and other structural features within the Colorado lane, and especially along its margins, are interpreted by Warner as elements of a middle Precambrian wrench fault system. The Colorado

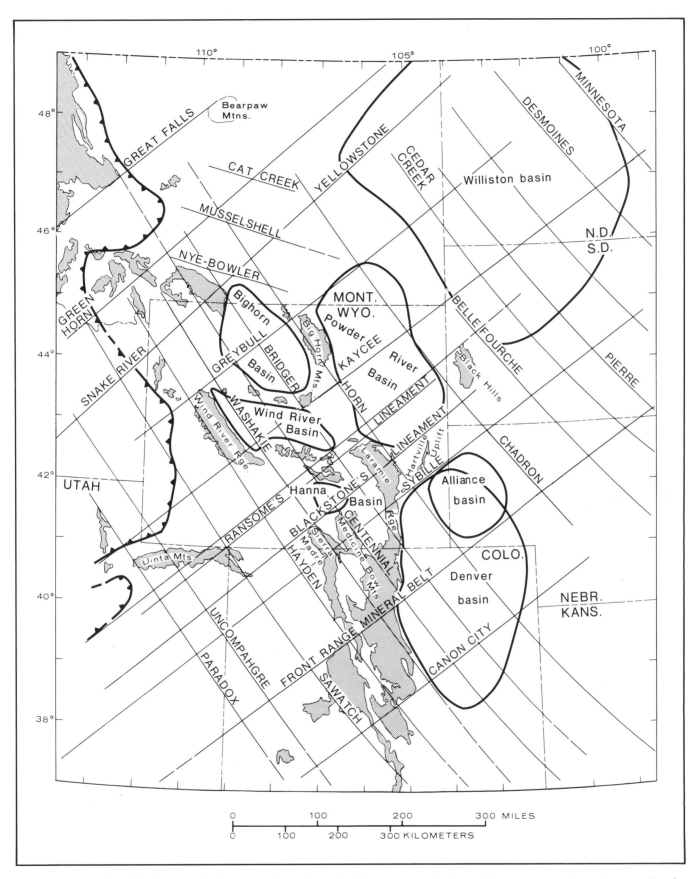

Figure 2—Principal middle Phanerozoic lineaments shown in relationship to present-day Precambrian exposures and to places mentioned in the text. Barbed lines are leading edges of Cordilleran thrusts; barbs point into allochthonous terrane. Lineaments and faults are dashed where inferred. From Maughan (1983). Used with permission of the Denver Region Exploration Geologists Society.

lane is shown to extend southwestward to the Grand Canyon region in Arizona and northeastward along the projection of the Sybille lineament probably to the Lake Superior region (Warner, 1978).

Other northeasterly linear structural trends on the Wyoming shelf are indicated by the spatial and temporal alignment of a variety of topographic and geologic features. On Figure 2 we identify the Kaycee lineament (designated as a "structural trend" by Maughan, 1966), the Greybull lineament here named for the approximate coincidence of the apparent structural zone aligned with the northeast-flowing Greybull River near the center of the Big Horn basin, and the northeast-trending Snake River–Yellowstone lineament (Maughan, 1983). Several additional northeast-trending lineaments occur southeast of the Front Range mineral belt lineament, but only one of these, the Canon City lineament, is shown in Figure 2.

The somewhat regular spacing of the major northeast lineaments may reflect fundamental properties of the crystalline basement complex. The NE fracture system apparently is not restricted to the Wyoming shelf. The NE lineaments seem to extend across the Colorado Plateaus west of the Rocky Mountain region. Others extend across the ancient continental shelf in the central United States as far east as the Appalachian orogenic belt as indicated by somewhat similar northeasterly linear topographic and geologic trends in that region. Conspicuous in the central United States is the northeasterly trend of the Mississippi Valley graben and related faults (Hildenbrand, 1982).

In western Montana, two major lineaments have been identified that have an approximately northeast trend. The Greenhorn lineament (Maughan and Perry, 1982) trends N 45° E, as does the Great Falls lineament (O'Neill and Lopez, 1983), which extends from north-central Idaho to the vicinity of the Bearpaw Mountains in north-central Montana.

A set of northwest-trending lineaments on the Wyoming shelf is approximately perpendicular to the northeasterly set described above. The northwest oriented set (identified in Figure 2) is inferred from depositional and erosional patterns in the middle Phanerozoic rocks similar to those observed along the northeast set. Some of the northwest lineaments are coincident with faults in Precambrian rocks and seem to be related to them, but their origin is more obscure than that of the other set. The northwest lineaments are more sharply defined by pronounced differences in thickness in some of the middle Phanerozoic strata. Furthermore, changes in facies are commonly abrupt along these lineaments, in contrast to the subtle changes in facies noted in equivalent strata along the northeast trends. Uplifts bounded by apparent normal- to high-angle reverse faults of the Ancestral Rocky Mountain system are conspicuously controlled by the northwest set of lineaments and in part define these lineaments. Recurrent movements along some of these lineaments are indicated by their coincidence with both older Precambrian faults and younger Laramide faults.

The lineaments shown in Figure 2 are composites of shorter linear trends evident on the lithofacies and isopach maps of each of the stratigraphic intervals on the U. S. Geological Survey paleotectonic maps of the Mississippian–Triassic Systems (McKee et al., 1959; McKee et al., 1967;

McKee et al., 1975; Craig et al., 1979). Figure 3 is a simplified version of one of these maps (Maughan, 1983) showing facies, thicknesses, and preservation of rocks representing a single interval of the Middle Pennsylvanian (Des Moinesian). The segments of the lineaments evident for each interval are seen to be either superimposed or laterally continuous with segments evident for other intervals of the paleotectonic maps. Segments of the lineaments on the interval maps indicate either syn-depositional or later structural movement that affected the thickness or facies of the rocks in each of these intervals. The accumulation of detrital sediments was commonly controlled by the topography that had developed along the lineaments. Conglomerates were deposited immediately adjacent to steeply uplifted areas (De Voto, 1980). Other clastic sediments were derived from moderately to slightly uplifted areas, and they accumulated as clastic aprons in adjacent marine littoral settings, or they were transported and commonly accumulated in sags or linear depressions that formed as topographic or structural features along some lineaments. Some eolian transport and deposition of sands may have occurred in extensive "sand seas," in which topographic differences due to differential uplift or subsidence localized deposition and created differences in thickness (Saperstone and Ethridge, 1984). Other eolian sand deposits seem to have been localized by orographic wind currents in which the sand transport capacity of the wind was lost against topographic highs (Fryberger, 1979; Maughan, 1980).

North–South and East–West Secondary System

Another system of lineaments secondary to the northeast and northwest system described above trends approximately north–south (N 5–15° E) and east–west (N 75–85° W). This secondary system, which diverges about 45° from the primary system, also seems to have originated in the Precambrian. The Lewis and Clark lane, the Uinta axis, and the Texas lane (Figure 1) are related to aulacogens (Harrison et al., 1974) that extend into the paleocontinental shelf along west–northwest to east–west trends. These approximately east–west lineaments and nearly perpendicular, approximately north–south linear trends are believed to have been influential in shaping the deposition of older Phanerozoic rocks, but evidence for them in pre-Upper Devonian rocks is not readily seen. No presently available maps of Middle Devonian and older Phanerozoic sequences of the Western Interior show adequate details of either thickness or lithofacies to define possible lineaments in these rocks. Subsurface data for these older rocks are sparse and some of the older Phanerozoic stratigraphic units, especially those of the Lower Devonian and Silurian, are absent over large parts of the region.

The Nye–Bowler, the Musselshell, and the Cat Creek lineaments (Figure 2) are conspicuous east–west structural elements in central Montana that are divergent from the principal system described above. These lineaments seem to extend the Lewis and Clark lane (Billingsley and Locke, 1939; Wallace et al., 1960) from the structurally complex areas associated with the Belt basin in northwestern Montana eastward into the craton. These lineaments seem to diminish and come to an end where they impinge on the Wyoming shelf in southeastern Montana, but there is a zone

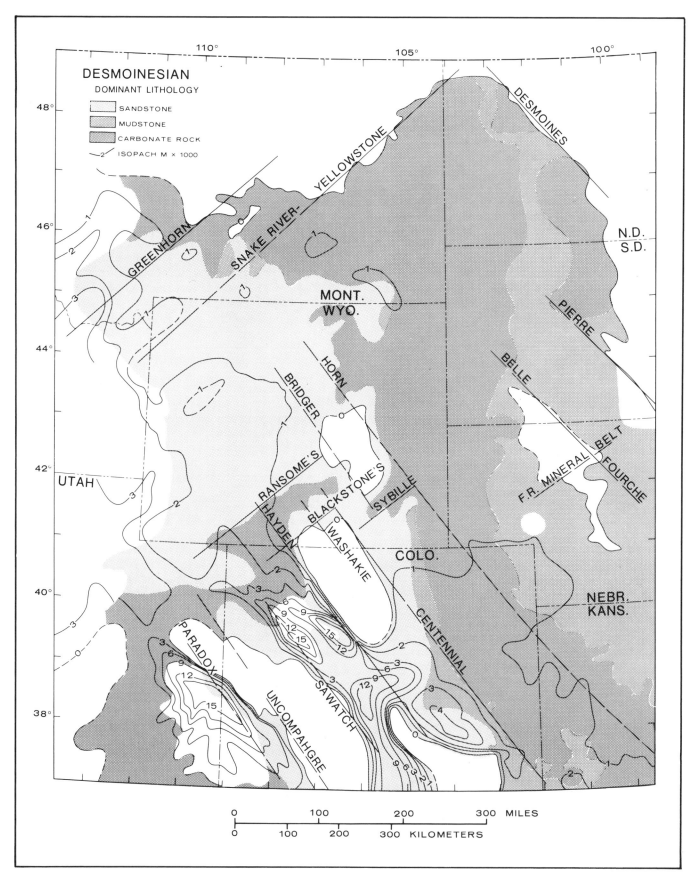

Figure 3—Generalized thickness and lithofacies of upper Middle Pennsylvanian (Des Moinesian) rocks and their relationship to the orthogonal lineaments. From Maughan (1983); modified in part from McKee et al. (1975). Used with permission of the Denver Region Exploration Geologists Society.

in southern Montana where the approximately east–west lineaments and the northeast and northwest fabric of the Wyoming shelf are superimposed. The east–west lineaments seem to have had little effect on depositional patterns in middle Phanerozoic rocks, except for the Devonian central Montana uplift and Yellowstone arch (Sandberg and Mapel, 1967) and the downwarping during Mississippian–Pennsylvanian time that formed the Big Snowy trough in approximately the same area as the central Montana uplift (Peterson, 1981; Maughan, 1984). Approximately east–west lineaments that are subparallel with the Nye–Bowler lineament have been noted in Upper Cretaceous and Cenozoic rocks and in surface features in the Bighorn Mountains and adjacent areas of northeastern Wyoming by Hoppin (1974), but there is no convincing evidence south of Montana to indicate any tectonic influence on lithofacies or thicknesses of the middle Phanerozoic rocks along these lineaments prior to the Laramide orogeny.

Approximately north–south lineaments are evidenced by marked topographic differences along major structural displacements on the geologic map (Figure 4). The roughly north–south-trending eastern flank of the Rocky Mountains in north–central Colorado seems continuous along a lineament that may extend northward along the northwestern edge of the Denver basin. A continuation of this lineament coincides farther north with the Fanny Peak lineament (Hoppin, 1974) from the Hartville uplift to the southern segment of the western flank of the Black Hills. A similar parallel lineament, unnamed on Figure 4, is defined by the eastern termini of several ranges of the Rocky Mountains in south–central Colorado and northern New Mexico. This lineament may continue northward across the Denver basin and define the southern segment of the eastern flank of the Black Hills.

Laramide Effects

Many of the same lineaments evident in the middle Phanerozoic rocks seem to have localized younger structures, especially those of the Laramide orogeny. Superimposing the orthogonal northwest and northeast system of lineaments, which were determined by analysis of the middle Phanerozoic data, onto the geologic map of the United States (Figure 4) shows that the lineaments coincide with the boundaries between many Laramide uplifts and basins, especially in Wyoming. Those that align with the northwest-trending lineaments include the Wind River Range, the Sierra Madre, the eastern and western flanks of the Wind River basin, parts of the Bighorn and Laramie mountains, and the Powder River basin. The Washakie lineament, which had localized the northeast flank of the late Paleozoic ancestral Front Range uplift, seems also to have localized the northeast flanks of the Laramide Sierra Madre and the Wind River Range. Other lineaments, parallel with the Washakie lineament, show similar recurrent vertical displacements. The Cedar Creek lineament in eastern Montana separates coherent blocks that have each been higher or lower than the other at various times. The block northeast of Cedar Creek apparently rose above the southwest block in Late Devonian time (Sandberg and Mapel, 1967); the northeast block was lower in Pennsylvanian and Permian time and was again the higher block in Late Cretaceous time.

Folds and faults bound many of the northwest-oriented Laramide mountain blocks and basins along the major northeast-trending lineaments. For example, the Black Hills are bounded on the southeast by the Sybille lineament and on the northwest by the Kaycee. The Kaycee lineament also coincides with the southeastern flank of the Bighorn Mountains and crosses the southeastern part of the Wind River Mountains at a place where the orientation of many structures changes. The overall trend of the Wind River Mountains also changes at the Kaycee lineament from north-northwest on the northwest side of the lineament to west-northwest on the other side. Other northeast-trending lineaments similarly separate blocks that differ in either orientation or style of structures.

Also notable in reference to Laramide structures is the secondary set of approximately east–west and north–south lineaments. These lineaments are evident in Upper Mississippian and Pennsylvanian rocks in central Montana (Maughan, 1983), and cursory study of older Phanerozoic and Precambrian rocks suggest that there was a structural influence on thicknesses, depositional and erosional limits, and lithofacies of these older rocks by this secondary system, as discussed above.

Many modern structural and topographic features also coincide with the lineaments defined here. The Horn and Cedar Creek lineaments, for example, have been named for their respective coincidence with Laramide structures, the Horn fault on the southeastern flank of the Bighorn Mountains and the Cedar Creek anticline (and fault) in eastern Montana. Tonal contrasts on Landsat images across nearly featureless plains coincide with segments of the Fanny Peak (Hoppin, 1974) and Kaycee lineaments (Figure 5).

The two systems of lineaments define nearly rectangular domains of Precambrian rocks for which the principal sense of Paleozoic movement of associated faults seems to have been vertical. Limited horizontal rotation of some domains at times produced small, lateral faults along some lineaments. Significant shear movements, however, or the incorporation of these faults into a wrench fault system have been suggested by others, and the possibility of major lateral translocation along some of the lineaments should be considered. Extensive shear movements would require the lateral offset of lineaments wherever they cross lineaments that trend normal to them—a relationship that we have not observed in our regional study. Detailed studies of smaller areas, however, such as the one by Stevenson and Baars (this volume), suggest some minor lateral offset of the lineaments. Further study may show significant lateral movement along some lineaments, but we presently conclude that the dominant movements during the Phanerozoic have been vertical. Major zones of crustal weakness such as the Lewis and Clark and Texas lanes have clearly accommodated varying amounts of Phanerozoic strike-slip.

Reversal of movement along faults on lineaments that project east–southeastward from the Lewis and Clark lane in northwestern Montana has been illustrated by Cooper (1956). In Lower Pennsylvanian strata, several normal faults that are downfaulted to the north suggest north–south extension. In contrast, Cretaceous (Laramide) tectonism was compressional and caused reverse slip along these same

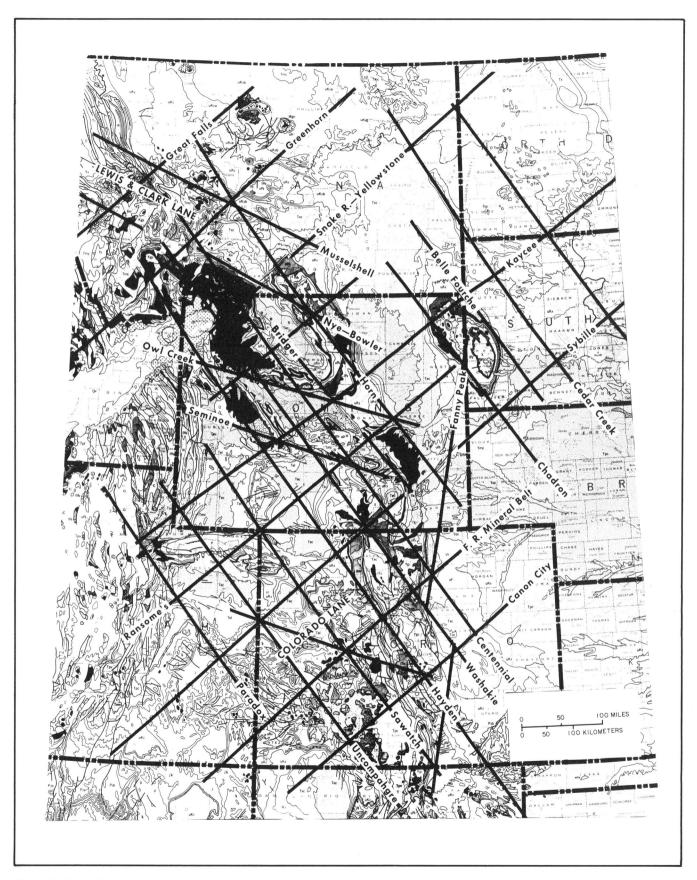

Figure 4—Major lineaments in relationship to the surface geology of Colorado, Wyoming, and parts of adjacent states. Base map from King and Beikman (1974).

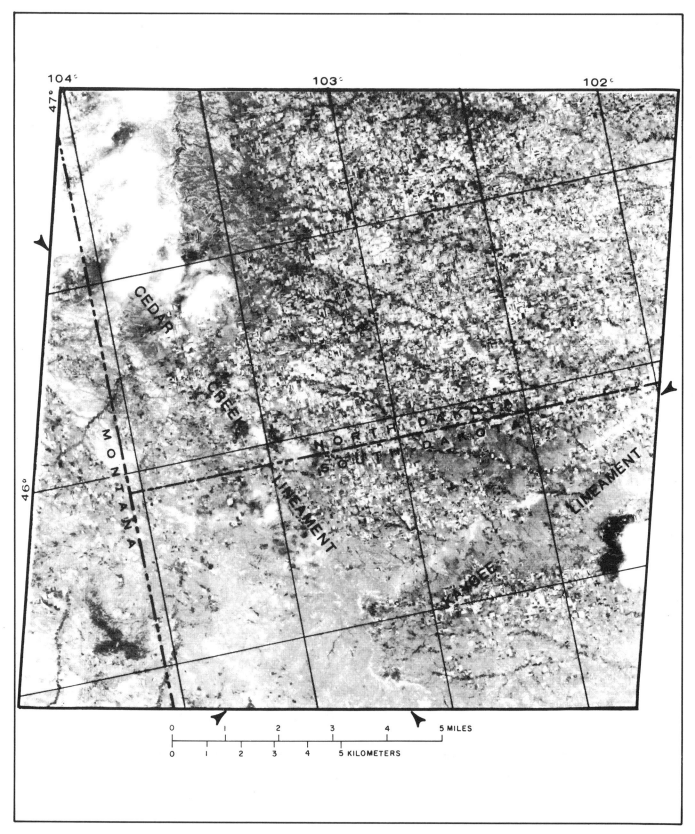

Figure 5—Landsat image of area around the intersection of North Dakota, South Dakota, and Montana showing Kaycee and Cedar Creek lineaments (see arrows). Image was recorded on frequency band 5, June 22, 1973.

faults. As illustrated by Cooper (1956), Laramide movements along many lineament-related faults were the reverse of the late Paleozoic movements.

CONCLUSIONS

Of the many unanswered questions regarding the lineaments described above, perhaps the most important is how they originated. Some important characteristics that may point to an answer include the following.

1. They extend well beyond the region described here, and cursory inspection suggests they they underlie most of North America. In fact, linear, nearly orthogonal fracture systems may be characteristic of continental plates in general.
2. Fragmentary documentation indicates many recurrent structural movements have been concentrated along the lineaments.
3. Motion on the lineaments, whether in response to horizontal extension or compression, seems consistent with apparent stress directions associated with major tectonic events that have resulted from the collisions of continental plates, continental fragmentation, and plate motion.
4. Many if not all of the lineaments represent fracture zones within Precambrian basement rocks.

Is it probable that these fracture zones represent ancient Precambrian fault zones? Might the crossing lineaments have been margins between bands of accretionary crust that formed at a Precambrian oceanic rift and that subsequently became incorporated in the continental plate? P. Hoffman of the Canadian Geological Survey (personal communication, 1982) has noted a remarkably similar pattern of faults and fractures in Precambrian terrain in the Great Slave Lake area, and he has speculated on just such an origin related to crustal accretion and transform faults. Why are there two nearly orthogonal systems in the Rocky Mountains, the northeast and northwest system and the secondary north–south and east–west system?

As we document the times, directions, and places of movements, we attempt to develop a better understanding of the tectonics of plate interiors. Our brief discussion here merely highlights the paleotectonic data available and concurs with many observations that many major lineaments in the Rocky Mountains region have a Precambrian ancestry.

REFERENCES CITED

Baars, D. L., and G. M. Stevenson, 1981, Tectonic evolution of the Paradox basin, Utah and Colorado, *in* D. L. Wiegand, ed., Geology of the Paradox basin: Denver, Rocky Mountain Association of Geologists, p. 23–31.

Billingsley, P. R., and A. Locke, 1939, Structure of ore districts in the continental framework: New York, American Institute of Mining and Metallurgical Engineers, 51 p.

Blackstone, D. L., Jr., 1956, Introduction to the tectonics of the Rocky Mountains: AAPG Rocky Mountain Section, Geological Record, February, p. 3–19.

Brown, D. L., 1978, Wrench-style deformational patterns associated with a meridional stress axis recognized in Paleozoic rocks in parts of Montana, South Dakota, and Wyoming, *in* 24th Annual Conference, The economic geology of the Williston basin, Montana, North Dakota, South Dakota, Saskatchewan, Manitoba: Billings, Montana Geological Society, p. 17–31.

Chamberlin, R. T., 1945, Basement control in Rocky Mountain deformation: American Journal of Science, v. 243-A, p. 98–116.

Condie, K. C., 1969, Petrology and geochemistry of the Laramie batholith and related metamorphic rocks of Precambrian age, eastern Wyoming: Geological Society of America Bulletin, v. 80, p. 57–82.

Cooper, G. G., 1956, The Tyler Formation and basement faulting, *in* Guidebook 7th Annual Field Conference: Billings, Montana, Billings Geological Society, p. 82–85.

Craig, L. C., and C. W. Connor, coordinators, 1979, Paleotectonic investigations of the Mississippian System in the United States: U. S. Geological Survey Professional Paper 1010, pts. I and II, 559 p.; pt. III, 15 plates in pocket.

De Voto, R. H., 1980, Pennsylvanian stratigraphy and history of Colorado, *in* H. C. Kent and K. W. Porter, eds., Colorado geology: Denver, Rocky Mountain Association of Geologists, p. 71–101.

Erslev, E. A., 1982, The Madison mylonite zone; a major shear zone in the Archean basement of southwestern Montana, *in* S. G. Reid, ed., Geology of Yellowstone Park area, 33rd Guidebook: Casper, Wyoming Geological Association, p. 213–221.

Fryberger, S. G., 1979, Eolian-fluviatile (continental) origin of ancient stratigraphic trap for petroleum in Weber Sandstone, Rangely oil field, Colorado: Mountain Geologist, v. 16, n. 1, p. 1–36.

Harrison, J. E., A. B. Griggs, and J. D. Wells, 1974, Tectonic features of the Precambrian Belt basin and their influence on post-Belt structures: U. S. Geological Survey Professional Paper 866, 15 p.

Hildenbrand, T. G., 1982, Model of the southeastern margin of the Mississippi Valley graben near Memphis, Tennessee, from interpretation of truck-magnetometer data: Geology, v. 10, p. 476–480.

Hobbs, W. H., 1904, Lineaments of the Atlantic border region: Geological Society of America Bulletin, v. 15, p. 483–506.

————, 1912, Earth features and their meaning: New York, Macmillan Co., 506 p.

Hoppin, R. A., 1974, Lineaments—their role in tectonics of central Rocky Mountains: AAPG Bulletin, v. 58, n. 11, p. 2260–2273.

Houston, R. S., et al., 1968, A regional study of rocks of Precambrian age in that part of the Medicine Bow Mountains lying in southeastern Wyoming—with a chapter on the relationship between Precambrian and Laramide structure: Wyoming Geological Survey Memoir 1, 167 p.

King, P. B., and H. M. Beikman, 1974, Geologic map of the

United States: U.S. Geological Survey Map, scale 1:2,500,000.

Kistler, R. W., and Z. E. Peterman, 1973, Variations in Sr, Rb, K, Na, and initial Sr^{87}/Sr^{86} in Mesozoic granite rocks and intruded wall rocks in central California: Geological Society of America Bulletin, v. 84, p. 3489–3512.

Locke A., P. R. Billingsley, and E. B. Mayo, 1940, Sierra Nevada tectonic patterns: Geological Society of America Bulletin, v. 51, p. 513–540.

Mallory, W. W., 1972a, Continental setting of the region, *in* W. W. Mallory, ed., Geological atlas of the Rocky Mountain region: Denver, Rocky Mountain Association of Geologists, p. 32–33.

———, 1972b, Regional synthesis of the Pennsylvanian System, *in* W. W. Mallory, ed., Geological atlas of the Rocky Mountain region: Denver, Rocky Mountain Association of Geologists, p. 111–127.

Matthews, V., III, ed., 1978, Laramide folding associated with basement block faulting in the western United States: Geological Society of America Memoir 151, 370 p.

Maughan, E. K., 1966, Environment of deposition of Permian salt in the Williston and Alliance basins, *in* Rau, J. L., ed., Second symposium on salt, v. 1, Geology, geochemistry, and mining: Cleveland, Northern Ohio Geological Society, p. 35–47.

———, 1967, Eastern Wyoming, eastern Montana, and the Dakotas, Chapter G, *in* E. D. McKee, S. S. Oriel, et al., Paleotectonic investigations of the Permian System in the United States: U.S. Geological Survey Professional Paper 515, p. 127–152.

———, 1980, Permian and Lower Triassic geology of Colorado, *in* H. C. Kent and K. W. Porter, eds., Colorado geology: Denver, Rocky Mountain Association of Geologists, p. 103–110.

———, 1983, Tectonic setting of the Rocky Mountain region during the late Paleozoic and the early Mesozoic, *in* Proceedings of symposium on the genesis of Rocky Mountain ore deposits: changes with time and tectonics: Denver Region Exploration Geologists Society, p. 39–50.

———, 1984, Paleogeographic setting of Pennsylvanian Tyler Formation and relation to underlying Mississippian rocks in Montana and North Dakota: AAPG Bulletin, v. 68, n. 2, p. 178–195.

Maughan, E. K., and W. J. Perry, Jr., 1982, Paleozoic tectonism in southwestern Montana (abs.): Geological Society of America Abstracts with Programs, v. 14, n. 6, p. 341.

McKee, E. D., S. S. Oriel, K. B. Kettner, M. E. MacLachlan, J. W. Goldsmith, J. C. MacLachlan, and M. R. Mudge, 1959, Paleotectonic maps of the Triassic System: U.S. Geological Survey Miscellaneous Geological Investigations Map I-300, 33 p.

McKee, E. D., S. S. Oriel, et al., 1967, Paleotectonic maps of the Permian System: U.S. Geological Survey Miscellaneous Geological Investigations Map I-450.

McKee, E. D., E. J. Crosby, et al., 1975, Paleotectonic investigations of the Pennsylvanian System in the United States: U.S. Geological Survey Professional Paper 853, pt. I, 349 p.; pt. II, 192 p.; pt. III, 17 plates in pocket.

Moore, V. A., and R. B. Nelson, 1974, Effect of Cambridge–Chadron structural trend on Paleozoic and Mesozoic thicknesses, western Nebraska: AAPG Bulletin, v. 58, n. 2, p. 260–268.

O'Leary, D. W., J. D. Friedman, and H. A. Pohn, 1976, Lineament and linear, a terminological reappraisal, *in* M. H. Podwysocki and J. L. Earle, eds., Proceedings of the Second International Conference on Basement Tectonics: Denver, Basement Tectonics Committee, Inc., p. 571–577.

O'Neill, J. M., and D. A. Lopez, 1983, Great Falls lineament, Idaho and Montana (abs.): AAPG Bulletin, v. 67, n. 8, p. 1350–1351.

Peterson, J. A., 1981, General stratigraphy and regional paleostructure of the western Montana overthrust belt, *in* T. E. Tucker, ed., Field Conference and Symposium Guidebook, Southwest Montana: Billings, Montana Geological Society, p. 5–35.

———, 1985, Regional stratigraphy and general petroleum geology of Montana and adjacent areas, *in* J. J. Tonnsen, ed., Montana oil and gas symposium 1985: Billings, Montana Geological Society, p. 5–45.

Prucha, J. J., J. A. Graham, and R. P. Nickelsen, 1965, Basement-controlled deformation in Wyoming province of Rocky Mountains foreland: AAPG Bulletin, v. 49, n. 7, p. 966–992.

Ransome, F. L., 1915, The Tertiary orogeny of the North American cordillera and its problems, *in* W. N. Rice, et al., Problems of American geology: New Haven, Yale University Press, p. 287–376.

Roberts, R. J., and M. R. Thomasson, 1964, Comparison of late Paleozoic tectonic history of northern Nevada and central Idaho, *in* Short papers in geology and hydrology: U.S. Geological Survey Professional Paper 475-D, p. D1–D6.

Sandberg, C. A., and W. J. Mapel, 1967, Devonian of the northern Rocky Mountains and plains, *in* International Symposium on the Devonian System: Calgary, Alberta Society of Petroleum Geologists, p. 843–877.

Saperstone, H. E., and F. G. Ethridge, 1984, Origin and paleotectonic setting of the Pennsylvanian Quadrant Sandstone, southwestern Montana, *in* J. Goolsby, and D. Morton, eds., The Permian and Pennsylvanian geology of Wyoming, 35th Annual Field Conference Guidebook: Casper, Wyoming Geological Association, p. 309–331.

Schmidt, C. J., and T. E. Hendrix, 1981, Tectonic controls for thrust belt and Rocky Mountain foreland structures in the northern Tobacco Root Mountains–Jefferson Canyon area, southwestern Montana, *in* T. E. Tucker, ed., Field Conference and Symposium Guidebook, southwest Montana: Billings, Montana Geological Society, p. 167–180.

Shurr, G. W., 1979, Upper Cretaceous tectonic activity on lineaments in western North Dakota: U.S. Geological Survey Open-File Report 79-1374, 24 p.

Sonnenberg, S. A., and R. J. Weimer, 1981, Tectonics sedimentation, and petroleum potential, northern Denver basin, Colorado, Wyoming, and Nebraska: Colorado School of Mines Quarterly, v. 76, n. 2, 45 p.

Thom, W. T., Jr., 1952, Structural features of the Big Horn basin rim, *in* Guidebook 7th Annual Field Conference: Casper, Wyoming Geological Association, p. 15–17.

Thomas, G. E., 1974, Lineament-block tectonics, Williston–Blood Creek basin: AAPG Bulletin, v. 58, n. 7, p. 1305–1322.

Tweto, O., and P. K. Sims, 1963, Precambrian ancestry of the Colorado Mineral Belt: Geological Society of America Bulletin, v. 74, p. 991–1014.

Wallace, R. E., A. B. Griggs, and S. W. Hobbs, 1960, Tectonic setting of the Coeur d'Alene District, Idaho: U. S. Geological Survey Professional Paper 440-B, p. B25–B27.

Warner, L. A., 1978, The Colorado lineament: a middle Precambrian wrench fault system: Geological Society of America Bulletin, v. 89, p. 161–171.

Weimer, R. J., 1980, Recurrent movement on basement faults, a tectonic style for Colorado and adjacent areas, *in* H. C. Kent and K. W. Porter, eds., Colorado Geology, Denver, Rocky Mountain Association of Geologists, p. 23–35.

Wise, D. U., 1982, Linesmanship and the practice of linear geo-art: Geological Society of America Bulletin, v. 93, p. 886–888.

Part II
NORTHERN ROCKY MOUNTAINS

General Stratigraphy and Regional Paleotectonics of the Western Montana Overthrust Belt [1]

J. A. Peterson
University of Montana
Missoula, Montana

A composite stratigraphic section ranging in age from late Precambrian to Holocene is present in the western Montana overthrust belt. Paleozoic rocks are 2000 to 5000 m thick and are dominated by shallow marine shelf limestone and dolomite facies, much of which is of carbonate bank or reefal origin in the lower and middle Paleozoic, and by shelf marine sandstone and carbonate in the upper Paleozoic. Source areas for Paleozoic and early Mesozoic clastic facies were in south–central Canada and north–central United States. Mesozoic rocks between 300 m and 7000 m thick are dominated by shallow water marine clastic facies in the pre-Cretaceous and by continental and near shore marine facies in the Cretaceous section, which becomes progressively coarser, volcanic-rich, and more continental in the younger beds. The source area for Mesozoic clastics was in east–central and northern Idaho and westernmost Montana which, beginning in Late Jurassic time, became the site of increasingly intense tectonic growth that culminated in the development of the thrust and fold belt and associated igneous activity in Middle to Late Cretaceous and early Tertiary time.

The main structural framework of western Montana was developed during late Precambrian Belt Supergroup deposition, and many of the prominent paleostructural elements persisted through most of the remainder of geologic time. The more important of these that influenced the nature and distribution of sedimentary facies during Paleozoic and Mesozoic time are the Lemhi arch, Alberta shelf, Beartooth shelf, Belt Island complex, Boulder high, Coeur D'Alene–central Montana trough, Big Snowy trough, and Snowcrest trough. In east–central Idaho, the Muldoon trough (north segment of the Sublett basin), which in Paleozoic time lay between the Antler orogenic belt and the Montana shelf province, was a regional area of active subsidence that received a great thickness of Paleozoic shallow to deep water marine sediments.

The lower and middle Paleozoic rock facies in western Montana includes a large thickness of porous dolomite, and the upper Paleozoic contains a large volume of clean shelf sandstone, some of which has good porosity. Potential petroleum source rocks are present in the Devonian, Mississippian, and Permian beds, some of which, despite deep burial of most of the Paleozoic section, are not highly altered thermally. The Mesozoic rock facies is primarily continental in origin, but a broad belt of intertonguing near shore marine and continental facies is present. The sandstone bodies in this facies offer reasonable potential for significant biogenic and thermal gas accumulations under adequate trapping conditions. Tertiary continental and lacustrine beds, as much as 3000 m or more thick, were deposited in localized downwarped basins. These beds contain substantial thicknesses of discontinuous alluvial sandstone and some carbonaceous to coaly beds and are of interest for their biogenic gas potential.

INTRODUCTION

This paper reviews the regional tectonic and stratigraphic history and the regional depositional facies relationships of Precambrian through Holocene rocks in western Montana (Figure 1). It is not intended to be a comprehensive or detailed description of the complete stratigraphic section. The reader is referred to other publications listed in the references for details of specific stratigraphic units, faunal data, age indications, regional correlations, and reviews of previous work.

The maps and charts included in this paper were prepared on a palinspastic base designed to restore stratigraphic data to approximate depositional positions. These palinspastic reconstructions use modifications of major thrust displacements estimated by several authors, including Mudge (1970), Skipp and Hait (1977), Ruppel (1978), Ruppel and Lopez (1984), and W. J. Perry (personal communication, 1980). Positions of the major thrust zones and the displacements used are shown in Figures 2 and 3. Figures 4, 5, 6, and 7 are palinspastically restored lithofacies cross sections.

---

[1] Reprinted with minor modification and updating of references from Symposium Guidebook to Southwest Montana Geological Society, 1981, Billings, Montana, p. 5–35, with permission of the Montana Geological Society.

Stratigraphic data were compiled from a variety of sources, including personal files; numerous unpublished theses; publications of the U.S. Geological Survey, the Montana Bureau of Mines and Geology, and the Montana Geological Society; and borehole data from geophysical and lithologic logs of the American Stratigraphic Company. Most of the important data sources are listed in the references. Paleozoic stratigraphic intervals for map construction were selected primarily on the basis of the predominance of carbonate versus clastic facies. Mesozoic intervals, which are primarily clastic deposits, were selected on the basis of major differences in the nature and depositional environments of the clastic sediments.

REGIONAL STRUCTURAL SETTING

The surface geology of western Montana is dominated by the western overthrust and disturbed belt, a northwestern area of primarily Belt Supergroup exposures, a west-central region of Mesozoic intrusives, a large east-central area of late Mesozoic and early Tertiary volcanic and coarse conglomeratic deposits, and a south-central area of exposed older Precambrian rocks. McMannis (1965) separated the region into three general provinces on the basis of differences in the tectonic and stratigraphic character of the rocks. First, the *Belt province* is characterized by exposures of the thick Precambrian Belt Supergroup. No exposures of older Precambrian rocks are known in this province, and several northwest–southeast-trending vertically faulted troughs filled with middle and late Tertiary sediments are present within this province. Second, the *batholithic province* is mainly confined between the Montana lineament and the west–east "Perry line" of faults north and west of Bozeman. The Boulder and Idaho batholith complexes and related intrusives and the "Sapphire block" (Hyndman et al., 1975) are within this province, as are several sharply defined Tertiary basins, many containing thick nonmarine deposits. Exposures of older Precambrian rocks are absent or rare within this province. Third, the *basement province* corresponds closely to the position of the Beartooth shelf (Figures 1, 2, and 3) and is dominated by exposures of older Precambrian metamorphic rocks, with Belt Supergroup rocks generally being absent. Paleozoic and Mesozoic sequences are thinner here than in the other provinces. It contains several well-defined block-faulted Tertiary basins and a network of high-angle faults.

The Montana plains province of eastern and central Montana, which is characterized by less complex structure and more uniform stratigraphy, merges westward into the northern part of the disturbed belt, the eastern part of the basement province, and the eastern flank of the Crazy Mountain basin.

These subdivisions of this complex area are reasonably well defined, although some characteristics of any given province are also found within the others.

REGIONAL PALEOGEOGRAPHY AND PALEOSTRUCTURE

Paleozoic rocks (Figure 2) range in thickness from 3000 ft (900 m) or less in the Montana plains province to over 15,000 ft (4500 m) in southwestern Montana (Figure 2). In the central parts of the Muldoon trough or Sublett basin of Idaho, they are at least 50,000 ft (15,000 m) thick (Peterson, 1977). The influence of several persistent Paleozoic plaeostructural features is reflected in thickness patterns as well as facies patterns of the Paleozoic sequence.

Mesozoic rocks (Figure 3) are absent over much of western Montana, which during most of this time was an actively emergent source area for clastic debris. Thickness patterns of Mesozoic rocks, when restored, indicate that they were probably more than 20,000 ft (6000 m) thick in parts of westernmost Montana but thinned eastward to less than 5000 ft (1500 m) in parts of the Montana plains (Figures 3 and 6). Several of the main Paleozoic paleostructural elements persisted throughout Mesozoic time, although development of the overthrust belt and intrusive activity changed the nature of paleostructural activity from one of relatively gentle uplifts and downwarps to one of greatly increased intensity during Mesozoic time.

During Paleozoic and early Mesozoic time, western Montana was a part of the broad craton–miogeosyncline border zone that extended the length of the western North American continent (Figure 1). Transgressive marine carbonate deposits dominated sedimentary facies during most of this time, but there were frequent regressive interruptions when clastic marine sediments, primarily originating from source terranes to the east and northeast, spread over previously deposited shallow water carbonate beds. Major clastic source areas in the western United States during Paleozoic time were located in and near the Canadian shield and the Transcontinental arch to the east. After Middle Devonian time, the Antler orogenic belt to the west became an active source area, although rarely were clastic materials from this belt of uplifts transported eastward far enough to invade the Rocky Mountain shelf.

Stratigraphic facies and thickness studies document the presence of several positive and negative paleostructural areas with varying degress of persistence that affected the depositional patterns of the major stratigraphic units. Some of these were present during late Precambrian Belt deposition and many continued throughout most or all of Phanerozoic time (Figures 1, 2, and 3). Among the more persistent regional features were the following (refer to the thickness and lithofacies maps of late Precambrian through Cretaceous (Montanan) sedimentary rocks in Figures 8 through 18):

The *Wyoming shelf* (Figure 1) occupied most of the Wyoming and southernmost Montana area during Paleozoic and early Mesozoic time.

The *central Montana trough* (Figures 1, 2, 3, and 7) was a subsiding trough area that was active during Precambrian Belt time and persisted with varying degrees of intensity throughout the remainder of geologic history. This trough area was dominated by carbonate, evaporite, and fine-grained clastic deposition during Paleozoic time and by thick clastic sediments, partly intertongued with volcanic sediments in Mesozoic time.

The *Alberta shelf* (Figures 1, 2, and 3) bordered the central Montana trough on the north and occupied the site of present-day southern Alberta, western Saskatchewan, and north-central Montana. This province was the site of predominantly carbonate and fine-grained clastic deposition

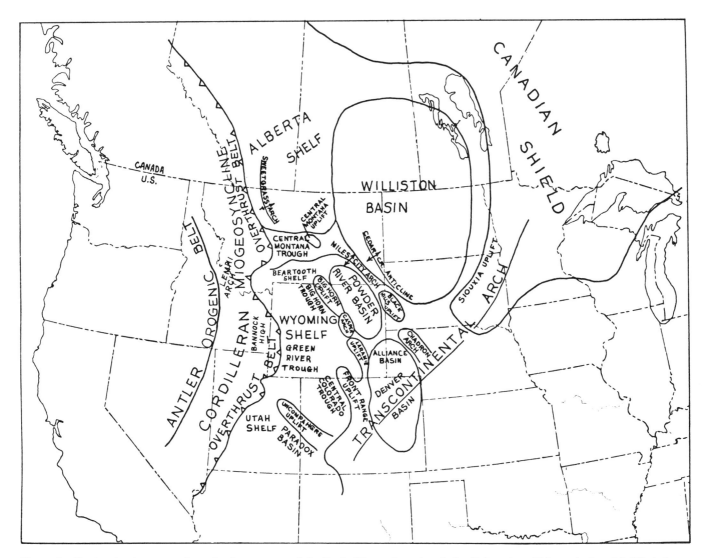

Figure 1—Regional paleogeography and paleostructure of the Rocky Mountain region during Paleozoic and Mesozoic time. Modified after Peterson (1981a).

in Paleozoic time and shelf sand, clay, and minor carbonate sediment in Mesozoic time. The Sweetgrass arch is a paleostructural element of the Alberta shelf.

The *Coeur D' Alene trough* (Figures 2, 8, and 9) was described by Harrison et al. (1974) as an elongate depositional trough during late Precambrian Belt deposition. This feature is part of a west-central downwarp that includes the central Montana trough. Restored thickness maps record evidence of its persistence through at least Paleozoic time and perhaps part of Mesozoic time.

The *Muldoon trough* (Figure 2) was a strongly subsiding Paleozoic trough area that formed the northern segment of the Sublett basin of southeastern Idaho and was particularly prominent in Devonian and Mississippian time (Rose, 1976).

The *Lemhi arch* (Figures 2 and 9) was an active positive area located in southwestern Montana and east-central Idaho. This feature was originally defined by Sloss (1954) as a Devonian arch, but it has since been documented in several other parts of the geologic column. In late Precambrian, Cambrian, and part of Devonian time it was a source area

fro clastic sediments and it formed a part of the cratonic shelf margin (Figures 2, 9, and 10) during the remainder of Paleozoic and early Mesozoic time. Several other names have been applied to various parts or phases of paleostructural growth in this area (Beaverhead arch, Scholten, 1967; Salmon River arch, Armstrong, 1975; Tendoy arch, Tysdal, 1976). The Lemhi arch may have extended toward the northwest to include the Precambrian source area referred to as "Belt Island" (Figures 2 and 8) by Harrison et al. (1974).

Other paleostructural features that were more local and less persistent include the following.

The *Beartooth shelf* (Figures 1, 2, and 3) was a northern part of the Wyoming shelf that bordered the central Montana trough on the south and was bounded on its west flank by the Greenhorn fault, which formed the east border of the Snowcrest trough (Figures 2 and 3). The Greenhorn fault underwent several phases of growth during Paleozoic and later time (Hadley, 1980).

The *Big Snowy trough* (Figures 2 and 3) was an elongate west–east belt of active subsidence during Mississippian and Pennsylvanian deposition and probably persisted well into

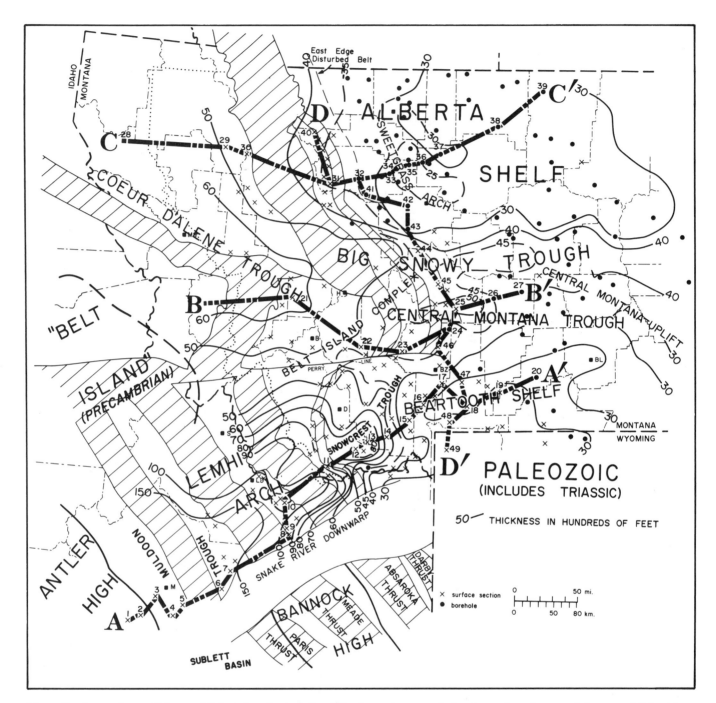

Figure 2—Approximate thickness of Paleozoic rocks, including Triassic, partly restored, showing main paleotectonic features of Paleozoic time. Cross-hatching indicates distances used for palinspastic reconstruction in main thrust area of western Montana and Idaho. Present-day outline of western Montana boundary is shown by dotted line within thrust belt. Lines of cross sections A–A′ through D–D′ in Figures 4, 5, 6, and 7 are shown. Cities shown for Montana are Butte (B), Billings (BL), Bozeman (BZ), Dillon (D), Great Falls (GF), Helena (H), Missoula (MS), and Shelby (SH); and for Idaho are Leadore (L), McKay (M), and Salmon (S). Palinspastic reconstruction of thrust belt features in southeastern Idaho is shown for comparison with those of western Montana. Modified after Peterson (1977, 1980). Used with permission of the Society of Economic Paleontologists and Mineralogists.

Mesozoic time. It formed the more active northern segment of the central Montana trough, which tended to separate into two parallel trough features in late Paleozoic time.

The *central Montana uplift* (Figure 1) was particularly active in Devonian time, but evidence also indicates its persistence during several times in the Paleozoic. Ancestral growth of this feature may have occurred in Cambrian time,

as shown by thickness patterns based on relatively limited control (Figure 9).

The *Belt island* system (Figures 3 and 15) was a complex of shallow marine emergent areas identified by Imlay et al. (1948) to explain variations in thickness patterns of marine Jurassic cyclic units. Earlier phases of this complex are evident for late Paleozoic and early Mesozoic time, although

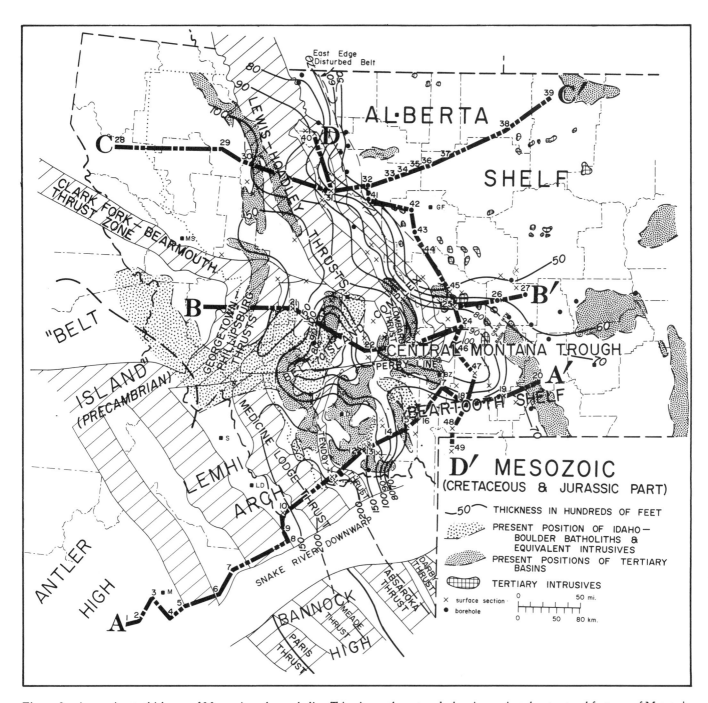

Figure 3—Approximate thickness of Mesozoic rocks, excluding Triassic, partly restored, showing main paleostructural features of Mesozoic time. Cross-hatching indicates distances used for palinspastic reconstruction in main thrust belts of western Montana and Idaho. See Figure 2 legend for further explanation.

removal of a substantial part of the older section by pre-Middle Jurassic erosion prevents an accurate appraisal of its early history. Parts of the Belt island complex may have been involved with the initial phases of structural activity that culminated in some Laramide structures.

The *Boulder high* (Figures 3 and 15) was noted by Mutch (1961) as an area of thinning in Early Cretaceous time. Evidence of its effect is also found in post-Cambrian thickness patterns. This feature formed a part of the Jurassic Belt island complex, and ultimately it became the site of emplacement of the Boulder batholith in late Mesozoic time. The northeasterly orientation of the northwest border of the Boulder high coincides closely with the trend of the Great Falls tectonic zone of O'Neill and Lopez (1985).

The *Snowcrest trough* (Figure 2) lies west of the Gravelly Range in southwestern Montana and is evident on several thickness maps (Figures 2, 11, 12, 13, and 14).

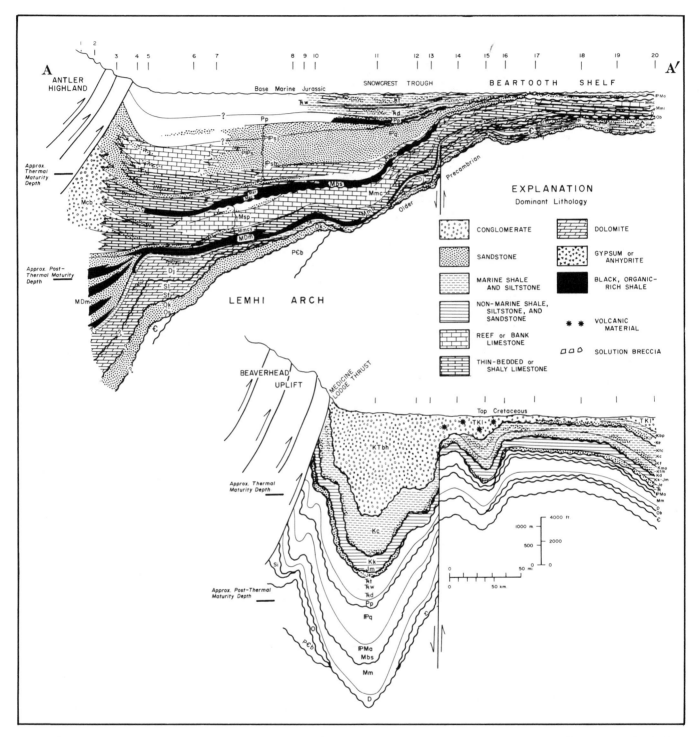

Figure 4—Southwest–northeast lithofacies cross section A–A', from east-central Idaho to central Montana. Datum for upper section is the base of the Jurassic marine sedimentary rocks; datum for lower section is estimated sea level at the end of Cretaceous time. Section is palinspastically restored in thrust belt. Line of cross section is shown in Figures 2 and 3. Abbreviations for formation names used in Figures 4-7 are given on p. 63. Also given are the references for the numbered sections used in Figure 4-7.

TKl—Livingston Fm
KTbh—Beaverhead Fm
Tklh—Hoppers Fm
Tklb—Billman Cr. Fm
Tklmc—Miner Cr. Fm
TKLc—Cokedale Fm
TKs—Sedan Fm
TKlm—Maudlow Fm
TKw—Willow Cr. Fm
Klc—Landslide Fm
Kev—Everts Fm
Kem—Elkhorn Mts. Volcanics
Kgs—Golden Spike Cgl
Kcc—Carter Cr. Fm
Kj—Jens Fm
Kcb—Coberly Fm
Kbd—Dunkelberg Fm
Ksm—Slim Sam Fm
Kl—Lennep Ss
Kfg—Fox Hills Ss
Khc—Hell Cr. Fm
Kjr—Judith R. = River Fm
Kcl—Claggett Fm
Kn—Niobrara Fm
Kcl—Carlile Fm
Kg—Greenhorn Fm

Kmd—Muddy Fm
Kt—Thermopolis Sh
Ksc—Skull Cr. = Creek Fm
Kd—Dakota Fm
Ktc—Telegraph Cr. = Creek Fm
Kf—Frontier Fm
Kc—Colorado Sh
Kco—Cody Sh
Km—Montana Gp
Kmr—Marias R. Fm
Ksm—St. Mary R. Fm
Kh—Horsethief Ss.
Kbp—Bearpaw Sh
Ktm—Two Medicine Fm
Kvi—Virgelle Ss
Keu—Eagle Ss
Kmk—Kevin Sh
Kmf—Ferdig Mbr
Kmcc—Cone Calc. Mbr
Kb—Blackleaf Fm
Kbv—Vaughan Mbr
Kbt—Taft Hill Mbr
Kbf—Flood Mbr
Kk—Kootenai Fm
Jm—Morrison Fm
Je—Ellis Gp

Js—Swift Fm
Jr—Rierdon Fm
Jsa—Sawtooth Fm
Jp—Piper m
Trc—Chugwater Fm
Trt—Thaynes Fm
Trw—Woodside Fm
Trd—Dinwoody Fm
Pp—Phosphoria Fm
Pr—Retort Sh.
Pm—Meade Peak Sh
Pt—Tosi Chert
Ps—Shedhorn Ss
Pf—Franson Mbr
Pg—Grandeur Mbr
Pennsylvanian:
q—Quadrant Fm
a—Amsden Fm
Mbs—Big Snow Gp
Mh—Heath Fm
Mo—Otter Fm
Mk—Kibbey Fm
Mcb—Copper Basin Fm
Mmc—Mission Canyon Fm
Mml—Lodgepole Fm
Mc—Castle Reef Dol

Ma—Allan Mtn. Ls
Mcrs—Sun River Dol
Msr—Surrett Can. Fm
Msc—South Cr. Fm
Msp—Scott Peak Fm
Mmjc—Middle Canyon Fm
MDm—Milligen Can. Fm
Dt—Three Forks Fm
Dtp—Potlatch Mbr
Dj—Jefferson Fm
Db—Birdbear Fm
Dd—Duperow Fm
Dsr—Souris R. Fm
Dm—Maywood Fm
Sf—Fish Haven Fm
Sl—Laketown Fm
Ob—Big Horn Fm
Ok—Kinnikinnic Fm
Os—Summerhouse Fm
Csr—Snowy Range Fm
Cpi—Pilgrim Ls
Cpa—Park Sh
Cm—Meagher Ls
Cw—Wolsey Sh
Cf—Flathead Ss
PCb—Belt Supergroup

1. Central Pioneer Mtns., Idaho T. 4 N., R. 21, 22 E., Skipp et al., 1979b
2. Iron Bog Cr. area, T. 6 N., R. 23 E., Idaho, Skipp et al., 1979b
3. Cabin Cr. area, T. 6 N. - R. 22 E., Idaho, Skipp et al., 1979b
4. Timbered Dome area, T. 3 N. - R. 25 E., Idaho, Skipp et al., 1979a
5. Arco Hills, T. 4 N., R. 26, 27 E, Idaho, Skipp et al., 1979a
6. Howe Peak area, T. 4, 5 N., R. 28, 29 E., Idaho, Skipp et al., 1979a, b
7. East Canyon-Box Canyon, T. 6, 7 N., R. 29 E., Idaho, Skipp et al., 1979a, b
8. So. Beaverhead Mtns., T. 9 N. - R. 32 E., Idaho, Skipp et al., 1979a, b
9. Copper Mtn-Blue Dome area, T. 10 N., R. 30 E., Idaho, Skipp et al., 1979a; Scholten and Hait, 1962
10. Deadman Lake area, T. 16 S., R. 10, 11 W., Montana, Skipp et al., 1979b; Scholten and Hait, 1962
11. Lima Peaks-Red Peaks area, T. 15 S., R. 8, 9 W., Mont., Scholten, 1950; Moritz, 1960
12. Blacktail Cr., T. 12 S., R. 6 W. Mont., Keenmon, 1950
13. Snowcrest Range, T. 11, 12 S., R. 5 W., Mont., Gealy, 1953; Kummel, 1960; Cressman and Swanson, 1964
14. Gravelly Range, T. 10 S., R. 1, 2 W., Mont., Hadley, 1980; Mann, 1954; Cressman and Swanson, 1964
15. Sphinx Mtn. area., T. 8 S., R. 2 E., Mont., Beck, 1960; Gardner et al., 1946
16. Garnet Mtn. area, T. 6, 7 S., R. 4 E., Mont., McMannis and Chadwick, 1964; Schwartz, 1972
17. Hyalite Canyon, T. 4 S., R. 6 E., Mont., McMannis, 1962
18. Mill Cr., T. 6 S., R. 9, 10 E. Mont., Wilson, 1936; McMannis, 1962
19. Picket Pen Cr., T. 4 S., R. 14 E., Mont., Gardner et al., 1946
20. Continental, No. 1 Govt., Sec. 33, T. 2 S., R. 19 E., Mont., American Strat. Co.
21. NE Flint Cr. Range, T. 8, 9 N., R. 11 W., Mutch, 1960; Gwinn, 1965
22. Jefferson Canyon, T. 1, 2 N., R. 2 W., Mont, Alexander, 1955; Gardner et al., 1946
23. Logan Area, T. 2 N., R. 2 E., Mont., Robinson, 1963; Hanson, 1960; McMannis, 1962
24. Sixteen Mine-Maudlow area, T. 4, 5 N., R. 6, 7, 8 E., Mont., Gardner et al., 1946; Skipp and McGrew, 1977
25. SW Castle Mtns. area, T. 7, 8 N., R. 7, 8 E., Mont., Gardner et al., 1946; Tanner, 1949

26. Amerada Russell No. 1, Sec. 1, T. 8 N., R. 13 E., Mont, American Strat. Co.
27. Continental NP No. 1, T. 9 N., R. 17 E., Mont, American Strat. Co.
28. Cabinet Mtns., T. 22-29 N., R. 28-30 W., Mont., Aadland, 1979
29. Spotted Bear, T. 25 N., R. 14 W., Mont., Ross, 1959; Theodosis, 1955
30. Prairie Reef area, T. 23 N., R. 11 W., Mont., Mudge, 1972; Wilson, 1955; Theodosis, 1955
31. Sun River Can., T. 22, 23 N., R. 9, 10 W., Mont., Mudge, 1972; Wilson, 1955
32. Phillips, Yeager-1, Sec. 6, T. 23 N., R. 6 W., Mont., American Strat. Co.
33. Brit. Amer., Severson-1, Sec. 7, 24 N., 2 W., Mont., American Strat. Co.
34. Gen. Pet., Holt-1, Sec. 30, 25 N., 1W., Mont., American Strat. Co.
35. Continental-1, State, 25 N., 1 E., Mont, American Strat. Co.
36. Huntley-1, Rossmiller, 27-26 N, 3 E., Mont., American Strat. Co.
37. Chamberlain, 1-Higgins, 30-27 N, 6E., Mont., American Strat. Co.
38. Texas Co., Kiemele-1, 26-31 N, 13 E., Mont., American Strat. Co.
39. So. Union, N. Chinook-1, 16-35 N., 19E., Mont., American Strat. Co.
40. Feather Woman Mtn. area, T. 29 N., R. 10, 11, W., Mont, Wilson, 1955
31. See Figure 6
32. See Figure 6
41. Phillips-1, Randall, 6-21 N., 5 W., Mont., American Strat. Co.
42. Schoonmaker-1, Stephan, 12-20N, 1E., Mont.
43. Riverdale, Murphy-1, 8-17N., 2E., Mont., American Strat. Co.
44. Smith R. Canyon, 13, 14, 15 N., 3, 4 E., Mont., Walker, 1974; Dahl, 1971
45. White Sulphur Spgs. area, 9, 10 N., 7E., Mont., Tanner, 1949; Dahl, 1971; Billings Geol. Soc., 1962
25. See Figure 5
24. See Figure 5
46. N. Bridger Mts., 3, 4 N., 6 E., Mont., Klemme, 1949; McMannis, 1955; Skipp and McGrew, 1977
47. Livingston area, 2, 3S., 9E., Mont., Gardner et al., 1946; Richards, 1957; Sando, 1972; Brown, 1957
18. See Figure 4
48. Cinnabar Mtn., 8S, 8E., Mont., Wilson, 1934; Kummel, 1960; McMannis, 1962
49. NW Yellowstone Park, Wyo. and Mont., Ruppel, 1972; Brown, 1957; McMannis, 1962

Figure 4 Legend

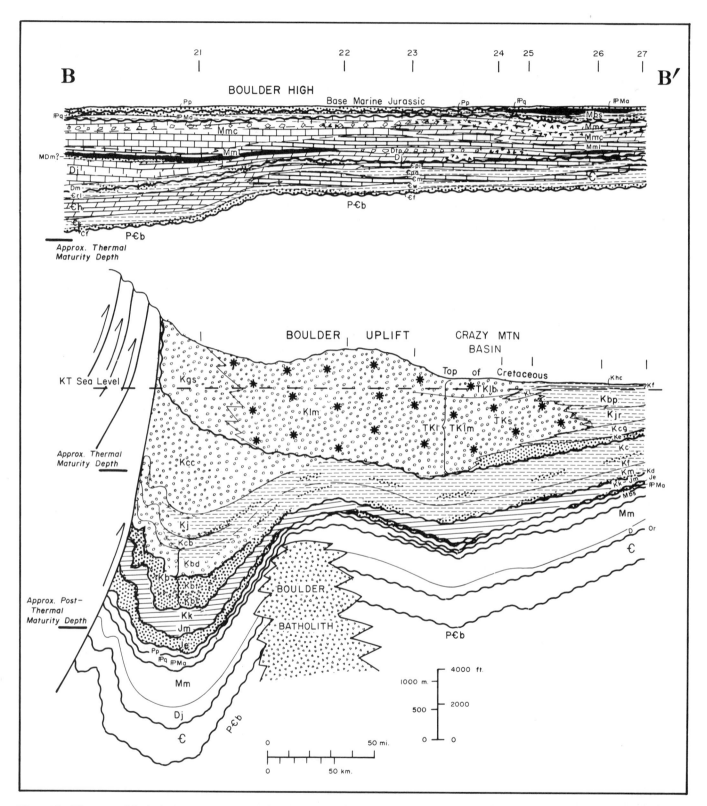

Figure 5—West–east lithofacies cross section B–B′, from west-central to east-central Montana. Datum for upper section is the base of the Jurassic marine sedimentary rocks; datum for lower section is estimated sea level at the end of Cretaceous time. Section is palinspastically restored in thrust belt. Line of cross section is shown in Figures 2 and 3. Lithologic explanation is shown in Figure 4, and explanation of formation abbreviations and numbered sections is given in Figure 4 legend.

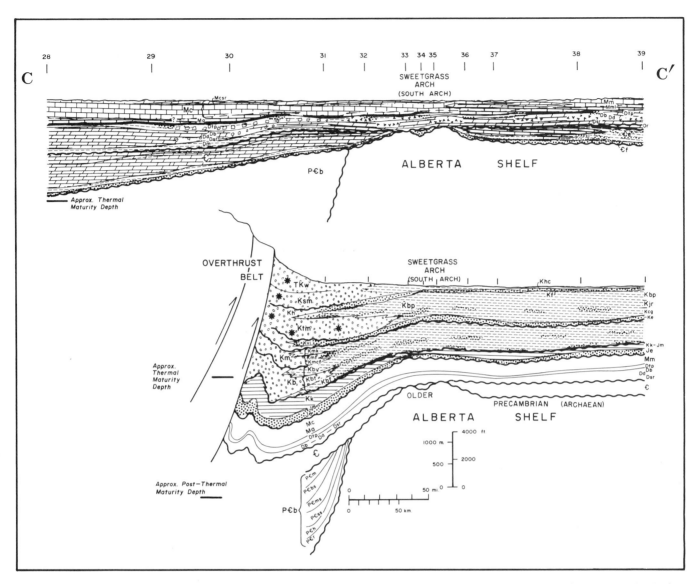

Figure 6—West–east lithofacies cross section C–C′, from northwestern Montana to north-central Montana. Datum for upper section is the base of the Jurassic marine sedimentary rocks; datum for lower section is estimated sea level at the end of Cretaceous time. Section is palinspastically restored in thrust belt. Line of cross section is shown in Figures 2 and 3; Lithologic explanation is shown in Figure 4, and explanation of formation abbreviations and numbered sections is given on Figure 4 legend.

STRATIGRAPHY AND SEDIMENTARY FACIES

Precambrian

Metamorphosed older Precambrian rocks are exposed in several mountain ranges north and west of Yellowstone National Park. These rocks have been studied by several authors, including Reid (1957), Foose et al. (1961), Stewart (1972), James and Hedge (1980), and others.

Proterozoic rocks of the Belt Supergroup make up most of the pre-Tertiary exposures in northwestern Montana and are present in some mountain ranges east of the overthrust belt. They are greater than 50,000 ft (15,000 m) thick in the vicinity of the Coeur D'Alene trough (Figure 8) and thin to zero roughly at the leading edge of the thrust belt in northwestern and southwestern Montana and along the "Perry line" near Bozeman, Montana (Figures 3, 4, 5, and 6). Belt equivalent rocks have been identified in boreholes east of Bozeman in the central Montana trough, or "Belt

embayment," area at least as far east as Billings. As pointed out by Harrison et al. (1974), much of the paleostructural grain of western Montana was developed at least as early as late Precambrian time and most of it was retained through most of the remainder of geologic time. Among the more persistent of the older features are the Coeur D'Alene trough, Lemhi arch, central Montana trough, Beartooth shelf, Alberta shelf, and perhaps the Boulder high or ancestral Belt Island (Jurassic) complex. Much of the Belt Supergroup sequence is composed of relatively fine-grained clastic rocks, although a narrow band of boulder conglomerate debris, the Lahood facies (McMannis, 1963), is present along the northwest border of the Beartooth shelf, and some coarse sandstone and pebbly beds occur in south-westernmost Montana in the vicinity of the Lemhi arch. These beds were apparently deposited along a fault-controlled shoreline in the vicinity of the present-day Horse Prairie fault (D. A. Lopez, personal communication, 1984).

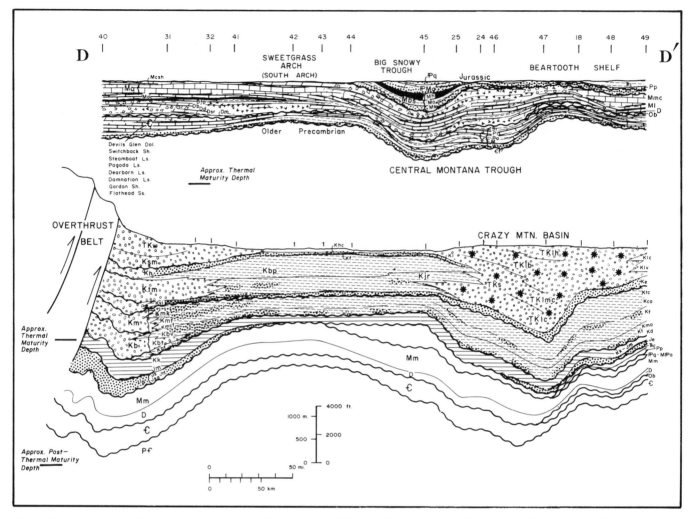

Figure 7—North–south lithofacies cross section D–D', from south of Glacier National Park to Yellowstone National Park. Datum for upper section is the base of the Jurassic marine sedimentary rocks; datum for lower section is estimated sea level at the end of Cretaceous time. Section is palinspastically restored in thrust belt. Line of cross section is shown in Figures 2 and 3. Lithologic explanation is shown in Figure 4, and explanation of formation abbreviations and numbered sections is given in Figure 4 legend.

A broad band of thick stromatolitic algal carbonate facies (Helena Formation and equivalents) is present on the west flank of the Alberta shelf and extends from near the Canadian border southward at least as far as the Helena vicinity (Figure 8; Theodosis, 1955; McGill and Sommers, 1967).

Cambrian

A transgressive sequence of Middle and Upper Cambrian rocks over 1000 ft (300 m) thick is present in west-central Montana and thickens to more than 2000 ft (600 m) in scattered exposures west of Butte (Figures 5, 6, and 9). The main trend of thickening follows the Coeur D'Alene–central Montana trough. The section is less than 150 m thick in the vicinity of the Sweetgrass arch area and thins to zero in southwestern Montana where a coarser clastic facies occurs near the Lemhi arch source area. Cambrian rocks are somewhat thicker in the area of the Snowcrest trough (Figure 4).

The main clastic source areas during Cambrian time were located in the continental interior on the Canadian shield and the Transcontinental arch (Figure 1). The widely

transgressive Flathead Sandstone is present at the base of the sequence almost everywhere and ranges in age from Middle to Upper Cambrian. In the Williston basin of eastern Montana and North Dakota, Cambrian rocks comprise a sandstone and shale facies that grades westward to green shale and limestone facies in central Montana. Approximately in the position of the central and northern overthrust belt, the Cambrian sequence becomes dominated by a thick section of sucrosic dolomite (Hasmark Formation and equivalents) (Hanson, 1952), which is more than 2000 ft (600 m) thick west of Helena and Butte, Montana (Figures 5 and 9). The dolomite facies is generally unfossiliferous, but careful outcrop and microscopic studies reveal evidence of bioclastic, algal, and oolitic fabric that probably originated as carbonate bank buildups in the boundary belt between the early Paleozoic shelf and miogeosyncline.

Ordovician–Silurian

Ordovician rocks are absent in most of western Montana except for the Beartooth shelf area where the Bighorn Dolomite, as thick as 300 ft (90 m), is present. The zero

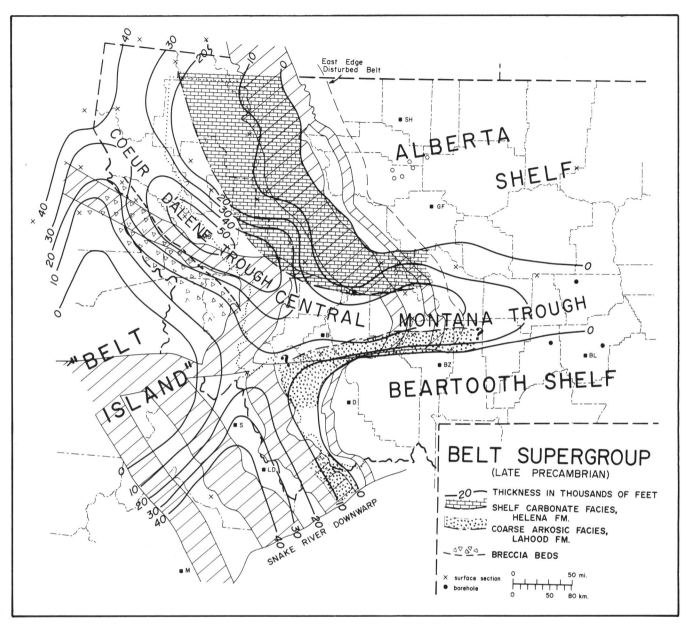

Figure 8—Thickness and general lithofacies of the upper Precambrian Belt Supergroup, partly restored. Data are palinspastically restored in thrust belt. Modified in part after Harrison (1972).

edge of the Bighorn extends along the north and west borders of the Beartooth shelf, projects slightly eastward into the area of the central Montana trough, and continues northward across the Alberta shelf into Canada (Figure 10). To the east, Ordovician rocks thicken uniformly into the Williston basin, where they reach thicknesses of more than 1000 ft (300 m). Ordovician rocks are very thick in the Muldoon trough of east-central Idaho (Kinnikinnic Quartzite) but they wedge out rapidly eastward and are truncated by Upper Devonian erosion approximately along the Montana–Idaho border (Figures 4 and 10) (James and Oaks, 1977; Scholten, 1967).

Silurian rocks have not been identified in western Montana, although the thick section of Silurian dolomite (Laketown Formation) in east-central Idaho pinches out eastward very close to the border between southwestern Montana and east-central Idaho a short distance west of the truncated edge of Ordovician rocks in the vicinity of the Lemhi arch (Figures 4 and 10). This erosional offlap relationship between the Bighorn–Interlake and the Kinnikinnic–Laketown formations and the overlying Upper Devonian beds, plus the general absence of sandstone and siltstone in the Silurian and Ordovician carbonate beds, suggest that the erosional edges of both units are some distance from the Ordovician or Silurian shorelines and that a broad area of originally deposited carbonate facies was removed by pre-Middle or Upper Devonian erosion.

Devonian

During Late Silurian through Middle Devonian time, most of the Rocky Mountain shelf was apparently emergent, and Lower and Middle Devonian rocks are absent over all of Montana except for scattered deposits in southern Montana

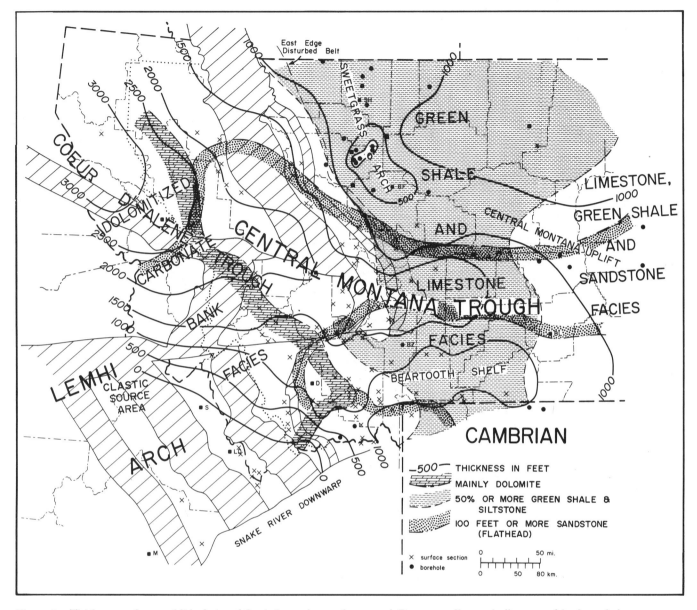

Figure 9—Thickness and general lithofacies of Cambrian rocks, partly restored. Data are palinspastically restored in thrust belt.

and a section that is increasingly older eastward into the Williston basin. Widespread but gentle erosion of the Silurian and Ordovician beds occurred at this time and resulted in the regional truncation offlap relationship between Upper Devonian rocks and the underlying units (Figure 4).

The worldwide Upper Devonian transgression appears to have spread completely across the Rocky Mountain shelf and covered all of Montana. Upper Devonian rocks comprise a transgressive sequence of fine- to medium-grained, red to green, clastic marine beds (Maywood or lower Souris River formations) overlain by a prominent sequence of generally porous stromatoporoid and coral-bearing dolomite, fossiliferous limestone, evaporite beds, and shaly limestone (Jefferson Formation) that form the bulk of the Devonian section. The strongly cyclic nature of this sequence has been described by several authors (Wilson, 1955; Sandberg and

Hammond, 1958; Rose, 1976). The Jefferson is overlain by regressive green and red shale, siltstone, evaporite beds, and shaly limestone beds of the Three Forks Formation, which grade upward into the Upper Devonian–Mississippian dark organic shale and siltstone beds of the Bakken and Sappington formations. Lithologic aspects and the facies distribution of these beds have been described in detail by Rau (1962), McMannis (1962), and Gutschick (1962).

The Upper Devonian rock sequence is more than 1000 ft (300 m) thick in most of western Montana and more than 2000 ft (600 m) thick in much of the western thrust belt area. It thickens rapidly westward into the Muldoon trough along the southwestern Montana–Idaho border (Figures 4 and 10). The western facies is dominated by dolomitized stromatoporoid and tabulate coral–bearing carbonate bank deposits commonly containing thick beds of highly porous sucrosic dolomite. The continued influence of the central

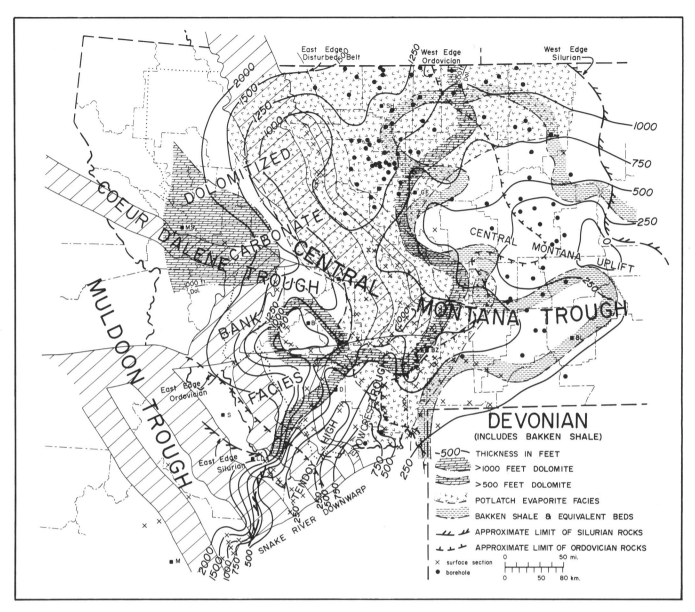

Figure 10—Thickness and general lithofacies of Devonian rocks, partly restored. Data are palinspastically restored in thrust belt. Approximate erosional limits of Silurian and Ordovician rocks are shown.

Montana trough is evident from the thick west–east trend in about the same position as that of the late Precambrian and Cambrian intervals. A relatively prominent area of thinning is present in the vicinity of the Lemhi arch in southwestern Montana ("Tendoy high"), and an accompanying thick southwest–northeast trend is present just to the east in the Snowcrest trough. The relatively prominent thin northwest–southeast area that coincides closely with the northwestern overthrust belt (Figure 10) may be related to outcrop leaching of gypsum or anhydrite beds of the Potlatch facies and similar beds in the underlying Jefferson Formation. Marked thinning or absence of Upper Devonian beds occurs along the central Montana uplift and extends westward along the Little Belt Mountains area south of Great Falls. The western part of this belt may also be related to evaporite solution of outcrop sections, although the

absence of lower Three Forks Potlatch breccias in the Little Belt Mountains suggests otherwise. The accompanying absence of the Devonian–Mississippian Bakken Formation beds in this area suggests that broad, gentle uplift occurred here at this time. The distribution of the Bakken and equivalent facies in south central Montana also demonstrates the continued influence of the central Montana trough and the Beartooth shelf (Figure 10).

The regional distribution of evaporite beds and related solution breccias in the Jefferson and Three Forks formations suggests that evaporitic conditions occurred in a back bank carbonate environment immediately east of the north–south-trending belt of carbonate bank facies buildup in westernmost Montana (Figure 10). This paleogeographic relationship may have resulted in the establishment of a westward-directed hydrodynamic gradient that enabled high

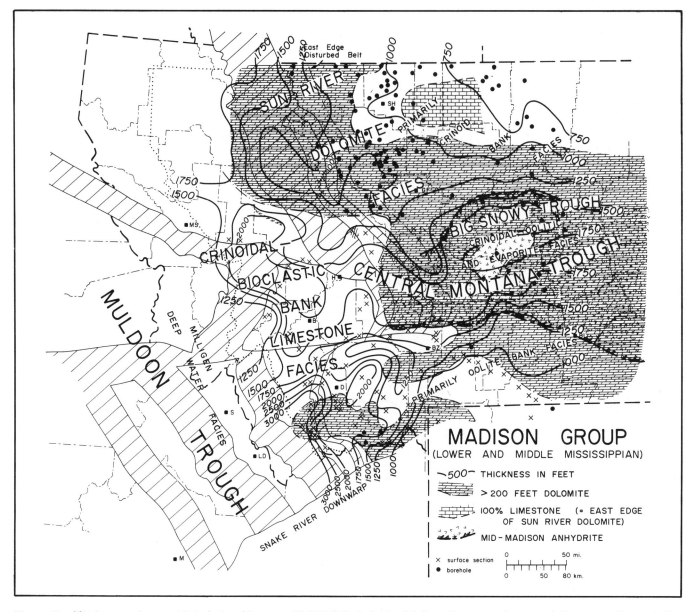

Figure 11—Thickness and general lithofacies of Lower and Middle Mississippian Madison Group, partly restored. Data are palinspastically restored in thrust belt.

magnesium waters of the shelf environment to move westward during low sea level evaporitic stages of the cycles, which caused early dolomitization of the thick carbonate facies by seepage reflux processes. Such a mechanism may have been in effect during much of the early to middle Paleozoic, as suggested by the similarity in gross dolomitic facies distribution of the Cambrian sequence. Evidence for shelf back bank evaporitic environments, however, is lacking in the earlier Paleozoic sequences.

Detailed lithologic and stratigraphic aspects of Devonian rocks in western Montana are described by Sloss and Laird (1945, 1946, 1947), Wilson (1955), Rau (1962), Sandberg (1962, 1965), Sandberg and McMannis (1964), Benson (1966), and Sandberg and Mapel (1967).

Lower-Middle Mississippian: Madison Group

The paleogeographic setting characteristic of the Devonian persisted into the Mississippian, and cyclic carbonate deposition predominated throughout the Lower and Middle Mississippian. These carbonate beds are not nearly as highly dolomitized as those of the Devonian, and evaporite deposits are not as widespread in the shelf area. Partly for these reasons, cycles in Mississippian rocks are more difficult to recognize. A thick crinoidal, bioclastic bank facies is present in approximately the same position as the thick Devonian stromatoporoid, coral bank facies (Figure 11). Evaporite beds east of the bank facies, however, are generally restricted to the central Montana trough where a prominent anhydrite or gypsum unit (middle Madison in age) is widely recognized in the subsurface and is represented by a prominent breccia interval in outcrop sections ("lower solution zone" of Sando, 1972, 1976). A thick trend of dolomite and anhydrite facies is generally prevalent in the area of the central Montana trough, which indicates continued relatively greater subsidence in that area. Thinning of the Madison section occurs on the Alberta shelf

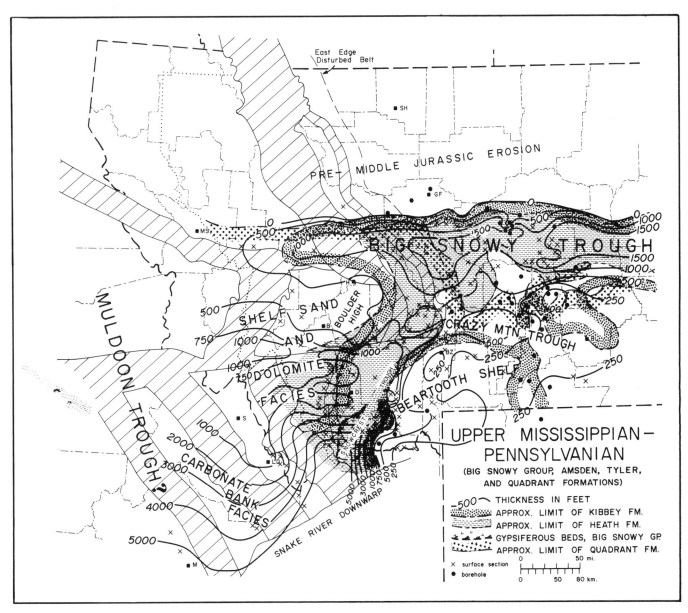

Figure 12—Thickness and general lithofacies of Upper Mississippian and Pennsylvanian rocks, partly restored. Data are palinspastically restored in thrust belt.

(Figure 11), which is characterized by crinoidal bank limestone facies. Thinning also occurs in the Beartooth shelf area south of the evaporitic trough belt where the carbonate beds are dominated by oolitic bank limestone facies.

The thick Madison carbonate bank facies in western Montana contains much less dolomite than the bank facies of the Devonian, which is almost entirely dolomite. The reasons for this may be related partly to generally diminished evaporitic conditions on the shelf during Madison deposition, which would have decreased the potential for development of westward-gradient seepage reflux systems. An additional reason may be that the Madison crinoid and oolite bank sediments contained considerably less original porosity than did the stromatoporoid and coral reefoid bank Devonian facies.

Detailed descriptions and analyses of the Madison rocks in the northern Rocky Mountains are published by several authors, including Sloss and Hamblin (1942), Sloss and

Laird (1945), Nordquist (1953), Andrichuk (1955), Mudge et al. (1962), Roberts (1966, 1979), Huh (1967), Sando (1967, 1972, 1976), Sando et al. (1969), Smith (1972), Rose (1976), Skipp et al. (1979), Smith and Gilmour (1979), and Peterson (1984b).

Upper Mississippian–Pennsylvanian

Upper Mississippian and Pennsylvanian beds (Big Snowy Group, Amsden and Quadrant Formations) are dominated by clastic sediments and represent a marked change in regional depositional and tectonic conditions of the western North American continental shelf at this time. These beds contrast sharply with the underlying stable shelf carbonate sediments that dominated the Rocky Mountain shelf and the adjacent border of the Cordilleran miogeosyncline during Ordovician, Silurian, Devonian, and Early–Middle Mississippian time. Tectonic activity in the middle and southern Rocky Mountains area that began in Late

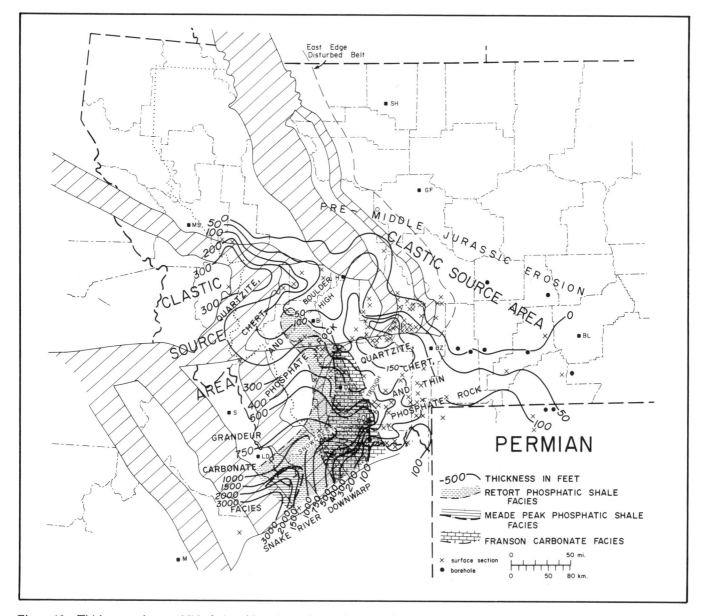

Figure 13—Thickness and general lithofacies of Permian rocks, partly restored. Data are palinspastically restored in thrust belt.

Mississippian time resulted in active growth of several prominent local uplifts and accompanying basins (Ancestral Rocky Mountains) and the rise of many new clastic source terranes. At the same time, the supply of clastic material from the Canadian Shield and Transcontinental arch increased and spread across the northern Rocky Mountain shelf. The regional paleogeographic and paleotectonic patterns in Montana appear to have remained essentially the same, but carbonate deposition was greatly diminished because of the increased supply of clastic material into the shallow water marine environment. Most of the organic carbonate bank growth that was characteristic of the middle Paleozoic time was eliminated on the shelf or craton edge during Late Mississippian–Pennsylvanian time. Farther west, however, at greater distances from clastic source areas along the western border of the shelf and the eastern flank of the Muldoon trough in east-central Idaho, a prominent carbonate bank facies is present (Figures 4, 11, and 12).

A substantial amount of the Upper Mississippian–Pennsylvanian section was removed by pre-Middle Jurassic erosion, but in general the major paleostructural elements of central to western Montana are reflected by the thickness and facies distribution patterns of this interval, particularly in the Upper Mississippian Big Snowy Group. Marked thickening of the section continues in the area of the central Montana trough, although this feature appears to separate into two west–east trough elements, the Crazy Mountain trough and the more sharply defined Big Snowy trough where the Big Snowy beds are more than 1500 ft (450 m) thick (Figures 7 and 12) and contain the thickest facies of the highly organic-rich Heath Formation. This thickness pattern is accompanied by more complex sedimentary facies relationships, and it is thought to reflect increased tectonic activity along the southern border of the Alberta shelf (Fanshawe, 1978; Maughan and Roberts, 1967). The Crazy Mountain trough merges westward with the north extension

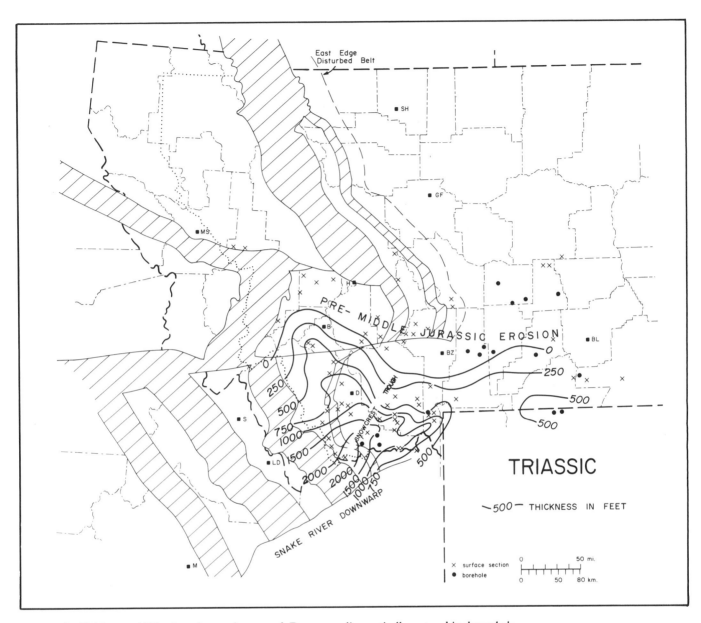

Figure 14—Thickness of Triassic rocks, partly restored. Data are palinspastically restored in thrust belt.

of the Snowcrest trough, which contains a substantially thicker section of Upper Mississippian and Pennsylvanian beds and tends to confine the southern extension of the Big Snowy facies. A markedly thinner sequence is present on the Beartooth shelf, and the Upper Mississippian–Pennsylvanian section is absent on the Alberta shelf. Thinning in both these areas is partly caused by pre-Middle Jurassic erosion, but they probably were also mildly elevated submarine shelf areas at this time. Restored thickness patterns show probable thinning in the vicinity of the Boulder high and the general area of the Lemhi arch (Figure 12).

A steadily increasing supply of quartzose sand from the northern source terrane is demonstrated by the almost complete dominance of clean shallow water marine and eolian sandstone beds in the upper part of the Upper Mississippian–Pennsylvanian clastic sequence (Quadrant Formation) (Figures 4, 5, 6, 7, and 12). This facies intertongues southwestward with dolomite and limestone beds of the carbonate bank facies along the east flank of the Muldoon trough and adjacent shelf margin. Much of the thick quartzose sand section is quartzitic or dolomitic, but substantial parts of it are composed of clean, well-sorted, porous sandstone in parts of southwestern Montana (D. Saperstone, personal communication, 1984).

Detailed studies of Upper Mississippian–Pennsylvanian rocks have been published by Sloss and Laird (1946), Easton (1962), Maughan and Roberts (1967), Harris (1972), Breuninger (1976), Rose (1976), Skipp et al. (1979), Smith and Gilmour (1979), and Maughan (1984).

Permian

The Permian section has a truncation offlap relationship with the underlying Pennsylvanian beds as the result of pre-Middle Jurassic erosion. The Permian sedimentary facies in southwestern Montana is characterized by phosphate, carbonate, and chert beds that represent the northern facies

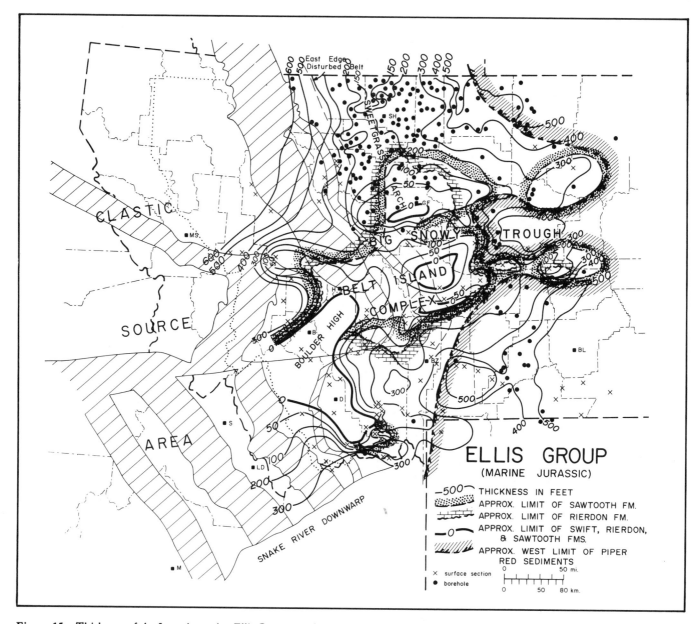

Figure 15—Thickness of the Jurassic marine Ellis Group, partly restored. Data are palinspastically restored in thrust belt.

of the Phosphoria basin of southeastern Idaho and western Wyoming. The northern clastic source area remained active during this time, as demonstrated by the presence of a quartzitic facies of Late Permian age (Shedhorn Sandstone) bordering the area of truncation (Figures 7 and 13). Evidence for a rejuvenated clastic source area in the vicinity of the Paleozoic Lemhi arch (Figures 2 and 13) is indicated by increased quartzose sand content of Permian beds west and northwest of Dillon. In southwesternmost Montana and adjacent Idaho, the Lower Permian section (Wolfcampian) contains a great thickness of carbonate beds (Grandeur Member), which represent an uninterrupted continuation of the depositional environments of Late Pennsylvanian time in that area (Figure 4). The major change in environmental conditions occurred at the close of the Grandeur cycle with the deposition of the overlying phosphate (Meade Peak and

Retort), chert (Tosi), and carbonate (Franson) beds of the Phosphoria Formation facies, which dominate the Permian section in southwestern Montana. The thickest post-Grandeur Permian deposits occur in the Snowcrest trough and in the adjacent area of subsidence immediately to the west. This belt falls generally between the western and eastern quartzitic facies and south of the Boulder high (Figure 13).

Because of the presence of economic phosphorite deposits, the Permian beds have been thoroughly studied. Many publications are available that contain details of the stratigraphy, facies distribution, depositional environments and biostratigrapahy of this section in the western Rocky Mountain region (Cressman, 1955; McKelvey et al., 1959; Cressman and Swanson, 1964; Sheldon, 1963; Yochelson, 1968; Peterson, 1972a, b, 1980, 1984a, b; Maughan, 1975).

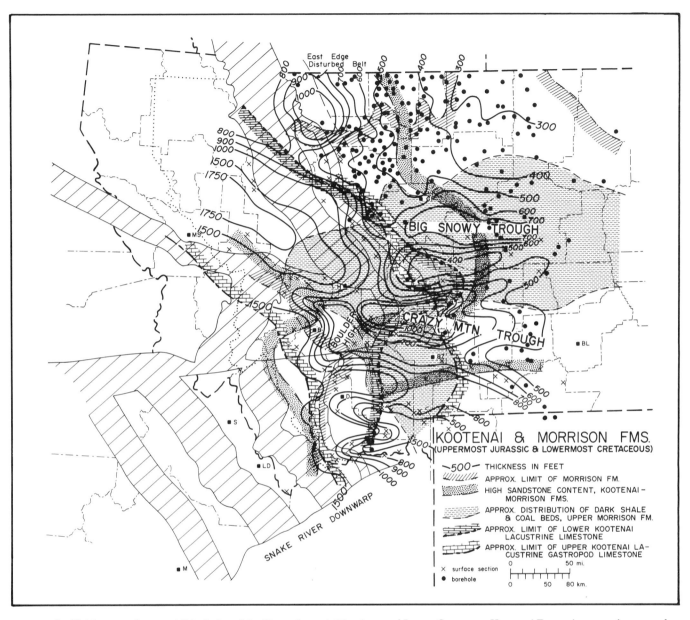

Figure 16—Thickness and general lithofacies of the Upper Jurassic Morrison and Lower Cretaceous Kootenai Formations, partly restored. Data are palinspastically restored in thrust belt.

Triassic

Rocks of Triassic age occur in a northward-thinning wedge that has a truncation offlap relationship with the underlying Permian as a result of pre-Middle Jurassic erosion (Figures 4, 7, and 14). The paleogeographic framework and source terrane characteristics of the Permian continued relatively unchanged into the Triassic, and deposition continued without evidence of marked physical interruption. Diminished influx of clastic material is indicated by the presence of finer grained clastic sediments in the Triassic beds than in the underlying Permian and Pennsylvanian facies. Triassic sediments, however, are in marked contrast to those of the underlying Permian, primarily with respect to the cessation of phosphatic, chert, and skeletal carbonate deposition that characterized the upper Paleozoic depositional facies. A primary reason for this is the apparent

rapid decline of marine rock-building faunas at the close of the Permian, a phenomenon that is particularly evident in the Permian–Triassic beds of southeastern Idaho and western Wyoming (Yochelson, 1968; Peterson, 1980, 1984a, b).

Previous publications on Triassic stratigraphy and facies in and near western Montana include those of Moritz (1951) and Kummel (1954).

Jurassic: Ellis Group

Ellis Group rocks are present in the Williston basin and in almost all of the northern Rocky Mountains, as well as the entire Rocky Mountain shelf area. These beds were deposited in three well-documented transgressive–regressive cycles of Middle and Late Jurassic age (Sawtooth–Piper, Rierdon, and Swift cycles) that are a product of worldwide

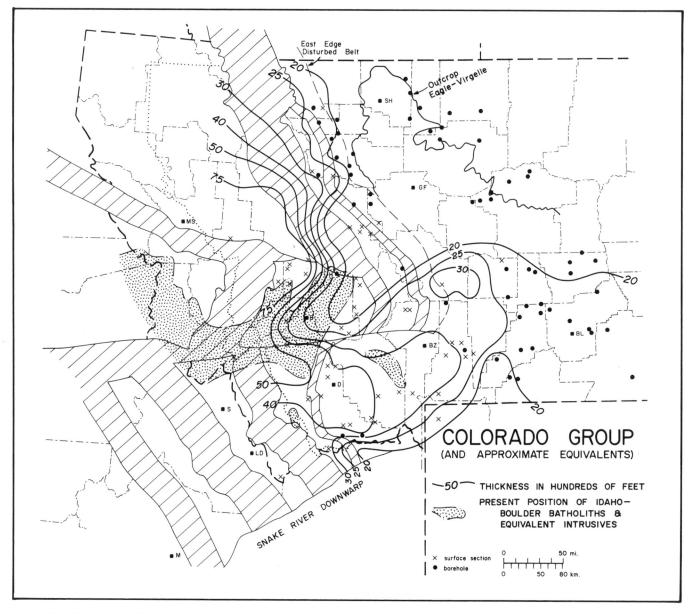

Figure 17—Thickness of Cretaceous Colorado Group and equivalent rocks, partly restored. Data are palinspastically restored in thrust belt. Present-day position (unrestored palinspastically) of Idaho and Boulder batholiths is shown.

transgressive events, which in the northern hemisphere represent incursion of northern (boreal) seas onto the continental shelf areas of that time. Rocks of Late Triassic and Early Jurassic age are not present in the northern Rocky Mountains. In Montana, Middle Jurassic basal transgressive units truncate underlying rocks ranging in age from Mississippian to Early Triassic.

Ellis paleogeography in the northern Rocky Mountains (Figure 15) was apparently dominated by the Belt Island complex of gentle uplifts. Evidence for this comes from various parts of the Ellis Group being thinned or absent because of erosion or nondeposition; this occurred during all or part of the three main depositional cycles. Vestiges of the Paleozoic and early Mesozoic paleostructural features were apparently present during Jurassic time but were much subdued. These include the Big Snowy trough and possibly

the Crazy Mountain and Snowcrest troughs, all of which show minor thickening of Ellis deposits (Figure 15). Rejuvenated growth of the Sweetgrass arch is reflected by the thinning and absence of some Ellis units coinciding with the trend of the arch. A major clastic source area appeared in southwestern Montana and adjacent Idaho at this time approximately in the position of the Paleozoic Lemhi arch (Figures 2 and 15). This event probably reflected the beginning of tectonic activity that ultimately resulted in development of the western North American thrust and fold belt. The Boulder high, which at this time may have represented initiation of the igneous and tectonic event that culminated with emplacement of the Boulder batholith, retained its expression as part of an eastward projection off the Lemhi arch and formed a part of the Belt Island complex.

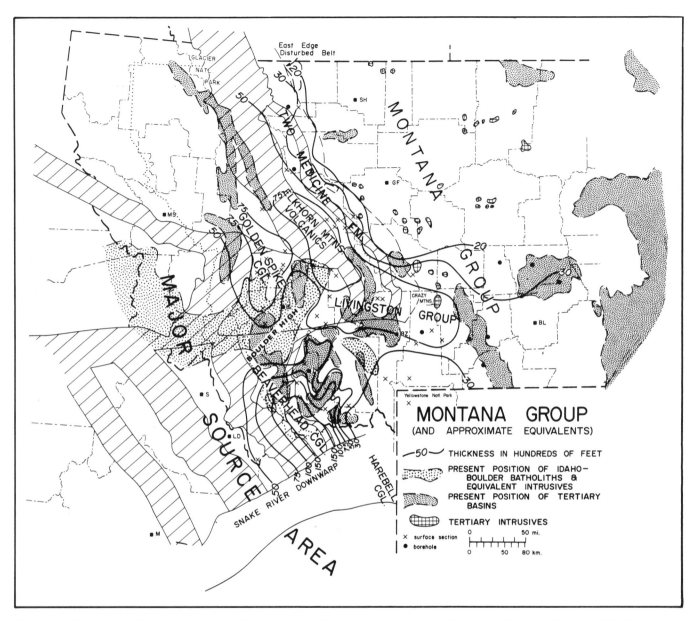

Figure 18—Thickness of Cretaceous Montana Group and equivalent rocks, partly restored. Data are palinspastically restored in thrust belt. Present-day positions (unrestored palinspastically) of Idaho and Boulder batholiths and Tertiary basins are shown.

Marine Jurassic rocks are dominated by fine- to medium-grained clastic sediments that originated mainly from the western source area. Significant carbonate deposits are present in the lower part of the Upper Jurassic sequence (Rierdon Formation), which are a northern extension of the thick Twin Creek Limestone facies of southeastern Idaho. These beds represent the final deposition of important marine carbonate sediments on the Rocky Mountain shelf. Marine clastic deposits of the Swift Formation grade into the basal beds of the overlying nonmarine Morrison Formation without evidence of unconformity.

Detailed analyses of the marine Jurassic sequences of Montana include those by Cobban (1945), Imlay (1945, 1957, 1980), Imlay et al. (1948), Moritz (1951, 1960), Schmitt (1953), Nordquist (1955), and Peterson (1957, 1972a, b).

Upper Jurassic–Lower Cretaceous: Kootenai and Morrison Formations

The Upper Jurassic Morrison Formation, the Lower Cretaceous Kootenai Formation and their equivalents are 500 ft (150 m) to more than 1500 ft (450 m) thick in western Montana (Figure 16). These beds represent an interruption of marine deposition in the western interior United States that affected the entire Rocky Mountain shelf. During this time, the shelf was covered with a blanket of varicolored clays, silts, sands, and continental lacustrine deposits. Much of the fine-grained fraction of this sedimentary blanket may have originated from fallout of fine-grained volcanic debris generated in the western North American volcanic field where important batholith emplacement occurred at this time (Stokes, 1950; Moberly,

1960; Peterson, 1966, 1972a, b; Suttner, 1969; Suttner et al., 1981). Prominent fluvial channel sand and gravel deposits are present in several parts of the Morrison and Kootenai section and their equivalents in all areas of the Rocky Mountains, and a widespread blanket deposit of boulder, gravel, and coarse sand material is commonly present at the base of the Kootenai and its equivalents on the Rocky Mountain shelf. A western source for this material is well documented, and its depositional environment has been the subject of much discussion (Stokes, 1950; Moberly, 1960; Peterson, 1966, 1972a, b; Suttner, 1969; Walker, 1974; Suttner et al., 1981).

Lacustrine limestone beds are frequently found interbedded with Morrison clastic beds, and evidence has been presented for relatively widespread, probable lacustrine dark shale and coaly deposits in the uppermost Morrison Formation, which generally cover the area of the central Montana trough (Peterson, 1966, 1972a, b; Walker, 1974). Relatively thick, widespread lacustrine limestone beds are also present in the lower and upper parts of the Kootenai Formation and equivalent beds in western Montana, southeastern Idaho, western Wyoming, and central Utah (Holm, et al., 1977). These deposits probably represent widespread north–south-oriented lake systems that formed in the foredeep area of the western tectonic belt that was progressively increasing in activity by Late Jurassic–Early Cretaceous time. At times of lower clastic influx or higher precipitation, the subsiding foredeep filled with nonmarine waters that persisted long enough to allow deposition of thick limestones containing ostracods, charaphytes, and gastropods. The upper part of the Kootenai grades upward from the lacustrine gastropod limestone beds and equivalent clastics into the lower part of the Blackleaf Formation, which contains the continental and nearshore marine sediments of the initial Cretaceous marine transgression.

Detailed studies of Morrison and Kootenai stratigraphy and facies distribution in western Montana include those by Peck (1941, 1956), Yen (1949, 1951), Moberly (1960), Suttner (1968, 1969), Paine (1970), Walker (1974), James and Oaks (1977), Holm et al. (1977, and Suttner et al. (1981).

Cretaceous: Colorado Group and Equivalent Rocks

Post-Kootenai Cretaceous rocks were deposited in a sequence of at least five transgressive–regressive cycles: the Dakota–Mowry and Frontier–Niobrara cycles, which make up the Colorado Group, and the Eagle–Claggett, Judith River–Bearpaw, and Fox Hills–Lance–Hell Creek cycles, which make up the Montana Group. In western Montana, all of these marine cycles grade westward into coarse-grained nonmarine facies that progressively increase in coarseness and content of volcanic material in the younger Cretaceous units (Figures 4, 5, 6, 7, and 17). Major tectonic elements affecting these sedimentary distribution patterns were the progressive growth and eastward spreading of the western Montana thrust and fold belt and the emplacement of the Upper Cretaceous Idaho and Boulder batholith systems and their satellite intrusives.

The Colorado Group comprises a sequence of well-defined, fossiliferous, marine transgressive–regressive cyclic clastic deposits (Dakota through Niobrara formations). These units grade westward into a complex, intertonguing

marine and nonmarine facies and finally a nonmarine, clastic, nonfossiliferous, partly volcanic sequence. The main belt of marine to nonmarine transition occurs roughly in the vicinity of the central and northern parts of the western Montana disturbed belt where the eastern open marine section grades westward into the Blackleaf and Marias River formations and their equivalents (Figures 5 and 6). Volcanic debris is common in both these units, which as pointed out by McMannis (1965), may be related to the early stages of emplacement of the Idaho batholith and associated volcanic activity. In southwestern Montana, the marine units of the Colorado Group extend somewhat farther west than in the west-central and northern areas.

Thickness of the Colorado Group and equivalents increases westward from less than 2000 ft (600 m) east of the overthrust belt in north-central Montana to more than 7500 ft (2300 m) in west-central Montana (Figure 17). Part of the section is removed by Cenozoic erosion along a broad belt roughly corresponding with the trend of the Sweetgrass arch and Little Belt Mountains. This thinning trend extends southwestward for some distance across the approximate position of the Boulder high, indicating continued growth of this paleostructural feature. The section is entirely removed by erosion or nondeposition in northwestern Montana, which in large part became a major clastic source area during Cretaceous time. A general thickening trend extends eastward from the central overthrust belt approximately in the position of the central Montana trough. Continued tectonic activity in westernmost Montana and adjacent Idaho expanded the size and elevation of the western source terrane and thus increased the influx of coarse clastic material by high gradient streams off the western highland. Some foredeep lacustrine deposits are also reported in part of the Colorado Group section in west-central Montana. Volcanic debris is common in parts of the sequence.

Detailed studies of the Colorado Group include those by Cobban (1951), Gwinn (1965), Cannon (1966), Schwartz (1972), Mudge (1972), McGookey et al. (1972), and Suttner et al. (1981)

Cretaceous: Montana Group and Equivalent Rocks

Five major facies of thick, coarse-grained nonmarine facies dominate Montana Group stratigraphy in western Montana (Figures 4, 5, 6, 7, and 18).

The *Beaverhead Conglomerate facies* is a complexly intertonguing facies, more than 15,000 ft (4500 m) thick, of quartzite and limestone-boulder conglomerate and sandstone, which according to Ryder and Scholten (1973) probably ranges in age from middle Colorado to Paleocene or Eocene. The source of some of the material was the rapidly expanding highland to the west associated with the evolving overthrust belt (Figures 4 and 18), but local sources were also present east of the overthrust belt.

The *Golden Spike facies* is at least 7500 ft (2300 m) thick and is similar in nature to the Beaverhead Conglomerate facies. Part of the Golden Spike could originally have been a northern continuation of the Beaverhead Conglomerate facies, and most of it may have been removed by subsequent erosion (Figures 5 and 18) (Gwinn and Mutch, 1965). A substantial portion of the Golden Spike facies, however, was probably derived from the Boulder high (Figure 18).

The *Elkhorn Mountains volcanics* occupy the frontal position of the overthrust belt in the vicinity of Helena and Butte and intertongue westward with part of the Golden Spike facies (Figures 5 and 18). This facies is a complex pile of volcanic flows, tuffs, breccias, and tuffaceous sandstone and mudstone that intertongues with and grades eastward into units of the Livingston Group (McMannis, 1955, 1965; Roberts, 1965; Skipp and McGrew, 1977). The origin of the material is closely related to the emplacement of the Boulder batholith complex (Smedes, 1966; Robinson et al., 1969).

The *Livingston Group* is more than 5000 ft (1500 m) thick and occupies the Crazy Mountain basin and adjacent area. The Livingston is a complex of mostly nonmarine volcaniclastic sandstone and mudstone containing a few volcanic flows and tuffs. The lower part intertongues rapidly eastward with marine units of the Montana Group on the east flank of the Crazy Mountain basin (Figures 5 and 18). The upper two-thirds of the Livingston intertongues with nonmarine beds of the Hell Creek and Lance formations (Roberts, 1965).

The *Two Medicine Formation* is preserved in outcrop remnants, some of which are at least 5000 ft (1500 m) thick, in the frontal zone of the northern disturbed belt near and south of Glacier National Park. Eastward in the subsurface, the nonmarine Two Medicine thins to 2000 ft (600 m) or less and grades into marine units of the Montana Group along the west flank of the Sweetgrass arch (Schwartz, 1972). To the south it intertongues with the Elkhorn Mountains volcanic facies (Figures 6 and 7).

The overall pattern of these facies relationships indicates that during Montana time, coarse boulder conglomerate debris that was derived primarily from erosion of Precambrian and Paleozoic rocks was transported eastward from the rising highland belt associated with the rapidly expanding western thrust and fold belt. Volcanic debris and flows were concentrated around the periphery of the Boulder high, which during Cretaceous (Montanan) time was the main site of emplacement of the Boulder batholith and its associated intrusive rocks. The marine Cretaceous seaway apparently transgressed westward into this complex of volcanic and conglomeratic debris with frequent rapid fluctuations of the western shoreline. At the same time, much volcanic ash and tuff fallout was incorporated with normal shallow water marine deposits to the east, which resulted in the common layers of bentonite and bentonitic shale characteristic of the Colorado and Montana Group beds in central and eastern Montana.

Cretaceous sedimentation in western Montana closed with the deposition of the nonmarine St. Mary's River Formation in the Glacier Park area and the Adel Mountains volcanic rocks to the south. In southwestern and west-central Montana, rocks of this age have been almost entirely removed by subsequent erosion, but age equivalents are probably preserved in the upper part of the Beaverhead Conglomerate and the Sphinx Conglomerate near Yellowstone National Park.

Detailed studies of Montana Group facies and correlative rocks in western Montana include those by McLaughlin and Johnson (1955), Klepper et al. (1957), Weimer (1960, 1961), Roberts (1963, 1965), Gwinn (1965), Gwinn and Mutch (1965), Viele and Harris (1965), Simms (1967), Wilson (1967), Ryder and Ames (1970), Ryder and Scholten (1973),

McGookey et al. (1972), and Skipp and McGrew (1977). Studies of the age and emplacement processes of the Boulder batholith complex include those by Weeks and Klepper (1954), Klepper et al. (1957), Chapman et al. (1955), Smedes (1966), Burfiend (1967), Robinson et al. (1969), Hamilton and Myers (1974), and Hyndman et al. (1975).

Tertiary

Sedimentation patterns of early Tertiary time reflect the continuation without significant interruption of the closing Cretaceous paleogeographic framework and depositional basins of the northern Rocky Mountains. The major change influencing depositional environments was the final withdrawal of Upper Cretaceous marine waters from the Rocky Mountain shelf. Marine deposition was replaced by widespread lacustrine and fluvial deposition across the emergent, gently sloping Late Cretaceous sea bottom, and extensive lacustrine and coastal swamp deposition of coaly beds and clastics occurred in the more active basinal areas of the shelf. These deposits make up the Hell Creek Formation and overlying Fort Union Group of the Montana plains (Figures 5 and 6). In western Montana, remnants of equivalent age rocks are scarce, but they are probably represented in the upper part of the Beaverhead and Sphinx conglomerates of southwestern Montana and the nonmarine Willow Creek Formation and Adel Mountains volcanics south of Glacier National Park.

By late Eocene time, the downfaulted continental Tertiary basins had been relatively well developed, and the remainder of Tertiary sedimentation was involved with basin-fill deposition in the many downwarped valley areas of present-day western Montana (Figure 18). Middle to Late Tertiary fluvial and lacustrine beds reach thicknesses of as much as 5000–10,000 ft (1500–3000 m) in several of the basins (C. A. Balster, personal communication, 1985). Burfiend (1967) reported gravity data that indicate that some of the valleys between Bozeman and Butte are filled with at least 5000 ft (1500 m) of unconsolidated material. According to Kuenzi and Fields (1971), the valley fill in the Jefferson basin east of Butte contains an upper Eocene or Oligocene to lower Miocene, fine-grained clastic sequence overlain by a Miocene and Pliocene sequence of coarser material that are separated by a marked unconformity. Similar stratigraphic sequences are reported in other Tertiary valleys of western Montana, and evidence suggests that the unconformity may be a regional feature common to valley fill history in many parts of western Montana (Robinson, 1961; Kuenzi and Richard, 1969). Many of the late Tertiary basins probably were controlled by rejuvenation of Precambrian tectonic features in western Montana (O'Neill and Lopez, 1985).

The age and mechanisms of thrusting and other tectonic activity that shaped the final form of western Montana are discussed in detail by many authors, including Alpha (1955, 1958), Poulter (1956, 1958), Scholten (1956, 1960, 1967, 1968), Reid (1957), Foose et al. (1961), Mutch (1961), Gwinn (1961), Childers (1963), Armstrong and Oriel (1965), Weidman (1965), McGill (1965), Smith and Barnes (1966), Scholten and Ramspott (1968), Mudge (1970), Klepper et al. (1971), Bregman (1971), Tysdal (1976), Skipp and Hait (1977), Ruppel (1978, 1982,), Ruppel and Lopez (1984), Hadley (1980), and Woodward (1980).

PETROLEUM GEOLOGY

Many of the components for petroleum generation, accumulation, and preservation are present in western Montana. Much of the geologic section has a shallow marine origin and contains variable lithologic facies that include substantial thicknesses of potential reservoir rock, source rock, and cap rock, which in many cases are closely associated under favorable stratigraphic trapping conditions. The presence of paleostructural elements and shelf marginal belts, which persisted throughout most of the depositional history here, provided a combination of favorable conditions for the localized development of good reservoir rock facies and the continuous growth of a stratigraphic and structural framework favorable to early migration and trapping of hydrocarbons at the time of generation. Factors that may negatively affect hydrocarbon accumulation include deep burial and tectonic alteration of clastic reservoir rocks; excessive thermal effects on source rocks; and destruction of early traps by stratigraphic and structural events in parts of the area, such as excessive burial depths and the complex structural and igneous activity during the Mesozoic and Tertiary. However, fracturing of potential reservoir rocks, particularly dolomites, by intense structural activity should be a positive factor.

Cambrian

The thick, porous Hasmark dolomite facies is of interest because of its potential for trapping petroleum, possibly nonindigenous, under suitable structural conditions. Evidence for adequate Cambrian source rocks is scarce, and burial depths beyond the postmature stage of organic diagenesis have affected these rocks in much of western Montana (Figures 4, 5, 6, and 7).

Devonian

Jefferson Formation dolomite beds are porous almost everywhere and commonly show indications of petroleum in the form of staining or odor. This thick reservoir facies is of major interest, although important petroleum accumulations have not yet been found in these beds in central to western Montana. Potentially good source rock beds are present in the Devonian–Mississippian Milligen Formation of east-central Idaho and southwestern Montana and in the thinner, partially equivalent, Bakken–Sappington facies in western Montana. Remnants of these facies may be present beneath thrusts in parts of central western or northwestern Montana. According to Perry et al. (1981), these beds have been thermally altered to postmaturity west of the Medicine Lodge thrust. In much of western Montana, however, these rocks probably have not been buried to depths beyond the mature stage (Figures 4, 5, 6, and 7).

Mississippian

Some porous dolomite units are present within the Madison bank carbonate facies in west-central to southwestern Montana. These beds show indications of petroleum at some outcrop localities. The Sun River Dolomite facies, the major Paleozoic reservoir in the Sweetgrass arch region, covers a broad area of northwestern

and west-central Montana (Figure 11) and is of major interest for future exploration. The Bakken–Sappington source rock facies is about 1000 ft (300 m) or more stratigraphically below the Sun River, but along with dark shale and argillaceous limestone beds in the lower Lodgepole and equivalents, it is potentially a good source rock facies in much of the area. Except for parts of southwestern and west-central Montana, these beds probably have not been buried to excessive depths, and in shallow burial areas they may never have been heated beyond the early stages of organic maturity (Figures 4, 5, 6, and 7). The highly organic-rich shale beds of the Heath Formation extend westward and southwestward from the Big Snowy trough and are present in a large part of the central and southwestern thrust belt and in east-central Idaho (Figure 12). This source rock facies has been deeply buried in east-central Idaho and in the Snowcrest trough, but in the remainder of western Montana burial depths probably did not reach the postmature stage.

Pennsylvanian–Permian

The Quadrant Formation and its equivalent shelf sandstone sequence are a major reservoir facies in much of the northern Rocky Mountains, particularly in Wyoming. These beds have been deeply buried and are quartzite or dolomite in most of southwestern and west-central Montana, but interbedded porous sandstones of probable eolian origin are present in the central and southwestern thrust belt and adjacent area (D. Saperstone, personal communication, 1984). The overlying Permian sandstone and carbonate beds generally are not very porous and are of only moderate interest as reservoirs. The highly organic-rich Retort and Meade Peak phosphatic shale beds have excellent source rock qualities in much of the northern Rocky Mountain region, including southwestern Montana. In some places these rocks are not thermally mature, such as in the Tendoy thrust area (Claypool et al., 1978; Perry et al., 1981, 1983), but in the Snowcrest trough and in some of the Tertiary basins, or under thrust sheets, they may have reached thermal maturity by late Mesozoic time (Figures 4 and 5).

Triassic–Jurassic

Reservoir and source rock quality of Triassic beds and of marine beds of the Jurassic appear to be low in west-central to southwestern Montana. However, in the thrust belt west of Great Falls, the Sawtooth Formation includes dark marine shale beds that are phosphatic in places; these are of interest as source rocks and may be the source for some of the Sweetgrass arch petroleum. Dark marine shale beds are a minor component of the Rierdon and Swift formations in the same area, and these, along with dark shale and coaly lacustrine beds in the upper Morrison Formation, deserve some attention for moderate to low potential.

Cretaceous

The intertonguing belt of marine and continental facies of the Colorado Group falls generally within and just east of the overthrust belt. Good clastic nearshore continental and marine reservoir sandstone beds and dark continental or marine shales are present within much of this facies. Within and adjacent to the thrust belt, these beds may have been buried deeply enough to reach thermal maturity (Figures 4, 5, 6, and 7). The similar continental to marine transition facies of the Montana Group generally occurs farther east. In most places, this sequence probably has not been buried to thermal maturity depths, but potential for biogenic gas generation may have been good if trapping conditions were favorable in these beds and in beds of the Colorado Group (Rice and Shurr, 1978; Clayton et al., 1982).

Tertiary

Discontinuous alluvial sandstone bodies, some with good porosity, are common in the early to middle Tertiary beds of the western Montana Tertiary basins, and some carbonaceous to coaly beds are also present. Except for the deeper parts of these basins, most of this section has not been buried to thermal maturity, but the possibility of small to moderate sized biogenic gas accumulations, mostly in stratigraphic traps, deserves consideration in future exploration programs.

Summary of Petroleum Geology

The western Montana overthrust belt contains a less complete and generally thinner stratigraphic section and less extensive potential source rock facies than that in the overthrust belt of western Wyoming, southeastern Idaho, and northern Utah where major petroleum accumulations have been discovered in recent years. The western Montana area is also characterized by extensive Mesozoic and Cenozoic igneous activity, in contrast with western Wyoming and southeastern Idaho. In addition, the high degree of structural complexity and surface exposure of much of the potentially favorable rock section and the detrimental effects of deep burial of Paleozoic rocks all detract from the petroleum potential of the region. The data presented by Claypool et al. (1978) and Perry et al. (1981, 1983), however, indicate that the Phosphoria and the highly organic-rich Big Snowy source rock facies have not reached the full maturity thermal stage in the Tendoy thrust region. Thus, windows of less rigid thermal effects may be preserved in places, particularly in the vicinity of persistent paleostructural highs. The palinspastically restored stratigraphic cross section in southwestern Montana (Figure 4) indicates that the middle and upper Paleozoic section had not been buried beyond the early mature stage by the end of Jurassic time. This was probably near the time of maximum burial depth of Paleozoic beds in this area. During Late Jurassic and later deposition, much of this area underwent relatively continuous uplifting as the Mesozoic fold and thrust system evolved. During this time, the Paleozoic section was relatively continuously uplifted and finally partly eroded, and the thermal gradient burial cycle may have either remained dormant or reversed in direction, resulting in the preservation of thermally less mature windows.

A large area of unexplored terrain remains in western Montana. Burial depths are within a favorable range in significant parts of the area, and a large volume of favorable reservoir facies is present. At this time, geologic appraisal and the drilling history of the area continues to point primarily toward its gas potential, with low probability for significant oil resources.

ACKNOWLEDGMENTS

The information used in this report was derived from many sources, most of which are acknowledged in the reference list. Revisions and improvements in the original manuscript were made after reviews by C. A. Balster, J. J. Tonnsen, R. G. Tysdal, T. E. Tucker, E. T. Ruppel, D. A. Lopez, and G. J. Verville.

REFERENCES CITED

Aadland, R. K., 1979, Cambrian stratigraphy of the Libby Trough, Montana: University of Idaho, Moscow, Idaho, Ph.D. thesis, 228 p.

Alexander, R. G., Jr., 1955, The geology of the Whitehall area, Montana: Yellowstone-Bighorn Research Association Contribution, 195, 111 p.

Alpha, A. G., 1955, Tectonic history of northeastern Montana: Billings Geological Society 6th Annual Field Conference Guidebook, p. 129–142.

———, 1958, Tectonic history of Montana: Billings Geological Society 9th Annual Field Conference Guidebook, p. 58–68.

Andrichuk, J. M., 1951, Regional stratigraphic analysis of the Devonian system in Wyoming, Montana, southern Saskatchewan and Alberta: AAPG Bulletin, v. 35, p. 2368–2408.

———, 1955, Mississippian–Madison group stratigraphy and sedimentation in Wyoming and southern Montana: AAPG Bulletin, v. 39, n. 11, p. 2170–2210.

Armstrong, F. C., and S. S. Oriel, 1965, Tectonic development of Idaho–Wyoming thrust belt: AAPG Bulletin, v. 49, p. 1847–1866.

Armstrong, R. L., 1975, Precambrian (1,500 m.y. old) rocks of central Idaho—the Salmon River arch and its role in Cordilleran sedimentation and tectonics: American Journal of Science, v. 275-A, p. 437–467.

Benson, A. L., 1966, Devonian stratigraphy of western Wyoming and adjacent areas: AAPG Bulletin, v. 50, p. 2566–2603.

Billings Geological Society, 1962, A. R. Hansen and J. H. McKeever, eds., Symposium, Devonian System of Montana and adjacent areas: Billings Geological Society, Billings, Montana, 152 p.

Bregman, M. L., 1971, Change in tectonic style along the Montana thrust belt: Geological Society of America, Geology, v. 4, p. 775–778.

Breuninger, R. H., 1976, *Palaeoaplysina* carbonate buildups from upper Paleozoic of Idaho: AAPG Bulletin, v. 60, p. 584–607.

Brown, C. W., 1957, Stratigraphic and structural geology of north-central northeast Yellowstone National Park, Wyoming and Montana: Ph.D. thesis, Princeton University, 287 p.

Burfiend, W. J., 1967, A gravity investigation of the Tobacco Root Mountains, Jefferson basin, Boulder batholith, and adjacent areas of southwestern Montana: Ph.D. thesis, Indiana University, Bloomington, Indiana, 146 p.

Cannon, J. L., 1955, Cretaceous rocks of northwestern Montana: Billings Geological Society Guidebook 6th Annual Field Conference, p. 107–119.

————, 1966, Outcrop examination and interpretation of paleocurrent patterns of the Blackleaf Formation near Great Falls, Montana: Billings Geological Society, 17th Annual Field Conference Guidebook, p. 71–111.

Chapman, R. W., D. Gottfried, and C. L. Waring, 1955, Age determinations on some rock from the Boulder batholith and other batholiths of western Montana: Geological Society of America Bulletin, v. 66, p. 607–610.

Childers, M. W., 1963, Structure and stratigraphy of the southwest Marias Pass area, Flathead, County, Montana: Geological Society of America Bulletin, v. 74, p. 141–164.

Claypool, G. E., A. H. Love, and E. K. Maughan, 1978, Organic geochemistry, incipient metamorphism, and oil generation in black shale members of Phosphoria Formation, western interior United States: AAPG Bulletin, v. 62, p. 98–120.

Clayton, J., M. R. Mudge, C. Lubeck, and T. A. Daws, 1982, Hydrocarbon source rock evaluation of the disturbed belt, northwestern Montana, *in* Powers, R. B., ed., Geologic Studies of the Cordilleran Thrust Belt: Rocky Mountain Association of Geologists, p. 777–804.

Cobban, W. A., 1945, Marine Jurassic formations of Sweetgrass Arch, Montana: AAPG Bulletin, v. 29, p. 1262–1303.

————, 1951, Colorado shale of central and northwestern Montana and equivalent rocks of the Black Hills: AAPG Bulletin, v. 35, p. 2170–2206.

Cressman, E. R., 1955, Physical stratigraphy of the Phosphoria Formation in part of southwestern Montana: U.S. Geological Survey Bulletin, 1027-A, p. 1–31.

Cressman, E. R., and R. W. Swanson, 1964, Stratigraphy and petrology of the Permian rocks of southwestern Montana: U.S. Geological Survey Professional Paper 313-C, p. 275–569.

Dahl, G. D., Jr., 1971, General geology of the area drained by the north fork of the Smith River, Meagher Co., Montana: M. S. thesis, Montana, College of Mineral Science and Technology, 56 p.

Easton, W. H., 1962, Carboniferous formations and faunas of central Montana: U.S. Geological Survey Professional Paper 348, 126 p.

Fanshawe, J. R., 1978, Central Montana tectonics and the Tyler Formation, *in* The economic geology of the Williston Basin: Montana Geological Society 24th Annual Conference, Williston Basin Symposium, p. 239–248.

Foose, R. M., D. U. Wise, and G. S. Garbarini, 1961, Structural geology of the Beartooth Mountains, Montana and Wyoming: Geological Society of America Bulletin, v. 72, p. 1143–1172.

Gardner, L. S., T. A. Hendricks, H. D. Hadley, and C. P. Rogers, Jr., 1946, Stratigraphic sections of upper Paleozoic and Mesozoic rocks in south-central Montana: Montana Bureau of Mines and Geology Memoir 24, 100 p.

Gealy, W. J., 1953, Geology of the Antone Peak Quadrangle, southwestern Montana: Ph.D. thesis, Harvard University, 143 p.

Gutschick, R. C., L. J. Suttner, and M. J. Switek, 1962, Biostratigraphy of transitional Devonian–Mississippian Sappington Formation of southwest Montana: Billings Geological Society Devonian System of Montana, 13th Annual Field Conference, p. 79–89.

Gwinn, V., 1952, Summary of Jurassic history in the western interior of the United States, *in* Billings Geological Society Guidebook, 3rd Annual Field Conference, p. 79–85.

————, 1961, Geologic map of the Drummond area, Granite and Powell Counties, Montana: Montana Bureau of Mines and Geology, Special Publication 19 (Map no. 3).

————, 1965, Cretaceous rocks of the Clark Fork Valley, central-western Montana: Billings Geological Society 16th Annual Field Conference Guidebook, p. 34–47.

Gwinn, V. E., and T. A. Mutch, 1965, Intertongued Upper Cretaceous volcanic and non-volcanic rocks, central-western Montana: Geological Society of America Bulletin, v. 76, p. 1125–1144.

Hadley, J. B., 1980, Geology of the Varney and Cameron Quadrangle, Madison County, Montana: U.S. Geological Survey Bulletin, 1459, 108 p.

Hamilton, W., and W. B. Myers, 1974, Nature of the Boulder batholith of Montana: Geological Society of America Bulletin, v. 85, p. 365–378.

Hanson, A. M., 1952, Cambrian stratigraphy in southwestern Montana: Montana Bureau of Mines and Geology Memoir 33, 32 p.

Harris, W. L., 1972, Upper Mississippian and Pennsylvanian sediments of central Montana: Ph.D thesis, University of Montana, Missoula, Montana, 251 p.

Harrison, J. E., 1972, Precambrian belt basin of northwestern United States, its geometry, sedimentation, and copper resources: Geological Society of America Bulletin, v. 83, p. 1215–1240.

Harrison, J. E., A. B. Griggs, and J. D. Wells, 1974, Tectonic features of the Precambrian belt basin and their influence on post-Belt structures: U.S. Geological Survey Professional Paper 866, 15 p.

Holm, M. R., W. C. James, and L. J. Suttner, 1977, Comparison of the Peterson and Draney limestones, Idaho and Wyoming, and the calcareous members of the Kootenai Formation, western Montana, *in* Rocky Mountain Thrust Belt, Geology and Resources: Wyoming Geological Association 19th Annual Field Conference, p. 259–270.

Huh, O., 1967, The Mississippian System across the Wasatch line, east-central Idaho, extreme southwestern Montana: Montana Geological Society 18th Annual Field Conference, p. 31–62.

Hyndman, D. W., J. L. Talbot, and R. B. Chase, 1975, Boulder batholith–a result of emplacement of a block detached from the Idaho batholith infrastructure: Geology, v. 3, p. 401–404.

Imlay, R. W., 1945, Occurrence of Middle Jurassic rocks in western interior of the United States: AAPG Bulletin, v. 29, p. 1019–1027.

————, 1948, Characteristic marine Jurassic fossils from the western interior of the United States: U.S. Geological Survey Professional Paper 214-B, p. 13–33.

————, 1957, Paleoecology of Jurassic seas in the western

interior of the United States: Geological Society of America Memoir 67, v. 2, p. 469–504.

_____, 1980, Jurassic paleobiogeography of the conterminous United States in its continental setting: U.S. Geological Survey Professional Paper 1062, 134 p.

Imlay, R. W., L. S. Gardner, C. P. Rogers, and H. D. Hadley, 1948, Marine Jurassic Formations of Montana: U.S. Geological Survey Oil and Gas Investigations Preliminary Chart 32.

James, H. L., and C. E. Hedge, 1980, Age of the basement rocks of southwest Montana: Geological Society of America Bulletin, v. 91, p. 11–15.

James, W. C., and R. Q. Oaks, Jr., 1977, Petrology of the Kinnikinic Quartzite, east-central Idaho: Journal Sedimentary Petrology, v. 47, p. 1491–1511.

Keenmon, K. A., 1950, Geology of the Blacktail-Snowcrest region, Beaverhead County, Montana: Ph.D. thesis, University of Michigan, 206 p.

Klepper, M. R., R. A. Weeks, and E. T. Ruppel, 1957, Geology of the southern Elkhorn Mountains, Jefferson and Broadwater Counties, Montana: U.S. Geological Survey Professional Paper 292, 82 p.

Klepper, M. R., G. D. Robinson, and H. W. Smedes, 1971, On the nature of the Boulder batholith of Montana: Geological Society of America Bulletin, v. 82, p. 1563–1580.

Kuenzi, W. D., and B. H. Richard, 1969, Middle Tertiary unconformity, North Boulder and Jefferson basins, southwestern Montana: Geological Society of America Bulletin, v. 80, p. 111–120.

Kuenzi, W. D., and R. W. Fields, 1971, Tertiary stratigraphy, structure, and geologic history, Jefferson Basin, Montana: Geological Society of America Bulletin, v. 82, p. 3374–3394.

Kummel, B., 1954, Triassic stratigraphy of southwestern Montana: Billings Geological Society 11th Annual Field Conference Guidebook, p. 239–243.

Lochman-Balk, C. 1972, Cambrian system, *in* Geologic Atlas of the Rocky Mountain region: Denver, Colorado, Rocky Mountain Association Geologists, p. 60–75.

McGill, G. E., 1965, Geologic map, northwest flank Flint Creek Range, Montana: Montana Bureau of Mines and Geology Special Publication 18 (Map no. 3).

McGill, G. E., and D. A. Sommers, 1967, Stratigraphy and correlation of the Precambrian Belt Supergroup of the southern Lewis and Clark Range, Montana: Geological Society of America Bulletin, v. 78, p. 343–352.

McGookey, D. P., W. A. Cobban, J. R. Gill, H. G. Goodell, L. A. Hale, J. D. Haun, D. G. MCubbin, R. J. Weimer, and G. R. Wulf, 1972, Cretaceous System, *in* Geologic Atlas of the Rocky Mountain Region: Denver, Colorado, Rocky Mountain Association Geologists, p. 190–228.

McKelvey, V. E., J. S. Williams, R. P. Sheldon, E. R. Cressman, T. M. Cheney, and R. W. Swanson, 1959, The Phosphoria, Park City, and Shedhorn Formations in the western phosphate field: U.S. Geological Survey Professional Paper 313-A, p. 1–47.

McLaughlin, K. P., and D. M. Johnson, 1955, Upper Cretaceous and Paleocene strata in Montana west of the continental divide, *in* Billings Geological Society, 6th Annual Field Conference, p. 4–12.

McMannis, W. J., 1955, Geology of the Bridger Range, Montana: Geological Society of America Bulletin, v. 66, p. 1385–1430.

_____, 1962, Devonian stratigraphy between Three Forks, Montana and Yellowstone Park: Billings Geological Society Guidebook 13th Annual Field Conference, p. 4–12.

_____, 1963, Lahood Formation—a coarse facies of the Belt Series in southwestern Montana: Geological Society of America Bulletin, v. 74, p. 407–436.

_____, 1965, Resume of depositional and structural history of western Montana: AAPG Bulletin, v. 49, p. 1801–1823.

McMannis, W. J., and R. A. Chadwick, 1964, Geology of the Garnet Mountain quadrangle, Gallatin County, Montana: Montana Bureau of Mines and Geology Bulletin, v. 43, 47 p.

Mann, J. A., 1954, Geology of part of the Gravelly Range, Montana: Yellowstone-Bighorn Research Association Contribution 190, 92 p.

Maughan, E. K., 1975, Montana, North Dakota, northeastern Wyoming, and northern South Dakota, Chapter O, *in* E. D. McKee, and E. J., Crosby, eds., Paleotectonic investigations of the Pennsylvanian system in the United States—Part I; Introduction and regional analyses of the Pennsylvanian System: U.S. Geological Survey Professional Paper 853-0, p. 279–293.

_____, 1984, Paleogeographic setting of Pennsylvanian Tyler Formation and relation to underlying Mississippian rocks in Montana and North Dakota: AAPG Bulletin, v. 68, p. 1778–1795.

Maughan, E. K., and A. E. Roberts, 1967, Big Snowy and Amsden Groups and the Mississippian–Pennsylvanian boundary in Montana: U.S. Geological Survey Professional Paper 554-B, 27 p.

Moberly, R., Jr., 1960, Morrison, Cloverly, and Sykes Mountain Formations, northern Big Horn Basin, Wyoming and Montana: Geological Society of America Bulletin, v. 71, p. 1137–1176.

Moritz, C. A., 1951, Triassic and Jurassic stratigraphy of southwestern Montana: AAPG Bulletin, v. 35, p. 1781–1814.

_____, 1960, Summary of Jurassic stratigraphy of southwestern Montana: Billings Geological Society Guidebook 11th Annual Field Conference, p. 239–243.

Mudge, M. R., 1970, Origin of the disturbed belt in northwestern Montana: Geological Society of America Bulletin, v. 81, p. 377–392.

_____, 1972, Pre-Quaternary rocks of the Sun River Canyon area, northwestern Montana: U.S. Geological Survey Professional Paper 663-A, 138 p.

Mudge, M. R., W. J. Sando, and J. T. Dutro, Jr., 1962, Mississippian rocks of Sun River Canyon area, Sawtooth Range, Montana: AAPG Bulletin, v. 46, p. 2003–2018.

Mutch, T. A., 1961, Geologic map of the northwest flank of the Flint Creek Range, Montana: Montana Bureau of Mines and Geology Special Publication, n. 22.

Nordquist, J. W., 1953, Mississippian stratigraphy of northern Montana: Billings Geological Society Guidebook 4th Annual Field Conference, p. 68–82.

_____, 1955, Pre-Rierdon Jurassic stratigraphy in northern

Montana and Williston basin: Billings Geological Society Guidebook 6th Annual Field Conference, p. 96–106.

O'Neill, J. M., and D. A. Lopez, 1985, Character and regional significance of Great Falls tectonic zone, east-central Idaho and west-central Montana: AAPG Bulletin, v. 69, p. 437–447.

Paine, M. H., 1970, A reconnaissance study of the gastropod limestone (Lower Cretaceous) in southwest Montana: Master's Thesis, Indiana University, Bloomington, Indiana, 243 p.

Peck, R. E., 1941, Lower Cretaceous Rocky Mountain non-marine microfossils: Journal of Paleontology, v. 15, p. 185–304.

——, 1956, North American Mesozoic Charophyta: U.S. Geological Survey Professional Paper 294-A, p. 1–44.

Perry, W. J., Jr., R. T. Ryder, and E. K. Maughan, 1981, The southern part of the southwest Montana thrust belt—a preliminary re-evaluation of structure, thermal maturation, and petroleum potential: Montana Geological Society Field Conference, southwest Montana, p. 261–273.

Perry, W. J., Jr., B. R. Wardlaw, N. H. Bostick, and E. K. Maughan, 1983, Structure, burial history, and petroleum potential of frontal thrust belt and adjacent foreland, southwest Montana: AAPG Bulletin, v. 67, p. 725–743.

Peterson, J. A., 1957, Marine Jurassic of northern Rocky Mountains and Williston basin: AAPG Bulletin, v. 41, p. 399–440.

——, 1966, Sedimentary history of the Sweetgrass Arch: Billings Geological Society Proceedings 17th Annual Field Conference and Symposium, p. 112–133.

——, 1972a, Permian sedimentary facies, southwestern Montana: Montana Geological Society 21st Annual Field Conference Guidebook, p. 69–74.

——, 1972b, Jurassic System, *in* Geologic Atlas of the Rocky Mountain Region: Denver, Colorado, Rocky Mountain Association Geologists, p. 177–189.

——, 1977, Paleozoic shelf margins and marginal basins, western Rocky Mountains–Great Basin, United States: Wyoming Geological Association 29th Annual Field Conference Guidebook, p. 69–74.

——, 1980, Permian paleogeography and sedimentary provinces, west-central United States, *in* Symposium on Paleozoic Paleogeography of west-central United States: Society Economic Paleontologists and Mineralogists, p. 271–292.

——, 1981a, General stratigraphy and regional paleostructure of the western Montana overthrust belt: Montana Geological Society Field Conference and Symposium, southwest Montana, p. 5–35.

——, 1981b, Stratigraphy and sedimentary facies of the Madison Limestone and associated rocks in parts of Montana, North Dakota, South Dakota, Wyoming, and Nebraska: U.S. Geological Survey Open-File Report 81-642, 96 p.

——, 1984a, Permian stratigraphy, sedimentary facies, and general petroleum geology, Wyoming and adjacent area, *in* Symposium of Permian and Pennsylvanian of Wyoming: Wyoming Geological Association, p. 25–64.

——, 1984b, Stratigraphy and sedimentary facies of the Madison Limestone and associated rocks in parts of Montana, Nebraska, North Dakota, South Dakota. and Wyoming: U.S. Geological Survey Professional Paper 1373-A, p. A1–A34.

Poulter, G. J., 1956, Geological map and sections of the Georgetown thrust area, Granite and Deer Lodge Counties, Montana: Montana Bureau of Mines and Geology, Geological Investigations Map, n. 3.

——, 1958, Geology of the Georgetown thrust area southwest of Phillipsburg, Montana: Montana Bureau of Mines and Geology, Geological Investigations Map n. 1.

Rau, J. L., 1962, The stratigraphy of the Three Forks Formation: Billings Geological Society 7th Annual Field Conference Guidebook, p. 35–45.

Reid, R. R., 1957, Bedrock geology of the north end of the Tobacco Root Mountains, Madison County, Montana: Montana Bureau of Mines and Geology, Memoir 36, 25 p.

Rice, D. D., and G. W. Shurr, 1978, Potential for major natural gas resources in shallow, low-permeability reservoirs of the Northern Great Plains, *in* The economic geology of the Williston basin: Montana Geological Society 24th Annual Conference, Williston Basin Symposium, p. 265–281.

Richards, P. W., 1957, Geology of the area east and southeast of Livingston, Park County, Montana: U.S. Geological Survey Bulletin, 1021-L, p. 385–438.

Roberts, A. E., 1963, The Livingston Group of south-central Montana: U.S. Geological Survey Professional Paper, 475-B, p. B86–B92.

——, 1965, Correlation of Cretaceous and lower Tertiary rocks near Livingston, Montana, *in* Geological Survey Research 1965: U.S. Geological Survey Professional Paper 525-B.

——, 1966, Stratigraphy of Madison Group near Livingston, Montana, and discussion of karst and solution-breccia features: U.S. Geological Survey Professional Paper 526-B, p. B1–B23.

——, 1979, Northern Rocky Mountain and adjacent plains region, *in* L. C. Craig and C. W. Connor, eds., Paleotectonic investigations of the Mississippian System in the United States—Part I. Introduction and regional analyses of the Mississippian System: U.S. Geological Survey Professional Paper 1010, p. 221–248.

Robinson, G. D., 1961, Origin and development of the Three Forks basin, Montana: Geological Society of America Bulletin, v. 72, p. 1003–1014.

Robinson, G. D., M. R. Klepper, and J. D. Obradovich, 1969, Overlapping plutonism, volcanism, and tectonism in the Boulder batholith region, western Montana: Geological Society of America Memoir 116, p. 557–576.

Rose, P. R., 1976, Mississippian carbonate shelf margins, western United States: U.S. Geological Survey Journal of Research, v. 4, n. 4, p. 449–466.

Ruppel, E. T., 1972, Geology of pre-Tertiary rocks in the northern part of Yellowstone National Park, Wyoming: U.S. Geological Survey Professional Paper 729-A, 66 p.

——, 1978, The Medicine Lodge thrust system, east-central Idaho and southwest Montana: U.S. Geological Survey Professional Paper 1031, 23 p.

_____, 1982, Cenozoic block uplifts in southwest Montana and east-central Idaho: U.S. Geological Survey Professional Paper 1224, 24 p.

Ruppel, E. T., and D. A. Lopez, 1984, The thrust belt in southwest Montana and east-central Idaho: U.S. Geological Survey Professional Paper 1278, 41 p.

Ruppel, E. T., R. J. Ross, Jr., and D. L. Schleicher, 1975, Precambrian Z and Lower Ordovician rocks in east-central Idaho: U.S. Geological Survey Professional Paper 889-B. p. 25-34.

Ryder, R. T., and H. T. Ames, 1970, Palynology and age of Beaverhead Formation and their paleotectonic implications in Lima region, Montana–Idaho: AAPG Bulletin, v. 54, p. 1155-1171.

Ryder, R. T., and R. Scholten, 1973, Synetectonic conglomerates in southwestern Montana—their nature, origin, and tectonic significance: Geological Society of America Bulletin, v. 84, p. 773-796.

Sandberg, C. A., 1962, Stratigraphic section of type Three Forks and Jefferson Formations at Logan, Montana, *in* Billings Geological Society Guidebook 13th Annual Field Conference, p. 47-50.

_____, 1965, Nomenclature and correlation of lithologic subdivisions of the Jefferson and Three Forks Formations of southern Montana and northern Wyoming: U.S. Geological Survey Bulletin 1994-N, p. N1-N18

Sandberg, C. A., and C. R. Hammond, 1958, Devonian System in Williston basin and central Montana: AAPG Bulletin, v. 42, n. 10, p. 2293-2334.

Sandberg, C. A., and W. J. McMannis, 1964, Occurrence and paleogeographic significance of the Maywood Formation of Late Devonian age in the Gallatin Range, southwestern Montana: U.S. Geological Survey Professional Paper 501-C, p. C50-C54.

Sandberg, C. A., and W. J. Mapel, 1967, Devonian of the northern Rocky Mountains Plains, *in* D. H. Oswald, ed., International Symposium on the Devonian System: Alberta Society of Petroleum Geologists, v. 1, p. 843-877.

Sando, W. J., 1967, Mississippian depositional provinces in the northern Cordilleran region: U.S. Geological Survey Professional Paper 575-D, p. D29-D38.

_____, 1972, Madison Group (Mississippian) and Amsden Formation (Mississippian and Pennsylvanian) in the Beartooth Mountains, northern Wyoming and southern Montana: Montana Geological Society 21st Annual Geological Conference Guidebook, p. 57-63.

_____, 1976, Mississippian history of the northern Rocky Mountain Region: U.S. Geological Survey Journal of Research, v. 4, n. 3, p. 317-338.

Sando, W. J., B. L. Mamet, and J. T. Dutro, Jr., 1969, Carboniferous megafaunal and microfaunal zonation in the northern Cordilleran of the United States: U.S. Geological Survey Professional Paper 613-E, 29 p.

Schmidt, C. J., 1975, An analysis of folding and faulting in the northern Tobacco Root Mountains, Montana: Ph.D. thesis, Indiana University, 480 p.

Schmitt, G. T., 1953, Regional stratigraphic analysis of Middle and Upper Jurassic in northern Rocky Mountains-Great Plains: AAPG Bulletin, v. 37, p. 355-393.

Scholten, R., 1950, Geology of the Lima Peaks area, Clark County, Idaho and Beaverhead County, Montana: Ph.D. thesis, University of Michigan, 300 p.

_____, 1956, Deformation of the geosynclinal margin in northern Rocky Mountains: Resumes, 20th International Geological Congress, p. 75-76.

_____, 1960, Sedimentation and tectonism in the thrust belt of southwestern Montana and east-central Idaho: Wyoming Geological Association 15th Annual Field Conference Guidebook, p. 77-84.

_____, 1967, Structural framework and oil potential of extreme southwestern Montana: Montana Geological Society Guidebook, 18th Annual Field Conference, p. 7-19.

_____, 1968, Model for evolution of Rocky Mountains east of Idaho batholith: Tectonophysics, v. 6, p. 109-126.

Scholten, R., and M. H. Hait, Jr., 1962, Devonian System from shelf-edge to geosyncline, southwestern Montana-central Idaho, *in* Montana Geological Society, 13th Annual Field Conference Guidebook, p. 13-22.

Scholten, R., and L. D. Ramspott, 1968, Tectonic mechanisms indicated by structural framework of central Beaverhead Range, Idaho–Montana: Geological Society of America Special Paper 104, 71 p.

Schwartz, R. K., 1972, Stratigraphic and petrographic analysis of the Lower Cretaceous Blackleaf Formation, southwestern Montana: Ph.D. thesis, Indiana University, 268 p.

Sheldon, R. P., 1963, Physical stratigraphy and mineral resources of Permian rocks in western Wyoming: U.S. Geological Survey Professional Paper 313-B, 273 p.

Skipp, B., and M. H. Hait, Jr., 1977, Allochthons along the northeast margin of the Snake River Plain, Idaho: Wyoming Geological Association 29th Annual Field Conference Guidebook, p. 499-516.

Skipp, B., and L. W., McGrew, 1977, The Maudlow and Sedan Formations of the Upper Cretaceous Livingston Group on the west edge of the Crazy Mountain basin, Montana: U.S. Geological Survey Bulletin 1422-B, 68 p.

Skipp, B., R. D. Hoggan, D. L. Schlercher, and R. C. Douglass, 1979a, Upper Paleozoic carbonate bank in east-central Idaho: U.S. Geological Survey Bulletin 1486, 78 p.

Skipp, B., W. J. Sando, and W. E. Hall, 1979b, The Mississippian and Pennsylvanian (Carboniferous) System in the United States—Idaho: U.S. Geological Survey Professional Paper 1110-AA, p. AA1-AA42.

Sloss, L. L., 1954, Lemhi arch, a mid-Paleozoic positive element in south-central Idaho: Geological Society of America Bulletin, v. 65, p. 365-368.

Sloss, L. L., and R. H. Hamblin, 1942, Stratigraphy and insoluble residues of Madison Group of Montana: AAPG Bulletin, v. 26, p. 305-333.

Sloss, L. L., and W. M. Laird, 1945, Mississippian and Devonian stratigraphy of northwestern Montana: U.S. Geological Survey Oil and Gas Investigations Preliminary Chart 15.

_____, 1946, Devonian stratigraphy of central and

northwestern Montana: U.S. Geological Survey Oil and Gas Investigations Preliminary Chart 25.

———, 1947, Devonian System in central and northwestern Montana: AAPG Bulletin, v. 31, p. 1404–1430.

Smedes, H. W., 1966, Geology and igneous petrology of the northern Elkhorn Mountains, Jefferson and Broadwater Counties, Montana: U.S. Geological Survey Professional Paper 510, 116 p.

Smith, A. G., and W. C. Barnes, 1966, Correlation of and facies changes in the carbonaceous, calcareous, and dolomitic formations of the Precambrian Belt–Purcell Supergroup: Geological Society of America Bulletin, v. 77, p. 1399–1426.

Smith, D. L., 1972, Depositional cycles of the Lodgepole Formation (Mississippian) in central Montana: Montana Geological Society 21st Annual Field Conference Guidebook, p. 29–36.

Smith, D. L., and E. H. Gilmour, 1979, The Mississippian and Pennsylvanian (Carboniferous) Systems in the United States—Montana: U.S. Geological Survey Professional Paper 1110-X, p. X1–X31.

Stewart, J. H., 1972, Initial deposits in the Cordilleran geosyncline: Evidence of a Late Precambrian continental separation: Geological Society of America Bulletin, v. 83, p. 1345–1360.

Stokes, W. L., 1950, Pediment concept applied to Shinarump and similar conglomerates: Geological Society of America Bulletin, v. 61, p. 91–98.

Suttner, L. J., 1968, Clay minerals in the Upper Jurassic–Lower Cretaceous Morrison and Kootenai Formations, southwest Montana: Earth Sciences Bulletin, v. 1, p. 5–14.

———, 1969, Stratigraphic and petrographic analysis of Upper Jurassic–Lower Cretaceous Morrison and Kootenai Formations, southwest Montana: AAPG Bulletin, v. 53, p. 1391–1410.

Suttner, L. J., R. K. Schwartz, and W. C. James, 1981, Late Mesozoic to Early Cenozoic foreland sedimentation in southwest Montana: Montana Geological Society Field Conference, southwest Montana, p. 93–103.

Tanner, J. J., 1949, Geology of the Castle Mountain area, Montana: Ph.D. thesis, Princeton University, 163 p.

Theodosis, S. D., 1955, Belt series of northwestern Montana: Billings Geological Society 6th Annual Field Conference Guidebook, p. 58–63.

Tysdal, R. G., 1976, Paleozoic and Mesozoic stratigraphy of the northern part of the Ruby Range, southwestern Montana: U.S. Geological Survey Bulletin, 1405-I, 126 p.

Viele, G. W., 1960, and F. G. Harris, 1965, Montana Group stratigraphy, Lewis and Clark County, Montana: AAPG Bulletin, v. 49, p. 379–417.

Walker, T. F., 1974, Stratigraphy and depositional environments of the Morrison and Kootenai Formations in the Great Falls area, central Montana: Ph.D. Thesis, University of Montana, Missoula, Montana, 195 p.

Weeks, R. A., and M. R. Klepper, 1954, Tectonic history of the northern Boulder batholith region, Montana: Geological Society of America Bulletin, v. 65, p. 1320–1331.

Weidman, R. M., 1965, The Montana lineament: Billings Geological Society 16th Annual Field Conference Guidebook: 137–143.

Weimer, R. J., 1960, Upper Cretaceous stratigraphy, Rocky Mountain area: AAPG Bulletin, v. 44, p. 1–20.

———, 1961, Spatial dimensions of Upper Cretaceous sandstone, Rocky Mountain area, *in* J. A. Peterson, and J. C. Osmond, eds., Geometry of sandstone bodies: AAPG Special Publication, p. 82–97.

Wilson, C. W., Jr., 1934, Geology of the thrust fault near Gardiner, Montana: Journal of Geology, v. 42, p. 649–663.

Wilson, J. L., 1955, Devonian correlations in northwestern Montana: Billings Geological Society 6th Annual Field Conference Guidebook, p. 70–77.

Wilson, M. D., 1967, Upper Cretaceous–Paleocene synorogenic conglomerates of southwestern Montana: AAPG Bulletin, v. 54, p. 1843–1867.

Woodward, L. A., 1980, Tectonic framework of disturbed belt of west-central Montana: AAPG Bulletin, v. 65, p. 291–302.

Yen, T. C., 1949, Review of the Lower Cretaceous fresh-water molluscan faunas of North America: Journal of Paleontology v. 23, p. 465–470.

———, 1951, Fresh-water mollusks of Cretaceous age from Montana and Wyoming: U.S. Geological Survey Professional Paper 233-A, p. 1–20.

Yochelson, E. L., 1968, Biostratigraphy of the Phosphoria, Park City and Shedhorn Formations: U.S. Geological Survey Professional Paper 313-D, p. 571–660.

Sedimentation and Tectonics of the Middle Proterozoic Belt Basin and Their Influence on Phanerozoic Compression and Extension in Western Montana and Northern Idaho

D. Winston
University of Montana
Missoula, Montana

The Belt Supergroup was deposited in a Middle Proterozoic intracratonic basin, occupied during much of its history by alluvial aprons that sloped down to a landlocked sea. Rocks from the Ravalli Group, middle Belt carbonate and Missoula Group have been classified into thirteen sediment types that are arranged in long-lived facies tract patterns. These sediment types and their sedimentologic interpretations are as follows: (1) gravel—braided stream gravel bars high on alluvial aprons; (2) cross-bedded sand—braided stream channels on upper and middle alluvial aprons; (3) flat-laminated sand—sheetfloods on sandflats on middle and lower alluvial aprons; (4) tabular silt—storm deposits on submerged mudflats; (5) even couple—distal sheetfloods on sandflat surfaces; (6) pinch-and-swell couple—subwavebase turbidite deposition in the Belt sea; (7) pinch-and-swell couplet—subwavebase underflow and interflow settle-out; (8) even couplet—traction accumulation followed by suspension settle-out from episodic floods that crossed exposed and submerged mudflats; (9) lenticular couplet—submerged mudflat surfaces reworked by waves; (10) microlamina—subaqueous surfaces of minimal accumulation, locally coated with organic material; (11) coarse sand and intraclast—small beaches and shoals; (12) discontinuous layer—combined traction and suspension settle-out on the sandflats or on alluvial apron surfaces during flood wane; and (13) carbonate mud—precipitated carbonate mud and terrigenous sand and mud accumulation mostly on shallow, submerged bottoms.

The lower Belt records maximum transgression of the Belt sea. Turbidite sand and pelagic mud were deposited across the central part of the basin. Carbonate mud was precipitated at times on its eastern side, and coarse conglomerate accumulated along its fault-bounded southern margin. The Ravalli Group records progradation of mudflats and alluvial aprons from the south and west across much of the basin. The middle Belt carbonate was deposited during a second large transgressive period, when terrigenous-to-carbonate cycles formed across the eastern part of the basin and turbidite sand and mud derived from the west was deposited in the deeper, locally slumped, western part of the basin. The Missoula Group represents a series of alluvial aprons that prograded northward into the basin, separated by transgressive mudflat and shallow water deposits. The Garnet Range Formation, near the top of the Missoula Group, represents incursion of open marine waters into the Belt basin.

A tectonic hypothesis proposes that the Belt basin was cut by at least four major growth fault zones which are based on linear trends of abrupt stratigraphic thickness changes that coincide with local patches of soft sediment deformation. Three nearly east-west fault lines are, in northward sequence, the Perry, Garnet, and Jocko lines. The fourth line, the Townsend line, trends northwestward from the Perry to the Jocko line. The inferred fault lines partly enclose at least four major crustal blocks that subsided during Middle Proterozoic time, forming the Belt basin. The crystalline Dillon block lies south of the Perry line, the southern basin boundary. Between the Perry line and the Garnet line are the Helena embayment block on the east and the Deer Lodge block on the west, separated by the Townsend line. The Ovando block lies north of the Deer Lodge block. North of the Ovando block is the Charlo block (or blocks). Cretaceous to Eocene folds and thrusts formed western and eastern thrust belts, separated by the central part of the Belt basin, which was not intensely deformed. Thrusts and folds are long and continuous within the Precambrian crustal blocks, but are broken locally along the east-west lines. Later Cenozoic extensional faulting produced distinctive domains over the Precambrian blocks and dislocations along the Precambrian lines. Because the central part of the basin is only mildly deformed, its relatively intact stratigraphic framework provides the basis for much of the sedimentologic interpretation.

INTRODUCTION

Although the Belt Supergroup has long been considered a classic North American Middle Proterozoic series, sedimentologic and tectonic understanding of the Belt basin (Figure 1) has been hampered by the immense scale of the Belt; formations are hundreds to thousands of meters thick and extend for hundreds of kilometers. Furthermore, Belt strata have been deformed and tectonically transported by Phanerozoic compression and extension, and are cut by igneous intrusions. Consequently, most of the effort in the Belt basin to date, carried on largely by geologists of the U.S. Geological Survey, has been devoted to surface mapping to establish the geometric relationships of the rocks. Contiguous mapping at a scale of 1:250,000 (Griggs, 1973; Mudge et al., 1983; Harrison et al., 1981, 1983; Ruppel et al., 1983; Wallace et al., 1981) has cleared up many of the stratigraphic problems, and there is now general agreement on stratigraphic framework of the Belt (Figure 2). In addition, delineation of the stratigraphic framework has made it possible to identify broad facies changes, at a scale of hundreds of kilometers, across the basin. Only more recently have geologists been able to analyze the sedimentology and paleotectonics of the Belt from this reasonably complete, broad geologic basis. This paper summarizes in overview fashion one interpretation of the sedimentology and tectonics of the Belt basin and proposes a fault block configuration for the basin that also appears to be reflected in Phanerozoic tectonic patterns. Although sedimentation in the Belt basin clearly developed in response to tectonic uplift along parts of the basin margin, sedimentology and the tectonic framework of the basin are discussed in two separate parts of this paper, because the analysis leading to the sedimentologic interpretation differs substantially from the lines of evidence that support the more tentative tectonic hypothesis.

PART I: BELT SEDIMENTATION AND STRATIGRAPHY

Sedimentologic Framework

The Belt basin (Figure 1) is here interpreted to have been intracratonic; bordered on the northeast by the stable North American craton, on the south by the rising Dillon block, and on the west by an inferred, unidentified cratonic block of continental proportions. As a result of tectonic uplift along the southern and western basin margins (Winston et al., 1984), or of prolonged shrinking of the Belt sea, great alluvial aprons and sandflats at times built far into the basin. At other times, the Belt sea expanded, burying the sandflat and alluvial sediments under subaqueous sediments of the sea margin and deeper parts of the Belt sea. These sedimentary environments produced suites of facies, each of which retained its individual character throughout most of Belt deposition. Transgressions formed similar sequences of facies at several stratigraphic levels within the Belt that are difficult to distinguish one from another in isolated outcrops. Regressions also produced sequences at several stratigraphic levels that are difficult to differentiate in faulted areas. For example, alluvial apron and sandflat facies

within the quartzite units of the Revett, Snowslip, Mount Shields, Bonner, and McNamara formations (Figure 2) show similar lateral changes in grain size, composition, and sedimentary structures. Red and green argillite units of the Ravalli and Missoula groups also have similar suites of sedimentary structures.

Diagenesis greatly altered the Belt sediments. Under oxidizing conditions, hematite formed, turning some Belt rocks red. Partial reduction in other sediments formed green chlorite and phengite, while greater reduction in still others produced darker green to very dark gray colors. In addition, clay in Belt sediments has been altered by diagenesis and burial metamorphism to illite (Maxwell and Hower, 1967), sericite and, low in the sequence, to biotite (Norwick, 1977). As a consequence of burial metamorphism, most Belt rocks are argillites and quartzites. Because formal units in the Belt are defined on the basis of grain size, color, and mineralogy, many formation boundaries are products of diagenesis and metamorphism. Although oxidizing and reducing diagenetic processes in places generally coincide with sedimentary environments, in many places they do not. In a previous study (Winston, 1978) I may have done a disservice to Belt sedimentology by proposing that sedimentary structures and color go hand in hand. Where rock packages based on sedimentary structures and packages based on color fail to coincide, formal stratigraphic units based on color commonly mask the continuity of sedimentologic packages based on sedimentary structures, detrital grain size, and composition. For this reason, cross sections depicting sedimentary facies generally differ from cross sections of formal stratigraphic units.

Belt Sediment Types

Because of the necessity to separate sedimentary characteristics of Belt rocks from diagenetically and metamorphically formed lithic characteristics, a new *sediment type* terminology is here proposed. Sediment types are defined on the basis of texture, sedimentary structures, and original composition. They represent empiric sedimentary elements which recur through much of the Belt. To emphasize their sedimentary aspect, they are given sediment status and described as *gravel, sand,* and *mud.* The thirteen sediment types defined here (Figure 3) apply to rocks studied within the Ravalli Group, middle Belt carbonate and Missoula Group. Additional, undefined sediment types occur in the Prichard and LaHood formations of the lower Belt, and in time others will probably be identified in the rest of the Belt. Coarse-grained, thick-bedded sediment types are wholly terrigenous, but fine-grained, thin-bedded types range from wholly terrigenous, to calcareous, to carbonaceous varieties (see Figure 3). The fine-grained types, for the most part, were deposited subaqueously in the Belt sea. When carbonate production and carbon incorporation were low, the terrigenous sediments formed an array or tract of sediment types identifiable on the basis of grain size and sedimentary structures. During periods of carbonate mud production or high organic carbon production, however, carbonate or carbon was added to the sediments, forming calcareous or carbonaceous varieties of the fine-grained sediment types. In the carbonate mud type, fine-grained calcite and dolomite

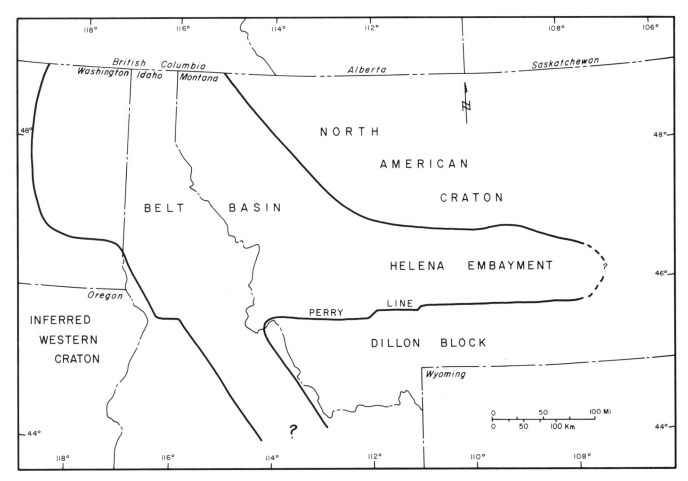

Figure 1—Postulated outline of the intracratonic Belt basin in the United States, partly restored along its northeastern margin to its pre-thrust position, but not restored with respect to Phanerozoic crustal shortening within the basin. Shown are the stable continental crust of the North American craton northeast of the basin, the Helena embayment, the tectonically positive Dillon block, and the inferred crustal block of continental proportions bordering the basin on the west.

mostly obscure terrigenous sedimentary structures. Sediment types presently recognized are illustrated in Figure 3 and are briefly defined as follows. See Winston (in press, b) for a more detailed treatment of sediment types.

Gravel Sediment Type

The gravel sediment type consists of quartz, quartzite, granite, feldspar, and chert granules, pebbles, and cobbles in a sand matrix. In some places the clasts are supported in flat-laminated or cross-bedded sand; in other places they are grain supported, with coarse sand infill. Clasts are crudely oriented parallel to bedding and in some places are imbricate. This sediment type forms lenses up to 1 m thick and tens of meters long in coarse, cross-bedded sand units. It is interpreted to represent channels and longitudinal bars of braided streams that crossed the upper and middle parts of great alluvial aprons.

Cross-Bedded Sand Sediment Type

The cross-bedded sand sediment type consists of coarse to fine, cross-bedded, commonly feldspathic, quartz sand in

beds as thick as 1 m. Trough, planar, and epsilon (Allen, 1965) cross beds occur in this type. Trough and planar cross beds are commonly unimodal, and form lenses 1–2 m thick in outcrops cut perpendicular to flow direction, but appear as tabular beds where the exposure is cut parallel to flow. Scale of the cross beds and grain size commonly decrease upward through intervals 1 m or more thick. The cross-bedded sand sediment type is interpreted to record migrating large-scale bedforms and laterally shifting edges of episodic, braided stream channels in the upper and middle parts of large alluvial aprons.

Flat-Laminated Sand Sediment Type

The flat-laminated sand sediment type consists of even, continuous, medium to very fine sand beds, 15 cm to 2 m thick, with even, flat internal laminae interstratified with occasional layers of climbing ripple cross laminae. Bases of beds are sharp and commonly rest on muddier layers with either depositional or slightly scoured contacts. Mudchips are commonly incorporated in the lower parts of the sand beds. Tops of some sand beds grade up into muddier strata of the discontinuous layer sediment type (described below), while

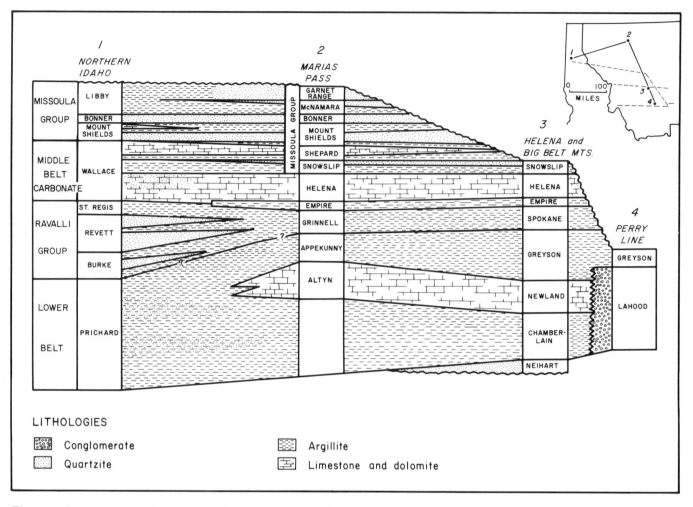

Figure 2—Stratigraphic section showing principal formations of the Belt Supergroup (not to scale).

others are marked by oscillation ripples. The flat-laminated sand sediment type is interpreted to have been deposited from the upper flow regime by episodic sheetfloods that crossed the middle and lower parts of alluvial aprons. As flow waned, it shifted to the lower part of the lower regime, depositing climbing ripples.

Tabular Silt Sediment Type

The tabular silt sediment type consists of even beds, 3–20 cm thick, of flat-laminated and climbing ripple cross-laminated silt and very fine sand. It commonly occurs interstratified with beds of the lenticular couplet sediment type (described below). Beds are characteristically not graded and have sharp tops and bottoms. The tabular silt sediment type may represent storm deposits in shallow waters of the Belt sea.

Even Couple Sediment Type

This sediment type is characterized by even, continuous beds, 3–15 cm thick, that grade from fine sand above sharp bases to silt mixed with clay in their upper parts. The term

couple here refers to graded beds 3–15 cm thick and is distinguished from the term *couplet*, which refers to thinner layers ranging from 0.3 cm to 3 cm thick. Bases of the couples are flat. The lower parts of some couples contain mud chips, and the upper parts are commonly cut by mud cracks or are marked by oscillation ripples. The even couple sediment type is interpreted to have been deposited by episodic sheetfloods that flowed in the upper regime from the toes of alluvial aprons across sandflats. As floods waned or were engulfed by rising water on the sandflat, flow shifted to the lower regime, depositing fine sand and mud.

Pinch-and-Swell Couple Sediment Type

The pinch-and-swell couple sediment type is characterized by graded beds, 3–15 cm thick, of fine sand to dark clay. The sharp sandy bases of the couples sag down into underlying dark mud as a result of loading or of cut and fill. In most places the lower sand lacks obvious stratification, but in other places laminae have been warped downward, indicating loading during sediment accumulation. Contorted, compressed, sand-filled cracks or small dikes cut some clay layers. This sediment type and the pinch-and-swell couplet

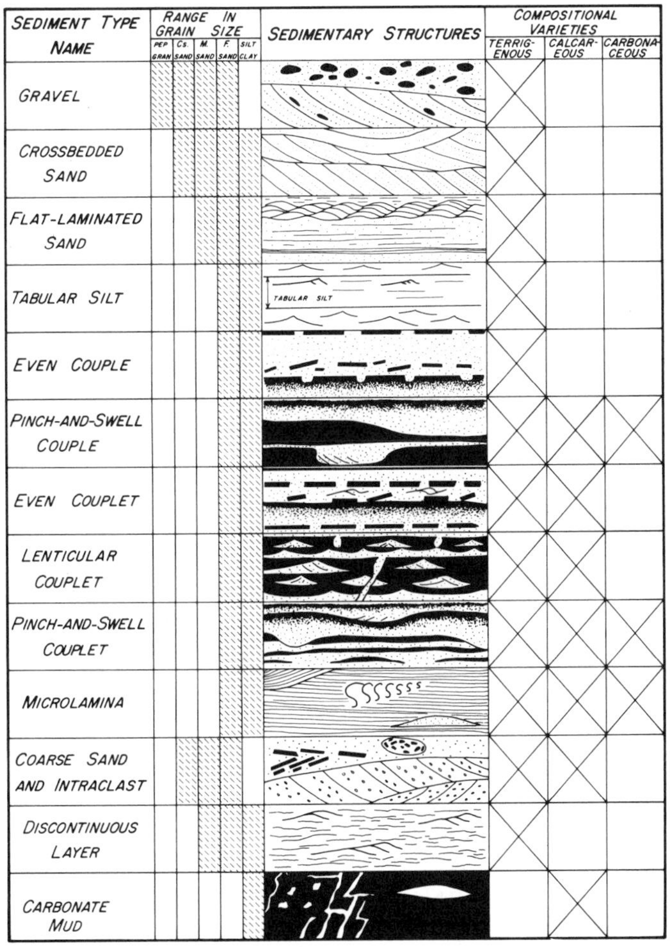

Figure 3—Sediment types of the Ravalli Group, the middle Belt carbonate, and the Missoula Group.

sediment type (described below) characterize the Wallace Formation. Both have wholly terrigenous, calcareous, and carbonaceous varieties. The graded beds of the pinch-and-swell couple sediment type are interpreted to record deposition on subwavebase surfaces of the Belt sea floor from turbidite underflows or from settle-out of interflows above a density boundary.

Pinch-and-Swell Couplet Sediment Type

The pinch-and-swell couplet sediment type is a thin, couplet-scale (0.3–3 cm) relative of the pinch-and-swell couple sediment type. Medium gray to light gray fine sand and silt grade to dark clay through layers 0.3–3 cm thick. Lower bounding surfaces of the fine sand and silt bow down into the underlying dark mud. Compacted sand-filled cracks are common, but mud chips are absent. Wholly terrigenous, calcareous, and carbonaceous varieties are common in the Wallace Formation. This sediment type appears to record the same turbidite underflow and interflow processes as those interpreted for the pinch-and-swell couple sediment type, but the thinner, finer grained layers represent a more distal setting.

Even Couplet Sediment Type

The even couplet sediment type consists of sharply based, very fine sand and silt layers that grade upward into clay, forming parallel-laminated to slightly wavy bedded, laterally continuous couplets 0.3–3 cm thick. Couplets range from wholly terrigenous to calcareous, and there are two distinctive subtypes. One subtype is moderately to intensely mud cracked and commonly contains mud chips in the lower sandy part of the couplet. Beds are commonly disrupted by penecontemporaneous deformation, probably as a result of desiccation. The other subtype lacks desiccation cracks, mud chips, and disruption. Consequently, the layers are more even and continuous. The even couplet sediment type is interpreted to record sheetflood deposition across exposed or submerged mudflats. The intensely mud cracked variety was deposited on continually exposed flats, whereas the more clay-rich, sparsely desiccated variety was deposited on occasionally exposed mudflats.

Lenticular Couplet Sediment Type

This sediment type contains couplet-scale (0.3–3 cm) layers of oscillation ripple-marked silt and fine sand, sharply overlain by clay. Bases of the rippled silt and sand layers are flat, and the clay layers in some intervals are cut by desiccation cracks. As in the even couplet sediment type, terrigenous and calcareous varieties can be identified. The lenticular couplet sediment type probably represents wave-reworked sediment that may originally have accumulated as even couplets on submerged mudflats.

Microlamina Sediment Type

This sediment type is characterized by millimeter-scale alternating laminae of quartzitic or dolomitic silt and clay. In some intervals the laminae are even and continuous, but in others they are commonly wrapped into tiny, tight, intrastratal folds, giving them the distinctive appearance of some Phanerozoic oil shale beds. The characteristic black color results from a small amount of carbon, but calcareous, non-carbonaceous green and even red varieties are known to occur. This sediment type may record episodic deposition of suspended sediment on protected surfaces far from major terrigenous sources. Some surfaces may have been covered with organic mats, and some were also occasionally exposed.

Coarse Sand and Intraclast Sediment Type

The coarse sand and intraclast sediment type is characteristically composed of coarse, well-rounded quartz sand, cross-bedded in a variety of directions, containing mud chips imbricated in a variety of directions. It commonly lies on flat to undulating, scoured surfaces and forms discontinuous lenses that extend across outcrops at a single stratum. In some beds the coarse grains are mostly ooliths; in others they consist mostly of intraclasts and "molar tooth" fragments (blades and spheres of fine-grained blocky calcite) that may have filled voids and cracks during early diagenesis. (See O'Connor, 1972, for a description of molar tooth structure.) Beds and lenses of the coarse sand and intraclast sediment type are interpreted to represent ephemeral beaches and shoals.

Discontinuous Layer Sediment Type

The discontinuous layer sediment type is characterized by decimeter- to meter-scale beds of discontinuous interlaminations of wavy and climbing ripple cross-laminated fine sand, silt, and clay. It is the absence of clearly definable graded couplets and couples that sets this sediment type apart. Instead, very fine sand and silt cross laminae form thin, discontinuous lenses that break up the continuity of the muddy laminae. Suspended clay drape is rare; thus, mud cracks are also rare. This sediment type appears to represent deposition from distal parts of floods or during waning flow, when silt and clay settled out simultaneously in small rivulets on alluvial apron or sandflat surfaces.

Carbonate Mud Sediment Type

The carbonate mud sediment type consists of decimeter- to meter-scale beds containing more than 50% micrite or dolomicrite, so that terrigenous sedimentary structures such as couples and couplets are generally indiscernible. Instead, the sediment type appears massive in outcrop, is commonly cut by molar tooth structures, and weathers tan. This sediment type records precipitation of carbonate mud, typically mixed with some silt and clay, on shallow subaqueous surfaces.

Facies Model

Stratigraphic cross sections demonstrate that the individual sediment types described above characteristically maintain regular stratigraphic positions with respect to each other throughout most of the Ravalli Group, the middle Belt carbonate, and the Missoula Group. This stratigraphic interrelationship, combined with the sedimentary processes inferred from grain size and sedimentary structures, is the basis of the above environments of deposition interpreted for each sediment type. Furthermore, integrating these environments laterally across the Belt basin leads to the generalized facies tract models of the Ravalli Group, the middle Belt carbonate, and the Missoula Group, which are illustrated in Figures 6, 8, and 10. The models are necessarily simplified. One reason for simplification is that in some places sediment types are interstratified. For instance, the cross-bedded sand, flat-laminated sand, and discontinuous layer types commonly form meter-scale fining-upward sequences deposited on the mid parts of alluvial aprons. Another reason for simplification is that the models are necessarily foreshortened, because at no time did all the sediment types accumulate simultaneously. As the facies tract shifted in response to tectonism or changes of Belt sea level, some sediment types migrated out of the area of Belt preservation as others entered it. Nonetheless, the models adequately conceptualize Belt deposition and have proven to be powerful tools in analyzing Belt stratigraphy and sedimentation, and in providing clues regarding Belt tectonics.

Based on the sedimentary processes interpreted from each sediment type and the constraints placed on them by their mutual facies arrangements (Winston, in press, b), the following generalized environmental interpretation of the facies tract is proposed (see Figures 6, 8, and 10). The gravel sediment type (see Figures 3, 9, and 10) was probably deposited high on alluvial aprons in longitudinal bars and in channels of braided streams of the Missoula Group flowing northward into the basin from the uplifted Dillon block or the inferred continent to the southwest. The gravel type compares closely with gravel in longitudinal bars of the Platte River (Smith, 1970) and many other braided stream deposits (Boothroyd and Ashley, 1975; Miall, 1978; Rust, 1979). However, in the Belt, where gravel locally forms the bases of fining-upward sequences, each sequence is more continuous, reflecting deposition from single flood events. Slope decreased down the alluvial aprons, where streams deposited the cross-bedded sand sediment type as bars and dunes from lower regime flow in broad, braided channels 1 m or more deep. These tabular cross beds are similar to those deposited by floods in Australia (Williams, 1971) and the Karoo region of South Africa (Stear, 1985). In places channel shapes are outlined by epsilon cross beds. In some fine sand channel deposits, such as those in the Revett Formation, flow was in the upper regime, and internal laminae of epsilon cross beds conform to the slope of the channel margins.

Farther down the alluvial apron, slope and flow velocity decreased, so that floods transported only fine to medium sand and mud. Channels broadened to shallow sheetflood surfaces, and flow shifted to the upper regime, depositing tabular beds of the flat-laminated sand sediment type across the fan toes. As floods waned, velocities decreased, and climbing ripple cross laminae formed in the lower flow regime, building upward to finer-grained layers. Clay draped some of the flood deposits, and small ephemeral ponds became floored with oscillation ripplemarks. These sediments compare closely with Bijou Creek flood deposits (McKee et al., 1967). Smoot (1983) has described similar beds from fan toes in the Wilkins Peak Member of the Eocene Green River Formation.

Beyond the fan toes, flood waters spread across sandflats, depositing either the even couple sediment type or the discontinuous layer type. Hydrologic factors determining which of these two sediment types accumulated are not yet understood. The even couple type simply records continued upper regime sheetflood from the fan toes farther out onto the sandflats, where finer-grained, thinner, graded sand beds, deposited by sheetflood, were capped by suspension settle-out mud. The discontinuous layer type, which lacks regular grading, was also probably deposited in very shallow water, where sandy current ripples formed in small rivulets between irregular patches of mud. Perhaps it represents lingering flow after a main flood event.

Individual floods formed fining-upward sequences at a meter scale or greater in the more proximal, coarse sand, decreasing to decimeter scale in the more distal, fine sand. As floods on the alluvial aprons waned, and flow velocities decreased, the facies tract at times shifted up the apron; thus, in places the flat-laminated sand sediment type overlies the cross-bedded sand type and the discontinuous layer type locally overlies them both.

As flood waters flowed across exposed mudflats, they picked up dried, broken mud polygons and deposited them as mud chips along with fine sand and silt in centimeter scale, flat-laminated and ripple cross-laminated layers. As flood waters ponded, and as standing water spread across the flats, mud settled from suspension, gradationally capping the fine sand and silt layers and forming the even couplet sediment type. The upper parts of the flats again became dry and deeply desiccated, producing the mud cracked variety of the even couplet sediment type. Farther out on the mudflats, where shallow water covered the flood deposits for longer periods of time, waves reworked the fine sediments, forming the lenticular couplet sediment type, characterized by centimeter-scale, oscillation rippled fine sand and silt, sharply overlain by mud drape. Thin lenses of the coarse sand and intraclast sediment type within intervals of the even and lenticular couplet types record small ephemeral beaches and shoals formed by waves breaking on the mudflat surfaces. Decimeter-scale beds of the tabular silt sediment type, mostly within intervals of the lenticular couplet sediment type, probably represent storm concentrations of fine sand and silt in shallow water.

Seaward from the abundantly rippled surfaces that produced the lenticular couplet sediment type, several other types accumulated. Each depended on the amount of terrigenous influx and the chemical and biological processes in the sea waters. Where floods brought large amounts of suspended silt and clay over subwavebase surfaces, even, graded couplets accumulated. Accordingly, these intervals of the even couplet sediment type tend to be finer-grained and much less disturbed than the mud cracked, mud chip-bearing variety. In some stratigraphic intervals the even couplet and lenticular couplet types are intimately interstratified, but in other intervals, single forms of couplets dominate sequences for tens of meters. At times, perhaps during periods of low sediment influx, pronounced sea expansion, or high nutrient production, the microlamina sediment type was widely deposited across both shallow and deeper parts of the basin. Its marked similarity to Eocene Green River oil shale strongly suggests a similar origin. At least some of it appears to have been organically bound, possibly by photosynthetic organisms that coated the bottom within the photic zone. During periods of sea expansion, Belt sea waters may have been brackish, but during some periods of inferred sea contraction, waters became supersaturated with respect to calcium carbonate. Calcareous and dolomitic varieties of the even couplet, lenticular couplet, and microlamina sediment types were then deposited, in many places abruptly above terrigenous varieties, forming terrigenous-to-carbonate sedimentary cycles several meters thick (O'Connor, 1972; Eby, 1977; Grotzinger, 1981a, b, in press; Winston et al., 1984). At times, carbonate production overwhelmed the periodic influx of suspended terrigenous sediments, and the carbonate mud sediment type accumulated.

During deposition of the middle Belt carbonate, when terrigenous-to-carbonate cycles spread across the eastern part of the basin, two additional sediment types were deposited in the deeper parts of the basin, particularly on the western side. They are the pinch-and-swell couple and pinch-and-swell couplet sediment types, which characterize the Wallace Formation. Like the other couple and couplet layers of the Belt, they represent single depositional events, but their characteristic load structures reflect abrupt sand deposition of water-saturated, subaqueous mud bottoms.

Their stratigraphic position, basinward from shallow water terrigenous-to-carbonate cycles, and the tendency of couplets in the central part of the basin to coarsen and thicken to couples in the western part of the basin indicate that the sediment came primarily from the west, transported as underflows, or possibly as interflows along salinity-generated, density boundaries.

Depositional History

The Belt stratigraphic sequence can be understood conceptually as resulting from transgressions and regressions of parts of the facies tract models described above. Initial rifting, about 1.5 b.y. ago (Elston, 1984; Harrison, 1984b), broke continental crust underlying the Belt basin into blocks, some of which are described in Part II of this paper. Subsiding blocks produced the Belt basin, in which formed the landlocked Belt sea. The sea spread over what may have been an extensive sand blanket covering the crystalline basement, now preserved only in isolated remnants such as the Neihart Quartzite.

Lower Belt

The initial transgression appears to have brought the Belt sea to its greatest areal extent, depositing debris flow conglomerate, turbidite sand, dark mud, and carbonate that are now represented by the LaHood, Chamberlain, Newland, Prichard, Altyn, and probably the Greyson and Appekunny formations (Figure 2). I have not studied these formations in sufficient detail to propose a sediment type classification or a first hand sedimentologic analysis of these units. However, other Belt workers have outlined well the sedimentological configuration of the lower Belt (Figure 4). McMannis (1963), Bonnet (1979), and Hawley et al. (1982) have substantiated a turbidite and debris flow origin for the spectacular LaHood conglomerate, reflecting uplift of the Dillon block and subsidence of the Belt basin north of the Perry line. These coarse-grained deposits along the southern margin of the basin interfinger abruptly with basinal mud of the Helena embayment, a large graben probably underlain by a subsiding block of continental crust. The block was bounded on the north by the Volcano Valley fault (Godlewski and Zieg, 1984), which appears to represent the eastern segment of the Garnet line (described in Part II). It moved down to the south, but did not have so much displacement as faults along the Perry line. North of Garnet line, the crystalline basement was very stable (McMannis, 1963, 1965) and was possibly covered with a coarse sand blanket during much of Middle Proterozoic time. Adjacent to this more stable basin margin, carbonate of the Newland, Altyn and Waterton formations was deposited (McMannis, 1963, 1965; Price, 1964; Fermor and Price, 1984; Hill and Mountjoy, 1984). Newland carbonate extends southwestward across the Helena embayment, whereas the Altyn and Waterton formations of the northeastern basin margin pass westward to the dark terrigenous sequence of the upper Prichard Formation (Aldrich Formation of Canada) (McMechan and Price, 1982). Cressman (1984) proposed that Prichard sediments in the west-central part of the basin were deposited as a large deltaic sequence and were derived from a land mass of continental proportions that lay south and west of the basin. Based on facies changes through

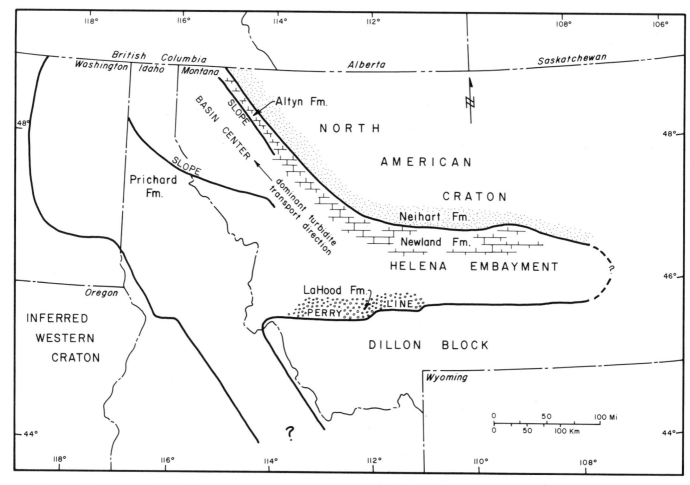

Figure 4—Generalized facies map of the lower Belt in the United States. Carbonate and terrigenous mud were deposited along the tectonically stable northeast boundary of the basin and in the Helena embayment (McMannis, 1963; Godlewski and Zieg, 1984; Fermor and Price, 1984; Hill and Mountjoy, 1984; White, 1984), while boulder conglomerate accumulated along the Perry line north of the uplifted Dillon block (McMannis, 1963; Hawley et al., 1982). The thick deltaic sequence of the western part of the basin received sediments from an inferred continent to the southwest (Cressman, 1985), and turbidite sediments were funneled northwestward down the basin axis (Godlewski and Zieg, 1984), shown here in its present, post-thrust position.

transgressive and regressive cycles, Finch and Baldwin (1984) supported Cressman's interpretation by inferring a northwest-trending basin axis through the east-central part of the basin (Figure 4).

Ravalli Group

The initial transgression of the lower Belt was followed by basinwide regression, recorded in the western part of the basin by upward transition from even, laminated black argillite of the Prichard Formation into lenticular, mud cracked couplets of the Ravalli Group. This regression is recorded through a stratigraphic interval of up to 500 m thick. Stratigraphic relations through this interval are complicated by the fact that some authors in the west (Mauk, 1983a, b; Cressman, 1985) place the Prichard-Burke boundary at the base of the transition interval, whereas farther east other authors (Finch and Baldwin, 1984; Van Loenen, 1984) place the Prichard-Burke boundary at the top of the interval, more closely coinciding with the original definition of the boundary (Ransome and Calkins, 1908). The original definition is depicted on Figure 5, in which the

base of the Burke is marked by the lowest mud cracked lenticular couplets that are here tentatively correlated lithostratigraphically with the Appekunny-Grinnell and Greyson-Spokane boundaries of the eastern part of the basin, a still controversial correlation (Harrison, 1972; Godlewski and Zieg, 1984; Whipple et al., 1984; Winston, in press, a). Figure 5, which is based on this correlation, illustrates the general sediment type facies changes from the Ravalli Group in the Coeur d'Alene district of Idaho northeastward to the Grinnell Formation of Glacier National Park. These facies changes provide the basis for the hypothetical Ravalli facies tract diagram of Figure 6; shifts in the facies tract recorded the depositional history of the Ravalli Group.

In the western part of the basin, subwavebase, evenly laminated argillite of the Prichard passes upward through bedded siltite to mud cracked, lenticular couplets of the Burke, recording northeastward progradation of mudflats into the basin. Sediment was probably derived from uplifted continental terranes west and southwest of the basin. The quartzite wedge of the lower informal member of the Revett Formation (White and Winston, 1982) is composed mostly of the cross-bedded sand and flat-laminated sand sediment

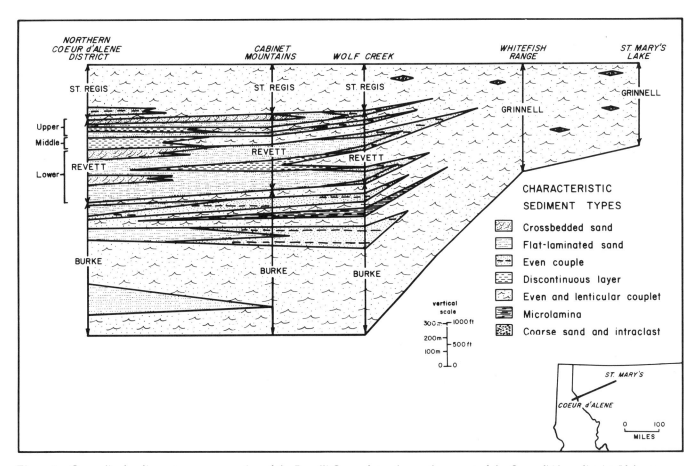

Figure 5—Generalized sediment type cross section of the Ravalli Group from the northern part of the Coeur d'Alene district, Idaho, to Glacier National Park, Montana. The Burke Formation is composed mostly of couplets interleaved with basinward-pinching tongues of the flat-laminated sand and even couple sediment types. One microlamina interval lies high in the Burke (Mauk, 1983a, b). In the west the Revett Formation consists of lower and upper quartzite wedges of the cross-bedded sand, flat-laminated sand, and even couple sediment types, separated by the middle Revett tongue of the discontinuous layer and lenticular couplet sediment types. The St. Regis Formation is marked at its base by mud cracked lenticular and even couplets, which are overlain by a wedge of the even couple sediment type. It passes up into even and lenticular couplets that characterize the body of the St. Regis. The Burke, Revett, and St. Regis are shown here and in Figure 2 to pass northeastward into the even and lenticular couplet types of the Grinnell Formation, which also contains lenses of the coarse sand and intraclast sediment type. Sources: Coeur d'Alene district—Hobbs et al., 1965; Winston, unpublished data. Cabinet Mountains—Wells et al., 1981. Wolf Creek—Johns, 1970. Whitefish Range—Johns, 1970. Glacier Park—Ross, 1959.

types and records continued progradation of a large alluvial apron eastward and northeastward into the basin. The top of the lower Revett marks a halt in progradation, and the overlying middle Revett member, composed of couplets and discontinuous layers, records the transgression of sandflats and mudflats westward over the alluvial apron. In the upper Revett, intervals of the cross-bedded sand and flat-laminated sand sediment types, interstratified with intervals of discontinuous layers and couplets, reflect alternating progradation and retreat of the alluvial aprons. Overlying the Revett Formation in the western part of the basin is the St. Regis Formation, composed mostly of even and lenticular couplets with interbeds of even couples and discontinuous layers. These record a major transgression of the Belt sea westward across the alluvial aprons—a transgression that continued into the overlying Wallace Formation.

The Revett quartzite wedges in the western part of the basin, above the argillite of the Burke and below the St. Regis argillite, impart the tripartite subdivision of the Ravalli Group. Where the quartzite wedges pinch out to the east, only a single argillite and siltite unit has been

recognized. In Glacier National Park this unit is called the Grinnell Formation; to the south near Helena, it is called the Spokane Formation. The lower Ravalli regression is represented in Glacier National Park by the upward transition from black argillite of the Appekunny Formation into mud cracked even and lenticular couplets of the Grinnell Formation. A similar transition through four shoaling-upward cycles occurs at the Greyson-Spokane boundary to the south (Bloomfield, 1983). Mud cracked even and lenticular couplets continue upward in the Grinnell and Spokane formations, indicating the persistence of exposed mudflats throughout Ravalli deposition across the eastern part of the basin. Even, tabular beds of cross-stratified, coarse-grained quartzite in the upper Grinnell may record floods that drained the North American craton and ponded in the Belt basin. Southward along the eastern part of the basin, coarse-grained quartzite beds diminish. However, an interval of even, fine-grained couples in the Spokane Formation (Connor, 1982, 1984) may be a distal tongue of the Revett, in which case sheetfloods from the west may have built a sandflat that extended for a period of time

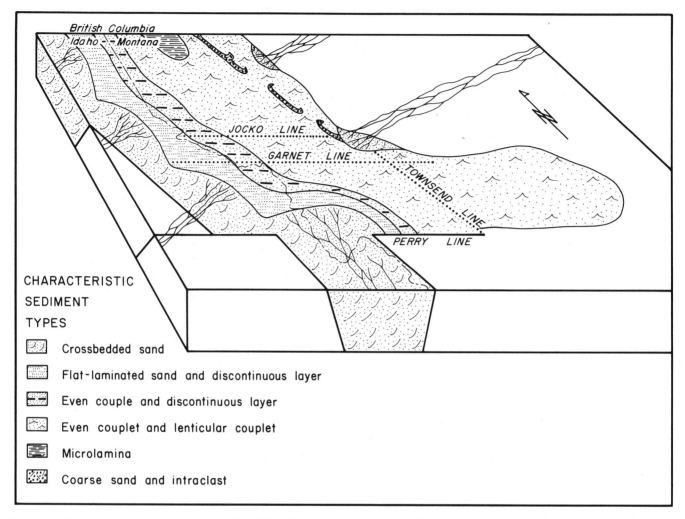

CHARACTERISTIC SEDIMENT TYPES

- Crossbedded sand
- Flat-laminated sand and discontinuous layer
- Even couple and discontinuous layer
- Even couplet and lenticular couplet
- Microlamina
- Coarse sand and intraclast

Figure 6—Hypothetical facies tract block diagram of Ravalli Group sediment types. The cross-bedded sand sediment type, deposited on alluvial aprons by braided streams, passes downslope to the flat-laminated sand and discontinuous layer sediment types, deposited by sheetfloods on the lower parts of the aprons. Flood deposits thinned and fined downslope to the even couple sediment type, deposited on the sandflats, passing in turn to the even couplet sediment type, deposited on the sea margin flats. Shallow, submerged surfaces were reworked by waves into the lenticular couplet sediment type. During times of minimal sediment influx, the microlamina sediment type accumulated in the basin center. Deposition along the eastern part of the basin was slow, and coarse sand, derived from streams draining the craton, was worked into ephemeral beaches of the coarse sand and intraclast sediment type.

nearly across the basin. The basinwide transgression, recorded in the St. Regis Formation in the western part of the basin, is represented in the eastern part by the transition from the Grinnell and Spokane formations up into the Empire Formation through beds of the coarse sand and intraclast sediment type, representing small transgressive beaches. This transgression culminated in the overlying middle Belt carbonate.

Middle Belt Carbonate

The middle Belt carbonate records the second great transgressive-regressive megasequence of the Belt. Although stratigraphic and sedimentologic details of the western part of the middle Belt carbonate are less well known than those of the eastern part, a generalized sediment type cross section of the middle Belt carbonate can be constructed. Figure 7

depicts how the lower part of the Wallace Formation, characterized by beds of the even and lenticular couplet sediment types, passes northeastward into calcareous even and lenticular couplets of the Empire Formation (uppermost Ravalli Group). This cross section also shows how the middle Wallace, characterized by sandy beds of the pinch-and-swell couple and couplet sediment types, passes into terrigenous-to-carbonate cycles (described below) that characterize the Helena Formation (Smith and Barnes, 1966; Harrison, 1984a). The thick, sandy couples of the middle Wallace are mostly restricted to the western part of the basin and probably reflect influx from the inferred western continental block (Winston et al., 1984; Grotzinger, in press). The couples grade eastward into thinner pinch-and-swell couplets, which in turn pass to even couplets in the lower terrigenous parts of some terrigenous-to-carbonate cycles of the Helena Formation. The cycles are marked by scoured bases overlain by thin lag deposits of the coarse sand

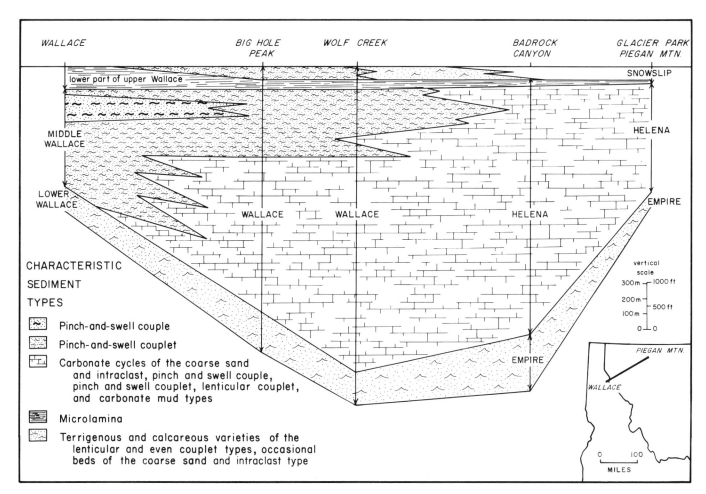

Figure 7—Generalized sediment type cross section of the middle Belt carbonate from Wallace, Idaho to Glacier National Park, Montana. Pinch-and-swell couples were deposited in the deeper, western part of the basin, near the inferred western continent. They pass eastward to pinch-and-swell couplets that become interstratified with even couplets in the lower parts of terrigenous-to-carbonate cycles that formed in shallow waters of the eastern part of the basin. The lower part of the upper Wallace is marked by an interval of the microlamina sediment type. Sources: Wallace section—Hobbs et al., 1965; Vance, 1981. Big Hole Peak—Grotzinger, 1981b, in press. Wolf Creek—Johns, 1970. Bad Rock Canyon—Johns, 1970. Glacier Park—Ross, 1959.

and intraclast sediment type reworked from the sediments below. Terrigenous even couplets lie above the basal lag deposits in most cycles. These are abruptly overlain by the upper carbonate parts of the cycles in which terrigenous couplets are mixed with, or are overwhelmed by, very fine-grained dolomite, forming the carbonate mud sediment type. The cycles are capped by scoured surfaces.

At the top of the middle Belt carbonate in the eastern part of the basin, calcareous microlaminae of the uppermost Helena Formation pass upward into terrigenous couplets of the lower Snowslip Formation. At about the same stratigraphic level, dark microlaminae and pinch-and-swell couplets of the upper part of the middle Wallace Formation extend across the central and western parts of the basin. Figure 8, which is based on the general middle Belt carbonate cross section (Figure 7), illustrates a hypothetical facies tract model for the middle Belt carbonate. Although development of the microlamina sediment type was locally only short-lived, its general position is illustrated. Shifts in the facies tract recorded the history of the middle Belt carbonate.

The transgression which began during deposition of the upper Ravalli Group continued, initiating deposition of the middle Belt carbonate. Periodically exposed and desiccated mudflats became progressively submerged, and even and lenticular couplets of the lower Wallace and Empire were deposited. As deeper water spread across the basin, terrigenous sediment from the west was trapped in the western Wallace facies, and carbonate cycles of the Helena developed in shallow water in the eastern and central parts of the basin. The cycles have been interpreted by some (O'Connor, 1972; Eby, 1977) as shoaling-upward marine cycles. Grotzinger (in press), however, concludes that the cycles may be either marine or lacustrine. The continuity of similar scale couplets from the terrigenous half-cycles into the carbonate half-cycles suggests to me that environments of deposition did not change substantially across the sharply defined half-cycle contacts. Instead, I interpret the terrigenous-carbonate contacts to represent chemical changes in Belt sea water from undersaturation with respect to calcium carbonate to supersaturation, similar to cycles of the Eocene Green River Formation (Surdam and Stanley,

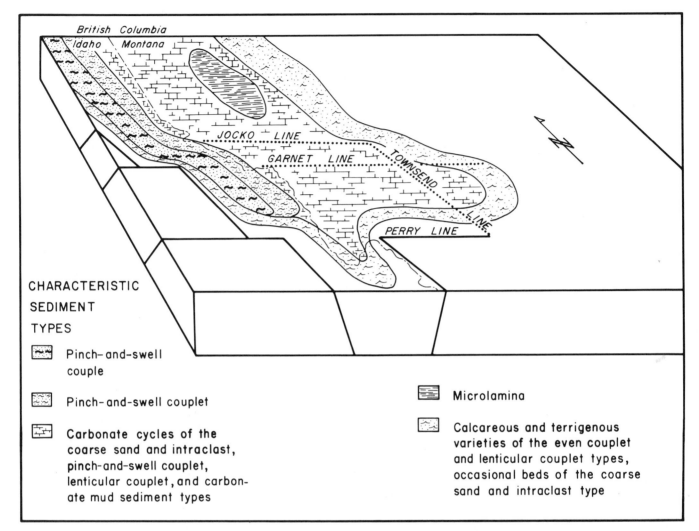

Figure 8—Hypothetical facies tract block diagram of middle Belt carbonate sediment types. The terrigenous variety of the lenticular couplet sediment type is inferred to have bordered the eastern margin of the basin, deposited during high stands of the Belt sea. Lower stands were accompanied by carbonate deposition, thus forming the terrigenous-to-carbonate cycles that stretched across the eastern and central shallow waters of the basin (Grotzinger, 1981a, b, in press). The inferred western continent contributed sediment to the western part of the basin. Narrow shelf deposits of the lenticular couplet sediment type rimmed the southwestern uplifted block and were bordered on the north and east by slumped slope deposits (see Figure 15) and deeper water sediments of the pinch-and-swell couple and couplet sediment types (Wallace et al., 1976; Godlewski, 1980).

1979). I interpret these widespread chemical changes to have resulted from shrinking of an intracratonic, landlocked Belt sea that was not connected to the open Proterozoic ocean. By this interpretation, the Belt cycles are more akin to Phanerozoic lacustrine cycles than they are to marine cycles.

Carbonate cycles of the Helena blend westward into pinch-and-swell couples and couplets of the Wallace, recording westward deepening of the basin. Laminated pinch-and-swell couples with scoured bases were probably deposited by eastward-directed turbidity flows. Non-laminated couplets (Lemoine, 1979) may represent fallout from interflows that traveled along salinocline surfaces. Halite molds (Grotzinger, 1981b) and shortite pseudomorphs (Harrison, 1984a) in the Wallace indicate that the western, deep-water portions of the Belt sea were at times hypersaline, perhaps having a salinity-stratified water

column. The southwestern margin of the basin appears to be reflected in argillite of the western Wallace Formation (Wallace et al., 1976). Mudflats that probably bordered the western continental terrane prograded eastward into the deeper part of the basin, flanked by slope deposits containing large slump folds and breccias (Wallace et al., 1976; Godlewski, 1977, 1980). Black microlamina beds at the top of the middle Wallace may be analogous to Phanerozoic oil shale deposits and may record freshening of the Belt sea before onset of regression, recorded in the lower Missoula Group.

Missoula Group

Basin configuration changed during deposition of the Missoula Group. The continental-scale source terranes west and southwest of the Belt basin, which had earlier provided sediment to the basin, may have become more subdued or cut off by intervening rift valleys during deposition of the Missoula Group. The predominant distal, seaward direction, which had for so long been generally northward and eastward, now became northwestward. The generalized cross section of the Missoula Group and upper Wallace Formation (Figure 9) illustrates the same general facies relationships as those of the underlying Ravalli Group and parts of the middle Belt carbonate. Figure 10 depicts a hypothetical Missoula Group facies tract based on facies changes in the Missoula Group. Couplets and couples of the Snowslip in the central part of the basin record northward progradation of exposed mudflats and sandflats from the southwest. Sediment influx along the eastern margin of the basin was limited, so that mudflat surfaces in that region remained submerged. Fine sand and silt were continually reworked by waves into lenticular layers occasionally interstratified with coarse sand carried in from the inferred sand blanket covering the eastern stable craton. Couplets of the Snowslip also became finer grained and thinner westward toward the Coeur d'Alene district and northwestward toward Clark Fork, Idaho, reflecting distal, probably submerged environments. Progradation of the Snowslip alluvial apron finally brought the flat-laminated and cross-bedded sand sediment types as far north as the Garnet line (Figure 10) before they were inundated by the Shepard transgression.

Stromatolites and thin lenses of coarse-grained beach quartzite (coarse sand and intraclast sediment type) mark the transgressive base of the Shepard. These beds are overlain by dark beds of the microlamina and lenticular couplet sediment types, which record submergence, soon followed by extrusion of the Purcell lavas of northern Montana and southern British Columbia (Willis, 1902; Price, 1964; McMechan, 1981). The dolomitic upper Shepard may record supersaturation of the Belt sea with respect to calcium carbonate, followed by progradation of exposed mudflats and sandflats of the Mount Shields Formation member 1, represented by terrigenous couples and couplets. Regression culminated in deposition of the Mount Shields member 2 sand wedge, which represents another major alluvial apron. In the south, Mount Shields 2 contains the gravel and cross-bedded sand sediment types, which record deposition in braided channels on the upper parts of the alluvial apron. Northward, the gravel and coarse sand pinch out, and Mount Shields member 2 becomes mostly fine-grained, tabular quartzite beds of the flat-laminated sand sediment type, deposited by sheetfloods across the lower part of the alluvial apron.

Coarse-grained oolitic quartzite and stromatolitic beds (coarse sand and intraclast sediment type), deposited in a transgressive beach sequence, mark the base of Mount Shields member 3. Continued transgression led to deposition of salt cast-bearing lenticular couplets, which characterize Mount Shields member 3 in the central part of the basin, and brought carbonate sediments into the northern part of the basin. Transgression peaked in the northern part of the basin with deposition of black, carbonaceous microlaminae, which grade progressively southward to green and then red varieties of the microlamina sediment type, deposited in more oxidized, but still submerged environments. Alluvial aprons again prograded northward and northwestward,

depositing the Bonner quartzite wedge. Again the alluvial facies tract (Figures 6 and 10) was established in the Bonner, with the downslope progression from gravel, to cross-bedded sand, to flat-laminated sand, to discontinuous layer sediment types. During the waning parts of each flood the facies tract shifted up the apron, producing fining-upward sequences 1 m or more thick. The McNamara Formation, although not extensively studied, appears to record yet another northwest-facing, transgressive-regressive megasequence. In contrast, the Garnet Range Formation, with its tidal channel deposits and hummocky cross-bedding (McGroder, 1984), appears to mark the introduction of truly marine conditions into the Belt basin (McGroder, 1984; Winston et al., 1984). Thus, the counter clockwise rotation of the paleoslope direction, first recorded in the lower Missoula Group, appears to culminate in the uppermost Missoula Group with marine inundation. The Garnet Range grades up into cross-bedded quartzite of the Pilcher Formation, which represents regressive marine deposits, overlain by fluvial quartzite (McGroder, 1984) below the Middle Cambrian marine transgressive deposits.

Sedimentologic Comparisons and Conclusions

Throughout most of its history, the Belt Supergroup appears to have been deposited in a mostly enclosed basin interpreted by Winston et al. (1984) to have been intracratonic, formed by incipient rifting (Winston et al., 1984). Southwestern and western source terranes of continental proportions supplied sediment for the lower Belt, the Ravalli Group, and the middle Belt carbonate. During Missoula Group deposition, the western source terrane receded, and sediments, derived from the south and southwest, were deposited on a northwest-facing regional slope that was eventually engulfed by marine transgression, recorded in the Garnet Range Formation.

Comparison of Belt sediments below the Garnet Range with those of modern and Phanerozoic alluvial fan, sandflat, and playa margin sediments reveals close similarities in depositional processes and facies distributions (Winston et al., 1984). Although sandflats and mudflats surrounding the Belt sea appear to be extensive, they do not appear to have been flooded by astronomic tides. The Belt lacks beach and barrier island sequences characteristic of modern and ancient macrotidal, mesotidal, and even microtidal coastlines. Although White (1984) built a case for intertidal deposition of the Altyn Formation, the channels he cites are inferred to have decimeter-scale relief, which is more characteristic of small channels on playas than tidal channels and creeks in marine intertidal zones. Above the lower Belt, reported local bimodal crossbeds (McMechan, 1981; Horodyski, 1983; Mauk, 1983b) are the only indication of possible tidal currents. These have not been supported by delineation of tidal channels and could represent tributary mouth bars of very broad and shallow, low gradient streams (Alam et al., 1985). Instead, the downslope progression from modern alluvial fans to saline lake margins, as described by Hardie et al. (1978), does compare closely with the descriptive facies tract model of the Belt.

Hardie et al. (1978, p. 17) described how modern alluvial fans, built largely of sand and gravel lenses deposited in

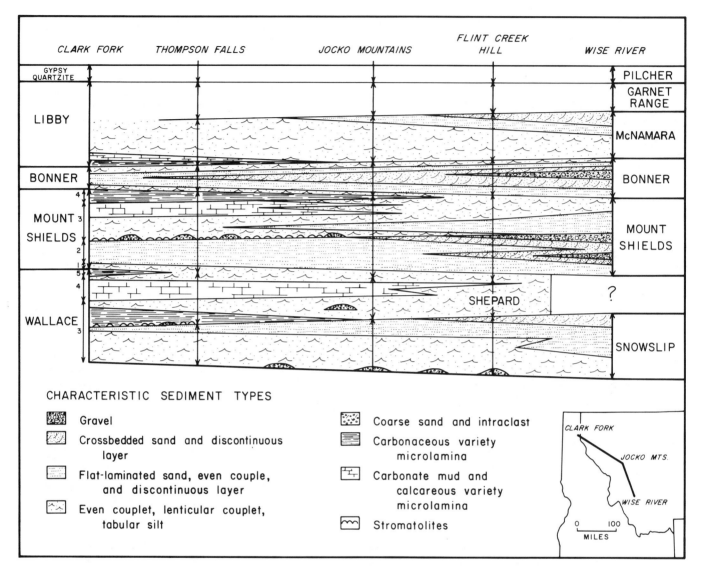

CHARACTERISTIC SEDIMENT TYPES

Gravel

Crossbedded sand and discontinuous layer

Flat-laminated sand, even couple, and discontinuous layer

Even couplet, lenticular couplet, tabular silt

Coarse sand and intraclast

Carbonaceous variety microlamina

Carbonate mud and calcareous variety microlamina

Stromatolites

Figure 9—Generalized sediment type cross section (not to scale) of the Missoula Group from the Snowslip through the McNamara formations. Cross section depicts downslope sediment type facies changes from Wise River, Montana, on the south to Clark Fork, Idaho, on the north. Quartzite wedges of the upper Snowslip, Mount Shields member 2, Bonner, and upper McNamara formations pinch basinward and illustrate the distal facies change from gravel, to cross-bedded sand, to flat-laminated sand, and even couple sediment types. The wedges are separated by southward-pinching tongues of the even and lenticular couplet sediment types, with beds of tabular silt and lenses of the coarse sand and intraclast sediment type. The tongues interfinger with northward-thickening intervals of the carbonate mud and microlamina sediment types. Sources: Wise River section—Calbeck, 1975. Flint Creek Hill—Winston, 1973; Winston and Wallace, 1983. Jocko Mountains—Winston, 1977. Thompson Falls—Winston, 1977. Clark Fork—Harrison and Jobin, 1963; Winston, 1977.

braided channels, pass at their toes to the

> sandflat subenvironment where the braid channels lose their identity and the floodwaters disperse as unchannelled, unconfined sheetfloods across a narrow flat (<<1% slope) sand plain.... Sand is the main component of the sediment and characteristically occurs as planar-parallel horizontal laminae and wavy laminated beds.... Shallow ponding of water on the sandflat as the adjoining saline lake expands with the flood may result in wind wave reworking of the surface of the sandflat, producing a variety of wave-ripple bedforms.... Deposition by each flood event may range from a package of laminated sand tens of centimeters thick to a thin sheet one or two centimeters thick....

In this passage Hardie and co-workers describe vividly how sediments analogous to the flat-laminated sand, even couple,

even couplet, and lenticular couplet sediment types are formed today. They (1978, p. 20) also describe how sandflats pass lakeward to the dry mudflat environment, whose sedimentary features are not well known, but can be subdivided into three basic types: (1) sheetwash mudflats whose deposits "would be graded, with traction load, flat lenticular sandy or silty laminae capped by a...mud drape deposited when the flow waned" (even couplet sediment type); (2) ponded water mudflats in which "sediment-charged sheetwash off the sandflats should rapidly decelerate as it enters the expanded lake and quickly deposit its load...in the form of a graded thin bed or thick lamina [which] could be reworked by wind waves to produce...coarse silt-fine sand lenses and muddy drapes... (lenticular couplet sediment type); and (3) exposed lake-

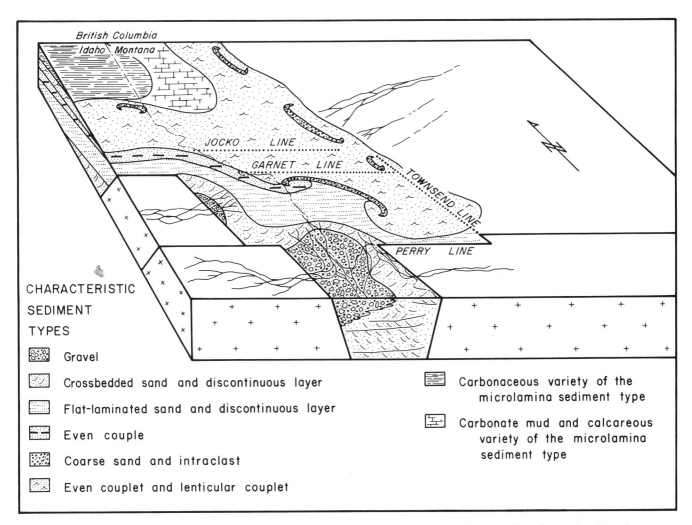

Figure 10—Hypothetical facies tract block diagram of Missoula Group sediment types. Braided streams from the south and southwest deposited the gravel and cross-bedded sand sediment types during floods high on alluvial aprons. Channeled flow broadened and blended into sheetfloods that crossed the lower apron and sandflats, depositing the flat-laminated sand, discontinuous layer, and even couple sediment types. Floods spread out across mudflats and ponded, depositing the even couplet sediment types which was desiccated between floods. Beaches of the coarse sand and intraclast sediment type formed on the mudflats, seaward from which waves worked mud and fine sand into the lenticular couplet sediment type. The microlamina sediment type formed in the basin center at times of low sediment influx, and the carbonate mud sediment type developed in areas of rapid carbonate mud precipitation.

bottom mudflats with a "highly churned soil-like zone" (churned dolomitic subfacies of O'Connor, 1972, at the tops of carbonate cycles).

Smoot (1983) described similar sequences from the Wilkins Peak Member of the Eocene Green River Formation, which correspond to the general Belt facies tract model on nearly a one-to-one basis (Winston et al., 1984). Other closely comparable graded, mud cracked redbeds have been described from the Upper Triassic Blomidon of Nova Scotia by Hubert and Hyde (1982), who interpreted that sequence to represent sandflat and playa deposition. The Triassic lacustrine Lockatong Formation of New Jersey and Pennsylvania (Van Houten, 1962) and the Triassic Edderfugedal of East Greenland (Clemmensen, 1978) both contain terrigenous-to-carbonate transgressive-regressive cycles very similar to those of the Helena Formation. Collectively, the initial rift configuration, the alluvial fan, sandflat, and lacustrine sedimentary sequences, and the

basaltic dikes and extrusions of the Triassic basins of eastern North America and Greenland (similar to diabase sills and the Purcell lavas of the Belt) probably represent the closest Phanerozoic tectonic and sedimentologic analog to the Proterozoic Belt basin.

PART II: PROTEROZOIC BLOCK FAULT PATTERNS WITHIN THE BELT BASIN AND THEIR INFLUENCE ON PHANEROZOIC STRUCTURES

Introduction

In Part I the Belt basin is interpreted as having been intracratonic, surrounded on the northeast, south and west by continental terranes. Its isolation from Proterozoic open oceans is based on sedimentologic analysis of the Belt Supergroup. In this part I propose, as a speculative hypothesis, that the Belt Supergroup was deposited on a cluster of subsiding crustal blocks (Figure 11), which are reflected not only in Belt strata, but are also evident in the patterns of Cretaceous to Eocene thrusts and later Cenozoic extensional faults. Both Proterozoic and Phanerozoic tectonic elements must be considered in analyzing Belt sedimentology. Not only did Proterozoic tectonics produce the Belt basin, but all Belt measured sections west of the Helena embayment are allochthonous, that is, tectonically transported by Phanerozoic compression and extension. Reliable estimates of tectonic disruption are crucial in reordering the Belt stratigraphic framework, upon which the sedimentologic and Proterozoic structural conclusions are largely based.

Analysis of Belt stratigraphy reveals in some places abrupt changes in stratigraphic thicknesses that are localized along linear trends. Furthermore, along these linear trends are patchy exposures of soft sediment deformation in Belt rocks. I interpret the linear thickness changes that coincide with patches of soft sediment deformation to reflect Proterozoic growth faults within the basin. These growth faults outline large quadrilaterals, here interpreted to be downdropped blocks of continental crust. In places, the rocks containing stratigraphic evidence for the block boundaries have been tectonically transported, so that the thickness changes no longer overlie the basement block boundaries. Fortunately, the blocks also appear to be outlined by major Phanerozoic faults, which thus serve as a guide for identifying the Proterozoic blocks. The danger of circular reasoning is obvious in proposing, on the one hand, that Proterozoic block faults affected Phanerozoic structure and, on the other hand, that in places the Proterozoic structures can be located by the discontinuities in Phanerozoic faults. The difficulty in separating Proterozoic structures from Phanerozoic overprint stands as the principal reason why this hypothesis is still so speculative. It can probably best be tested by painstakingly interweaving stratigraphic and structural analyses. It is with the hope that others will test the hypothesis that I present it here in its formulative stage, separated from the sedimentologic analysis, which I believe stands on firmer ground.

Hypothesis

I hypothesize that Proterozoic growth faults in the Belt basin form at least four major linear trends, which partly outline or border at least five crustal blocks, and that differential subsidence of four of them produced the Belt basin (Figure 11) (Winston, 1981, 1982). The block boundaries have been sites of Phanerozoic faults, and I refer to the hypothetical Proterozoic fault trends as *lines* to separate them conceptually from Phanerozoic faults that can be mapped on the ground. The southernmost Proterozoic line is the Perry line, long recognized as a major tectonic feature in Montana (Thom, 1923; Harris, 1957). It is expressed as a series of en echelon faults that extend from Camp Creek in the southern Highland Mountains in the west, to the Pass fault in the southern Bridger Range in the east. Each fault segment of the line trends east, but their en echelon arrangement gives an east-northeast trend to the line. The Perry line may extend as far east as Billings, Montana (Peterson, 1981). It may extend westward into Idaho, where west-northwest-trending structures reflect a

"major transverse lineament" (Desmarais, 1983, p. 125) in the vicinity of Lost Trail Pass. Faults along the Perry line moved down to the north during the Middle Proterozoic and separated the uplifted crystalline Dillon block to the south from the subsiding Belt basin to the north (Figure 11). North of the Perry line and west of the Bridger Range lies the Deer Lodge block (Winston, in press, c) (Figure 11), which subsided with respect to the Dillon block. North of the Perry line and east of the Bridger Range is the block underlying the Helena embayment (Harrison et al., 1974a). It also subsided with respect to the Dillon block and is separated from the Deer Lodge block by the proposed northwest-trending Townsend line (Figure 11). Faults along the southern part of the Townsend line probably moved down to the east, at least during early Belt deposition, so that the Helena embayment block subsided more deeply than the Deer Lodge block. Bounding the Deer Lodge and Helena embayment blocks on the north is the Garnet line (Figure 11). It extends from the Lolo fault in the west, eastward to the north end of the Helena Valley, where it intersects the Townsend line, and continues eastward through the southern part of the Little Belt Mountains. The segment of the Garnet line north of the Deer Lodge block moved down to the north. Conversely, the segment east of the Townsend line moved down to the south, forming the northern edge of the Helena embayment graben block. North of the Deer Lodge block and west of the Townsend line is the Ovando block (Figure 11). It subsided with respect to both the Deer Lodge block on the south and the stable crystalline basement east of the Townsend line. The Ovando block is bounded on the north by the Jocko line (Figure 11), which extends from the Coeur d'Alene district east-southeastward, through the lower reaches of the Flathead and Jocko rivers, along the northern edge of the Ovando Valley, to Rogers Pass, where it meets the Townsend line. Like faults along the Perry line, faults along the Jocko line may have been an en echelon system of east–west faults, giving the Jocko line its west-northwest trend. North of the Jocko line is a block, or set of blocks, that at least in places, moved up with respect to the Ovando block, thus making the Ovando block a graben. This region north of the Jocko line is tentatively called the Charlo block (Figure 11), with the suspicion that future study may reveal more than one block.

Thus, the Perry, Garnet, Jocko, and Townsend lines partly border five blocks of continental crust: the Dillon, Deer Lodge, Helena embayment, Ovando, and Charlo blocks. The Dillon, Deer Lodge, and Ovando blocks are arranged in a down-to-the-north, stairstep fashion, so that the Ovando block represents a graben faulted downward against the Charlo block (Figure 12). East of the Townsend line, the Helena embayment block offsets the position of the graben to the southeast. As so defined, none of the blocks are entirely surrounded by faults. The Miner Lake-Beaverhead Divide fault, described by Ruppel in this volume, could represent the west-bounding fault of the Dillon block, as diagrammed in Figures 6, 8, and 10. Furthermore, it could have projected northwestward to the position of the present Bitterroot Valley, thus separating the thin Belt section of the Deer Lodge block from the thicker Belt section deposited to the west, now represented by the Sapphire allochthon

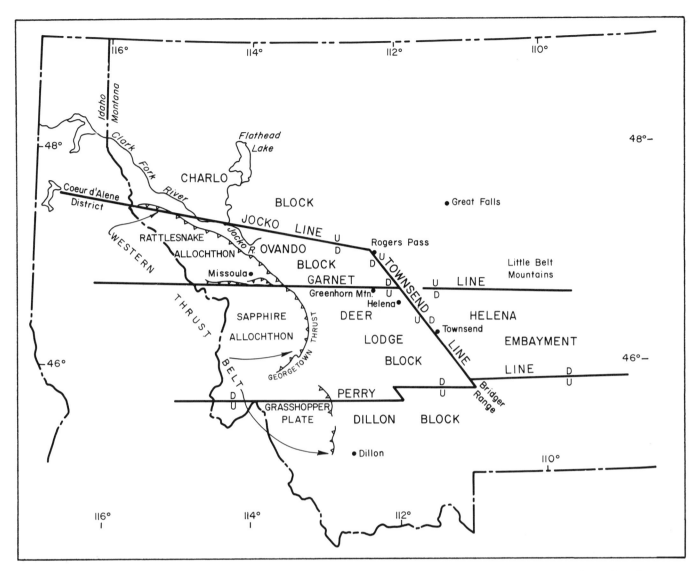

Figure 11—Hypothetical crustal blocks of the Belt basin. The Perry, Garnet, Jocko, and Townsend lines cut continental crust into the Dillon, Deer Lodge, Helena embayment, Ovando, and Charlo blocks. The western edges of the Dillon, Deer Lodge, and Ovando blocks are tentatively marked by the western thrust belt.

(Figures 11 and 12). Therefore, the west edge of the Deer Lodge block is tentatively placed at the eastern limit of the Sapphire allochthon, marked by the Georgetown thrust. The western edge of the Ovando block is likewise conditionally placed at the eastern limit of Rattlesnake allochthon (Figure 11). The region north of the Jocko line, collectively placed in the Charlo block, is not further delimited by proposed Proterozoic faults. Westward thickening of the Belt section from the eastern thrust belt to about the Mission Range (Mudge, 1972a) indicates that this part of the Charlo block tilted westward, whereas northeastward thickening of the Belt section from the Coeur d'Alene district to the Libby thrust suggests that the western part of the Charlo block tilted eastward. Perhaps the Charlo block was indeed broken by Proterozoic faults, in which case the Hope fault possibly overlies a deep seated Proterozoic fault (Harrison et al., 1974a). Further speculation is premature, and the region is consequently treated here as a single block. Finally, the

eastern limit of the Helena embayment remains undefined; probable Belt rocks are reported in the subsurface as far east as Billings, Montana (Peterson, 1981).

Proterozoic Evidence

Evidence supporting Proterozoic fault movement along the proposed lines (Figures 11 and 12) ranges from convincing to speculative, and in many places is based on limited control. Nonetheless, I judge the following evidence sufficient for presenting the above block fault pattern as a viable hypothesis.

Perry Line

Proterozoic faulting along the Perry line is most clearly illustrated in the lower Belt LaHood Formation where great thicknesses of boulder conglomerate border the Dillon block. Although thrust sheets of LaHood have been tectonically

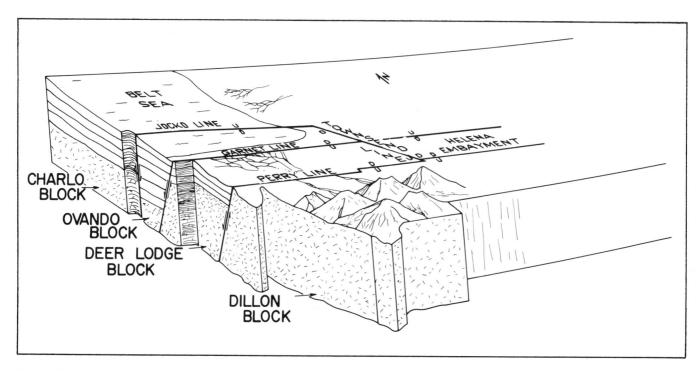

Figure 12—Block diagram of the Belt basin during deposition of the Missoula Group, showing the nearly east-west Perry line, Garnet line, and Jocko line and the northwest-trending Townsend line. The fault lines cut continental crust into at least five blocks: the Dillon, Deer Lodge, Helena embayment, Ovando, and Charlo blocks.

transported along the Perry line, McMannis (1963) concluded that geometric relationships of the LaHood to the Dillon block suggest (particularly in the eastern and western parts of the Perry line) strong fault movement during LaHood deposition. Hawley et al. (1982, p. 5) believe "components of the LaHood to have been eroded from a rugged land area of high relief and deposited in a deep basin north of a fault-shoreline." McMannis (1963) pointed out that the upward transition from the conglomeratic LaHood through fine-grained calcareous argillite of the Newland to terrigenous argillite of the Greyson reflects less fault movement along the Perry line and possible southward onlap of Belt sediments across the line. Faulting along the Perry line probably continued sporadically after LaHood deposition. Just north of the Perry line in the Highland Mountains, Thorson (1984, p. 11) reported that above the LaHood, there is a massive sulfide "deposited on the down-dropped blocks of a horst and graben terrane created by local faulting just before, or contemporaneous with, the deposition of the sulfide." Evidence reflecting behavior of the Perry line during deposition of the later Belt has been mostly removed by pre-Middle Cambrian erosion in the southern part of the basin. However, the large southward-thickening quartzite wedges of the Missoula Group farther north in the basin may reflect renewed uplift of the Dillon block or of blocks to the southwest.

Garnet Line

Down-to-the-south movement of faults along the northern edge of the Helena embayment has been analyzed most fully by Godlewski and Zieg (1984). They reported a

linear trend of debris flows in the Newland Formation just south of the Garnet line and parallel to it. The flows contain carbonate clasts that Godlewski and Zieg believe were transported from shallow water environments on an uplifted northern block southward across the Garnet line into deeper water carbonate sediments of a subsiding southern block (Helena embayment). Down-to-the-north movement along the Garnet line west of the Townsend line is reflected in abrupt northward thickening of the Helena, Snowslip, and Shepard formations, which span the line. The line is best constrained in the vicinity of Greenhorn Mountain (Figure 13), where the Helena Formation thickens from 1220 m near Helena (Knopf, 1963) to 2290 m north of the Garnet line (Bierwagen, 1964). The Snowslip thickens from 233 m at Birdseye (Winston, unpublished section) to 966 m north of the Garnet line (Bierwagen, 1964), and the Shepard thickens from 88 m at Birdseye (Winston, unpublished section) to 1029 m north of the Garnet line (Bierwagen, 1964).

Slump breccia occurs in Mount Shields member 3 within the Garnet line (Figures 13 and 14) and pre-Middle Cambrian faulting has brought the McNamara on the north downward against Mount Shields member 2 west of Greenhorn Mountain and has brought Mount Shields member 3 on the north downward against Mount Shields member 2 east of Greenhorn Mountain. To the west, stratigraphic data reflecting the Garnet line become less constrained because Belt outcrops on either side become more widely separated by Phanerozoic rocks. However, Mount Shields member 2 in the southeastern Flint Creek Range measures 160 m (Winston, unpublished section), and is thinner than the 255 m in the nearest section of Mount

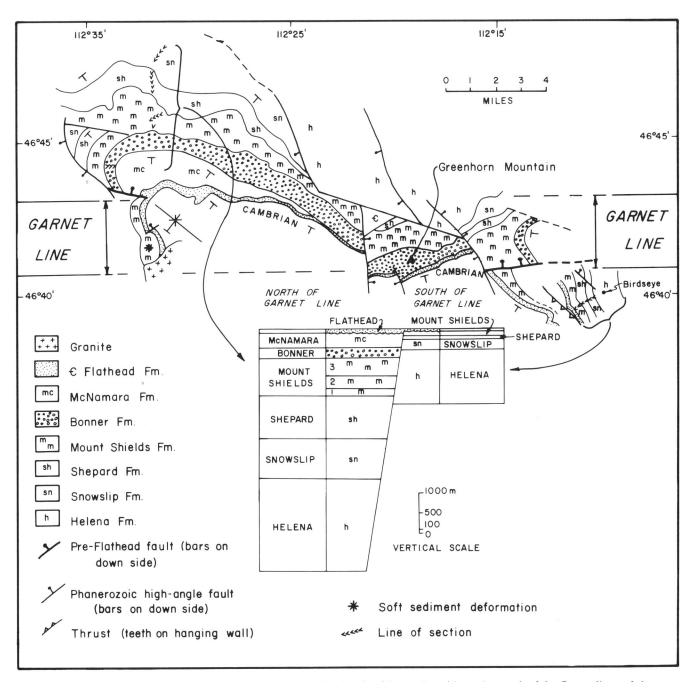

Figure 13—Map of Greenhorn Mountain area (see Figure 11), showing the thin stratigraphic section south of the Garnet line and the contrasting thick stratigraphic section north of the line, recording down-to-the-north growth faulting during deposition of the Missoula Group. Faulting continued into Late Proterozoic time, and the unfaulted Flathead Formation shows that pre-Middle Cambrian erosion cut down to the McNamara Formation north of the Garnet line, but down to the Mount Shields south of the line. Sources: Bierwagen, 1964; Wallace et al., 1981; Winston and Baldwin, 1977, unpublished data.

Shields member 2 north of the Garnet line near Ovando (Slover and Winston, in press). West of the Deer Lodge and Ovando blocks the Garnet line may extend between the Sapphire allochthon and the Rattlesnake allochthon, where Wallace et al. (1976) projected an approximately east-west belt of soft sediment slumps in the Wallace Formation

(Figure 14) north of the Idaho batholith. Still farther west Armstrong et al. (1977), on the basis of 87Sr/86Sr ratios, plotted a sharp east-west offset of the western limit of North American continental crust along the projection of the Garnet line, possibly indicating that the line played a part in Late Proterozoic continental separation.

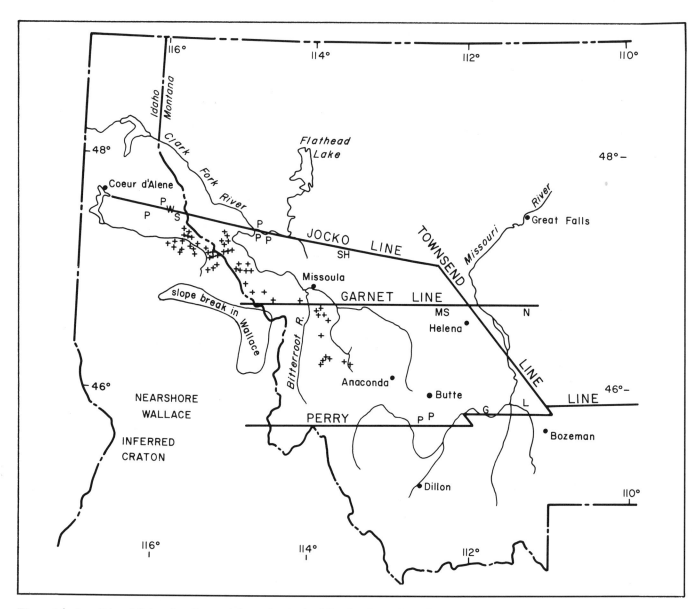

Figure 14—Localities of Belt soft sediment deformation in the following formations: Greyson (G), LaHood (L), Mount Shields (MS), Newland (N), Prichard (P), Shepard (SH), and Wallace (W). Crosses (+) are locations of slump breccia in the Wallace Formation reported by Wallace et al. (1976), who also located the western source terrain, as well as nearshore and slope break environments in the Wallace Formation. Sources: Ransome and Calkins, 1908; Chevillon, 1977; Webster, 1981; Hawley et al., 1982; Godlewski and Zieg, 1984; Winston, unpublished data.

Jocko Line

Proterozoic down-to-the-south movement along the Jocko line is most clearly evidenced in the Coeur d'Alene district on the basis of both stratigraphic thickness changes and soft sediment deformation. The Jocko line (Figures 12 and 15) is marked by the Osburn fault, and the growth faulting is best illustrated in the Revett and St. Regis formations (Figure 15). Composite sections north of the Jocko line show that the total Revett thickens from 580 m north of the Jocko line (Winston, unpublished section) to more than 1160 m (White and Winston, 1982) south of it; that the St. Regis thickens from 314 m north of the Jocko line at Military Gulch (Winston, unpublished section) to approximately 1500 m along Silver Creek south of the Jocko line (Wallace

and Hosterman, 1956); and the Wallace thickens from an estimated 460 m along the North Fork of the Coeur d'Alene River (Winston, unpublished observation) north of the Jocko line to 1067 m south of Wallace, Idaho (Vance, 1981). Immediately south of the Osburn fault, the St. Regis contains clasts (Ransome and Calkins, 1908; Hobbs et al., 1965), which Chevillon (1977) attributed to intrastratal brecciation by Proterozoic faulting along the Osburn fault (Jocko line). Spectacular folds in the middle Wallace (Ransome and Calkins, 1908), which appear to have formed intrastratally in soft sediment, also coincide with the Jocko line. Thus, from a variety of evidence taken from several stratigraphic levels, down-to-the-south movement along the Jocko line in the Coeur d'Alene district seems well substantiated. Cretaceous thrusts and east-facing folds north

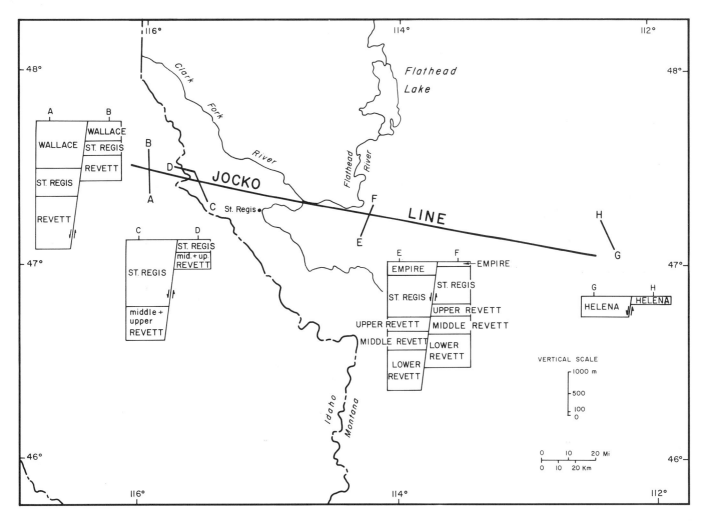

Figure 15—Map and cross sections showing thickness changes across the Jocko line, revealing down-to-the-south growth faulting along the line during deposition of the Ravalli Group and middle Belt carbonate. Sources: Section A-B (Wallace, Idaho and North Fork of the Coeur d'Alene River to Kellogg, Idaho)—Hobbs et al., 1965; Vance, 1981; Winston, unpublished data. Section C-D (Silver Creek to Revett Lakes and Military Gulch)—Alleman, 1983. Section E-F (Flathead Reservation Divide to National Bison Range)—Mauk, 1983b; Winston, unpublished data. Section G-H (Rogers Pass to Falls Creek Ridge)—Eby, 1977.

of the Jocko line, and north-facing folds south of the line do not appear to have enough displacement to alter the regional evidence for southward thickening. Nor is it substantially altered by the 26 km of Cenozoic right lateral movement along the Osburn fault (Hobbs et al., 1965).

From the Coeur d'Alene district east to St. Regis, Montana, the Revett Formation continues to thicken southward across the Jocko line (Alleman, 1983), probably recording down-to-the-south growth faulting along the line (Figure 15). East of the confluence of the Flathead River with the Clark Fork River, the Ravalli Group and middle Belt carbonate thicken on both sides of the Jocko line, and down-to-the-south movement is not so sharply delineated (Figure 15). In the eastern thrust belt, rocks of the Belt Supergroup are brought to the surface along the Silver King, Hoadley, Eldorado, and Steinbach thrusts. McDonough (1985) interpreted abrupt stratigraphic thickening of the Helena and Snowslip formations within the Silver King thrust plate to reflect down-to-the-south Proterozoic growth faulting along the Jocko line. He estimated about 30 km of northeastward tectonic transport along the Silver King and

Hoadley thrusts, placing the Jocko line about where east-trending Tertiary normal faults border the northern edge of the Ovando Valley. In a similar way, Eby (1977) has shown that on the Eldorado plate the Helena Formation thickens southward from 176 m at Falls Creek Ridge to 456 m at Rogers Pass, a distance of 11 km. This abrupt thickening is also interpreted to reflect down-to-the-south Proterozoic faulting along the Jocko line. In the same area, pre-Middle Cambrian erosion has beveled the Belt section northward, so that Middle Cambrian rocks rest on progressively older units from Mount Shields member 1 north of Rogers Pass to the Helena at Falls Creek Ridge. This beveling may reflect Late Proterozoic down-to-the-south warping across the Jocko line.

Townsend Line

Evidence for the Townsend line differs from that for the east-west Proterozoic lines. Because the proposed Townsend line trends at a high angle to the direction of Cretaceous to Eocene thrusting, little stratigraphic evidence remains in close proximity to the line. However, Harrison (1972, figure

3) showed that the lower Belt section thickens eastward from 560 m on the Deer Lodge block to 7200 m in the Helena embayment. From another point of view, the Townsend line seems necessary to explain the offset of the Helena graben on the east with respect to the Ovando graben on the west (Figures 11 and 12).

Phanerozoic Response to the Proterozoic Blocks

Some Proterozoic blocks appear to be reflected by Phanerozoic structural patterns, and for that reason they are discussed here. Paleozoic carbonate rocks generally thicken westward across the central and western parts of the Belt basin, reflecting deposition on a trailing continental margin, but evidence in them which might support the proposed Proterozoic blocks is unreported. However, the Helena embayment apparently continued to subside in the Paleozoic and received a thick Paleozoic section in a segment of the central Montana trough (Peterson, 1981).

Cretaceous to Eocene Compression

Patterns of Cretaceous to Eocene thrusts and folds tend to outline the Proterozoic blocks (Figures 16 and 17). In a general way, folds and thrusts are long and continuous above the Proterozoic blocks, but become broken and segmented over the Proterozoic east-west lines. The northwest-trending Townsend line is reflected by differences in styles of ramping. Compression produced both a western thrust belt in Montana and Idaho and an eastern thrust belt in Montana. The western thrust belt is broken by the Jocko, Garnet, and Perry lines into three segments. The northernmost segment is a collection of thrust plates, here termed the Rattlesnake allochthon (expanded Rattlesnake plate of Wallace and Lidke, 1980). Its northeastern edge extends from the east-trending, north-facing folds immediately south of the Jocko line in the Coeur d'Alene district eastward to the Savenac syncline, where it curves southeastward (Lonn, 1984) through the west side of the Squaw Peak Range, to the Rattlesnake plate on the southwest side of the Jocko Mountains and Garnet Range. At its juncture with the Garnet line, the Rattlesnake allochthon meets the torn, northern edge of the Sapphire allochthon, which is the central segment of the western thrust belt. The arcuate form of the Sapphire allochthon, marked on its leading edge by the Georgetown thrust, suggests that its northern and southern extremities were dragged and torn along the Garnet and Perry lines. Likewise, the Grasshopper plate, representing the southern segment of the western thrust belt (Ruppel et al., 1981), also appears to have been dragged along its northern edge where the Johnson thrust curves northwestward across the Perry line.

The southern limit of the eastern thrust belt may be represented in part by the frontal fold and thrust zone west of the Dillon block (Ruppel et al., 1981) and also by Laramide-style compressional faults within the Dillon block. In the Belt basin, the eastern thrust belt is sharply offset eastward along the Perry line by a series of strike-slip tear faults (Schmidt and Hendrix, 1981; Werkema et al., 1981; Woodward, 1981; Lageson et al., 1983). Woodward (1981) described that segment of the eastern thrust belt that curves northeastward into the thick Belt stratigraphic pile of the Helena embayment and swings northwestward near the Garnet line. The arcuate trace of this segment of the eastern thrust belt suggests that it was dragged by crystalline basement along the Perry and Garnet lines. North of the Garnet line the Scout Camp and Moors Mountain thrusts (Birkholtz, 1967; Bregman, 1976; Woodward, 1981) of the eastern belt ramped across the Townsend line onto the buttress of the crystalline basement. From north of Helena to the vicinity of Rogers Pass the northwest-trending Steinbach, Eldorado, and Hoadley thrusts of the eastern thrust belt repeat the Belt section in steeply dipping sheets, also interpreted here to reflect ramping onto crystalline basement across the Townsend line. Near Rogers Pass the Townsend and Jocko lines are inferred to intersect, and, although the thrust sheets continue northeast of Rogers Pass, evidence for projection of the Townsend line becomes tenuous on stratigraphic grounds. Instead, structural evidence reflecting the Jocko line becomes more apparent in the divergent west-northwest trend of the Silver King thrust (McDonough, 1985), which overrides the Hoadley and more closely approximates the trace of the Jocko line. A west-northwest-trending ramp anticline in the Silver King thrust plate parallels the Jocko line and coincides with abruptly southward-thickened Helena and Snowslip sections that are interpreted to reflect growth faulting along the Jocko line (McDonough, 1985). Thus, northeastward thrusting up a ramp along the Jocko line appears to be recorded as a faulted hanging wall anticline in the Silver King plate, transported a distance of approximately 30 km. Consequently, Cretaceous to Eocene compression along both the eastern and western segments of the Jocko line formed thrusts that moved northeastward across the line and diverge from the more northerly trend of folds and thrusts of the line.

Cenozoic Extension

From latest Eocene to Holocene time, the Belt basin has undergone crustal extension, and once more the Proterozoic blocks have been revealed, this time by patterns of normal faults that formed distinctive domains over each of the Proterozoic blocks (Figures 18 and 19). Normal faults cutting the Dillon block trend mostly northeastward and northwestward, possibly responding to Early Proterozoic structural trends (Schmidt and Garihan, 1979). Those that cut the Deer Lodge block are concentrated in deep, north-trending valleys that separate large batholiths. The Townsend line marks the eastern limit of most basin-and-range type faults; only a few extensional faults cut Belt rocks of the Helena embayment to the east. As pointed out by Reynolds (1979), faults focused in the Townsend Valley curve toward the west where they approach the Garnet line to the north and pass into right lateral strike-slip faults that appear to have transferred extensional strain more broadly to the many northwest-trending high-angle faults of the Ovando block. Where these faults approach the Jocko line, they again curve toward the west into the right lateral strike-slip St. Marys fault. Like faults of the Garnet line, the St. Marys fault also probably functioned as an extensional tear fault, reflecting greater westward movement of the Ovando block with respect to the Charlo block(s). High-angle extensional faults splay from the St. Marys fault northward and northwestward across the Charlo block(s). West of the Ovando block are the Clark Fork-Ninemile and

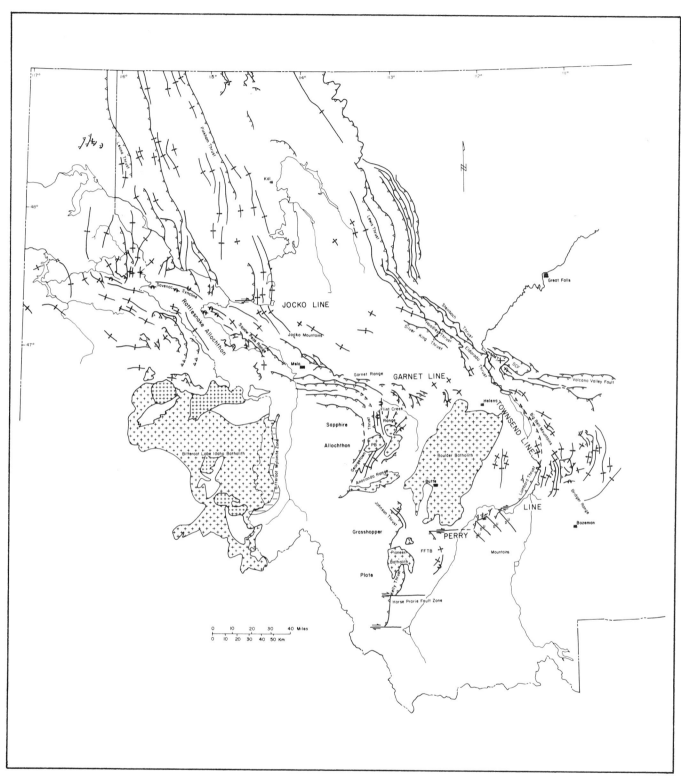

Figure 16—Map of Cretaceous to Eocene folds, thrusts, and tear faults resulting from crustal compression. Shown are the western and eastern thrust belts, offset along the Perry, Garnet, and Jocko lines and ramped along the Townsend line east of the Ovando block. Abbreviations are as follows: frontal fold and thrust belt (FFTB), Moors Mountain thrust (MMT), Philipsburg batholith (PB), Rattlesnake thrust system (RTS), and Scout Camp thrust (SCT). Sources: Weed, 1899; Calkins and Emmons, 1915; Mertie et al., 1951; Myers, 1952; Ross et al., 1955; Poulter, 1958; McGill, 1959; Ross, 1959, 1963; Csejtey, 1962; Mutch, 1960; Nelson and Dobell, 1961; Childers, 1963; Harrison and Jobin, 1963; Knopf, 1963; McMannis, 1963; Nelson, 1963; Robinson, 1963, 1967; Bierwagen, 1964; Skipp and Peterson, 1965; Groff, 1967; Hall, 1968; Johns, 1970; Fraser and Waldrop, 1972; Kleinkopf and Mudge, 1972; Mudge, 1972a, b; Griggs, 1973; Mudge et al., 1974, 1983; Harrison et al., 1974b, 1980; Wells, 1974; Calbeck, 1975; Harrison, 1977; McGrew, 1977a, b, c; Aadland and Bennett, 1979; Rember and Bennett, 1979; Whipple, 1979; Hyndman, 1980; Wallace and Lidke, 1980; Ruppel et al., 1981, 1983; Schmidt and Hendrix, 1981; Wallace et al., 1981; Woodward, 1981.

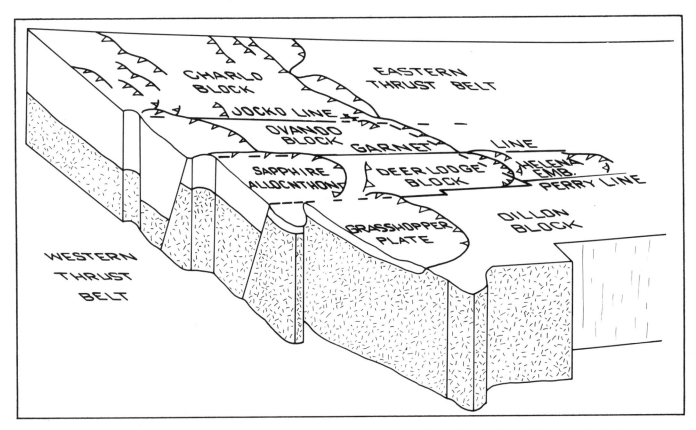

Figure 17—Block diagram of Cretaceous to Eocene western and eastern thrust belts, offset along the Perry, Garnet, and Jocko lines and ramped along the Townsend line east of the Ovando block.

Boyd Mountain faults, which extend from the Garnet line northwestward to the Jocko line. Down-to-the-southwest movement with rotation of their hanging walls indicates that they are probably listric normal faults. Where they curve westward along the Jocko line, movement becomes mostly strike slip, and they merge with the Osburn fault, which has had 26 km of right lateral displacement in the Coeur d'Alene district. Normal faults in the eastern part of the Charlo block trend northward, whereas those of the western part appear to splay out from the St. Marys fault and trend northwestward. Thus, each Proterozoic block seems to have its own style of Cenozoic extensional faults.

Tectonic Discussion

The idea that Mesozoic and Cenozoic structures in the Belt basin were controlled in part by Precambrian structures is not new. Peale (1896) recognized that the Perry line marked the edge of a Precambrian buttress around which wrapped the Laramide fold belt. This discovery was further developed by Thom (1923), Harris (1957), Robinson (1959), McMannis (1963, 1965), Harrison et al. (1974a), Schmidt and Hendrix (1981), Hawley et al. (1982), and Lageson et al. (1983). The major difference between the present proposal and preceding interpretations involves the concept of the "Montana lineament" or "Lewis and Clark line," which is an apparent belt of northwest-trending structures that crosses western Montana from north of Helena through the Coeur

d'Alene district of Idaho (Billingsley and Locke, 1939; Blackstone, 1956; Weidman, 1965; Harrison, 1972; Kleinkopf, 1984). Some authors have suggested that it is a transform fault zone (Smith, 1965; Talbot and Hyndman, 1973; Sheriff et al., 1984); others believe it represents a zone of deep-seated Precambrian crustal weakness or faulting (Wallace et al., 1960; Hobbs et al., 1965; Harrison et al., 1974a; Reynolds, 1977, 1984a; Reynolds and Kleinkopf, 1977). Aside from Maughan's objection in this volume that the Lewis and Clark is not a line, but instead a "lane," the concept of the Lewis and Clark line appears to project Proterozoic fault evidence from the Garnet line north of Helena along trends of Cenozoic extensional faults to Proterozoic faults of the Jocko line in the Coeur d'Alene district. Therefore, in terms of the proposal presented here, the so-called "Lewis and Clark line" represents parts of two Proterozoic lines, and its continued usage distorts geologic thinking.

Assessment of Phanerozoic tectonic displacement is crucial in reconstructing the Belt stratigraphic framework, upon which the sedimentologic interpretations so heavily rest. During compression, the greatest crustal shortening was absorbed in the western and eastern thrust belts (Figure 16). The central part of the basin, although transported, does not appear to have been severely disrupted. For this reason, Belt sections appear to have retained most of their original positions relative to each other, and the stratigraphic framework appears to be relatively intact.

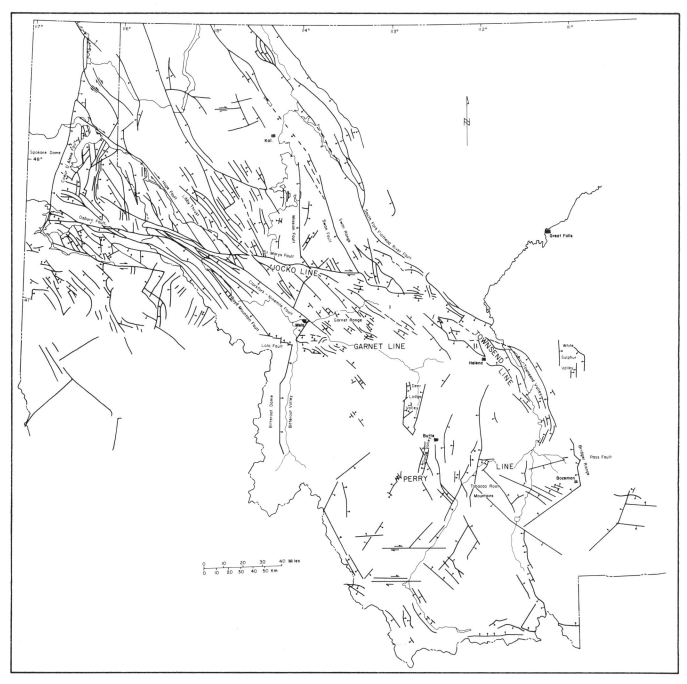

Figure 18—Map of Cenozoic high angle, listric normal and right lateral faults resulting from extension. Proterozoic blocks are reflected as Cenozoic fault domains: the Dillon block by northwest- and northeast-trending faults; the Deer Lodge block by fault concentrations in few valleys; the Helena embayment by few extensional faults; and the Ovando block by northwest-trending normal faults, separated from the Charlo block by right lateral extensional faults along the St. Marys fault, overprinting the Jocko line. Sources: Calkins and Emmons, 1915; Mertie et al., 1951; Myers, 1952; Ross et al., 1955; Nelson and Dobell, 1961; Knopf, 1963; Nelson, 1963; Robinson, 1963; Ross, 1963; Bierwagen, 1964; Hobbs et al., 1965; Cremer, 1966; Groff, 1967; Hall, 1968; Johns, 1970; Kuenzi and Fields, 1971; Mudge, 1972b; Griggs, 1973; Mudge et al., 1974, 1977, 1983; Harrison, 1977; Aadland and Bennett, 1979; Rember and Bennett, 1979; Reynolds, 1979; Whipple, 1979; Schmidt and Hendrix, 1981; Wallace et al., 1981; Woodward, 1981; Ruppel et al., 1983.

High-angle extensional faults are numerous within the basin, but do not appear to have transported sections across great distances within the basin. The 12 km of extension attributed by Constenious (1981) to the Flathead fault and the 26 km of right lateral offset along the Osburn fault are exceptional, and reasonable stratigraphic adjustments can easily be made. Thus, neither compression nor extension

appears to have severely disrupted the stratigraphic framework of the central basin between the thrust belts, and I have confidence in the sedimentologic conclusions based on these sections. In addition, thicknesses of Belt rocks are so great and facies changes so gradual that, proportional to the stratigraphic scale, the scale of compressional and extensional tectonic transport does not obliterate all general

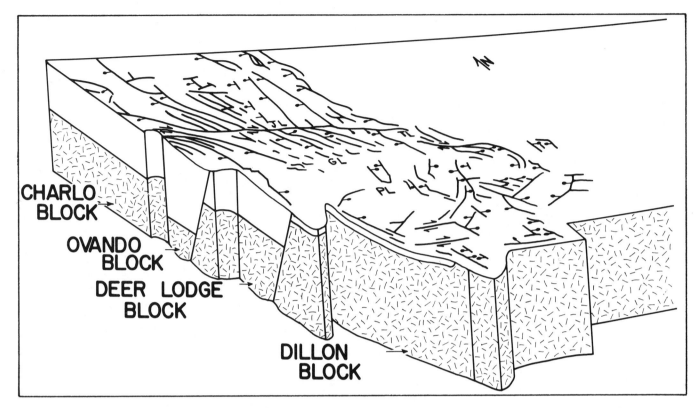

Figure 19—Block diagram of Cenozoic high angle, listric normal and right lateral extensional faults reflecting the influence of Proterozoic lines; Perry line (PL), Garnet line (GL), Jocko line (JL), and Townsend line (TL).

facies changes, even in the complex thrust belts, where at least the direction of thrusting can be identified. Tectonic problems may not be so insurmountable as Reynolds (1984b) infers, and others (Whipple et al., 1984; Harrison, 1984a; Wallace et al., 1984) have recently succeeded in constructing reasonable Belt stratigraphic cross sections despite the Phanerozoic tectonic problems.

One reason that sedimentation and tectonics can be analyzed relatively independently in Belt studies is that, although stratigraphic thicknesses change across growth fault lines, facies do not, except along the Perry line. The independence of facies changes and growth faults in the alluvial apron and sea margin sediments probably indicates that accumulation rates exceeded relative fault motion rates, and that the deep-seated faults were not reflected in the topography of the basin except along the southern margin. If accumulation rates were similar to Phanerozoic rates (Grotzinger, in press), then block faulting and extension were relatively slow and prolonged, so that continental separation was not achieved until Late Proterozoic time (Stewart, 1972).

Finally, I reemphasize the speculative and hypothetical basis for this proposal. Proposing a system of Proterozoic faults along lines that were later reactivated by Phanerozoic faulting is tenuous. The Perry and Garnet lines seem most fully supported where they bound the Helena embayment east of the eastern thrust belt. The lines within the eastern and western fold and thrust belts have received, and should continue to receive critical appraisal.

ACKNOWLEDGMENTS

I wish to thank Phil Jacob, who worked with me in the field intermittently from 1975 to 1977 and helped develop much of the stratigraphic framework; Dave Baldwin, who identified Proterozoic faults in the Greenhorn Mountain area; Jim Reid, who assisted me in the field from 1978 to 1980 and who compiled much of the data; and Marvin Woods, who helped edit and prepare this paper. Bob Horodyski, Margot McMechan, Ed Ruppel, and Jon Thorson read an early version of the manuscript, and their very thoughtful criticism has resulted in considerable revision. John Cuplin, Laurie Emmart, and Kay Kendrik drafted figures. The research was supported by National Science Foundation Grants PRM-801149 and EAR-08409507.

REFERENCES CITED

Aadland, R. K., and E. A. Bennett, 1979, Geologic map of the Sandpoint quadrangle, Idaho and Washington: Geologic Map Series, Idaho Bureau of Mines and Geology, Moscow. Scale 1:250,000.

Alam, M. M., K. A. W. Crook, and G. Taylor, 1985, Fluvial herring-bone cross-stratification in a modern tributary mouth bar, Coonamble, New South Wales, Australia: Sedimentology, v. 32, p. 235–244.

Alleman, D. G., 1983, Stratigraphy and sedimentation of the Precambrian Revett Formation, northwest Montana and northern Idaho: Master's thesis, University of Montana, Missoula, 103 p.

Allen, J. R. L., 1965, A review of the origin and characteristics of recent alluvial sediments: Sedimentology, v. 5, p. 91–191.

Armstrong, R. L., W. H. Taubeneck, and P. O. Hales, 1977, Rb-Sr and K-Ar geochronology of Mesozoic granitic rocks and their Sr isotopic composition, Oregon, Washington, and Idaho: Geological Society of America Bulletin, v. 88, p. 397–411.

Bierwagen, E. E., 1964, Geology of the Black Mountain area, Lewis and Clark and Powell counties, Montana: Ph.D. thesis, Princeton University, Princeton, New Jersey, 166 p.

Billingsley, P. R., and A. Locke, 1939, Structure of ore deposits in the continental framework: American Institute of Mining and Metallurgy Engineers Transactions, v. 144, p. 9–64.

Birkholz, D. O., 1967, Geology of the Camas Creek area, Meagher County, Montana: Master's thesis, Montana College of Mineral Science and Technology, Butte, 68 p.

Blackstone, D. L., Jr., 1956, Introduction to the tectonics of the Rocky Mountains, *in* 1956 Geological Record, Rocky Mountain Section, AAPG: Denver, Petroleum Information, p. 3–19.

Bloomfield, S. L., 1983, The Proterozoic Greyson-Spokane transition sequence: a stratigraphic and gravity study, west-central Montana: Master's thesis, University of Montana, Missoula, 96 p.

Bonnet, A., 1979, Depositional sequences and sedimentary-tectonic implications of the limestone-rich interval of the LaHood Formation (Precambrian Y), southwestern Montana (abs.): Geological Society of America Abstracts with Programs, v. 11, n. 6, p. 266.

Boothroyd, J. C., and G. M. Ashley, 1975, Process, bar morphology, and sedimentary structures on braided outwash fans, northeastern Gulf of Alaska, *in* A. V. Jopling, and B. C. McDonald, eds., Glaciofluvial and Glaciolacustrine Sedimentation: Society of Economic Paleontologists and Mineralogists Special Publication 23, p. 193–222.

Bregman, M. L., 1976, Change in tectonic style along the Montana thrust belt: Geology, v. 4, p. 775–778.

Calbeck, J. M., 1975, Geology of the central Wise River Valley, Pioneer Mountains, Beaverhead County, Montana: Master's thesis, University of Montana, Missoula, 89 p.

Calkins, F. C., and W. H. Emmons, 1915, Description of the Philipsburg quadrangle, Montana: U.S. Geological Survey Geology Atlas Folio 196, 25 p.

Chevillon, C. V., 1977, Tectonically induced Proterozoic soft sediment deformation at the Atlas property, Coeur d'Alene mining district, Idaho (abs.): Geological Society of America Abstracts with Programs, v. 9, n. 6, p. 716.

Childers, M. O., 1963, Structure and stratigraphy of the southwest Marias Pass area, Flathead County, Montana: Geological Society of America Bulletin, v. 74, n. 2, p. 141–164.

Clemmensen, L. B., 1978, Lacustrine facies and stromatolites from the Middle Triassic of East Greenland: Journal of Sedimentary Petrology, v. 48, p. 1111–1128.

Connor, J. J., 1982, The Spokane Formation (Belt Supergroup) in the Helena salient, western Montana (abs.): Geological Society of America Abstracts with Programs, v. 14, n. 6, p. 307.

———, 1984, Geochemistry of the Middle Proterozoic Spokane, Grinnell, and St. Regis formations of the Belt Supergroup (abs.), *in* S. W. Hobbs, ed., The Belt: Abstracts with Summaries, Belt Symposium II, 1983: Butte, Montana Bureau of Mines and Geology Special Publication 90, p. 102–103.

Constenious, K. N., 1981, Stratigraphy, sedimentation, and tectonic history of the Kishenehn basin, northwestern Montana: Master's thesis, University of Wyoming, Laramie, 116 p.

Cremer, E. A., III, 1966, Gravity determination of basement structure configuration, southern Deer Lodge Valley, Montana: Master's thesis, University of Montana, Missoula, 23 p.

Cressman, E. R., 1984, Paleogeography and paleotectonic setting of the Prichard Formation—a preliminary interpretation (abs.), *in* S. W. Hobbs, ed., The Belt: Abstracts with Summaries, Belt Symposium II, 1983: Butte, Montana Bureau of Mines and Geology Special Publication 90, p. 8–9.

———, 1985, The Prichard Formation of the lower part of the Belt Supergroup (Middle Proterozoic), near Plains, Sanders County, Montana: U.S. Geological Survey Bulletin 1553, 64 p.

Csejtey, B., 1962, Geology of the southeast flank of the Flint Creek Range, western Montana: Ph.D. thesis, Princeton University, Princeton, New Jersey, 175 p.

Desmarais, N. R., 1983, Geology and geochronology of the Chief Joseph plutonic-metamorphic complex, Idaho: Ph.D. thesis, University of Washington, Seattle, 150 p.

Eby, D. E., 1977, Sedimentation and early diagenesis within eastern portions of the "Middle Belt Carbonate Interval" (Helena Formation), Belt Supergroup (Precambrian Y), western Montana: Ph.D. thesis, State University of New York at Stony Brook, Stony Brook, 712 p.

Elston, D. P., 1984, Magnetostratigraphy of the Belt Supergroup—a synopsis: *in* S. W. Hobbs, ed., The Belt: Abstracts with Summaries, Belt Symposium II, 1983: Butte, Montana Bureau of Mines and Geology Special Publication 90, p. 88–90.

Fermor, P. R., and R. A. Price, 1984, Stratigraphy of the lower part of the Belt-Purcell Supergroup (Middle Proterozoic) in the Lewis thrust sheet of southern Alberta and British Columbia, *in* J. D. McBane and P. B. Garrison, eds., Northwest Montana and Adjacent Canada: Montana Geology Society 1984 Field Conference Guidebook, p. 73–89.

Finch, J. C., and D. O. Baldwin, 1984, Stratigraphy of the Prichard Formation, Belt Supergroup (abs.), *in* S. W. Hobbs, ed., The Belt: Abstracts with Summaries, Belt Symposium II, 1983: Butte, Montana Bureau of Mines and Geology Special Publication 90, p. 5–7.

Fraser, G. D., and H. A. Waldrop, 1972, Geologic map of the Wise River quadrangle, Silver Bow and Beaverhead counties, Montana: U.S. Geological Survey Geologic Quadrangle Map GQ-988.

Godlewski, D. W., 1977, The origin of the middle Wallace breccias (abs.): Geological Society of America Abstracts with Programs, v. 9, n. 6, p. 727.

———, 1980, Origin and classification of the middle Wallace breccias: Master's thesis, University of Montana, Missoula, 74 p.

Godlewski, D. W., and G. A. Zieg, 1984, Stratigraphy and depositional setting of the Precambrian Newland Limestone (abs.), *in* S. W. Hobbs, ed., The Belt: Abstracts with Summaries, Belt Symposium II, 1983: Butte, Montana Bureau of Mines and Geology Special Publication 90, p. 2–4.

Griggs, A. B., 1973, Geologic map of the Spokane quadrangle, Washington, Idaho, and Montana: U.S. Geological Survey Miscellaneous Geologic Investigations Map I-768.

Groff, S. L., 1967, A summary description of the geology in the Smith River Valley, Meagher County, Montana, *in* L. B. Henderson, ed., Centennial basin of southwestern Montana: Montana Geological Society 18th Annual Field Conference Guidebook, p. 93–102.

Grotzinger, J. P., 1981a, Late Precambrian Belt cyclic lacustrine sedimentation in the Wallace Formation, northwestern Montana and northern Idaho, U.S.A. (abs.): Geological Association of Canada, Joint Annual Meeting, Calgary, Abstracts, v. 6, p. A-23.

———, 1981b, The stratigraphy and sedimentation of the Wallace Formation, northwest Montana and northern Idaho: Master's thesis, University of Montana, Missoula, 153 p.

———, in press, Shallowing-upward cycles of the Wallace Formation (Belt Supergroup), northwestern Montana and northern Idaho, *in* S. Roberts, ed., Belt Supergroup: A guide to Proterozoic Rocks of Western Montana and Adjacent Areas: Butte, Montana Bureau of Mines and Geology Special Publication 94.

Hall, F. B., II, 1968, Bedrock geology, north half of Missoula 30′ quadrangle, Montana: Ph.D. thesis, University of Montana, Missoula, 253 p.

Hardie, L. A., J. P. Smoot, and H. P. Eugster, 1978, Saline lakes and their deposits: a sedimentological approach, *in* A. Matter, and M. E. Tucker, eds., Modern and Ancient Lake Sediments: International Association of Sedimentologists Special Publication 2, p. 7–41.

Harris, S. A., 1957, Tectonics of Montana as related to the Belt Series, *in* R. W. Graves, ed., Crazy Mountain basin: Billings Geology Society 8th Annual Field Conference Guidebook, p. 22–33.

Harrison, J. E., 1972, Precambrian Belt basin of northwestern United States—its geometry, sedimentation, and copper occurrences: Geological Society of America Bulletin, v. 83, n. 5, p. 1215–1240.

———, 1977, Preliminary geologic map of the southwestern part of the Wallace 1:250,000 sheet: U.S. Geological Survey Open-File Report OF-77-33.

———, 1984a, Stratigraphy and lithocorrelation of the Wallace Formation (abs.), *in* S. W. Hobbs, ed., The Belt: Abstracts with Summaries, Belt Symposium II, 1983: Butte, Montana Bureau of Mines and Geology Special Publication 90, p. 24–26.

———, 1984b [Summary of] Session on geochemistry and geophysics: *in* S. W. Hobbs, ed., The Belt: Abstracts with Summaries, Belt Symposium II, 1983: Butte, Montana Bureau of Mines and Geology Special Publication 90, p. 98–100.

Harrison, J. E., and D. A. Jobin, 1963, Geology of the Clark Fork quadrangle, Idaho-Montana: U.S. Geological Survey Bulletin 1141-K, 38 p.

Harrison, J. E., M. D. Kleinkopf, and J. D. Obradovich, 1972, Tectonic events at the intersection between the Hope fault and the Purcell Trench, northern Idaho: U.S. Geological Survey Professional Paper 719, 24 p.

Harrison, J. E., A. B. Griggs, and J. D. Wells, 1974a, Tectonic features of the Precambrian Belt basin and their influence on post-Belt structures: U.S. Geological Survey Professional Paper 866, 15 p.

———, 1974b, Preliminary geologic map of part of the Wallace 1:250,000 sheet, Idaho–Montana: U.S. Geological Survey Open-File Report OF-74-37.

———, 1981, Generalized geologic map of the Wallace 1 × 2 quadrangle, Montana and Idaho: U.S. Geological Survey Miscellaneous Field Studies Map MF-1354-A.

Harrison, J. E., M. D. Kleinkopf, and J. D. Wells, 1980, Phanerozoic thrusting in Proterozoic Belt rocks, northwestern United States: Geology, v. 8, p. 407–411.

Harrison, J. E., E. R. Cressman, and J. W. Whipple, 1983, Preliminary geologic and structural maps of part of the Kalispell 1 × 2 quadrangle, Montana: U.S. Geological Survey Open-File Report OF-83-502.

Hawley, D., A. Bonnet-Nicolaysen, and W. Coppinger, 1982, Stratigraphy, depositional environments, and paleotectonics of the LaHood Formation, *in* Rocky Mountain Section Field Trip Guidebook, 35th Annual Meeting: Bozeman, Montana, Geological Society of America, 20 p.

Hill, R., and E. W. Mountjoy, 1984, Stratigraphy and sedimentology of the Waterton Formation, Belt Purcell Supergroup, Waterton Lakes National Park, southwest Alberta, *in* J. D. McBane and P. B. Garrison, eds., Northwest Montana and Adjacent Canada: Montana Geological Society Field Conference Guidebook, p. 91–100.

Hobbs, S. W., A. B. Griggs, R. E. Wallace, and A. B. Campbell, 1965, Geology of the Coeur d'Alene district, Shoshone County, Idaho: U.S. Geological Survey Professional Paper 478, 139 p.

Horodyski, R. J., 1983, Sedimentary geology and stromatolites of the Middle Proterozoic Belt Supergroup, Glacier National Park, Montana: Precambrian Research, v. 20, p. 391–425.

Hubert, J. F., and M. G. Hyde, 1982, Sheet-flow deposits of graded beds and mudstones on an alluvial sandflat-playa system: Upper Triassic Blomidon redbeds, St. Mary's Bay, Nova Scotia: Sedimentology, v. 29, p. 457–474.

Hughes, G. C., 1981, Tertiary stratigraphy and depositional history of the Cry Creek Valley, Montana, *in* T. E. Tucker, ed., Southwest Montana: Montana Geological Society Field Conference Guidebook, p. 111–119.

Hyndman, D. W., 1980, Bitterroot dome-Sapphire tectonic block, an example of a plutonic-core gneiss-dome complex with its detached suprastructure: Geological Society of America Memoir 153, p. 427–443.

Johns, W. M., 1970, Geology and mineral deposits of Lincoln and Flathead counties, Montana: Montana Bureau of

Mines and Geology Bulletin 79, 182 p.

Kleinkopf, M. D., 1984, Gravity and magnetic anomalies of the Belt basin, United States and Canada (abs.), *in* S. W. Hobbs, ed., The Belt: Abstracts with Summaries, Belt Symposium II, 1983: Butte, Montana Bureau of Mines and Geology Special Publication 90, p. 85–87.

Kleinkopf, M. D., and M. R. Mudge, 1972, Aeromagnetic, Bouguer gravity, and generalized geologic studies of the Great Falls-Mission Range area, northwestern Montana: U.S. Geological Survey Professional Paper 726-A, 19 p.

Knopf, A., 1963, Geology of the northern part of the Boulder batholith and adjacent area, Montana: U.S. Geological Survey Miscellaneous Geologic Investigations Map I-381.

Kuenzi, W. D., and R. W. Fields, 1971, Tertiary stratigraphy, structure, and geologic history, Jefferson basin, Montana: Geological Society of America Bulletin, v. 82, p. 3373–3394.

Lageson, D. R., C. J. Schmidt, H. W. Dresser, M. Walker, R. B. Berg, and H. L. James, 1983, Road logs 1 and 2, *in* D. L. Smith, compiler, Guidebook of the Fold and Thrust Belt, West-Central Montana: Butte, Montana Bureau of Mines and Geology Special Publication 86, 98 p.

Lemoine, S. R., 1979, Correlation of the upper Wallace with the lower Missoula Group and resulting facies interpretations, Cabinet and Coeur d'Alene Mountains, Montana: Master's thesis, University of Montana, Missoula, 162 p.

Lonn, J., 1984, Structural geology of the Tarkio area, Mineral County, Montana: Master's thesis, University of Montana, Missoula, 51 p.

Mauk, J. L., 1983a, Stratigraphy and sedimentation of the Proterozoic Burke and Revett formations, Belt Supergroup, Flathead Reservation, western Montana (abs.): Geological Society of America Abstracts with Programs, v. 15, n. 5, p. 424–425.

———, 1983b, Stratigraphy and sedimentation of the Proterozoic Burke and Revett formations, Flathead Reservation, western Montana: Master's thesis, University of Montana, Missoula, 106 p.

Maxwell, D. T., and J. Hower, 1967, High-grade diagenesis and low-grade metamorphism of illite in the Precambrian Belt Series: American Mineralogist, v. 52, p. 843–857.

McDonough, D. T., 1985, Structural evolution of the southeastern Scapegoat Wilderness, west-central Montana: Master's thesis, University of Montana, Missoula, 125 p.

McGill, G. E., 1959, Geologic map of the northwestern flank of the Flint Creek Range, western Montana: Butte, Montana Bureau of Mines and Geology Special Publication 18.

McGrew, L. W., 1977a, Geologic map of the Sixteen Mile quadrangle, Gallatin and Meagher counties, Montana: U.S. Geological Survey Geologic Quadrangle Map GQ-1383.

———, 1977b, Geologic map of the Black Butte Mountain quadrangle, Meagher County, Montana: U.S. Geological Survey Geologic Quadrangle Map GQ-1381.

———, 1977c, Geologic map of the Ringling quadrangle, Meagher County, Montana: United States Geological Survey Geologic Quadrangle Map GQ-1382.

McGroder, M. F., 1984, Stratigraphy and sedimentation of the Proterozoic Garnet Range Formation, Belt Supergroup, Montana: Master's thesis, University of Montana, Missoula, 107 p.

McKee, E. D., E. J. Crosby, and H. L. Berryhill, 1967, Flood deposits, Bijou Creek, Colorado: Journal of Sedimentary Petrology, v. 37, p. 829–851.

McMannis, W. J., 1963, LaHood Formation—a coarse facies of the Belt Series in southwestern Montana: Geological Society of America Bulletin, v. 74, n. 4, p. 407–436.

———, 1965, Résumé of depositional and structural history of western Montana: AAPG Bulletin, v. 49, p. 1801–1823.

McMechan, M. E., 1981, The Middle Proterozoic Purcell Supergroup in the southwestern Purcell Mountains, British Columbia and the initiation of the Cordilleran miogeocline, southern Canada and adjacent United States: Bulletin of Canadian Petroleum Geologists, v. 29, p. 583–621.

McMechan, M. E., and R. A. Price, 1982, Transverse folding and superposed deformation, Mount Fisher area, southern Canadian Rocky Mountain thrust and fold belt: Canadian Journal of Earth Sciences, v. 19, n. 5, p. 1011–1024.

Mertie, J. B., Jr., R. P. Fischer, and S. W. Hobbs, 1951, Geology of the Canyon Ferry quadrangle, Montana: U.S. Geological Survey Bulletin 972, 97 p.

Miall, A. D., 1978, Lithofacies types and vertical profile models in braided river deposits: a summary, *in* A. D. Miall, ed., Fluvial Sedimentology: Canadian Society of Petroleum Geologists Memoir 5, p. 579–604.

Mudge, M. R., 1972a, Pre-Quaternary rocks of the Sun River Canyon area, northwestern Montana: U.S. Geological Survey Professional Paper 663-A, 138 p.

———, 1972b, Structural geology of the Sun River Canyon and adjacent areas, northwestern Montana: United States Geological Survey Professional Paper 663-B, 52 p.

Mudge, M. R., R. L. Earhart, K. C. Watts, Jr., E. T. Tuchek, and W. L. Rice, 1974, Mineral resources of the Scapegoat Wilderness, Powell and Lewis and Clark counties, Montana, with a section on geophysical surveys by D. L. Peterson: U.S. Geological Survey Bulletin 1385-B, 82 p.

Mudge, M. R., R. L. Earhart, and D. D. Rice, 1977, Preliminary bedrock geologic map of part of the northern disturbed belt, Lewis and Clark, Teton, Pondera, Glacier, Flathead, and Powell counties, Montana: United States Geological Survey Open-File Report OF-77-25.

Mudge, M. R., R. L. Earhart, J. W. Whipple, and J. E. Harrison, 1983, Geologic and structure maps of the Choteau 1 × 2 quadrangle, northwestern Montana: Butte, Montana Bureau of Mines and Geology, Montana Atlas 3-A.

Mutch, T. A., 1960, Geologic map of the northeast flank of the Flint Creek Range, western Montana: Butte, Montana Bureau of Mines and Geology Special Publication 22.

Myers, W. B., 1952, Geology and mineral deposits of the northwest quarter of Willis quadrangle and adjacent

Browns Lake area, Beaverhead County, Montana: U.S. Geological Survey Trace Elements Investigations Report 259, 46 p.

Nelson, W. H., 1963, Geology of the Duck Creek Pass quadrangle, Montana: U.S. Geological Survey Bulletin 1121-J, 56 p.

Nelson, W. H., and J. P. Dobell, 1961, Geology of the Bonner quadrangle, Montana: U.S. Geological Survey Bulletin 1111-F, p. 189–235.

Norwick, S. A., 1977, The regional Precambrian metamorphic facies of the Prichard Formation of western Montana and northern Idaho: Ph.D. thesis, University of Montana, Missoula, 129 p.

O'Connor, M. P., 1972, Classification and environmental interpretation of the cryptalgal organosedimentary "molar tooth" structure from the Late Precambrian Belt-Purcell Supergroup: Journal of Geology, v. 80, p. 592–610.

Peale, A. C., 1896, Description of the Three Forks quadrangle [Montana]: U.S. Geological Survey Geological Atlas Folio 24, 5 p.

Peterson, J. A., 1981, General stratigraphy and regional paleostructure of the western Montana overthrust belt, *in* T. E. Tucker, ed., Southwest Montana: Montana Geological Society Field Conference Guidebook, p. 5–35.

Poulter, G. J., 1958, Geology of the Georgetown thrust area southwest of Philipsburg, Montana: Butte, Montana Bureau of Mines and Geology Geologic Investigation Map No. 1.

Price, R. A., 1964, The Precambrian Purcell System in the Rocky Mountains of southern Alberta and British Columbia: Bulletin of Canadian Petroleum Geologists, Guidebook, v. 12, p. 399–426.

Ransome, F. L., and F. C. Calkins, 1908, The geology and ore deposits of the Coeur d'Alene district, Idaho: U.S. Geological Survey Professional Paper 62, 203 p.

Rember, W. C., and E. H. Bennett, 1979, Geologic map of the Hamilton quadrangle, Idaho: Moscow, Idaho Bureau of Mines and Geology Geologic Map Series.

Reynolds, M. W., 1977, Character and significance of deformation at the east end of the Lewis and Clark line, Montana (abs.): Geological Society of America Abstracts with Programs, v. 9, n. 6, p. 758–759.

――――, 1979, Character and extent of basin-range faulting, western Montana and east-central Idaho, *in* G. W. Newman and H. D. Goode, eds., Basin and Range Symposium and Great Basin Field Conference Guidebook: Rocky Mountain Association of Geologists and Utah Geological Association, p. 185–193.

――――, 1984a, Tectonic setting and development of the Belt basin, northwestern United States (abs.), *in* S. W. Hobbs, ed., The Belt: Abstracts with Summaries, Belt Symposium II, 1983: Butte, Montana Bureau of Mines and Geology Special Publication 90, p. 44–46.

――――, 1984b, [Summary of] Session on structure and tectonics, *in* S. W. Hobbs, ed., The Belt: Abstracts with Summaries, Belt Symposium II, 1983: Butte, Montana Bureau of Mines and Geology Special Publication 90, p. 53–56.

Reynolds, M. W., and M. D. Kleinkopf, 1977, The Lewis and Clark line, Montana-Idaho: a major intraplate tectonic boundary (abs.): Geological Society of America Abstracts with Programs, v. 9, n. 7, p. 1140–1141.

Robinson, G. D., 1959, The disturbed belt in the Sixteen Mile area, Montana, *in* C. R. Hammond and H. Trapp, Jr., eds., Sawtooth-Disturbed Belt Area: Billings Geological Society 10th Annual Field Conference Guidebook, p. 34–40.

――――, 1963, Geology of the Three Forks quadrangle, Montana: U.S. Geological Survey Professional Paper 370, 143 p.

――――, 1967, Geologic map of the Toston quadrangle, southwestern Montana: U.S. Geological Survey Miscellaneous Geologic Investigations Map I-486.

Ross, C. P., 1959, Geology of Glacier National Park and the Flathead region, northwestern Montana: U.S. Geological Survey Professional Paper 296, 125 p.

――――, 1963, The Belt Series in Montana: U.S. Geological Survey Professional Paper 346, 122 p.

Ross, C. P., D. A. Andrews, and I. J. Witkind, 1955, Geologic map of Montana: U.S. Geological Survey. Scale 1:500,000.

Ruppel, E. T., C. A. Wallace, R. G. Schmidt, and D. A. Lopez, 1981, Preliminary interpretation of the thrust belt in southwest Montana and east central Idaho, *in* T. E. Tucker, ed., Southwest Montana: Montana Geological Society Field Conference Guidebook, p. 139–159.

Ruppel, E. T., J. M. O'Neill, and D. A. Lopez, 1983, Preliminary geologic map of the Dillon 1 × 2 quadrangle, Montana: U.S. Geological Survey Open-File Report OF-83-168.

Rust, B. R., 1979, Coarse alluvial deposits, *in* R. G. Walker, ed., Facies Models: Geological Association of Canada, Geoscience Canada, Reprint Series 1, p. 9–21.

Schmidt, C. J., and J. M. Garihan, 1979, A summary of Laramide basement faulting in the Ruby, Tobacco Root, and Madison Ranges and its possible relationship to Precambrian continental rifting (abs.): Geological Society of America Abstracts with Programs, v. 11, n. 6, p. 301.

Schmidt, C. J., and T. E. Hendrix, 1981, Tectonic controls for thrust belt and Rocky Mountain foreland structures in the northern Tobacco Root Mountains-Jefferson Canyon area, southwestern Montana, *in* T. E. Tucker, ed., Southwest Montana: Montana Geological Society Field Conference Guidebook, p. 167–180.

Sheriff, S. D., J. W. Sears, and J. N. Moore, 1984, Montana's Lewis and Clark fault zone: an intracratonic transform fault system: Geological Society of America Abstracts with Programs, v. 16, n. 6, p. 653–654.

Skipp, B., and A. D. Peterson, 1965, Geologic map of the Mauldow quadrangle, southwestern Montana: United States Geological Survey Miscellaneous Investigations Map I-452.

Slover, S. M., and D. Winston, in press, Fining-upward sequences in Mount Shields members one and two, central Belt basin, Montana, *in* S. Roberts, ed., Belt Supergroup: A Guide to Proterozoic Rocks of Western Montana and Adjacent Areas: Butte, Montana Bureau of Mines and Geology Special Publication 94.

Smith, A. G., and W. C. Barnes, 1966, Correlation of and facies changes in the carbonaceous, calcareous, and dolomitic formations of the Precambrian Belt-Purcell Supergroup: Geological Society of America Bulletin, v.

77, n. 12, p. 1399–1426.

Smith, J. G., 1965, Fundamental transcurrent faulting in northern Rocky Mountains: AAPG Bulletin, v. 49, p. 1398–1409.

Smith, N. D., 1970, The braided stream depositional environment: comparison of the Platte River with some Silurian clastic rocks, north-central Appalachians: Geological Society of America Bulletin, v. 81, p. 2293–3014.

Smoot, J. P., 1983, Depositional subenvironments in an arid closed basin; the Wilkins Peak Member of the Green River Formation (Eocene), Wyoming, U.S.A.: Sedimentology, v. 30, p. 801–827.

Stear, W. M., 1985, Comparison of the bedform distribution and dynamics of modern and ancient sandy ephemeral flood deposits in the southwestern Karoo region, South Africa: Sedimentary Geology, v. 45, p. 209–230.

Stewart, J. H., 1972, Initial deposits in the Cordilleran geosyncline: evidence of a late Precambrian (<850 m.y.) continental separation: Geological Society of America Bulletin, v. 83, p. 1345–1360.

Surdam, R. C., and K. O. Stanley, 1979, Lacustrine sedimentation during the culminating phase of Eocene Lake Gosiute, Wyoming (Green River Formation): Geological Society of America Bulletin, v. 90, pt. 1, p. 93–110.

Talbot, J. L., and D. W. Hyndman, 1973, Relationship of the Idaho batholith structures to the Montana lineament: Northwest Geology, v. 2, p. 48–52.

Thom, W. T., Jr., 1923, The relation of deep-seated faults to the surface structural features of central Montana: AAPG Bulletin, v. 7, p. 1–13.

Thorson, J. P., 1984, Suggested revisions of the lower Belt Supergroup stratigraphy of the Highland Mountains, southwestern Montana (abs.), *in* S. W. Hobbs, ed., The Belt: Abstracts with Summaries, Belt Symposium II, 1983: Butte, Montana Bureau of Mines and Geology Special Publication 90, p. 10–12.

Vance, R. B., 1981, Geology of the NW1/4 of the Wallace 15′ quadrangle, Shoshone County, Idaho: Master's thesis, University of Idaho, Moscow, 102 p.

Van Houten, F. B., 1962, Cyclic sedimentation and the origin of analcime-rich Upper Triassic Lockatong Formation, west-central New Jersey and adjacent Pennsylvania: American Journal of Science, v. 260, p. 561–576.

Van Loenen, R. E., 1984, Geologic map of the Mount Henry roadless area, Lincoln County, Montana: U.S. Geological Survey Miscellaneous Field Studies Map MF-1534-A.

Wallace, C. A., and D. J. Lidke, 1980, Geologic map of the Rattlesnake Wilderness study area, Montana: U.S. Geological Survey Miscellaneous Field Studies Map MF-1235-A.

Wallace, C. A., J. E. Harrison, M. R. Klepper, and J. D. Wells, 1976, Carbonate sedimentary breccias in the Wallace Formation (Belt Supergroup), Idaho and Montana, and their paleogeographic significance (abs.): Geological Society of America Abstracts with Programs, v. 8, n. 6, p. 1159.

Wallace, C. A., R. G. Schmidt, M. R. Waters, D. J. Lidke, and A. B. French, 1981, Preliminary geologic map of parts of the Butte 1 × 2 quadrangle, central Montana: U.S. Geological Survey Open-File Report OF-81-1030.

Wallace, C. A., J. E. Harrison, J. W. Whipple, E. T. Ruppel, and R. G. Schmidt, 1984, A summary of the stratigraphy of the Missoula Group, Belt Supergroup, and a preliminary interpretation of basin-subsidence characteristics, western Montana, northern Idaho, and eastern Washington (abs.), *in* S. W. Hobbs, ed., The Belt: Abstracts with Summaries, Belt Symposium II, 1983: Butte, Montana Bureau of Mines and Geology Special Publication 90, p. 27–29.

Wallace, R. E., and J. W. Hosterman, 1956, Reconnaissance geology of western Mineral County, Montana: U.S. Geological Survey Bulletin 1027-M, p. 575–612.

Wallace, R. E., A. E. Griggs, A. B. Campbell, and S. W. Hobbs, 1960, Tectonic setting of the Coeur d'Alene district, Idaho: U.S. Geological Survey Professional Paper 400, B25–27.

Webster, T. A., 1981, Faulting and slumping during deposition of the Precambrian Prichard Formation, near Quinns, Montana: a model for sulfide deposition: Master's thesis, University of Montana, Missoula, 71 p.

Weed, W. H., 1899, Description of the Little Belt Mountains quadrangle (Montana): U.S. Geological Survey Geologic Atlas Folio 56, 9 p.

Weidman, R. M., 1965, The Montana lineament, *in* R. W. Fields and W. Shepard, eds., Geology of the Flint Creek Range, Montana: Billings Geological Society 16th Annual Field Conference Guidebook, p. 137–143.

Wells, J. D., 1974, Geologic map of the Alberton quadrangle, Missoula, Sanders, and Mineral counties, Montana: U.S. Geological Survey Geologic Quadrangle Map GQ-1157.

Wells, J. D., D. A. Lindsey, and R. E. Van Loenen, 1981, Geology of the Cabinet Mountain Wilderness, Lincoln and Sanders counties, Montana: U.S. Geological Survey Bulletin 1501-A, 24 p.

Werkema, M. A., T. E. Hendrix, and C. J. Schmidt, 1981, Analysis of the Sappington fault in the "J" structure, Jefferson County, Montana, *in* T. E. Tucker, ed., Southwest Montana: Montana Geological Society Field Conference Guidebook, p. 181–189.

Whipple, J. W., 1979, Preliminary geologic map of the Rogers Pass area, Lewis and Clark County, Montana: U.S. Geological Survey Open-File Report OF-79-710.

Whipple, J. W., J. J. Connor, O. B. Raup, and R. G. McGimsey, 1984, Preliminary report on the stratigraphy of the Belt Supergroup, Glacier National Park and adjacent Whitefish Range, Montana, *in* J. D. McBane and P. B. Garrison, eds., Northwest Montana and Adjacent Canada: Montana Geological Society Field Conference Guidebook, p. 33–50.

White, B., 1984, Stromatolites and associated facies in shallowing-upward cycles from the Middle Proterozoic Altyn Formation of Glacier National Park, Montana: Precambrian Research, v. 24, p. 1–26.

White, B. G., and D. Winston, 1982, The Revett-St. Regis "transition zone" near the Bunker Hill Mine, Coeur d'Alene district, Idaho, *in* R. R. Reid and G. A. Williams, eds., Society of Economic Geologists, Coeur d'Alene Field Conference, Idaho—1977: Moscow, Idaho Bureau of Mines and Geology Bulletin 24, p. 25–30.

Williams, G. E., 1971, Flood deposits of the sand-bed ephemeral streams of central Australia: Sedimentology, v. 17, p. 1–40.

Willis, B., 1902, Stratigraphy and structure, Lewis and Livingston ranges, Montana: Geological Society of America Bulletin, v. 13, p. 305–352.

Winston, D., 1973, The Precambrian Missoula Group of Montana as a braided stream and sea-margin sequence, *in* Belt Symposium, v. 1: Moscow, University of Idaho, Department of Geology, and Idaho Bureau of Mines and Geology, p. 208–220.

————, 1977, Alluvial fan, shallow water and sub-wave base deposits of the Belt Supergroup near Missoula, Montana: Geological Society of America Rocky Mountain Section, 30th Annual Meeting Field Guide No. 5, 41 p.

————, 1978, Fluvial systems of the Precambrian Belt Supergroup, Montana and Idaho, U.S.A., *in* A. D. Miall, ed., Fluvial Sedimentology: Canadian Society of Petroleum Geologists Memoir 5, p. 343–359.

————, 1981, Late Precambrian faults in the Belt basin and their influence on Cretaceous thrusting (abs.): Geological Association of Canada, Joint Annual Meeting, Calgary, Abstracts, v. 6, p. A-62.

————, 1982, The effect of Precambrian Belt basin blocks on Cretaceous to Eocene compression and middle to late Cenozoic extension, Montana and Idaho (abs.): Geological Society of America Abstracts with Programs, v. 14, n. 6, p. 354.

————, in press, a, Stratigraphic correlation and nomenclature of the Middle Proterozoic Belt Supergroup, Montana, Idaho, and Washington, *in* S. Roberts, ed., Belt Supergroup: A Guide to Proterozoic Rocks of Western Montana and Adjacent Areas: Butte, Montana Bureau of Mines and Geology Special Publication 94.

————, in press, b, Sedimentology of the Ravalli Group, middle Belt carbonate and Missoula Group, Middle Proterozoic Belt Supergroup, Montana, Idaho and Washington, *in* S. Roberts, ed., Belt Supergroup: A Guide to Proterozoic Rocks of Western Montana and Adjacent Areas: Butte, Montana Bureau of Mines and Geology Special Publication 94.

————, in press, c, Middle Proterozoic tectonics of the Belt basin, western Montana and northern Idaho, *in* S. Roberts, ed., Belt Supergroup: A Guide to Proterozoic Rocks of Western Montana and Adjacent Areas: Butte, Montana Bureau of Mines and Geology Special Publication 94.

Winston, D., and C. A. Wallace, 1983, Field guide for trip no. 3: The Helena Formation and the Missoula Group at Flint Creek Hill, near Georgetown Lake, western Montana, *in* S. W. Hobbs, ed., Guide to Field Trips, Belt Symposium II: Missoula, Department of Geology, University of Montana, p. 66–81.

Winston, D., M. Woods, and G. Byer, 1984, The case for an intracratonic Middle Proterozoic Belt-Purcell basin: tectonic, stratigraphic and stable isotopic considerations, *in* J. D. McBane and P. B. Garrison, eds., Northwest Montana and Adjacent Canada: Montana Geological Society Field Conference Guidebook, p. 103–118.

Woodward, L. A., 1981, Tectonic framework of the Disturbed Belt of west-central Montana: AAPG Bulletin, v. 65, p. 291–302.

The Lemhi Arch: A Late Proterozoic and Early Paleozoic Landmass in Central Idaho

E. T. Ruppel
U.S. Geological Survey
Denver, Colorado

The Lemhi arch was a northwest-trending landmass in central Idaho during late Proterozoic and early to middle Paleozoic time. The arch first formed in late middle Proterozoic or early late Proterozoic time, when middle Proterozoic miogeosynclinal sedimentary rocks were arched into an elongate dome. It was deeply eroded in late Proterozoic time, and as much as 4500 m of clastic rocks were stripped away. The eroded edges of the middle Proterozoic rocks on the west flank of the arch were partly covered in late Proterozoic(?) and Cambrian time by the onlapping Wilbert Formation, but sedimentation apparently did not continue into later Cambrian time. On the east flank of the arch, westward-thinning marine sedimentation began with deposition of the Middle Cambrian Flathead Formation and continued through Late Cambrian time.

During Ordovician and Silurian time the east flank of the arch was dry. The west flank was partly submerged in Early Ordovician time, and the onlapping nearshore clastic and carbonate rocks of the Summerhouse Formation were deposited. The Wilbert and Summerhouse formations are successively overlain by the eastward-thinning marine rocks of the Kinnikinic Quartzite (Middle Ordovician) and the Saturday Mountain Formation (Late Ordovician and younger). The west flank of the arch was briefly exposed to erosion after deposition of the Saturday Mountain Formation, but it was again partly submerged in Middle and Late Silurian time when the eastward-thinning Laketown Dolomite was deposited.

Both flanks of the arch were exposed in Early Devonian time, but in Middle Devonian time deposition was renewed on the west flank, as fresh and brackish water sandstone was deposited in channels cut deeply into the Ordovician rocks. Later Middle and Late Devonian age rocks of the Jefferson Formation on the west flank of the arch indicate eastward transgression, followed by Late Devonian marine sedimentation. The east flank of the arch was exposed until late in the Devonian, when a thin sequence of marine carbonate rocks of the Jefferson and Three Forks formations was deposited across the top of the arch and marine sedimentation in this region was continuous from the miogeosyncline far onto the craton. The Lemhi arch continued to influence marine deposition even after it was submerged, separating the region of shelf deposition in southwestern Montana and east-central Idaho from the region of miogeosynclinal deposition in central Idaho. The arch was a landmass again through much of Mesozoic time and was overridden by the Medicine Lodge thrust plate in late Early and Late Cretaceous time.

INTRODUCTION

The eastward changes in thicknesses and lithologies of the lower Paleozoic sedimentary rocks in the southern parts of the Lemhi Range and Beaverhead Mountains, east-central Idaho (Figure 1), were recognized fairly early in geologic studies of this region. Sloss (1954) first described these changes in the Devonian Jefferson Formation at the south end of the Lemhi Range and suggested that they resulted from depositional onlap onto a middle Paleozoic high area, which he named the Lemhi arch. Sloss also suggested that the Lemhi arch had an earlier, pre-Devonian history, but the older sedimentary rocks of the region were poorly known then. Scholten (1957) subsequently described three other high areas: the Skull Canyon uplift in the Beaverhead Mountains, a pre-Middle Ordovician uplift; the Tendoy

dome in the Tendoy Mountains, a Devonian uplift; and the Beaverhead arch in the south parts of the Lemhi Range and Beaverhead Mountains, a Mississippian uplift. Each of these uplifts was inferred from regional distribution and thickness patterns of the Paleozoic formations, which were interpreted to have been deposited in a geosynclinal basin in central Idaho and on a more stable cratonic shelf in southwestern Montana. These two were apparently separated by a tectonic and stratigraphic hinge zone approximately located in the Beaverhead Mountains. Neither regional thrust faulting nor underlying Proterozoic sedimentary rocks had yet been defined, so that their implications to the uplifts could not be considered by Sloss or Scholten.

A fifth high area, the Salmon River arch, was named by Armstrong (1975). Armstrong suggested, on the basis of radiometric data, that the Precambrian rocks exposed in

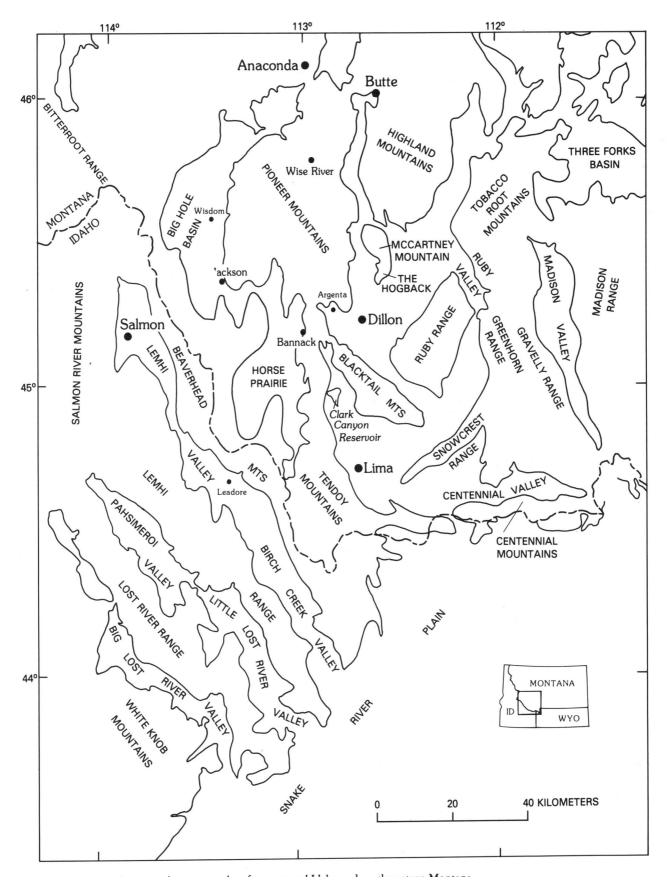

Figure 1—Index map of present-day topography of east-central Idaho and southwestern Montana.

east-central Idaho were older than the Belt Supergroup of
west-central Montana and that they had been arched into a
broad uplift that formed the south side of the Belt basin.
The Salmon River arch was thought to have persisted as a
positive region until middle or late Paleozoic time. It
incorporated the Belt island of Harrison et al. (1974) at its
western end and extended to the southeast to include the
regions of the Lemhi arch of Sloss (1954) and the early
Paleozoic uplifts described by Scholten (1957).

The different arches, domes, and uplifts named in east-
central Idaho and southwest Montana have some features in
common, and some of the names have been used inter-
changeably, particularly those of the Lemhi arch and the
Salmon River arch. Evidence now available on the structural
and stratigraphic history of the region suggests, however,
that most of the high areas are different parts of a single
arch that was formed in late middle Proterozoic time. This
arch persisted as a positive region until Late Devonian time
and continued to influence marine depositional patterns
throughout most of the rest of the Paleozoic. The first name
proposed, the Lemhi arch (Sloss, 1954), is appropriate for
this persistent uplift, but the arch was located farther west
than Sloss suggested, because the rocks documenting it are
allochthonous and have been displaced eastward, perhaps as
much as 150–170 km, by Medicine Lodge thrusting (Ruppel,
1978; Ruppel and Lopez, 1984). The Lemhi arch and the
Salmon River arch seem to be the same feature because all
of the stratigraphic changes in lower Paleozoic rocks
attributed to the Lemhi arch by Sloss (1954) were
incorporated into the description of the Salmon River arch
(Armstrong, 1975). Changes in Proterozoic rocks ascribed by
Armstrong to the Salmon River arch are related instead to
an older, early middle Proterozoic uplift called the Belt uplift
in this report. The name Salmon River arch is not used
here. The Belt uplift persisted through most of middle
Proterozoic time and separated the Belt basin in west-central
Montana from the Proterozoic miogeosyncline in central
Idaho, before uplift of the Lemhi arch.

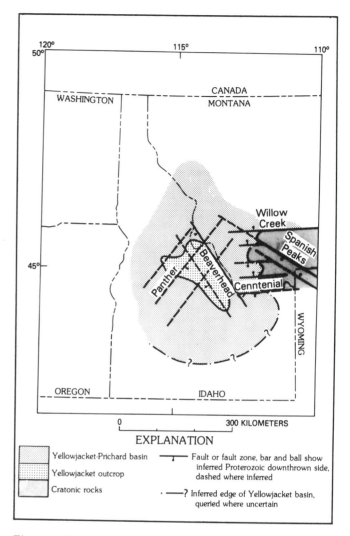

Figure 2—Sketch map of the early middle Proterozoic Yellowjacket
depositional basin.

GEOLOGIC FRAMEWORK BEFORE UPLIFT OF THE LEMHI ARCH

The oldest Proterozoic sedimentary rocks exposed in east-
central Idaho are included in the early middle Proterozoic
Yellowjacket Formation, a sequence of fine to very fine
grained feldspathic quartzite, siltstone, and argillite perhaps
as much as 8000 m thick (Lopez, 1981). The base of the
formation is not known to be exposed, and its top is the sole
zone of the Medicine Lodge thrust system, so its stratigraphic relationships to other Precambrian rocks are not
known. The formation is thought to be autochthonous
(Ruppel and Lopez, 1984) and equivalent to the Prichard
Formation of the Belt Supergroup in west-central Montana
(Ruppel, 1975), which it closely resembles. The Yellowjacket
is interpreted to be a deep marine turbidite sequence (Lopez,
1981) (Figure 2), which Hahn and Hughes (1984)
considered to have been deposited in an intracratonic basin.

The formation is intruded by several large porphyritic
quartz monzonite plutons that have yielded U-Th-Pb ages of

about 1.4 b.y., the minimum age of the Yellowjacket
Formation (Lopez, 1981; Evans, 1981; Evans and Zartman,
1981). Its maximum possible age is about 1.7 b.y. (see
Harrison, 1972). The formation is cut off on its eastern edge
by a zone of steep faults, the Miner Lake–Beaverhead Divide
fault zone, and that part of the formation now preserved in
east-central Idaho probably was deposited in the central part
of the Yellowjacket depositional basin (Figure 3) (Lopez,
1981; Ruppel and Lopez, 1984; Ruppel et al., 1983). The
movement on the Miner Lake–Beaverhead Divide fault zone
appears to have taken place in an episode of regional
tectonism that preceded or accompanied the emplacement of
the granitic plutons west of the fault zone about 1.3–1.4 b.y.
ago. The Yellowjacket depositional basin was destroyed at
this time, and the cratonic region of southwestern Montana
was uplifted and exposed as the eastern facies of the
Yellowjacket Formation was eroded (Ruppel and Lopez,
1984).

This cratonic uplift is the eastern part of the cratonic
region south of the Belt basin, which includes Belt island at

its western end (Harrison et al., 1974). The uplift has been described by Harrison et al. as a large cratonic prong or island that was a persistent source of detritus shed northward into the Belt basin throughout much of middle Proterozoic time (Figure 4). The evidence that it was also a source for the sediments now included in the middle Proterozoic miogeosynclinal rocks of the Lemhi Group, Swauger Formation, and Lawson Creek Formation in central Idaho is less clear (Ruppel, 1975; Hobbs, 1980). These units are dominantly fine- to medium-grained feldspathic quartzites that are now exposed in a narrow band of allochthonous rocks, principally in the Lemhi Range and Beaverhead Mountains in east-central Idaho. Their coarser grained, nearer shore equivalents, if any were deposited, were either destroyed by late Proterozoic erosion or are hidden beneath the Medicine Lodge thrust plate (Ruppel, 1978). Their finer grained distal equivalents, presumably deposited farther west in the Proterozoic miogeosyncline, have not been recognized. They may be hidden beneath other major thrust plates or may have been destroyed by late Proterozoic erosion or by emplacement of the Idaho batholith. The limited evidence on source directions in the preserved quartzitic rocks, however, does suggest a landmass toward the northeast. The sandstones in the Big Creek Formation in the Lemhi Group (Ruppel, 1975) coarsen toward the northeast, from fine to medium grained. Also, the clean quartzite of the Swauger Formation closely resembles part of the Missoula Group, which is its temporal equivalent in the Belt Supergroup, determined on the basis of paleomagnetic data (D. P. Elston and S. L. Bressler, personal communication, 1979). This suggests that they probably were derived from a similar cratonic source. Hobbs (1980) suggested that the Swauger and Lawson Creek formations are so like the upper part of the Missoula Group as to suggest deposition in proximate or interconnected basins.

The uplift that formed in early middle Proterozoic time when the Yellowjacket depositional basin was destroyed separated the Belt basin in west-central Montana from the Proterozoic miogeosyncline in central Idaho during the rest of middle Proterozoic time. Sediments derived from the uplift were distributed northward into the Belt basin to form part of the Belt Supergroup and probably southward into the Proterozoic miogeosyncline to form the middle Proterozoic rocks of central Idaho. The eastern part of the uplift was largely bounded by faults (Figure 4), including the Willow Creek fault zone (McMannis, 1963, 1965; Robinson, 1963), the Miner Lake–Beaverhead Divide fault zone (Ruppel et al., 1983), and probably the Centennial fault or other fault zones at the southern edge of the cratonic block (Ruppel, 1982). The uplift probably was cut near its center by northeast-trending faults that earlier had partly controlled the Yellowjacket depositional basin (O'Neill et al., 1982; Hahn and Hughes, 1984). Its westernmost known extension was Belt island in west-central Idaho (Harrison et al., 1974). Taken together, these features form a west–northwest-trending uplift parallel to the Coeur d'Alene trough in the axial part of the Belt basin (Harrison, 1972; Harrison et al., 1974) and about parallel to the Salmon River arch of Armstrong (1975), but farther north. This uplift seems most appropriately called the Belt uplift, to recognize the earlier

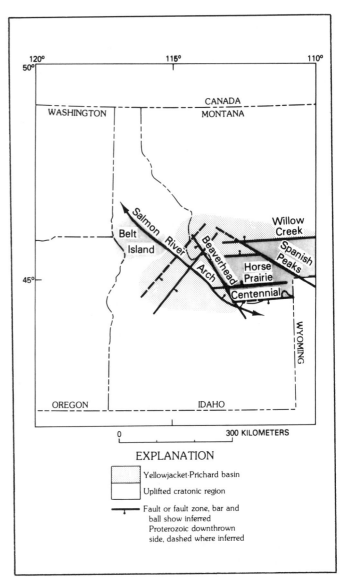

Figure 3—Sketch map of the Yellowjacket basin after early middle Proterozoic faulting and cratonic uplift.

descriptions of parts of it by Harrison et al. (1974). The uplift was formed after deposition of the Yellowjacket Formation, about 1.3–1.4 b.y. ago (Evans and Zartman, 1981; Evans, 1981; Lopez, 1982) and continued to shed sediments through most of the rest of middle Proterozoic time.

The Belt Uplift and the Salmon River Arch

The name Belt uplift is more appropriate than the name Salmon River arch for the uplift that formed in early middle Proterozoic time. The definition of the Salmon River arch (Armstrong, 1975) was based on less complete information than is available now, particularly in terms of regional structural relationships and the relationships of middle Proterozoic rocks. The Precambrian rocks of east-central Idaho were interpreted to be older than the Belt Supergroup (Armstrong, 1975), but later studies have shown them to be a thick sequence of miogeosynclinal rocks that are

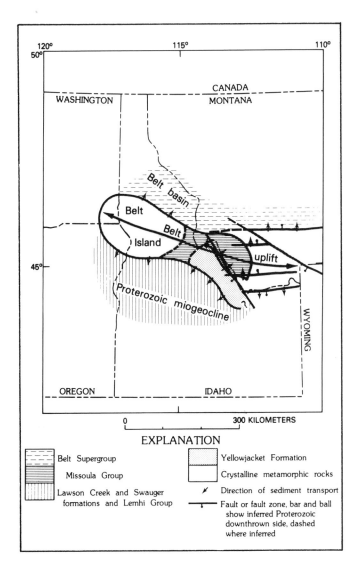

Figure 4—Sketch map showing location and relationships of the Belt uplift.

combined two very different uplifts into one: the Belt uplift of early middle Proterozoic age in central Idaho and southwest Montana and the Lemhi arch of late Proterozoic and early Paleozoic age in central Idaho. Because two features were incorporated into one, the axis of the Salmon River arch was placed well to the south of the cratonic region of southwestern Montana, about where the Lemhi arch was then presumed to be (Figure 3).

Summary: The Belt Uplift

In summary, the Belt uplift combines the Belt island of Harrison et al. (1974) and the southwestern Montana cratonic region into a single west–northwest-trending uplift that separated the Belt basin of west-central Montana from the middle Proterozoic miogeosyncline in central Idaho. It was uplifted after deposition of the Yellowjacket Formation, in an episode of regional tectonism and associated plutonism about 1.3–1.4 b.y. ago and controlled regional depositional patterns throughout middle Proterozoic time. Erosion of the uplifted eastern facies of the Yellowjacket Formation and of underlying Archean crystalline metamorphic rocks provided a major southern source of fine-grained sediments for the Belt Supergroup and a northern source for equivalent rocks deposited in the Proterozoic miogeosyncline, those of the Lemhi Group, Swauger Formation, and Lawson Creek Formation.

The east end of the Belt uplift was fragmented by renewed faulting in late middle Proterozoic time, and a small, fault-bounded embayment was formed at this time on the north flank of the Belt uplift, about in the region of the present Big Hole basin and reaching as far south as the Horse Prairie fault zone (Figures 1 and 4). This embayment was subsequently filled with Missoula Group sediments that may have joined with the temporally equivalent rocks of the Swauger and Lawson Creek formations across the central part of the Belt uplift, but no autochthonous Proterozoic rocks other than the Yellowjacket Formation are now known in this area. The influence of the Belt uplift on deposition ended about 800 m.y. ago when sedimentation in the miogeosynclinal region in central Idaho and the Belt basin was ended by uplift.

THE LEMHI ARCH

The Lemhi arch was a major landmass in central Idaho that separated the miogeosyncline farther west from an intermittent marine embayment on the east during much of late Proterozoic and early Paleozoic time. It influenced late Proterozoic, early Paleozoic, and probably late Paleozoic–early Mesozoic marine sedimentation patterns; provided a source for later Mesozoic detrital rocks to the east; and was overridden by the Medicine Lodge thrust in Late Cretaceous time (Sloss, 1954; Ruppel, 1978) (Figure 5). The arch is evident in shoaling patterns in rocks of many ages and in the many arches and uplifts that have been described in this region to explain anomalous early Paleozoic marine sedimentation patterns (Sloss, 1954; Scholten, 1957; Armstrong, 1975). Rocks from the western, miogeosynclinal side of the arch have been thrust eastward across the arch and concealing it and overlap rocks deposited on the eastern,

lithologically distinct from the Belt Supergroup, but equivalent in age (Ruppel, 1975; Hobbs, 1980; Lopez, 1981, 1982; Evans, 1981; Evans and Lund, 1981; Evans and Zartman, 1981; D. P. Elston and S. L. Bressler, personal communication, 1979). The effects of regional thrust faulting were largely unknown, and only small thrusts were thought to be present in east-central Idaho (Armstrong, 1975). Later mapping, however, has shown that the Yellowjacket Formation in east-central Idaho is everywhere separated from younger middle Proterozoic rocks by the sole zone of the Medicine Lodge thrust and that all of the post-Yellowjacket Proterozoic and Paleozoic rocks are miogeo-synclinal rocks tectonically displaced eastward perhaps as much as 150–170 km (Ruppel, 1975, 1978; Ruppel et al., 1981; Ruppel and Lopez, 1984). Furthermore, the effects of late Proterozoic deformation and deep erosion in the allochthonous middle Proterozoic rocks of east-central Idaho had not been recognized. As a result, the Salmon River arch

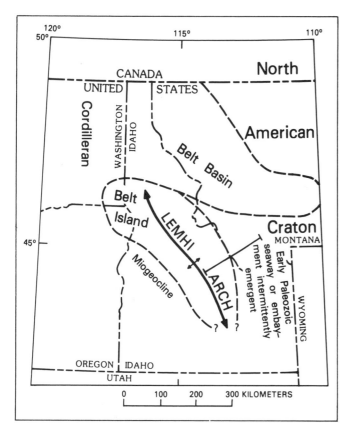

Figure 5—The location of the Lemhi arch, as defined in this paper.

1975) and by Lower or Middle Ordovician rocks elsewhere in these ranges.

Stratigraphic Relationships on the West Flank of the Lemhi Arch

Evidence for the Lemhi arch is found principally in depositional patterns of lower Paleozoic stratigraphic units in east-central Idaho on the west flank of the Lemhi arch, which thin eastward to disappear between the Lemhi Range and the Beaverhead Mountains against a highland underlain by deformed and deeply eroded middle Proterozoic rocks (Table 1 and Figure 6).

Lower Cambrian rocks are known only in the southern part of the Lemhi Range, where they are included in the Wilbert Formation. The Wilbert Formation was originally and tentatively thought to be late Proterozoic in age (Ruppel et al., 1975). Derstler and McCandless (1981) and McCandless (1982) have since reported fossils indicating that at least the upper part of the formation in the southernmost part of the Lemhi Range is of Early Cambrian age. The age of the lower part of the formation, in which no fossils have been found, remains uncertain, but is tentatively assigned as late Proterozoic and perhaps partly Cambrian.

The Lower Ordovician Summerhouse Formation (Ruppel et al., 1975) is more widely distributed in the central and southern parts of the Lemhi Range than the Wilbert Formation, but both formations thin and disappear to the east and north. The Summerhouse Formation was apparently deposited in a nearshore, perhaps lagoonal, depositional environment. The Middle Ordovician Kinnikinic Quartzite is the lowest Paleozoic formation that is widely distributed in east-central Idaho. It is a nearly white, clean, vitreous quartzite that conformably overlies the Summerhouse Formation in a few places, but more commonly overlies middle Proterozoic rocks with strong angular unconformity. It wedges out in the southern part of the Beaverhead Mountains (Scholten, 1957; Ramspott, 1962). The overlying Middle and Upper Ordovician and younger Saturday Mountain Formation, which is a thick sequence of dolomite in the central part of the Lemhi Range, thins and becomes sandy toward the southeast in the Lemhi Range, with abundant thin interbeds and partings of sandstone and conglomeratic sandstone and pyritic sandstone in a few places (Ross, 1961; Ruppel and Lopez, 1981). It is absent in the Beaverhead Mountains except in a small, thrust-faulted sliver near Leadore, Idaho (Ruppel, 1968).

No Silurian rocks are present in the Beaverhead Mountains or in the southern part of the Lemhi Range where the uppermost part of the Saturday Mountain Formation (which in places is as young as Silurian) has been removed by pre-Devonian erosion. The Middle and Upper Silurian Laketown Dolomite is also absent in these areas, but is present in places in the central Lemhi Range where it was deposited in a discontinuous, northward- and eastward-thinning wedge across a weathered and irregularly eroded surface cut into the Saturday Mountain Formation (Ruppel, 1968; Ruppel and Lopez, 1981).

The Devonian Jefferson Formation, described by Sloss (1954), is thickest in the Lost River Range and in the central part of the Lemhi Range, where it includes both

seaway side of the arch. Thus, rocks originally deposited on opposite shores of the arch are now placed nearly together.

The Lemhi arch formed by regional uplift and folding that ended sedimentation in the middle Proterozoic miogeosyncline. This deformation is documented in east-central Idaho by the nearly complete erosion of some middle Proterozoic rocks and by the angular relationships between middle Proterozoic rocks and overlying early Paleozoic rocks. The youngest known middle Proterozoic rocks, those of the Lawson Creek Formation (Hobbs, 1980), are preserved only in the northern part of the Lost River Range. They have been completely removed by erosion elsewhere in east-central Idaho, as has part, or in places all, of the underlying Swauger Formation (Ruppel, 1980). As a result, upper Proterozoic and lower Paleozoic rocks overlie the eroded edges of rocks as old as those in the upper part of the Lemhi Group. As much as 4500 m of strongly folded middle Proterozoic sedimentary rocks were eroded before deposition of any upper Proterozoic or lower Paleozoic rocks. Such deep erosion suggests that the Lemhi arch was uplifted in late middle Proterozoic or early late Proterozoic time. The uplift may have been related to the East Kootenay orogeny in Canada about 800 m.y. ago (Gabrielese, 1972) or perhaps earlier (McMechan and Price, 1982). The middle Proterozoic rocks are overlain with strong angular unconformity by upper Proterozoic(?) and Cambrian rocks in the southern part of the Lemhi Range and in the central part of the Lost River Range (Ross, 1947; Baldwin, 1951; Ruppel et al.,

Table 1—Generalized thicknesses and eastward thinning of lower and middle Paleozoic sedimentary rocks in east-central Idaho.

AGE		FORMATION	THICKNESS (METERS)[1]		
			Beaverhead Mountains	Lemhi Range[2]	Lost River Range
Devonian	Late and Middle	Three Forks and Jefferson Formations	30-90	90-825	900 (includes Grand View Dolomite)
	Early				
Silurian		Laketown Formation	0	0-100	400
Ordovician	Late	Saturday Mountain Formation	0	85-370	370-400
	Middle	Kinnikinic Quartzite	0-90	150-400	610
	Early	Summerhouse Formation	0	0-305	180
Cambrian and Late Proterozoic(?)		Wilbert Formation	unknown	0-300	uncertain
ANGULAR UNCONFORMITY					
Middle Proterozoic		Lawson Creek Formation and older rocks	unknown	Swauger and Gunsight Formations	Lawson Creek and Swauger Formations

1. Principal references:
   Beaverhead Mountains: Scholten, 1957; Scholten and Hait, 1962; Scholten and Ramspott, 1968; Ruppel, 1968; Luchitta, 1966; Smith, 1961; Ramspott, 1961.
   Lemhi Range: Sloss, 1954; Ross, 1961; Beutner, 1968; Ross, R. J., Jr., 1959; McCandless, 1982; Ruppel, 1968, 1980; Rupel, Ross, and Schleicher, 1975; Ruppel, Watts, and Peterson, 1970; Ruppel and Lopez, 1981.
   Lost River Range: Ross, 1947; Mapel, Read, and Smith, 1965; Mapel and Shropshire, 1973; Hays, McIntyre, and Hobbs, 1978; McIntyre and Hobbs, 1978.
2. In Lemhi Range, maximum thicknesses are in central part of range, and thin to south and east.

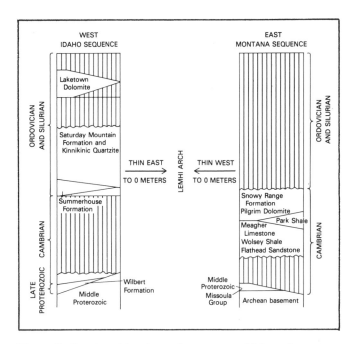

Figure 6—Stratigraphic columns from western Idaho and eastern Montana showing thinning of lower Paleozoic formations against the Lemhi arch.

Formation in southwestern Montana, which also is of Late Devonian age. The dolomite is part of a nearly continuous blanket of marine sedimentary rocks that extended in Late Devonian time from the Cordilleran miogeosyncline far onto the craton, for the first time since deposition of the early middle Proterozoic Yellowjacket Formation.

Stratigraphic Relationships on the East Flank of the Lemhi Arch

The stratigraphic record in southwestern Montana on the east flank of the Lemhi arch is far less complete than that in east-central Idaho (Figures 6 and 7). Only Cambrian rocks are present beneath the continuous blanket of Upper Devonian marine carbonate rocks in most of this region, and these are of Middle and Late Cambrian age, rather than of Early Cambrian age like those on the west flank of the arch (Sloss and Moritz, 1951).

The westernmost complete Cambrian sequences are near Camp Creek, in the southern part of the Highland Mountains, where the section is 400 m thick (Hanson, 1952), and in the Blacktail Mountains east of Dillon, Montana, where it is 270 m thick (Pecora, 1981) (Figure 1). (See isopachous maps in McMannis, 1965; Scholten, 1957.) These sequences include all the Middle and Upper Cambrian formations of the Montana shelf sequence (see Deiss, 1936; Robinson, 1963; Grant, 1965; Ruppel, 1972) except, locally, the Snowy Range Formation. The section thickens eastward to about 500 m in the Ruby Range (Tysdal, 1976), thins westward to about 230 m near Jackson, Montana (Ruppel et al., 1981), and thins southward to about 12 m in the Tendoy Range and is absent in the Beaverhead Mountains (Scholten et al., 1955; Scholten, 1957).

Middle and Upper Devonian strata (Figure 7). In these areas it is underlain in several places by partly conglomeratic sandstone beds that were deposited in steep-walled channels cut deeply into underlying Silurian and Ordovician rocks (Churkin, 1962; Hait, 1965; Mapel and Shropshire, 1973; Hoggan, 1981; Ruppel and Lopez, 1981). These sandstones contain Middle Devonian fossils, principally fishes, that suggest deposition in fresh or brackish water, perhaps in a marginal estuary or on a tidal flat (Hait, 1965; Denison, 1968; Hoggan, 1981). The overlying sequence of marine carbonate rocks includes sandy dolomite and quartz sandstone interbeds that are abundant in the lower part of the sequence in the central part of the Lemhi Range, but are much less so in the Lost River Range to the west (Churkin, 1962). The Jefferson Formation is thinner in the southern part of the Lemhi Range, and most of it there is probably of Late Devonian age (Sloss, 1954; Ross, 1961).

In the Beaverhead Mountains, the Jefferson Formation is thin, of Late Devonian age, and overlies either Kinnikinic Quartzite or Proterozoic quartzites. In places, the basal part of the formation includes gravels accumulated on the crest of the Lemhi arch, evaporites, and marine carbonate rocks (Ramspott, 1962). The basal beds are overlain by a thin sequence of dark-colored dolomite like that of the Jefferson

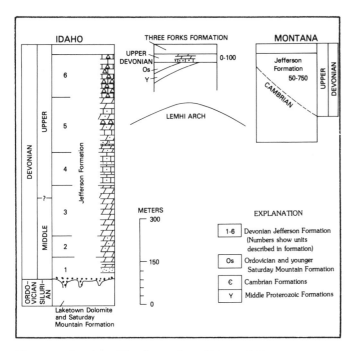

Figure 7—Stratigraphic columns from Idaho and Montana showing thinning of the Jefferson Formation against the Lemhi arch.

Part of the westward thinning of the Cambrian sequence is a reflection of late Middle Cambrian erosion, which locally removed part or all of the Park Shale and in a few places removed all of the Cambrian rocks (Myers, 1952). Pre-Devonian erosion removed much of the Snowy Range Formation (Sloss and Moritz, 1951; Lowell, 1965), but overall westward depositional thinning is evident as well. The Flathead Sandstone and Meagher Limestone are each only a few meters thick in the Tendoy Mountains, and the Wolsey Shale is absent, all apparently a result of depositional thinning (Scholten et al., 1955; Scholten, 1957; McMannis, 1965). Younger Cambrian rocks are absent in the Tendoy Mountains. Cambrian rocks near Jackson, Montana, also seem to be depositionally thinned (Ruppel et al., 1981; Zimbelman, 1981). The Meagher Limestone and Pilgrim Dolomite at Jackson each are only about 13 m thick and are separated by only a few thin beds of Park Shale. The Meagher also is partly sandy and silty here, and the dolomite is partly cross bedded and detrital. In the Blacktail Mountains part of the Meagher is argillaceous, and Pecora (1981) suggested that the Meagher Formation coarsens and thins westward toward Jackson, Montana, and laps onto a western source of the detritus, the Lemhi arch.

Ordovician and Silurian rocks are absent in most of southwestern Montana, suggesting that the east flank of the Lemhi arch probably was being eroded during Ordovician and Silurian time. Isolated exposures of Middle Ordovician Kinnikinic Quartzite in the vicinity of the Beaverhead Mountains are remnants of the Medicine Lodge thrust plate, tectonically carried in from farther west (Sloss and Moritz, 1951; Scholten et al., 1955; McMannis, 1965). The Ordovician Bighorn Dolomite to the east in the Madison Range (Sloss and Moritz, 1951; Hanson, 1952; Hadley,

1980) may be close to the original depositional edge, which was destroyed by erosion before deposition of the Upper Devonian Jefferson Formation (McMannis, 1965).

All of the Jefferson Formation in southwestern Montana is of Late Devonian age (Sloss and Moritz, 1951; Scholten and Hait, 1962; Hadley, 1980). It thins westward from about 80 m thick in the Ruby Range (Tysdal, 1976) to about 35 m thick in the Blacktail Mountains (Pecora, 1981). The Jefferson is absent locally in the Tendoy Mountains and the southernmost Beaverhead Mountains (Scholten and Hait, 1962), the area of the Tendoy dome of Scholten (1957; McMannis, 1965), which is interpreted as the last remnant of the Lemhi arch to have been submerged (Scholten, 1957). The Upper Devonian Three Forks Formation extends across the entire region and ranges from about 30–35 m thick in southwestern Montana to about 90 m thick in east-central Idaho. In general, the Three Forks Formation thickens westward from the cratonic shelf to the miogeosyncline, across the crest of the Lemhi arch.

Summarized History of the Lemhi Arch

The Lemhi arch was formed in central Idaho in late middle or early late Proterozoic time when middle Proterozoic miogeosynclinal rocks were uplifted and folded into a broad, northwest-trending arch. This was a period of deformation that ended middle Proterozoic sedimentation in the northern part of the Proterozoic miogeosyncline. The arch persisted as a landmass into late Proterozoic and early Paleozoic time, until it finally was submerged in Late Devonian time when marine sedimentary rocks of the Jefferson and Three Forks formations were deposited across it in a continuous blanket that reached from the Cordilleran miogeosyncline onto the cratonic shelf of southwestern Montana. The arch continued to influence marine sedimentation patterns even after it was submerged, because different sequences of marine rocks of various ages were deposited on its opposite sides (Sloss and Moritz, 1951; Scholten, 1957; Ruppel, 1978). In effect, its crest was a hinge zone that separated the region of later Paleozoic miogeosynclinal sedimentation in central and western Idaho from the region of marine deposition on the cratonic shelf.

During most of Mesozoic time, the Lemhi arch probably was a persistent high area, supplying sediments rather than receiving them. The flood of detritus into western Montana in Jurassic and Early Cretaceous time was partly from a western source (McMannis, 1965; Peterson, 1981) and indicates renewed uplift in the region of the Lemhi arch. In late Early and Late Cretaceous time, the Lemhi arch was overridden by major thrust plates from the west. The Medicine Lodge thrust system shifted together or overlapped middle Proterozoic rocks scalped from the crest of the arch and Paleozoic sedimentary rocks from the Paleozoic-Cordilleran miogeosyncline farther west, across sedimentary rocks partly of different ages deposited on the east flank of the arch, as shown in Figure 8.

Erosion of the Lemhi Arch in Late Proterozoic and Early Paleozoic Time

The volume of detritus eroded from the Lemhi arch is uncertain, because the original spatial relationships of rocks that were at one time more widely separated but that are

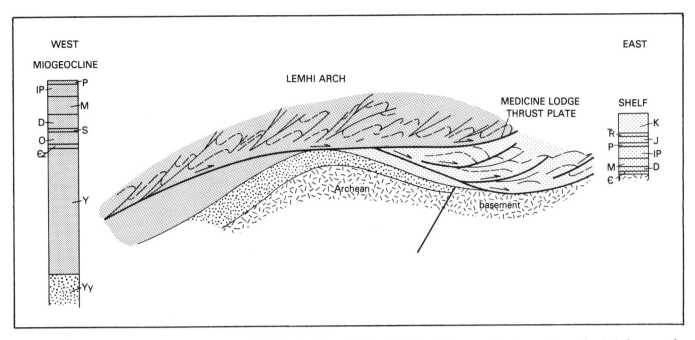

Figure 8—Schematic cross section across the Lemhi arch after thrust faulting. Relative thicknesses of units are shown. Symbols for ages of units: K, Cretaceous; J, Jurassic, ℞, Triassic; P, Permian; ℙ, Pennsylvanian; M, Mississippian; D, Devonian; S, Silurian; O, Ordovician; Є, Cambrian; Z, late Proterozoic; Y, middle Proterozoic; Yy, middle Proterozoic Yellowjacket Formation.

now placed nearly together by thrust faulting are only partly understood. Most of the Lawson Creek Formation (about 1300 m thick), most or in places all of the underlying Swauger Formation (about 3100 m thick), and part of the Gunsight Formation at the top of the Lemhi Group were eroded before deposition of succeeding upper Proterozoic or lower Paleozoic rocks. This represents a total eroded thickness of perhaps as much as 4500 m in a region about 150 km long and about 60 km wide. That area represents only part of the crest and west flank of the Lemhi arch; the rest of the arch is now concealed beneath thrust plates, and the extent of erosion is unknown. A conservative guess at the volume of detritus removed by late Proterozoic and early Paleozoic erosion is about 30,000 km³, assuming a minimum eroded prism about 200 km long, 50 km wide, and 3 km thick. But because the observed eroded block nearly approaches the size of the assumed minimum prism, the actual volume of eroded sediments could be several times the minimum volume, perhaps as much as 100,000 km³

Regardless of the actual volume of eroded sediments, it seems clear that a great amount of detritus was eroded from the Lemhi arch. A small part of this detritus was redeposited and preserved in the small areas of upper Proterozoic and lower Paleozoic formations of east-central Idaho, but the greater part of it was deposited elsewhere. The upper Proterozoic(?) and Cambrian Wilbert Formation (Ruppel et al., 1975; McCandless, 1982), which consists of sandstone, conglomeratic sandstone, shale, and a few beds of sandy limestone and dolomite, seems to be composed mainly of detritus eroded from the Lawson Creek and Swauger formations. The Wilbert is known only in the central part of the Lost River Range and in the southern parts of the Lemhi Range and Beaverhead Mountains (Ross, 1947, 1961;

Baldwin, 1951; Ruppel et al., 1975). The sandstone of the Lower Ordovician Summerhouse Formation (Ruppel et al., 1975), which lies above the Wilbert Formation, apparently represents a nearshore deposit derived mainly from the Lawson and Swauger formations, but perhaps also from the underlying Wilbert Formation, which it depositionally overlaps. The Summerhouse is present only in a few places in the Lemhi and Lost River ranges.

The Middle Ordovician or older Clayton Mine Quartzite of central Idaho (Hobbs et al., 1968) includes conglomerate and beds containing coarse fragments of detrital feldspar that suggest a nearby source in crystalline rocks, perhaps on the northern part of the Lemhi arch. The Lower or Middle Cambrian Cash Creek Quartzite of the same region (Hobbs et al., 1968) is a moderately to poorly sorted, pebbly quartzite that suggests nearness to a source area, perhaps the Lemhi arch (see Stewart and Suczek, 1977). It overlies a sequence of poorly sorted, partly pebbly, largely detrital rocks of uncertain age and limited areal extent, which probably represent a local deposit on the southwestern flank of the Lemhi arch.

The much more widespread Kinnikinic Quartzite, which is the only thick, widespread blanket of sandstone in this region, is composed of supermature quartz sand that seems most likely to have been derived at least partly from erosion of quartzites like those of the Swauger Formation. A similar source is suggested for equivalent, clean quartzites of Middle Ordovician age farther south in the Paleozoic miogeo-syncline (James and Oaks, 1977). Some of the supermature sand in the Kinnikinic, however, was probably derived from erosion and reworking of upper Proterozoic and lower Paleozoic rocks like those in the Wilbert and Summerhouse formations, because the Kinnikinic depositionally overlaps the eroded edges of these formations, as well as the upturned edges of the deformed and deeply eroded middle Proterozoic rocks of the Lemhi arch.

Other lower or middle Paleozoic formations in both southwestern Montana and east-central Idaho contain abundant beds of sandstone and sandy dolomite where they lap against the Lemhi arch, which suggests that this sandstone was eroded from middle and upper Proterozoic and lower Paleozoic rocks exposed on the arch. However, the total amount of sandstone in the upper Proterozoic and lower and middle Paleozoic rocks in east-central Idaho and southwestern Montana is small compared with the great volume thought to have been removed from the Lemhi arch. The only known rocks that are sufficiently thick and widespread to account for the volume of detritus eroded from the arch are those of the upper Proterozoic and lower Paleozoic sequences in southeastern Idaho. These rocks suggest that a large part of the detritus eroded from the Lemhi arch, as well as that eroded from the exposed cratonic region in southwestern Montana, was carried toward the south and southeast to form the clastic rocks that are now included in the upper Precambrian Pocatello Formation and younger upper Proterozoic and Lower to Middle Cambrian formations in the vicinity of Pocatello, Idaho (Crittenden et al., 1971; Link, 1983).

SUMMARY AND CONCLUSIONS

The many arches, domes, and uplifts of middle Proterozoic to middle Paleozoic age that have been described in central Idaho and southwestern Montana seem to be different parts of only two major uplifts. The older one, the Belt uplift, combines the Belt island of Harrison et al. (1974) and the southwestern Montana cratonic region. It trended about west-northwest from southwestern Montana into west-central Idaho. It was uplifted early in middle Proterozoic time, about 1.4 b.y. ago, and separated the Belt basin in west-central Montana from the middle Proterozoic miogeosyncline in central Idaho. It was a major source of detritus for the Belt Supergroup deposited in the Belt basin and for equivalent rocks deposited in the middle Proterozoic miogeosyncline—those of the Lemhi Group, Swauger Formation, and Lawson Creek Formation. The uplift controlled depositional patterns until late middle or early late Proterozoic time when uplift and folding ended sedimentation in both the Belt basin and in the miogeosyncline.

The Lemhi arch, a northwest-trending arch, formed in central Idaho in late middle or early late Proterozoic time, and it separated the late Proterozoic and Paleozoic miogeosyncline in western Idaho from the cratonic shelf in southwestern Montana. It was most likely uplifted at the same time that uplift and folding ended sedimentation in the middle Proterozoic miogeosyncline and in the Belt basin. It was underlain by the strongly folded middle Proterozoic quartzitic rocks of central Idaho, which were deeply eroded in late Proterozoic and early Paleozoic time and thus supplied detritus for the marine sequences that overlap it and probably for the Proterozoic rocks of the Pocatello

Formation in southeastern Idaho. The arch remained a major source of sediments until Late Devonian time when it finally was overlapped by the marine Jefferson and Three Forks formations. It continued, however, to influence marine depositional patterns even after it was overlapped, because it remained a zone of stratigraphic transition between the Cordilleran miogeosyncline and the cratonic shelf through middle and late Paleozoic time. It was uplifted again in the early Mesozoic and became a source of sediments for Jurassic and younger Mesozoic rocks in southwestern Montana, rocks that also lapped northward against the Belt island of Imlay et al. (1948). In late Early and Late Cretaceous time, the Lemhi arch was overridden by major thrust plates from the west. Both the crest of the arch, underlain by deeply eroded middle Proterozoic rocks, and upper Proterozoic and Paleozoic miogeosynclinal rocks deposited on its western flank were transported eastward as part of the Medicine Lodge thrust plate, perhaps as much as 150–170 km into east-central Idaho. Here they tectonically overlap a different sequence of rocks deposited on the east flank of the arch and on the cratonic shelf in southwestern Montana.

ACKNOWLEDGMENTS

I am deeply indebted to Prof. L. L. Sloss of Northwestern University, who first described and named the Lemhi arch, and to Prof. Robert Scholten of Pennsylvania State University, who frequently and generously has shared his time and his knowledge of southwest Montana geology. Similarly, many colleagues in the U.S. Geological Survey have helped me. I am particularly indebted to Jack E. Harrison and Chester A. Wallace, who have shared their knowledge of the Belt Supergroup, and to S. Warren Hobbs, for his help in understanding the rocks and structure of central Idaho.

REFERENCES CITED

Armstrong, R. L., 1975, Precambrian (1500 million years old) rocks of central Idaho—the Salmon River arch and its role in Cordilleran sedimentation and tectonics: American Journal of Science, v. 275-A, p. 437–467.

Baldwin, E. M., 1951, Faulting in the Lost River Range area of Idaho: American Journal of Science, v. 249, p. 884–902.

Beutner, E. C., 1968, Structure and tectonics of the southern Lemhi Range, Idaho: Ph.D. thesis, Pennsylvania State University, University Park, 106 p.

Churkin, M., Jr., 1962, Facies across Paleozoic miogeosynclinal margin of central Idaho: AAPG Bulletin, v. 46, p. 569–591.

Crittenden, M. D., F. E. Schaeffer, D. E. Trimble, and L. A. Woodward, 1971, Nomenclature and correlation of some Upper Precambrian and basal Cambrian sequences in western Utah and southeastern Idaho: Geological Society of America Bulletin, v. 82, p. 581–602.

Deiss, C., 1936, Revision of type Cambrian formations and sections of Montana and Yellowstone National Park:

Geological Society of America Bulletin, v. 47, p. 1257–1342.

Denison, R. H., 1968, Middle Devonian fishes from the Lemhi Range of Idaho: Fieldiana Geology, v. 16, p. 269–288.

Derstler, K., and D. O. McCandless, 1981, Cambrian trilobites and trace fossils from the southern Lemhi Range, Idaho: their stratigraphic and paleotectonic significance (abs.): Geological Society of America Abstracts with Programs, v. 13, n. 4, p. 194.

Evans, K. V., 1981, Geology and geochronology of the eastern Salmon River Mountains, Idaho, and implications for regional Precambrian tectonics: Ph.D. thesis, Pennsylvania State University, University Park, 222 p.

Evans, K. V., and K. Lund, 1981, The Salmon River "arch"? (abs.): Geological Society of America Abstracts with Programs, v. 13, n. 5, p. 448.

Evans, K. V., and R. E. Zartman, 1981, U-Th-Pb zircon geochronology of Proterozoic Y granitic intrusions in the Salmon area, east-central Idaho (abs.): Geological Society of America Abstracts with Programs, v. 13, n. 4, p. 195.

Gabrielese, H., 1972, Younger Precambrian of the Canadian Cordillera: American Journal of Science, v. 272, p. 521–536.

Grant, R. E., 1965, Faunas and stratigraphy of the Snowy Range Formation (Upper Cambrian) in southwestern Montana and northwestern Wyoming: Geological Society of America Memoir 96, 171 p.

Hadley, J. B., 1980, Geology of the Cameron and Varney quadrangles, Madison County, Montana: U.S. Geological Survey Bulletin 1459, 108 p.

Hahn, G. A., and G. J. Hughes, Jr., 1984, Sedimentation, tectonism, and associated magmatism of the Yellowjacket Formation in the Idaho cobalt belt, Lemhi County, Idaho: Montana Bureau of Mines and Geology Special Publication 90, p. 65–67.

Hait, M. H., Jr., 1965, Structure of the Gilmore area, Lemhi Range, Idaho: Ph.D. thesis, Pennsylvania State University, University Park, 134 p.

Hanson, A. M., 1952, Cambrian stratigraphy in southwestern Montana: Montana Bureau of Mines and Geology Memoir 33, 46 p.

Harrison, J. E., 1972, Precambrian Belt basin of northwestern United States: its geometry, sedimentation, and copper occurrences: Geological Society of America Bulletin, v. 83, p. 1215–1240.

Harrison, J. E., A. B. Griggs, and J. D. Wells, 1974, Tectonic features of the Precambrian Belt basin and their influence on post-Belt structures: U.S. Geological Survey Professional Paper 866, 15 p.

Hays, W. H., D. H. McIntyre, and S. W. Hobbs, 1978, Geologic map of the Lone Pine Peak quadrangle, Custer County, Idaho: U.S. Geological Survey Open-File Report 78-1060.

Hobbs, S. W., 1980, The Lawson Creek Formation of middle Proterozoic age in east-central Idaho: U.S. Geological Survey Bulletin 1482-E, 12 p.

Hobbs, S. W., W. H. Hays, and R. J. Ross, Jr., 1968, The Kinnikinic Quartzite of central Idaho—redefinition and subdivision: U.S. Geological Survey Bulletin 1254-J, 22 p.

Hoggan, R., 1981, Devonian channels in the southern Lemhi Range, Idaho: Montana Geological Society, Southwest Montana Field Conference Guidebook, p. 45–47.

Imlay, R. W., L. S. Gardner, C. P. Rogers, Jr., and H. D. Hadley, 1948, Marine Jurassic formations of Montana: U.S. Geological Survey Oil and Gas Investigations Preliminary Chart 32.

James, W. C., and R. Q. Oaks, Jr., 1977, Petrology of the Kinnikinic Quartzite (Middle Ordovician), east-central Idaho: Journal of Sedimentary Petrology, v. 47, p. 1491–1511.

Link, P. K., 1983, Glacial and tectonically influenced sedimentation in the upper Proterozoic Pocatello Formation, southeastern Idaho: Geological Society of America Memoir 157, p. 165–181.

Lopez, D. A., 1981, Stratigraphy of the Yellowjacket Formation of east-central Idaho: Ph.D. thesis, Colorado School of Mines, Golden, 252 p.

———, 1982, Constraints on the shape and position of the Yellowjacket (Proterozoic Y) depositional basin (abs.): Geological Society of America Abstracts with Programs, v. 14, n. 6, p. 320.

Lowell, W. R., 1965, Geologic map of the Bannack-Grayling area, Beaverhead County, Montana: U.S. Geological Survey Miscellaneous Geologic Investigations Map I-433.

Luchitta, B. K., 1966, Structure of the Hawley Creek area, Idaho–Montana: Ph.D. thesis, Pennsylvania State University, University Park, 204 p.

Mapel, W. J., and K. L. Shropshire, 1973, Preliminary geologic map and section of the Hawley Mountain quadrangle, Custer, Butte, and Lemhi counties, Idaho: U.S. Geological Survey Miscellaneous Field Studies Map MF-546.

Mapel, W. J., W. H. Read, and R. K. Smith, 1965, Geologic map and sections of the Doublespring quadrangle, Custer and Lemhi counties, Idaho: U.S. Geological Survey Geologic Quadrangle Map GQ-464.

McCandless, D. O., 1982, A reevaluation of Cambrian through Middle Ordovician stratigraphy of the southern Lemhi Range: Master's thesis, Pennsylvania State University, University Park, 157 p.

McIntyre, D. H., and S. W. Hobbs, 1978, Geologic map of the Challis quadrangle, Custer County, Idaho: U.S. Geological Survey Open-File Report 78-1059.

McMannis, W. J., 1963, LaHood Formation—a coarse facies of the Belt Series in southwestern Montana: Geological Society of America Bulletin, v. 74, p. 407–436.

———, 1965, Resumé of depositional and structural history of western Montana: AAPG Bulletin, v. 49, p. 1801–1823.

McMechan, M. E., and R. A. Price, 1982, Superimposed low-grade metamorphism in the Mount Fisher area, southeastern British Columbia: implications for the East Kootenay orogeny: Canadian Journal of Earth Sciences, v. 19, p. 476–489.

Myers, W. B., 1952, Geology and mineral deposits of the northwest quarter Willis quadrangle and adjacent Browns Lake area, Beaverhead County, Montana: U.S.

Geological Survey Open-File Report, 46 p.

O'Neill, J. M., D. A. Lopez, and N. R. Desmarais, 1982, Recurrent movement along, and characteristics of, northeast-trending faults in part of east-central Idaho and west-central Montana (abs.): Geological Society of America Abstracts with Programs, v. 14, n. 6, p. 345.

Pecora, W. C., 1981, Bedrock geology of the Blacktail Mountains, southwestern Montana: Master's thesis, Wesleyan University, Middletown, Connecticut, 203 p.

Peterson, J. A., 1981, General stratigraphy and regional paleostructure of the western Montana overthrust belt: Montana Geological Society, Southwest Montana, Field Conference and Symposium Guidebook, p. 5–35.

Ramspott, L. D., 1962, Geology of the Eighteenmile Peak area, and petrology of the Beaverhead pluton: Ph.D. thesis, Pennsylvania State University, University Park, 215 p.

Robinson, G. D., 1963, Geology of the Three Forks quadrangle, Montana: U.S. Geological Survey Professional Paper 370, 143 p.

Ross, C. P., 1947, Geology of the Borah Peak quadrangle, Idaho: Geological Society of America Bulletin, v. 58, p. 1085–1160.

———, 1961, Geology of the southern part of the Lemhi Range, Idaho: U.S. Geological Survey Bulletin 1081-F, p. 189–260.

Ross, R. J., Jr., 1959, Brachiopod fauna of the Saturday Mountain Formation, southern Lemhi Range, Idaho: U.S. Geological Survey Professional Paper 294-L, p. 441–461.

Ruppel, E. T., 1968, Geologic map of the Leadore quadrangle, Lemhi County, Idaho: U.S. Geological Survey Geologic Quadrangle Map GQ-733.

———, 1972, Geology of pre-Tertiary rocks in the northern part of Yellowstone National Park, Wyoming: U.S. Geological Survey Professional Paper 729-A, p. 1–66.

———, 1975, Precambrian Y sedimentary rocks in east-central Idaho: U.S. Geological Survey Professional Paper 889-A, 23 p.

———, 1978, Medicine Lodge thrust system, east-central Idaho and southwest Montana: U.S. Geological Survey Professional Paper 1031, 23 p.

———, 1980, Geologic map of the Patterson quadrangle, Lemhi County, Idaho: U.S. Geological Survey Geologic Quadrangle Map GQ-1529.

———, 1982, Cenozoic block uplifts in southwest Montana and east-central Idaho: U.S. Geological Survey Professional Paper 1224, 24 p.

Ruppel, E. T., and D. A. Lopez, 1981, Geologic map of the Gilmore quadrangle, Lemhi County, Idaho: U.S. Geological Survey Geologic Quadrangle Map GQ-1543.

———, 1984, The thrust belt in southwest Montana and east-central Idaho: U.S. Geological Survey Professional Paper 1278, 41 p.

Ruppel, E. T., K. C. Watts, and D. L. Peterson, 1970, Geologic, geochemical, and geophysical investigations in the northern part of the Gilmore mining district,

Lemhi County, Idaho: U.S. Geological Survey Open-File Report, 56 p.

Ruppel, E. T., R. J. Ross, Jr., and D. Schleicher, 1975, Precambrian Z and Lower Ordovician rocks in east-central Idaho: U.S. Geological Survey Professional Paper 889-B, p. 25–34.

Ruppel, E. T., C. A. Wallace, R. G. Schmidt, and D. A. Lopez, 1981, Preliminary interpretation of the thrust belt in southwest and west-central Montana and east-central Idaho: Montana Geological Society, Southwest Montana Field Conference and Symposium Guidebook, p. 139–159.

Ruppel, E. T., J. M. O'Neill, and D. A. Lopez, 1983, Preliminary geologic map of the Dillon 2° quadrangle, Montana: U.S. Geological Survey Open-File Report 83-168.

Scholten, R., 1957, Paleozoic evolution of the geosynclinal margin north of the Snake River Plain, Idaho–Montana: Geological Society of America Bulletin, v 68, p. 151–170.

Scholten, R., and M. H. Hait, 1962, Devonian system from shelf edge to geosyncline, southwestern Montana–central Idaho: Billings Geological Society, Thirteenth Annual Field Conference Guidebook, p. 13–22.

Scholten, R., and L. D. Ramspott, 1968, Tectonic mechanisms indicated by structural framework of central Beaverhead Range, Idaho–Montana: Geological Society of America Special Paper 104, 70 p.

Scholten, R., K. A. Keenmon, and W. D. Kupsch, 1955, Geology of the Lima region, southwestern Montana, and adjacent Idaho: Geological Society of America Bulletin, v. 66, p. 345–404.

Sloss, L. L., 1954, Lemhi arch, a mid-Paleozoic positive element in south-central Idaho: Geological Society of America Bulletin, v. 65, p. 365–368.

Sloss, L. L., and C. A. Moritz, 1951, Paleozoic stratigraphy of southwestern Montana: AAPG Bulletin, v. 35, p. 2135–2169.

Smith, J. G., 1961, The geology of the Clear Creek area, Montana–Idaho: Master's thesis, Pennsylvania State University, University Park, 75 p.

Stewart, J. H., and C. A. Suczek, 1977, Cambrian and latest Precambrian paleogeography and tectonics in the western United States, *in* J. H. Stewart, C. H. Stevens, and A. E. Fritsche, eds., Paleozoic paleogeography in the western United States: Society of Economic Paleontologists and Mineralogists, Pacific Section, Pacific Coast Paleogeography Symposium 1.

Tysdal, R. G., 1976, Paleozoic and Mesozoic stratigraphy of the northern part of the Ruby Range, southwestern Montana: U.S. Geological Survey Bulletin 1405-I, 26 p.

Zimbelman, D. R., 1981, Stratigraphy of Precambrian and Cambrian sedimentary rocks, Polaris 1 SE quadrangle, Beaverhead County, Montana (abs.): Geological Society of America Abstracts with Programs, v. 13, n. 4, p. 231.

Syntectonic Conglomerates in Southwestern Montana: Their Nature, Origin, and Tectonic Significance[1] (with an Update)

R. T. Ryder
U.S. Geological Survey
Reston, Virginia

R. Scholten
Pennsylvania State University
University Park, Pennsylvania

During Late Cretaceous (late Albian–Cenomanian) through Paleocene and probably early Eocene time, southwestern Montana and adjacent Idaho were the depocenter for thick accumulations of syntectonic quartzite conglomerate, limestone conglomerate, and minor amounts of sandstone collectively known as the Beaverhead Formation. In this study, the depositional and deformational history of the Beaverhead is documented in detail in an attempt to understand the tectonic and topographic development of southwestern Montana and east-central Idaho.

Clast imbrication and composition measurements suggest that two fundamentally different source areas existed for the Beaverhead. The limestone conglomerate clasts were derived primarily from Mississippian and Triassic carbonate rocks exposed by uplift of the Blacktail–Snowcrest basement arch and the Ancestral Beaverhead Range along the present-day Montana–Idaho border. These deposits, which derived their detritus from an area less than 25 km away, probably represent coalesced alluvial fans. Although the quartzite conglomerate beds were deposited simultaneously with the limestone clasts, the quartzite was derived from a more distant source. The vast quantity of Belt quartzite clasts that initially reached what is now southwestern Montana in late Albian time apparently originated from the large area of Belt Group exposure on the northeast side of the Idaho batholith approximately 80 km to the west and northwest of the study area. Gradual expansion of this source area by active uplift continuously provided steep slopes needed for the transport of gravel by braided stream systems to the adjacent subsiding alluvial plain. Recycling of the Belt clasts owing to Late Cretaceous uplift in the present eastern Snake River plain carried them as far east as western Wyoming, where they became incorporated in the Harebell and Pinyon conglomerates of western Wyoming.

The Beaverhead Formation exhibits two distinct structural patterns: (1) an earlier cratonic pattern composed of northeast-trending, gently plunging, open folds and associated high-angle faults, and (2) a northwest-trending "geosynclinal" pattern characterized by major upthrusts such as the Tendoy, Cabin, and Fritz Creek faults, and large, gently plunging, open folds. The older structural pattern seems to be related to the tectonism of the Late Cretaceous to early Paleocene that created the Blacktail–Snowcrest basement arch. The younger, more pervasive northwest trend may have a causal connection with the large uplift in the region of the Idaho batholith. It is plausible that gravitative energy created by this uplift was available not only for eastward fluvial transport of Belt clasts, but also for northeastward downslope tectonic transport. According to this hypothesis, large sheets of bedded rock slid eastward along detachment surfaces within the Belt, and as these surfaces were folded and locked, new ones developed to the east. This eastward migrating tectonic front reached the Lima, Montana, region soon after major Beaverhead sedimentation had ceased during middle to late Paleocene or early Eocene time. At this time, the Beaverhead rocks, which in part had inherited the earlier northeast trends, were cut by large upthrusts, folded into broad anticlinal structures, and overridden by low-angle sheets of Carboniferous limestone.

[1]Reprinted from the Geological Society of America Bulletin, v. 84, p. 773–796, March 1973.

INTRODUCTION

The thick sequence of conglomerate, sandstone, and minor carbonate rocks that crops out in southwestern Beaverhead County, Montana, and adjacent Clark County, Idaho, was named the Beaverhead Formation by Lowell and Klepper (1953). The detritus is dominated by conglomerate units and exhibits marked changes in facies and thickness. Locally it may be as much as 15,000 ft (4550 m) thick. These strata rest with angular unconformity on rocks as old as Precambrian and as young as Early Cretaceous (Albian) and are themselves deformed into structures whose trends roughly coincide with those found in underlying rocks. The unit is unconformably overlain by mildly deformed rocks not older than late Eocene, whose structures bear little relationship to earlier trends. The Beaverhead Formation is thus a syntectonic deposit in which there is recorded major tectogenic and morphogenic events of Late Cretaceous and early Tertiary age in southwestern Montana and adjacent Idaho. Together with rocks partially equivalent in age farther west in Idaho and farther east in Montana, an area of 10,000 mi^2 (26,000 km^2) is involved.

In this study, we attempt to reconstruct the tectonic and topographic development of southwestern Montana and east-central Idaho on the basis of the interpreted depositional and deformational history of the Beaverhead Formation. To this end, a detailed stratigraphic, sedimentologic, and structural study of the Beaverhead was conducted in the region around Lima, Montana. The Lima region (Figure 1), centrally located within the limits of Beaverhead exposures, was chosen as the study area because the major lithofacies of the Beaverhead are well exposed in this region. Moreover, this area contains complex deformational structures whose development in most cases has involved the Beaverhead. This makes it possible to relate the Beaverhead to other evidence of the local structural history as worked out by previous authors.[2]

REGIONAL SETTING OF THE BEAVERHEAD FORMATION

Structural Framework

Patterns and Phases of Deformation

Southwestern Montana and east-central Idaho straddle a gradational boundary between two distinct styles of deformation (Figure 1). Tectonism east of the Tendoy Range is manifested largely by uplifted cratonic areas that include Precambrian basement rocks, have diverse trends, and are partly bordered by high-angle upthrusts. Three major uplifted areas, plunging southwest to southeastward, can be recognized between the Tendoy Range and the Yellowstone Plateau. The structure in the area of the former geosyncline,

west of the Beaverhead Range, is characterized by major low-angle thrust faults that trend north to northwest and involve complexly deformed Beltian and Paleozoic rocks, but which are not known to involve basement rocks. The thrust belt is bordered on the west by the Idaho batholith and plunges southeastward beneath the Snake River lava plain, reappearing in southern Idaho and western Wyoming. The Beaverhead and Tendoy ranges, situated along the former geosynclinal hinge zone, appear to be the meeting ground between the cratonic and geosynclinal provinces where both types of structures are present (Scholten, 1967). These two patterns are also reflected in the Beaverhead Formation, which exhibits (1) an early northeast-trending cratonic pattern composed of gently plunging, open folds and high-angle faults associated with the southwest-plunging Blacktail–Snowcrest arch; and (2) a northwest-trending "geosynclinal" pattern characterized by major thrust faults and related folds. Northwest-trending basin-and-range normal faults further complicate the picture. The major structures are shown in Figures 1, 2, and 3.[3]

Two major faults, the Tendoy upthrust and the Medicine Lodge overthrust, were interpreted by Tanner (1963) as being major left-lateral wrench faults on the basis of shear-fracture orientation in Beaverhead quartzite clasts. Objections to this interpretation were noted by Scholten (1964). The objections are strengthened by the fact that 250 fracture orientation measurements taken in various places along the Tendoy fault show a predominance of shear planes dipping 50° SW and 40° NE. This does *not* suggest a subvertical orientation of the intermediate principal stress as required by the wrench concept, but, if anything, a subhorizontal one.

The truncation of northeast trends in the Beaverhead Formation by the Tendoy thrust south of Lima and south and west of Dell (Figure 2) indicates that in this instance the northwest-trending structure was superimposed on the northeast-trending one. Other relationships suggest that this conclusion can be generalized for the entire region. Both field and well data (Gorder, 1960; Scholten, 1967) show that the northwest-plunging Lima anticline, as expressed in the Beaverhead–Aspen unconformity, was superimposed on the preexisting steep southeastern flank of the Blacktail–Snowcrest arch. If the timing had been the reverse, the fold in the unconformity would plunge steeply southeastward. The overturned syncline north of the Lima anticline near Monida evidently resulted from refolding of an earlier syncline whose initial northeast trend can still be seen farther north. Similar overprinting effects can be seen in the Mesozoic rocks south and southwest of Lima and west of Dell (Scholten et al., 1955).

The Medicine Lodge overthrust mass overrode northwest folds and basement upthrusts, and consequently it appears to be the youngest tectogenic feature, although minor subsequent renewal of movement seems to have occurred locally along the upthrusts. The large basin-and-range normal faults are clearly post-tectogenic.

---

[2]Editor's Note: original text has been omitted at this point.

[3]Editor's Note: Figure 2 of the original text has been omitted at this point and figures renumbered; original Table 1 deleted.

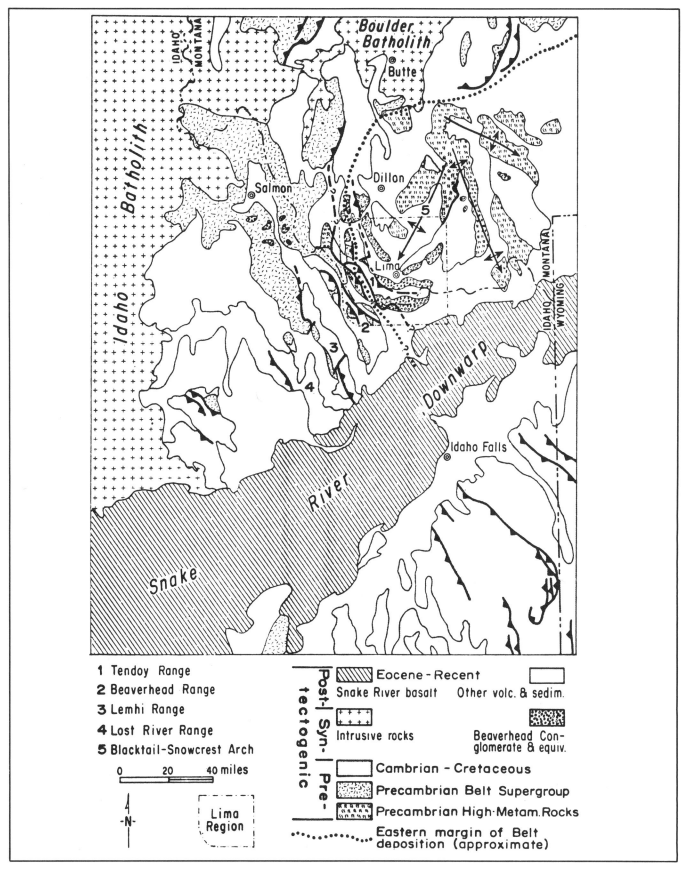

1 Tendoy Range
2 Beaverhead Range
3 Lemhi Range
4 Lost River Range
5 Blacktail-Snowcrest Arch

0   20   40 miles

-N-

Lima Region

Post- | Syn- | Pre- tectogenic

Eocene - Recent
Snake River basalt

Other volc. & sedim.

Intrusive rocks

Beaverhead Con-glomerate & equiv.

Cambrian - Cretaceous

Precambrian Belt Supergroup

Precambrian High-Metam. Rocks

Eastern margin of Belt deposition (approximate)

Figure 1—Generalized geologic map of southwestern Montana and central and southern Idaho, showing location of area studied (Lima region).

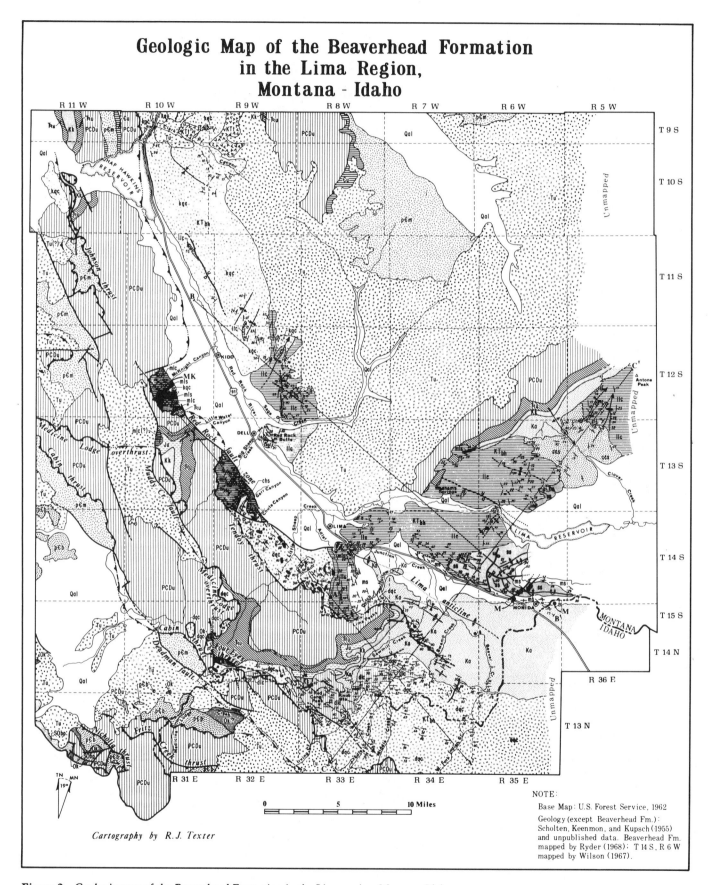

Figure 2—Geologic map of the Beaverhead Formation in the Lima region, Montana-Idaho.

Explanation

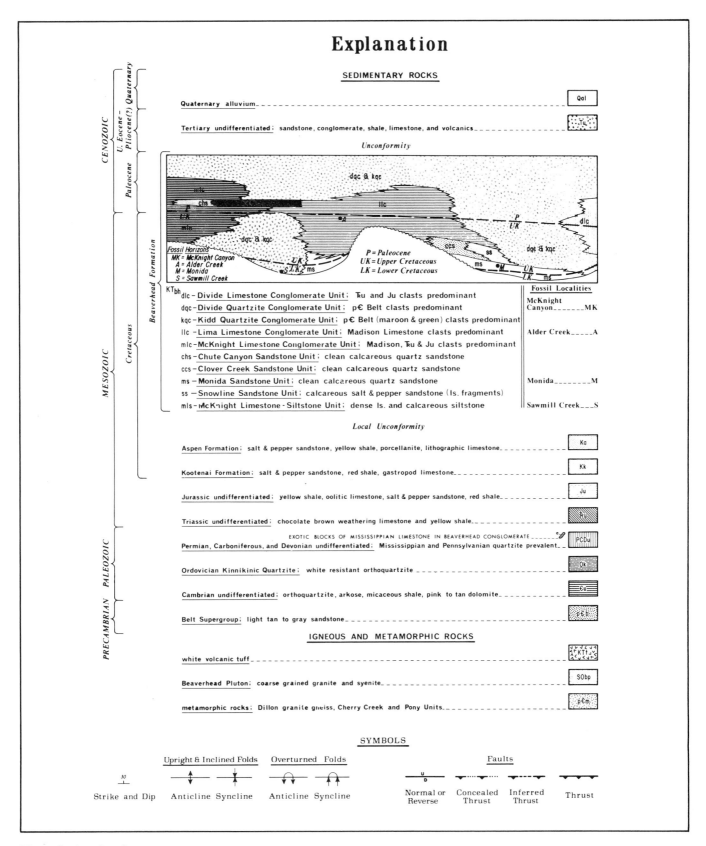

SEDIMENTARY ROCKS

Quaternary alluvium _ `Qal`

Tertiary undifferentiated: sandstone, conglomerate, shale, limestone, and volcanics _ _ _ _ _ _ _ _ _ `Tu`

Unconformity

Fossil Horizons
MK = McKnight Canyon
A = Alder Creek
M = Monida
S = Sawmill Creek

P = Paleocene
UK = Upper Cretaceous
LK = Lower Cretaceous

KTbh
dlc – Divide Limestone Conglomerate Unit; Tu and Ju clasts predominant

dqc – Divide Quartzite Conglomerate Unit; pE Belt clasts predominant

kqc – Kidd Quartzite Conglomerate Unit; pE Belt (maroon & green) clasts predominant

llc – Lima Limestone Conglomerate Unit; Madison Limestone clasts predominant

mlc – McKnight Limestone Conglomerate Unit; Madison, Tu & Ju clasts predominant

chs – Chute Canyon Sandstone Unit; clean calcareous quartz sandstone

ccs – Clover Creek Sandstone Unit; clean calcareous quartz sandstone

ms – Monida Sandstone Unit; clean calcareous quartz sandstone

ss – Snowline Sandstone Unit; calcareous salt & pepper sandstone (ls. fragments)

mls – McKnight Limestone - Siltstone Unit; dense ls. and calcareous siltstone

Fossil Localities
McKnight Canyon _ _ _ _ _ _ _ MK

Alder Creek _ _ _ _ _ _ A

Monida _ _ _ _ _ _ _ _ M

Sawmill Creek _ _ _ S

Local Unconformity

Aspen Formation; salt & pepper sandstone, yellow shale, porcellanite, lithographic limestone _ _ _ _ _ _ _ `Ka`

Kootenai Formation; salt & pepper sandstone, red shale, gastropod limestone _ _ _ _ _ _ _ _ _ _ _ _ _ `Kk`

Jurassic undifferentiated; yellow shale, oolitic limestone, salt & pepper sandstone, red shale _ _ _ _ _ _ _ `Ju`

Triassic undifferentiated; chocolate brown weathering limestone and yellow shale _ _ _ _ _ _ _ _ _ _ _ _ `Ru`

EXOTIC BLOCKS OF MISSISSIPPIAN LIMESTONE IN BEAVERHEAD CONGLOMERATE _ _ _ _ _ _
Permian, Carboniferous, and Devonian undifferentiated; Mississippian and Pennsylvanian quartzite prevalent _ _ `PCDu`

Ordovician Kinnikinic Quartzite; white resistant orthoquartzite _ _ _ _ _ _ _ _ _ _ _ _ _ _ _ _ _ _ _ `Ok`

Cambrian undifferentiated; orthoquartzite, arkose, micaceous shale, pink to tan dolomite _ _ _ _ _ _ _ _ _ `Eu`

Belt Supergroup; light tan to gray sandstone _ `pEb`

IGNEOUS AND METAMORPHIC ROCKS

white volcanic tuff _ `KTt`

Beaverhead Pluton; coarse grained granite and syenite _ `SObp`

metamorphic rocks; Dillon granite gneiss, Cherry Creek and Pony Units _ _ _ _ _ _ _ _ _ _ _ _ _ _ _ `pEm`

SYMBOLS

	Upright & Inclined Folds		Overturned Folds			Faults		
30	Anticline	Syncline	Anticline	Syncline	Normal or Reverse	Concealed Thrust	Inferred Thrust	Thrust
Strike and Dip								

Figure 2—(continued).

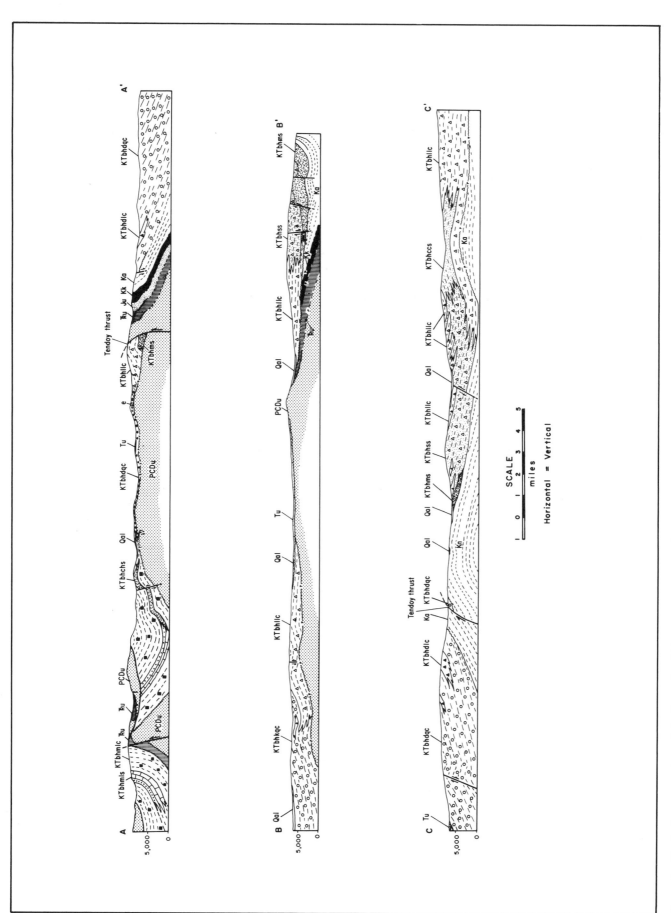

Figure 3—Geologic cross sections of the Beaverhead Formation. Locations shown on Figure 2.

STRATIGRAPHY OF THE BEAVERHEAD FORMATION

Type Section

The type section of the Beaverhead, described by Lowell and Klepper (1953), is located near the mouth of McKnight Canyon in T.12S.,R.10W. (Figure 2). Here the exposed section, approximately 9000 ft (2750 m) thick, is divided into the following members: an upper conglomerate member, a limestone–siltstone member, and a lower conglomerate member. Unfortunately, this section is bounded by faults and thus its temporal relationship with the rest of the Beaverhead cannot be readily documented.

Nomenclature Applied to the Mappable Units

The limestone conglomerate and limestone–siltstone lithologies situated at McKnight Canyon represent only a small part of the Beaverhead Formation. The entire Beaverhead Formation is a complex mosaic comprised of five distinct lithofacies: quartzite conglomerate, limestone conglomerate, sandstone, limestone and siltstone, and exotic limestone blocks.

Several of the major lithofacies are divided into additional compositional entities, thus creating a total of ten distinctly mappable units within the Beaverhead Formation, to which may be added the individual mappable exotic limestone blocks. These subdivisions within the major lithofacies are mutually intertongued and lack significant lateral continuity (1–20 mi [1.5–30 km]). Earlier (Ryder, 1967), the term *lithosome* was applied to them. This terminology does not seem appropriate for subdivisions of a formation and is therefore not used in this paper. The mappable lithologies in the Beaverhead are referred to informally as "units" and have been given specific names. The rock unit boundaries shown on the geologic map (Figure 2) are not everywhere parallel to stratification.

We prefer not to follow Wilson's (1970) assignment of formational status to the Monida Sandstone separate from the Beaverhead Formation, or his earlier practice (Wilson, 1967) of elevating the Beaverhead rocks, including the Monida Sandstone, to group status. There are not merely two, but ten mappable units within the Beaverhead, including several interbedded sandstones rather than just one. To assign formational names to each of these ten lithic variants seems undesirable to us, partly because the formal nomenclature would become complex and partly because on maps the arbitrary cutoffs generally used for formal formations would obscure the true time–space relationships between the lithic units, which interfinger over short distances.

The ten mapped lithic units of the Beaverhead Formation include the three members defined by Lowell and Klepper in McKnight Canyon, as well as the Lima and Monida Formations recognized by Wilson. In order to avoid confusion, the names Lima, Monida, and McKnight are retained. The lithologic units are given the following names: 1. Divide quartzite conglomerate unit; 2. Kidd quartzite conglomerate unit; 3. Divide limestone conglomerate unit; 4. Lima limestone conglomerate unit; 5. McKnight limestone conglomerate unit; 6. Chute Canyon sandstone unit; 7. Clover Creek sandstone unit; 8. Monida sandstone unit; 9. Snowline sandstone unit; and 10. McKnight limestone-siltstone unit. Concise descriptions of the individual units are given in Table 1.

Distribution, Stratigraphic Position, and Thickness of the Units

The map pattern of the lithologic units illustrates the complex intertonguing between adjacent units and their limited geographic range (Figure 2). In general, the contacts between tongues represent the actual locations of the units rather than diagrammatic sketches. Structural symbols are included to distinguish deformational features from depositional tongues and lithic unit contacts.

The stratigraphic position of the lithologic units on each side of the Tendoy thrust is well documented; however, the mutual correlation between these two isolated rock assemblages is difficult. In the footwall block east of Lima, the Lima limestone conglomerate intertongues with the Clover Creek, Snowline, and Monida sandstone units. South of Lima, a quartzite conglomerate unit overlies it, but also intertongues with the Monida sandstone. The limestone conglomerate east and north of Dell is correlated with the Lima unit on lithologic grounds and intertongues with the Kidd quartzite conglomerate. The limestone conglomerate southwest of Dell is correlated with the McKnight limestone conglomerate unit northwest of Dell on lithologic grounds and on the assumption that the Chute Canyon sandstone in the Big Sheep Creek area is equivalent to the limestone-siltstone unit in the McKnight Canyon area. The hanging wall assemblage consists of the intertongued Divide limestone and Divide quartzite conglomerate exposed along the state border and the Continental Divide. The key to the relationship between these structurally isolated sections is that the quartzite conglomerate units in the vicinity of Sawmill Creek and Dutch Hollow, just south of Lima, are identical to those exposed nearby along the Continental Divide. They appear to belong to the same unit. Thus, the thick Divide conglomerate unit in Idaho is probably in part equivalent to the Monida sandstone and its correlatives, including the Lima limestone conglomerate. A generalized interpretive stratigraphic section of the Beaverhead Formation across southwestern Montana and adjacent Idaho is represented in the explanation for Figure 2. For simplicity, the Kidd and Divide quartzite conglomerate units are grouped together.

Regional relationships of the Beaverhead with volcanic rocks near Clark Canyon, in the northwestern part of the study area (Figure 2) remain obscure. Lowell (1965) stated that these white, quartzite clast-bearing tuffs overlie the Beaverhead and older rocks with angular conformity. However, it appears to us that at least some of the volcanic rocks are intertongued with the Beaverhead, a relationship similar to that of syntectonic clastic rocks with the Elkhorn Mountain volcanics to the north (Gwinn and Mutch, 1965) and Livingston volcanics to the east (Roberts, 1963).

The total thickness of the Beaverhead Formation is unknown because of poor exposures and truncated sections. However, complete sections of the individual lithic units were measured in several localities. The range in thickness

Table 1—Intertongued lithologic units of the Beaverhead Formation.[a]

Lithofacies	Unit name		Composition of framework constituents and clast or grain size	Bedding and sedimentary structures
Quartzite conglomerate	Divide quartzite conglomerate	East of meridian through Lima	Belt dominant: pink quartzite with red specks, and blue-black coarse-grained quartzite. Also black chert breccia and dacite porphyry clasts of unknown age. Clasts up to 1.5 m in diameter	Crude horizontal stratification, beds up 1.8 m thick, channels common; interbedded sandstone displays planarlaminae and some low-angle cross-beds; festoons near base; clasts well rounded and imbricated
		West of meridian through Lima	Belt dominant; multicolored quartzites. Also white Kinnikinic quartzite and a few clasts of Dillon granite gneiss and Beaverhead granite-syenite. Clasts up to 1.8 m in diameter	
	Kidd quartzite conglomerate		Belt dominant: coarse, red to purple quartzite, dark maroon quartzite, and light green, siliceous argillite. Clasts up to 2 m in diameter	
Limestone conglomerate	Divide limestone conglomerate		Triassic limestone dominant; Mississippian limestone increases toward top. Clasts up to 2.4 m in diameter bedded sandstone	Crude horizontal stratification, beds up to 6 m thick; inter-displays faint planar laminae; local large-scale foreset bedding, clast imrbication and graded beds; channels common; clasts subangular to rounded
	Lima limestone conglomerate	Dell area	Mississippian limestone dominant; Belt quartzite common. Some Triassic limestone, Phosphoria chert and Quadrant quartzite	
		Lima area	Mississippian limestone dominant, Flathead quartzite in minor amount. Clasts up to 1.6 m in diameter	
	McKnight limestone conglomerate		Mississippian limestone dominant in most places; Triassic limestone locally dominant. Red-maroon Belt quartzite and light-green siliceous argillite at top. Clasts up to 0.8 m in diameter	
Sandstone	Chute Canyon sandstone		Clean, fine-grained, friable, calcareous quartz sandstone, white buff colored with red bands	Planar bedforms; medium-scale low-angle cross-beds predominant some medium-scale festoon cross-beds
	Clover Creek		Clean, medium- to coarse-grained calcareous quartz sandstone, with local lenses rich in Phosphoria chert grains	
	Monida sandstone		Clean, medium- to coarse-grained calcareous quartz sandstone	
	Snowline sandstone		Salt-and-pepper sandstone composed of quartz and limestone grains	
Limestone and Siltstone	McKnight limestone		Lithographic limestone, light tan to pinkish gray, containing some gastropods and oncolites. Light-tan calcareous siltstone with gastropods	Thick limestone beds up to 3 m; massive and medium-bedded siltstone
Exotic limestone blocks			Huge slabs of Mississippian limestone up to 400 m in length	

[a]The arrangement in this table does not imply a stratigraphic succession (compare with Figure 2).

of each unit is shown in Table 2. There appears to be a westward thickening of the Beaverhead to possibly 15,000 ft (4550 m) in Clark County, Idaho, and 10,000 ft (3050 m) near McKnight Canyon.

Nature of the Basal Beaverhead Contact

In most localities, the Beaverhead rests with distinct angular unconformity on rocks as young as the Aspen Formation (Figure 2), which is Lower Cretaceous (Albian). For example, at Clover Creek, the map pattern of the Beaverhead clearly indicates the partial truncation of the Aspen strata. In the northern part of the area, the conclusion is inescapable that the Beaverhead rests on rocks ranging in age from Precambrian to Cretaceous beneath a cover of younger Tertiary sediment. Near the Red Conglomerate Peaks, the evidence is more spectacular in outcrop. Here, the Divide limestone conglomerate, which dips about 40° south, rests on vertical to overturned beds of the Kootenai Formation and on a small overturned anticline that involves the Aspen Formation. A well (Cities Services No. 1 Emerick) drilled in Aspen strata on the axis of the Lima Anticline (Figure 2, T.14S.,R.7W.) also suggests a strong angular unconformity. The trace of the Beaverhead–Aspen contact records a gentle northwest plunge of the axis, in contrast to the 60° dips measured from cores in the Aspen; the regional structural pattern suggests the Aspen dips southeastward (Scholten et al., 1955; Scholten, 1967).

Despite these examples of obvious Beaverhead–Aspen discordance, some localities suggest conformity between these units. Near Sawmill Creek, the quartzite conglomerate of the Divide unit is interbedded with yellow shale and chert-bearing sandstone that are identical to those of the underlying Aspen Formation and that gradually decrease upsection. Where a prominent physical break is absent, the Beaverhead–Aspen contact is arbitrarily drawn below the first upsection appearance of Belt quartzite clasts. The thick Divide quartzite conglomerate unit north of the Montana–Idaho border also appears conformable with the Aspen Formation. Along this 8 mi (13 km) stretch, both the Divide unit and the Aspen dip gently southward and maintain the same structural trends. The argument for conformity in this area is further supported by the transitional zone of Belt quartzite conglomerate, shale, and "salt and pepper" sandstone between the two formations. The arbitrary Beaverhead–Aspen contact was not mapped eastward into Idaho.

The absence of quartzite clasts in the Monida area makes it difficult to distinguish the Beaverhead and Aspen sandstones. Wilson (1967, 1970) mapped the Beaverhead and Aspen on the basis of the high chert content and angular grains in the latter. On the strength of supposedly mid-Campanian pollen dates, he proposed a major unconformity between them, spanning the time period from Cenomanian through Santonian. This evidence is permissive but not conclusive; on the basis of reevaluated pollen assemblages, the sample date could be equally well placed in the mid-Turonian to late Coniacian interval (Ryder and Ames, 1970), with the 300 m of Beaverhead strata beneath the sample accounting for the remaining Cenomanian and early Turonian time. Physical evidence suggests the

Beaverhead–Aspen strata are separated at most by a minor disconformity.

The general picture shows an increasing hiatus at the Beaverhead–Aspen unconformity as the Blacktail–Snowcrest arch is approached. The strong angular unconformity near the flanks of the uplift gradually disappears toward the southeast.

Age and Regional Correlations

A Late Cretaceous to Paleocene age was assigned to the Beaverhead Formation by Yen (*in* Lowell and Klepper, 1953) and Wilson (1967, 1970). An analysis made by Ryder and Ames (1970) of palynomorph assemblages from three localities (Sawmill Creek, Monida, and Alder Creek, marked S, M, and A, respectively, in map and section, Figure 2) suggests that the lower Divide quartzite conglomerate is as old as late Albian and probably no younger than late Cenomanian. The upper Monida sandstone appears to fall in the mid-Turonian to late Coniacian interval; near Monida it forms the flanks of a syncline. The core of this syncline is occupied by the Snowline sandstone (Figure 2) indicating that the latter is younger. The Lima limestone conglomerate at Alder Creek is latest Cretaceous in age.[4] The considerable thickness of quartzite conglomerate overlying the Lima unit in this area suggests that continuation of Beaverhead deposition into early Eocene time cannot be ruled out.

The Beaverhead Formation thus appears to be correlative or partly correlative with a host of clastic rocks in south-central Montana and northwestern Wyoming, including the Frontier Formation, Cody Shale, Eagle Sandstone, Livingston Group, Fort Union Formation, Bacon Ridge Sandstone, Mesa Verde Formation, Meeteetse Formation, and the Harebell and Pinyon conglomerates (McMannis, 1955, 1965; Love, 1956; Hale, 1960; Weimer, 1960). The Beaverhead is unique by comparison to these units because it represents nearly continuous coarse clastic sedimentation from late Albian well into Paleocene or perhaps early Eocene and because Precambrian detritus appears much earlier. Although the Pinyon and Harebell conglomerates contain much Precambrian (Belt) detritus, its first appearance is in late Late Cretaceous time, in contrast to their appearance in late Early Cretaceous time in the Beaverhead.

SEDIMENTOLOGY OF THE BEAVERHEAD FORMATION

The appearance of both well-rounded Belt quartzite clasts and more angular Madison–Thaynes limestone conglomerates within the Beaverhead is suggestive of two fundamentally different source areas: one area that supplied detritus from a distant Precambrian terrane and another that shed limestone detritus from nearby outcrops of Paleozoic

---

[4]Ryder and Ames (1970) assigned a latest Cretaceous to Paleocene age to the Alder Creek assemblage primarily on the basis of the restricted range of *Aquilapollenites* sp. However, K. R. Newman (1969, personal communication), who studied this genus extensively in several Rocky Mountain localities, suggests that the Alder Creek forms probably are latest Cretaceous in age. Tschudy and Leopold (1970) also restrict the range of this genus to the Campanian and Maestrichtian.

Table 2—Type sections of Beaverhead units.[a]

Unit Name	Location of Section	Measured Thickness (m)	Completeness of the Section
Divide quartzite conglomerate	Lima Peaks sec. 4, T. 15 S., R. 8 W.	45	Very incomplete, may attain >4550 m in Clark Co., Idaho
Kidd quartzite conglomerate	McKnight Canyon	300	Very incomplete, may attain >1500 m near Kidd, Montana
Divide limestone conglomerate	Knob Mountain Montana: sec. 26, 35, T. 15 S., R. 8 W. Idaho: sec. 2, 11, T. 13 N. R. 33 E.	1400	Complete section; however, it may be about 1000 m, thicker toward the west
Lima limestone conglomerate	Lima Peaks sec. 4, T. 15 S., R. 8 W.	600	Complete section; however, it is >1500 m north of Snowline, Montana
	Red Rock Butte sec. 10, T. 13 S., R. 9 W.	200	Incomplete, top and base absent
McKnight limestone conglomerate	McKnight Canyon sec. 21, 28, 34, T. 12 S., R. 10 W.	1950	Incomplete, top and base absent; conglomerate probably replaces 420 m McKnight limestone toward the west
	Chute Canyon sec. 4, T. 14 S., R. 9 W.	200	Incomplete, top absent
Chute Canyon sandstone	Chute Canyon sec. 4, T. 14 S., R. 9 W.	90	Incomplete, base absent
Clover Creek sandstone	Clover Divide sec. 13, T. 13 S., R. 6 W.	1100	Incomplete, base absent
Monida sandstone	Dutch Hollow sec. 34, 35, 36, T. 14 S., R. 8 W.	1450	Complete section
	Monida sec. 34, T. 14 S., R. 6 W.	720	Complete section
Snowline sandstone	Snowline sec. 11, 14, 15, T. 14 S., R. 7 W.	1740	Complete section
McKnight limestone and siltstone	McKnight Canyon sec. 28, T. 12 S., R. 10 W.	420	Complete section

[a]The arrangement in this table does not imply a stratigraphic succession.

and Mesozoic formations. Clast imbrication and composition measurements were used to delineate the major directions of conglomerate transport.

Clast Imbrication Measurements

The oblate ellipsoidal clasts within many of the conglomerate units are well suited for imbrication measurements and have been so used, except where there was reason to think that the depositional fabric was altered by soft sediment deformation.

The results of about 1500 imbrication measurements are shown in Figure 4. The generalized transport directions interpreted on this basis are extremely variable (Figure 5).

However, within individual lithologic units, the directions are consistent. These measurements suggest that the various conglomerate units were derived from the following directions: 1. Divide quartzite conglomerate, southwest to south; 2. Kidd quartzite conglomerate, north to northwest; 3. Lima limestone conglomerate, radially off the Blacktail-Snowcrest arch; and 4. McKnight limestone conglomerate, west to northwest.

Clast Composition Measurements

Knowledge of regional outcrop patterns of pre-Beaverhead rocks, of the tectonic evolution of southwestern Montana and central Idaho, and of the clast composition of

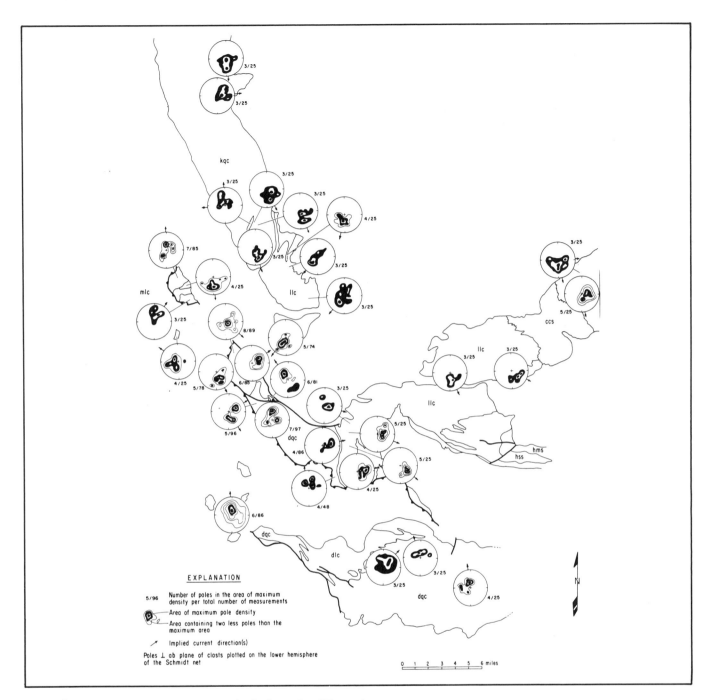

Figure 4—Clast imbrication measurements in the Beaverhead Formation.

the various lithic units within the Beaverhead, restricts the range of possible source areas for these units. To this end, composition data were gathered from measured sections (109 sample sites), local outcrops (255 sample sites), and float-covered slopes (7 sample sites). Fifty clasts were counted at each sample site. Clast percentages at each sample site are shown in Figure 6, and Table 3 summarizes the ranges of clast percentages within each of the Beaverhead units.

Discussion of Source Areas

On the whole, the interpretations as to source areas based on composition counts are consistent with those based on

paleocurrent data. The critical clast lithologies are especially the pre-Dillon amphibolite, the quartzites of the Precambrian Belt Supergroup and the Ordovician Kinnikinic Formation and to a lesser extent the Cambrian Flathead Quartzite and the Ordovician Beaverhead syenite. The various limestone clasts offer little information as to their source area, because bedrock outcrops of these limestones are widespread throughout the region.

The Belt clasts could not have been derived from the east where Cambrian strata rest directly on pre-Beltian crystalline rocks. A southerly source, now buried beneath the Snake River lava, was implied by Tanner (1963). Wilson (1967, 1970), on the basis of clast size and composition data

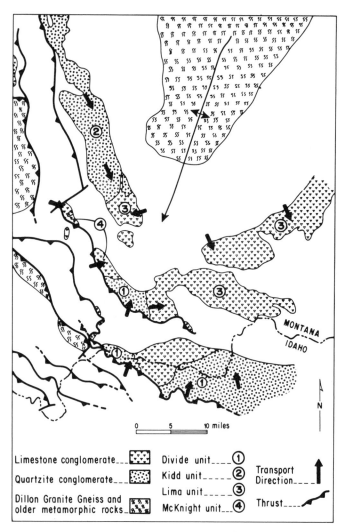

Figure 5—Interpreted transport directions for different units of the Beaverhead Formation, based on clast imbrication data (Figure 4).

the eastern Snake River lava plain, ending with Pennsylvanian and Permian at the southern tips of the ranges; Mesozoic rocks predominate along the projected trend on the opposite side of the plain. To assume both such an abrupt loop in the Belt isopachs and such a major reversal in the regional tectonic plunge to the southeast exactly beneath the lava cover seems fortuitous and strains credulity. The width of the eastern lava plain hardly leaves room for a Belt subcrop large enough to account for so much Belt quartzite conglomerate. As stated by Love and Reed (1968), it would have been necessary for at least 15,000 ft (4550 m) of Mesozoic and Paleozoic rocks to have been eroded before the top of the Belt was exposed.

For these reasons, we do not favor a southerly or southeasterly bedrock source for the Belt clasts in the Beaverhead. On the other hand, a less spectacular Late Cretaceous uplift in the present eastern Snake River plain is not implausible, causing not the exposure of Belt bedrock, but simply the erosion of previously deposited Divide quartzite conglomerate and its transportation eastward into Wyoming and northward into southwestern Montana. This would explain similar clast lithologies found in the Beaverhead, Harebell, and Pinyon conglomerates (D. A. Lindsey, 1969, personal communication), and also the observation by J. D. Love (1968, personal communication) that Triassic and Permian shale fragments are mixed with Belt clasts, since the Divide quartzite conglomerate almost certainly overlies these rocks beneath the eastern Snake River lava. Such a history would also be in accord with the generally northerly paleocurrent data in the Divide unit (Figure 5) and the southeasterly paleocurrent data in the Pinyon and Harebell units (Lindsey, 1970). Axelrod (1968) showed a moderate topographic high in this location in Eocene time on paleobotanical grounds (Figure 7). In general, the concept of recycling of clasts off uplifts of various ages allows us to look for much more distant ultimate sources of Belt cobbles and boulders than if we assumed that the conglomerate units were all first-generation deposits. Each new uplift would renew the energy required to transport such clasts, and the total distance of transport of quartzite clasts could thus be great.

For the ultimate source of the enormous amount of Belt quartzite clasts in the Divide unit, as well as in the Harebell and Pinyon formations, we are forced to look toward the southwest, west, or northwest, that is, toward some part nearer the axis of the former geosyncline. Abundant Kinnikinic and a few Beaverhead syenite clasts in the Divide unit point in the same direction. Kinnikinic bedrock does not occur east of the Tendoy Range, and the syenite intruded the central Beaverhead Range. The nearest Belt outcrops occur in the Beaverhead Range, locally in the northern Tendoy Range, and in the basin between these ranges (Figures 1 and 2). Belt also undoubtedly subcrops beneath the mid-Tertiary to late Tertiary beds in parts of this basin. The basin area, however, was probably denuded tectonically by gliding of Paleozoic carbonate rocks northeastward across the Divide unit and thus could not have been the source of Belt clasts in the Divide unit. The Belt exposures in the Tendoy and Beaverhead ranges are too small to have supplied such quantities of Belt clasts, and the rock types exposed there are different from many of the clasts found in the Divide unit. Moreover, if this nearby area had been a

from the area between Lima and Monida, concludes that the conglomerate clasts were transported to the south and northeast. Love and Reed (1968) postulated a Late Cretaceous northwest extension of the ancestral Tetons into eastern Idaho, referred to as the Targhee uplift and likewise buried beneath the Snake River lava to account for ubiquitous Belt clasts in the Harebell and Pinyon conglomerates of western Wyoming. If some major uplift had indeed existed south or southeast of the Beaverhead area and been eroded down deeply into Belt, it could conceivably have been the source of Belt clasts in the Beaverhead also. Aside from the fact that any hypothesis of large subcrops of Belt beneath the lava cannot readily be tested, two facts militate against such a hypothesis. First, the pre-Paleozoic wedge-out of the Belt, which trends southeastward through southwesternmost Montana (Figure 1), would have to loop abruptly eastward approximately along the Idaho–Montana border, swinging wide around the center of the postulated uplift; several thousand meters of Belt would have to occur within this embayment. Second, at present, ever younger rocks predominate toward the south in the ranges north of

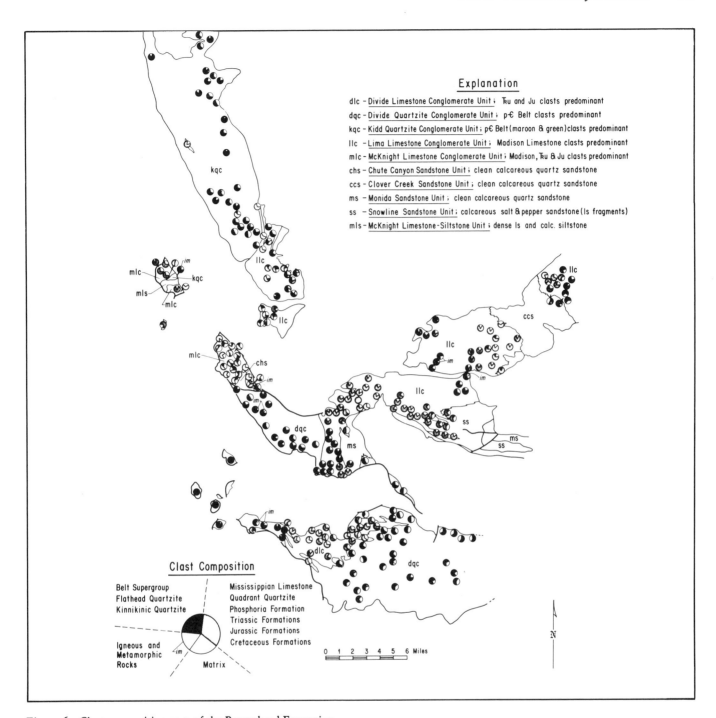

Explanation

dlc – Divide Limestone Conglomerate Unit ; ₸u and Ju clasts predominant
dqc – Divide Quartzite Conglomerate Unit ; p-€ Belt clasts predominant
kqc – Kidd Quartzite Conglomerate Unit ; p€ Belt (maroon & green) clasts predominant
llc – Lima Limestone Conglomerate Unit ; Madison Limestone clasts predominant
mlc – McKnight Limestone Conglomerate Unit ; Madison, ₸u & Ju clasts predominant
chs – Chute Canyon Sandstone Unit ; clean calcareous quartz sandstone
ccs – Clover Creek Sandstone Unit ; clean calcareous quartz sandstone
ms – Monida Sandstone Unit ; clean calcareous quartz sandstone
ss – Snowline Sandstone Unit ; calcareous salt & pepper sandstone (ls fragments)
mls – McKnight Limestone-Siltstone Unit ; dense ls and calc. siltstone

Clast Composition

Belt Supergroup
Flathead Quartzite
Kinnikinic Quartzite

Mississippian Limestone
Quadrant Quartzite
Phosphoria Formation
Triassic Formations
Jurassic Formations
Cretaceous Formations

Igneous and
Metamorphic
Rocks

Matrix

0 1 2 3 4 5 6 Miles

N

Figure 6—Clast composition map of the Beaverhead Formation.

major source, the unit should be far more contaminated with other clasts, especially since much of the present exposure of Belt probably did not occur until post-Laramide time, as a result of erosion following block-faulting of the ranges. Farther to the southwest, Belt rocks crop out in a strip along the west flank of the southern Lemhi Range (Beutner, 1968; see also Figure 1). Again, however, much of the exposure is probably due to erosion in post-Paleocene or at least post-Cretaceous time, and again one would expect a much greater admixture of clasts from the wide belt of Paleozoic rocks, which any streams carrying Belt clasts from this source area would have traversed. For all of the above reasons, we

disagree with Wilson (1967, 1970) who maintains that the quartzite conglomerate units near Lima were primarily derived from nearby uplifts to the southwest.

These considerations leave only one major plausible source for the enormous quantities of Belt quartzite clasts: the large area of Belt exposure on the east flank of the Idaho Batholith, approximately 50 mi (80 km) to the west and northwest (Figure 1). As for the Kidd quartzite conglomerate, paleocurrent and clast composition measurements are mutually consistent and likewise suggest transport from the northwest, presumably from the same source. Maroon quartzite and green argillite clasts in the

Table 3—Clast percentages (%) within Beaverhead Units.

Conglomerate Units	No. of Sample Localities 50 clasts per Locality	Cretaceous, Jurassic and Triassic	Phosphoria and Quadrant	Carboniferous Limestone	Kinnikinic Quartzite	Flathead Quartzite	Belt Quartzite	Igneous and Metamorphic	Matrix
Divide limestone conglomerate	55	25–80[a]	0–12	0–25	0	0	0–2	0	10–30
Divide quartzite conglomerate									
East Idaho	46	0–10	0	0–5	0–20	0	50–85[a]	0–5[b]	15–50
Little Sheep Creek	19	0–5	0	0–5	0–20	0	50–85[a]	0–5[c,d]	15–50
Kidd quartzite conglomerate	31	0	0	0–10	0	0	50–75[a,e]	0	20–50
Lima limestone conglomerate									
Dell area	30	0–20	0–15	30–75[a]	0	0	25–50[a,e]	0	15–20
Lima area	90	0–10	0–30	50–85[a]	0	0–12	0	0–5[f]	15–50
McKnight limestone conglomerate	80	0–30	0–15	10–80[a]	0–2	0	0–30[e]	0	20–50

[a] Dominant lithology (ies).
[b] Dacite porphyry
[c] Contains Dillon Granite Gneiss.
[d] Contains Beaverhead granite and syenite.
[e] Green siliceous argillite from Belt Supergroup.
[f] Contains pre-Dillon metamorphic rocks.

Kidd unit crop out northwest of Dillon, Montana, at least 80 km to the north-northwest. On the basis of literature data and personal observations of the Belt Supergroup in Montana and Idaho (Salmon River Range, Glacier National Park, Missoula area, Beaverhead Range, and Lemhi Range) we conclude that the green argillite is exposed only to the north and northwest of the area of Beaverhead deposition.

A westerly to northwesterly source for much of the Beaverhead is consistent with conclusions expressed in the literature regarding the existence of highlands in central Idaho during the Cretaceous and Tertiary. Thick Cretaceous sediments in Wyoming, Idaho, and Montana require the existence of uplifts to the west (Eardley, 1962; Gilluly, 1963). According to Axelrod (1968), this situation remained essentially the same through Eocene time. On the basis of floral remains from Eocene deposits, he reconstructed a topographic divide in central Idaho rising over 4000 ft (1200 m) above sea level, with local peaks as much as 6000 ft (1800 m) high and an eastern flank that stood about 2000–2500 ft (600–750 m) above the area of Beaverhead deposition (Figure 7).

The conclusion that the source area for the quartzite conglomerates was 50 mi (80 km) or more away is not necessarily inconsistent with the abundance of cobble- and boulder-sized clasts. Cobbles in the Swiss Molasse (Nagelfluh) conglomerates spread 15 km north of the front of the nappes from which they were derived; some that were eroded from the cores of the nappes must have traveled considerably farther. Molasse conglomerate units near Chiasso, northern Italy, contain enormous boulders of Bergell granite and tonalite that were transported southward over a distance of 50 km or more (Gansser, 1962; Scholten, personal observation). In light of such data, a distant source of quartzite clasts need not be rejected on the basis of their sizes, particularly when it is considered that steep slopes could be continuously maintained between contiguous areas of active uplift and subsidence, allowing repeated recycling of the conglomerate.

The limestone conglomerate clasts, on the other hand, appear to have been more locally derived. The angularity and chemical instability of the clasts suggest a transport distance of less than 30 mi (50 km) (Plumley, 1948). Clast imbrication data from the Lima unit suggest that its detritus was radially distributed off the Blacktail–Snowcrest arch. This is consistent with the clast composition, for the Flathead quartzite and Pony (?) amphibolite clasts that occur in it are restricted geographically in bedrock exposure. There is no Flathead in the southern Tendoy Range (Scholten,

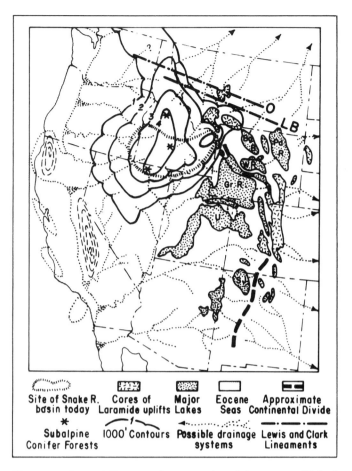

Figure 7—General Eocene paleogeography of the western United States interpreted from paleobotanical data. Redrawn from Axelrod (1968, Figure 10). The dashed lines labeled LB (Lake Basin) and O (Osburn) have been added by us and mark the two most prominent lineaments in the Lewis and Clark zone. Note major topographic highland in central Idaho and moderate local high in eastern Idaho.

1957). A feather-edge of suspected Flathead equivalent occurs locally in the southern Beaverhead Range, thickening westward to a few tens of meters in the southern Lemhi Range (Beutner and Scholten, 1967). However, the nearest and only definitely known Flathead quartzite exposures are in the northwest flank of the Blacktail–Snowcrest arch near Clark County Dam and Ashbough Canyon. They are now less than 15 mi (25 km) north of the nearest occurrence of the Lima unit, and there must have been much closer outcrops at the time of Beaverhead deposition in both flanks of the arch. The amphibolite clasts in the Lima unit probably came from the metamorphic core of the arch, where they are widely exposed (Scholten et al., 1955). The partial contamination of the Lima unit near Dell by clasts of Belt quartzite signifies a zone of mixing between distantly and locally derived debris. The McKnight and Divide limestone conglomerates were derived from local source areas to the southwest and west that have long since been obscured by younger tectonism. Some of the boulders in the Divide limestone conglomerate consist of paper-thin shale that could not have withstood prolonged transport.

Material in the Clover Creek, Snowline, and Monida sandstones probably came from the same source area as that of the Lima limestone conglomerate unit. This is suggested by the intimate intertonguing with the Lima unit and by the presence of limestone fragments in the sandstone. The quartz undoubtedly was largely derived from nearby exposures of the friable Pennsylvanian Quadrant Quartzite, which were also in all likelihood the sources for the Chute Canyon sandstone unit. The small number of "salt and pepper" sandstones (low-rank graywacke) interbedded with the Monida sandstone probably were derived from the same source as that of the Divide quartzite conglomerate; it is similar to the low-rank graywackes interbedded with this conglomerate.

The exotic slabs of Mississippian limestone embedded in the conglomerate units (Figure 2) probably came from nearby exposures of this formation, presumably on the crest and flanks of the Blacktail–Snowcrest arch. Their preponderance southwest of Lima suggests transport more or less down the axial plunge of this arch.

Environment and Mechanism of Deposition

In view of the conclusions regarding the locations and elevations of the source areas and in light of the great thickness of the Beaverhead Formation, it appears that the conglomerates were deposited in a rapidly subsiding basin that lay east of the main central Idaho topographic axis and that was dotted or bordered by local uplifts. Brown (1962) and Axelrod (1968) maintain that Cretaceous and Paleocene floras in the western Great Plains and the vicinity of the present Continental Divide, which is near the area of Beaverhead deposition, lived in a warm, humid lowland within 1000 ft (300 m) of sea level. Paleoclimatic studies by Millison (1964) indicate that there was abundant rainfall in southwestern Montana during Late Cretaceous time.

The association of well-rounded, percussion-marked, imbricated clasts, channeling, and generally well-stratified sandstone indicates that the quartzite conglomerate was deposited from strong turbulent traction currents within high-gradient braided channels. Most of the braided rivers that drained the central Idaho upland were competent enough to transport coarse gravel tens of miles beyond the mountain front, constructing enormous fan complexes on the subsiding alluvial plain. The presence of medium-scale festoon cross-beds in the sandstone of the Divide conglomerate unit near Sawmill Creek indicates that migrating sand waves were at least locally an important part of bedload transport. They probably occupied an earlier, more distal phase of braided-stream development, which was eventually buried by the advancing higher gradient channel deposits.

Similar environments are found today near the foothills of the Himalayas where the Kosi River flows for 75 mi (120 km) over a gently sloping alluvial fan of its own construction in a broad channel. Since 1763, the Kosi has slowly migrated 60 mi (100 km) to the west with concurrent deposition of gravel and sand (Holmes, 1965). L. D. Meckel (1968, personal communication) reports major braided alluviating rivers carrying gravel and sand in southern California, including the Santa Clara River which derives detritus from more than 30 mi (50 km) away.

The sediments composing the limestone conglomerate units were probably deposited as smaller alluvial fans at the base of nearby highlands, extending out of canyon mouths and coalescing to form an integral part of the alluvial plain, much as they do in the present intermontane basins. Perennial braided streams migrating across the fans, in a manner similar to that described by Boothroyd and Ashley (1972) in southeastern Alaska, probably supplied most of the detritus. Like the Beaverhead limestone conglomerate, the Alaskan fans also contain numerous imbricated clasts. The steep, large-scale, foreset bedding observed at the head of Chute Canyon suggests that some of the detritus was deposited in a lacustrine environment and built up Gilbert deltas similar to those in the Pleistocene Lake Bonneville. J. H. Elison (1970, personal communication) so interpreted similar deposits from the Tertiary North Horn conglomerates in central Utah.

The bulk of the sandstone units was probably deposited in braided channel complexes at the distal ends of the alluvial fans, whereas the freshwater limestone near McKnight Canyon accumulated in a lake that resulted from stream ponding. For the exotic blocks of Madison limestone, a mechanism of gravity sliding during periods of torrential rainfall seems the most plausible.

TECTONIC IMPLICATIONS

The conclusions regarding the time and manner of deposition of the Beaverhead Formation and the source areas for the clasts carry major implications with respect to the tectonic development of the Rocky Mountains north of the eastern Snake River plain and south of the Lewis and Clark Lineament. In many places, the structural pattern in this region is still not known in detail, but in general it is characterized by: (1) major refolded overthrusts involving rocks as old as the Belt Supergroup, but so far as known, not the pre-Belt basement rocks; (2) smaller sheets of middle to upper Paleozoic carbonate and shale that rest on rocks ranging in age from Precambrian to Paleocene; (3) cascade-type and recumbent folds that are particularly complex in the carbonates; and (4) range front block faults that caused eastward tilting of the ranges.

Scholten (1968) suggested that the energy for the eastward tectonic transport was gravitative in nature; that is, that the structures east of the Idaho batholith are primarily the result of gliding and cascading off a regional uplift in central Idaho, which had created a temporary reversal of the regional slope of the décollement surfaces. The same mechanism had been suggested for southeastern Idaho and western Wyoming by Eardley (1967) and Crosby (1968). Mudge (1970) applied the principle of downslope gravity gliding to northwestern Montana, but the structural framework there appears to be different in character, resembling rather the framework of the Canadian Rockies, where flat bedding plane thrust faults and mildly folded strata are characteristic and structural complexities, though present, are less ubiquitous. Price and Mountjoy (1970; see also Price, 1969) allege that the thrust surfaces in the Canadian Rockies always had a regional westward gradient.

They think that uplift in the geosyncline created only a topographic highland, which then caused lateral spreading and thrusting against the west-sloping surfaces under the force of gravity. If one accepts this "continental glacier" mechanism for the northern sector, the mechanism in the southern sector may, by analogy, be likened to that of a valley glacier, in which downslope gliding and internal deformation accompany and overshadow lateral spreading.

It is commonly difficult to demonstrate that the slope of a given surface was at one time temporarily the reverse of what it is today, as is here postulated for the region east of the Idaho batholith during Cretaceous and early Tertiary time. In this instance, however, the Beaverhead conglomerates provide at least a strong suggestion. The conclusion that deposition of thousands of meters of coarse quartzite conglomerate in southwestern Montana and adjacent Idaho had already begun in late Early Cretaceous time, and that these conglomerate units were mainly derived from an area some 50 mi (~80 km) to the west or northwest, implies the existence of a high and large uplift at that time, located in the region of the Idaho batholith. At least the northern part of this uplift was already eroded down to the Belt over a wide area by Albian time, and both vertical rise and vigorous erosion must have continued through most or all of the Paleocene and perhaps early Eocene, as evidenced by the enormous volumes of Belt cobbles that continued to pour into southwestern Montana. North of the batholith area, by contrast, conglomerate in the Lower Cretaceous Kootenai Formation of northwestern Montana (Mudge and Sheppard, 1968) and the Upper Jurassic to Lower Cretaceous Kootenay and Blairmore formations of southern Canada (Rapson, 1965) are minor compared to the Beaverhead, especially with regard to the volume of Precambrian clasts. The conglomerate units south of the Snake River plain differ from the Beaverhead in that "Eocambrian" clasts first appear in early Late Cretaceous time and do not become abundant until the late Late Cretaceous (Armstrong, 1968). Thus, the Idaho batholith area may be unique in the eastern Sevier belt and its northern continuation into Canada in that high uplift was in part earlier and in part greater than elsewhere.

This makes it appear plausible that the basal décollement slope from central Idaho into western Montana was indeed temporarily reversed during Cretaceous and early Tertiary time. The geosyncline of central Idaho was intruded by a giant batholith and became the locus of a "meso-undation" in the sense of Van Bemmelen (1966, 1967). Gravitative energy was available not only for eastward fluvial transport of sand- to boulder-sized clasts, but for northeastward downslope tectonic transport. The contrast in the overall structural style between this area and the Rockies of northwest Montana and Canada may find its explanation in the different degrees of uplift in the center of the geosyncline (Scholten, 1970, 1973). Large sheets of bedded rock slid eastward across detachment surfaces within the Belt, possibly aided by high pore pressures related to hydrothermal activity in the intruded area (Goguel, 1969). Regional isostatic subsidence due to the piling up of thrust sheets caused the Beaverhead foredeep to endure as a "sediment trap" (Price, 1969), and it maintained for a long time the topographic slope required for distant fluvial

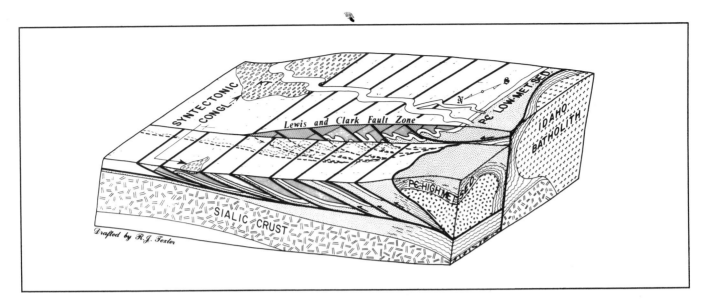

Figure 8—Schematic diagram, looking south, showing interpreted differential uplift north and south of the Lewis and Clark fault zone (compare Figure 7), leading to gravity gliding and cascading down a temporarily reversed décollement slope south of the zone (as postulated by Scholten, 1968) and to gravitative spreading against a west-sloping décollement surface north of the zone (as postulated by Price, 1969; Price and Mountjoy, 1970). The Beaverhead quartzite conglomerates (KTbh) were derived from the uplifted region south of the zone.

transport and the structural slope needed for eastward progression of tectonic gliding. As older "thrusts" were folded and locked, new ones developed to their east. Shallower and more localized sheets of Paleozoic carbonate rock glided or cascaded off local highs into adjacent lows (Scholten and Ramspott, 1968). Finally, in the course of Tertiary time, the undation subsided and the décollement slopes reverted to their original westward gradient.

Scholten (1970, 1973) points out that the line separating the two tectonic provinces in the northern Rockies is the west–northwest-striking zone of transverse faults known as the Lewis and Clark lineament. The western end of this lineament coincides closely with the northern boundary of an Eocene high in central Idaho as contoured topographically by Axelrod (1968) on paleobotanical evidence (Figure 7). This high would seem to be the aftermath of the Cretaceous to Paleocene undation in the geosyncline, and the lineament would seem to have acted as an important hinge in the geosyncline, with the south side raised relative to the north side. Figure 8 illustrates this situation diagrammatically and also shows the different gravitative mechanisms that are thought to have been responsible for the folding and thrusting in the Rockies of the northern United States and southern Canada: gliding south of the lineament as postulated by Scholten (1968), and spreading north of the lineament as postulated by Price (1969).

One further conclusion of a tectonic nature may be drawn from this study of the Beaverhead Formation. In the Tendoy Range, the subhorizontal Medicine Lodge gravity glide sheet, composed of slices of contorted Mississippian limestone and remarkably undeformed shale, overrides Divide quartzite conglomerate containing large slabs of Mississippian limestone (Figure 2). This sheet can have come only from the subsequently downfaulted area between the Tendoy and

Beaverhead ranges, where late Tertiary rocks rest on the Belt Group. It is unlikely that such a small, thin, and locally derived sheet could have pushed conglomerate thousands of meters thick out of the way, gliding across a surface that sloped only a few degrees. The sheet must have ridden out essentially across a gravel-covered land surface, perhaps shoveling only a veneer of gravel out of the way. Furthermore, the Beaverhead horizon, which contains similar exotic slabs near Lima and Dell, appears to be Paleocene, as it is stratigraphically higher than the fossil horizon at Alder Creek. This suggests that the conglomerate beneath the gliding sheet represents only the upper part of the Beaverhead Formation and that it was not the bottom part of a pile of conglomerate originally thousands of meters thick.

Overthrusts or gliding sheets that moved across or not far below a subaerial surface (epiglyptic thrusts) have been described in other areas (Ampferer, 1928; Longwell, 1949; Accordi, 1955; Pierce, 1957; Debelmas and Kerckhove, 1973). This thrust across the fluvial Beaverhead conglomerate appears to be another well-documented example.[5]

ACKNOWLEDGMENTS

The National Science Foundation (Project Grant GP-449) and the Penrose Bequest of the Geological Society of America (Grant No. 1097-66) provided financial support.

Joseph T. Ryder provided invaluable assistance and encouragement during the field seasons 1964, 1965, and

---

[5]Editor's Note: Original text has been omitted at this point.

1966. The assistance of James LeAnderson during the summer of 1965 is appreciated. Discussions with Michael D. Wilson, who worked concurrently on certain aspects of the Beaverhead Formation, were most beneficial. Special thanks are due Eugene G. Williams and Russell R. Dutcher who visited the study area and supplied useful suggestions. The authors are grateful for the valuable critique of this manuscript by D. A. Lindsey and W. K. Hamblin.

REFERENCES CITED

Accordi, B., 1955, Le dislocazioni delle cime (Gipfelfaltungen) delle Dolomiti: Ferrara University Annals series 9, n. 2, p. 65–189.

Ampferer, O., 1928, Die Reliefüberschiebung des Karwendelgebirges: Geologisches Bundesanstalt Jahrbuch, Band 78, p. 241–256.

Armstrong, R. L., 1968, Sevier orogenic belt in Nevada and Utah: Geological Society of America Bulletin, v. 79, p. 429–458.

Axelrod, D. I., 1968, Tertiary floras and topographic history of the Snake River Basin, Idaho: Geological Society of America Bulletin, v. 79, p. 713–734.

Beutner, E. C., 1968, Structure and tectonics of the southern Lemhi Range, Idaho: Ph.D. thesis, Pennsylvania State University, University Park, 106 p.

Beutner, E. C., and R. Scholten, 1967, Probable Cambrian strata in east-central Idaho and their paleotectonic significance: AAPG Bulletin, v. 51, p. 2305–2311.

Boothroyd, J. C., and G. M. Ashley, 1972, Continental sedimentation on tectonically active geosynclinal basin, glacial outwash plain of northeastern Gulf of Alaska: (abs.) AAPG Bulletin, v. 56, p. 604–605.

Brown, R. W., 1962, Paleocene flora of the Rocky Mountains and Great Plains: U.S. Geological Survey Professional Paper 375, 119 p.

Crosby, G. W., 1968, Western Wyoming overthrust belt: AAPG Bulletin, v. 52, p. 2000–2015.

Debelmas, J., and C. Kerckhove, 1973, Large gravity nappes in the French-Italian and French-Swiss Alps, *in* K. A. De Jong, and R. Scholten, eds., Gravity and tectonics: New York, Interscience, p. 189–200.

Eardley, A. J., 1962, Structural geology of North America: New York, Harper and Row, 743 p.

————, 1967, Idaho–Wyoming thrust belt: Its divisions and an analysis of its origin: Intermountain Association of Geologists 15th Annual Field Conference, p. 35–44.

Gansser, A., 1962, The central and southeastern Alps: Guidebook for the International Field Institute: American Geological Institute, p. 47–86.

Gilluly, J., 1963, Tectonic evolution of the western United States: Geological Society London Quarterly Journal, v. 199, p. 133–174.

Goguel, J., 1969, Le rôle de l'eau et de la chaleur dans les phénomènes tectoniques: Revue Géographie Physique et Géologie Dynamique, v. 11, p. 153–164.

Gorder, J. D., 1960, The Lima anticline in the West Yellowstone earthquake area: Billings Geological Society 11th Annual Field Conference Guidebook, p. 265–267.

Gwinn, V. E., and T. A. Mutch, 1965, Intertongued Cretaceous volcanic and non-volcanic strata, Montana: Geological Society of America Bulletin, v. 76, p. 1125–1144.

Hale, L. A., 1960, Annotations to accompany Cretaceous correlation chart: Wyoming Geological Society Guidebook 15th Annual Field Conference, p. 131–136.

Holmes, A., 1965, Principles of physical geology: New York, Ronald Press Company, 1288 p.

Lindsey, D. A., 1970, Facies and paleocurrents in conglomerates of the Harebell Formation and Pinyon Conglomerate, northwestern Wyoming (abs.): Geological Society of America, Abstracts with Programs, Rocky Mountain Section, v. 2, n. 5, p. 341.

Longwell, C. R., 1949, Structure of the northern Muddy Mountain area, Nevada: Geological Society of America Bulletin, v. 60, p. 923–967.

Love, J. D., 1956, Cretaceous and Tertiary stratigraphy of the Jackson Hole area, NW. Wyoming: Wyoming Geological Association Guidebook 11th Annual Conference, p. 76–93.

Love, J. D., and J. C. Reed, 1968, Creation of the Teton landscape; the geologic story of Grand Teton National Park: Grand Teton Natural History Association, 120 p.

Lowell, W. R., 1965, Geologic map of the Bannack-Grayling area, Beaverhead County, Montana: U.S. Geological Survey Miscellaneous Geologic Investigations Map I-433.

Lowell, W. R., and M. R. Klepper, 1953, The Beaverhead Formation, a Laramide deposit in Beaverhead County, Montana: Geological Society of America Bulletin, v. 64, p. 235–244.

McMannis, W. J., 1955, Geology of the Bridger Range, Montana: Geological Society of America Bulletin, v. 66, p. 1385–1430.

————, 1965, Résumé of depositional and structural history of western Montana: AAPG Bulletin, v. 49, p. 1801–1823.

Millison, C., 1964, Paleoclimatology during Mesozoic time in the Rocky Mountain area: Mountain Geologist, v. 1, p. 79–88.

Mudge, M. R., 1970, Origin of the Disturbed Belt in northwestern Montana: Geological Society of America Bulletin, v. 81, p. 377–392.

Mudge, M. R., and R. A. Sheppard, 1968, Provenance of igneous rocks in Cretaceous conglomerates in northwestern Montana, *in* Geological Survey research 1968: U.S. Geological Survey Professional Paper 600-D, p. D137–D146.

Norris, D. K., R. D. Stevens, and R. K. Wanless, 1965, K–Ar age of igneous pebbles in the McDougall–Segur conglomerate, southeastern Canadian Cordillera: Canada Geological Survey Paper 65-25, 11 p.

Pierce, W. G., 1957, Heart Mountain and South Fork thrusts of Wyoming: AAPG Bulletin, v. 41, p. 591–626.

Plumley, W. J., 1948, Black Hills terrace gravels: A study in sediment transport: Journal of Geology, v. 56, p. 526–577.

Price, R. A., 1969, The southern Canadian Rockies and the role of gravity in low-angle thrusting, foreland folding, and the evolution of migrating foredeeps (abs.):

Geological Society of America, Abstracts with Programs for 1969, Pt. 5, Rocky Mountain Section, p. 284–286.

Price, R. A., and E. W. Mountjoy, 1970, Geologic structure of the Canadian Rocky Mountains between Bow and Athabasca Rivers—A progress report, *in* Wheeler, J. O., ed., Structure of the southern Canadian Cordillera: Geological Association Canada Special Paper 6, p. 7–25.

Rapson, J. E., 1965, Petrography and derivation of Jurassic–Cretaceous clastic rocks, southern Rocky Mountains, Canada: AAPG Bulletin, v. 49, p. 1426–1452.

Roberts, A. E., 1963, The Livingston Group of south-central Montana: U.S. Geological Survey Professional Paper 475, p. B86–B92.

Ryder, R. T., 1967, Lithosomes in the Beaverhead Formation, Montana–Idaho: A preliminary report: Montana Geological Society Guidebook 18th Annual Field Conference, p. 63–70.

Ryder, R. T., and H. T. Ames, 1970, The palynology and age of the Beaverhead Formation and their paleotectonic implications in the Lima region, Montana–Idaho: AAPG Bulletin, v. 54, p. 1155–1171.

Scholten, R., 1957, Paleozoic evolution of the geosynclinal margin north of the Snake River Plain, Idaho–Montana: Geological Society of America Bulletin, v. 68, p. 151–170.

———, 1964, Crushed pebble conglomerates: A discussion: Journal of Geology, v. 72, p. 486–489.

———, 1967, Structural framework and oil potential of extreme southwestern Montana: Montana Geological Society Guidebook 18th Annual Field Conference, p. 7–20.

———, 1968, Model for evolution of Rocky Mountains east of Idaho Batholith: Tectonophysics, v. 6, p. 109–126.

———, 1970, Origin of the Disturbed Belt in northwestern Montana: Discussion: Geological Society of America Bulletin, v. 81, p. 3789–3791.

———, 1973, Gravitational mechanisms in the northern Rocky Mountains of the United States, *in* K. A. De Jong and R. Scholten, eds., Gravity and tectonics: New York, Interscience, p. 473–489.

Scholten, R., and L. D. Ramspott, 1968, Tectonic mechanisms indicated by structural framework of central Beaverhead Range, Idaho–Montana: Geological Society of America Special Paper 104, 70 p.

Scholten, R., K. A. Keenmon, and W. O. Kupsch, 1955, Geology of the Lima region, southwestern Montana and adjacent Idaho: Geological Society of America Bulletin, v. 66, p. 345–404.

Tanner, W. F., 1963, Crushed pebble conglomerate of southwestern Montana: Journal of Geology, v. 71, p. 637–641.

Tschudy, B. D., and E. B. Leopold, 1970, Aquilapollenites (Rouse) Funkhouser-selected Rocky Mountain taxa and their stratigraphic ranges, *in* R. M. Kosanke, and A. T. Cross, eds., Symposium on palynology of the Late Cretaceous and early Tertiary: Geological Society of America Special Paper 127, p. 113–168.

Van Bemmelen, R. W., 1966, The structural evolution of the southern Alps: Geological Mijnbouw, v. 45, p. 405–444.

———, 1967, The importance of the geonomic dimensions for geodynamic concepts: Earth Science Review, v. 3, p. 79–110.

Weimer, R. J., 1960, Upper Cretaceous stratigraphy, Rocky Mountain area: AAPG Bulletin, v. 44, n. 1, p. 1–20.

Wilson, M. D., 1967, The stratigraphy and origin of the Beaverhead Group in the Lima area southwestern Montana: Ph.D. thesis, Northwestern University, Evanston, Illinois, 171 p.

———, 1970, Upper Cretaceous–Paleocene synorogenic conglomerates of southwestern Montana: AAPG Bulletin, v. 54, p. 1843–1867.

UPDATE, January 1985

INTRODUCTION

In the 12 years since the publication of the Ryder and Scholten (1973) paper, the geology of the Lima region has been further refined by numerous outcrop investigations and by five holes drilled into the deep subsurface. Many of these outcrop and subsurface investigations have led to the collection and interpretation of new data that are highly relevant to the origin of the Beaverhead Formation. This short note is an attempt to bring our 1973 study of the Beaverhead up to date by integrating it with new geologic investigations of the Lima region and with new concepts on the evolution of the southwest Montana and east-central Idaho thrust belt. The update is not intended to be an exhaustive treatment of post-1973 geologic studies in the Lima region, and thus we cite only those studies thought to be pertinent to the Beaverhead Formation.

STRUCTURAL FRAMEWORK

The overlapping of northwest-trending geosynclinal structures with older northeast-trending cratonic structures in the Beaverhead Formation (Ryder and Scholten, 1973; Scholten, 1967) seems to be consistent with current concepts regarding the tectonic evolution of southwestern Montana. However, the terms geosynclinal and cratonic structures have been modernized and replaced, respectively, by the terms Sevier, Sevier-type, or foreland thrust belt structures and Laramide structures (Beutner, 1977; Hamilton, 1978; Perry et al., 1983). Based on our conclusions regarding the timing of these structural features, Burchfiel and Davis (1975) state that the Lima region appears to be one of the few places in the western Cordillera where the Laramide structures are synchronous with the foreland thrust belt structures.

More serious modifications to the structural framework proposed by Ryder and Scholten (1973) involve the character and origin of the faults identified by us as the Tendoy upthrust and Medicine Lodge overthrust (Figure 2). There is now good evidence to indicate that the Tendoy fault, which we had interpreted as a basement upthrust, has a listric geometry and thus is similar to other low-angle faults of thin-skinned origin in the foreland thrust belt (Skipp and Hait, 1977; Hammons, 1981; Perry et al., 1981, 1983). The Four Eyes, Cabin, Johnson, Fritz Creek, and Nicholia thrusts, located west of the Tendoy thrust and identified on Figure 2, have also been reinterpreted to be listric faults rather than steepening-downward faults (Skipp and Hait, 1977; Dubois, 1982; Scholten, 1982; Skipp, 1984). (Skipp [1984] reinterpreted the Nicholia thrust to be the core of a ramp anticline within the Fritz Creek thrust plate.) According to Perry et al. (1983), the slices of crystalline early Proterozoic basement rocks composing the Cabin and Johnson thrusts have been skimmed off the southwest-plunging nose of the Blacktail–Snowcrest uplift in a manner similar to that proposed by Armstrong and Dick (1974). Scholten (1982) considers all of the listric faults to be ultimately rooted in the basement with the exception of the

Medicine Lodge overthrust[1]. This fault was interpreted by Ryder and Scholten (1973) as a gravity glide sheet that postdated the Tendoy thrust and other northwest-trending thrust faults of the region. Most investigators consider it to be no different in geometry and timing from the rest of the thrusts in the southwest Montana thrust belt (Beutner, 1977; Skipp and Hait, 1977; Ruppel, 1978; Perry et al., 1981, 1983; Skipp, 1984). One of us (Scholten), however, maintains that it overrides the Four Eyes Canyon thrust and that a gravity gliding origin for at least the final emplacement of the Medicine Lodge thrust remains likely.

Deep drilling approximately 8 mi (13 km) southeast of Lima has revealed that the Lima anticline is rootless (Skipp and Hait, 1977). The blind thrusts beneath the Lima anticline are named the Lima thrust system by Perry et al. (1981, 1983) and are considered to be forward imbricate slices of the southwest Montana thrust belt.

Another significant modification to the structural framework of the Lima region as viewed by Ryder and Scholten (1973) is the interpretation by Perry et al. (1981, 1983) that calls for a major low-angle, northwest-dipping Laramide thrust beneath the basement-cored Blacktail–Snowcrest uplift. Perry et al. (1981, 1983) name this fault the sub-Snowcrest Range thrust. Abrupt changes in the thickness of upper Paleozoic rocks, the character of aeromagnetic data, and the depth to crystalline basement between the Blacktail–Snowcrest uplift and the adjacent Centennial basin support this interpretation. The Snowcrest and Greenhorn basement thrusts (Klepper, 1950; Hadley, 1960) are subparallel to and probably merge with the sub-Snowcrest Range thrust at depth (Perry et al., 1983). Perry and co-workers (1981, 1983) further propose that the sub-Snowcrest Range thrust extends southwestward beneath the Tendoy thrust and possibly the Medicine Lodge thrust. Geologic and geophysical studies by Skipp et al. (1983) suggest that the sub-Snowcrest Range thrust extends as far west as the central Beaverhead Mountains.

Except for Scholten's (1982) opinion that final emplacement of the Medicine Lodge thrust sheet occurred by gravity gliding, and that the Tendoy and Four Eyes Canyon thrusts are as deeply rooted in the basement as the Cabin and Johnson thrusts, we agree with the new structural interpretations of the Lima region as presented in this section.

STRATIGRAPHY OF THE BEAVERHEAD FORMATION

Ryder and Scholten (1973) used the name Beaverhead Formation for the complexly intertongued quartzite and limestone conglomerates and equivalent and underlying sandstone units in southwest Montana and east-central Idaho. The most widespread lithologic components of the

---

[1]The term Medicine Lodge overthrust is here restricted to the local, shallow feature initially described by Kirkham (1927) and is not to be confused with Ruppel's (1978) usage of the term "Medicine Lodge thrust system," a regional and more complex feature affecting deeper seated rocks.

Beaverhead were subdivided into informal units. We again use that nomenclature in this paper. Wilson (1970) removed the sandstone-dominated Monida Formation from the Beaverhead and assigned informal names to the numerous intervening conglomerate and sandstone tongues.

Current studies in Beaverhead stratigraphy are in step with Wilson's (1970) approach. Nichols et al. (1985) elevate the Lima limestone conglomerate of Ryder and Scholten (1973) to formation status, calling it the Lima Conglomerate and raised the Beaverhead Formation to group status, a practice followed earlier by Wilson (1967). The McKnight, Divide, and Kidd conglomerate units and the Chute Canyon sandstone unit of Ryder and Scholten (1973) (Figure 2) are combined into one unit by Nichols et al. (1985) and referred to as unnamed conglomerates and sandstones, undivided. Moreover, Nichols et al. (1985) remove the remaining sandstone units (Monida, Snowline, and Clover Creek) of Ryder and Scholten (1973) from the Beaverhead Formation and refer to them as unnamed Cretaceous sandstones, undivided. Sandstone and conglomeratic sandstone beds known locally near Sawmill Creek (Figure 2), which Ryder and Scholten (1973) assign to the oldest part of the Beaverhead, have also been removed from the Beaverhead by Nichols et al. (1985) and referred to as the "beds at Shine Hill" probably equivalent to the Frontier Formation. Perry et al. (1983) also suggest that the beds near Sawmill Creek should be removed from the Beaverhead and assigned to the Frontier Formation. Thus, Nichols and co-workers (1985) apply the term Beaverhead Group mainly to the conglomerate units of Ryder and Scholten (1973).

New palynological data from the Beaverhead Formation, reevaluation of old data, and recent advances in Cretaceous palynomorph biostratigraphy have caused Nichols et al. (1985) to modify the Albian–Paleocene age range assigned by Ryder and Scholten (1973) to the Beaverhead Formation. They find no evidence for rocks of Albian and Paleocene age[2] in the Beaverhead Formation recognized by Ryder and Scholten (1973). Furthermore, they suggest that the Beaverhead of Ryder and Scholten (1973) is interrupted by a major hiatus, as first proposed by Wilson (1967, 1970). Palynomorph assemblages identified by Nichols et al. (1985) indicate a Campanian–Maestrichtian age for the Beaverhead Group as redefined by them and an early–middle Campanian age for their Lima Conglomerate, which is at the base of the Beaverhead. Sandstone units that underlie or are equivalent to the Lima Conglomerate (Monida sandstone of Wilson,

1970; Snowline, Clover Creek, and Monida sandstone units of Ryder and Scholten, 1973) are assigned a Coniacian–Campanian age. The sandstone and conglomeratic sandstone that compose the beds at Shine Hill and that represent the oldest beds in the Beaverhead of Ryder and Scholten (1973) are tentatively assigned a Cenomanian–Turonian age. The unconformity recognized by Nichols et al. (1985), which possibly spans Coniacian–Santonian time, is situated between the undivided sandstones and the beds at Shine Hill.

Additional information on the age of the Beaverhead comes from whole rock K-Ar dates obtained from igneous rocks southwest of Dillon, Montana (Figure 1), as reported by de la Tour-du-Pin (1983). Intrusive rocks dated 75.2–77.6 m.y. old are said to cross cut the formation, implying a pre-late Campanian age for part of the Beaverhead exposed at that locality. Moreover, volcanic rocks were found to lie within the Beaverhead, confirming our earlier opinion (Ryder and Scholten, 1973). Volcanics dated as old as 90.6–92.8 m.y. indicate that the oldest Beaverhead at that locality is pre-late Cenomanian, according to the geologic time scales of Obradovich and Cobban (1975) and the Geological Society of America (1983). If these results are confirmed, a late Albian age for the earliest part of the Beaverhead cannot be excluded.

The Aspen Formation (Albian), the term which Ryder and Scholten (1973) assigned to the chiefly nonmarine clastic units beneath the Beaverhead Formation, has also been subdivided, redated in part, and tentatively renamed. The volcaniclastic-bearing sequence composing the lower half of the Aspen Formation in the outcrop and subsurface of southwest Montana has been reassigned by Schwartz (1982), Perry et al. (1983), and Dyman et al. (1984) to the Blackleaf Formation after a lithologically similar stratigraphic unit of Albian age in north-central Montana (Cobban et al., 1976) and west-central Montana (Gwinn, 1965). V. E. Ames (in Skipp and Hait, 1977) recognizes the bentonitic Vaughn Member of the Blackleaf in deep drill holes southeast of Lima, a unit which Perry et al. (1983) refer to as the upper member of the Blackleaf Formation. Moreover, Perry et al. (1983) assign a black shale unit in the lower part of the Blackleaf Formation to the Flood(?) Member called the Thermopolis(?) Formation by Witkind and Prostka (1980). Turonian to Cenomanian palynomorphs have been identified by Nichols (in Perry et al., 1983) in beds assigned by Ryder and Scholten (1973) to the upper half of the Aspen Formation. These beds of sandstone, siltstone, yellow-weathering shale, and thin limestone situated below the beds at Shine Hill are probable equivalents of the Frontier Formation and have been called "Upper Cretaceous rocks, undivided" by Perry et al. (1983).

The palynologic and radiometric age dates, coupled with deep drilling data and regional stratigraphic investigations, have provided valuable additions to the information reported by us in 1973. At the same time, it is clear that conflicting evidence continues to exist concerning the age of the oldest Beaverhead Formation as used by Ryder and Scholten (1973) and the relationship of the Beaverhead to the underlying rocks. If the reported age assignments are all accepted, the oldest Beaverhead of Ryder and Scholten (1973) in the north is coeval with the Frontier-equivalent

---

[2]Nichols et al. (1985) report that the late Albian–late Cenomanian age assigned by Ryder and Ames (1970) and Ryder and Scholten (1973) to the Beaverhead at the Sawmill Creek locality ("beds at Shine Hill" locality) was based on the presence of palynomorph species, such as *Eucommidites minor* Groot and Penny, which have little biostratigraphic significance. The palynomorph *Tricolpites* sp. identified by Ryder and Ames (1970) from this locality was reidentified by Nichols et al. as *Tricolpites interangulus* Newman, a species having a middle to upper Campanian stratigraphic range. A Paleocene age for the uppermost Beaverhead is ruled out by Nichols et al. (1985) because they find no palynomorphs characteristic of the Maestrichtian *Wodehouseia spinata* Assemblage Zone of Nichols et al. (1982) or younger zones.

beds at Shine Hill of Nichols et al. (1985) (our lowermost part of the Divide quartzite conglomerate unit), and it is also coeval with the unnamed sandstone, siltstone, shale, and limestone unit of Perry et al. (1983) above the Blackleaf Formation in the Lima area.

If Nichols et al. (1985) and Wilson (1967) are correct in interpreting an unconformity between their Beaverhead Group and the beds at Shine Hill, this unconformity should also be present in the equivalent lowermost beds of the Divide quartzite conglomerate unit between the headwaters of Swamp Creek and the eastern edge of the study area in T. 13 N., R. 35 E. (Figure 2). Moreover, this interpretation would require an unconformity near the base of the eastward-tapering tongue of the Divide limestone conglomerate unit (secs. 29, 30, T. 14 N., R. 34 W.) rather than at the base of the Divide quartzite conglomerate unit as shown by Ryder and Scholten (1973) (Figure 3, cross sections A–A' and C–C'). If present, this unconformity must merge with the profound angular unconformity between the Divide limestone conglomerate unit and the upturned Mesozoic–Paleozoic rocks forming the southeast margin of the Blacktail–Snowcrest uplift. Detailed mapping and dating of strata between the southeast part of T. 14 N., R. 33 E. and the southwest part of T. 14 N., R. 34 E. are needed to help resolve whether or not an intraformational unconformity exists in the Divide quartzite conglomerate unit of Ryder and Scholten (1973).

SEDIMENTOLOGY OF THE BEAVERHEAD FORMATION OF RYDER AND SCHOLTEN (1973)

Recent advances in the stratigraphy of lower Paleozoic and Precambrian sedimentary rocks in southwest Montana and central Idaho (Ruppel, 1975; Ruppel et al., 1975; Hobbs, 1980; Dover, 1981; Lopez, 1982; McCandless, 1982) have made it possible for quartzite clasts in the Beaverhead (identified only as Belt quartzites by us in 1973) to be assigned to specific source terranes. For example, Ruppel (in Perry and Sando, 1982) has identified clasts of the Lemhi Group (middle Proterozoic) in the Little Sheep Creek part of the Divide quartzite conglomerate unit, which probably originated in the northern Lemhi Range. Clasts of pink quartzite with red specks described by us in the Idaho part of the Divide quartzite conglomerate unit were probably derived from the southern Beaverhead Mountains and Lemhi Range (Figure 1), specifically the Swauger Formation (middle Proterozoic) (Ruppel, 1975) and/or the upper Proterozoic Wilbert Formation (Ruppel et al., 1975), which was recently assigned on the basis of fossil data to the Lower Cambrian (Derstler and McCandless, 1981; McCandless, 1982). Clasts described as light-green, siliceous argillite by Ryder and Scholten (1973) in the Kidd quartzite conglomerate unit probably were derived from the Apple Creek Formation (Middle Proterozoic) of Ruppel (1975) now exposed in the northern Lemhi Range. According to Skipp et al. (1979), clasts of black chert pebble conglomerate (described as black chert breccia by us in 1973), which locally compose as much as 10% of the Idaho part of the Divide quartzite conglomerate unit, may have been derived from the Mississippian Copper Basin Formation now exposed in

the Pioneer and White Knob mountains of central Idaho (Paull et al., 1972; Skipp and Hait, 1977).

These random observations of quartzite and other resistant clasts in the quartzite conglomerates of the Beaverhead corroborate our earlier interpretation that the ultimate source of many of the quartzite conglomerates was along the east flank of the Idaho batholith. These data also suggest that uplifts in the vicinity of both the Atlanta and Bitterroot lobes of the Idaho batholith (Armstrong et al., 1977) may have been active source terranes for the Beaverhead. More systematic studies that match quartzite clasts to specific source terranes are needed to establish detailed dispersal patterns for the Beaverhead quartzite conglomerates and to further test the long-distance source concept of Ryder and Scholten (1973).

We adhere to our belief that a Late Cretaceous–Paleocene uplift of Beaverhead conglomerate in the vicinity of the eastern Snake River plain caused Precambrian quartzite clasts to be recycled into the Harebell Formation (Upper Cretaceous) and Pinyon Conglomerate (Upper Cretaceous–Paleocene) of northwest Wyoming (Figure 9). Love (1973, 1982; Love and Keefer, 1975) maintains that a much higher uplift in the same general area, named the Targhee uplift (Love and Reed, 1968), caused local removal of all the Mesozoic and Paleozoic rocks thus exposing Precambrian that became the source of the Harebell and Pinyon clasts. He dismisses our interpretation on the grounds that lithologic differences exist between the Beaverhead and Harebell–Pinyon deposits and that large boulders in the latter require a nearby bedrock source. Wiltschko and Dorr (1983) also recognize the existence of the Targhee uplift. In reply, we repeat our previous arguments against it (Ryder and Scholten, 1973) and point out that lithologic differences also exist *within* the Beaverhead and that its large boulders clearly had a distant origin. Ruppel and Lopez (1984) suggest that conglomerates in the Harebell and Pinyon were derived locally from the leading edge of the thrust belt, which according to them swings abruptly eastward through the Centennial Range into the region of the proposed Targhee uplift. This hypothesis is no more appealing to us than the Targhee uplift hypothesis for explaining the origin of conglomerates in the Harebell and Pinyon, because there is no evidence for Precambrian quartzite (allochthonous or autochthonous) in the vicinity of the eastern Snake River plain. We agree with Lindsey (1972), Beutner (1977), and Kraus (1983) that, ultimately, the conglomerates of the Harebell and Pinyon must have had the same distant source as the conglomerates of the Beaverhead.

A detailed examination of the lithology and paleontology of the exotic limestone blocks in the Divide quartzite conglomerate unit near Little Sheep Creek (Figure 2) has enabled Perry and Sando (1982) to correlate them with exposures of the Madison Group in the Tendoy Range. They argue for a derivation of the blocks from a leading thrust edge and their incorporation in the underlying and peripheral Beaverhead Formation prior to the eastward propagation of the Tendoy thrust. We concede that analogy with the setting of exotic blocks in other regions makes it more plausible to relate these limestone blocks to a thrust belt source than to a Blacktail–Snowcrest uplift source

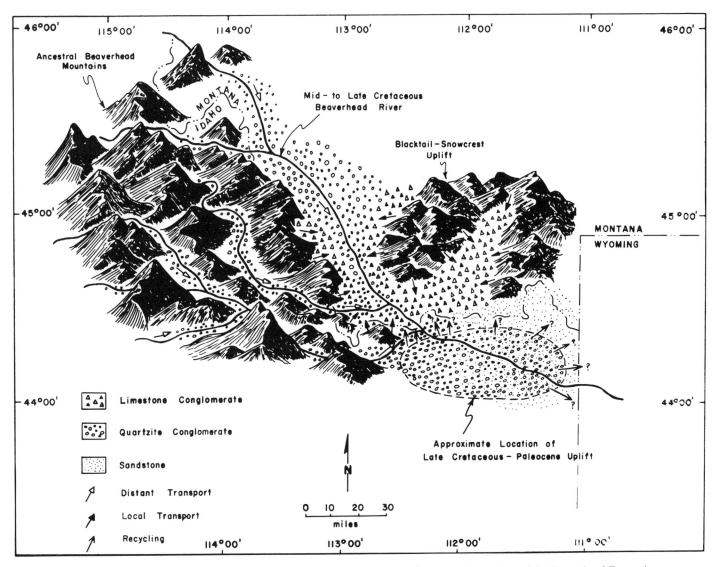

Figure 9—Interpreted paleotopography of southwestern Montana and eastern Idaho during deposition of the Beaverhead Formation.

(Ryder and Scholten, 1973). We still believe, however, that the exotic blocks in the Lima and McKnight conglomerate units were derived from the Blacktail–Snowcrest uplift.

Interpretations regarding the source terranes of the limestone conglomerates have remained basically intact except for recent changes by Haley (1985). Haley (1983a) suggested that the source of the polymict conglomerates composed of Precambrian quartzite and Paleozoic and Mesozoic carbonates and exposed near Dell lies exclusively in the thrust belt to the west, whereas we reasoned in favor of a mixed contribution from the Blacktail–Snowcrest uplift and the thrust belt.

We may have underestimated the influence of the Blacktail–Snowcrest uplift on Beaverhead sedimentation in the area of the Red Conglomerate Peaks (Figures 2 and 9). We considered the Divide limestone conglomerate unit in this area to have been derived from the northwest-trending thrust belt. In view of recent suggestions, however, that the Blacktail–Snowcrest uplift is bounded on the southeast by a major basement thrust, probably continuing southwestward beneath the Tendoy thrust (Perry et al., 1981, 1983), a

derivation from the leading edge of this sub-Snowcrest Range thrust seems possible. However, the Divide quartzite conglomerate unit in this same area clearly has not been derived from that source in view of the absence of Precambrian quartzite in the Tendoy Range and the Blacktail–Snowcrest uplift.

Ryder and Scholten (1973) concluded that the Beaverhead Formation was the product of coalescing alluvial fans and associated braided streams. Haley (1983b) finds evidence of alluvial fans dominated by debris flows and by braided streams in the Beaverhead Formation.

TECTONIC IMPLICATIONS

Ryder and Scholten (1973) link the origin of the voluminous quartzite conglomerates of the Beaverhead Formation with the emplacement and subsequent uplift of the Idaho batholith. Scholten (1982) again points to evidence provided by the Beaverhead for great mid-Cretaceous uplift in the region of the Idaho batholith. He relates this uplift to the rising, in the core of the orogen, of previously subducted thrust slices of continental crust, once they became dissociated from the sinking upper mantle that had initially dragged them downward and westward.

In the area of uplift, rapid erosion into the initially deeply buried middle Proterozoic quartzites apparently generated the boulders and cobbles of the Beaverhead. Expansion of the uplift continued to provide the steep slopes needed for their transport into the foredeep of southwest Montana and east-central Idaho. When thrusting began to affect the area of Beaverhead deposition, eastward-migrating tectonic fronts were created that were capable of recycling clasts in conveyor-belt fashion from one thrust sheet to another. Each of the major thrust sheets contributed some of its own detritus so that the volume of quartzite conglomerate at the final depositional site was at least 100 mi³ (~400 km³). The original volume of quartzite debris was much greater if one accepts our proposal calling for the partial uplift of this depocenter and recycling of the clasts into northwest Wyoming (Figure 9).

Beutner (1977) recognizes the distant source for the quartzite conglomerates in the Beaverhead, but attributes their origin to the stacking of thrust sheets in the southern Lemhi Range and southern Beaverhead Mountains against a westward protruding buttress of the craton in southwest Montana, which includes the Blacktail–Snowcrest uplift. The absence of Beaverhead in these thrusts, however, argues against this, nor do these ranges expose a sufficiently large region of Proterozoic quartzite, even after deep Cenozoic erosion of fault blocks, to account for the enormous volume of Upper Cretaceous quartzite detritus. The only major exposure of quartzite lies to the west-northwest on the flank of the Idaho batholith, and sedimentologic data reported by Ryder and Scholten (1973) tend to support a derivation from this source.

Ruppel and Lopez (1984) have suggested that the Beaverhead quartzite clasts were deposited in an eastward-migrating peripheral "moat" in front of the advancing "Medicine Lodge thrust system" and then rode "piggy back" on smaller thrust sheets into southwest Montana. The Ruppel and Lopez (1984) hypothesis requires extensive tectonic recycling of quartzite clasts and the overriding of these clasts by the advancing Medicine Lodge thrust sheet. Although possible, it is significantly weakened by the absence of conglomerate of the Beaverhead beneath this thrust sheet in the northern Lemhi Range.

Our suggestion of large-scale gravity gliding (Figure 8) off the Idaho batholith (see also Scholten, 1968, 1973) has not been widely accepted, although it has been supported by Chase and Talbot (1973) and Hyndman (1980). It was later abandoned by Scholten (1982). Hamilton (1978) opts for the gravity spreading model of Price and Mountjoy (1970) and Price (1973) to explain the genesis of the foreland thrust belt in the western United States. According to the Price and Mountjoy (1970) and Price (1973) hypothesis, foreland thrusting in the Canadian Rockies was controlled by lateral gravitational spreading of a high mobile core of supracrustal rocks against a westward-dipping basement. Batholithic intrusions such as the Idaho batholith are a major component of the mobile core (Hamilton, 1978). The gravitational spreading hypothesis is also supported by the quantitative model studies of Elliott (1976).

Over the years, various authors have argued in favor of westward continental underthrusting (Scholten, 1957; Misch, 1960; Bally et al., 1966; Burchfiel and Davis, 1972, 1975; Coney, 1972, 1973; Dickinson, 1976; Blackstone, 1977; Lowell, 1977; Bally and Snelson, 1980; Price, 1981). The extension of this concept and its application to the northern Rocky Mountains of the United States by Scholten (1982) may account for the great Cretaceous uplift in central Idaho necessitated by our interpretation of the conglomerate in the Beaverhead Formation.

ACKNOWLEDGMENTS

The writers thank James A. Peterson, the editor of this volume, for his endorsement of our 1973 paper and for providing us with the opportunity to revise some of our earlier thoughts. William J. Perry, Jr. has repeatedly encouraged the senior author (Ryder) to write an update of the Ryder and Scholten (1973) paper and has openly shared with him recent ideas and developments concerning the stratigraphy and structure of the southwest Montana thrust belt.

REFERENCES CITED

Armstrong, R. L., and H. J. B. Dick, 1974, A model for the development of thin overthrust sheets of crystalline rock: Geology, v. 2, n. 1, p. 35–40.

Armstrong, R. L., W. H. Taubeneck, and P. O. Hales, 1977, Rb–Sr and K–Ar geochronology of Mesozoic granitic rocks and their Sr isotopic composition, Oregon, Washington, and Idaho: Geological Society of America Bulletin, v. 88, n. 3, p. 397–411.

Bally, A. W., and S. Snelson, 1980, Realms of subsidence, *in* A. D. Miall, ed., Facts and principles of world petroleum occurrence: Canadian Society of Petroleum Geologists Memoir 6, p. 9–94.

Bally, A. W., P. L. Gordy, and G. A. Stewart, 1966, Structure, seismic data and orogenic evolution of southern Canadian Rocky Mountains: Bulletin of Canadian Petroleum Geology, v. 14, p. 337–381.

Beutner, E. C., 1977, Cause and consequences of curvature in the Sevier orogenic belt, Utah to Montana, *in* E. L. Heisey, D. E. Lawson, E. R. Norwood, P. H. Wach, and L. A. Hale, eds., Rocky Mountain thrust belt geology and resources: Wyoming Geological Association, Montana Geological Society, and Utah Geological Society 29th Field Conference Guidebook, p. 353–365.

Blackstone, D. L., Jr., 1977, The overthrust belt salient of the Cordilleran fold belt—western Wyoming–southeastern Idaho–northeastern Utah, *in* E. L. Heisey, D. E. Lawson, E. R. Norwood, P. H. Wach, and L. A. Hale, eds., Rocky Mountain thrust belt geology and resources: Wyoming Geological Association, Montana Geological Society, and Utah Geological Society 29th Field Conference Guidebook, p. 367–384.

Burchfiel, B. C., and G. A. Davis, 1972, Structural framework and evolution of the Cordilleran orogen, western United States: American Journal of Science, v. 272, p. 97–118.

———, 1975, Nature and controls of Cordilleran orogenesis,

western United States: extension of an earlier hypothesis: American Journal of Science, v. 275-A, p. 363–396.

Chase, R. B., and J. L. Talbot, 1973, Structural evolution of the northeastern border zone of the Idaho batholith, western Montana (abs.): Geological Society of America Abstracts with Programs, v. 5, n. 6, p. 470–471.

Cobban, W. A., C. E. Erdmann, R. W. Lemke, and E. K. Maughan, 1976, Type sections and stratigraphy of the Members of the Blackleaf and Marias River Formations (Cretaceous) of the Sweetgrass Arch, Montana: U.S. Geological Survey Professional Paper 974, 66 p.

Coney, P. J., 1972, Cordilleran tectonics and North America plate motion: American Journal of Science, v. 272, p. 603–628.

———, 1975, Plate tectonics of marginal fold–thrust belts: Geology, v. 1, p. 131–134.

de la Tour-du-Pin, H., 1983, Contribution à l'étude géologique de l'Overthrust Belt du Montana: Ph.D. Thesis, Université du Bretagne Occidentale, Brest, France, 167 p.

Derstler, K., and D. O. McCandless, 1981, Cambrian trilobites from the southern Lemhi Range, Idaho: their stratigraphic and paleotectonic significance (abs.): Geological Society of America Abstracts with Programs, v. 13, n. 4, p. 194.

Dickinson, W. R., 1976, Sedimentary basins developed during the evolution of Mesozoic–Cenozoic arc–trench system in western North America: Canadian Journal of Earth Science, v. 13, p. 1268–1287.

Dover, J. H., 1981, Geology of the Boulder–Pioneer Wilderness study area, Blaine and Custer Counties, Idaho, *in* U.S. Geological Survey and U.S. Bureau of Mines, Mineral Resources of the Boulder–Pioneer Wilderness study area, Blaine and Custer Counties, Idaho: U.S. Geological Survey Bulletin 1497, p. 21–75.

Dubois, D. P., 1982, Tectonic framework of basement thrust terrane, northern Tendoy Range, southwest Montana, *in* R. B. Powers, ed., Geologic studies of the Cordilleran thrust belt, volume 1: Denver, Rocky Mountain Association of Geologists, p. 145–158.

Dyman, T. S., R. Niblack, and J. E. Platt, 1984, Measured stratigraphic section of Lower Cretaceous Blackleaf Formation and lower Upper Cretaceous Frontier Formation (lower part) near Lima, in southwestern Montana: U.S. Geological Survey Open-File Report 84-838, 25 p.

Elliott, D., 1976, The motion of thrust sheets: Journal of Geophysical Research, v. 81, n. 5, p. 949–963.

Geological Society of America, 1983, Decade of North American Geology 1983 Time Scale: Geological Society of America, Boulder, Colorado.

Gwinn, V. E., 1965, Cretaceous rocks of the Clark Fork Valley, central western Montana: Billings Geological Society 16th Annual Field Conference Guidebook, p. 34–57.

Hadley, J. B., 1960, Geology of the northern part of the Gravelly Range, Madison County, Montana: Billings Geological Society 11th Annual Field Conference Guidebook, p. 149–153.

Haley, J. C., 1983a, Depositional processes in Beaverhead Formation, southwestern Montana and northeastern Idaho and their tectonic significance (abs.): AAPG Bulletin, v. 67, n. 8, p. 1340.

———, 1983b, The sedimentology of a synorogenic deposit: the Beaverhead Formation of Montana and Idaho (abs.): Geological Society of America Abstracts with Programs, v. 15, n. 6, p. 589.

———, 1985, Upper Cretaceous (Beaverhead) synorogenic Montana–Idaho thrust belt and adjacent foreland: relationships between sedimentation and tectonism: Ph.D. thesis, Johns Hopkins University, Baltimore, Maryland, 530 p.

Hamilton, W., 1978, Mesozoic tectonics of the western United States, *in* D. G. Howell, and K. A. McDougall, eds., Mesozoic paleogeography of the western United States, Pacific Coast Paleogeography Symposium 2, Society of Economic Paleontologists and Mineralogists, Pacific Section, p. 33–70.

Hammons, P. E., 1981, Structural observations along the southern trace of the Tendoy fault, southern Beaverhead County, Montana, *in* T. E. Tucker, ed., Montana Geological Society field conference and symposium guidebook to southwest Montana: Billings, Montana Geological Society, p. 253–260.

Hobbs, S. W., 1980, The Lawson Cres–dome complex with its detached suprastructure: Geological Society of America Memoir 153, p. 427–443.

Hyndman, D. W., 1980, Bitterroot dome–Sapphire tectonic block, an example of a plutonic core gneiss–dome complex with its detached suprastructure: Geological Society of America Memoir 153, p. 427–443.

Kirkham, V. R. D., 1927, A geologic reconnaissance of Clark and Jefferson and parts of Butte, Custer, Fremont, Lemhi, and Madison counties, Idaho: Idaho Bureau of Mines and Geology Pamphlet 19, 47 p.

Klepper, M. R., 1950, A geologic reconnaissance of parts of Beaverhead and Madison Counties, Montana: U.S. Geological Survey Bulletin 969-C, p. 55–85.

Kraus, M. J., 1983, Late Cretaceous–Early Tertiary exotic quartzite dispersal in northwest Wyoming (abs.): Geological Society of America Abstracts with Programs, v. 15, n. 5, p. 332.

Lindsey, D. A., 1972, Sedimentary petrology and paleocurrents of the Harebell Formation, Pinyon Conglomerate, and associated coarse clastic deposits, northwestern Wyoming: U.S. Geological Survey Professional Paper 734-B, 68 p.

Lopez, D. A., 1982, Stratigraphy of the Yellowjacket Formation of east-central Idaho: U.S. Geological Survey Open-File Report OF81-1088, 218 p.

Love, J. D., 1973, Harebell Formation (Upper Cretaceous) and Pinyon Conglomerate (uppermost Cretaceous and Paleocene), northwestern Wyoming: U.S. Geological Survey Professional Paper 734-A, 54 p.

———, 1982, A possible gap in the western thrust belt in Idaho and Wyoming, *in* R. B. Powers, ed., Geologic studies of the Cordilleran thrust belt, volume 1: Denver, Rocky Mountain Association of Geologists, p. 247–259.

Love, J. D., and J. C. Reed, 1968, Creation of the Teton landscape; the geologic story of Grand Teton National Park: Grand Teton Natural History Association, 120 p.

Love, J. D., and W. R. Keefer, 1975, Geology of sedimentary rocks in southern Yellowstone National Park,

Wyoming: U.S. Geological Survey Professional Paper 729-D, 60 p.

Lowell, J. D., 1977, Underthrusting origin for thrust-foldbelts with application for the Idaho–Wyoming belt, *in* E. L. Heisey, D. E. Lawson, E. R. Norwood, P. H. Wach, and L. A. Hale, eds., Rocky Mountain thrust belt geology and resources: Wyoming Geological Association, Montana Geological Society, and Utah Geological Society 29th Field Conference Guidebook, p. 449–455.

McCandless, D. O., 1982, A reevaluation of Cambrian through Middle Ordovician stratigraphy of the southern Lemhi Range: Master's thesis, Pennsylvania State University, University Park, 157 p.

Misch, P., 1960, Regional structural reconnaissance in central-northeast Nevada and some adjacent areas—observations and interpretations: International Association of Petroleum Geologists, 11th Annual Field Guidebook, p. 17–42.

Nichols, D. J., S. R. Jacobson, and R. H. Tschudy, 1982, Cretaceous palynomorph biozones for the central and northern Rocky Mountain region of the United States, *in* R. B. Powers, ed., Geologic studies of the Cordilleran thrust belt, volume 1: Denver, Rocky Mountain Association of Geologists, p. 721–733.

Nichols, D. J., W. J. Perry, Jr., and J. C. Haley, 1985, Reinterpretation of the palynology and age of Laramide syntectonic deposits, southwestern Montana, and revision of the Beaverhead Group: Geology, v. 13, n. 2, p. 149–153.

Obradovich, J. D., and W. A. Cobban, 1975, A time-scale for the Late Cretaceous of the Western Interior of North America: The Geological Association of Canada Special Paper 13, p. 31–54.

Paull, R. A., M. A. Wolbrink, R. G. Volkmann, and R. L. Grover, 1972, Stratigraphy of Copper Basin Group, Pioneer Mountains, south-central Idaho: AAPG Bulletin, v. 56, n. 8, p. 1370–1401.

Perry, W. J., Jr., and W. J. Sando, 1982, Sequence of deformation of Cordilleran thrust belt in Lima, Montana region, *in* R. B. Powers, ed., Geologic studies of the Cordilleran thrust belt, volume 1: Denver, Rocky Mountain Association of Geologists, p. 137–144.

Perry, W. J., Jr., R. T. Ryder, and E. K. Maughan, 1981, The southern part of the southwest Montana thrust belt: a preliminary re-evaluation of structure, thermal maturation and petroleum potential, *in* T. E. Tucker, ed., Montana Geological Society field conference and symposium guidebook to southwest Montana: Billings, Montana Geological Society, p. 261–273.

Perry, W. J., Jr., B. R. Wardlaw, N. H. Bostick, and E. K. Maughan, 1983, Structure, burial history, and petroleum potential of frontal thrust belt in adjacent foreland, southwest Montana: AAPG Bulletin, v. 67, n. 5, p. 725–743.

Price, R. A., 1973, Large-scale gravitational flow of supracrustal rocks, southern Canadian Rockies, *in* K. A. De Jong, and R. Scholten, eds., Gravity and Tectonics: New York, Wiley-Interscience, p. 491–502.

———, 1981, The Cordilleran fold and thrust belt in the southern Canadian Rocky Mountains, *in* K. R. McClay,

and N. J. Price, eds., Thrust and nappe tectonics: Geological Society of London, Symposium volume, p. 427–448.

Price, R. A., and E. W. Mountjoy, 1970, Geologic structure of the Canadian Rocky Mountains between Bow and Athabasca Rivers—a progress report, *in* J. O. Wheeler, ed., Structure of the southern Canadian Cordillera: Geological Association of Canada Special Paper 6, p. 7–25.

Ruppel, E. T., 1975, Precambrian Y sedimentary rocks in east-central Idaho: U. S. Geological Survey Professional Paper 889-A, p. 1–23.

———, 1978, Medicine Lodge thrust system, east-central Idaho and southwest Montana: U.S. Geological Survey Professional Paper 1031, 23 p.

Ruppel, E. T., and D. A. Lopez, 1984, The thrust belt in southwest Montana and east-central Idaho: U. S. Geological Survey Professional Paper 1278, 41 p.

Ruppel, E. T., R. J. Ross, Jr., and D. Schleicher, 1975, Precambrian Z and Lower Ordovician rocks in east-central Idaho: U. S. Geological Survey Professional Paper 889-B, p. 25–34.

Ryder, R. T., and H. T. Ames, 1970, Palynology and age of Beaverhead Formation and their paleotectonic implications in Lima region Montana–Idaho: AAPG Bulletin, v. 54, n. 7, p. 1155–1171.

Ryder, R. T., and R. Scholten, 1973, Syntectonic conglomerates in southwestern Montana: their nature, origin, and tectonic significance: Geological Society of America Bulletin, v. 84, n. 3, p. 773–796.

Scholten, R., 1957, Deformation of the geosynclinal margin north of the Snake River Plain: Resumenes, 20th International Geologic Congress, p. 75–76.

———, 1967, Structural framework and oil potential of extreme southwestern Montana: Billings, Montana Geological Society Guidebook 18th Annual Field Conference, p. 7–20.

———, 1968, Model for evolution of Rocky Mountains east of Idaho batholith: Tectonophysics, v. 6, p. 108–126.

———, 1973, Gravitational mechanisms in the northern Rocky Mountains of the United States, *in* K. A. De Jong, and R. Scholten, eds., Gravity and Tectonics: New York, Wiley-Interscience, p. 473–489.

———, 1982, Continental subduction in the northern Rockies—a model for back-arc thrusting in the western Cordillera, *in* R. B. Powers, ed., Geologic studies of the Cordilleran thrust belt, volume 1: Denver, Rocky Mountain Association of Geologists, p. 123–136.

Schwartz, R. K., 1982, Broken Early Cretaceous foreland basin in southwestern Montana: sedimentation related to tectonism, *in* R. B. Powers, ed., Geologic studies of the Cordilleran thrust belt, volume 1: Denver, Rocky Mountain Association of Geologists, p. 159–183.

Skipp, B., 1984, Geologic map and cross sections of the Italian Peak and Italian Peak Middle Roadless Areas, Beaverhead County, Montana and Clark and Lemhi Counties, Idaho: U. S. Geological Survey Miscellaneous Field Studies Map MF-1601-B, scale 1:62,500.

Skipp, B., and M. H. Hait, Jr., 1977, Allochthons along the northeast margin of the Snake River Plain, Idaho, *in* E. L. Heisey, D. E. Lawson, E. R. Norwood, P. H. Wach, and L. A. Hale, eds., Rocky Mountain thrust belt geology and resources: Wyoming Geological Association, Montana Geological Society, and Utah Geological Society 29th Field Conference Guidebook, p. 499–515.

Skipp, B., H. J. Prostka, and D. L. Schleicher, 1979, Preliminary geologic map of the Edie Ranch quadrangle, Clark County, Idaho and Beaverhead County, Montana: U.S. Geological Survey Open-File Report 79-845, scale 1:62,500.

Skipp, B., B. K. Lucchitta, and D. M. Kulik, 1983, West-to-east transported Sevier-style thrust plates and north-to-south transported foreland-style reverse faults in the Beaverhead Mountains (abs.): Geological Society of America Abstracts with Programs, v. 15, n. 5, p. 318.

Witkind, I. J., and H. J. Prostka, 1980, Geologic map of the southern part of the Lower Red Rock Lake quadrangle, Beaverhead and Madison Counties, Montana, and Clark County, Idaho: U.S. Geological Survey Miscellaneous Investigations Series Map I-1216.

Wilson, M. D., 1967, The stratigraphy and origin of the Beaverhead Group in the Lima area, southwestern Montana: Ph.D. thesis, Northwestern University, Evanston, Illinois, 171 p.

————, 1970, Upper Cretaceous–Paleocene synorogenic conglomerates of southwestern Montana: AAPG Bulletin, v. 54, n. 10, p. 1843–1867.

Wiltschko, D. V., and J. A. Dorr, Jr., 1983, Timing of deformation in overthrust belt and foreland of Idaho, Wyoming, and Utah: AAPG Bulletin, v. 67, n. 8, p. 1304–1322.

Paleotectonic Implications of Arkose Beds in Park Shale (Middle Cambrian), Bridger Range, South-Central Montana

J. C. Fryxell
ARCO Exploration and Technology Co.
Plano, Texas

D. L. Smith[1]
Montana State University
Bozeman, Montana

The Cambrian System in the Bridger Range of south-central Montana is represented by the Sauk Sequence, transgressive–regressive package of fine-grained clastic and carbonate rocks. The sequence includes the transgressive Flathead Sandstone, followed by three shale–limestone couplets. The tectonic framework, which strongly influenced Cambrian and later Paleozoic sedimentation, was inherited from Precambrian time. The three couplets may be the products of sea level fluctuation and gentle tectonism along Precambrian structural elements.

In south-central Montana, the Park Shale is micaceous shale with siltstone at the base and limestone at the top. In the northern Bridger Range, however, the lower 30 m is interbedded arkosic sandstone and shale. The arkosic sandstone has two major facies, one rich in quartz and orthoclase and the other in arkose, glauconite, and grainstone intraclasts. It is interpreted to have been deposited in a shallow water, near shore, point source (island) environment. Variations in depositional energy and tectonic stability resulted in the two major facies.

The occurrence of Park sandstone beds containing grains of orthoclase and plagioclase and gneissic quartzo-feldspathic pebbles requires the presence of (1) a localized island of Precambrian crystalline rock, which was an erosional remnant that was exposed above the depositional interface through most of the Middle Cambrian, or (2) an island of Precambrian crystalline rock that was exposed by late Middle Cambrian tectonism.

The paucity of conspicuous feldspar grains and quartzo-feldspathic pebbles in the upper Wolsey Shale and Meagher Limestone, and the abundance of basement generated grains in the Park arkosic interval favor the second hypothesis. Stratigraphic and paleocurrent data do not define a source area proximal to the northern Bridger Range. Mineralogy of the arkosic sandstones suggests that the source terrain was located in the basement-cored Precambrian Dillon block.

The northern margin of the Dillon block is defined by the Willow Creek fault zone, which bisects the Bridger Range into northern and southern segments. Rocks north of the fault, which include the Park arkose, appear to be allochthonous. Thus, the original location of the source and depositional environment of the arkosic interval may be to the south and west of the present-day Bridger Range. This may help explain the lack of a readily apparent source area. If a source area for the Park arkose can be defined in the Dillon Block, these sandstones may prove useful in refining estimates of displacement along the Willow Creek fault zone.

INTRODUCTION

In the northern Bridger Range, the lower 30 m of the Park Shale consists of an interbedded arkosic sandstone and shale sequence (Figure 1). These arkosic sandstone beds represent an interesting stratigraphic and sedimentologic problem that has not been addressed until recently (Fryxell, 1982). Middle Cambrian rocks in the study area are dominated by fine-grained clastic and carbonate sediments totaling 498 m in thickness; feldspar-rich intervals occur in rocks of the Flathead, Wolsey, and Park formations. The stratigraphically "high" occurrence of arkose in this sequence and the areally restricted distribution of the sandstones represent local tectonic instability and disruption of stable, marginal shelf conditions during early Park time.

The purpose of this paper is (1) to describe the stratigraphy and areal distribution of the arkose, (2) to develop a depositional model that explains the stratigraphically high and areally restricted occurrence of the arkosic sandstones, and (3) to discuss the Cambrian and Laramide tectonic implications of the arkose.

[1]Deceased July 31, 1983.

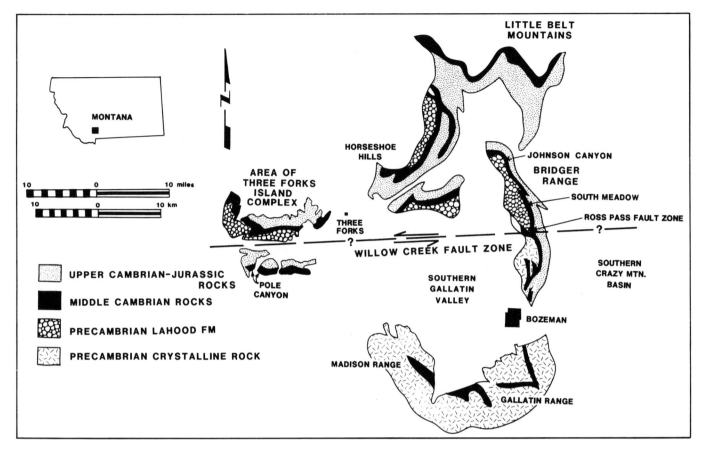

Figure 1—Distribution of Middle Cambrian rocks referred to in the text and their locations relative to the Willow Creek fault zone. Park arkosic sandstones are found in the northern Bridger Range north of the Pass fault.

The study area encompasses the Bridger Range and the Horseshoe Hills to the west (Figure 1). The Park Shale, which forms part of the Middle Cambrian sequence, crops out along the western flank of the Bridger Range and in the Horseshoe Hills. To the south, immediately adjacent to the study area, the Park Shale crops out in the Madison–Gallatin Range complex.

REGIONAL TECTONIC FRAMEWORK AND CONTROLS ON SEDIMENTATION

The primary structural and stratigraphic framework of western Montana evolved in Precambrian Y time (800–1700 m.y. ago) during Belt sedimentation (McMannis, 1965). Evidence suggests that sediments were deposited in the Belt basin, a large epicratonic trough that formed a southeast-trending seaway along the eastern margin of an unidentified cratonic block of continental proportions (Figure 2) (Winston, this volume). The southeasternmost part of the basin appears to have swung inland, forming the east–west-trending Helena embayment. The embayment was bounded to the north by the North American craton and the Garnet line, and to the south by the Precambrian Dillon block and the Perry line (Winston, this volume). The Dillon block is essentially the same tectonic province as the Paleozoic Wyoming shelf of Sloss (1950).

The northern margin of the Dillon block was coincident with the Paleozoic shelf break and is presently defined by

the east–west-trending Perry line. This lineament is thought to represent a major Precambrian structural discontinuity that is still active today (Sloss, 1950; Bonnet, 1979). It is comprised of several en echelon fault zones, including the Willow Creek fault zone (Figure 2). The term "Willow Creek fault zone" is the phrase that is more widely used in the literature for discussing the southern margin of the Helena embayment and its relationship to tectonics and sedimentation within the study area. For this reason, the term "Willow Creek fault zone" and not "Perry line" will be used throughout the rest of this paper.

Arkosic sandstones and conglomerates of the LaHood Formation, which form part of the Belt Supergroup, were deposited immediately north of the Willow Creek fault zone, along the southern margin of the Helena embayment. South of the Willow Creek fault zone Beltian rocks are not present (Sloss, 1950; Bonnet, 1979). Paleozoic formations are thicker in east–west zones that parallel the trend of the Helena embayment and the Willow Creek fault zone (Figure 2). Jurassic and parts of Cretaceous sections are also thicker in east–west zones that parallel the trend of the embayment. These trends suggest that the tectonic framework influencing Paleozoic deposition was inherited from Precambrian time and continued to modify sedimentation into Mesozoic time (Sloss, 1950; McMannis, 1965; Bonnet, 1979). These Precambrian structural elements later influenced the style and location of Laramide deformation and associated igneous activity (Winston, 1983).

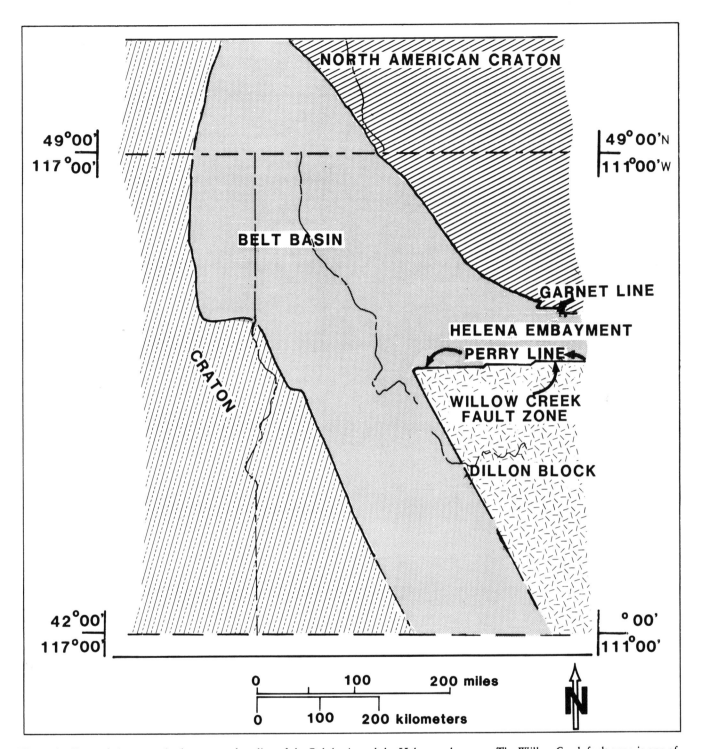

Figure 2—Precambrian tectonic elements and outline of the Belt basin and the Helena embayment. The Willow Creek fault zone is one of the en echelon fault zones comprising the Perry Line. Taken from Winston, this volume. Used with permission of the author.

REGIONAL STRATIGRAPHY AND SEDIMENTATION

The Sauk Sequence (Sloss, 1963) was deposited on the Wyoming shelf and the western edge of the North American craton during the eastward transgression of the Cambrian sea. The sequence is a 450–500 m thick transgressive-regressive package of rocks consisting of a basal sandstone and three shale–limestone repetitions with minor interbedded siltstone and sandstone (Figure 3). These three shale–limestone couplets may reflect the complex interaction of sea level fluctuation and subtle, differential epeirogenic upwarping and downwarping of the Helena embayment and Wyoming shelf.

In the Bridger Range, the Sauk sequence is represented by the Flathead Sandstone, Wolsey Shale, Meagher Limestone,

Park Shale, Pilgrim Limestone, and Snowy Range Formation. The Snowy Range Formation consists of the Dry Creek Shale and Sag Pebble Conglomerate. Intervals rich in quartz and feldspar occur in the basal Flathead Sandstone and middle Wolsey Shale. The most prominent quartzo-feldspathic interval is located in the Park Shale (Figure 3).

Prior to the initial transgression of the Cambrian sea, an irregular post-Belt/pre-Flathead erosional surface developed. Topographic "highs" on this surface consisted of resistant Precambrian metamorphic and Belt sedimentary rocks (Deiss, 1935; Lochman-Balk, 1972; Kennedy, 1980). As the sea transgressed eastward during Flathead time, these topographic highs became "islands," generating point sources of coarse grained-clastic detritus and thus greatly modifying local depositional environments. The influence of the islands on sedimentation is best seen in Flathead or Wolsey strata as abbreviated sections or lithologic variations, such as an arkosic sandstone sequence occurring within a shale. Graham and Suttner (1974) have documented such lithologic and thickness variations in the Flathead Sandstone and Wolsey Shale, near Three Forks Montana, about 64 km to the west of the study area. Hanson (1952), Alexander (1951), Mann (1950), McMannis (1955), and LeBauer (1964) have also noted the influence of Precambrian topographic highs on Cambrian sedimentation.

STRATIGRAPHY OF THE PARK SHALE

Regionally, the Park Shale is a green to maroon, fissile, micaceous shale that averages 66 m in thickness (Hanson, 1952). Occasionally, intercalated siltstone beds occur at its base. The lower contact of the Park Shale is conformable with the underlying Meagher Limestone and the upper contact is conformable with the overlying Pilgrim Limestone. The upper third of the Park contains interbedded limestone that becomes the dominant lithology upward toward the base of the Pilgrim.

In the northern Bridger Range, however, the lower 30 m of the Park consists of interbedded arkosic sandstone, siltstone, and shale (Figure 4). The arkosic interval extends from Morrison Canyon in the north to Dry Creek Canyon in the south (Figure 5). Individual arkosic sandstone beds are grossly lenticular, laterally discontinuous, and characterized by sharp contacts with the enclosing shale. These beds range in thickness from 5 to 17 cm and occur sporadically within the interval. The stratigraphic separation between beds increases upsection (Figure 4).

The arkosic interval of the Park Shale consists of two major facies: a quartz- and orthoclase-rich sandstone and a calcareous, fossiliferous, arkosic, glauconitic sandstone. Both facies are characterized by bright salmon-pink feldspar grains. Major beds (5–17 cm thick) of the quartz and orthoclase-rich facies occur at North Cottonwood Creek, Mill Creek, Hardscrabble Peak, and Sacajawea Peak. There is little or no occurrence of outcrop or float between any of these locations of this facies. In contrast, the glauconite-rich facies occurs only between Hardscrabble and Sacajawea peaks and is restricted to the lower 2.5–3 m of the Park Shale. This facies appears to be more continuous than the quartz- and orthoclase-rich facies. Other glauconite-rich

intervals, lacking the clastic and fossiliferous components of the facies in the Bridger Range, are found to the south in the Gallatin Range and to the west in the Horseshoe Hills (Verral, 1952; McMannis and Chadwick, 1964).

LITHOLOGY OF THE PARK SHALE ARKOSIC INTERVAL

The interpretation of field data was hampered by the sporadic distribution of outcrop along strike in the Bridger Range, the total lack of Cambrian outcrops to the east in the Crazy Mountain basin (Figure 1), and the lack of outcrop for 11 km to the west. Exposures of the Park Shale within the Bridger Range are poor because the shale slumps badly. As a result of limitations, the major characteristics of each facies must be viewed collectively to support the interpretation for that facies.

Quartz- and Orthoclase-Rich Facies

Description

The major constituents of this facies are quartz and orthoclase (Table 1). Minor amounts of microcline and plagioclase are also present; recognizable mafic minerals are rare. Locally developed flat-pebble conglomerate is comprised of laminated to cross-stratified calcareous siltstone. The facies is cemented with sparry calcite. Individual grains range from sand to granules in size. Sorting is poor and grains are angular to rounded.

Orthoclase grains show the greatest amount of alteration and are characterized by a strongly developed iron oxide clay "coating," resulting in the characteristic salmon-pink color of the Park arkose. Plagioclase is characterized by weakly developed, light gray clay coatings. Iron oxide coatings occur either as a halo around a grain edge or as a patch on a fracture or cleavage plane. Microcline is generally unaltered.

Interpretation

Seven major characteristics summarize the petrologic and stratigraphic relationships of the quartz-orthoclase-rich facies: (1) quartzo-feldspathic mineralogy; (2) trace amounts of glauconite and finely comminuted fossil debris; (3) absence of sedimentary structures except for ripple marks on the *tops* of individual beds, local soft sediment deformation due to loading at the base of a bed, and imbrication of lithoclasts; (4) graded bedding at the base of individual beds; (5) abrupt contacts with the enclosing shale; (6) lateral discontinuity of beds; and, (7) abrupt, sporadic, vertical distribution of the arkose.

All of these above characteristics except for the first suggest deposition in a relatively shallow water, wave agitated, near shore point source or island environment (Figure 6). The abrupt contacts of the arkosic sandstone with the enclosing shale, the presence of graded beds, the lateral discontinuity, and the sporadic occurrence of individual sandstone beds all indicate sharp pulses of rapid, coarse-grained sedimentation followed by an abrupt return to conditions favoring mud sedimentation.

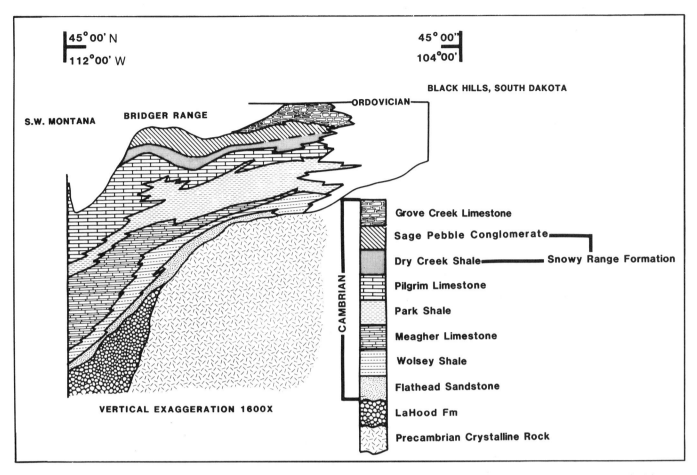

Figure 3—Reconstructed cross-section of the Sauk sequence from southwestern Montana to the Black Hills, South Dakota. Modified from Lochman-Balk, 1972.

Glauconite-Rich Facies

Description

The major constituents of the glauconite-rich facies are glauconite, large grainstone intraclasts, quartz, orthoclase, and calcite cement. Minor amounts of plagioclase and microcline are also present. Recognizable lithoclasts occur occasionally. The grainstone intraclasts are composed mainly of the green algae *Girvanella* and inarticulate brachiopod and trilobite hash. The clastic component of the facies is made up of grains ranging from fine sand to granules in size and is poorly sorted. Grains are very angular to subrounded, indicating a limited distance of transport. The alteration of feldspar in this facies is the same as that in the quartz- and orthoclase-rich facies.

An important, but minor, constituent of the glauconite-rich facies is small, composite pebbles ranging in size from 2 mm to 2 cm. These pebbles are comparable in mineralogy to the clastic component of both facies. In thin section, the pebbles exhibit the same style of alteration as the clastic component of both facies. The grain-to-grain contacts in the pebbles show varying amounts of distintegration, ranging from incipient to complete.

Interpretation

There are seven major characteristics of the glauconite-rich facies that are important to the interpretation of its depositional environment: (1) abundant glauconite and fossil debris, often concentrated in lag deposits; (2) grainstone intraclasts containing inarticulate brachiopod and trilobite debris and the green algae *Girvanella*; (3) greater areal distribution and lateral continuity than the quartz- and orthoclase-rich facies; (4) weakly developed cross stratification; (5) detrital grains; and (6) abundant scour-and-fill sequences.

The abundance of glauconite, fossil debris, and greater carbonate content than in the quartz- and orthoclase-rich facies suggests deposition in a low-energy, shallow water, stable, marginal shelf environment (Figure 7).

The absence of glauconite, large grainstone intraclasts, and abundant fossil debris from the rest of the Park Shale indicate that the intrabasinal factors controlling sedimentation during deposition of the glauconite-rich facies were unique. This facies probably reflects a period of tectonic stability and a decrease in the rate and amount of clastic sediment being introduced into the depositional environment. These changes enhanced glauconite formation,

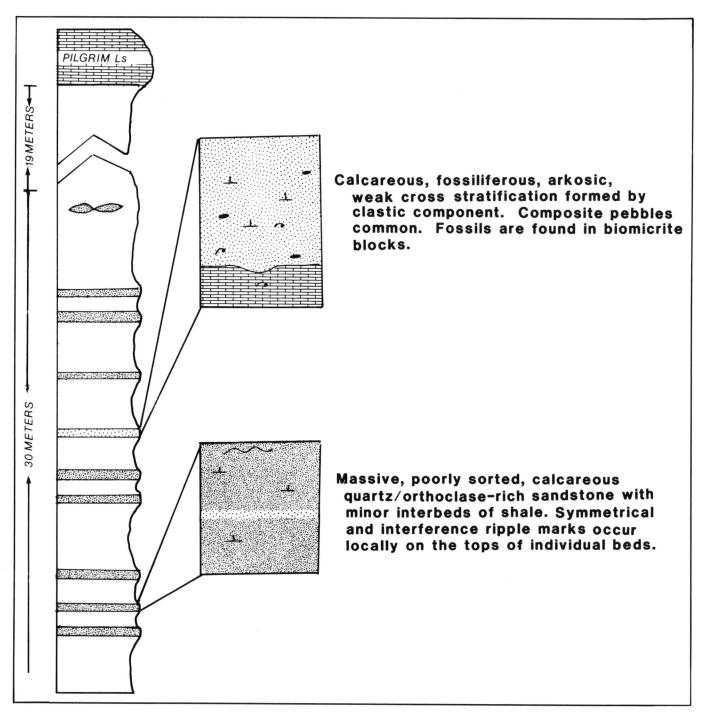

Figure 4—Composite stratigraphic profile of the Park Shale arkosic interval. Unpatterned areas represent shale.

and the development of carbonate buildups, which were apparently the source of the grainstone intraclasts. Disruption of stable, marginal shelf conditions is indicated by the clastic component of the facies, locally developed shelly lag deposits, scour-and-fill sequences, and weakly developed cross stratification. The detrital quartz and feldspar indicate a renewed influence of the clastic point source.

PROVENANCE OF THE PARK SHALE ARKOSIC INTERVAL

Petrographic data suggest that the arkosic sandstones of the Park Shale were derived from a weathered zone of granitic Precambrian basement rock. The paleosol may have developed during periods of subaerial exposure and concomitant shallowing of the surrounding depositional

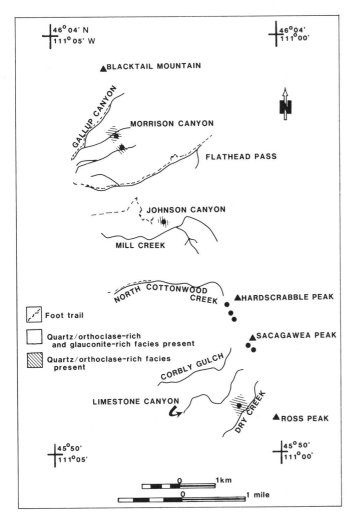

Figure 5—Distribution of the quartz- and orthoclase-rich and glauconite-rich sandstones in the northern Bridger Range. Dots represent measured sections or areas of arkosic sandstone float.

environment. We infer the presence of a weathered zone from the following: (1) an iron–clay–chlorite association; (2) the presence of composite grains and pebbles, which may have been derived from the lower part of a paleosol; and (3) the wide range of grain sizes within the composite pebbles, which is a characteristic of decomposed granite (Pettijohn, 1978).

The arkosic sandstones of the Park were apparently derived from a quartz- and orthoclase-rich granite as indicated by the average mineralogy (Table 1). Quartz: orthoclase ratios vary from 36:41.5 to 45:36. Plagioclase, microcline, chloritized mafics and mica are present in amounts less than 10%. The presence of occasional relict metamorphic textures, such as quartz ribboning and weakly developed schistosity, suggest that the source rock was slightly metamorphosed.

The estimated percentages of quartz, orthoclase, plagioclase, and microcline in the Park arkosic sandstones, compared with those of the LaHood, reveal some significant differences in mineralogy. The differences between the two lie in the sharply contrasting percentages of microcline, orthoclase, and plagioclase. The average mineralogy of the

LaHood at four major outcrops within and adjacent to the study area are listed in Table 1 (McMannis, 1963). Note that orthoclase is much less abundant in the LaHood (0–14.9%) than in the Park. Microcline (6.5–23%) and plagioclase (14.5–40%) are much more abundant in the LaHood. In comparison, the Park arkosic sandstones average 6% microcline and 3–9% plagioclase. These differences in mineralogy suggest that the Precambrian sedimentary rock was *not* the source for the Cambrian sandstones.

PALEOTECTONIC IMPLICATIONS OF THE PARK ARKOSIC INTERVAL

The sedimentologic and stratigraphic data discussed in this paper can be summarized as follows.

1. A major arkosic interval is located in the upper part of a transgressive–regressive package of fine-grained clastic and carbonate sedimentary rocks. Quartz and orthoclase components of the interval were derived from a granitic basement source rock.
2. The Park arkosic interval has an extremely restricted areal distribution, and vertical distribution of arkosic sandstone beds within that interval is abrupt and sporadic.
3. A suite of sedimentary structures and stratigraphic relationships indicates sporadic pulses of rapid sedimentation in a shallow water, marginal shelf environment (usually dominated by fine-grained clastic and carbonate sedimentation), alternating with periods of quiescence.

Any model generated to explain the arkosic interval of the Park Shale must take into account these three major observations. The presence in the Park of sandstone beds derived from a granitic source indicates a period of tectonic instability and disruption of stable, marginal shelf conditions during the deposition of the lower Park. There are two possible hypotheses that may explain the occurrence of arkosic sandstone in the Park Shale: (1) the presence of a tectonically stable island of Precambrian crystalline rock that influenced sedimentation throughout Middle Cambrian time; and (2) exposure of Precambrian source rock in late Middle Cambrian time resulting from movement along Precambrian zones of structural weakness or from the development of a local fault zone.

The first hypothesis requires the presence of a localized, erosional remnant of Precambrian crystalline rock that must have risen at least 200 m above the surrounding Precambrian–Cambrian erosional surface and that was exposed above the depositional interface throughout most of Middle Cambrian time (Figure 8). A height of 200 m represents the combined thickness of Flathead through Meagher rocks. An island with this amount of relief would have produced significant variations in the thickness of Flathead through Meagher strata and in the lithologic types adjacent to the island or islands. Feldspathic sandstone or arkose would have been generated from Flathead to Park time, resulting in a predominantly clastic sequence. However, lithologic and stratigraphic evidence in the study

Table 1—Comparison of average mineral composition (%) of the Park Shale and LaHood Formation arkosic sandstones.[a]

	Quartz	Orthoclase	Microcline	Plagioclase	Chloritized Mafics and Micas
Park arkosic sandstone					
Composite grains	35.0	41.5	6.5	9.5	5.0
Composite pebbles	45.0	36.0	6.0	3.0	7.0
LaHood arkosic sandstone[b]					
Bridger Range	45.0	1.9	13.8	35.6	3.4
Horseshoe Hills	46.5	0.0	12.2	40.0	1.3
Lewis and Clark Caverns area	46.7	14.9	23.0	14.5	8.0
Jefferson Canyon	35.2	2.2	6.5	33.0	11.0

[a]From McMannis (1963).
[b]Average mineral composition of the LaHood arkosic sandstone readjusted to show only percentages of detrital minerals (McMannis, 1963).

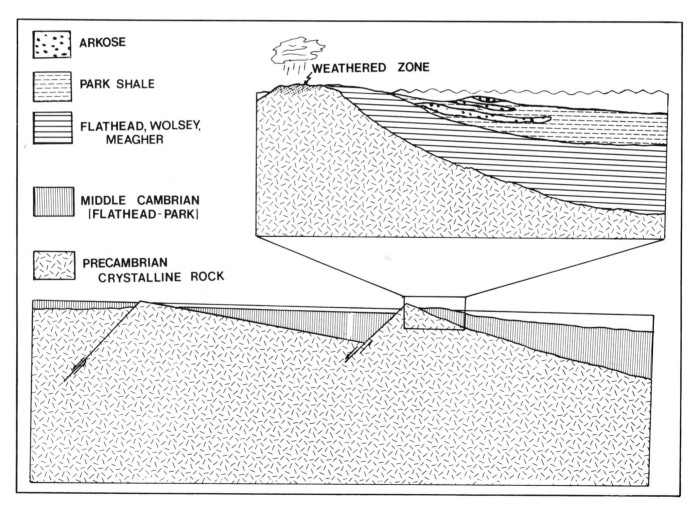

Figure 6—Hypothesized depositional environment of the quartz- and orthoclase-rich facies.

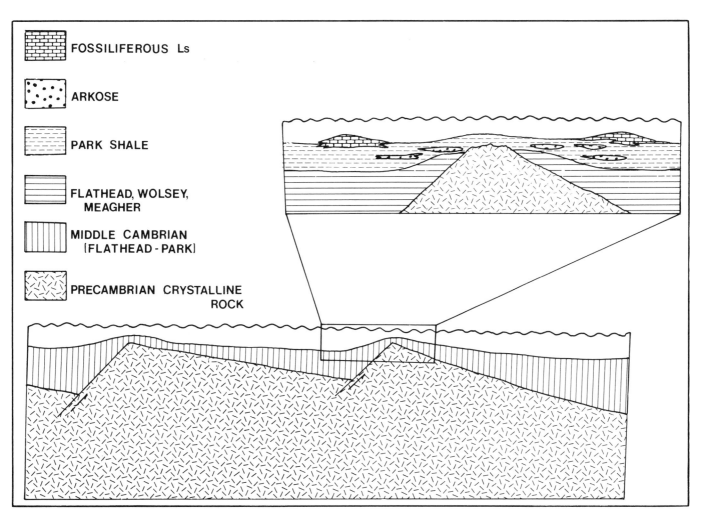

FOSSILIFEROUS Ls

ARKOSE

PARK SHALE

FLATHEAD, WOLSEY, MEAGHER

MIDDLE CAMBRIAN (FLATHEAD - PARK)

PRECAMBRIAN CRYSTALLINE ROCK

Figure 7—Hypothesized depositional environment of the glauconite-rich facies.

area indicates that deposition alternated between fine-grained clastics and carbonates and that the island or islands were apparently not usually within the range of effective wave base. No abbreviated or thinned section is known in the Bridger Range, Horseshoe Hills, or the Gallatin–Madison Range complex to the south (Figure 1).

The erosional remnant hypothesis does not account for the restricted areal occurrence of the arkose nor for the absence of arkose within the Meagher Limestone. Local sea level fluctuations would most likely have produced similar arkosic sandstone sequences in the Horseshoe Hills and Gallatin–Madison Range if Precambrian islands were present. If islands were not present in these areas, then the question must be asked, Why did an island with 200 m relief occur *only* near the northern Bridger Range?

The second hypothesis requires that an island of Precambrian crystalline rock was exposed by late Middle Cambrian reactivation of Precambrian structural zones of weakness, such as the Willow Creek fault zone, or by development of a Cambrian fault or fault zone. The development of such a fault zone may have resulted in a small scale horst-and-graben setting. Repeated differential uplift of these small blocks would be analogous to the floor boards on an old porch, differentially rising and lowering

and thus generating a local clastic source and sediment trap complex (Figure 9).

During periods of "uplift," these islands would form local areas of shoaling. Subaerial weathering of the Precambrian crystalline rock would result in an easily erodable source of coarse-grained clastics in a depositional environment otherwise dominated by fine-grained clastic and carbonate sedimentation. The very fine to granule-sized clastic debris derived from the weathered zone would subsequently be deposited in the small surrounding basins or "grabens" associated with this uplift(s). As the result of intervening periods of subsidence and possible local sea level rise, the islands would be submerged. This would eliminate them as a source of coarse-grained clastics and would allow normal shelf sedimentation to resume. The repeated differential uplift and subsidence of the Precambrian crystalline blocks, possibly combined with local sea level fluctuations, could result in the sporadic, vertical distribution of sandstone beds seen in the Park arkosic interval. The dominance of the quartz- and orthoclase-rich facies throughout the arkosic interval of the Park suggests that "floorboard" tectonism persisted throughout the time of lower Park deposition. The occurrence of the glauconite-rich facies in the lower third of this interval could reflect a period of tectonic stability,

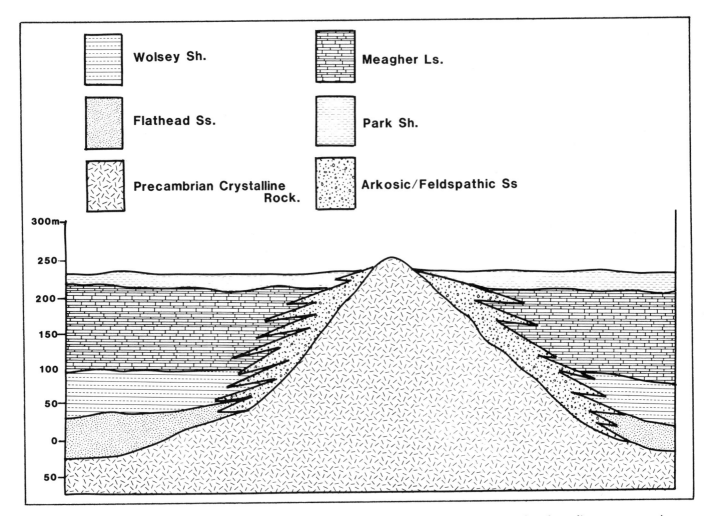

Figure 8—Hypothetical facies relationships between an erosional remnant of Precambrian basement rock and an adjacent transgressive sequence of shale and carbonate sediments in a tectonically stable, marginal shelf setting.

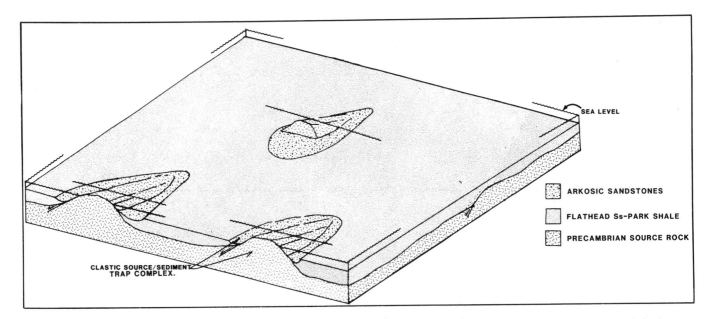

Figure 9—Diagrammatic sketch of the "floor board" tectonic model. Tectonically rejuvenated, tilted fault blocks generated local clastic source and sediment trap complexes. Crests of fault blocks would be subaerially exposed, dip slopes would be at or below sea level.

possible sea level rise, shelf subsidence, or any combination of the three.

The strongest arguments supporting the hypothesis of a tectonically rejuvenated island are (1) the notable absence of quartzo-feldspathic detritus in the upper Wolsey Shale and Meagher Limestone; (2) the occurrence of a major arkosic interval derived from basement rock stratigraphically high in the Sauk Sequence; and (3) the areally restricted occurrence of the arkose.

These lines of evidence strongly suggest that the Park arkosic interval is the product of late Middle Cambrian tectonic rejuvenation of Precambrian crystalline rock. Subsequent weathering and erosion would have provided the coarse-grained clastic detritus.

The word "uplift" is used here only to define a block of crystalline rock that has been uplifted just enough to permit the exposure of basement rock. When Precambrian islands of crystalline rock influenced Cambrian sedimentation during the deposition of the lower Wolsey Shale, the amount of uplift along proposed bounding horst and graben faults must have been at least 140 m, which is the combined thickness of middle and upper Wolsey Shale and Meagher Limestone. The uplift of such blocks would have also resulted in weathering and erosion of Flathead through Meagher rocks. Lithoclasts derived from these formations are rare. This may be due in part to our difficulty in recognizing lithoclasts derived from lower Wolsey and Flathead sandstones. Lithoclasts derived from the Meagher would not have withstood significant distance of transport. These factors bias any estimates of the contribution of these rocks to the resulting influx of clastic detritus.

The idea that Precambrian islands in southwest Montana were structurally controlled during Cambrian time is not new. Graham and Suttner (1974) noted the parallel trend of the Three Forks island complex with the Willow Creek fault zone and the restriction of the islands to the north side of the fault zone. Work by Beutner and Scholten (1967) and Tysdal (personal communication to Lochman-Balk, 1972) document Cambrian tectonism just east of the Idaho and Montana border.

The location, size, and shape of the island or islands from which the arkose was derived is unknown. No abbreviated Cambrian section is known in the area and no paleocurrent or paleosource direction data are available because of present-day slumping of outcrops. However, because the arkose is restricted to the northern Bridger Range and a relatively proximal source area is required for this type of sediment, the sediment-producing island must have been close by.

Several lines of evidence indicate that the source terrane for the Park arkosic sandstones was located in the basement-cored Precambrian Dillon block. The mineralogy of the arkose and the presence of relict, weakly developed metamorphic textures suggest that the source area was predominantly granitic in character and slightly metamorphosed. In positive Paleozoic areas, such as the Wyoming shelf, the basement complex is dominated by granite; however, in negative Paleozoic areas, such as the Helena embayment, metasediments and ultrabasics are dominant (Sloss, 1950). The location of the source terrane in

the Dillon block is further supported by the fact that Belt rocks north of the Willow Creek fault zone are 3000–5000 m thick (McMannis, 1963). Such a thickness precludes the exposure of any Precambrian granitic rock either as an erosional remnant or by faulting within the Helena embayment. Stratigraphic relationships within the arkosic interval place some restrictions as to where in the Dillon block the source may have been located.

The arkosic sandstones of the Park, which are most abundant between Hardscrabble and Sacajawea peaks, thin and pinch-out toward the south and toward the Willow Creek fault zone (Figure 1 and 6). In the southern Bridger Range, which is cored by Precambrian metamorphic rocks, there is no abbreviated section that might indicate the presence of a clastic point source. Abbreviated sections are unknown to the south in the Gallatin–Madison Range complex. Also, Cambrian sections in these areas lack arkosic intervals. These trends are contrary to what would be expected if the source area lay immediately to the south. Other possible locations for the source terrane within the study area would be fortuitous. Because of the lack of Cambrian outcrops to the east and a paucity of subsurface data in the southern Crazy Mountain basin, a southeasterly source can not be ruled out (Figure 1).

An overview of the structural relationships between rocks to the north of the Willow Creek fault zone and those to the south in the foreland suggest an alternative explanation for the lack of a readily apparent source rock for the Park arkosic sandstones. We would like to present the following new model for the source terrane.

The trend of the easternmost disturbed belt reflects the zero isopach line of the Helena embayment (Figure 10). At the southern margin of the embayment, the trend of the disturbed belt is displaced westward along the fault zone. In the vicinity of the Pioneer batholith, south of the Willow Creek fault, the trend continues as the frontal fold and thrust belt of Ruppel et al. (1981).

Robinson (1959) and Woodward (1981), among others, proposed that the rocks of the Belt embayment are subtended by one or more décollement surfaces located at or near the base of the Belt Supergroup. It is thought that these décollement surfaces, which formed during Laramide time, transported Belt and younger rocks eastward into their present-day geographic locations. Right lateral tear thrusting apparently occurred along the Willow Creek fault zone as successive thrust sheets were emplaced in the Helena embayment (Schmidt and Hendrix, 1981). Thus, rocks north of the fault zone, which bisects the Bridger Range, would be allochthonous with respect to those south of the fault. The original geographic location of the source terrane and depositional environment of the Park arkosic sandstones would then be to the south and west of the present-day Bridger Range.

To substantiate or refute this hypothesis, a detailed study comparing the mineralogy of the Park arkose to that of granitic areas in the Dillon block is needed (Figure 10). If a source area for the Park arkosic interval is defined in the block, then these sandstones may prove to be useful stratigraphic marker beds for refining estimates of displacement along the Willow Creek fault zone.

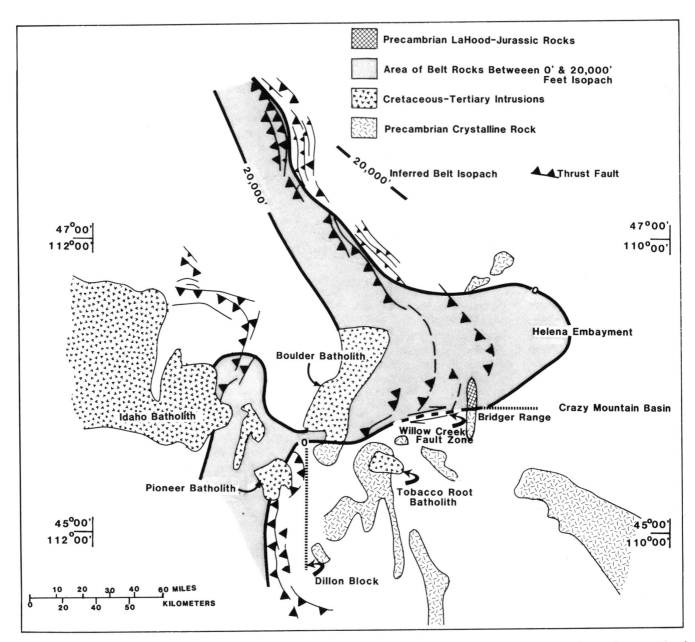

Figure 10—Location map of fold and thrust belt and foreland structures at the end of Laramide deformation. Along the southern margin of the Precambrian Helena embayment, the zero isopach of Belt strata is coincident with the Willow Creek fault zone, as shown by the dashed isopach line. The hachured lines represent the outline of the Dillon Block.

CONCLUSIONS

The arkosic sandstones of the Park Shale represent a period of tectonic instability and possible concomitant fluctuation in sea level during lower Park deposition. The sporadic vertical distribution of beds within the arkosic interval, its limited areal distribution and the occurrence of a major quartzo-feldspathic zone high in the Sauk Sequence can best be explained by the exposure of Precambrian granitic rock resulting from the development of a small-scale horst and graben system. This system would have developed as the result of local tectonic instability.

Evidence suggests that differential uplift and subsidence of these blocks (or "floor board" tectonism) resulted in the generation of local clastic source and sediment trap complexes. Exposure and weathering of the source rock produced an easily erodable source of coarse-grained detritus, the characteristic salmon-pink color of the potassium feldspar, and the composite pebbles of the glauconite-rich facies.

The dominance of the quartz- and orthoclase-rich facies in the arkosic interval of the Park Shale indicates that floor board tectonism dominated during the time of lower Park deposition. The glauconite-rich facies represents a unique period of tectonic stability and possible local sea level rise. The lack of arkose and the increasing amount of intercalated limestone in the upper third of the Park reflects increasing tectonic stability, continued subsidence of the shelf, and transgression of the Cambrian sea, all culminating in deposition of the Pilgrim Limestone.

The Park arkosic interval demonstrates that the long-standing picture of a stable, slowly subsiding shelf throughout the Cambrian is not totally accurate. This interval represents a dramatic facies change and a period of tectonism that has not previously been recognized.

The quartzo-feldspathic mineralogy of the Park arkosic sandstones, the presence of relict, weakly developed metamorphic textures, the evidence for a granitic nature for the positive Paleozoic areas, and the presence of Belt rocks in excess of 4500 m thick all support the hypothesis that the source terrane for the arkosic interval was located in the Dillon block. If a source area for the arkose can be found in the Dillon block, then the Park arkosic sandstones may prove to be useful in refining estimates of displacement along the Willow Creek fault zone.

ACKNOWLEDGMENTS

The senior author greatly appreciates the review of early manuscript drafts by R. J. Woodward. Discussions on regional structure and stratigraphy in south-central Montana with Don Winston, Ed Ruppel, and Christina Lochman-Balk were very helpful and enlightening. Geraldine Sinegal helped type the manuscript. Special appreciation and thanks to Anita Moore-Nall, Sarah Anderson, and Dave Love for their support and encouragement in presenting this paper after Don Smith's death.

REFERENCES CITED

Alexander, R. G., 1951, Geology of the Whitehall area, Montana: Yellowstone–Bighorn Research Project, Contribution 195, 107 p.

Beutner, E. C., and R. Scholten, 1967, Probable Cambrian strata in east-central Idaho and their paleotectonic significance: AAPG Bulletin, v. 55, p. 2305–2309.

Bonnet, A. T., 1979, Lithostratigraphy and depositional setting of the limestone-rich interval of the LaHood Formation (Belt Supergroup), southwestern Montana: Master's thesis, Montana State University, Bozeman, 87 p.

Deiss, C. F., 1935, Cambrian–Algonkian unconformity in western Montana: Geological Society of America Bulletin, v. 46, p. 95–124.

Fryxell, J. C., 1982, Depositional environments and provenance of arkosic sandstone, Park Shale, Middle Cambrian, Bridger Range, southwestern Montana: Master's thesis, Montana State University, Bozeman, 95 p.

Graham, S. A., and L. J. Suttner, 1974, Occurrence of Cambrian islands in southwestern Montana: Mountain Geologist, v. 11, p. 71–84.

Hanson, A. M., 1952, Cambrian stratigraphy in southwestern Montana: Montana Bureau of Mines and Geology Memoir, n. 33, 46 p.

Kennedy, W. B., 1980, The pre-Flathead weathered zone in southwestern Montana and northwestern Wyoming

(abs.): Geological Society of America Abstracts with Programs, v. 6, n. 6, p. 276.

LeBauer, L. R., 1964, Petrology of the Middle Cambrian Wolsey Shale of southwestern Montana: Journal of Sedimentary Petrology, v. 34, p. 503–511.

Lochman-Balk, C., 1972, Cambrian System, *in* Geologic Atlas of the Rocky Mountain Region: Rocky Mountain Association of Geologists, p. 60–75.

McMannis, W. J., 1955, Geology of the Bridger Range area, Montana: Ph.D. thesis, Princeton University, Princeton, New Jersey, 182 p.

———, 1963, LaHood Formation—a coarse facies of the Belt Series in southwestern Montana: Geological Society of America Bulletin, v. 74, p. 407–436.

———, 1965, Resumé of depositional and structural history of western Montana: AAPG Bulletin, v. 49, p. 1801–1823.

McMannis, W. J., and R. A. Chadwick, 1964, Geology of the Garnet Mountain Quadrangle: Montana Bureau of Mines and Geology Bulletin, n. 34, 47 p.

Mann, J. A., 1950, Geology of part of the Gravelly Range area, Montana: Ph.D. thesis Princeton University, Princeton, New Jersey, 170 p.

Pettijohn, F. J., 1975, Sedimentary rocks: New York, Harper Row Publishers, 628 p.

Robinson, G. D., 1959, The disturbed belt in the Sixteen Mile area, Montana: Billings Geological Society Annual Field Conference Guidebook 10, p. 34–40.

Ruppel, E. T., C. A. Wallace, R. G. Schmidt, and D. A. Lopez, 1981, Preliminary interpretation of the thrust belt in southwest and west-central Montana and east-central Idaho: Montana Geological Society Field Conference and Symposium Guidebook 25, p. 139–160.

Schmidt, C. J., and T. E. Hendrix, 1981, Tectonic controls for thrust belt and Rocky Mountain foreland structures in the northern Tobacco Root Mountains–Jefferson Canyon area, Montana: Montana Geological Society Field Conference and Symposium Guidebook 25, p. 167–180.

Sloss, L. L., 1950, Paleozoic sedimentation in Montana area: AAPG Bulletin, v. 34, p. 423–451.

———, 1963, Sequences in the Cratonic Interior of North America: Geological Society of America Bulletin, v. 74, p. 93–114.

Verral, P., 1955, Geology of the Horseshoe Hills area, Montana: Ph.D. thesis, Princeton University, Princeton, New Jersey, 261 p.

Winston, D., 1983, Middle Proterozoic Belt basin syndepositional faults and their influence on Phanerozoic thrusting and extension (abs.): AAPG Bulletin with Abstracts, v. 67, n. 8, p. 291–302.

Woodward, L. A., 1981, Tectonic framework of disturbed belt of west-central Montana: AAPG Bulletin, v. 65, n. 2, p. 291–302.

Structural Influences on Cretaceous Sedimentation, Northern Great Plains

L. O. Anna
Consultant
Lakewood, Colorado

Cretaceous rocks in the northern Great Plains can be divided into four chronostratigraphic intervals. The intervals consist of marine and nonmarine shale and sandstone. Thickness patterns for each interval show subtle and distinct linear trends that have been interpreted to reflect movement of basement faults and fault blocks. Recurrent movement has occurred throughout Paleozoic and Mesozoic time. Reactivated faults are propagated upward, deforming overlying sediments by faulting and probably by drape or force folding. Thickness trends of four intervals and trends of sandstone thicknesses are generally oriented northeastward and, more rarely, northwestward. A northeast-southwest pure and/or simple shear stress system during Cretaceous time created a structural style in the northern Great Plains that consisted of a series of grabens, half-grabens, and horsts. The strike of these features was predominantly northeast-southwest. Relief on these structures was small, ranging from a few feet to tens of feet. Translated stress from continental margins to the northern Great Plains during Cretaceous time created a definite structural style for the area. That style was later enhanced as part of the Laramide orogeny by reactivated Precambrian (basement) faults and fault blocks.

INTRODUCTION

Structural influence on sedimentation has long been a recognized process. Tectonism, vertical and lateral distribution of sediment, structure, and stratigraphic position are all interconnected and intimately associated. Most literature deals with these relationships near orogenic belts or near large structural features. As a result, there is a paucity of information dealing with these relationships during anorogenic time in a relatively stable and quiescent area, such as the northern Great Plains. The dynamics of these relationships are not spectacular and are difficult to document, yet subtle changes and events are an important part of the total geologic history.

The Northern Great Plains physiographic province is thought by many authors to have been part of a cratonic shelf during most of Mesozoic and Cenozoic time, except during the Laramide orogeny. Some authors, however, believe that during anorogenic periods, the craton was tectonically active and not passive (Thom, 1923; Sonnenberg, 1956; Smith, 1965; Sales, 1968; Stone, 1969; Thomas, 1974, 1976; Shurr, 1976; Brown, 1978; Weimer, 1978, 1980).

Recurrent movement on basement faults, initially developed in Precambrian time and the effect of that movement on the vertical and lateral distribution of Cretaceous sediment in the northern Great Plains is the focus of this paper. The regional aspect allows liberal interpretations; however, detailed studies such as Shurr

(1976), Weimer et al. (1982), and Shurr and Rice (this volume) document basement-controlled, fault block movement during Cretaceous time. Other regional studies that associate structural adjustment with sediment distribution include those by Chamberlin (1945), Smith (1965), Thomas (1974), Merewether and Cobban (1981), Gerhard et al. (1982), and Maughan (1983).

GENERAL CRETACEOUS PALEOGEOGRAPHY

The initial advance of the Early Cretaceous sea in North America was during early–middle Albian time. The sea extended from the arctic to the Gulf of Mexico and was approximately 1000 mi (1600 km) wide. During Late Cretaceous time, the sea was at its widest extent, but marine deposition was interrupted by frequent regression. The regressions were controlled partly by worldwide eustatic changes in sea level and partly by regional tectonism. As a result, nonmarine sediments intertongue with thick sequences of marine shale (Weimer, 1960).

The eastern margin of the Western Interior basin was formed by the central part of the North American craton, the Canadian shield, and the southwestern extension of the Transcontinental arch (Figure 1). The western margin was formed by the repeatedly rising Cordilleran highland. This highland was a main source of the clastics that ultimately filled the epicontinental seas (Gill and Cobban, 1973). Within the basin, sedimentary processes were influenced to

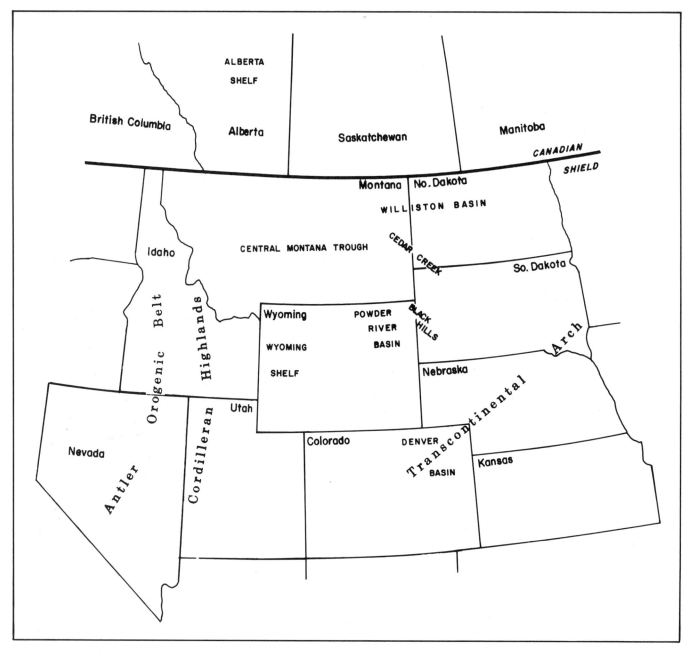

Figure 1—Regional paleogeography and paleostructure of the Western Interior of the United States during Cretaceous time.

a varying degree by the quantity and quality of sediment, sea level changes, regional tectonism, and elevation and depression of paleostructural elements associated with structural development, and growth of the Western Interior.

The major Cretaceous paleostructural element in the northern Great Plains was the Williston basin; also important were the Powder River basin, the central Montana trough, the Cedar Creek anticline, the Transcontinental arch, and the Alberta shelf (Figure 1).

STRATIGRAPHY

Lower Cretaceous

Lower Cretaceous rocks (Figure 2) in the study area consist of sequences of both marine and nonmarine clastic

sedimentary rocks. Thicknesses in the northern Great Plains range from a feather edge in eastern North Dakota and South Dakota to more than 1400 ft (430 m) in west-central Montana.

The Lower Cretaceous Lakota and Fuson formations and equivalents (Figure 2) are generally fluvial sandstone, siltstone, and shale. The Lakota consists mostly of sandstone and occasional conglomerate. Locally, the Lakota has scoured into the underlying Morrison Formation, and where the Morrison is thin or absent, into or through the Swift Formation. The Lakota ranges in thickness from 0 to 100 ft (0 to 30 m), except on the south flank of the Black Hills uplift, where the formation is a few hundred feet thick. Generally, the Lakota is a channel and valley-fill deposit, but in the subsurface, it is often difficult to distinguish from the

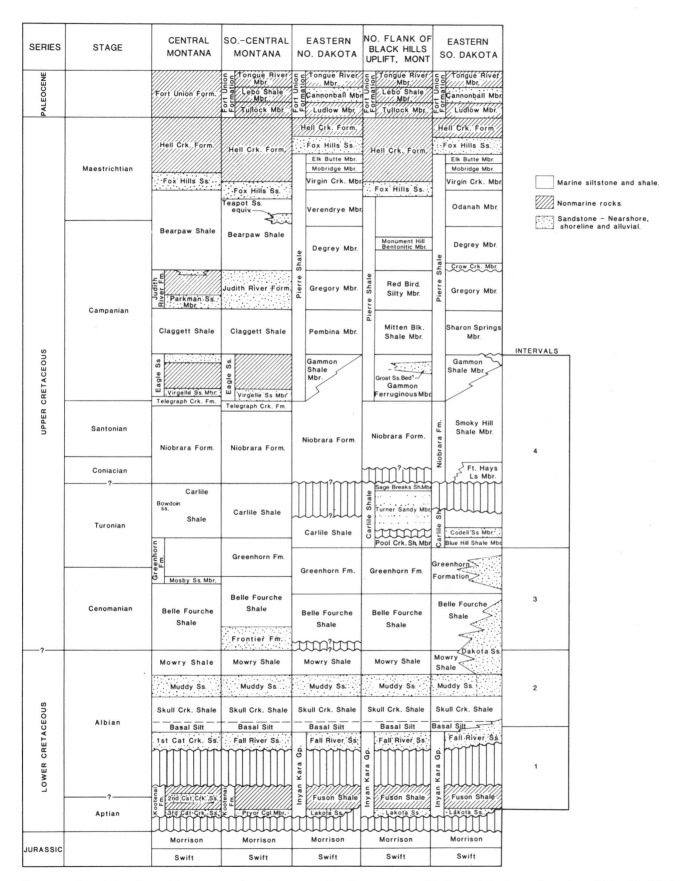

Figure 2—Correlation chart of Cretaceous rocks and corresponding chronostratigraphic intervals in the northern Great Plains. Modified from Rice (1976).

valley-fill of the overlying Fuson Formation. From geophysical log characteristics, the Fuson Formation appears to consist mostly of valley-fill and channel margin silty shale with occasional thick, well-developed, channel-fill sandstone beds. The Fuson ranges in thickness from approximately 400 ft (120 m) in central Montana to less than a few feet in eastern North Dakota and South Dakota, and finally to a zero edge in eastern South Dakota.

The Fall River Sandstone represents the initial advance of the Early Cretaceous sea, which rapidly deposited fine sand, silt, and clay in marginal marine, tidal flat, coastal swamp, and deltaic environments (Waage, 1959). Waage (1959) and Mettler (1966) have noted nonmarine point bar and channel deposits in northeastern Wyoming. Silt and shale deposits in central Montana and Wyoming suggest a deeper water shelf environment or the distal end of a large fluvial system. The Lakota and Fuson formations and the Fall River Sandstone thin eastward and are truncated by pre–Skull Creek erosion. Total thickness of the three formations ranges from about 700 ft (210 m) in central Montana to a zero edge in eastern North Dakota and South Dakota.

The Skull Creek Shale consists of two informal lithologies: (1) a lower, slightly glauconitic siltstone, informally termed the "basal silt," and (2) an upper shale. The silt lithology is of regional extent and contains minor amounts of sand that increase in central and southcentral Montana. A regional marker bed consisting of siltstone (identified from geophysical log characteristics) is present in the upper part. The shale lithology was apparently deposited under strong reducing conditions (Wulf, 1962), and consists mainly of black shale with associated pyrite and considerable organic matter. The formation ranges in thickness from 0 ft in eastern South Dakota to more than 250 ft (75 m) in parts of Montana, Wyoming, and western South Dakota and North Dakota.

Withdrawal of the Skull Creek sea created an unconformity at the base of the Muddy/Newcastle Sandstone in eastern Wyoming and southeastern Montana. Incisement into the Skull Creek during the hiatus was subsequently filled with channel-fill and valley-fill of the Muddy/Newcastle Sandstone (Baker, 1962; W. D. Stone, 1971; Waring, 1975; Weimer et al., 1982). As sea level rose, marginal marine deposits formed over a broader area than the valley-fill deposits; later, the sea transgressed from west to east, causing development of extensive delta systems in eastern Montana and northeastern Wyoming and, later, in southeastern South Dakota. Sediment supply to the deltas originated in eastern and south-central South Dakota, and the deltas supplied sediment to the shelf areas in central Montana and Wyoming. In northwestern Montana, a delta system from the west was also supplying sediment to the shelf areas in central Montana.

Thickness of the Muddy/Newcastle Sandstone is variable, ranging from 0 ft in large areas of North Dakota to tens of feet in central Montana and Wyoming, and increasing abruptly to a few hundred feet in southeastern North Dakota and eastern and south-central South Dakota. Where the formation is a few hundred feet thick, the Muddy/Newcastle is referred to as the Dakota Formation (Figure 2).

As the sea encroached farther eastward during the time of late Muddy/Newcastle Sandstone deposition, the Mowry Shale was deposited over the central and western part of the northern Great Plains. The Mowry (Figure 2) consists of dark to light gray, siliceous shale containing disarticulated fish scales and bones. Davis (1970) and Cluff (1976) reported the source of the silica to be biogenic. In the northern Great Plains, the Clay Spur bentonite marks the top of the Mowry and also divides the Lower from the Upper Cretaceous. This bentonite is used as a regional time marker. Thickness of the Mowry ranges from a zero edge in eastern and southeastern South Dakota and eastern North Dakota to more than 700 ft (200 m) in south-central Montana.

Upper Cretaceous

The sedimentary pattern for Upper Cretaceous rocks in the northern Great Plains can be described in terms of four main transgressions and four main regressions (Weimer, 1960).

Transgression 1—The Belle Fourche Shale and Greenhorn formations (Figure 2) were deposited as a continuation of the Mowry transgression and extend over most of the northern Great Plains region. The Belle Fourche Shale is about 500 ft (150 m) thick (although thicknesses vary greatly) and consists of gray to black shale with numerous bentonite beds. The Greenhorn Formation is about 200 ft (60 m) thick and consists of a thin upper limestone; a thin, tight sandstone; and a lower, chalky shale. Rice and Shurr (1980) divided the two formations into three main lithologies from east to west: (1) chalk and shaly chalk, (2) shale, and (3) shelf sandstone.

The Carlile Shale (Figure 2) is about 300 ft (90 m) thick and consists of gray marine shale with a thin, tight sandstone (Bowdoin Sandstone in Montana and Turner Sandstone in eastern Powder River basin, Wyoming). Rice and Shurr (1980) divide the Carlile into three lithologies from east to west: (1) shale and shaly chalk, (2) shelf sandstone, and (3) shale.

Regression 1—The Frontier Formation (Figure 2) is the westward equivalent of the Belle Fourche Shale and the Greenhorn and Carlile formations. Although the Frontier is areally restricted, the formation is 500–2000 ft (150–600 m) thick and consists of alternating beds of deltaic sandstone and shale (Barlow and Haun, 1966).

Transgression 2—The Niobrara Formation (Figure 2) is about 350 ft (100 m) thick and consists of gray shale with lenticular chalky beds characterized by small, white calcareous lenses or "white specks." Lithologic variations range from dominantly chalk in the east to mostly shale in the west.

Regression 2—In eastern Montana, the Telegraph Creek Formation (Figure 2) is about 300 ft (90 m) thick and consists mainly of sandy shale with thin beds of concretionary sandstone in the middle and upper parts, which often forms capping escarpments.

The Eagle Formation (Figure 2), including the Virgelle Sandstone, is about 600 ft (180 m) thick and consists of light colored, friable, massive, nonmarine and marine sandstone. Marine sandstone units that extend into the Powder River basin are the Shannon, Sussex, and Groat sandstone beds of the Gammon Member of the Pierre Shale (Rice and Shurr, 1980).

Transgression 3—The Claggett Shale (Figure 2) in

Montana is about 300 ft (90 m) thick and consists of dark marine shale and siltstone. Numerous thick and persistent bentonite beds occur in the lower part, including the Ardmore bentonite bed (Figure 2) which is a regional time datum easily recognized on geophysical logs.

Regression 3—The Judith River Formation (Figure 2) in Montana is about 400 ft (120 m) thick and thins eastward. The formation consists of nonmarine, light-colored sandstone with abundant coarse volcanic detritus to the west. Finer grained marine equivalents lie to the east. The Judith River is stratigraphically equivalent to the Parkman Sandstone Member and an unnamed shale member of the Steele Shale in the western Powder River basin, Wyoming.

The Mesaverde Formation (Figure 2) in the Powder River basin consists of the lower Parkman Sandstone Member, an unnamed shale member, and the upper Teapot Sandstone Member. A major unconformity exists at the base of the Teapot Sandstone.

Transgression 4—The Bearpaw Shale (Figure 2) in Montana consists of about 800 ft (250 m) of dark marine shale with numerous bentonite beds. The Bearpaw is equivalent to the Lewis Shale in the western Powder River basin and to the upper Pierre Shale of the eastern Powder River basin, North Dakota and South Dakota. The Pierre Shale, which consists of over 2000 ft (600 m) of dark marine shale, is also the eastern equivalent of the Eagle Sandstone, Claggett Shale, Judith River Formation, Mesaverde Formation, Lewis Shale, and Bearpaw Shale.

Regression 4—The final regression of the Late Cretaceous sea deposited the Fox Hills Sandstone (Figure 2) and equivalents. The overlying Hell Creek Formation was being deposited as the continental part of the regressive system (Frye, 1969). Both formations and their equivalents are areally extensive in the subsurface and crop out over sizeable areas in southern and central North Dakota.

STRUCTURAL INFLUENCE ON STRATIGRAPHY

Introduction

The degree of structural influence on stratigraphy depends in part on the scale of viewing. Large-scale patterns of sedimentation are influenced by the rate and extent of orogenesis and basin development. Smaller scale patterns can be influenced by syndepositional structural adjustments. In turn, lithology, facies, texture, porosity, permeability, quality and quantity of source material, subsidence rates, and diagenetic history are controlled by both large- and small-scale tectonism and structure.

The physical framework in areas having little or no data can be more clearly defined if certain structural adjustments produce specific or predictable sedimentation patterns. If only part of a rock unit can be described, certain parameters of other parts of the system can be predicted when: (1) the paleostructural history of the area is known, and (2) a depositional model is established that takes structural adjustment into account.

Recurrent movement of Precambrian basement blocks in the northern Great Plains (although subtle compared with orogenic events) has affected lithofacies and the distribution of primary porosity and permeability. Fault block movement controls topography or bathymetry (Thomas, 1974; Shurr, 1976; Weimer et al., 1982); these in turn control facies and

physical characteristics of the sediment, depending on the quality and quantity of available sediment. The following sections describe the lineations expressed by sedimentary geometries and textures believed to be the result of recurrent movement of Precambrian basement faults.

Most of the units described herein are chronostratigraphic intervals; a chronostratigraphic interval is defined as everywhere representing the same horizon in geologic time (North American Stratigraphic Code, 1983). Boundaries of the intervals are generally based on approximate time surfaces that were mapped from geophysical well logs (Figure 3). This approach affects not only the accuracy of the maps, but also an understanding of stratigraphic relationships and the reconstructions of geologic events inferred from these relationships (Oriel, 1959).

Thickness patterns are shown for each of four time intervals (Intervals 1, 2, 3, and 4) in Figures 4–7, respectively, and for two sandstone accumulations in Figures 8 and 9 (for part of Intervals 1 and 2). These parameters are then related to possible structural controls on their distribution. Sand thicknesses were determined from geophysical logs. Thickness changes of sediments are controlled by several factors (Sonnenberg and Weimer, 1981): (1) differing rates of sedimentation, such as onlap and offlap; (2) differing rates of uplift or subsidence; (3) erosion or fill; (4) differential compaction; and (5) faulting. By mapping chronostratigraphic intervals, effects of erosion and fill can be eliminated. At the scale and contour intervals used in preparing maps for this report, thickness changes from differential compaction are also thought to have been eliminated. Care was used to isolate thickness changes that were thought to result from local faulting. Interpretations in this report were made from more detailed maps than could be published here. Detailed maps will soon be published (Anna, in press).

Lineaments, as used in this report, are based on interpretations of subsurface data (O'Leary et al., 1976). The lineaments are derived from linear trends of thickness and from sedimentation patterns. When two or more linear trends are superimposed or nearly so, then that trend is mapped as a lineament zone (Figure 10). Several of the lineament zones have been similarly mapped and named on the basis of analysis of Paleozoic rock distributions (Maughan, 1983). Maughan's work reinforces the hypothesis that basement faults of Precambrian age were reactivated at various times during Phanerozoic time. The paleostructural elements shown on Figure 10 are classified as lineaments, although their present-day structural configuration may be well documented. For example, the Cedar Creek anticline is now an anticline, but in the geologic past, it may have operated as a fault, a hinge line, or a negative area or it may have remained quiescent.

Interval 1

The regional unconformity at the base of Interval 1 marks the base of the Lower Cretaceous (Figures 2 and 3). The top of the interval is a regional marker bed, identified on geophysical logs, located near the top of the Marine basal silt of the Skull Creek Shale. Interval 1 is more than 600 ft (180 m) thick in central Montana, with a conspicuous east–west trending tongue of thick sediment extending into eastern Montana (Figure 4). The interval is also thick on the south flank of the Black Hills uplift, and thins to 61 m or less along the eastern flank of the Williston basin and in the central part of the Powder River basin. A rose diagram of all linear thick and thin trends shows a strong preferred northeast direction and a less prominent northwest direction (Figure 11).

In eastern and north-central South Dakota, Interval 1 thins greatly over a large northeast-trending elevated structural block that includes the Sioux uplift. In approximately the same area, Merewether and Cobban (1981) show a northeast-trending paleostructure developed during Carlile time, although the structure was then depressed, in contrast to its elevated position in Interval 1 time. The Powder River basin was relatively positive during deposition of Interval 1, receiving little sediment except for a few linear belts adjacent to the Black Hills uplift. In eastern South Dakota, it is not known whether there was substantial deposition in Inyan Kara sediments and subsequent uplift and erosion, or little original deposition, or both.

Comparison of thick and thin isopach patterns of various Intervals show in several localities the axes of thin linear trends generally overlie the axes of thick linear trends. For example, the Great Falls lineament shows an altering thick and thin pattern for Intervals 1, 2 and 3. Comparing the Des Moines, Powder, Mondak, and Kaycee lineaments for Intervals 2 and 3 show thick linear trends overlie a thin linear trend or vice versa. Some linears show a scissor effect; that is, thick along one part of the trend and thin along the other part (Kaycee and Mondak linears for Interval 3). Comparing net sand thickness maps with underlying Interval maps show similar thick over thin (and vice versa) patterns. Thin Muddy/Newcastle (Figure 8) overlies thick Interval 1 (Figure 4) along the Great Falls linear; whereas, a thick Muddy/Newcastle overlies a thin Interval 1 along the Kaycee, Greybull, and Des Moines linears.

Admittedly, linear trends are not always obvious nor even existent. Several factors may account for this: little or no recurrent movement for an area for a particular time, no sediment to distribute, bathymetry or topography not affected by fault movement, or energy levels not affected by a bathymetric or topographic change. Yet, there is enough evidence to show that sedimentation patterns are being influenced by linear structural features or blocks on a regional and local scale.

Interval 2

Interval 2 consists of units from the regional marker bed in the basal silt of the Skull Creek Shale up to the top of the Mowry Shale, including the Muddy/Newcastle Formation (Figure 2). The top of the Mowry is picked at the Clay Spur bentonite, a regional marker at the top of the Lower Cretaceous.

An isopach map of Interval 2 (Figure 5) shows an overall thickening in west-central and south-central Montana. Most of the thickening in these areas is attributed to the Mowry Shale. Substantial thickening is also shown in southeastern South Dakota and northern Nebraska, although this thickening is mostly attributed to the Muddy/Newcastle Sandstone. Thin areas in northeastern North Dakota are attributed to thinning of all the formations within Interval 2.

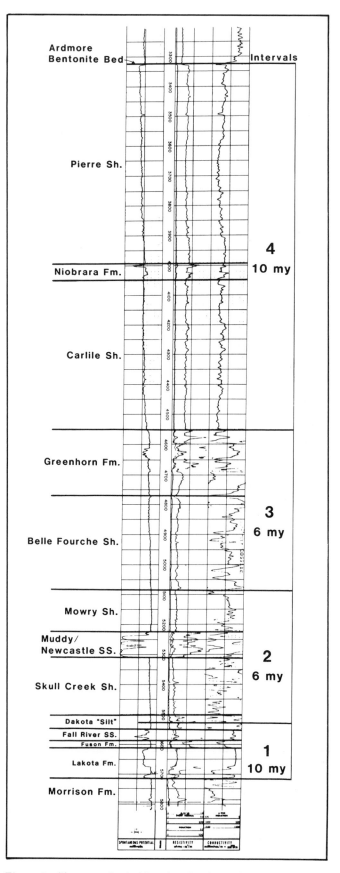

Figure 3—Type geophysical log showing formations and chronostratigraphic intervals with respective time increments for the northern Great Plains.

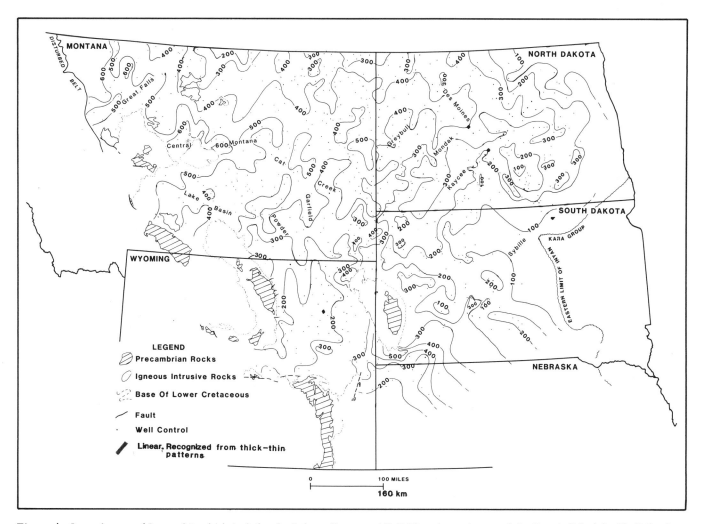

Figure 4—Isopach map of Interval 1, which includes the Lakota, Fuson, and Fall River formations, and the "basal silt" of the Skull Creek Shale and equivalents.

A net sandstone thickness map of the Muddy/Newcastle Sandstone (Figure 8) within Interval 2 shows a broad northeast-trending band of thick sandstone that was apparently deposited in a deltaic environment in southeastern and south-central South Dakota and northern Nebraska (Witzke et al., 1983). In other areas of the northern Great Plains, the internal and external geometry of the sandstone indicates a variety of depositional environments, ranging from upper meander belt (fluvial), to deltaic, to marginal marine (Baker, 1962; Wulf, 1962; Waring, 1975; Weimer et al., 1982). Most of the sediment deposited in northeastern Wyoming, eastern Montana, North Dakota, and South Dakota originated east and southeast of these areas; in north-central and northwestern Montana, sediment originated in southwestern Saskatchewan, southeastern Alberta, Canada, and in northwestern Montana. This interpretation is supported by the wide area of thin sediment in central Montana. The overall thickness pattern has a northeasterly trend or grain, especially east of the Cedar Creek anticline in northeastern Montana and western North Dakota. The northeasterly trend is also present west of the Black Hills uplift, except for a strong northwesterly trend in western South Dakota paralleling the Cedar Creek anticline. The northeasterly

grain of Muddy/Newcastle sedimentation contrasts with the northwesterly grain of Inyan Kara Group sedimentation (Figure 9).

The top of Interval 2 marks the top of the Lower Cretaceous. Vail et al. (1977) and Hancock (1974) show a substantial worldwide increase in sea level starting in Late Cretaceous time. That rise may, in part, be tectonically controlled (Vail et al., 1977).

Interval 3

Interval 3 consists of formations between the top of the Mowry Shale and the top of the Greenhorn Formation, including the Belle Fourche Shale and equivalents (Figure 2). This interval was apparently deposited in a marine environment and consists of shale, calcareous shale, and limestone with occasional thin, low-permeability sandstone or siltstone beds.

Depocenters for Interval 3 (Figure 6) are the Powder River basin and the northern Black Hills (Figure 9). There was significant downwarping in the Powder River basin, possibly related to incipient movement to the Laramide orogeny, although the main orogenic event started after Interval 4 deposition. Interval 3 is thin in central and northern Montana because of major unconformities; it is

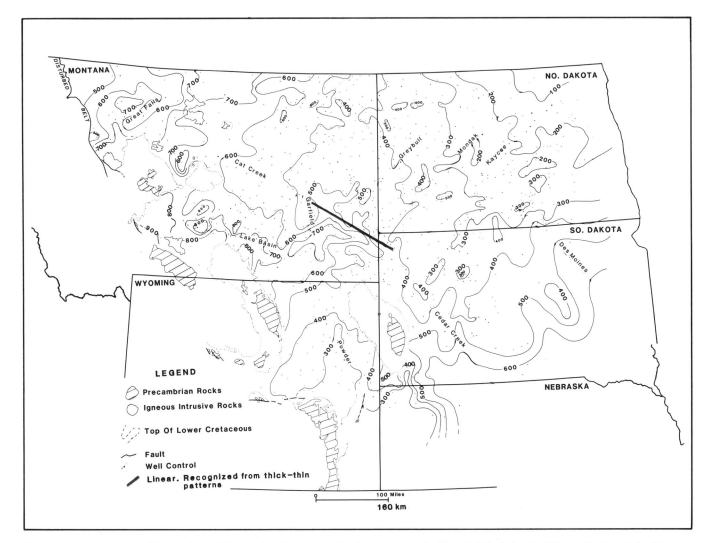

Figure 5—Isopach map of Interval 2, which includes the base of the silt marker in the "basal silt" of the Skull Creek Shale and the Mowry Shale and equivalents.

also thin in eastern North Dakota and South Dakota and moderately thick in the Williston basin. South of the Central Montana uplift, the interval thickens abruptly across a feature that may have been a hingeline between the Alberta shelf and the Wyoming shelf. A rose diagram of thick and thin linear trends of Interval 3 indicates that northwest and northeast directions dominate (Figure 11).

Thick and thin patterns coincide for both Intervals 2 and 3 (Figures 5 and 6) near the periphery of the Williston basin or near present-day uplifts, especially to the west. A rose diagram (Figure 11) illustrating directions of thick and thin trends shows the northeast direction that is dominant in Interval 2 time (Early Cretaceous) and the northwest direction dominant in Interval 3 time (Late Cretaceous).

Interval 4

Interval 4 extends from the top of the Greenhorn Formation through the regionally extensive Ardmore bentonite bed of the Pierre Shale (base of the Claggett Shale in Montana) (Figure 2). The Ardmore is approximately equivalent to the top of the Eagle Sandstone in Montana. Interval 4 consists mostly of marine shale, but thick

prograding sandstone units of the Eagle Sandstone and equivalents in the western part form significant sandstone intervals. Within Interval 4 the Niobrara Formation consists of chalk in North Dakota and South Dakota (Rice and Shurr, 1980), and thin, low-permeability, marine sandstone, siltstone, and shale in Montana, Wyoming, and North Dakota.

Thickness patterns of this interval (Figure 7) generally follow those of Interval 3. The Powder River basin was significantly downwarped; however, the area north of the Black Hills was a structural and topographic low. Downwarping seems to have increased progressively north and west toward the disturbed belt in west-central Montana. There was thinning of Interval 4 starting between the Powder River basin and a small depression in southeastern Montana. This may reflect initial stages of the final separation of the Williston basin from the Powder River basin. The area of thickening in Custer and Carter counties in southeastern Montana was slightly downwarped, possibly due to a pulling apart of these basins. During the Laramide orogeny, this area was upwarped into its present configuration, the Miles City arch. The area near the Cedar

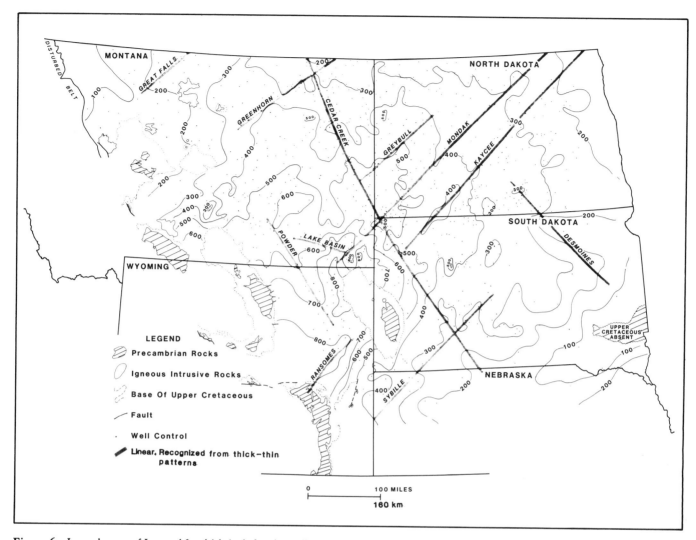

Figure 6—Isopach map of Interval 3, which includes the Belle Fourche and Greenhorn formations and equivalents.

Creek anticline was structurally low during Interval 4 time, as indicated by the linear thickening trend paralleling its axis. The Hartville fault, southwest of the Black Hills, was very active during this interval; within about 10 mi (15 km) along the trace of the present-day Hartville fault, there is an approximately 800 ft (250 m) increase in the thickness of Interval 4. A rose diagram of Interval 4 shows a dominant northeast trend for thick and thin linear features (Figure 11).

A composite rose diagram of thick and thin linear trends for Intervals 1–4 is shown in Figure 12. The frequency of the linear trends is taken from Anna (in press). All diagrams show distinct northeast–northwest preferred trends, with the northeast direction often dominant. The importance of these trends is discussed in the following section.

STRUCTURAL AND STRATIGRAPHIC MODEL

The influence of paleostructure on sedimentation and stratigraphic relationships in the northern Great Plains is reflected by lithofacies distribution and thickness patterns resulting from the presence of grabens, half-grabens, and

horsts. Recurrent movement of basement blocks in the study area occurred periodically, with blocks being elevated at certain times and depressed at other times. Reactivated faults in Precambrian rocks appear to have propagated upward, deforming overlying sediments by faulting or by force or drape folding. These faults and folds are manifested either as linear features on a paleosurface of deposition or on the present-day surface.

The response of the rock column to stress in the study area is regionally repetitive and is expressed by structural patterns of faults, fractures, and folds that are also repetitive. Wrench or strike-slip faults exist as simple shears and are commonly associated with folds and with thrust and reverse faults (Thomas, 1974). Scissor-type faults are also common and are characterized by reversal of apparent dip-slip displacement along strike. Folds, thrusts, and reverse faults are commonly recognized as Laramide features, but are not usually directly observed as pre-Laramide (paleostructure) features. Drape or force folds responding to basement faulting are probably the most common type of paleostructure (see Stearns, 1971, for a review of drape folds). The change of topographic or bathymetric relief from one fault block to another can be in the form of a drape fold and not necessarily in the form of a fault block.

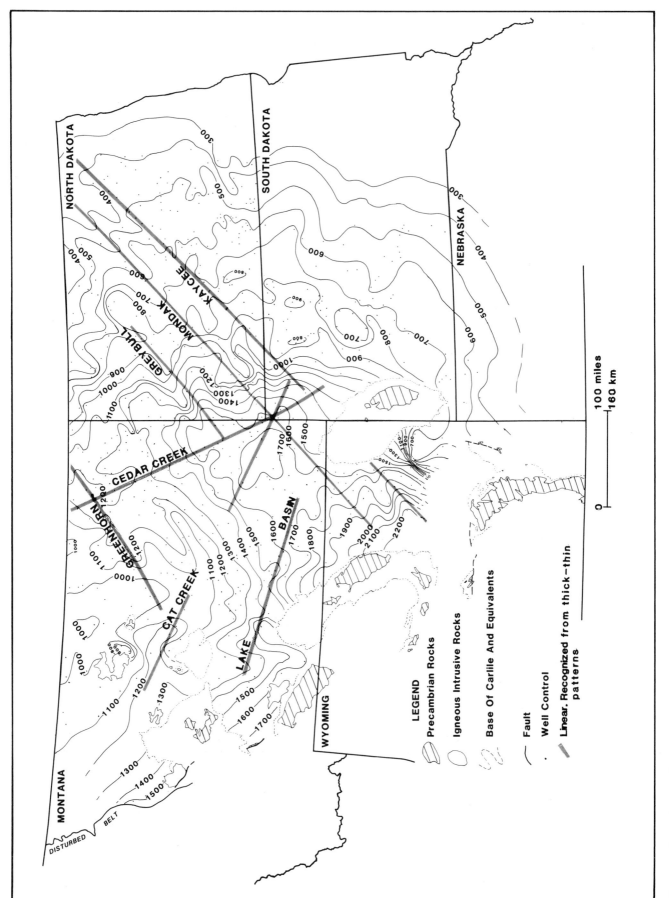

Figure 7—Isopach map of Interval 4, which includes the Carlile and Niobrara formations and the Pierre Shale to the top of the Ardmore bentonite bed and equivalents.

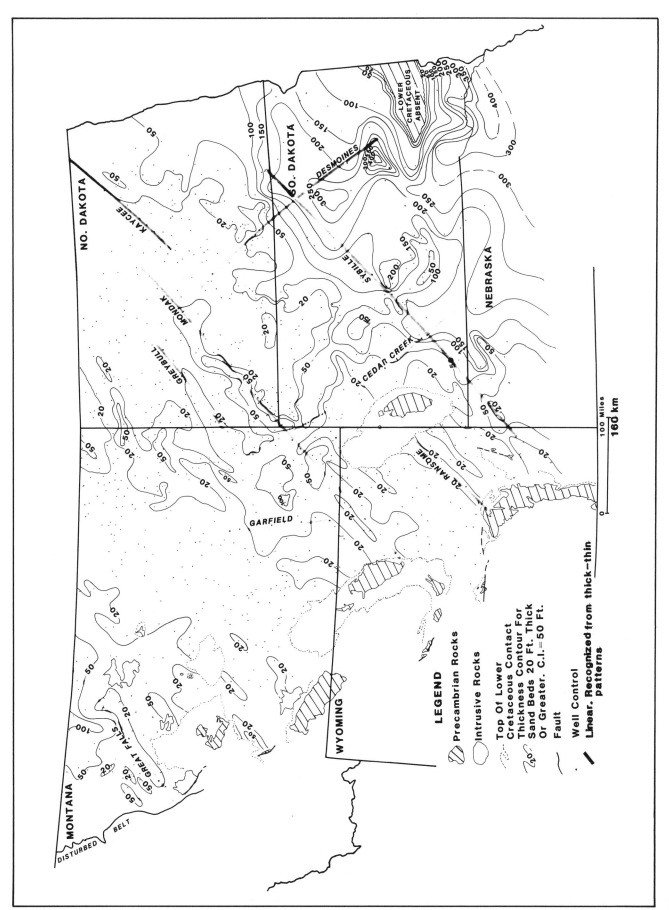

Figure 8—Net sandstone thickness of the Muddy/Newcastle Sandstone and equivalents.

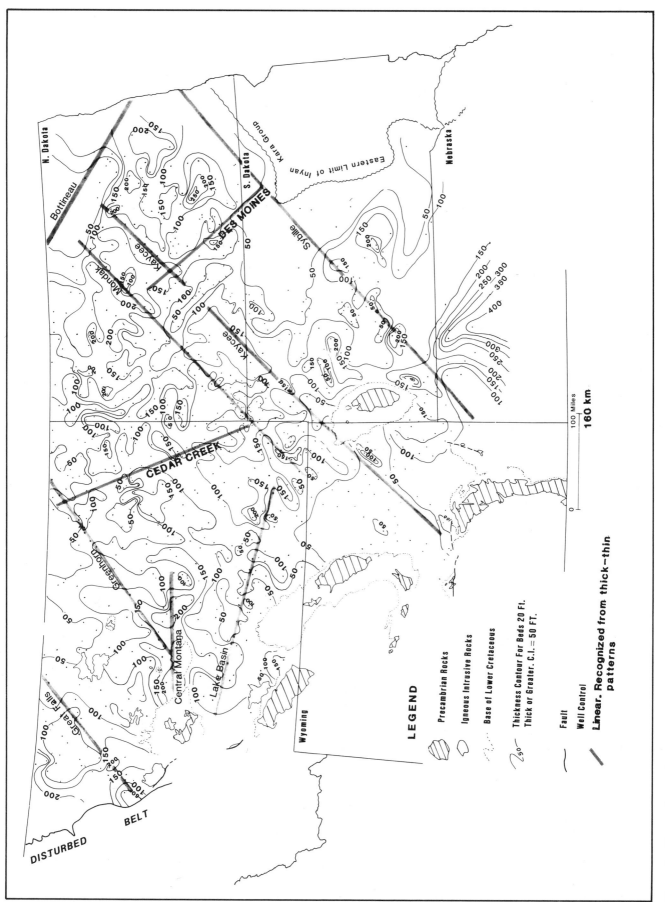

Figure 9—Net sandstone thickness of the Inyan Kara Group and equivalents.

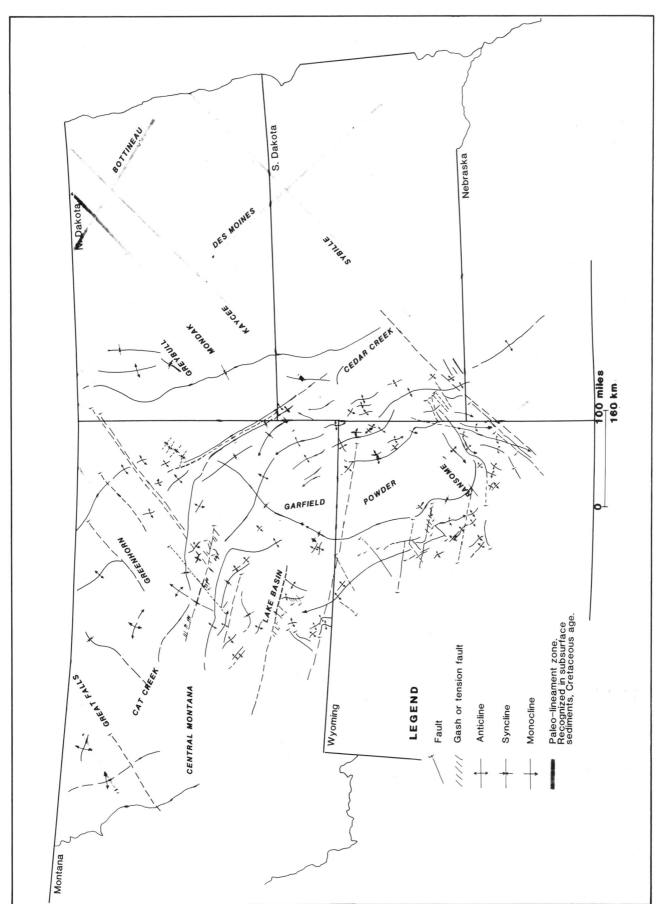

Figure 10—Present-day and ancient lineament zones of the northern Great Plains. Lineament zones are based on coinciding lineaments shown in Figures 4–9 and on detailed work by Anna (in press). Strike of faults and folds is shown on Figure 15. Structure modified from D. S. Stone (1971). Used with permission of the Wyoming Geological Association.

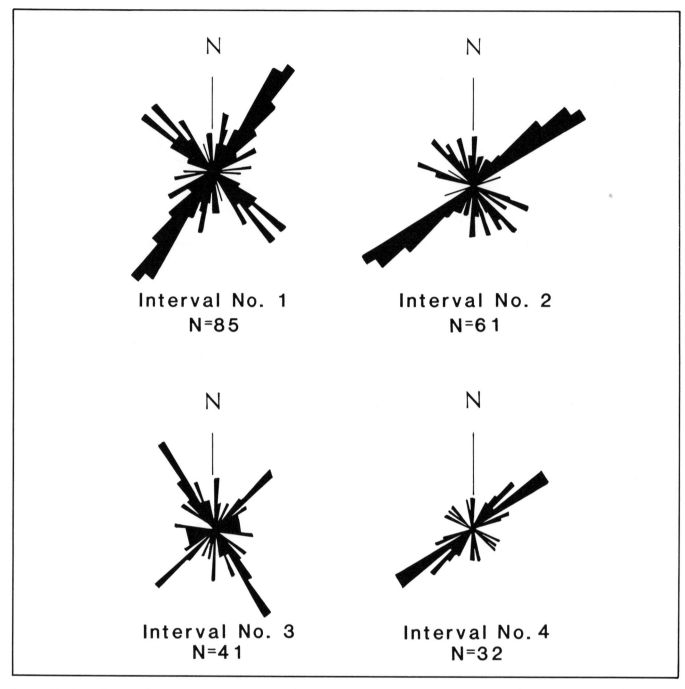

Figure 11—Rose diagrams for isopachs of Intervals 1–4 showing the direction of thick and thin linear trends. *N*, number of lineaments used in each diagram. Frequency was the only criterion used. Modified from more detailed work by Anna (in press).

The observed distribution and thickness of sediments in this detrital depositional system apparently stem from recurrent movement of Precambrian blocks as horsts, grabens, or half-grabens, from eustatic changes in sea level, and from the quantity and quality of available sediment. A series of diagrams summarizing possible depositional environments that existed in the northern Great Plains resulting from structural adjustment and relative changes in sea level are shown in Figure 13 (for a more complete summary, see Weimer et al., 1982). Initial conditions prior to a change in sea level involving a horst–graben sequence

in the marine environment are shown in Figure 13A. Offshore bars develop over the horst because of the slightly shallower water (higher energy). After a drop in sea level, incisement occurs on the graben block (Figure 13B), because streams are channeled into topographic or structural lows (higher energy). As sea level rises, valley-fill and channel-fill deposition occurs in the incised channel (Figure 13C). Continued rise in sea level results in deposition over the entire graben, not only in the incised channel (Figure 13D). With continued rise in sea level, thick nearshore, shoreface, and well-drained swamp deposits are deposited over horst or

Figure 12—Composite rose diagram for Intervals 1–4 showing the direction of thick and thin linear trends. Frequency ($N = 219$) was the only criterion used. Modified from more detailed work by Anna (in press).

stress analysis (McKinstry, 1953) can be applied to analyze the stress history causing the fault block movement. The best fit seems to be a pure shear/simple shear-type stress system. Maximum horizontal stress in the study area is oriented in a general northeast–southwest direction.

There is other evidence supporting this type of stress system. For example, Warner (1978) described the Colorado lineament as a major northeast-trending wrench fault developed in Precambrian time (Figure 14). Gerhard et al. (1982) ascribe the northeast-trending Colorado lineament and the Fromberg fault zone (Figure 14) as major structural features in Precambrian, Paleozoic, and Mesozoic time. Maughan (1983) also shows a lineament mosaic derived from Paleozoic rock distributions. Also, major normal faults in the northern Great Plains are oriented northeast–southwest, whereas major thrusts and folds are oriented northwest–southeast (Figure 15). Thus, the faults parallel a tensional stress direction and the folds parallel a compressional stress direction. Because the thick–thin alignment of Cretaceous rocks is dominantly northeasterly (Figure 11), a northeast–southwest tensional structural style appears to have prevailed during Cretaceous time. Therefore, a series of grabens, half-grabens, and associated horsts was the dominant structural style for the northern Great Plains during Cretaceous time.

Wicks (1980) mapped primary and secondary fractures in the Fall River Formation in the Black Hills area. He demonstrated that primary fractures were generally oriented northeast–southwest. Wicks also postulated that the fractures developed prior to the Black Hills uplift in Laramide time. This coincides with the northeast–southwest thick–thin alignment that dominates in the subsurface in the northern Great Plains.

CONCLUSIONS

The vertical and lateral distribution of Cretaceous sediments was influenced by a basement fault and fracture system originally developed in Precambrian time and having recurrent movement in Paleozoic and Mesozoic time. Reactivated faults in basement rocks are propagated upward, deforming overlying sediment by faulting and probably by drape or force folding. The faults are manifested as linear features on a paleosurface (Shurr and Rice, this volume).

Present-day thick–thin sedimentation patterns and primary fractures are dominantly oriented northeast–southwest, and the primary stress direction in Cretaceous time was generally northeast–southwest as well. Thus, the following conclusions have been drawn.

1. The structural style in the northern Great Plains was a series of grabens, half-grabens, and horsts, whose strike was predominantly aligned northeast–southwest. Relief on these structures was probably small, ranging from one to a few meters. There was, however, apparently enough relief to initiate incisement or to change energy levels to distribute sediment in a linear pattern.

2. Most structural features originated from a compressional pure or simple shear stress system or both. The primary stress direction had a general orientation of northeast–southwest. Therefore, tensional structural features

upthrown blocks (Figure 13E); estuarine, tidal flat, and poorly drained swamp deposits form over graben or downthrown blocks. The final step in the process is a complete transgression resulting in marine deposits at the base of the next cycle (Figure 13F). To summarize, in a nonmarine environment, rivers channel into topographically or structurally low areas, resulting in thicker, coarser grained deposits. In contrast, interchannel areas develop in topographically or structurally high areas, resulting in finer grained deposits. In a marine environment, sand deposits generally develop over structurally high areas (due to shoaling effects), and shale deposits accumulate in structurally low areas. The same principles apply to areas of half-grabens, but the geometry of the deposit is modified because the configuration of the structural block is modified.

Structural adjustments or recurrent movement of basement blocks may result from changes in magnitude of the local stress field or, less frequently, the regional stress field. Basement blocks might also experience structural adjustment or recurrent movement when the resistance to movement in one direction is greater than another; therefore, the new direction of movement along a fault plane would be in the direction of least resistance. Thus, a particular block may be elevated at one time and depressed at another, depending on the change of direction of the stress and the amount of change in the resistance.

If one assumes that the thick and thin linear trends are a result of basement fault block movement, a McKinstry-type

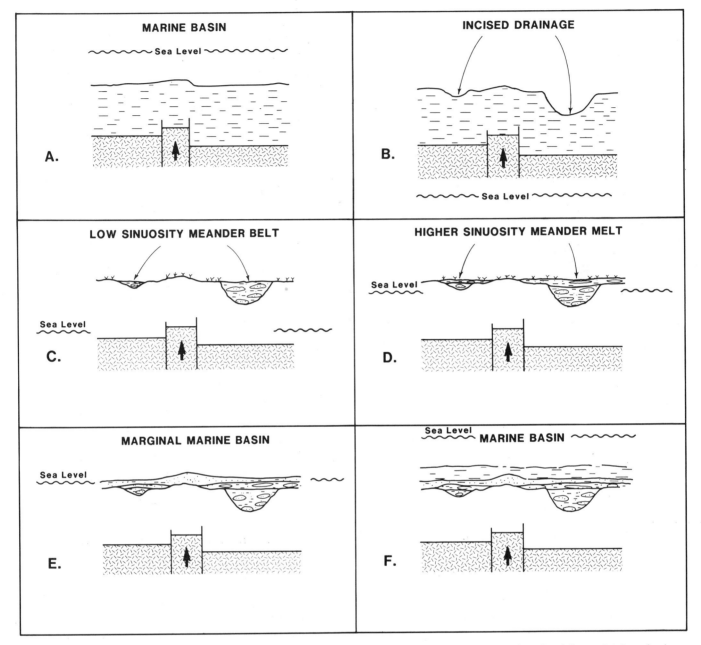

Figure 13—Diagrams showing postulated sedimentary response to structural adjustment for one cycle of sea level fluctuation in a clastic environment. Modified from Weimer et al. (1982).

were aligned northeast–southwest and compressional structural features were aligned northwest–southeast. This style was manifested during the Laramide orogeny.

3. Movement on basement fault blocks can create relief for distributing sediments and can also create a mechanism for listric normal faulting (Figure 16). Therefore, orientation, distribution and timing of listric faulting in the study area may be predictable because of the orientation and distribution of thick and thin patterns of Cretaceous rocks. Listric faulting can also create substantial fracturing from associated antithetic or keystone faulting. Davis (1983) demonstrated that listric normal faulting is a significant structural style for the Denver basin, which has before been little recognized in the basin. Other Rocky Mountain basins may have areas of listric normal faulting as well.

ACKNOWLEDGMENTS

The author gratefully acknowledges George Shurr and Roger Miller for their constructive review. Special appreciation is extended to Bill Thetford, Will Baca, Bob Fourney, and Sherri Thetford for drafting, and to Janet Hall for typing.

REFERENCES CITED

Anna, L., in press, Geologic framework of the ground-water flow system in Jurassic and Cretaceous rocks, Northern Great Plains: U.S. Geological Survey Professional Paper 1402-B.

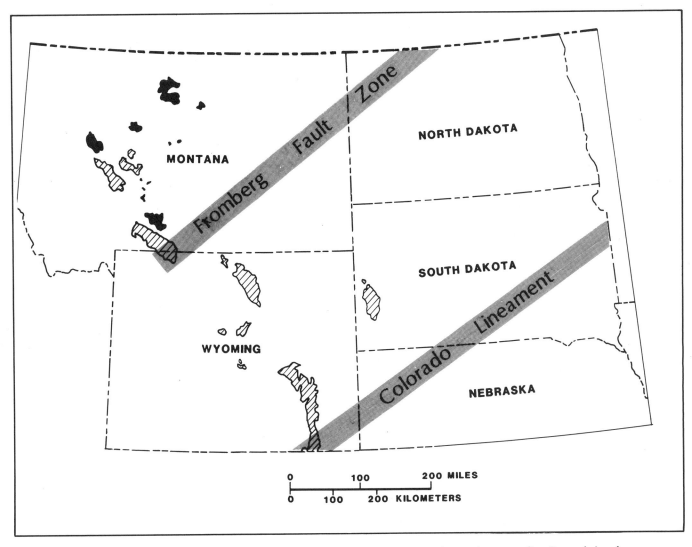

Figure 14—Orientation of the Colorado lineament and the Fromberg fault zone, two major northeast-trending Precambrian shear zones. After Gerhard et al. (1982) and Warner (1980).

Baker, D. R., 1962, The Newcastle Formation in Weston County, Wyoming—A non-marine (alluvial) plain deposit: Wyoming Geological Association 17th Annual Field Conference Guidebook, p. 148–162.

Barlow, J. A., Jr., and J. D. Haun, 1966, Regional stratigraphy of Frontier Formation and relation to Salt Creek field, Wyoming: AAPG Bulletin, v. 50, n. 10., p. 2185–2196.

Brown, D. L., 1978, Wrench-style deformational patterns associated with a meridional stress axis recognized in Paleozoic rocks in parts of Montana, South Dakota, and Wyoming: Billings, Montana Geological Society, Williston Basin Symposium Guidebook, p. 17–31.

Chamberlin, R. T., 1945, Basement control in Rocky Mountain deformation: American Journal of Science, v. 243-A, p. 98–116.

Cluff, R. M., 1976, Paleoecology and depositional environments of the Mowry Shale (Albian) Black Hills region: Master's thesis, University of Wisconsin, Madison, 104 p.

Coney, P. J., 1975, Overview of Late Cretaceous through Cenozoic Cordilleran plate tectonics (abs.): Geological Society of America Abstracts with Programs, v. 7, n. 7, p. 1035.

Davis, J. C., 1970, Petrology of Cretaceous Mowry Shale of Wyoming: AAPG Bulletin, v. 54, n. 3, p. 487–502.

Davis, T. L., 1983, Influence of Transcontinental arch on Cretaceous listric-normal faulting, west flank, Denver basin (abs.): AAPG Bulletin, v. 67, p. 1333.

Dickinson, W. R., 1974, Plate tectonics and sedimentation, *in* Tectonics and Sedimentation: Society of Economic Paleontologists and Mineralogists Special Publication 22, p. 1–27.

Frye, C. I., 1969, Stratigraphy of the Hell Creek Formation in North Dakota: North Dakota Geological Survey Bulletin 54, 65 p.

Gerhard, L. C., S. B. Anderson, J. A. LeFever, and C. G. Carlson, 1982, Geologic development origin, and energy mineral resources of Williston basin, North Dakota: AAPG Bulletin, v. 66, p. 989–1020.

Gill, J. R., and W. A. Cobban, 1973, Stratigraphy and geologic history of the Montana group and equivalent rocks, Montana, Wyoming and North and South

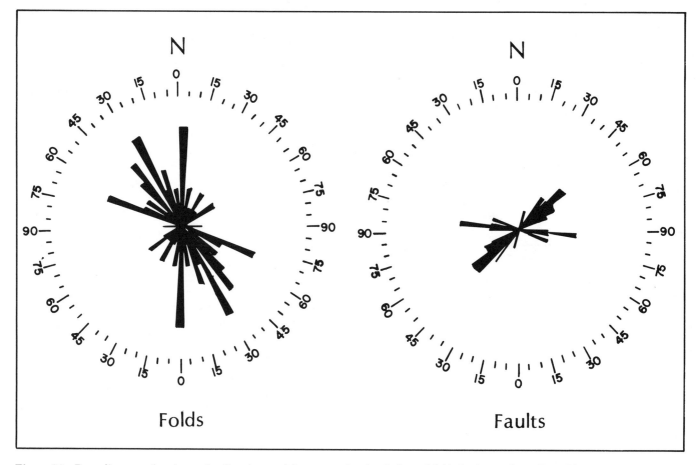

Figure 15—Rose diagrams showing strike directions and frequency of major faults and folds in the northern Great Plains (see Figure 10). Most directions are taken from D. S. Stone (1971). Used with permission of the Wyoming Geological Association.

Dakota: U.S. Geological Survey Professional Paper 776, 37 p.

Hancock, J. A., 1974, The sequence of facies in the Upper Cretaceous of northern Europe compared with that in the Western Interior: Special Publication of the Geological Association of Canada 13, p. 83–118.

Haun, J. D., and H. C. Kent, 1965, Geologic history of the Rocky Mountain region: AAPG Bulletin, v. 49.

Hodgson, R. A., 1965, Genetic and geometric relations between structures in basement and overlying sedimentary rocks, with examples from Colorado plateau and Wyoming: AAPG Bulletin, v. 49, n. 7, p. 935–949.

Maughan, E. K., 1983, Tectonic setting of the Rocky Mountain region during the late Paleozoic and the early Mesozoic, *in* Proceedings of the Symposium on the genesis of Rocky Mountain ore deposits: Changes with time and tectonics, Denver Region of Exploration Geologists Society, p. 39–50.

McKinstry, H. E., 1953, Shears of the second order: American Journal of Science, v. 251, p. 401–414.

Merewether, E. A., and W. A. Cobban, 1981, Mid-Cretaceous formations in eastern South Dakota and adjoining areas—stratigraphic, paleontologic, and structural interpretations, *in* Cretaceous stratigraphy and sedimentation in northwest Iowa, northeast Nebraska, and southeast South Dakota: Iowa Geological Survey Guidebook Series n. 4, p. 43–56.

Mettler, D. E., 1966, West Moorcroft Dakota field, Crook County, Wyoming: The Mountain Geologist, v. 3, p. 89–92.

North American Stratigraphic Code, 1983, North American Commission on Stratigraphic Nomenclature: AAPG Bulletin, v. 67, p. 841–887.

O'Leary, D. W., J. D. Friedman, and H. A. Pohn, 1976, Lineament and linear, a terminological reappraisal, *in* M. H. Podwysocki, and J. L. Earle, eds., Second International Conference on Basement Tectonics Proceedings: Denver, Basement Tectonic Committee, Inc., p. 571–577.

Oriel, S. S., 1959, Lithofacies–thickness maps and subdivisions of the Triassic system, *in* E. D. McKee, et al., eds., Paleotectonic maps of the Triassic system: U.S. Geological Survey Miscellaneous Geologic Investigations Map I-300.

Rice, D. D., 1976, Correlation chart of Cretaceous and Paleocene rocks of the Northern Great Plains: U.S. Geological Survey Oil and Gas Investigations Chart OC-70.

Rice, D. D., and G. W. Shurr, 1980, Shallow, low-permeability reservoirs of Northern Great Plains—Assessment of their natural gas reservoirs: AAPG Bulletin, v. 67, n. 7, p. 969–987.

Sales, J. K., 1968, Crustal mechanics of Cordilleran foreland deformations, A regional and scale model approach: AAPG Bulletin, v. 52, n. 10, p. 2016–2044.

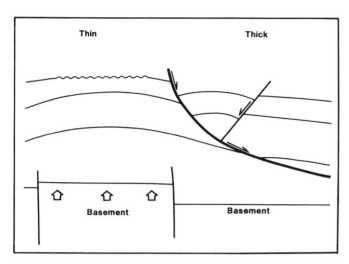

Figure 16—Possible mechanism for initiating listric faults. Orientations and locations may be predictable by mapping thick and thin sediment distributions.

Shurr, G. W., 1976, Lineament control of sedimentary facies in the Northern Great Plains, U.S., *in* M. H. Podwysocki, and J. L. Earle, eds., Second International Conference on Basement Tectonics Proceedings: Denver, Basement Tectonic Committee, Inc., p. 413–422.

Smith, J. G., 1965, Fundamental transcurrent faulting in northern Rocky Mountains: AAPG Bulletin, v. 49, n. 9, p. 1398–1409.

Sonnenberg, F. P., 1956, Tectonic patterns of central Montana: Billings Geological Society 7th Annual Field Conference Guidebook, p. 73–78.

Sonnenberg, S. A., and R. J. Weimer, 1981, Tectonics, sedimentation, and petroleum potential, northern Denver basin, Colorado, Wyoming, and Nebraska: Colorado School of Mines Quarterly, v. 76, n. 2, 45 p.

Stearns, D. W., 1971, Mechanisms of drape folding in the Wyoming province: Wyoming Geological Association 23rd Annual Field Conference Guidebook, p. 125–143.

Stone, D. S., 1969, Wrench faulting and Rocky Mountain tectonics: Wyoming Geological Association Earth Science Bulletin, v. 2, n. 2, p. 27–41.

————, 1971, Tectonic sketch map of the central Rocky Mountains: Wyoming Geological Association 23rd Annual Field Conference Guidebook, 1 sheet.

Stone, W. D., 1971, Stratigraphy and exploration of the Lower Cretaceous Muddy Formation northern Powder River basin, Wyoming and Montana: The Mountain Geologist, v. 9, n. 4, p. 355–378.

Thom, W. T., Jr., 1923, The relation of deep-seated faults to the surface structural features of central Montana: AAPG Bulletin, v. 7, n. 1, p. 1–13.

Thomas, G. E., 1974, Lineament-block tectonics— Williston–Blood Creek basin: AAPG Bulletin, v. 58, n. 7, p. 1305–1322.

————, 1976, Lineament-block tectonics—North America–Cordilleran orogen, *in* M. H. Podwysocki and J. L. Earle, eds., Second International Conference on Basement Tectonics Proceedings: Denver, Basement Tectonics Committee, Inc., p. 361–370.

Vail, P. R., R. M. Mitchum, Jr., and S. Thompson III, 1977, Seismic stratigraphy and global changes of sea level, Part 4, *in* Seismic stratigraphy, Applications to hydrocarbon exploration: AAPG Memoir 26, p. 83–97.

Waage, K. M., 1959, Stratigraphy of the Inyan Kara Group in the Black Hills: U.S. Geological Survey Bulletin 1081-B, p. 11–90.

Waring, J., 1975, Depositional environments of the Lower Cretaceous muddy sandstone, southeastern Montana: Ph.D. thesis, Texas A&M University, College Station, 195 p.

Warner, L. A., 1978, The Colorado Lineament: A middle Precambrian wrench fault system: Geological Society of America Bulletin, v. 89, n. 2, p. 161–171.

Weimer, R. J., 1960, Upper Cretaceous stratigraphy, Rocky Mountain area: AAPG Bulletin, v. 44, n. 1, 20 p.

————, 1978, Influence of transcontinental arch on Cretaceous marine sedimentation: A preliminary report, *in* Energy resources of the Denver basin: Denver, Rocky Mountain Association of Geologists, p. 211–222.

————, 1980, Recurrent movement on basement faults, a tectonic style for Colorado and adjacent areas, *in* Colorado geology: Denver, Rocky Mountain Association of Geologists, p. 23–35.

Weimer, R. J., J. J. Emme, C. L. Farmer, L. O. Anna, T. L. Davis, and R. L. Kidney, 1982, Tectonic influences of sedimentation, Early Cretaceous, east flank Powder River basin, Wyoming and South Dakota: Colorado School of Mines Quarterly, v. 77, n. 4, 61 p.

Wicks, J., 1980, A regional study of joints in the Fall River Formation (Cretaceous), Black Hills of South Dakota and Wyoming: Master's thesis, University of Toledo, Toledo, Ohio, 74 p.

Witzke, B. J., G. A. Ludvigson, J. R. Poppe, and R. L. Ravn, 1983, Cretaceous paleogeography along the eastern margin of the Western Interior seaway, Iowa, southern Minnesota, and eastern Nebraska and South Dakota, *in* M. W. Reynolds and E. D. Dolly, eds., Mesozoic paleogeography of the west-central United States: Rocky Mountain Paleogeography Symposium 2, Rocky Mountain Section, Society of Economic Paleontologists and Mineralogists, p. 225–252.

Wulf, G. R., 1962, Lower Cretaceous Albian rocks in Northern Great Plains: AAPG Bulletin, v. 46, n. 8, p. 1371–1415.

Paleotectonic Controls on Deposition of the Niobrara Formation, Eagle Sandstone, and Equivalent Rocks (Upper Cretaceous), Montana and South Dakota

G. W. Shurr
U.S. Geological Survey
St. Cloud State University
St. Cloud, Minnesota

D. D. Rice
U.S. Geological Survey
Denver, Colorado

Deposition of the Niobrara Formation, Eagle Sandstone, and equivalent Upper Cretaceous rocks was controlled by paleotectonic activity on lineament-bound basement blocks in Montana and South Dakota. Linear features observed on Landsat images provide an interpretation of lineament geometry that is independent of stratigraphic data. Paleotectonism on lineament-bound blocks is documented in three areas that were located in distinctly different depositional environments.

In central Montana, coastal and inner shelf sandstone and nonmarine coastal plain and wave-dominated deltaic deposits reflect paleotectonic control by lineaments trending north–south, east–west, northwest, and northeast. In the northern Black Hills, chalk and outer shelf sandstone reflect control by lineaments trending north–south, northwest, and northeast. In central South Dakota, erosion and deposition of chalk and calcareous shale on a west-sloping carbonate ramp were controlled by lineaments that generally trend northeast and northwest.

Paleotectonism on lineament-bound blocks characterized four tectonic zones located in the Late Cretaceous seaway: the western foredeep, the west-median trough, the east-median hinge, and the eastern platform. The regional geometry of all four tectonic zones appears to be related to the geometry of the convergent plate margin on the west. Paleotectonic activity on lineament-bound blocks may have been the result of horizontal forces related to the convergent margin and to vertical forces related to the movement of the North American plate.

INTRODUCTION

How important is paleotectonism in the development of the stratigraphic record? That question has been raised in a variety of stratigraphic and tectonic settings, but it has rarely been fully answered. The Cretaceous of the Western Interior of North America has recently been the subject of work on paleotectonic controls (e.g., Weimer, 1983). However, the relative importance of paleotectonism and other controls on deposition, such as eustatic changes, is difficult to assess. Indeed, it may not be possible to completely separate the relative influence of tectonism and eustacy by using only the information available in the stratigraphic record. However, paleotectonic controls produce distributions and trends that are predictable and hence can be assessed.

Paleotectonic controls can be most easily evaluated if eustatic influences are minimized by focusing on a single cycle of transgression and regression. Depositional cycles resulting from sea level changes are well documented in the Cretaceous (Weimer, 1960; Hattin, 1964; Kauffman, 1969). In general, the cycles are lithogenetic packages often bounded at the top and bottom by regional unconformities; the cycles are thus depositional sequences in the sense of Mitchum et al. (1977). By examining a single depositional sequence, paleotectonic controls can be assessed in several different depositional settings.

The Niobrara cyclothem of Kauffman (1969) is examined in this study. It includes the Niobrara Formation deposited during an initial transgression and the Eagle Sandstone deposited during a subsequent regression (T_2 and R_2, respectively, of Weimer, 1960). Abundant outcrops in the Western Interior have yielded fossils used to build a detailed biostratigraphy for the cycle. Radiometric dating of bentonites and volcanic rocks intertonguing with sedimentary rocks has calibrated the biostratigraphy to produce an exceptionally well-documented time-stratigraphic framework (Gill and Cobban, 1973; Obradovich and Cobban, 1975; Cobban and Merewether, 1983). The depositional sequence is Santonian to early Campanian in age and encompasses the time span from 78 to 86 m.y. before present. Rock stratigraphic units included in the depositional sequence are shown in Figure 1. The Eagle Sandstone, the Gammon Shale (including the Shannon Sandstone Member), and the Niobrara Formation have been targets for hydrocarbon exploration, and consequently, extensive subsurface data are available in Montana and the Dakotas.

Subsurface data and outcrop information have been integrated to produce an interpretation of regional paleogeography (Rice and Shurr, 1983). That paleogeography (Figure 2) refines the well-known picture (e.g., Reeside, 1957) of broad facies belts aligned north and south in the Cretaceous seaway. It is now apparent that different depositional environments characterized the different tectonic settings that existed from west to east across the seaway. Coastal and inner shelf sandstone, including shoreface and nearshore facies, was apparently deposited on a broad, tectonically active shelf along the western side of the seaway in central Montana. Thick, noncalcareous shale accumulated by eastward progradation of shelf and slope sediments into a subsiding basin situated near the geographic center of the seaway; subsequently, outer shelf sandstone was deposited above the shale that filled the basin in eastern Montana and western South Dakota. Shale and chalk were deposited on a west-sloping ramp, and submarine erosion occurred on the eastern margin of the seaway in central South Dakota.

In this paper we describe the stratigraphy and structural geology in study areas in each of these three depositional and tectonic settings to arrive at two main conclusions. First, paleotectonism exerted a fundamental control on sedimentation throughout the area of marine deposition, including the cratonic eastern side. Second, the paleotectonic features influencing deposition had different orientations in the western, central, and eastern parts of the seaway. Our study relies heavily on the concept of lineament-block tectonics and on the use of satellite images to map surface expression of lineaments. Lineament blocks interpreted from Landsat images represent geometric patterns that can be tested for paleotectonic significance using stratigraphic data.

Basement Blocks and Landsat Linear Features

Lineaments, as defined here, are zones of structural weakness and/or fault zones that bound discrete tectonic blocks in the Precambrian basement. Lineament-block tectonics assumes that major tectonic elements, such as basins and arches, are mosaics of lineament-bound blocks. The concept has been used to describe the development of the Williston basin in terms of wrench-fault tectonics (Thomas, 1974). Laramide structures along block boundaries

have also been related to vertical displacement on basement blocks (Stearns, 1978), and monoclines surrounding the Black Hills provide specific examples of structures controlled by block movement (Lisenbee, 1978). In addition to explaining Laramide structures, lineament-block tectonics has been extensively employed in interpretations of paleotectonism.

In the Williston basin, lineament-bound blocks are suggested to have influenced Paleozoic (Thomas, 1974; Brown, 1978; Maughan, 1983) and Mesozoic (Shurr, 1979; Anna, 1983) sedimentation. Cretaceous movement of basement blocks has been documented in the Powder River basin by Slack (1981) and by Weimer et al. (1982). In the Denver basin, basement blocks controlled deposition throughout Paleozoic and Mesozoic time (Weimer, 1980; Sonnenberg and Weimer, 1981). Only two of these eight studies employed satellite images to map the surface expression of block boundaries.

Lineaments can be interpreted from observations made on Landsat images, and consequently, basement-block geometries can be mapped. Lineaments acting as block boundaries usually have surface expression in landscape components, such as drainage, topography, and vegetation. These landscape components are represented by linear patterns of tone, texture, and color on Landsat images. *Linear features* observed on Landsat are the surface expression of block boundaries and provide the empirical basis for interpretation of lineaments, as shown in Figure 3. Some individual linear features probably reflect individual basement faults and associated folds located along block boundaries. Not all observed linear features, however, mark a subsurface structure. Zones of concentrated short linear features and individual long linear features at the surface are used to interpret regional lineaments that outline basement blocks. This procedure represents an application of lineament-block tectonics to Landsat observations, and these observations can be integrated with stratigraphic data to interpret paleotectonism.

The control exerted by lineament-bound blocks on sedimentation was recognized early in central Montana where lineaments have distinct structural expression (Thom, 1923; Sonnenberg, 1956). To the east in the Williston basin, where lineaments have more subtle landform expression, Landsat images have greatly facilitated the identification and mapping of lineaments in the northern Black Hills and in central South Dakota.

STUDY AREAS

Overview

In this paper, we assess the influence of paleotectonism in different depositional environments in three specific study areas (Figure 4). Each of the three study areas represents a different sector of the Cretaceous seaway. The first area is a part of central Montana that includes the Bearpaw Mountains and Central Montana uplift as major tectonic features. The second area includes the northern Black Hills and adjacent parts of the Williston and Powder River basins. The third area is on the southeastern flank of the Williston basin in central South Dakota. This paper describes the

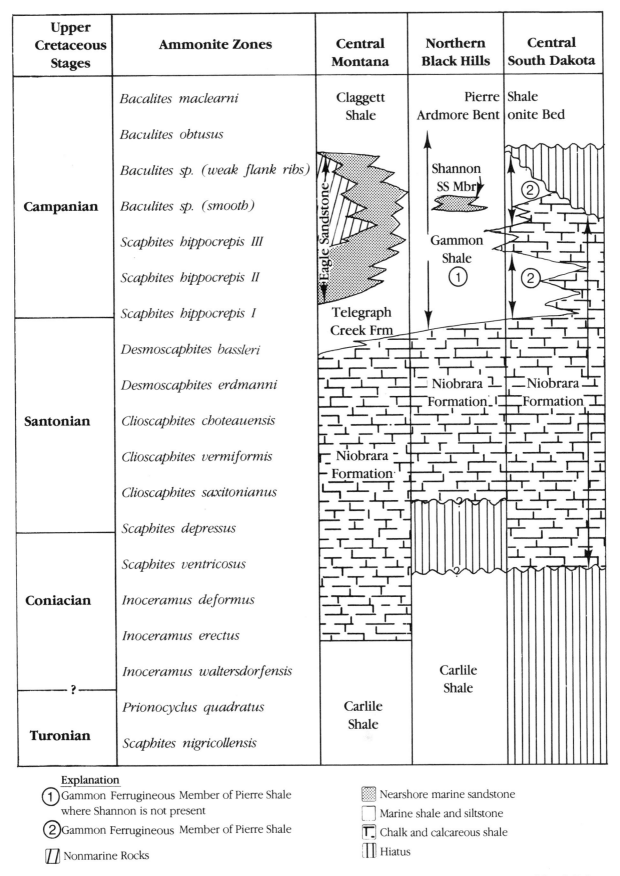

Upper Cretaceous Stages	Ammonite Zones	Central Montana	Northern Black Hills	Central South Dakota
Campanian	*Bacalites maclearni*	Claggett Shale	Pierre Shale	
	Baculites obtusus		Ardmore Bent	onite Bed
	Baculites sp. (weak flank ribs)	Eagle Sandstone	Shannon SS Mbr	②
	Baculites sp. (smooth)			
	Scaphites hippocrepis III		Gammon Shale ①	②
	Scaphites hippocrepis II			
	Scaphites hippocrepis I	Telegraph Creek Frm		
Santonian	*Desmoscaphites bassleri*			
	Desmoscaphites erdmanni		Niobrara Formation	Niobrara Formation
	Clioscaphites choteauensis			
	Clioscaphites vermiformis	Niobrara Formation		
	Clioscaphites saxitonianus			
	Scaphites depressus			
Coniacian	*Scaphites ventricosus*			
	Inoceramus deformus			
	Inoceramus erectus			
	Inoceramus waltersdorfensis		Carlile Shale	
? Turonian	*Prionocyclus quadratus*	Carlile Shale		
	Scaphites nigricollensis			

Explanation

① Gammon Ferrugineous Member of Pierre Shale where Shannon is not present

② Gammon Ferrugineous Member of Pierre Shale

▧ Nonmarine Rocks

▨ Nearshore marine sandstone

▢ Marine shale and siltstone

⊤ Chalk and calcareous shale

▥ Hiatus

Figure 1—Stratigraphic framework of the Eagle Sandstone, Niobrara Formation, and equivalent rocks in Montana and South Dakota. Modified from Rice (1976), Rice and Shurr (1983), and Cobban and Merewether (1983). Used with permission of the Society of Economic Paleontologists and Mineralogists, Rocky Mountain Section.

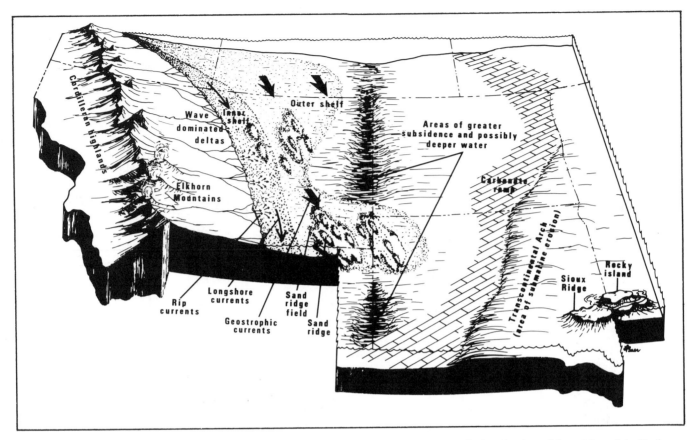

Figure 2—Regional paleogeographic reconstruction of a sector in the Late Cretaceous seaway during the time of deposition of the Eagle Sandstone and Niobrara Formation. From Rice and Shurr (1983). Used with permission of the Rocky Mountain Association of Economic Paleontologists and Mineralogists, Rocky Mountain Section.

Figure 3—Individual linear features observed on Landsat images are components of a wide lineament zone that bounds basement blocks and appear to be the surface expression of structural features associated with the block boundary. No horizontal or vertical scale. From Shurr (1982). Used with permission of the Saskatchewan Geological Society.

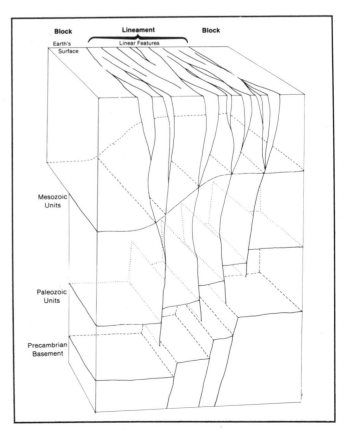

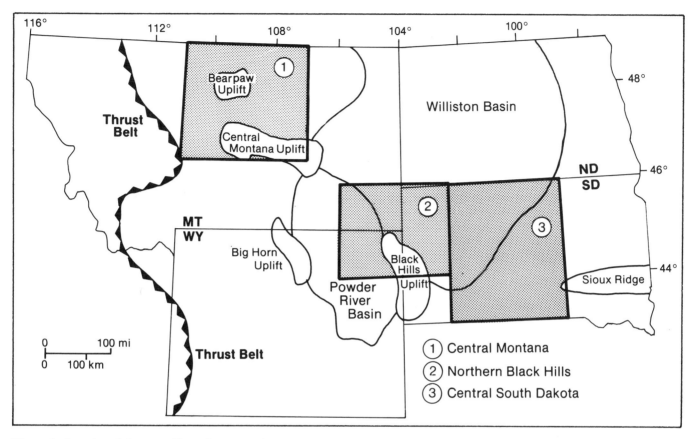

Figure 4—Location of three specific study areas in this paper and major tectonic elements.

Niobrara Formation, Telegraph Creek Formation, Eagle Sandstone, Gammon Shale, and Gammon Ferruginous Member of the Pierre Shale (Figure 1). These units constitute the Niobrara cyclothem that is a depositional sequence genetically distinct from the underlying Carlile Shale and overlying Claggett and Pierre shales.

Although stratigraphic units and rock types are different in each of the study areas (Figure 1), a record of overall progradation persists across the seaway. The transgressions that initiated and terminated the Niobrara cyclothem were relatively rapid, and as a consequence, the stratigraphic record in these rocks is dominated by the effects of progradation and regression. Thus, the influence of eustacy is minimized and the interpretation of paleotectonism is made possible.

Rocks in the central Montana study area document environments of deposition along the western margin of the Cretaceous seaway (Figure 2). After the initial transgression, progradation of offshore calcareous shale and siliciclastic siltstone and shale is recorded, respectively, in the Niobrara and Telegraph Creek formations (Figure 1). As rising Cordilleran highlands and volcanic sources, such as the Elkhorn Mountains, shed clastics to the east, the offshore siltstone and shale were overlain by coastal sandstone of the lower Virgelle Member of the Eagle Sandstone. This sandstone was apparently deposited as shoreface sand and as inner shelf sand ridges. Continued progradation carried nonmarine coastal plain and wave-dominated deltaic

deposits eastward across the coastal sand. Marine sandstone in the upper part of the Eagle indicates short progradational events in an overall transgression that led to initiation of the Claggett cyclothem. In summary, the central Montana study area was located on the nearshore and inner shelf margin adjacent to the source of siliciclastic sediments that were being dispersed into the Cretaceous seaway.

Rocks in the northern Black Hills study area document environments of deposition in the central part of the seaway (Figure 2). After the initial transgression, calcareous shale and chalk of the Niobrara Formation were deposited in areas of deeper water and/or greater subsidence. In the overlying Gammon Shale, offshore marine siltstone and shale formed thick progradational prisms that filled the northern area of deeper water and/or greater subsidence. Later in the overall progradation, outer shelf sandstone of the Shannon Sandstone Member was deposited as sand ridge fields, while progradation of the shelf margin continued to the southeast. Marine shale that overlies the Shannon was deposited on a relatively flat shelf during the early stages of the transgression that initiated the Claggett cyclothem. Thus, in the vicinity of the northern Black Hills, areas of deeper water and/or greater subsidence acted as sediment sinks that controlled the eastward distribution of siliciclastics deposited on the western shelf.

Rocks in the central South Dakota study area document environments on the eastern seaway margin (Figure 2). Calcareous shale and chalk of the Niobrara were deposited on a series of carbonate ramps that sloped northwestward into areas of deeper water. The ramps represented the northwestern flank of the Transcontinental arch and generally sloped away from a rocky island located on the Sioux ridge. The initial transgression of the cyclothem is marked by an unconformity at the base of the Niobrara. This unconformity is overlain by a chalk tongue equivalent to the coastal sandstone of the Virgelle Sandstone Member and to thick progradational shale prisms of the Gammon. An upper chalk tongue of the Niobrara is equivalent to sand ridges in the Shannon and to thick progradational shale prisms that filled the area of deeper water on the Wyoming–South Dakota border. The final transgression correlates with an unconformity at the top of the Niobrara. Deposition of calcareous sediments characterized the eastern margin of the seaway because the siliciclastics derived from the west were largely trapped in the central part of the seaway.

Central Montana Study Area

The central Montana study area (Figure 4) has numerous local domes and igneous centers. These uplifts are located east of the main mountain front and have been collectively referred to as the Central Montana Rockies (Eardley, 1962). In the southern part of the study area, the Little Belt and Big Snowy mountains (Figure 5) are structural domes with relief exceeding 10,000 ft (3000 m) (Dobbin and Eardmann, 1955). These two domes and the Cat Creek fault zone (Figure 5) constitute the western and northern limits of the Central Montana uplift as recognized by Sonnenberg (1956). In the central part of the study area the Highwood, Moccasin, and Judith mountains (Figure 5) are igneous centers and the latter two have doming associated with igneous activity. In the northern part of the study area, the Bearpaw and Little Rocky mountains (Figure 5) are domed igneous centers where Precambrian rocks are also elevated and exposed. Early interpretations by Thom (1923) and Sonnenberg (1956) have related many of the structures to basement blocks that have been the sites of recurrent paleotectonic activity. In a review of earlier literature, Johnson and Smith (1964, p. 64-65) summarize by saying

...Central Montana has been structurally active during much of Paleozoic and Mesozoic time and deposition of most Paleozoic and Mesozoic sedimentary rocks...was influenced by ancestral structural features that had the positions and trends of the present surface features.

It has been suggested by Mahr (1969) and Schorning (1972) that a large uplift coinciding with the present-day Bearpaw Mountains influenced deposition of the Eagle Sandstone. These studies, however, do not emphasize the lineament-block model.

Lineaments mapped on Landsat images greatly aid in the interpretation of paleotectonic controls on sedimentation. Parts of at least eight regional lineaments have been mapped in the central Montana study area (lineaments A–H in Figure 5). Lineaments at this scale are shown as broad areas bounded in places by long linear features and generally enclose areas where shorter linear features are concentrated. The lineament zones have expression on aeromagnetic and gravity maps and are believed to represent the boundaries of

basement blocks (Shurr, 1983). The geologic significance of lineaments is further demonstrated by their relationship to specific structural features (Figure 5A). The Highwood, Little Belt, Big Snowy, Moccasin/Judith, and Little Rocky mountains lie along lineaments A, B, and D. The Cat Creek fault zone approximates lineament F. The maximum extent of thrust faults around the southern flank of the Bearpaw Mountains is limited by lineaments A, G, and H.

The record of progradation shown by the Niobrara, Telegraph Creek, and Eagle formations in central Montana is reflected in the position of strandlines shown in Figure 5. In general, the shape of strandline 1 appears to be related to the locations of the Highwood and Little Belt mountains, and the positions of strandlines 2 and 3 to the locations of the Big Snowy, Moccasin/Judith, and Bearpaw mountains (Figure 5B). Strandline shapes and positions, however, are even more clearly tied to lineaments interpreted from Landsat (Figure 5C): lineament A coincides with a local bulge on strandline 2: lineaments A and B outline large bulges in strandlines 1, 2, and 3; lineament C parallels strandline 4; lineaments D and E parallel and enclose strandline 3; and lineaments G and H parallel and enclose strandlines 2 and 3. We suggest that the grid of lineaments outlines subtle topographic features that controlled the position of strandlines. In general, blocks in the western part of the study area were apparently topographically higher and were the sites of nonmarine deposition; marine deposition in the east occurred on topographically lower and submerged blocks.

The Eagle Sandstone has been the subject of detailed investigation (Rice, 1980) in an area south of the Bearpaw Mountains (dashed box in Figure 5C). Numerous subsurface control points produced by hydrocarbon exploration have been integrated with outcrop information. In this area, the Eagle consists of three members. The lower member, the Virgelle Sandstone Member, is an 80–120-ft (24–40-m) thick unit that coarsens upward with bioturbated and parallel-bedded sandstone in the lower part and cross-bedded sandstone in the upper part; it is interpreted to have been deposited in an aggrading coastal and interdeltaic setting associated with strandlines 1 and 2 (Figure 5). The informal middle member of the Eagle is 70–170 ft (20–50 m) thick and is composed of two lithofacies that are in part laterally equivalent: (1) a mudstone unit that includes some siltstone, sandstone, and coal beds; and (2) a sheetlike sandstone unit that has scour surfaces with coarse grains concentrated in the lower part and some trough cross bedding in the upper part. These lithofacies of the middle member are interpreted to represent deposition in coastal plain (mud) and wave-dominated delta (sand) environments at the time of maximum progradation shown by strandline 3. The informal upper member of the Eagle is 30–95 ft (10–30 m) thick. It consists of interbedded sandstone, siltstone, and shale deposited in tidal flat environments and of coarsening-upward, shoreface sandstone. These lithofacies are the result of overall transgression (strandline 4, Figure 5) that ultimately terminated Eagle deposition.

Isopach patterns for the total Eagle interval (Figure 6) are related to the distribution of lithofacies in the three members. Delta front sandstone of the middle member is present in the central part of the study area where the Eagle

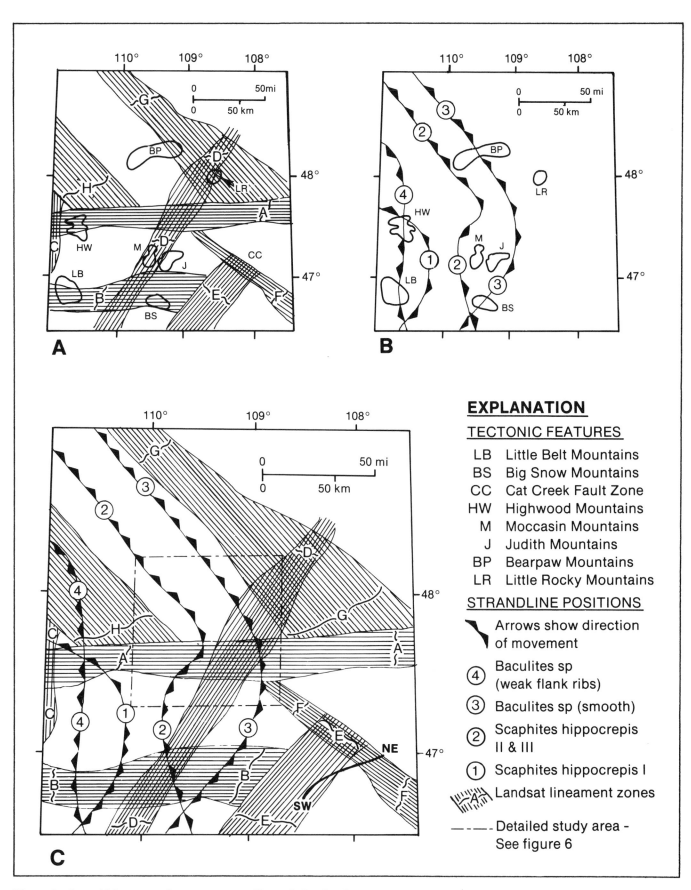

Figure 5—Central Montana study area (area 1 on Figure 4). Landsat lineaments appear to be closely related to tectonic elements and to Late Cretaceous strandline positions. From Rice (1980). A. Relationship of lineaments to structural features. B. Location of strandlines relative to tectonic elements (see text for discussion). C. Relationship of lineaments to strandlines. (SW–NE line is location of stratigraphic section in Figure 7.) From Rice (1980).

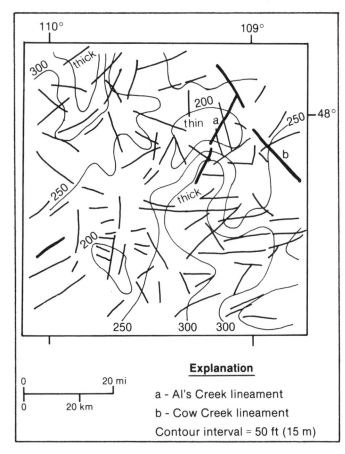

Explanation

a - Al's Creek lineament
b - Cow Creek lineament
Contour interval = 50 ft (15 m)

Figure 6—Map showing isopach patterns in Eagle Sandstone (Rice, 1980), Landsat linear features, and named structural lineaments. See Figure 5C for location.

is greater than 300 ft (90 m) thick. Thinning occurs in two general directions: toward the west where nonmarine rocks represent the middle member and toward the east where coastal sandstone grades into marine shale.

Isopach patterns in the Eagle Sandstone seem to be related to the orientation and distribution of individual linear features mapped on Landsat images (Figure 6). Linear features parallel contours, mark indentations on overall contour trends, and are concentrated in areas of maximum thickness change. Named structural and fault zones are associated with several linear features. The Al's Creek and Cow Creek lineaments, mapped and named by Hearn et al. (1964), are of particular interest (Figure 6). These structural lineaments are evident as linear features on Landsat images and correspond approximately with an area of thin Eagle deposition. The area of thin deposition has paleotectonic significance because it coincides with a reentrant on the facies boundary between the nearshore marine sandstone and the marine siltstone and shale in the Virgelle Sandstone Member (Rice, 1980). East–west linear features in the central part of the area of detailed study are components of lineament A (Figure 5). Linear features trending northeast are elements of lineament D. Linear features in the northern half of the area that trend northwest represent lineaments G and H. The area of thick delta front sandstone in the central part of the area coincides with the intersection of lineaments A and D.

Lineaments also apparently influenced deposition on the inner shelf as well as near the strandlines. Northeastward across the Cat Creek fault zone, which defines the north side of the Central Montana uplift and approximates lineament F (Figure 5), the Eagle Sandstone grades laterally into the Gammon Shale (Figure 7). Near the town of Winnett, excellent exposures of the lower member show imbricated sandstone lenses (Rice and Shurr, 1983). These lenses are interpreted to be inner shelf sand ridges that prograded toward the southwest away from the Cat Creek fault zone (lineament F, Figure 5). Marine shale was deposited on the topographically lower block northeast of lineament F while sand ridges prograded across the block toward the southwest. Any influence on sedimentation by lineament E has not been established.

In summary, paleotectonism on lineament-bound blocks in central Montana appears to have controlled deposition of the Eagle Sandstone. Generally, nonmarine rocks grade eastward into coastal sandstone that was deposited directly on the shoreface and on sand ridges located farther seaward on the inner shelf. This eastward gradation of sedimentary environments reflecting paleobathymetry suggests that the net movement on the mosaic of basement blocks was generally down to the basin.

Northern Black Hills Study Area

The northern Black Hills study area (Figure 4) has a relatively simple tectonic setting: the Black Hills uplift and the Miles City arch to the northwest separate the Williston basin on the northeast from the Powder River basin on the southwest. Local domes and igneous centers are relatively small and isolated. Monoclines are the most conspicuous structural features in the study area, and these monoclines are forced folds developed over differentially uplifted basement blocks (Lisenbee, 1978). Lineaments interpreted from linear features mapped on Landsat images have been related to monoclines and to other geologic features (Shurr, 1982). Lineaments, which coincide with monoclines, were apparently active during both Paleozoic and Mesozoic time and outline regional fracture domains. Linear features mark specific subsidiary structures along monoclines, and some individual linear features have expression as faults on seismic records. Lineament-bound blocks influenced sedimentation on the outer western shelf and in the basin of the Cretaceous seaway.

The Niobrara cyclothem in the northern Black Hills consists of the 200-ft (60-m) thick Niobrara Formation and the more than 1000-ft (300-m) thick Gammon Shale. The Gammon Shale has formation status in subsurface and in areas where the Shannon Sandstone Member is present; elsewhere, the interval is referred to as the Gammon Ferruginous Member of the Pierre Shale (Rice et al., 1982). Chalk of the Niobrara grades laterally to calcareous shale (Figure 8) and noncalcareous shale. In a few places pre-Niobrara faults are associated with these facies changes. Similar paleofaults have been identified on the southwestern flank of the Black Hills; however, faulting was post-Niobrara and pre-Gammon (Asquith, 1970). The lower part of the Gammon is offshore marine siltstone and shale units that display sigmoidal geometry, apparently resulting from southeastward progradation of the shelf margin. The upper

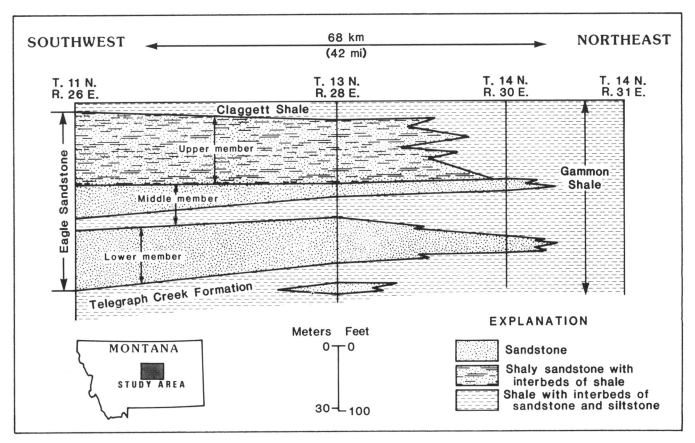

SOUTHWEST ←——— 68 km (42 mi) ———→ NORTHEAST

T. 11 N.
R. 26 E.
T. 13 N.
R. 28 E.
T. 14 N.
R. 30 E.
T. 14 N.
R. 31 E.

Claggett Shale

Eagle Sandstone

Upper member

Middle member

Lower member

Gammon Shale

Telegraph Creek Formation

MONTANA
STUDY AREA

Meters Feet
0 — 0

30 — 100

EXPLANATION

Sandstone

Shaly sandstone with
interbeds of shale

Shale with interbeds of
sandstone and siltstone

Figure 7—Facies changes in Eagle Sandstone (Rice and Shurr, 1983) across Cat Creek fault zone. See Figure 5C for location.

part of the Gammon is a complex lithosome that includes two types of shelf sandstone and a variety of sand body geometries (Shurr, 1983). The majority of the lithosome is made up of thin sandstone beds or laminae interbedded with shale and siltstone. The second type of shelf sandstone forms a variety of distinctive sand bodies in the Shannon Sandstone Member. The Shannon is stratigraphically equivalent to the middle member of the Eagle in central Montana and is generally 150–200 ft (45–60 m) below the top of the Gammon. Gammon strata above the Shannon record the end of progradation and the beginning of the transgression that terminated the depositional sequence and continued into the Claggett cyclothem. Effects of paleotectonism on deposition of thick Gammon Shale are difficult to assess. The thin Niobrara and Shannon with their distinctive chalk and sandstone facies provide two stratigraphic units that were more sensitive to paleotectonic controls.

Niobrara thickness and facies changes have been described in the Great Plains east of the Black Hills (Shurr, 1984b). In general, Niobrara chalks thin and pass laterally into calcareous and noncalcareous shale in the same areas where Gammon Shale is very thick. Chalks have expression on electric logs as units of high resistivity and large departures on the spontaneous potential (SP) log (Figure 8). Calcareous shale has lower resistivity and small SP departures, but it is not easily distinguished from noncalcareous shale, which has low resistivity and SP curves with little character. The map

of these lithologies (Figure 9) is based on approximately one well per township in Montana and Wyoming, but well control is more sparse in South Dakota where hydrocarbon exploration has not been as extensive. Reconstruction of the facies pattern is highly conjectural over the area of the Black Hills (Figure 9) where the Niobrara has been removed by erosion.

Paleotectonic controls on Niobrara deposition can be assessed by comparing facies patterns with linear features mapped on Landsat images (Figure 9). Linear features shown in Figure 9 are the primary components of lineaments interpreted to outline a basement-block mosaic in the northern Black Hills (Shurr, 1982). In general, northwest-oriented features appear to control the overall facies patterns, and linear features trending northeast and north mark small incursions in the regional trends. Lineament control appears to be more clearly defined in Montana and Wyoming than in the South Dakota part of the study area. This apparent difference in lineament control may be a real expression of the degree of basement-block control or it may be a result of differences in Landsat mapping or of differences in subsurface control.

A northwest-trending block, which was the site of chalk deposition in Wyoming and Montana (area 1 on Figure 9), is particularly noteworthy because it seems to support an interpretation of Niobrara deposition in areas to the east in the Dakotas. Specifically, the deposition of chalk seems to have been restricted to uplifted blocks adjacent to

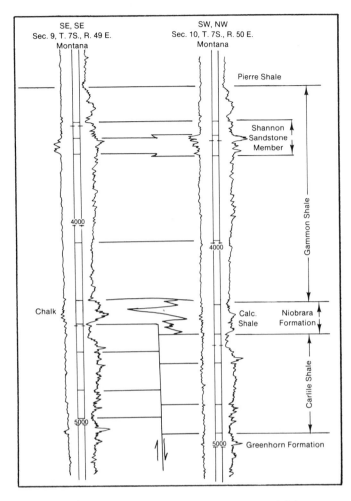

Figure 8—Correlation of electric (spontaneous potential) logs showing thickness changes in the Shannon Sandstone Member and facies changes in the Niobrara Formation across a paleofault marked by a Landsat linear feature. See Figure 9 for location.

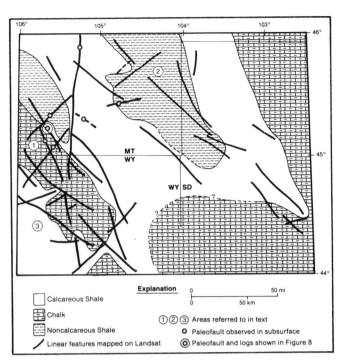

Figure 9—Lithologic variations in the Niobrara Formation around the northern Black Hills appear to be associated with Landsat linear features noted by Shurr (1982). This is study area 2 on Figure 4. Used with permission of the Saskatchewan Geological Society.

topographically lower blocks where noncalcareous mud was deposited (Shurr, 1984b). The northeastern margin of the block on which chalk was deposited is sharply marked by linear features and pre-Niobrara faulting (Figure 8); shale was deposited to the northeast on a down-dropped block. Chalk deposited on the uplifted block grades to calcareous shale toward the southwest, suggesting that the block was tilted and that relief was lower on the southwestern flank than on the northeastern flank. Paleofaults recognized in the subsurface support the interpretation of difference in relief on a sloping block. At the northern side of the sloping block in Montana, pre-Niobrara faults are more numerous (Figure 9). At the southern and eastern margins of the sloping block in Wyoming, offsets are not readily evident, and relative block movement may have produced monoclinal flexures rather than faults. Widely spaced control points in South Dakota probably preclude identification of paleofaults.

The Shannon Sandstone Member of the Gammon Shale provides an additional opportunity to assess paleotectonism. This shelf sandstone coarsens from bioturbated fine-grained sandstone with interlaminated siltstone and shale upward to cross-bedded medium- to coarse-grained sandstone.

Sandstone bodies that make up the Shannon fall into a hierarchy of sizes (Shurr, 1984a): facies packages as much as 10–20 ft (3–6 m) thick and covering approximately 0.05 mi^2 (0.12 km^2) probably represent submarine sand waves; elongate lenses as much as 40–60 ft (12–18 m) thick and covering approximately 20 mi^2 (50 km^2) are interpreted to be sand ridges; regional lentils as much as 50–75 ft (15–23 m) thick and covering approximately 580 mi^2 (1500 km^2) are the record of sand ridge fields; and a sandstone sheet about 100 ft (30 m) thick and covering approximately 7000 mi^2 (18,200 km^2) comprises coalesced sand ridge fields bounded by linear features. The outline of the sandstone sheet shown in Figure 10 includes three coalesced lentils (A, B, and C in Figure 10) and shows expression of individual sand ridges (e.g., area 1 in Figure 10) near the margins of the sand ridge fields. Sandstone in the southwestern part of the study area may be at a slightly different stratigraphic position and may represent a lentil in a sandstone sheet that extends south and west into Wyoming. As with the Niobrara, reconstruction of facies patterns across the Black Hills uplift is highly conjectural because the Shannon has been removed there by erosion.

Northwest linear features were important controls on sandstone geometry in South Dakota and Wyoming, and in addition, north–south and northeast linear features were important in Montana. In general, the shelf sand of the Shannon accumulated on the same blocks that had been depositional sites for calcareous and noncalcareous mud during the time of Niobrara deposition. Linear features that bound blocks containing Niobrara shale are essentially the same as those that controlled Shannon sandstone

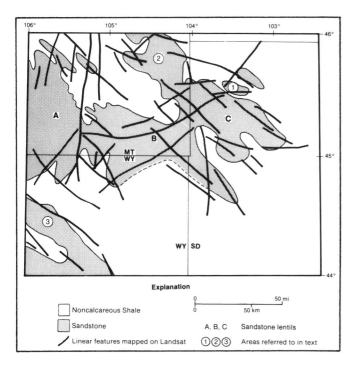

Figure 10—Distribution of the Shannon Sandstone Member of the Gammon Shale in the area of the northern Black Hills. Landsat linear features noted by Shurr (1982) appear to control sandstone distribution. This is study area 2 on Figure 4. Used with permission of the Saskatchewan Geological Society.

distribution (compare areas 2 and 3 in Figure 9 with areas 2 and 3 in Figure 10). The northwest-trending block of chalk deposition in Wyoming (area 1 in Figure 9) is bounded by blocks that later received sand during the time of Shannon deposition (A in Figure 10). The generalization that sand accumulated on blocks that earlier received mud is not clearly applicable to the part of the block (area 1 in Figure 9) along the Montana border where lentil A (Figure 10) appears to have been deposited in an area of earlier chalk deposition. However, the sandstone lentil is thicker northeast of the chalk block in the area of the downdropped block where Niobrara mud accumulated. Figure 8 shows the increase in sandstone thickness that takes place across a paleofault controlling Niobrara facies change.

Shelf sandstone appears to have accumulated, or rather was preserved, in slight topographic depressions that remained on the shelf surface after major shale infilling. As a consequence, blocks containing Shannon sandstone are also the blocks that received Niobrara shale. This generalization seems to lend support to the hypothesis that Cretaceous shelf sandstone bodies are parts of "relict" sand masses reworked and "frozen" in areas of slightly deeper water (Hobson et al., 1982). Hobson et al., however, interpret the relict masses to be the remains of earlier shoreline or even nonmarine sediments, and we do not believe that that can be demonstrated in the Shannon. Shelf sand preserved in topographic depressions does not seem compatible with an interpretation that sand had been concentrated on subtle topographic highs on the shelf surface (Rice and Shurr, 1983). Further stratigraphic studies on the Gammon above the Shannon are needed to test the hypothesis that sand accumulated in topographic lows.

Linear features visible on Landsat images define blocks that controlled sedimentation during the time of Niobrara and Shannon deposition in the northern Black Hills. This conclusion is broadly similar to interpretations of paleotectonism made by Slack (1981) for the northern Powder River basin, but differs in several important details. Slack shows a dominance of northeast-trending features; however, we note that linear features trending north–south and northwest are as conspicuous as those trending northeast (Figures 9 and 10). He also postulates a simple northeast-trending "Belle Fourche arch" in Wyoming; however, we believe that mosaic blocks showing differential uplift are complex. Furthermore, paleofaults in Wyoming are interpreted by Slack from indirect evidence such as facies patterns; in contrast, paleofaults are clearly observed in the subsurface in southeastern Montana (Figure 8).

Central South Dakota Study Area

The structural setting of central South Dakota is the simplest of the three study areas. It is situated on the southeastern flank of the Williston basin adjacent to the Sioux ridge (Figure 4). Although few folds or faults are known to exist in the area, linear features observed on Landsat have been interpreted as lineaments that bound basement blocks (Shurr, 1981b). The present-day structure is a mosaic of blocks that step down to the north and west into the Williston basin. Lineaments marking block margins generally correspond with major streams that flow northeastward into the Missouri River. The lineament-block mosaic also constituted paleotectonic elements that apparently controlled deposition of the Niobrara cyclothem.

Three chalk tongues are present in the Niobrara Formation of central South Dakota. The lowest tongue is distinctive in Nebraska but is difficult to distinguish in South Dakota. It correlates with the unconformity at the base of the Niobrara in the northern Black Hills (Figure 1). Western limits of the middle chalk tongue east of the Black Hills are shown in Figure 9. The middle chalk tongue in central South Dakota is equivalent to thick Gammon shales in the northern Black Hills, and the upper chalk in central South Dakota is equivalent to the Shannon in the northern Black Hills (Figure 1). Chalk tongues coalesce in eastern South Dakota where intervening calcareous shale units thin and the Niobrara is a single chalk unit about 200 ft (60 m) thick. Eastward from the Black Hills, the upper part of the Gammon thins and the top of the Niobrara is marked by an unconformity throughout most of central and eastern South Dakota. These stratigraphic relationships are interpreted to reflect deposition on a series of carbonate ramps (Shurr, 1984b) that gently sloped northwestward from the Transcontinental arch and down into deeper water in the central part of the seaway. The highest part of the Sioux ridge apparently stood as a rocky island on the crest of the Transcontinental arch (Figure 2). Clastic components in the Niobrara increase adjacent to the ridge, and a shallow water facies equivalent to the Niobrara has been described as a localized basin-filling unit on the ridge (Witzke et al., 1983). In eastern North Dakota, channels at the base and top of the Niobrara trend northwestward and the Niobrara consists of a lower calcareous shale and an upper chalk (Reiskind, 1982). These observations are consistent with a northwest paleoslope off the Transcontinental arch, which was the site

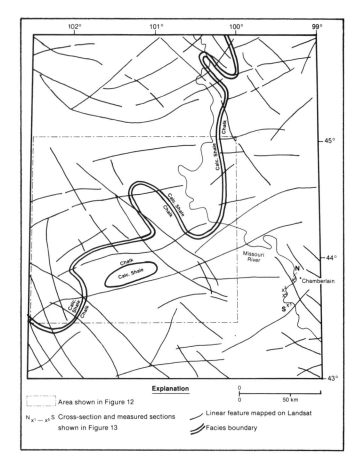

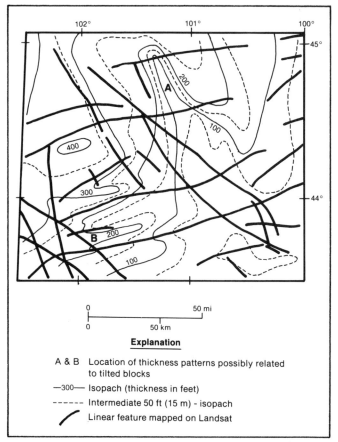

Figure 11—Lateral facies change of chalk (Niobrara Formation) to calcareous shale (Gammon Shale) relative to Landsat linear features in central South Dakota. This is study area 3 on Figure 4. Modified from Shurr (1978). Used with permission of the Montana Geological Society.

Figure 12—Isopach map of the Gammon Shale in central South Dakota. See Figure 11 for location of map area. Modified from Shurr (1978). Used with permission of the Montana Geological Society.

of major chalk deposition. Chalks mainly accumulated in shallow water on the crest of the arch, which was apparently a submarine sill. Terrigenous material diluted the carbonate sediments in slightly deeper water on the north flank of the arch and in the much deeper water of the basin to the northwest (Shurr, 1984b).

Linear features mapped on Landsat images in central South Dakota have been ranked according to frequency of observation and have been integrated with a map of vertical magnetic intensity. Subparallel pairs of long linear features are interpreted to bound basement blocks, and some specific linear features seem to have affected Niobrara and Gammon facies (Figure 11). The general northeast–southwest trend of the facies line between the upper chalk tongue and calcareous shale is interrupted by northwest-trending perturbations. The general northeasterly trend and northwesterly perturbations of this facies line seem to correspond with specific linear features. Along the Missouri River the correspondence of facies and linear features is not as distinct as in the northern part of the study area, where the relationship between the facies line and linear features is possibly clearer. In the southwestern part of the study area the upper chalk is not developed locally in an area that is outlined by linear features. Blocks, which were the site of chalk accumulation in the eastern and southern part of the

study area, are interpreted to have stood higher relative to blocks where calcareous shale was deposited in the basin to the northwest. This block geometry represents westward down-to-the-basin displacements that mirror block displacements in central Montana.

The Gammon in central South Dakota is the lowest member of the Pierre Shale, and thickness patterns generally parallel facies changes within the Niobrara Formation. Shales of the Gammon are thin where Niobrara chalks are well-developed (Figures 11 and 12). Thickness patterns in the Gammon (Figure 12) commonly show a close relationship with specific linear features and seem to indicate deposition on tilted basement blocks similar to those northwest of the Black Hills. For example, the 200-ft (60-m) isopach in the northern part of the area (A in Figure 12) corresponds with a northwestward extension of chalk (Figure 11). These relationships suggest that an underlying block was uplifted on the western side and sloped gently to the east. Another possible tilted block may be present near location B (Figure 12). A block in which chalk deposition was dominant sloped gently westward toward location B. At this location there was an abrupt decrease in elevation across the linear feature into an area where shale was deposited.

The Gammon thins by facies change and by erosion eastward until the Sharon Springs Member of the Pierre

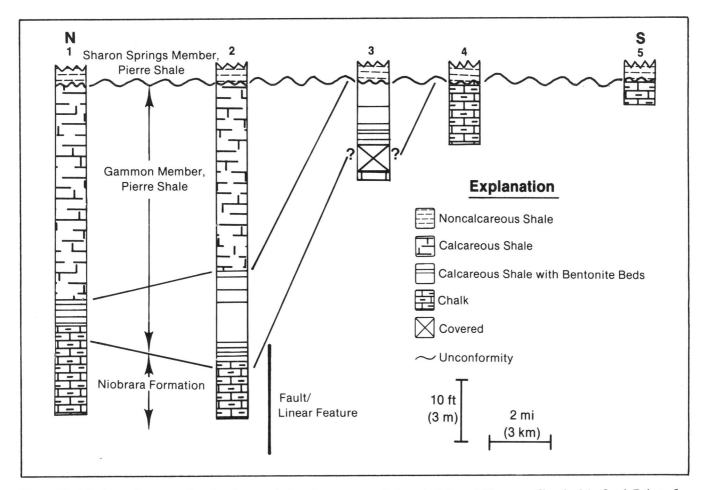

Figure 13—North–south cross section showing correlation of strata exposed along the Missouri River near Chamberlain, South Dakota. See Figure 11 for location.

Shale directly overlies the Niobrara. The top of the Gammon has been eroded south and east of the area of abrupt thinning and approximately coincident with the 150-ft (45-m) isopach on Figure 12.

Exposures along the Missouri River show the Niobrara unconformably overlain by the Sharon Springs Member of the Pierre Shale (Sections 4 and 5, Figure 13). Calcareous shale below the Sharon Springs may be equivalent to the Gammon, but no biostratigraphic control is available. Although the amount of erosion at the unconformity is difficult to evaluate, erosional cutout can be demonstrated by correlation of units defined by bentonite beds. Tracing units in outcrop suggests that calcareous shale thins southward because of erosion rather than a facies change. Phosphatic nodules and concentrations of abraded vertebrate material, which commonly characterize unconformities, are locally incorporated at the base of the Sharon Springs. This is only one of a series of anastomosing unconformities observed in exposures of the lower Pierre along the Missouri River (Schultz, 1965). An abrupt change in thickness of units below the unconformity at the top of the Niobrara between sections 2 and 3 (Figure 13) appears to be coincident with a Landsat linear feature trending northwestward. The sense of movement on this fault is down on the northeast side of the

linear feature, assuming greater erosion on the uplifted block. However, the sense of displacement on a fault exposed in a roadcut near section 2 is opposite to that indicated by the strata; specifically, the north block has moved up relative to the south block. Overlying units of the Pierre have been removed by recent erosion along the Missouri so that the fault can be dated only as post-Niobrara. Other paleofaults associated with linear features south of Chamberlain can be dated as post-Niobrara and pre-Sharon Springs; these faults have upthrown blocks to the northwest and correspond with linear features trending northeast (Shurr, 1979). All of the faults are near vertical.

In summary, on the eastern side of the Late Cretaceous seaway, carbonate ramps sloped away from the Transcontinental arch. Margins of these ramps were apparently controlled by lineament-bound blocks within deeper water parts of the seaway. Although tilted blocks are not as readily identified in central South Dakota as in the northern Black Hills, the regional northwesterly slope from the arch into the seaway implies tilting. Reentrants in the facies line at the margin of the carbonate ramp probably represent submarine channels lying along block boundaries oriented northwestward, perpendicular to the axis of the Transcontinental arch. Erosion on the shallower parts of the ramp in the southern and eastern part of the study area seems to have been controlled by varying amounts of uplift on fault-bounded blocks.

OTHER EXAMPLES OF PALEOTECTONISM

In the northern Great Plains sector of the Late Cretaceous seaway there is evidence for paleotectonism recorded in depositional sequences below and above the Niobrara cyclothem. Anna (1983) described lineament patterns in several Cretaceous stratigraphic units below the Niobrara. Rice (1984) demonstrated that the Mosby Sandstone Member of the Belle Fourche Shale reflects higher energy depositional environments on Bowdoin dome and the Central Montana uplift than in intervening areas of low energy.

The Parkman Sandstone lies 200 ft (60 m) above the Gammon on Porcupine dome at the east end of the Central Montana uplift. Here, thickness and facies patterns parallel lineaments, and paleocurrents interpreted from sedimentary structures trend parallel with and perpendicular to faults in the Cat Creek fault zone (Shurr, 1979). In the northern Black Hills the thickness of the Ardmore Bentonite Beds, which lie directly above the Gammon Shale, appears to be related to lineaments (Shurr, 1979). Thickness and facies patterns for several members in the lower part of the Pierre Shale in central South Dakota have been interpreted to be controlled by lineament-bound basement blocks (Shurr, 1978). In addition, paleofaults exposed along the Missouri River mark lineaments and perhaps influenced deposition of lower members of the Pierre Shale above the Niobrara (Shurr, 1979).

REGIONAL TECTONIC SETTING

Relationship to Depositional Environments

The Late Cretaceous seaway of North America flooded a variety of tectonic provinces, and the three study areas described in this report provide an insight into the different structural styles of these provinces. Four longitudinal tectonic zones (Table 1) have been named by McNeil and Caldwell (1981) by modifying tectonic provinces originally proposed by Kauffman (1977). From west to east these zones are the western foredeep, the west-median trough, the east-median hinge, and the eastern platform. A map of these zones during deposition of the Niobrara cyclothem is shown in Figure 14; boundaries are based on differences in lithology and thickness patterns (see Table 1).

Different depositional environments were present in each of the four tectonic zones (Table 1). Subareal and coastal environments were located in the western foredeep; continental shelf and slope environments were located on the west-median trough; a carbonate ramp was located in the east-median hinge; and erosion characterized shallow water areas of the ramp on the eastern platform. Water depths in each of the four tectonic zones were controlled by the relative rates of subsidence and sedimentation (McNeil and Caldwell, 1981). In the western foredeep, subsidence and sedimentation rates were maximum and were approximately

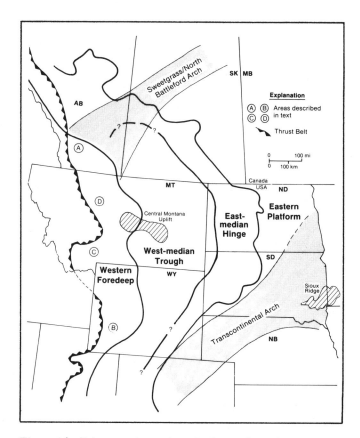

Figure 14—Paleotectonic provinces in the northern Great Plains during time of deposition of the Eagle Sandstone, Niobrara Formation, and equivalent rocks. Tectonic zones are described in Table 1.

equal; consequently, water depths were probably minimal. In the west-median trough the rates were high, but not equal. Initially, subsidence rates apparently exceeded sedimentation rates and thus basin water resulted. Subsequently, subsidence rates were about equal to sedimentation rates, and water depths decreased as the basin filled by eastward progradation of shelf and slope sediments. In the east-median hinge, rates of subsidence and sedimentation were moderate to low and were not equal. Again, when initial subsidence rates exceeded sedimentation rates, deep water was apparently maintained over the carbonate ramp. When subsequent subsidence rates were approximately equal to sedimentation rates, water depths decreased as terrigenous clastics from the west engulfed the carbonate ramp. On the eastern platform, subsidence and sedimentation rates were low and were approximately equal; consequently water depths were minimal.

Shallow water environments have only recently been recognized in tectonic zones on the eastern side of the seaway (Witzke et al., 1983). As a consequence, previous estimates of water depths in areas of chalk deposition have been controversial and range from 200 to 1600 ft (60 to 490 m) (Weimer, 1983). In addition, interpretation of an eastern carbonate ramp modifies earlier paleogeographic models for the seaway. We accept Asquith's (1970) model of basin infilling by westward progradation of shelf and slope

Table 1—Longitudinal tectonic zones of the Late Cretaceous seaway of North America.[a]

Tectonic Zones	Subsidence and Sedimentation Rates	Sedimentary Environments	Water Depth
Western foredeep	Maximum	Coarse terrigenous clastics of shallow and marginal marine and coastal plain origin.	<50 m or subaeral
West-median trough	High	Fine terrigenous shelf and slope clastics deposited as basin fill; subsequent deposition of sandstone.	200–300 m
East-median hinge	Moderate to low	Fine terrigenous clastics and fine carbonates deposited on basinward part of ramp.	100–200 m
Eastern platform	Low	Fine terrigenous clastics and fine carbonates deposited on shallow water part of ramp; erosion.	<100 m

[a]From McNeil and Caldwell (1981), as modified from Kauffman (1977).

environments. However, we believe that the model must be modified to include the ramp and shallow water environments on the eastern side of the seaway (Rice and Shurr, 1983). The modified paleogeographic model appears to be compatible with the paleotectonic model shown in Figure 14.

The most striking aspect of the map of tectonic zones (Figure 14) is the eastward bulge of the western foredeep and the parallel shape of tectonic zones on the eastern side of the seaway. We have previously interpreted the parallel patterns of terrigenous clastics and carbonates to represent dilution of bioclastic sediment by terrigenous material from the west (Rice and Shurr, 1983). Structural control of these patterns, however, has been demonstrated in the three study areas, and we now believe that these parallel relationships also reflect paleotectonism.

Tectonic Zones

In addition to the eastward bulge of longitudinal tectonic zones, there are several distinctive trends along the axis of the seaway in each of the four tectonic zones.

The western foredeep zone narrows as the west-median trough approaches the thrust belt in Alberta and Wyoming (locations A and B in Figure 14). The stratigraphic record in these areas has been successfully modeled by assuming that foredeep subsidence is the result of crustal loading by thrust sheets (Beaumont, 1981; Jordan, 1981). However, preexisting structural features in the foreland area have also influenced the subsidence (Beaumont, 1981; Wiltschko and Dorr, 1983). Similar interaction of thrust belt and foreland trends are present in southwestern (Perry et al., 1983) and northwestern (Winston, 1983) Montana (locations C and D, respectively, in Figure 14). The central part of the foredeep zone lying in northwestern Wyoming and Montana could

better be termed a "foreland," applying the distinctions between foreland and foredeep discussed by Wiltschko and Dorr (1983).

In the west-median trough, areas of maximum subsidence change orientation along the trend of the zone. Thickness patterns indicate an elongate area of maximum subsidence in southwestern Saskatchewan trending northwestward (Williams and Burke, 1964). Along the North Dakota–Montana border, thick areas trend north–south (Shurr, 1979). In Wyoming, depocenters are elongate northeastward (Asquith, 1970). Areas of maximum subsidence are approximately at the center of the west-median trough where it trends northwestward and southwestward, but in Montana, the areas of maximum subsidence are found near the eastern boundary of the zone.

Widths of the east-median hinge vary from north to south. In the Dakotas the wide hinge zone is defined by the western limit of the upper and lower chalk tongues of the Niobrara and by the associated change in Gammon Shale thickness. In Saskatchewan the zone is narrow and is based on thickness patterns and distribution of erosion to the east (Williams and Burke, 1964). In Wyoming the hinge zone is also narrow and is defined by the western limits of carbonate units in the Niobrara (McGookey et al., 1972).

The eastern platform also has north–south variations, although the differences are not conspicuous in Figure 14. In Manitoba and eastern North Dakota the Niobrara contains calcareous shale, in central and eastern South Dakota it is dominantly chalk, and to the south in Nebraska it again contains calcareous shale. In addition, the hiatuses represented by the unconformities at the base and at the top of the Niobrara are greatest over the Transcontinental arch in western Nebraska and central South Dakota (Shurr, 1984b).

Lineament-Block Tectonics

There is a range of structural styles in the four longitudinal tectonic zones of the Cretaceous seaway. Viscoelastic subsidence appears to have been dominant in the western foredeep adjacent to well-defined thrust sheets. Lineament-block tectonics was the dominant style eastward from the margin of the foredeep through the trough, hinge, and platform. The cratonic areas may have experienced little paleotectonism on the far eastern side of the platform.

The three study areas described in this report are all characterized by lineament-bound blocks that apparently influenced deposition of sediments now constituting the Niobrara cyclothem. This tectonic style is in direct contrast to a viscoelastic lithosphere response to thrust sheet loading postulated in Alberta by Beaumont (1981) and in Wyoming by Jordan (1981). Perhaps the lithosphere in the central "foreland" portion of the western tectonic zone did not deform in a viscoelastic manner or perhaps the response to thrust sheet loading was localized very near the margin of the well-defined thrust belt. In any case, block movements in the west-median trough, east-median hinge, and eastern platform appear to have been far beyond the influence of subsidence induced by crustal loading in the thrust belt. On the crest of the Transcontinental arch in extreme eastern South Dakota, there are indications that parts of the eastern platform experienced essentially no tectonism during Late Cretaceous time (Shurr, 1981a). The evidence consists of Niobrara occurrences on Precambrian rocks that presently have elevations identical with the elevation of apparent Late Cretaceous sea levels during the time of deposition of the Niobrara (Pitman, 1978). This implies no paleotectonic movements.

Individual lineament-bound blocks in the three study areas are the constituent parts of larger tectonic elements in the tectonic zones (Figure 14). For example, east–west lineaments in central Montana are part of the Central Montana uplift (Figure 5) and northeast lineaments in central South Dakota compose the flank of the Transcontinental arch (Figure 9). In northern Montana, northwest-trending Landsat linear features (Shurr, 1983) parallel the northwesterly trend of the northern part of the west-median trough.

From west to east across Montana and the Dakotas, there are differences in the trend of lineaments that influenced sedimentation. The lineaments trend northwest and northeast in all three areas. In addition, however, there are north–south and east–west trending features in central Montana (Figures 5 and 6) and north–south trending features in the northern Black Hills (Figures 9 and 10). Linear features trending north–south and east–west have also been observed in central South Dakota (Shurr, 1981b), but none of the north–south and east–west features appear to have influenced depositional patterns.

The pattern of the tectonic zones, the predominance of lineament-block tectonics, and the variation of orientation of linear features across the seaway provide the basis for speculation on the nature of the forces that produced the paleotectonism.

Driving Forces

Paleotectonic evaluation of Landsat linear features based on stratigraphic data emphasizes vertical movements. Although horizontal forces have been used in some lineament tectonic models for the northern Great Plains (e.g., Thomas, 1974; Brown, 1978), we have examined the vertical forces in our work.

Forces related to the convergent plate margin west of the study area were dominantly horizontal, but several mechanisms have been proposed that would have transformed the forces to vertical. For example, it has been suggested (Swift and Rice, 1984) that subsidence induced by thrust sheet loading extended far to the east where it influenced paleotectonism on blocks that were sites of sand ridge fields on the western shelf. However, calculations made for the areas adjacent to the Alberta (Beaumont, 1981) and Wyoming (Jordan, 1981) thrust margins do not warrant this hypothesis. A second example of vertical forces possibly resulting from activity at the plate margin is interaction of a shallow subduction slab with lithosphere blocks in the North American plate. A change from steep to shallow angle of plate descent has been proposed to have occurred at the approximate time of Gammon deposition (Dickinson and Snyder, 1978). If the subducting slab were large enough, this effect could have extended far eastward into the west-median trough. It is doubtful, however, that subduction influences could extend all the way to the east-median hinge and eastern platform. In addition, paleotectonic consequences of this subduction model have not been sufficiently deduced to provide specific predictions that could be tested.

Vertical forces not related to plate interactions could also have been the source of paleotectonic displacements. Mantle plumes or "hotspots," such as the well-documented one currently beneath Yellowstone Park, could jostle blocks in thick, overlying continental lithosphere. Mesozoic hotspot traces have been described in eastern North America (Crough, 1981) and in northern Africa (Van Houten, 1983); Cenozoic hotspot positions have been located in the northern Great Plains (Crough et al., 1980). No Mesozoic hotspots, however, have as yet been recognized in the area shown in Figure 14. Crustal loading by water and sediments has recently been employed in a model describing carbonate to terrigenous clastic sequences (Walker et al., 1983). Tests of this model, however, require descriptions of depositional patterns above the Niobrara cyclothem that are beyond the scope of this paper.

Whatever the source for forces driving lineament-block tectonism, one conclusion seems clear: block movements far to the east show variations along the seaway similar to north–south variations along the convergent margin. Specifically, thrust belt and foredeep couples were apparently active in Alberta and Wyoming, while in Montana preexisting foreland structures were very important (Figure 14). The Montana portion of the margin also coincides with the westernmost position of the batholith belt reconstructed by Hamilton (1969). In the seaway to the east, the broad pattern of tectonic zones bulges eastward from Montana (Figure 14). In addition, north–south and east–west linear features affected sedimentation in central Montana and the northern Black Hills, but not in central South Dakota. Kluth and Coney (1981) made similar observations for the late Paleozoic margin of the southern United States. Although that margin was the site of a continent–continent collision, the geometry of the Mesozoic western margin of North America may have influenced tectonism in the seaway in an analogous manner. Thomas (1983) has emphasized the importance of preexisting structures along a convergent margin by noting the correspondence of promontories and embayments on the preexisting continental margin with resulting recesses and salients in the Paleozoic orogenic belt of eastern North America. The Montana sector of the western margin appears generally to have been a salient adjacent to the Central Montana uplift. This distinctive basement-controlled feature appears to correspond with an embayment in the continental margin, has had a long paleotectonic history, and extends far eastward. Forces driving block movements in the trough, median, and hinge may have been influenced by the geometry of this embayment and associated salient in the orogenic belt.

Although north–south and east–west linear features do not seem to be important throughout the area shown in Figure 14, northwest and northeast linear features are ubiquitous. In addition, large-scale tectonic elements composed of northwest and northeast block mosaics are also present (e.g., the Transcontinental arch) (Figure 14). Upper Cretaceous movement of the North American plate was to the northwest (Coney, 1978), perpendicular and parallel to the northwest and northeast features, respectively. If movements of the asthenosphere were coupled to the lithosphere base, horizontal forces might have been transformed to vertical forces to produce movements of the lineament-bound blocks. Such transformations could have been the result of lateral differences in buoyancy in the lithosphere or the shape of the bottom lithosphere surface. Interactions of this nature between lithosphere and asthenosphere have been postulated (Artyushkov et al., 1980).

In summary, we speculate that the driving forces that produced lineament-block tectonics in the northern Great Plains sector of the Cretaceous seaway were the result of two force fields. The first was a nonuniform field directed from west to east that was related to the convergent margin and influenced by the irregular geometry of the margin. The second was a more uniform northwest-directed field related to movement of the North American plate. Although it is not clear how these horizontal forces were transformed to vertical forces that produced lineament-block tectonism, mechanisms for that transformation have been discussed in the literature. The fundamental conclusion of our study is that preexisting zones of weakness in the lithosphere plate responded to these driving forces and the resulting paleotectonism influenced Late Cretaceous sedimentation.

CONCLUSIONS

Zones of lithospheric weakness that bound discrete lithosphere blocks are visible on Landsat images. The patterns can be compared with stratigraphic data to assess paleotectonism. By focusing on a single depositional sequence generated by a single cycle of transgression and regression, the effects of sea level change are minimized and paleotectonism in a variety of depositional settings can be investigated.

The stratigraphic record for a sequence composed of rocks equivalent to the Niobrara Formation, Telegraph Creek Formation, and Eagle Sandstone was described for three areas of the northern Great Plains. In all three areas lineament-bound blocks appear to be reflected in patterns of sedimentation. In central Montana, coastal and inner shelf sandstone and nonmarine coastal plain and wave-dominated delta deposits were apparently influenced by lineaments trending north–south, east–west, northwest, and northeast. In the northern Black Hills, chalk and outer shelf sandstone show controls by lineaments trending north–south, northwest, and northeast. In central South Dakota calcareous shale, chalk, and areas of erosion reflect the geometry of lineaments that generally trend northeast and northwest.

Lineament-block tectonics thus appears to characterize the structural style of structural zones extending far to the east of the Late Cretaceous continental margin. Tectonic variations within the seaway are similar to variations along the trend of the orogenic belt. Driving forces for the lineament blocks may have been related to the geometry of the western margin of the North American plate and to movement of the plate itself.

REFERENCES CITED

Anna, L. O., 1983, Structure influence on Lower and Middle Cretaceous sedimentation—northern Great Plains (abs.): AAPG Bulletin, v. 67, p. 1329.

Artyushkov, E. V., A. E. Shlesinger, and A. L. Yanshin, 1980, The origin of vertical crustal movements within lithospheric plates, *in* A. W. Bally, P. L. Bender, T. R. McGetchin, and R. I. Walcott, eds., Dynamics of plate interiors: American Geophysical Union Geodynamic Series, v. 1, p. 37–51.

Asquith, D. O., 1970, Depositional topography and major marine environments, Late Cretaceous, Wyoming: AAPG Bulletin, v. 54, p. 1184–1224.

Beaumont, C., 1981, Foreland basins: Geophysical Journal of the Royal Astronomical Society, v. 65, p. 291–329.

Brown, D. L., 1978, Wrench-style deformation patterns associated with a meridional stress axis recognized in Paleozoic rocks in parts of Montana, South Dakota, and Wyoming: Montana Geological Society 24th Annual Conference Proceedings, Williston Basin Symposium, p. 17–31.

Cobban, W. A., and E. A. Merewether, 1983, Stratigraphy and paleontology of Mid-Cretaceous rocks in Minnesota and contiguous areas: U.S. Geological Survey Professional Paper 1253, 52 p.

Coney, P. J., 1978, Mesozoic-Cenozoic Cordilleran plate tectonics, *in* R. B. Smith and G. P. Eaton, eds., Cenozoic tectonics and regional geophysics of the western Cordillera: Geological Society of America Memoir 152, p. 33–50.

Crough, S. T., 1981, Mesozoic hotspot epeirogeny in eastern North America: Geology, v. 9, p. 2–6.

Crough, S. T., W. J. Morgan, and R. B. Hargraves, 1980, Kimberlites—their relation to mantle hotspots: Earth and Planetary Science Letters, v. 50, p. 260–274.

Dickinson, W. R., and W. S. Snyder, 1978, Plate tectonics of the Laramide orogeny, *in* V. Matthews, ed., Laramide folding associated with basement block faulting in the western United States: Geological Society of America Memoir 151, p. 355–366.

Dobbin, C. E., and C. E. Erdmann, 1955, Structure contour map of the Montana Plains: U.S. Geological Survey Map OM-178B.

Eardley, A. J., 1962, Structural geology of North America: New York, Harper and Row, 743 p.

Gill, J. R., and W. A. Cobban, 1973, Stratigraphy and geologic history of the Montana Group and equivalent rocks, Montana, Wyoming, and North and South Dakota: U.S. Geological Survey Professional Paper 776, 37 p.

Hamilton, W., 1969, Mesozoic California and the underflow of Pacific mantle: Geological Society of America Bulletin, v. 80, p. 2409–2430.

Hattin, D. E., 1964, Cyclic sedimentation in the Colorado Group of west-central Kansas, *in* D. F. Merriam, ed., Symposium on cyclic sedimentation: Kansas Geological Survey Bulletin 169, p. 205–217.

Hearn, B. C., Jr., W. T. Pecora, and W. C. Swadley, 1964, Geology of the Rattlesnake quadrangle, Bearpaw Mountains, Blaine County, Montana: U.S. Geological Survey Bulletin 1181-B, 66 p.

Hobson, J. P., Jr., M. L. Fowler, and E. A. Beaumont, 1982, Depositional and statistical exploration models, Upper Cretaceous offshore sandstone complex, Sussex Member, House Creek field, Wyoming: AAPG Bulletin, v. 66, p. 649–688.

Johnson, W. D., and H. R. Smith, 1964, Geology of the Winnett-Mosby Area, Petroleum, Garfield, Rosebud, and Fergus counties, Montana: U.S. Geological Survey Bulletin 1149, 91 p.

Jordan, T. E., 1981, Thrust loads and foreland basin evolution, Cretaceous, western United States: AAPG Bulletin, v. 65, p. 2506–2520.

Kauffman, E. G., 1969, Cretaceous marine cycles of the Western Interior: Mountain Geologist, v. 6, p. 227–245.

———, 1977, Geological and biological overview—Western Interior Cretaceous basin: Mountain Geologist, v. 14, p. 75–99.

Kluth, C. F., and P. J. Coney, 1981, Plate tectonics of the ancestral Rocky Mountains: Geology, v. 9, p. 10–15.

Lisenbee, A. L., 1978, Laramide structure of the Black Hills uplift, South Dakota-Wyoming-Montana, *in* V. Matthews, ed., Laramide folding associated with basement block faulting in the western United States: Geological Society of America Memoir 151, p. 165–196.

Mahr, P. D., 1969, Eagle gas accumulations of the Bearpaw uplift area, Montana: Montana Geological Society 20th Annual Field Conference Guidebook, Eastern Montana Symposium, p. 121–127.

Maughan, E. K., 1983, Tectonic setting of the Rocky Mountain region during the late Paleozoic and the early Mesozoic: Proceedings of Symposium on the Genesis of Rocky Mountain Ore Deposits—Changes with Time and Tectonics, p. 39–50.

McGookey, D. P., J. D. Haun, L. A. Hale, H. G. Goodell, D. G. McCubbin, R. J. Weimer, and G. R. Wulf, 1972,

Cretaceous System, *in* W. M. Mallory, ed., Geologic atlas of the Rocky Mountain region: Rocky Mountain Association of Geologists, p. 190–228.

McNeil, D. H., and W. G. E. Caldwell, 1981, Cretaceous rocks and their foraminifera in the Manitoba escarpment: Geological Association of Canada Special Paper n. 21, 439 p.

Mitchum, R. M., Jr., P. R. Vail, and S. Thompson, 1977, The depositional sequence as a basic unit for stratigraphic analysis, *in* C. E. Payton, ed., Seismic stratigraphy—applications to hydrocarbon exploration: AAPG Memoir 26, p. 53–62.

Obradovich, J. D., and W. A. Cobban, 1975, A time-scale for Late Cretaceous of the Western Interior of North America, *in* W. G. E. Caldwell, ed., The Cretaceous System in the Western Interior of North America: Geological Association of Canada Special Paper n. 13, p. 31–54.

Perry, W. J., Jr., B. R. Wardlaw, N. H. Bostick, and E. K. Maughan, 1983, Structure, burial history, and petroleum potential of frontal thrust belt and adjacent foreland, southwest Montana: AAPG Bulletin, v. 67, p. 725–743.

Pitman, W. C., 1978, Relationships between eustacy and stratigraphic sequences of passive margins: Geological Society of America Bulletin, v. 89, p. 1389–1403.

Reeside, J. B., Jr., 1957, Paleoecology of Cretaceous seas of the Western Interior of the United States, *in* H. S. Ladd, ed., Treatise on marine ecology and paleoecology: Geological Society of America Memoir 67, v. 2, p. 505–541.

Reiskind, J., 1982, Niobrara Formation (Upper Cretaceous), eastern North Dakota (abs.): AAPG Bulletin, v. 66, p. 622.

Rice, D. D., 1976, Correlation chart of Cretaceous and Paleocene rocks of the northern Great Plains: U.S. Geological Survey Chart OC-70.

———, 1980, Coastal and deltaic sedimentation of Upper Cretaceous Eagle Sandstone—relation to shallow gas accumulations, north-central Montana: AAPG Bulletin, v. 64, p. 316–338.

———, 1984, Widespread, shallow marine, storm generated sandstone units in the Upper Cretaceous Mosby Sandstone, central Montana, *in* R. W. Tillman and C. T. Siemers, eds., Siliciclastic shelf sediments: Society of Economic Paleontologists and Mineralogists Special Publication No. 34.

Rice, D. D., and G. W. Shurr, 1983, Patterns of sedimentation and paleogeography across the Western Interior seaway during time of deposition of Upper Cretaceous Eagle Sandstone and equivalent rocks, northern Great Plains, *in* M. W. Reynolds and E. D. Dolly, eds., Mesozoic paleogeography of the west-central United States: Rocky Mountain Section Society of Economic Paleontologists and Mineralogists, Rocky Mountain Paleogeography Symposium 2, p. 337–358.

Rice, D. D., G. W. Shurr, and D. L. Gautier, 1982, Revision of Upper Cretaceous nomenclature in Montana and South Dakota: U.S. Geological Survey Bulletin 1529-H, p. 99–104.

Schorning, F., 1972, Reservoir geology of the Tiger Ridge

area: Montana Geological Society 21st Annual Field Conference Guidebook, Crazy Mountains basin, p. 149–154.

Schultz, L. G., 1965, Mineralogy and stratigraphy of the lower part of the Pierre Shale, South Dakota and Nebraska: U.S. Geological Survey Professional Paper 392-B, p. B1–B19.

Shurr, G. W., 1978, Paleotectonic controls on Cretaceous sedimentation and potential gas occurrences in western South Dakota: Montana Geological Society 24th Annual Conference Proceedings, Williston Basin Symposium, p. 265–281.

———, 1979, Lineament control of sedimentary facies in the northern Great Plains, United States: Proceedings of Second International Conference on Basement Tectonics, p. 413–422.

———, 1981a, Cretaceous sea cliffs and structural blocks on the flanks of the Sioux Ridge, South Dakota and Minnesota, *in* Cretaceous stratigraphy and sedimentation in northwest Iowa, northeast Nebraska, and southeast South Dakota: Iowa Geological Survey Guidebook Series No. 4, p. 25–41.

———, 1981b, Lineaments as basement-block boundaries in western South Dakota: Proceedings of Third International Conference of Basement Tectonics, p. 177–184.

———, 1982, Geological significance of lineaments interpreted from Landsat images near the northern Black Hills, *in* J. E. Christopher and J. Kaldi, eds., Fourth international Williston basin symposium: Saskatchewan Geological Society Special Publication No. 6, p. 313–320.

———, 1983, Landsat linear features in Montana plains: AAPG Bulletin, v. 67, p. 1355.

———, 1984a, Geometry of shelf sandstone bodies in the Shannon Sandstone of southeastern Montana, *in* R. W. Tillman and C. T. Siemers, eds., Siliciclastic shelf sediments: Society of Economic Paleontologists and Mineralogists Special Publication No. 34, p. 63–83.

———, 1984b, Regional setting of the Niobrara Formation in the northern Great Plains: AAPG Bulletin, v. 68, p. 598–609.

Slack, P. B., 1981, Paleotectonics and hydrocarbon accumulation, Powder River basin, Wyoming: AAPG Bulletin, v. 65, p. 730–743.

Sonnenberg, F. P., 1956, Tectonic patterns of central Montana: Billings Geological Society 7th Annual Field Conference Guidebook, p. 73–81.

Sonnenberg, S. A., and R. J. Weimer, 1981, Tectonics, sedimentation, and petroleum potential, north Denver basin, Colorado, Wyoming, and Nebraska: Colorado School of Mines Quarterly, v. 76, n. 2, 45 p.

Stearns, D. W., 1978, Faulting and forced folding in the Rocky Mountains foreland, *in* V. Matthews, ed., Laramide folding associated with basement block faulting in the western United States: Geological Society of America Memoir 151, p. 1–38.

Swift, D. J. P., and D. D. Rice, 1984, Sand bodies on muddy shelves—a model for sedimentation in the Western

Interior seaway, North America, *in* R. W. Tillman and C. T. Siemers, eds., Siliciclastic shelf sediments: Society of Economic Paleontologists and Mineralogists Special Publication No. 34, p. 43–62.

Thom, W. T., Jr., 1923, The relation of deep-seated faults to the surface structural features of central Montana: AAPG Bulletin, v. 7, p. 1–13.

Thomas, G. E., 1974, Lineament-block tectonics: Williston–Blood Creek basin: AAPG Bulletin, v. 58, p. 1305–1322.

Thomas, W. A., 1983, Continental margins, orogenic belts, and intracratonic structures: Geology, v. 11, p. 270–272.

Van Houten, F. B., 1983, Sirte basin, north-central Libya—Cretaceous rifting above a fixed mantle hotspot? Geology, v. 11, p. 115–118.

Walker, K. R. O., G. Shanmugam, and S. C. Ruppel, 1983, A model for carbonate to terrigenous clastic sequences: Geological Society of America Bulletin, v. 94, p. 700–712.

Weimer, R. J., 1960, Upper Cretaceous stratigraphy, Rocky Mountain area: AAPG Bulletin, v. 44, p. 1–20.

———, 1980, Recurrent movement on basement faults, a tectonic style for Colorado and adjacent areas, *in* H. C. Kent and K. W. Porters, eds., 1980 Symposium, Colorado Geology: Rocky Mountain Association of Geologists, p. 23–35.

———, 1983, Relation of unconformities, tectonics, and sea level changes, Cretaceous of the Denver basin and adjacent areas, *in* M. W. Reynolds and E. D. Dolly, eds., Mesozoic paleogeography of the west-central United States: Rocky Mountain Section Society of Economic Paleontologists and Mineralogists, Rocky Mountain Paleogeography Symposium 2, p. 359–376.

Weimer, R. J., J. J. Emme, C. L. Farmer, L. O. Anna, T. L. Davies, and R. L. Kidney, 1982, Tectonic influence on sedimentation, Early Cretaceous, east flank Powder River basin, Wyoming and South Dakota: Colorado School of Mines Quarterly, v. 73, 62 p.

Williams, G. D., and C. F. Burke, Jr., 1964, Upper Cretaceous, *in* R. G. McCrossan and R. P. Glaister, eds., Geological history of western Canada: Alberta Society of Petroleum Geologists, p. 169–189.

Wiltschko, D. V., and J. A. Dorr, Jr., 1983, Timing of deformation in overthrust belt and foreland of Idaho, Wyoming, and Utah: AAPG Bulletin, v. 67, p. 1304–1322.

Winston, D., 1983, Middle Proterozoic Belt basin syndepositional faults and their influence on Phanerozoic thrusting and extension (abs.): AAPG Bulletin, v. 67, p. 1361.

Witzke, B. J., G. A. Ludvigson, J. R. Poppe, and R. L. Ravn, 1983, Cretaceous paleogeography along the eastern margin of the Western Interior seaway, Iowa, southern Minnesota, eastern Nebraska and South Dakota, *in* M. W. Reynolds and E. D. Dolly, eds., Mesozoic paleogeography of the west-central United States: Rocky Mountain Section Society of Economic Paleontologists and Mineralogists, Rocky Mountain Paleogeography Symposium 2, p. 225–252.

Cedar Creek: A Significant Paleotectonic Feature of the Williston Basin

J. H. Clement[1]
Shell Western Exploration and Production Inc.
Houston, Texas

Cedar Creek is the major anticlinal structure marking the southwestern flank of the Williston basin. More than 327 million bbl of oil have been produced from Paleozoic carbonate reservoirs in 15 fields along the feature. This pronounced fold developed through recurrent tectonic movements along a northwest-striking fault zone. There are four major periods of tectonism documentable in the Cedar Creek area from early Paleozoic to middle Tertiary times.

Uplift and fault movement accompanied northward and eastward tilting of the major Cedar Creek block in Early Devonian time. Silurian strata were eroded prior to Middle Devonian time, and a karst plain developed. Middle and Upper Devonian sediments progressively onlapped and infilled the uplifted, northwest-plunging element. During latest Late Devonian time, fault movement occurred along the main fault zone. The block was uplifted and tilted northward and eastward. Extensive erosion resulted in the near peneplanation of the structure and in truncation of Upper Devonian strata. Positive paleostructural influence continued during Mississippian time. In late Mississippian and Early Pennsylvanian time, the central and northern part of the Cedar Creek area was gently downwarped with relative down-to-the-east fault movement along most of the faults. Similar faulting and subsidence influenced the deposition and preservation of Permian and Triassic evaporite-rich red bed sequences. Relative tectonic stability was attained by the Middle Jurassic and maintained until post-Paleocene time.

The main uplift of the Cedar Creek block occurred during post-Paleocene time and was accompanied by major flexuring and deep fault adjustment. Northwestward plunge along the anticline crest was increased. The entire area was subsequently uplifted during epeirogenic phases in middle Tertiary time, and Paleocene and Upper Cretaceous strata were eroded along the axis of the present structure.

INTRODUCTION

Paleozoic sedimentation patterns, paleotectonics, and paleogeography of the Northern Great Plains and Rocky Mountain regions are related to the geologic history of the western border of the North American craton, presently the stable interior of the continent. During Paleozoic time, accumulation, preservation, and erosion of sediments were affected by recurrent tectonic events associated with several regional structural elements including the Transcontinental arch, the Alberta and Wyoming shelves, the Central Montana trough, and the Williston basin, the latter being the major Paleozoic paleostructural feature of the Northern Great Plains. Semiregional features that exerted important influence on the Williston basin area of Montana and South Dakota include the Black Hills uplift, the Central Montana uplift, the Weldon fault system, and the Cedar Creek anticline, the subject of this paper.

GENERAL GEOLOGY

The Cedar Creek anticline is the major anticlinal structure demarcating the southwestern flank of the Williston basin. This pronounced surface feature extends linearly about 233 km (145 mi), striking S 30° E from northwest of Glendive, Montana, to just west of Buffalo, South Dakota (Figure 1). It displays a width varying from 9 to 32 km (6 to 20 mi) and encompasses an area in excess of 5180 km² (2000 mi²).

Surface geology and structural maps of the anticline were first prepared in 1921 by geologists of the Northern Pacific Railway and in the early 1930s by the U.S. Geological Survey (Dobbin and Larsen, 1936). New detailed maps of the bedrock geology of the region are being published by the Montana Bureau of Mines and Geology (Vuke-Foster et al., in press, 1986).

The surface fold is strongly asymmetric. The southwestern flank dips generally S 60° W at angles varying from 4 to 40°. The northeastern flank dips generally N 60° E at angles of less than 1°. Upper Cretaceous Bearpaw (Pierre) shales are exposed along the erosionally breached axis, and uppermost Cretaceous Hell Creek and Fox Hills strata crop out on the flanks, as do beds of the

---

[1]Present Address: Shell Oil Company, P. O. Box 481, Houston, TX 77001.

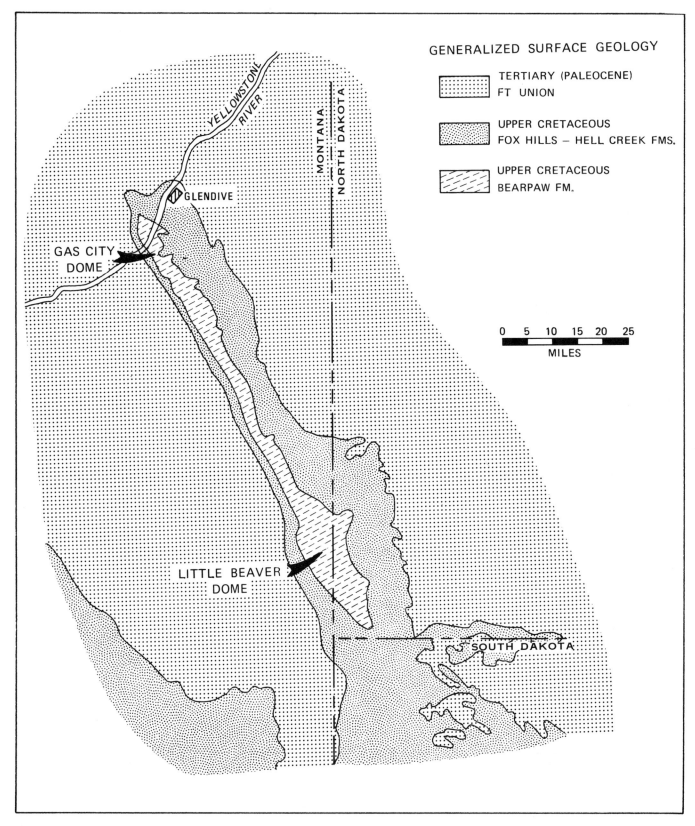

Figure 1—Generalized map of surface geology, Cedar Creek anticline, Montana, North Dakota, and South Dakota.

Paleocene Fort Union Formation, the youngest strata involved in the structural deformation.

From Little Beaver dome, the structurally highest part of the present crestal area, the major surface axis plunges to the northwest at less than 3.7 m/km (20 ft/mi) and to the southeast at approximately 7.5 m/km (40 ft/mi). Surface faults and fracture systems mapped in the exposed Paleocene and Upper Cretaceous strata show varying, but generally small, magnitudes of displacement. However, their patterns reflect post-Paleocene uplift, extension, and probable minor strike-slip (wrench) movement along this ancient structure.

ECONOMIC GEOLOGY

Hydrocarbons (gas) were first discovered along the Cedar Creek anticline in 1912 in Upper Cretaceous Judith River and Eagle sandstones at a depth of about 230 m (750 ft) on Gas City dome. The area of production from these shallow gas horizons was extended during the 1920s and 1930s over a large part of the anticline (Figure 2). About 140 billion cubic feet of gas was produced from over 200 wells. Most of the shallow Cretaceous gas sands are used today for gas storage.

Most significantly, Cedar Creek is the most prolific oil-productive structure in the state of Montana (Figures 3, 4, and 5). The initial discovery of oil was made at Little Beaver dome in 1936, but it was not until late 1951 that the discovery of significant commercial accumulations in the Pine and Glendive fields stimulated the exploration and development of the extensive producing trend. Over 327 million bbl of oil have been produced from Paleozoic (Ordovician, Silurian, and Mississippian) carbonate reservoirs in 15 fields. Ultimate primary and secondary (water flood) phases of production will probably exceed 400 million bbl of oil and 65 billion cubic feet of solution gas.

The most significant fields are Pine, with nearly 100 million bbl produced (153,616 bbl in March 1984); Cabin Creek with over 85 million bbl produced (120,486 bbl in March 1984); and Pennel–Lookout Butte (combined), with nearly 80 million bbl produced (323,274 bbl in March 1984). Today, Cedar Creek remains the leading oil-producing trend in Montana, contributing more than 700,000 bbl per month or about 8.4 million bbl per year.

TECTONIC HISTORY

Subsurface data from more than 1200 wells drilled on and around Cedar Creek during the past 35 years document significant tectonic activity from early Paleozoic through middle Tertiary time. Recurrent faulting occurred along a generally linear, principal fault zone or zones that segmented the feature into two or more major crustal blocks. The relative displacements along the fault zone(s) and the structural attitudes of the major blocks, principally the northeastern block (which is herein called the Cedar Creek block), have significantly affected the geology of sedimentary strata since Ordovician time. This is apparent in both local and regional thickness patterns, uncon-formities, and depositional facies. Consequently, the

migration and accumulation of hydrocarbons, although not the subject of this paper, were significantly influenced by this geologic history (Clement, 1976).

The sequence of paleotectonic events in the Cedar Creek area was first cited in publications by McCabe (1954) and Strickland (1954). It has been referred to by several authors since then and has been of interest to petroleum explorationists in the Williston basin. The evolution of the Cedar Creek area is best demonstrated by studying local documentable movements along the principal fault zone(s) and then relating them to more regional geologic conditions. Because critical subsurface control exists on both flanks of the structure in proximity to Gas City field (Figures 3 and 6), a sequence of paleostructural sections is shown for this locale (Figure 7). Similar analyses can be made at several localities along the structure.

Precambrian–Ordovician

The Williston basin is a structural and sedimentary intracratonic basin with an area of more than 130,000 km² (50,000 mi²). It began to take shape as a distinctive area of crustal subsidence and sediment accumulation during Early Ordovician or possibly Late Cambrian time (Lochman-Balk and Wilson, 1967). Regionally, the Upper Cambrian–Lower Ordovician Deadwood Formation thins uniformly eastward, but is interrupted in several areas where the strata are somewhat anomalously thin; for example, this occurs over the Nesson anticline and in the southern part of the Cedar Creek anticline. However, it is uncertain whether or not these areas are related to depositional thinning over Early Cambrian–Precambrian structural highs or over buried paleotopography on the Precambrian surface (Peterson, 1981).

The sparsity of pre–Middle Ordovician well penetrations in the Cedar Creek area precludes any documentable conclusions that Cedar Creek was "predestined" by a northwest-southeast-striking Precambrian fault zone, although it probably was. Gravity and magnetic information and basement age dates (Peterman et al., 1983) suggest but do not confirm this relationship.

Major porosity zones in the Upper Ordovician Red River and Stony Mountain formations constitute primary hydrocarbon reservoirs in the Cedar Creek fields; hundreds of subsurface penetrations have been made. Isopach and facies maps of these strata do not indicate significant crustal instability in southeastern Montana during Ordovician time (Figure 8). The Williston area was a shallow sedimentary basin, and widespread areas periodically emerged to create broad tidal flat and intertidal regions. The Cedar Creek area occupied an intermediate position between the shallow basin area to the east and an oceanic platform to the west–southwest. Therefore, the area was an optimum site for deposition of the cyclic marine and peritidal carbonate sequences typical of the Ordovician (Roehl, 1967; Clement, 1985).

Post-Silurian to Pre-Middle Devonian

In Late Silurian time, probably during late Niagaran (about 400 m.y. ago), the entire cratonic region including the Williston basin was affected by uplift and erosion, and the Cedar Creek evolved as a northwest striking structure

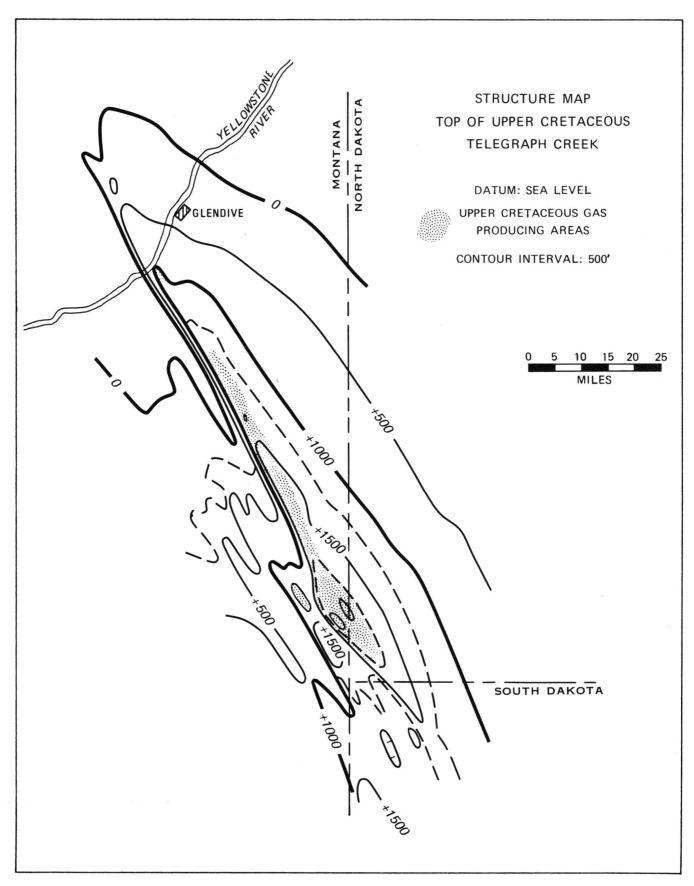

Figure 2—Structural contour map of the top of the Upper Cretaceous Telegraph Creek Formation and location of shallow Upper Cretaceous Judith River and Eagle gas producing areas.

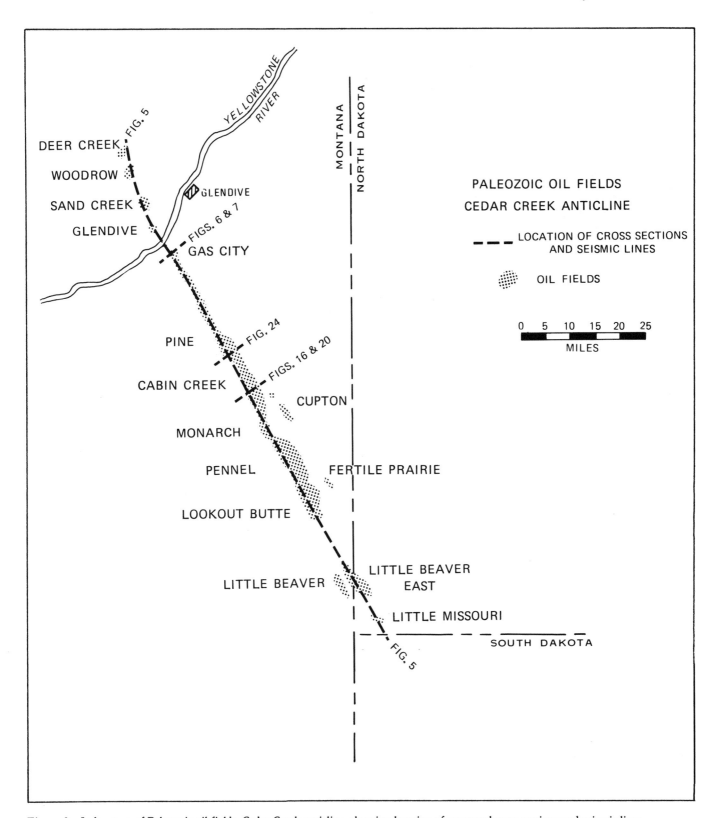

Figure 3—Index map of Paleozoic oil fields, Cedar Creek anticline, showing location of structural cross sections and seismic lines.

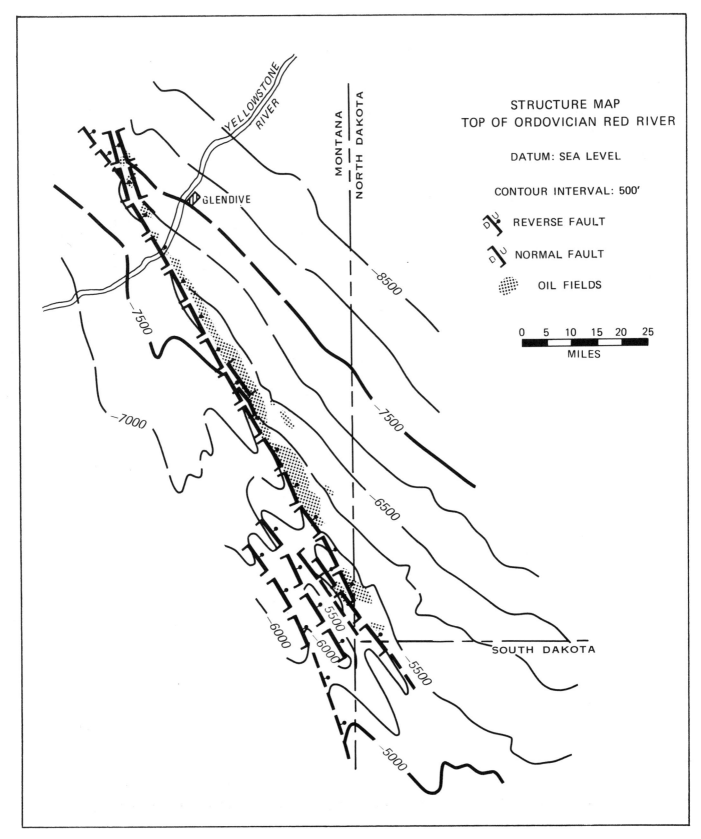

Figure 4—Structural contour map of the top of the Ordovician Red River Formation. Names of fields are shown in Figure 3.

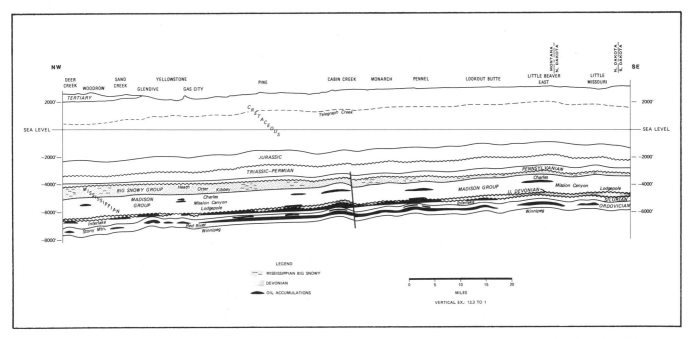

Figure 5—Northwest–southeast structural cross section along the axis of Cedar Creek anticline showing structural and stratigraphic positions of major oil accumulations. Line of section is shown in Figure 3.

(Figure 9). Displacement along the principal fault is conjectural because the original thickness of Silurian strata was greatly obscured by extensive erosion during and following Devonian time. The Silurian sequence is composed mainly of light colored, very finely crystalline dolomite mudstones and wackestones. It is partly algal and coralline rich and is composed dominantly of shallow marine and peritidal sediments.

On the basis of the greater thickness of Silurian rocks basinward and the readily correlated erosional thinning toward and locally on the Cedar Creek feature, it is apparent that several hundred feet of Silurian sediments were eroded prior to Middle and Late Devonian deposition. Topography was apparently developed on a Silurian karst plain (Roehl, 1967). Channels were incised in this surface up to 30 m (100 ft) deep in the Cabin Creek and adjacent areas. These channels were subsequently filled with impermeable Devonian shales and erosional detritus. The Silurian carbonate "hills" between channels today constitute small but effective "paleogeomorphic" hydrocarbon traps in the Wills Creek area of the Cabin Creek field.

Middle Devonian seas transgressed this karst surface and deposited successively younger carbonate and carbonate-rich clastic sediments higher on the east flank and northwest plunge area of the uplifted, slightly northwest-plunging, east-dipping Cedar Creek block (Figure 10). There was intermittent erosion of Middle Devonian units on the uplift, and the distribution shown in Figure 10 is due to both depositional onlap and erosional cycles. Again, post–Upper Devonian erosion obscures conclusive evidence of fault displacement along the trend, although a broad, northwest-plunging east-dipping structure is subtly delineated.

Upper Devonian seas further transgressed the area and deposited a few hundred meters of carbonate sediments that

infilled and onlapped Middle Devonian and Silurian strata (Figure 11). Easily correlated strata within the Duperow and Souris River formations thin gradually onto the feature, indicating continued syndepositional uplift and eastward tilting of the Cedar Creek block.

Late Devonian–Pre-Mississippian

The first pronounced, *clearly documentable* fault movement along the principal fault zone(s) occurred during latest Late Devonian time and possibly earliest Mississippian time (about 350 m.y. ago), which is based on the distribution of the early Kinderhookian upper Bakken and Lodgepole formations. The Cedar Creek block was uplifted relative to the southwest block and tilted significantly northward and eastward (Figure 7a). A maximum displacement of over 245 m (800 ft) occurred along the southwestern flank of Glendive field (Davis and Hunt, 1956). Where encountered by wells at several localities, the zone is a high-angle reverse system with an estimated dip of 72–80°. Rupture appears to have commenced near the Deer Creek area and to have extended as far south as Little Beaver.

Extensive erosion during the latest Devonian–early Mississippian hiatus resulted in the near peneplanation of the structure and truncation of Upper Devonian strata over a large area (Figure 12). All Devonian strata were removed from locally uplifted structures such as Glendive, Gas City, Pine, and Pennel, and the Silurian surface was again exposed. A nearly normal Upper Devonian section was preserved along the downthrown southwestern block. The result was a broad, northwest-striking, faulted anticlinal structure (Figure 13). A large area of closure extended linearly for over 145 km (90 mi), and significant local structural closures as deep as 90 m (300 ft) existed.

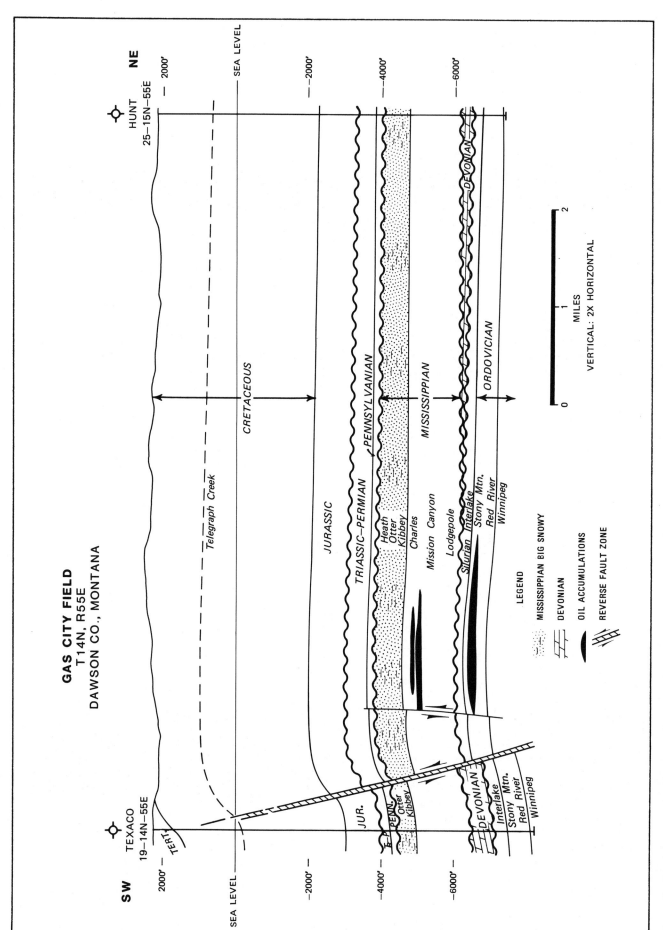

Figure 6—Southwest-northeast structural cross section across Gas City field. Line of section is shown in Figure 3.

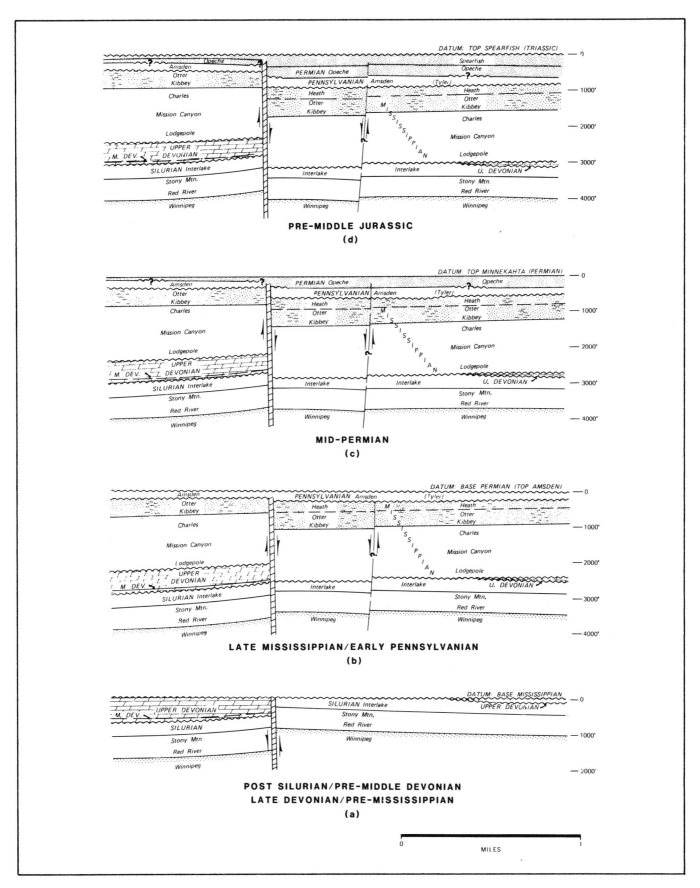

Figure 7—Paleostructural cross sections across Gas City field illustrating major fault displacements and significant stratigraphic changes across the principal and subsidiary faults. Line of sections is shown in Figure 3.

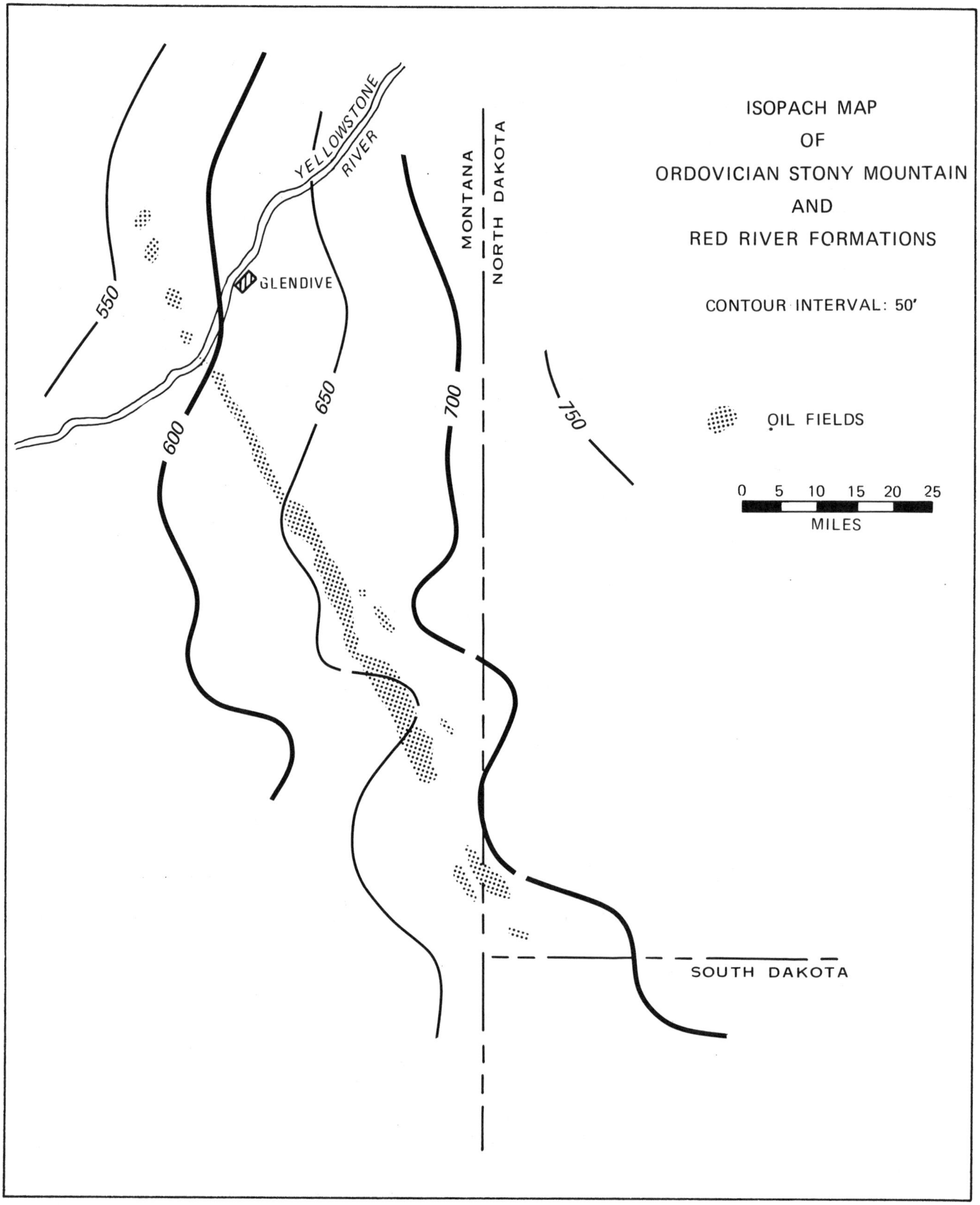

Figure 8—Distribution and thickness of combined Ordovician Stony Mountain and Red River formations.

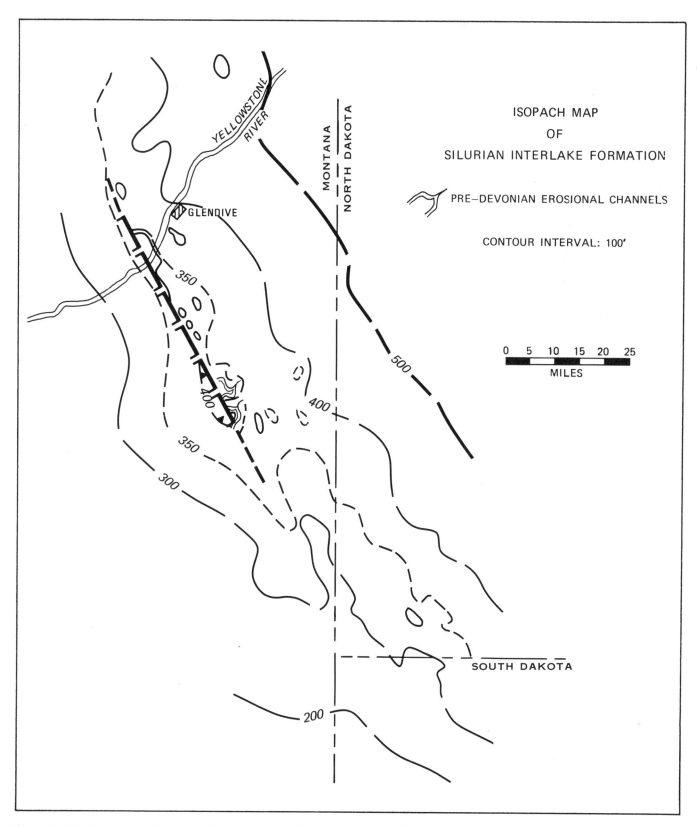

Figure 9—Distribution and thickness of Silurian Interlake Formation. Note significant pre-Devonian erosional channels in Cabin Creek field area.

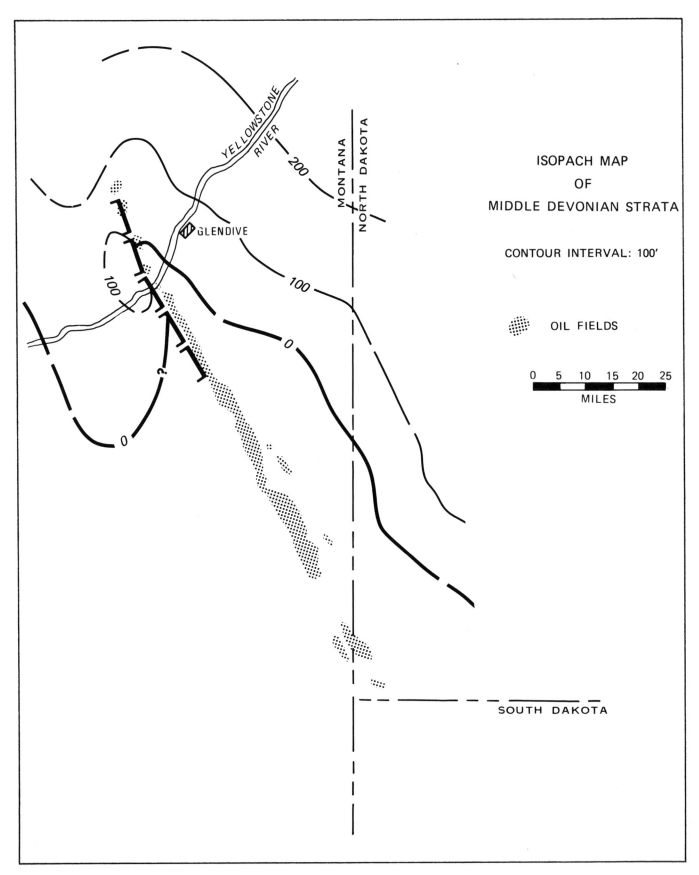

Figure 10—Distribution and thickness of Middle Devonian strata.

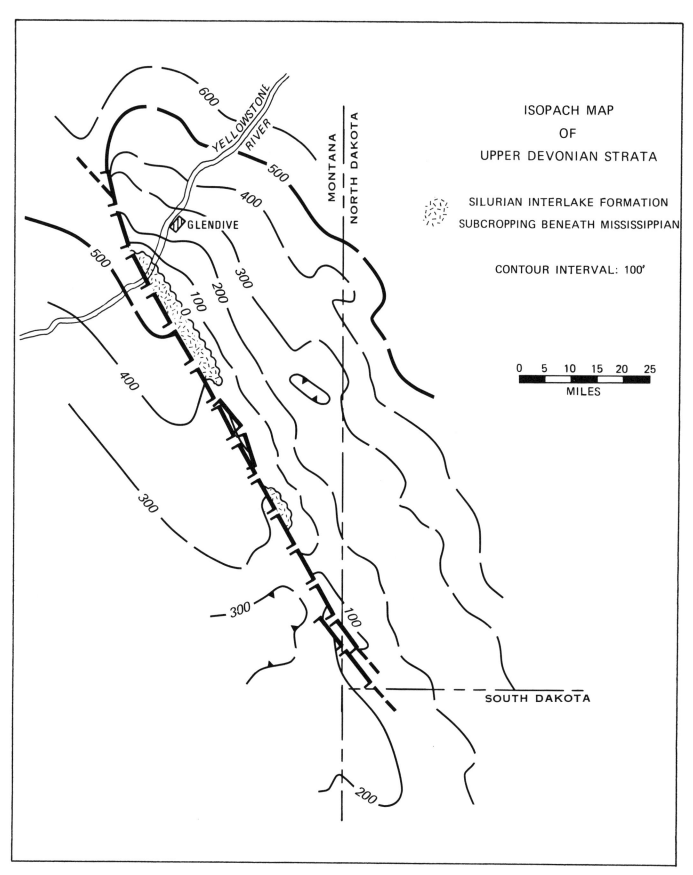

Figure 11—Distribution and thickness of Upper Devonian strata.

Late Mississippian–Early Pennsylvanian

Stratigraphic thinning and minor onlapping of Lower Mississippian Lodgepole strata over the Cedar Creek block and subtle thickening in the subsided southwestern block indicate that the pre-Mississippian unconformity surface was slightly upwarped (Figure 14). Devonian–Mississippian Bakken strata are present on the downthrown, southwestern block in the northern part of the anticline and are absent on the crest. Crinoidal grainstone and packstone "banks" occur in the basal Kinderhookian Lodgepole Formation along Cedar Creek. They produce oil in the Pine and Pennel fields, usually where they lie directly on Silurian strata.

Peterson (1984) illustrates that the lower half of the Kinderhook contains "Waulsortian" crinoidal carbonate mound facies on the western, southwestern, and eastern flanks of the Williston basin. He states that "the main crinoidal mound belt appears to follow the general trends of the Central Montana uplift and the Cedar Creek anticline, probably reflecting (pls. 5 and 12) continued paleostructural influence of these features into earliest Mississippian time."

For the most part, the Mississippian Period was characterized by tectonic stability and relatively continuous deposition over the Cedar Creek area. Continued positive influence of the feature is indicated by (1) the semiregional thinning of the Madison Group in the area (Figure 15), (2) a general tendency for carbonate facies within the group to be more bioclastic or grain dominated (Peterson, 1981), and (3) the gradual termination of upper Madison Charles salts along the northwestward plunge from the Deer Creek field to north Pine field. Part of this salt termination, however, may be due to postdepositional salt solution.

Contemporaneous with the late phases of Chesterian Heath deposition (about 310 m.y. ago), the central and northern part of the area underwent gentle downwarping as the Cedar Creek block tilted northwestward and the southwestern block uplifted and tilted (shifted?) northward along the ancestral principal fault zone (Figure 7b). Normal fault displacement (down to the northeast) occurred along or aligned with the ancestral zone, as well as along newly developed subsidiary faults, such as the small one shown at Gas City field on Figure 7b and the significant east-bounding fault at Cabin Creek field (Figure 16). This is not a unique phenomenon in the Montana region because similar reversals of fault movement of similar age are known along the Cat Creek and Delphia zones of central Montana and the Weldon zone of western Williston basin. All are probably genetically related to major adjustments in the regional tectonic framework that formed the dominantly east–west-striking Big Snowy basin. It is notable that Cedar Creek strikes at nearly right angles to the Big Snowy patterns.

Late Chesterian Heath and Early Pennsylvanian Tyler rocks in southeastern Montana appear to be thickest along Cedar Creek (Figure 17). These strata terminate or abruptly thin across the principal and subsidiary fault zones. Only a thin Tyler–Heath section, for instance, is preserved on the Cabin Creek horst block; the strata thickening abruptly across the normal fault bounding the horst on the east.

The unconformable and/or time–facies relationships of the Heath, Tyler, and lower Amsden strata in the area are not well understood. Study of the available data, however, indicates that significant erosion and/or nondeposition probably occurred during latest Mississippian or more likely,

Early Pennsylvanian time. Early Pennsylvanian regional uplift and erosion over the broad northern Wyoming shelf and western Williston basin areas resulted in the removal of Chesterian Big Snowy sediments over much of the region (Figure 17).

Figure 18 is a paleostructural interpretation of the Red River Formation showing the effects of the Late Mississippian–Early Pennsylvanian structural adjustments. Although several large local areas of closure remained, the overall extent of closure was segmented and diminished, except at Cabin Creek where the large horst block was created. Relative maximum displacement at the Red River level along the principal fault zone was reduced to about 120 m (400 ft), and the entire Cedar Creek block acquired a stronger northwestern plunge.

Permian–Triassic

Widespread, uniform carbonate strata in the Pennsylvanian lower Amsden and Minnelusa formations indicate a period of relative structural stability and marine transgression during Late Pennsylvanian time. However, recurrent downwarping and continued displacement or "growth" occurred along the active Cedar Creek zones during Permian and Triassic time (185–280 m.y. ago) and was partly contemporaneous with deposition of these evaporite-rich red bed sequences (Figure 19). Abrupt thickening of Permian and Triassic beds occurs across the principal and subsidiary faults and is particularly notable along the Cabin Creek horst where massive salt members of the Opeche Formation are present and abrupt thickening of strata of the Spearfish Formation occur on the downthrown, northeastern side of the fault (Figure 16). Salt solution may account for some of this change, but there are no significant local structural anomalies or thickness changes in overlying Jurassic or Cretaceous sedimentary rocks of the horst block to document this mechanism.

The combined effects of the Late Mississippian through Triassic tectonism resulted in a grossly reversed dis-placement of the Red River reservoirs of as much as 230 m (750 ft). This caused readjustment of the early Paleozoic strata within the two mobile blocks to nearly their original pre-Silurian positions as shown in the cross-sectional interpretation of the Gas City area (Figures 7c and d).

When calibrated carefully with synthetic seismograms obtained from modern log suites, high resolution CDP data can provide an innovative technique to aid in the interpretation of structural history that is not generally available from subsurface well data or conventional representation of seismic data. For example, Figure 20 is a seismic line across the Cabin Creek field that, when flattened on the Jurassic Piper Limestone event (Figure 21), dramatically displays the pre-Jurassic structure. Although the characteristics of the major paleostructure at Cabin Creek are well known from subsurface data, this section indicates that the faulting sequence on the western flank is probably more complex than was previously evident from the limited well data.

Jurassic–Early Paleocene

Relative tectonic stability existed by Middle Jurassic time and persisted throughout Jurassic, Cretaceous, and Paleocene time. Continued gentle subsidence along the fault zone and

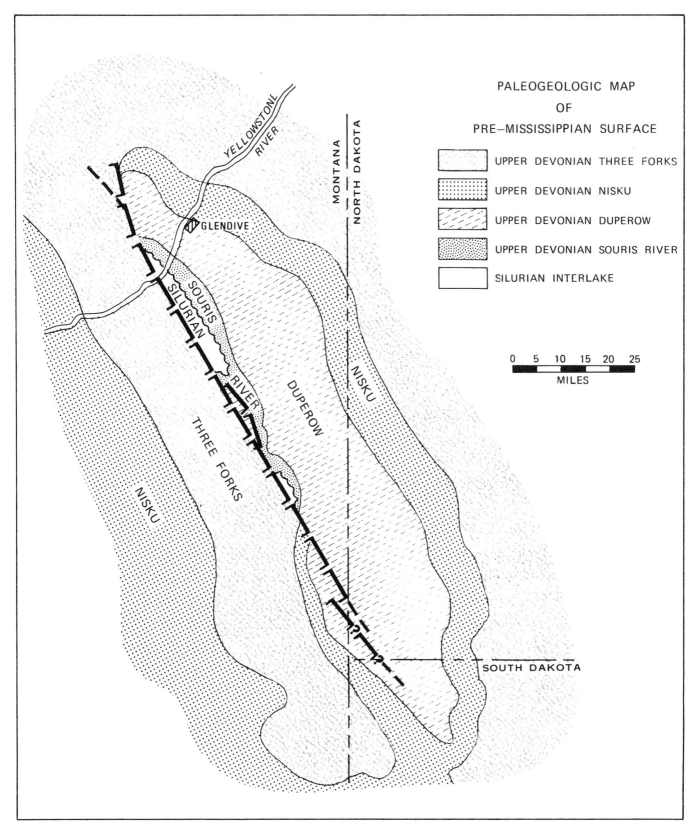

Figure 12—Paleogeologic map of pre-Mississippian surface depicting the prominent ancestral Cedar Creek feature.

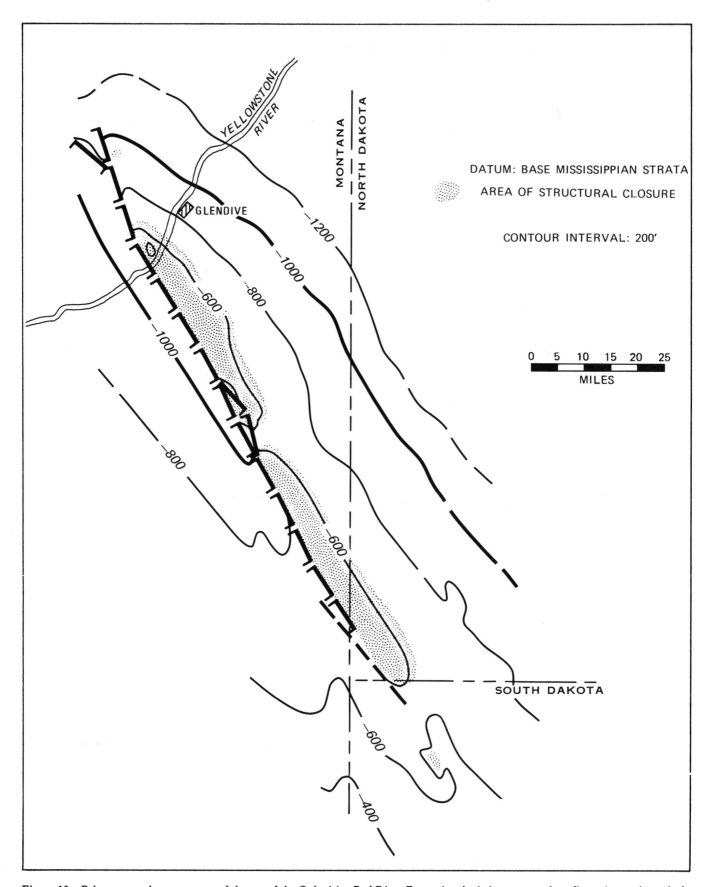

Figure 13—Paleostructural contour map of the top of the Ordovician Red River Formation depicting structural configuration at the end of Devonian time. Note extensive areas of structural fault closure.

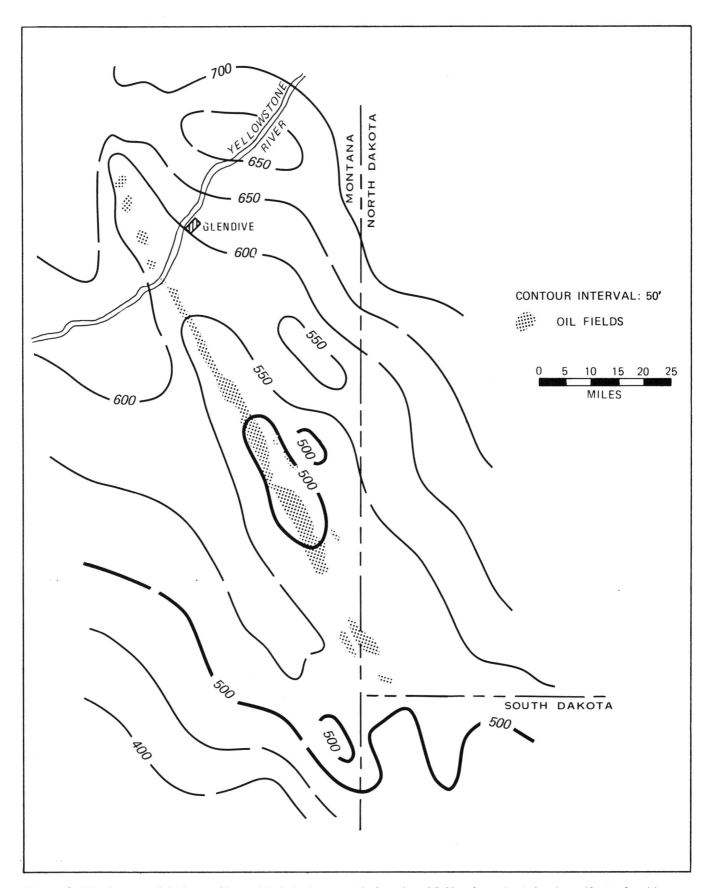

Figure 14—Distribution and thickness of Lower Mississippian strata (Lodgepole and Bakken formations) showing evidence of positive influence of the ancestral Cedar Creek structure.

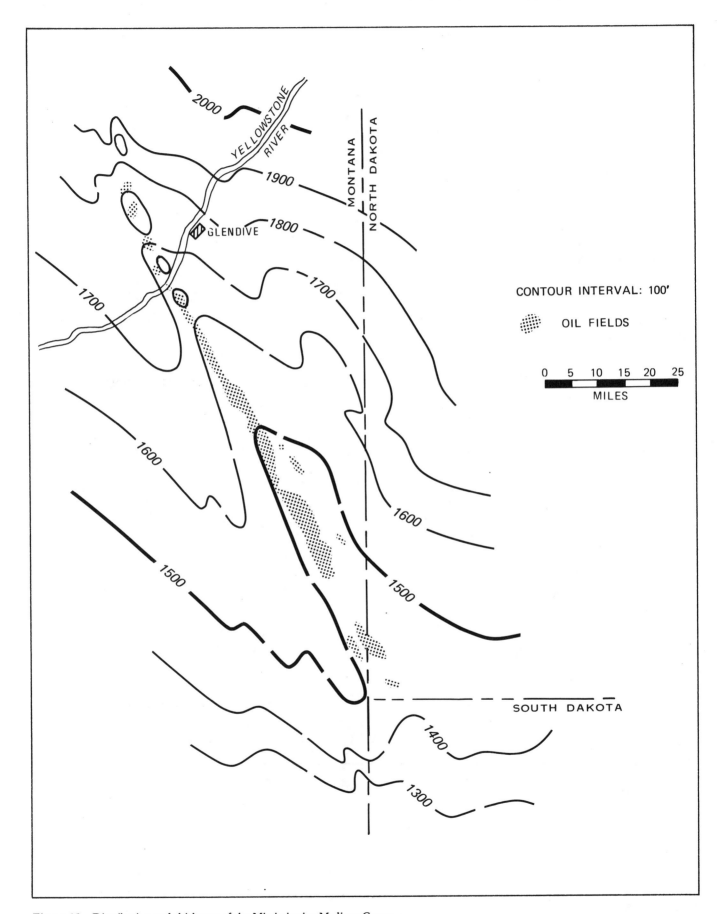

Figure 15—Distribution and thickness of the Mississippian Madison Group.

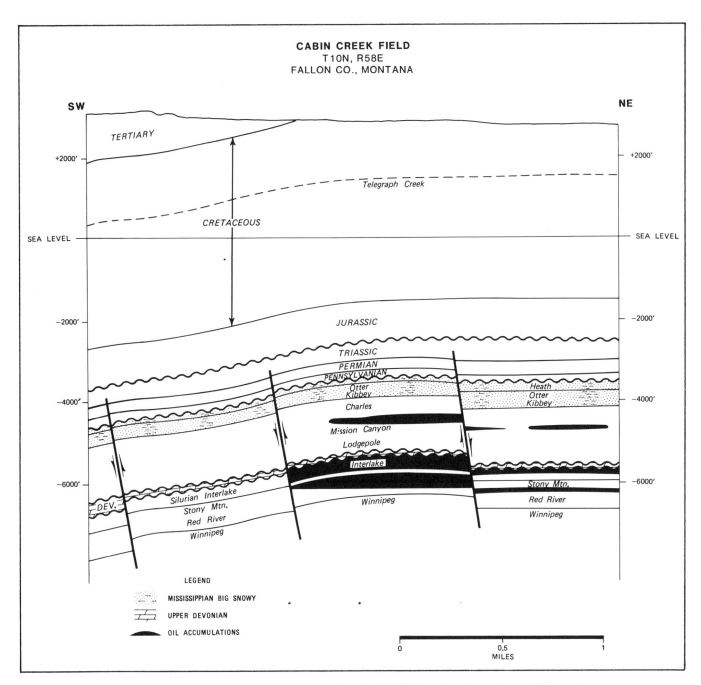

CABIN CREEK FIELD
T10N, R58E
FALLON CO., MONTANA

SW

NE

TERTIARY

+2000'

+2000'

Telegraph Creek

CRETACEOUS

SEA LEVEL

SEA LEVEL

-2000'

-2000'

JURASSIC

TRIASSIC

PERMIAN

PENNSYLVANIAN

Otter
Kibbey

Heath
Otter
Kibbey

-4000'

Charles

-4000'

Mission Canyon

Lodgepole

Interlake

-6000'

Stony Mtn.

-6000'

Silurian Interlake
Stony Mtn.
Red River

Winnipeg

Red River

Winnipeg

DEV.

Winnipeg

LEGEND

MISSISSIPPIAN BIG SNOWY

UPPER DEVONIAN

OIL ACCUMULATIONS

0   0.5   1
MILES

Figure 16—Southwest–northeast structural cross section across Cabin Creek field. Line of section is shown in Figure 3.

crestal portion of the present-day anticline is suggested by the thickening of the Upper Cretaceous Telegraph Creek to Middle Jurassic Piper Limestone interval in the area (Figure 22). Thickness patterns of smaller stratigraphic intervals are similar; none suggest significant fault movements.

Near the end of late Santonian time (about 70 m.y. ago), the Red River Formation had acquired the structural configuration shown in Figure 23. The main area of closure was shifted northward because of increased northwestward plunge; an eastward dip of about 5.6 m/km (30 ft/mi) prevailed as did significant areas of local closure at Deer Creek, Glendive, and Cabin Creek.

Slight thinning of the uppermost Cretaceous sequences in the Cedar Creek area suggests renewed positive movement in latest Cretaceous (Maestrichtian) time. Shoaling depositional conditions are suggested by the apparent localization of the gas-productive marine sandstones and siltstones in the Upper Cretaceous Judith River and Eagle formations.

Post-Paleocene

Terrestrial deposits of the Fort Union (Tongue River) Formation of Paleocene age (about 55 m.y. old) are the youngest Tertiary beds present in the anticline area.

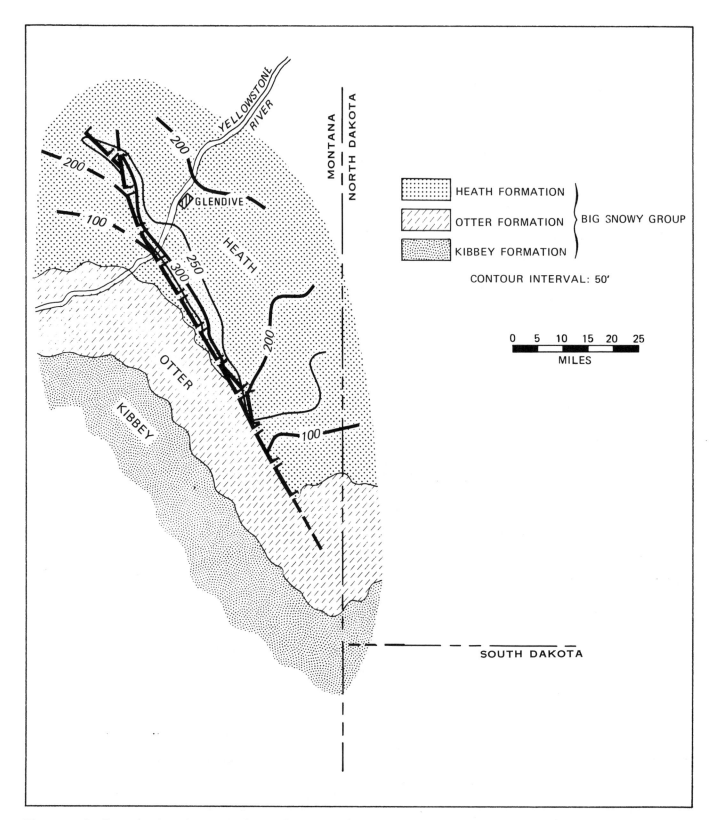

Figure 17—Pre-Pennsylvanian subcrop map showing distribution of the Upper Mississippian Big Snowy Group. Note significant thicknesses of Heath–Tyler strata on the Cedar Creek block east of the principal fault zones.

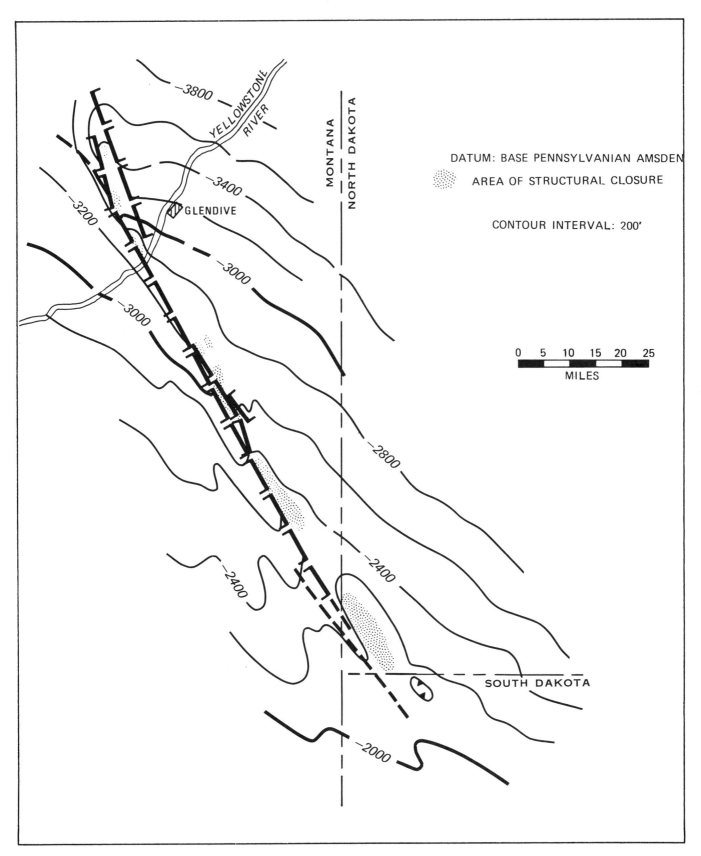

Figure 18—Paleostructural contour map of the top of the Ordovician Red River Formation depicting structural configuration during Early Pennsylvanian time.

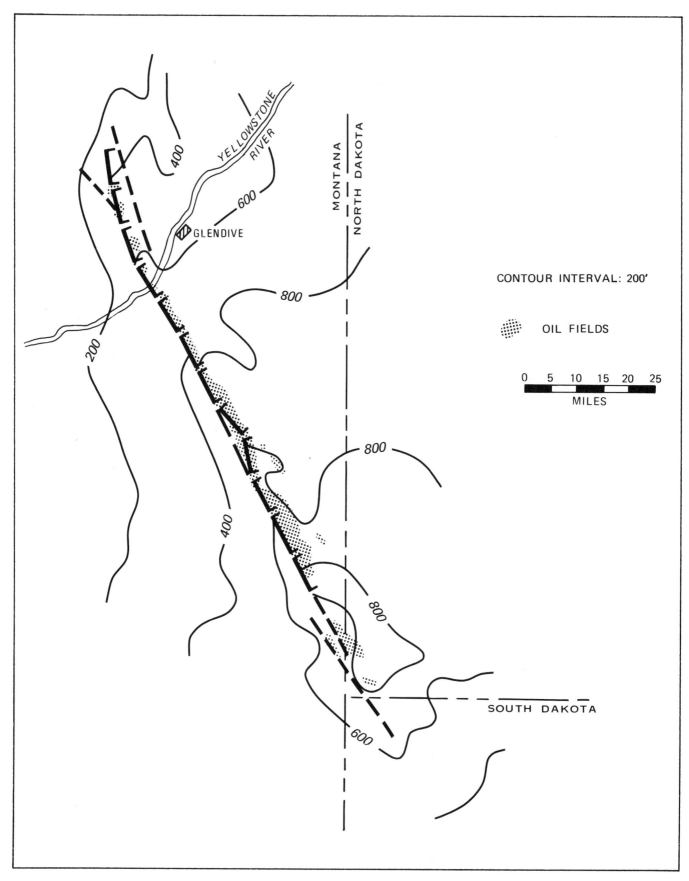

Figure 19—Distribution and thickness of combined Triassic and Permian strata. Note significant thicknesses on Cedar Creek block east of the principal fault zones.

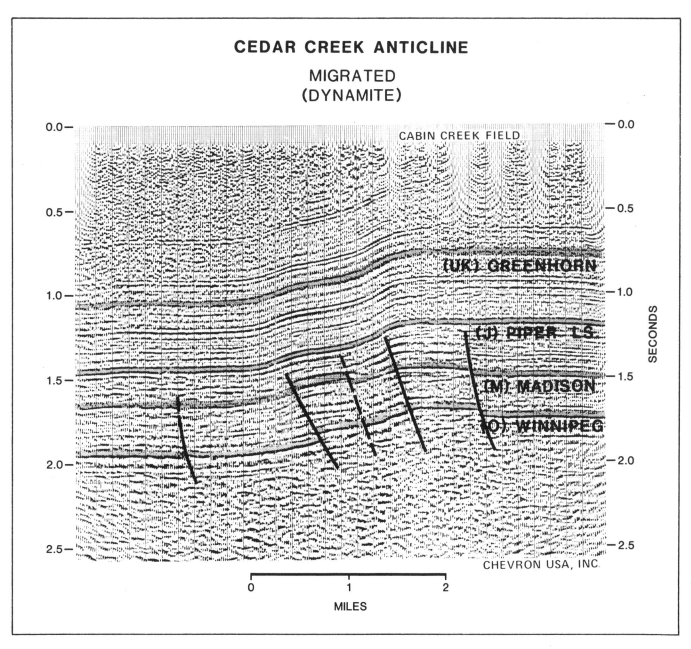

CEDAR CREEK ANTICLINE

MIGRATED
(DYNAMITE)

CABIN CREEK FIELD

(UK) GREENHORN

(J) PIPER LS.

(M) MADISON

(O) WINNIPEG

SECONDS

CHEVRON USA, INC.

0   1   2

MILES

Figure 20—Seismic line across Cabin Creek field. Line of section is shown in Figure 3.

Following their deposition, the Cedar Creek block underwent its greatest uplift, ranging from about 370 to 490 m (1200–1600 ft) between Glendive and Pennel fields. A linear belt of asymmetric drape folding was created nearly coincident and generally in line with the ancestral fault zone(s). Northwestward regional plunge was markedly increased, and eastward dip was more than doubled along most of the feature.

Major areas of structural closure on the deeper Paleozoic horizons were diminished or eliminated. Except for the long maintained local closures at Deer Creek, Glendive, and Cabin Creek, only local, relatively low-relief closures remained (Figures 3 and 4). Significant structural closures, however, were created in the Little Beaver and Little Beaver East

areas where little significant closure was apparent during earlier phases of tectonism. Subsidiary structures were developed southwest of Cedar Creek, namely the Plevna and Westmoreland anticlines. Although it is speculated that these features probably had earlier periods of structural development, documentation is lacking.

Surface faults are present along Cedar Creek anticline, but they generally have displacements of small magnitudes. In the densely drilled crestal portion of the feature, relatively few faults with significant displacement (greater than 4.5 m [15 ft]) are detectable in subsurface Mesozoic strata.

Because of the sparsity of wells drilled in the steeply dipping, drape-folded western flank, major subsurface faulting in Triassic and younger beds is rarely documentable.

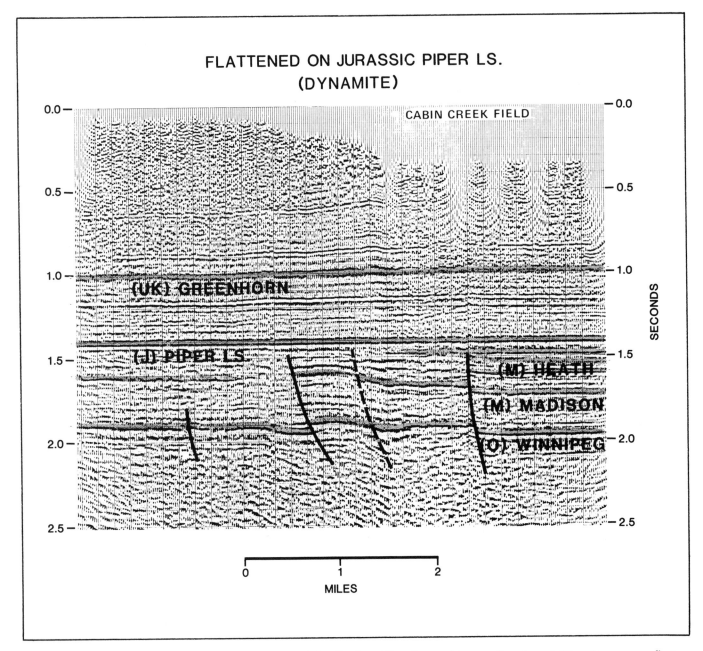

FLATTENED ON JURASSIC PIPER LS.
(DYNAMITE)

CABIN CREEK FIELD

(UK) GREENHORN

(J) PIPER LS

(M) HEATH

(M) MADISON

(O) WINNIPEG

SECONDS

MILES

Figure 21—Seismic line across Cabin Creek field, showing Cedar Creek anticline, flattened (restored) on Jurassic Piper Limestone reflector to illustrate pre-Jurassic fault movements and significant stratal thickening across faults. Line of section is shown in Figure 3.

Although faults in Paleozoic strata have been encountered in the few wells drilled on this flank, resolving the time of major movements on such faults—whether during Paleozoic episodes of deformation or during post-Paleocene time—is difficult.

Modern CDP reflection seismic data across the west flank of Cedar Creek conclusively show that significant post-Paleocene fault movement occurred that displaced (up to the east) Paleozoic, Jurassic, and Cretaceous strata (Figure 24). The fault zones are high-angle reverse systems that disrupt strata as young as the Upper Cretaceous Greenhorn. Major fault displacement diminishes upward through the Cretaceous strata and terminates in the drape folding of the

highly ductile Upper Cretaceous shales. No faults with *major* displacements have been mapped at the surface along the western flank of the feature; however, recent detailed mapping indicates that faulting is probably observable at the surface along the western flank in the southern part of the anticline (Vuke-Foster et al., in press; personal communication, 1984). The sequence of faulting on the western flank may be even more complicated than indicated by subsurface well data and/or the available high-resolution seismic data.

The present Cedar Creek axis was breached during epeirogenic uplift of middle Tertiary time that removed at least 460 m (1500 ft) of Paleocene and Upper Cretaceous

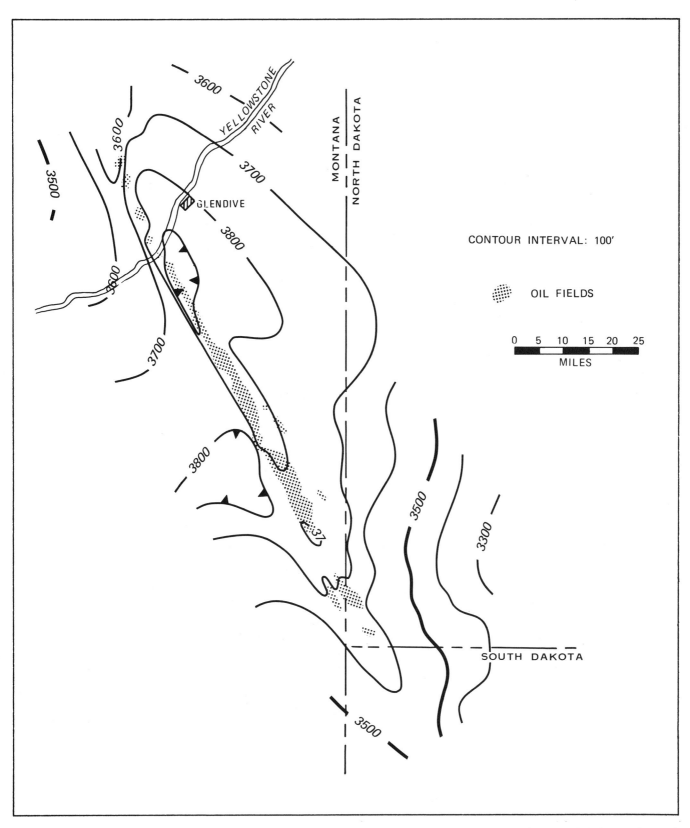

Figure 22—Isopach map of interval between top of Telegraph Creek and Jurassic Piper showing evidence of continued slight subsidence of the Cedar Creek block during Late Jurassic to middle Late Cretaceous.

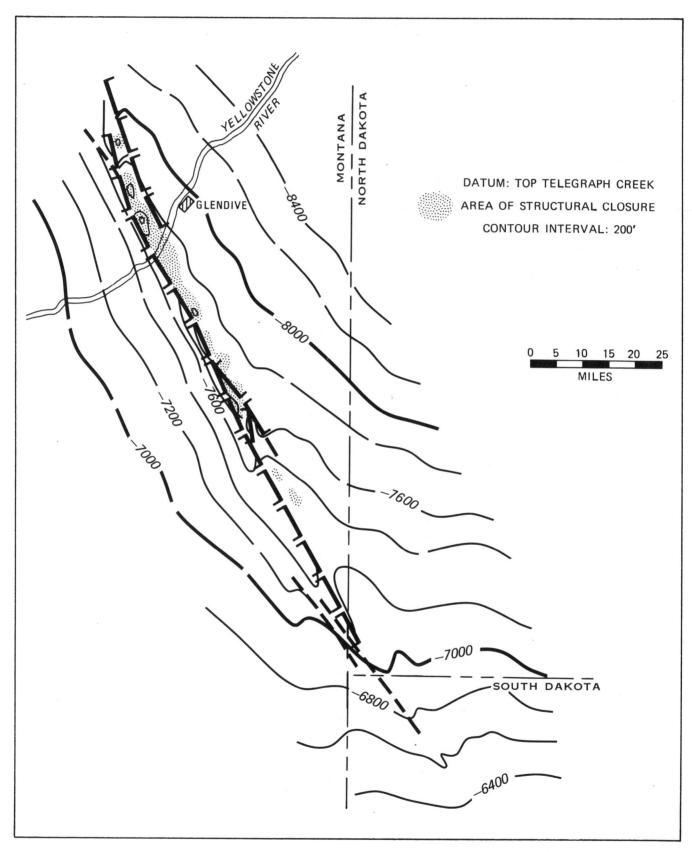

Figure 23—Paleostructural contour map of the top of the Ordovician Red River Formation depicting the structural configuration in Late Cretaceous time.

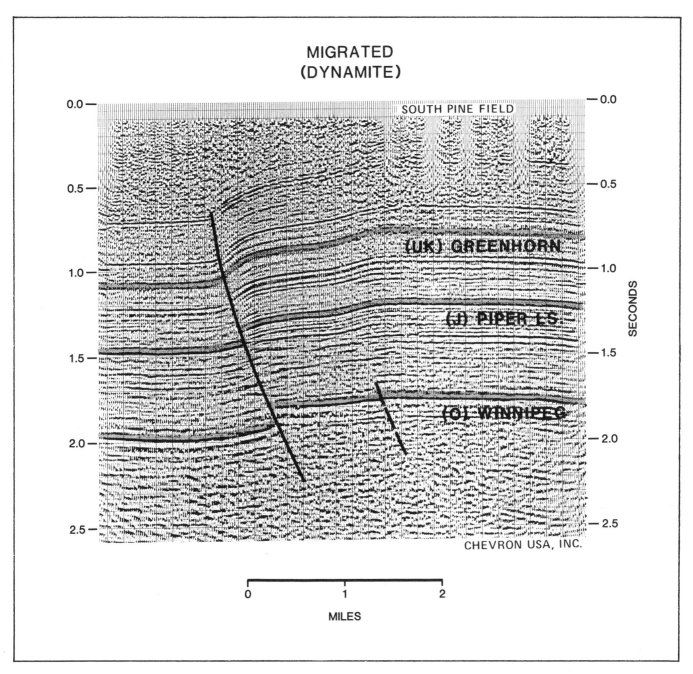

**MIGRATED
(DYNAMITE)**

SOUTH PINE FIELD

(UK) GREENHORN

(J) PIPER LS.

(O) WINNIPEG

CHEVRON USA, INC.

0   1   2

MILES

Figure 24—Seismic line across the southern part of Pine field, showing Cedar Creek anticline. Line of section is shown in Figure 3.

strata. Absence of younger Tertiary sediments in the area precludes dating the structure more definitively than post-Paleocene.

CONCLUSIONS

The fundamental stress system responsible for the regional basement tectonics creating Cedar Creek is believed to have been dominantly compressional. This stress caused episodic, differential, nearly vertical basement uplift and strike-slip (wrench) faulting, as evidenced by the following: high-angle reverse attitude of major fault zones; the subtle, almost en echelon pattern of structural closures; and the patterns of mappable, small magnitude, surface faults and fractures.

There remain many unanswered questions about the Cedar Creek structural feature. An awareness of the significant, recurrent paleotectonic history of the feature, however, may provide oil explorationists, stratigraphers, and structural geologists with new ideas in the search for more difficult to locate hydrocarbon accumulations.

ACKNOWLEDGMENTS

Special recognition is due to the late Thomas Reed Barnes, a distinguished exploration geologist, for his early vision of the giant hydrocarbon potential of the Cedar Creek anticline and for his many contributions during the formative phases of Exploration and Development by Shell Oil Company. The writer thanks Shell Western E&P Inc. for permission to publish this paper and for providing basic information, facilities, and services to prepare it. Permission to publish the seismic data herein was generously granted by Chevron USA Inc. Studies by the Exploration and Production staffs of Shell Oil Company's former Billings Division and Denver area offices, particularly by F. P. Van West and P. O. Roehl, were freely drawn upon, and their contributions are gratefully acknowledged. Sincere thanks are extended to Ruth V. Haag for typing the manuscript and to Jerry J. Baker, Mary E. Nelson, and Stacy L. Smith for the initial preparation and final drafting of the illustrations.

REFERENCES CITED

Clement, J. H., 1976, Geologic history—key to accumulation at Cedar Creek (abs.): AAPG Bulletin, v. 60, p. 2067–2068.

———, 1985, Depositional sequences and characteristics of Ordovician Red River reservoirs, Pennel field, Fallon County, Montana, *in* P. O. Roehl and P. W. Choquette, eds., Carbonate Petroleum Reservoirs: New York, Springer-Verlag, p. 71–84.

Davis, C. E., and R. G. Hunt, 1956, Geology and oil production on the northern portion of the Cedar Creek anticline, Dawson County, Montana: Bismarck, First International Williston Basin Symposium, p. 121–129.

Dobbin, C. E., and R. M. Larsen, 1936, Geologic and structure contour map of the Cedar Creek anticline, Dawson, Prairie, Wibaux, and Fallon counties, Montana, and Bowman County, North Dakota: U.S. Geological Survey Map, scale 1:63,360, 2 sheets.

Lochman-Balk, C., and J. L. Wilson, 1967, Stratigraphy of Upper Cambrian–Lower Ordovician subsurface sequences in Williston basin: AAPG Bulletin, v. 51, n. 6, p. 883–917.

McCabe, W. S., 1954, Williston basin Paleozoic unconformities: AAPG Bulletin, v. 38, n. 9, p. 1997–2100.

Peterman, Z. E., S. S. Goldich, and R. E. Zartman, 1983, High-grade reworked Archean rocks in the basement of the Williston basin, North Dakota (abs.): Geological Society of America Abstracts with Programs, 96th Annual Meeting, p. 660.

Peterson, J. A., 1984, Stratigraphy and sedimentary facies of the Madison Limestone and associated rocks in parts of Montana, North Dakota, South Dakota, and Wyoming: U.S. Geological Survey, Professional Paper 1273A, p. A1–A34.

Roehl, P. O., 1967, Stony Mountain (Ordovician) and Silurian (Interlake) facies analogs of Recent low energy marine and subaerial carbonates, Bahamas: AAPG Bulletin, v. 51, n. 10, p. 1979–2032.

Strickland, J. W., 1954, Cedar Creek anticline, eastern Montana (abs.): AAPG Bulletin, v. 38, n. 5, p. 947–948.

Vuke-Foster, S. M., R. B. Colton, M. C. Stickney, J. E. Robocker, E. M. Wilde, and K. C. Christensen, 1986, Geology of the Baker and Wibaux 30' × 60' quadrangles, eastern Montana and adjacent North Dakota: Butte, Montana Bureau of Mines and Geology, Map Series No. 41.

Part III
MIDDLE ROCKY MOUNTAINS

Tectonic Development of the Idaho–Wyoming Thrust Belt[1]

F. C. Armstrong
S. S. Oriel
U.S. Geological Survey
Denver, Colorado

Three stages are evident in the tectonic development of southeastern Idaho and western Wyoming. First are the changing patterns of tectonic elements during deposition; second, development of northward-trending folds and thrust faults; and third, development of block faults that produced horst ranges and graben valleys.

During Paleozoic time about 50,000 ft (15,000 m) of marine sediments, mostly limestone and dolomite, were deposited in a miogeosyncline and about 6000 ft (1800 m) of mixed marine sediments were deposited on the shelf to the east. Detritus came from both east and west from Cambrian time onward. Starting in Mississippian time, the belt between shelf and miogeosyncline, where thicknesses increase markedly, shifted progressively eastward.

During Mesozoic time about 35,000 ft (10,500 m) of marine and continental sediments were deposited in the western part of the region and about 15,000 ft (4500 m) in the eastern part, with terrestrial deposits becoming increasingly dominant. Western positive areas became the chief source of detritus. The belt of maximum thickening and the site of maximum deposition were relocated progressively eastward; maximum thicknesses of succeeding geologic systems are not superposed. In Late Triassic a belt on the west rose and the miogeosyncline started to break up. As Mesozoic time progressed the western high spread eastward, until by the end of the Jurassic the miogeosyncline gave way to intracratonic geosynclinal basins that received thick deposits, particularly in Cretaceous time. Cenozoic sedimentary rocks are products of orogeny in the region.

The second stage, which overlapped the first, produced folds overturned toward the east and thrust faults dipping gently west in a zone, convex to the east, 200 mi (320 km) long and 60 mi (~100 km) wide. Stratigraphic throw on many larger faults is about 20,000 ft (6000 m); horizontal displacement is at least 10–15 mi (16–24 km). Lack of metamorphism and mylonite along the faults is striking. From west to east, the thrust faults cut progressively younger beds, have progressively younger rocks in their upper plates, and are estimated to be successively younger. Thrusting started in the west in latest Jurassic and ended in the east perhaps as late as early Eocene time; detritus shed from emergent upper plates is preserved in coarse terrestrial strata of corresponding ages.

West of the thrust belt is a northwestward-trending area underlain mostly by lower Paleozoic rocks and flanked on the east and west by upper Paleozoic and Mesozoic rocks. Scattered pieces of eastward-dipping thrust faults have been reported west of the older rocks. This central area of old rocks has been interpreted as: (1) part of a large continuous thrust sheet moved scores of kilometers from the west, or (2) an uplifted segment of the earth's crust from which thrust sheets on the east and west were derived. Both interpretations have defects; relative thrust ages are difficult to explain under the first; a large positive gravity anomaly, expectable under the second, is apparently absent.

Block faulting, the third stage of tectonic development, started in Eocene time. Faulting has continued to Holocene time, as indicated by broken alluvial fans, displaced basalt flows less than 27,000 years old, and earthquakes. North-trending and east-trending fault sets are recognized. Old east-trending steep faults in the Bear River Range may be tear faults genetically related to thrusting. Movement along many faults has been recurrent. Patches of coarse Tertiary gravel on the flanks and crests of ranges, for which there is no provenance with present topography, may record reversed vertical movement along some north-trending faults. Present topographic relief of basins and ranges is tectonic.

[1] Adapted from AAPG Bulletin, v. 49, n. 11, p. 1847–1866.

INTRODUCTION

Many hundreds of man-years have been devoted to geologic studies of the thrust belt of southeastern Idaho and western Wyoming. The stimulus has been economic.

Classic reports resulted from the search for coal, oil, and gas (Veatch, 1907; Schultz, 1914), and phosphate (Mansfield, 1927). The region contains the best deposits of the western phosphate field, one of the large phosphate fields of the world, and the phosphate is being mined at an increasing rate for fertilizer and chemicals. Along its eastern margin the region contains oil and gas fields that likely are but a small sample of potential reserves. Coal continues to contribute to the region's economy in the new steam-generating plants constructed by Utah Power and Light Company, near strip mines a few kilometers south of Kemmerer, Wyoming.

Geologic complexities of the region have made a sound scientific approach necessary for intelligent exploration. It is thus no accident that economically motivated studies have resulted in important contributions to geologic science. The region contains some of the earliest thrust faults described in North America; the Bannock and Absaroka thrust faults are widely known and discussed in many textbooks on structural geology.

Despite the amount of work already done, parts of the region have not been studied since Peale's (1879) reconnaissance for the Hayden Territorial Survey, and answers to fundamental questions continue to be elusive. Accordingly, this paper is a progress report that summarizes some of the information and ideas that have emerged in recent years and some of the problems that remain.

SCOPE OF PAPER

The region discussed is southeast of the Snake River Plain, west of the Green River basin, and north of the Utah state line (Figure 1). The discussion is divided into three parts. The first deals with the changing patterns of deposition during the tectonic evolution in Paleozoic and early Mesozoic time of a miogeosyncline on the west and a shelf on the east; the second deals with regional deformation that formed north-trending folds and large thrust faults in late Mesozoic and early Tertiary time; and the third deals with block faults, active since Eocene time, that have produced the present horst ranges and graben valleys.

PART 1. DEPOSITIONAL HISTORY

Aggregate thicknesses of sedimentary rocks deposited in southeastern Idaho and adjacent western Wyoming during Paleozoic and Mesozoic time are about 100,000 ft (30,000 m). The entire thickness was not deposited at any one place, however; rather, the site of maximum deposition changed with time. The changing patterns of deposition are shown in isopachous maps and diagrammatic cross sections for each of the geologic systems of the Paleozoic and Mesozoic eras. Elements emphasized are the platform on the east where the sedimentary rocks are thin, the miogeosyncline on the

west where the rocks are thick, and a transitional area between. The position of this transitional area changed from system to system. Where data are available the sites of maximum deposition are shown; the positions of these also changed from system to system. The isopachous maps show the present distribution of thicknesses; the region is not yet sufficiently well understood to construct worthwhile palinspastic maps.

Cambrian

The isopachous map of the Cambrian System excludes the basal quartzite (Figure 2). A meaningful isopachous map of the quartzite is difficult to construct because different geologists have placed the Cambrian–Precambrian contact at different places in the thick sequence of conformable quartzite and argillite in the western part of the area. Accurate placement of this contact is one of the problems yet to be solved. The basal quartzite in the diagrammatic cross section is transgressive and is called Brigham on the west and Flathead on the east. Early Cambrian trilobites are present near the top of the Brigham Quartzite in the Portneuf Range (Oriel, 1964). The Flathead is Middle Cambrian.

Above the quartzite is a thick sequence of carbonate rocks, shale, and some arkosic quartzite; the sequence thins eastward. The isopachous map includes essentially the rocks deposited during Middle and Late Cambrian time. Westward thickening appears about 15 mi (24 km) east of the Idaho–Wyoming state line, and the site of maximum deposition was somewhere on the west. The Worm Creek Quartzite Member of the St. Charles Limestone contains abundant detrital cleavage fragments of potassium feldspar and some of the rock is arkose. The Worm Creek is not present in Wyoming; in Idaho it thickens westward and its feldspar content increases. This evidence indicates that during Late Cambrian time there was a source area at the west underlain by granite or granitic metamorphic rocks. During most of Paleozoic time major source areas both on the east and on the west (Rubey, 1955; Ross, 1962) shed detritus recurrently into the miogeosyncline.

Ordovician

The transitional area during Ordovician time was about 20 mi (32 km) east of the Idaho–Wyoming state line; as during Cambrian time, the site of maximum deposition was on the west (Figure 3). In Idaho, Lower and Middle Ordovician rocks are assigned to the Garden City Limestone and Swan Peak Quartzite, respectively. The entire thickness of Ordovician rocks in western Wyoming is included in the Upper Ordovician Bighorn Dolomite, which is correlative with the Fish Haven Dolomite in Idaho.

The Swan Peak is white pure quartzite composed of uniformly medium-sized, frosted, well-rounded quartz grains of high sphericity. The grains are probably of second or higher cycle. In this part of Idaho, most of the sand may have been derived from the north, as suggested by Ross (1964), but it is difficult to visualize only a northern source for the large volume of quartzose detritus in the Ordovician quartzite that extends southwest through Utah and into Nevada.

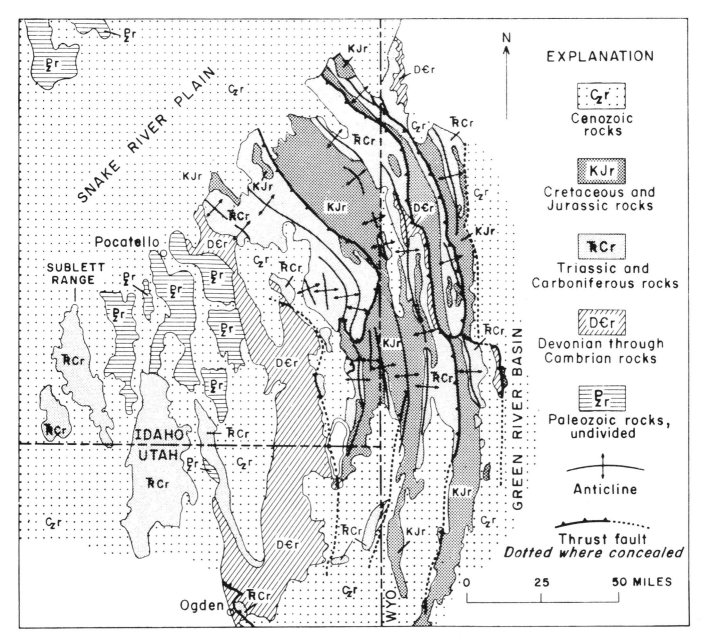

Figure 1—Generalized geologic map of Idaho–Wyoming thrust belt. Modified from Stose (1932).

Silurian

Silurian rocks, represented by the Laketown Dolomite in southeastern Idaho and adjacent Utah (Figure 4), have not been found in western Wyoming (Berdan and Duncan, 1955). The absence of Lower and Middle Ordovician and Silurian rocks in Wyoming was formerly interpreted as suggesting nondeposition during those times. The recent discovery of fossiliferous exposures of Middle(?) and Upper Ordovician and Silurian rocks among the Precambrian crystalline rocks 22 mi (35 km) southeast of Laramie, Wyoming (Chronic and Ferris, 1961), however, suggests a different interpretation. Consequently, the zero lines on the Ordovician and Silurian isopachous maps (Figures 3 and 4) probably are erosional rather than depositional limits.

Devonian

The isopachous map of the Devonian System shows local basins west of the Idaho–Wyoming state line in which thick deposits accumulated (Figure 5).

Devonian rocks in western Wyoming have been mapped as the Darby Formation, which is divisible into a lower dolomite, comparable and correlative with the Jefferson Dolomite, and an upper siltstone and limestone unit that is correlated with the Three Forks Limestone. In southeasternmost Idaho, the names Jefferson and Three Forks have been used. In north-central Utah the Jefferson Formation is divided (Williams, 1948) into a lower member, the Hyrum Dolomite Member that correlates with the Jefferson Dolomite, and an upper member, the Beirdneau

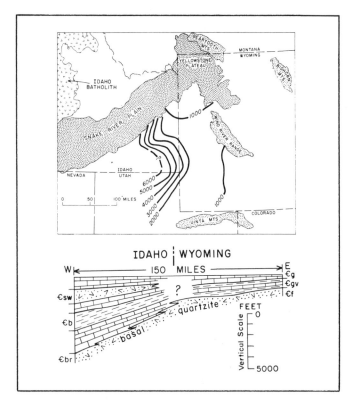

Figure 2—Isopachous map of the Cambrian System and generalized section. Contoured thicknesses are of Cambrian rocks above basal quartzite. Thicknesses are drawn in part from Lochman-Balk (1960) and Trimble and Carr (1962). Stratigraphic units shown in section: €sw, Worm Creek Quartzite Member of St. Charles Limestone; €b, Bloomington Formation; €br, Brigham Quartzite; €g, Gallatin Limestone; €gv, Gros Ventre Formation; €f, Flathead Quartzite.

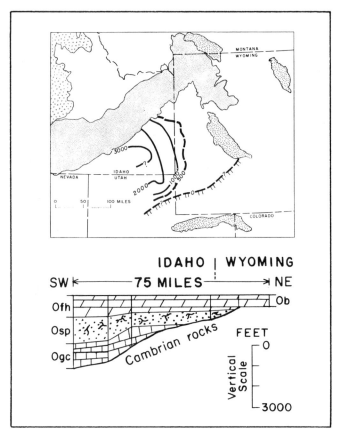

Figure 3—Isopachous map of the Ordovician System and generalized section. Map drawn in part from Hintze (1960). Stratigraphic units shown in section: Ofh, Fish Haven Dolomite; Osp, Swan Peak Quartzite; Ogc, Garden City Formation; Ob, Bighorn Dolomite.

Sandstone Member that correlates with the Three Forks Limestone. Because fine-grained quartz sandstone is present in the top of the Jefferson in part of southeastern Idaho, and because farther northeastward the sandstone of the Beirdneau grades laterally into the siltstone and limestone of the upper division of the Darby, the sandstone in the Beirdneau likely was derived from the west or southwest.

Below the Hyrum in Utah (Williams, 1948; Williams and Taylor, 1964) and in extreme southeastern Idaho (Armstrong, 1953; Coulter, 1956) is the Lower Devonian Water Canyon Formation, which consists of interbedded dolomite, argillaceous and silty dolomite, with cross-bedded quartz sandstone and calcareous quartz sandstone. Lower Devonian rocks of this type have not been recognized in western Wyoming. The presence of the Bighorn Dolomite on the east in Wyoming, blanketing underlying possible sources of quartz sand, requires that Osmond's (1962) interpretation of an eastern provenance for the quartz sand in the Water Canyon Formation and Sevy Dolomite in Utah and Nevada be modified for the sand in the Water Canyon in Idaho. We interpret the quartz sand in the Water Canyon in southeastern Idaho as having been derived from the southwest, transported by long-shore currents flowing northeastward along the eastern margin of the miogeosyncline.

Mississippian

Mississippian rocks thicken markedly westward from about 40 mi (64 km) west of the Idaho–Wyoming state line (Figure 6). During Cambrian through Devonian time this transitional area between miogeosyncline and shelf was east of its position in Mississippian time. From Mississippian time on, however, the position of the transitional area reversed with time and from then on it lay progressively farther to the east.

Because of facies changes and other stratigraphic complexities, Mississippian rocks in the region have been mapped in different ways and assigned different formational names. In the westernmost part of the region (not shown on Figure 6) Mississippian rocks have been called, in ascending order, Lodgepole Limestone, Deep Creek Formation, Great Blue Limestone, and Manning Canyon Shale (Carr and Trimble, 1961). In most of southeastern Idaho, however, Mississippian rocks have been mapped as two formations, the Madison Limestone and the overlying Brazer Limestone (Mansfield, 1927). In parts of adjacent western Wyoming the Mississippian consists of a basal carbonate unit and the lower part of the overlying Amsden Formation. In surface mapping the Madison and Brazer have not been differentiated; instead they have been mapped together as the basal carbonate unit, Madison and Brazer undivided

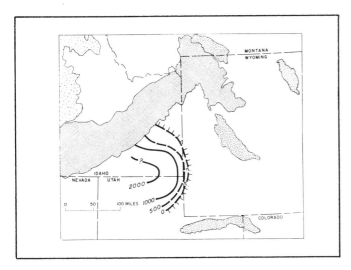

Figure 4—Isopachous map of the Silurian System. Thicknesses in Utah from Hintze (*in* Rush, 1963).

(Rubey, 1958).

Mississippian nomenclature in the region has been modified in recent years. The name Brazer was restricted in 1959 to its occurrence in the Crawford Mountains, Utah (Sando et al., 1959), and the Brazer Limestone of former usage in southeastern Idaho was subsequently redescribed as the Chesterfield Range Group, consisting of the Little Flat Formation and the overlying Monroe Canyon Limestone (Dutro and Sando, 1963). In the same article (Dutro and Sando, 1963) the name of the unit previously mapped as Madison Limestone (Mansfield, 1929) was changed to Lodgepole Limestone. Fine-grained sandstone in the lower part of the Brazer, now named the sandstone member of the Little Flat Formation, is the stratigraphic equivalent of the Humbug Formation to the southwest. The Humbug is slightly thicker and coarser grained than the sandstone member of the Little Flat and contains detrital grains of microcline and plagioclase (Gilluly, 1932); both minerals have been recognized recently in the Little Flat Formation in the Fish Creek Range. Accordingly, the sand in the Little Flat is interpreted as having been derived from a source area on the southwest.

The presence in Mississippian time of land on the northwest also is indicated by coal beds in the Milligen Formation (Umpleby et al., 1930) in the Wood River region of Idaho north of the Snake River Plain (Figure 6).

Pennsylvanian

A large amount of westward thickening of Pennsylvanian rocks is apparent about 20 mi (32 km) west of the Idaho–Wyoming state line (Figure 7), and the site of maximum deposition is farther west, as shown. The upper part of the Amsden Formation and the lower and middle parts of the Wells Formation make up the Pennsylvanian in western Wyoming. In southeasternmost Idaho the lower and middle parts of the Wells are the only Pennsylvanian rocks. Farther west in Idaho, Pennsylvanian rocks are assigned to the Oquirrh Formation (Carr and Trimble, 1961). The proportion of sandstone in Pennsylvanian rocks increases eastward, which suggests a source on the east. Westward

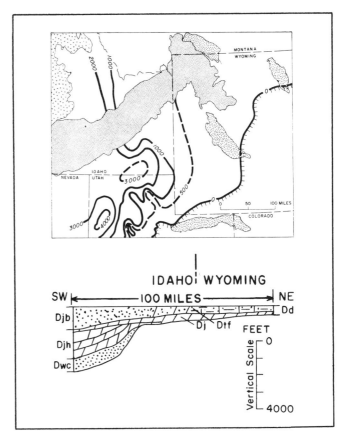

Figure 5—Isopachous map of the Devonian System and generalized section. Thicknesses in Utah are from Rigby (1963); in Wyoming, partly from C. A. Sandberg (unpublished data). Stratigraphic units shown in section: Djb, Beirdneau Sandstone Member of Jefferson Dolomite; Djh, Hyrum Dolomite Member of Jefferson Dolomite; Dwc, Water Canyon Formation; Dtf, Three Forks Limestone; Dj, Jefferson Dolomite; Dd, Darby Formation.

from southeasternmost Idaho, however, Pennsylvanian rocks contain a large amount of sandstone and, although the writers do not have supporting quantitative data, their impression is that the total amount of the quartz sand fraction in the thick western sequence exceeds that in the thin eastern sequence. The large amount of sandstone in the Pennsylvanian rocks on the west, the presence of thick, coarse conglomerate beds in the Wood River Formation north of the Snake River Plain and east of the Idaho batholith (Umpleby et al., 1930; Ross, 1937) near the 8000-ft (2400-m) isopachous line (Figure 7), indicate a western source. Detritus in Pennsylvanian rocks of southeastern Idaho and adjacent Wyoming probably was derived both from the east and the west.

Permian

By Permian time, the eastern margin of the miogeosyncline was still farther east, close to the Idaho–Wyoming state line. The axis of maximum deposition lies between the 9000-ft (2700-m) isopachous lines and trends northeastward (Figure 8). A high area of similar trend was present farther to the west (McKee et al., 1967).

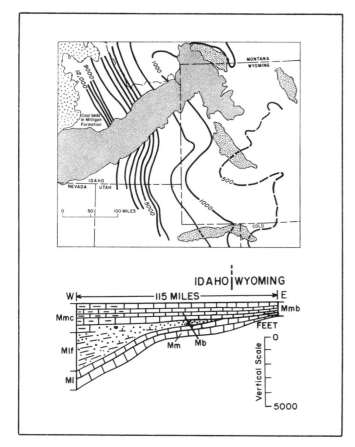

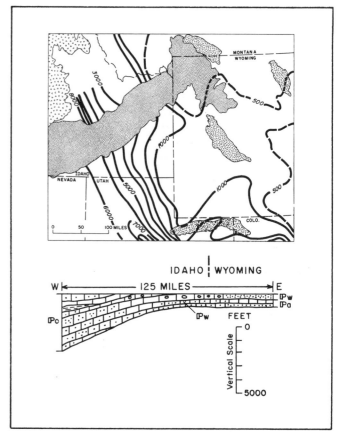

Figure 6—Isopachous map of the Mississippian System and generalized section. Thicknesses are partly from Strickland (1956) and A. E. Roberts (unpublished data). Stratigraphic units shown in section: Mmc, Monroe Canyon Formation; Mlf, Little Flat Formation; Ml, Lodgepole Limestone; Mb, Brazer Limestone; Mm, Madison Limestone; and Mmb, Madison and Brazer Limestones, undivided. Lower part of Amsden Formation not shown in section.

Figure 7—Isopachous map of the Pennsylvanian System and generalized section. Thicknesses are partly from Williams (1962) and W. W. Mallory (unpublished data). Stratigraphic units shown in section: ℙo, Oquirrh Formation; ℙw, Wells Formation, lower and middle parts; and ℙa, Amsden Formation, upper part.

The upper part of the Wells Formation, now assigned to the Grandeur Tongue of the Park City Formation (McKelvey et al., 1959), and the Phosphoria Formation were deposited in Permian time. Pronounced thickening westward resulted almost wholly from thickening of the lower unit that is roughly equivalent to the top of the Wells. The Wells is interbedded sandstone and limestone that probably were deposited in shallow water. The sand may have been derived from both east and west. The Phosphoria is mudstone, chert, phosphorite, and limestone and was deposited in deep water.

Triassic

All Paleozoic rocks in southeastern Idaho were deposited in a marine environment. Lower Triassic rocks are also marine, but the rest of the Triassic is nonmarine.

Triassic sediments thicken markedly westward from about 30 mi (48 km) east of the Idaho–Wyoming border, and the center of a deep basin is shown by the 8000-ft (2400-m) isopach line (Figure 9). In central Idaho west of the basin, a northeast-trending ridge rose early in Late Triassic time and shed coarse detritus into the basin on the southeast to form

the Higham Grit, which probably was deposited as coalescing fans along a mountain front (McKee et al., 1959). The rise of this ridge, perhaps foreshadowed in the Permian, indicates the start of the breakup of the miogeosyncline.

Jurassic

During Jurassic time, the familiar pattern of shelf on the east and geosyncline on the west, although evident, is not as sharply defined as in older systems, and thick deposits of sedimentary rocks are present in western Wyoming (Figure 10). The location of the western margin of the basin is not known, but it probably was not very far west. The nonmarine conditions of Late Triassic persisted into Early Jurassic time. The Late Triassic northeast-trending high in central Idaho persisted into the Early Jurassic and supplied detritus for the Nugget Sandstone. During this time the high appears to have spread eastward and perhaps the relief was less. Subsequently, the sea returned, and the Twin Creek Limestone, Preuss Sandstone, and Stump Sandstone were deposited under marine conditions. In early Late Jurassic time the high on the west rose, again spread eastward, and shed detritus now found in the Preuss and Stump.

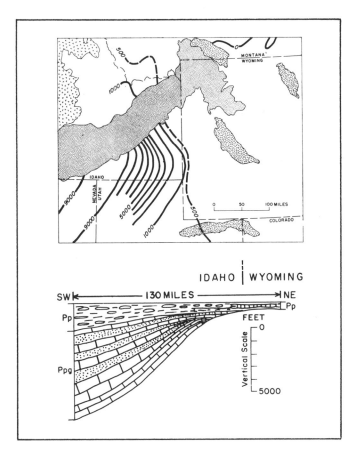

Figure 8—Isopachous map of the Permian System and generalized section. Map generalized after McKee et al. (1967). Stratigraphic units shown in section: Pp, Phosphoria Formation; and Ppg, Grandeur Tongue of Park City Formation which occurs at top of, and has been mapped with, upper part of Wells Formation.

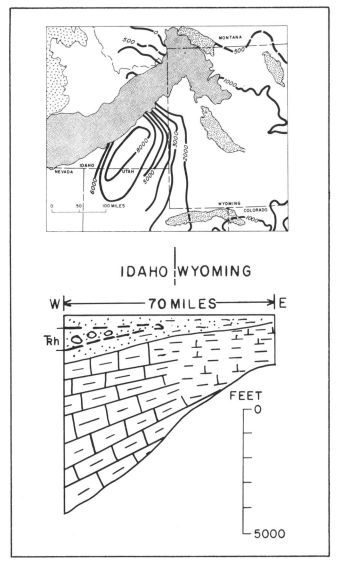

Figure 9—Isopachous map of the Triassic System and generalized section. Map generalized from McKee et al. (1959). Stratigraphic unit designated on section is Higham Grit (Trh).

Lower Cretaceous

During Early Cretaceous time the axis of a deep, north-trending basin was 5–10 mi (8–16 km) west of the Idaho–Wyoming border (Figure 11). In latest Jurassic or earliest Cretaceous time, parts of southeastern Idaho were raised and mountains were formed. Coarse detritus was deposited in coalescing alluvial fans to form the Gannett Group, and orogenic sediments, such as the Wayan Formation[2], continued to be deposited throughout Early Cretaceous time. Because Lower Cretaceous sediments probably were not deposited west of the Bear River Range (Armstrong and Cressman, 1963), the western parts of the isopachous lines may be shown incorrectly on the map. The Gannett and Wayan thin markedly eastward and become finer grained.

---

[2] Although the Wayan Formation has previously been assigned to the Lower(?) and Upper Cretaceous, it is here regarded as Lower Cretaceous. Studies in the type area, particularly by W. W. Rubey, have shown that many of the rocks mapped as Wayan belong to the Gannett Group. The younger rocks present are western facies of the Bear River and Aspen formations of Early Cretaceous age and are arbitrarily depicted in the Figure 11 cross section to be about 900 m thick.

Upper Cretaceous

In Late Cretaceous time an axis of maximum deposition was 10–15 mi (8–16 km) east of the Idaho–Wyoming border (Figure 12). Farther east are other basins in which thick prisms of sediments were deposited. In the western basin probably more than two-thirds of the rocks were deposited in early Late Cretaceous (Colorado) time, whereas in the eastern basins most of the rocks were deposited in late Late Cretaceous (Montana) time. In a sense this reflects continued eastward movement with time of the site of maximum deposition. Detritus in sandy and partly conglomeratic tongues in the Frontier, Hilliard, and Adaville formations was derived from the west and northwest.

Patterns of Deposition

The systemic isopachous maps show gross patterns in addition to the features already mentioned. Because of insufficient information these gross patterns, however, may

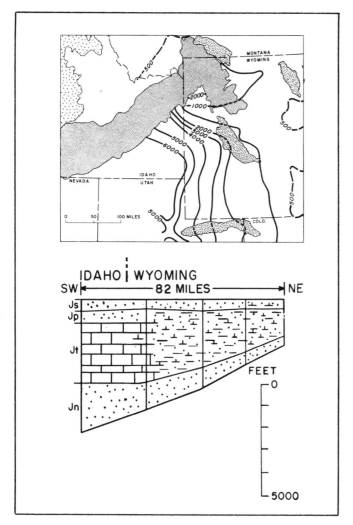

Figure 10—Isopachous map of the Jurassic System and generalized section. Map generalized from McKee et al. (1956). Stratigraphic units shown in section: Js, Stump Sandstone; Jp, Preuss Sandstone; Jt, Twin Creek Limestone; Jn, Nugget Sandstone.

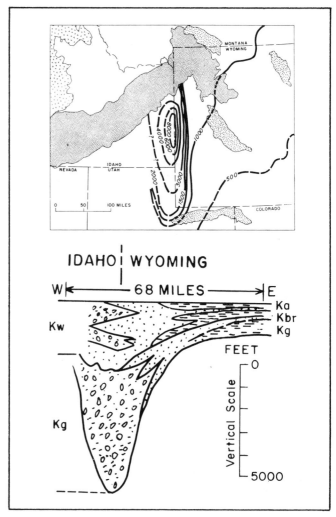

Figure 11—Isopachous map of the Lower Cretaceous Series and generalized section. Map modified from Haun and Barlow (1962), used by permission of the Wyoming Geological Association. Stratigraphic units shown in section: Kw, Wayan Formation; Kg, Gannett Group; Ka, Aspen Formation; Kbr, Bear River Formation.

be more apparent than real. Isopachous lines on the Cambrian–Silurian maps are strongly concave westward, and the axis of concavity trends slightly north of east. The Devonian map is characterized by a subdued westward concavity and a northward trend for the eastern margin of the miogeosyncline. After Devonian time the eastern margin of the miogeosyncline had a well-defined northward to northwestward trend, and the axis of concavity had a northeastward trend. These changes in pattern in the transitional Devonian Period may be related to tectonic activity—perhaps the Antler orogeny (Roberts et al., 1958)—on the west. These changes in pattern also may help explain the greater continuity northward, across the eastern part of the Snake River Plain, for upper than for lower Paleozoic stratigraphic units.

The northerly trend of the eastern margin of the miogeosyncline persisted into Cretaceous time, and the isopachous lines lost their former westward concavity. During Jurassic time, however, the eastern margin of the miogeosyncline is not well defined. Can the Jurassic Period be compared with the Devonian as an unsettled, transitional

period after which a different pattern emerged in response to new forces? Were the deep, north-trending Cretaceous basins true descendants of the miogeosyncline? They may not have been, because these deep basins developed on the edge of a cratonic area that had been moderately stable for the previous 450 m.y. Certainly during Late Cretaceous time a different pattern developed in Wyoming. During latest Cretaceous (Lance) time, two deep east-trending basins began to form on the shelf in Wyoming, one across northern Wyoming and the other across southern Wyoming (Love et al., 1963). The northern basin extended from the vicinity of Yellowstone Park eastward to the Black Hills and later developed into the Wind River and related basins; the southern one extended across south-central Wyoming and later developed into the Great Divide and related basins.

The sedimentary record thus indicates that during Paleozoic time there was a miogeosyncline on the west and a platform on the east. A thick accumulation of marine strata, mostly limestone and dolomite, was deposited across the area. Throughout Paleozoic time detritus came recurrently from both the east and the west. Beginning in Mississippian

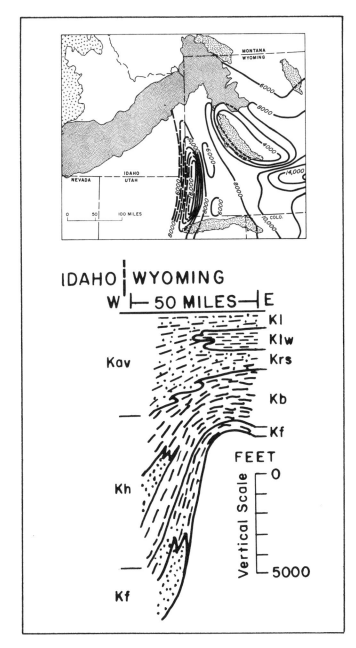

Figure 12—Isopachous map of the Upper Cretaceous Series and generalized section. Map generalized from Weimer (1961). Stratigraphic units shown in section: Kav, Adaville Formation; Kh, Hilliard Shale; Kf, Frontier Formation; Kl, Lance Formation; Klw, Lewis Shale; Krs, Rock Springs Formation; Kb, Baxter Shale. Relationship of western sequence to eastern sequence, shown as now commonly accepted, has not been established.

time the miogeosyncline was shifted eastward; at the same time the site of maximum deposition may also have shifted eastward, a trend which, with temporary reversal in the Permian, continued into Mesozoic time.

These shifts in the position of the miogeosyncline continued during the Mesozoic when marine and continental sedimentary rocks were deposited across the area. A "high" on the west became the principal source of detritus, and with the passage of time increasing proportions of continental strata were deposited, finally to the exclusion of

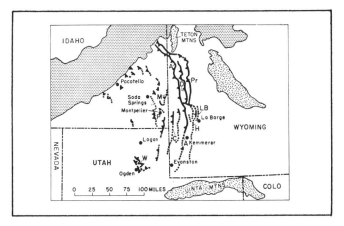

Figure 13—Mapped traces of major thrust faults. Solid line indicates exposed fault trace; broken line indicates concealed fault trace. Major faults shown: P, Paris; M, Meade; C, Crawford; A, Absaroka; D, Darby; Pr, Prospect; H, Hogsback; LB, La Barge; W, Willard. Modified from Rubey (1958) and Armstrong and Cressman (1963).

marine sediments. As Mesozoic time progressed, the high on the west spread eastward until the miogeosyncline was destroyed in Late Jurassic time. In Late Cretaceous time basins developed on what had been the stable, cratonic platform on the east.

Destruction of the miogeosyncline in late Mesozoic time was accompanied by the development of large north-trending folds and by eastward movement on thrust faults along the eastern flank of the miogeosyncline. Thus, the second chapter of tectonic history began.

PART 2. THRUST FAULTING

The thrust belt forms an eastwardly convex, arcuate zone about 200 mi (320 km) long and 60 mi (~100 km) wide, extending from the Snake River Plain southward into Utah (Figure 13). It has been said (Rubey, 1955) to consist of parts of Idaho and Utah that were pushed, or slipped, into Wyoming between the Teton Mountain and Uinta Mountain buttresses of Precambrian rock. The major thrusts to be discussed are, from west to east, the Paris, the Absaroka, the Darby, the Prospect, and the Hogsback.[3] The probable ages of the Meade, Crawford, and La Barge thrusts are shown in figures that follow.

Structural Relationships of Faults

The thrust faults of the Idaho–Wyoming thrust belt are unlike many in the world in that strata involved are unmetamorphosed and no major fault breccia or mylonite is present. In structural sections (Figure 14) prepared by Rubey (Rubey and Hubbert, 1959), the Absaroka, Darby, and Prospect thrust faults are shown dipping slightly to moderately westward. Drilling has confirmed low westward dip for the Hogsback thrust (not shown in Figure 14) about

---

[3]Hogsback is a new name for the thrust fault south of Snider basin called Darby by Schultz (1914) and subsequent workers and Darby(?) by Oriel (1962). The fault is named for its excellent exposures along Hogsback Ridge west of La Barge, Wyoming.

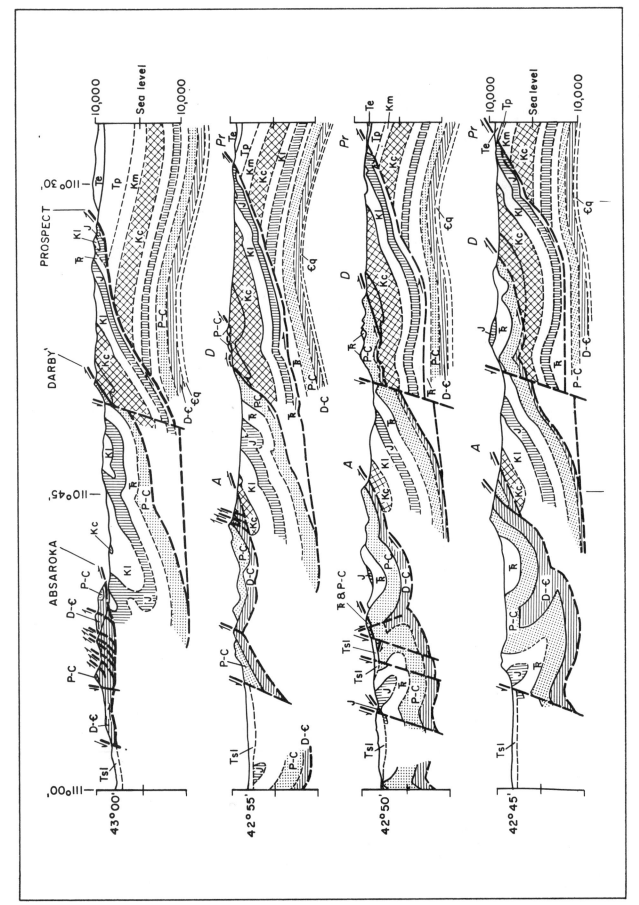

Figure 14—West-east structure sections, somewhat generalized, across Bedford, Blind Bull Creek, and part of Merna Butte Quadrangles, Wyoming. Horizontal and vertical scales same. Cq, Cambrian quartzite; D-Є, Devonian to Cambrian Gros Ventre Formation, inclusive; P-Є, Permian to Carboniferous, inclusive; Tr, Triassic; J, Jurassic; Kl, Lower Cretaceous; Kc, Cretaceous (Colorado); Km, Cretaceous (Montana); Tp, Paleocene; Te, Eocene; Tsl, Pliocene Salt Lake Formation. A, Absaroka thrust fault; D, Darby thrust fault; Pr, Prospect thrust fault. From Rubey and Hubbert (1959, p. 188).

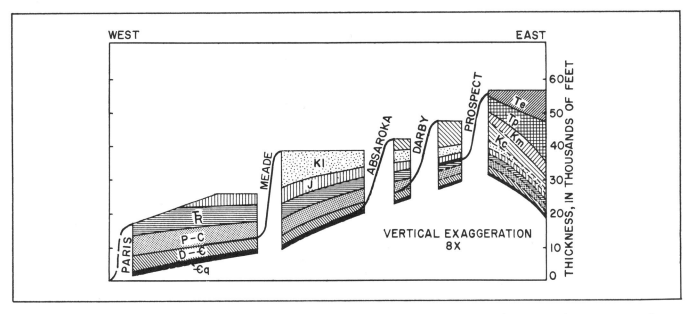

Figure 15—Stratigraphic sections in four large overthrust sheets and in Green River Basin. Diagram shows approximate constancy of stratigraphic throw along different faults and progressively younger age of involved rocks from west to east. Formational symbols as in Figure 14. Modified from Rubey and Hubbert (1959).

40 mi (65 km) south of the cross sections; there the average dip of the Hogsback thrust, which has been penetrated by several wells across a downdip direction of 10 mi (~16 km) is about 15° W. The Darby and Prospect faults mostly parallel bedding in the upper plates of the thrusts, although they locally cut across bedding. In the upper two cross sections the Absaroka fault cuts across bedding both above and below the fault surface, whereas in the lower two cross sections the Absaroka is like the Darby and Prospect faults and parallels bedding in the upper plate. Strata above and below all the faults are folded. Although folding formerly was thought to have preceded thrusting, it is now believed to have accompanied the faulting. Minimum horizontal displacements along the faults were 10–15 mi (8–16 km) and may have been considerably more.

Stratigraphic throw on each of the major thrust faults is about 20,000 ft (6000 m) as indicated by the vertical scale in Figure 15. Also illustrated is the fact that the upper plate of each successively more eastward thrust fault contains successively younger strata.

Dating Major Movements

Few of the thrust faults can be dated within narrow limits by usual geologic methods. One that can be, however, is the Prospect thrust (Figure 16). The fault cuts the Hoback Formation of Paleocene and earliest Eocene age; the Eocene part of the Hoback is folded with the underlying strata, but only the Paleocene part is cut by the fault. South of where the fault cuts the Hoback, its trace is overlain by Wasatch strata of late early Eocene age. Major movement on the fault thus occurred in middle early Eocene time.

Somewhat similar evidence is used to date major movement on the Absaroka thrust north of Kemmerer. A short distance south of the area illustrated in Figure 17, the youngest unit cut by the Absaroka is the Adaville Formation, whose topmost part is of latest Cretaceous (Lance or late Montana) age. The Evanston Formation, which is of latest

Cretaceous (Lance) and Paleocene age (Rubey et al., 1961), crops out in a north-trending belt close to the trace of the Absaroka; and locally the basal beds of the Evanston, the Lance part, have been cut by late small movement on the Absaroka. Although the Evanston has not been found directly overlying the fault, it overlies rocks above and below the fault with angular unconformity. The Evanston thus was deposited in its present site after most of the movement; major thrust movement on the Absaroka is thereby dated as latest Cretaceous.

Another type of evidence was used to date first movement on the Paris thrust (Armstrong, 1962; Armstrong and Cressman, 1963). At Red Mountain in the Gannett Hills (Figure 18), coarse conglomeratic redbeds, 5000 ft (1500 m) thick, are a coarse facies of the lower part of the Gannett Group of earliest Cretaceous age whose basal beds probably are of latest Jurassic age. Examination of nearby Gannett stratigraphic sections shows that the conglomeratic unit coarsens westward and more than doubles in thickness in a distance of 8 mi (12 km), thus indicating a western source for the detritus. A source area only 25–30 mi (40–50 km) to the west is suggested by the rate of increase in pebble size (Rubey, *in* Moritz, 1953) and by the survival of nonresistant rocks in the conglomerate (Plumley, 1948). Pebbles of Paleozoic rocks in the conglomerate were derived from formations now exposed in the upper plate of the Paris thrust 20–30 mi (30–50 km) west of Red Mountain. If the conglomeratic unit on Red Mountain is a synorogenic deposit (Eardley, 1960) as the evidence suggests, then first movement on the Paris thrust is dated as latest Jurassic and earliest Cretaceous.

Evidence for accurately dating the other thrust faults shown on Figure 13 is meager. Nonetheless, in Figure 19 an attempt is made to date principal movement on each fault, with the realization that some dates may be slightly in error. Comparison of Figures 13 and 19 shows that the oldest faults are on the west and the youngest on the east, a fact

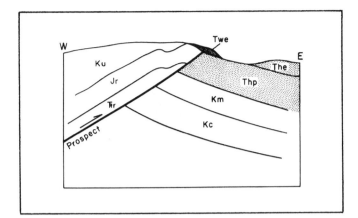

Figure 16—Schematic structure section of Prospect thrust fault west of Daniel, Wyoming, showing basis for dating major movement on fault. Twe, Wasatch Formation strata of latest early Eocene age; The, Hoback Formation of Eardley and others (1944), upper part of earliest Eocene age; Thp, Hoback Formation, lower part of Paleocene age; Km, Cretaceous (Montana); Kc, Cretaceous (Colorado); Ku, Cretaceous undivided; Jr, Jurassic rocks; Trr, Triassic rocks.

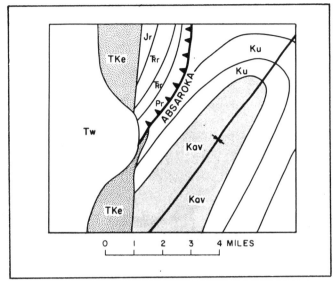

Figure 17—Schematic map of part of Absaroka thrust fault, showing the basis for dating major movement on fault in northern part of Kemmerer Quadrangle. Tw, Wasatch Formation (Eocene); TKe, Evanston Formation (Paleocene and uppermost Cretaceous); Kav, Adaville Formation (Upper Cretaceous); Ku, Cretaceous undivided; Jr, Jurassic rocks; Trr, Triassic rocks; Pr, Permian rocks. Modified from unpublished map by W. W. Rubey, J. I. Tracey, Jr., and S. S. Oriel.

which has been recognized before. Because the Darby and Prospect thrusts probably are not the northward continuations of the Hogsback and LaBarge faults, these two pairs of faults have been shown one above the other on the right side of Figure 19. Not shown on Figure 19 are times and durations of recurrent movements known or interpreted to have occurred on some of the faults.

Data from Figures 13 and 19 are plotted as a graph in Figure 20; the ordinate is time in millions of years, and the abscissa is distance in miles, measured westward from the trace of the easternmost thrust about normal to the trends of the faults and fold axes. The graph was conceived initially to illustrate the futility of applying the term Laramide to deformations in the thrust belt. Use of the term masks, rather than reveals, temporal relationships discovered in the region. Surprisingly, both the well-supported and indirectly inferred dates cluster along a straight line. Perhaps even more disconcerting is an implication of the westward projection of this straight line. The Antler orogenic belt of Nevada has been inferred to project northward into Idaho (Churkin, 1962; Roberts and Thomasson, 1964). The distance between the northward projection of the belt and the easternmost fault is plotted in Figure 20; it intersects the line determined by the Idaho–Wyoming thrusts at about 380 m.y. or at an age of about Middle Devonian (Holmes, 1959). The result is not greatly discrepant from the Late Devonian or Early Mississippian to Early Pennsylvanian age deduced by Roberts et al. (1958) from other evidence. The graph may or may not have real meaning. An inference that could be drawn from the graph, for example, by projecting the line toward the right, would be the location of the next thrust faulting on the east.

Gross Structure West of Thrusts

A question of some concern is, What is the gross crustal structure of the area of mainly lower Paleozoic rocks that extends northward from Ogden, Utah, and lies west of the Idaho–Wyoming thrust belt (Figure 1)? The contrast in

detail shown on the east and on the west in Figure 1 reflects contrasts in what is known about the area. Most of the area between Ogden and Pocatello is underlain by rocks of the Brigham through Ordovician units; upper Paleozoic and Mesozoic units are present on both sides of the area. Triassic rocks occur in the Sublett Range on the west and all Mesozoic systems are present in the thrust belt on the east.

East-dipping thrust faults (Figure 13) near the western margin of the area of Paleozoic rocks have been reported by Blackwelder (1910) and Eardley (1944) in the Ogden area, by Murdock (1961) just north of the Utah–Idaho state line, by Anderson (1928) and Ludlum (1943) southeast of Pocatello and by W. J. Carr and D. E. Trimble (personal communication, 1963) southwest of Pocatello.

Many geologists (Richards and Mansfield, 1912; Richardson, 1941; Eardley, 1944; Crittenden, 1961) have inferred that the Willard thrust near Ogden connects with the Paris thrust to the east. On the basis of this interpretation (Figure 21) the area of Paleozoic rocks from Ogden northward is regarded as a remnant of the upper plate of a large thrust that moved scores of kilometers from the west and is therefore allochthonous. In addition to other objections (Armstrong and Cressman, 1963), this interpretation is difficult to reconcile with the progressively younger ages of the thrust faults eastward from the Paris thrust, particularly if there is more than one major eastward-dipping thrust on the west. Moreover, if this block is allochthonous and did move scores of kilometers eastward, it is surprising that nowhere in southeastern Idaho or western Wyoming has a western facies of any stratigraphic unit yet been recognized east of an eastern facies. On the south in Utah, however, the Willard thrust fault separates a western thick sequence of the east from an eastern thin sequence on the west (Blackwelder, 1925; Crittenden, 1961).

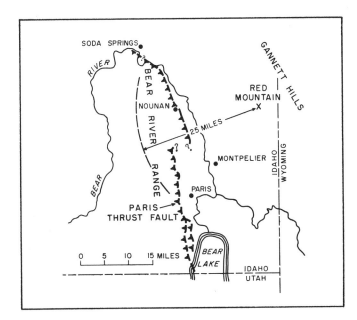

Figure 18—Map of localities cited in dating first movement on Paris thrust fault.

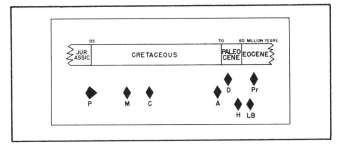

Figure 19—Estimated ages of principal movement on major thrust faults. Not shown are times of recurrent movements on some faults. Numerical ages from Holmes (1959). Letter symbols are same as on Figure 13, with which this figure should be compared.

Juxtaposition of western and eastern sequences is also evident still farther south, east of Provo, along the Charleston–Nebo thrust (Baker, 1959; Crittenden, 1959, 1961).

An alternate interpretation for the area of older Paleozoic rocks has been suggested (Eardley, 1944). According to this interpretation (Figure 22) the area was raised as a large wedge-shaped segment of the earth's crust from which the thrust plates were pushed or slid, and the area is thus autochthonous. If a segment of the earth's crust were raised as postulated, one might expect crystalline rocks of the basement to have been brought close enough to the surface so that their presence would be indicated, because of their greater density, by a gravity high. The gravity map of Idaho (Bonini, 1963), however, does not show a gravity high southeast of Pocatello suitable to support this wedge hypothesis.

These two interpretations are based on the assumption that the westward-dipping eastern thrusts and the eastward-dipping western thrusts are connected—if not physically, then at least temporally and genetically. As yet, however, data are inadequate to date accurately the western thrusts and to demonstrate that they are the same ages as the eastern thrusts. Accordingly, a third possibility is that the faults are in fact unrelated, at least in the ways suggested thus far.

PART 3. BLOCK FAULTING

Block faulting, the third and latest chapter in the tectonic evolution of the Idaho–Wyoming thrust belt, began during the Eocene and has continued to the present.

Northeast of Kemmerer, Wyoming, several steeply dipping faults cut the plate of Mesozoic rocks (Figure 23) that lies above the late Paleocene Hogsback thrust fault (Figure 19). Some of the high-angle faults also cut the Wasatch Formation of late early Eocene age that

unconformably overlaps the rocks above the thrust. Other steep faults, probably of early Eocene age, do not offset the Wasatch.

Eocene block faults have not been recognized in the western part of the region because Eocene rocks, used to date the faults, are absent where needed.

On the west, many block faults cannot be dated within narrow limits. In a part of the Portneuf Range within the Bancroft Quadrangle, Idaho, three sets of block faults have been mapped (Figure 24). The oldest trends northwest and does not cut the Salt Lake Formation; the next set trends northeast and cuts the Salt Lake; and the youngest set trends north to northwest, cuts the Salt Lake, and parallels present mountain fronts. Fossils have not been found in the Salt Lake in this area, and the formation is assumed to be Pliocene as it is at many places nearby.

In the next range to the east, the Bear River Range, similarly trending fault sets have been mapped (Figure 25). The oldest set, which trends east and northeast, may be tear faults related to thrusting (Armstrong, 1964).

The next set trends northwest, does not cut the Salt Lake Formation, and is cut by a younger northeast-trending set, which in turn is cut by a younger northwest-trending set that cuts the Salt Lake and parallels the mountain front.

In the western part of the region, the late northwest-trending faults served as conduits along which basalt flows rose during Pliocene–Holocene time.

Some block faults have clearly moved more than once. Along one fault Tertiary strata have been offset about 1000 ft (300 m) vertically, whereas elsewhere along the same fault a modern alluvial fan has been offset only 50 ft (15 m) (W. W. Rubey, J. I. Tracey, and S. S. Oriel, unpublished data). Vertical movement along some block faults may have been in opposite directions at different times. This conclusion is suggested by the presence, on the crests and flanks of present ranges, of coarse Tertiary gravel for which there is no possible source with the present topography. These gravels were derived from the west, an area that is now low, at a time when the source was higher than the site of deposition.

Movement on the northwestward-trending block faults was principally vertical. Gem Valley is a graben that separates the Fish Creek and Portneuf Ranges from the ranges on the east (Figure 26). Gravimetric surveys by Mabey (Mabey and Armstrong, 1962) reveal marked negative anomalies (Figure 26) that are interpreted to indicate a relief of 7000–10,000 ft (2100–3000 m) between

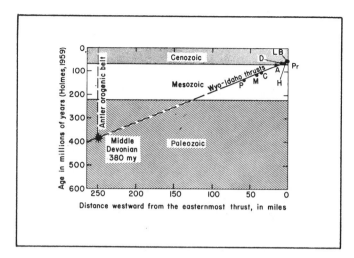

Figure 20—Relationship of dates of principal movement to positions of present traces of major thrust faults. Curiously, extension of line in graph, based on Idaho–Wyoming thrust dates, yields possible age for northward projection of Antler orogenic belt on west surprisingly close to age inferred by Roberts et al. (1958). Based on data in Figures 13 and 19. Letter symbols same as in Figure 13.

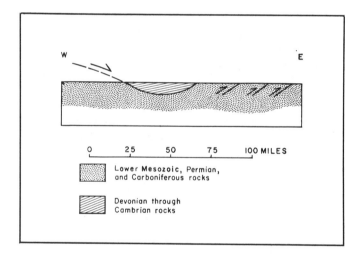

Figure 21—Schematic structural section illustrating overthrust origin inferred for area of Paleozoic rocks. Area extends north-northwestward from Ogden, Utah. See Figure 1.

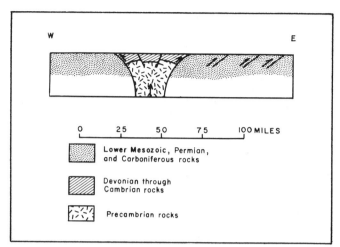

Figure 22—Schematic structure section illustrating uplifted wedge origin inferred for area of Paleozoic rocks.

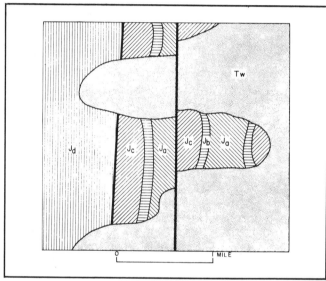

Figure 23—Schematic geologic map of part of Fort Hill Quadrangle, Wyoming, showing that Wasatch Formation, Tw, is cut by some steeply dipping faults but not by others. Ja to Jd, Jurassic units, from oldest to youngest, in plate above Hogsback thrust fault.

the crests of the flanking ranges and the bottom of the Tertiary fill in the valley. This relief is the result of movement on the block faults.

The young age of some block faulting hardly needs emphasis. Fault scarps are abundant, and modern alluvial fans have been cut by Holocene faults. Basalt in Gem Valley dated as younger than 27,000 years (Bright, 1963) has been cut by faults. Earthquakes continue in the area.

RELATIONSHIP OF GEOLOGIC HISTORY TO PETROLEUM EXPLORATION

We have attempted to show the sequence of geologic events in part of an orogenic belt. Gross generalizations, as well as small geologic features, provide clues for petroleum

exploration. Only a few can be mentioned here, for most lie beyond the scope of this paper.

The first broad episode in late Paleozoic and Mesozoic time was a progressive eastward shift of a miogeosyncline encroaching on the craton that resulted in conditions favorable for the development of hydrocarbons and stratigraphic traps. Recurrently throughout deposition, both western and eastern sources shed detritus into the miogeosyncline to form eastward-thinning and westward-thinning detrital wedges, respectively. Precise delineation of regional facies changes and of pinch-outs of detrital wedges is critical to exploration.

The second broad episode, which overlapped late stages of the first, produced the folded and thrust belt along the eastern flank of the miogeosyncline; movement along the thrusts, measured in at least tens of kilometers, was progressively younger eastward. The times of hydrocarbon migration resulting from deformation may also have been progressively younger eastward. Thrust plates cover and conceal oil and gas fields, only a few of which have yet been found. Although some surface and near-surface allochthonous rocks may seem to be unfavorable prospects for drilling, they may conceal distinctly different rocks in underlying autochthonous sequences that are potential oil reservoirs. The structural history of stratigraphic traps is also important, because tilts and rotations accompanying folding during thrusting may have caused hydrocarbons to escape from some and to be concentrated in others.

The latest broad episode involves block faulting that apparently increased in intensity westward and is still active. Some minor dislocations and modifications of hydrocarbon fields are apparent in the east. Block faulting also resulted in the deposition of detrital wedges in Tertiary sequences that have yielded some gas and oil. Some areas in the western part of the region have been so intensely shattered by several sets of block faults that hydrocarbon concentrations accumulated prior to block faulting may have escaped late in the tectonic history. These same sets of faults, however, have provided channels for hydrothermal solutions that locally have deposited base metal minerals.

ACKNOWLEDGMENTS

Material for the paper has been assembled from many sources and we gratefully acknowledge this debt. We also thank many of our colleagues on the U.S. Geological Survey for their constructive discussion and criticism. Special thanks are due E. D. McKee, W. W. Mallory, A. E. Roberts, A. L. Benson, C. A. Sandberg, W. W. Rubey, and J. I. Tracey for use of unpublished material. Our greatest debt is owed William W. Rubey. Much of the information and many of the ideas on which this paper is based were originally set forth by him (Rubey, 1955; Rubey and Hubbert, 1959), and the paper was greatly improved as the result of his criticism. The paper also profited from the suggestions of G. D. Robinson and W. H. Hays.

REFERENCES CITED

Anderson, A. L., 1928, Portland cement materials near Pocatello, Idaho: Idaho Bureau of Mines and Geology Pamphlet 28, 14 p.

Armstrong, F. C., 1953, Generalized composite stratigraphic section for the Soda Springs Quadrangle and adjacent areas in southeastern Idaho: Intermontane Association of Petroleum Geologists 4th Annual Field Conference Guidebook, Guide to the geology of northern Utah and southeastern Idaho.

———, 1962, Indirect dating of the Paris thrust fault, southeastern Idaho (abs.): Geological Society of America Special Paper 73, p. 21.

———, 1965, Northwestward projection of the Paris thrust fault, southeastern Idaho (abs): Geological Society of America Special Paper 82, p. 317.

Armstrong, F. C., and E. R. Cressman, 1963, The Bannock thrust zone, southeastern Idaho: U.S. Geological Survey Professional Paper 374-J, p. J1–J22.

Baker, A. A., 1959, Faults in the Wasatch Range near Provo, Utah: Intermontane Association of Petroleum Geologists 10th Annual Field Conference Guidebook, N. C. Williams, ed., Geology of the Wasatch and Uinta Mountains, transition area, p. 153–158.

Berdan, J. M., and H. Duncan, 1955, Ordovician age of rocks mapped as Silurian in western Wyoming: Wyoming Geological Association 10th Annual Field Conference Guidebook, N. C. Williams, ed., Green River basin, p. 48.

Blackwelder, E., 1910, New light on the geology of the Wasatch Mountains, Utah: Geological Society of America Bulletin, v. 21, p. 517–542.

———, 1925, Wasatch Mountains revisited (abs.): Geological Society of America Bulletin, v. 36, p. 132–133.

Bonini, W. E., 1963, Gravity anomalies in Idaho: Idaho Bureau of Mines and Geology Pamphlet 132, 10 p.

Bright, R. C., 1963, Pleistocene Lakes Thatcher and Bonneville, southeastern Idaho: Ph.D. thesis, Minneapolis University, Minneapolis, Minnesota.

Carr, W. J., and D. E. Trimble, 1961, Upper Paleozoic rocks in the Deep Creek Mountains, Idaho, *in* Short papers in the geologic and hydrologic sciences: U.S. Geological Survey Professional Paper 424-C, p. C181–C184.

Chronic, J., and C. S. Ferris, Jr., 1961, Early Paleozoic outlier in southeastern Wyoming: Rocky Mountain Association of Geologists 12th Annual Field Conference Guidebook, R. R. Berg, ed., Lower and middle Paleozoic rocks of Colorado, p. 143–146.

Churkin, M., Jr., 1962, Facies across Paleozoic miogeosynclinal margin of central Idaho: AAPG Bulletin, v. 46, n. 5, p. 569–591.

Coulter, H. W., 1956, Geology of the southeast portion of the Preston Quadrangle, Idaho: Idaho Bureau of Mines and Geology Pamphlet 107, 48 p.

Crittenden, M. D., Jr., 1959, Mississippian stratigraphy of the central Wasatch and western Uinta Mountains, Utah: Intermontane Association of Petroleum Geologists 10th Annual Field Conference Guidebook, N. C. Williams, ed., Geology of the Wasatch and Uinta Mountains, transition area, p. 63–74.

———, 1961, Magnitude of thrust faulting in northern Utah: U.S. Geological Survey Professional Paper 424-D, p. D128–D133.

Dutro, J. T., and W. J. Sando, 1963, New Mississippian formations and faunal zones in Chesterfield Range, Portneuf Quadrangle, southeast Idaho: AAPG Bulletin, v. 47, n. 11, p. 1963–1986.

Eardley, A. J., 1944, Geology of the north-central Wasatch Mountains, Utah: Geological Society of America Bulletin, v. 55, p. 819–894.

———, 1960, Phases of orogeny in the fold belt of western Wyoming and southeastern Idaho: Wyoming Geological Association 15th Annual Field Conference Guidebook, D. P. McGookey, ed., Overthrust belt of southwestern Wyoming and adjacent areas, p. 37–40.

Eardley, A. J., et al., 1944, Hoback–Gros Ventre–Teton [Range, Wyo.], field conference: Michigan University geologic map, tectonic map, with sections, 2 sheets.

Gilluly, J., 1932, Geology and ore deposits of the Stockton

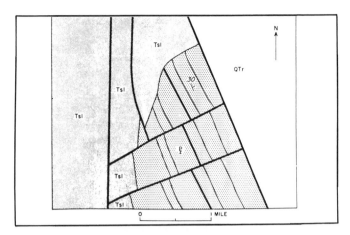

Figure 24—Schematic geologic map of part of Bancroft Quadrangle, Idaho, showing sets of faults mapped and their relationships to rock units: QTr, Quaternary and uppermost Tertiary rocks; Tsl, Pliocene Salt Lake Formation; Pz, Paleozoic rocks.

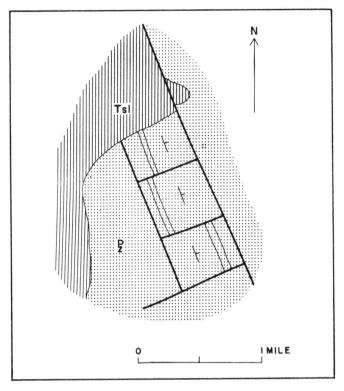

Figure 25—Schematic geologic map of part of Soda Springs Quadrangle, Idaho, showing sets of faults mapped and their relationships to rock units: Tsl, Pliocene Salt Lake Formation; Pz, Paleozoic rocks.

and Fairfield Quadrangles, Utah: U.S. Geological Survey Professional Paper 173, 171 p.

Haun, J. D., and J. A. Barlow, Jr., 1962, Lower Cretaceous stratigraphy of Wyoming: Wyoming Geological Association 17th Annual Field Conference Guidebook, R. L. Enyert, ed., Early Cretaceous rocks of Wyoming and adjacent areas, p. 15–22.

Hintze, L. F., 1960, Ordovician tectonics of western Wyoming and vicinity: Wyoming Geological Association 15th Annual Field Conference Guidebook, D. P. McGookey, ed., Overthrust belt of southwestern Wyoming and adjacent areas, p. 110–115.

Holmes, A., 1959, A revised geological time-scale: Edinburgh Geological Society Transactions, v. 17, pt. 3, p. 183–216.

Lochman-Balk, C., 1960, The Cambrian section of western Wyoming: Wyoming Geological Association 15th Annual Field Conference Guidebook, D. P. McGookey, ed., Overthrust belt of southwestern Wyoming and adjacent areas, p. 98–108.

Love, J. D., P. O. McGrew, and H. D. Thomas, 1963, Relationship of latest Cretaceous and Tertiary deposition and deformation to oil and gas in Wyoming, *in* The backbone of the Americas—Tectonic history from Pole to Pole, a symposium: AAPG Memoir 2, p. 196–208.

Ludlum, J. C., 1943, Structure and stratigraphy of part of the Bannock Range, Idaho: Geological Society of America Bulletin, v. 54, p. 973–986.

Mabey, D. R., and F. C. Armstrong, 1962, Gravity and magnetic anomalies in Gem Valley, Caribou County, Idaho, *in* Short papers in geology, hydrology, and topography: U.S. Geological Survey Professional Paper 450-D, p. D73–D75.

Mansfield, G. R., 1927, Geography, geology, and mineral resources of part of southeastern Idaho: U.S. Geological Survey Professional Paper 152, 453 p.

———, 1929, Geography, geology, and mineral resources of the Portneuf Quadrangle, Idaho: U.S. Geological Survey Bulletin 803, 110 p.

McKee, E. D., S. S. Oriel, V. E. Swanson, M. E. MacLachlan, J. C. MacLachlan, K. B. Ketner, J. W. Goldsmith, R. Y. Bell, and D. J. Jameson, 1956, Paleotectonic maps of the Jurassic system, with a separate section on Paleogeography by R. W. Imlay: U.S. Geological Survey Miscellaneous Geological Investigations Map I-175, 6 p.

McKee, E. D., S. S. Oriel, K. B. Ketner, M. E. MacLachlan, J. W. Goldsmith, J. C. MacLachlan, and M. R. Mudge, 1959, Paleotectonic maps of the Triassic system: U.S. Geological Survey Miscellaneous Geological Investigations Map I-300.

McKee, E. D., S. S. Oriel, et al., 1967, Paleotectonic maps of the Permian system: U.S. Geological Survey Miscellaneous Geological Investigations Map I-450.

McKelvey, V. E., et al., 1959, The Phosphoria, Park City, and Shedhorn formations in the western phosphate field: U.S. Geological Survey Professional Paper 313-A, p. 1–47.

Moritz, C. A., 1953, Summary of the Cretaceous stratigraphy of southeastern Idaho and western Wyoming: Intermontane Association of Petroleum Geologists Guidebook 4th Annual Field Conference, p. 63–72.

Murdock, C. N., 1961, Geology of the Weston Canyon area, Bannock Range, Idaho: Master's thesis, Utah State University, Logan, Utah.

Oriel, S. S., 1962, Main body of Wasatch Formation near La Barge, Wyoming: AAPG Bulletin, v. 46, n. 12, p. 2161–2173.

———, 1965, Brigham, Langston, and Ute formations in Portneuf Range, southeastern Idaho (abs.): Geological Society of America Special Paper 82, p. 341.

Osmond, J. C., 1962, Stratigraphy of Devonian Sevy

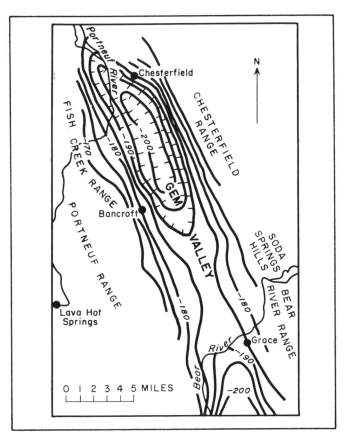

Figure 26—Bouguer anomaly map of Gem Valley, Caribou County, Idaho. (Gravity contour interval 5 mGal.) After Mabey and Armstrong (1962).

Dolomite in Utah and Nevada: AAPG Bulletin, v. 46, n. 11, p. 2033–2056.

Peale, A. C., 1879, Report on the geology of the Green River district: U.S. Geological and Geographical Survey Territorial 11th Annual Report, p. 511–646.

Plumley, W. J., 1948, Black Hills terrace gravels; a study in sediment transport: Journal of Geology, v. 56, n. 6, p. 526–577.

Richards, R. W., and G. R. Mansfield, 1912, The Bannock overthrust; a major fault in southeastern Idaho and northeastern Utah: Journal of Geology, v. 20, p. 681–707.

Richardson, G. B., 1941, Geology and mineral resources of the Randolph Quadrangle, Utah–Wyoming: U.S. Geological Survey Bulletin 923, 54 p.

Rigby, J. K., 1963, Devonian System of Utah: Utah Geological and Mineralogical Survey Bulletin 54, p. 75–88.

Roberts, R. J., and M. R. Thomasson, 1964, Comparison of late Paleozoic depositional history of northern Nevada and central Idaho, *in* Short papers in geology and hydrology: U.S. Geological Survey Professional Paper 475-D, p. D1–D6.

Roberts, R. J., P. E. Hotz, J. Gilluly, and H. G. Ferguson, 1958, Paleozoic rocks of north-central Nevada: AAPG Bulletin, v. 42, p. 2813–2857.

Ross, C. P., 1937, Geology and ore deposits of the Bayhorse region, Custer County, Idaho: U.S. Geological Survey Bulletin 877, 161 p.

——, 1962, Paleozoic seas, central Idaho: Geological Society of America Bulletin, v. 73, p. 769–794.

Ross, R. J., Jr., 1964, Relations of Middle Ordovician time and rock units in Basin Ranges, western United States: AAPG Bulletin, v. 48, n. 9, p. 1526–1554.

Rubey, W. W., 1955, Early structural history of the overthrust belt of western Wyoming and adjacent states: Wyoming Geological Association 10th Annual Field Conference Guidebook, G. G. Anderson, ed., Green River basin, p. 125–126.

——, 1958, Geologic map of the Bedford Quadrangle, Wyoming: U.S. Geological Survey Geological Quadrangle Map GQ-109.

Rubey, W. W., and M. K. Hubbert, 1959, Overthrust belt in geosynclinal area of western Wyoming in light of fluid-pressure hypothesis, Pt. 2 *of* Role of fluid pressure in mechanics of overthrust faulting: Geological Society of America Bulletin, v. 70, p. 167–206.

Rubey, W. W., S. S. Oriel, and J. I. Tracey, 1961, Age of the Evanston Formation, western Wyoming, *in* Short papers in the geologic and hydrologic sciences: U.S. Geological Survey Professional Paper 424-B, p. B153–B154.

Rush, R. W., 1963, Silurian strata in Utah: Utah Geological and Mineralogical Survey Bulletin 54, p. 63–73.

Sando, W. J., J. T. Dutro, and W. C. Gere, 1959, Brazer Dolomite (Mississippian), Randolph Quadrangle, northeast Utah: AAPG Bulletin, v. 43, p. 2741–2769.

Schultz, A. R., 1914, Geology and geography of a portion of Lincoln County, Wyoming: U.S. Geological Survey Bulletin 543, 141 p.

Stose, G. W., 1932, Geologic map of the United States: U.S. Geological Survey.

Strickland, J. W., 1956, Mississippian stratigraphy, western Wyoming: Wyoming Geological Association 11th Annual Field Conference Guidebook, R. R. Berg, ed., Jackson Hole, p. 51–57.

Trimble, D. E., and W. J. Carr, 1962, Preliminary report on the stratigraphy of the Paleozoic rocks southwest of Pocatello, Idaho: U.S. Geological Survey Open-File Report, 18 p.

Umpleby, J. B., L. G. Westgate, and C. P. Ross, 1930, Geology and ore deposits of the Wood River region, Idaho: U.S. Geological Survey Bulletin 814, 250 p.

Veatch, A. C., 1907, Geography and geology of a portion of southwestern Wyoming: U.S. Geological Survey Professional Paper 56, 178 p.

Weimer, R. J., 1961, Uppermost Cretaceous rocks in central and southern Wyoming and northwest Colorado: Wyoming Geological Association 16th Annual Field Conference Guidebook, G. J. Wiloth, ed., Late Cretaceous rocks Wyoming and adjacent areas, p. 17–28.

Williams, J. Stewart, 1948, Geology of the Paleozoic rocks, Logan Quadrangle, Utah: Geological Society of America Bulletin, v. 59, p. 1121–1164.

——, 1962, Pennsylvanian System in central and northern Rocky Mountains, *in* C. C. Branson, ed., Pennsylvanian System in the United States: AAPG, p. 159–187.

Williams, J. Stewart, and M. E. Taylor, 1964. The Lower Devonian Water Canyon Formation of northern Utah, *in* Contributions to geology: Wyoming University, v. 3, n. 2, p. 38–53.

TIME OF THRUSTING IN IDAHO-WYOMING THRUST BELT—DISCUSSION

E. W. Mountjoy
McGill University
Montreal, Quebec

The American Association of Petroleum Geologists and the editors and authors of the recent issue[4] concerning the Rocky Mountain sedimentary basins (AAPG *Bulletin*, v. 49, n. 11) are to be commended for publishing the papers of this symposium so soon after they were presented orally.

One of the papers of considerable interest is the excellent summary of the geology and tectonics of the Idaho-Wyoming thrust belt by Armstrong and Oriel (1965). Armstrong and Oriel have presented some important data concerning the dating and temporal relationships of the thrust belt faults in this region in a clear and well-illustrated paper based not only on their own work but also on data summarized from several sources, especially the work of Rubey (1958; Rubey and Hubbert, 1959). They concluded mainly on the basis of stratigraphic evidence that thrusting began in latest Jurassic time in the west and ended in the east about early Eocene time.

The purpose of this note is to consider carefully the evidence used to date thrusting in this region. Armstrong and Oriel (p. 1857) reported that "Few of the thrust faults can be dated within narrow limits by usual geologic methods." Nevertheless, they stated that three of the thrusts, the Prospect, Absaroka, and Paris thrusts, can be dated. There is no doubt of the age of the Prospect thrust, the most easterly major thrust, because the trace of the thrust fault is overlain by late early Eocene Wasatch strata and cuts the Paleocene and earliest Eocene Hoback Formation. However, the evidence for the Absaroka and Paris thrusts is not nearly so convincing, and this writer suggests that they can not be dated other than generally.

Armstrong and Oriel (1965) concluded that the Evanston Formation of latest Cretaceous and Paleocene age was deposited after most of the movement along the Absaroka thrust, and the movement is thereby dated as latest Cretaceous. The significant relationships are illustrated in their Figure 17. In addition they state that "... the Evanston has not been found directly overlying the fault...." (p. 1857). However, as shown in their Figure 17, the actual relationships are covered by the Eocene Wasatch Formation so that other interpretations are possible without more critical evidence to the contrary. Armstrong and Oriel suggested that only the basal beds of the Evanston Formation have been cut by the Absaroka thrust, but it could equally well cut through *all* of the Evanston Formation beneath the cover of the Eocene Wasatch Formation. The unconformable base of the Evanston Formation need not be related to the thrusting. Consequently, the Absaroka thrust might be as young as early Eocene and may have formed at the *same* time or nearly the same time as the Prospect thrust.

The Paris thrust has been dated as latest Jurassic and earliest Cretaceous on the basis of the presence of coarsely conglomeratic redbeds of the Lower Cretaceous Gannett Group which contain Paleozoic pebbles of a nonresistant character. Armstrong and Oriel infer that the source of the Gannett Group pebbles of the Gannett Hills was about 25–30 mi (40–50 km) west of their present location and suggested that these pebbles were derived from erosion of Paleozoic rocks of the Paris thrust sheet. A more complete discussion is given in the report by Armstrong and Cressman (1963). Again the evidence is not conclusive, however, because these conglomerates could equally well have been derived before thrusting from some other uplifted source areas presently buried beneath the Meade or Paris thrust sheet. If the sequence of thrust development suggested by Armstrong and Oriel is correct, then surely this conglomeratic unit should also be found in the Meade thrust sheet. Armstrong and Cressman (1963) earlier concluded that the evidence regarding the Paris thrust sheet being the source of the conglomerate of the Gannett Group was not conclusive. Again the evidence is not compelling that the Paris thrust is Early Cretaceous and latest Jurassic; it could well be much younger, possibly as young as early Eocene. Consequently, much of the thrusting in the Idaho-Wyoming thrust belt may have taken place during the earliest Eocene or slightly earlier. Moreover, the sequence of thrusting from west to east cannot be positively demonstrated, although it appears to be a reasonable inference from the present structural and stratigraphic information. Also, applying the term Laramide to deformations in the thrust belt may not be so futile as suggested by Armstrong and Oriel.

Having worked in another well-known thrust belt for several years, the central part of the Alberta Rocky Mountains, I am interested in methods used for dating times of thrusting. In the Alberta Rockies near Jasper, the only evidence available is that the youngest beds (unnamed unit above the Brazeau Formation) cut by thrust faults or deformed in associated folds are of Late Cretaceous or Paleocene age. Geometric and fold relationships suggest an order of development from west to east (Fox, 1959; Mountjoy, 1960, 1962), which indicates that all of the thrust sheets did not form at the same time. Two scraps of information help date the thrusting of the southern Alberta Rockies. First, the upper limit of thrusting can be reasonably well dated near the International Boundary in the Flathead valley south of Crowsnest Pass on the basis of late Eocene and early Oligocene Kishenehn strata (Russell, 1954; Price, 1962). These strata were derived from progressive denudation of a local area of high relief and were deposited in a basin bordered in part by normal faults that offset the Lewis thrust (Price, 1959, 1962). The second bit of information comes from the eastern Foothills where thrust faults appear to be covered by the undeformed Paleocene Paskapoo and Porcupine Hills formations (Bossart, 1957; Fox, 1959). However, this may be a situation in which thrust faults do not reach the present erosion surface as happens elsewhere in the Rockies (Fox, 1959) and also in the Appalachians (Gwinn, 1964). The stratigraphic evidence

---

[4]Adapted from AAPG Bulletin, v. 50, n. 12, p. 2612-2621.

from western Canada in general indicates that the Canadian Rocky Mountain uplift and deformation culminated at about the mid-point of Eocene time (Taylor et al., 1964). By this statement I do not wish to imply that all the folding and thrusting of the Alberta Rockies took place simultaneously in middle Eocene time, but that most of the thrust movements and uplifts appear to have occurred at this time. The available stratigraphic evidence does not rule out the possibility of somewhat earlier (especially Paleocene) uplift and thrust movements. It is interesting that the most accurate dating of thrusting in both the Alberta and Idaho–Wyoming thrust belts is approximately the same—Eocene.

The writer is not personally familiar with the structure of the Idaho–Wyoming thrust belt. This discussion has been written with the anticipation that Armstrong and Oriel will provide further evidence for time of thrusting and a more complete discussion of other ways in which the basic information may be interpreted, which they may not have had space to do in their excellent summary paper.

References Cited

Armstrong, F. C., and E. R. Cressman, 1963, The Bannock thrust zone, southeastern Idaho: U.S. Geological Survey Professional Paper 374-J, 22 p.

Armstrong, F. C., and S. S. Oriel, 1965, Tectonic development of Idaho–Wyoming thrust belt: AAPG Bulletin, v. 49, p. 1847–1866.

Bossart, D. O., 1957, Relationship of the Porcupine Hills to early Laramide movements: Alberta Society of Petroleum Geologists 7th Annual Field Conference Guidebook, E. W. Jennings, ed., p. 46–51.

Fox, F. G., 1959, Structure and accumulation of hydrocarbons in southern Foothills, Alberta, Canada: AAPG Bulletin, v. 48, p. 992–1025.

Gwinn, V. E., 1964, Thin-skinned tectonics in the Plateau and northwestern Valley and Ridge provinces of the central Appalachians: Geological Society of America Bulletin, v. 75, p. 863–900.

Mountjoy, E. W., 1960, Structure and stratigraphy of the Miette and adjacent areas, eastern Jasper National Park, Alberta: Ph.D. thesis, University of Toronto, Toronto, Ontario.

———, 1962, Mount Robson (southeast) map-area, Rocky Mountains of Alberta and British Columbia: Geological Survey of Canada Paper 61-31, 144 p.

———, in press, Significance of rotated folds and thrust faults, Alberta Rocky Mountains.

Price, R. A., 1959, Geology, Flathead, British Columbia and Alberta: Geological Survey of Canada Map 1-1959.

———, 1962, Fernie map-area, east half, Alberta and British Columbia: Geological Survey Canada Paper 61-24, 65 p.

Rubey, W. W., 1958, Geology of Bedford quadrangle, Wyoming: U.S. Geological Survey Quadrangle Map GQ-109.

Rubey, W. W., and M. K. Hubbert, 1959, Role of fluid pressure in mechanics of overthrust faulting, II. Overthrust belt in geosynclinal area of western Wyoming in light of fluid-pressure hypothesis: Geological Society of America Bulletin, v. 70, p. 167–206.

Russell, L. S., 1954, The Eocene–Oligocene transition as a time of major orogeny in western North America: Transactions of the Royal Society of Canada, Series 3, v. 48, sec. 4, p. 65–69.

Taylor, R. S., W. H. Mathews, and W. O. Kupsch, 1964, Tertiary, geological history of western Canada: Alberta Society of Petroleum Geologists, p. 190–194.

TIMES OF THRUSTING IN IDAHO–WYOMING THRUST BELT: REPLY[5]

S. S. Oriel
U.S. Geological Survey
Denver, Colorado

F. C. Armstrong
U.S. Geological Survey
Spokane, Washington

Professor Mountjoy's kind discussion, on the preceding pages, of our summary paper (Armstrong and Oriel, 1965) on the tectonic development of the Idaho–Wyoming thrust belt is most welcome for its own sake and because it has led the editors of the AAPG *Bulletin* to invite us to reply. We welcome the opportunity for more thorough discussion of the times and durations of thrust movements than was possible in the original paper.

[5]Reply received March 26, 1966; accepted May 10, 1966.

Mountjoy points out, quite correctly, that in our 1965 paper unequivocal evidence is presented for dating the frontal Prospect thrust as Eocene but not for the greater age of thrusts farther west; he goes on to suggest that perhaps all the thrusts are Eocene and to note that thrusting in the Alberta Rockies also is Eocene. The plain, though no doubt unintended, implication of his suggestion is that folding and thrusting occurred during a very brief interval simultaneously throughout the Cordillera. Such a view is at variance with the experience of most modern structural geologists. It does not fit the evidence available from the Idaho–Wyoming thrust belt, which indicates that principal movements were not simultaneous on all the thrusts and that some western thrusts have been inactive, or nearly so, since well before Eocene time. Before data supporting this firm statement are offered, it may be worthwhile to consider some general problems in dating thrust faults.

Problems in Dating Thrust Faults

Paucity of Precise Dates

Thrust movements are difficult to date precisely in most of the world's thrust belts. A reason is that well-dated bracketing strata are preserved in very few places within orogenically active mountain systems. Along the mountain front and beyond, however, deposition predominates over erosion. Synorogenic and postorogenic debris preserved along the mountain front may provide the critical stratigraphic record of major geologic events and thus facilitate dating of frontal faults. Precise dates become far more scarce with increasing distance from the thrust belt front. Of course, a precise date for a frontal fault does not establish the age for an entire mountain system.

Accordingly, coarse detrital units are being used increasingly to date tectonic events, but proving that a sediment is the product of a specific tectonic event is difficult. Does a particular conglomerate bed record movement along one thrust fault or along another or some other orogenic event?

Paucity of precise dates raises many questions, a few of which are discussed here.

Duration of Thrust Movement

Discussions of times of thrusting commonly imply that the movement is brief. Yet movements on thrust faults that transport enormous plates of rock scores of kilometers laterally and that result in stratigraphic throws of many kilometers cannot take place instantaneously. One estimate of the possible average rate of movement of a thrust sheet, based on ages of the youngest rocks exposed below successive thrust plates (Rubey and Hubbert, 1959) is about 1 mi (1.5 km) per 1 m.y. (Rubey and Hubbert, 1959) or 1.5 mm per year. This estimate may be low, because it envisions exceedingly slow, successive, eastward movements of the several thrust plates and is based on conservative estimates of minimum horizontal displacement. Far more rapid movement, about 1 in. (2.5 cm) per year or 16 mi (25 km) per 1 m.y., has been measured briefly along an active thrust fault that has cut oil well casings in the Buena Vista oil field of California (McMasters, 1943; Gilluly, 1949); this figure may be excessive over a long span of time. In the absence of better information, let us assume that the rate of thrust movement is between the two figures and is several kilometers per 1 m.y.

The time involved in the movement of a thrust sheet such as the Absaroka, then, is significant. If the sheet moved only about 10–15 mi (15–25 km) (Rubey and Hubbert, 1959), then approximately 3–5 m.y. may have been required; if the sheet moved as far as 30 mi (50 km), then as much as 10 m.y. may have elapsed during transport, a time comparable

with the estimated 11 or 12 m.y. duration of Paleocene time (Geological Society of London, 1964).

If the estimated rates of thrusting previously mentioned are not greatly in error, then the span of time required for movement exceeds intervals of geologic time that we are now able to resolve on the basis of stratigraphic paleontology.

Recurrent Thrust Movements

Recurrent movements may seriously complicate the dating of thrust movements. Once a glide surface or thrust fault develops, continued application of forces that produce other thrust faults can, under some circumstances, result in renewed movement along the first surface of rupture. Not only do large thrust displacements require significant spans of time, but movement may have occurred during several such spans.

Thus, recognition that two distinct conglomerate units of different ages may have been derived from the upper plate of the Paris fault has led to the suspicion that this thrust plate has moved more than once. Some of the fragments in both conglomerate units are of rock types found only in the Paris thrust plate.

Movement of Parts of a Single Thrust Plate

Still another point not discussed in the original paper is the possibility that different parts of a single thrust sheet have moved at different times. Localities at which precise dates for fault movement are available are far too few to establish that a thrust sheet moved as a unit. A thrust plate as large as the Absaroka, which has been traced about 200 mi (320 km), could have moved by small increments at different times in different places without failing and breaking into separate plates.

The point is not entirely hypothetical, for it may explain apparent conflicts in time of thrusting of different parts of the Hogsback fault. Observations near La Barge, Wyoming, suggest one date for movement. Strata that overlie the fault here are believed to be involved in the faulting farther north (Michael, 1960). One possible explanation is that different parts of the thrust plate moved at different times. Other possible explanations are that the critical strata at the two places may not be identical or that structure at one of the localities may be more complex than now believed.

Precision in Discussing Dates of Thrusting

The long time probably required for movement of large thrust sheets imposes the need for precision in discussions of times of thrusting. By "time" or "date" of thrusting do we mean when thrusting began, when most of the movement took place, or when thrusting ceased?

Possible recurrent movement must also be considered in discussions of thrust dates. Times of recurrent movement were specifically excluded from diagrams in the original paper (Armstrong and Oriel, 1965, Figures 19 and 20) illustrating temporal relationships among the thrust faults. Precise dates are insufficient to enable us to decipher the entire life history of any one thrust fault.

The absence of proof that all parts of a large thrust plate moved synchronously requires that a hard-earned precise date be applied only to a relatively small area. A precise date for Absaroka fault movement near Kemmerer, Wyoming, may not apply to the time of movement on the Absaroka 150 mi (240 km) away along the edge of the Snake River Plain.

These considerations bear on the discussions to follow, and great care in analysis is necessary if we hope to gain an understanding of a complex assemblage of orogenic events. Now we can return to data that bear on faults in the Idaho–Wyoming thrust belt and answer points raised in Mountjoy's discussion.

Idaho–Wyoming Thrust Belt

Hogsback Thrust Fault

In illustrating the diversity of information used to date thrust faults, we ignored the Hogsback fault in the original paper. The basis for dating the Hogsback fault is the same as that for dating the Prospect fault.

On Hogsback Ridge near La Barge, Wyoming, movement along the Hogsback thrust ended before latest Paleocene time. The Hogsback fault lies west of and above the easternmost thrust known at this latitude, the La Barge or Hilliard, which was active in early Eocene time (Murray, 1960). The La Barge or Hilliard fault cuts rocks that Murray assigned to the Evanston Formation. The upper part of Murray's Evanston is here assigned to the Chappo Member of the Wasatch Formation of latest Paleocene and earliest Eocene age. The westward-dipping Hogsback fault, Cambrian rocks above the fault, and Upper Cretaceous rocks below it are overlain with angular unconformity by nearly horizontal beds along the crest of Hogsback Ridge (Oriel, 1963). The horizontal beds are distinctive and are assigned to the Chappo Member of the Wasatch Formation (Oriel, 1962), in which vertebrate remains of latest Paleocene (Clarkforkian) age have been found (Gazin, 1942, 1956a). The Hogsback fault ceased to move before the upper Paleocene strata were deposited. Major movement along the Hogsback fault at this locality, therefore, took place before latest Paleocene time. Even the very last stage of movement cannot have been as recent as Eocene.

Absaroka Thrust Fault

The relationships of the Absaroka fault, whose age Mountjoy questions, are similar near Kemmerer, Wyoming, to those of the Hogsback fault, although proof is more difficult. Critical evidence here is the relationship of the Lazeart syncline, a part of which is shown on Figure 17 of Armstrong and Oriel (1965), to the Absaroka thrust and the age and areal distribution of the Evanston Formation relative to these two features.

The Lazeart syncline (Veatch, 1907) is a long, slightly sinuous, asymmetric chevron syncline with a steep west limb and a gentle east limb; at places along the west limb beds are overturned toward the east and consequently dip westward. The trace of the axial plane of the syncline roughly parallels the trace of the Absaroka fault for about 70 mi (110 km) and lies about 1–4 mi (1.5–6.5 km) east of it

(Veatch, 1907. Fractures of the west limb of the syncline extend in places for some distance east of the Absaroka fault (Veatch, 1907). Strata above the fault are folded similarly in both extent and degree of overturning and parallel with those below the fault. These relationships suggest that the syncline and the thrust fault were formed by the same forces and at the same time.

The youngest unit east of the Absaroka that is involved in the gross folding of the syncline and that was presumably cut by the fault is the Adaville Formation, whose upper part is of Late Cretaceous (late Montana or early Lance) age.

The Evanston Formation is divisible into a lower and an upper part. The lower part consists of conglomerate and sandstone with some interbedded mudstone; it contains fossils of latest Cretaceous (late Lance, Hell Creek) age (Rubey et al., 1961). The upper part of the Evanston consists mainly of siltstone, mudstone, and sandstone; it contains fossils of early late Paleocene (Tiffany) age (Gazin, 1956b). Basal strata of the Evanston dip moderately westward, but the dips decrease upward in the sequence; the upper beds of the formation are nearly flat.

The lower (Cretaceous) part of the Evanston lies with marked angular unconformity on upturned and eroded Adaville and older Cretaceous beds in the west limb of the Lazeart syncline about 9 mi (14.5 km) north-northwest of Kemmerer and 1 mi (1.5 km) northeast of Hams Fork (Veatch, 1907). Farther south, however, basal beds of the Evanston lie parallel with strata in the east limb of the syncline along the east flank of the Fossil basin. Basal Evanston strata have been folded at one locality along the axis of the Lazeart syncline on the southwest side of Hams Fork. At another place 4.5 mi (7 km) north and along the trace of the Absaroka, basal Evanston beds are slightly overridden by the fault. Basal Evanston strata overlie units both above and below the Absaroka fault (Armstrong and Oriel, 1965), although they have not been observed to overlie the fault trace itself. Moreover, the linear belt of basal Evanston is not offset by the Absaroka fault.

The upper (Paleocene) part of the Evanston, in contrast, is mainly flat and only gently folded locally. In holes drilled through the Absaroka fault neither lower nor upper parts of the Evanston have been found beneath the fault. No faults have been observed during mapping by Rubey, Oriel, and Tracey in the upper part of the Evanston where, on the basis of subsurface information, it is believed to overlie the buried trace of the Absaroka fault about 6.5 mi (10.5 km) west of Kemmerer.

These relationships are consistent with the inference that folding of the Lazeart syncline and movement along the Absaroka fault were essentially contemporaneous. The Absaroka plate near Kemmerer apparently reached essentially its present site as the syncline formed and before the basal part of the Evanston was deposited. Major movement on the Absaroka thrust is thus post-Adaville and pre–basal Evanston, or late Late Cretaceous.

Minor involvement of basal Evanston beds in the deformation indicates that a small orogenic pulse occurred later. This pulse involved lateral transport of only a few hundred meters in contrast to the probable 15 mi (25 km) or more of movement late in Cretaceous time. The absence of faulting in the upper part of the Evanston where it covers the Absaroka indicates that movement on the fault had ceased before deposition of these beds and that the late pulse took place between latest Cretaceous and early late Paleocene time.

If the inference that major movement on the Absaroka thrust and folding of the Lazeart syncline took place at the same time is not accepted, then perhaps most of the Absaroka movement took place after deposition of the basal Evanston. This alternative interpretation, however, calls for such special and fortuitous geologic events as to render it unlikely.

According to this alternative, the parallelism of folds above and below the Absaroka fault may reflect long persistence of a single stress field. The part of the basal Evanston now resting on the upper plate of the fault would have been transported piggyback to its present position. The amount of lateral transport must have been sufficient to move a western belt of basal Evanston (above the fault) just exactly the right distance to be aligned with an eastern belt of basal Evanston (below the fault). The credibility of this alternative is strained still further by absence of basal Evanston strata along most exposures of the Absaroka fault trace and in boreholes through the fault. This absence can be explained, under the assumption that basal Evanston deposits are older than the thrusting, only by spotty and patchy preservation of these beds. Thus, a nearly precise alignment of erratic patches of basal Evanston requires an even greater coincidence. The problems raised by the alternative make the first interpretation more palatable.

Paris Thrust Fault

The Paris fault lies considerably west of the thrust belt front and is more difficult to date precisely. The youngest rocks cut by the Paris fault in southeastern Idaho, where we have studied the relationships in the field, are Early Triassic in age. The oldest rocks directly above the fault are not Eocene as previously reported (Mansfield, 1927) but more likely Pliocene where we have examined them; that is, we assign to the Salt Lake Formation patches of strata previously mapped as Wasatch Formation. The age span for Paris fault movement by the bracket method—Early Triassic–Pliocene—is so long as to be virtually meaningless.

However, if conglomerates in the Lower Cretaceous Gannett Group are the product of uplift and erosion related to movement along the Paris fault, they provide a means for dating some of this movement more closely. Mountjoy recognizes that evidence for dating movement on the Paris fault is not conclusive and suggests that Gannett conglomerates could have been derived from some other uplifted source areas now buried beneath the Meade or Paris thrust sheets. Pebbles in the conglomerates probably were not derived from uplifted source areas now concealed by the Meade and Paris thrust sheets because exposed rocks beneath both thrust sheets are thrust plates of uppermost Paleozoic and Mesozoic units. No middle to lower Paleozoic units are evident there now which would have provided a source for lithologic types recognized in the Gannett pebbles. These lithologic types are exposed only above and west of the Paris fault. The presence of an unknown source for these pebbles beneath the Paris or Meade faults is unsupported by present knowledge of the structure and is not required by any fact known to us.

That Gannett conglomerates record early movement on the Paris thrust may be unconvincing if considered alone, but other considerations lend support to this conclusion. First, no evidence has been found of prethrust block faulting. All of the uplift-bounding normal faults that we have been able to date formed after thrusting. Second, the Idaho–Wyoming thrust belt is characterized by an eastward progression of belts of coarse detritus. Conglomerates are thick, extensive, and rather coarse in units ranging from Upper Triassic to lower Eocene. The westernmost known occurrence and coarsest facies of each unit lie in a belt east of the coarsest facies of the next older unit.

A tenable inference that may be drawn is that the detrital units record tectonic events and that the progressively more eastward positions of successively younger coarse detrital units record progressively more eastward movements on the faults. This inference is supported by evidence of eastward progression in the development and eventual destruction of the Cordilleran miogeosyncline.

Additional support for the greater antiquity of the Paris fault is found along tectonic strike on the south in Utah, where Spieker (1946) and Eardley (1944) recognized evidence of a Late Jurassic or Early Cretaceous orogeny that involved folding and thrusting; neither they nor other geologists have recognized Eocene thrusting in that area.

Because the Paris thrust may have moved more than once (Armstrong and Cressman, 1963), we have discussed only first movement on this fault in contrast to last or major movement on other faults. Lack of upper Mesozoic and lower Cenozoic strata in the area prevents accurate dating of major and last movements. Recurrent movement is suggested by the apparent absence of pebbles of distinctive, resistant Cambrian and upper Precambrian quartzites in the Lower Cretaceous Gannett Group and their abundance in the uppermost Cretaceous basal part of the Evanston Formation farther east. Although the middle part, stratigraphically, of the Paris thrust plate was exposed during Gannett deposition, the lower part of it apparently was not (Armstrong and Cressman, 1963) because the resistant rock types are not preserved. No source other than the lower part of the Paris thrust plate in the Bear River Range is known for the Cambrian and upper Precambrian quartzite pebbles found in the Evanston Formation near Kemmerer. Thus, the sole of the Paris thrust plate had not broken out to the land surface during Gannett deposition, but the front of the plate composed of younger formations (Rubey and Hubbert, 1959) had; the basal part of the plate was exposed during Evanston deposition. The Evanston pebbles probably record recurrent movement along the Paris fault, or they may possibly suggest block fault uplift of the ancestral Bear River Range; they probably were not recycled out of an older conglomerate because no such conglomerate has been found.

Several independent lines of evidence, none of which alone is conclusive, thus point to a Late Jurassic and Early Cretaceous age for the first, and perhaps the major, movement on the Paris thrust.

Meade Thrust Plate

Another point raised by Mountjoy is that, if the thrusting sequence is progressively younger eastward, then the Gannett Group should be found in the Meade thrust sheet.

The Meade thrust zone extends eastward around the south end of Snowdrift Mountain into Crow Creek and thence north and northwestward to the southern edge of the Snake River Plain (Armstrong and Cressman, 1963). In the northern part of this zone Gannett rocks do occur in the upper plates of thrust slices (Mansfield, 1927, 1952). Mapping by Oriel and L. B. Platt in the Crow Creek part of the zone shows that Gannett rocks also occur in previously unrecognized thrust slices that likely are parts of the upper plate of the Meade thrust zone.

The absence of the Gannett Group in the western part of the Meade thrust plate is expectable. The longitude of the eastern part of the Meade thrust zone is the approximate western limit of known Lower Cretaceous rocks; because the source of the Gannett conglomerates was nearby, Lower Cretaceous strata probably were not deposited as far west as the area from which much of the Meade sheet was transported (Armstrong and Cressman, 1963; Armstrong and Oriel, 1965). Moreover, the Meade thrust plate has been subjected to erosion during most of the time since the Late Cretaceous, and Lower Cretaceous rocks, if present earlier, have been stripped. In addition, Cenozoic, particularly late Cenozoic, block faults have raised the western part of Meade sheet exposures so that not only probable Cretaceous, but also thousands of feet of possible Jurassic, strata have been stripped away.

The Meade fault cuts the Gannett Group, indicating that movement on this fault was in latest Early Cretaceous time or later. If the date for first movement on the Paris fault is accepted, then here is another illustration of the eastward progression of tectonic events.

Some Generalizations

The foregoing evidence for dating the several thrust faults accomplishes two things. First, it establishes the pre-Eocene age of at least two major thrust faults. Second, it suggests, but does not prove, increasing ages for principal movements of the thrust faults from east to west.

Strata bracketing the Hogsback fault near La Barge and the Absaroka fault near Kemmerer establish, beyond doubt in our view, the pre-Eocene age of principal movements along these faults at these localities. Mountjoy's statement that the most accurately dated faults in the Idaho–Wyoming thrust belt are of Eocene age is incorrect. Only the faults along the front of the Idaho–Wyoming thrust belt, the youngest, are of this age.

The fairly well-dated thrust faults along the latitude of La Barge and Kemmerer also indicate westward-increasing ages for principal movements. Major movement occurred in the early Eocene on the La Barge or Hilliard fault on the east, in the Paleocene on the Hogsback fault, and in latest Cretaceous on the Absaroka fault. The earliest movement on the Paris fault, on the west side of the Idaho–Wyoming thrust belt, was probably latest Jurassic and earliest Cretaceous in age.

Eastward progression of orogeny in the Cordilleran region has long been recognized. It is supported by the foregoing dates of thrust movements. It is supported by the progressively younger ages eastward of the youngest rocks present below the various thrust plates (Rubey and Hubbert, 1959). It is also supported by information on the

development and later the destruction of the miogeosyncline in which sediments of the Cordillera were deposited. Still other support comes from the distribution and ages of coarse detrital rock units along the eastern margin of the Cordilleran region. As orogeny in the thrust belt consisted of folding and thrust faulting, corroborative evidence of eastward progression of orogeny supports an eastward progression with time of thrust faulting.

The pattern of progressively younger faults eastward is apparently contradicted locally by geometric relationships of some structures. In the Meade thrust plate, for example, along the Left Fork of Twin Creek northeast of Georgetown, Idaho, a younger western thrust slice overlies an older eastern thrust slice (Armstrong and Cressman, 1963). There are other localities in the Idaho–Wyoming thrust belt (Staatz and Albee, 1963; Jobin and Soister, 1964; Albee, 1964) where folds and steeply dipping faults in eastern underlying thrust sheets do not affect western overlying thrust sheets; some eastern folds and faults apparently predate the western thrust fault. Fault traces at all such localities are closely spaced and no independent basis has been found for dating these faults except relative to one another. We interpret them as slices of the same major thrust fault. If they are, then the western slice might well have moved last, as inferred in the development of back-limb thrusts (Douglas, 1950; Irish, 1965). Late or recurrent movement of a western fault slice, late forceful squeezing and shattering of a small slice between major thrust faults, and different times of movement at different places along the same major thrust may explain local relationships that apparently contradict the regional pattern. Whatever the explanation, the fact is that the few available precise dates indicate progressively younger movement eastward along the major thrust faults regionally.

Canadian Disturbed Belt

Our knowledge of the literature on the Canadian Rocky Mountain thrust belt is meager, and we are, regrettably, even less familiar with the rocks themselves. It would be risky to attempt to apply ideas gained in one region to another not known to us. However, we are puzzled by statements that the Canadian Rocky Mountain deformation culminated in middle Eocene time. Information that contributes to our perplexity includes available evidence for Paleocene and older orogenic events (Bossart, 1957; Taylor et al., 1964). Are these simply "precursors" of Laramide orogeny or the main event? Also of interest are the well-reasoned inferences from structural data (Mountjoy, 1962) that major thrust sheets were developed from southwest to northeast. If there was a progression in the development of the thrust faults, then how could thrusting have taken place almost everywhere in middle Eocene time, particularly in view of the enormous amounts of time required for transport of each thrust plate?

Our intent in asking these questions is not to suggest that thrusting in the Canadian disturbed belt occurred at precisely the same times as thrusting in the Idaho–Wyoming thrust belt. Individual events no doubt overlapped, but precise coincidence of major events in the two widely separated regions is unlikely. A general impression, that we are not prepared to defend, is that some of the latest events along the front of the Canadian belt may have occurred somewhat later than the last events along the front of the Idaho–Wyoming belt.

Dating of Orogenies and the Laramide

Despite the common interest that Mountjoy and we share in thrust faults, we may be very much apart on one issue. Neither we nor our colleagues working in the Idaho–Wyoming thrust belt (Rubey and Hubbert, 1959; Eardley, 1960; Armstrong and Cressman, 1963) have believed it necessary to have orogeny or deformation culminate in a rousing crescendo for a region or mountain system. Precise dates for frontal faults of a mountain system do not date the system, but merely the last of a whole sequence of tectonic events that include equally large or larger earlier episodes of tectonism that are more difficult to date. The local field evidence in the Idaho–Wyoming thrust belt, as in central Utah (Spieker, 1946), supports the concept of multiple, nonsynchronous tectonic events in orogeny (Shepard, 1923; Spieker, 1946; Gilluly, 1949, 1963, 1965; Aubouin, 1965).

Acceptance of the concept of multiple tectonic events in orogeny may explain, in part, our not using the term Laramide, with which Mountjoy takes issue. The term Laramide was for decades a keystone in the theory of periodic worldwide diastrophism. The Laramide revolution formerly was used to mark the end of the Mesozoic and beginning of the Cenozoic eras. The Laramide revolution has also been defined as: "A period of mountain building and erosion in Rocky Mountain region that began in late Cret. time and ended in early Tert. time" (Wilmarth, 1938, p. 1149). Mountjoy applies the term to middle Eocene events. Our colleagues working in other parts of the continent apply the term to events ranging from middle Mesozoic to middle Cenozoic in age. It is difficult to tell what is meant by use of the term.

No doubt the term Laramide is just as useful and just as much needed in some discussions as the term Paleozoic. However, assignment of a stratigraphic unit to the lower Middle Devonian seems more meaningful to us than assignment to the Paleozoic. Moreover, however great their shortcomings, criteria are available for recognition of the beginning and the end of the Paleozoic. Tectonic events in many parts of the continent cannot be dated more precisely than Laramide. If they can be, use of the term conceals the precision, for Laramide has been applied to episodes of orogeny extending through at least 60 m.y. (Gilluly, 1963).

Semantic quibbling over the term Laramide, however, is not the real issue. The danger of use of the term is that it may imply synchroneity of tectonic events within a region or between regions. Mounting evidence in the western United States, in contrast, demonstrates that crustal deformation proceeds ". . . in a quasicontinuous, although doubtless episodic way" (Gilluly, 1965, p. 23). The concept has also been described as ". . . long lasting periods of tectonic deformation, at a most variable speed, with paroxysms, which may occur at different times in different places along the length of the range" (Goguel, 1965, p. 195).

The title of this reply, therefore, for all of the foregoing reasons, is *"Times"*—not "Time"—"of thrusting in the Idaho–Wyoming thrust belt."

Acknowledgments

We are indebted to our colleagues W. W. Rubey, J. I. Tracey, Jr., and L. B. Platt for critical review of this manuscript and for their permission to make use of unpublished information. The manuscript also profited greatly from suggestions by G. D. Robinson.

The origin of working ideas is commonly difficult to recall. The previous discussion no doubt makes use of many ideas first enunciated by our colleagues as well as a few of our own. Our debt to our colleagues, particularly Rubey and Tracey, is therefore considerable.

References Cited

Albee, H. F., 1964, Preliminary geologic map of the Garns Mountain NE quadrangle, Teton County, Idaho: U.S. Geological Survey Mineral Investigations Field Studies Map MF-274.

Armstrong, F. C., and E. R. Cressman, 1963, The Bannock thrust zone, southeastern Idaho: U.S. Geological Survey Professional Paper 347-J, p. J1–J22.

Armstrong, F. C., and S. S. Oriel, 1965, Tectonic development of Idaho–Wyoming thrust belt: AAPG Bulletin, v. 49, n. 11, p. 1847–1866.

Aubouin, J., 1965, Geosynclines, developments in geotectonics 1: Amsterdam, Elsevier, 335 p.

Bossart, D. O., 1957, Relationship of the Porcupine Hills to early Laramide movements, *in* Alberta Society of Petroleum Geologists 7th Annual Field Conference Guidebook, E. W. Jennings, ed., p. 46–51.

Douglas, R. J. W., 1950, Callum Creek, Langford Creek, and Gap map-areas, Alberta: Geological Survey of Canada Memoir 255, 124 p.

Eardley, A. J., 1944, Geology of the north-central Wasatch Mountains, Utah: Geological Society of America Bulletin, v. 55, p. 819–894.

———, 1960, Phases of orogeny in the fold belt of western Wyoming and southeastern Idaho, *in* D. P. McGookey, ed., Overthrust belt of southwestern Wyoming and adjacent areas: Wyoming Geological Association 15th Annual Field Conference Guidebook, p. 37–40.

Gazin, C. L., 1942, Fossil Mammalia from the Almy Formation in western Wyoming: Washington Academy of Science Journal, v. 32, n. 7, p. 217–220.

———, 1956a, The upper Paleocene Mammalia from the Almy Formation in western Wyoming: Smithsonian Miscellaneous Collection, v. 131, n. 7, p. 1–18.

———, 1956b, The occurrence of Paleocene mammalian remains in the Fossil basin of southwestern Wyoming: Journal of Paleontology, v. 30, n. 3, p. 707–711.

Geological Society of London, 1964, The Phanerozoic time-scale, a symposium: Quarterly Journal of the Geological Society of London, v. 120, supplement, 458 p.

Gilluly, J., 1949, Distribution of mountain building in geologic time: Geological Society of America Bulletin, v. 60, p. 561–590.

———, 1963, The tectonic evolution of the western United States: Quarterly Journal of the Geological Society of London, v. 119, p. 133–174.

———, 1965, Volcanism, tectonism, and plutonism in the western United States: Geological Society of America

Special Paper 80, 69 p.

Goguel, J., 1965, Tectonics and continental drift, *in* A symposium on continental drift: Philosophical Transactions of the Royal Society of London, n. 1088, p. 194–198.

Irish, E. J. W., 1965, Geology of the Rocky Mountain Foothills, Alberta: Geological Survey of Canada Memoir 334, 241 p.

Jobin, D. A., and P. E. Soister, 1964, Geologic map of the Thompson Peak quadrangle, Bonneville Co., Idaho: U.S. Geological Survey Mineral Investigations Field Studies Map MF-284.

Mansfield, G. R., 1927, Geography, geology, and mineral resources of part of southeastern Idaho: U.S. Geological Survey Professional Paper 152, 453 p.

———, 1952, Geography, geology, and mineral resources of the Ammon and Paradise Valley quadrangles, Idaho: U.S. Geological Survey Professional Paper 238, 92 p.

McMasters, J. H., 1943, Buena Vista Hills area of the Midway–Sunset oil field: California State Division of Mines Bulletin 118, p. 517–518.

Michael, R. H., 1960, Hogsback and Tip Top units, Sublette and Lincoln counties, Wyoming, *in* D. P. McGookey, ed., Overthrust belt of southwestern Wyoming and adjacent areas: Wyoming Geological Association 15th Annual Field Conference Guidebook, p. 210–216.

Mountjoy, E. W., 1962, Mount Robson (southeast) map area, Rocky Mountains of Alberta and British Columbia: Geological Survey of Canada Paper 61-31, 114 p.

———, 1966, Time of thrusting in the Idaho–Wyoming thrust belt: discussion: AAPG Bulletin, v. 50, n. 12, p. 2612–2614.

Murray, F. E., 1960, An interpretation of the Hilliard thrust fault, Lincoln and Sublette counties, Wyoming, *in* D. P. McGookey, ed., Overthrust belt of southwestern Wyoming and adjacent areas: Wyoming Geological Association 15th Annual Field Conference Guidebook, p. 181–186.

Oriel, S. S., 1962, Main body of the Wasatch Formation near La Barge, Wyoming: AAPG Bulletin, v. 46, n. 12, p. 2161–2173.

———, 1963, Preliminary map of the Fort Hill quadrangle, Lincoln Co., Wyoming: U.S. Geological Survey Oil and Gas Investigations Map OM-212.

Rubey, W. W., and M. K. Hubbert, 1959, Overthrust belt in geosynclinal area of western Wyoming in light of fluid-pressure hypothesis, Pt. 2 *of* Role of fluid pressure in mechanics of overthrust faulting: Geological Society of America Bulletin, v. 70, p. 167–206.

Rubey, W. W., S. S. Oriel, and J. I. Tracey, Jr., 1961, Age of the Evanston Formation, western Wyoming, *in* Short papers in the geologic and hydrologic sciences: U.S. Geological Survey Professional Paper 424-B, p. B153–154.

Shepard, F. P., 1923, To question the theory of periodic diastrophism: Journal of Geology, v. 31, p. 599–613.

Spieker, E. M., 1946, Late Mesozoic and early Tertiary history of central Utah: U.S. Geological Survey Professional Paper 205-D, p. 117–161.

Staatz, M. H., and H. F. Albee, 1963, Preliminary geologic map of the Garns Mountain SE quadrangle, Bonneville and Teton counties, Idaho: U.S. Geological Survey Mineral Investigations Field Studies Map MF-262.

Taylor, R. S., W. H. Mathews, and W. O. Kupsch, 1964, Tertiary, *in* Geological history of western Canada: Alberta Society of Petroleum Geologists, p. 190–194.

Veatch, A. C., 1907, Geography and geology of a portion of southwestern Wyoming: U.S. Geological Survey Professional Paper 56, 178 p.

Wilmarth, M. G., 1938, Lexicon of geologic names of the United States: United States Geological Survey Bulletin 896, 2396 p.

TECTONIC DEVELOPMENT OF THE IDAHO–WYOMING THRUST BELT: AUTHORS' COMMENTARY

S. S. Oriel

U.S. Geological Survey
Denver, Colorado

F. C. Armstrong

U.S. Geological Survey
Spokane, Washington

Republication of our 1965 and 1966 papers here in this volume and in the volume compiled by Perry et al. (1984) is gratifying, because these papers continue to be among the most frequently cited products of our work. We have not thanked editor J. A. Peterson adequately for his insistence that we participate in the 1964 Rocky Mountain Section meeting of AAPG at Durango, Colorado, for which our original paper was written, after his preferred participant and our mentor, W. W. Rubey, firmly declined. This brief commentary, written in response to Peterson's request, is intended to put the papers into perspective. Our purpose here is to review some pertinent contributions that seem most significant since our reports were published and some older ones sometimes overlooked.

Our original reports are now antiquated. We would now delete most references to geosynclines, especially to their western margins, and would instead emphasize miogeoclines, foreland moats, and continental shelves and slopes. The title would perhaps now be "The Idaho–Wyoming salient of the Cordilleran foreland thrust belt (Figure 27)," to place our work into a modern continental and plate tectonic framework (Burchfiel and Davis, 1975; Davis et al., 1978; Hamilton, 1978). (Some of our colleagues, concerned with petroleum rather than tectonics, would prefer the designation "Wyoming–Utah–Idaho salient.") A modern version might include such terms as duplex, horse, ramp, listric, blind thrust, ramp anticline (Boyer and Elliott, 1982), and of course, "balanced sections" (Dahlstrom, 1969). (Unbalanced sections generally are "yours"; the balanced ones are "ours," even if a few do not illustrate "conservation of volume.")

Preparation of our 1965 report was a labor of love because each of us had worked in his respective area for about a decade and our colleagues had worked there even longer. The myriad observations, documented in subsequent U.S. Geological Survey publications, were just beginning to form exciting cohesive patterns. (Yes, there was a time in olden days when comprehensive and thorough investigations were not only tolerated, but encouraged and expected.)

We are indebted to E. W. Mountjoy for allowing us to reprint his discussion of our paper. After that published exchange, we exchanged week-long field conferences in the summer of 1967 to familiarize ourselves with each other's regions and concepts. A happy consequence was the development of enduring friendships with Canadian colleagues, including Eric Mountjoy, Raymond Price, and John Wheeler.

Relationship of Geosynclinal Stages to Newer Concepts

In our initial paper (Armstrong and Oriel, 1965), three stages were recognized in the accumulation of the thick strata in the Idaho–Wyoming salient. Although linked then to shifts in the belt or hinge between shelf and miogeosyncline, these are now commonly described in plate tectonic terms.

The initial stage is represented by a miogeoclinal wedge of late Proterozoic to middle Paleozoic strata deposited along the passive, rifted, and attenuated western margin of the North American protocontinent (Stewart, 1972; Aitken, 1981). The margin, which now trends northerly, truncates southwest-trending Archean to middle Proterozoic terranes and tectonic provinces, presumably bounded in part by ancient sutures. The basal part of the miogeoclinal wedge consists of detrital strata derived from the craton on the east; it ranges from argillite to quartzitic conglomerate, thickening westward to more than 25,000 ft (7000 m) (Crittenden et al., 1971; Stewart, 1972). The sequence is assigned the same age as the Windermere Group in southern British Columbia, which it resembles. Both include a lower part of glaciomarine diamictite (Christie-Blick, 1982; Link, 1983) and basalt. The uppermost part is quartzitic sandstone, which transgressed onto the craton and is

younger eastward. The basal detrital sequence is overlain by more than 14,000 ft (4000 m) of eastwardly thinning Cambrian–Devonian limestone and dolomite interbedded with some quartzite and shale. The northwestern source of detrital potassium feldspar (nonperthitic microcline) in the Upper Cambrian basal Worm Creek Quartzite Member of the St. Charles Limestone and in the Park Shale Member of the Gros Ventre Formation (Deboer and Middleton, 1983) remains unexplained, although it could have been gneiss or recycled from Proterozoic strata along the Lemhi arch (Ruppel, in press; Stewart and Suczek, 1977).

The second stage of stratal accumulation was characterized by an eastward shift (Armstrong and Oriel, 1965) of the miogeoclinal wedge from Mississippian time on. This eastward shift may have been an isostatic response to crustal loading by the obduction during Late Devonian and Early Mississippian Antler deformation of oceanic rocks (including some from island arc settings) of the Roberts Mountain allochthon (Poole and Sandberg, 1977), and during Permian–Triassic Sonoman deformation of the Golconda allochthon (Silberling and Roberts, 1962; Speed, 1977; Stevens, 1977; Stewart et al., 1977). Deformation accompanying accretion of both of these obducted oceanic masses (Dickinson, 1977) produced turbidite detritus deposited to the east in troughs or foreland basins (Nilsen and Stewart, 1980) flanking the eastward-shifted progradational carbonate continental shelf margin (Rose, 1976). The second stage also included the development of the large foreland-like Oquirrh basin associated with Pennsylvanian and Permian deformation of the Ancestral Rocky Mountains (Jordan and Douglass, 1980).

The third and greater change in stratal accumulations during Mesozoic and early Paleogene time constitutes the synorogenic record of compression within the thrust belt. Diverse terms have been applied to the tectonic setting in which these very thick, mainly terrigenous strata accumulated from Early Cretaceous to early Eocene time. *Exogeosyncline* was defined by Kay (1947, 1951) to distinguish the foredeeps that developed within continental cratons and received detritus from adjoining mountain belts, from the *orthogeosynclines* (continental shelves and slopes) that flanked the craton. The strata within the foredeep, the clastic wedge of King (1959), have long been recognized as analogous to the molasse (Trümpy, 1960; Van Houten, 1973) in the external (or foreland) part of the Alpine mountain system and to the Siwalik Group in the Himalayan foredeep. But the genesis of these features remained unexplained until linked to other ideas by our Canadian colleagues in an elegant model. The foredeep is attributed (Price, 1973) to isostatic subsidence of a flexurally rigid lithosphere resulting from loading by supracrustal rocks tectonically thickened by thrusting. The thrusting, in turn, is a response of brittle supracrustal rocks (Price and Mountjoy, 1970) to the upwelling and lateral spreading of a hot, mobile infrastructure (Bucher, 1955; Rodgers, 1964), represented in the Canadian Cordillera by the Shuswap Metamorphic Complex, or an associated magmatic arc, which was initiated by either crustal shortening or convergence (Price, 1981). The crustal rheologic implications of the isostatic development of a foredeep or

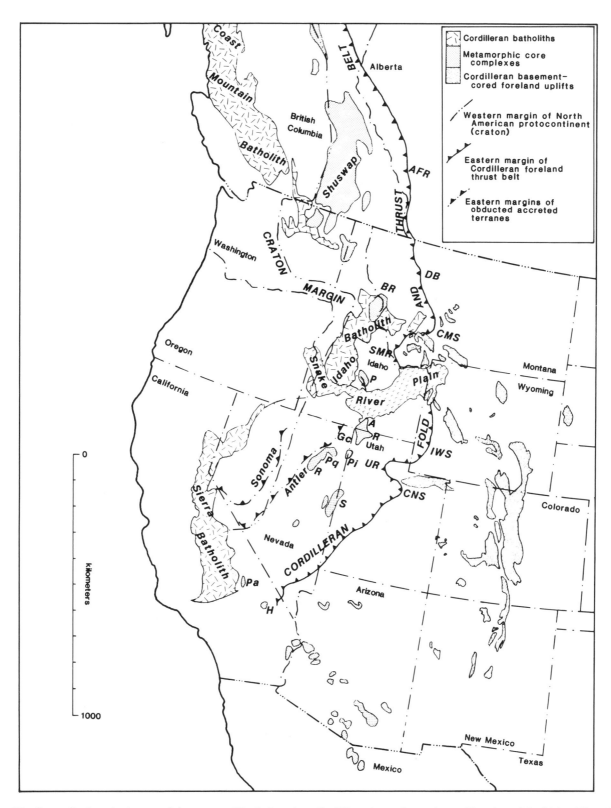

Figure 27—Generalized geologic map of the western North American Cordillera shows the setting and location of the Idaho–Wyoming salient of the Cordilleran foreland thrust belt, and the distributions of basement-cored foreland uplifts to the east, and of metamorphic core-complexes and magmatic arcs to the west. The eastern margin of the thrust belt includes the following parts: AFR, Alberta Front Range; DB, Disturbed belt in western Montana; CMS, Central Montana salient; SMR, Southwestern Montana reentrant (around Lemhi uplift); IWS, Idaho–Wyoming salient; UR, Uinta reentrant; CNS, Charleston-Nebo salient. Metamorphic core-complexes include: The Shuswap terrane; BR, Bitterroot; P, Pioneer; A, Albion; R, Raft River; GC, Grouse Creek; PI, Pilot; Pq, Pequop; R, Ruby Range; S, Snake Range; Pa, Panamint; H, Halloran. (Modified from Hamilton, 1978; Armstrong, 1982; Beutner, 1977; and Davis and others, 1978).

moat filled with synorogenic detritus have been considered by Jordan (1981) and Beaumont et al. (1982).

Thrust Belt Geometry

Major advances in our understanding of the Idaho–Wyoming salient of the Cordilleran thrust belt have come from industry data on subsurface geometry. Those of our generation were implored, both by early training and apprenticeship mentors, to avoid inferring or defining a fault unless field relationships absolutely required one. The emphasis was on conservatism in structural interpretations. As advances were made in understanding regional stratigraphic relationships, a basis was developed for discriminating episodes of deformation (Armstrong and Cressman, 1963) of diverse structures viewed earlier as linked (Mansfield, 1927). Rapid advances in aerial photographic methods and their applications have also contributed to our knowledge of structural details, particularly when accompanied by field checks. Structural field relationships are much more complex than initially suspected.

Several major petroleum companies, particularly Chevron, have been generous in releasing proprietary data to improve industry and scientific knowledge of the thrust belt's geometry. Their geologists have published not only interpretations of structural complexities evident in the region and at rich and prolific hydrocarbon fields, but also the basic borehole and seismic reflection data on which the interpretations are based (e.g., Royse et al., 1975), thereby enabling others to confirm the soundness of their conclusions. Borehole, dipmeter, and numerous other data are also presented in the comprehensive regional paper by Lamerson (1982), which includes numerous structural sections controlled by seismic reflection surveys that regrettably are not reproduced. Still another contribution by a Chevron geologist (Lamb, 1980) defines and describes the Painter Reservoir giant oil field in southwestern Wyoming, using numerous subsurface data and seismic reflection surveys that are not reproduced in his papers.

Primary seismic reflection surveys for parts of the Idaho–Wyoming belt have been released for publication not only by Chevron (Royse et al., 1975), but also by Arco (Norton, 1983), Conoco (McClellan and Storrusten, 1983), and Champlin (Dixon, 1982).

The success of exploration efforts in the Idaho–Wyoming salient has been properly attributed to advances in methods used to gather seismic reflection data and to breakthroughs in seismic data processing. These tools have undeniably promoted efficiency in exploration. However, the asymmetry of folds within the region was recognized decades ago by surface mappers, but nevertheless was not always taken into account in the siting of early exploration tests. Unfortunately, modern seismic reflection data remain almost inaccessible to scientists responsible for ascertaining the mineral resource potential of public lands in which industry members and other citizens have a considerable stake.

Information based on diverse subsurface data of generally unspecified quality illustrating the structural complexities of producing oil and gas fields is being released by many companies, as in papers by Evans and Barrett (1977), Kelly and Hine (1977), Conner and Covlin (1977), Frank and Gavlin (1981), Hoffman and Kelly (1981), Petroleum Information Corporation (1981), Bishop (1982), Frank et al. (1982), Hoffman and Bakells-Baldwin (1982), Walker (1982), and West and Lewis (1982).

The new structural information establishes that the top surface of the basement generally slopes gently westward throughout the thrust belt foreland, as demonstrated by Bally et al. (1966) and that for most of the belt, basement clearly is not involved in the "thin-skinned" deformation (Rich, 1934; Rodgers, 1949, 1964). Pieces of basement involved locally in thrusting, as in northern and central Utah, could represent decapitated tops of basement highs formed during the late Paleozoic Ancestral Rocky Mountain deformation.

The near absence of basement involvement in the Idaho–Wyoming salient contrasts markedly with the abundance of basement massifs within the nappes of the Alps (Trümpy, 1960, 1980). This contrast is attributed by Dewey and Bird (1970) to the distinction between continental edge (ocean–continent) and intracontinental (continent–continent) collisions. The presence of a thick anisotropic miogeoclinal wedge in the North American Cordillera apparently provided a supracrustal cover deep enough to absorb brittle deformation. Sedimentary rock sequences within the Alps are far thinner (Trümpy, 1960) and accumulated mainly well within the Pangean continent (Dietz and Holden, 1970) rather than along a long-lasting trailing continental margin. Basement is involved, however, in compressive structures of Laramide-type foreland uplifts east of the miogeoclinal wedge (Armstrong, 1974; Hamilton, 1981; Gries, 1983; Lowell, 1983) where overlying strata are far thinner than in the thrust belt.

Industry subsurface data and regional geologic mapping confirm that thrust surfaces are parallel to incompetent rock layers but cut steeply across more competent rock units (Rich, 1934) to form flats, ramps, and ramp anticlines. Such data also provided a basis for preparing well-constrained structural sections (Royse et al., 1975) like those prepared earlier in Canada, as well as regional restored sections (Bally et al., 1966; Price and Mountjoy, 1970). Industry data also proved that normal faults within the thrust belt, long considered to maintain their steep dips to considerable depths, are listric, flattening downward.

The role of salt flowage within the Jurassic Preuss redbeds in modifying initial structures within the fold and thrust belt (Lelek, 1982) and in forming ductile layers bounding disharmonic folds had not been recognized earlier.

Times of Thrusting

The eastward progression of the times of transport of the major thrust sheets, from the western hinterland to the eastern foreland, has been substantiated by new data having greater precision than was possible two decades ago in the Idaho–Wyoming salient (Figure 28) and has also been defined now in Utah (Villien and Kligfield, in press).

The relative youthfulness of the easternmost thrust faults is well established. On the north, the Jackson thrust of Horberg (1938) and the Prospect thrust of Rubey and

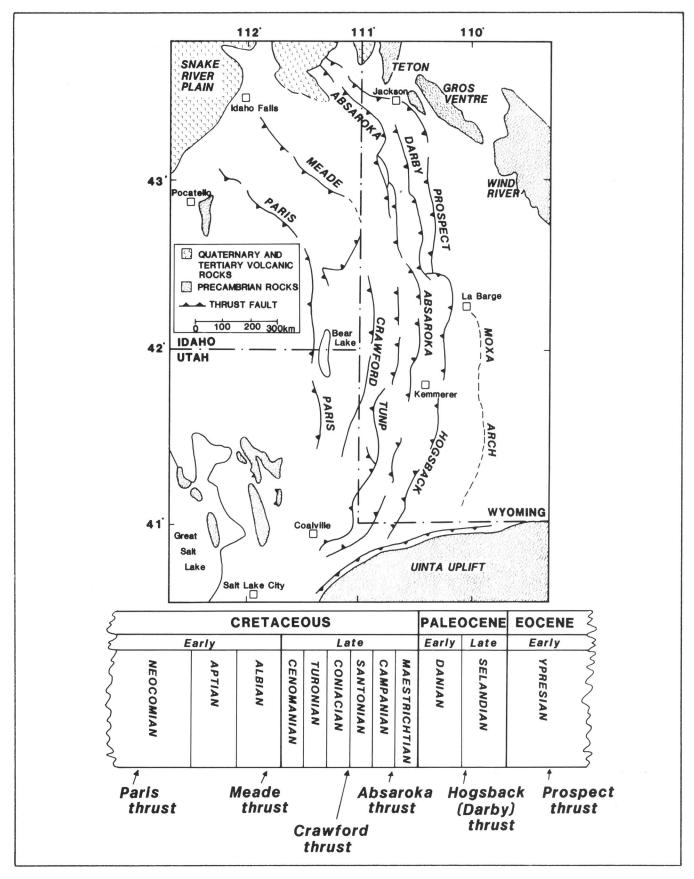

Figure 28—Generalized tectonic map of the major thrust faults of the Idaho–Wyoming salient shows the locations and extents of the principal thrust faults and the times of their major motions.

Hubbert (1959), or the Cliff Creek thrust of Dorr et al. (1977), moved in latest Paleocene or earliest Eocene time (Dorr et al., 1977). The easternmost fault on the south, the La Barge thrust, is of early (possibly middle early) Eocene age (Oriel, 1969). The foregoing ages are based on terrestrial vertebrate fossils whose utility in ascertaining the Paleocene–Eocene boundary is questioned by some palynologists (Wing, 1984).

The middle Paleocene age inferred by Oriel (1969) for the Hogsback or Darby thrust fault has been confirmed by Dorr and Gingerich (1980). Nevertheless, the Hogsback fault south of Snider basin may be linked to both the Darby and Prospect thrusts on the north (Royse et al., 1975; Blackstone, 1977), although Royse et al. (1975) recognized that the Darby fault north of Snider basin is older than the Prospect thrust. The Hogsback thrust sheet may have moved during both Paleocene and early Eocene time (Royse, 1985).

The Late Cretaceous major movement of the Absaroka thrust sheet (Oriel and Tracey, 1970) has been established more precisely by Chevron investigators (Royse et al., 1975; Lamerson, 1982). Palynologic calibration of Cretaceous units in the northern Fossil basin (Nichols and Jacobson, 1982; Jacobson and Nichols, 1982) enables us to date major Absaroka fault movement as late Campanian or earliest Maestrichtian. To the south, a western slice of the Absaroka thrust sheet (the "Early Absaroka fault" of Royse et al., 1975) is dated by them on the basis of its associated synorogenic product, the Little Muddy Creek Conglomerate, as middle Santonian (Jacobson and Nichols, 1982).

Movement of the Crawford thrust fault, which overlies and marks the western boundary of the Absaroka thrust sheet, is presumed to have produced the Echo Canyon Conglomerate (Royse et al., 1975) of Coniacian or earliest Santonian age (Jacobson and Nichols, 1982).

The Crawford thrust fault has been linked on the basis of regional relationships and geometry to the Meade thrust fault, exposed north and west of the Crawford thrust (Oriel and Platt, 1980; Blackstone, 1980); together they were designated the Meade–Crawford thrust zone by Royse et al. (1975). We believe, but cannot prove, that on the contrary, movement of the Meade sheet preceded that of the Crawford. Although some frontal imbricate thrust slices of the Meade thrust fault locally cut the Wayan Formation (as restricted by Rubey, 1973), we regard the Wayan Formation as the synorogenic product of major movement of the Meade thrust sheet, although we have not proved it to be on the basis of clast compositions. The Early Cretaceous age of the Wayan (Rubey, 1973) has been confirmed by vertebrates and pollen collected by J. A. Dorr, Jr., and dated as middle Albian (Wiltschko and Dorr, 1983), although palynomorphs suggest a late Albian age as more probable (S. N. Nelson, personal communication, 1982). Thus, we view the Meade thrust sheet as having been emplaced considerably before the Crawford sheet.

We know of no new information comparable in quality to that of Armstrong and Cressman (1963) to alter or date more precisely the major movement of the Paris thrust sheet during the very early Cretaceous.

An apparent exception to the eastward progression in times of thrusting of the major thrust faults, or "the dogma

of west to east younging" faults (Steidtmann et al., 1984), is represented by the Medicine Butte thrust fault (Veatch, 1907). This minor fault has minimal stratigraphic throw in the subsurface (Lamerson, 1982), and its movements were during Paleocene and early Eocene time (Lamerson, 1982). We infer from geometric relationships that this fault represents late movement of a hanging wall anticline cored by salt in the Jurassic Preuss redbeds.

The times of movement of minor imbricate thrust slices, with respect to movements along related major thrust faults, are poorly dated in the Idaho–Wyoming salient. Dating of movement along the Commissary thrust slice of the Absaroka thrust sheet remains imprecise (Rubey et al., 1975; Lamerson, 1982). The Bridger Hill thrust slice of the Absaroka sheet (Lamerson, 1982) moved before deposition of the Maestrichtian Hams Fork Conglomerate, as did the Absaroka thrust. The inference that thrust slices young both eastward and westward away from a major thrust fault (Douglas, 1950; Dahlstrom, 1970), although challenged by Jones (1984) and unproven by our data, is apparently confirmed by Royse (1985). A continuing problem is the basis for distinguishing between major thrusts and their imbricate slices in a region where most thrust faults emerge from the same lithostratigraphic décollement surface.

Foreland, Hinterland, Metamorphic Core Complexes, and Magmatic Arc

Despite widespread use of the term *foreland* for thrust and fold belts of mountain systems, their distal (from the craton) boundaries cannot be defined unambiguously. Yet many earth scientists distinguish between foreland (external or cratonward) and hinterland (internal or oceanward) parts of mountain systems to contrast rock compositions, metamorphic facies, megastructures, and strain features recording the degree of ductility. The Idaho–Wyoming salient's western transition into the hinterland is obscured by younger sedimentary and volcanic strata and by superimposed extensional deformation, in contrast to the southern Canadian Cordilleran segment (Price, 1981), which is less disrupted by early Paleogene extension (Fox and Beck, in press). Because of this, differences continue on placement of the boundary. A resource-based boundary was placed by Powers (1977) at the trace of the Paris thrust fault. In a recent unpublished manuscript, a boundary based on contrasts in strain features was placed at the trace of the Meade thrust fault. In past talks, we have contrasted structural styles west and east of the Paris thrust trace along the east side of the Bear River Range. But the boundary clearly lies between the Bear River and Bannock ranges on the east and the Albion and Raft River ranges on the west (Armstrong, 1972; Armstrong and Hansen, 1966; Allmendinger and Jordan, 1981) in a region now complicated by extensive denudation faults (Armstrong, 1972) that place allochthons of younger rocks on older strata. These were incorrectly designated as thrust plates by Oriel and Platt (1979).

Exposed within the Albion and Raft River ranges are metamorphic core complexes, such as in the Shuswap terrane in southern Canada. The tectonic significance of core complexes and their relationship to the foreland thrust belt have been much debated partly because of the diversity of tectonic features designated as core complexes (Crittenden et al., 1980). Ambiguities arise because the region between the complexes and the foreland fold and thrust belt is concealed by diverse deformational, volcanic, and detrital products of extension, whose magnitude in the Basin and Range province may be half of the present width of the province (Hamilton and Myers, 1966; Hamilton, 1978).

Compounding the problem is that the buoyant upwelling and lateral spreading of the hot infrastructure hypothesized in the Price and Mountjoy model (1970) requires extension over the mass grading eastward into the compression produced by the spreading. The products of this extension are not easily distinguished from those of the late Paleogene and Neogene extensional Basin and Range deformation of southern Idaho and western Utah. Indeed, in places even the products of Mesozoic compressional deformation have been confused with those of Neogene extension (Oriel and Moore, 1985), as is evident from misinterpretations of the structure of Teton Pass.

The core complexes form two distinct belts: One extends southward from the Shuswap terrane in southern British Columbia to southeastern California and lies between (and near) the magmatic arc on the west and the fold and thrust belt on the east (Figure 27); this belt is polygenetic (Armstrong, 1982). The other belt extends southeastward from the foreland thrust belt and is considered to be monogenetic. The former belt had been interpreted (Misch, 1960; Armstrong and Hansen, 1966; Price and Mountjoy, 1970) as an infrastructural or diapiric orogenic core zone in the hinterland of the Cordilleran thrust belt, probably resulting from lithospheric heating (Armstrong, 1974) along a long-lived and eastward-dipping (Hamilton, 1969, 1978) middle to late Mesozoic subduction zone (Armstrong, 1982) which also produced the magmatic arc. The core complexes consist of formerly deeper crustal levels that have been raised by thermal diapirism or by crustal shortening and thickening produced by collisional compression. The core complexes in this belt are polygenetic in that they localized the much later and very different extensional deformation that modified them (Armstrong, 1982; Coney and Harms, 1984).

A foremost problem in the western part of the thrust belt has been in distinguishing and dating the various allochthons produced by the several episodes during each contrasting deformation. The contrasting structures are recognized by geometry (Platt and Coward, 1982) and radiometric dates (Armstrong, 1968; Allmendinger and Jordan, 1981, 1984; Allmendinger et al., 1984; Miller, in press). Compressive deformation and magmatism span much of Jurassic, Cretaceous, and early Paleogene time.

Mechanics of Thrusting

Acceptance of the thrust fault concept, like that of continental drift, was long delayed because earth scientists could not explain the lateral forces required and the ways in which they could be transmitted for great distances in materials of finite strength.

Field relationships recognized in Switzerland and Scotland late in the nineteenth century, however, made acceptance of the thrust fault concept inescapable (Hubbert and Rubey, 1959). The paradox of thrust faulting remained: What was the origin of the lateral forces, and how could these forces be transmitted through relatively weak rocks, especially anisotropic strata?

Numerous problems remain, but current concepts have overcome the major obstacles to understanding. The role of high fluid pressures in reducing the stresses required to transport thrust sheets great distances laterally over nearly horizontal surfaces was demonstrated by Hubbert and Rubey (1959). Plate tectonics and plate collisions demonstrate the reality of enormous lateral forces within the earth's crust. The analogy of wedge-shaped masses of soil or snow being pushed by bulldozers (Davis et al., 1983; Dahlen et al., 1984) is readily applied to accretionary wedges at plate boundaries. The lateral spreading model of Price and Mountjoy (1970) provided for lateral forces closer to the miogeoclinal wedge, which are analogous to those forces at accretionary wedges at plate margins, although much stronger. Spreading was a product of crustal thickening by tectonic progradation (Price, 1981), by crustal shortening of previously attenuated crust at the continental margin (Hellwig, 1976; Okulitch, 1984), or by heat (and possibly material) transfer along (Hamilton, 1978) or east of (Armstrong, 1982) the magmatic arc. The inferred spreading site along southern British Columbia is part of the Omineca Crystalline belt, a crustal welt marking the Jurassic boundary between the ancient North American continent and a large composite accreted terrane that includes Stikinia (Monger et al., 1982).

Continuing arguments regarding the respective roles of underthrusting versus overthrusting seem to us, as they did half a century ago to James Gilluly, to be fruitless exercises in semantic quibbling: it depends on whether one is standing on the allochthon or the autochthon. Thrust deformation requires a force couple.

Implications of Triassic Detritus

The contention that thrusting in the Idaho–Wyoming belt was underway during Jurassic time (as previously discussed) and the long-known Jurassic dates for parts of the magmatic arc have implications for the Triassic detrital strata in southeastern Idaho. The Lower(?) Triassic Timothy Sandstone and the Upper Triassic Higham Grit contain potassium feldspar, enough in places to make them arkosic, as well as clasts of quartzite that presumably came from Proterozoic or Paleozoic strata. Both feldspathic and volcanogenic detritus are present in Triassic Chinle strata southward in the Cordilleran region and on the Colorado Plateau. Their sources have been attributed (Blakey and Gubitosa, 1983) to volcanic arcs to the west and volcanic complexes along the Mogollon highlands to the south. Despite the presence of Triassic plutons on the east side of the Sierra Nevada (Chen and Moore, 1982), Anderson et al. (1984) cite evidence for unidentified southern, rather than western, sources. The source of Chinle detritus need not, of course, be similar to that for units in southeastern Idaho. Conceivably, some of the detritus was derived from microcontinents on the west colliding with the continent (Jones et al., 1978; Coney et al., 1980) although times of accretion or docking onto the continent remain uncertain (Davis et al., 1978).

Extensional Tectonics

The third part of our initial paper, concerned with block faulting, would now be entitled "Extensional tectonics," the scope of which is too great to be addressed here. We have previously stressed here that Basin and Range structures are overprinted on parts of the thrust belt.

Some steep extensional or relaxation faults in the eastern part of the region cut Eocene Wasatch strata and some do not (Oriel, 1969). As described by Oriel, some would now be designated as listric and some as growth faults.

In the west, too few Paleogene rocks are present to date the inception and progression of extensional faulting at different times and places. We had believed some extension may have been as old as the Eocene. The presence of several sets of normal faults in the Bear River, Portneuf, and other ranges to the west establishes that there were several stages of extensional faulting. The youngest sets cut not only Neogene strata of the Salt Lake Formation, but also valley basalts and alluvial fans, and parallel present mountain fronts, indicating that they are still active. The abundance of volcanic material in such older Tertiary units as the andesitic Bridger Formation, east of the belt, and the rhyolitic Fowkes Formation, Norwood Tuff, and Keetley Volcanics, which are as old as middle Eocene, had suggested early phases of extension. These units and associated structures, however, were probably products of deformation linked to Paleogene arc volcanism (Dickinson, 1979) that produced the widespread Eocene Challis and Absaroka volcanics and intrusive rocks at Caribou Mountain and in the Snake River Range. Basin and Range extension was probably not initiated here until Neogene time, as indicated by voluminous rhyolitic Salt Lake strata. This extension apparently resulted in thinning of the crust to about 30 km and caused eruptions in intermontane valleys of alkali olive basalts with anomalous oceanic affinities, which according to Armstrong (U.S. Geological Survey, 1962) were probably derived from partial melting of mantle peridotite at depths greater than those of associated olivine tholeiites of the Snake River Plain (Leeman, 1982).

Hydrocarbon Resources

In view of the great development of significant oil and gas fields in the Idaho–Wyoming salient, as indicated by foregoing citations, it is difficult now to recall that the province was condemned by many for hydrocarbon exploration for decades, despite the conviction among those who knew the rocks that geologic factors were favorable.

The major field discoveries, especially along the leading edge of the Absaroka thrust fault, were anticipated about 80 years ago by Veatch (1907) who noted oil occurrences in rocks assigned by him to the Jurassic System and in Cretaceous source beds. Veatch's other noteworthy achievements have been admired elsewhere (Rubey et al., 1975; Oriel, 1979; Parker, 1983). Factors favorable for the occurrence of hydrocarbons were summarized by Sheldon (1963), Armstrong and Oriel (1965), Oriel (1969), and Monley (1971). Despite these factors, the overthrust belt was classified as having ". . . a fair to poor potential to become a significant petroleum province" (Lyth, 1971, p. 412; see also Miller et al., 1975). Our response was to present talks before the Rocky Mountain Association of

Geologists in January 1972 and in April 1976 and before the Wyoming Geological Association in September 1976, challenging those low estimates (Oriel, 1976).

The history of petroleum exploration and development in the thrust belt is summarized by Hogden and McDonald (1977), Petroleum Information Corporation (1981), Ver Ploeg and De Bruin (1982), and Powers (1983). Current estimates of undiscovered recoverable oil and gas for the region are by Coury and Powers (*in* Dolton et al., 1981, and *in* Petroleum Information Corporation, 1981).

The recent period of very active and successful exploration in the Idaho–Wyoming thrust belt is attributable to several factors worth citing. Not the least was the nation's energy crisis precipitated by the actions of OPEC nations in 1973. The discovery of the Pineview field in northern Utah in 1975 by the American Quasar Petroleum Company, on a farm-out agreement from Amoco, demonstrated beyond question that prolific hydrocarbon fields were indeed present in the province. The Amoco Production Company previously had the foresight to negotiate an agreement with the Union Pacific Railroad to aggressively explore federal land grants to the railroad late in the last century. Another factor was the conviction among many geologists of Chevron USA, Inc. (several of whom had been students of D. L. Blackstone) that the province was favorable for oil exploration, which led them to maintain leases and continue research in the region. Finally, advancing concepts on favorable source beds, reservoirs, trapping mechanisms, and hydrocarbon thermal maturation, as well as enhanced seismic techniques, have increased the probability of success in exploration in this area.

Some Gratuitous Concluding Remarks

One prerogative of maturity is to express some observations candidly. Dismaying to us have been the declining (almost vanishing) skills in, and support for, regional surface geologic mapping, both within the profession at large and in our own organization (Gilluly, 1977). Geologic mapping is a powerful tool because its unique discipline requires widespread observations (for completeness), precision and accuracy (for reproducibility of primary data), and geometric consistency of spatial and temporal relationships (for logic and meaning.) The final product, the geologic map, is a succinct graphic summary of knowledge of the compositions, distributions, and interrelationships of earth materials as interpreted by geologists for specified purposes on the basis of prevailing concepts. A complete geologic map is a multipurpose tool, although no geologic map can ever serve all unanticipated needs. Despite the considerable investments now required for such work, most geologic maps demonstrate their value by repaying society several-fold within a few years after completion. Many currently accepted concepts of the thrust belt and its broad regional framework emerged from painstaking quadrangle geologic mapping.

Some current "authorities" on the regional geology cannot distinguish Cambrian Worm Creek quartzitic arkose from Pennsylvanian Wells orthoquartzite, or cannot express the considerable distinction between the Jurassic Preuss redbeds and Cambrian Gros Ventre shales. Authors of several widely cited papers summarizing the geology have barely set foot in the region. On the other hand, some of our own field geologist colleagues have failed to utilize abundant subsurface data, much of which is no longer proprietary. Much misinformation has been published.

Also dismaying has been the decline in scholarship in published and unpublished reports both by recent students and their advisors. Citations are frequently of recent speculative summaries published in periodicals devoted to newsworthiness rather than of the original reports in which information and concepts are supported by firm data. Too many attributions to previous work are erroneous. Parochialism in citing references in some published reports reflects a lack of awareness, or an unwillingness by some authors to concede, that other institutions have contributed to our understanding of the region.

We have also been puzzled by the apparent reluctance of authors of many manuscripts we have reviewed to acknowledge adequately the scientists who have influenced them and the institutions that have supported their work.

No doubt these factors arise from the publish-or-perish syndrome and the grantsmanship malaise afflicting all the sciences.

We are privileged to have participated in the increasing understanding of one of the best known thrust and fold belts in the world. No other such belt is as well-defined geometrically by detailed mapping and petroleum exploration or is as well dated, although strain analyses of deformational structures have only just begun in this region, in contrast to some other parts of the world.

Acknowledgments

A paper such as this is based on the contributions of many investigators in industry, government, and academia, although we have undoubtedly overlooked some in an epoch of proliferating publication. We are indebted to all who have shared their knowledge with us, especially our colleagues in Canada and in industry. We have profited from the manuscript reviews by Warren Hamilton, Wallace R. Hansen, Frank Royse, Jr., and Karl Kellogg and from helpful discussions with Warren Hamilton and Frank Royse. Publication authorized by the Director of the U.S. Geological Survey.

References Cited

Aitken, J. E., 1981, The Cambrian System in the southern Canadian Rocky Mountains, Alberta and British Columbia: Second International Symposium on the Cambrian System, Guidebook for Field Trip 2, 61 p.

Allmendinger, R. W., and T. E. Jordan, 1981, Mesozoic evolution, hinterland of the Sevier orogenic belt: Geology, v. 9, n. 7, p. 308–313.

———, 1984, Mesozoic structure of the Newfoundland Mountains, Utah: horizontal shortening and subsequent extension in the hinterland of the Sevier belt: Geological Society of America Bulletin, v. 95, n. 11, p. 1280–1292.

Allmendinger, R. W., D. M. Miller, and T. E. Jordan, 1984, Known and inferred Mesozoic deformation in the hinterland of the Sevier belt, northwest Utah, *in* Geology of northwest Utah, southern Idaho and northeast Nevada: Utah Geological Association

Publication 13, p. 21–34.

Anderson, T. H., G. B. Haxel, L. T. Silver, J. H. Stewart, and J. E. Wright, 1984, Late Triassic paleogeography of the southern Cordilleran: the problem of a source for voluminous volcanic detritus in the Chinle Formation of the Colorado Plateau region (abs.): Geological Society of America Abstracts with Programs, v. 16, n. 6, p. 430.

Armstrong, F. C., and E. R. Cressman, 1963, The Bannock thrust zone, southeastern Idaho: U.S. Geological Survey Professional Paper 374J, 22 p.

Armstrong, F. C., and S. S. Oriel, 1965, Tectonic development of Idaho–Wyoming thrust belt: AAPG Bulletin, v. 49, n. 11, p. 1847–1866.

Armstrong, R. L., 1972, Low-angle (denudation) faults, hinterland of the Sevier orogenic belt, eastern Nevada and western Utah: Geological Society of America Bulletin, v. 83, n. 6, p. 1729–1754.

———, 1968, Mantled gneiss domes in the Albion Range, southern Idaho: Geological Society of America Bulletin, v. 79, n. 10, p. 1295–1314.

———, 1974, Magmatism, orogenic timing, and orogenic diachronism in the Cordillera from Mexico to Canada: Nature, v. 247, p. 344–351.

———, 1982, Cordilleran metamorphic core complexes— from Arizona to southern Canada: Annual Review of Earth and Planetary Sciences, v. 10, p. 129–154.

Armstrong, R. L., and E. Hansen, 1966, Cordilleran infrastructure in the eastern Great basin: American Journal of Science, v. 264, n. 2, p. 112–127.

Bally, A. W., P. L. Gordy, and G. A. Stewart, 1966, Structure, seismic data and orogenic evolution of southern Canadian Rocky Mountains: Bulletin of Canadian Petroleum Geology, v. 14, n. 3, p. 337–381.

Beaumont, C., C. E. Keen, and R. Boutilier, 1982, A comparison of foreland and rift margin sedimentary basins: Philosophical Transactions of the Royal Society of London, Series A, v. A305, p. 295–317.

Bishop, R. A., 1982, Whitney Canyon–Carter Creek gas field, southwest Wyoming, *in* E. L. Heisey, ed., Geologic studies of the Cordilleran thrust belt: Rocky Mountain Association of Geologists Symposium, v. II, p. 591–599.

Blackstone, D. L., Jr., 1977, The overthrust belt salient of the Cordilleran fold belt, western Wyoming–southeastern Idaho–northeastern Utah, *in* E. L. Heisey, ed., Rocky Mountain thrust belt: Joint Wyoming–Montana–Utah Geological Associations Field Conference Guidebook, p. 367–384.

———, 1980, Tectonic map of the Overthrust belt, western Wyoming, southeastern Idaho and northeastern Utah, 4th ed.: Wyoming Geological Survey Map Series 8A.

Blakey, R. C., and R. Gubitosa, 1983, Late Triassic paleogeography and depositional history of the Chinle Formation, southern Utah and northern Arizona, *in* M. W. Reynolds, and E. D. Dolly, eds., Mesozoic paleogeography of the west-central United States: Rocky Mountain Section, Society of Economic Paleontologists and Mineralogists Symposium, v. 2, p. 57–76.

Boyer, S. E., and D. Elliot, 1982, Thrust systems: AAAB Bulletin, v. 66, p. 1196–1230.

Bucher, W. H., 1955, Deformation in orogenic belts: Geological Society of America Special Paper 62, p. 343–368.

Burchfiel, B. C., and G. A. Davis, 1975, Nature and controls of Cordilleran orogenesis, western United States; extensions of an earlier synthesis: American Journal of Science, v. 275-A, p. 363–396.

Chen, J. H., and J. G. Moore, 1982, Uranium–lead isotopic ages from the Sierra Nevada batholith, California: Journal of Geophysical Research, v. 87, n. B6, p. 4761–4784.

Christie-Blick, N., 1982, Upper Proterozoic and Lower Cambrian rocks of the Sheeprock Mountains, Utah: regional correlation and significance: Geological Society of America Bulletin, v. 93, n. 8, p. 735–750.

Coney, P. J., and T. A. Harms, 1984, Cordilleran metamorphic core complexes: Cenozoic extensional relics of Mesozoic compression: Geology, v. 12, n. 9, p. 550–554.

Coney, P. J., D. L. Jones, and J. W. H. Monger, 1980, Cordilleran suspect terranes: Nature, v. 288, p. 329–333.

Conner, D. C., and R. J. Covlin, 1977, Development geology of Pineview field, Summit Co., Utah: Wyoming, Montana, and Utah Geological Association Guidebook, E. L. Heisey, ed., Rocky Mountain and thrust belt geology and resources, p. 639–650.

Crittenden, M. D., Jr., F. E. Schaeffer, D. E. Trimble, and L. A. Woodward, 1971, Nomenclature and correlation of some upper Precambrian and basal Cambrian sequences in western Utah and southeastern Idaho: Geological Society of America Bulletin, v. 82, n. 3, p. 581–602.

Crittenden, M. D., Jr., P. J. Coney, and G. H. Davis, eds., 1980, Cordilleran metamorphic core complexes: Geological Society of America Memoir 153, 490 p.

Dahlen, F. A., J. Suppe, and D. Davis, 1984, Mechanics of fold-and-thrust belts and accretionary wedges: Cohesive Coulomb theory: Journal of Geophysical Research, v. 89, n. B12, p. 10,087–10,101.

Dahlstrom, C. D. A., 1969, Balanced cross sections: Canadian Journal of Earth Sciences, v. 6, p. 743–757.

———, 1970, Structural geology in the eastern margin of the Canadian Rocky Mountains: Bulletin of Canadian Petroleum Geology, v. 18, n. 3, p. 332–406.

Davis, D., J. Suppe, and F. A. Dahlen, 1983, Mechanics of fold-and-thrust belts and accretionary wedges: Journal of Geophysical Research, v. 88, n. B2, p. 1153–1172.

Davis, G. A., J. W. H. Monger, and B. C. Burchfiel, 1978, Mesozoic construction of the Cordilleran "collage," central British Columbia to central California, *in* Pacific coast paleogeography: Pacific Section, Society of Economic Paleontologists and Mineralogists Symposium, v. II, p. 1–32.

Deboer, D., and L. Middleton, 1983, Sedimentology of a shallowing-upward sequence in Middle Cambrian carbonate–siliciclastic associations, western Wyoming (abs.): AAPG Bulletin, v. 67, n. 3, p. 448–449.

Dewey, J. F., and J. M. Bird, 1970, Mountain belts and the new global tectonics: Journal of Geophysical Research, v. 75, n. 14, p. 2625–2547.

Dickinson, W. R., 1977, Paleozoic plate tectonics and the evolution of the Cordilleran continental margin, *in* Pacific coast paleogeography: Pacific Section, Society of Economic Paleontologists and Mineralogists Symposium, v. I, p. 137–155.

———, 1979, Cenozoic plate tectonic setting of the Cordilleran region in the United States, *in* Pacific coast paleogeography: Pacific Section, Society of Economic Paleontologists and Mineralogists Symposium, v. 3, p. 1–13.

Dietz, R. S., and J. C. Holden, 1970, Reconstruction of Pangaea: break up and dispersion of continents, Permian to present: Journal of Geophysical Research, v. 75, n. 26, p. 4939–4956.

Dixon, J. S., 1982, Regional structural synthesis, Wyoming salient of western Wyoming overthrust belt: AAPG Bulletin, v. 66, n. 10, p. 1560–1580.

Dolton, G. L., K. H. Carlson, R. R. Charpentier, A. B. Coury, R. A. Crovelli, S. E. Frezon, A. S. Khan, J. H. Lister, R. H. McMullin, R. S. Pike, R. B. Powers, E. W. Scott, and K. L. Varnes, 1981, Estimates of undiscovered recoverable conventional resources of oil and gas in the United States: U.S. Geological Survey Circular 860, 87 p.

Dorr, J. A., Jr., and P. D. Gingerich, 1980, Early Cenozoic mammalian paleontology, geologic structure, and tectonic history in the overthrust belt near La Barge, western Wyoming: University of Wyoming, Contributions to Geology, v. 18, n. 2, p. 101–115.

Dorr, J. A., Jr., D. R. Spearing, and J. R. Steidtmann, 1977, Deformation and deposition between a foreland uplift and an impinging thrust belt: Hoback basin, Wyoming: Geological Society of America Special Paper 177, 82 p.

Douglas, R. J. W., 1950, Callum Creek, Langford Creek, and Gap map-areas, Alberta: Geological Survey of Canada Memoir 255, 124 p.

Evans, S. B., and W. J. Barrett, 1977, The Craven Creek field, Lincoln County, Wyoming, *in* E. L. Heisey, ed., Rocky Mountain thrust belt geology and resources: Wyoming, Montana, and Utah Geological Associations Guidebook, p. 611–616.

Fox, K. F., and M. E. Beck, Jr., in press, Paleomagnetic results for Eocene rocks from northeastern Washington and the Tertiary tectonics of the Pacific Northwest: Tectonics.

Frank, J. R., and S. Gavlin, 1981, Painter Reservoir, East Painter Reservoir, and Clear Creek fields, Uinta County, Wyoming, *in* Energy resources of Wyoming: Wyoming Geological Association 32nd Annual Field Conference Guidebook, S. G. Reid, ed., p. 83–93.

Frank, J. R., S. Cluff, and J. M. Bauman, 1982, Painter Reservoir, East Painter Reservoir, and Clear Creek fields, Uinta County, Wyoming, *in* Geologic studies of the Cordilleran thrust belt: Rocky Mountain Association of Geologists Symposium, v. II, p. 601–611.

Gilluly, J., 1977, American geology since 1910—a personal appraisal, *in* Annual Review of Earth Planetary Sciences, v. 5, p. 1–12.

Gries, R., 1983, Oil and gas prospecting beneath the Precambrian of foreland thrust plates in the Rocky Mountains: AAPG Bulletin, v. 67, n. 1, p. 1–26.

Hamilton, W., 1969, Mesozoic California and the underflow of Pacific mantle: Geological Society of America Bulletin, v. 81, n. 12, p. 2409–2430.

————, 1978, Mesozoic tectonics of the western United States, *in* Pacific coast paleogeography: Pacific Section, Society of Economic Paleontologists and Mineralogists Symposium, v. II, p. 33–70.

————, 1981, Plate-tectonic mechanism of Laramide deformation: University of Wyoming, Contributions to Geology, v. 19, n. 2, p. 87–92.

Hamilton, W., and W. B. Myers, 1966, Cenozoic tectonics of the western United States: Review of Geophysics, v. 5, p. 509–549.

Helwig, J., 1976, Shortening of continental crust in orogenic belts and plate tectonics: Nature, v. 260, p. 768–770.

Hoffman, M. E., and J. M. Kelly, 1981, Whitney Canyon-Carter Creek field, Uinta and Lincoln Counties, Wyoming, *in* Energy resources of Wyoming: Wyoming Geological Association 32nd Annual Field Conference Guidebook, S. G. Reid, ed., p. 99–107.

Hoffman, M. E., and R. N. Bakells-Baldwin, 1982, Gas giant of the Wyoming thrust belt: Whitney Canyon–Carter Creek field, *in* Geologic studies of the Cordilleran thrust belt: Rocky Mountain Association of Geologists Symposium, v. II, p. 613–618.

Hogden, H. J., and R. E. McDonald, 1977, History of oil and gas exploration in the overthrust belt of Wyoming, Idaho, and Utah, *in* E. L. Heisey, ed., Rocky Mountain thrust belt geology and resources: Wyoming, Montana, and Utah Geological Associations Guidebook, p. 37–69.

Horberg, L., 1938, The structural geology and physiography of the Teton Pass area, Wyoming: Augustana Library Publications, n. 16, 85 p.

Hubbert, M. K., and W. W. Rubey, 1959, Role of fluid pressure in mechanics of overthrust faulting: I. Mechanics of fluid-filled porous solids and its application to overthrust faulting: Geological Society of America Bulletin, v. 70, n. 2, p. 115–166.

Jacobson, S. R., and D. J. Nichols, 1982, Palynological dating of syntectonic units in the Utah–Wyoming thrust belt: The Evanston Formation, Echo Canyon Conglomerate, and Little Muddy Creek Conglomerate, *in* Geologic studies of the Cordilleran thrust belt: Rocky Mountain Association of Geologists Symposium, v. II, p. 735–750.

Jones, D. L., N. J. Silberling, and S. W. Hillhouse, 1978, Microplate tectonics of Alaska—significance for the Mesozoic history of the Pacific Coast of North America, *in* Pacific coast paleogeography: Pacific Section, Society of Economic Paleontologists and Mineralogists Symposium, v. 2, p. 71–74.

Jones, P. R., 1984, Sequence of formation of back-limb thrusts and imbrications: implications for development of Idaho–Wyoming thrust belt: AAPG Bulletin, v. 68, n. 7, p. 816–818.

Jordan, T. E., 1981, Thrust loads and foreland basin evolution, Cretaceous, western United States: AAPG Bulletin, v. 65, n. 12, p. 2506–2520.

Jordan, T. E., and R. C. Douglass, 1980, Paleogeography and structural development of the Late Pennsylvanian to early Permian Oquirrh basin, northwestern Utah, *in* Rocky Mountain paleogeography, Rocky Mountain Section, Society of Economic Paleontologists and Mineralogists Symposium, v. 1, p. 217–238.

Kay, G. M., 1947, Geosynclinal nomenclature and the craton:

AAPG Bulletin, v. 29, p. 1287–1293.

Kay, M., 1951, North American geosynclines: Geological Society of America Memoir 48, 143 p.

Kelly, J. M., and F. O. Hine, 1977, Ryckman Creek field, Uinta County, Wyoming, *in* E. L. Heisey, ed., Rocky Mountain thrust belt geology and resources: Wyoming, Montana, and Utah Geological Associations Guidebook, p. 618–628.

King, P. B., 1959, The evolution of North America: Princeton, Princeton University Press, 190 p.

Lamb, C. F., 1980, Painter Reservoir field—giant in the Wyoming thrust belt, *in* M. T. Halbouty, ed., Giant oil and gas fields of the decade 1968–1978: AAPG Memoir 30, p. 281–288 (also *in* AAPG Bulletin, v. 64, n. 5, p. 638–644).

Lamerson, P. R., 1982, The Fossil basin and its relationship to the Absaroka thrust system, Wyoming and Utah, *in* Geologic studies of the Cordilleran thrust belt: Rocky Mountain Association of Geologists Symposium, v. I, p. 279–340.

Leeman, W. P., 1982, Olivine tholeiitic basalts of the Snake River Plain, Idaho, *in* B. Bonnichsen and R. M. Breckenridge, eds., Cenozoic geology of Idaho: Idaho Bureau of Mines and Geology Bulletin 26, p. 181–191.

Lelek, J. J., 1982, Anschutz Ranch East field, northeast Utah and southwest Wyoming, *in* Geologic studies of the Cordilleran thrust belt: Rocky Mountain Association of Geologists Symposium v. II, p. 619–631.

Link, P. K., 1983, Glacial and tectonically influenced sedimentation in the upper Proterozoic Pocatello Formation, southeastern Idaho, *in* D. M. Miller, V. R. Todd, and K. A. Howard, Tectonic and stratigraphic studies in the eastern Great basin: Geological Society of America Memoir 157, p. 165–181.

Lowell, J. D., ed., 1983, Rocky Mountain foreland basins and uplifts: Rocky Mountain Association of Geologists Symposium Guidebook, 392 p.

Lyth, A. L., 1971, Summary of possible future petroleum potential, Region 3, western Rocky Mountains, *in* I. H. Cram, ed., Future petroleum provinces of the United States—their geology and potential: AAPG Memoir 15, v. 1, p. 406–412.

Mansfield, G. R., 1927, Geography, geology and mineral resources of part of southeastern Idaho: U.S. Geological Survey Professional Paper 152, 453 p.

McClellan, B. D., and J. A. Storrusten, 1983, Utah–Wyoming overthrust line, *in* A. W. Bally, ed., Seismic expression of structural styles: AAPG Studies in Geology Series, n. 15, v. 3, p. 4.1-39–4.1-44.

Miller, B. M., H. L. Thomsen, G. L. Dalton, A. B. Coury, T. A. Hendricks, F. E. Lennartz, R. B. Powers, E. G. Sable, and K. L. Varnes, 1975, Geological estimates of undiscovered recoverable oil and gas resources in the United States: U.S. Geological Survey Circular 725, 78 p.

Miller, D. M., in press, Sedimentary and igneous rocks of the Pilot Range and vicinity, Utah and Nevada, *in* Orogenic patterns and stratigraphy of north-central Utah and southeastern Idaho: Utah Geological Association Field Conference and Symposium.

Misch, P., 1960, Regional structural reconnaissance in central-northeast Nevada and some adjacent areas:

observations and interpretations: Intermountain Association of Petroleum Geologists 11th Annual Field Conference Guidebook, p. 17–42.

Monger, J. W. H., R. A. Price, and D. J. Templeman-Kluit, 1982, Tectonic accretion and the origin of the two metamorphic and plutonic welts in the Canadian Cordillera: Geology, v. 10, n. 2, p. 70–75.

Monley, L. E., 1971, Petroleum potential of Idaho–Wyoming overthrust belt, *in* I. H. Cram, ed., Future petroleum provinces of the United States—their geology and potential: AAPG Memoir 15, v. 1, p. 509–530.

Nichols, D. J., and S. R. Jacobson, 1982, Cretaceous biostratigraphy in the Wyoming thrust belt: Mountain Geologist, v. 19, n. 3, p. 73–78.

Nilsen, T. H., and J. H. Stewart, 1980, The Antler orogeny— mid-Paleozoic tectonism in western North America (Penrose Conference Report): Geology, v. 8, n. 6, p. 298–302.

Norton, M. A., 1983, Kemmerer area, Lincoln County, Wyoming, *in* A. W. Bally, ed., Seismic expression of structural styles: AAPG Studies in Geology Series, n. 15, v. 3, p. 4.1-45–4.1-47.

Okulitch, A. V., 1984, The role of the Suswap Metamorphic Complex in Cordilleran tectonics, a review: Canadian Journal of Earth Science, v. 21, n. 10, p. 1171–1193.

Oriel, S. S., 1969, Geology of the Fort Hill quadrangle, Lincoln County, Wyoming: U. S. Geological Survey Professional Paper 594-M, 40 p.

———, 1976, Oil in the Idaho–Wyoming thrust belt (abs.): Rocky Mountain Association of Geologists and Wyoming Geological Association Newsletters, July.

———, 1979, Concepts of the Idaho–Wyoming thrust belt since the Hayden Survey (abs.): Geological Society of America Abstracts with Programs, v. 11, n. 6, p. 298.

Oriel, S. S., and J. I. Tracey, Jr., 1970, Uppermost Cretaceous and Tertiary stratigraphy of Fossil basin, southwestern Wyoming: U.S. Geological Survey Professional Paper 635, 53 p.

Oriel, S. S., and L. B. Platt, 1979, Younger-over-older thrust plates in southeastern Idaho (abs.): Geological Society of America Abstracts with Programs, v. 11, n. 6, p. 298.

———, 1980, Geologic map of the Preston 1° X 2° quadrangle, southeastern Idaho and western Wyoming: U.S. Geological Survey Miscellaneous Investigations Series Map I-1127.

Oriel, S. S., and D. W. Moore, 1985, Geologic map of the West and East Palisades Roadless areas, Idaho and Wyoming: U.S. Geological Survey Miscellaneous Field Studies Map MF-1619B, scale 1:50,000.

Parker, J. M., 1983, Concepts and creativity (AAPG Presidential Address): AAPG Explorer, v. 4, n. 10, p. 12–13.

Perry, W. J., D. H. Roeder, and D. R. Lageson, compilers, 1984, North American thrust-faulted terranes: AAPG Reprint Series, n. 27, 466 p.

Petroleum Information Corporation, 1981, The Overthrust Belt—1981: Denver, 251 p.

Platt, L. B., and R. I. Coward, 1982, The Sheep Creek window through the Oquirrh Formation, southeastern Idaho, *in* Geologic studies of the Cordilleran thrust belt: Rocky Mountain Association of Geologists Symposium, v. II, p. 845–849.

Poole, F. G., and C. A. Sandberg, 1977, Mississippian paleogeography and tectonics of the western United States, *in* Pacific coast paleogeography: Pacific Section, Society of Economic Paleontologists and Mineralogists Symposium, v. 1, p. 67–85.

Powers, R. B., 1977, Assessment of oil and gas resources in the Idaho–Wyoming thrust belt, *in* E. L. Heisey, ed., Rocky Mountain thrust belt geology and resources: Wyoming, Montana, and Utah Geological Associations 29th Annual Field Conference Guidebook, p. 629–637.

———, 1983, Petroleum potential of Wilderness lands in Wyoming–Utah–Idaho thrust belt: U.S. Geological Survey Circular 902N, 14 p.

Price, R. A., 1973, Large-scale gravitational flow of supracrustal rocks, southern Canadian Rockies, *in* K. A. DeJong, and R. A. Scholten, eds., Gravity and tectonics: New York, Wiley-Interscience, p. 491–502.

———, 1981, The Cordilleran foreland thrust and fold belt in the southern Canadian Rockies, *in* Thrust and nappe tectonics: Geological Society of London, p. 427–448.

Price, R. A., and E. W. Mountjoy, 1970, Geologic structure of the Canadian Rocky Mountains between Bow and Athabasca rivers—a progress report, *in* J. O. Wheeler, ed., Structure of the southern Canadian Cordilleran: Geological Association of Canada Special Paper, n. 6, p. 7–25.

Rich, R. L., 1934, Mechanics of low-angle overthrust faulting as illustrated by Cumberland thrust block, Virginia, Kentucky, and Tennessee: AAPG Bulletin, v. 18, n. 12, p. 1584–1596.

Rodgers, J., 1949, Evolution of thought on structure of middle and southern Appalachians: AAPG Bulletin, v. 33, n. 10, p. 1643–1654.

———, 1964, Basement and no-basement hypotheses in the Jura and the Appalachian Valley and Ridge, *in* Tectonics of the Southern Appalachians: Virginia Polytechnical Institute, Department of Geological Sciences, Memoir 1, p. 71–80.

Rose, P. R., 1976, Mississippian carbonate shelf margins, western United States: U.S. Geological Survey Journal Research, v. 4, n. 4, p. 449–466.

Royse, F., Jr., 1985, Geometry and timing of the Darby–Prospect–Hogsback thrust fault system, Wyoming (abs.): Geological Society of America Abstracts with Programs, v. 17, n. 4.

Royse, F., Jr., M. A. Warner, and D. L. Reese, 1975, Thrust belt structural geometry and related stratigraphic problems, Wyoming–Idaho–northern Utah, *in* Deep Drilling Frontiers—1975: Rocky Mountain Association of Geologists Symposium, p. 41–54.

Rubey, W. W., 1973, New Cretaceous formations in western Wyoming thrust belt: U.S. Geological Survey Bulletin 1372-I, 35 p.

Rubey, W. W., and M. K. Hubbert, 1959, Role of fluid pressure in mechanics of overthrust faulting. II. Overthrust belt in geosynclinal area of western Wyoming in light of fluid-pressure hypothesis: Geological Society of America Bulletin, v. 70, p. 167–205.

Rubey, W. W., S. S. Oriel, and J. I. Tracey, Jr., 1975, Geology of the Sage and Kemmerer quadrangles, Lincoln County, Wyoming: U.S. Geological Survey Professional

Paper 855, 18 p.

Ruppel, E. T., 1986, The Lemhi arch, a major late Precambrian–early Paleozoic paleotectonic feature, Montana–Idaho: this volume.

Sheldon, R. P., 1963, Physical stratigraphy and mineral resources of Permian rocks in western Wyoming: U.S. Geological Survey Professional Paper 313-B, 273 p.

Silberling, N. J., and R. J. Roberts, 1962, Pre-Tertiary stratigraphy and structure of northwestern Nevada: Geological Society of America Special Paper 72, 58 p.

Speed, R. C., 1977, Island-arc and other paleogeographic terranes of late Paleozoic age in Western Great basin, *in* Pacific coast paleogeography: Pacific Section, Society of Economic Paleontologists and Mineralogists Symposium, v. 1, p. 349–362.

Steidtmann, J. R., D. J. Hurst, and M. W. Shuster, 1984, Structural evolution of part of the Wyoming thrust belt and foreland: a sedimentary–tectonic approach (abs.): Geological Society of America Abstracts with Programs, v. 16, n. 6, p. 667.

Stevens, C. H., 1977, Permian depositional provinces and tectonics, western United States, *in* Pacific coast paleogeography: Pacific Section, Society of Economic Paleontologists and Mineralogists Symposium, v. 1, p. 113–135.

Stewart, J. H., 1972, Initial deposits in the Cordilleran geosyncline: evidence of a late Precambrian (<850 m.y.) continental separation: Geological Society of America Bulletin, v. 83, n. 5, p. 1345–1360.

Stewart, J. H., and C. A. Suczek, 1977, Cambrian and latest Precambrian paleogeography and tectonics in the western United States, *in* Pacific coast paleogeography: Pacific Section, Society of Economic Paleontologists and Mineralogists Symposium, v. 1, p. 1–17.

Stewart, J. H., J. R. MacMillan, K. M. Nichols, and C. H. Stevens, 1977, Deep-water upper Paleozoic rocks in north-central Nevada—a study of the type area of the Havallah Formation, *in* Pacific coast paleogeography: Pacific Section, Society of Economic Paleontologists and

Mineralogists Symposium, v. 1, p. 337–347.

Trümpy, R., 1960, Paleotectonic evolution of the central and western Alps: Geological Society of America Bulletin, v. 71, n. 6, p. 843–908.

———, 1980, An outline of the geology of Switzerland: Twenty-sixth International Geology Congress Guidebook G 10, pt. A, 104 p.

U.S. Geological Survey, 1962, Synopsis of geologic, hydrologic, and topographic results: U.S. Geological Survey Professional Paper 450A, 257 p.

Van Houten, F. B., 1973, Meaning of molasse: Geological Society of America Bulletin, v. 84, n. 6, p. 1973–1976.

Veatch, A. C., 1907, Geography and geology of a portion of southwestern Wyoming: U.S. Geological Survey Professional Paper 56, 178 p.

Ver Ploeg, A. J., and R. H. De Bruin, 1982, The search for oil and gas in the Idaho–Wyoming–Utah salient of the overthrust belt: Geological Survey of Wyoming Report of Investigations, n. 21, 108 p.

Villien, A., and R. Kligfield, 1986, Thrusting and synorogenic sedimentation in central Utah: this volume.

Walker, J. P., 1982, Hogback Ridge field, Rich County, Utah: thrust-belt anomaly or harbinger of further discoveries?, *in* Geologic studies of the Cordilleran thrust belt: Rocky Mountain Association of Geologists Symposium, v. II, p. 581–590.

West, J., and H. Lewis, 1982, Structure and palinspastic reconstruction of the Absaroka thrust, Anshutz Ranch area, Utah and Wyoming, *in* Geologic studies of the Cordilleran thrust belt: Rocky Mountain Association of Geologists Symposium, v. II, p. 633–639.

Wiltschko, D. V., and J. A. Dorr, Jr., 1983, Timing of deformation in overthrust belt and foreland of Idaho, Wyoming and Utah: AAPG Bulletin, v. 67, n. 8, p. 1304–1322.

Wing, S. L., 1984, A new basis for recognizing the Paleocene/Eocene boundary in Western Interior North America: Science, v. 226, p. 439–441.

Thrusting and Synorogenic Sedimentation in Central Utah

A. Villien[1]
R. M. Kligfield
University of Colorado
Boulder, Colorado

New paleontologic data on detrital lithostratigraphic units in central Utah, which are inferred to represent the synorogenic products of compressive deformation, provide a basis for dating the times of transport of each of several thrust sheets. The thrust belt in central Utah can be divided geometrically into four major thrust systems from west to east: the Canyon Range, Pavant, Gunnison, and Wasatch thrust systems. Biostratigraphic correlations and constraints imposed by the geometry of thrust systems suggest the following ages for thrusting events: late Albian for the Pavant 1 thrust, late Santonian–early Campanian for the Pavant 2 thrust, middle to late Campanian for the late Canyon Range thrust, late Maestrichtian for the Gunnison thrust system, and late Paleocene for the Wasatch thrust system.

In the hinterland, a combination of structural, stratigraphic, and chronologic evidence suggests that shortening was accommodated by the development of a backward-breaking thrust sequence: Pavant 1 thrust, Pavant 2 thrust, and (late) Canyon thrust. This led to the formation of successive overlapping unconformities of late Cenomanian, early to middle Campanian, and late Campanian age. In the foreland, the Gunnison thrust system has a ramp-flat geometry; a series of blind, splay, imbricate faults are associated with a major ramp beneath Sevier and Sanpete valleys. Late Cretaceous and Paleocene unconformities coincide with the development of an imbricate fan structure, which was subsequently deformed during late Paleocene time by formation of a deeper duplex structure within the Wasatch thrust system. Associated back thrusts accommodated shortening toward the surface at the west side of the Wasatch Plateau.

Tectonic loading produced a foreland basin and a correlative forebulge located east of the toes of the major thrust sheets. The times of superimposed thrusting phases, when compared with eustatic episodes recorded in the Cretaceous seaway, indicate that episodes of continental tectonism were approximately synchronous with eustatic rises in central Utah. Overall structural zonation and sedimentation patterns in central Utah can be correlated in a general sense with similar features in the Wyoming thrust belt salient farther north, but some significant differences are that some of the compressive stresses in central Utah were accommodated by such backward-breaking thrusts as the Pavant 2 fault.

INTRODUCTION

The Mesozoic to early Tertiary evolution of central Utah was dominated by the effects of large-scale thrust faulting, accompanied by syntectonic erosion and deposition. Subsequent extensional tectonics during the Tertiary has severely dissected the geometry produced during that time and led to discontinuous exposures of thrust belt structures. Nonetheless, a sufficient amount of subsurface information from seismic profiling and well log analyses is now available to attempt to link remnants of individual thrusts into a single, overall thrust system network.

Pieces of this framework are well known from numerous published papers. In the western part of the study area

(Figures 1 and 2), the Canyon and Pavant thrusts have been described by Christiansen (1952) and Crosby (1959) and their regional significance outlined by Armstrong (1968) and Burchfiel and Hickcox (1972) as part of the Sevier orogenic deformational events (Early to Late Cretaceous). More recently, detailed mapping in the Pavant Range synthesized by Baer et al. (1982) shows the surface thrust geometry, whereas Allmendinger et al. (1983) have summarized results from COCORP seismic reflection data that illuminate deep relationships between Mesozoic and Cenozoic structures within and west of the Sevier Desert. In the Leamington area between the Gilson Mountains and the Canyon Range, the surface geology has been summarized by Higgins (1982) and Morris (1983).

The tectonic evolution of the eastern margin of central Utah (Figures 1 and 2) has long been a subject of attention. A complex deformation model involving 16 episodes of crustal movement (some of regional importance and others

[1]Société Nationale Elf Aquitaine (Production), St. Martory, France.

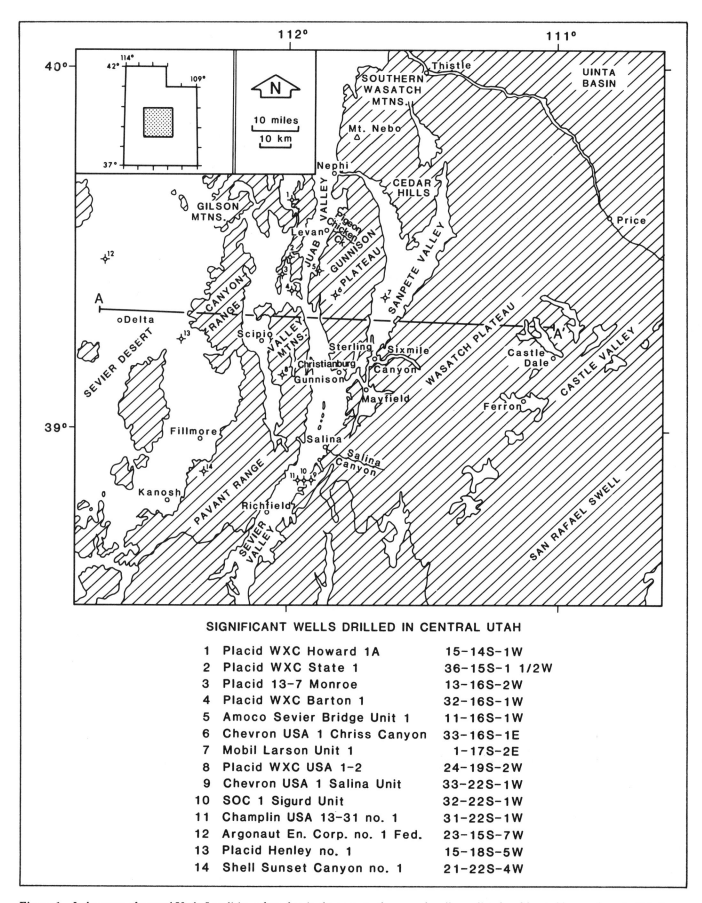

SIGNIFICANT WELLS DRILLED IN CENTRAL UTAH

1	Placid WXC Howard 1A	15–14S–1W
2	Placid WXC State 1	36–15S–1 1/2W
3	Placid 13–7 Monroe	13–16S–2W
4	Placid WXC Barton 1	32–16S–1W
5	Amoco Sevier Bridge Unit 1	11–16S–1W
6	Chevron USA 1 Chriss Canyon	33–16S–1E
7	Mobil Larson Unit 1	1–17S–2E
8	Placid WXC USA 1–2	24–19S–2W
9	Chevron USA 1 Salina Unit	33–22S–1W
10	SOC 1 Sigurd Unit	32–22S–1W
11	Champlin USA 13–31 no. 1	31–22S–1W
12	Argonaut En. Corp. no. 1 Fed.	23–15S–7W
13	Placid Henley no. 1	15–18S–5W
14	Shell Sunset Canyon no. 1	21–22S–4W

Figure 1—Index map of central Utah. Localities referred to in the text are shown and wells are listed and located by number. A–A′ is line of cross section in Figure 8. Diagonal lines indicate areas of pre-alluvium exposures.

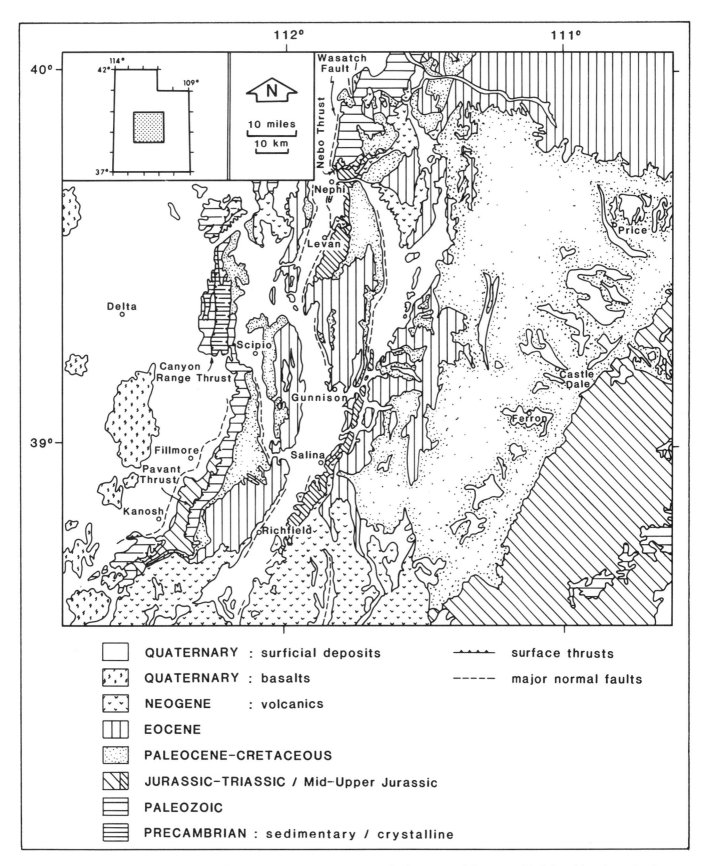

QUATERNARY : surficial deposits

QUATERNARY : basalts

NEOGENE : volcanics

EOCENE

PALEOCENE-CRETACEOUS

JURASSIC-TRIASSIC / Mid–Upper Jurassic

PALEOZOIC

PRECAMBRIAN : sedimentary / crystalline

surface thrusts

major normal faults

Figure 2—Geologic map of central Utah showing same areas as in Figure 1. Surface traces of thrusts are labeled, and locations of major Basin and Range normal faults are represented. Paleocene–Cretaceous strata are shown by a single symbol. If late Tertiary extension is omitted, such Paleocene–Cretaceous exposures from the Pavant Range, Valley Mountains, Canyon Range, Gunnison Plateau, and Cedar Hills form a northeast-trending clastic wedge that parallels Wasatch Plateau exposures. Map adapted from Hintze (1973b). Used by permission of the Brigham Young University Department of Geology.

very localized) lasting from Late Jurassic to Pleistocene time was invoked by Spieker (1946, 1949) and his students. Subsequently, controversy developed regarding the roles and magnitude of salt tectonics. Stokes (1952, 1956, 1982), Moulton (1975), Baer (1976), and Witkind (1982) have argued that diapirism of the Jurassic Arapien Shale played a dominant role in producing the structures observed within the Gunnison Plateau and Sevier and Sanpete valleys. This problem has also been discussed by Weiss (1982).

In contrast, Lawton et al. (1982) and particularly Royse (1983) and Standlee (1982, 1983) interpret subsurface data as indicating that eastward-directed thrusting affected central Utah east of the Gunnison Plateau in a classic pattern of deformation involving foreland clastic deposition, with some westward-directed backthrusting in the Sanpete Valley. Our data support the latter interpretation, although thrust relationships undoubtedly have been modified locally by episodes of Cenozoic diapiric movements.

Central Utah is also studied from the point of view of synorogenic sedimentation. Deposition of conglomerates and clastic sediments during the Sevier–Laramide deformational events (Late Jurassic–Eocene) has been discussed by Hintze (1973a), Lawton (1982), and Fouch et al. (1982, 1983).

Our purpose here is to provide a stratigraphic and tectonic framework for central Utah based on a combination of surface and subsurface data. The results of additional surface mapping, available drill hole records and general depth interpretations help to define a structural overview of thrusting and sedimentation, although seismic reflection profiles are still proprietary and cannot be published here. We proceed by examining paleontologic ages that date syntectonic conglomerates and clastic sediments. We then describe the geometric relationships between these synorogenic sediments and the individually studied major thrusts. Together, these data provide constraints for a unified thrust system model for tectonics and sedimentation in central Utah. More detailed descriptions of the localities and their paleontologic assemblages have been given in Villien (1984).

STRATIGRAPHY

Post-Cretaceous sedimentary and volcanic rocks (Figure 2) overlie and mask much of the lithostratigraphic and biostratigraphic relationships of Cretaceous strata in central Utah. This is in contrast to the situation farther north where Upper Jurassic–Cretaceous stratigraphic units are exposed and easily studied (Peterson, 1972; Imlay, 1980; Nichols and Jacobson, 1982a, b). Rock units such as the Twist Gulch, North Horn, and Flagstaff formations exhibit rapid facies changes whose spatial patterns are not yet well understood. As a result, the units cannot be traced laterally over the entire region with assurance on the basis of lithology. To ascertain biostratigraphic correlations for the Upper Jurassic–Cretaceous rocks of central Utah, we have reexamined the ages of exposed surface sections and supplemented these data with information from wells whose locations are indicated in Figure 1. The ages of critical parts of surface sections and drill holes have been reexamined by

Elf Aquitaine paleontologists, who obtained new dates from palynomorphs and microfauna.

The correlations proposed in this paper are largely biostratigraphic correlations, depending on the accuracy of the reported age results. In the sections that follow, we examine the stratigraphic record for four intervals: Middle–Late Jurassic, Early Cretaceous, Late Cretaceous, and latest Cretaceous–Eocene (Figure 3). In each discussion, we begin by identifying the widespread occurrence of rocks of similar lithofacies in both surface sections and in wells; the ages obtained from palynology and faunal assemblages are then used to ascertain the relationship of these lithocorrelations into biostratigraphic correlations (see Figures 4–6). This information is then used in a later section to identify the signature of thrusting phases in the stratigraphic record and to discuss the timing of thrusting phases.

Middle to Late Jurassic

Rocks of Middle to Late Jurassic age were studied in the Chicken Creek section of the Gunnison Plateau and in the Placid WXC Howard #1A (sec. 5, T. 14 S., R. 1 W.) (locations given in Figure 1 and stratigraphic columns shown in Figure 4). Samples from both areas contain palynomorphs and calcareous nannofossils whose ages are indicated in Figure 4.

Spieker (1946) assigned the Arapien Formation to the Middle and Late Jurassic and divided it into the Twist Gulch Member and Twelvemile Canyon Member. This formation was subsequently studied by Hardy (1952) who wanted to elevate the rank of the Twist Gulch to a formation. More recently, Sprinkel (1982) and Sprinkel and Waanders (in press) have assigned basal Arapien Shale strata on the basis of electric and lithologic logs and palynologic assemblages to the upper two members of the Twin Creek Limestone of northern Utah—the Leeds Creek and Giraffe Creek members. They indicated that the Arapien Shale is underlain by the lower five members of the Twin Creek. Three dinoflagellate assemblages were recognized: the Bajocian assemblage is associated with the Sliderock and Rich members, the Bathonian assemblage characterizes the Boundary Ridge and Watton Canyon members and the lowermost part of the Arapien Shale, and the Callovian assemblage has been recognized in the Arapien Shale.

From 0 to 300 m (1000 ft) in the Chicken Creek section (Figure 4), the lithologic units encountered (mainly limestone) comprise type 1 of Spieker (1946) and have been assigned by Sprinkel and Witkind (1982) to the Twin Creek Limestone (Leeds Creek Member?). Thin-bedded limestone and shale with interlayered gypsum are present between 0 and 150 m (500 ft). Palynomorphs indicate that this section is of Bathonian age. These strata are overlain between 150 and 300 m (500 and 1000 ft) by interbedded limestone and mudstone with gypsum. At a level of approximately 300 m (1000 ft), late Callovian palynomorphs were found in argillaceous limestone, suggesting that the interval between 150 m and 300 m is probably of Callovian age.

The overlying strata from 300 to 1200 m (1000 to 4000 ft) (Figure 4) in the Chicken Creek section (mainly mudstone, gypsiferous shale, and sandstone) resemble those in types 2, 3, and 4 of Spieker (1946), and are assigned by

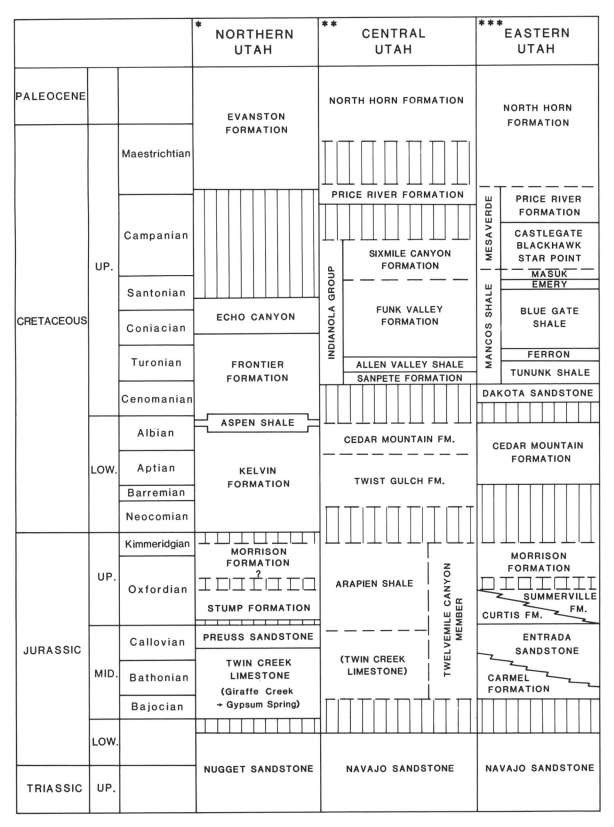

Figure 3—Regional correlation chart for Jurassic, Cretaceous, and lower Tertiary units in central Utah and adjacent areas. An Early Cretaceous age for the Twist Gulch Formation is documented by dating; late Maestrichtian–Paleocene attribution for the North Horn Formation is hypothetical. This informal classification for central Utah is discussed in text.

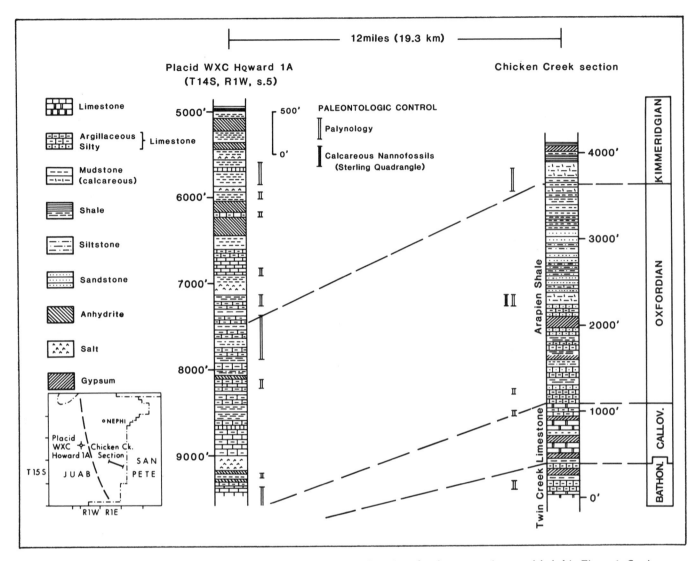

Figure 4—Biostratigraphic correlations for the Middle and Late Jurassic. Locations for the two sections are labeled in Figure 1. Cuttings from the Placid WXC Howard #1A borehole have been described by D. A. Sprinkel. Dashed line on the small index map represents a boundary between open and restricted marine environments for the early Oxfordian stage.

Sprinkel and Witkind (1982) to the Arapien Shale. Interbedded argillaceous limestone and silty mudstone present between 300 and 540 m (1000 and 1800 ft) have yielded early Oxfordian palynomorphs. Calcareous mudstone and siltstone containing some coarse-grained conglomeratic sandstone are present between 660 and 1050 m (2200 and 3500 ft) and contain palynomorphs and calcareous nannofossils of Oxfordian age. This unit, which forms the upper part of Spieker's Twelvemile Canyon Member in central Utah, cannot be temporally correlative with the Preuss and Entrada sandstones because of their age disparities. Instead, they must be temporally equivalent to the Stump and Curtis formations of northern Utah and eastern Utah, respectively (Figure 3).

Similar sequences, with some facies variations, are found in the Placid WXC Howard #1A well (Figure 4). The lower part of the section (below ~2800 m, or 9400 ft) contains mainly carbonates and was assigned by Sprinkel (1982) to the Twin Creek Limestone. The upper part (up to 1500 m,

or 5000 ft), however, probably represents the Arapien shale including argillaceous limestone and calcareous mudstone with salt and anhydrite. Palynomorphs indicate that the strata below ~2800 m are Callovian and older, whereas between 2800 and 2250 m (7500 ft), they are of Oxfordian age. These are overlain by red, salt-bearing shale, mudstone, and siltstone that form a distinctive unit of Kimmeridgian age.

The lithologic differences between the Howard #1A well and the Chicken Creek section for the early Oxfordian stage are probably related to differences in local depositional environments. Before thrusting, the Howard #1A sequence was west of the palinspastic position of the Chicken Creek section. The early Oxfordian strata of the Howard #1A sequence probably were deposited in a basin, while the Chicken Creek section of the same age in the Gunnison Plateau represents nearshore deposits. During late Oxfordian–early Kimmeridgian time, both sections record a transgression represented by calcareous mudstone. At both places the red, gypsum-bearing shale, siltstone, and sandstone of Kimmeridgian age represent the youngest Jurassic strata in central Utah and are equivalent to the Morrison Formation in eastern Utah (Figure 3).

The apparent continuity of Middle to Late Jurassic sedimentation in central Utah thus provides no evidence of major thrusting phases. If there was deformation of Middle to Late Jurassic age, then it must have occurred farther to the west.

Early Cretaceous

Strata of Early Cretaceous age were studied at the Chicken Creek and Christianburg sections at the northwestern and southern ends of the Gunnison Plateau, respectively (Figures 1 and 5). In addition, cuttings from the Placid WXC State #1 well located west of the Juab Valley (Figures 1, 5, and 8) were examined.

At Chicken Creek, interbedded gray-green and red shale and sandstone, some conglomeratic and some calcareous, are found in strata that Hardy and Zeller (1953) mapped to the south as the Twist Gulch Member (Figure 5). Samples from gray-green sandstone with plant fossils have yielded nonmarine palynomorphs suggesting an Aptian–Albian age. The nature of the underlying contact with Kimmeridgian strata is concealed by Quaternary deposits along Chicken Creek and the neighboring Pigeon Creek sections. North of Nephi (Figure 1), however, a minor unconformity between similar mudstone beds and underlying Arapien Shale has been reported (Le Vot, 1984). These data suggest that the Twist Gulch Member, as defined by Spieker (1946) and mapped by Hardy and Zeller (1953), is thus of Early Cretaceous age (Figure 3).

The beds of the Twist Gulch Formation at Chicken Creek are overlain by red-orange siltstone, conglomerate (labeled A″ in Figure 5), and freshwater limestone. Thin-bedded, gray mudstone beds found between conglomeratic sandstone beds contain a flora of late Albian age (Standlee, 1982). These beds have been previously mapped by Hunt (1950) as the Upper Cretaceous Indianola Group. Standlee (1982) has assigned this unit to the Cedar Mountain Formation, which must now be considered as late Albian age on the Gunnison Plateau.

Similar lithofacies tentatively assigned to the Cedar Mountain Formation are found in the Christianburg section to the south (Figures 1 and 5). The section described contains abundant conglomerate beds (A′ in Figure 5) containing upper Paleozoic chert pebbles and freshwater limestone beds. Algal stromatolites and nodular oncolites as well as gastropods and charophytes occur on a distinctive ledge on the east front of the San Pitch Mountains. Samples from dark gray calcareous mudstone (NW¼, sec. 18, T. 19 S., R. 2 E.) have yielded nonmarine palynomorphs of Early Cretaceous age (Figure 5). Late Albian nonmarine palynomorphs have also been found by Placid (D. Sprinkel, personal communication, 1983). The sequence of rocks at Christianburg has previously received different attributions: Spieker (1946) assigned these rocks to the Morrison(?) Formation, Gilliland (1963) chose the Sanpete Formation, and Taylor (1980) the North Horn Formation. On the basis of the above evidence, we conclude that the section is of Aptian–Albian age, as is the previously described Chicken

Creek section, and that it is equivalent to the Cedar Mountain Formation.

Cuttings from the Placid WXC State #1 well (sec. 36, T. 15 S., R. 1½ W.) (Figures 1, 5, and 8) within a sequence of shale, siltstone, and partly conglomeratic sandstone (A‴ in Figure 5) contain Early Cretaceous palynomorphs. This age is in excellent agreement with results found in the nearby Placid WXC Barton #1 well (Figure 1) (D. Sprinkel, personal communication, 1983).

Thus, widespread conglomerate (A′, A″, and A‴ in Figure 5) beds in central Utah appear to be of late Albian age. Thickness variations between the Chicken Creek–Christianburg surface sections and the WXC State #1 well can probably be attributed to differences in the dispersal directions of the alluvial fans that existed at that time (map on Figure 5).

Late Cretaceous

Fouch et al. (1983) have described temporal and depositional relationships among Upper Cretaceous rocks within and east of the Sanpete Valley and along the southern margin of the Uinta basin. In the Gunnison Plateau, Valley Mountains, and Pavant Range (Figure 1) regions of central Utah, overlying Tertiary deposits, structural complexities, and sparsity of subsurface information have hindered the study of Upper Cretaceous rocks. Data obtained from well-exposed strata in the Chicken Creek section and the Sixmile Canyon section east of Sterling (Figure 1) are discussed below, followed by their implications for units represented by cuttings from the Placid WXC USA #1-2 well (sec. 24, T. 19 S., R. 2 W.).

The lower parts of the Sixmile Canyon and Chicken Creek sections contain rocks of the Sanpete Formation and the Indianola Group, respectively (Figures 3 and 6). No Cenomanian fossils have been recovered in either section. At Sixmile Canyon, the Sanpete Formation is characterized by sandstone that grades upward into littoral marine beds. In the Placid WXC Barton #1 well (sec. 32, T. 16 S., R. 1 W.) beneath the Juab Valley (Figure 1), marginal marine strata with palynomorphs of Cenomanian–Turonian age unconformably overlie Lower Cretaceous continental rocks (D. Sprinkel, personal communication, 1983). The presence of coarse conglomerate locally at the base of the Sanpete Formation in Sixmile Canyon and its absence in the Barton #1 well suggest that it may be a fan delta of limited areal extent, rather than a major synorogenic conglomerate.

Fouch et al. (1983) have suggested that the Sanpete may be a lateral equivalent of and analogous to the Dakota (Figure 3) and therefore of Cenomanian age. Strata in the Barton #1 well are similar to those found in the lowest parts of both the Chicken Creek and Sixmile Canyon sections shown in Figure 6. All three sections record a continental to marine facies transition, suggesting that the region was uplifted during early Cenomanian time prior to marine deposition. Evidence of this event in the form of an unconformity or continental to marine transition should have been preserved at the top of the WXC State #1 well (Figure 5), but the highest part of this section has been eroded and only Tertiary strata are now found.

In the Sixmile Canyon area, the upper marine beds of the Sanpete Formation contain *Inoceramus labiatus* of early

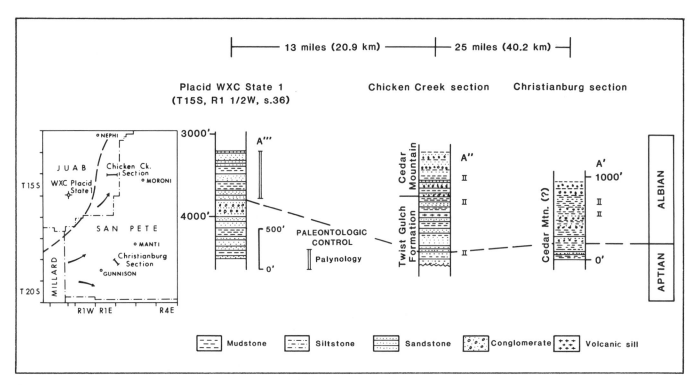

Figure 5—Surface to subsurface biostratigraphic correlations for part of the Early Cretaceous. Locations for the three sections are also labeled on Figure 1. Borehole cuttings were described by D. A. Sprinkel. The Aptian–Albian boundary is approximate and is more representative of the base of synorogenic deposits of late Albian age. Arrows on the small index map represent directions of development of alluvial fans which are similar to Early Cretaceous paleodrainage directions, mentioned by McCubbin (1961).

Turonian age. These beds intertongue with sandy marine shales of the Allen Valley Shale (Figures 3 and 6), which have yielded calcareous nannofossils of early Turonian age. This confirms the age equivalence between the Tununk Member of the Mancos Shale and the Allen Valley Shale.

The open marine environments inferred for these strata may represent one of the global transgressive maxima (T_6, latest Cenomanian–early Turonian, of Kauffman, 1973, 1977).

A 180 m (600 ft) sequence of variegated shale, sandstone, and limestone containing nonmarine palynomorphs of Cenomanian–Turonian(?) age is found at the bottom of the Placid WXC USA #1-2 section (Figure 6). These strata are unconformably(?) overlain by conglomerate (probably Santonian in age, as discussed later). Ryer (1983) has pointed out that the thickness of the Tununk increases from the Wasatch Plateau toward the southwest from 90 to 240 m (300 to 800 ft). The absence of strata equivalent to the Tununk in the USA #1-2 well may be interpreted as evidence of local uplift in this region. Tununk–Allen Valley strata may extend westward beneath the Sevier Desert.

In the Sixmile Canyon region, Allen Valley sandy shale grades upward into interbedded sandstone and shale of marine, nearshore facies, which are in turn overlain by sandy shale and bioturbated sandstone representing the Funk Valley Formation (Figure 6, section from 450 to 1200 m, or 1500 to 4000 ft). The lowest sandy beds of the Funk Valley near Salina contain *Collignoniceras woolgari* of early middle Turonian age (Ryer, 1983). The middle part of the Funk

Valley Formation at Sixmile Canyon (Figure 6, 780-m, or 2600-ft level) contains *Inoceramus deformis* of Coniacian age (Fisher et al., 1960).

The situation in Chicken Creek and Pigeon Creek areas of the Gunnison Plateau is not as clear because of the lack of paleontologic control. Lawton (1982) reports lithologic correlations between the Sanpete and Funk Valley formations and a sequence of the following rocks (in ascending order): (1) conglomeratic, braided, and fluvial facies; (2) alluvial fan facies; (3) facies (1) and (2) repeated; and finally, (4) lagoonal–deltaic facies. In our view, the lowest part of Lawton's sequence is temporally equivalent to the uppermost parts of the Cedar Mountain Formation of late Albian and perhaps as young as early Cenomanian age (Figure 5, A″). We believe that the overlying upward-coarsening sandstone units can be best assigned late Cenomanian–Turonian and possibly Coniacian ages (Figure 6). Such a correlation is consistent with the presence of marginal marine beds of similar age found in the Placid Barton WXC #1 well nearby.

Marine Funk Valley strata in the Sixmile Canyon section are overlain by continental deposits of the lower member of the Sixmile Canyon Formation at about the 1200 m (4000 ft) level (Figure 6). The nature of this contact has been interpreted in different ways: Pinnell (1972) reported the presence of an unconformity near Thistle (Figure 1) at the base of beds assigned to the Sixmile Canyon Formation; Lawton (1982) proposed that the rapid change from upper Funk Valley to Sixmile Canyon Formation can be best

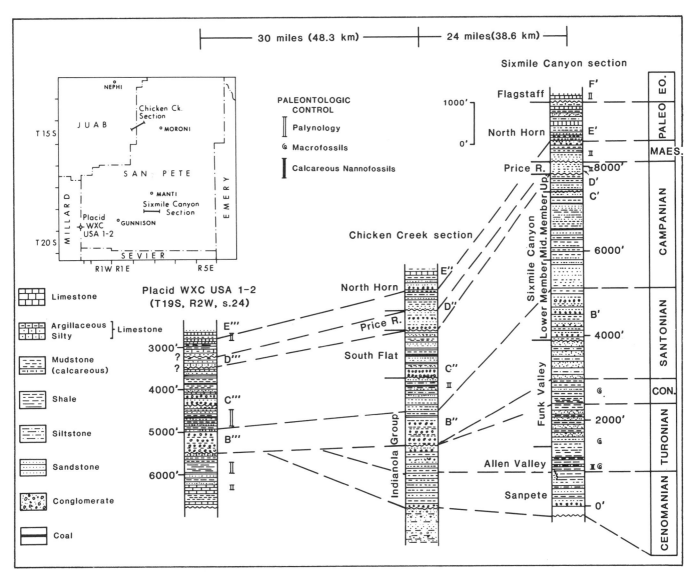

Figure 6—Biostratigraphic correlations for the Late Cretaceous and Early Tertiary. Locations for the three sections are labeled on Figure 1. Unconformities, with or without angular relationships, are represented. Absence of Coniacian strata in the Chicken Creek section and in the Placid WXC USA #1-2 wells hypothetical, and additional dates might change such correlation. The presence of Santonian beds in the Chicken Creek section has not yet been proven. Letter symbols refer to synorogenic deposits discussed in text.

interpreted as a widespread hiatus; and Fouch et al. (1983) identified *Clioscaphites* zones of middle Santonian age in sandy, littoral marine sandstone and cobble conglomerate of the Indianola(?) Group. They consider this sequence of rocks to represent the youngest marine beds in the northern Wasatch Plateau and to be a part of the upper Funk Valley Formation (Figure 3).

We have examined samples of calcareous shale at two different localities (NW¼ sec. 8, T. 19 S., R. 2 E.) in beds previously mapped as Arapien Shale by Hardy (1952) near Sterling (Figure 1). Both localities yielded an abundant microflora of Late Jurassic age, but also contained calcareous foraminifera like *Globotruncana arca* (Cushman) of Santonian–Campanian age. The best interpretation is that the palynologic assemblages have been redeposited and that the youngest occurrence of marine beds in the Sanpete Valley may be as young as late Santonian. This is temporally

equivalent to the former Masuk Member at the top of the Mancos Shale (Figure 3).

The age of the detrital beds and conglomerates of the lower member of the Sixmile Canyon Formation (Figure 6, 1200–1500 m, or 4000–5000 ft, B′) has been ascertained. No distinctive fossils have been recovered from the upper Funk Valley and the lower Sixmile Canyon formations (Fouch et al., 1983). However, samples from Sixmile Canyon (Figure 6) of carbonaceous shales of the Sixmile Canyon Formation (coal-bearing member at 1920 m, or 6400 ft) yielded palynomorphs of Campanian age, the age of the Blackhawk Formation (Figure 3). Therefore, the alluvial fan conglomerates (B′) must have been deposited in late Santonian or possibly early Campanian time.

The above relationships suggest that during middle Santonian time, while claystone of the Blue Gate Member and the regressive marine Emery Sandstone were deposited along Castle Valley (Figures 1 and 3), nearshore, bioturbated sands of the upper Funk Valley Formation accumulated along the western edge of the Wasatch Plateau. All of these regions are overlain by alluvial fan deposits represented by sandstone with conglomerate lenses and coal-bearing beds of Campanian age in the Sixmile Canyon area. However, in the Sanpete Valley, marine claystone and siltstone representing an upper tongue of the Blue Gate Member ("Masuk Member") of late Santonian appears to be present. This documents that the marine transgression represented could extend this far west during late Santonian time.

Similar facies relationships are found in the Valley Mountains and in the Chicken Creek section. In the WXC USA #1-2 well, a thick succession of conglomerate containing pebbles of various lithologies, such as carbonate and quartzite (B''' in Figure 6), is overlain by sandstone containing conglomeratic and coal-bearing beds. Cuttings containing palynomorphs of Campanian age were found above the thick conglomerate unit (Figure 6). The entire section rests unconformably on Cenomanian–Turonian(?) shale, sandstone, and limestone. The lithologic similarities among the Sixmile Canyon section and related regions, the Chicken Creek section, and the Placid USA #1-2 well indicate the presence of a widespread major conglomerate (Figure 6, B) of late Santonian age throughout central Utah.

The lowest coal-bearing member can be traced from the Wasatch Plateau (Blackhawk Formation, middle member of the Sixmile Canyon Formation), westward to the Gunnison Plateau (lower coal unit below the South Flat Formation), and to the Valley Mountains (coal-bearing units at 1500–1350 m, or 5000–4500 ft, in the WXC USA #1-2 well) (Figure 6). In both the Sixmile Canyon section (C' in Figure 6) and the Chicken Creek section (C''), overlying conglomeratic sandstone containing quartzite pebbles was deposited and partially eroded. Fouch et al. (1983) have suggested that the upper beds of the Sixmile Canyon Formation (quartzose rocks) in the Sixmile Canyon area may be the erosional remnants of the Castlegate Sandstone (Figure 3). The Castlegate Sandstone contains the fossil mollusk *Baculites asperiformis* of early late Campanian age at its base in Price Canyon in Castle Valley (Figure 1). These conglomerate and sandstone beds (C', C'', and C''' in Figure 6) (Castlegate remnants) directly overlie lower Campanian coal-bearing strata and grade from quartzose beds (in the Sixmile Canyon section and the WXC USA #1-2 well) northward into alluvial fan conglomerate containing mostly quartzite pebbles (in the Chicken Creek section). The widespread occurrence of these rocks throughout central Utah indicates that the event producing them was regional in scope and occurred early in late Campanian time (Fouch et al., 1983).

Along the western edge of the Wasatch Plateau, the uppermost beds of the Sixmile Canyon Formation are overlain by conglomerate of the Price River Formation (Figure 3 and D' in Figure 6). These have yielded palynomorphs of latest Campanian age from carbonaceous siltstone in Sixmile Canyon (Fouch et al., 1983). Although the pebbly sandstone beds there appear to be concordant with the underlying Sixmile Canyon Formation, there may be a hiatus between the two formations. However, westward near Christianburg (Figure 1), quartzite conglomerate

(NE¼, sec. 13, T. 19 S., R. 1 E.) unconformably overlies the Lower Cretaceous conglomerates shown in Figure 5 (A'). This quartzite conglomerate has been identified as the Price River Formation (Witkind, 1982).

A sequence of dominant sandstone with minor shale beds (D'') is found at the 1050 m (3500 ft) level of the Placid WXC USA #1-2 well (Figure 6). The lower conglomerate unit in the Canyon Range (Figure 1) (Unit A of Stolle, 1978) may be of the same age and similar in composition. (This possible correlation is inferred from the dominant quartzitic clasts.)

Fouch et al. (1983) have proposed that upper Campanian synorogenic deposits developed during thrusting in the west. Paleontologic data are lacking for the above-named deposits, but tentative lithocorrelations would link the quartzite, sandstone, and conglomerate in the USA #1-2 well, the Chicken Creek section (D''), and the Price River section of the Sixmile Canyon section (D') (Figure 6) and relate them to an early to late Campanian synorogenic episode.

The widespread occurrence of three Late Cretaceous synorogenic deposits is summarized in the correlations of Figure 6: late Santonian (B), early to late Campanian (C), and latest Campanian (D).

Latest Cretaceous to Early Tertiary

Uppermost Cretaceous and basal Tertiary strata are present in both the Sixmile Canyon and Chicken Creek sections; only a few hundred feet of strata are found in the top of the Placid WXC USA #1-2 well (Figure 6). The North Horn Formation consists largely of conglomerate in the Canyon and Pavant ranges. It grades eastward into sandstone, shale, and limestone on the Gunnison Plateau. The upper part of the Placid WXC USA #1-2 well penetrated the North Horn Formation; cuttings of carbonaceous siltstone yielded palynomorphs of latest Cretaceous–early Tertiary age (Figure 6, E'''). On the Gunnison Plateau, Standlee (1982) reported Maestrichtian palynomorphs from samples collected directly above the unconformity between the North Horn and Price River formations in the Sixmile Canyon region (below E' in the Sixmile Canyon section of Figure 6). In the region between the Sanpete Valley and the San Rafael Swell, Maestrichtian lacustrine beds form the lower part of the North Horn Formation, whereas palynomorphs of early Eocene age occur in limestone overlying the intertonguing boundary between the Flagstaff and North Horn formations (Fouch et al., 1983). In summary the North Horn Formation, a syntectonic unit, is probably of late Maestrichtian to middle or late Paleocene age (Figure 6).

Figure 6 indicates a probable late Paleocene–early Eocene age for the Flagstaff Limestone (F'). Late Paleocene palynomorphs were reported by Fouch et al. (1983) for conglomeratic beds of the Flagstaff Formation at Dark Canyon in easternmost Utah. In Sixmile Canyon, the lacustrine limestone that unconformably overlies Upper Cretaceous beds have yielded palynomorphs of early Eocene age. Farther to the west in the Canyon and Pavant ranges, the North Horn and Flagstaff form a continuous interfingering depositional sequence. Stolle (1978) reported a Paleocene–Eocene age for this sequence (his Unit B).

THRUST SYSTEM GEOMETRY

The simplified structural map of central Utah (Figure 7) can be divided into two major areas—the foreland and the hinterland. The foreland part of the thrust belt extends from the Wasatch Plateau into the Sevier and Juab valleys, whereas the hinterland part includes the western Valley Mountains, the area covered by the Canyon and Pavant ranges, and the far western mountains.

A generalized cross section from the Sevier Desert region in the west to the Wasatch Plateau in the east is shown on Figure 8. The upper cross section shows the relationships between thrusts, stratigraphy, and normal faults. The lower cross section synthesizes the Cretaceous–Paleocene thrust geometry. For ease in following the discussions of the structural geology given in the next sections, we briefly preview the nomenclature used in this paper to describe the thrust system geometry.

The following structures are found in the cross section of Figure 8: ramps, flats, imbricate splay faults, emergent thrusts, blind thrusts, horses, duplex structures, and backthrusts. Our usage is in general agreement with that reviewed by Boyer and Elliott (1982) and Butler (1982). Individual named thrusts (Figure 8, bottom) that are parallel to formation contacts in incompetent strata are known as *flats*; where they cut steeply upward across competent horizons in the stratigraphy they are called *ramps*. The overall geometry resembles the staircase thrust topography described in the Canadian Rocky Mountains (Bally et al., 1966; Dahlstrom, 1970; Price, 1981), the Wyoming portion of the Cordilleran thrust belt (Armstrong and Oriel, 1965; Royse et al., 1975; Dixon, 1982), the Moine thrust system of Scotland (Elliott and Johnson, 1980), and the Helvetic Nappes of Switzerland (Ramsay et al., 1983). Where the thrusts climb upsection and break the erosion surface, they form *imbricate fan structures* (Boyer and Elliott, 1982) in either *forward-breaking* or *backward-breaking* sequences. If the thrust displacement ceases before reaching the surface, it is known as a *blind thrust* (Thompson, 1981). A *horse* is a body of rock bounded on all sides by splay faults; it often contains a hanging-wall anticline and footwall syncline pair of folds. A package of horses, bounded by roof and floor thrusts, is known as a *duplex structure* (Elliott and Johnson, 1980; Boyer and Elliott, 1982); duplexes form in either forward- or backward-breaking sequences. *Backthrusts* refer here to eastward-dipping thrusts whose sense of motion (east-to-west vergence) is opposite to, but coeval with, the large majority of thrusts in this part of Utah.

Four major thrust systems are shown in the bottom cross section of Figure 8. Their structural geometries are described in detail in the following sections and in Figures 9, 10, and 11. From west to east (from the hinterland toward the foreland), these thrust systems are the Canyon Range, the Pavant (I and II), the Gunnison, and the Wasatch. We refer to the package of rock contained between two of these thrust systems as a *plate*. Thus, the rocks overlying the Gunnison thrust system but underlying the Pavant thrust system form the Gunnison plate.

As illustrated in Figure 8, the hinterland part of the thrust belt contains major horizontal faults in the subsurface that form parts of the Canyon, Pavant, and Gunnison thrust systems. Total shortening is on the order of 200 km and may

be more, although these allochthonous sheets have been disrupted by major extension along westward-dipping normal faults. The best known example of these normal fault systems is the Sevier Desert detachment. Its geometry, outlined originally by McDonald (1976), has been further elucidated on the basis of the west-central Utah COCORP line (Allmendinger et al., 1983).

The foreland part of the thrust belt is characterized by thrust faulting that has only recently been recognized, largely on the basis of seismic and drill hole data (Standlee, 1982, 1983; Royse, 1983). On the west flank of the Gunnison Plateau, the Gunnison thrust system may sole into a flatlying or moderately westward-dipping décollement zone at about 7–8 km depth (Figure 8). This could represent the level at which Cambrian shale beds act as detachment horizons; alternatively, it may sole into a brittle–ductile transition zone of the basement at about 10–12 km depth, as shown in Figure 8. The Gunnison plate (shown lying beneath the Gunnison Plateau in Figure 8) is itself underlain by a duplex structure belonging to the Wasatch thrust system.

The Wasatch plate duplex structure is overlain by a complex series of blind, imbricate splay faults of the Gunnison system at its western end. The eastern edge of deformation has propagated within the incompetent Arapien Shale, which led to the structural decoupling between the strata above and below the Middle to Upper Jurassic strata. Some backthrusting in the Sanpete and Sevier valleys appears to be geometrically linked to duplex formation beneath the Gunnison Plateau. Late Tertiary extension and normal faults that appear to affect the basement, possibly along weak zones inherited from late Precambrian rifting, and late Paleozoic Ancestral Rocky Mountain deformation further complicate the picture. Total shortening in the foreland province appears to be on the order of 20 km.

To illustrate key tectonic and sedimentologic relationships, we discuss three areas in detail. These are the Sevier Valley near Salina, the Sanpete Valley near Sterling, and the western Pavant Range near Fillmore (Figure 1).

Hinterland

Parts of three thrust systems are found in the hinterland: the Canyon and Pavant thrust systems are mostly exposed at the surface, and the western part of the Gunnison thrust system is found at depth (Figure 8). The Pavant thrust system is further subdivided into two plates based on the presence of the Pavant 1 thrust (lower) and the Pavant 2 thrust (upper); both are overlain by the Canyon Range thrust (Figure 8).

The deeper Pavant plate is bounded by the Pavant 1 and 2 thrusts. The flatlying Pavant 1 thrust may lie directly above Middle to Upper Jurassic strata in its footwall, which apparently were drilled in the frontal part (Placid WXC USA #1-2). The presence of Arapien beds to the west is likely, as they represent ideal gliding surfaces and might explain the evaporites found in the Sevier Desert (Argonaut Energy Corp. #1, sec. 23, T. 15 S., R. 7 W.) of Tertiary(?) age (Mitchell, 1979). To the west, where this Pavant 1 thrust is displaced by the Sevier Desert detachment, salt flowing along the detachment from an Arapien source could have led

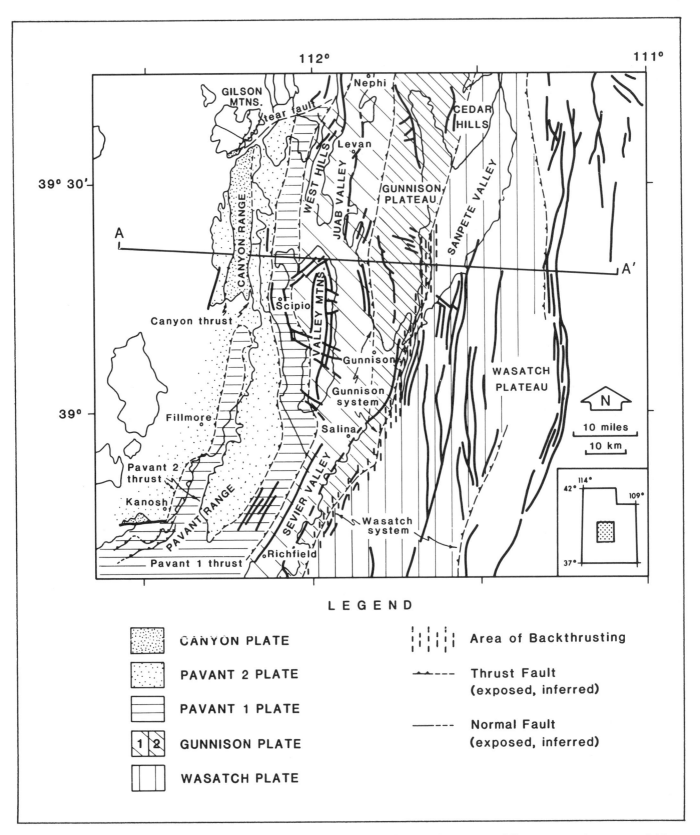

Figure 7—Generalized structural map of central Utah. Surface traces of thrust faults in the Pavant and Canyon ranges integrate available mapping (Christiansen, 1952; Lautenschlager, 1952; Crosby, 1959; Callaghan and Parker, 1962; Hickcox, 1971; Villien, 1980; Millard, 1982; Davis, 1982; Holladay, 1983; George, 1983), and normal faults are from the geologic map of Utah (compiled by Hintze, 1980). Deflections along the Pavant and Gunnison thrusts are related to facies changes and prethrusting deformation within basement and Paleozoic rocks; the eastern extent of the Gunnison thrust is linked to the paleogeographic pattern of the Arapien basin (Sprinkel and Waanders, in press). Strike-slip faults which occurred during late Tertiary extension are omitted.

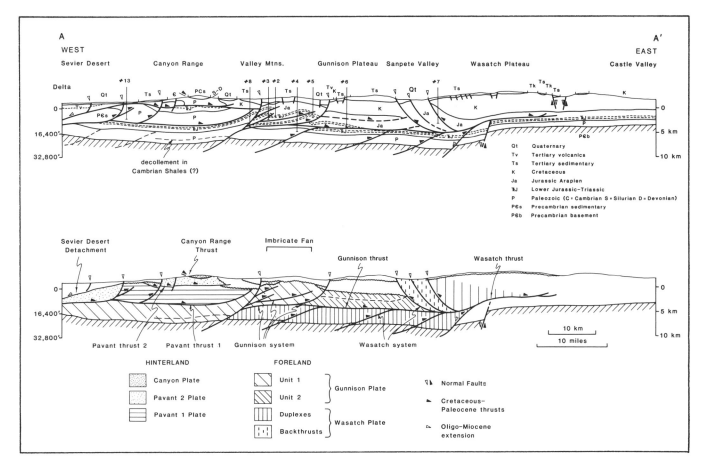

Figure 8—Structural cross sections from the Sevier Desert to Castle Valley, showing subsurface structural geometry based on surface geology, borehole and seismic control (including COCORP line on the western end), and gravity surveys. The upper cross section shows the stratigraphic units involved, and the lower cross section emphasizes tectonic elements (no vertical exaggeration). Locations of cross sections and wells are in Figure 1. Evident is the geometry of major thrust sheets and their relationships to the younger extensional detachments. Basement configuration is based on inferred regional relationships rather than on seismic profiling, and a deep-seated interpretation for the Wasatch Fault system is preferred (Zoback, 1982, 1983). Uncertainties remain for the western extent of the Gunnison plate and the nature of the deformation occurring within the basement below the eastern side of the Sevier Desert. The fault at the east end of the cross section is also interpreted as having two senses of fault displacement (Standlee, 1982), with compression preceding extension.

to diapiric structures found in the Tertiary section of the Sevier Desert basin fill.

The overlying Pavant 2 plate is bounded by the Pavant 2 and Canyon thrusts. Its lower Paleozoic rocks form a major hanging-wall anticline. The Pavant 2 thrust merges with the Pavant 1 thrust at depth, within the eastern part of the Sevier Desert. The Canyon plate, whose remnants are found in the House Range and westward, represents the overlying dominant thrust sheet (Blanchet et al., 1981, 1983), which has controlled the structural configuration of western Utah.

Pavant Range, Near Fillmore

Near Fillmore, between the central and southern Pavant Range (Figures 1 and 2), lower Paleozoic rocks (mostly Cambrian) override Paleozoic and lower Mesozoic strata found in a tectonic window underlain by deeper thrusts (Figure 9). Late Tertiary extensional faults further complicate the thrust geometry on the western side of the region. On the eastern side, the Pavant 2 thrust places Cambrian–Ordovician quartzite and carbonate over the Jurassic Navajo Sandstone and Twin Creek Limestone, and

dips 15 to 20° eastward. The easterly dip may be the result of late extension in the Sevier Valley to the east. The structure section on Figure 9 integrates surface mapping (Hickcox, 1971; Baer et al., 1982; George, 1983) and subsurface information provided by wells and seismic profiling. A gently westward-dipping thrust (labeled 2 in Figure 9) further complicates the Pavant 1 body. This intermediate thrust has been encountered in the Shell Sunset Canyon #1 well (sec. 21, T. 22 S., R. 4 W., not shown on Figure 9) at 2520 m (8265 ft) and brings Ordovician strata over Devonian Sevy Dolomite. It probably merges with the Pavant 1 thrust at depth on the western side of the Pavant Range. The timing of thrusting as well as structural development of thrusts in the Pavant region is discussed later, after the geometry of thrusts in the foreland is outlined.

Foreland

Two thrust systems are recognizable in the foreland: the Gunnison and the Wasatch thrust systems (Figure 8). The rather complex geometry is illustrated in more detail on Figures 10 and 11, which have been constructed from a

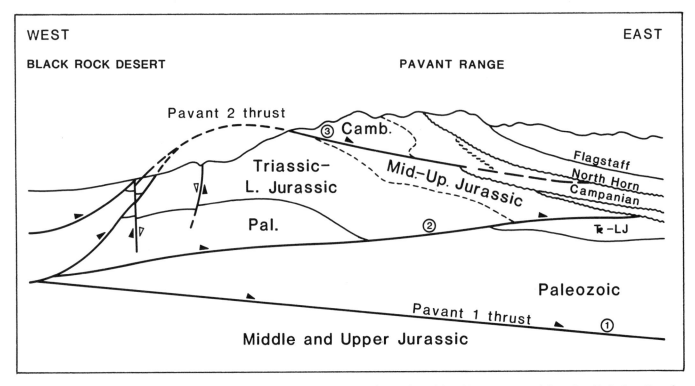

Figure 9—Structural cross section across the Pavant Range near Fillmore. The backward-breaking sequence of thrusting (1, 2, then 3) and folding of the Pavant 2 thrust are discussed in detail in the text. Major extension in the Sevier Valley may have caused eastward tilting of the Pavant 1 thrust and overlying rocks.

combination of surface mapping, new field work in critical areas, and from interpretation of seismic lines and drill hole data.

The Wasatch plate consists of two pieces: a western, geometrically and structurally lower piece that is found at a depth of 3000–6000 m (10,000-20,000 ft) beneath the Gunnison Plateau; and an eastern, upper piece that overrode the lower one westward and lies above the Arapien Shale, forming the Wasatch Plateau in a morphologic sense. The rocks of the lower, western piece (Figures 8 and 10) form a duplex structure whose floor is localized either in the Cambrian shales immediately above the basement or in the basement itself. We are not able to choose conclusively between these two alternatives on the basis of the existing data. The successive imbrication of each horse within this Wasatch duplex may have been the cause of folding of the overlying Arapien Shale and younger rocks to form the observed antiformal structure along the western edge of the Wasatch Plateau (Figure 10). The eastern (upper) Wasatch plate contains folds of small amplitude related to the propagation of blind thrusts within the Arapien Shale below the Wasatch Plateau. Backthrusts (westward-directed overthrusts) are found at the western edge of the Wasatch Plateau (Figure 11) (Standlee, 1982). The backthrusts form the westernmost boundary of the upper Wasatch plate. The shortening in the lower duplex structure occurs within Cambrian to Lower Jurassic rocks. This shortening was accommodated higher in the section by the formation of backthrusts in the eastern Wasatch plate.

The region between the Wasatch fault (normal fault) to the west, the Wasatch plate backthrusts in the east, and the underlying Wasatch duplex structure presently forms a

triangle zone directly beneath the Gunnison Plateau; the rocks overlying the duplex structure belong to the Gunnison plate.

The Gunnison plate consists of all rocks lying above the Gunnison thrust, which forms the roof for the underlying Wasatch duplex structure (Figures 8, 10, and 11). Its western boundary is defined by the presence of the overlying Pavant 1 thrust (see above) in the hinterland. From west to east, the Gunnison thrust system changes stratigraphic level and geometric character: beneath the hinterland it is traceable between basement and Cambrian shale as a flat; possibly it ramps upward out of the basement in this region into the Cambrian shales. A major ramp in the Gunnison thrust occurs where it changes stratigraphic level from the Cambrian shale upward into the Arapien Shale beneath the Gunnison Plateau. A series of imbricate splay faults branch upward from their major ramp in the Gunnison system. These splay thrusts accommodate shortening of hanging-wall strata required by displacement across the ramp. (This geometry is very similar to that shown by the Absaroka thrust in the Alpine area of the Idaho–Wyoming thrust belt as it climbs upsection from the lower Paleozoic into the Jurassic [figure 6 in Royse et al., 1975].) The Gunnison thrust system continues to the east in the Arapien Shale until its displacement dies out beneath the Wasatch Plateau (Figures 8, 10, and 11), most likely in a series of blind splay thrusts. The region where the Gunnison thrust dies out is indicated by the change from solid to dashed lines indicating the thrust in Figures 10 and 11.

The imbricate fan structures of the Gunnison system below the Sevier Valley, from the surface to a depth of ~3000 m (10,000 ft), are separated from the main body of

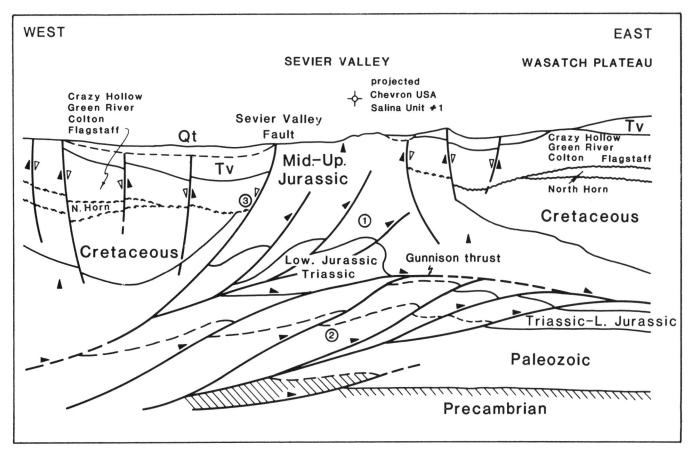

Figure 10—Structural cross section across the Sevier Valley, south of Salina. Tertiary volcanic rocks are downdropped west of the major normal fault (3), recognized along the eastern edge of the Sevier Valley. The unconformity beneath the Flagstaff truncates the unconformity beneath the North Horn above both ends of the deep antiformal stack of horses forming a duplex (2), which has also deformed the overlying imbricate fan (1) (which in turn has been further deformed by salt diapirism).

the thrusted plate by a late Tertiary extensional fault (labeled 3 in Figure 10). On the surface, this boundary (extensional fault) forms a pattern of northeast- and northwest-trending normal faults on the east side of the Sevier Valley between Richfield and Salina (Figure 1). East of the extensional fault, the Standard Oil of California #1 Sigurd Unit (sec. 32, T. 22 S., R. 1 W., not shown on Figure 10) spudded in the Arapien Shale and encountered the Navajo at 2766 m (9070 ft), while west of the fault, the Champlin #13-31 USA well (sec. 31, T. 22 S., R. 1 W.) went through an upper and lower Tertiary section down to 1380 m (4600 ft) (base of the Flagstaff Limestone) and then into the Twist Gulch Formation and Arapien Shale. Therefore, movement on the fault has produced a western, downthrown side that is probably not related to a diapiric movement.

The cross section of Figure 10 passes about 20 km to the south of a gravity profile made by Brown and Cook (1982) (their profile A–A'). This profile extends from Jacks Peak on the Pavant Range divide to east of Salina; the western part of their profile obliquely intersects the northwest-trending normal faults on the west side of the Sevier Valley. The Placid WXC USA #1-2 well (sec. 24, T. 19 S., R. 2 W., located near the west end of their B–B' profile) encountered Tertiary and Upper Cretaceous strata underlain by Paleozoic rocks; these are in turn underlain by Arapien and pre-Arapian strata. The Phillips United States #D-1 well (sec.

20, T. 22 S., R. 3 E.), on which the density distribution for the lithologic units is based, was drilled near the east end of the A–A' profile of Brown and Cook (1982) where there are no such major thrusts at depth. The geometry of Figure 10 displays an offset along the fault labeled 3, predicted on the basis of projection across the Sevier Valley at the base of the Oligocene strata, which is not seen on the Brown and Cook A–A' profile.

The gravity survey supports the general interpretation of the cross section geometry (Figure 10). If gravity modeling is adjusted in the western area where pre-Arapien age beds are not shown faulted, then the gravity data would more closely coincide with the geometry of Figure 10.

The extensional fault trajectory below 6000 m (20,000 ft) is speculative. Discontinuous low-angle, westward-dipping reflectors at 2.5 to 3.0 sec may represent the normal fault soling into an older thrust fault; alternatively, the reflectors may represent a thrust surface that may be cut by a more steeply dipping normal fault. The quality of seismic data on each side of the projected fault is equivocal. The Wasatch fault near Nephi, however, has been interpreted by Zoback (1982) as dipping steeply and, therefore, offsetting the older thrusts. Smith (1983) favors the dying out of the fault onto an older thrust 15 km to the south, near Levon.

The lower boundary of the Gunnison plate below the Pavant 1 thrust is not well defined. The interpretation presented on Figure 8 for a gentle, westward-dipping

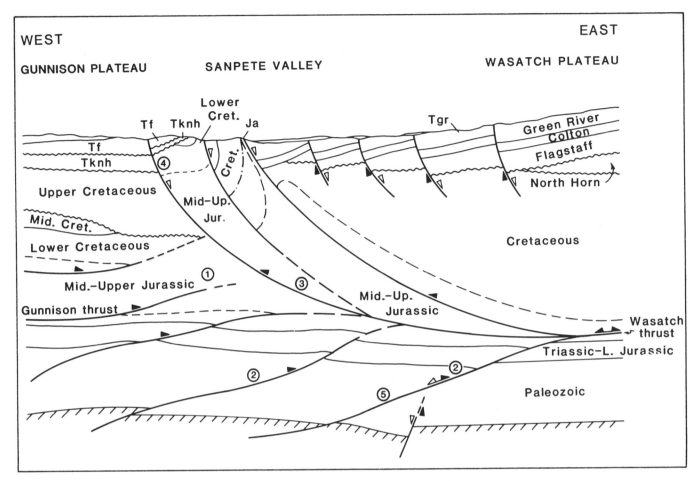

Figure 11—Structural cross section across the Sanpete Valley, near Sterling. The Flagstaff unconformity truncates the North Horn unconformity only within the Wasatch Plateau, while both unconformities appear to be parallel along the eastern edge of the Gunnison Plateau. This reflects the successive development of the imbricate fan (1), deformed by the deep duplex structure (2) with subsequent formation of backthrusts (3 and 4), which are antithetic to the major sense of tectonic transport. These east-dipping thrusts bring to the surface slices of Cretaceous strata above Jurassic Arapien Shale; however, this Jurassic–Cretaceous contact is 3600 m (12,000 ft) deep below the Gunnison Plateau (Standlee, 1982). Listric normal faults along these surfaces explain the formation of Sanpete Valley and are the product of extension that also affected the deepest frontal thrust (5).

Gunnison thrust (dashed line, bottom cross section) is based only on prominent but discontinuous reflections that tie in with similar reflections near the east end of the west-central COCORP line at about 2–2.5 sec (Allmendinger et al., 1983). Because the position of crystalline basement is uncertain, these reflections may represent an intrabedding décollement within lower Paleozoic strata representing a minor amount of displacement. Whether or not the basement is offset by faults as suggested in Figures 8, 10, and 11 needs to be conclusively demonstrated and is presently uncertain.

Sevier Valley, Near Salina

The Arapien Shale and Twist Gulch Formation are complexly deformed south of Salina (Figures 1 and 2). They are overlain to the east by conglomerate and sandstone of possible Early Cretaceous age (by comparison with a similar section located to the north near Christianburg) and also unconformably overlain by North Horn as well as Flagstaff beds that grade upward into the Colton and Green River

formations. Oligocene and younger volcanic and sedimentary rocks are present on both sides of the Sevier Valley, related to volcanic events in the Marysvale area where radiometric ages indicate late Oligocene–Miocene activity (30–18 m.y. ago) (Steven et al., 1979). The structural cross section shown in Figure 10, located south of Salina in the Sevier Valley (Figure 1), integrates critical surface relationships, drill hole data, and available seismic control.

Surface mapping by Hardy (1952) in Arapien strata shows at least two sets of fold axes. One strikes north–northeast and is deformed by a locally developed southeast-trending alignment. North of the section on Figure 10, in Salina Canyon (Figure 1), the double unconformity (Flagstaff–North Horn) is the result of two deformational events. Subsurface data are available from boreholes: the Champlin USA13-31 #1 and Standard Oil of California #1 Sigurd Unit and the more recent Chevron USA #1 Salina Unit. The latter penetrated a repeated section of Arapien and the Navajo at 2700 m (9000 ft), continued through Triassic strata into Permian rocks, and bottomed at

5314 m (17,423 ft), possibly in Navajo Sandstone (Standlee, 1982). Therefore, at least two thrusts were encountered: one in the Arapien section and a west-dipping one below the Permian. Projection at depth of known thicknesses and geometric reconstruction suggest that the moderately west-dipping Triassic sequence is doubled and that the Permian rocks form only a thin layer. One can regard the thrust within the Arapien as part of the imbricate fan structure (labeled 1 in Figure 10) and the lower one as part of the antiformal stack of horses of the underlying Wasatch plate (labeled 2 in Figure 10).

Although the general antiform that involves Paleozoic–Triassic and Lower Jurassic strata is evident on the A–A' gravity profile of Brown and Cook (1982), the nature of basement involvement in thrusting is less clear. Seismic reflections in some areas display the broken nature of the Precambrian crystalline basement. A prominent north-striking normal fault affects pre-Arapien strata along the eastern side of the Sanpete and Sevier valleys. Standlee (1982) has pointed out that the dip-slip component of displacement increases southward along this fault; a west-dipping surface (45°) overlies a similar but smaller normal fault, which probably offsets basement and is downthrown to the west (labeled 5 in Figure 11). The steep basement fault could have influenced several phases of deformation: first a ramp in the thrust developed above a basement step, which was subsequently reactivated during a late Tertiary phase of normal faulting.

To determine whether basement was involved in thrusting (thick-skin geometry) or whether thrusting occurred within Cambrian shale above the basement (thin-skin geometry), it is necessary to calibrate and identify seismic reflection data with drill hole data. Data from drill holes on the Wasatch Plateau have been used to identify reflections down into Cambrian shale; the significance of deeper reflections in the basement is still unresolved in central Utah, although it is clear that basement is involved in thrusting in the neighboring Uinta Mountains and northern Wasatch Range (Crittenden, 1972; Bryant, 1980; Bruhn and Beck, 1981).

The basement warps that control Jurassic and Cretaceous paleogeography and thrust belt geometry have also been discussed by Picha and Gibson (1983). He has emphasized the north-directed and east to northeast-directed trends, but one can add the northwest-trending axis (troughs and uplifts) related to the Ancestral Rockies (Kluth and Coney, 1981), which has favored their involvement in thrusting by local elevation of the top of the crystalline basement along the highs. Such an interpretation suggested by S. S. Oriel (personal communication, 1983) is supported by surface observations along the western side of Mount Nebo near Nephi (Figure 1) where exposures of Precambrian gneiss with amphibolite cannot be explained solely by movement along the Wasatch fault (Le Vot, 1984).

Sanpete Valley, Near Sterling

Northeast of the Salina area, the Sanpete Valley near Sterling (Figures 1 and 2) forms the boundary between the Wasatch and Gunnison plateaus. Facies changes in Cretaceous rocks with opposite vergence occur over short distances across the Sanpete Valley where deformed Arapien beds are partly overlain by Tertiary strata forming the western edge of the Wasatch monocline. North Horn and Flagstaff unconformities are present on both sides of the valley, and late Tertiary deformation is documented by normal faults (trending north and northeast where the Sanpete Valley intersects the Sevier Valley), while local diapirism along some of these faults is as young as Quaternary. The structure section in Figure 11 represents the deformation that affects the western edge of the Wasatch Plateau and is based on surface mapping (Taylor, 1980; Villien, 1980) and seismic data cited by Standlee (1982).

Most of the shallow relationships within the Sanpete Valley are obscured by late Tertiary fill and faults, but detailed work conducted by Hardy (1952) shows that folds in Arapien beds verge mainly westward. He recognized westward-dipping thrusts (trending north–northeast) bringing younger Arapien rocks over older Arapien west of Mayfield (Figure 1). Standlee (1982) has suggested that the faults are backthrusts that dip eastward as shown in Figure 11. These thrusts are poorly exposed, but abrupt thickening and thinning of cargneule gypsum beds support their existence. Their dip is still a problem because several phases of deformation took place since Late Cretaceous time, with two deformational events recorded during late Maestrichtian and late Paleocene time (double North Horn–Flagstaff unconformities). The imbricate fan structure (1) shown in the left part of Figure 11, involving Middle to Upper Jurassic and pre–North Horn rocks, forms the eastern edge of the Gunnison thrust system and represents unit 2 of the Gunnison plate (Figure 8). This imbricate fan is underlain and deformed by the northeastern part of the deep antiformal stack of horses whose general geometry is apparent on gravity profiles (B–B' of Brown and Cook, 1982), but the main difference between this and the Salina area is the importance of backthrusting. Upper Santonian marine beds appear to be present east of the Ninemile Reservoir; their presence west of Cenomanian–Turonian strata (Sanpete and Allen valleys) indicates that they may have undergone westward-directed thrusting that has accommodated the deeper shortening within the duplex. Thrusts of Indianola over North Horn along the eastern side of the Gunnison Plateau (Burma and Hardy, 1953; Weiss, 1982) may also represent backthrusting. Farther to the north in the Sanpete Valley (Figure 1), the Mobil #1 Larson Unit (sec. 1, T. 17 S., R. 2 E.) seems to have encountered a tectonic thickening of east-dipping Arapien Shale. This is also seen on the seismic profiles mentioned by Standlee (1982).

THRUSTING PHASES

Preceding parts of this report have considered a preliminary biostratigraphic correlation of Cretaceous–lower Tertiary rocks of central Utah and the large-scale geometry of its thrust belt. It is clear from the sedimentary records that a number of discrete thrusting phases of regional importance took place in central Utah between Albian and early Tertiary time. In the following section, we present the stratigraphic and structural evidence that we believe allows us to assign specific thrusting events to discrete age brackets. We conclude by reviewing the sequence of development of each thrust in the larger context of the thrust belt geometry as a whole.

The recognition in the sedimentary record of a thrusting phase depends on the complex interplay of the geometry of thrusting, the sea level at the time of the thrusting event, and the particular facies present when thrusting was initiated. In their now classic papers on thrusting phases in the Wyoming, Idaho, and Utah portion of the Cordilleran fold and thrust belt, Armstrong and Oriel (1965) and Oriel and Armstrong (1966) were able to identify a series of thrusting events through both geometric brackets and the sedimentary record and to link them to motions of the Paris–Willard, Meade–Crawford, Absaroka, Darby, and Hogsback thrust systems.

An emergent thrust carries strata of its hanging wall to the surface where they are subject to erosion. Near the thrust toe, synorogenic sediments may be deposited as alluvial fan conglomerates whose clasts have compositions reflecting the strata in the hanging wall. Farther from the thrust, the thrusting event can sometimes be recognized by an abrupt change in facies, for example from marine to nonmarine.

A thrusting event also produces a clastic wedge (Price and Mountjoy, 1970) or foreland basin (Beaumont, 1981) at considerable distances from the locus of thrusting. The sudden onset of renewed deposition in the adjacent foreland can lead to changes in the stratigraphic record that can be used to identify phases of thrusting (Jordan, 1981), as long as the effects from the rise and fall of sea level can be isolated (Kauffman, 1981).

In the case of blind thrusts, no synorogenic conglomerates need to be produced. The recognition of the thrusting phase is less obvious but still possible. As in the case of emergent thrusts, the uplift of strata on the hanging wall leads to erosion, and sedimentation after uplift and erosion produces unconformities, either angular unconformities where folding and/or tilting was involved or disconformities.

It is therefore apparent that the age of a thrusting event must be documented by one or all of the following criteria: (1) the age of a synorogenic conglomerate reflects the approximate age of thrusting; (2) the age of a prominent facies transition from, for example, marine to continental may reflect the age of thrusting; (3) the age of an overlying unconformity must be at least as early as and, more likely, may mark the end of a thrusting event; and (4) the age of renewed sedimentation in a foreland basin setting must reflect the elastic or viscoelastic response of the lithosphere to thrusting in the adjacent thrust belt (Beaumont, 1981).

The thrusting phases we infer from the stratigraphic record are identified in Figure 12. Synorogenic conglomerates are labeled A through F on both Figure 12 and in the stratigraphic columns of Figures 5 and 6. The wavy lines indicate the inferred latest movements of a thrusting event. In some places a thrust is sealed by sediments whose age provides an upper limit; in other places this relationship is derived by indirect means.

Late Albian Thrusting Phase A

We have identified an alluvial fan deposit of late Albian age in the Christianburg and Chicken Creek sections (A in

Figure 5). In the Placid WXC State #1 well, a coarse-grained sandstone is present that may be correlative with these presumed synorogenic conglomerates. The slightly different facies could have been produced within an alluvial fan setting. Although we suggest that this sandstone records the same event as in the two surface sections, it is also possible that its age is older; in that case, it would record a pre–late Albian event.

The synorogenic conglomerate (Phase A) contains clasts of Paleozoic rocks including quartzite, chert, and limestone. These rocks now are found directly east of the Pavant 1 thrust (see A in Figure 13). The USA #1-2 well (Figure 1) cuts the hanging-wall strata of the Pavant 1 thrust (well #8 in Figure 8 cross section) before piercing the Pavant 1 thrust at depth. At approximately the 1800-m (6000-ft) level, Placid has identified Cenomanian–Turonian palynomorphs in Cretaceous strata that rest unconformably on the hanging wall of Paleozoic rocks. These Cenomanian–Turonian strata are equivalent in age to the Sanpete Formation farther east (Figure 12) and suggest a latest time for thrusting age along the Pavant 1 thrust.

We conclude that the thrusting on the Pavant 1 thrust began as early as late Albian and that deformation associated with thrusting ended by Cenomanian–Turonian time. This interpretation depends on the identification by Placid of Cenomanian–Turonian palynomorphs. In cuttings from the same well, Elf Aquitaine palynologists found only Early Cretaceous palynomorphs, which we think may have been reworked. If, however, these are not reworked and are of Early Cretaceous age, then we have little control for the end of Pavant 1 thrusting. Pavant 1 thrusting in this latter case could be any time younger than late Albian. This point is of significance in deciding whether the thrusting sequence in the hinterland was backward- or forward-breaking, and we will return to it subsequently.

Late Santonian–Early Campanian Thrusting Phase B

A thick sequence of alluvial fan conglomerates (Phase B) is found in the Sixmile Canyon section, the Chicken Creek section, and the Placid USA #1-2 well (Figure 6). The clasts consist of Paleozoic rocks including limestone, dolomite, quartzite, and chert, all of which are present in the hanging wall of the Pavant 2 thrust (Figure 8). The conglomerates are found to the east of the Pavant 2 thrust but *above* the Pavant 1 thrust in the USA #1-2 well. They are therefore associated with thrusting along the Pavant 2 thrust.

All conglomerates are overlain by Campanian coal beds that are continental and not related to thrusting phases. The conglomerates do not directly overlie the hanging wall of the Pavant 2 thrust, but do unconformably overlie younger strata. The change in facies from a marine sequence (middle to upper Santonian) to thick continental deposits may be the synorogenic product of Pavant 2 thrusting during late Santonian–early Campanian time.

Middle and Late Campanian Thrusting Phases C and D

In the Sixmile Canyon region, Fouch et al. (1983) identified quartzose sandstone overlying the Campanian coal-bearing sequences. This was interpreted as a distal product of thrusting events that occurred at the early to late Campanian boundary. At Chicken Creek, conglomeratic sandstone unconformably overlies coal-bearing beds at the

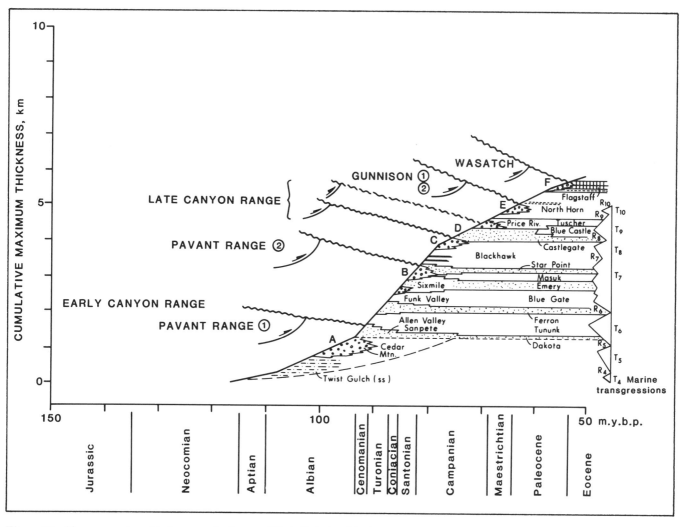

Figure 12—Tectonostratigraphic framework of central Utah. Relationships between thrust and sedimentation with correlative eustatic episodes show that thrust movements in the orogenic belt occur at the same time as transgressions in the foreland basin. Eustatic standstills, in turn, produced progradation events, and normal regressive sequences follow times of active continental tectonism (Kauffman, 1981). The volume of clastic wedges is linked to the thrust belt structural geometry (emergent or blind thrusts).

base of the South Flat Formation (Figure 6). In the Placid WXC USA #1-2 well to the west, conglomerates unconformably overlie Campanian coal beds (the lowest members) and consist mostly of quartzitic pebbles. The Price River Formation (Phase D) is found elsewhere above the youngest coal-bearing beds, but an earlier conglomerate (Phase C) occurs between the lowest and the highest coal beds (Figure 6). The dominantly quartzose rocks in these conglomerate sequences are found throughout central Utah and undoubtedly record a major thrusting event or events.

The Canyon Range thrust crops out in the Canyon Range (Figure 1) where latest Cretaceous conglomerate (Stolle, 1978) is present in both the hanging wall and the footwall. The clasts in the footwall conglomerates are of boulder size, but only small amounts of conglomerate are found in the hanging wall. Furthermore, the footwall conglomerate is folded beneath the Canyon Range thrust; the entire package (hanging wall, thrust, and footwall) is overlain by Paleocene sandstone and limestone. Hanging-wall strata consist of Precambrian quartzite, while footwall strata consist of Ordovician and Devonian strata. These geometric and

stratigraphic relationships suggest the following sequence: pre-latest Cretaceous thrusting along the Canyon Range thrust (Campanian thrust of Figure 12), deposition of an alluvial fan conglomerate in front of the emergent thrust, renewed movement (also in Campanian) along the thrust causing the slight folding of the footwall conglomerate, and subsequent sealing of the thrust in Paleocene time by continued erosion from the hanging-wall strata.

The Placid WXC USA #1-2 well and the Chicken Creek section are east of the Canyon Range. We interpret these spatial patterns as indicating the presence throughout central Utah of synorogenic conglomerates that result from motion on the Canyon Range thrust during middle and late Campanian time.

Late Maestrichtian Thrusting Phase E

The North Horn Formation is composed of sandstone, limestone, and locally, conglomerate at Sixmile Canyon, Chicken Creek, and the USA #1-2 well (Figure 6). At Chicken Creek and Sixmile Canyon and in the Canyon Range, the unit is underlain by an angular unconformity only; in other places the base is a disconformity. Only Maestrichtian palynomorphs were found at Sixmile Canyon, but toward the west, palynomorphs span from late Maestrichtian to early Tertiary in age. The unconformity at the base of the North Horn is therefore mainly of late Maestrichtian age and represents the latest episode of thrusting.

The North Horn Formation or equivalents (E) is found throughout central Utah: In the Canyon Range, conglomerates sealing the late Canyon Range thrusting phases are overlain unconformably by North Horn strata; it is likely that the Pavant 2 thrust is also overlain by North Horn strata (Figure 9). North Horn rocks are found on both sides of the Sevier Valley (Figure 10); at Sixmile Canyon (Figures 6 and 11) they lie in angular unconformity over Cretaceous strata.

Detritus for the North Horn apparently was shed by moving thrust systems of complex geometry. Where uplift and deformation occurred, the North Horn was laid down in unconformity above underlying strata; where there was a lack of such uplift, the North Horn was deposited conformably on underlying strata. The geometry of the Gunnison thrust system is complex throughout central Utah. The irregular topography created by movement on such a system (containing ramps, flats, and blind splay thrusts) would have produced regions of localized deformation and uplift. We therefore conclude that the Gunnison thrust system was active during late Maestrichtian time.

The late Maestrichtian timing of movement on the Gunnison thrust system helps to explain some of the anomalous geometries known from seismic reflection surveys across the Sevier Valley near Salina (Figure 10). The development of an imbricate fan structure, reflected by north- to northeast-trending folds in the Arapien on the surface and by salt pillowing at depth, apparently bowed up the pre–North Horn beds on the eastern end. This Cretaceous section is overlain unconformably by the North Horn, prominent on seismic profiles, which supports a late Maestrichtian age for the development of the imbricate fan.

Late Paleocene Thrusting Phase F

The Flagstaff Formation (F in Figures 6 and 12) contains palynomorphs of early Eocene age in limestone lying directly above an unconformity in Sixmile Canyon (Figure 6). Spieker (1949) also reported the presence of a similar unconformity in Salina Canyon near Salina, at the south end of the Wasatch Plateau. Farther to the east, Fouch et al. (1983) have reported the presence of localized conglomerates of middle Paleocene age east of the San Rafael Swell.

We interpret the local unconformity at the base of the Flagstaff Formation as evidence of renewed deposition at the end of a late Paleocene thrusting phase that interrupted sedimentation patterns across the Wasatch Plateau and the Sanpete and Sevier valleys regions.

The Wasatch thrust system, which formed as a result of this late Paleocene shortening, produced two contrasting structural styles. Duplex formation accommodated shortening at deepest levels beneath the Sanpete and Sevier valleys to the west and backthrusts accommodated shortening in the cover at the west end of the Wasatch Plateau to the east (Figure 8). Such reverse faults also occur

east of the San Rafael Swell (R. W. Allmendinger, personal communication, 1983). Viewed in this way, the eastward propagation of thrusting during Eocene time is a continuation of "Sevier" type thrusting events and is unrelated to classic concepts of a "Laramide" style orogeny.

In both the Sanpete Valley (Figure 11) and the Sevier Valley near Salina (Figure 10), late Paleocene deformation is recorded by the Flagstaff unconformity which truncates the earlier North Horn unconformity. The Flagstaff unconformity also appears to be restricted to the eastern and western edges of the Gunnison Plateau. Both unconformities are easily recognized in seismic profiles used to prepare Figures 8, 10, and 11. Late Paleocene shortening and formation of the duplex structure has consequently deformed the overlying, older Gunnison thrust system imbricate fan structure. This may have been responsible for producing southeast-trending folds in the Arapien Shale on the surface and the associated small-scale thrust faults seen in Lost Creek Canyon (~6 km south of Salina). Because of the incompetence of the Arapien Shale and the complexity of the structures observed, it is not prudent to assign any fold axis set to a particular thrusting phase, especially following late Tertiary extensional deformation.

At higher structural levels, backthrusting resulted in the truncation of the North Horn unconformity by the Flagstaff unconformity, especially on the eastern side of the Sanpete Valley (Figure 11). There are several sedimentologic consequences of backthrusting. On each side of the Sanpete Valley, uplifted regions produced by backthrusting apparently caused the development of both eastward- and westward-directed sediment dispersal in late Paleocene lithic quartz sandstone, recognized by Stanley and Collinson (1979). An eastward to northeastward change in sediment dispersal direction of similar sandstone units north of Salina is also related to thrust geometry, whereas northwest-directed sediment dispersal in feldspathic sandstone of the overlying Colton Formation (Eocene) is related to uplifted source areas to the east and southeast. The northwest-trending highs of the Ancestral Rocky Mountains may have also influenced structures produced by the compressive deformation phases (S. S. Oriel, personal communication, 1983) and affected the trends of sediment dispersal directions.

SEQUENCE OF THRUSTING

The structural evolution of central Utah during Late Cretaceous–early Tertiary time is shown in sequential stages of development in Figure 13. Deformation was largely confined to the hinterland areas (Canyon Range, Sevier Desert region, and Pavant Range) during late Albian–late Campanian time, but it migrated into the foreland region (Gunnison Plateau, Sanpete Valley, and Wasatch Plateau) beginning in late Maestrichtian time.

This scenario has many elements in common with development of the Overthrust Belt farther to the north (Oriel and Armstrong, 1966; Royse et al., 1975) where a classic forward-breaking pattern of thrusting has been documented: deformation migrates in space and time from west to east as each younger thrust forms beneath and in front of its predecessor. However, there are some significant differences with the sequence of thrusting to the north. The sequence of thrusting in the hinterland region in central Utah appears to have formed by a combination of forward-breaking and backward-breaking thrusting. The evidence for such a sequence of events is reviewed below.

The structural development of thrusts in the hinterland appears to have taken place as follows. Motion on the Canyon thrust (early Canyon phase) (not shown on Figures 12 and 13) carried the Canyon plate onto what is now the hinterland. This was synchronous with or followed by formation of the underlying Pavant 1 thrust, which formed in a classic forward-breaking sequence. A backward-breaking thrust sequence then developed; formation of the Pavant 2 thrust and emplacement of the Pavant 2 sheet further transported and deformed the overlying early Canyon plate. Finally, renewed motion occurred on the already existing Canyon thrust (late Canyon phase).

As previously shown, the Pavant 1 thrust produced a late Albian synorogenic conglomerate (Figure 5). Strata of both the hanging wall and footwall of the Pavant 1 thrust are overlapped by Cenomanian strata; Paleozoic in the hanging wall and Lower Cretaceous in the footwall (B in Figure 13).

During late Santonian time, the Pavant 2 thrust formed west of the Pavant 1 thrust (B in Figure 13). The Pavant 1 thrust is sealed by the Cenomanian–Turonian unconformity (Figure 12), and to the west Campanian strata overlap Cenomanian–Turonian rocks (Figure 9). (These ages come from the Placid WXC USA #1-2 and WXC Barton #1 wells located farther east [Figure 1], but projections of the encountered depths combined with seismic reflections support the interpretation.) We have previously inferred that the Campanian conglomerate is related to movement on the Pavant 2 thrust (Figure 9). Therefore, the Pavant 2 thrust is younger than the Pavant 1 thrust and must have developed as a younger slice of the Pavant 1 thrust in a backward-breaking sequence. Further evidence of this interpretation comes from the geometry of folded rocks in the Pavant 2 thrust footwall. Along the western boundary of the Pavant Range (e.g., just east of Fillmore), the hanging wall of the Pavant 1 thrust contains kink-band folds in Triassic and Lower Jurassic rocks. The Pavant 2 thrust truncates these folds in a way that is incompatible with their having been produced by deformation in the footwall of the Pavant 2 thrust (i.e., they are not footwall syncline-type folds). The Pavant 2 thrust has brought Cambrian rocks over Lower Jurassic rocks. Therefore, a significant amount of displacement must have occurred on the Pavant 2 thrust. We interpret these geometric relationships as further evidence that the Pavant 2 thrust formed to the west of, and is younger than, the Pavant 1 thrust.

We should note that an intermediate thrust, labeled 2 in Figure 11, is found between the Pavant 1 and Pavant 2 thrusts; it is recognizable in both seismic sections and drill hole records. Very little stratigraphic separation occurs along this thrust. If it moved in post Cenomanian–Turonian(?) and pre-Campanian time, then a late Coniacian age is possible, which could explain the eastern synorogenic deposits at the Coniacian–Santonian boundary.

The exact location of the previously developed Canyon Range plate at this time is not well known (Figure 13B). However, if it had been overlying or even nearby, it would have undergone the structural effects of underthrusting

along the Pavant 1 and/or Pavant 2 thrusts. This would have resulted in deformation of the Canyon Range thrust at this time, an argument suggested by Burchfiel and Hickcox (1972). The major role played by the Canyon plate in the development of the Cordilleran fold and thrust belt has been outlined by Blanchet et al. (1983). The events in the hinterland area can be summarized as follows: initiation of thrusting along the Pavant 1 thrust (late Albian); development of an intermediate thrust shown on Figures 13B and 9 (late Coniacian?) and the Pavant 2 thrust (late Santonian) in a backward-breaking sequence; and finally, renewed movement along the Canyon Range thrust (late Campanian), also in a backward-breaking sense with respect to the underlying Pavant 1 thrust.

Deformation and thrusting migrated into the foreland region during late Maestrichtian time with eastward transport along the Gunnison thrust system. This was followed in late Paleocene time by duplex formation at depth and backthrusting (westward transport directions) on the Wasatch thrust system (F in Figure 13).

An Alternative Thrusting Sequence

The backward-breaking sequence of thrust development in the hinterland that we have just outlined depends critically on the Cenomanian–Turonian age that we have assigned to strata in the Placid USA #1-2 well. As discussed earlier, if this age is incorrect and we can only identify it as Early Cretaceous, then the timing of motion along the Pavant 1 thrust is less well constrained: thrusting must have been between Early Cretaceous and Campanian time. If this premise is accepted, then it is possible that thrusting phases in the hinterland developed in an eastward progressing sequence. The reasoning is closely analogous to that used by Oriel and Armstrong (1966) in discussing the sequence of thrusting in the Overthrust Belt to the north.

In this alternative scenario, late Albian motion along the Canyon Range and/or Pavant 2 thrust would have produced a synorogenic conglomerate of this age east of the thrust toe. A period of uplift and erosion could have removed a large portion of this conglomerate, leaving only minor amounts in the frontal part. Development of the Pavant 1 thrust east of an older Pavant 2 thrust would have cut across the remnants of the late Albian conglomerate produced previously. During late Campanian time, continued uplift and erosion would have covered the entire region, in particular sealing the Pavant 1 thrust. The final geometry produced by this sequence of events would closely resemble that shown in C–D in Figure 13.

There are two major objections to accepting the forward-breaking alternative thrust sequence as just outlined. First, there is very little evidence for an extensive late Albian conglomerate that ought to be found above and west of the Pavant 1 thrust. We believe that it is improbable that this conglomerate was eroded everywhere *except* at the site where the Pavant 1 thrust would subsequently form. Second, the Pavant 1 hanging-wall kink-band folds are truncated by the Pavant 2 thrust. This would be difficult to explain if the sequence of thrusting had been Pavant 2 followed by Pavant 1.

Taken together, the evidence for the Cenomanian–Turonian age of the strata in the Placid USA #1-2 well and the geometric arguments presented above seem to us to rule

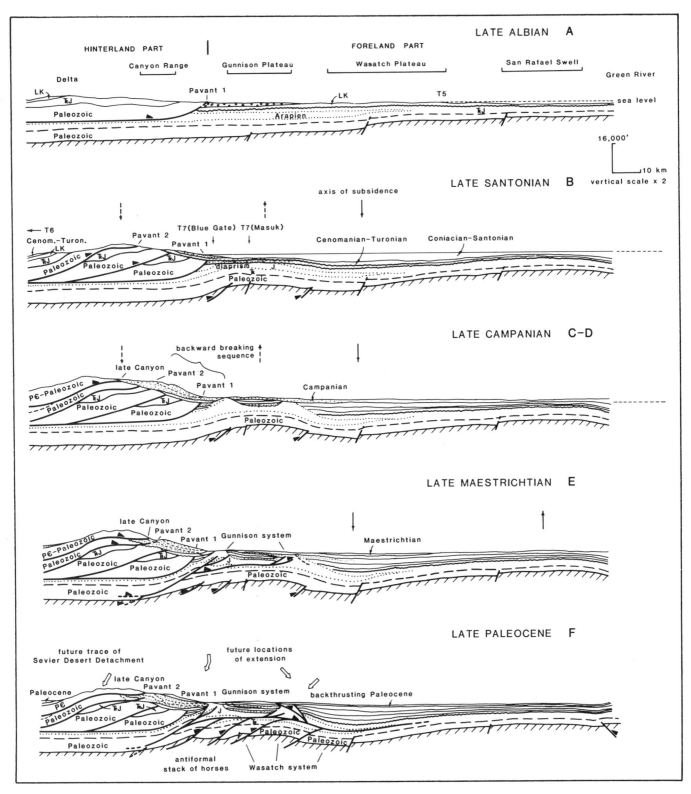

Figure 13—Structural development of central Utah. Each stage is represented by a cross section showing paleostructural and stratigraphic relationships along the same location as cross section A–A′ shown in Figure 1. Vertical exaggeration (2:1) allows detailed graphic representation. Arrows indicate movements occurring during each interval. Tectonic loading within the hinterland has produced foreland deformation that controls diapiric movements, subsidence that exceeds sediment loading, and "forebulge" development. The clastic wedges, developed at the toes of the major thrust plates, may not have been the locus of highest subsidence. Erosion of part of the uplifted thrust plate by late Albian time redistributed the load causing subsidence eastward, and the paleotopography after this major Middle Cretaceous thrusting was mainly controlled by isostatic subsidence. The backward-breaking sequence occurring in the hinterland during the Campanian did not produce a major shift in the subsidence axis, and the sedimentation in this basin during Maestrichtian was also related to the high developed on its eastern margin. Paleogeography during Paleocene time was inherited from a previous stage, and locations for late Tertiary extension are linked to structural discontinuities of the thrust belt.

out the forward-breaking thrust sequence in favor of the backward-breaking sequence suggested previously.

STRUCTURAL EVOLUTION AND RELATED PROBLEMS

Effect of Thrusting on Basin Development

Late Santonian and late Campanian detrital strata record the effects of structural loading on the adjacent foreland by thrust emplacement in the hinterland. A foreland basin or clastic wedge was formed (Figure 13) of the type documented by Price and Mountjoy (1970) and Beaumont (1981) using a theory of viscoelastic flexure of the lithosphere. The loading resulted in the deposition of major clastic wedges and the initiation of a subsidence axis 50 km away from the toes of the thrusted sheets with a correlative uplifted "bulge" area 120 km eastward in the San Rafael Swell. Nevertheless, the San Rafael area appears to have deformed earlier, analogous to the Wyoming Moxa arch, which can be interpreted as a forebulge; both arches were well expressed in Maestrichtian time. In summary, evolution of the foreland basins is linked to structural loading in the adjacent thrust belt; the greatest subsidence seems to have been located east of the clastic wedges.

The relationships illustrated in Figure 12 are very similar to those of Royse et al. (1975) and Jordan (1981) for the Wyoming fold and thrust belt. In the Wyoming segment of the thrust belt, however, the thickness versus age graph (Jordan, 1981) shows a steeper gradient for early Tertiary time than in central Utah (Figure 12). Emergent thrusts there produced a greater amount of detritus not furnished by the blind structures present in central Utah.

Relationship between Tectonic and Eustatic Episodes

Transgressions and regressions recognized in the interior Cretaceous seaway (Kauffman, 1977) suggest relationships between tectonic and eustatic episodes. As noted by Kauffman (1981), times of active tectonism along the western margin of the seaway correlate roughly with eustatic changes in sea level, which produce transgression. This is documented for the late Albian (T_5), the late Santonian (upper T_7), the middle Campanian (T_8), the late Campanian (T_9), and the late Maestrichtian (T_{10}). Two major transgressions do not seem to have thrust counterparts—the lower Turonian (T_6) and the Coniacian–Santonian boundary (lower T_7)—but may be related to the initiation of basement deformation in the foreland (Figure 13B). The lower Turonian deposits are recorded far to the west and indicate that the eastern margin of the hinterland area was actively subsiding to compensate for the warping that had occurred in front of it. Nevertheless, our data suggest that major transgressive events do correlate with deformation events (horizontal and vertical scales) in central Utah. Figure 12 also illustrates that broad regressions produce upward-coarsening, regressive sedimentary sequences that do not record specific thrusting events (Sanpete, Ferron, Star Point, and Castlegate).

Role of Basement during Thrusting

In the hinterland, lower Mesozoic and Paleozoic strata form the backbone of primary thrust sheets. We do not have enough evidence to determine if the Cambrian shale forms a décollement horizon or if a major décollement is located deeper, in the upper Precambrian. The crystalline basement in the hinterland had to have been involved in some way in shortening, but in many places extension has obscured the previous relationships. In the foreland, the Middle Jurassic Arapien represents a disharmonic horizon that resulted in a decoupling between pre- and post-Arapien strata and did not allow the deeper thrusts to cut across higher sedimentary beds. Lower Jurassic and Triassic strata are deformed into kink folds and imbricate fans with little stratigraphic separation. Paleozoic strata and possibly some basement rocks form the main body of the deep duplex structure.

We are unable to ascertain a cause of the backward-breaking sequence that developed in the hinterland. It is possible, however, that basement warping in the foreland during Santonian and Campanian time (Figure 13B and C-D) may have influenced stress regimes in the hinterland.

The palinspastic position of the Arapien basin was most likely controlled by basement structure. The position of this basin and reactivation of the related basement structure during both compression and extension phases probably influenced both structure and sedimentation.

As a result of tectonic loading and reactivation of basement structures in the thrust belt, diapiric movements affected sedimentation of late Santonian marine beds containing Mid–Upper Jurassic palynomorphs.

The main expression of deformation in Phanerozoic strata of the foreland started in late Maestrichtian time and was followed by a late Paleocene episode. It is likely that deformation within the foreland area also affected the basement during early Late Cretaceous time as shown in Figure 13. The region may have been affected by post-Paleocene deformation, but its record must lie at depth along the western margin of the Colorado Plateau. The observed continuity in space and time between events in the hinterland and foreland and their deformational surface expressions suggest that they are the result of a continuous process of crustal shortening that has affected both the sedimentary cover and the basement. In the absence of "foreland type" uplifts in central Utah, it is not fruitful to separate deformation phases according to the classic concepts of Sevier versus Laramide orogenic styles.

All the major structures of central Utah were apparently emplaced by late Paleocene time. The Late Cretaceous to Eocene structural geometry clearly influenced the locations of late Tertiary extensional structures. For example, the extensional fault now on the east side of the Sevier Valley formed on the rear end of the foreland duplex structure of the lower Wasatch plate. Similarly, the backthrusts of the western edge of the Wasatch Plateau accommodated extension. These are interpreted here as examples of reactivation of earlier thrust faults during Basin and Range extension.

ACKNOWLEDGMENTS

This study is part of a Ph.D. dissertation by A. Villien at the University of Colorado under the direction of E. Kauffman, R. Kligfield, and S. Oriel (U.S. Geological Survey,

Denver). The writers are grateful to Elf Aquitaine
Petroleum for permission to publish this paper. Discussions
toward understanding the larger regional aspects of thrust
belt geology with M. Zimmermann and B. Wilson in the
Denver office have been extremely beneficial, as have
paleontologic reports from the Centre de Recherches de Pau
(Societe Nationale Elf Aquitaine Production, or SNEA-P) by
J. Aubert, J. P. Blanc, J. Le Fevre, G. Peniguel, C. Poumot, and
C. Seyve. All chronostratigraphic assignments not attributed
to others in this report are based on SNEA-P paleontologic
reports. Drafting by P. Hutson and typing by C. Rogers of
Elf Aquitaine petroleum and by P. Franz and K. Fox of the
Geology Department are also appreciated.

Invaluable assistance and continuous support by Societe
Nationale Elf Aquitaine (SNEA) through a student program
under French–American management with R. Blanchet
(Universite de Bretagne Occidentale), B. Plauchut
(SNEA-P), S. Oriel (USGS), M. Tardy (Universite de
Savoie), and R. Vernet and M. Zimmermann (Elf Aquitaine
Petroleum) made possible completion of an earlier third-
cycle thesis directed by R. Blanchet and M. Tardy (Villien,
1980).

Also, D. Sprinkel generously provided his knowledge, and
thanks are extended to Placid Oil Company for well cuttings.
Finally, field work with M. Le Vot in central Utah yielded
many contributary observations, and earlier drafts of this
manuscript were improved by the critical reviews of S. Oriel
who suggested many corrections and gave his assistance at
all stages.

REFERENCES CITED

Allmendinger, R. W., J. W. Sharp, D. Von Tish, L. Serpa, L.
    Brown, S. Kauffman, J. Oliver, and R. B. Smith, 1983,
    Cenozoic and Mesozoic structure of the eastern Basin
    and Range province, Utah, from COCORP seismic
    reflection data: Geology, v. 11, p. 532–536.

Armstrong, F. C., and S. S. Oriel, 1965, Tectonic
    development of Idaho–Wyoming thrust belt: AAPG
    Bulletin, v. 49, n. 11, p. 1847–1866.

Armstrong, R. L., 1968, Sevier orogenic belt in Nevada and
    Utah: Geological Society of America Bulletin, v. 79, p.
    429–458.

Baer, J. L., 1976, Structural evolution of central Utah—late
    Permian to Recent, *in* J. G. Hill, ed., Symposium on
    geology of the Cordilleran Hingeline: Rocky Mountain
    Association of Geologists, p. 37–45.

Baer, J. L., R. L. Davis, and S. E. George, 1982, Structure and
    stratigraphy of the Pavant Range, central Utah, *in* D. L.
    Nielson, ed., Overthrust Belt of Utah: Utah Geological
    Association Publications 10, p. 31–48.

Bally, A. W., P. L. Gordy, and G. A. Stewart, 1966, Structure,
    seismic data and orogenic evolution of the Southern
    Canadian Rocky Mountains: Canadian Petroleum
    Geology Bulletin, v. 14, n. 3, p. 337–381.

Beaumont, C., 1981, Foreland basins: Geophysical Journal of
    the Royal Astronomical Society, v. 65, p. 192–329.

Blanchet, R., M. Tardy, and A. Villien, 1981, Un profile
    tectonique des Cordilleres de l'Amerique du Nord
    (Sierra Nevada–Grand Bassin–Overthrust Belt)—
    charriages majeurs et structures polyphasees: Comptes

Rendus des Seances de l'Academie des Sciences, t. 292,
    ser. 11, p. 1299–1304.

Blanchet, R., H. de la Tour-du-Pin, M. Le Vot, F. Roure, M.
    Tardy, and A. Villien, 1983, Structures majeures de la
    Cordillere Ouest Americaine a l'Est de la Sierra
    Nevada—implications paleogeographiques: Compte
    Rendus des Seances de l'Academie des Sciences, v. 296,
    ser. II, p. 863–868.

Boyer, S., and D. Elliott, 1982, Thrust systems: AAPG
    Bulletin, v. 66, n. 9, p. 1196–1230.

Brown, R. P., and K. L. Cook, 1982, A regional gravity
    survey of the Sanpete–Sevier valleys and adjacent areas
    in Utah, *in* D. L. Nielson, ed., Overthrust Belt of Utah:
    Utah Geological Association Publication 10, p. 121–135.

Bruhn, R., and S. L. Beck, 1981, Mechanics of thrust faulting
    in crystalline basement, Sevier orogenic belt, Utah:
    Geology, v. 9, p. 200–204.

Bryant, B., 1980, Metamorphic and structural history of the
    Farmington Canyon Complex, Wasatch Mountains,
    Utah (abs.): Rocky Mountain Section of the Geological
    Society of America, 33rd Annual Meeting Abstracts
    with Programs, p. 269.

Burchfiel, B. D., and C. W. Hickcox, 1972, Structural
    development of central Utah, *in* J. L. Baer and E.
    Callaghan, eds., Plateau–Basin and Range transition
    zone, central Utah: Utah Geological Association
    Publication 2, p. 55–66.

Burma, B. H., and C. T. Hardy, 1953, Pre–North Horn
    orogeny in Gunnison Plateau, Utah: AAPG Bulletin, v.
    37, n. 3, p. 549–553.

Butler, R. W. H., 1982, The terminology of structures in
    thrust belts: Journal of Structural Geology, v. 4, p.
    239–246.

Callaghan, E., and R. L. Parker, 1962, Geology of the Sevier
    quadrangle, Utah: U.S. Geological Survey Geology
    Quadrangle Map, GQ-156.

Christiansen, F. W., 1952, Structure and stratigraphy of the
    Canyon Range, Utah: Geological Society of America
    Bulletin, v. 63, p. 717–740.

Crosby, G. W., 1959, Geology of the South Pavant Range,
    Millard and Sevier counties, Utah: Brigham Young
    University of Geology Studies, v. 6, n. 3, 59 p.

Crittenden, M. D., Jr., 1972, Willard thrust and the Cache
    allochthon, Utah: Geological Society of America
    Bulletin, v. 83, p. 2871–2880.

Dahlstrom, C. D. A., 1970, Structural geology in the eastern
    margin of the Canadian Rocky Mountains: Canadian
    Petroleum Geology Bulletin, v. 18, n. 3, p. 332–406.

Davis, R. L., 1982, Geology of the Dog Valley–Red Ridge
    area, southern Pavant Mountains, Millard County, Utah:
    Master's thesis, Brigham Young University, Provo,
    Utah.

Dixon, J. S., 1982, Regional structural synthesis, Wyoming
    salient of the western Overthrust Belt: AAPG Bulletin,
    v. 66, n. 10, p. 1560–1580.

Elliott, D., and M. R. W. Johnson, 1980, Structural evolution
    in the northern part of the Moine thrust zone:
    Transactions of the Royal Society of Edinburgh, Earth
    Sciences, v. 71, p. 69–96.

Fisher, D. J., C. E. Erdman, and J. B. Reeside, Jr., 1960,
    Cretaceous and Tertiary formations of the Book Cliffs,

Carbon, Emery, and Grand counties, Utah and Garfield and Mesa counties, Colorado: U.S. Geological Survey Professional Paper 332, 80 p.

Fouch, T. D., T. F. Lawton, D. J. Nichols, W. B. Cashion, and W. A. Cobban, 1982, Chart showing preliminary correlation of major Albian to middle Eocene rock units from the Sanpete Valley in central Utah to the Book Cliffs in eastern Utah, *in* D. L. Nielson, ed., Overthrust Belt of Utah: Utah Geological Association Publication 10, p. 267–272.

———, 1983, Patterns and timing of synorogenic sedimentation in Upper Cretaceous rocks of central and northeast Utah, *in* M. W. Reynolds and E. D. Dolly, eds., Mesozoic paleogeography of west-central United States: Rocky Mountain Section, Society of Economic Paleontologists and Mineralogists, p. 305–336.

George, S. E., 1983, The geology of the Fillmore and Kanosh quadrangles, Millard County, Utah: Master's thesis, Brigham Young University, Provo, Utah.

Gilliland, W. N., 1963, Sanpete–Sevier Valley anticline of central Utah: Geological Society of America Bulletin, v. 74, p. 115–124.

Hardy, C. T., 1952, Eastern Sevier Valley, Sevier and Sanpete counties, Utah: Utah Geological and Mineralogical Survey Bulletin 43, 98 p.

Hardy, C. T., and H. D. Zeller, 1953, Geology of the west-central part of the Gunnison Plateau, Utah: Geological Society of America Bulletin, v. 64, p. 1261–1278.

Hickcox, C. W., 1971, The geology of a portion of the Pavant Range allochthon, Millard County, Utah: Ph.D. thesis, Rice University, Houston, Texas, 67 p.

Higgins, J., 1982, Geology of the Champlin Peak quadrangle, Juab and Millard counties, Utah: Brigham Young University Geological Studies, v. 29, pt. 2, p. 40–58.

Hintze, L. F., 1973a, Geologic history of Utah: Brigham Young University Geological Studies, v. 20, pt. 3, 181 p.

———, 1973b, Utah geological highway map: Brigham Young University Geological Studies Special Publications 3.

———, 1980, Geological map of Utah: Utah Geological and Mineralogical Survey, scale 1:500,000.

Holladay, J. D., 1983, Geology of the northern Canyon Range, Millard and Juab counties: Master's thesis, Brigham Young University, Provo, Utah.

Hunt, R. E., 1950, Geology of the northern part of the Gunnison Plateau: Ph.D. thesis, Ohio State University, Columbus.

Imlay, R. W., 1980, Jurassic paleogeography of the conterminous United States in its continental setting: U.S. Geological Survey Professional Paper 1062, 134 p.

Jordan, T. E., 1981, Thrust loads and foreland basin evolution, Cretaceous, western United States: AAPG Bulletin, v. 65, p. 2506–2520.

Kauffman, E. G., 1973, Stratigraphic evidence for Cretaceous eustatic changes (abs.): Geological Society of America Abstracts with Programs, Annual Meeting, p. 687.

———, 1977, Geological and biological overview: Western Interior Cretaceous basin: The Mountain Geologist, v. 14, n. 3-4, p. 75–99.

———, 1981, Interaction of tectonic, sedimentologic, and eustatic history in the Western Interior Cretaceous

seaway of North America (abs.), *in* J. R. Steidtmann, ed., Sedimentary tectonics—principles and applications: University of Wyoming, Wyoming Geological Association, Geological Survey of Wyoming Joint Meeting, p. 15.

Kluth, C. F., and P. J. Coney, 1981, Plate tectonics of the Ancestral Rocky Mountains: Geology, v. 9, p. 10–15.

Lautenschlager, H. K., 1952, The geology of the central part of the Pavant Range, Utah: Ph.D. thesis, Ohio State University, Columbus, 188 p.

Lawton, T. F., 1982, Lithofacies correlations within the Upper Cretaceous Indianola Group, central Utah, *in* D. L. Nielson, ed., Overthrust Belt of Utah: Utah Geological Association Publication 10, p. 199–213.

Lawton, T. F., W. R. Dickinson, and W. S. Jefferson, 1982, Inferred eastern extent of Overthrust Belt in central Utah (abs.): AAPG Bulletin, v. 66, p. 592.

Le Vot, M., 1984, Etude géologique d'un segment de l'Overthrust Belt—le Nord-Centre Utah, USA: Thèse 3ᵉᵐᵉ Cycle, Université de Bretagne Occidentale, Brest, Societe Nationale Elf Aquitaine Production.

McCubbin, D. G., 1961, Basal Cretaceous of southwestern Colorado and southeastern Utah: Ph.D. thesis, Harvard University, Cambridge, Massachusetts, 172 p.

McDonald, R. E., 1976, Tertiary tectonics and sedimentary rocks along the transition: Basin and Range province to Plateau and Thrust Belt province, *in* J. G. Hill, ed., Symposium on Geology of the Cordilleran Hingeline: Rocky Mountain Association of Geologists, p. 281–317.

Millard, A. W., 1982, Geology of the southwestern quarter of the Scipio North (15′) quadrangle, Millard and Juab counties, Utah: Master's thesis, Brigham Young University, Provo, Utah.

Mitchell, G. C., 1979, Stratigraphy and regional implications of the Argonaut Energy no. 1 Federal, Millard County, Utah, *in* G. W. Newman and H. D. Goode, eds., Basin and Range Symposium: Rocky Mountain Association of Geologists and Utah Geological Association, p. 503–514.

Morris, H., 1983, Interrelations of thrust and transcurrent faults in the central Sevier orogenic belt near Leamington, Utah: Geological Society of America Memoir 157, p. 75–81.

Moulton, F. C., 1975, Lower Mesozoic and upper Paleozoic petroleum potential of the hingeline area, central Utah, *in* D. W. Bolyard, ed., Symposium on Deep Drilling Frontiers in the Central Rocky Mountains: Rocky Mountain Association of Geologists, p. 87–97.

Nichols, D. J., and S. R. Jacobson, 1982a, Palynological dating of syntectonic units in the Utah–Wyoming thrust belt: the Evanston Formation, Echo Canyon Conglomerate, and Little Muddy Creek Conglomerate, *in* R. B. Powers, ed., Geologic Studies of the Cordilleran Thrust Belt: Rocky Mountain Association of Geologists, p. 735–750.

———, 1982b, Palynostratigraphic framework for the Cretaceous (Albian–Maestrichtian) of the Overthrust Belt of Utah and Wyoming: Palynology, v. 6, p. 119–147.

Oriel, S. S., 1969, Geology of the Fort Hill quadrangle, Lincoln County, Wyoming: U.S. Geological Survey

Professional Paper 594-M, 40 p.

Oriel, S. S., and F. C. Armstrong, 1966, Times of thrusting in Idaho–Wyoming thrust belt: Reply: AAPG Bulletin, v. 50, p. 2614–2621.

Peterson, J. A., 1972, Jurassic system, *in* W. W. Mallory, ed., Geologic Atlas of the Rocky Mountain Region: Rocky Mountain Association of Geologists, p. 177–189.

Picha, F., and R. I. Gibson, 1983, Basement control of the Sevier Orogenic Belt in Utah (abs.): Geological Society of America Abstracts with Programs, v. 15, n. 5, p. 378.

Pinnell, M. L., 1972, Geology of the Thistle quadrangle, Utah: Brigham Young University Geological Studies, v. 19, p. 89–130.

Price, R. A., 1981, The Cordilleran foreland thrust and fold belt in the Southern Canadian Rocky Mountains, *in* K. R. McClay and N. J. Price, eds., Thrust and Nappe Tectonics: Geological Society of London Special Publication 9, p. 427–448.

Price, R. A., and E. W. Mountjoy, 1970, Geologic structure of the Canadian Rocky Mountains between Bow and Athabasca Rivers—a progress report, *in* J. O. Wheeler, ed., Structure of the southern Canadian Cordillera: Geological Association of Canada Special Paper 6, p. 7–25.

Ramsay, J. G., M. Casey, and R. Kligfield, 1983, Role of shear in development of the Helvetic fold-thrust belt of Switzerland: Geology, v. 11, p. 439–442.

Royse, F., Jr., 1983, Extensional faults and folds in the foreland thrust belt, Utah, Wyoming, Idaho (abs.): Geological Society of America Abstracts with Programs, v. 15, p. 295.

Royse, F., Jr., M. A. Warner, and D. L. Reese, 1975, Thrust belt structural geometry and related stratigraphic problems, Wyoming–Idaho–Northern Utah, *in* D. W. Bolyard, ed., Symposium on Deep Drilling Frontiers in the Central Rocky Mountains: Rocky Mountain Association of Geologists, p. 41–54.

Ryer, T. A., 1983, Early Late Cretaceous paleogeography of east-central Utah, *in* M. W. Reynolds and E. D. Dolly, eds., Mesozoic Paleogeography of West-Central United States: Rocky Mountain Section, Society of Economic Paleontologists and Mineralogists, p. 253–272.

Smith, R. B., 1983, Cenozoic tectonics of the eastern Basin–Range: inferences on the origin and mechanism from seismic reflection and earthquake data (abs.): Geological Society of America Abstracts with Programs, v. 15, n. 5, p. 287.

Spieker, E. M., 1946, Late Mesozoic and early Cenozoic history of central Utah: U.S. Geological Survey Professional Paper 205-D, 161 p.

———, 1949, The transition between the Colorado Plateau and the Great basin in central Utah: Utah Geological Society Guidebook to the Geology of Utah, n. 4, 106 p.

Sprinkel, D. A., 1982, Twin Creek Limestone–Arapien Shale relations in central Utah, *in* D. L. Nielson, ed., Overthrust Belt of Utah: Utah Geological Association Publication 10, p. 169–179.

Sprinkel, D. A., and I. J. Witkind, 1982, Road log—relations between overthrusts and salt diapirs, central Utah, *in* D. L. Nielson, ed., Overthrust Belt of Utah: Utah

Geological Association Publication 10, p. 331–334.

Sprinkel, D. A., and G. L. Waanders, in press, Twin Creek Limestone–Arapien Shale stratigraphic and paleontologic relations, Arapien basin, Utah—A preliminary appraisal: AAPG Bulletin, v. 68, p. 950.

Standlee, L. A., 1982, Structure and stratigraphy of Jurassic rocks in central Utah: their influence on tectonic development of the Cordilleran foreland thrust belt, *in* R. B. Powers, ed., Geologic Studies of the Cordilleran Thrust Belt: Rocky Mountain Association of Geologists, p. 357–382.

———, 1983, Structural controls on the Mesozoic–Tertiary tectonic evolution of central Utah (abs.): Geological Society of America Abstracts with Programs, v. 15, p. 295.

Stanley, K. O., and J. W. Collinson, 1979, Depositional history of Paleocene–lower Eocene Flagstaff Limestone and coeval rocks, central Utah: AAPG Bulletin, v. 63, p. 311–323.

Steven, T. A., C. G. Cunningham, C. W. Naeser, and H. H. Mehnert, 1979, Revised stratigraphy and radiometric ages of volcanic rocks and mineral deposits in the Marysvale area, west-central Utah: U.S. Geological Survey Bulletin 1469, 40 p.

Stokes, W. L., 1952, Salt-generated structures of the Colorado Plateau and possible analogies (abs.): AAPG Bulletin, v. 36, p. 961.

———, 1956, Tectonics of Wasatch Plateau and nearby areas (abs.): AAPG Bulletin, v. 40, p. 790.

———, 1982, Geologic comparisons and contrasts, Paradox and Arapien basins, *in* D. L. Nielson, ed., Overthrust Belt of Utah: Utah Geological Association Publication 10, p. 1–11.

Stolle, J. M., 1978, Stratigraphy of the Lower Tertiary and Upper Cretaceous(?) continental strata in the Canyon Range, Juab County, Utah: Brigham Young University Geological Studies, v. 25, pt. 3, p. 117–139.

Taylor, J. M., 1980, Geology of the Sterling quadrangle, Sanpete County, Utah: Brigham Young University Geological Studies, v. 27, pt. 1, p. 117–135.

Thompson, R. I., 1981, The nature and significance of large "blind" thrusts within the northern Rocky Mountains of Canada, *in* K. R. McClay and N. J. Price, eds., Thrust and Nappe Tectonics: Geological Society of London Special Publication 9, p. 449–462.

Villien, A., 1980, Etude geologique d'un segment de l'Overthrust Belt—l'Utah centro-meridional, USA: Thèse 3ᵉᵐᵉ Cycle, Université de Bretagne Occidentale, Brest, Societe Nationale Elf Aquitaine Production, 251 p.

———, 1984, Central Utah deformation belt: Ph.D. thesis, University of Colorado, Boulder, 345 p.

Weiss, M. P., 1982, Structural variety on east front of the Gunnison Plateau, central Utah, *in* D. L. Nielson, ed., Overthrust Belt of Utah: Utah Geological Association Publication 10, p. 49–63.

Witkind, I. J., 1982, Salt diapirism in central Utah, *in* D. L. Nielson, ed., Overthrust Belt of Utah: Utah Geological Association Publication 10, p. 13–30.

Zoback, M. L., 1982, Preliminary interpretation of a 30-km-long seismic reflection profile in the hingeline near Nephi, Utah (abs.), *in* T. L. Britt, ed., Program and Abstracts for the 1982 Symposium on the Overthrust Belt of Utah: Utah Geological Association Publication 11, p. 16.

———, 1983, Structural style along the Sevier frontal thrust zone in central Utah (abs.): Geological Society of America Abstracts with Programs, v. 15, n. 5, p. 377.

The Influence of Preexisting Tectonic Trends On Geometries of the Sevier Orogenic Belt and Its Foreland in Utah

F. Picha
Gulf Oil Exploration and Production Co.
Houston, Texas

Tectonic style of the late Mesozoic to early Cenozoic(?) Sevier orogenic belt in Utah was greatly affected by preexisting structural trends that date from late Precambrian rifting and fragmentation of the North American continent. The frontal zone of the thrust belt was superimposed on the edge of the late Precambrian craton (Cordilleran hingeline) that was marked by a system of prominent faults and structural highs, such as the Fillmore arch, the north–south-trending ancestral Wasatch and Ancient Ephraim faults, and the northeast-trending Leamington, Scipio, Cove Fort, and Paragonah lineaments. The renewed activity of these structures affected the geometries of the late Paleozoic Ancestral Rocky Mountain uplifts and basins, the extent of the Jurassic evaporitic Arapien basin, and the sedimentary pattern of the Cretaceous foreland basin. During the compressional Sevier tectonism, some of these fault zones were reactivated as tectonic ramps (e.g., the Ancient Ephraim fault) and tear faults (e.g., the Leamington fault). The Fillmore arch and other structural highs along the edge of the late Precambrian craton caused ramping of the inner Keystone, Pavant, Canyon, Paris, and Willard thrust sheets and telescoping of the frontal thrust sheets.

During the Basin and Range extension, a major low-angle, normal fault (the Sevier Desert detachment) developed at the western side of the uplifted Fillmore arch. The northeast-trending Leamington, Cove Fort, and Paragonah lineaments were reactivated as left-lateral strike-slip faults. Mobilization of major fault zones that were inherited from late Precambrian rifting may account for some crustal shortening during the compressional Sevier orogeny and, conversely, for some crustal extension during the Basin and Range extensional event.

INTRODUCTION

The Sevier orogenic belt in western North America is a structurally complex terrane of Precambrian, Paleozoic, and Mesozoic rocks that were folded, faulted, and thrust onto the stable foreland. The compressional tectonism related to the development of an active continental margin along the Pacific side of North America lasted from Late Jurassic to early Tertiary time (Armstrong and Oriel, 1965), although some uncertainty exists about the timing of the earliest and latest events in various regions. The frontal zone of the Sevier belt is typically a thin-skinned structural complex consisting of numerous allochthonous and parautochthonous units detached at various stratigraphic levels and bound by thrusts dipping generally westward. This style is now clearly distinguished from the Late Cretaceous to early Tertiary Laramide foreland tectonism that typically involved the crystalline basement in the Rocky Mountains and Colorado Plateau areas (Gries, 1983).

The presence of low-angle thrusting in southeastern Nevada and Utah has been established by the work of Misch (1960), Armstrong (1968), Crittenden (1974), Burchfiel and Davis (1975), Allmendinger and Jordan (1981), and others. The internal geometries, the eastern extent of deformation, and the influence of basement tectonics, however, remain uncertain and are the subject of much discussion.

The classic model of thin-skinned thrust tectonics developed in Alberta (Bally et al., 1966; Dahlstrom, 1969; Price and Mountjoy, 1970) and in Wyoming (Royse et al., 1975; Dixon, 1982) suggests that the lowest level of deformation is a décollement that soles out at or above the crystalline basement, which slopes gently westward under the orogenic belt. The basement in those areas is not involved in thrusting and has little or no impact on the overlying thrust sheets. The significant tectonic shortening of the supracrustal rocks in the foreland thrust belts, however, has to be compensated at deeper crustal levels. Two concepts, which are often adjusted to specific conditions but are similar in principle, have been proposed to account for the shortening of the lower crust: (1) subduction of the deeper crust combined with decoupling of the supracrustal rocks (e.g., Bally et al., 1966; Bally, 1975;

309

Burchfiel and Davis, 1975; Price, 1981; Scholten, 1982) and (2) tectonic thickening of the original geosynclinal basement, which consisted of previously rifted attenuated continental crust. The latter model proposed by Helwig (1976) for the Alpine belt and adopted by Okulitch (1984) for the Canadian Rockies eliminates the need for the subduction of large volumes of continental crust, a concept that defies the principles of buoyancy. The occurrence of crystalline basement rocks in the inner zones of orogenic belts, such as the western Cordillera (see Scholten, 1982, for review), European Alps, and Carpathians, suggests that some major thrusts are rooted in the crystalline basement and that the basement rocks have moved into upper crustal levels. A significant amount of shortening can thus be accommodated by tectonic stacking of the deeper crust. Tectonic movements involving the basement are not necessarily confined to inner cores of orogenic belts. Significant interactions between the basement and progressing thrust belt, such as buttressing by foreland structures and reactivation of ancient transverse faults, have been reported elsewhere (see Beutner, 1977).

The Sevier thrust belt south of the Uinta Mountains in Utah provides an interesting opportunity for studying the relationship between the thrust belt and underlying basement structures. Unlike the Wyoming thrust salient north of the Uinta Mountains or the Canadian Rocky Mountain Foothills, which evolved on a gently westward-dipping platform east of the edge of the craton, the frontal zone of the Sevier belt in Utah overlies a complexly faulted margin of the North American craton (Cordilleran hingeline) that dates from late Precambrian rifting and continental separation.

In this paper we identify significant late Precambrian structures in western Utah that were reactivated during Phanerozoic time. We examine their influence on the geometries of the Mesozoic–Cenozoic Sevier orogenic belt and the superimposed Basin and Range extensional structures. The study is based on interpretation of surface and subsurface geology and published and unpublished geophysical data. During an earlier stage of the study, R. I. Gibson collaborated in joint papers on the Cordilleran hingeline in Utah and Nevada (Picha and Gibson, 1983, 1985).

LATE PRECAMBRIAN STRUCTURAL TRENDS

Late Precambrian rifting and the development of a passive continental margin in western North America set a structural pattern that persisted throughout the entire region during Phanerozoic time. The single most significant structural element of this late Precambrian tectonism in western Utah is the Cordilleran hingeline (or the Wasatch line of Kay's [1951] definition, which is a linear feature that separated the stable craton on the east during late Precambrian and early Paleozoic time from the attenuated, thinner crust of the miogeocline on the west. It roughly parallels the front of the Sevier thrust belt with which it sometimes has been confused. Stewart (1972, 1980), Burchfiel and Davis (1975), Stokes (1976), and other workers have related the origin of the hingeline to the fragmentation of the North American continent during late Precambrian rifting.

The hingeline is marked at the surface by a system of Tertiary normal faults. In northern Utah, the hingeline has been identified with the Wasatch fault, but south of Mount Nebo, the continuation of this fault is uncertain. Stokes (1976) connected the Wasatch fault with the Las Vegas line of Welsh (1959) in southern Nevada and with the Garlock fault in California. However, a complex feature like the Cordilleran hingeline cannot be tied to a single surface fault. Some Tertiary faults probably do reflect the existing weak zones along the Precambrian hingeline, but individually they should not be visualized as the hingeline proper. Picha and Gibson (1985) interpreted the hingeline as a broad and variable edge of the craton that is simply faulted in some places, but complexly faulted and less distinctive in other places.

Several significant structural features, such as the Fillmore arch, the Ancient Ephraim fault, and the northeast–southwest-trending lineaments identified in the hingeline area, document the complexity of the late Precambrian cratonic margin.

Fillmore Arch and Other Basement Highs

The tectonic window present in the southern Pavant Range and the eastward dip of the Pavant thrust sheet on the east side of the range suggest the presence of north–south-trending uplift along the western side of the Pavant and Canyon ranges (Figure 1). Hansen et al. (1980) described the structure as a part of a loosely defined Fillmore–Maricopa arch. They suggested that the arch represents the margin of the late Precambrian craton and extends the length of the western Cordillera from Idaho to Mexico.

The existence of the Fillmore arch is indicated by seismic data including the COCORP line (Allmendinger et al., 1983). The value of Bouguer gravity for the Fillmore arch is approximately 50 mGal higher than values in the Uinta Mountains and the Wasatch Plateau (Cook et al., 1975). The strong, broad Bouguer gravity maximum at the western side of the Canyon and Pavant ranges is unrelated to the topography of the ranges.

Picha and Gibson (1983) interpreted the Fillmore arch to be a thicker crustal block separated from the Wasatch Plateau by a weaker and thinner zone made up of the Sevier and Sanpete valleys (Figures 2 and 3). As a result of late Precambrian rifting, the Fillmore arch was positioned at the western edge of the thicker crust of the North American craton. West of the Fillmore arch, a thinner, attenuated crust lay under the Cordilleran miogeocline.

Similar structures are probably located in the Wasatch Front in northern Utah and in the Blue Mountains, Muddy Mountains and Keystone thrust area in southern Utah and southeastern Nevada (Figure 2). Like the Fillmore arch the other structures display tectonic windows through the overlying Sevier thrust sheets. Gravity data indicate highs similar to, but less conspicious than, the Fillmore arch. The Fillmore arch is thus a prominent segment of a discontinuous chain of structural highs marking the edge of the craton. The hypothetical line separating the late Precambrian craton from the miogeocline probably lies west of these uplifts.

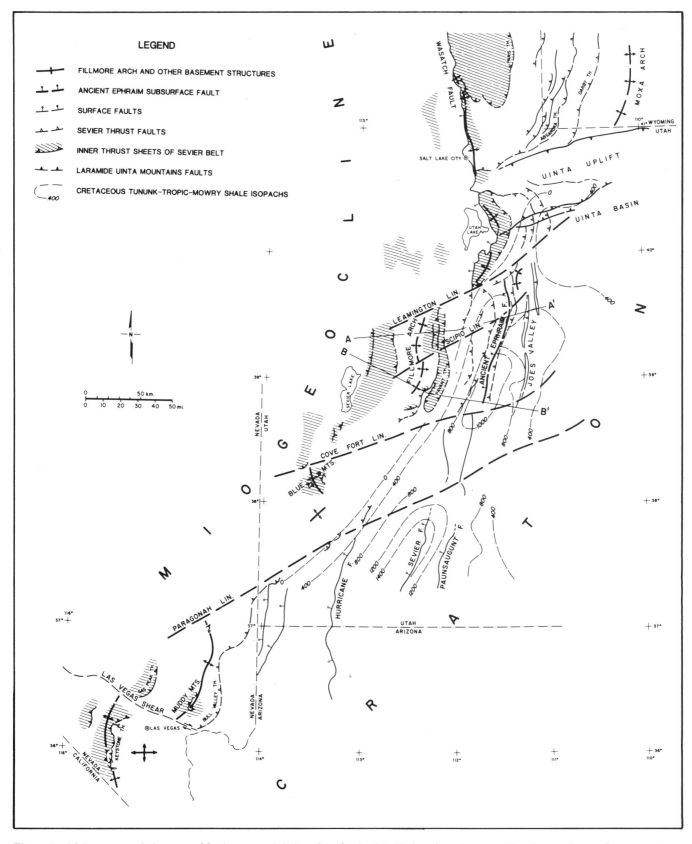

Figure 1—Major structural elements of Sevier orogenic belt and its foreland in Utah and southeastern Nevada superimposed on tectonic framework of late Precambrian rifted cratonic margin. Tectonic windows in inner Paris–Willard, Pavant, and Keystone thrust sheets coincide with old basement highs. The Wasatch syncline is bounded at depth by the Ancient Ephraim fault, which is interpreted as the easternmost tectonic ramp. Leamington and other northeast-trending lineaments offset the front of Sevier belt. Tununk, Tropic, and Mowry shales (Upper Cretaceous) isopachs (based on well data) show basement related depocenters within the foreland basin.

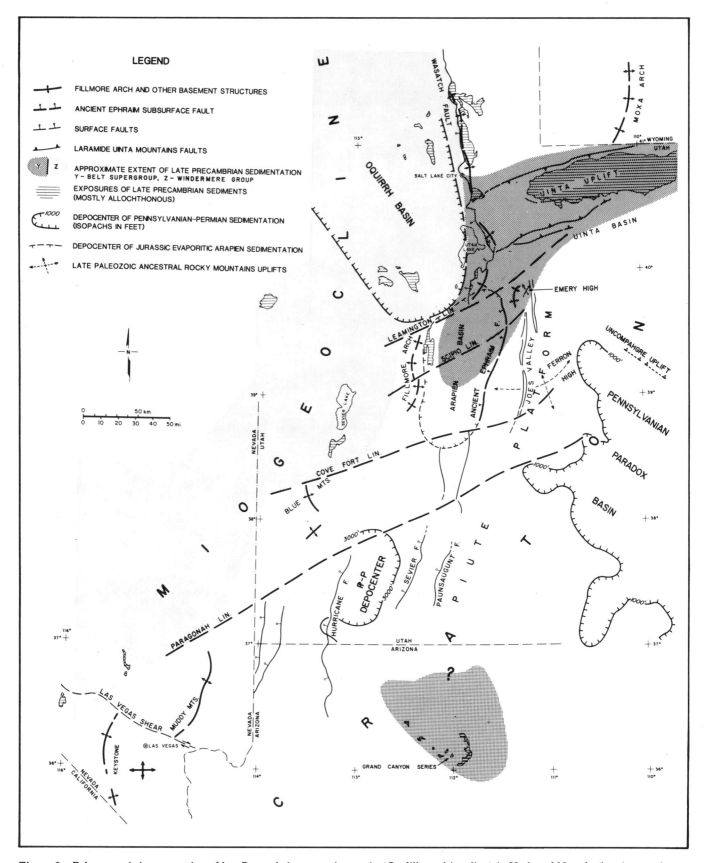

Figure 2—Paleotectonic interpretation of late Precambrian cratonic margin (Cordilleran hingeline) in Utah and Nevada showing northeast-trending lineaments, Ancient Ephraim fault, and structural highs such as Fillmore arch. Superimposed are tectonically bounded Pennsylvanian–Permian Oquirrh basin, Pennsylvanian Paradox basin (1000 ft [300 m] isopach), Pennsylvanian–Permian depocenter in southwestern Utah (3000 ft [900 m] isopach), and Jurassic evaporitic Arapien basin.

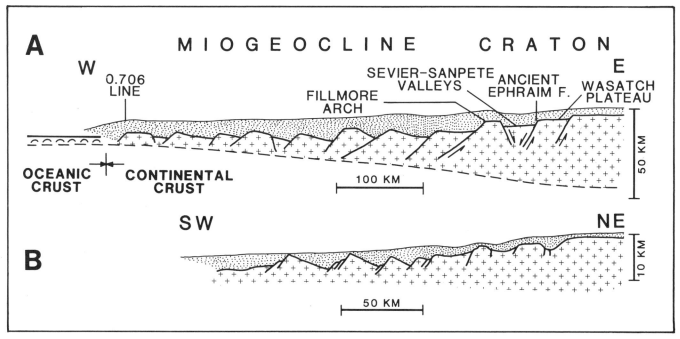

Figure 3—A. Hypothetical position of various crustal blocks at the edge of late Precambrian rifted margin of western North America. Arrows indicate possible fault zones activated during compressional Sevier orogeny and extensional Basin and Range tectonism. B. Example of a modern continental margin from the eastern Atlantic Ocean for comparison. From Roberts et al. (1981). Used with permission of the Institute of Petroleum.

Ancient Ephraim Fault

The north-south-trending Ancient Ephraim fault (Moulton, 1976), which separates at depth the Wasatch Plateau from the Sevier and Sanpete valleys depression, is another prominent basement feature whose origin might be related to late Precambrian fragmentation of the North American continent. This fault is concealed under the detached Wasatch syncline (Figures 2, 4, and 5) and directly affects only the underlying Paleozoic strata. It is apparent, however, that the surface fault zone bounding the eastern side of the Wasatch syncline is related to the Ancient Ephraim fault, which during Sevier thrusting may have acted as the easternmost ramp. This preexisting zone of weakness was reactivated again during the extensional Basin and Range regime.

The Ancient Ephraim fault is parallel to the Fillmore arch and represents a major break within the late Precambrian cratonic margin. The Scofield Reservoir–Joes Valley surface fault system on the eastern side of the Wasatch Plateau may reflect yet another late Precambrian zone of weakness within the cratonic margin.

Northeast-Trending Lineaments

Several northeast- and east-trending lineaments and mineral belts have been described in the transitional zone between the Colorado Plateau and the Great Basin province (e.g., Morris and Shepard, 1964; Stewart et al., 1971; Shawe and Stewart, 1976; Rowley et al., 1978). Some of these features are related to extensional Basin and Range tectonics, whereas others are much older and have been repeatedly active through time. On the basis of interpretation of geology and geophysical data, Picha and Gibson (1985) identified four northeast-trending lineaments whose geologic history can be traced back to the late

Precambrian rifting and fragmentation of the North American continent:

1. The Leamington lineament terminates the Fillmore arch on the north and marks the southern limit of the Oquirrh Formation in the East Tintic and Gilson mountains. It roughly delineates the southern edge of the Charleston–Nebo thrust salient, and it follows the northern edge of the late Paleozoic Emery High (Figure 2).
2. The Scipio lineament offsets the Fillmore arch, the Canyon Mountains thrust, and the eastern side of the Gunnison Plateau, and it may affect the deeper structure of the Wasatch Plateau (Figure 2).
3. The Cove Fort lineament probably terminates the Fillmore Arch and the Pavant Range on the south and marks the northern margin of the Marysvale volcanic field (Figure 5).
4. The Paragonah lineament passes through the attenuated zone north of the Beaver Dam Mountains, appears to terminate the Hurricane fault system north of Cedar City, and roughly follows the southeastern side of the Marysvale volcanic field (Figure 5).

These lineaments, not unlike the transform faults of modern continental margins, offset the edge of the late Precambrian craton between Las Vegas and Salt Lake City (Figure 2). The southern side of each fault is upthrown and offsets the craton edge westward. Rather than following an imaginary straight line, the cratonic margin thus trends southwest in a series of steps (Figure 2).

Similar northeast-trending lineaments, possibly of Precambrian age, have been recognized in the Canadian Rockies (Price, 1981) and in the Northern Rockies in the United States (see Scholten, 1982, for review).

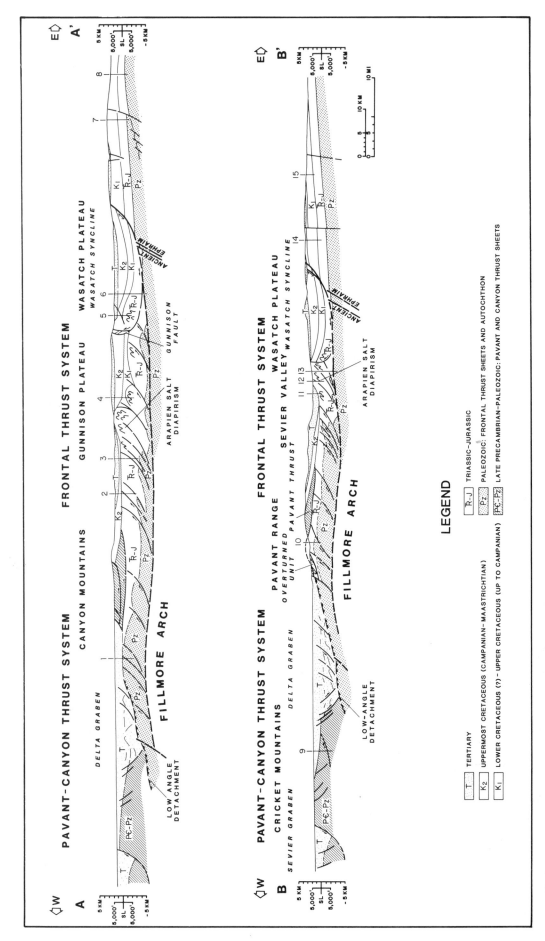

Figure 4—Regional cross sections (A–A' and B–B') through the Sevier orogenic belt and its foreland in central Utah (see Figure 1 for location) based on surface and subsurface geology and numerous seismic data. Pavant-Canyon thrust system (striped pattern) ramps over the Fillmore arch; frontal thrust system (dotted pattern on Paleozoic strata) consists of numerous imbricate sheets, including an overturned unit exposed on the western side of Pavant Range. Low-angle detachment underneath Sevier Desert is intersected by at least one normal fault associated with late Tertiary and Quaternary volcanic activity. Frontal thrusts were reactivated as extensional listric faults; Wasatch syncline slid westward against Gunnison Plateau. Numbers refer to following oil wells: 1. Placid #1 Henley; 2. Placid 13-7 Monroe; 3. Placid WXC-1 Barton; 4. Dixel #1 Gunnison State; 5. Phillips #1 Price; 6. Hanson 1A-X Moroni; 7. Tenneco #1 Clear Creek; 8. Oxydental #1 Gordon Creek; 9. Cominco #2 Beaver River; 10. Shell #1 Sunset Canyon; 11. Champlin 13-31 USA; 12. Standard California #1 Sigurt; 13. Chevron #1 Salina; 14. Gulf #1 Johnson Livestock; and 15. Moore Hiram #1 Coral Federal.

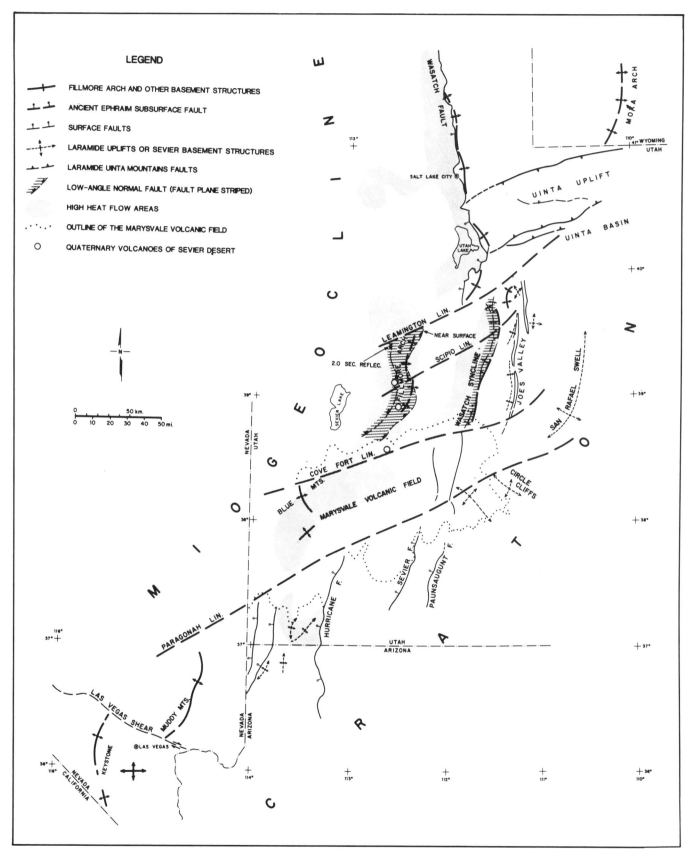

Figure 5—Basin and Range low-angle extensional faulting superimposed on late Precambrian tectonic pattern in Utah and southeastern Nevada. Present thermal activity (Utah Geological and Mineral Survey, 1983) outlines zones of significant crustal extension, such as the edge of the late Precambrian craton and the structurally weak zone of the Sevier and Sanpete valleys. Sevier Desert detachment (striped) surfaces at western side of Fillmore arch. Detached Wasatch syncline slid westward from tectonic ramp bounded by Ancient Ephraim fault; San Rafael Swell and Circle Cliffs are Laramide uplifts.

THE INFLUENCE OF LATE PRECAMBRIAN TECTONICS ON DEPOSITIONAL AND STRUCTURAL HISTORY

The overall structural and depositional pattern of the Cordilleran hingeline resembles the modern rifted continental margins as portrayed, for example, by Roberts et al. (1981) in the eastern Atlantic. The schematic cross section shown in Figure 3 which extends from the Wasatch Plateau into the Great Basin, indicates the hypothetical position of various crustal blocks at the edge of the North American platform in late Precambrian time.

The Fillmore arch, Ancient Ephraim fault, northeast-trending lineaments, and other structural elements of the late Precambrian cratonic margin are reflected in the depositional and structural history of the region. When subjected to stress these features repeatedly reactivated as tectonic uplifts, buttresses, ramps, tear faults, extensional detachments, and strike-slip faults. Their subsequent activity affected both the distribution of sediments in various basins and the geometries of the compressional Sevier orogenic belt and the extensional Basin and Range tectonism. The present surface expression of these features may be misleading because it often reflects only the last episode in their long history.

Late Precambrian and Paleozoic History

The structural pattern set by late Precambrian rifting and continental separation in western Utah is best reflected by the distribution of upper Precambrian sedimentary rocks. The older series (equivalent of the Belt Supergroup) exposed in the Uinta Mountains has been interpreted as a fill of an aborted east–west rift (Crittenden and Wallace, 1973; Stokes, 1976). Our seismic data (unpublished) indicate the presence of equivalent upper Precambrian strata in the northwestern corner of the Wasatch Plateau (Figure 2). These rocks are separated from the overlying Cambrian strata by an angular unconformity and were apparently deposited on a tectonically strained block adjacent to the Uinta aulacogen. Another grabenlike structure has been suggested by Bayley and Muehlberger (1968) to connect the Grand Canyon Series of northern Arizona with the main north-south-trending rift farther west. All these deposits probably represent an early stage of rifting and continental separation characterized by development of intrasialic transverse grabens (aulacogens) that cut deeply into the North American continent.

The younger late Precambrian series (equivalent of the Windermere Group, <850 m.y. old) has closer affinities with Early Cambrian sedimentary patterns and was deposited unconformably on the older series (Stewart, 1972). The younger rocks probably represent the development of a continental terrace on the older Belt–Purcel series (Burchfiel and Davis, 1975).

As indicated by the presence of the late Precambrian rocks (<850 m.y. old) in the Paris-Willard, Canyon Mountains, and other inner thrust sheets and by their absence in the frontal imbricate thrusts (Figures 1, 2, and 4), the eastern extent of the younger Precambrian series hypothetically can be located west of the Fillmore arch and other structural highs marking the edge of the craton.

In Paleozoic time the late Precambrian structural trends in the hingeline area affected, among other things, the geometries of the Ancestral Rocky Mountain uplifts and basins. The Pennsylvanian–Permian Oquirrh basin formed on a weak crust west of the Precambrian cratonic margin. The pre-Sevier extent of the thick Oquirrh deposits to the south and east was apparently constrained by the precursors of the modern Wasatch fault and the Leamington lineament (Figure 2).

The Piute platform (Hintze, 1973), a Pennsylvanian uplift underlying the present Wasatch Plateau, is bounded on the western side by the Ancient Ephraim fault. Detailed isopach maps and seismic data (unpublished) indicate that the northeast-trending lineaments may have had some influence on segmentation of the Piute platform into such smaller structural units as the Emery and Ferron highs (Figure 2).

In Permian time the Leamington lineament separated the Phosphoria–Park City facies in northwestern Utah from the time equivalent Kaibab carbonate facies in the Wasatch Plateau.

Sevier Orogeny

The old structural pattern of the late Precambrian cratonic margin is most clearly shown in the geometries of the Sevier orogenic belt. On the basis of stratigraphy and structural position, two principal groups of thrust sheets, inner and frontal, can be distinguished in the Utah–Nevada segment of the thrust belt. The inner Paris-Willard, Mount Nebo, Canyon, Tintic, Pavant, and Keystone thrust sheets are typically detached at the upper Precambrian and Cambrian levels and are characterized by a thick upper Precambrian and Paleozoic geoclinal sequence. They ramped over the Wasatch Front, Fillmore arch, Muddy Mountains, and the other basement highs of the cratonic margin (Figure 1). Thick piles of late Cretaceous synorogenic conglomerates on the eastern side of the Pavant, Canyon, and Mount Nebo ranges probably mark the eastern edge of these plates.

The frontal system consists of numerous imbricate sheets of upper Paleozoic and Mesozoic rocks having only small displacements (a few kilometers) relative to one another. Their sedimentary sequence more closely resembles the platform facies of the Colorado Plateau. The presence of an overturned unit in the southern Pavant Range (Crosby, 1959), as well as the stacking of thrust sheets on the western side of the Fillmore arch, indicate possible duplex structures on the top of the arch.

The postthrust uplift of the basement structures and extensional relaxation caused the inner thrust sheets to be eroded and tectonically denuded by backsliding along the low-angle detachment on the western side of the Fillmore arch. The underlying frontal thrust system was exposed in a series of tectonic windows, such as those in the Wasatch Front, southern Pavant Range, Blue Mountains, and Muddy Mountains. The uplift and/or subsidence of basement structures may also account for gravitational gliding in the frontal zone of the thrust belt.

The salt-bearing anticlinal structures in the Sevier and Sanpete valleys have been traditionally interpreted as salt diapirs (Stokes, 1952; Baer, 1976; Witkind, 1982). However, new data from field observations, deep wells, and seismic surveys collected mostly by private industry (and including my work) and data published (Standlee, 1982) or presented at the Utah Geological Association Field Trip in 1982 revealed that the salt diapirism is only a secondary phenomenon superimposed on primary compressional Sevier structures.

The Jurassic evaporitic basin (Arapien Formation) formed in the weak zone between the Fillmore arch to the west and the Ancient Ephraim fault to the east (Figure 2). The northern limit of thick Arapien deposition is marked by the Leamington lineament and the southern limit under the Marysvale volcanic field is near the Cove Fort lineament. Hansen et al. (1980) interpreted the Arapien basin as a rift valley extending from the Gulf of Mexico to Canada, but there is no remaining evidence of regional thermal uplift or volcanic activity. An extensional event of such magnitude is unlikely to have occurred during the generally compressional regime of the Sevier orogeny. I believe that the Arapien salt basin is an early foreland basin that formed in response to an early stage of the Sevier orogeny. The extent and internal geometry of the basin was partly determined by the preexisting tectonic framework whose elements were reactivated by the advancing Sevier tectonism.

Seismic data have revealed the existence of a major synclinal structure in the western part of the Wasatch Plateau. This structure, which I call the Wasatch syncline, is detached at a lower Mesozoic level from the underlying autochthonous strata of the Wasatch Plateau (Figure 4). I interpret the Wasatch syncline as a Sevier compressional structure, which during the subsequent Basin and Range extension, slid back into the Sevier and Sanpete valleys. The fault bounding the eastern side of the syncline is tentatively considered to be the leading edge of the Sevier thrust belt. This fault follows the eastern side of the Ancient Ephraim fault, which acted as a ramp (Figures 1 and 4). There is a remote possibility that some minor displacement could have reached as far east as the Joes Valley graben and the Scofield Reservoir fault system.

The northeast-trending lineaments were apparently reactivated as right-lateral tear faults that offset the frontal zone of the eastward-propagating thrust belt. The most prominent shift occurred on the Leamington lineament. North of the Leamington lineament, the Sevier thrusting progressed far into a structurally weak zone, a precursor of the present Uinta Basin, and formed the Charleston–Nebo thrust salient. South of the Leamington fault, the high-standing basement block of the Wasatch Plateau hindered the eastward progress of thrusting. Other right-lateral shifts apparently occurred on the Cove Fort and Paragonah lineaments (Figure 1). The front of the thrust belt thus gradually retreats southwestward reflecting the gradually higher structural position of cratonic basement blocks.

The thin-skinned structures in the frontal zone of the Sevier thrust belt were locally affected by the Late Cretaceous to early Tertiary Laramide deformation that typically involved the crystalline basement. The east-west-trending faults on the southern side of the Laramide Uinta Mountains uplift, which overprint the thin-skinned Sevier structure of the Charleston–Nebo thrust salient, are a good example. It is also possible that the Fillmore arch and other basement highs along the cratonic margin were uplifted during Laramide tectonism. Such uplifting may actually have started the Basin and Range extension. According to Armstrong (1974) and Burchfiel and Davis (1975) the

Laramide deformation is an expression of relative eastward propagation of crustal compression. The kinematic and dynamic relationship of the Laramide deformation to the thin-skinned Sevier tectonism in the Hingeline area is little understood, partly because both structural trends are overprinted by the Basin and Range extensional event. Scattered pieces of information, however, indicate that the intensity of the thin-skinned Sevier tectonism along the Cordilleran hingeline in Utah decreased southward. On the other hand, the Laramide tectonic style seems to be more visible in southern Utah and Nevada and prevails in southern Arizona. This general trend indicates that south of the Uinta Mountains the crystalline basement was mobilized during the late Mesozoic to early Tertiary orogenies. The thin-skinned model of foreland thrusting, well documented in the Northern Rockies, is thus less applicable in the southern part of the western Cordillera.

Cretaceous Foreland Basin

The late Precambrian structural trends also had an impact on the distribution of sediments in the Cretaceous foreland basin that formed on the depressed cratonic crust in front of the advancing Sevier orogenic belt. The western margin of Cretaceous deposition followed the eastern side of the Fillmore arch and other basement highs that were situated on the edge of the late Precambrian platform. Although the overall geometry of the foreland basin was set by the Sevier tectonism, the deposition was locally affected by differential subsidence of preexisting crustal blocks.

The deepest segment of the foreland basin is located just north and south of the Uinta Mountains on the weak, fragmented crust adjacent to the late Precambrian triple junction between the north-south-trending main rift and the east-west-trending Uinta Mountains aulacogen. More subtle structural trends are reflected in the sedimentary pattern of marine shales, which because of slow deposition, reflect the motion on basement faults better than rapidly deposited sandstones. The isopach map of the Upper Cretaceous Tununk, Tropic, and Mowry shales (Figure 1), shows two depositional centers, one between the Leamington and Cove Fort lineaments and one located south of the Paragonah lineament. Another example higher in the stratigraphic succession is the Blue Gate Shale, which is generally confined to the area between the Leamington lineament on the north and the Cove Fort lineament on the south.

Basin and Range Extensional Tectonism

In early Tertiary time, the dominantly compressional Sevier and Laramide orogenies were replaced by the Basin and Range extension. During the early stage of extension, a major, low-angle, normal fault developed at the western side of the uplifted Fillmore arch. It was recognized by McDonald (1976) and is called the Sevier Desert detachment by Allmendinger et al. (1983). On the basis of the analysis of a large amount of unpublished seismic data, a remarkable detachment plane has been mapped under most of the Sevier Desert (Figure 5). This fault plane surfaces near the top of the Fillmore arch at the west side of the Pavant and Canyon ranges and dips gently (10–15°) to the west below the deep Delta graben, which has more than 3700 m of Tertiary sedimentary fill (Figure 4). The detachment on the western side of the Fillmore arch follows, at least locally, the preexisting Pavant thrust.

Structural reconstruction of the Delta graben and the upper thrust plates on the Fillmore arch suggests at least 20–30 km of extension in the area just west of the town of Fillmore. Allmendinger et al. (1983) calculated 30–60 km of extensional displacement along this fault detachment plane as far west as the Snake Range. They also pointed out that over its entire length, the low-angle fault is not significantly cut by any high-angle normal fault. Other seismic data indicate that the Sevier Desert detachment may not be perfectly featureless. A vertical offset of the detachment plane recognized in the deep central zone of the Delta graben (Figure 4) shows that at least some tectonic disruption of the detachment plane might have occurred here as a result of rejuvenation of faults bounding the Fillmore arch on the west. Moreover, the alignment of the late Tertiary and Quaternary basaltic volcanic cones and lava flows along this fault zone (Figure 5) suggests that deep fractures reach into the lower crust or the upper mantle. Both the regional detachment and the young volcanism are located along the old cratonic margin, which is marked by structural highs such as the Fillmore arch.

The development of the low-angle fault on the western side of the Fillmore arch is possibly related to the existence of a major tectonic discontinuity between the thicker crustal block of the arch and the thinner attenuated crust to the west. If so, similar detachments could have formed on the edges of other late Precambrian crustal blocks as well. The most probable places where such phenomena could occur are the Wasatch Front and the adjacent Salt Lake Desert, the Escalante Desert in southwestern Utah, and the Las Vegas area in southeastern Nevada. Similar but less significant extension probably occurred at the Ancient Ephraim fault, as indicated by the detachment and westward sliding of the Wasatch syncline (Figure 4). The frontal thrust faults that bound this syncline were reactivated as normal listric faults.

These analyses are strongly supported by the distribution of heat flow in western Utah (Utah Geological and Mineral Survey, 1983). The most significant anomaly follows the western side of the Wasatch fault to the Leamington lineament where it shifts west of the Fillmore arch into the Sevier Desert (Figure 5). It apparently marks the active zone of crustal extension on the Fillmore arch and other structural highs associated with the cratonic margin. Another narrow anomalous geothermal zone in the Sevier and Sanpete valleys roughly follows the downthrown side of the Ancient Ephraim fault.

In response to Basin and Range extension, the northeast-trending Leamington lineament was reactivated as a left-lateral strike-slip fault, which further affected the Charleston–Nebo thrust salient. The measured net slip on the Leamington lineament thus resulted from a combined affect of the right-lateral motion during the compressional Sevier orogeny and the left-lateral motion during the Basin and Range extension.

CONCLUSIONS

The classic model of thin-skinned thrust tectonics developed in Alberta and Wyoming suggests that the thrusting soles out at or above the crystalline basement and that the basement has little or no impact on the geometry of the thrust belt proper. This concept can be applied to many segments of foreland thrust belts, but it should not be viewed as a universal truth, because there are numerous exceptions.

Whether or not the frontal zone of any thrust belt is significantly affected by basement tectonics may depend primarily on the existence of tectonically active zones under the thrust belt. Unlike the southern Alberta Foothills and the Wyoming salient, where the thrusting progressed on a gently dipping stable platform and was thus uninhibited by basement tectonics, the frontal zone of the Sevier belt in western Utah and southwestern Nevada was superimposed on a structurally complex margin of the North American craton (Cordilleran hingeline) that dates from late Precambrian rifting and continental separation. During the compressional Sevier orogeny, some of the ancient faults were reactivated as tectonic ramps and tear faults. The high-standing blocks along the edge of the craton caused telescoping of imbricate thrusts, and their subsequent uplift probably enhanced some gravity gliding and development of major, low-angle extensional detachments. Whether or not some of the major thrust faults are rooted in the crystalline basement remains open to further investigation.

Tectonic mobilization of the deeper crustal pattern inherited from the late Precambrian rifting may account for some crustal shortening during the Sevier and Laramide compressional orogenies and some crustal extension during the subsequent Basin and Range extensional event.

ACKNOWLEDGMENTS

The author thanks Gulf Oil Company for supporting this study and granting permission to publish this paper, and R. I. Coward, and C. A. Ross for comments on the manuscript. M. D. Picard, L. F. Hintze, and G. W. Crosby critically reviewed a preliminary version of the manuscript and suggested modifications that significantly improved the final draft of the publication.

REFERENCES CITED

Allmendinger, R. W., and T. E. Jordan, 1981, Mesozoic evolution, hinterland of the Sevier orogenic belt: Geology, v. 9, p. 308–313.

Allmendinger, R. W., J. W. Sharp, D. Von Tish, L. Serpa, L. Brown, S. Kaufman, J. Oliver, and R. B. Smith, 1983, Cenozoic and Mesozoic structure of the eastern Basin and Range province, Utah from COCORP seismic-reflection data: Geology, v. 11, p. 532–536.

Armstrong, R. L., 1968, Sevier orogenic belt in Nevada and Utah: GSA Bulletin, v. 79, p. 429–458.

————, 1974, Magmatism, orogenic timing, and orogenic diachronism in the Cordillera from Mexico to Canada: Nature, v. 247, p. 348–351.

Armstrong, F. C., and S. S. Oriel, 1965, Tectonic development of the Idaho–Wyoming thrust belt: AAPG Bulletin, v. 49, p. 1847–1866.

Baer, J. L., 1976, Structural evolution of Central Utah—Late Permian to Recent, *in* J. G. Gilmore, ed., Geology of the Cordillera Hingeline: Denver, Rocky Mountain Association of Geologists Symposium, p. 37–46.

Bally, A. W., 1975, A geodynamic scenario for hydrocarbon occurrences: Tokyo, 9th World Petroleum Congress, v. 2, p. 23–44.

Bally, A. W., P. L. Gordy, and G. A. Stewart, 1966, Structure, seismic data and orogenic evolution of southern Canadian Rockies: Bulletin of Canadian Petroleum Geologists, v. 14, p. 337–381.

Bayley, R. W., and W. R. Muehlberger, 1968, Basement rock map of the United States: U.S. Geological Survey, scale 1:2,500,000.

Beutner, E. C., 1977, Causes and consequences of curvature in the Sevier orogenic belt, Utah to Montana, *in* E. L. Heisey, D. E. Lawson, E. R. Norwood, P. H. Wach, and L. A. Hale, eds., Rocky Mountain Thrust Belt Geology and Resources: Wyoming Geological Association 29th Annual Field Conference Guidebook, p. 353–365.

Burchfiel, B. C., and G. A. Davis, 1975, Nature and controls of Cordilleran orogenesis, western United States: Extensions of an earlier synthesis: American Journal of Science, v. 275-A, p. 363–396.

Cook, K. L., J. R. Montgomery, J. T. Smith, and E. F. Gray, 1975, Simple Bouguer gravity anomaly map of Utah: Utah Geological and Mineral Survey Map 37, scale 1:1,000,000.

Crittenden, M. D., Jr., 1974, Regional extent and age of thrusts near Rockport Reservoir and relation to possible exploration targets in Northern Utah: AAPG Bulletin, v. 58, p. 2428–2435.

Crittenden, M. D., and C. A. Wallace, 1973, Possible equivalents of the Belt Supergroup in Utah: Belt Symposium, Idaho Bureau of Mines and Geology, v. 1, p. 116–138.

Crosby, G. W., 1959, Geology of the South Pavant Range, Millard and Sevier Counties, Utah: Brigham Young University Geological Studies, v. 6, n. 3, 59 p.

Dahlstrom, C. D. A., 1969, Balanced cross sections: Canadian Journal of Earth Science, v. 6, p. 743–757.

Dixon, J. S., 1982, Regional structural synthesis Wyoming salient of Western overthrust belt: AAPG Bulletin, v. 66, p. 1560–1580.

Gries, R., 1983, Oil and gas prospecting beneath Precambrian of foreland thrust plates in Rocky Mountains: AAPG Bulletin, v. 67, p. 1–28.

Hansen, A. R., F. C. Moulton, and B. F. Owings, 1980, Utah–Arizona overthrust–hingeline belt: Oil and Gas Journal, v. 78, n. 47, p. 188–199.

Helwig, T., 1976, Shortening of the continental crust in orogenic belts and plate tectonics: Nature, v. 260, p. 768–770.

Hintze, L. T., 1973, Geologic history of Utah: Brigham Young University Geological Studies, v. 20, pt. 3, p. 181.

Kay, M., 1951, North American geosynclines: Geological Society of America Memoir 48, 143 p.

McDonald, R. E., 1976, Tertiary tectonics and sedimentary rocks along the transition: Basin and Range province to Plateau and Thrust Belt province, Utah, *in* J. G. Hill, ed., Geology of the Cordilleran Hingeline: Denver,

Rocky Mountain Association of Geologists Symposium, p. 281–317.

Misch, P., 1960, Regional structural reconnaissance in central northeast Nevada and some adjacent areas—observations and interpretations, *in* Geology of East Central Nevada: Intermountain Association of Petroleum Geologists and Eastern Nevada Geological Society, 11th Annual Field Conference Guidebook, p. 17–42.

Morris, H. T., and W. M. Shepard, 1964, Evidence for a concealed tear fault with large displacement in the Central East Tintic Mountains, Utah: U. S. Geological Survey Professional Paper, 501-C, p. 19–21.

Moulton, F. C., 1976, Lower Mesozoic and Upper Paleozoic petroleum potential of the Hingeline area, Central Utah, *in* J. G. Hill, ed., Geology of the Cordilleran Hingeline: Denver, Rocky Mountain Association of Geologists Symposium, p. 219–229.

Okulitch, A. V., 1984, The role of the Shuswap Metamorphic Complex in Cordilleran tectonism: a review: Canadian Journal of Earth Sciences, v. 21, p. 1171–1193.

Picha, F., and R. I. Gibson, 1983, Basement control of the Sevier Orogenic Belt in Utah (abs.): 36th Annual Meeting Rocky Mountain Section and 79th Annual Meeting Cordilleran Section, Geological Society of America Programs with Abstracts, p. 378.

————, 1985, Cordilleran hingeline: Late Precambrian rifted margin of the North American craton and its impact on the depositional and structural history, Utah and Nevada: Geology, v. 13, p. 465–468.

Price, R. A., 1981, The Cordilleran foreland thrust and fold belt in the southern Canadian Rocky Mountains, *in* K. R. McClay, and N. J. Price, eds., Special Publication No. 9: Geological Society of London, p. 427–448.

Price, R. A., and E. W. Mountjoy, 1970, Geologic structure of the Canadian Rocky Mountains between Bow and Athabasca rivers—a progress report, *in* J. O. Wheeler, ed., Structure of the southern Canadian Cordillera: Geological Association of Canada Special Paper 6, p. 7–39.

Roberts, D. G., D. G. Masson, L. Montadert, and O. de Charpal, 1981, Continental margin from the Porcupine Seabight to the Armorican marginal basin, *in* L. V. Illing, and G. D. Hobson, eds., Petroleum geology of the continental shelf of north-west Europe: London, Institute of Petroleum, p. 455–473.

Rowley, P. D., P. W. Lipman, H. H. Mehnert, E. A. Lindsey, and J. J. Anderson, 1978, Blue Ribbon Lineament, an east-trending structural zone within the Pioche mineral belt of southwestern Utah and eastern Nevada: U. S. Geological Survey Journal of Research, v. 6, n. 2, p. 175–192.

Royse, F., Jr., M. A. Warner, and D. L. Reese, 1975, Thrust belt structural geometry and related stratigraphic problems Wyoming–Idaho–northern Utah, *in* Deep Drilling Frontiers of the Rocky Mountains: Denver, Rocky Mountain Association of Geologists Symposium, p. 41–54.

Scholten, R., 1982, Continental subduction in the Northern U.S. Rockies—a model for back-arc thrusting in the western Cordillera, *in* R. B. Powers, ed., Geologic studies of the Cordilleran thrust belt: Rocky Mountain Association of Geologists, v. 1, p. 123–1136.

Shawe, D. R., and J. H. Stewart, 1976, Ore deposits as related to tectonics and magnetism, Nevada and Utah: American Institute of Mining, Metallurgical, and Petroleum Engineers, Economics Transactions, v. 260, p. 225–232.

Standlee, L. A., 1982, Structure and stratigraphy of Jurassic rocks in central Utah: their influence on tectonic development of the Cordilleran foreland thrust belt, *in* R. B. Powers, ed., Geologic studies of the Cordilleran thrust belt: Rocky Mountain Association of Geologists, v. 1, p. 357–382.

Stewart, J. H., 1972, Initial deposits in the Cordilleran geosyncline; evidence of a late Precambrian (<850 m.y.) continental separation: Geological Society of America Bulletin, v. 83, p. 1345–1360.

———, 1980, Geology of Nevada—a discussion to accompany the Geologic Map of Nevada: Nevada Bureau of Mines and Geology Special Publication 4, 136 p.

Stewart, J. H., W. J. Moore, and I. Zietz, 1971, East–west pattern of Cenozoic igneous rocks, aeromagnetic anomalies, and mineral deposits, Nevada and Utah: Geological Society of America Bulletin, v. 88, p. 67–77.

Stokes, W. L., 1952, Salt-generated structures of the Colorado Plateau and possible analogies (abs.): AAPG Bulletin, v. 36, p. 961.

———, 1976, What is the Wasatch Line?, *in* J. G. Hill, ed., Geology of the Cordilleran Hingeline: Denver, Rocky Mountain Association of Geologists Symposium, p. 11–25.

Utah Geological and Mineral Survey, 1983, Energy resources map of Utah, Salt Lake City, Utah Geological and Mineral Survey Map 68.

Welsh, J. E., 1959, Biostratigraphy of the Pennsylvanian and Permian Systems in southern Nevada: Ph.D. dissertation, University of Utah, Salt Lake City, Utah.

Witkind, I. J., 1982, Salt diapirism in Central Utah, *in* D. L. Nielson, ed., Overthrust Belt of Utah: Utah Geological Association Publication 10, p. 13–30.

Geometry, Distribution, and Provenance of Tectogenic Conglomerates Along the Southern Margin of the Wind River Range, Wyoming

J. R. Steidtmann
University of Wyoming
Laramie, Wyoming

L. T. Middleton
Northern Arizona University
Flagstaff, Arizona

R. J. Bottjer[1]
K. E. Jackson[2]
L. C. McGee[3]
E. H. Southwell
University of Wyoming
Laramie, Wyoming

S. Lieblang[4]
Northern Arizona University
Flagstaff, Arizona

Surface and subsurface studies of the tectogenic conglomerates along the southern margin of the Wind River Range show that they are of three general types with regard to size, geometry, and depositional environment and that each of these types can be related to a specific structural setting. The Fort Union Formation (Paleocene) and the main body of the Wasatch Formation (early Eocene) make up a large clastic apron and were generated by uplift of the range as a whole along the Wind River fault zone. The South Pass Formation (late Oligocene–early Miocene), conglomerate in the White River Formation (Oligocene), and the recently identified Leckie Beds (post–South Pass) are each intramontane alluvial deposits that indicate faulting in the core of the range. Fault scarp clastic wedges in the Cathedral Bluffs Tongue of the Wasatch Formation (early Eocene) and the newly recognized Circle Bar beds (Miocene) document up-to-the-north motion along the Continental fault during Eocene time and later postcompressional collapse (down-to-the-north) in Pliocene time.

Conclusions from this study suggest that a steep tear fault may have uncoupled the Wind River Range from basins to the south as it was thrust toward the southwest. Furthermore, there may have been significant uplift in the core of the range after the main motion on the Wind River fault, and therefore, Laramide deformation may have lasted long after early Eocene time.

INTRODUCTION

The Wind River Range is the largest uplift in the Wyoming foreland province, exposing Precambrian crystalline rocks along a core 125 mi (200 km) long and 25 mi (40 km) wide in west-central Wyoming (Figure 1). It was uplifted by motion on the Wind River fault which shortened the crust by at least 13 mi (20 km) (Brewer et al., 1980) and vertically offset the crystalline basement by at least 7 mi (11 km). During and after uplift in late Cretaceous and Tertiary time, the Paleozoic and Mesozoic sedimentary cover and the crystalline core of this range were eroded and large volumes of sediment were shed into adjacent basins. Thus, the history of the uplift and subsequent collapse of the range are recorded in these syntectonic and posttectonic sedimentary rocks, and specific structural and compositional aspects of the uplift are reflected in sediment composition, texture, geometry, and distribution.

Presumably, a similar relationship between uplifted source rock and derived sediments existed for most of the ranges

[1]Present Address: Amoco Production Company, Denver, CO.
[2]Present Address: Tenneco Oil Company, Lafayette, LA.
[3]Present Address: U.S. Geological Survey, Denver, CO.
[4]Present Address: Union Oil Company of California, Los Angeles, CA.

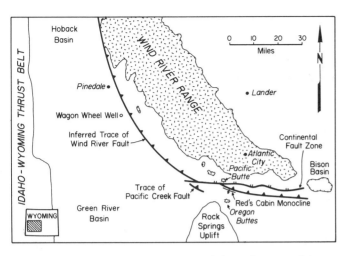

Figure 1—Index map showing location of the study area and the physiographic and geologic features discussed in the text.

throughout the Rocky Mountain foreland, but along the south end of the Wind River Range, postcompressional collapse has been sufficient to lower these sediments and thus preserve them from erosion. In this area, a relatively complete record of latest Cretaceous and Tertiary mountain flank sediments, some possibly as young as Pliocene, are preserved.

This paper reports preliminary results of our studies to determine the distribution and stratigraphic relationships among the conglomeratic facies of these deposits and to infer from their composition, texture, and sedimentary structures their relationship to the uplift and collapse of the range. Our interpretations are, to a large degree, based on the assumption that large accumulations of coarse clastics require steep gradients and topographic relief, which in turn may be related to faulting. Data for our study come from our own surface and subsurface observations as well as the published works of others.

Our findings indicate that the main conglomerate units are of three general types with regard to size, geometry, and depositional environment and, furthermore, that each of these types can be related to a specific structural setting. Large (subregional) basin margin aprons of proximal and distal alluvial fan deposits flank the entire west and south sides of the range (Berg, 1963) and are apparently related to the main uplift of the range by movement on the Wind River fault. Within the range, more separate and distinct proximal alluvial deposits were generated by subsequent faulting that uplifted the Precambrian core over much of the length of the range. Finally, very local, extremely coarse prisms of proximal alluvial fan sediments were deposited immediately adjacent to steeply dipping tear faults that were later the sites of postcompressional collapse of the range.

Stratigraphic Setting

Sedimentary rocks associated with uplift and subsequent collapse of the Wind River Mountains range in age from latest Cretaceous (Lance) through Tertiary (Figure 2), but only those that are Eocene and younger are well exposed at the surface. The stratigraphic relationships and zircon fission-track ages for several of these units are discussed by Steidtmann and Middleton (1986).

Those conglomerate units that we feel are a direct response to motion on faults along or within the southern end of the Wind River Range include the Fort Union Formation (Paleocene), main body of the Wasatch Formation (early Eocene), Cathedral Bluffs Tongue of the Wasatch Formation (early Eocene), South Pass Formation (late Oligocene–early Miocene), conglomerate beds in the White River Formation (Oligocene), the Circle Bar beds (Miocene), and the Leckie beds (post–South Pass). The distribution of these units and general paleocurrent directions for each are summarized in Figure 3.

SUBREGIONAL CLASTIC APRON

Introduction

The Fort Union (Paleocene) and main body of the Wasatch (early Eocene) in this area make up the large clastic apron that flanks the Wind River Range along its western and southern margins. Most of these sediments are of alluvial plain or distal alluvial fan origin. In general, proximal facies of these deposits are not observable because they have been overrun by the Wind River fault and few cores record them at depth. Thus, neither the structural relationships between these deposits and the range itself, nor their coarseness and its implications concerning stream gradients, are known.

Fort Union Formation

The Fort Union is present only in the subsurface in the immediate area. To the south it is exposed in the Rock Springs uplift (Figure 1) where it is as much as 1800 ft (575 m) thick and composed mainly of interbedded sandstone, conglomerate, mudstone, and coal (Winterfeld, 1979). To the east it is exposed in the Bison basin (Figure 1) where it is about 900 ft (275 m) thick. Here it consists of a sequence of thin, lenticular conglomerate beds and medium- to coarse-grained planar- and cross-bedded sandstone enveloped in carbonaceous overbank siltstone and claystone (Southwell, in prep.). These overbank deposits constitute more than 90% of the Fort Union in the Bison basin and, in places, show pedogenic differentiation and both floral and faunal bioturbation. Although the great amount of overbank facies suggests a meandering river environment, the association of large-scale trough cross-stratification, planar-bedding, and scour-and-fill structures in the sandstone and conglomerate indicate mid-channel bars and bedforms of a braided stream (Figure 4). This association of overbank and braided channel deposition is typical of an anastomosing stream as described by Smith and Putnam (1980).

The conglomerate consists of subangular to well-rounded clasts 0.25–4 in. (0.6–10 cm) in diameter. Clasts are composed of gray, black, green, and banded chert and siliceous sandstone and shale fragments. For the most part, the chert appears to be derived from the Madison Limestone (Mississippian) and Phosphoria Formation (Permian); the sandstone fragments from the Tensleep Sandstone (Pennsylvanian), Nugget Sandstone (Jurassic), and Cloverly Formation (Cretaceous); and the shale clasts from the Mowry Shale (Cretaceous). These channel-fill sequences average approximately 8 ft (2.5 m) in thickness and 160 ft (48 m) in lateral extent. Cross-bedding indicates a sediment

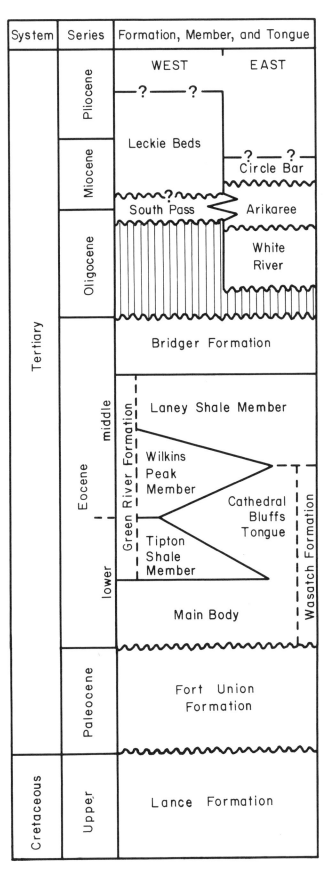

Figure 2—Composite stratigraphic section showing the relationships among Tertiary units exposed along the south flank of the Wind River Range. Adapted in part from Zeller and Stephens (1969).

source to the northwest (Figure 3). The association of mechanically more labile shale clasts with the ultradurable chert clasts suggests a short transport history and therefore a nearby source (Abbott and Peterson, 1978).

Along the south and west flanks of the range our information about the Fort Union comes entirely from the subsurface. Our examination of cores from the Wagon Wheel well (Figure 1) indicates that about 1200 ft (360 m) of sediment, palynologically dated as Paleocene (Law, 1981), consists mainly of K-feldspar granules floating in a silt matrix and suggests that proximal fan deposition (mudflow?) during the Paleocene occurred at least 12 mi (20 km) west of the buried trace of the Wind River fault. Furthermore, these deposits indicate that, at least in this area, erosion had reached the crystalline core by early Paleocene time. Whether or not these beds should be called Fort Union is debatable. Law (1981) referred to them as the "unnamed unit" and indicated that this unit is conformably overlain by the Fort Union in the deeper part of the basin. Near the perimeter of the basin this contact is unconformable. This unnamed unit is truncated and does not appear at the surface.

In his discussion of Laramide sediments along the Wind River fault, Berg (1963) described a Paleocene "basinal facies" composed of nonmarine shale, coal, siltstone, and sandstone that he designated as the Hoback Formation because of its similarity to this unit exposed at the surface in the Hoback basin 80 mi (130 km) to the northwest. These are most likely the same sedimentary rocks identified as Fort Union by Law (1981).

Unfortunately, Law's (1981) correlation does not differentiate the Fort Union from the overlying early Eocene Wasatch Formation, and it is therefore difficult to determine the respective thicknesses. Furthermore, because of the complex lateral and vertical facies changes between the two units, it is likely that their thicknesses change rapidly from place to place. Estimates from Berg's (1963, his figure 8) schematic diagram indicate that the Fort Union may be as thick as 7000 ft (2100 m) and the Wasatch as much as 6000 ft (1800 m). The log from the Wagon Wheel well (Martin and Shaughnessy, 1969) shows thicknesses of about 3300 ft (1000 m) for the Fort Union and 7000 ft (2100 m) for the Wasatch. Zeller and Stephens (1969) stated that the average thickness of the Wasatch in wells drilled south of Oregon Buttes (Figure 1) is just over 3000 ft (900 m). This rather wide range of thicknesses for the Fort Union and Wasatch is probably a function of the range of true thicknesses, errors in estimates from the subsurface because of difficulties in picking Tertiary tops, and structural complexities.

Local thickness variations of the Fort Union and the underlying Lance Formation may indicate time of motion on the Pacific Creek fault (Figure 1). Seismic sections indicate that units identified as Lance and Fort Union thin locally. MacLeod (1981) interpreted this to indicate control of local depositional patterns by motion on the Pacific Creek fault in latest Cretaceous and Paleocene time.

Main Body of the Wasatch Formation

The upper part of the main body of the Wasatch Formation is exposed in the southwestern part of the area (Figure 3), but here too it is likely that most of the proximal

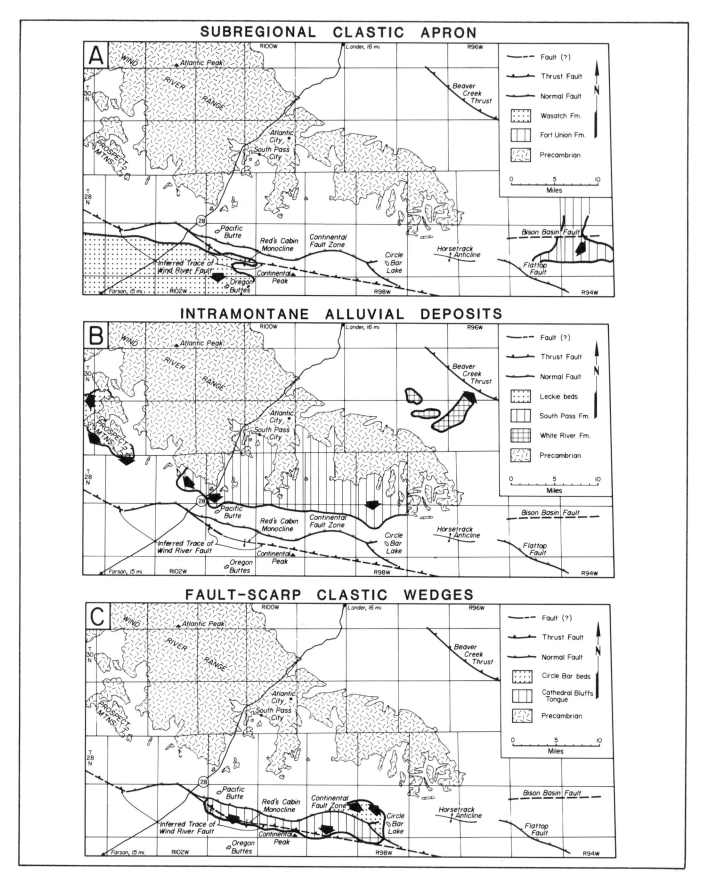

Figure 3—Distribution and general paleocurrent directions of tectogenic conglomerate units along the south flank of the Wind River Range. A. Subregional clastic apron deposits (Fort Union and main body of the Wasatch). B. Intramontane alluvial deposits (South Pass, White River, and Leckie). C. Fault scarp clastic wedges (Cathedral Bluffs and Circle Bar).

Figure 4—Sandstone and conglomerate facies of the Fort Union Formation in the Bison basin. Outcrop shows stacks of stringerlike fills of trough cross-bedded sandstone and conglomerate composed of angular clasts. Hammer shows scale.

deposits have been overrun by the Wind River fault and are not available to direct observation. The main body consists of variegated mudstone, sandstone, and feldspathic pebble conglomerate. These beds can be distinguished from the overlying, lithologically identical Cathedral Bluffs Tongue of the Wasatch Formation only where the Tipton Tongue of the Green River Formation separates the two. Near Pacific Butte (Figure 1) the upper 180 ft (50 m) of the main body is exposed in the core of Red's Cabin monocline (McGee, 1983). The upper part of the main body of the Wasatch is composed of beds of red and gray conglomerate 2–4 ft (0.6–1.2 m) thick that fine upward into variegated drab, maroon, tan, and greenish-blue sandy, arkosic siltstone, mudstone, and claystone. Here, only about 7% of the main body consists of conglomerate beds that are generally only moderately well sorted and in some places contain up to 40% silt and sand matrix. Clasts are subrounded, as much as 3 in. (8 cm) in diameter, and include schist, vein quartz, chert, and gneiss. Crude horizontal or very low-angle cross-bedding is outlined by pebble imbrication, which, along with rare high-angle cross-bedding, indicates a southerly transport direction (Figure 3).

The distance from probable source and the sedimentary texture and structures of the main body indicate deposition on an alluvial braid plain. The low-angle and horizontal stratification represents high velocity flow in braided channels with deposition occurring on longitudinal bars. The fining-upward sequences from conglomerate to sandstone attest to waning flow following major flood events. The lateral impersistence of the coarse facies is consistent with sheet flooding both within and proximal to incised braided channels on the medial to distal parts of an alluvial fan system (Bluck, 1967; Steel, 1974). Presence of matrix-supported beds indicates that this depositional style was periodically interrupted by debris flows. Larsen and Steel

(1978) described a similar conglomerate and explain the abundance of matrix by repeated inundation of fan slopes by water and fine-grained sediment in a tectonically unstable area resulting in the addition of fines to previously deposited coarse clastics.

INTRAMONTANE ALLUVIAL DEPOSITS

Introduction

The alluvial deposits that we relate to faulting in the core of the range include the South Pass Formation (late Oligocene–early Miocene), conglomerate in the White River Formation (Oligocene), and the Leckie beds (post–South Pass).

South Pass Formation

The South Pass Formation is a granule to boulder conglomerate, 0–350 ft (0–105 m) thick exposed along the south flank of the Wind River Range (Figure 3). It overlies rocks of the Bridger Formation (middle Eocene) and locally overlies Precambrian crystalline rocks.

The conglomerate consists of pebbles, cobbles, and boulders in a matrix of coarse sand and pebbles (Bottjer, 1984). Clast composition of the South Pass is variable, reflecting both the complex source rocks of the Precambrian and sediment supply from proximal and distal sources. Presence of well-rounded and very angular grains of the same size and composition in the matrix supports the interpretation of both proximal and distal sources. Clast types include granite, diabase, basalt, gneiss, granodiorite, and epidotized and sheared crystalline rocks.

Although the conglomerate is generally matrix supported, deposition by fluvial processes is indicated by the coarseness of the matrix (sand to pebbles) and the fact that it is well stratified. Gross lenticular bedding, scour surfaces, cross-stratification, channel fills as much as 100 ft (30 m) wide, and clast imbrication all suggest deposition in braided streams with low bed relief. Shallow flows are indicated by the predominance of planar bedding (Allen, 1967), which for coarse sediments (i.e., grains more than 0.6 mm in diameter) is a stable bed configuration in both the upper and lower flow regimes (Southard, 1971). Both downstream and laterally accreting mid-channel bars are indicated by large-scale planar cross-stratification that dips both parallel and oblique to the mean flow direction. Figure 5 shows a transverse cross section of a gravelly mid-channel bar that grew by vertical aggradation and lateral accretion.

Weak distributary paleocurrent trends (Figure 3), proximity to major uplifts, and the coarseness of the deposits all indicate an alluvial fan setting adjacent to highlands within the core of the Wind River Range.

Conglomerate in the White River Formation

Conglomerate beds similar to those of the South Pass occur in the White River Formation near the southeastern terminus of the Wind River Range (Figure 3). The conglomerate and coarse sandstone beds are laterally discontinuous and enveloped in white, ashy siltstone characteristic of the White River. The conglomerate is mainly clast supported but contains rare matrix-supported beds.

In the eastern part of the deposit clasts of igneous and metamorphic rocks dominate, the most common rock types being granite, quartz monzonite, and mafic plutonic rocks. Sedimentary rock fragments dominate the clast population in the western part of the area and include sandstone from the Flathead (Cambrian), Tensleep (Pennsylvanian), and Chugwater (Triassic) formations; siltstone from the Gros Ventre Formation (Cambrian); intraformational conglomerate and limestone from the Gallatin Limestone (Cambrian); dolomite from the Bighorn Dolomite (Ordovician) and Amsden Formation (Pennsylvanian); and chert from the Phosphoria Formation (Permian).

The abundance of clast-supported conglomerate, clast imbrication, high gravel:sand ratio, poor sorting, and scoured surfaces suggests a braided stream environment (Lieblang, 1983). Laterally discontinuous conglomerates are typical of modern braided systems (Smith, 1970; Collinson, 1978), and longitudinal, diagonal, and transverse bars, similar to those in the South Pass Formation, were identified by association of sedimentary structures and unit geometries (Lieblang, 1983).

Longitudinal bars were recognized in the field by a basal deposit of massive, clast-supported conglomerate and/or horizontally stratified clast- and matrix-supported conglomerate overlying a scoured base (Figure 6). These units were apparently emplaced during major flooding events as diffuse gravel sheets (Hein and Walker, 1977). The remainder of the bar is composed of trough cross-stratified conglomerate and sandstone reflecting lower stage cut and fill during emergence. Finer grained facies representing settling during low energy stages compose the bar tops.

Although not as abundant, facies associations representing transverse bar formation and migration also occur. These consist of basal units of large-scale planar tabular cross-bedding (avalanche deposition on bar fronts) overlain by horizontally stratified beds (upper flow regime sheet flow across bar tops). Dune migration in channels is indicated by stacked sequences of trough cross-stratified sandstone and numerous cut-and-fill structures.

An eastward paleocurrent direction is indicated by the analysis of clast imbrication and is supported by a west-to-east decrease in labile clast components and clast size (Figure 3). Furthermore, there is an eastward change from proximal (longitudinal) to distal (transverse) bar types similar to that described by Smith (1970), Hein and Walker (1977), and Collinson (1978).

Leckie Beds

The deposits we here informally designate as the Leckie beds have only recently been identified as tectogenic in origin. Thus, much remains to be learned about their composition, provenance, and structural relationships. Until 1983 these coarse clastics were considered to be of glacial origin following the interpretation of Moss and Holmes (1955). Mapping described by Richmond (1983), however, shows these deposits to be discrete, bouldery alluvial fans immediately adjacent to topographic highs along the west and southwest side of the Wind River Range (Figure 3). The depositional surface and shape of these fanglomerates is preserved at several localities and according to Richmond (1983), the beds are offset by normal faults, some possibly of Quaternary age. Our field observations generally corroborate those of Richmond, but we favor the more cautious age designation of post–South Pass rather than his Miocene until more data have been collected.

FAULT SCARP CLASTIC WEDGES

Introduction

Wedges of proximal fan deposits formed along steeply dipping faults that were probably tear faults during uplift of the range and were later sites of postcompressional collapse. Those deposits along the tear faults are represented by the Cathedral Bluffs Tongue of the Wasatch Formation (early Eocene) and those along the collapse faults by the Circle Bar beds (Miocene), recently identified by Jackson (1984). Although the Cathedral Bluffs is significantly coarser than the Circle Bar, both represent extremely local deposition along fault-controlled topographic highs. Textural and compositional differences between the two units are related to the source rock.

Cathedral Bluffs Tongue of the Wasatch Formation

Where the Tipton Tongue of the Green River Formation is present, an upper unit of the Wasatch Formation, the Cathedral Bluffs Tongue, is recognized in the eastern Green River basin. It consists of arkosic siltstone, sandstone, and conglomerate generally quite similar to, although coarser than, the main body of the Wasatch (McGee, 1983). It changes thickness markedly over short distances, ranging from less than 100 ft (30 m) on the southeast end of Red's Cabin monocline to more than 700 ft (210 m) just 1 mi (~1.5 km) northwest on Pacific Butte (Figure 3).

In the vicinity of the monocline, boulder–cobble conglomerate is interbedded with variegated siltstone and

Figure 5—Transverse cross section of a gravelly mid-channel bar in the South Pass Formation. Bedding shows that the bar grew both by lateral accretion (A) and vertical aggradation (B). Lens cap shows scale.

claystone. The cobble-sized clasts are as much as 10 in. (25 cm) in diameter, consist of dark schist and gneiss with some chert and quartz, and show both isolate and contact imbrication. Boulders as large as 15 ft (4.5 m) in diameter are strewn over the surface of Pacific Butte (Figure 7), and discontinuous patches of similar sized boulders are present at several other locations immediately adjacent to the Continental fault (Figure 3). Apparently these boulders are a weathering lag reworked out of the Cathedral Bluffs, and they decrease in both size and number away from the Continental fault until none are present only 1 mi (~1.5 km) to the south.

Most of the largest clasts are composed of granite, gneiss, schist, pegmatite, and pegmatitic quartz. The relative amount of these clast types varies markedly over very short distances, indicating distinctive source rock types and transport distances short enough to inhibit mixing. Rocks of the greenschist–amphibolite metamorphic facies near Atlantic City (Figure 1) 12 mi (~20 km) to the northeast make up most of the pebble- and cobble-sized clasts, whereas the boulders are mainly pegmatite and granite derived from just north of the fault. At several locations these conglomerates contain gold that also had a nearby source (Love et al., 1978).

The decrease in maximum clast size southward from the Continental fault and the general gradation from conglomerate to sandstone and siltstone toward the south suggest deposition in a south-flowing alluvial system characterized by discrete, low-sinuosity channels such as might occur on the proximal and medial parts of alluvial fans (Figure 3). Sedimentary structures consist of low-angle to horizontal stratification, well-developed clast imbrication, and cut-and-fill sequences. These features are common on modern longitudinal bars in braided systems (Ore, 1963; Hein and Walker, 1977; Miall, 1977).

Thin beds of arkosic, conglomeratic sandstone are interbedded with the conglomerate units. These sands probably were deposited on the downstream reaches of

longitudinal bars during low-flow stages (Bluck, 1979; Steel and Thompson, 1983) and also filled topographically low areas oblique to the bar axes that were scoured out during low-stage bar dissection.

The proportion of mudstone increases to the south and reflects more distal fan and upper alluvial plain sedimentation. Roehler (1969) also considered the sandstone and conglomerate in the Cathedral Bluffs to represent alluvial fan deposits that grade into mudstone and fine sandstone on an alluvial plain.

Circle Bar Beds

A previously unrecognized sequence of tuffaceous sandstone, volcanic ash, and conglomerate was informally designated the Circle Bar beds by Jackson (1984) for exposures along the north side of the Continental fault ~17 mi (27 km) southeast of Atlantic City (Figure 3). Previously, these rocks had been mapped as part of the Arikaree Formation (Oligocene–Miocene) by Denson and Pipiringos (1974) and Zeller and Stephens (1969). The Circle Bar unit varies in thickness from 260 to 400 ft (80 to 120 m) over a distance of only 2 mi (~3 km) and rests with angular unconformity on the Bridger (middle Eocene) and Arikaree (Oligocene–Miocene) formations where they have been folded along the Continental fault. There is no doubt that the Circle Bar is younger than early Miocene because it overlies and contains reworked fragments of the Arikaree Formation. Jackson (1984) assigned it an age of late Miocene or Pliocene and Steidtmann and Middleton (1986) obtained a middle Miocene zircon fission-tracks age from ash beds in this unit.

The lower part of the Circle Bar beds consists of fine-grained, gray, tuffaceous sandstone containing scattered pebbles of Precambrian crystalline rocks, broken and rounded fragments of white, silicified root casts, and boulders of sandstone from the Arikaree Formation. The Circle Bar coarsens upward and toward the southwest where lenses and continuous beds of pebble, cobble, and boulder

Figure 6—Cross section of longitudinal bar in conglomerate of the White River Formation. Bar consists of massive, structureless clast- and matrix-supported conglomerate (A) interbedded with finely laminated siltstone (B) scoured into massive, tuffaceous siltstone (C). Hammer shows scale.

conglomerate are interbedded with, and channeled into, cross-bedded tuffaceous sandstone. Fragments of metagraywacke and other mafic metamorphic rocks constitute most of the clasts of pebble and cobble size and were probably derived from the basal conglomerate of the adjacent Arikaree Formation where only these sizes are available. Rounded fragments of Arikaree sandstone make up most of the boulder-sized clasts, some as large as 9 ft (3 m) in diameter.

Paleocurrent directions, shown by imbrication of the tabular metamorphic clasts, indicate transport toward the north and northeast (Figure 3). These directions, along with the clast size trends and the reworked fragments of Arikaree sandstone indicate that sediments of the Circle Bar beds were shed off a high-standing southern block along the Continental fault. Deposition was restricted to the immediate vicinity of the topographic high on alluvial fans where, from time to time, ash-rich sediment accumulated in small ponds.

DISCUSSION

The schematic diagram in Figure 8 shows the general distribution of the three types of conglomeratic units and their relationships to major structural elements. Although there are other conglomerates in the Tertiary record of this area, most are thin and discontinuous and cannot be related to identifiable events associated with the structural development of the Wind River Range. The widespread, range-encircling nature of the Fort Union and main body of the Wasatch clastic apron indicates that these sediments recorded the uplift of the range as a whole. This is not to say, however, that the sedimentary record indicates that the range was uplifted as one intact block along its entire length. On the contrary, interpretations by Berg (1983) and Steidtmann et al. (1983) suggest that the Wind River fault is a zone consisting of numerous segments that may have

moved independently. It is unlikely, however, that additional source area and current direction studies will have the resolution to identify separate uplifted blocks. The data are simply not available at the surface or in the scant subsurface record and subtle source area and current direction differences between individual, but coalescing, fans of various sizes would be difficult to recognize. Furthermore, much of the evidence has been overrun by syndepositional movement on the Wind River fault.

Because of the separate and distinct nature of the intramontane alluvial deposits (Figure 8), they can be used as evidence for uplift in specific areas. Analysis of the South Pass Formation indicates uplift in the core of the range that postdates main motion on the Wind River fault. This interpretation is indicated by several lines of evidence including paleocurrent indicators (Figure 3), position of the deposit on the toe of the Wind River fault, textural evidence for both distal and proximal sources, and lack of compositional control by local source rocks.

The nature of this uplift in the core is not certain but it is quite likely that it occurred along faults or fault zones that now occupy the obvious break in slope immediately west of the highest part of the range (Figure 9). Faults and fault zones similar to those envisioned for this area have been mapped northward along the range by Richmond (1945), Berg (1961, 1963), and Frost (personal communication, 1984). Couples and Stearns (1978, their figure 9) indicate a fault at this position bounding the east side of their "shattered lobe." If volume of derived sediment is an indicator of displacement along the fault, it is clear that uplift along this proposed fault was significantly less than that on the Wind River fault, probably about 3300 ft (1000 m).

The tectonic significance of conglomerate in the White River Formation is not as clear cut. This debris was derived mainly from the Wind River Range to the west and south and was deposited in a basin flanked by the dip slopes of the Wind River Range and the hanging wall of the Beaver

Figure 7—Surface of Pacific Butte showing boulders as large as 15 ft (4.5 m) in diameter left as a lag from weathering of the Cathedral Bluffs Tongue of the Wasatch Formation. Trace of the Continental fault (which uplifted the source for these clasts) is immediately behind the ridge (**X**). Pack on boulder shows scale.

Creek thrust (Figure 3). Whether deposition of this Oligocene conglomerate resulted from continued denudation of the original uplift or from renewed uplift is not entirely clear, but the disconformity between the White River and the subjacent Wasatch suggests that uplift was renewed. If this was the case, the White River conglomerate, together with the South Pass, provides one line of evidence that, in the Wind River Range, Laramide uplift lasted long after early Eocene time.

Tectonic implications of the Leckie beds are, at this time, necessarily sketchy. Little is known of them, but it seems clear that distinct alluvial fans were shed from locally uplifted blocks in the core of the range (Figure 3). The geomorphic expression of the fans can still be easily recognized on topographic maps and from the air. A more specific determination of their significance awaits further study.

The recognition of the northern side of the Continental fault as a source for Cathedral Bluffs sediments during early Eocene time is a particularly significant finding. Prior to this study, the Continental fault was recognized as a down-to-the-north, postcompressional collapse fault. The fact that the north side was apparently up during part of the time when the Wind River Range was thrust southwestward suggests that the early Eocene Continental fault served as a tear that uncoupled the range on the north from the basins to the south. Coarse latest Cretaceous and Paleocene deposits were probably also deposited along this zone of tearing during earlier phases of thrusting, but except for a Paleocene boulder bed in the subsurface south of the Bison basin, rocks of this age are not exposed in this area.

Finally, the Circle Bar beds represent the first sedimentologic evidence of the collapse of the Wind River Range along the Continental fault. There has never been much doubt that collapse did occur because Eocene sediments on the south are juxtaposed against Miocene sediments on the north. The Circle Bar beds, however, were

shed by the high-standing south side of the fault during collapse in mid-Miocene time.

CONCLUSIONS

Data from the tectogenic sediments shed by the Wind River Range can be used to determine the details of its uplift and subsequent collapse. The subregional clastic apron that encircles the range records general uplift of the range along the Wind River fault, and intramontane alluvial deposits indicate significant uplift in the core after major movement on the Wind River fault ceased. Fault scarp clastic wedges identify both a steep tear fault that uncoupled the range from basins to the south and later collapse of the range along this same steep fault.

ACKNOWLEDGMENTS

This project was supported by National Science Foundation Grant no. EAR-8108939 to Steidtmann and Middleton and by the Amoco Production Company, Denver, Colorado.

REFERENCES CITED

Abbott, P. L., and G. L. Peterson, 1978, Effects of abrasion durability on conglomerate clast populations: examples from Cretaceous and Eocene conglomerates of the San Diego area, California: Journal of Sedimentary Petrology, v. 48, p. 31–42.

Allen, J. R. L., 1967, Depth indicators of clastic sequences: Marine Geology, v. 5, p. 429–446.

Berg, R. R., 1961, Laramide tectonics of the Wind River Mountains: Wyoming Geological Association 16th

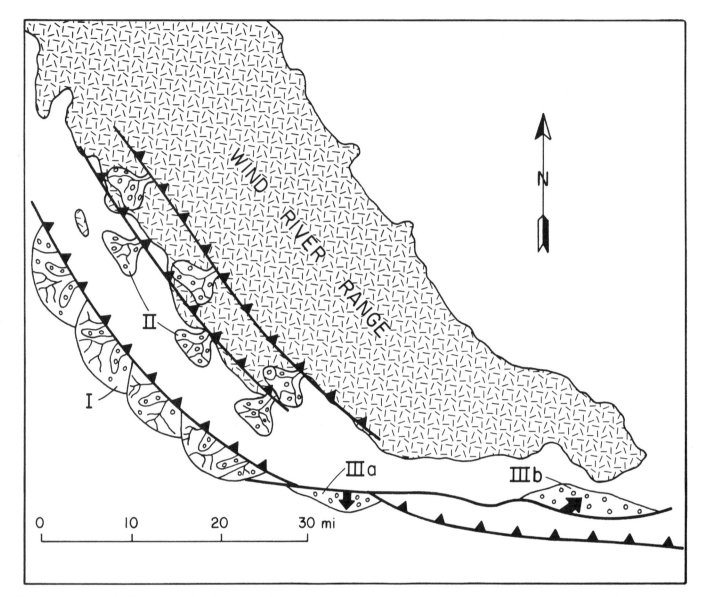

Figure 8—Schematic diagram showing the general distribution of the three types of conglomeratic units and their relationships to major structural elements. Type I deposits are the subregional clastic aprons along the main range-bounding fault zone. Type II are intramontane alluvial deposits associated with faulting in the core of the range. Type III are fault scarp clastic wedges. Type IIIa is related to tear faulting during compression and was transported southward. Type IIIb is related to postcompressional collapse of the range along the zone of tearing and was transported to the north.

Annual Conference Guidebook, p. 70–80.

———, 1963, Geometry of the Wind River thrust, *in* J. D. Lowell and R. Gries, eds., Rocky Mountain foreland basins and uplifts: Rocky Mountain Association of Geologists Guidebook, p 257–262.

———, 1983, Laramide sediments along Wind River thrust, Wyoming, *in* O. E. Childs and B. W. Beebe, eds., Backbone of the Americas: AAPG Memoir 2, p. 220–230.

Bluck, B. J., 1967, Deposition of some upper Old Red Sandstone conglomerates in the Clyde area: a study in the significance of bedding: Scottish Journal of Geology, v. 3, p. 139–167.

———, 1979, Structure of coarse grained braided stream alluvium: Transactions of the Royal Society of Edinburgh, v. 70, p. 181–221.

Bottjer, R. J., 1984, Depositional environments and tectonic significance of the South Pass Formation, southern Wind River Range, Wyoming: Master's thesis, University of Wyoming, Laramie, 229 p.

Brewer, J. A., S. B. Smithson, J. E. Oliver, S. Kaufman, and L. D. Brown, 1980, The Laramide orogeny: evidence from COCORP deep crustal seismic profiles in the Wind River Mountains, Wyoming: Tectonophysics, v. 62, p. 165–189.

Collinson, J. D., 1978, Alluvial sediments, *in* H. G. Reading, ed., Sedimentary environments and facies: New York, Elsevier, p. 15–60.

Couples, G., and D. W. Stearns, 1978, Analytical solutions applied to structures of the Rocky Mountains foreland on local and regional scales, *in* V. Matthews, ed., Laramide folding associated with basement block

Figure 9—Oblique view of plastic relief map (2° sheets) of the northwestern quarter of Wyoming. Arrow points to the break in slope just west of the crest in the range that is probably the surface expression of a zone of faulting. Uplift along these faults supplied clasts to the South Pass conglomerate.

faulting in the western United States: Geological Society of America Memoir 151, p. 313–336.

Denson, N. M., and G. N. Pipiringos, 1974, Geologic map and sections showing areal distribution of Tertiary rocks near the southeastern terminus of the Wind River Range, Fremont and Sweetwater counties, Wyoming: U.S. Geological Survey Miscellaneous Investigations Series Map I-835.

Hein, F. J., and R. G. Walker, 1977, Bar evolution and development of stratification in the gravelly, braided, Kicking Horse River, British Columbia: Canadian Journal of Earth Science, v. 14, p. 562–570.

Jackson, K. E., 1984, Geology of the Circle Bar Lake quadrangle, Fremont and Sweetwater counties, Wyoming: Master's thesis, University of Wyoming, Laramie, 77 p.

Larsen, V., and R. J. Steel, 1978, The sedimentary history of a debris-flow dominated, Devonian alluvial fan: a study of textural inversion: Sedimentology, v. 25, p. 37–39.

Law, B. E., 1981, Section C–C': subsurface and surface correlations of some Upper Cretaceous and Tertiary rocks, northern Green River basin, Wyoming: U.S. Geological Survey Open-File Report 81-663.

Lieblang, S., 1983, Sedimentology and petrology of conglomerate and sandstone in the Oligocene White River Formation, central Wyoming: Master's thesis, Northern Arizona University, Flagstaff, 177 p.

Love, J. D., J. C. Anteweiler, and E. L. Mosier, 1978, A new look at the origin and volume of the Dickie Springs–Oregon Gulch placer gold at the south end of the Wind River Mountains: Wyoming Geological Association 30th Annual Field Conference Guidebook, p. 379–391.

MacLeod, M. K., 1981, The Pacific Creek anticline: buckling above a basement thrust fault: Wyoming University Contributions to Geology, v. 19, p. 143–160.

Martin, W. B., and J. Shaughnessy, 1969, Project Wagon Wheel: Wyoming Geological Association 21st Annual Field Conference Guidebook, p. 145–153.

McGee, L. C., 1983, Tectonics and sedimentation at Red's Cabin monocline, south end of the Wind River Range, Wyoming: Master's thesis, University of Wyoming, Laramie, 107 p.

Miall, A. D., 1977, A review of the braided-river depositional environment: Earth Science Review, v. 13, p. 1–62.

Moss, J. H., and W. G. Holmes, 1955, Pleistocene geology of the southwestern Wind River Mountains, Wyoming: Geological Society of America Bulletin, v. 66, p. 629–654.

Ore, H. T., 1963, The braided stream depositional environment: Ph.D. thesis, University of Wyoming, Laramie, 182 p.

Richmond, G. M., 1945, Geology of northwest end of the Wind River Mountains, Sublette County, Wyoming: U.S. Geological Survey Oil and Gas Investigation Series Map 31.

———, 1983, Modification of glacial sequence along Big Sandy River, southern Wind River Range, Wyoming (abs.): Geological Society of America Abstracts with Programs, Rocky Mountain–Cordilleran Sections, p. 431.

Roehler, H. W., 1969, Stratigraphy and oil-shale deposits of Eocene rocks in the Washakie basin: Wyoming Geological Association 21st Annual Field Conference Guidebook, p. 197–206.

Smith, D. G., and P. E. Putnam, 1980, Anastomosed river deposits: modern and ancient examples in Alberta, Canada: Canadian Journal of Earth Science, v. 17, p. 1396–1406.

Smith, N. D., 1970, The braided stream depositional environment: comparison of the Platte River with some Silurian clastic rocks, north-central Appalachians: Geological Society of America Bulletin, v. 81, p. 2993–3014.

Southard, J. B., 1971, Representations of bed configurations in depth–velocity size diagrams: Journal of Sedimentary Petrology, v. 41, p. 903–915.

Southwell, E. H., in prep., Depositional and pedogenic processes of the Fort Union Formation, Bison basin, central Wyoming: Master's thesis, University of Wyoming, Laramie.

Steel, R. J., 1974, New Red Sandstone floodplain and piedmont sedimentation in the Hebridean Province, Scotland: Journal of Sedimentary Petrology, v. 44, p. 336–357.

Steel, R. J., and D. B. Thompson, 1983, Structures and textures in Triassic braided stream conglomerates ("Bunter" Pebble Beds) in the Sherwood Sandstone Group, North Staffordshire, England: Sedimentology, v. 30, p. 341–368.

Steidtmann, J. R., and L. T. Middleton, 1986, Eocene–Pliocene stratigraphy along the southern margin of the Wind River Range, Wyoming: Revisions and implications from field and fission-track studies: The Mountain Geologist, v. 23, p. 19–25.

Steidtmann, J. R., L. C. McGee, and L. T. Middleton, 1983, Laramide sedimentation, folding and faulting in the southern Wind River Range, Wyoming, *in* J. D. Lowell and R. Gries, eds., Rocky Mountain foreland basins and uplifts: Rocky Mountain Association of Geologists Guidebook, p. 161–168.

Winterfeld, G. F., 1979, Geology and mammalian paleontology of the Fort Union Formation, eastern Rock Springs uplift, Sweetwater County, Wyoming: Master's thesis, University of Wyoming, Laramie, 181 p.

Zeller, H. D., and E. V. Stephens, 1969, Geology of the Oregon Buttes area, Sweetwater, Sublette, and Fremont counties, southwestern Wyoming: U.S. Geological Survey Bulletin 1256, 60 p.

Tectonics and Sedimentology of Uinta Arch, Western Uinta Mountains, and Uinta Basin

R. L. Bruhn
M. D. Picard
University of Utah
Salt Lake City, Utah

J. S. Isby
Sohio Petroleum Company
Gulf Coast Division
Houston, Texas

The Wasatch Mountains mark a major zone of structural transition in the Sevier–Laramide orogenic belt. A regional system of frontal and sidewall thrust-fault ramps developed in the overthrust belt beneath the western edge of the Wasatch Mountains during the Late Cretaceous. The basal décollement stepped upsection to the east, from a depth of about 15 km in Proterozoic strata and crystalline basement in the region west of the Wasatch Mountains, to a level in lower Paleozoic and Mesozoic strata toward the east. This ramp system strongly influenced the geometry of the overlying allochthons as they were transported eastward toward the foreland. Large hanging wall folds formed during formation of frontal and sidewall ramps (fault propagation folds) and during subsequent transport of thrust sheets over the underlying ramps. Lateral boundaries of the Absaroka and Charleston allochthons formed above major sidewall ramp systems that extended into the crystalline basement. The Uinta arch is a large, north-vergent anticlinorium in the Wasatch Mountains that forms the westward continuation of the Uinta Mountain structure. Structurally, the arch lies in the southernmost part of the hanging wall of the early Paleocene Hogsback thrust system. The arch began to form during latest Cretaceous and early Paleocene time as eastward movement on the Hogsback décollement was transferred onto a south-dipping thrust ramp beneath the Uinta arch. The anticlinal structure of the Uinta arch formed in response to distortion of the hanging wall above the fault ramp and involved a component of northward-directed crustal shortening in addition to the sinistral slip. Uplift of the Uinta Mountains occurred at the same time and may reflect regional displacement of the crust along the south-dipping fault system. Uplift and faulting in the Uinta Mountains continued throughout the Paleocene and Eocene, postdating movement on the Hogsback thrust.

Fluvial strata of the Currant Creek, Uinta, and Duchesne River formations (Late Cretaceous–Paleocene in age) record tectonism in the central Rocky Mountains. They are related to separate episodes of regional uplift in the Wasatch Mountains and Uinta Mountains and in western and southwestern Colorado. The Currant Creek Formation of Maestrichtian–Paleocene(?) age documents major tectonism in the north-central Utah overthrust belt by its thick conglomerate and sandstone units deposited in coalescing alluvial fans. Conglomerate clast composition and southerly paleocurrent directions in the upper part of the formation indicate uplift of the western end of the Uinta arch. The Uinta Formation was derived from volcanic terrains exposed to the east and deposited within the fluvial and lacustrine facies of Lake Uinta. Sediment of the Eocene–Oligocene Duchesne River Formation was derived from the Uinta Mountains. Conglomerate and sandstone is rich in sedimentary and metasedimentary rock fragments. Southward-directed fluvial deposition occurred within the Uinta Basin in response to Eocene–early Oligocene uplift and faulting of the Uinta Mountains.

INTRODUCTION

The Uinta structural arch is a large, east-trending anticlinorium in the Wasatch Mountains of north-central Utah (Figure 1). The anticlinorium extends westward from the Uinta Mountains into the eastern Great Basin, where it has been disrupted by Tertiary faulting and plutonism (Crittenden, 1976). About 5 km of Precambrian (Y age) sandstone and shale were deposited over metamorphic basement rock in the Uinta arch (Eardley, 1944, 1951; Crittenden, 1976). This thick section of sedimentary rock is interpreted as the fill of an east-trending aulacogen that extended eastward through the area of the Uinta Mountains and may have continued west of the Wasatch Mountains into

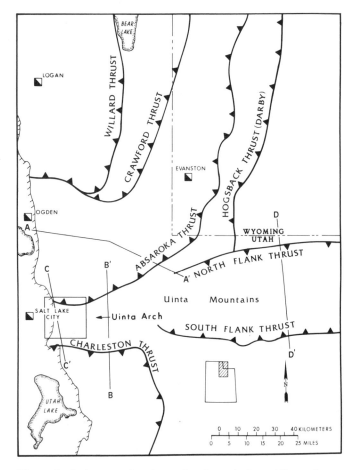

Figure 1—Index map showing major thrust faults and lines of cross sections discussed in this paper. Box shown near Salt Lake City is area of Figure 5.

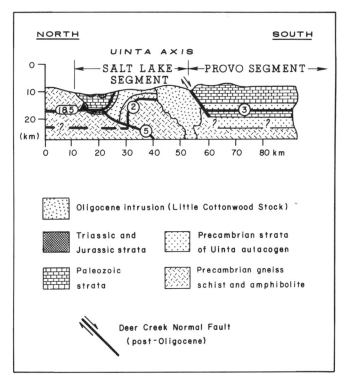

Figure 2—Present structure of the central Wasatch Mountains showing a generalized, north–south structural cross section of the Uinta arch at Salt Lake City, Utah. Major thrust faults #1–5 same as in Figures 11 and 12. See text for discussion. (See line C–C′ in Figure 1 for location.)

the eastern Great Basin. Latest Proterozoic through Jurassic strata of the Cordilleran miogeosyncline were subsequently deposited over the top of the Uinta aulacogen.

The Uinta arch (Figures 1 and 2) is a significant regional structure. It segments the overthrust belt of northern Utah into two structural salients that contain different allochthonous thrust sheets and have different deformational histories (Figure 1) (Royse et al., 1975; Crittenden, 1976; Blackstone, 1977; T. Lawton, personal communication, 1984). Several regional structural hypotheses have been proposed for the Uinta arch and Uinta Mountains. Sales (1968) suggested that they were part of a large sinistral shear zone that formed a regional strike-slip boundary in the Rocky Mountain foreland, which was ultimately related to subduction along the western edge of the continent. Beutner (1977) viewed the Uinta arch and westernmost part of the Uinta Mountains as a rigid structural buttress to eastward-advancing thrust sheets during the Late Cretaceous. Hamilton (1981) suggested that the Uinta arch–Uinta Mountains structure formed along the northern margin of the Colorado subplate, which was a large, essentially coherent fragment of the North American continent that rotated a few degrees clockwise relative to the region north of the Uinta Mountains during the Late Cretaceous and early Tertiary. The result of this rotation was

sinistral–oblique convergence across the Uinta arch and Uinta Mountains, with northeast compression transmitted northward into the area of the Laramide uplifts in Montana, Wyoming, and Colorado. He envisioned lesser amounts of compression in the Colorado Plateau. Each of these tectonic interpretations requires extensive structural reactivation and modification of the Precambrian Uinta aulacogen. The hypotheses of Sales (1968) and Hamilton (1981) predict an important component of sinistral slip along the Uinta arch and Uinta Mountains during deformation.

Crittenden (1976) cited evidence for multiple phases of deformation in the Uinta arch. He proposed that the structure evolved during several periods of alternating north–south and east–west oriented compression. He inferred an intimate relationship between eastward-directed thrusting on the Willard, Absaroka, and Charleston thrust systems (Figure 1) and north–south shortening in the Uinta arch, but remained uncertain about the origin of the large-scale rotations in the stress field.

We present here a new hypothesis concerning the structural history of the Uinta arch and its relationship to the evolution of the Uinta Mountains and the foreland fold and thrust belt in Utah. Structural relationships between the east-trending Uinta arch and more north-trending thrust belt structures are clarified by interpreting vertical and down-plunge cross sections. The chronology of deformation in the Uinta arch and Uinta Mountains and the relationship of these deformational events to the thrust belt are based on our studies of several synorogenic sedimentary deposits

(Picard et al., 1983). Thrusting within the Sevier–Laramide orogenic belt largely controlled the tectonic history of Upper Cretaceous and lower Tertiary rocks of north-central Utah. The Sevier highland provided a constantly available source of clastic material to the north-central foreland basin throughout Late Cretaceous time. The Currant Creek Formation (Table 1) is one of several synorogenic sequences that was deposited in continental settings east of the ancestral Wasatch Range. The green, gray, and maroon claystone and minor sandstone of the eastern Uinta Formation represent channel and flood plain deposits of a low-gradient stream system flowing westward across a nearly flat alluvial plain into a receding remnant of Lake Uinta (Stagner, 1941; Andersen and Picard, 1974). The saline facies and the sandstone and limestone facies of the formation represent the youngest deposits of Lake Uinta remnant in the west (Dane, 1954; Picard, 1955).

Paleozoic rocks in northeastern Utah are dominantly marine, but Mesozoic and Tertiary rocks are dominantly nonmarine (Table 1). The order of abundance of nonmarine rocks by stratigraphic thickness is as follows: sandstone (most abundant), fine-grained rocks (siltstone and claystone), carbonate rocks (dominantly in the lower Tertiary Green River and Flagstaff formations), and conglomerate. Conglomerate is volumetrically concentrated in Late Cretaceous–early Tertiary synorogenic deposits of the Price River, North Horn, Currant Creek, and Duchesne River formations. In contrast, carbonate rocks dominate the Paleozoic strata, followed by sandstone and fine-grained rocks; conglomerate is scarce.

Throughout the entire succession quartzarenite and subarkose are the principal sandstone types. The Mesozoic and Tertiary sandstones contain more feldspar than do the Paleozoic sandstones. There are much fewer detailed petrographic data for the Paleozoic clastics; they likely contain slightly more feldspar than we indicate (Table 1). When plotted on a QFL (quartz–feldspar–lithic "triangle") diagram, nearly all of the sandstone falls in the source field of continental blocks—quartzo-feldspathic sandstone that is low in lithic fragments. Sediment sources are generally considered to be either on stable shields and platforms or in uplifts that mark plate boundaries and trends of intraplate deformation that transect the continental blocks (Dickinson et al., 1983). Some of the sandstone that plots in the craton interior subfield is multicycle sandstone whose grains were dominantly derived from sedimentary rocks very low in lithic fragments. In other sandstone that is low in lithic fragments, grains were recycled from sedimentary sources, but there was also a first-cycle contribution of grains from igneous source rocks.

STRUCTURAL EVOLUTION OF UINTA ARCH AND RELATIONSHIP TO OVERTHRUST BELT AND UINTA MOUNTAINS

Understanding the structural relationships between the Uinta Mountains, Uinta arch, and the overthrust belt has long been recognized as an important problem in the tectonics of Utah and Wyoming (Roberts et al., 1965; Sales, 1968; Crittenden, 1976; Beutner, 1977; Hansen, 1965, 1984;

Bruhn et al., 1983a). The structural geometry, locations, and timing of uplifts of major mountain ranges are particularly important in deciphering the provenance of synorogenic sedimentary deposits. In the following sections we review geologic data, interpret geologic mapping, and propose new hypotheses that bear on this problem.

Thrust Belt Structural Transition

The western edge of the Wasatch Mountains is the site of a major system of thrust-fault ramps in the overthrust belt (Royse et al., 1975; Bruhn et al., 1983a; Smith and Bruhn, 1984). These ramps are important in that they mark a major step upward in the original level of the Cretaceous and Paleocene basal décollement: from a depth of about 12–15 km in Precambrian strata and crystalline basement west of the Wasatch Mountains to a depth of 5–10 km in Paleozoic and Mesozoic strata in the overthrust belt east of the mountains.

Major frontal and sidewall ramp anticlines occur in several thrust sheets in the Wasatch Mountains (Figure 3) (Royse et al., 1975; Smith and Bruhn, 1984). The large anticlines in the Charleston thrust sheet (see RA2 and RA3 on Figure 3) mark a rise in décollement from a level in Proterozoic strata and crystalline basement west of the Wasatch Mountain front to a level in the Jurassic Arapien and Twin Creek formations beneath the mountains to the east.

There is direct geologic evidence for a rise in the level of décollement from the Precambrian to the Mesozoic level crossing the position of the Wasatch Mountain front at the northern and southern edges of the Charleston allochthon. The Charleston thrust ramp has been partly exhumed along the southern edge of the Uinta arch where it occurs in the backlimb of the large, east-vergent Cottonwood fold (Riess, 1985). Here the Charleston thrust has been partly reactivated as a low- to moderate-angle normal fault during uplift and extension of the Wasatch Mountains during late Cenozoic time (Hopkins and Bruhn, 1983; Royse, 1983; Riess, 1985). Geologic mapping (Baker and Crittenden, 1961; Baker, 1964; Riess, 1985) demonstrates that the thrust fault cut up-section toward the east, rising through the Paleozoic section to a flat in the Twin Creek Formation. Farther south, in the southernmost part of the Charleston allochthon, the Mount Nebo thrust faulted Precambrian through Mesozoic strata in the hanging wall (RA3) over the Jurassic Arapien Formation (Black, 1965).

There is evidence, however, suggesting that thrust fault flats exist at levels deeper than the Mesozoic strata in the footwall of the Charleston allochthon. For example, the inferred blind décollement beneath the Cottonwood district (Figures 1 and 2) could continue southward and extend beneath all or part of the southern Wasatch Mountains. Also, Baker and Crittenden (1961) and Baker (1964) mapped a décollement, the Deer Creek thrust, in the Mississippian Manning Canyon Formation in the allochthon. Perhaps this stratigraphic unit is also faulted in the footwall.

Royse et al. (1975) first recognized the large basement-cored culmination of the northern Wasatch Mountains as a hanging wall anticline (RA1) that developed above a footwall ramp in the Absaroka thrust system (Figure 3). The fold is a doubly plunging anticline cored by highly

Table 1—Stratigraphic sequence of the western Uinta Mountains and Uinta basin.

Age	Formation	Depositional Environment	Rock Type[a]	QFL[b] (Sandstone Type)	Thickness (ft)
Eocene–Oligocene(?)	Duchesne River	Alluvial	Ss, Cgl, F, V, R	87-2-11	3,800
Cretaceous–Paleocene	Currant Creek	Alluvial	Cgl, Ss, F, R, V	94-1-5	4,800
Cretaceous	Mesaverde Group	Deltaic and marine (clastic shoreline)	Ss, F, Cgl, C	69-8-23, 84-2-14	1,000
	Mancos	Marine (offshore)	F, V		2,500
	Frontier	Deltaic and marine (clastic shoreline and offshore)	Ss, F, C, V	86-6-6	450
	Mowry	Marine	F, V, C		250
	Dakota	Alluvial	Ss, F, R, V	95-3-2	180
Jurassic	Cedar Mountain	Alluvial	F, Ss, R, V		130
	Morrison	Alluvial and lacustrine	F, Ss, Cgl, Carb, V, R		1,425
	Curtis–Stump	Marine (clastic shoreline)	Ss, F, Carb		165
	Entrada–Preuss	Evaporitic, marine (shallow), and alluvial	Ss, F, R	89-8-3	700
	Twin Creek	Marine (shallow)	Carb, F, Ss, E	Subarkose	740
	Nugget	Evaporitic and lacustrine	Ss, F, Carb, R, E	82-15-3	1,310
Triassic	Popo Agie–Chinle	Alluvial and lacustrine	F, Ss, V, R	Subarkose	410
	Gartra	Alluvial	Ss, Cgl, F, R	91-8-1	50
	Ankareh	Marine (shallow)	F, Ss, E, R	Subarkose, arkose	720
	Thaynes	Marine (shallow)	F, Carb, Ss	Subarkose, arkose	375
	Woodside	Marine (tide dominated)	F, Ss, E, R	73-22-5	800
	Dinwoody	Marine	F, Ss, Carb		80
Permian	Park City–Phosphoria	Marine (shallow)	Carb, F, Ss		650
Pennsylvanian	Weber	Evaporitic and lacustrine	Ss, Carb, F, R, E	Quartzarenite	1,500
	Morgan	Marine (shallow) and evaporitic	Carb, Ss, F, R, E	Quartzarenite	325
	Round Valley	Marine	Carb, F		350
Mississippian	Manning Canyon	Marine (offshore)	F, Carb, Ss	Quartzarenite	225
	Humbug	Marine (shallow)	Carb, Ss	Quartzarenite	350
	Deseret	Marine (shallow)	Carb, Ss	Quartzarenite	650
	Madison	Marine (shallow)	Carb		250
Cambrian	Tintic Quartzite	Marine (shallow)	Ss	Quartzarenite, subarkose	375
Precambrian	Uinta Mountain Group	Alluvial, deltaic, and marine (shallow)	Ss, F, R	Subarkose, quartzarenite	>20,000

[a]Rock Types: Cgl = conglomerate, Ss = sandstone, F = siltstone and claystone, Carb = carbonate rocks, E = evaporites, R = red beds or varicolored rocks, C = Coal, V = volcanic or volcaniclastic rocks.
[b]QFL means quartz–feldspar–lithic diagram percentages.

imbricated Archean basement gneiss and lower Paleozoic strata (Figure 4) (see cross sections in Royse et al., 1975; Bruhn et al., 1983a; Smith and Bruhn, 1984). The overall geometry and internal structure of the anticline suggests that it developed by progressive imbrication of a large basement "high" or protuberance that extended westward in the footwall (see Figure 8). The basal décollement stepped upward across this protuberance from a level in crystalline basement in the west to the lower Paleozoic section beneath the Wasatch Mountains to the east. Progressive, eastward-directed imbrication of the footwall was underway by Maestrichtian time and must have continued into Paleocene time during movement on the Hogsback thrust system.

The Cottonwood fold is a large, east-vergent anticline exposed in the interior of the Uinta arch southeast of Salt Lake City (Figures 5 and 6). This anticline was refolded about an east-trending hinge line during development of the Uinta arch and plunges northward beneath the upper sheet of the Mount Raymond thrust fault. In the southernmost part of the arch the anticline plunges southward beneath the Charleston allochthon (Riess, 1985). Down-plunge views of geologic maps of the Cottonwood fold demonstrate the original geometric relationships between the anticline and the Charleston and Mount Raymond thrust faults—namely, the faults emerged from the back limb of the anticline (Figures 6 and 7) (Riess, 1985). A particularly important observation is that both faults cut across the same

stratigraphic units in the fold's back limb and merge into a thrust fault "flat" within the Twin Creek Formation in the region east of the Wasatch Mountain front, which indicates that the fold was an important structure in the transition zone between the Charleston and Absaroka thrust systems. We infer that the Alta–Grizzly thrust (Crittenden, 1976), which crops out in the eastern part of the Cottonwood fold, represents splays of the Absaroka or Charleston thrust systems, or both.

The blind thrust fault inferred in the interior of the Cottonwood fold presumably represented the easternmost tip of the regional basal décollement that extended throughout the area west of the Wasatch Mountains. This décollement was located at a depth of 12–15 km in Precambrian strata and crystalline basement.

The geometry of major thrust faults during middle and Late Cretaceous can be inferred using the present positions of the large, frontal, and sidewall anticlines in hanging wall thrust sheets. Drill hole and seismic reflection data and published information on hanging and footwall cutoffs on major thrust systems (Royse et al., 1975; Dixon, 1982; Bruhn et al., 1983a; Lamerson, 1982) are also useful in reconstructing the geometry. The stratigraphic level of the thrust faults beneath the Wasatch Mountains is known from geologic mapping, seismic reflection, and well data. The structural relief of fault ramps and the stratigraphic level of the basal décollement in the region west of the Wasatch

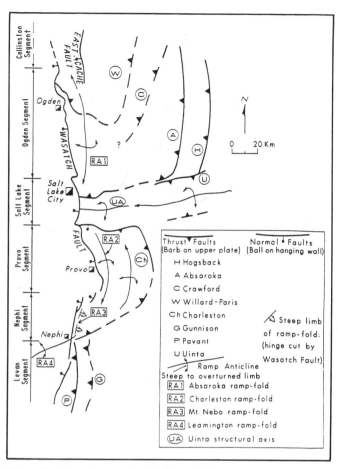

Figure 3—Tectonic map of the Wasatch Mountains and adjacent areas showing major hanging wall folds in thrust sheets and Late Cenozoic structural segmentation of the Wasatch normal fault zone. After Smith and Bruhn (1984), copyright ©1984 by the American Geophysical Union.

Mountains are estimated by noting the oldest rocks preserved in the hanging wall thrust sheets. The basal décollement west of the fault ramp must have been located deeper than the oldest rocks preserved in the hanging wall sheet after it had moved over the ramp. The ramp height can thus be estimated as the difference in stratigraphic thickness between the oldest strata in the hanging wall sheet and the youngest strata in the underlying footwall. Applying this approach to thrust sheets preserved in the Wasatch Mountains indicates that several large frontal and sidewall fault ramps with reliefs of 5–10 km were located along the western edge of the mountains (Figure 8). The positions of these ramps were apparently controlled by inherited crustal structures in the crystalline basement, such as the boundaries of the Pennsylvanian–Permian Oquirrh basin and the Precambrian Uinta aulacogen (Bruhn et al., 1983b; Smith and Bruhn, 1984).

Structure of Uinta Mountains

Structurally, the Uinta Mountains are a north-verging anticlinorium that trends eastward (Figure 1). On their eastern end, the mountains die out into a series of northwest–striking folds and thrust faults in the Axial arch of Colorado, which in turn connects to the Grand Hogback

monocline in the White River uplift. The Uinta Mountains continue westward into the Uinta arch, where late Mesozoic structures of the thrust belt are folded into a north-verging anticlinorium (Crittenden, 1976).

Discussions of Uinta Mountain structure often center on the large, partly en echelon series of reverse and thrust faults that extend along the northern flank of the range. Significant displacements have occurred along some of these faults with estimates of vertical stratigraphic separation as large as 10 km (Hansen, 1965; Ritzma, 1969; Gries, 1983). Hansen (1965) suggested that as much as 6 km of sinistral slip may have occurred along part of the north flank. Large thrust and reverse faults also occur along most of the south flank of the mountains where they are buried by sedimentary deposits in the Uinta basin (Gries, 1983). These faults apparently have displacements of several kilometers, but there are few published data on them.

Although the exposed faults are large, we consider the anticlinal geometry of the mountains to be even more significant because the fold geometry provides insights into the deep crustal structure (Figure 8). Our preliminary modeling of the deep structure is based on the following assumptions: (1) deformation in the crust occurred on discrete fault zones at depths above the brittle–ductile transition; (2) the long, gently dipping limb of the large anticline reflects the presence of a fault ramp that extends from the brittle, sedimentary section downward to the brittle–ductile transition in the basement; (3) the ratio of vertical to horizontal displacement perpendicular to the mountain range is a reasonable estimate for the tangent of the ramp's dip; and (4) we assume that the cross-sectional geometry is the result of crustal shortening across the mountain range.

Crustal displacements can be estimated from a cross section (Figure 9) constructed on the basis of well data that are available for the western Uinta Mountains. The vertical uplift of strata in the core of the Uinta fold compared with the same strata in the adjacent Green River basin is about 12 km, and the amount of horizontal shortening across the mountain range is at least 13 km. We obtained this latter figure by summing a minimum of 6 km of horizontal translation on the thrust fault encountered in the Shell Dahlgreen Creek well with 7 km of shortening estimated from the geometry of the fold. The amount of movement on the thrust fault is probably greater than 6 km, because the fault could extend at a low angle for some distance south of the well. An additional 1–2 km of horizontal shortening, perhaps more, apparently occurred across the reverse fault on the southern flank of the mountains (Figure 9).

Figure 10 is a speculative drawing depicting our model for the evolution of the Uinta Mountains in the region just east of the Hogsback thrust system. We infer that the thick Uinta Mountain Group represents the fill of east-trending trough or aulacogen that extended into the craton in late Precambrian time (Wallace and Crittenden, 1969; Sears et al., 1982). A thin, discontinuous section of strata accumulated in the Uinta Mountains region during the Paleozoic and Mesozoic (Herr et al., 1982) as deposition was interrupted by marine regressions.

Formation of the Uinta Mountains was underway in early Paleocene time and continued throughout the Eocene. We

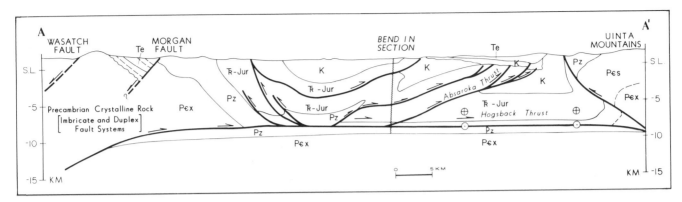

Figure 4—Regional cross section along line A–A' (Figure 1) of the thrust belt from the Wasatch fault zone to the northern edge of the Uinta Mountains. The cross section is based partly on work by Smith and Bruhn (1984) and by Royse and Lamerson (1984). Evidence for the position of the Hogsback thrust beneath the western Uinta Mountains is discussed in the text. Rock units: Te = Tertiary strata, K = Cretaceous, TR–Jur = Triassic-Jurassic, Pz = Paleozoic, PЄs = Precambrian strata, and PЄx =Archaean crystalline basement.

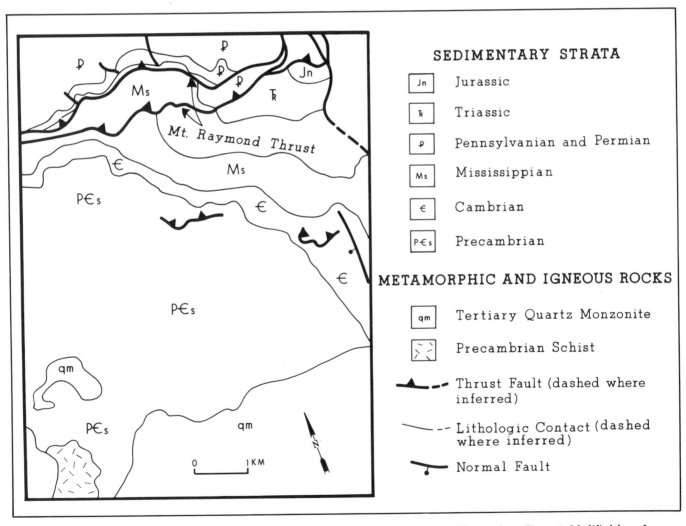

Figure 5—Geologic map of part of the Cottonwood area. Map area is indicated by box in the Uinta arch on Figure 1. Modified from James (1979). Used by permission of the Utah Geological and Mineralogical Survey.

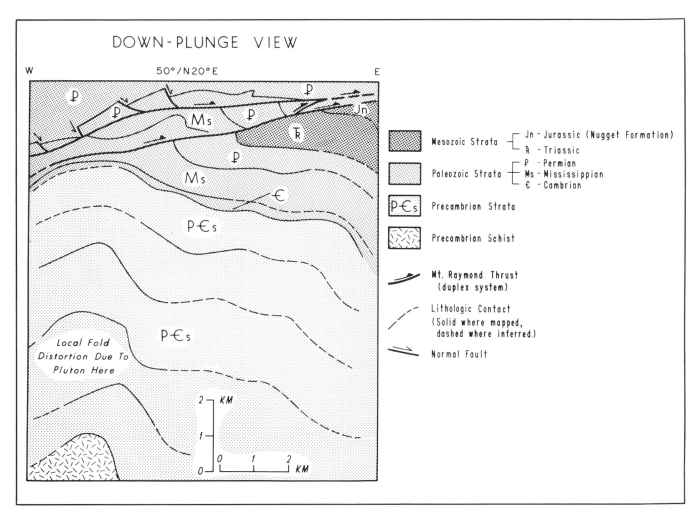

Figure 6—Down-plunge cross section of area shown in Figure 5 showing large, north–northeast plunging anticline in the Cottonwood area. Projection is viewed looking along the fold hinge, which plunges 50° toward N 20° E.

infer a minimum of 13 km of northward displacement on a major thrust fault system that dips southward to a depth of at least 15–20 km. Below this depth, the lower crust may have deformed ductilely, either by broadly distributed flow or perhaps by displacements on a complex system of ductile shear zones. We infer, as did Sears et al. (1982) and Hansen (1984), that some of the Precambrian faults along the edge of the aulacogen may have been reactivated during formation of the Uinta Mountains. The thick section of Tertiary strata in the Uinta basin may reflect crustal loading and isostatic subsidence caused by movement on the south flank thrust system. If so, then the amount of displacement on these faults may be much larger than we show in our cross section.

Structural Relationship of Uinta Mountains to Uinta Arch

The geometrical relationships of the Uinta Mountains and Uinta arch are partly obscured by extensive outcrops of Oligocene volcanic rocks along the western end of the Uinta Mountains. However, published cross sections on USGS quadrangle maps of the western Uinta Mountains and eastern part of the Uinta arch show that the north-vergent anticlinal structure of the Uinta mountains extended

westward into the arch prior to extrusion of the Oligocene volcanic cover (Bromfield et al., 1970; Bromfield and Crittenden, 1971). Two tectonic events have apparently disrupted this original structural continuity. First, Eocene–early Oligocene uplift and faulting were super-imposed on earlier structures in the Uinta Mountains (Hansen, 1965, 1984). Hansen (1965) has estimated that latest Cretaceous–early Paleocene uplift along the northern flank of the Uinta Mountains was about 3–4 km on the basis of the composition of clasts in conglomerate of the Fort Union Formation. Subsequent uplift along the north and south flank thrust fault systems resulted in as much as 8–10 km of vertical displacement during Eocene–early Oligocene time. Most of the uplift was concentrated in fault zones along the northern flank of the mountains. Notably, these faults truncate the Fort Union Formation and underlying strata, clearly demonstrating that at least two periods of uplift occurred in the Uinta Mountains (Hansen, 1984). We speculate that this latter period of deformation may have been responsible for the westward-plunging structural closure that separates the Uinta Mountains from the Uinta arch. The north and south flank thrust fault systems terminate near the western end of the Uinta Mountains as

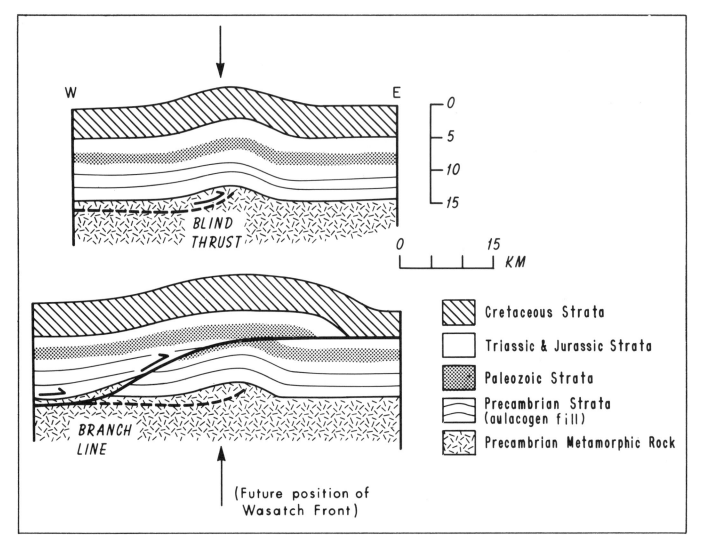

Figure 7—Diagram showing the inferred evolution of the Cottonwood fold and Mount Raymond thrust fault. Top—blind detachment propagates eastward beneath the Cottonwood area during Late Cretaceous time, forming a large, east-vergent anticline in the hanging wall. Bottom—continued crustal shortening is accommodated by formation of the Mount Raymond thrust fault on the back limb of the anticline as décollement steps up-section to the east.

vertical separation across the faults decreases toward the west. This westward decrease in separation on the thrust faults would naturally lead to the development of a westward plunging structural closure in the Uinta Mountain anticlinorium. We do not know the extent of Eocene deformation in the Uinta arch. The north and south flank thrust fault systems in the Uinta Mountains do not continue westward into the arch, therefore Eocene deformation may have been mostly concentrated in the Uinta Mountains and may have died out to the west. This suggestion, however, remains speculative and is based solely on structural geometry.

The second important tectonic event leading to disruption of the Uinta arch–Uinta Mountain structural system was the intrusion of Oligocene plutons and widespread normal faulting during the Neogene (Figure 2; Crittenden, 1976; Hopkins and Bruhn, 1983; Royse, 1983). Broad doming and low- to high-angle normal faulting in the interior of the arch developed during igneous intrusion; this assumption is

based on mapped relationships between plutons and country rocks on geologic maps of the Cottonwood area (James, 1979). Subsequent extension led to the structural reactivation and localized "back sliding" of thrusts in the Charleston allochthon (Hopkins and Bruhn, 1983; Royce, 1983; Reiss, 1985). Preliminary work on the Neogene fault systems indicates that as much as 11 km of vertical uplift may have occurred on the southwestern edge of the Uinta arch along the Wasatch normal fault zone, resulting in regional northeasterly tilting of the arch (Figure 2) (W. T. Parry and R. L. Bruhn, unpublished data). This tilting and associated normal faulting in the interior of the arch has clearly disrupted the continuity of Cretaceous and lower Tertiary structures, but not so greatly as to completely obscure the original, north-vergent anticlinorium that extended continuously between the present Uinta Mountains and Uinta arch.

Our interpretation of the evolution of the Uinta arch during Late Cretaceous and early Paleocene time is depicted

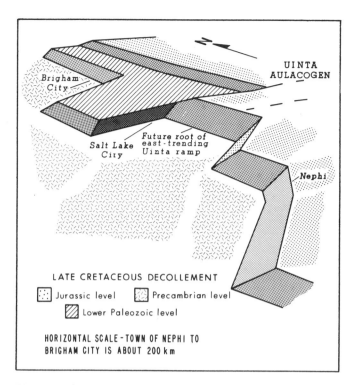

Figure 8—Generalized model of footwall structural topography along the western edge of the Wasatch Mountains during Late Cretaceous time. The view is toward the northeast, and hanging wall strata have been removed to show the inferred structural topography of the footwall. Blind thrusts and thrust fault flats in the footwall of the Charleston allochthon are also inferred, but not indicated on the diagram. See text for discussion.

in Figures 11 and 12. Campanian–Maestrichtian time was marked by eastward transport of the Absaroka ① and Charleston ③ thrust sheets (see circled numbers on Figure 11). The Mount Raymond thrust ② represented the major fault linking the two sheets. We presume that differences in thicknesses of strata and amounts of displacement between the Charleston and Absaroka thrust sheets led to development of a tear fault, or faults, between the two sheets on the basis of variations in stratigraphic thicknesses and east-trending folds in the Mount Raymond and Charleston thrust sheets. West-to-east ramping of thrusts from Precambrian to Mesozoic rocks resulted in development of the large, north-trending Cottonwood fold beneath the Mount Raymond thrust and northern part of the Charleston thrusts during Campanian–Maestrichtian time, resulting in about 1–2 km of uplift in the proto-Uinta arch (Figure 7). The trace of the blind detachment ④ associated with development of this fold in the proto-Uinta arch is illustrated in Figure 11. The extent of this blind fault in the region to the north and south of the arch is not known.

During Maestrichtian–Paleocene time displacement on the Hogsback thrust system ⑤ north of the Uinta arch led to reactivation and eastward displacement on the Absaroka décollement ① beneath the northern Wasatch Mountains (see circled numbers on Figure 12). The Hogsback thrust (reactivated Absaroka décollement) did not terminate into a tear fault in the upper sheet along its southern margin

(Bruhn et al., 1983a), but rather, the fault must have plunged southward into an east-striking, south-dipping sidewall fault ramp ⑤ that cut deeply into the crystalline basement beneath the Uinta arch. Evidence for this structure includes the following. First, as much as 20 km of eastward displacement occurred on the Hogsback thrust system along the northern edge of the Uinta arch and western Uinta Mountains during Paleocene time according to published estimates of displacement (Royse et al., 1975; Dixon, 1982; Lamerson, 1982). Second, uplift of the Uinta Mountains began in latest Cretaceous or Paleocene time, but culminated during the Eocene (Blackstone, 1977; Gries, 1983; Hansen, 1984). Consequently, the southern edge of the Hogsback thrust was truncated by the North Flank thrust fault in the western Uinta Mountains. However, the North Flank thrust does not cut any of the exposed rocks in the core of the Uinta arch in the region west of the Uinta Mountains. That is, the rocks in the Uinta arch form the hanging wall of the Hogsback thrust in this area. This means that (1) eastward displacement of the Hogsback thrust during Paleocene must have been transferred onto a fault system beneath the Uinta arch, and (2) that as much as 20 km of sinistral motion may have occurred along this deeply seated fault system as the crust south of the axis moved eastward with respect to the footwall of the Hogsback thrust in the region to the north (Figure 12). Presumably, sinistral slip occurred at the same time in the Uinta Mountains. This early phase of deformation in the Uinta Mountains may be recorded by the coarse conglomerate in the top of the Currant Creek Formation, which contain boulders of Cretaceous rocks.

AGE, PETROGRAPHY, AND PROVENANCE OF CURRANT CREEK, UINTA, AND DUCHESNE RIVER FORMATIONS

Age

The Currant Creek Formation is unconformably overlain by the Duchesne River Formation and unconformably overlies the Cretaceous Mesaverde Formation. On the basis of rock types and stratigraphic position, which are similar to those of the North Horn Formation of the Wasatch Plateau, the formation is assigned a Late Cretaceous–early Tertiary age (Figure 13). Bryant (cited in Bruhn et al., 1983a) has collected pollen and freshwater ostracodes from near the base that are Maestrichtian in age. The age of the upper part of the formation is uncertain, but is probably late Paleocene.

The Uinta Formation is conformably overlain by the Duchesne River Formation and, in turn, conformably overlies the Green River Formation of which it is a partial time equivalent. Three vertebrate faunal zones, which indicate a middle–late Eocene age, are recognized in the Uinta Formation (Osborn, 1895; Peterson, 1932).

The Duchesne River Formation overlies both the Currant Creek and Uinta formations in the Uinta basin and is overlain only locally by younger Tertiary gravel surfaces and Quaternary alluvium. A late Eocene–early Oligocene age for the formation is assigned on the basis of vertebrate remains in the lower part of it (Simpson, 1933; Wood et al., 1941) and K-Ar dating of volcanic ash in the upper part (Dawson, 1966; B. Bryant, personal communication, 1982).

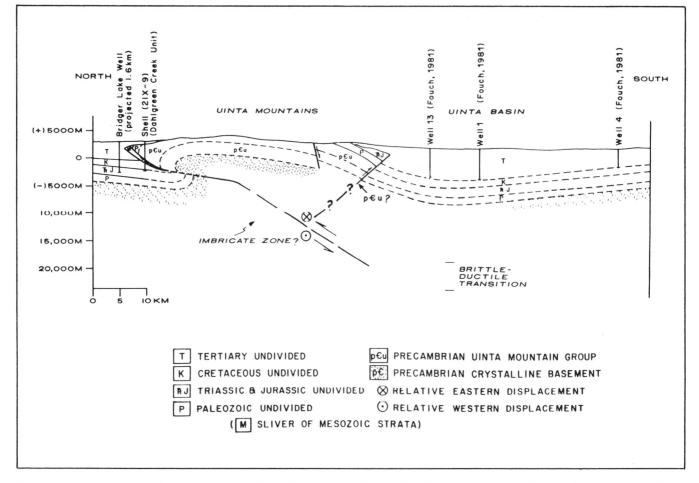

Figure 9—Regional cross sections of the western Uinta Mountains. See line D–D' in Figure 1 for location. After Bruhn et al. (1983a). Used by permission of the Utah Geological and Mineralogical Survey.

General Stratigraphy, Sedimentary Structures, and Bedforms

The Currant Creek Formation is composed of conglomerate, sandstone, siltstone, mudstone, and claystone. A west-to-east thinning from about 1500 m thick at its western margin to where the formation is overlapped by the overlying Duchesne River Formation to the east is accompanied by a decrease in grain size (Figure 14). Conglomerate beds are commonly graded, lenticularly stratified, and imbricated and display cut-and-fill structures. Upward-fining conglomerate–sandstone units are scoured into underlying finer-grained rocks. Sandstone is trough and planar cross-stratified and occurs both as separate beds and as lenses within conglomerate. Planar bedding is also common in sandstone, as is graded bedding. Parting lineation is rare. Fine-grained sequences are poorly exposed, but where seen, are commonly planar bedded.

Coarse- to fine-grained sandstone, siltstone, mudstone, claystone, and carbonate rocks characterize the Uinta Formation. Maximum thickness is approximately 600 m. Sandstone is trough and planar cross-stratified. Planar-bedded sandstone also occurs and is often massive in appearance. Sandstone alternates with beds of mudstone, claystone, and calcareous claystone.

The Duschesne River Formation is composed of conglomerate, sandstone, siltstone, mudstone, and claystone.

Maximum thickness of the formation is about 1200 m (Isby and Picard, 1983). A stratigraphic thickness of 878 m was measured near Vernal, Utah (Andersen and Picard, 1972). Stratigraphic thickness of the formation increases markedly to the south. Simultaneously, the grain size decreases greatly southward. Conglomerate beds display indistinct horizontal stratification or form broad shallow troughs. Isolated sandstone lenses within conglomerate beds are common. Conglomerate and sandstone beds often disconformably overlie fine-grained rocks. Sandstone is commonly trough and planar cross-stratified. Horizontal and inclined planar stratification are also abundant. Fine-grained rocks are horizontally stratified or laminated. Ripple stratification is rare.

Petrography

Conglomerate in the Currant Creek Formation is petromict and is clast- and matrix-supported. Sandstone is moderately to well sorted and ranges in composition from quartzarenite to litharenite (classification of Folk, 1980). Most quartz grains are monocrystalline and nonundulatory. Feldspar is rare, constituting less than 1% of most samples (Figure 15). Chert and other sedimentary rocks are the most abundant rock fragments.

Sandstone of the Uinta Formation ranges from poorly to well sorted. The degree of sorting increases with an increase

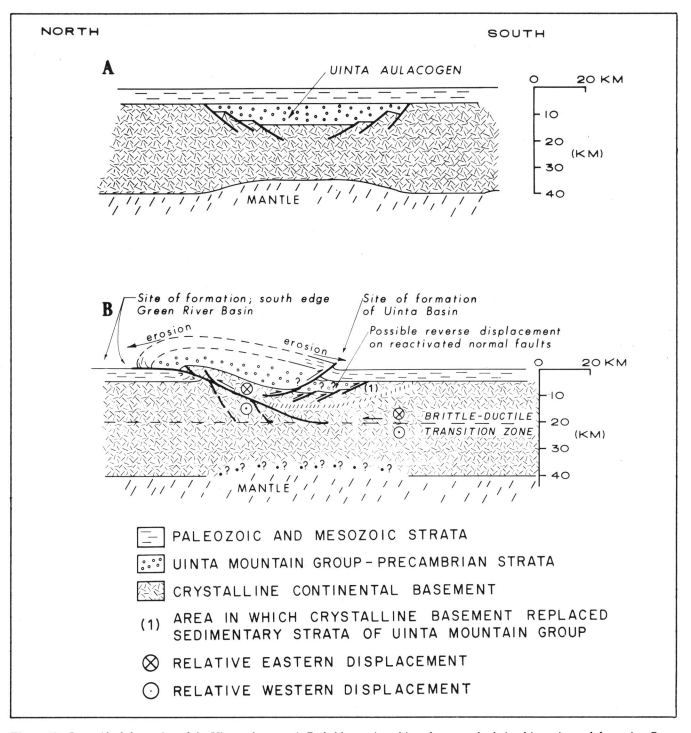

Figure 10—Laramide deformation of the Uinta aulacogen. A. Probable stratigraphic and structural relationships prior to deformation. B. Relationships following northward thrusting of aulacogen. After Bruhn et al. (1983a). Used by permission of the Utah Geological and Mineralogical Survey.

in grain size. The sandstone is feldspar-rich (Stagner, 1941; Picard and Dideriksen, 1978) and is classified as arkose and subarkose (Figure 15).

Duchesne River conglomerate is pectromictic, both clast- and matrix-supported and generally very poorly sorted. Sandstone in this formation is also poorly sorted and ranges in composition from quartz-rich to lithic-rich varieties. The great majority of quartz grains are monocrystalline. Feldspar

is rare and generally constitutes less than 5% of the grains in a sample (Figure 15). Rock fragments consist mainly of sedimentary types. Carbonate and clastic fragments are more abundant than chert.

Depositional Environments

Sedimentary evidence from the Currant Creek Formation suggests that it was deposited largely within braided streams

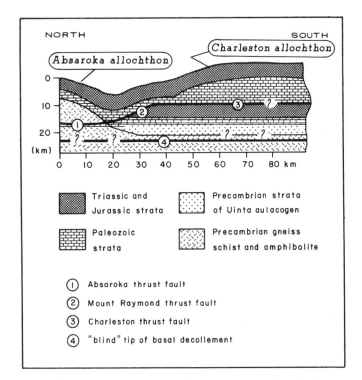

Figure 11—Structural geometry of the Uinta arch region during latest Cretaceous (Campanian–Maestrichtian) time. Cross section extends along line B–B′ in Figure 1. See text for further details.

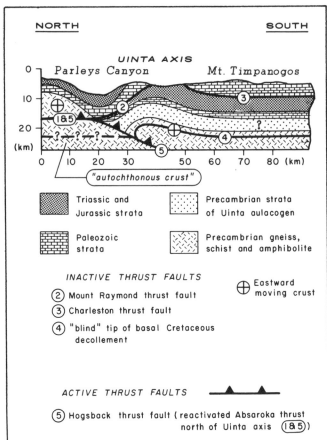

Figure 12—Formation of the Uinta arch and relationship of the arch to thrust faulting during the Paleocene. Cross section extends along line B–B′ in Figure 1. See text for further details.

as part of an alluvial fan complex in a humid environment (Isby and Picard, 1985). The proximal fan facies is represented by conglomerate and lesser amounts of sandstone that formed as channel fills, longitudinal bars, and sieve deposits. Distal facies consist of sandstone and fine-grained clastic strata that were deposited as part of an intricate braided stream network. Numerous sandstone channels, often having conglomeratic bases, divide and rejoin.

The Uinta Formation was largely deposited in lacustrine and fluvial settings. The saline facies and the sandstone/limestone facies of this formation record the regression and last several million years of deposition within Lake Uinta. Fluvial facies composed of medium- to coarse-grained, cross-stratified sandstone are well developed in the eastern Uinta basin. Deposition likely occurred within wide meandering rivers.

A fluvial environment of deposition is indicated for the Duchesne River Formation (Andersen and Picard, 1974; Picard and Anderson, 1975; Maxwell and Picard, 1976). Lenticular conglomerate and sandstone bodies are interpreted as stream channel deposits. In the more proximal parts of the formation, deposition occurred within high-gradient braided streams. In the more distal regions, lower gradient meandering streams prevailed. Interspersed with channel deposits are fine-grained rocks probably representing floodplain deposits.

Paleocurrents

Paleocurrent directions measured in sandstone of the Currant Creek Formation at 17 separate stratigraphic intervals indicate southerly and southeasterly transport of sediment (Figure 16). Directions derived from 23 separate imbricated conglomerate units are dominantly toward the south–southeast. Paleocurrent directions derived from cross-stratified sandstone in the Uinta Formation are dominantly toward the west–northwest (Figure 17). Duchesne River Formation paleocurrent directions, as determined from trough axes of cross stratification, are strongly toward the south–southwest (Figure 17).

Provenance of Currant Creek Formation

Two principal sources have been proposed for sediment in the Currant Creek Formation: (1) the Strawberry Valley (Charleston) thrust sheet (Walton, 1957) and (2) the area of junction of the Uinta and Wasatch mountains (Garvin, 1969). According to Garvin (1969), the Strawberry Valley thrust was an unreasonable suggestion since little Weber Sandstone and no Precambrian Mutual Formation are exposed there. Similarly, Beutner (1977) suggested that sediment was derived from the western end of the Uinta reentrant where the greatest uplift in the thrust belt in east–central Utah occurred.

In the Wasatch Range there are two distinct areas where strata of the Mutual Formation, Tintic Quartzite, Big Cottonwood Formation, and cherty carbonate beds are exposed. These are (1) Big and Little Cottonwood canyons near Salt Lake City and adjoining areas to the south and

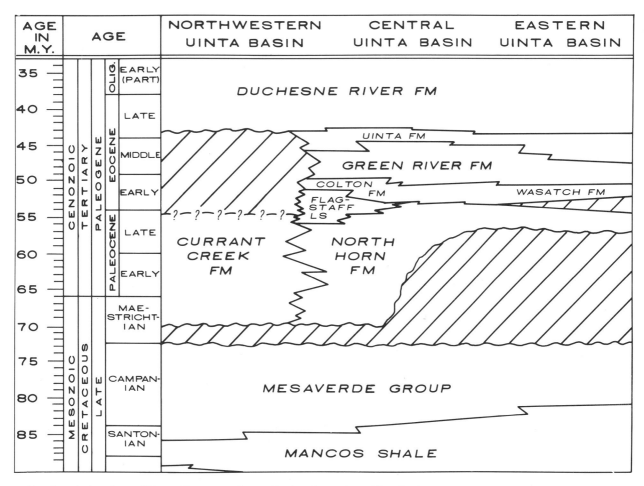

Figure 13—Correlation chart of Upper Cretaceous–Lower Tertiary formations, Uinta Basin.

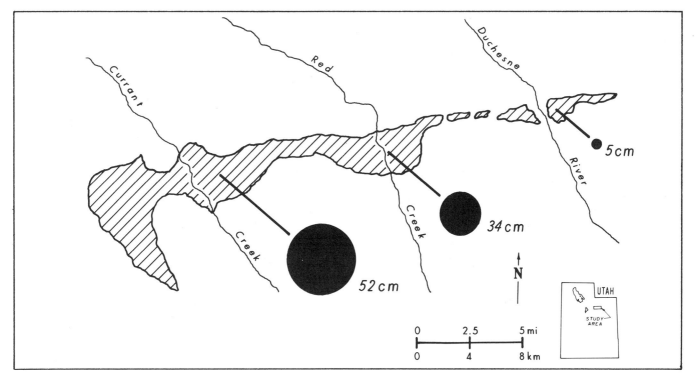

Figure 14—Distribution of mean maximum clast size in the Currant Creek Formation.

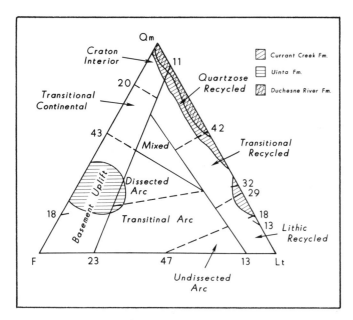

Figure 15—Quartz-Feldspar-Lithic plot of sandstone in Currant Creek, Uinta, and Duchesne River formations. Qm = monocrystalline quartz, F = feldspar, Lt = total polycrystalline lithic fragments.

southeast; and (2) northeast of Ogden, where extensive outcrops of Precambrian and Paleozoic strata are exposed.

Charleston Thrust

Exposures near Salt Lake City occur within or adjacent to the Charleston thrust sheet. The upper plate of the Charleston–Nebo thrust contains Cambrian and Precambrian sedimentary rocks. Near Deer Creek reservoir, the exposed upper plate is composed of the Cambrian Tintic Quartzite and Ophir Formation, as well as the Precambrian Big Cottonwood Formation (Baker, 1964). In addition to their exposure in the Cottonwood area, the Mutual, Big Cottonwood, and Tintic formations are exposed in the upper plate of the Charleston thrust in the Mount Timpanogos area (Baker and Crittenden, 1961). The basal décollement of the Charleston sheet immediately west of this region is likely rooted in stratigraphically lower Precambrian sedimentary rocks or crystalline basement.

The age of the major tectonic episode along the Charleston thrust sheet is constrained by the fact that in the western Uinta basin the Charleston–Nebo thrust displaces the Late Cretaceous (Santonian–Campanian) Mesaverde Group and is in turn onlapped by the Maestrichtian–Paleocene North Horn Formation (Crittenden, 1976). Jefferson (1982) has interpreted the North Horn Formation as Paleocene(?) in age where it overlaps the Nebo thrust in the Cedar Hills. As suggested by Bruhn et al. (1983a), the age of emplacement of the Charleston thrust is likely late Late Cretaceous.

The majority of clasts in the Currant Creek Formation were probably not derived from the Charleston–Nebo thrust sheet. Clasts at the base of the Currant Creek include Precambrian and Cambrian quartzite. Figure 18 shows the

distribution of conglomerate (Type A) that is rich in clasts of Precambrian and Cambrian age quartzite and chert derived from Mississippian carbonate. If indeed detritus were shed from the Charleston plate, we would expect evidence of systematic un-roofing of the thrust sheet and an inverse-stratigraphy in the conglomerate, not seen (Figure 18).

Willard Thrust Sheet

The Willard thrust sheet also contains possible Currant Creek source rocks. Mullens (1969) and Crittenden (1972a, b) show Precambrian and Cambrian clastic rock as well as Mississippian carbonate cropping out in the upper sheet of the Willard thrust. These include the following units or their lateral equivalents: Mineral Fork Formation (contains dark grey quartzite), Mutual Quartzite, Tintic Quartzite, and Mississippian–Permian cherty limestone.

The majority of vector resultants calculated for paleocurrent measurements in the Currant Creek Formation are oriented toward the south–southeast. This configuration favors a more northerly than westerly source. Figure 14 illustrates the drastic decrease in maximum clast size from west to east for this formation. This may appear to favor a more westerly source, but the present outcrop extent of the formation is part of a wide lobe that extended to the south.

The Willard thrust can be structurally correlated with the Paris thrust in Idaho. Armstrong and Oriel (1965) believe that the Paris thrust was finally emplaced by the beginning of Late Cretaceous time. Because of structural complexities and extensive Cenozoic cover in north-central Utah, age of emplacement is unknown but may be as late as middle Cretaceous (Bruhn et al., 1983a). Crawford's (1979) provenance studies of the Maestrichtian–Paleocene Evanston Formation show it contains most of the same clast types as the Currant Creek Formation. The Evanston Formation also contains clasts of carbonate rock and crystalline basement rock of the Farmington Canyon Complex. These less resistant rock types decrease greatly in size and abundance farther from the source area (Crawford, 1979). Crawford maintains that clasts in the Evanston Formation were largely shed from the Willard allochthon to the northwest. The conglomerate of the Currant Creek Formation is genetically related to that of the Evanston Formation. Yet, differences in maximum clast sizes prohibit the Currant Creek conglomerate from being a simple distal equivalent of the Evanston. Northeast of Ogden the Willard thrust sheet is folded about the large basement-cored anticline in the younger Absaroka plate (Royse et al., 1975). This uplift and folding of the Willard sheet is recorded by the Maestrichtian–Paleocene Evanston Formation. The bulk of the conglomerate in the Currant Creek Formation also records this uplift but was derived from a different part of the thrust sheet. The Willard plate extends to the south and west of its westernmost exposure near Ogden (Crittenden, 1972b). The Currant Creek source may thus have been a more southerly or southwesterly extension of the Willard thrust sheet that was uplifted within the Absaroka plate but was erosionally removed from the upper levels of the allochthon.

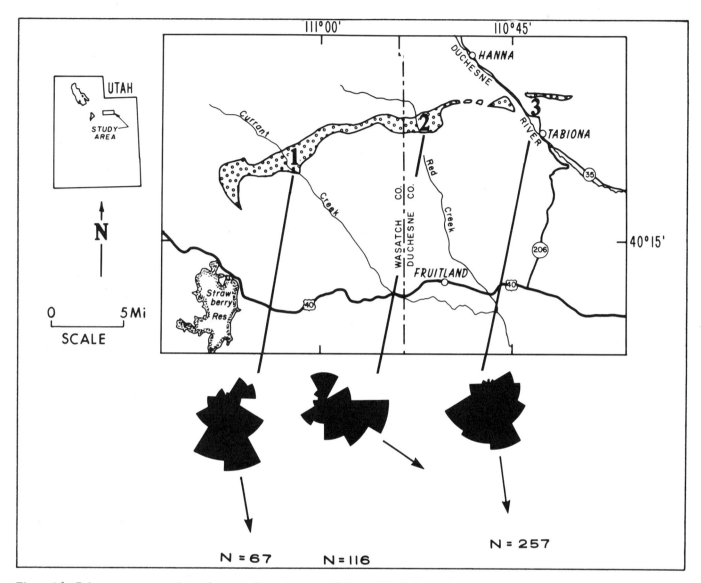

Figure 16—Paleocurrent pattern in sandstone and conglomerate of Currant Creek Formation.

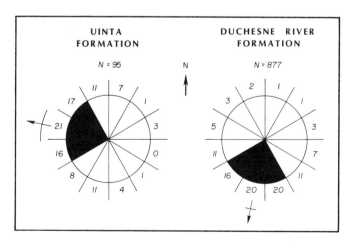

Figure 17—Compass diagrams of paleocurrent measurements from Uinta and Duchesne formations. Numbers around circles represent the percentage of measurements in each 30° sector. Shaded sectors are modes. Vector resultant directions and 95% confidence limits are shown.

Provenance of Uinta and Duchesne River Formations

The source area of the Uinta Formation was much more distant than that of the Currant Creek and Duchesne River formations. The high feldspar content of the sandstone and the types of feldspar in it indicate a volcanic rock source. Strong westerly and northwesterly paleocurrent directions make the deeply rooted Laramide uplifts of western and southwestern Colorado probable source terranes.

The Duchesne River Formation is composed of recycled sedimentary and low-grade metasedimentary rocks derived from the Uinta Mountains to the north. Conglomerate units show a roughly inverse stratigraphy. Mesozoic and Paleozoic age clasts are concentrated toward the base, while upper beds show a concentration of Precambrian age clasts. The formation records the final and major episode of uplift in the Uinta Range. Rocks of the Duchesne River Formation lap over Uinta Mountain structures and are not deformed.

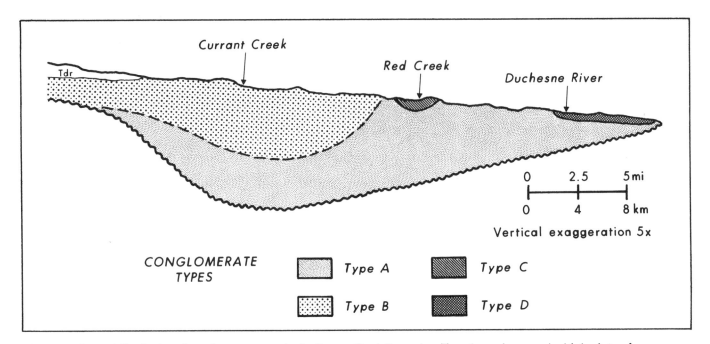

Figure 18—Lateral distribution of conglomerate types in the Currant Creek Formation. Type A conglomerate is rich in clasts of Precambrian and Cambrian quartzite and chert derived from Mississippian carbonate. Type B conglomerate is rich in clasts of Pennsylvanian and Permian sandstone. Types C and D conglomerate are mainly composed of Mesozoic clasts representing younger synorogenic deposits.

PALEOGEOGRAPHY

By Maestrichtian time the Cretaceous seaway had retreated from what is now central and eastern Utah (Fouch et al., 1983), and nonmarine environments were dominant (Figure 19). Thrusting at the western boundary of the foreland basin occurred during the Late Cretaceous and resulted in the deposition of extensive coalescing alluvial fans that prograded eastward from the active Sevier–Laramide orogenic belt. Alluvial fan facies broadly graded into an extensive alluvial floodplain.

By early Tertiary time this configuration had changed (Figure 20). Lakes occupied much of the former marine depositional basin. This depositional pattern continued through at least late Eocene time and is represented by alluvial plain and lacustrine facies of varying lateral extent (Figures 21 and 22). During Paleocene time, the sedimentary body of the Currant Creek Formation became bounded on the north by the newly uplifted Uinta arch.

The Currant Creek Formation is a remnant of a once much more laterally extensive clastic wedge that protruded into the north-central Utah foreland basin and extended all along the Cordilleran thrust front. The major part of this projection possibly lay northeast of the Currant Creek Formation, but it has been removed by erosion. It was probably located where older, upturned strata now flank the Uinta Mountains. The younger Duchesne River Formation, which formed largely in response to the last major mountain building events of the Uintas during Eocene (Andersen and Picard, 1972, 1974), also onlaps older strata and may cover other remaining portions of the thick, laterally extensive, coalescing alluvial fan deposit of the Currant Creek Formation.

CONCLUSIONS

The Wasatch Mountains mark the site of two regional structural features in the Sevier–Laramide orogenic belt. These are (1) a regional system of frontal and sidewall thrust fault ramps that developed during the Late Cretaceous and (2) the Uinta arch, which is the zone of structural interaction between the Uinta Mountains and the foreland thrust belt.

The basal décollement of the Sevier–Laramide orogen obtained a steplike geometry during crustal shortening in the Late Cretaceous, stepping up-section to the east. This regional system of frontal and sidewall ramps was located along the western edge of the Wasatch Mountains where it strongly influenced the geometrical evolution of the thrust belt. It marked the site at which the basal décollement rooted in crystalline basement toward the west, or "hinterland" of the orogen, and delimited the along-strike extent of the Absaroka and Charleston allochthons.

The Uinta arch began to form during latest Cretaceous–early Paleocene time as the basal décollement of the Hogsback thrust system rooted southward into the crust beneath the east-trending Uinta aulacogen. Eastward transport on the Hogsback thrust system resulted in about 20 km of sinistral slip along the south-dipping thrust ramp beneath the Uinta arch, thus forming a large, north-vergent anticlinorium in the southernmost part of the Hogsback allochthon. Subsequent deformation may have occurred in the arch during the late Paleocene and Eocene. Uplift of the Uinta Mountains during the late Paleocene and Eocene resulted in truncation of the Hogsback thrust system by the North Flank thrust in the westernmost Uinta Mountains. However, the North Flank thrust system did not continue to

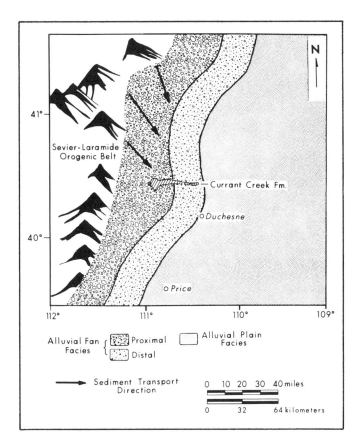

Figure 19—Proposed paleogeographic map of north-central and northeastern Utah in early Maestrichtian time.

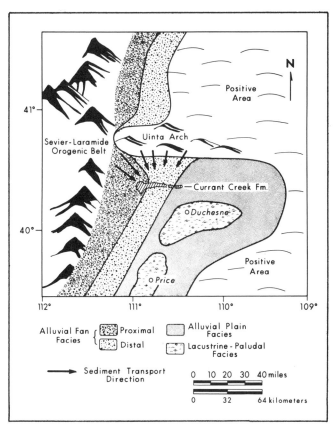

Figure 20—Proposed paleogeographic map of north-central and northeastern Utah in early Paleocene time.

the west into the area of the Uinta arch. The early Tertiary structure of the Uinta arch was disrupted by the intrusion of Oligocene plutons and the development of large normal fault systems, including the Wasatch normal fault zone.

The Maestrichtian–late Paleocene(?) Currant Creek Formation was deposited largely by braided streams within a humid alluvial fan complex. These fans extended into the north-central Utah foreland basin from the active Sevier–Laramide orogenic belt. Proximal facies are composed of conglomerate and sandstone that formed as longitudinal bars, channel fills, and sieve deposits. Distal facies are made up of sandstone and finer grained clastic units that were deposited within channels or overbank regions. In the northwestern Uinta basin the Currant Creek Formation is succeeded by the fluvial Duchesne River Formation, which was deposited mainly by relatively small, rapidly aggrading, southward-flowing streams. Most stream channels probably were braided and had gradients and had high gradients and high flow velocities (Andersen and Picard, 1972, 1974). Farther east, lacustrine and fluvial deposits of the Uinta Formation preceded deposition of the Duchesne River Formation.

Clast types in Currant Creek conglomerates are principally quartzite of Precambrian and Cambrian age and chert that originated from Mississippian carbonates. These are likely derived from rocks within the southern part of the Willard allochthon that were uplifted by folding in the underlying Absaroka allochthon in latest Cretaceous time. Where the

conglomerate contains abundant clasts of Pennsylvanian and Permian age sandstone some detritus may have been locally derived from the Charleston thrust sheet. Separate conglomerate units that are composed primarily of clasts of Mesozoic age rock are present near the top of the Currant Creek Formation. Paleocurrent directions measured from imbricated clasts and associated sandstone indicate a southerly and southwesterly transport direction. These conglomerates likely represent younger synorogenic deposits associated with uplift on the north—the western end of the Uinta arch.

The Currant Creek Formation is a remnant of a once much more laterally extensive clastic wedge. Early in its depositional history, sediment was mainly derived from the Sevier–Laramide overthrust belt to the northwest. By early Tertiary time (likely late Paleocene), however, both the source area and the depositional basin of the Currant Creek Formation had been modified by the initial uplift of the Uinta Mountains.

ACKNOWLEDGMENTS

D. L. Blackstone, Jr., Michael DePanger, and Frank Royse, Jr., critically reviewed the manuscript and contributed many helpful suggestions that improved it. Bruhn acknowledges discussions of thrust belt structure and sedimentology with S. L. Beck, M. Bradley, M. D. Crittenden, Jr., D. L. Hopkins,

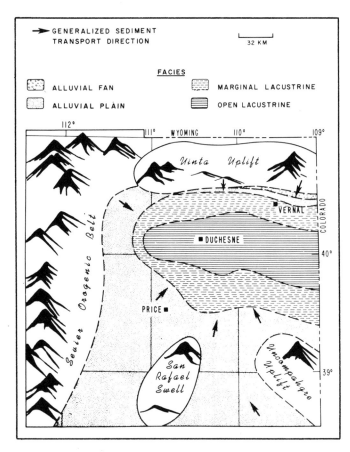

Figure 21—Proposed middle Eocene paleogeography of the upper Parachute Creek Member of the Green River Formation.

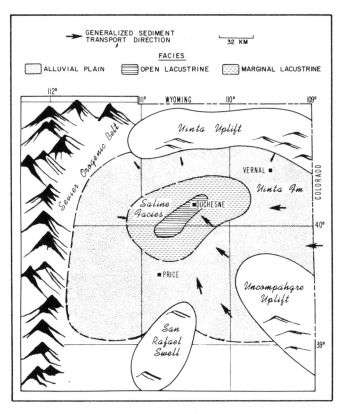

Figure 22—Proposed late Eocene paleogeography of northeastern Utah.

and Frank Royse, Jr. Picard and Isby thank Helene Greenberg and Paul Onstott for drafting assistance and S. J. Lewis, G. D. Lord, and Tom Toronto for assistance in the field. Steve Russo and Suzanne Zink typed several versions of the manuscript. Picard presented an earlier version of some of our ideas as a keynote address to the AAPG Rocky Mountain Section Meeting in Billings, Montana, September 1983. This work was partly supported by a Petroleum Research Fund grant from the American Chemical Society (to Bruhn) and by the Mineral Leasing Fund, University of Utah.

REFERENCES CITED

Andersen, D. W., and M. D. Picard, 1972, Stratigraphy of the Duchesne River Formation (Eocene–Oligocene?), northern Uinta Basin, Northeastern Utah: Utah Geological and Mineralogical Survey Bulletin 97, 29 p.
———, 1974, Evolution of synorogenic clastic deposits in the intermontane Uinta basin of Utah, *in* W. R. Dickinson, ed., Tectonics and sedimentation: SEPM Special Publication, n. 22, p. 167–189.

Armstrong, R. L., and S. S. Oriel, 1965, Tectonic development of the Idaho–Wyoming thrust belt: AAPG Bulletin, v. 49, p. 1847–1866.

Baker, A. A., 1964, Geologic map and sections of the Aspen Grove Quadrangle, Utah: U.S. Geological Survey, Geologic Quadrangle Map GQ-239, scale 1:24,000.

Baker, A. A., and M. D. Crittenden, Jr., 1961, Geologic map of the Timpanogos Cave quadrangle, Utah: U.S. Geological Survey, Geologic Quadrangle Map GQ-132, scale 1:24,000.

Beutner, E. C., 1977, Causes and consequences of curvature in the Sevier orogenic belt, Utah to Montana, *in* E. L. Heisey et al., eds., Rocky Mountain thrust belt geology and resources: Wyoming Geological Association, 29th Field Conference Guidebook, p. 353–365.

Black, B. A., 1965, Nebo overthrust, southern Wasatch Mountains, Utah: Brigham Young University Geologic Studies, v. 12, p. 55–89.

Blackstone, D. L., Jr., 1977, The overthrust belt salient of the Cordilleran fold belt, western Wyoming–southeastern Idaho–northeastern Utah, *in* E. L. Heisey et al., eds., Rocky Mountain thrust belt geology and resources: Wyoming Geological Association, 29th Field Conference Guidebook, p. 367–384.

Bromfield, C. S., A. Baker, and M. D. Crittenden, 1970, Geologic map of the Heber Quadrangle, Wasatch and Summit Counties, Utah: U.S. Geological Survey, Geologic Quadrangle Map GQ-864, scale 1:24,000.

Bromfield, C. S., and M. D. Crittenden, 1971, Geologic map of the Park City East Quadrangle, Summit and Wasatch Counties, Utah: U.S. Geological Survey, Geologic Quadrangle Map GQ-852, scale 1:24,000.

Bruhn, R. L., M. D. Picard, and S. L. Beck, 1983, Mesozoic and early Tertiary structure and sedimentology of the central Wasatch Mountains, Uinta Mountains and Uinta basin: Utah Geological and Mineralogical Survey Special Studies, n. 59, p. 63–105.

Bruhn, R. L., M. D. Picard, and B. Griffey, 1983, Influence of inherited basement structure on thrust sheet geometry, Sevier orogenic belt, Utah (abs.): Geological Society of America Abstracts with Programs, v. 15, p. 377.

Crawford, K. A., 1979, Sedimentology and tectonic significance of the Late Cretaceous–Paleocene Echo Canyon and Evanston synorogenic conglomerates of the north-central Utah thrust belt: Master's thesis, University of Wisconsin, Madison, 143 p.

Crittenden, M. D., Jr., 1961, Magnitude of thrust faulting in northern Utah: U. S. Geological Survey Professional Paper 424-D, p. D128–D131.

———, 1972a, Geologic map of the Brown's Hole quadrangle, Utah: U.S. Geological Survey, Geologic Quadrangle Map GQ-968, scale 1:24,000.

———, 1972b, Willard thrust and Cache allochthon, Utah: Geological Society of America Bulletin, v. 83, p. 2871–2880.

———, 1976, Stratigraphic and structural setting of the Cottonwood area, Utah, *in* J. G. Hill ed., Symposium on geology of the Cordilleran Hingeline: Rocky Mountain Association of Geologists, p. 363–379.

Dane, C. H., 1954, Stratigraphic and facies relationships of upper part of Green River Formation and lower part of Uinta Formation in Duchesne, Uinta, and Wasatch counties, Utah: AAPG Bulletin, v. 38, p. 405–425.

Dawson, M. R., 1966, Additional late Eocene rodents (Mammalia) from the Uinta Basin: Carnegie Museum Annals, v. 38, p. 97–114.

Dickinson, W. R., L. S. Beard, G. R. Brakenridge, et al., 1983, Provenance of North American Phanerozoic sandstones in relation to tectonic setting: Geological Society of America Bulletin, v. 94, p. 222–235.

Dixon, J. S., 1982, Regional structural synthesis, Wyoming salient of western overthrust belt: AAPG Bulletin, v. 66, p. 1560–1580.

Eardley, A. J., 1944, Geology of the north-central Wasatch Mountains, Utah: Geological Society of America Bulletin, v. 55, p. 819–894.

———, 1951, Tectonic divisions of North America: AAPG Bulletin, v. 35, p. 2229–2237.

Folk, R. L., 1980, Petrology of sedimentary rocks: Austin, Texas, Hemphills, 184 p.

Fouch, T. D., 1981, Distribution of rock types, lithologic groups, and interpreted depositional environments for some lower Tertiary and Upper Cretaceous rock from outcrops at Willow Creek–Indian Canyon through the subsurface of Duchesne and Altamont oil fields, southwest to north-central parts of the Uinta Basin, Utah: U. S. Geological Survey, Oil and Gas Investigation Map, Chart OC-81.

Fouch, T. D., T. F. Lawton, D. J. Nichols, W. B. Cashion, and W. A. Cobban, 1983, Patterns and timing of synorogenic sedimentation in Upper Cretaceous rocks of central and northeast Utah, *in* M. W. Reynolds, and E. D. Dolly, eds., Mesozoic paleogeography of west-central United States: Society of Economic Paleontologists and Mineralogists, Rocky Mountain Section Symposium 2, p. 305–336.

Garvin, R. F., 1969, Stratigraphy and economic significance of the Currant Creek Formation, northwest Uinta basin, Utah: Utah Geological and Mineralogical Survey Special Studies, n. 27, 62 p.

Gries, R., 1983, Oil and gas prospecting beneath Precambrian of foreland thrust plates in Rocky Mountains: AAPG Bulletin, v. 67, p. 1–28.

Hamilton, W., 1981, Plate tectonic mechanism of Laramide deformation: University of Wyoming Contributions to Geology, v. 19, p. 87–92.

Hansen, W. R., 1965, Geology of the Flaming Gorge area, Utah–Colorado–Wyoming: U.S. Geological Survey Professional Paper 490, 196 p.

———, 1984, Post-Laramide tectonic history of the eastern Uinta Mountains, Utah, Colorado, Wyoming: Mountain Geologist, v. 21, p. 5–29.

Herr, R. G., M. D. Picard, and S. H. Evans, Jr., 1982, Age and depth of burial, Cambrian Lodore formation, northeastern Utah and northwestern Colorado: University of Wyoming Contributions to Geology, v. 21, p. 115–121.

Hopkins, D. L., and R. L. Bruhn, 1983, Extensional faulting in the Wasatch Mountains, Utah (abs.): Geological Society of America Abstracts with Programs, v. 15, p. 402.

Isby, J. S., and M. D. Picard, 1983, Currant Creek Formation: record of tectonism in Sevier–Laramide orogenic belt, north-central Utah: University of Wyoming Contributions to Geology, v. 22, p. 91–108.

———, 1985, Depositional setting of Upper Cretaceous–Lower Tertiary Currant Creek Formation, north-central Utah, *in* M. D. Picard, ed., Geology and energy resources, Uinta basin of Utah: Utah Geological Association, p. 39–49.

James, L. P., 1979, Geology, ore deposits, and history of the Big Cottonwood mining district, Salt Lake County, Utah: Utah Geological and Mineralogical Survey Bulletin 114, 98 p.

Jefferson, W. S., 1982, Structural and stratigraphic relations of Upper Cretaceous to Lower Tertiary orogenic sediments of the Cedar Hills, Utah, *in* D. L. Nielson, ed., Overthrust belt of Utah: Utah Geological Association Publication, n. 10, p. 65–80.

Lamerson, P. R., 1982, The Fossil basin and its relationship to the Absaroka thrust system, Wyoming and Utah, *in* R. B. Powers, ed., Geologic Studies of the Cordilleran thrust belt: Rocky Mountain Association of Geologists, Denver, Colorado, p. 279–340.

Maxwell, T. A., and M. D. Picard, 1976, Small channel-fill sequences in the Duchesne River Formation near Vernal, Utah: possible examples of transitions from meandering to braided stream deposits: Utah Geology, v. 3, p. 61–66.

Mullens, T. E., 1969, Geologic map of the Causey Dam quadrangle, Weber County, Utah: U.S. Geological Survey, Geologic Quadrangle Map GQ-790, scale 1:24,000.

Osborn, H. F., 1895, Fossil mammals of the Uinta Basin: American Museum of Natural History Bulletin, v. 7, p. 72–105.

Picard, M. D., 1955, Subsurface stratigraphy and lithology of

Green River Formation in Uinta basin, Utah: AAPG Bulletin, v. 39, p. 75–102.

Picard, M. D., and D. W. Andersen, 1975, Paleocurrent analysis and orientation of sandstone bodies in the Duchesne River Formation (Eocene–Oligocene?), northern Uinta Basin, northeastern Utah: Utah Geology, v. 2, p. 1–15.

Picard M. D., and D. J. Dideriksen, 1978, Volcanic sandstone in the Green River and Uinta formations, Uinta basin, Utah (abs.): AAPG Bulletin, v. 62, p. 890–891.

Picard, M. D., R. L. Bruhn, and S. L. Beck, 1983a, Mesozoic and early Tertiary paleostructure and sedimentology of central Wasatch Mountains, Uinta Mountains, and Uinta Basin (abs): AAPG Bulletin, v. 67, p. 1351–1352.

Peterson, O. A., 1932, New species from the Oligocene of the Uinta: Carnegie Museum Annals, v. 21, p. 61–78.

Riess, S. K., 1985, Structural geometry of the Charleston thrust fault, central Wasatch Mountains, Utah: Master's thesis, University of Utah, Salt Lake City, 73 p.

Ritzma, H. R., 1969, Tectonic resume, Uinta Mountains, *in* J. B. Lindsay, ed., Geologic guidebook of the Uinta Mountains: Intermountain Association of Geologists, 16th Annual Field Conference, p. 57–63.

Roberts, R. J., M. D. Crittenden, E. W. Tooker, et al., 1965, Pennsylvanian and Permian basins in northwestern Utah, northeastern Nevada and south-central Idaho: AAPG Bulletin, v. 49, p. 1926–1956.

Royse, F., Jr., 1983, Extensional faults and folds in the foreland thrust belt, Utah, Wyoming, Idaho (abs.): Geological Society of America Abstracts with Programs, v. 15, p. 295.

Royse, F., Jr., and P. Lamerson, 1984, Field trip guide, northern Utah thrust belt and north Cottonwood uplift: AAPG Structural Geology School, Park City, Utah.

Royse, F., Jr., M. A. Warner, and D. L. Reese, 1975, Thrust belt structural geometry and related stratigraphic problems, Wyoming–Idaho–northern Utah, *in* Bolyard, D. W., ed., Symposium on deep drilling frontiers in the central Rocky Mountains: Rocky Mountain Association of Geologists, Denver, Colorado, p. 41–54.

Sales, J. K., 1968, Crustal mechanics of Cordilleran foreland deformation—a regional and scale-model approach: AAPG Bulletin, v. 52, p. 2016–2044.

Sears, J. W., P. J. Graff, and G. S. Holden, 1982, Tectonic evolution of lower Proterozoic rocks, Uinta Mountains, Utah and Colorado: Geological Society of America Bulletin, v. 93, p. 990–997.

Simpson, G. G., 1933, Glossary and correlation charts of North American Tertiary mammal–bearing formations: American Museum of Natural History Bulletin, v. 67, p. 79–121.

Smith, R. B., and R. L. Bruhn, 1984, Intraplate extensional tectonics of the eastern Basin Range: inferences on structural style from seismic reflection data, regional tectonics, and thermal–mechanical models of brittle–ductile deformation: Journal of Geophysical Research, v. 89, p. 5733–5762.

Stagner, W. L., 1941, The paleogeography of the eastern part of the Uinta basin during Uinta B (Eocene) time: Carnegie Museum Annals, v. 28, p. 273–308.

Wallace, C. A., and M. D. Crittenden, Jr., 1969, The stratigraphy, depositional environment, and correlation of the Precambrian Uinta Mountain Group, western Uinta Mountains, Utah, *in* J. B. Lindsay, ed., Geologic Guidebook of the Uinta Mountains: Intermountain Association of Geologists, 16th Annual Field Conference, p. 127–141.

Walton, P. T., 1957, Cretaceous stratigraphy of the Uinta Basin, *in* O. G. Seal, ed., Guidebook to the geology of the Uinta basin: Intermountain Association of Petroleum Geologists, 8th Annual Field Conference Guidebook, p. 97–101.

Wood, H. E. (chairman), et al., 1941, Nomenclature and correlation of the North American continental Tertiary: Geological Society of America Bulletin, v. 52, p. 1–48.

Plate Tectonics of the Ancestral Rocky Mountains

C. F. Kluth
Chevron U.S.A. Inc.
Denver, Colorado

The Ancestral Rocky Mountains were intracratonic block uplifts that formed in Colorado and the surrounding region during Pennsylvanian time. Their development was related to the collision suturing of North America with South America–Africa, which also resulted in the Ouachita–Marathon orogeny. In Early Pennsylvanian time, suturing was taking place only in the Ouachita region, and foreland deformation took place largely in the midcontinent. By Middle Pennsylvanian time, the length of the active suture zone had increased, extending from the Ouachita region to the Marathon region. At this same time, deformation of the craton also increased in intensity and in areal extent, culminating in the Ancestral Rocky Mountains. In Late Pennsylvanian time, suturing was taking place only in the Marathon region, and cratonic deformation decreased areally and spread southward into New Mexico and western Texas and westward into the Cordillera miogeocline. The Ancestral Rocky Mountains and related features in a broad area of the western United States were formed while an irregularly bounded peninsula of the craton (including the transcontinental arch) was pushed northward and northwestward by the progressive collision suturing of North America and South America–Africa. This intraplate deformation was, in some respects, like the Cenozoic deformation of Asia in response to the collision with India.

INTRODUCTION

The Ancestral Rocky Mountain uplifts and basins formed in the area that is now Colorado and neighboring states during Pennsylvanian time (Lee, 1918; Melton, 1925; Ver Wiebe, 1930; Heaton, 1933; Eardley, 1951; Curtis, 1958; Mallory, 1972a, b, 1975) (Figures 1 and 2). These features are enigmatic in terms of plate tectonics because they occurred in an intraplate located as far as 1500 km from any active plate margin. The late Paleozoic deformation of the craton disrupted Cordilleran sedimentation patterns that had existed for approximtely 200 m.y. and that were reestablished after the uplifts were eroded. The uplifts were not intensely folded mountain chains, but apparently were broad block uplifts of considerable relief bounded by narrow, complex fault zones. Furthermore, this deformational event was curiously amagmatic (Curtis, 1958).

The existence of the Ancestral Rocky Mountains is largely inferred from unconformities, from isopachous patterns and sedimentation rates (Figure 3), and from coarse arkosic sediments that were apparently shed from the uplifts and deposited in adjacent basins (Mallory, 1958, 1972a, b, 1975, Howard, 1966; Mack et al., 1979). In some areas, there is additional evidence of the deformation along fault zones that were active during Pennsylvanian time and that are exposed today (Mallory, 1958; DeVoto et al., 1971; Szabo and Wengard, 1975; Stone, 1977).

A plate tectonic model for the Ancestral Rocky Mountains previously proposed (Kluth and Coney, 1980, 1981) suggested a close relationship between the Ancestral Rocky Mountains and other uplifts of nearly the same age in the midcontinent and in the western United States. The model suggested a close relationship between these intracratonic features and the deformation of the Ouachita and Marathon regions along the southern margin of the North American craton. Our purpose in this paper is to expand the statement of that model beyond the previous limited format and to update it with information published since its introduction.

In this paper, we briefly summarize stratigraphy and structural evidence related to late Paleozoic deformation in several regions. The relationship of one region to another is illustrated on a series of paleotectonic maps (see Figures 4–8). We compiled these maps using the thickness and facies relationships for specific time intervals to interpret magnitude, rate, and timing of local tectonic events (see Figure 3). Uncertainties in the geologic information make the insights gained by these maps qualitative. This type of analysis, however, does show general regional relationships. These maps support the interpretation of the Ancestral Rocky Mountains as part of a much larger area of deformation, which we must study in both space and time to understand its tectonic development. Finally, we propose a plate tectonics model that attempts to explain this enigmatic late Paleozoic orogenic event.

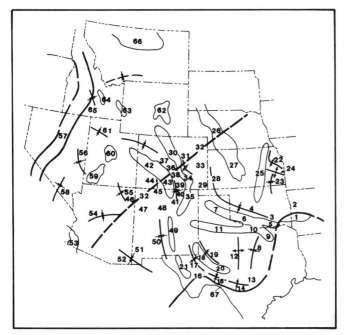

Figure 1—Location map of features in the text. 1. Ouachita Mountains; 2. Arkoma (McAlester) basin; 3. Arbuckle Mountains; 4. Anadarko basin; 5. Criner Hills uplift; 6. Wichita Mountains; 7. Amarillo Mountains; 8. Fort Worth basin; 9. Muenster arch; 10. Red River uplift; 11. Matador arch; 12. Bend arch; 13. Llano uplift; 14. Kerr basin; 15. Val Verde basin; 16. Marathon uplift; 17. Delaware basin; 18. Central Basin platform; 19. Midland basin (Delaware and Midland basins and the Central Basin platform outline the earlier Tobosa basin); 20. Fort Stockton–Ozona "high"; 21. Diablo platform; 22. Forest City basin; 23. Cherokee basin; 24. Bourbon arch; 25. Nemaha ridge, 26. Cambridge arch; 27. Central Kansas uplift; 28. Hugoton embayment; 29. Keyes dome; 30. Frontrange uplift; 31. Denver basin; 32. Transcontinental arch; 33. Las Animas arch; 34. Apishapa uplift; 35. Sierra Grande uplift; 36. Central Colorado trough; 37. Sawatch uplift; 38. Hartsel uplift; 39. Sangre de Cristo uplift; 40. Cimarron arch; 41. Rowe-Mora (Taos Trough) basin; 42. Uncompahgre uplift; 43. San Luis highland; 44. Paradox basin; 45. Sneffels horst; 46. Defiance uplift; 47. Zuni uplift; 48. Joyita uplift; 49. Pedernal uplift; 50. Orogrande basin; 51. Florida uplift; 52. Pedregosa basin; 53. Ensenada land; 54. Arizona sag; 55. Black Mesa sag; 56. Ely basin; 57. Antler orogenic belt; 58. Bird Springs basin; 59. Piute uplift; 60. Emery uplift; 61. Oquirrh basin; 62. Pathfinder uplift; 63. Bannock highland; 64. Copper basin uplift; 65. Wood River basin; 66. Milk River uplift; 67. Devils River uplift.

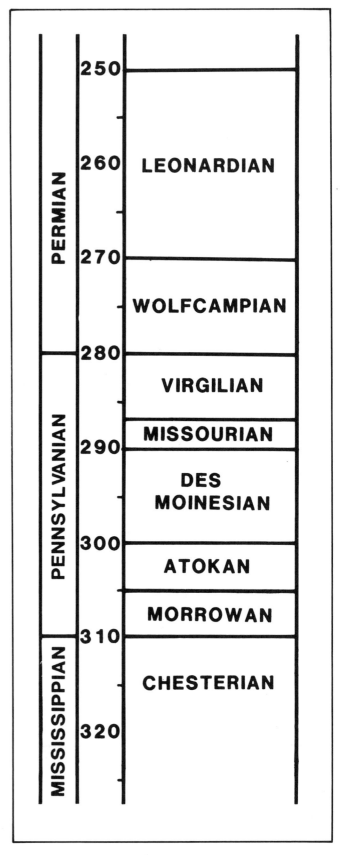

Figure 2—Pennsylvanian time scale as used in this paper.

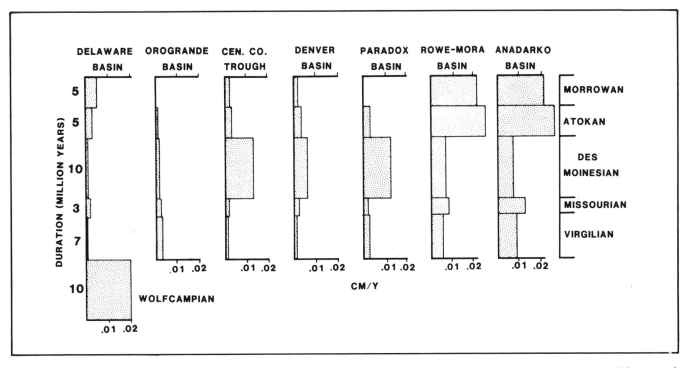

Figure 3—Sedimentation rate histogram of various basins in the Ancestral Rocky Mountains. Although these graphs oversimplify many of the complexities of basin geometry and changes in sedimentation rates, they do show the general pattern and timing of tectonic events discussed in the text.

SUMMARY OF REGIONAL TECTONIC PATTERNS

Late Mississippian Time

The tectonic pattern of uplifts and basins that dominated Pennsylvanian time in the west-central United States had already begun to emerge sometime before earliest Pennsylvanian time. (Flawn et al., 1961; Roberts et al., 1965; Baars, 1966; Ham and Wilson, 1967; Galley, 1958; Gerhard, 1972; Wilson, 1975b; Mallory, 1975; Bachman, 1975; Gorham, 1975; Spoelhof, 1976; DeVoto, 1980b).

In many areas, the late Paleozoic tectonic fabric was determined by much older structures (Galley, 1958; Flawn et al., 1961; Baars, 1966; Baars and See, 1968; Hoffman et al., 1974; Stevenson and Baars, 1977; Baars and Stevenson, 1981; Budnick, 1983). In general, the pre-Pennsylvanian precursory patterns in the Ancestral Rocky Mountains were subdued and vaguely outlined. They only hinted at the tectonic developments that followed.

In contrast to the subdued tectonic activity in the region of the Ancestral Rocky Mountains, the Ouachita region was the site of subsidence and deposition of approximately 4 km of sediments (Stanley Group) during Mississippian time (Osagean–Chesterian) (Goldstein and Hendricks, 1962; Morris, 1974; Graham et al., 1975; Glick, 1975); it also received volcanics from a source to the south (Hass, 1950; Mose, 1969; Graham et al., 1975; Niem, 1977). North of the Ouachita region movement apparently occurred along the eastern flank of the Nemaha uplift during Late Mississippian time (Lee, 1943; Merriam, 1963; Prichard, 1975).

Morrowan Time

Deposition of thick sequences of sediments, some of which contained large exotic blocks (Shideler, 1970), continued in the Ouachita region (Flawn et al., 1961; Ham and Wilson, 1967; Morris, 1974; Graham et al., 1975) during Morrowan time (Figure 4). These thick sedimentary sections deposited in this region during Late Mississippian–Early Pennsylvanian time reflect the development of a foredeep basin and the northward advance of the Ouachita thrust system (Viele, 1973). North of the Ouachita region the Nemaha uplift was a low positive area (Merriam, 1963; Frezon and Dixon, 1975). Major deformation of the Anadarko basin region, west of the Ouachita trough, began in Morrowan time (van der Gracht, 1931; Tomlinson and McBee, 1959; Ham and Wilson, 1967; Frezon and Dixon, 1975; W. G. Brown, 1984, personal communication). By late Morrowan time, the eastern (Criner Hills) part of the Amarillo Mountains–Wichita Mountains–Criner Hills trend had approximately 2 km of structural relief (Tomlinson and McBee, 1959; Frezon and Dixon, 1975).

Evidence suggests that central Texas was a shallow marine shelf, deforming gently through most of Pennsylvanian time. Movement patterns are inferred largely from thickness variations over much of the region (Tomlinson and McBee, 1959; Flawn et al., 1961; Ham and Wilson, 1967; Frezon and Dixon, 1975; Crosby and Mapel, 1975; Erxleben, 1975). The shelf deepened to the south and received sediments from a southerly source (Van Watterschoot Van der Gracht, 1931; King, 1937; Crosby and Mapel, 1975; Ross, 1978a). To the north, the tectonic

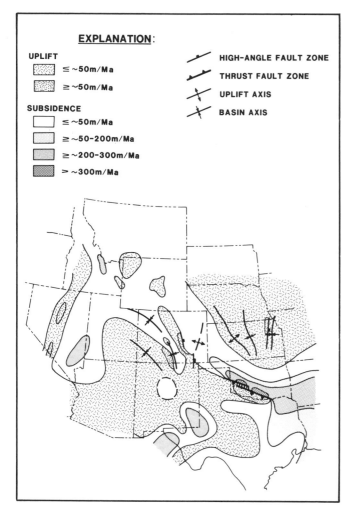

Figure 4—Late Mississippian and Morrowan tectonic features of the Ancestral Rocky Mountains. There was rapid subsidence in the Ouachita trough and Anadarko basin and an uplift of the Amarillo–Wichita–Criner Hills trend. Other features were more subdued. Modified after McKee and Crosby (1975).

The first major influx of arkosic conglomerates into the Anadarko basin apparently reflects accelerated subsidence of the Anardarko basin and uplift of the Amarillo–Wichita–Criner Hills trend during Atokan time (Rascoe, 1962; Frezon and Dixon, 1975; Stewart, 1975). In central Texas, the east flank of the Bend arch subsided into the Fort Worth basin through Atokan time (Cheney, 1940; Crosby and Mapel, 1975). The Llano uplift was formed during Atokan and early Des Moinesian time (Plummer, 1950; Ham and Wilson, 1967), while the Kerr basin just to the southwest received a thick sequence of clastic sediments (Flawn et al., 1961; Crosby and Mapel, 1975). During this same time, the Central Basin platform appears to have uplifted, initiating the segmentation of the Tobosa basin into the Midland and Delaware basins (Galley, 1958; Harrington, 1963).

During Atokan time, the tectonic features in the Ancestral Rocky Mountain region became well defined. Movement on faults that bounded the uplifts accelerated, and coarse, arkosic sediment was shed from the upthrown blocks into the adjacent basins (Sutherland, 1963; DeVoto, 1971, 1972, 1980b; DeVoto et al., 1971; Wilson, 1975b; Mallory, 1972b; Casey, 1980a, b). In the regions surrounding the Ancestral Rocky Mountains, Atokan age deformation was generally in the form of broad, gentle folding or minor faulting (Wengard and Matheny, 1958; Adams, 1962; Mallory, 1975; Bachman, 1975; Kottlowski, 1975; Maughan, 1975; Blakey, 1980). Farther to the south and west, tectonic patterns appear to have been largely inherited from those established during the Devonian–Mississippian Antler orogeny (Bissell, 1960, 1974; Brill, 1963; Roberts et al., 1965; Skipp and Hall, 1980), or even earlier (Ross, 1973; Greenwood et al., 1977).

Des Moinesian Time

Tectonic activity appears to have waned in the Ouachita region during Des Moinesian time. (Goldstein and Hendricks, 1962; Woods and Addington, 1973; Frezon and Dixon, 1975). The dominant style of tectonism in the region was broad upwarping (Ham and Wilson, 1967; Morris, 1974; Frezon and Dixon, 1975; Glick, 1975) (Figure 6).

Evidence shows that in central Texas, the Muenster, Red River, and Matador arches were leveled and buried by the end of Des Moinesian time and did not affect subsequent sedimentation (Crosby and Mapel, 1975). Continued eastward downwarping of the east flank of the Bend "arch" resulted in a series of horsts and grabens along the Fort Chadbourne fault system on the "crest" of the monoclinal flexure during Des Moinesian time (Cheney, 1940; Crosby and Mapel, 1975).

The greatest development of the Marathon basin apparently occurred during Des Moinesian time (King, 1931, 1958; Ham and Wilson, 1967; Crosby and Mapel, 1975; Thomson and McBride, 1978). Subsidence of the Val Verde basin and the differential subsidence of the southern Tobosa basin resulted in continued development of the southern part of the Central Basin platform (Galley, 1958; Crosby and Mapel, 1975). To the north in New Mexico, slow subsidence of the Orogrande basin and uplift of the Pedernals and Florida uplifts began in Des Moinesian time as structural patterns became aligned north–south (Kottlowski, 1969; Kottlowski and Stewart, 1970; Greenwood et al., 1977; Seager, 1981).

features of the Ancestral Rocky Mountains region were becoming more clearly outlined in Morrowan time than they had been earlier (Maher, 1953; Adams, 1962; Ellis, 1966; DeVoto et al., 1971; DeVoto, 1972, 1980b; Mallory, 1975; Kottlowski, 1975; Wilson, 1975a, b; Maughan, 1975; Bachman, 1975; Chronic and Williams, 1978; Casey, 1980a, b).

Atokan Time

Rapid subsidence and deposition apparently continued in the Ouachita region at least through early Atokan time, but slowed by late Atokan (Morris, 1974) (Figure 5). The depocenter in the region moved northward during Early Pennsylvanian time (Koinm and Dickey, 1967), suggesting progressive northward development of the foredeep as the thrust system advanced from the south. Maximum intensity of deformation along the eastern margin of the Nemaha uplift and elsewhere in Kansas and Nebraska was during late Atokan or early Des Moinesian time (Stewart, 1975; Prichard, 1975).

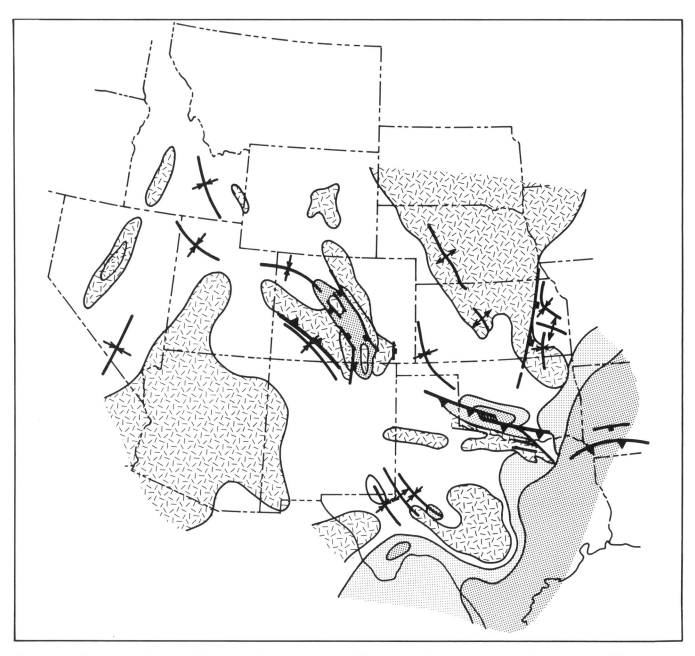

Figure 5—Atokan tectonic features of the Ancestral Rocky Mountains. Thrusting had begun to disrupt the Ouachita region, deformation intensified in the area of the Anadarko basin and Wichita Mountains, and the features of the Ancestral Rocky Mountains became more clearly defined. (See Figure 4 for explanation of symbols.) Modified after McKee and Crosby, (1975).

The Des Moinesian was the time of maximum development of the Ancestral Rocky Mountains. The western side of the Front range uplift (Mallory, 1971, 1972b, 1975; Wilson, 1975b; DeVoto, 1980b;), the northern side of the Apishapa uplift (Wilson, 1975b; DeVoto, 1980b), and the Uncompahgre uplift (Mallory 1972a, 1975; Stone, 1977; DeVoto, 1980b; Lindsey and Schaefer, 1984) probably achieved their maximum uplift at this time. The Central Colorado trough subsided and formed a deep, narrow, segmented basin between the Frontrange and Uncompahgre uplifts (Miller et al., 1963; Sutherland, 1963, 1972; Baltz, 1965; DeVoto, 1972, 1980b; DeVoto et al., 1971; DeVoto and

Peel, 1972; Walker, 1972; Mallory, 1975; Bachman, 1975; Szabo and Wengard, 1975; Casey, 1980a, b). By late Des Moinesian time, movement on some parts of the Uncompahgre uplift had ended while other parts of the uplift were still very active (Spoelhof, 1976; Frahme and Vaughn, 1983).

Tectonic features elsewhere in the region were active, but deformation was less intense than in the Colorado part of the Ancestral Rocky Mountains (Wengard and Malheny, 1958; Mallory, 1967, 1975; Maughan, 1975, 1979; Greenwood et al., 1977; Skipp and Hall, 1980).

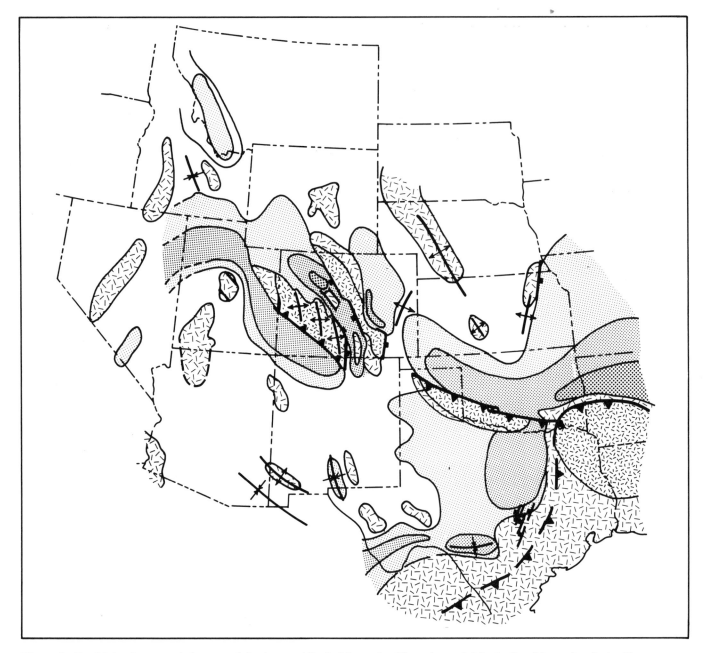

Figure 6—Des Moinesian tectonic features of the Ancestral Rocky Mountains. Thrusting ended in the Ouachita region during Des Moinesian time, reflecting decreased activity of the suture zone to the south of that area. Rapid subsidence of the Marathon region reflected the beginning of activity of the suture zone to the south of that area. The Ancestral Rocky Mountains achieved their most rapid development during the Des Moinesian. (See Figure 4 for explanation of symbols.) Modified after McKee and Crosby (1975).

Missourian Time

Uplift in the Ouachita region and on the Nemaha ridge is interpreted to have been relatively minor during Missourian time (Merriam, 1963; Ham and Wilson, 1967; Stewart, 1975) (Figure 7). The western (Amarillo–Wichita) part of the Amarillo–Wichita–Criner Hills trend was uplifted (Frezon and Dixon, 1975), while the Red River and Muenster arches were the sites of subsidence during Late Pennsylvanian time (Crosby and Mapel, 1975).

Vigorous uplift continued throughout Missourian time in the Ancestral Rocky Mountains but at a gradually diminishing rate (Wilson, 1975b; Mallory, 1975; Welsh,

1979; DeVoto, 1980b; Frahme and Vaughn, 1983). In the surrounding regions, features smaller than those in Colorado were active (Maughan, 1975; Maughan and Roberts, 1967).

Virgilian Time

Sporadic minor movement appears to have continued in the southern midcontinent through Virgilian time. The exception to this general condition of slow tectonic activity was the Arbuckle Mountains, which were uplifted and provided a source for arkosic sediments during latest Pennsylvanian time (Ham and Wilson, 1967; Frezon and Dixon, 1975) (Figure 8).

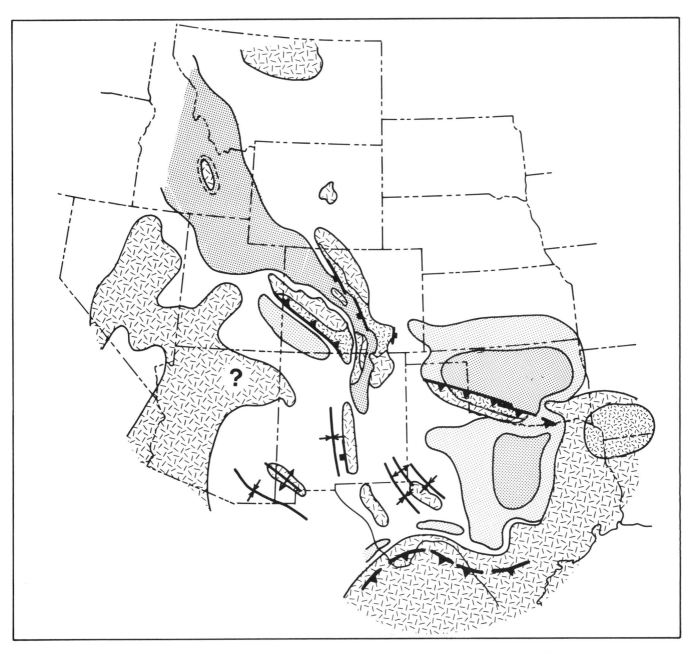

Figure 7—Missourian tectonic features discussed in the text. Tectonic activity in the Ouachita region and the midcontinent began to wane, reflecting an end of the continental collision to the south of that area. Deformation intensified in the Marathon region and in the southwestern United States, reflecting continued activity along the collision suture south of that area. The deformation in the Ancestral Rocky Mountains began to gradually decrease in intensity. (See Figure 4 for explanation of symbols.) Modified after McKee and Crosby (1975).

Southwest of the Ouachita region, however, tectonic activity was apparently intensifying. Subsidence to the west in the Midland basin formed the west flank of the composite Bend arch (Cheney, 1940; Crosby and Mapel, 1975). Over 3000 m of clastic sediments accumulated in the Val Verde basin during Virgilian and earliest Permian time (Flawn et al., 1961; Crosby and Mapel, 1975). At that same time, northward thrusting began in the Marathon region and the Central Basin platform continued to rise between the Delaware basin on the west and the Midland basin on the east (Galley, 1958; Flawn et al., 1961; Harrington, 1963; Hills, 1970; Crosby and Mapel, 1975; Ross, 1978a). The

Diablo platform in westernmost Texas has a tectonic history similar to that of the Central Basin platform. The rate of tectonic activity increased to a maximum during Virgilian time in this region, as well as to the north in New Mexico (Galley, 1958; Otte, 1959; King, 1965; Kottlowski, 1965a, b, 1969; Kottlowski and Stewart, 1970; Crosby and Mapel, 1975; Bachman, 1975; Greenwood et al., 1977; Ross, 1978a, b).

In most of the Ancestral Rocky Mountain region, tectonic activity was apparently diminishing in Virgillian time, although local areas of vigorous uplift continued, especially in the southern part of the region (Sutherland, 1963, 1972;

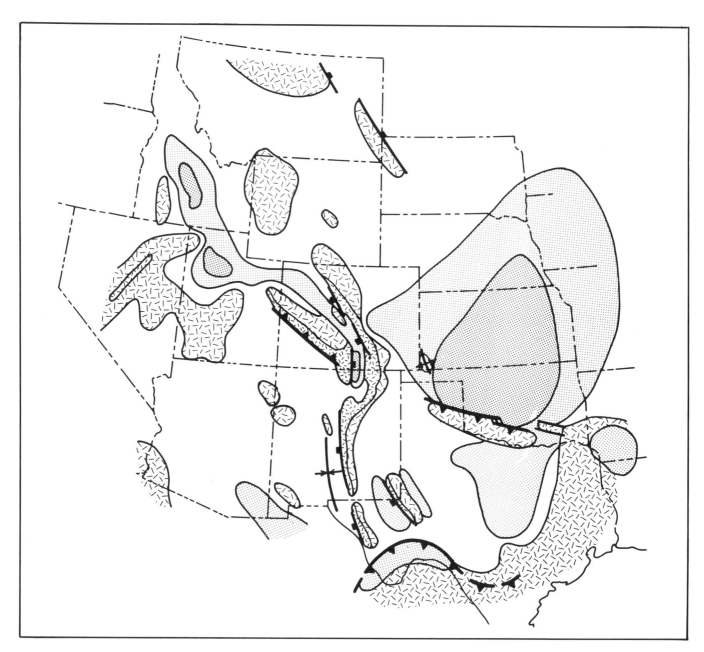

Figure 8—Virgilian and Early Permian tectonic features of the Ancestral Rocky Mountains. Thrusting continued in the Marathon region, reflecting continued activity of the suture zone to the south of that area, but activity was greatly reduced farther to the east. Foreland deformation gradually decreased and affected areas in southwestern United States. Thrusting and foreland deformation ended synchronously in the Early Permian. (See Figure 4 for explanation of symbols.) Modified after McKee and Crosby (1975).

Baltz, 1965; DeVoto, 1972, 1980b; DeVoto and Peel, 1972; Mallory, 1975; Wilson, 1975b; Bachman, 1975; Tweto, 1980; Casey, 1980a, b).

West of the Ancestral Rocky Mountains, tectonic activity increased in intensity in Virgilian time (Roberts et al., 1965; Ross, 1978a; Wilson, 1975a; Maughan, 1975; Ketner, 1977; Jordan and Douglas, 1980; Skipp and Hall, 1980). Stone et al. (1980) and Stevens and Stanton (1980) have reported that farther to the west a widespread Late Pennsylvanian– Early Permian deformational event occurred in eastern California that included tilted horst and graben blocks bounded by high-angle faults. It is believed that these events are related to similar contemporaneous events farther to the east.

Early Permian Time

Tectonic activity in the southern midcontinent continued to slow during Early Permian time as the Anadarko basin was filled with sediments and the trend of the Amarillo and Wichita Mountains and the Arbuckle Mountains were covered (Roscoe, 1962; Ham and Wilson, 1967). To the southwest, however, in the Marathon region and the foreland to the north, tectonic activity was still vigorous in the early part of Wolfcampian time. The activity decreased through the Wolfcampian and ended during the Leondardian (Galley, 1958; Crosby and Mapel, 1975; Ross, 1978a, b).

In the Ancestral Rocky Mountain region, there was less tectonic activity in Early Permian time than in mid-Pennsylvanian time. The uplifts and basins appear to have had sporadic movements through Early Permian time, but the basins were filling and sediment was overlapping the adjacent uplifts (DeVoto, 1972; Freeman and Bryant, 1977; Maughan, 1980; Peterson, 1980). Some areas of significant movement on the Uncompahgre uplift (Leudke and Burbank, 1962; Stone, 1977; Weimer, 1980; Maughan, 1980) apparently continued until the Leonardian (Rascoe and Baars, 1972). In regions adjacent to the Ancestral Rocky Mountains tectonic activity decreased and ended in Wolfcampian or Leonardian time (Ross, 1973, 1978a; Maughan, 1975; Wengard and Matheny, 1975; Jordan and Douglass, 1980).

INTERPRETATION

The pattern and timing of events summarized above suggest that the Ancestral Rocky Mountains were part of a complex, intraplate response of the foreland to events that produced the Ouachita–Marathon orogeny. The general spread from east to west of deformation in the foreland and the correlation in timing of deformation in the Ouachita–Marathon orogen with that of the Ancestral Rocky Mountains is too close to be considered coincidental. The Antler orogeny ended before the development of the Ancestral Rocky Mountains began, and the Ancestral Rocky Mountains were tectonically quiet before the Sonoma orogeny began along the western margin of North America (Gilluly, 1967; Burchfiel and Davis, 1972; Speed, 1979; Speed and Sleep, 1982). Thus, we prefer our interpretation presented here to any that might relate the Ancestral Rocky Mountains to events along the western plate margin, because the Pennsylvanian was a period of relative quiescence there.

Although not critical to the interpretation that the Ancestral Rocky Mountains and the Ouachita–Marathon orogeny are tectonically closely related, it is assumed that both phenomena were caused by a continent–continent collision along the southern margin of North America. This model was proposed by others and is supported by diverse data (Dewey and Bird, 1970; Van der Voo and French, 1974; Graham et al., 1975; Wickham et al., 1976; Ross, 1978a, 1979; Scotese et al., 1979; Bambach et al., 1980; Pindell and Dewey, 1982; Hallam, 1983; Van der Voo, 1983; cf. Goldstein, 1981, 1984).

There has been some discussion about the positions of the continents during Pennsylvanian time (cf. Irving, 1977; Morel and Irving, 1980; Van der Voo and French, 1974; Scotese et al., 1979; Hallam, 1983). The strength of the diverse evidence, however, appears to support a plate configuration in which South America was on the south side of the Pennsylvanian suture zone (Pangea A) in the Ouachita–Marathon region. We use that configuration here, but we recognize the possibility that microplates or suspect terranes (Coney et al., 1980), as yet unidentified, might also have been involved between Laurasia and Gondwana (cf. reconstruction of Pindell and Dewey, 1982).

The Early Pennsylvanian deformation of the craton is related to the early deformational phases of the Ouachita orogeny. The progressive, southwestward migration of the center of rapid sedimentation (Graham et al., 1975) (Black

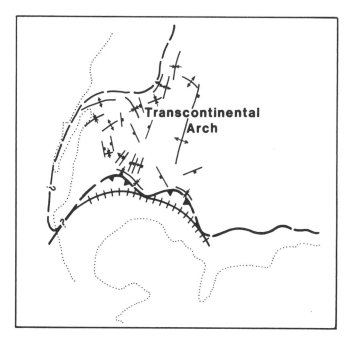

Figure 9—Schematic map of Middle Pennsylvanian to Early Permian tectonic features in North America. During this time, the southwest trending peninsula of North America, which included the Transcontinental arch, was deformed across its entire width from the Ouachita–Marathon region through the Cordilleran miogeocline and accreted terranes. This deformation was apparently related to the progressive southwestward collision between North America and South America–Africa.

Warrior, Arkoma, Fort Worth, Val Verde–Kerr, and Marathon basins) and the progressive southwestward migration of deformation of the orogen are interpreted as a response of the southwestward migration throughout Pennsylvanian time of the suture zone between North America and South America–Africa.

The suture zone between North America and South America–Africa apparently lengthened in a southwestward direction, as reflected in increased deformation of the Ouachitas and areas along the continental margin to the southwest of the Ouachitas. As this occurred, the deformation of the craton increased in intensity and areal extent and gradually moved westward. By Middle Pennsylvanian time, the collision zone was active from the Ouachita region to the Marathon region. The large-scale thrusting seen in the Ouachitas suggests a culmination of continental suturing to the south. Subsidence in the Marathon region, by analogy with earlier phases of development in the Ouachita region, reflects an earlier stage of suturing collision. At the same time, deformation of the cratonic region had spread northwestward, thus forming the Ancestral Rocky Mountains. The Ancestral Rocky Mountains developed rapidly and shed coarse arkosic sandstones into adjacent basins. During Middle and Late Pennsylvanian time, the craton of western North America was being deformed from its southeastern margin to the Cordilleran margin (Figure 9).

Later in Pennsylvanian and Early Permian time, activity generally waned in the Ouachita region, which reflected a slowing of the collision suturing there. At the same time the

Marathon region was experiencing intense folding and thrusting, suggesting an increase of activity along the suture to the south of the Marathon region. As the length of the actively closing suture zone decreased and moved southwestward along the margin, the intensity and areal extent of deformation of the craton decreased and spread southward and westward.

A relationship between the Ouachita orogeny and deformation in the Anadarko basin has been suggested by Ham and Wilson (1967), between the Wichita orogeny and the Ancestral Rocky Mountains by Tomlinson and McBee (1959), and between the Marathon region and New Mexico and western Texas by Ross (1978a, 1979). A general relationship has been suggested between the Ancestral Rocky Mountains and the Ouachita–Marathon orogeny by Coney (1978), Burchfiel (1978), Kluth and Coney (1980, 1981), and Casey (1980b). The interpretation presented above indicates that deformation over an area of approximately 30×10^6 km^2 of the craton, from Texas and Oklahoma to Montana, Idaho, Nevada, and perhaps eastern California, can be related to this progressive suturing of the two continents as part of the formation of Pangea.

In general, the foreland uplifts appear to be block uplifts of great structural relief having minor en echelon folding within them (Curtis, 1958; Tomlinson and McBee, 1959; Kottlowski, 1969; Mallory, 1975). In some areas, the fault zones that bound the blocks have characteristics of both normal and reverse faults (Merriam, 1963). In other areas the faults are unequivocally normal faults (DeVoto et al., 1971), while in still other areas the faults are reverse or thrust faults (Frahme and Vaughan, 1983). Geometry and kinematics of other less well documented faults have been interpreted in a wide variety of ways. Blocks within the fault zone have complex histories of reversal of relative movements (Merriam, 1963; Baars, 1966; Stevenson and Baars, 1977; Frezon and Dixon, 1975). In some areas, whole uplifts show a reversal of movement relative to surrounding areas (Tomlinson and McBee, 1959; Coogan, 1964; Crosby and Mapel, 1975; Frezon and Dixon, 1975; Wilson and Laule, 1979). Horizontal movement on specific zones has been interpreted to have occurred in many areas (Otte, 1959; Flawn et al., 1961; Harrington, 1963; Hills, 1970; Burgess, 1976; Walper, 1977). Because of the large structural relief on these blocks, it seems reasonable to interpret the largest component of movement along the faults bounding the uplifts as vertical. Because many of these zones also have characteristics common to strike-slip faults, it suggests that many of the zones may have had at least some component of strike-slip movement.

Model

The inferred component of transcurrent movement suggested on the faults bounding the foreland block uplifts probably reflects that the North American craton was undergoing internal translation (Coney, 1973, 1978; Burchfiel, 1978; Casey, 1980) resulting from the collision of North America with South America–Africa. The large fault block mountains formed when the southwestern peninsular projection of the North American craton (including the Transcontinental arch), between the Cordilleran miogeocline and the Ouachita–Marathon miogeocline, was pushed northward and northwestward by the continental collision

(Figure 9). The wrenching and translation appear to have been caused by distributive shear of a large area of the craton, such as that suggested by Hills (1970) for the Delaware basin region of western Texas, because there is no evidence for large-scale throughgoing megashears with great displacement in the Pennsylvanian or pre-Pennsylvanian rocks.

An additional significant factor that influenced the timing and the pattern of deformation of the craton was the inferred irregular southern margin of North America during the collision (Thomas, 1977). Such promontories as the Llano region of central Texas and such embayments as the Anadarko and Arkoma basins and the Marathon region probably had an effect on the stress–strain patterns and the timing of deformation within the craton. The general effect that such irregularities in the continental margin would have on the strain pattern produced during collisions has been discussed by Tapponier and Molnar (1976) and Molnar and Tapponier (1975), and their effect on the structural and sedimentary patterns has been covered by Dewey and Burke (1974). More specific data on the configuration of the continental margins of North and South America–Africa are required before this effect can be fully evaluated.

No large-scale foreland block uplifts such as those associated with the Ancestral Rocky Mountains have been found that represent episodes of suturing farther northeastward along the Appalachian Mountain trend (Ham and Wilson, 1967; Thomas, 1983). There is evidence, however, of deformation of the craton in the eastern midcontinent in the form of arches and basins associated with suturing in the Appalachians (Thomas, 1983). The collision along the Appalachian trend involved a relatively straight continental margin (Thomas, 1977, 1983) and involved the main mass of the North American Precambrian craton. The absence of large-scale block uplifts in the eastern midcontinent thus suggests that conditions there must have been unfavorable for their development. A major difference between that eastern region and the region of the Ancestral Rocky Mountains is the location of the western area on a narrow peninsula of the North American craton that had preexisting weaknesses and a more irregular margin.

Evidence suggests that the crust was fractured and weakened in late Precambrian time during a rifting event that formed the southern margin (Stewart, 1976) and probably during other pre-Pennsylvanian tectonic events. When suturing started to affect the area south of the Ouachita–Marathon region, the peninsula of Precambrian crust of the North American craton was too weak and narrow to resist the effects of the collision. Preexisting zones of weakness were reactivated by the collision, thus forming uplifts and basins that were controlled by the patterns of stress which changed as the suture progressed southwestward along the irregular margin.

The high angle between the uplifts and basins of the Ancestral Rocky Mountains and the suture is apparently the result of a strong northward component of stress set up in the craton by the collision. Preexisting zones of weakness having the proper orientation to the stress fields that existed at the time were reactivated to produce the uplifts and basins. An example of this changing pattern is in the Delaware basin of western Texas (Hills, 1970). Early faulting in that region probably resulted from the Early

Pennsylvanian collision in the Ouachita region. The Late Pennsylvanian fault patterns probably relate to the collision taking place in the Marathon region to the south.

The generally northerly component of the stress field that was set up in the craton by the collision resulted in large vertical components of movement on the fault zones that bound the uplifts. In some zones, such as the one between the Paradox basin and the Uncompahgre uplift (Frahme and Vaughn, 1983; White and Jacobson, 1983) and between the Anadarko basin and the Wichita–Amarillo mountains trend (Brewer, 1982; Brown, 1981; Brewer et al., 1983; cf. Tanner, 1967; Harding and Lowe, 1979), the geometry is that of a thrust fault. In other zones, such as along the eastern margin of the Central Colorado trough (DeVoto, 1972, 1980b), the faults were higher angle. It is inferred that Pennsylvanian kinematics of some of the preexisting zones had at least some component of strike-slip movement. The inferred amount and sense of strike-slip displacement on these zones varied within the region. During the Pennsylvanian and Permian deformation, the entire western United States that was situated on the narrow peninsula of the craton may have acted as a large-scale shear zone or zones. Blocks within the large-scale zone may have had a variety of movement patterns relative to one another, just as faults in smaller scale shear zones can show a variety of slip orientations.

One preexisting zone of weakness in which there appears to have been movement is the Anadarko basin, an aulacogen formed during a Precambrian rifting of southern North America (Hoffman et al., 1974). This zone of inherited structural discontinuity was weak enough to be deformed throughout Pennsylvanian time. The Arbuckle uplift and related features developed in Late Pennsylvanian time, after the culmination of tectonic activity in the Ouachita region. In our model, we view them as having been block-uplifted and folded by late movements along the Anadarko zone in response to the combined effects of wrenching and compression caused by the suturing taking place to the southwest in the Marathon area. Another inherited and reactivated zone of weakness is the fault zone along the southwest margin of the Uncompahgre uplift. This zone has a long history of complex movements, including the Pennsylvanian deformation (Baars, 1966; Stevenson and Baars, 1977; Stone, 1977).

The deformation of the craton in Pennsylvanian time, including the Ancestral Rocky Mountains, as interpreted here, resulted from the collision of North America and South America–Africa. This deformation is analogous in some respects to the deformation of Asia in response to the Cenozoic collision with India (Molnar and Taponnier, 1975). Both areas represent a similar plate tectonic setting. Differences in plate geometries resulted in differences in stress and strain patterns and thus in fault geometry and kinematics of the two areas. Nevertheless, both regions represent the intraplate deformation of a large area of the continental crust in response to a continent–continent collision. In both regions the continental crust is deforming to "escape" the collision.

We have pointed out that extrapolation of rock mechanics data suggests that intraplate deformation should not occur in response to plate boundary interaction (Warner, 1983). Despite this interpretation, examples of such intraplate

deformation occur in non-collisional plate tectonic settings, such as the Mesozoic–early Cenozoic deformation of the western United States (Burchfiel and Davis, 1972, 1975) and the present-day deformation of parts of South America (Jorden et al., 1983). Intraplate deformation also occurs in collisional settings, such as the Cenozoic deformation of Asia (Molnar and Tapponier, 1975; Tapponier and Molnar, 1976), the Cenozoic deformation of Europe (Celal-Sengor, 1976; Ziegler, 1982), and the Pennsylvanian deformation of the western United States (Burchfiel, 1978; Kluth and Coney, 1980, 1981; Casey, 1980). More detailed work is required before the mechanical aspects of these types of intraplate deformations are fully understood.

SUMMARY

The timing, rates, and distribution of tectonic events summarized here suggest that the Ancestral Rocky Mountains were part of a complex intraplate response to the collision of North America and South America–Africa. This same collision apparently also produced the Ouachita–Marathon orogeny. It is suggested that the areal extent of the cratonic deformation is related to the length and activity along the suture zone between the continents. In Early Pennsylvanian time, the collision was taking place only south of the Ouachita region. At that time, foreland deformation was restricted largely to the Anadarko basin and the midcontinent. By Middle Pennsylvanian time, the collision was taking place from south of the Ouachita region along the irregular margin into the area south of the Marathon region. The foreland deformation, in response to the collision, reached its greatest extent and intensity at this time, culminating in the Ancestral Rocky Mountains. During Late Pennsylvanian and into Early Permian time, the collision process generally slowed south of the Ouachitas, but it continued south of the Marathon region. As the length of the actively converging margin decreased, the areal extent of the deformation of the craton decreased and spread southward from the Ancestral Rocky Mountains into New Mexico and western Texas. Suturing, as reflected by thrusting in the Marathon region, and deformation of the craton ended synchronously in early Permian time.

The Ancestral Rocky Mountains and related foreland uplifts do not have analogs elsewhere along the Appalachian–Ouachita–Marathon trend. This suggests that their tectonic setting was unique. Their setting is interpreted to have been on the southwest-trending peninsula (including the Transcontinental arch) of the North America craton that was the deformed foreland during the continental collision. These foreland uplifts were a product of the continental collision along an irregular continental margin that tended to push the craton to the north and northwest. The Pennsylvanian deformation of North America is, in some general ways, analogous to the Cenozoic intraplate deformation of Asia in response to the collision with India.

ACKNOWLEDGMENTS

Discussions with many friends and colleagues at the University of Arizona (especially P. J. Coney, W. R.

Dickinson, and Larry Mayer), at Chevron U.S.A. Inc. (especially W. G. Brown, J. Carney, and E. L. DuFeu) and elsewhere were very important in helping me over the past several years to crystallize the ideas presented here. This paper was improved by comments from D. L. Baars, W. L. Bilodeau, R. Kligfield, D. G. Morse, J. A. Peterson, L.A. Standlee, J. S. Wickham, and an anonymous reviewer.

REFERENCES CITED

Adams J. E., compiler, 1962, Foreland Pennsylvanian rocks of Texas and eastern New Mexico: *in* Pennsylvanian System in the United States—a symposium: Tulsa, Oklahoma, AAPG, p. 372–384.

Baars, D. L., 1966, Pre-Pennsylvanian paleotectonics—key to basin evaluation and petroleum occurrences in Paradox Basin: AAPG Bulletin, v. 50, p. 2082–2111.

Baars, D. L., and P. D. See, 1968, Pre-Pennsylvanian stratigraphy and paleotectonics of the San Juan Mountains, southwestern Colorado: Geological Society of America Bulletin, v. 79, p. 333–350.

Baars, D. L., and G. M. Stevenson, 1981, Tectonic evolution of the Paradox Basin, Utah and Colorado, *in* Wiegand, D. L., ed., Geology of the Paradox Basin: Rocky Mountain Association of Geologists Guidebook, p. 23–31.

Bachman, G. O., 1975, New Mexico, *in* McKee, E. D., and Crosby, E. J., coordinators, Paleotectonic investigations of the Pennsylvanian System in the United States: U.S. Geological Survey Professional Paper 853, p. 233–243.

Baltz, E. H., 1965, Stratigraphy and history of Raton Basin and notes on San Luis Basin, Colorado–New Mexico: AAPG Bulletin, v. 49, p. 2041–2075.

Bambach, R. K., C. R. Scotese, and C. R. Ziegler, 1980, Before Pangea: the geographics of the Paleozoic World: American Scientist, v. 68, p. 26–38.

Bissell, H. J., 1960, Eastern Great Basin Permo-Pennsylvanian strata—Preliminary statement: AAPG Bulletin v. 44, p. 1424–1435.

———, 1974, Tectonic control of late Paleozoic and early Mesozoic sedimentation near the hinge line of the Cordilleran miogeosynclinal belt, *in* Dickinson, W. R., ed., Tectonics and Sedimentation: Society of Economic Paleontologists and Mineralogists Special Publication 22, p. 83–97.

Blakey, R. C., 1980, Pennsylvanian and Early Permian paleogeography, southern Colorado Plateau and vicinity, *in* Fouch, T. D., and E. R. Magathan, eds., Paleozoic paleogeography of west central United States: Rocky Mountain Section, Society of Economic Paleontologists and Mineralogists Symposium 1, p. 239–257.

Brewer, J. A., 1982, Study of southern Oklahoma aulacogen, using COCORP deep seismic-reflection profiles, *in* Gilbert, M. C., and R. N. Donovan, eds. Geology of the eastern Wichita Mountains, southwestern Oklahoma, Guidebook for Field Trip 1: South-Central Section of the Geological Society of America Meeting, Norman, Oklahoma, p. 31–39.

Brewer, J. A., R. Good, J. E. Oliver, L. D. Brown, and S.

Kaufman, 1983, COCORP profiling across the southern Oklahoma aulacogen: overthrusting of the Wichita Mountains and compression within the Anadarko Basin: Geology, v. 11, p. 109–114.

Brill, K. G., Jr., 1963, Permo-Pennsylvanian stratigraphy of western Colorado Plateau and eastern Great Basin region: AAPG Bulletin, v. 74, p. 307–330.

Brown, W. G., 1981, Wachita Valley fault system—a new look at an old fault (abs.): AAPG Bulletin, v. 65, p. 1496.

Budnick, R. T., 1983, Recurrent intraplate deformation on the Ancestral Rocky Mountain orogenic belt (abs.): Geological Society of America Abstracts with Programs, v. 15, p. 535.

Burchfiel, B. C., 1978, Geologic history of the central western United States: Proceedings of the Fifth Quadrennial International Association on the Genesis of Ore Deposits Symposium, p. 1–11.

Burchfiel, B. C., and G. A. Davis, 1972, Structural framework and evolution of the southern part of the Cordilleran orogen, western United States: American Journal of Science, v. 272, p. 97–118.

———, 1975, Nature and controls of Cordilleran orogenesis, western United States: an extension of an earlier synthesis: American Journal of Science, v. 275-A, p. 363–396.

Burgess, W. J., 1976, Geologic evolution of the mid-continent and Gulf Coast areas—a plate tectonics view: Gulf Coast Association of Geological Societies Transactions, v. 26, p. 132–143.

Casey, J. M., 1980a, Depositional systems and paleogeographic evolution of the late Paleozoic Taos Trough, northern New Mexico, *in* Fouch, T. D., and E. R. Magathan, eds., Paleozoic paleography of west-central United States: Rocky Mountain section, Society of Economic Paleontologists and Mineralogists Symposium 1, p. 181–196.

———, 1980b, Depositional systems and basin evolution of the late Paleozoic Taos Trough, northern New Mexico: Texas Petroleum Research Committee, University Division, Report 80-1, 236 p.

Celal-Sengor, A. M., 1976, Collision of irregular continental margins: Implications for foreland deformation of Alpine-type orogens: Geology, v. 4, p. 779–782.

Cheney, M. G., 1940, Geology of north-central Texas: AAPG Bulletin, v. 24, p. 65–118.

Chronic, J., and C. A. Williams, 1978, The Glen Eyrie Formation (Carboniferous) near Colorado Springs, *in* Pruitt, J. D., and P. E. Coffin, eds. Energy Resources of the Denver Basin: Denver, Colorado, Rocky Mountain Association of Geologists, p. 199–206.

Coney, P. J., 1978, Mesozoic–Cenozoic Cordilleran plate tectonics, *in* Smith, R. B., and G. P. Eaton, Cenozoic tectonics and regional geophysics of the western Cordillera: Geological Society of America Memoir 152, p. 33–49.

Coney, P. J., D. L. Jones, and J. W. H. Monger, 1980, Cordilleran suspect terranes: Nature, v. 288, p. 329–333.

Coogan, A. H., 1964, Early Pennsylvanian history of Ely Basin, Nevada: AAPG Bulletin, v. 48, p. 487–495.

Crosby, E. J., and J. W. Mapel, 1975, Central and West Texas, *in* McKee, E. D., and E. J. Crosby, coordinator,

Paleotectonic investigations of the Pennsylvanian System in the United States: U. S. Geological Survey Professional Paper, 853, p. 196–232.

Curtis, B. F., 1958, Pennsylvanian paleotectonics of Colorado and adjacent areas, *in* Symposium on Pennsylvanian rocks of Colorado and adjacent areas: Denver, Colorado, Rocky Mountain Association of Geologists, p. 9–12.

DeVoto, R. H., 1971, Geologic history of South Park and geology of the Antero Reservoir Quadrangle, Colorado: Colorado School of Mines Quarterly, v. 66, 90 p.

_____, 1972, Pennsylvanian and Permian stratigraphy and tectonism in central Colorado: Colorado School of Mines Quarterly, v. 67, n. 4, p. 139–185.

_____, 1980a, Mississippian stratigraphy and history of Colorado, *in* Kent, H. C., and K. W. Porter, eds., Colorado Geology: Denver, Colorado, Rocky Mountain Association of Geologists, p. 57–70.

_____, 1980b, Pennsylvanian stratigraphy and history of Colorado, *in* Kent, H. C., and K. W. Porter, eds., Colorado Geology: Denver, Colorado, Rocky Mountain Association of Geologists, p. 71–101.

DeVoto, R. H., and F. A. Peel, 1972, Pennsylvanian and Permian stratigraphy and structural history, northern Sangre de Cristo Range, Colorado: Colorado School of Mines Quarterly, v. 67, n. 4, p. 283–320.

DeVoto, R. H., F. A. Piel, and W. H. Pierce, 1971, Pennsylvanian and Permian stratigraphy, tectonism and history, northern Sangre de Cristo Range, Colorado, *in* James, H. L., ed., San Luis Basin: New Mexico Geological Society Guidebook, p. 141–163.

Dewey, J. F., and J. M. Bird, Mountain belts and the new global tectonics: Journal of Geophysical Research, v. 75, p. 2625–2647.

Dewey, J. F., and K. C. A. Burke, 1974, Hotspots and continental breakup: Implications for collision orogeny: Geology, v. 2, p. 57–60.

Eardley, A. J., 1951, Structural Geology of North America: New York, Harper and Brothers, 624 p.

Ellis, C. H., 1966, Paleontologic age of the Fountain Formation south of Denver, Colorado: The Mountain Geologist, v. 3, p. 155–160.

Erxleben, A. W., 1975, Depositional systems in Canyon Group (Pennsylvanian System), north-central Texas: University of Texas, Bureau of Economic Geology, Report of Investigations, n. 82, 76 p.

Flawn, P. T., A. Goldstein, Jr., P. B. King, and C. E. Weaver, 1961, The Ouachita System: The University of Texas, Bureau of Economic Geology Publication 6120, 401 p.

Frahme, C. W., and E. B. Vaughn, 1983, Paleozoic geology and seismic stratigraphy of the northern Uncompahgre front, Grande County, Utah, *in* Lowell, J. D., ed., Rocky Mountain foreland basins and uplifts: Denver, Colorado, Rocky Mountain Association of Geologists, p. 201–211.

Freeman, V. L., and B. Bryant, 1977, Red bed formations in the Aspen region, Colorado, *in* Veal, H. K., ed., Exploration frontiers of the central and southern Rockies: Rocky Mountain Association of Geologists, p. 181–189.

Frezon, S. E., and G. H. Dixon, 1975, Texas panhandle and Oklahoma, *in* McKee, E. D., and E. J. Crosby, coordinators, Paleotectonic investigations of the

Pennsylvanian System in the United States: U.S. Geological Survey Professional Paper 853, p. 177–195.

Galley, J. E., 1958, Oil and geology in the Permian Basin of Texas and New Mexico, *in* Weeks, L. G., ed., Habitat of Oil—A symposium: Tulsa, Oklahoma, AAPG, p. 395-446.

Gerhard, L. C., 1972, Canadian depositional environments and paleotectonics, central Colorado: Colorado School of Mines Quarterly, v. 67, n. 4, p. 1–36.

Gilluly, J., 1967, Chronology of tectonic movements in the western United States: American Journal of Science, v. 265, p. 306–331.

Glick, E. E., 1975, Arkansas and northern Louisiana, *in* McKee, E. D., and E. J. Crosby, coordinators, Paleotectonic investigations of the Pennsylvanian System in the United States: U.S. Geological Survey Professional Paper 853, p. 157–175.

Goldstein, A. G., 1981, Comment on plate tectonics of the Ancestral Rocky Mountains: Geology, v. 9, p. 387–388.

_____, 1984, Tectonic controls of Late Paleozoic subsidence in the south central United States: Journal of Geology, v. 92, p. 217-222.

Goldstein, August, Jr., and T. A. Hendricks, 1962, Late Mississippian and Pennsylvanian sediments of Ouachita facies, Oklahoma, Texas, and Arkansas, *in* Pennsylvanian System in the United States—a symposium: Tulsa, Oklahoma, AAPG, p. 385-430.

Gorham, F. D., Jr., 1975, Tectogenesis of the central Colorado Plateau aulacogen, *in* Fassett, J. E., ed., Canyonlands Country: Four Corners Geological Society Guidebook, p. 211–216.

Graham, S. A., W. R. Dickinson, and R. V. Ingersoll, 1975, Himalayan–Benal model for flysch dispersal in the Appalachian–Ouachita System: Geological Society of America Bulletin, v. 86, p. 273–286.

Greenwood, E., F. E. Kottlowski, Jr., and S. Thompson III, 1977, Petroleum potential and stratigraphy of the Pedregosa Basin, comparison with Permian and Orogrande basins: AAPG Bulletin, v. 61, p. 1448–1469.

Hallam, A., 1983, Supposed Permo-Triassic megashear between Laurasia and Gondwana: Nature, v. 301, p. 499–502.

Ham, W. E., and J. L. Wilson, 1967, Paleozoic epeirogeny and orogeny in the central United States: American Journal of Science, v. 265, p. 332–407.

Harding, T. P., and J. D. Lowe, 1979, Structural styles, their plate tectonic habitat and hydrocarbon traps in petroleum provinces: AAPG Bulletin, v. 63, p. 1016–1058.

Harrington, J. W., 1963, Opinion of structural mechanics of Central Basin Platform area, West Texas: AAPG Bulletin, v. 47, p. 2023–2038.

Hass, W. H., 1950, Age of the lower part of the Stanley Shale: AAPG Bulletin, v. 34, p. 1578–1584.

Heaton, R. L., 1933, Ancestral Rockies and Mesozoic and late Paleozoic stratigraphy of Rocky Mountain region: AAPG Bulletin, v. 17, p. 109–168.

Hills, J. M., 1970, Late Paleozoic structural directions in southern Permian Basin, West Texas and southeastern New Mexico: AAPG Bulletin, v. 54, p. 1809–1827.

Hoffman, P. F., J. F. Dewey, and K. Burke, 1974, Aulacogens and their genetic relationship to geosynclines with a

Proterozoic example from the Great Slave Lake, Canada: *in* Dott, R. H., Jr., and R. H. Shaner, eds., Modern and ancient geosynclinal sedimentation: Society of Economic Paleontologists and Mineralogists Special Publication 19, p. 38–55.

Howard, J. D., 1966, Patterns of sediment dispersal in the Fountain Formation of Colorado: The Mountain Geologist, v. 3, p. 147–153.

Irving, E., 1977, Drift of the major continental blocks since the Devonian: Nature (London), v. 270, p. 304–309.

Jordan, T. E., and R. C. Douglass, 1980, Paleogeography and structural development of the Late Pennsylvanian to Early Permian Oquirrh Basin, northwestern Utah, *in* Fouch, T. D., and E. R. Megathan, eds., Paleozoic paleogeography of west-central United States: Rocky Mountain Section, Society of Economic Paleontologists and Mineralogist Symposium 1, p. 217–238.

Jorden, T. E., B. L. Isacks, R. W. Allmendinger, J. A. Brewer, V. A. Ramos, and C. J. Ando, 1983, Andean tectonics related to geometry of subducted Nazea plate: Geological Society of America Bulletin, v. 94, p. 341–361.

Ketner, K. B., 1977, Late Paleozoic orogeny and sedimentation, southern California, Nevada, Idaho, and Montana, *in* Stewart, J. H., C. H. Stevens, and A. E. Fritsche, eds., Paleozoic paleogeography of the western United States: Society of Economic Paleontologists and Mineralogists Pacific Section, Pacific Coast Symposium 1, p. 363–369.

King, P. B., 1931, Descriptive geology, Part 1 of the geology of the Glass Mountains, Texas: Texas University Bulletin 3038, 167 p.

———, 1937, Geology of the Marathon region, Texas: U.S. Geologic Survey Professional Paper, 187, 148 p.

———, 1958, Problems of boulder beds of Haymond Formation, Marathon Basin, Texas: American Association of Petroleum Geologist Bulletin, v. 42, p. 1731–1734.

———, 1965, Geology of the Sierra Diablo region, Texas: U.S. Geological Survey Professional Paper 480, 185 p.

Kluth, C. F., and P. J. Coney, 1980, Plate tectonics of the Ancestral Rocky Mountains (abs.): Geological Society of America Abstracts with Programs, v. 12, p. 277.

———, 1981, Plate tectonics of the Ancestral Rocky Mountains: Geology, v. 9, p. 10–15.

Koinm, D. N. and P. A. Dickey, 1967, Growth faulting in McAlester Basin of Oklahoma: AAPG Bulletin, v. 51, p. 710–718.

Kottlowski, F. E., Jr., 1965a, Southwestern New Mexico basins: AAPG Bulletin, v. 49, p. 2120–2139.

———, 1965b, Facets of the late Paleozoic strata in southwestern New Mexico, *in* Fitzsimmons, J. and C. Lochman-Balk, eds., Southwestern New Mexico: New Mexico Geological Society Guidebook, p. 141–147.

———, 1969, Summary of the late Paleozoic in the El Paso border region, *in* Kottlowski, F. E., Jr., and D. Lemone, eds., New Mexico Bureau of Mines Circular 104, p. 38–50.

———, 1975, Stratigraphy of the San Andres Mountains in central New Mexico: New Mexico Geological Society Guidebook, p. 95–104.

Kottlowski, F. E., Jr., and W. Stewart, 1970, The Wolfcampian Joyita Uplift in central New Mexico: New Mexico State Bureau of Mines and Mineral Technology, Memoir 23, pt. 1, 82 p.

Lee, W. T., 1918, Early Mesozoic physiography of the southern Rocky Mountains: Smithsonian Miscellaneous Collections, v. 69, n. 4, 41 p.

Lee, Wallace, 1943, The stratigraphy and structural development of the Forest City Basin in Kansas: Kansas Geological Survey Bulletin 51, 142 p.

Lindsey, D. A., and R. A. Schaefer, 1984, Principal reference section for the Sangre de Cristo Formation (Pennsylvanian and Permian), northern Sangre de Cristo Range, Saguache County, Colorado: U.S. Geological Survey Miscellaneous Field Studies Map MF-1622-A.

Luedke, R. G., and R. S. Burbank, 1962, Geology of the Ouray Quadrangle, Colorado: U.S. Geological Survey Geologic Quadrangle Map GQ-152.

Mack, G. H., L. J. Suttner, and J. R. Jennings, 1979, Permo-Pennsylvanian climatic trends in the Ancestral Rocky Mountains, *in* Baar, D. L., ed., Permianland: Four Corners Geological Society Guidebook, p. 7–12.

Maher, J. C., 1953, Paleozoic history of southeastern Colorado: AAPG Bulletin, v. 37, p. 2475–2489.

Mallory, W. W., 1958, Pennsylvanian coarse arkosic redbeds and associated mountains in Colorado, *in* Symposium on Pennsylvanian rocks of Colorado and adjacent areas: Denver, Colorado, Rocky Mountain Association of Geologists, p. 17–20.

———, 1967, Pennsylvanian and associated rocks in Wyoming: U.S. Geological Survey Professional Paper 554-G, 31 p.

———, 1972a, Pennsylvanian arkose and the Ancestral Rocky Mountains, *in* Mallory, W. W., ed. Geologic Atlas of the Rocky Mountain Region: Denver, Colorado, Rocky Mountain Association of Geologists, p. 131–132.

———, 1972b, Regional synthesis of the Pennsylvanian system, *in* Mallory, W. W., ed., Geologic Atlas of the Rocky Mountain Region: Denver, Colorado, Rocky Mountain Association of Geologists, p. 111–127.

———, 1975, Middle and southern Rocky Mountains, northern Colorado Plateau and eastern Great Basin, *in* McKee, E. D., and E. J. Crosby, coordinators, Paleotectonic investigations of the Pennsylvanian System in the United States: U.S. Geological Survey Professional Paper 853, p. 265–278.

Maughan, E. K., 1975, Montana, North Dakota, northeastern Wyoming, and northern South Dakota, *in* McKee, E. D., and E. J. Crosby, coordinators, Paleotectonic investigations of the Pennsylvanian system in the United States: U.S. Geological Survey Professional Paper 853, p. 279–293.

Maughan, E. K., 1979, Pennsylvanian (Upper Carboniferous) System in Wyoming: U.S. Geological Survey Professional Paper 1110-M-DD, p. U16–U33.

Maughan, E. K., 1980, Permian and Lower Triassic Geology of Colorado, *in* Kent, H. C., and K. W. Porter, eds., Colorado Geology: Denver, Colorado, Rocky Mountain Association of Geologists, p. 103–110.

Maughan, E. K., and A. E. Roberts, 1967, Big Snowy and

Amsden Groups and the Mississippian Pennsylvanian boundary in Montana: U.S. Geological Survey Professional Paper 554-B, 27 p.

McKee, E. D., and E. J. Crosby, coordinators, 1975, Paleotectonic investigations of the Pennsylvania System in the United States: U.S. Geological Survey Professional Paper 853, pt. I, 349 p.; pt. II, 192 p.

Melton, F. A., 1925, The Ancestral Rocky Mountains of Colorado and New Mexico: Journal of Geology, v. 33, p. 84–89.

Merriam, D. F., 1963, The geologic history of Kansas: Kansas Geological Survey Bulletin, v. 162, 327 p.

Miller, J. P., A. Montgomery and P. K. Sutherland, 1963, Geology of part of the southern Sangre de Cristo Mountains, New Mexico: New Mexico Bureau of Mines and Mineral Resources Memoir 11, 106 p.

Molnar, P., and P. Tapponier, 1975, Cenozoic tectonics of Asia: Effects of a continental collision: Science, v. 189, p. 419–426.

Morel, P., and E. Irving, 1980, Late Paleozoic reconstruction of the continents based on paleomagnetism, *in* Pilger, R. H., Jr., ed., The origin of the Gulf of Mexico and the early opening of the central North Atlantic Ocean: Proceedings of a symposium on the Gulf of Mexico, Louisiana State University, Baton Rouge, Louisiana, p. 75–78.

Morris, R. C., 1974, Sedimentary and tectonic history of the Ouachita Mountains, *in* Dickinson, W. R., ed., Tectonics and sedimentation: Society of Economic Paleontologists and Mineralogists Special Publication 22, p. 120–142.

Mose, D., 1969, The age of the Hatlon Tuff of the Ouachita Mountains, Oklahoma: Geological Society of America Bulletin, v. 80, p. 2373–2378.

Otte, C., Jr., 1959, Late Pennsylvanian and early Permian stratigraphy of the northern Sacramento Mountains, Otero County, New Mexico: New Mexico State Bureau of Mines and Mineral Resources Bulletin 50, 111 p.

Niem, A. R., 1977, Mississippian pyroclastic flow and ash-fall deposits in the deep-marine Ouachita Flysch Basin, Oklahoma and Arkansas: Geological Society of America Bulletin, v. 88, p. 49–61.

Peterson, J. A., 1980, Permian paleogeography and sedimentary provinces, west-central United States, *in* Fouch, T. D., and E. R. Magathan, eds. Paleozoic paleogeography of west-central United States: Rocky Mountain Section, Society of Economic Paleontologists and Mineralogists Symposium 1, p. 271–292.

Pindell, J., and J. F. Dewey, 1982, Permo-Triassic reconstruction of western Pangea and the evolution of the Gulf of Mexico/Caribbean regions: Tectonics, v. 1, p. 179–211.

Plummer, F. B., 1950, Carboniferous rocks of the Llano region of central Texas: Texas University Publication 1329, 170 p.

Prichard, G. E., 1975, Nebraska and adjoining parts of South Dakota and Wyoming, *in* McKee, E. D., and E. J. Crosby, coordinators, Paleotectonic investigations of the Pennsylvanian System in the United States: U. S. Geological Survey Professional Paper 853, p. 115–126.

Rascoe, B., Jr., 1962, Regional stratigraphic analysis of Pennsylvanian and Permian rocks in western mid–continent, Colorado, Kansas, Oklahoma, Texas: AAPG Bulletin, v. 46, p. 1345–1370.

Rascoe, B., Jr., and D. L. Baars, 1972, Permian System, *in* Mallory, W. W. ed., Geologic Atlas of the Rocky Mountain Region: Denver, Colorado, Rocky Mountain Association of Geologists, p. 143–165.

Roberts, R. J., M. D. Crittenden, Jr., E. W. Tooker, H. T. Morris, R. K. Hose, and T. M. Cheney, 1965, Pennsylvanian and Permian basins in northeastern Nevada and south-central Idaho: AAPG Bulletin, v. 49, p. 1926–1956.

Ross, C. A., 1973, Pennsylvanian and Early Permian depositional history, southeastern Arizona: AAPG Bulletin, v. 57, p. 887–912.

———, 1978a, Pennsylvanian and Early Permian depositional framework, southeastern Arizona, *in* Callender, J. F., J. C. Wilt, and R. E. Clemons, eds., Land of Cochise: New Mexico Geological Society–Arizona Geological Society Guidebook, p. 193–200.

———, 1978b, Late Pennsylvanian and Early Permian sedimentary rocks and tectonic settling of the Marathon geosyncline, *in* Mazullo, S. J., ed., Tectonics and Paleozoic facies of the Marathon geosyncline, West Texas: Permian Basin Section, Society of Economic Paleontologists and Mineralogists Publication 78-17, p. 89–93.

———, 1979, Late Paleozoic collision of North and South America: Geology, v. 7, p. 41–44.

Scotese, C. R., R. K. Bambach, C. Barton, R. Van der Voo, and A. M. Ziegler, 1979, Paleozoic base maps: Journal of Geology, v. 87, p. 217–277.

Seager, W. R., 1981, Geology of Organ Mountains and southern San Andres Mountains, New Mexico: New Mexico Bureau of Mines and Mineral Resources Memoir 36, 97 p.

Shideler, G., 1970, Provenance of Johns Valley boulders in late Paleozoic Ouachita facies, southeastern Oklahoma and southwestern Arkansas: AAPG Bulletin, v. 54, p. 786–806.

Skipp, B., and W. E. Hall, 1980, Upper Paleozoic paleotectonics and paleogeography of Idaho, *in* Fouch, T. D., and E. R. Magathan, eds., Paleozoic paleography of west-central United States: Rocky Mountain Section, Society of Economic Mineralogists and Paleontologists Symposium 1, p. 387–422.

Speed, R. C., 1979, Collided Paleozoic microplate in the western United States: Journal of Geology, v. 87, p. 279–292.

Speed, R. C., and N. H. Sleep, 1982, Antler orogeny and foreland basin: a model: Geological Society of America Bulletin, v. 93, p. 815–828.

Spoelhof, R. W., 1976, Pennsylvanian stratigraphy and paleotectonics of the western San Juan Mountains, southwestern Colorado, *in* Epis, R. C., and R. J. Weimer, eds., Studies in Colorado field geology: Colorado School of Mines Professional Contributions, n. 8, p. 159–179.

Stevens, C. H., and W. Stanton, 1980, *Stylastraca*-bearing gravity flow sequence of Early Permian age in eastern California (abs.): Geological Society of America Abstracts with Programs, v. 12, p. 154.

Stevenson, G. M., and D. L. Baars, 1977, Pre-Carboniferous

paleotectonics of the San Juan Basin: New Mexico Geological Society Guidebook, 28th Field Conference, San Juan Basin, p. 99–110.

Stewart, G. F., 1975, Kansas, *in* McKee, E. D., and E. J. Crosby, coordinators, Paleotectonic investigation of the Pennsylvanian System in the United States: U. S. Geological Survey Professional Paper 853, p. 127–156.

Stewart, J. H., 1976, Late Precambrian evolution of North America—plate tectonic implication: Geology, v. 4, p. 11–15.

Stone, D. S., 1977, Tectonic history of the Uncompaghre Uplift, *in* Veal, H. K., ed., Exploration frontiers of the central and southern Rockies: Denver, Colorado, Rocky Mountain Association of Geologists Guidebook, p. 23–30.

Stone, P., C. H. Stevens, and C. D. Cavit, 1980, A regional Early Permian angular unconformity in eastern California (abs.): Geological Society of America Abstracts with Programs, v. 12, p. 154.

Sutherland, P. K., 1963, Paleozoic rocks, *in* Geology of the southern Sangre de Cristo Mountains, New Mexico: New Mexico Bureau of Mines and Mineral Resources Memoir 11, p. 22–46.

———, 1972, Pennsylvanian stratigraphy, southern Sangre de Cristo Mountains, New Mexico, *in* Mallory, W. W., ed., Geologic Atlas of the Rocky Mountain Region, Rocky Mountain Association of Geologists, p. 139–142.

Szabo, E., and S. A. Wengard, 1975, Stratigraphy and tectogenesis of the Paradox Basin, *in* Fassett, J. E., ed., Canyonland Country: Four Corners Geological Society Guidebook, p. 193–210.

Tanner, J. H., III, 1967, Wrench fault movements along Wachita Valley fault, Arbuckle Mountain area, Oklahoma: AAPG Bulletin, v. 51, p. 126–141.

Tapponier, P., and P. Molnar, 1976, Slip line field theory and large scale continental tectonics: Nature, v. 264, p. 319–324.

Thomas, W. A., 1977, Evolution of Appalachian–Ouachita salients and recesses from recesses and promontories in the continental margins: American Journal of Science, v. 277, p. 1233–1278.

———, 1983, Continental margins, orogenic belts, and intracratonic structures: Geology, v. 11, p. 270–272.

Thomson, A., and E. J. McBride, 1978, Summary of the geologic history of the Marathon geosyncline, *in* Mazullo, S. J., ed., Tectonics and Paleozoic facies of the Marathon geosyncline, West Texas: Permian Basin Section, Society of Economic Paleontologists and Mineralogists Publication 78-17, p. 79–88.

Tomlinson, C. W., and W. McBee, Jr., 1959, Pennsylvanian sediments and orogenies of Ardmore district, Oklahoma, *in* Petroleum geology of southern Oklahoma: Ardmore Geological Society, v. 2, p. 3–52.

Tweto, O., 1980, Tectonic history of Colorado, *in* Kent, H. C., and K. W. Porter, eds., Colorado Geology, Denver, Colorado, Rocky Mountain Association of Geologists, p. 5–9.

Van der Voo, R., 1983, A plate tectonics model for the Paleozoic assembly of Pangea based on paleomagnetic data, *in* Hatcher, R. D., Jr., H. Williams, and I. Zietz,

eds., Contributions to the tectonics and geophysics of mountain chains: Geological Society of America Memoir 158, p. 19–23.

Van der Voo, R., and R. B. French, 1974, Apparent polar wandering for the Atlantic bordering continents: Late Carboniferous to Eocene: Earth Science Reviews, v. 10, p. 99–119.

Van Waterrschoot Van der Gracht, W. A., 1931, Permo-Carboniferous orogeny in south-central United States: AAPG Bulletin, v. 15, n. 9, p. 991–1057.

Ver Wiebe, W. A., 1930, Ancestral Rocky Mountains: AAPG Bulletin, v. 14, p. 765–788.

Viele, G. W., 1973, Structure and tectonic history of the Ouachita Mountains, Arkansas, *in* DeJong, K. A., and R. Scholten, eds., Gravity and tectonics: New York, John Wiley and Sons, p. 361–377.

Walker, T. R., 1972, Bioherms in the Minturn Formation (Des Moines age), Vail-Minturn area, Eagle County, Colorado: Colorado School of Mines Quarterly, v. 67, n. 4, p. 249–277.

Walper, J. I., 1977, Paleozoic tectonics of the southern margin of North America: Gulf Coast Association of Geological Societies Transactions, v. 27, p. 230–241.

Warner, L. A., 1983, Comment on plate tectonics of the Ancestral Rocky Mountains: Geology, v. 11, p. 120–121.

Weimer, R. J., 1980, Recurrent movement on basement faults, a tectonic style for Colorado and adjacent areas, *in* Kent, H. C., and K. W. Porter, eds., Colorado Geology: Denver, Colorado, Rocky Mountain Association of Geologists, p. 23–35.

Welsh, J. E. 1979, Paleogeography and tectonic implications of the Mississippian and Pennsylvanian in Utah, *in* Newman, G. W., and H. D. Goode, eds., Basin and Range Symposium: Rocky Mountain Association of Geologists and Utah Geological Association, p. 93–106.

Wengard, S. A., and M. L. Matheny, 1958, Pennsylvanian System of the Four Corners Region: Bulletin, v. 42, p. 2048–2106.

White, M. A., and M. I. Jacobson, 1983, Structures associated with the southwest margin of the ancestral Uncompahgre Uplift, *in* Averett, W. R., ed., Northern Paradox Basin–Uncompaghre Uplift: Grand Junction Geological Society Guidebook, p. 33–39.

Wickham, J., D. Roeder, and G. Briggs, 1976, Plate tectonics models for the Ouachita foldbelt: Geology, v. 4, p. 173–176.

Wilson, B. R., and S. W. Laule, 1979, Tectonics and sedimentation along the Antler orogenic belt of central Nevada, *in* Newman, G. W., and H. D. Goode, eds., Basin and Range Symposium: Rocky Mountain Association of Geologists and Utah Geological Association, p. 81–92.

Wilson, R. F., 1975a, Nevada and southern California, *in* McKee, E. D., and E. J. Crosby, coordinators, Paleotectonic investigations of the Pennsylvanian System in the United States: U. S. Geological Survey Professional Paper 853, p. 311–328.

———, 1975b, Eastern Colorado, *in* McKee, E. D., and E. J. Crosby, coordinators, Paleotectonic investigations of the Pennsylvanian System in the United States: U. S.

Geological Survey Professional Paper 853, p. 245–264.

Woods, R. D., and J. W. Addington, 1973, Pre-Jurassic geologic framework, northern Gulf Basin: Gulf Coast Association of Geological Societies Transactions, v. 23, p. 92–108.

Ziegler, P. A., 1982, Faulting and graben formation in western and central Europe: Philosophical transactions of the Royal Society of London, v. 305, p. 113–143.

Post-Mississippian Paleotectonic, Stratigraphic, and Diagenetic History of the Weber Sandstone in the Rangely Field Area, Colorado

M. H. Koelmel
Chevron U.S.A. Inc.
Denver, Colorado

Rangely Field is situated in Rio Blanco County, Colorado, on a doubly plunging anticline of Laramide age. The Rangely structure is asymmetric with the steepest flank to the southwest. The Pennsylvanian–Permian Weber Sandstone is the primary producing formation, with cumulative production exceeding 670 million bbl. The Weber is a subarkosic arenite deposited in an eolian regime. It interfingers with the alluvial Maroon Formation in the southern and southeastern part of Rangely Field. Isopach maps of the Pennsylvanian formations suggest a paleotectonic platform in the Rangely area and a Pennsylvanian–Permian north–south-trending arch west of the Laramide Douglas Creek arch. Hydrocarbons migrated into the Rangely Field area prior to the Laramide orogeny and were stratigraphically trapped at the Weber–Maroon transition zone. Subsequent Laramide structure localized the hydrocarbon accumulation.

Diagenetic history of the Weber Sandstone differs between the Uinta and Piceance basins. Weber diagenesis in the Uinta basin is dominated by silica precipitation, and porosity appears to be residual primary. Weber diagenesis in the Piceance basin involved dissolution of detrital material and precipitation of a complex sequence of carbonate cements. Weber porosity in the Piceance basin is both residual primary and secondary. The boundary between these two diagenetic regimes coincides with a Pennsylvanian paleoarch. The diagenetic model proposed for the Rangely area assumes a paleotectonic basin geometry consisting of a gently dipping western limb and a steeply dipping eastern limb. Silica precipitation commenced after Weber deposition throughout the Rangely area. Pre-Laramide salt tectonics may have caused sufficient faulting to permit fluid communication between the Eagle Valley Evaporites and the Weber Sandstone. Saline solutions from the Eagle Valley Evaporites migrated into the Weber in the Piceance basin halting silica precipitation and initiating precipitation of carbonate cements. Precipitation of silica continued in the Uinta basin. Development of secondary porosity in the Piceance basin occurred prior to, or simultaneously with, oil migration.

INTRODUCTION

Rangely Field in Rio Blanco County, Colorado, has a cumulative production exceeding 670 million bbl. Hydrocarbon is structurally trapped in a Laramide doubly plunging anticline. The Rangely anticline is asymmetric with a gently dipping northeastern limb and a steeply dipping southwestern limb. The Pennsylvanian–Permian Weber Sandstone is the main producing interval. Minor hydrocarbon production has been established in the Morrison Formation, Dakota Formation, and Mancos Group. The objective of this study is to chronologically reconstruct the sequence of geologic events that affected the hydro-carbon accumulation at Rangely Field. The study area encompasses a 23,000 mi^2 (59,800 km^2) area in northwestern Colorado and northeastern Utah (Figure 1). The hydrocarbon potential of the study area has been previously discussed by Picard (1956), Jensen (1958), Wells (1958), Cheney and Sheldon (1959), Millison (1962), Hansen (1963), Porter (1963), Osmond et al. (1965), Folsom (1968), Dunn (1974), and Sanborn (1977). In addition to Rangely Field, Weber production has been established at Ashley Valley, Elk Springs, Iles Dome, Maudlin Gulch, Thornburg, and Winter Valley fields (Figure 1). Cumulative production totals for these fields are listed in Table 1. Weber production accounts for a small fraction of the cumulative totals for Iles Dome and Maudlin Gulch fields. The chronologic reconstruction of events begins prior to Weber deposition.

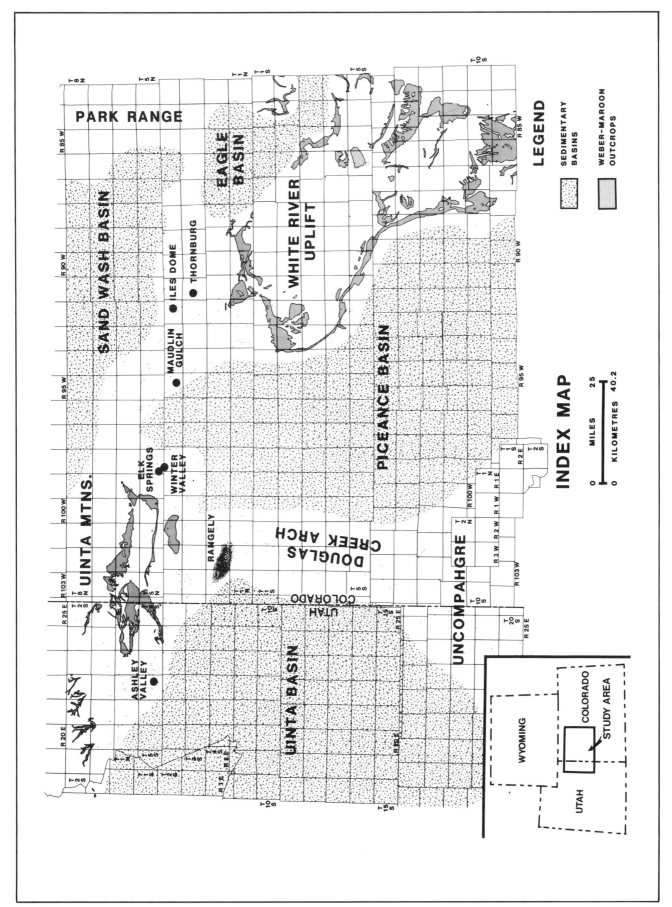

Figure 1—Index map of the study area indicating major Laramide basins and uplifts. Fields with hydrocarbon production established in the Weber Sandstone are also shown.

Table 1—Cumulative production totals for fields shown in Figure 1.[a]

Field Name	Year Discovered	Cumulative Production (bbl)
Ashley Valley	1948	18,960,388
Elk Springs	1926	545,672
Iles Dome	1924	18,476,314
Maudlin Gulch	1947	7,385,156
Rangely	1932	670,838,059
Thornburg	1925	753,686
Winter Valley	1958	303,708

[a]Source: Petroleum Information Production Report

LOWER PENNSYLVANIAN (MORROWAN–DESMOINESIAN) HISTORY

The Morrowan–Desmoinesian series of the Lower Pennsylvanian consists of the Round Valley Limestone, Morgan Formation, Maroon Formation, Belden Shale, and Minturn Formation (Figure 2). Associated with the Minturn Formation are the Eagle Valley Evaporites. The Morrowan and Atokan series are dominated by carbonate and shale deposition within the study area. Clastic deposition increased significantly during Desmoinesian time, although deposition of carbonates and evaporites persisted into early Desmoinesian time. Arkosic material shed from the rising Pennsylvanian highlands appeared during Atokan time (De Voto, 1972; Mallory, 1972; Walker, 1972) and is considered part of the Maroon Formation for the purpose of this study. De Voto (1972) and Walker (1972) include the arkosic material as a facies of the Minturn Formation and Belden Shale in the Eagle basin. The boundaries between the Morgan Formation, Minturn Formation, and Belden Shale are not identifiable facies changes based on the present subsurface control and are thus only areal changes in nomenclature.

The Morgan, Minturn, and Maroon lithofacies map (Figure 3) represents deposition during early Desmoinesian time. The lithofacies map is based on surface data from Boggs (1966), Mallory (1971), Tillman (1971), Walker (1972), and Driese and Dott (1984), as well as on subsurface control. The eastern part of the lithofacies map is modified from work by M. B. Cooper (1978, personal communication) and Walker (1972). The Morrowan–Desmoinesian paleotopography was dominated by two ancestral highlands in the vicinity of the Laramide Park range and Uncompahgre uplift. A third ancestral highland southeast of the study area has been proposed by De Voto (1972) and Walker (1972). This third ancestral highland would have occupied a position similar to the Laramide Sawatch range. The Maroon arkoses were shed from the ancestral Front range and ancestral Uncompahgre uplift and were probably deposited in an alluvial fan environment.

The Morgan Formation has been subdivided into an upper and lower member by Hansen (1977). The lower member consists of varigated shale and siltstone with interbedded sandstone and limestone. The upper member consists of cross-bedded and bioturbated sandstone, limestone, and dolomite. Sadlick (1955, 1957) suggested a shallow marine

environment of deposition for the Morgan Formation. Driese and Dott (1984) have proposed an alternating marine and eolian environment of deposition for the upper member of the Morgan in the Uinta Mountain area. Driese and Dott suggest that the environmental shifts resulted from eustatic changes. The Morgan Formation interfingers with the alluvial Maroon Formation in the vicinity of Rangely Field. The Morgan Formation is bounded on the east by algal limestone and dolomite of the Minturn Formation. Walker (1972) has speculated that the Minturn algal mounds restricted flow from the Morgan sea into the Eagle basin area creating an evaporitic environment. Alternatively, the evaporitic environment could have been created by the eustatic changes proposed by Driese and Dott (1984).

The Lower Pennsylvanian isopach map (Figure 4) includes the Round Valley Limestone, Morgan Formation, Maroon Formation (part), Belden Shale, Minturn Formation, and Eagle Valley Evaporites and is constructed from 68 subsurface well penetrations within the study area. Since the Maroon Formation was probably deposited from Atokan through Wolfcampian time, a criterion was necessary to distinguish Lower Pennsylvanian Maroon from Upper Pennsylvanian–Permian Maroon. Within the study area, the Morgan and Minturn formations can generally be distinguished from the Weber Sandstone by a higher resistivity on electric logs. The resistivity increase is seemingly correlatable into the Maroon Formation and is considered to be a reasonable approximation of the boundary between the Lower Pennsylvanian Maroon and the Upper Pennsylvanian–Permian Maroon.

The gross basin geometry during Morrowan–Desmoinesian time was asymmetric with a steeply dipping eastern limb. A broad paleoarch may have existed at the western margin of the study area. Rangely Field appears to be on the southern margin of a paleotectonic platform. The thick Maroon section immediately south of Rangely Field (Figure 4) is apparently due to an east–west-trending fault, between T. 2 S. and T. 3 S., which was active during deposition. The Vail–McCoy trough described by Boggs (1966), Tillman (1971), and Walker (1972) dominates the eastern portion of the study area. The steep isopach gradient on the northeast margin of the trough may be indicative of a fault that was active during deposition of the Lower Pennsylvanian system. The basinal asymmetry that developed during Morrowan–Desmoinesian time persisted throughout Weber–Maroon deposition into Wolfcampian time. The paleotectonic platform north of Rangely Field may have influenced Weber deposition by impeding the northward progradation of the Maroon arkoses.

UPPER PENNSYLVANIAN–LOWER PERMIAN (DESMOINESIAN–WOLFCAMPIAN) HISTORY

The Weber Sandstone was deposited primarily during Desmoinesian and Wolfcampian time (Figure 2) with a disconformity eliminating most of the Missourian and Virgilian section (Bissell and Childs, 1958; Bissell, 1964). The Weber is thought to have been deposited in an eolian environment throughout the northern part of the study area

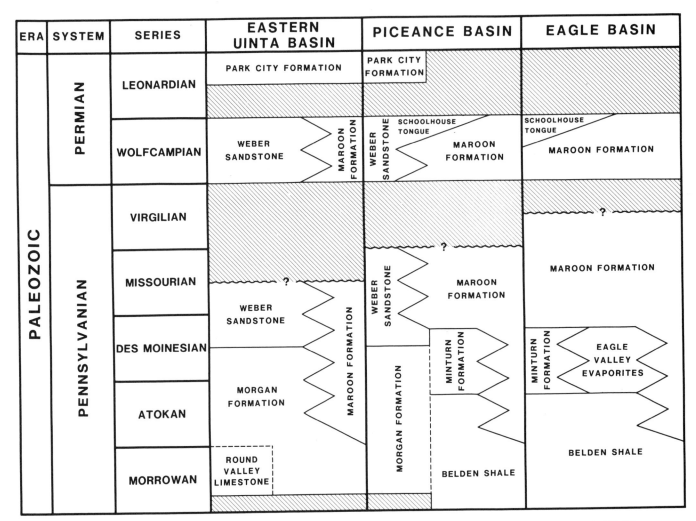

ERA	SYSTEM	SERIES	EASTERN UINTA BASIN	PICEANCE BASIN	EAGLE BASIN
PALEOZOIC	PERMIAN	LEONARDIAN	PARK CITY FORMATION	PARK CITY FORMATION	
		WOLFCAMPIAN	WEBER SANDSTONE / MAROON FORMATION	WEBER SANDSTONE / SCHOOLHOUSE TONGUE / MAROON FORMATION	SCHOOLHOUSE TONGUE / MAROON FORMATION
	PENNSYLVANIAN	VIRGILIAN			?
		MISSOURIAN	? WEBER SANDSTONE / MAROON FORMATION	? WEBER SANDSTONE / MAROON FORMATION	MAROON FORMATION
		DES MOINESIAN		MINTURN FORMATION	MINTURN FORMATION / EAGLE VALLEY EVAPORITES
		ATOKAN	MORGAN FORMATION	MORGAN FORMATION	
		MORROWAN	ROUND VALLEY LIMESTONE	BELDEN SHALE	BELDEN SHALE

Figure 2—Pennsylvanian and Lower Permian stratigraphic columns for the eastern Uinta, Piceance, and Eagle basins. Modified from Mallory (1975).

(Figure 5). At Whiterocks Canyon, the Weber section consists of an upper 1190 ft (363 m) of eolian facies and a basal 640 ft (195 m) of marine facies (Bissell and Childs, 1958). The contact between the Weber Sandstone and the Morgan Formation is ambiguous at this locality, and part of the marine facies measured by Bissell and Childs (1958) may belong to the Morgan Formation. The Whiterocks Canyon section measured by Bissell and Childs (1958) is incorporated into this study. The marine facies does not appear to extend very far east of Whiterocks Canyon. Extensive Weber outcrops are exposed north of Rangely in Dinosaur National Monument and have been described in detail by Doe (1973) and Fryberger (1979). The eolian Weber is a fine-grained, subangular to subrounded sandstone with large-scale cross bedding indicative of barchan and transverse dunes. Interdunal areas intermittently contain lenticular carbonate beds. Large-scale soft sediment deformation is observable in the Weber outcrops at Dinosaur National Monument (Doe, 1973) and in cores from Rangely Field.

The Maroon arkoses continued to be shed from the ancestral Front Range and ancestral Uncompahgre uplift during Desmoinesian and Wolfcampian time. The Maroon

Formation was probably deposited in an alluvial fan environment, but the Wolfcampian section may have been deposited in a fluvial fan environment similar to the one proposed for the Cutler Formation by Campbell (1979).

A broad transition zone existed between the alluvial Maroon arkoses and the eolian Weber at the close of Weber–Maroon deposition. This transition zone is particularly well developed in the eastern part of the study area and is identified as the Schoolhouse Tongue of the Weber Sandstone by Brill (1944, 1952) and Bissell and Childs (1958). The stratigraphic relationship between the Weber Sandstone, transition zone, and Maroon Formation is depicted in the diagrammatic cross sections (Figure 6). Brill (1944) and Bissell and Childs (1958) suggested that the Schoolhouse Tongue was an extension of the eolian Weber facies that had advanced southeastward as Maroon deposition subsided during early Wolfcampian time. The Schoolhouse Tongue, however, is less mineralogically mature than the eolian Weber facies and commonly exhibits a bimodal grain size distribution (Figure 7). The Schoolhouse Tongue may have been deposited in an expanding interfan region dominated by eolian processes with intermittent fluvial deposition. The eolian processes may have been

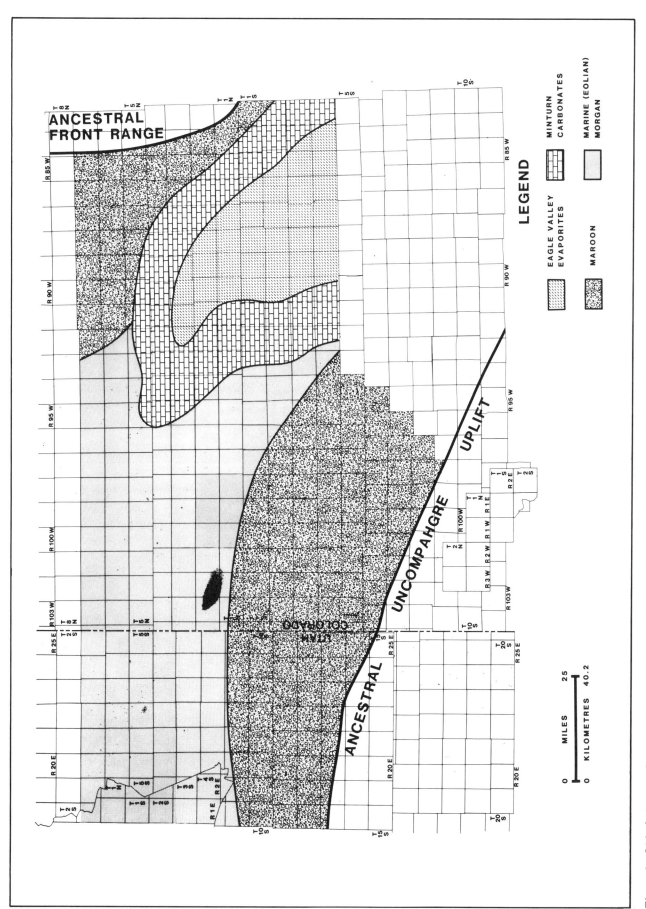

Figure 3—Lithofacies map depicting Morgan, Minturn, and Maroon depositional environments during early Desmoinesian time. The lithofacies map is based on surface data by Boggs (1966), Mallory (1971), Tillman (1971), Walker (1972), and Driese and Dott (1984), as well as on subsurface control. The eastern part of the lithofacies map is modified from work by M. B. Cooper (1978, personal communication) and Walker (1972).

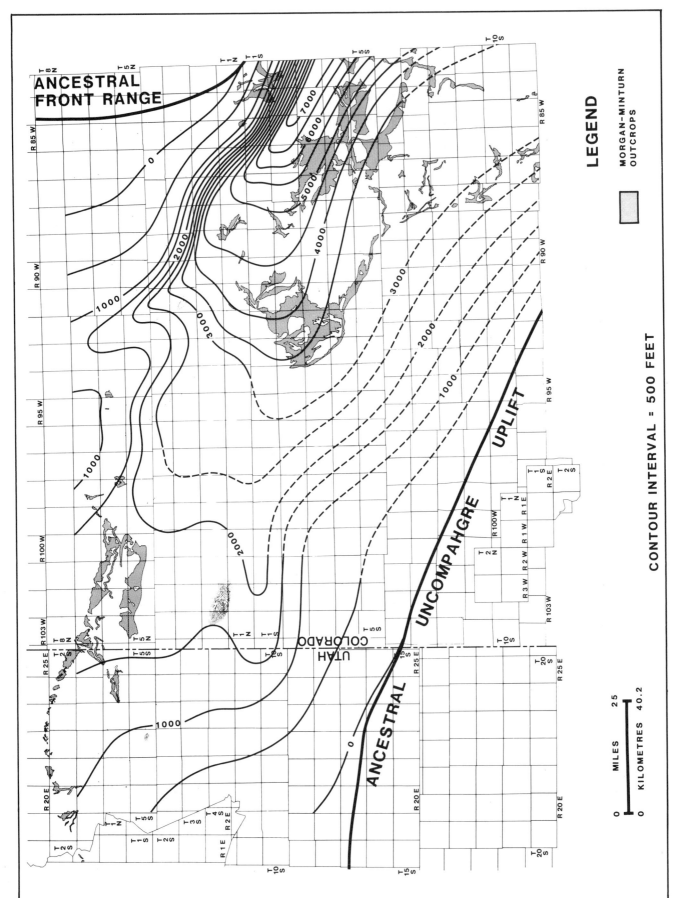

Figure 4—Lower Pennsylvanian isopach map based on 68 subsurface well penetrations within the study area.

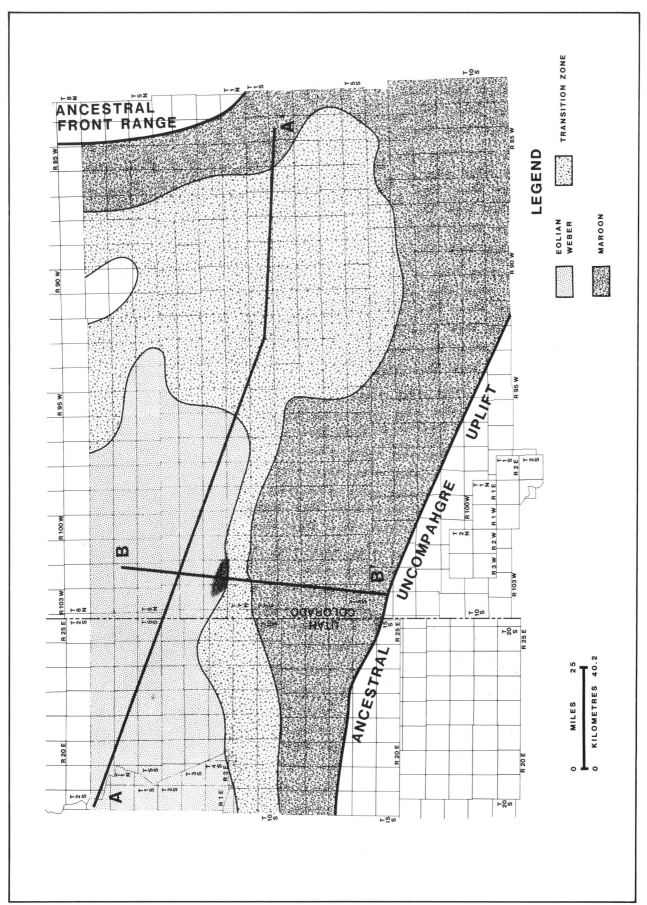

Figure 5—Lithofacies map depicting Weber and Maroon depositional environments during early Wolfcampian time. Stratigraphic cross sections A–A′ and B–B′ are shown in Figure 6.

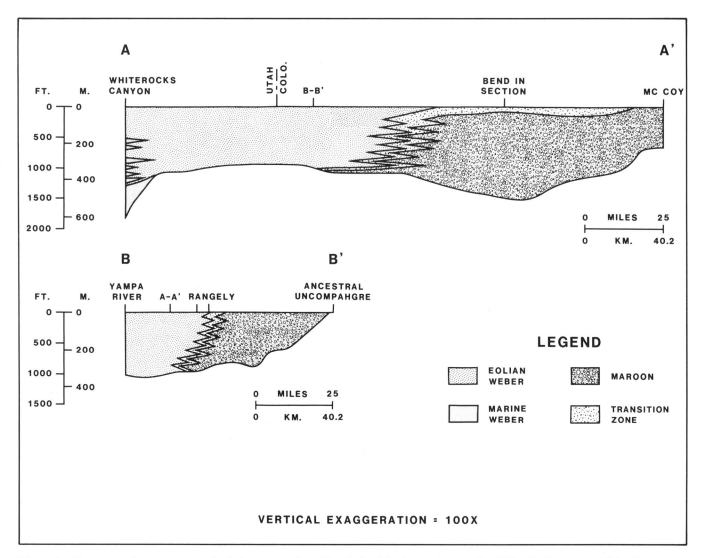

Figure 6—Diagrammatic cross sections depicting the stratigraphic relationships between the eolian Weber facies, marine Weber facies, transition zone, and Maroon arkoses. Location of stratigraphic cross sections A–A′ and B–B′ are shown in Figure 5.

reworking the distal facies material of the retreating Maroon alluvial (or fluvial) fans, which would account for the lower mineralogic maturity and bimodal grain size distribution. A transition zone facies similar to the Schoolhouse Tongue is observable in cores immediately south of Rangely Field.

The Weber–Maroon isopach map (Figure 8) is constructed from 84 subsurface well penetrations in addition to the surface sections measured by Fryberger (1979) and Bissell and Childs (1958). The Upper Pennsylvanian–Permian Maroon Formation is segregated from the Lower Pennsylvanian Maroon by the previously discussed resistivity criterion. The basin geometry during Weber and Maroon deposition was similar to the Lower Pennsylvanian basin geometry. The thickest section of the Weber–Maroon system occurs near Maroon Bells, Colorado, in the southeastern corner of the study area. The Maroon arkoses measure nearly 15,000 ft (4580 m) thick in this area because of a subsiding basin at the time of deposition. Tweto (1977) suggested that subsidence may have been partly related to flowage of the underlying Eagle Valley Evaporites in response to loading.

An area of nondeposition occurs 30 mi (48 km) west of the ancestral Front Range in T. 6 N., R. 90 W. (Figure 8) and apparently represents a paleohigh during Weber–Maroon deposition. The paleohigh apparently subsided during early Wolfcampian time, because Schoolhouse Tongue deposition is influenced less by the paleohigh than the underlying Maroon arkoses. As during Lower Pennsylvanian Maroon deposition, a thick Upper Pennsylvanian–Permian Maroon section occurs south of Rangely Field, because an east-west-trending fault between T. 2 S. and T. 3 S. was active during deposition. The Weber–Maroon isopach map (Figure 8) suggests the presence of a broad, north-south-trending paleoarch west of the Laramide Douglas Creek arch. The paleoarch may have prevented the marine facies of the Weber Sandstone from transgressing farther eastward. The rapid thickening of the Weber section between Dry Fork Canyon (T. 3 S., R. 20 E., USM), Utah, and Whiterocks Canyon (T. 2 N., R. 1 E., USM), Utah, is due to the additional 640 ft (195 m) of marine facies at Whiterocks Canyon, as reported by Bissell and Childs (1958).

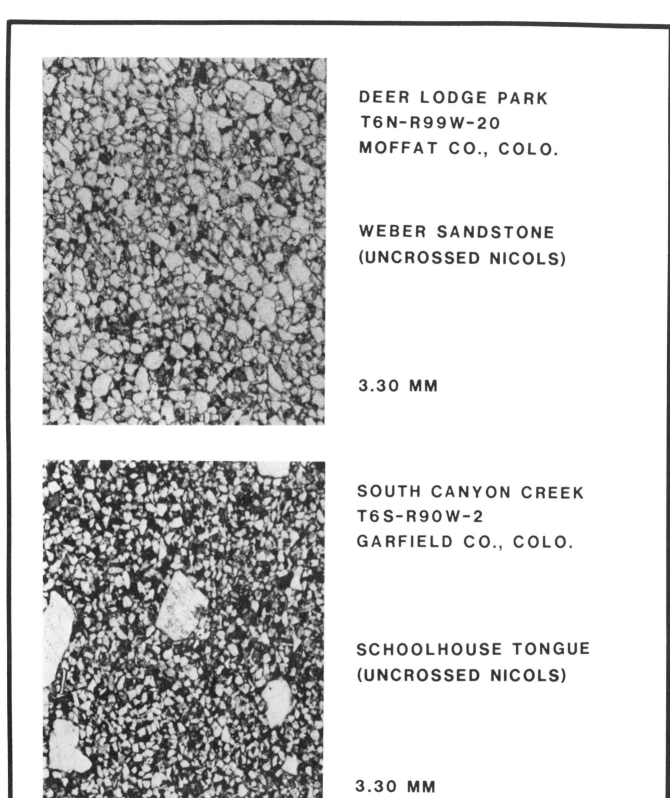

DEER LODGE PARK
T6N-R99W-20
MOFFAT CO., COLO.

WEBER SANDSTONE
(UNCROSSED NICOLS)

3.30 MM

SOUTH CANYON CREEK
T6S-R90W-2
GARFIELD CO., COLO.

SCHOOLHOUSE TONGUE
(UNCROSSED NICOLS)

3.30 MM

Figure 7—Photomicrographs of the eolian Weber facies at Deer Lodge Park (Sec. 20, T. 6 N., R. 99 W.) in Dinosaur National Monument and of the bimodal grain size distribution within the Schoolhouse Tongue at South Canyon Creek (Sec. 2, T. 6 S., R. 90 W.) near Glenwood Springs, Colorado. The vertical field of view is indicated for each photomicrograph.

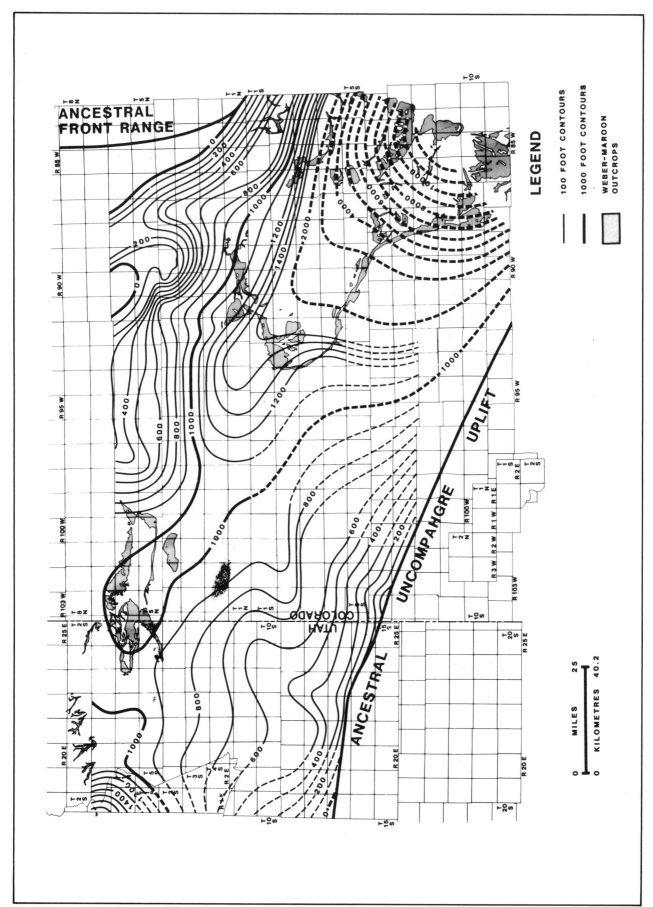

Figure 8—Weber and Upper Pennsylvanian–Permian Maroon isopach map constructed from 84 subsurface well penetrations and from surface sections measured by Fryberger (1979) and Bissell and Childs (1958).

The Weber isopach map (Figure 9) is constructed from surface sections measured by Fryberger (1979) and Bissell and Childs (1958) and from 96 subsurface penetrations. The transition zone (Schoolhouse Tongue) is included with the Weber Sandstone. The Weber sections within Dinosaur National Monument are 1000 ft (305 m) thick (Fryberger, 1979). The Shell Oil, Government 22X-17 well (Sec. 17, T. 2 N., R. 97 W.) penetrated 1080 ft (330 m) of Weber section, extending the 1000 ft (305 m) isopach contour from the Dinosaur National Monument 25 mi (40 km) to the southeast. Rangely Field is located on a steep, isopach gradient that is areally restricted. The steep isopach gradient reflects a rapid facies transition in the vicinity of Rangely Field and may be the combined result of three factors. First, the east-west-trending paleofault south of Rangely Field was active during Weber–Maroon deposition and locally steepened the paleogradient on the northern flank of the ancestral Uncompahgre uplift. The steepened paleogradient permitted alluvial (fluvial) Maroon sediments to be transported farther northward into the area south and southeast of Rangely Field. Second, the Lower Pennsylvanian isopach map (Figure 4) suggests that Rangely was positioned on the southern margin of a paleotectonic platform during Morrowan–Desmoinesian time. The paleotectonic platform may have persisted into Desmoinesian–Wolfcampian time and may have impeded the northward progradation of the alluvial (fluvial) Maroon sediments. Third, Fryberger (1979) indicates that the dominant dune transport direction for the eolian facies of the Weber Sandstone was to the south. The combined effect of these three factors could cause the steep isopach gradient that is restricted to the vicinity of Rangely Field. The rapid facies transition reflected by the steep isopach gradient was probably the key factor in stratigraphically trapping large quantities of hydrocarbon in the vicinity of Rangely Field prior to the Laramide orogeny.

A petrographic analysis of the uppermost Weber Sandstone was undertaken within the study area. Shown in Table 2 are the 48 samples collected from 27 localities. It is important to note that the sample quantity and distribution are insufficient to be statistically conclusive. Any study of this nature, however, will be severely limited by the sample and core material available from existing well control. Nevertheless, some significant patterns can be recognized from the petrographic data base. All but two of the samples used in the petrographic analyses were collected from the upper 200 ft (61 m) of the Weber Sandstone. The two exceptions are from the Frontier Refining Company, Colorow #1 well (Sec. 17, T. 2 N., R. 97 W.) and were collected approximately 300 ft (92 m) below the top of the Weber Sandstone. Numerous maturity indices are proposed in the literature. For the purpose of this analysis, a variation of the maturity index proposed by Pettijohn (1975) was used and is computed as follows:

$$MI = Q/DG \times 100$$

where *MI* is the maturity index, *Q* is the number of monocrystalline quartz grains, and *DG* is the number of detrital grains.

The maturity index map (Figure 10) indicates a systematic increase in maturity from Maroon sediments, through transition zone (Schoolhouse Tongue) sediments, to eolian Weber sediments. In general, Maroon sediments have maturity indices of less than 60, the transition zone sediments have maturity indices ranging from 60 to 75, and the eolian Weber sediments have maturity indices greater than 75. As previously discussed, the Schoolhouse Tongue may have been deposited in an expanding interfan region dominated by eolian processes with intermittent fluvial deposition. This type of depositional environment would explain the observed mineralogic maturity relationships between the Maroon Formation, Schoolhouse Tongue, and eolian facies of the Weber Sandstone.

PERMIAN-CRETACEOUS (WOLFCAMPIAN-CAMPANIAN) HISTORY

During this time period, the Weber Sandstone within the study area underwent a complex diagenetic history. Presently, there are no suitable benchmarks to date the various stages of Weber diagenesis. Weber diagenesis probably began contemporaneously with or shortly after deposition, and essentially ended with the migration of hydrocarbon into the study area prior to the Laramide orogeny. There is some local petrographic evidence for post-Laramide diagenesis (telediagenesis) in areas where the Weber Sandstone is exposed. A diagenetic sequence is proposed in Figure 11 for the Weber Sandstone in the Piceance and Uinta basins. The stage 3 carbonate cements are not listed in order of precipitation, although the textural relationship of the stage 3 carbonate cements does suggest a pattern of increasing formation fluid salinity with time. Precipitation of the stage 1 carbonate cements probably commenced during or shortly after Weber deposition in both the Piceance and Uinta basins. The stage 1 carbonate cements are generally localized in patches (Figure 12) that weather as spherical nodules on the surface outcrops. The stage 1 carbonate cements are contaminated with detrital and authigenic clays and can be readily distinguished from the stage 3 carbonate cements, which are characteristically devoid of clays.

Precipitation of the stage 1 carbonate cements was followed by silica precipitation in the form of quartz overgrowths. The stage 2 quartz overgrowths are present in both the Piceance and Uinta basins, although the stage 2 overgrowths are more prevalent in the Uinta basin. Weber porosity in the Uinta basin is residual primary and appears to be an inverse function of the quantity of silica precipitated as quartz overgrowths. Core material from the Equity Oil Company Ashley Valley Fee #1 well (Figure 12) and the Chevron Oil Company Red Wash Unit #219 well (Figure 13) demonstrates this relationship. Euhedral quartz overgrowths have decreased primary porosity in the Ashley Valley Fee #1 well, but a significant amount of primary porosity is still present. Core material from the Red Wash Unit #219 well has virtually no porosity. When a thin section from the Unit #219 well is viewed in polarized light, the loss of primary porosity appears to result from pressure solution of the detrital grains. When the same thin section is observed by cathodoluminescence, it becomes apparent that the loss of primary porosity is mainly due to silica precipitation in the form of stage 2 quartz overgrowths. Only localized pressure solution is evident. The stage 2

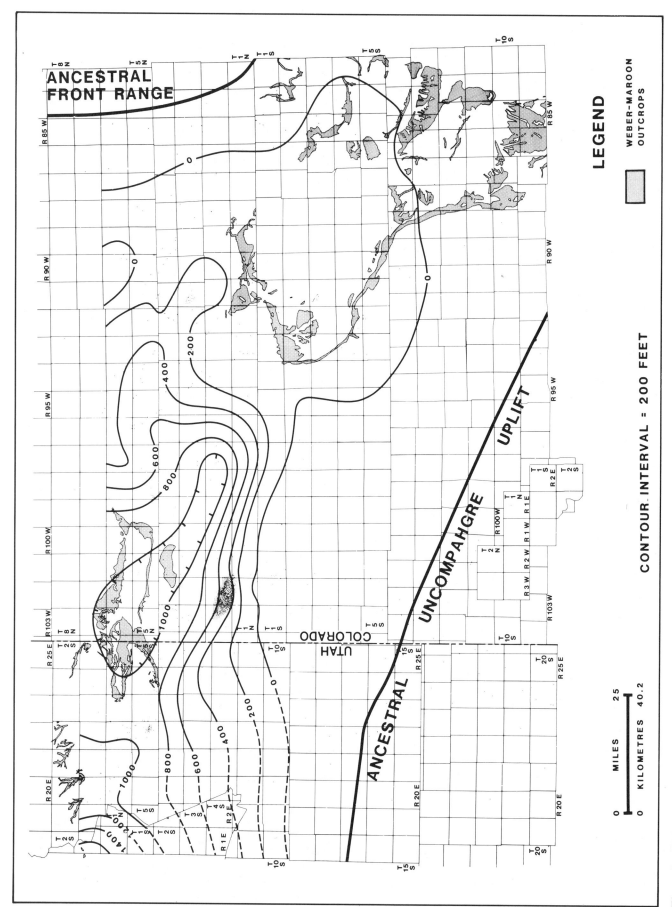

Figure 9—Weber isopach map constructed from 96 subsurface well penetrations and from surface sections measured by Fryberger (1979) and Bissell and Childs (1958).

Table 2—Names and locations of wells from which samples were taken for petrographic analysis of the upper Weber Sandstone.

Sample No.	Location	Well	Depth (ft)
1	T1N-R103W-15	Chevron Oil Co., Federal #14-15	10,323
2	T1N-R103W-15	Chevron Oil Co., Federal #14-15	10,337
3	T2N-R94W-4	Texaco, Inc., Wilson Creek #45	10,608
4	T2N-R97W-17	Frontier Refining, Colorow #1	14,545
5	T2N-R97W-17	Frontier Refining, Colorow #1	14,605
6	T2N-R102W-18	Chevron Oil Co., Gray 7B	6,370
7	T2N-R102W-18	Chevron Oil Co., Gray 7B	6,390
8	T2N-R102W-20	Chevron Oil Co., Fee 55	6,117
9	T2N-R102W-20	Chevron Oil Co., Fee 55	6,142
10	T2N-R102W-29	Chevron Oil Co., Fee 39	5,924
11	T2N-R102W-29	Chevron Oil Co., Fee 39	5,951
12	T2N-R102W-29	Chevron Oil Co., Fee 39	5,976
13	T2N-R102W-30	Chevron Oil Co., Emerald 46	6,443
14	T2N-R103W-10	Chevron Oil Co., McLaughlin 34	6,620
15	T2N-R103W-10	Chevron Oil Co., McLaughlin 34	6,669
16	T2N-R103W-29	Asamera Oil Co., Raven Ridge #1	13,237
17	T2N-R103W-29	Asamera Oil Co., Raven Ridge #1	13,253
18	T2N-R103W-29	Asamera Oil Co., Raven Ridge #1	13,263
19	T2N-R103W-36	Chevron Oil Co., Emerald 20	6,125
20	T2N-R103W-36	Chevron Oil Co., Emerald 20	6,164
21	T3N-R94W-18	Texaco, Inc., Wilson Creek #40	9,252
22	T3N-R94W-18	Texaco, Inc., Wilson Creek #40	9,264
23	T3N-R94W-34	Texaco, Inc., Wilson Creek #15	8,200
24	T3N-R104W-12	Union Texas Petroleum, Val Miller #1-12	9,334
25	T3N-R104W-12	Union Texas Petroleum, Val Miller #1-12	9,420
26	T4N-R92W-13	Argo Oil Co., Craig #1	5,502
27	T4N-R92W-13	Argo Oil Co., Craig #1	5,512
28	T4N-R95W-27	Texaco, Maudlin Gulch #8	8,294
29	T4N-R95W-27	Texaco, Maudlin Gulch #8	8,302
30	T5N-R95W-31	Texaco, Treleaven Government #7	8,195
31	T5N-R95W-31	Texaco, Treleaven Government #7	8,225
32	T6N-R99W-20	Deer Lodge Park, Colorado	Surface
33	T1S-R91W-18	Buford Road, Colorado	Surface
34	T1S-R92W-19	Buford Road, Colorado	Surface
35	T1S-R101W-10	Chevron Oil, Douglas Pass Unit #1	8,715
36	T1S-R101W-10	Chevron Oil, Douglas Pass Unit #1	8,792
37	T2S-R84W-9	McCoy, Colorado	Surface
38	T4S-R92W-27	Rifle Creek Fish Hatchery, Colorado	Surface
39	T6S-R90W-2	South Canyon Creek, Colorado	Surface
40	T2N-R1E-18	Whiterocks Canyon, Utah	Surface
41	T3S-R20E-8	Dry Fork Canyon, Utah	Surface
42	T3S-R20E-8	Dry Fork Canyon, Utah	Surface
43	T5S-R22E-23	Equity Oil Co., Ashley Valley Fee #1	4,173
44	T5S-R22E-23	Equity Oil Co., Ashley Valley Fee #1	4,255
45	T5S-R22E-23	Equity Oil Co., Ashley Valley Fee #1	4,274
46	T7S-R24E-24	Chevron Oil Co., Red Wash Unit #219	17,963
47	T7S-R24E-24	Chevron Oil Co., Red Wash Unit #219	18,100
48	T9S-R25E-34	Phillips Petroleum, Watson B-1	11,225

quartz overgrowths appear to be the dominant Weber diagenetic stage throughout that part of the Uinta basin included in the study area. Weber diagenesis in the Piceance basin, however, appears to be dominated by stage 3 carbonate cements. When stage 3 carbonate cements are observed in contact with stage 2 quartz overgrowths, the overgrowths are commonly etched and, in places, corroded by stage 3 carbonate cements. This textural relationship indicates that precipitation of the stage 3 carbonate cements followed precipitation of the stage 2 quartz overgrowths.

The stage 3 carbonate cements were precipitated by a solution that was apparently very corrosive and increasing in salinity with time. The stage 3 carbonate cements commonly replace stage 1 carbonate cements, framework feldspars

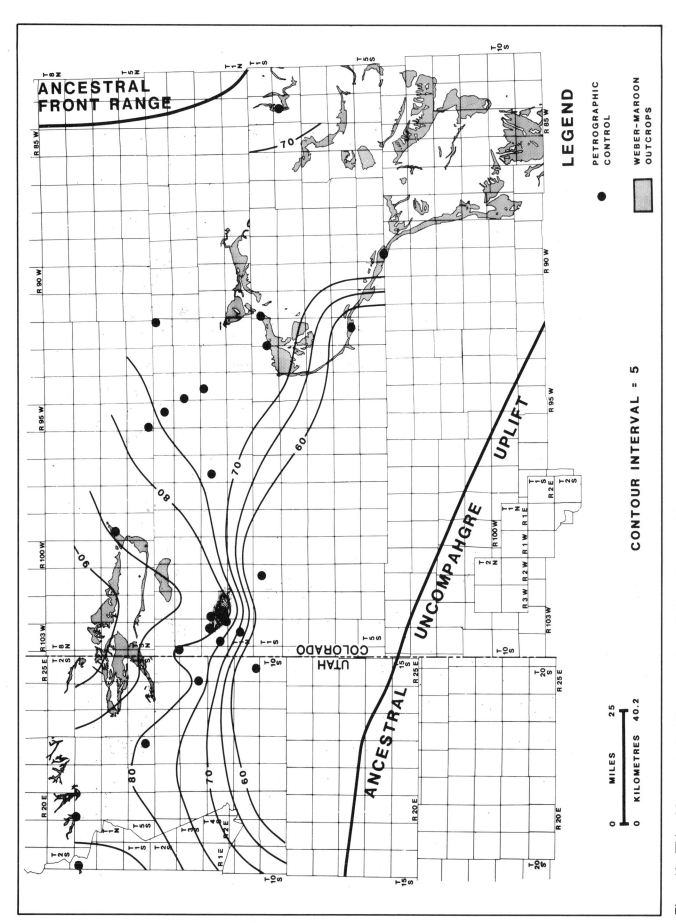

Figure 10—Weber–Maroon maturity index map made on the basis of petrographic control.

DIAGENETIC HISTORY

PICEANCE BASIN

STAGE 1   CARBONATE CEMENT AND
MINOR AUTHIGENIC CLAY

STAGE 2   QUARTZ OVERGROWTHS

STAGE 3   CARBONATE CEMENTS
1. CALCITE
2. FERROAN CALCITE
3. DOLOMITE
4. FERROAN DOLOMITE
5. ANHYDRITE
6. AUTHIGENIC ILLITE

STAGE 4   SECONDARY POROSITY
DEVELOPMENT

STAGE 5   HYDROCARBON MIGRATION

UINTA BASIN

STAGE 1   CARBONATE CEMENT AND
MINOR AUTHIGENIC CLAY

STAGE 2   QUARTZ OVERGROWTHS

STAGE 3   MINOR CARBONATE CEMENTS

STAGE 4   HYDROCARBON MIGRATION

Figure 11—Proposed diagenetic sequence for the Weber Sandstone in the Piceance and Uinta basins.

(particularly plagioclase), and biotite both in framework and in matrix material. The stage 3 carbonate cements sometimes exhibit a poikilotopic texture. Maroon arkoses in the vicinity of the facies transition with the Weber Sandstone also contain stage 3 carbonate cements. The precipitation sequence of the stage 3 carbonate cements generally begins with calcite and ferroan calcite, followed by dolomite and ferroan dolomite, and ending with very minor amounts of anhydrite. Core material from the Chevron Oil Company Fee #39 well at Rangely Field (Figure 14) demonstrates the paragenetic sequence. The euhedral habit of the stage 3 carbonate cement is evident in the polarized light photomicrograph. Stage 3 dolomites commonly occur as perfect rhombohedrons. Anhydrite is the last of the stage 3 cements to precipitate and occurs in residual primary porosity voids. When the carbonate patch (Figure 14) is viewed by cathodoluminescence, the paragenetic sequence can be reconstructed. Precipitation began with a ferroan calcite phase followed by dolomite, ferroan dolomite, dolomite, and ferroan dolomite. The iron source for the ferroan carbonate phases at Rangely Field appears to have been the biotite (now altered) in the Maroon arkoses within

the transition zone. The alteration of biotite and dissolution of feldspars also released alumina, which was apparently reprecipitated in authigenic illite (Figure 15). Authigenic illite precipitation seems to have been coincident with precipitation of the stage 3 carbonate cements.

The stage 3 carbonate cements appear to be the dominant Weber diagenetic event in the Piceance basin. The stage 3 carbonate cement map (Figure 16) indicates the bulk volume percentages of stage 3 cements based on the petrographic control. It is interesting to note that the highest bulk volume percentage of stage 3 carbonate cements occurs in the vicinity of the Eagle Valley Evaporites (Figure 3). On the basis of this observation, the asymmetric paleotectonic basin geometry, and the apparent paragenetic sequence of the stage 3 cements, I propose a diagenetic model for the Weber Sandstone in the study area. The stage 1 carbonate cements were precipitated throughout the study area prior to precipitation of the stage 2 quartz overgrowths. Following commencement of stage 2 silica precipitation in the Piceance basin, fluid communication occurred between the Eagle Valley Evaporites and the eolian facies of the Weber Sandstone. Pre-Laramide salt tectonics may have caused sufficient faulting to permit fluid communication between the Eagle Valley Evaporites and the Weber Sandstone in the eastern part of the study area. Once the Weber Sandstone in the eastern part of the study area became charged with saline brines, the brines apparently had sufficient hydrodynamic head to migrate westward up the gentle paleoslope. As the evaporitic solutions migrated westward, the Weber Sandstone fluid pH probably increased and effectively ended stage 2 silica precipitation in the Piceance basin. The increase in salinity and pH also tended to destabilize the feldspars and alter the biotite, which subsequently released alumina and iron into the formation water. Any fixed point on the migration path would thus register an increase in salinity with time, which is in agreement with the observed paragenetic sequence. Weber Sandstone water analyses at Rangely prior to the initiation of the current water flood program indicated that NaCl equivalent salinities approximated 120,000 ppm and that pH values ranged from 6.5 to 7.0. Stage 3 carbonate cements apparently do not dominate Weber diagenesis west of Rangely Field, with the exception of the two samples collected from the Chevron Oil Company Red Wash Unit #219 well (Sec. 24, T. 7 S., R. 24 E.). West of Rangely Field, the stage 2 quartz overgrowths appear to dominate Weber diagenesis.

In the Piceance basin, a dissolution stage apparently followed precipitation of the stage 3 carbonate cements. The dissolution of stage 3 cements resulted in secondary porosity development. The dissolution stage may have occurred simultaneously with hydrocarbon migration. Weber porosity in the Piceance basin is generally a combination of residual primary and secondary porosity. Both types of porosity can be observed in the Chevron Oil Company Fee #39 well (Figure 17). The top scanning electron micrograph (SEM) in Figure 17 shows areas of primary porosity (upper left) and areas of secondary porosity development (upper right). The bottom SEM shows "channel-like" porosity development characteristic of secondary porosity. Weber

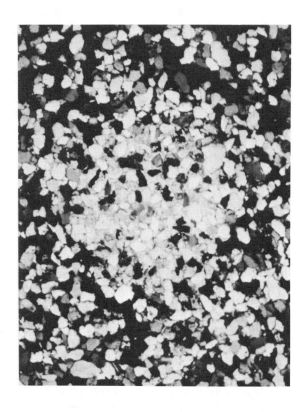

DRY FORK CANYON
T3S-R20E-6
UINTAH COUNTY, UTAH

STAGE 1 CARBONATE CEMENT
(CROSSED NICOLS)

3.00 MM

ASHLEY VALLEY FEE #1 (4173')
T5S-R22E-23
UINTA COUNTY, UTAH

QUARTZ OVERGROWTHS
(UNCROSSED NICOLS)

0.52 MM

Figure 12—Photomicrographs of the stage 1 carbonate cements and stage 2 quartz overgrowths (see Figure 11). The vertical field of view is indicated for each photomicrograph.

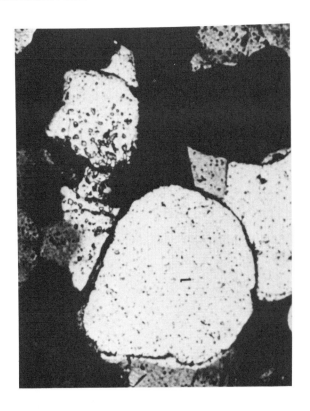

RED WASH UNIT #219 (17,963')
T7S-R24E-24
UINTAH COUNTY, UTAH

QUARTZ OVERGROWTHS
(CROSSED NICOLS)

0.34 MM

RED WASH UNIT #219 (17,963')
T7S-R24E-24
UINTAH COUNTY, UTAH

QUARTZ OVERGROWTHS
(CATHODOLUMINESCENCE)

0.34 MM

Figure 13—Photomicrograph and cathodoluminescence micrograph of stage 2 quartz overgrowths. The polarized light photomicrograph suggests extensive pressure solution, whereas the cathodoluminescence micrograph indicates extensive silica precipitation. The vertical field of view is indicated for each micrograph.

RANGELY FEE #39 (5975.5')
T2N-R102W-29
RIO BLANCO CO., COLO.

STAGE 3 CARBONATE CEMENTS
(CROSSED NICOLS)

0.20 MM

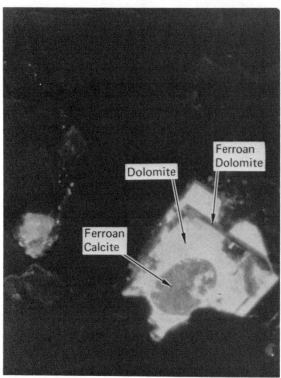

RANGELY FEE #39 (5975.5')
T2N-R102W-29
RIO BLANCO CO., COLO.

STAGE 3 CARBONATE CEMENTS
(CATHODOLUMINESCENCE)

0.20 MM

Figure 14—Photomicrograph and cathodoluminescence micrograph of the stage 3 carbonate cements. The paragenetic sequence of the carbonate patch can be observed in the cathodoluminescence micrograph. The vertical field of view is indicated for each micrograph.

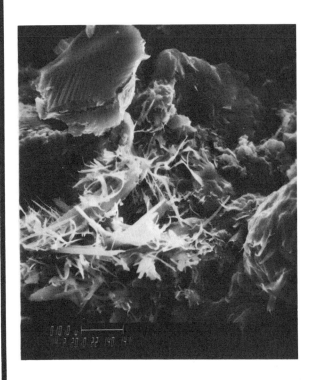

RANGELY FEE #39 (5975.5')
T2N-R102W-29
RIO BLANCO CO., COLO.

AUTHIGENIC ILLITE
(S.E.M.)

68 MICRONS

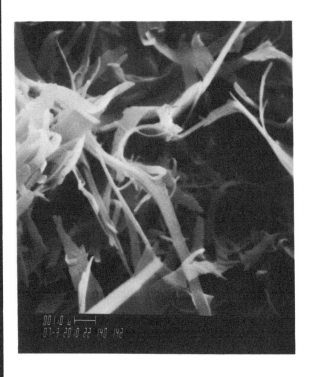

RANGELY FEE #39 (5975.5')
T2N-R102W-29
RIO BLANCO CO., COLO.

AUTHIGENIC ILLITE
(S.E.M.)

13 MICRONS

Figure 15—Authigenic illite precipitated in conjunction with the stage 3 carbonate cements. The vertical field of view is indicated for each scanning electron micrograph (SEM).

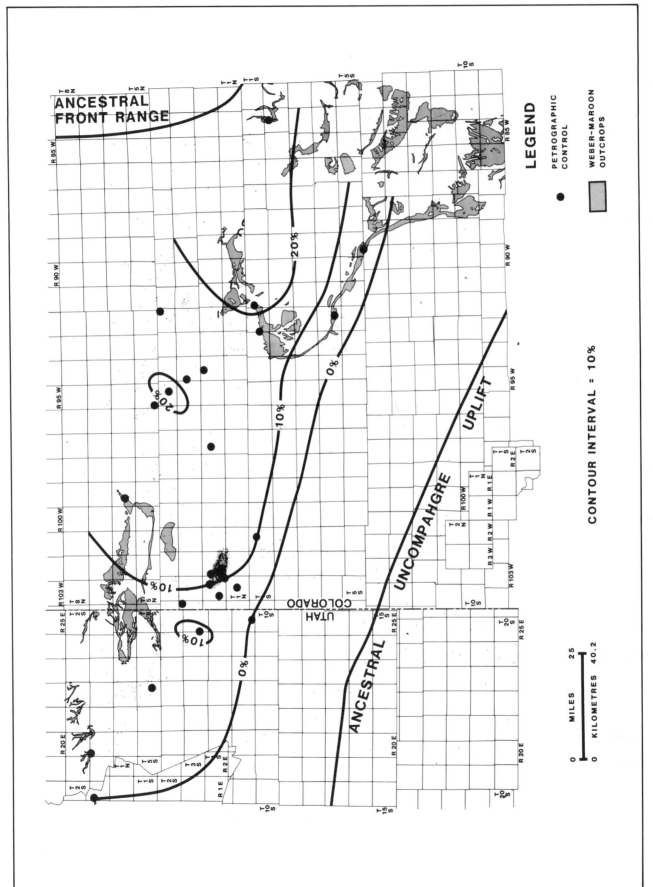

Figure 16—Bulk volume percentages of stage 3 carbonate cements based on the petrographic control. The highest bulk volume percentage occurs in the vicinity of the Eagle Valley Evaporites.

porosity in the Uinta basin is generally residual primary as observed in the Equity Oil Company Ashley Valley Fee #1 well (Figure 12).

Geochemical typing of the hydrocarbon produced from the Weber Sandstone at Rangely Field indicates a Pennsylvanian–Permian source rock. The only significant Pennsylvanian–Permian source rock in the study area is the Belden Shale (Figure 2). The Belden Shale does not appear to have sufficient oil-prone organic material and areal extent to have been the source for the large hydrocarbon accumulation at Rangely Field. Consequently, it is assumed that Permian oil from the Phosphoria basin migrated through the Weber and into the study area prior to the Laramide orogeny. Since Weber diagenesis was essentially complete prior to hydrocarbon migration, the Permian oil migration fairways presumably followed routes with the highest residual primary porosity and/or routes with the highest bulk volume percentage of stage 3 carbonate cements that could be dissolved to create secondary porosity. Stated otherwise, Permian oil migration fairways presumably followed routes with the lowest bulk volume percentage of framework grains unless otherwise influenced by structure. This hypothesis will hold only for clastic rocks with soluble cements and a relatively low bulk volume percentage of matrix material.

The oil migration routes map (Figure 18) is based on the bulk volume percentage of framework grains for the 27 localities used in the petrographic analyses. Quartz overgrowths were counted as framework material, since silica was effectively insoluble during hydrocarbon migration compared to stage 3 carbonate cements. Although this hypothesis is tenuous, the 80% contour may have significance in the case of the Weber Sandstone. Most of the established Weber production occurs in areas where the bulk volume percentage of framework grains is less than 80%. The Thornburg and Iles fields are in an area having slightly greater than 80% framework grains. The Schoolhouse Tongue outcrops in the southeastern part of the study area are apparently in a favorable area and are commonly oil stained. The Weber–Maroon isopach map (Figure 8), however, indicates that the Schoolhouse Tongue outcrops were positioned in a structurally low area during Pennsylvanian–Permian time. Since the general Pennsylvanian–Permian basin geometry persisted into Cretaceous time (Sanborn, 1977), Permian oil probably migrated into the Schoolhouse Tongue only after the onset of the Laramide orogeny when the White River uplift became a positive feature.

Oil produced from the Weber Sandstone at Ashley Valley Field does not correlate geochemically with oil produced from the Weber Sandstone at Rangley Field. Permian oil may have migrated into the Ashley Valley area (Figure 18) and subsequently into the Uinta Mountains before the Ashley Valley structure developed. Permian oil from the Phosphoria basin migrated into the vicinity of Rangely Field prior to the Laramide orogeny (perhaps during Campanian time). The Permian oil was stratigraphically trapped at the Weber–Maroon facies transition, as suggested by Campbell (1955), Hoffman (1957), and Bissell (1964). However, a diagenetic influence as indicated by the oil migration routes map (Figure 18) may have been necessary to prevent the

hydrocarbon accumulation from being stratigraphically trapped farther to the west, where the Laramide structure could not have concentrated the accumulation.

CRETACEOUS (CAMPANIAN) TO PRESENT HISTORY

During the Laramide orogeny, the Uncompahgre uplift and the Park range area again became positive elements. The White River uplift, Douglas Creek arch, and Uinta Mountains (Figure 1) are all Laramide features. The rise of the Uinta Mountains disrupted the Permian oil migration route from the Phosphoria basin into the study area. Rangely Field is located on a doubly plunging anticline of Laramide age (Figure 19). The Rangely structure is asymmetric with the steepest flank to the southwest and is positioned between the northern termination of the Douglas Creek arch and the southern margin of the Uinta Mountains. The culmination of the Rangely structure coincides with a transition zone between two major thrusts on the southern margin of the Uinta Mountains. The thrust faults on either side of the transition zone account for significant amounts of crustal shortening, and high-angle faults in the transition zone account for a minor amount of crustal shortening. Consequently, regional crustal shortening is maintained by shifting compression from the transition zone on the Uinta Mountain front to the Rangely structure. Formation of the Rangely anticline concentrated the Permian oil that had been stratigraphically trapped in the eolian Weber facies prior to the Laramide orogeny. Hydrocarbon production was established from the Mancos Group at Rangely Field as early as 1902. In 1932, the California Company (now Chevron U.S.A. Inc.) completed the Raven #1-A well in the Weber Sandstone and discovered Rangely Field. Cumulative Weber production at Rangely Field presently exceeds 670 million bbl of oil.

SUMMARY

The Weber Sandstone was deposited in an eolian environment throughout the northern part of the study area. The eolian Weber is a fine-grained, subarkosic arenite with large-scale cross bedding indicative of barchan and transverse dunes. The eolian Weber interfingers with the Maroon arkoses shed from the ancestral Front Range and ancestral Uncompahgre uplift. A rapid facies transition from eolian Weber sediments to alluvial Maroon sediments occurs immediately south and southeast of Rangely Field. Diagenesis probably commenced during or shortly after Weber deposition with precipitation of the stage 1 carbonate cements and minor authigenic clays. Precipitation of the stage 1 carbonate cements was followed by silica precipitation in the form of quartz overgrowths. The stage 2 quartz overgrowths are present in both the Piceance and Uinta basins, but dominate Weber diagenesis only in the Uinta basin. Weber porosity in the Uinta basin is residual primary and appears to be an inverse function of the quantity of silica precipitated as quartz overgrowths. Precipitation of the stage 2 quartz overgrowths was followed by precipitation of the stage 3 carbonate cements. The stage

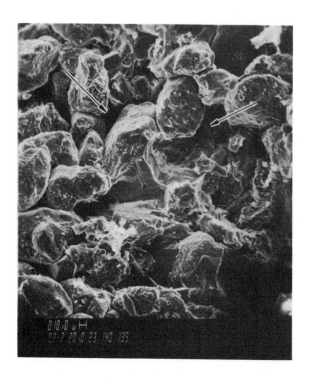

RANGELY FEE #39 (5975.5')
T2N-R102W-29
RIO BLANCO CO., COLO.

POROSITY DEVELOPMENT
(S.E.M.)

400 MICRONS

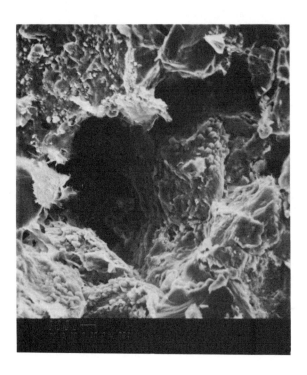

RANGELY FEE #39 (5975.5')
T2N-R102W-29
RIO BLANCO CO., COLO.

POROSITY DEVELOPMENT
(S.E.M.)

190 MICRONS

Figure 17—Scanning electron micrographs (SEM) of primary and secondary porosity in the Weber Sandstone at Rangely Field. The vertical field of view is indicated for each SEM.

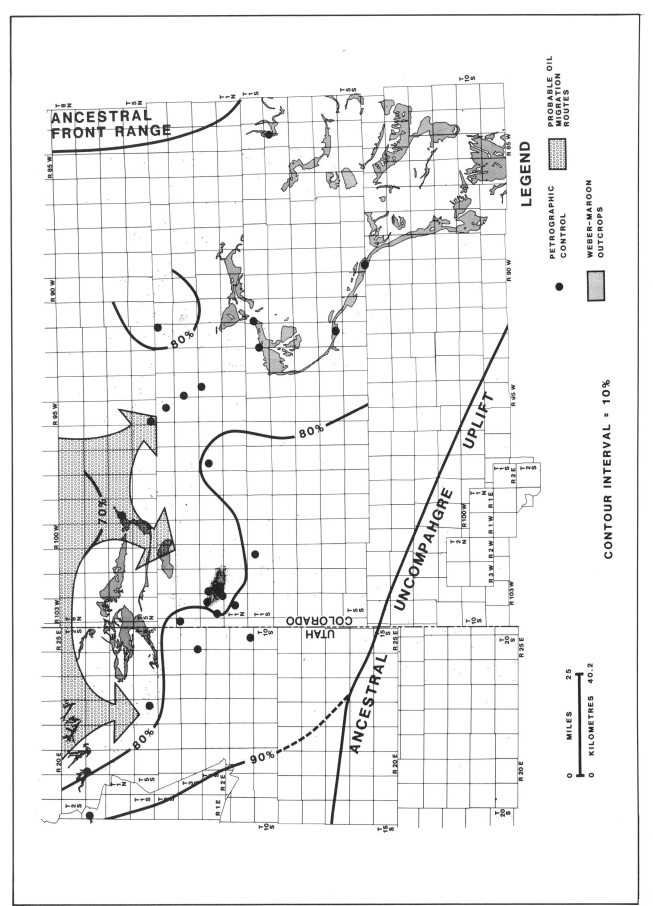

Figure 18—Oil migration routes map based on the bulk volume percentage of framework grains. Hydrocarbon migration fairways presumably followed routes with the lowest percentages of framework grains, unless otherwise influenced by structure.

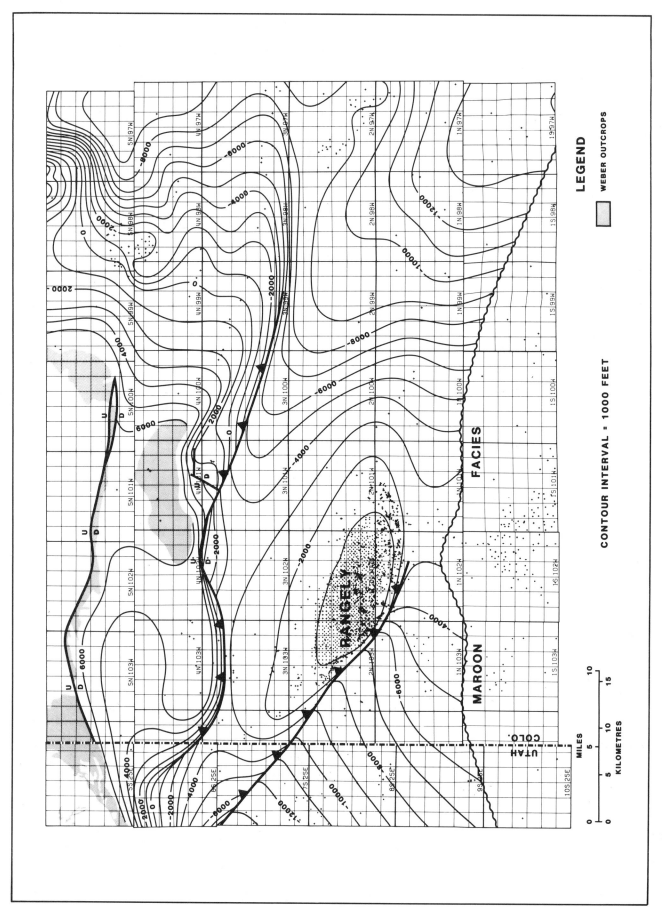

Figure 19—Top Weber structure map of the Rangely Field area based on surface outcrop data and subsurface control. The culmination of the Rangely anticline coincides with a transfer zone between two major thrusts on the southern margin of the Uinta Mountains.

3 carbonate cements may have precipitated from a saline solution that migrated into the Weber Sandstone from the Eagle Valley Evaporites. The stage 3 carbonate cements appear to be the dominant Weber diagenetic event in the Piceance basin. Weber porosity in the Piceance basin is both residual primary and secondary. Permian oil from the Phosphoria basin migrated into the study area prior to the Laramide orogeny and was stratigraphically trapped along the abrupt, Weber–Maroon facies transition south and southeast of Rangely Field. Formation of the Rangely structure during the Laramide orogeny concentrated the stratigraphically trapped hydrocarbon accumulation.

ACKNOWLEDGMENTS

I thank Chevron U.S.A. Inc. for granting me permission to publish this paper. I gratefully acknowledge the assistance of D. W. Richards during the petrographic analyses and the helpful suggestions of R. B. Christy and D. B. Givens during the interpretation of the data. I also thank Asamera Oil Company, Equity Oil Company, Texaco Inc., and Union Texas Petroleum for supplying core material for the petrographic analyses. This manuscript has benefitted from suggestions made by Kenneth G. Brill, Jr., John C. Osmond, Graham S. Campbell, James R. Wood, Manuel N. Bass, and Charles F. Kluth.

REFERENCES CITED

Bissell, H. J., 1964, Lithology and petrography of the Weber Formation in Utah and Colorado, *in* Guidebook of the geology and mineral resources of the Uinta Basin: Intermountain Association of Petroleum Geologists 13th Field Conference, p. 67–91.

Bissell, H. J., and O. E. Childs, 1958, The Weber Formation of Utah and Colorado, *in* Symposium on Pennsylvanian rocks of Colorado and adjacent areas: Denver, Rocky Mountain Association of Geologists, p. 26–30.

Boggs, S., Jr., 1966, Petrology of Minturn Formation, east-central Eagle County, Colorado: AAPG Bulletin, v. 50, p. 1399–1422.

Brill, K. G., Jr., 1944, Late Paleozoic stratigraphy, west-central and northwestern Colorado: Geological Society of America Bulletin, v. 55, p. 621–656.

———, 1952, Stratigraphy in the Permo-Pennsylvanian zeugogeosyncline of Colorado and Northern New Mexico: Geological Society of America Bulletin, v. 63, p. 809–880.

Campbell, G. S., 1955, Weber pool of Rangely field, Colorado, *in* Guidebook to the Geology of northwest Colorado: Intermountain Association of Petroleum Geologists 6th Field Conference, p. 99–100.

Campbell, J. A., 1979, Lower Permian depositional system, northern Uncompahgre Basin, *in* Permianland 1979: Four Corners Geological Society 9th Field Conference, p. 13–22.

Cheney, T. M., and R. P. Sheldon, 1959, Permian stratigraphy and oil potential, Wyoming and Utah, *in* Guidebook to the geology of the Wasatch and Uinta

Mountains: Intermountain Association of Petroleum Geologists 10th Field Conference, p. 90–100.

De Voto, R. H., 1972, Pennsylvanian and Permian stratigraphy and tectonism in central Colorado: Colorado School Mines Quarterly, v. 67, n. 4, p. 139–186.

Doe, T. W., 1973, Deformed cross-bedding from the Weber Formation (Pennsylvanian–Permian), northeastern Utah and northwestern Colorado: Master's Thesis, University of Wisconsin, Madison, Wisconsin, 87 p.

Driese, S. G., and R. H. Dott, Jr., 1984, Model for sandstone–carbonate "cyclothems" based on upper member of Morgan Formation (Middle Pennsylvanian) of northern Utah and Colorado: AAPG Bulletin, v. 68, p. 574–597.

Dunn, H. L., 1974, Geology of petroleum in the Piceance Creek basin, northwestern Colorado: Rocky Mountain Association of Geologists 25th Field Conference Guidebook, p. 217–224.

Folsom, L. W., 1968, Economic aspects of Uinta Basin gas development, *in* Natural gases of North America—a symposium: AAPG Memoir 9, v. 1, p. 199–208.

Fryberger, S. G., 1979, Eolian–fluviatile (continental) origin of ancient stratigraphic trap for petroleum in Weber sandstone, Rangely oil field, Colorado: The Mountain Geologist, v. 16, p. 1–36.

Hansen, A. R., 1963, The Uinta basin—structure, stratigraphy, and tectonic setting, in oil and gas possibilities of Utah, re-evaluated: Utah Geological and Mineralogical Survey Bulletin, v. 54, p. 175–176.

Hansen, W. R., 1977, Geologic map of the Jones Hole quadrangle, Uintah County, Utah, and Moffat County, Colorado: U. S. Geological Survey Quadrangle Map GQ-1401, scale 1:1:4000.

Hoffman, F. H., 1957, Possibilities of Weber stratigraphic traps, Rangely area, northwest Colorado: AAPG Bulletin, v. 41, p. 894–905.

Jensen, F. S., 1958, Oil and Gas in Permo-Pennsylvanian rocks of the Maroon basin, northwestern Colorado and northeastern Utah, *in* Symposium on Pennsylvanian rocks of Colorado and adjacent areas: Rocky Mountain Association of Geologists, p. 122–128.

Mallory, W. W., 1971, The Eagle Valley Evaporite, northwest Colorado—a regional synthesis: U.S. Geological Survey Bulletin, 1311-E, 37 p.

———, 1972, Regional synthesis of the Pennsylvanian System, *in* Geologic atlas of the Rocky Mountain region: Denver, Rocky Mountain Association of Geologists, p. 111–127.

———, 1975, Middle and southern Rocky Mountains, northern Colorado Plateau, and eastern Great Basin region, *in* E. D. McKee, and E. J. Crosby, coordinators, Introduction and regional analyses of the Pennsylvanian System, Pt. 1 of Paleotectonic investigations of the Pennsylvanian System in the United States: U. S. Geological Survey Professional Paper 853, p. 265–278.

Millison, C., 1962, Accumulation of oil and gas in northwestern Colorado controlled principally by stratigraphic variations, *in* Exploration for oil and gas in northwestern Colorado: Denver, Rocky Mountain Association of Geologists, p. 41–48.

Osmond, J. C., R. Locke, A. C. Dillé, W. Practorious, and J. C. Wilkins, 1968, Natural gas in Uinta Basin, Utah, *in* Natural gases of North America—a symposium: AAPG Memoir 9, v. 1, p. 174–198.

Pettijohn, F. J., 1975, Sedimentary Rocks (3rd edition): New York, Harper & Row, 628 p.

Picard, M. D., 1956, Summary of Tertiary oil and gas fields in Utah and Colorado: AAPG Bulletin, v. 40, p. 2956–2960.

Porter, L., Jr., 1963, Stratigraphy and oil possibilities of the Green River formation in the Uinta basin, *in* Oil and gas possibilities of Utah, re-evaluated: Utah Geological and Mineralogical Survey Bulletin, v. 54, p. 193–198.

Sadlick, W., 1955, Carboniferous formations of northeastern Uinta Mountains: Wyoming Geological Association 10th Annual Field Conference Guidebook, p. 49–59.

———, 1957, Regional relations of Carboniferous rocks of northeastern Utah, *in* Guidebook to the geology of the Uinta basin: Intermountain Association of Petroleum Geologists 8th Annual Field Conference Guidebook, p.

56–77.

Sanborn, A. F., 1977, Possible future petroleum of Uinta and Piceance basins and vicinity, northeast Utah and northwest Colorado, *in* Exploration frontiers of the central and southern Rockies: Rocky Mountain Association of Geologists, p. 151–166.

Tillman, R. W., 1971, Petrology and paleoenvironments, Robinson Member, Minturn Formation (Desmoinesian), Eagle Basin, Colorado: AAPG Bulletin, v. 55, p. 593–620.

Tweto, O., 1977, Tectonic history of west-central Colorado, *in* Exploration frontiers of the central and southern Rockies: Rocky Mountain Association of Geologists, p. 11–22.

Walker, T. R., 1972, Bioherms in the Minturn Formation (Des Moines age), Vail–Minturn area, Eagle County, Colorado: Colorado School Mines Quarterly, v. 67, n. 4, p. 249–277.

Wells, L. F., 1958, Petroleum occurrence in the Uinta basin, *in* Habitat of oil: AAPG, p. 344–365.

Relationship of Unconformities, Tectonics, and Sea Level Changes in the Cretaceous of the Western Interior, United States[1]

R. J. Weimer
Colorado School of Mines
Golden, Colorado

Intrabasin tectonics and sea level changes influenced patterns of deposition and geographic distribution of major unconformities within the Cretaceous of the Western Interior. Nine major regional to near regional unconformities have been identified. Previous workers have related five of these unconformities to sea level changes and to well known regressive–transgressive cycles. The origin of the other four unconformities may be related either to tectonic movement or sea level changes.

The approximate dates for unconformities are estimated as follows (formations involved are in parentheses): (1) late Neocomian to early Aptian, 112 m.y. (base lower Mannville, Lakota, Lytle); (2) late Aptian–early Albian, ~100 m.y. (upper Mannville, Fall River, Plainview); (3) Albian, ~97 m.y. (Viking, Muddy, Newcastle, or J Sandstone); (4) early Cenomanian, ~95 m.y. (lower Frontier–Peay, and D); (5) Turonian, ~90 m.y. (base upper Frontier or upper Carlile); (6) Coniacian, ~89 m.y. (base Niobrara or equivalents); (7) early Santonian, ~80 m.y. (Eagle, lower Pierre and upper Niobrara); (8) late Campanian, ~73 m.y. (mid-Mesaverde, Ericson, base Teapot); (9) late Maestrichtian, ~66 m.y. (top Lance or equivalents). Variations in the accuracy of the dating are probably within 1 m.y. because of problems in accurately defining the biostratigraphic level of the breaks and in the precision of radiometric dates. The unconformities are grouped into three types: those completely within nonmarine strata such as at the base and top of the Cretaceous, those involving both marine and nonmarine strata, and those within marine strata, as currently mapped.

Three examples are described as typical of the unconformities, all thought to be related primarily to drops in sea level, but with minor influence by tectonic movement. One is the ~97 m.y. unconformity, with which the petroleum-producing J and Muddy Sandstone is related. A second is ~90 m.y. unconformity which is recognized by relationships within the shelf, slope, and basin deposits of the Greenhorn, Carlile, and Frontier formations. The third is the ~80 m.y. unconformity within the basin and shoreline regression associated with the upper Niobrara, lower Pierre, Eagle, and Shannon formations.

Several billion barrels of oil were found in sandstones associated with unconformities in the Cretaceous of the Rocky Mountain region. Future stratigraphic trap exploration is guided by a knowledge of tectonic influence on sedimentation during sea level changes and how these factors control distribution of source rock, migration patterns, reservoir rock, and seal.

INTRODUCTION

Depositional systems on the margins of continents, both modern and ancient, have been related to sea level changes. Two simplified diagrams by Vail et al. (1977) illustrate highstand and lowstand conditions and associated unconformities (Figure 1). The highstand diagram represents a depositional system that might be observed in many coastal areas today or at times of highstand in the past. The four main components are the coastal plain, shelf, slope, and rise (or deep water basin). An unconformity related to coastal onlap during the rise of sea level to the highstand condition is shown. With a drop in sea level to the edge of the continental shelf (a lowstand condition), sediment bypasses the shelf and is deposited in deep water. The entire shelf is exposed to subaerial erosion and streams adjusting to the lower base level incise into the older shelf deposits. The depocenter shifts from the deltaic system under the highstand to marine subsea fans of the lowstand. Unconformities are present within the marine strata and

[1] Adapted from R. J. Weimer, 1984, Relation of Unconformities, Tectonics, and Sea Level Changes, Cretaceous of Western Interior, U.S.A., *in* J. S. Schlee, ed., Interregional Unconformities and Hydrocarbon Accumulation: AAPG Memoir 36, p. 7–35.

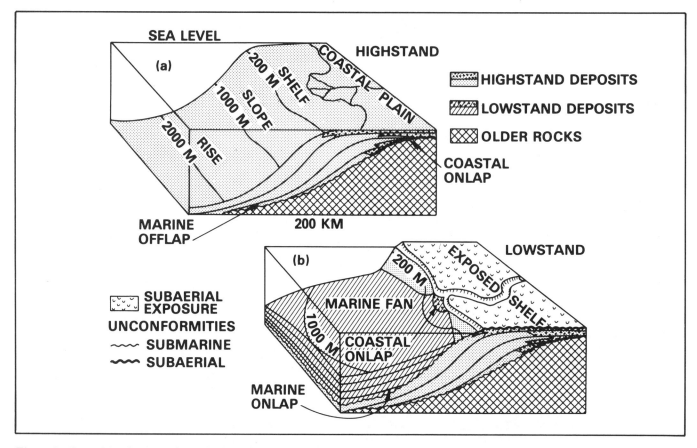

Figure 1—Depositional pattern during highstand (A) and lowstand (B) of sea level. From Vail et al. (1977).

also project into the coastal plain deposits. Several depositional models are possible in ancient strata between these two end members of sea level conditions.

If concepts illustrated by the diagrams from continental margins are applied to ancient interior cratonic basins, then marine and associated deposits represent sea level highstands and unconformities represent erosion during lowstands. Without considering tectonic influence on local depositional basins, a low sea level which exposed much of the shelf on the margin of the continent must have also exposed cratonic regions to subaerial exposure, thus developing interregional unconformities. If a time of significant sea level drop can be identified on a worldwide basis, then a predictive model could be formulated to search for an erosional surface in an ancient interior sequence. Conversely, an incised drainage system with a surface of erosion developed on an ancient marine sequence could be evidence for a significant drop in sea level. The purpose of this paper is to evaluate unconformities observed within the Cretaceous of the Western Interior basin and to determine if the associated surfaces of erosion are related to sea level changes and/or tectonics. Depositional models, previously constructed for the Cretaceous, are expanded to incorporate tectonic influence on sedimentation during sea level changes.

Recognition and Evaluation of Breaks in the Record

Because of the significance of recognizing and evaluating stratigraphic breaks in ancient sequences, how one defines an unconformity is important. I prefer the definition that an unconformity is a sedimentary structure in which two groups of rocks are separated by an erosional surface; the erosion may be by subaerial or submarine processes. Blackwelder (1909) described the factors that must be evaluated to understand the significance of a stratigraphic break. These factors are angular discordance, hiatus (what is missing), nature of the contact (for example, sharp, erosional, or irregular), aerial extent of break, duration of erosion, and cause.

Many types of breaks occur in the stratigraphic record. Unconformity is used for a major break that normally, but not always, can be traced over large areas. This structure records major changes that result from tectonic or sea level variations. A major surface of nondeposition, if recognized within strata, is referred to as a paraconformity. A minor break in the record associated with scour or nondeposition, normally found within a depositional environment, is called a diastem.

The use of unconformities for correlation is widespread. The erosional surface of an unconformity is useful in separating older from younger strata. However, to establish time relations among the strata, independent methods of correlation must be used. Otherwise, one cannot determine an accurate reconstruction of events recorded by the unconformity.

Petroleum Reservoir Rocks and Quaternary Deposits

Reservoir rocks for petroleum in ancient cratonic basins are most commonly associated with highstand conditions.

This concept is illustrated by observing the present high sea level and associated base level for deposition. Sand accumulates in shoreline, deltaic, fluvial (point bar), eolian, or lacustrine settings (Figure 2). During a stillstand or a slow increase in sea level in the Quaternary, the shoreline prograded by lateral accretion (Figure 3) in areas where rates of deposition exceeded rates of subsidence (or submergence). When sea level dropped and base level controlling stream deposition or erosion was lowered, the streams incised into older deposits and an erosional surface formed over large areas. Sand deposits may have developed in the deep water (basin) setting during the lowstand. With a rise in sea level, sediment was deposited in the incised drainage as valley-fill deposits.

Valley-fill deposits are illustrated for the modern Mississippi River valley (Figure 4) by Fisk (1947). The Mississippi and other Holocene fluvial systems have been studied to understand the origin and importance of point-bar sand deposits as a reservoir for petroleum. The lateral changes from point-bar sands to impermeable siltstones and claystones of floodplain or floodbasin deposits have been described as conditions favorable for stratigraphic trapping of petroleum. A more widespread sandstone of a fluvial braided channel complex was identified by Fisk at the base of the valley fill (Figure 4).

The relationship of valley-fill deposits to underlying bed rock and the surface of erosion separating older and younger strata have not been extensively studied in Quaternary deposits. Yet, the recognition of this type of unconformity is important in interpreting ancient strata. It is essential to recognize the difference between the scour surface at the base of a point bar sand within the valley fill (a diastem), and the surface of erosion associated with a sea level drop and drainage incisement (an unconformity) (Figure 4).

Similar problems exist in defining and evaluating breaks in the record of marine sequences in ancient cratonic basins. The most easily recognized breaks commonly occur within shelf sequences. The breaks associated with sedimentation on modern shelves are documented for the Gulf of Mexico. The record of Holocene deposition on the shelf of the northwest Gulf of Mexico is summarized by Curray (1960, 1975). Most of the shelf was subaerially exposed during the Wisconsin glaciation lowstand. With the Holocene rise in sea level, mud, sand, or shell were deposited over shelf areas where a sediment supply was available. Pleistocene sediments are exposed over a large area of the shelf, or are covered by a thin veneer of relict or palimpsest sands, reworked by waves during the transgression. Thus, an erosional surface modified by marine processes occurs at the base of the Holocene.

Incisement of a drainage system into Pleistocene shelf deposits, with a subsequent Holocene valley fill, is well documented by Nelson and Bray (1970). An area of the Texas shelf encompassing 2600 sq km (Figure 5) was studied in detail by using bottom samples, marine sonoprobe profiles, and cores. A paleovalley from 9.6 to 12.8 km wide was mapped for a distance of approximately 80 km (Figure 6). The valley was incised to an unknown depth by the combined flow of the Sabine and Calcasieu rivers (Figure 5) during the Wisconsin lowstand. The stratigraphic relations of three types of fill in the paleovalley are plotted on longitudinal and transverse sections (Figures 6 and 7). The lowest fill, older than 10,200 years, consists of fluvial–deltaic sands of unknown thickness because the base of the valley was not mapped. As the rising sea flooded the valley, estuarine and lagoonal clay and sandy mud were deposited in the valley. These deposits vary in thickness from 6 to 15 m and range in age from 10,200 years to approximately 5200 years. The final fill is composed of marine mud and sandy mud which varies in thickness from a wedge edge to 12 m. During the past 5000 years a marine sand bar complex known as the Sabine and Heald banks has formed approximately 30 km seaward from the present shoreline. The offshore bar is 10 to 13 km wide, 65 km long, and varies in thickness from a wedge edge to 6 m (Figures 6 and 7). The sand bodies may have a transitional base with underlying Holocene marine mud or may rest on an erosional surface on the Pleistocene (Figure 7). Although the sand is dominantly detrital quartz, local concentration of shells make up 100% of the banks. The offshore sand bar has a northwest trend parallel with the present shoreline and a portion of the paleovalley; it overlaps the southern margin of the paleovalley (Figures 6 and 7).

This study by Nelson and Bray (1970) illustrates three types of Holocene sand bodies associated with sea level changes that have potential as petroleum reservoirs. These are the fluvial–deltaic sands of the valley fill; the linear offshore marine bar and associated thin lag deposit; and the thin sands of the shoreline zone. Important stratigraphic breaks are associated with these sand bodies. Similar sand bodies have been recognized in ancient sequences but not always have their origin been related to eustatic changes.

Criteria for recognition of sea level changes in cratonic basins are listed here.

1. Regression of shoreline with incised drainage followed by overlying marine shale.
2. Valley-fill deposits (of incised drainage system) overlying marine shale:
   A. Root zones at or near base of valley-fill sequence.
   B. Paleosoil on scour surface.
3. Unconformities within basin:
   A. Missing faunal zone.
   B. Missing facies in a normal regressive sequence (e.g., shoreface or delta front sandstones).
   C. Paleokarst with regolith or paleosoil.
   D. Concentration of one or more of the following on a scour surface: phosphate nodules, glauconite, recrystallized shell debris to form thin lenticular limestone layers.
4. Thin widespread coal layer overlying marine regressive delta front sandstone deposits indicate rising sea level.
5. Correlation with the record of sea level changes from other continents.

When these criteria are used in conjunction with the factors listed above in evaluating unconformities, breaks which are the result of local tectonic influence can be separated from those caused by sea level changes.

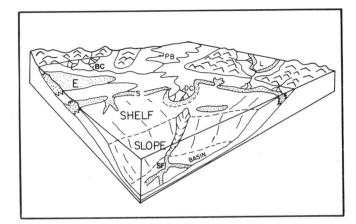

Figure 2—Environments of deposition for sandstone reservoirs under highstand conditions: S, shoreline; SH, shelf; DC, delta distributary channel; PB, fluvial point bar; E, eolian; L, lacustrine. For lowstand conditions: DW, deep water or basin.

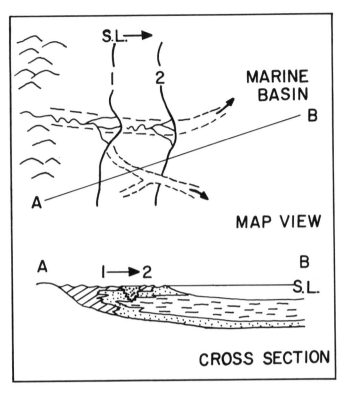

Figure 3—Shoreline regression from position 1 to position 2 during high sea level with high terrigenous influx. Drainage incised during sea level drop (dashed lines) and filled during subsequent sea level rise (section A–B).

REGIONAL SETTING OF CRETACEOUS BASIN

The Western Interior Cretaceous basin of the North American continent is one of the largest cratonic (foreland or back-arc) basins in the world. Because of economic products, mainly petroleum and coal, strata deposited in this basin have been thoroughly studied, both in outcrop and subsurface occurrences. The original basin was 800–1650 km wide and extended from the Arctic to the Gulf of Mexico (Figure 8). The relationship of the basin to other structural elements in the western part of North America is outlined by Dickinson (1976) (Figure 9). The foreland basin formed on a thick continental crust and was bordered on the west by a fold-thrust belt and on the east by the Canadian shield.

During the Early Cretaceous, sediments were derived from both sides of the basin, though the thickest strata are along the western margin. During the Late Cretaceous, the dominant source of sediment was along the western margin and lithofacies were controlled by changes in environments from coastal plain to shoreline to marine shelf and the deeper water of the basin. Intertonguing of nonmarine strata on the west with marine strata in the center of the basin is the dominant pattern of sedimentation (Figure 10). Thickness of the Cretaceous strata varies from 600 to 7000 m. The thinnest sections occur in the geographic center to the eastern margin of the basin because sedimentation rates were slower there. Total organic content is higher in these strata because of the slow sedimentation rates and because of anoxic conditions in deeper water that favored preservation of organic matter.

The Cretaceous basin was deformed during the Laramide orogeny and segmented into the present-day intermontane basins of the Rocky Mountain region. Areas between these basins were uplifted and subsequent erosion has removed Cretaceous strata from the structural high areas. Hence, reconstruction of the entire original depositional basin requires correlation among the intermontane Laramide structural basins. Because of widespread faunas and floras and closely spaced subsurface well control, accurate correlations are possible within much of the stratigraphic

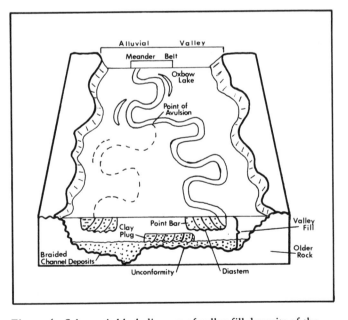

Figure 4—Schematic block diagram of valley-fill deposits of the Mississippi River (modified from Fisk, 1947). Breaks in the sequences are diastems at the base of meander-belt sands and the unconformity at the base of the valley fill. Braided channel deposits are widespread sands at base of valley fill; sand lenses in upper part are point-bar deposits with clay plugs of abandoned channel shown in black.

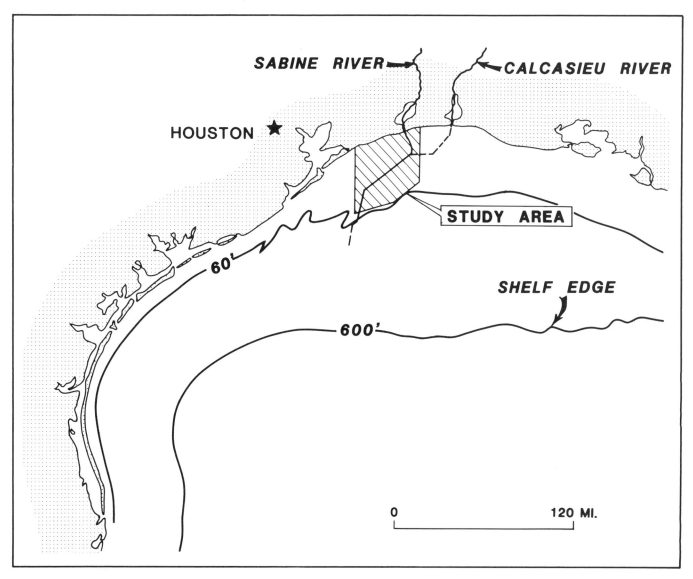

Figure 5—Index map with part of shelf, Gulf of Mexico Holocene sedimentation. Ruled area represents study area of Nelson and Bray (1970).

section. Therefore, time-stratigraphic units can be mapped, breaks accurately evaluated, and facies changes and depositional models reconstructed.

Stratigraphic concepts derived from studies of this Cretaceous depositional basin have widespread application to understanding detrital sequences in all ancient basins which had a structural setting on, or marginal to, cratonic regions of continental plates.

UNCONFORMITIES WITHIN THE CRETACEOUS BASIN

Unconformities within the Cretaceous basin have been recognized by many investigators. The best published synthesis of the Cretaceous system for the Rocky Mountain region in the United States is the Geologic Atlas published by the Rocky Mountain Association of Geologists. The

Cretaceous chapter, compiled by McGookey (1972), identifies many unconformities within the system, but only eight or nine within the different structural basins can be placed in a regional framework. A restored section (Figure 10) is plotted from the western margin of the Cretaceous basin (generally western Wyoming and western Montana) to the geographic center of the basin (eastern Colorado, Black Hills area, and eastern Alberta). Strata are dominantly nonmarine in the western portion of the basin and dominantly marine in the geographic center of the basin.

Unconformities are in three positions: those completely within nonmarine strata such as at the base and top of the Cretaceous; those involving both marine and nonmarine strata; and, those totally within the marine strata, as currently mapped. Wavy lines extending completely across the diagram represent times when the entire basin was subjected to subaerial or submarine erosion. In general, the amount of erosion of underlying strata associated with each

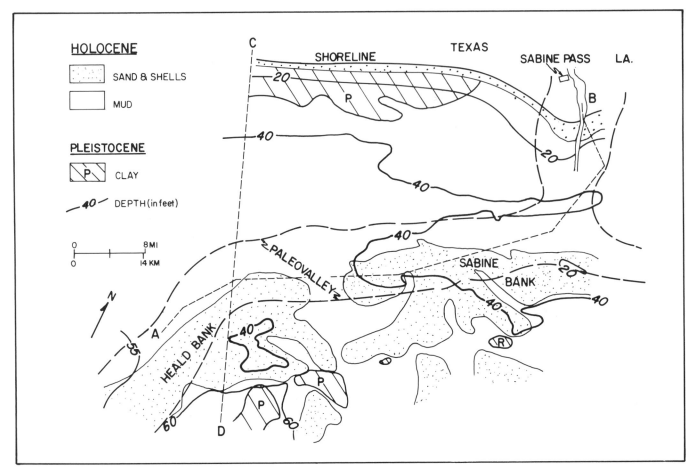

Figure 6—Map of offshore Texas (area shown on Figure 5) with distribution of Holocene and Pleistocene sediments on sea floor. Approximate locations of paleovalleys of Sabine and Calcasieu rivers are shown. After Nelson and Bray (1970).

break is less than one hundred meters. Where regional beveling occurs, the angularity is too small to recognize on a local basis.

On Figure 10 strata are plotted relative to age, and thickness of section is not considered. Uncertainty exists in the dating of many of the unconformities. The time scale and faunal zones of Obradovich and Cobban (1975) and modified by Fouch et al. (1983) were used to date the major unconformities. Their approximate dates (± 1 m.y.), together with associated formations (in parentheses), are estimated as follows: (1) late Neocomian to early Aptian, 112 m.y. (base lower Mannville, Lakota, or Lytle); (2) late Aptian–early Albian, 100 m.y. (upper Mannville, Fall River, Plainview); (3) Albian, 97 m.y. (Viking, Muddy, Newcastle, or J); (4) early Cenomanian, 95 m.y. (lower Frontier–Peay and D); (5) late Turonian, 90 m.y. (base upper Frontier, upper Carlile or Juana Lopez); (6) Coniacian, 89 m.y. (base Niobrara or Fort Hayes); (7) early Santonian, 80 m.y. (Eagle, lower Pierre–upper Niobrara); (8) late Campanian–early Maestrichtian, 73 m.y. (mid-Mesaverde, base Ericson, base Teapot); and (9) late Maestrichtian, 66 m.y. (top Lance or equivalents). The ages of stage boundaries are based on work by Obradovich and Cobban (1975) for the Western Interior Cretaceous, and modified by Lanphere and Jones

(1978) and Fouch (1983). The positions of some of the stage boundaries relative to this radiometric time scale are not in agreement with those published by other workers, for example Van Hinte (1976) or Kauffman (1977). These variations in the age of stage boundaries are plotted on Figure 11.

Many of the above unconformities can be related to sea level changes and to well known regressive and transgressive cycles. However, one major problem is determining if some of the breaks are associated with regional or local tectonics instead of sea level changes. The difficulties in relating transgressions and regressions to sea level changes or tectonics in the Canadian portion of the Cretaceous basin were discussed by Jeletzky (1978).

INFLUENCE OF SEA LEVEL CHANGES ON DEPOSITIONAL SYSTEMS

Changes in sea level have a direct influence on base levels of erosion and deposition, which control sedimentation. The influence varies among major environments of deposition but the most noticeable effect is in nonmarine and shallow-marine environments. Overall, drainages adjust to a lower

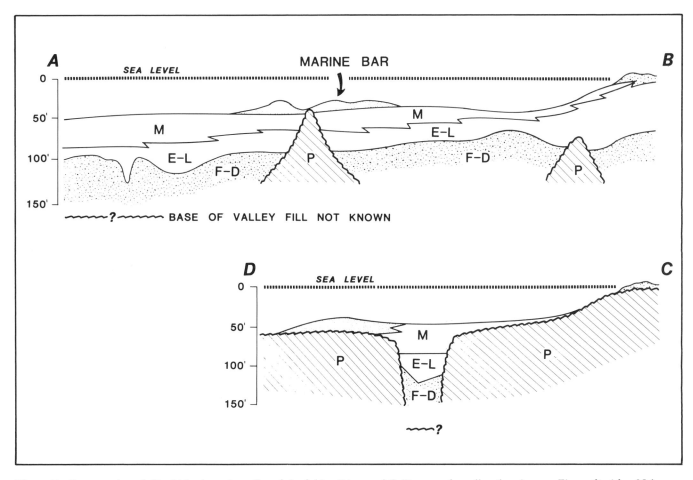

Figure 7—Cross sections A–B within the paleovalley of the Sabine River and C–D across the valley (locations on Figure 6). After Nelson and Bray (1970). M, marine; E-L, estuary-lagoon; F-D, fluvial-deltaic; P, Pleistocene.

base level (lowstand) by incisement (erosion) into underlying strata; when base level rises, streams aggrade and deposition resumes in valleys and marginal areas. Shorelines normally regress during lowstands and transgress during highstands. Depending on the magnitude of change, marine shelf sedimentation may be influenced only slightly, or widespread erosion may occur either in a subaerial or submarine setting. Generally, the deposits that are studied in cratonic basins are associated with highstands, whereas lowstands are represented by breaks that may or may not have been identified, especially in nonmarine strata.

Many of the unconformities in the Cretaceous basin are associated with overall regressive cycles of shoreline movement. The widespread regressive shoreline and shallow marine sandstones may be related to a stillstand or a slow rise or lowering of sea level where a high rate of sediment supply prevails. The Holocene Mississippi River delta complex is an example of a regressive event during a Holocene stillstand, or slightly rising sea level, because of a high sediment input to the basin. The shoreline has prograded seaward approximately 160 km during the last 5000 years. Thus, a shoreline regression in an ancient sequence need not be related to a sea level drop. The best single indicator of an ancient falling sea level is incised

drainage with root zones or paleosoils on the surface of erosion at the base of valley-fill deposits, as summarized in the list of criteria previously given. Other criteria that are useful in establishing eustatic changes are also listed.

Sea level curves for the Cretaceous have been published by Hancock (1975), Kauffman (1977), and Hancock and Kauffman (1979). These authors relate major transgressive and regressive cycles for the western United States with those of north Europe and relate these recorded events to sea level changes. Based on criteria discussed in this paper, a modified sea level curve for the Western Interior has been prepared (Figure 11) and compared with Hancock's curve for north Europe. The comparison of events between continents allows for discrimination of those shoreline movements caused by sea level changes. However, when the unconformities are added, a more accurate dating can be determined for the lowstands.

Transgressions in the Cretaceous are represented by widespread marine shale strata between sandstone units of regressive events, some of which are capped by unconformities (Figure 10). From oldest to youngest these shale or chalk formations are as follows: (1) Clearwater of Canada, an event represented by nonmarine strata in the United States (the Lakota or Lytle); (2) Skull Creek; (3)

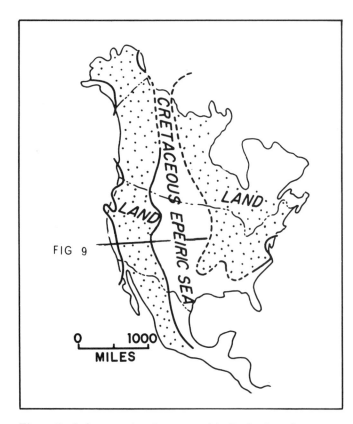

Figure 8—Index map showing geographic distribution of Cretaceous seaway in interior of continent. Location of cross section (Figure 9) with representative tectonic elements is indicated.

Mowry-Huntsman; (4) Greenhorn; (5) upper Carlile; (6) Niobrara; (7) Claggett; and (8) Bearpaw. Because of their wide distribution, these formations have been related to highstand conditions. However, underlying valley-fill deposits, where present, are also related to rising sea level and transgressive events. Because of the unconformities, in many areas regional sedimentation patterns cannot be easily related to symmetric cycles of transgression and regression, although locally this has been done (Kauffman, 1977). Because the breaks are generally on top of the regressive phases, cycles, if present, are asymmetric in favor of the regression event.

TECTONIC INFLUENCE ON SEDIMENTATION

Depositional models proposed for the Cretaceous do not generally consider whether or not syndepositional tectonic movement occurred. Detailed studies in eastern Colorado have clearly established that structural elements had periodic movement on the Cretaceous sea floor (Weimer, 1978), 1980). Major northeast-trending basement fault blocks in the northern Denver basin are mapped as extensions of well-documented Precambrian shear zones observed in Front Range outcrops (Figure 12). Recurrent movement occurred on several of these paleostructures during the Pennsylvanian, Permian, and Cretaceous (Sonnenberg and Weimer, 1981).

The shear zones, Precambrian in age, are "weak rock" and bound rigid fault blocks. At times during the Phanerozoic when the crust was highly stressed, the stress was relieved primarily by movement along these preexisting lines of weakness. Cretaceous strata clearly show that fault movement was sporadic, affecting some layers but not others. The strata which record movement are referred to as "tectonically sensitive intervals." They are the keys to reconstructing the size and distribution of paleotectonic elements. Recurrent movements on the basement fault blocks are normally in the same direction, but important reversals in movement (structural inversions) along faults have been recorded.

Fault block boundaries may be recognized in the sedimentary sequence overlying the basement by direct offset of sedimentary layers, abrupt change in strike and dip related to drape folding, and closely spaced fractures in competent layers. Strata overlying major basement blocks generally have uniform dip over the extent of the block. Different sized blocks are referred to as first, second, or third order features.

One of the most important premises in establishing a tectonic control on sedimentation is that fault block movement controls topography and bathymetry (referred to as structural topography). We can make the following generalities concerning sedimentation. In continental deposits rivers flow on topographic lows, whereas interchannel areas generally occur over higher structural areas. However, because of the leveeing process associated with channels, interchannel deposits may be deposited in areas topographically lower than the channel.

In shoreline deposits with a high sediment influx, deltas develop in structural and topographic low areas and inter-deltaic deposits occur over and around the more positive structural blocks. When sediment influx is low the structural and topographic low areas (deltas) may become estuaries. In marine deposits, topographically high blocks may be shoal areas and thus control the distribution of sand bodies or reefs. Moreover, in deeper water deposits, sand turbidites (both calciclastic and siliciclastic) are deposited in bathymetrically low areas, which may coincide with downthrown blocks. In summary, thin successions are associated with paleohighs and thick successions are associated with paleolows, but sand deposits may occur in either setting depending upon the depositional environment.

Depositional topography may have developed within the Cretaceous basin because of thickness variations in units related to rates of sedimentation (Asquith, 1970). Thick and thin sediment accumulations, associated with depositional topography, can be confused with thickness patterns related to tectonics, and a careful analysis of depositional environments, processes, and subtle breaks is needed to determine what controlled thickness variation.

Tectonic movements can influence the accuracy of time correlations. Older units can be elevated to the same or a higher stratigraphic level than younger strata. An example of this relation is demonstrated in the discussion of the J Sandstone.

By applying the above concepts, the cause of an unconformity can be evaluated—tectonic, sea level change,

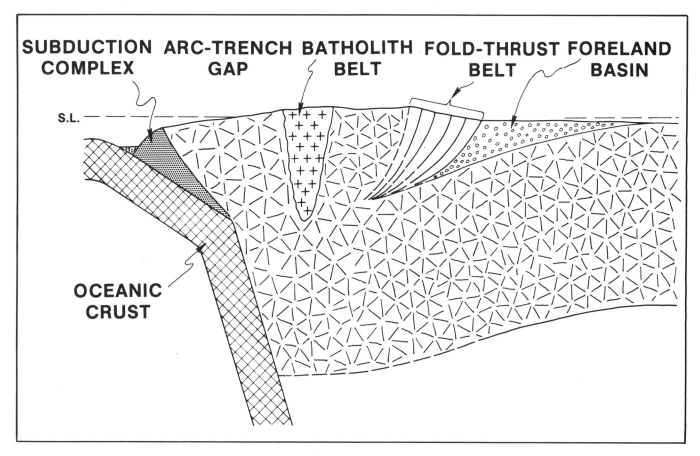

Figure 9—Schematic diagram to illustrate tectonic elements across western United States during the Cretaceous. S.L., sea level. After Dickinson (1976).

or a combination of both. Depositional models constructed for all ancient depositional basins should consider these factors to reconstruct accurately the recorded geologic events.

EXAMPLES OF UNCONFORMITIES ASSOCIATED WITH CRETACEOUS SEA LEVEL CHANGES

Although field, paleontologic, and subsurface data are available to support all of the unconformities in the Cretaceous shown on Figure 10, only three breaks, regarded as typical, are described in the following discussion. The 97 m.y. and 90 m.y. unconformities can be related to events on the continental margins and are, therefore, interregional in nature. The 80 m.y. break is now best defined within the marine basin deposits but probably extends to the margin of the basin and elsewhere on the continent.

BASINWIDE INCISEMENT OF DRAINAGE DURING SEA LEVEL CHANGE (97 M.Y. AGO)

An important transgressive–regressive–transgressive sequence is recorded in the Albian strata of the Western Interior Cretaceous basin. A widespread marine shale, mapped throughout the basin, is known in different areas as the Skull Creek, Kiowa, Thermopolis, and Joli Fou (Canada) shales (Figures 10 and 13). Equivalent strata in western Wyoming in the basin margin area are generally included in the Dakota Group or Bear River Formation. The shale deposits, which accumulated during a highstand of the Albian Sea, are correlated over large areas, either by contained faunas or by stratal continuity. The shales, generally 30 to 60 m thick, represent the first widespread transgression of the Cretaceous sea into the United States portion of the Western Interior basin. Overlying regressive sandstone units named the Muddy, J, or Viking (Canada) sandstones are widespread and productive of petroleum in stratigraphic or structural traps. Generally less than 30 m thick, these sandstones were deposited in a range of environments from freshwater to marine. They are generally regarded as deposits related to a lowering of sea level. The following transgression is recorded by the widespread marine Mowry Formation and other highstand deposits. When the history of these strata is related to radiometric dates from associated bentonite beds, the sequence spans the time interval of approximately 96 to 98 m.y. ago. The major event correlates with the worldwide sea level drop 97 m.y. ago, reported by Vail et al. (1977) and Hancock (1975).

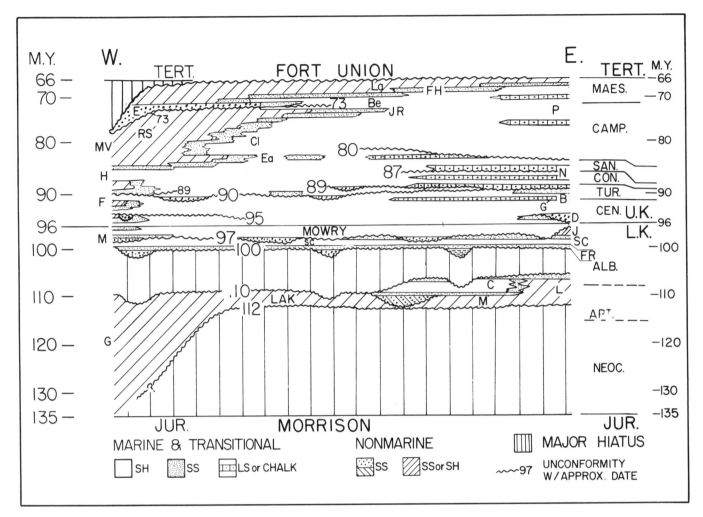

Figure 10—Diagrammatic east–west section across Cretaceous basin showing stratigraphic position and approximate dates of major intrabasin unconformities (modified after McGookey, 1972). Formations or groups to the west are G, Gannett; SC, Skull Creek; M, Mowry; F, Frontier; H, Hilliard; MV, Mesaverde; RS, Rock Springs; E, Ericson; Ea, Eagle; Cl, Claggett; JR, Judith River; Be, Bearpaw; FH, Fox Hills; La, Lance. To the east, formations are L, Lytle; LAK, Lakota; FR, Fall River; SC, Skull Creek; J and D, Sandstones of Denver basin; G, Greenhorn; B, Benton; N, Niobrara; P, Pierre; M and C, McMurray and Clearwater of Canada. The vertical ruled lines represent unconformities where a major hiatus is recognized. When the gap is a million years or less, vertical ruling is omitted.

The events described above are known largely from detailed stratigraphic studies in the Powder River basin, Wyoming, and the Denver basin, Colorado, Wyoming, and Nebraska. Only the Denver basin work is summarized in this paper but literature describing the Powder River basin is extensive (Weimer, 1982).

Denver Basin

Petroleum was discovered in upper sandstones (D and J) of the Dakota Group in the Denver basin in 1923. Since the discovery of the Wattenberg gas field in 1970, more than 1000 wells have been drilled to the J Sandstone in an area covering 600,000 acres across the deeper portion of the Denver basin (Figure 14). Production is about 24 km east of outcrop sections which are described by MacKenzie (1971), Clark (1978), and Suryanto (1979). Details of the Wattenberg field have been published by Matuszczak (1976) and Peterson and Janes (1978). Because of new data from the outcrop and subsurface, a reinterpretation of the depositional model for the J Sandstone is now possible. An

important new element is the recognition of the Wattenberg area as a paleostructure that influenced J Sandstone deposition (Weimer, 1980; Weimer and Sonnenberg, 1982).

The J Sandstone (Muddy) is present in outcrop sections from the Wyoming state line to the South Platte River southwest of Denver (Figure 14). Two types of sandstone bodies comprising the J Sandstone in the Denver basin were described by MacKenzie (1965) from outcrops in the Fort Collins area (Figure 15). One type, named the Fort Collins Member by MacKenzie, is a very fine-grained to fine-grained sandstone containing numerous marine trace fossils and is interpreted to be delta front, shoreline, and marine bar sandstone that was deposited during rapid regression of the shoreline of the Skull Creek sea. A second type is fine- to medium-grained, well-sorted, cross-stratified sandstone (channels) containing carbonized wood fragments and associated shales and siltstones. These lithologies were named the Horsetooth Member by MacKenzie. Valleys of an extensive drainage system were incised into the Fort Collins Member or the underlying Skull Creek Shale and contain a

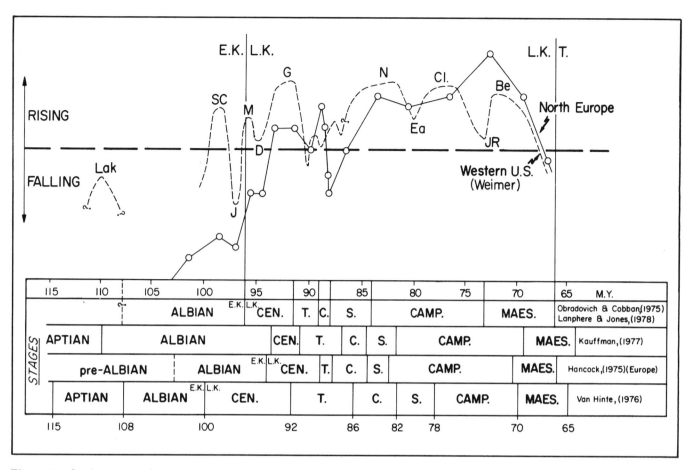

Figure 11—Sea level curves for the United States and Europe. Letters designate the same formations as on Figure 10. Abbreviations for stages: Cen., Cenomanian; T., Turonian; C., Coniacian; S., Santonian; Camp., Campanian; Maes., Maestrichtian; E.K., Early Cretaceous; L.K., Late Cretaceous; T., Tertiary.

fill of sandstone, siltstone, and claystone varying in origin from alluvial plain to shoreline deposits (Horsetooth Member). The J varies in thickness from 6 to 46 m (Figures 15 and 16).

The thick dominantly fresh to brackish water J Sandstone facies in the Golden area was interpreted by Waage (1953), Haun (1963), MacKenzie (1971), and Matuszczak (1976) to be laterally equivalent to the thin marine sandstone facies north of Boulder. This interpretation led to the concept of a northwest-trending marine basin in the area between Boulder and Fort Collins.

The interpretation shown on Figure 15 correlates the Golden area sections (Weimer and Land, 1972) with the Horsetooth Member. The Kassler Sandstone, the lower unit of the J, where present, rests on an erosional surface cut into the Skull Creek Shale. Sandstone of the Fort Collins Member is interpreted as having been removed by erosion prior to deposition of the Kassler. Root zones are found in the Kassler only a few feet above the base. In addition, conglomeratic sandstones with chert pebbles up to 1 cm in diameter are sporadically present in the Kassler (Poleschook, 1978). Thus, lower J Sandstone is interpreted as a valley-fill complex of a major drainage system. Cross strata in the Kassler Sandstone (lower J) indicate a dominantly southeast transport direction (Poleschook, 1978; Lindstrom, 1979) (Figure 17). The drainage patterns are

interpreted as tributary rather than distributary as previously described.

A north–south electric log section east of the outcrop across the Wattenberg field shows a similar interpretation of the J Sandstone (Figure 16). The widespread gas-bearing sandstone at Wattenberg is the Fort Collins Member of the J, which is transitional with the underlying Skull Creek. Major channel sandstones of the Horsetooth Member are present to the north and south of Wattenberg (Figures 16 and 17). In the Third Creek field (sec. 19, T. 2 S., R. 67 W.) (Figure 16), root zones are preserved below the J channel sandstone and above the erosional surface on top of the marine Fort Collins Member. Based on core interpretation of facies, at least 10 m of the Fort Collins Member was eroded prior to deposition of the fluvial channel sandstone that is oil-productive. Previous interpretations have shown the channel sandstone to be the lateral equivalent of the Fort Collins Member at Wattenberg. Since the channel sandstone can be demonstrated to be younger than the erosional surface cut on the Fort Collins Member, such a facies interpretation is in error.

Tectonics and Sedimentation Model for J Sandstone, Denver Basin

The following model for tectonic influence on J Sandstone sedimentation is proposed for the Denver basin. At the end

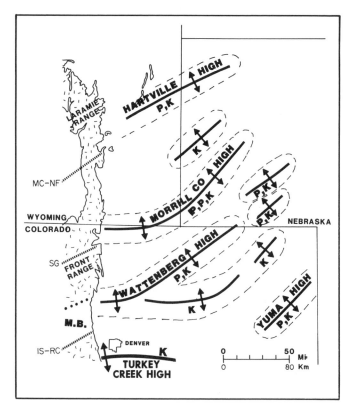

Figure 12—Summary diagram of east-northeast trending paleostructures (basement fault blocks) in northern Denver basin. Times of dominant movement are IP, Pennsylvanian; P, Permian; K, Cretaceous. Outcrop of Precambrian with major shear zones along left side of diagram. MC–NF, Mullen Creek–Nash Fork; SG, Skin Gulch; IS–RC, Idaho Springs–Ralston Creek; M.B., Colorado mineral belt.

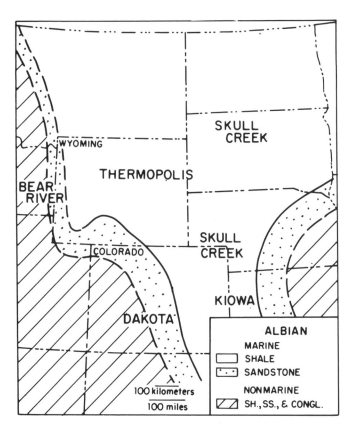

Figure 13—Outline of Cretaceous seaway during Albian time.

of Skull Creek deposition (T_1, Figure 18), a regressive event began that deposited shoreline and shallow-marine sandstones with a transitional contact with underlying Skull Creek Shale. Depositional patterns over basement fault blocks, where slight fault block movement influenced sedimentation, are illustrated. Rivers and associated deltas positioned themselves in structural and topographically low areas, grabens, whereas delta margin or interdeltaic sedimentation occurred along an embayed coast over structural horst blocks. Delta front and shoreface sands extended seaward from the shoreline to a distance controlled by effective wave base. The shoreline prograded seaward to position T_2 and a sheetlike sand body was deposited over a large area (Wattenberg pay sandstone or Fort Collins Member).

A drop in sea level occurred (T_3) during which all, or a large portion, of the depositional basin (Skull Creek seaway) was drained (Figure 19). River drainages were incised into marine shales and/or the regressive shoreline sandstones in topographic lows which generally correspond to the graben fault block area. Over much of the Denver basin the base of the incisement is on the Fort Collins Member (T_1 or T_2 sand complex). Only locally did the erosional surface cut into the Skull Creek Shale.

A rise in sea level occurred during which the incised valleys were probably modified and filled with fluvial and

estuarine sandstone, siltstone, and shale. Vertically, the valley fill has a wide variety of fluvial environments in the lowermost part and estuarine or deltaic environments in the upper part (Figure 20). With a rising sea level the earliest fluvial deposits were deposited in narrow valleys as upper meander belt sandstones. As sea level increased, the lower meander belt environments shifted landward and channel meandering widened the valley by scour. These sandstones may overlie either the upper meander belt sandstone or the Fort Collins Member. The final deposits of the J are transitional estuarine, deltaic, or shoreline deposits. After the valleys filled, deposition covered the interstream divide areas. With a continued rise in sea level (T_4, Figure 21) and minor renewed fault block movement, strata were eroded from the top of the horst blocks and an extensive thin transgressive-lag deposit of conglomeratic or coarser grained sandstone formed over the horst blocks on a surface of erosion. Following T_4 the entire region received marine siltstone and shale (Mowry or Graneros shales).

In the above model, an important unconformity separates T_1 and T_2 deposits from T_4 deposits. This basinwide unconformity (T_3, Figure 19) may be within sandstone deposits (i.e., valley-fill sandstones rest on older regressive sandstones), or between sandstone and marine shale deposits (i.e., valley-fill sandstones rest on Skull Creek Shale). In portions of the Wattenberg field, the

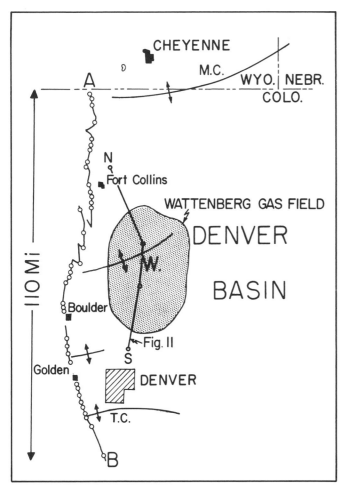

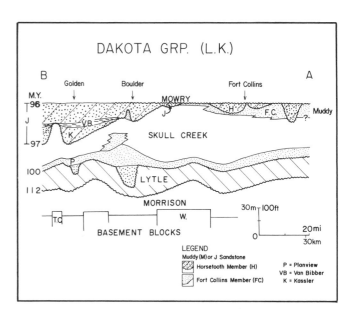

Figure 15—Stratigraphic restored section A–B through the Dakota Group (Lower Cretaceous) (location on Figure 14). Modified after MacKenzie (1971).

Figure 14—Map of Denver basin showing outcrop sections of Dakota Group and paleotectonic elements: M.C., Morrill Co. High; W., Wattenberg; T.C., Turkey Creek. Locations of measured surface sections indicated by O.

unconformity is at the base of the Mowry Shale or top of the regressive sandstone.

Previous correlations which show the J Sandstone to be deposited across the basin during one major regressive event need to be modified. Sandstones above the unconformity (Horsetooth Member) are younger than the regressive sandstone at the top of the Skull Creek (Fort Collins Member), although because of tectonic movement the older sandstones are now at a stratigraphic high position (Figures 15 and 16). This model has important implications for future petroleum exploration in the Denver basin.

A relation exists between major northeast-trending Precambrian shear zones mapped in the Front Range and paleostructure in the northern Denver basin (Figure 12) (Sonnenberg and Weimer, 1981). Recurrent movement on these old fault zones has controlled thickness variations and depositional facies in Paleozoic and Mesozoic strata. Five major east-northeast-trending paleostructures occur in the northern Denver basin which had recurrent movement during the Cretaceous and some have documented

movement during the Permian. The major paleostructures are the Wattenberg high, the Morrill County high, the Hartville high, the Turkey Creek high, and the Yuma high (Figure 12). These paleohighs vary in width from 32 to 40 km and in length from 80 to 290 km. Several important northwest- and north-trending paleostructures are omitted from Figure 12 (Sonnenberg and Weimer, 1981). Each structural paleohigh should be investigated to determine if recurrent movement influenced sedimentation as documented in the Wattenberg area.

Recurrent movement on paleostructural elements affected the Cretaceous seaway in a broader sense. The five paleostructural elements (Figure 12) collectively have been grouped together as a broad structural arch referred to as the Transcontinental arch (Weimer, 1978). Structural movement on this broad arch during the time of J or Muddy Sandstone deposition created a topographic high which divided the drainage in the Cretaceous basin during the low sea level stand (T₃; Figure 19). A general south-flowing drainage developed in southwest Wyoming and eastern Colorado (Figure 22), whereas a north-flowing system developed in northern Wyoming (Powder River basin; Weimer et al., 1982).

EROSION OF SHELF STRATA DURING SEA LEVEL CHANGE (90 M.Y. AGO)

Unconformities within marine strata have been described in the Cretaceous by Reeside (1944, 1957), McGookey (1972), and Merewether and Cobban (1972, 1973), Merewether et al. (1976), and Merewether et al. (1979). The stratigraphic positions of the documented unconformities are shown on Figure 10. Uncertainty exists as to the cause of

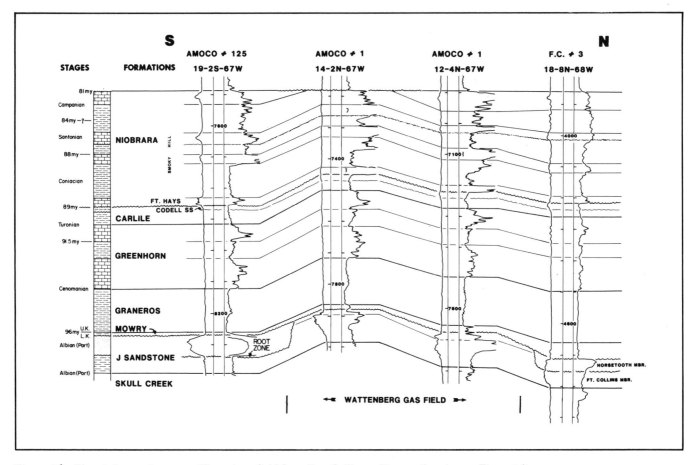

Figure 16—Electric log section across Wattenberg field from Fort Collins to Denver (location on Figure 14).

the marine unconformities: Are they the result of sea level change with subaerial or submarine erosion? Do they result from submarine erosion associated with tectonic movement, nondeposition, or a combination of these processes? To what extent is the geographic distribution of the unconformities influenced by depositional topography, which in turn is controlled by the interplay of rates of sedimentation and oceanic processes?

Model for Depositional Topography

Depositional topography commonly developed in ancient basins which had a high input of terrigenous sediment. Asquith (1970) described depositional topography in the Western Interior Cretaceous basin for Campanian and Maestrichtian strata in Wyoming. A modified model from Asquith for highstand sea level depositional topography includes environments of deposition related to coastal plain, shoreline, shallow marine/shelf, slope, and basin (Figure 23A). Because of high rate of sedimentation and lateral accretion (progradation), higher than normal depositional dip formed in two areas: the delta front–prodelta, and the slope. The primary dips (clinoforms) in these areas of dominantly clay and silt deposition are generally less than 1°, as determined from closely spaced well log correlations (Figure 23B). Because of their low dip, the clinoforms are not generally recognized on seismic sections or on outcrop.

The slowest rates of sedimentation existed in deeper water areas (basin, Figure 23A) where sedimentation of organic rich chalk and clay, related to pelagic sedimentation, was dominant. Water depths are difficult to reconstruct but, in general, shelf depths are estimated to vary from 30 to 90 m. Depths for chalk sedimentation in the basin are controversial but estimates range from 60 to 490 m (Kent, 1967; Eicher, 1969; Hattin, 1975a; Kauffman, 1977). On the basis of the physical evidence for the shelf, slope, and basin model (Figure 25), I favor basin water depths in the range of 180 to 300 m during sea level highstands.

Water depths greatly influenced sedimentation or erosion during sea level changes. During lowstand events erosion by wave energy or by subaerial processes may have removed shelf and slope deposits, depending on the magnitude of the sea level drop (Figure 23C). Sand normally confined to the coastal plain and shoreline areas was transported to the shelf area. During a subsequent sea level rise, the sand may have been reworked into marine shelf complexes, either as narrow linear marine bars, as thin transgressive sheet sands, or as a final fill of incised drainages. The shelf deposits of the Gulf of Mexico illustrate these types of sand bodies and were discussed previously (Figures 5, 6, and 7).

Although the above model applies generally to interpreting all strata in the Cretaceous basin, an example illustrating the components in the marine phase of the

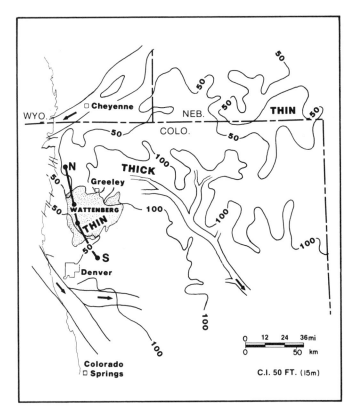

Figure 17—Isopach map of J Sandstone (includes Fort Collins and Horsetooth Members) with location of major incised valleys as indicated by lower J (equivalent of Kassler Sandstone of outcrop on Figure 15). Stippled pattern is area of gas production from J Sandstone. Modified from Haun (1963) and Matuszczak (1976).

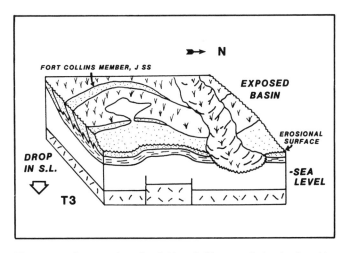

Figure 19—Lowstand sea level (time 3, T_3) recorded as basin-wide erosional surface resulting from subaerial exposure. Root zones form on exposed marine shales and sandstones.

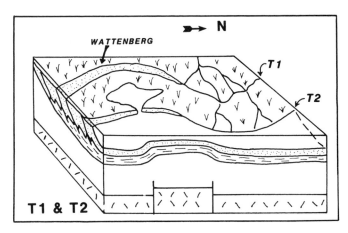

Figure 18—Depositional and tectonic model for Fort Collins Member of J Sandstone showing highstand regression over basement fault blocks (Wattenberg high) with penecontemporaneous movement. T_1, time 1; T_2, time 2. Rate of sediment supply exceeds rate of subsidence or submergence.

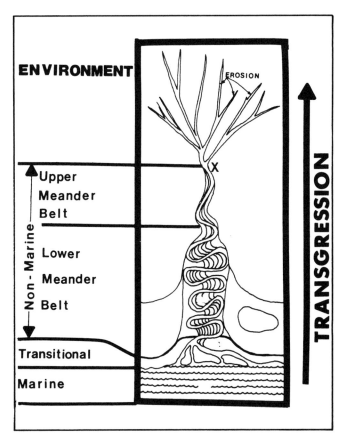

Figure 20—Environments of deposition in an alluvial valley and shoreline setting. X marks reference point for vertical changes in lithology in valley-fill deposits as sea level rises and environments shift landward during transgression.

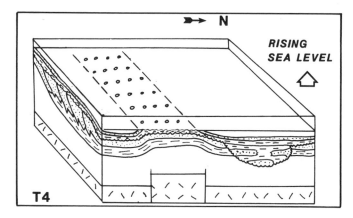

Figure 21—Rising sea level during Time 4 (T₄) with fill of incised valley and deposition of marine shale and sandstones. A thin transgressive lag (generally less than 0.3 m thick) of conglomeratic sandstone (indicated by circles) occurs in association with basement-controlled horst block.

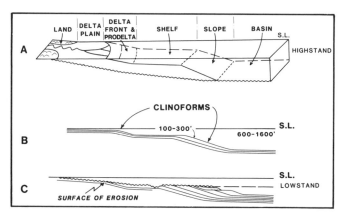

Figure 23—Cross section showing depositional topography in Cretaceous basin during highstand of sea level (A). Areas of depositional dip (clinoforms) develop by lateral accretion during deposition (B). Erosional surface develops across shelf during lowstand of sea. S.L., sea level; numbers give water depth in feet.

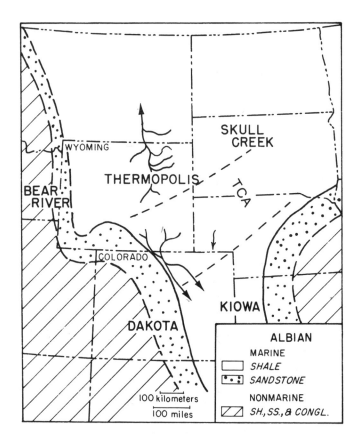

Figure 22—Map showing shoreline trends during high stand of sea level during Skull Creek deposition with direction of flow of incised drainage during subsequent sea level drop (lines within marine area). A drainage divide developed across the basin in the area of the Transcontinental arch (TCA).

model has been developed for Cenomanian and Turonian strata in the Denver basin and adjacent areas. A stratigraphic section from near Hays, Kansas, to central Wyoming incorporates surface and subsurface data (Figures 24 and 25). Ages of formations at each end of the section are based on ammonite correlations. Elsewhere subsurface correlations are by stratal continuity of bentonite layers in shale units, and thin limestones (chalks) identified in thousands of well logs in the Denver basin. Cobban and associates have developed faunal zones (Table 1) for the Cenomanian, Turonian, and Coniacian stages (Merewether et al., 1979). These faunal zones have been related to lithologic markers and traced over a large area as time-stratigraphic units (Weimer, 1978, 1983; Sonnenberg and Weimer, 1981). Shelf sedimentation for the Cenomanian and Turonian is well illustrated by the interbedded sandstones and shales of the Frontier Formation (Figure 25) in central Wyoming (Merewether et al., 1979). They estimate water depths on the shelf of less than 130 m. Synchronous basin sedimentation is represented by chalks of the Greenhorn and Fairport of Kansas (Figure 25). A major unconformity associated with shelf sedimentation (90 m.y. lowstand) is present within the Carlile and Frontier formations.

Carlile Formation

The Carlile Formation in the Great Plains province consists of four widespread members which in ascending order are the Fairport chalk, the Blue Hill Shale, the Codell Sandstone, and the Juana Lopez. A fifth member, the Sage Breaks Shale, is present as the upper Carlile in the northern Denver basin and Powder River basin area. The Juana Lopez member is a lenticular limestone unit that is locally conspicuous in outcrop along the west flank of the Denver basin but is generally too thin (less than 1 m) to map in the subsurface. The unit is ubiquitous in outcrop along the southern margin of the Denver basin and adjacent areas to the south. The unit is absent in outcrop (Figure 25) over most of central Kansas (Hattin, 1975b).

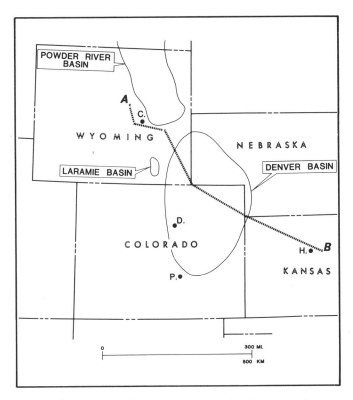

Figure 24—Index map for stratigraphic section from central Wyoming to central Kansas (Figure 25). C, Casper; D, Denver; P, Pueblo; H, Hays.

The Carlile contains one or more unconformities in the upper portion which play a significant role in the thickness and depositional patterns. The thickness of the Carlile varies from less than 15 m to more than 60 m (Figures 25 and 26). The Fairport and Blue Hill members in Kansas are progressively cut out westward by an unconformity at the base of the Codell Sandstone, or in its absence the unconformity at the base of the Fort Hays Limestone (Figure 25). A similar relationship was observed in north–south correlations from the more complete Pueblo, Colorado, section (similar to central Kansas) with the strata in the northern Denver basin (Weimer, 1978).

If a model for a shelf, slope, and basin is used for deposition of the Carlile (Figure 23), the associated regional unconformities can be easily explained by a combination of eustatic changes and tectonic movement. Several important stratigraphic relations in the area from central Wyoming to Kansas support this model.

The Fairport chalk of Kansas (basin deposit) thickens westward and changes across the Denver basin to siltstone and shale with the well-developed clinoforms of slope deposits (Figure 25), especially in southeastern Wyoming and western Nebraska.

The Codell Sandstone in portions of central Kansas and in the Pueblo, Colorado, area has a transitional contact with the underlying Blue Hill Shale (Hattin, 1962; Merriam, 1957, 1963; Pinel, 1977). However, over much of the Denver basin the Codell is sporadically developed, and where present, an unconformity is observed at the base, which has

a hiatus increasing in magnitude to the west (Figure 25). A subcrop map of formations beneath the surface of erosion illustrates these relationships (Figure 26). In the northern Front Range area and the Laramie basin, the Codell Sandstone rests on strata equivalent to the middle Greenhorn (faunal zone 8 or 9).

The regional subcrop pattern of the Fairport and Greenhorn beneath the unconformity shows a broad eastward bulge in southern Wyoming and northern Colorado (Figure 27). This feature is related to a structural doming with the beveling of faunal zones 9 through 13 from central Wyoming to Kansas. The regional distribution of regressive sandstone during faunal zone 14 (*Prionocyclus hyatti*) suggests a major sea level drop of short time duration (Table 1). The scour and fill pattern in central Wyoming (Figure 25), showing remnants of faunal zones 12 and 13 (Fairport equivalents) on faunal zone 8 (lower Greenhorn equivalents), suggests movement of local tectonic elements superimposed on the broad doming. Structural movement with erosion started during or after the deposition of faunal zones 12 and 13 (the unnamed member of the Frontier Formation of Merewether et al., 1979). Units 12 and 13 were either deposited over the entire area and subsequently removed by erosion, or they were deposited only in topographic (structural) low areas. Regional correlations suggest that sea level remained high during faunal zones 9 through 13 (Table 1) so the erosion is thought to be related to structural movement and submarine scour. The clinoform pattern of slope and basin deposits in the upper Greenhorn and Fairport formations (Figure 25) may be the result of rapid deposition because of sediment recycling by erosion on the top and deposition on the margin of the broad structural element.

Erosion by subaerial or submarine processes on the shelf during a sea level drop (faunal zone 14) removed strata and erosional depressions (valleys?) were cut into shelf deposits as base level was lowered. Chert pebbles and fine- to coarse-grained sand were transported across the shelf by streams and currents. These scours are observed mainly in the northernmost Denver basin.

With the subsequent rise in sea level (the time of either late faunal zone 14 or zone 15) three types of sandstone were deposited above the surface of erosion: (1) thick sand deposits of fresh or brackish water origin accumulated in the scour depressions, (2) thin widespread fine-grained bioturbated shelf sand, and (3) coarsening upward fine- to medium-grained sand reworked into marine bar complexes. These sands are best preserved in the northern Denver basin. These types of sand occurrences, deposited during a changing sea level, are believed similar to those observed on the modern shelf of the Gulf of Mexico (Figures 5, 6, and 7) as described by Nelson and Bray (1970).

In the southern Denver basin and Kansas, where sedimentation was continuous from the underlying Blue Hill through the Codell, the sand was deposited either as a shallow marine bar sequence or along a regressing shoreline during the lowstand of sea level. These sands are slightly older than the sands above the surface of erosion in the central and northern Denver basin (Figure 25).

The Juana Lopez in Colorado, a thin relict or palimpsest deposit, rests on an erosional surface with chert and

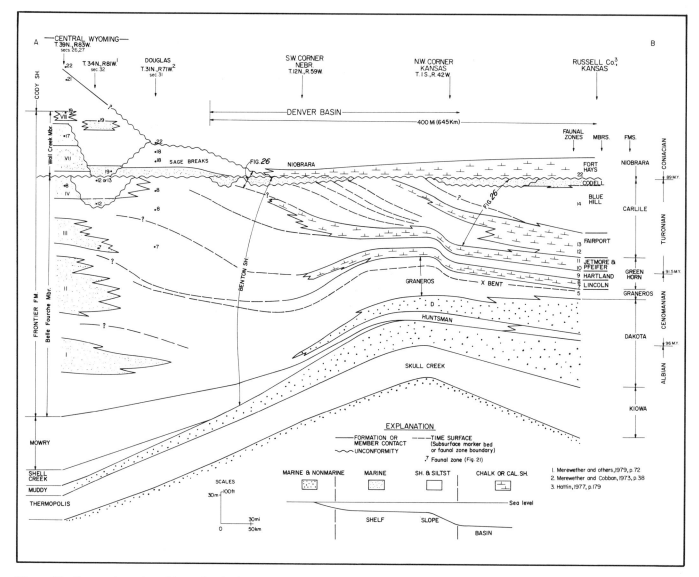

Figure 25—Restored stratigraphic section for lower part of Cretaceous from central Wyoming to central Kansas. Unconformities associated with J and D sandstones are not shown.

phosphate pebbles and coarse sand occurring as a lag in the lower portion. Bioclastic recrystallized limestone with shark teeth occurs in the upper portion. This unit, where well developed (for example, southern Colorado and northern New Mexico), contains faunal zones 15, 16, and 17 (Kauffman, 1977; Hook and Cobban, 1979). The Frontier units 6 and 7 (Wall Creek Member) (Figure 25) record this highstand in central Wyoming.

Erosion in central Wyoming described by Merewether et al. (1979) developed during or near the end of faunal zone 18 and removed sandstone and shale containing zones 16 and 17. The scour was filled by marine deposits containing faunal zone 19 (Figure 25). These relations are interpreted to result from tectonic movement and submarine scour similar to those previously described in the same area for faunal zones 8 through 13. However, an alternative

interpretation is for erosion to have occurred on the shelf during minor sea level drops at the times of faunal zones 10 or 11 and 18 or 19. Some regional evidence supports a possible drop during 18 or 19.

Over much of the central Denver basin and eastward into Kansas (Figure 25) a widespread surface of erosion is also present above the Codell Sandstone with the upper part of the Fort Hays Member of the Niobrara Formation (either faunal zone 21 or 22) resting on the surface of erosion. This unconformity has been related by Weimer (1978) to result from erosion and then marine onlap on a broad northeast-trending structural element called the Transcontinental arch. Sparse faunal evidence suggests that the hiatus represents the time of faunal zones 15 through 21 or 22. The erosion may have been associated with the possible sea level drops during faunal zones 18 or 19.

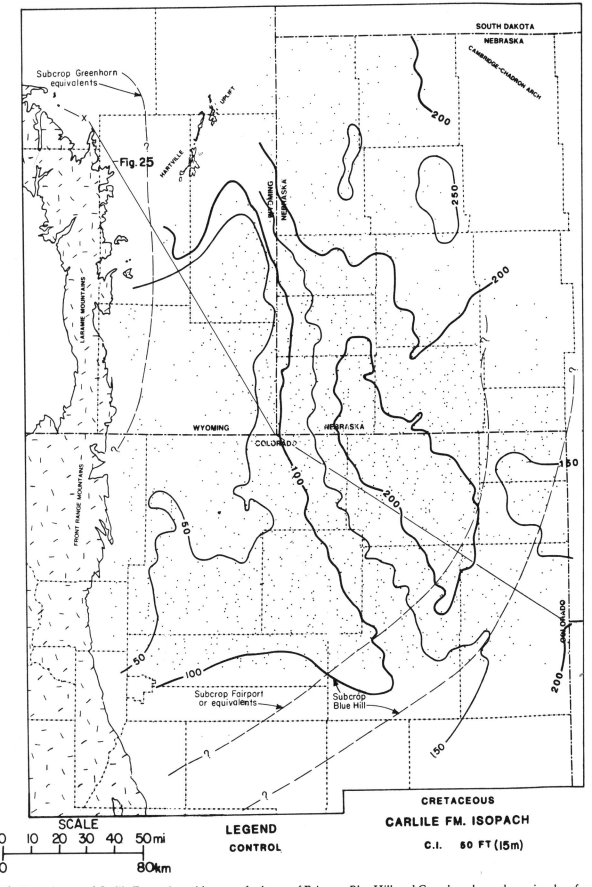

Figure 26—Isopach map of Carlile Formation with areas of subcrop of Fairport, Blue Hill, and Greenhorn beneath erosional surface at base of Codell Sandstone or Fort Hays Limestone (see Figure 25).

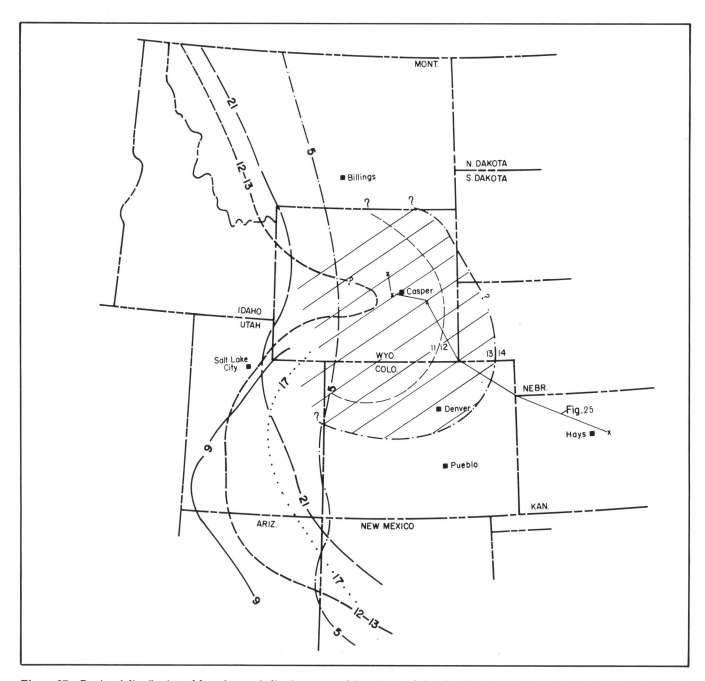

Figure 27—Regional distribution of faunal zones indicating geographic position of shoreline during highstands of sea level. Faunal zones 5 and 12-13, after Cobban and Hook (1979, 1980); faunal zones 9, 17, and 21, after Hook and Cobban (1977, 1979, 1980). Diagonal ruling indicates area of erosion because of structural movement and a low stand of sea level (faunal zone 14).

The approximate positions of the western shoreline of the basin for highstands during faunal zones 5, 9, 12, 13, 17, and 21 (Figure 27) have been mapped by Hook and Cobban (1977, 1979, 1981a) and Cobban and Hook (1979, 1981). Because of the lack of detailed faunal data on a regional basis, the pattern of deposition and erosion is preliminary in nature and no attempt has been made to change shoreline trends. Some of these shorelines must be significantly modified because of the erosion of faunal zones from the paleohigh (Figure 25), especially for faunal zones 9 through 14.

EXAMPLE OF TECTONIC MOVEMENT AND SEA LEVEL CHANGES AFFECTING BASIN DEPOSITS (80–81 M.Y. AGO)

Niobrara Formation

Following the late Turonian (Carlile and equivalents of previous section) depositional patterns changed significantly. During the Coniacian, the Niobrara Formation (composed of chalk and organic rich shale deposition, similar in lithology to the Greenhorn) was deposited over eastern Colorado, Kansas, and adjacent areas (Figures 10 and

Table 1—Faunal zones in lower part of Upper Cretaceous. After Merewether et al. (1979). Dots are faunal zones with radiometric dates. After Obradovich and Cobban (1975); modified by Fouch (1983).

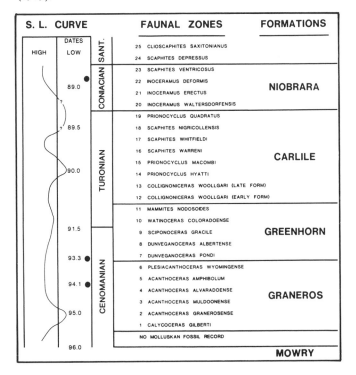

S. L. CURVE			FAUNAL ZONES	FORMATIONS
HIGH	DATES LOW	SANT.	25 CLIOSCAPHITES SAXITONIANUS	
		CONIACIAN	24 SCAPHITES DEPRESSUS	
	89.0 ●		23 SCAPHITES VENTRICOSUS	NIOBRARA
			22 INOCERAMUS DEFORMIS	
			21 INOCERAMUS ERECTUS	
	? 89.5		20 INOCERAMUS WALTERSDORFENSIS	
		TURONIAN	19 PRIONOCYCLUS QUADRATUS	
			18 SCAPHITES NIGRICOLLENSIS	
			17 SCAPHITES WHITFIELDI	
	90.0		16 SCAPHITES WARRENI	CARLILE
			15 PRIONOCYCLUS MACOMBI	
			14 PRIONOCYCLUS HYATTI	
			13 COLLIGNONICERAS WOOLLGARI (LATE FORM)	
			12 COLLIGNONICERAS WOOLLGARI (EARLY FORM)	
			11 MAMMITES NODOSOIDES	
	91.5		10 WATINOCERAS COLORADOENSE	
			9 SCIPONOCERAS GRACILE	GREENHORN
			8 DUNVEGANOCERAS ALBERTENSE	
	93.3 ●	CENOMANIAN	7 DUNVEGANOCERAS PONDI	
			6 PLESIACANTHOCERAS WYOMINGENSE	
	94.1 ●		5 ACANTHOCERAS AMPHIBOLUM	
			4 ACANTHOCERAS ALVARADOENSE	GRANEROS
	95.0		3 ACANTHOCERAS MULDOONENSE	
			2 ACANTHOCERAS GRANEROSENSE	
			1 CALYCOCERAS GILBERTI	
	96.0		NO MOLLUSKAN FOSSIL RECORD	
				MOWRY

28). This lithologic change from Carlile to Niobrara was the result of a significant increase in water depth as the shoreline zone shifted to western Wyoming and Utah (highstand 21, Figure 27). The shelf–slope break shifted westward in a corresponding manner to central Wyoming (Figure 29). Thus, the northern Denver basin, where shelf and slope deposits were dominant in the Carlile, became an area of chalk and shale deposits of the basin setting.

Whereas thickness changes associated with the Greenhorn chalks follow a regional pattern (Sonnenberg and Weimer, 1981), the Niobrara shows many thickness anomalies of a more local distribution (Figure 28). The Niobrara Formation in the northern Denver basin varies in thickness from 75 m to more than 150 m. The formation or equivalent strata are thicker to the south in the Pueblo area and to the north in the Powder River basin (Figure 28).

Within the Niobrara, four limestone intervals and three intervening shale intervals occur regionally and are easily recognized on geophysical logs (Figure 16). The lower limestone is named the Fort Hays and the overlying units are grouped together as the Smoky Hill Member (Scott and Cobban, 1964). The lower boundary is an erosional surface where the limestone is in contact with either Carlile Shale or, in some places, strata as old as the Greenhorn Formation. The upper boundary of the Niobrara is placed at the contact with the noncalcareous black shales of the Pierre Shale. Locally, the contact is an erosional surface; elsewhere the contact is transitional.

Regionally, the northern Denver basin has a thin Niobrara section that is related to structural movement on the Transcontinental arch (Weimer, 1978) and depositional thinning because of slow rates of deposition. Superimposed on the broad pattern are the major northeast–southwest axes of thinning (Figure 28), which are related to second-order paleotectonic elements (Sonnenberg and Weimer, 1981). Thinning over the paleohighs results from erosion of the upper chalk and underlying shale (Figure 16), whereas these strata are preserved in structural and topographic lows between the structural highs (Weimer, 1980).

The time of major movement on the paleohighs was near the end of Niobrara deposition or early in the deposition of the Pierre Shale. The maximum thinning occurs over the Wattenberg paleostructure where up to 30 m of Upper Niobrara was removed by sea floor erosion prior to deposition of the overlying Pierre (Figure 28).

Minor thinning also occurs at the base of each of the four limestones of the Niobrara in local areas. Regional unconformities can be mapped at the base of the Fort Hays and the base of the second chalk from the top of the Niobrara (Figure 16). Because these breaks occur within the deep water sediments generally during highstand conditions, the breaks may be associated with marine onlap processes as diagrammed by Vail et al. (1977).

A major question is what changed the pattern of structural movement from a broad regional uplift centered in Wyoming during Carlile deposition (Figures 26 and 27) to the movement on northeast structural trends centered in Colorado during late Niobrara and early Pierre deposition (Figure 28). The answer may be related to the direction and rate of movement of the American plate. Normally the plate had western movement to establish the tectonic framework shown on Figures 9 and 29. Either a more rapid rate of westward movement, or a more northwesterly movement may have caused vertical movement on northeast-trending basement faults with submarine erosion recurring over the higher standing blocks.

At approximately the same time as fault blocks were reactivated in the northern Denver basin, volcanic activity on the sea floor, during deposition of the Austin chalk, occurred in Texas (Figure 29) (Simmons, 1967). Ancient submarine volcanoes occur along a northeast trend across central Texas. The lithology of the igneous material is ultrabasic to basic in composition and the features are referred to as "serpentine plugs" altered from igneous rocks rich in olivine. Carbonate reefs are found fringing or capping the volcanoes. Depth of the sea during deposition of the upper Austin is estimated to have been between 60 to 245 m (Simmons, 1967).

The northeast trend of the volcanoes was related by Simmons (1967) to extrusion above basement fault systems. Tension along the Balcones fault zone opened the crust to allow mantle-derived material to be intruded in the strata or to be erupted on the sea floor as volcanoes.

The Austin chalk of Texas is correlated with the Niobrara Formation of eastern Colorado and Kansas. Thus, at the time of the volcanic events in Texas, basement fault block movement occurred on second order structures along the ancestral Transcontinental arch (Figure 29). Moreover, major thrusting was in progress along the Meade–Crawford thrust sheets in western Wyoming and Utah (Royse et al., 1975; Jordan, 1981). The timing of these events is approximately 80–81 m.y. ago.

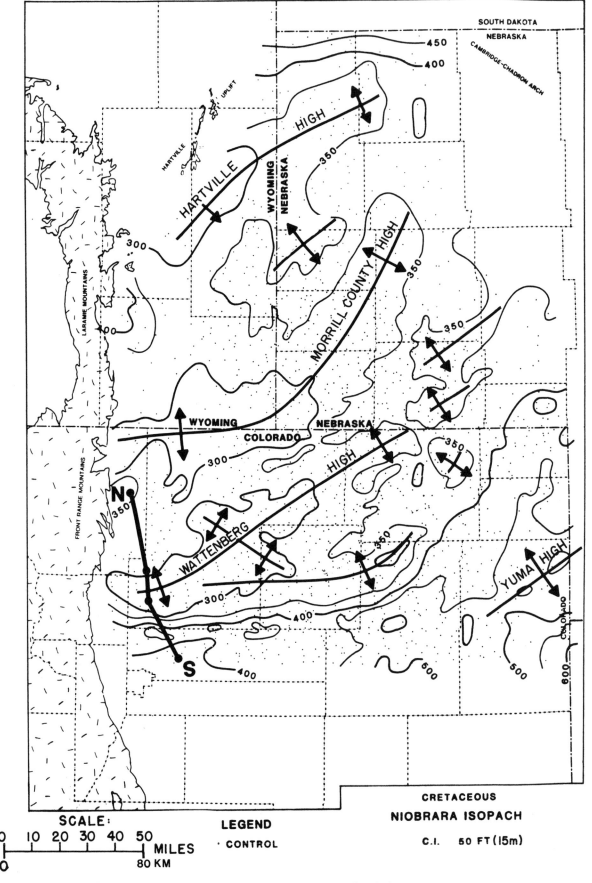

Figure 28—Isopach map of Niobrara Formation with tectonic elements associated with thinning.

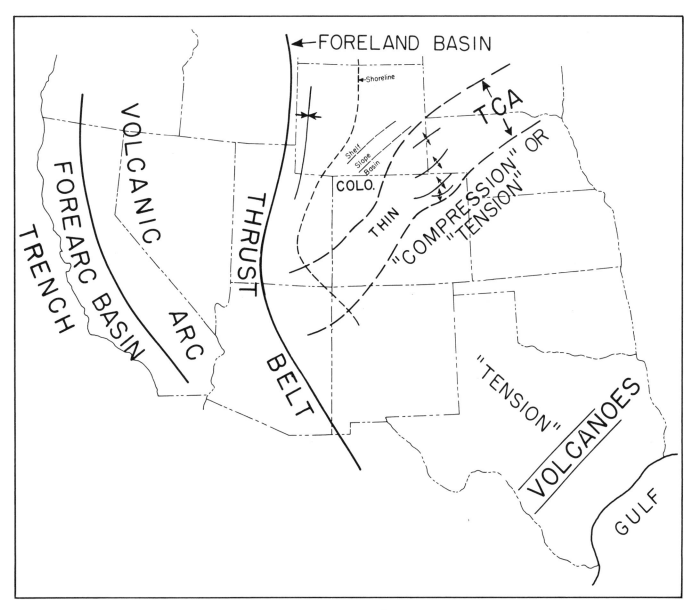

Figure 29—Map of western United States showing events in Cretaceous foreland basin during deposition of upper Niobrara and Austin strata (80–81 m.y. ago).

Uplift of the fault blocks on the sea floor in the northern Denver basin could have resulted from compression with the blocks forming topographic and structural highs; or, a broad arch related to compression may have been modified by tension with some blocks subsiding to lower structural positions than adjacent more positive blocks.

The structural movement was synchronous with a possible sea level drop. The upper Niobrara chalk is a basin facies equivalent of the Eagle and Shannon sandstones of Montana and Wyoming. The wide distribution of these shelf sandstones may have been related to events similar to the processes described previously for the Carlile Formation. A lower sea level would have affected storm wave base and associated scour. By a lowering of storm wave base, strata were removed by erosion over the topographic (structural) highs but not in the intervening structural and topographic

lows. Thus, the Niobrara isopach pattern and associated unconformities are interpreted to result from a northwest direction of plate motion, the development of compression to uplift parts of the sea floor, and submarine scour during a contemporaneous sea level drop. Following these events, a normal pattern of tectonics and sedimentation for the Western Interior Cretaceous was resumed for the time of lower Pierre sedimentation.

PETROLEUM OCCURRENCE

Modern petroleum exploration integrates geologic factors that relate to origin, migration, and accumulation of oil and gas. The factors are source beds, generation area, migration paths, reservoir, seal, time of formation, and preservation of trap.

The principal source beds in the Denver basin have been identified as the Benton Group (Graneros, Greenhorn, and Carlile) by Clayton and Swetland (1980). They believe that oil was generated in the deeper parts of the Denver basin and that migration occurred from these organically rich layers into the J (Muddy) Sandstone. In the proposed depositional model (Figures 23 and 25), the organic-rich source beds originated as basin and slope deposits. These are deeper water deposits where anoxic conditions favored preservation of organic matter.

Reservoir rocks in the J (Muddy) Sandstone are of two main types: one is the widespread marine delta front sandstone (Fort Collins Member) illustrated by the Wattenberg field. The second is the fluvial channel sandstone facies of the valley-fill deposits (Horsetooth Member). Because most of the individual sand bars within the channel complex are small in geographic distribution, the oil fields are small (generally less than 2 million bbl of reserves per field).

The trap for petroleum may be either facies changes within the valley fill from reservoir to nonreservoir rock, or the unconformity surface. In the latter case, wedge out of sandstone against an impermeable valley wall occurred, or erosion of porous sandstone with impermeable shales placed in the scour forming a trap.

Reservoir rocks in the Frontier Formation in central Wyoming are lenticular marine shelf sandstones which occur both above and below the major surface of unconformity (Figure 25). Petroleum occurs where offshore marine sandstones are in favorable structural condition for entrapment. Over 400 million bbl of oil have been produced from the Second Wall Creek Sandstone at the Salt Creek field, Wyoming (Barlow and Haun, 1966). This sandstone is shown as Unit 3 of the Frontier Formation on Figure 25 (Merewether et al., 1979). Many other fields in central Wyoming also produce from the Frontier Formation.

CONCLUSIONS

Stratigraphic evaluation of ancient sequences has been directed principally toward the construction of a depositional model to explain the origin and distribution of formations, facies, and time-stratigraphic units. For some sequences, little attention has been given to the possible influence of intrabasin deformation on sedimentation and the development of unconformities, especially during times of major eustatic changes.

This paper presents data and concepts that illustrate unconformities associated with the 97, 90, and 80–81 m.y. sea level changes. During these times, complex relationships existed among the following: sea level fluctuations and related changes in base level of erosion and deposition; tectonic movements and/or climatic changes in the source area that influenced the rate of sediment supply to the basin; unconformities within the basin; distribution of sandstone reservoirs related to environments of the shelf, shoreline, and alluvial valleys; and the influence of intrabasin recurrent movement of basement fault blocks. The stratigraphic record from studies of inland cratonic basins should be coordinated with studies of plate margin areas and of other continents.

By evaluating the above factors, a more complete geologic model can be constructed and used as a powerful predictive tool in stratigraphic trap exploration for petroleum. Improved modeling will aid significantly in the interpretation and use of seismic data both in the older mature areas of exploration and in frontier areas.

ACKNOWLEDGMENTS

I appreciate S. A. Sonnenberg's assistance in compiling and interpreting subsurface data, and am grateful to the Getty Oil Company for making the research possible by generous financial support. Barbara Brockman typed the manuscript and Craig Corbin drafted many of the illustrations. I thank M. Reynolds, T. D. Fouch, and P. C. Weimer for helpful suggestions in improving the manuscript.

REFERENCES CITED

Asquith, D. O., 1970, Depositional topography and major marine environments, Late Cretaceous, Wyoming: AAPG Bulletin, v. 54, n. 7, p. 1184–1224.

Barlow, J. A., Jr., and J. D. Haun, 1966, Regional stratigraphy of Frontier Formation and relation to Salt Creek field, Wyoming: AAPG Bulletin, v. 50, n. 10, p. 2185–2196.

Blackwelder, E., 1909, The valuation of unconformities: Journal of Geology, v. 17, p. 289–300.

Clark, B. A., 1978, Stratigraphy of the Lower Cretaceous J Sandstone, Boulder County, Colorado—a deltaic model, *in* J. D. Pruit, and P. E. Coffin, eds., Proceedings, Symposium on mineral resources of the Denver basin: Rocky Mountain Association of Geologists, p. 237–246.

Clayton, J. L., and P. J. Swetland, 1980, Petroleum generation and migration in Denver basin: AAPG Bulletin, v. 64, p. 1613–1634.

Cobban, W. A., and S. C. Hook, 1979, *Collignoniceras woollgari woollgari* (Mantell) ammonite fauna from Upper Cretaceous of Western Interior, United States: New Mexico State Bureau of Mines and Mineral Resources Memoir 37, p. 5–51.

———, 1980, Occurrence of *Ostrea beloiti* Logan in Cenomanian rocks of Trans-Pecos Texas: New Mexico Geological Society 31st Field Conference Guidebook, p. 169–172.

———, 1981, New turrilitid ammonite from mid-Cretaceous (Cenomanian) of southwest New Mexico: New Mexico State Bureau of Mines and Mineral Resources Circular 180, p. 22–35.

Curray, J. R., 1960, Sediments and history of Holocene transgression, continental shelf, northwest Gulf of Mexico, *in* F. P. Shepard, F. B. Phleger, and T. J. van Andel, eds., Recent sediments, northwest Gulf of Mexico: AAPG Special Publication, p. 221–266.

———, 1975, Marine sediments, geosynclines and orogeny *in* A. G. Fischer, and S. Judson, eds., Petroleum and global tectonics: Princeton University Press, Princeton, New Jersey, p. 157–217.

Dickinson, W. R., 1976, Plate tectonic evolution: AAPG Continuing Education Course Note Series 1, p. 46.

Eicher, D. L., 1969, Paleobathymetry of Cretaceous Greenhorn sea in eastern Colorado: AAPG Bulletin, v. 53, n. 5, p. 1075–1090.

Fisk, H. N., 1947, Fine-grained alluvial deposits and their effects on Mississippi River activity: Vicksburg, Mississippi, U. S. Corps of Engineers, Waterway Experiment Station, 98 p.

Fouch, T. D., 1983, Patterns of synorogenic sedimentation in Upper Cretaceous rocks of central and northeastern Utah, *in* M. Reynolds, and E. Dolly, eds., Mesozoic paleogeography of west-central United States: Denver, Society of Economic Paleontologists and Mineralogists, Rocky Mountain Section Special Publication, p. 305–336.

Hancock, J. M., 1975, The sequence of facies in the Upper Cretaceous of northern Europe compared with that in the Western Interior, *in* W. G. C. Caldwell, ed., The Cretaceous System in the Western Interior of North America: Geological Association of Canada Special Paper No. 13, p. 82–118.

———, and E. G. Kauffman, 1979, The great transgressions of the Late Cretaceous: Geological Society of London Journal, v. 136, p. 175–186.

Hattin, D. E., 1962, Stratigraphy of the Carlile Shale (Upper Cretaceous) in Kansas: Kansas Geological Survey Bulletin, n. 156, 155 p.

———, 1975a, Stratigraphy and depositional environment of Greenhorn Limestone (Upper Cretaceous) of Kansas: Kansas Geological Survey Bulletin, n. 209, 128 p.

———, 1975b, Stratigraphic study of the Carlile-Niobrara (Upper Cretaceous) unconformity in Kansas and northeastern Nebraska: Geological Association of Canada Special Paper No. 13, p. 195–210.

Haun, J. D., 1963, Stratigraphy of Dakota Group and relationship to petroleum occurrence, northern Denver basin, *in* P. J. Katich, and D. W. Bolyard, eds., Geology of the northern Denver basin and adjacent uplifts: Rocky Mountain Association of Geologists Guidebook, p. 119–134.

Hook, S. C., and W. A. Cobban, 1977, *Pycnodonte newberryi* (Stanton)—common guide fossil in Upper Cretaceous of New Mexico: New Mexico State Bureau of Mines and Mineral Resources Annual Report, p. 48–54.

———, 1979, *Prionocyclus novimexicanus* (Marcou)— common Upper Cretaceous guide fossil in New Mexico: New Mexico State Bureau of Mines and Mineral Resources Annual Report, p. 35–42.

———, 1980, Some guide fossils in Upper Cretaceous Juana Lopez member of Mancos and Carlile shales, New Mexico: New Mexico State Bureau of Mines and Mineral Resources Annual Report, p. 38–49.

———, 1981b, Late Greenhorn (mid-Cretaceous) discontinuity surfaces, southwest New Mexico, *in* S. C. Hook, compiler, Contributions to mid-Cretaceous paleontology and stratigraphy of New Mexico: New Mexico State Bureau of Mines and Mineral Resources Circular 180, p. 5–36.

Jeletzky, J. A., 1978, Causes of Cretaceous oscillations of sea level in western and arctic Canada and some general geotectonic implications: Geological Survey of Canada Paper 77–18, 38 p.

Jordan, T. E., 1981, Thrust loads and foreland basin evolution, Cretaceous, western United States: AAPG Bulletin, v. 65, p. 2506–2520.

Kauffman, E. G., 1977, Upper Cretaceous cyclothems, biotas, and environments, Rock Canyon anticline, Pueblo, Colorado: The Mountain Geologist, v. 14, n. 3 and 4, p. 129–152.

Kent, H. C., 1967, Microfossils from the Niobrara Formation (Cretaceous) and equivalent strata in northern and western Colorado: Journal of Paleontology, v. 41, n. 6, p. 1433–1456.

Lanphere, M. A., and D. L. Jones, 1978, Cretaceous time scale from North America, *in* G. V. Cohee, and M. F. Glaessner, eds., The geologic time scale: AAPG Studies in Geology, n. 6, p. 259–268.

Lindstrom, L. J., 1979, Stratigraphy of the South Platte Formation (Lower Cretaceous), Eldorado Springs to Golden, Colorado, and channel sandstone distribution of J Member: Master's thesis, Colorado School of Mines, Golden, Colorado, 142 p.

MacKenzie, D. B., 1965, Depositional environments of Muddy Sandstone, Western Denver basin, Colorado: AAPG Bulletin, v. 49, p. 186–206.

———, 1971, Post-Lytle Dakota Group on west flank of Denver basin, Colorado: The Mountain Geologist, v. 8, n. 3, p. 91–131.

Matuszczak, R. A., 1976, Wattenberg Field: a review, *in* R. C. Epis, and R. J. Weimer, eds., Studies in Colorado field geology: Golden, Colorado, Colorado School of Mines Professional Contribution 8, p. 275–279.

McGookey, D. P., 1972, Cretaceous system, *in* W. W. Mallory, ed., Geologic atlas Rocky Mountain region: Rocky Mountain Association of Geologists Special Publication, p. 190–228.

Merewether, E. A., and W. A. Cobban, 1972, Unconformities within the Frontier Formation, northwestern Carbon County, Wyoming: U.S. Geological Survey Professional Paper 800-D, p. D57–D66.

Merewether, E. A., and W. A. Cobban, 1973, Stratigraphic sections of the Upper Cretaceous Frontier Formation near Casper and Douglas, Wyoming: Wyoming Geological Association Earth Science Bulletin, v. 6, n. 4, p. 38–39.

Merewether, E. A., W. A. Cobban, and C. W. Spencer, 1976, The Upper Cretaceous Frontier Formation in the Kaycee-Tisdale Mountain area, Johnson County, Wyoming: Wyoming Geological Association 28th Annual Field Conference Guidebook, p. 33–44.

Merewether, E. A., W. A. Cobban, and E. T. Cavanaugh, 1979, Frontier Formation and equivalent rocks in eastern Wyoming: The Mountain Geologist, v. 16, n. 3, p. 67–101.

Merriam, D. F., 1957, Subsurface correlation and stratigraphic relation of rocks of Mesozoic age in Kansas: Kansas Geological Survey Oil and Gas Investigations n. 14, 25 p.

———, 1963, The geologic history of Kansas: Kansas Geological Survey Bulletin, n. 162, 309 p.

Nelson, H. F., and E. E. Bray, 1970, Stratigraphy and history of the Holocene sediment in the Sabine-High Island area, Gulf of Mexico, *in* J. P. Morgan, and R. H. Shaver, eds., Deltaic sedimentation modern and ancient: Society of Economic Paleontologists and Mineralogists Special Publication, n. 15, p. 48–77.

Obradovich, J. D., and W. A. Cobban, 1975, A time scale for the Late Cretaceous of the Western Interior of North America, *in* W. G. A. Caldwell, ed., The Cretaceous System in the Western Interior of North America: Geological Association of Canada Special Paper, n. 13, p. 31–54.

Peterson, W. L., and S. D. Janes, 1978, A refined interpretation of depositional environments of Wattenberg Field, Colorado, *in* J. D. Pruit, and P. E. Coffin, eds., Rocky Mountain Association of Geologists Symposium, Energy Resources of the Denver Basin: Denver, Rocky Mountain Association of Geologists, p. 141–147.

Pinel, M. J., 1977, Stratigraphy of the upper Carlile and lower Niobrara formations (Upper Cretaceous), Fremont and Pueblo counties, Colorado: Master's thesis, Colorado School of Mines, Golden, Colorado, 111 p.

Poleschook, D., Jr., 1978, Stratigraphy and channel discrimination, J Sandstone, Lower Cretaceous group, south and west of Denver, Colorado: Master's thesis, Colorado School of Mines, Golden, Colorado, 226 p.

Reeside, J. B., Jr., 1944, Map showing thickness and general character of the Cretaceous deposits in the Western Interior of the United States: U.S. Geological Survey Oil and Gas Investigations Map OM-10.

———, 1957, Paleoecology of the Cretaceous seas of the Western Interior of the United States: Geological Society of America Memoir 67, p. 505–542.

Royse, F., Jr., M. A. Warner, and D. L. Reese, 1975, Thrust belt structural geometry and related stratigraphic problems, Wyoming-Idaho-northern Utah, *in* D. W. Bolyard, ed., Deep drilling frontiers of the central Rocky Mountains: Denver, Rocky Mountain Association of Geologists, p. 41–55.

Scott, G. R., and W. A. Cobban, 1964, Stratigraphy of the Niobrara Formation at Pueblo, Colorado: U.S. Geological Survey Professional Paper 454-L, p. L1–L30.

Simmons, K. A., 1967, A primer on "serpentine plugs" in south Texas, *in* W. G. Ellis, ed., Contributions to the geology of south Texas: San Antonio, South Texas Geological Society, v. 7, n. 2, p. 125–132.

Sonnenberg, S. A., and R. J. Weimer, 1981, Tectonics, sedimentation, and petroleum potential, northern Denver basin, Colorado, Wyoming, and Nebraska:

Colorado School of Mines Quarterly, v. 76, n. 2, 45 p.

Suryanto, U., 1979, Stratigraphy and petroleum geology of the J Sandstone in portions of Boulder, Larimer, and Weld Counties, Colorado: Master's thesis, Colorado School of Mines, Golden, Colorado, 173 p.

Vail, P. R., R. M. Mitchum, Jr., and S. Thompson, III, 1977, Seismic stratigraphy and global sea-level changes, part 3: AAPG Memoir 26, p. 63–82.

Van Hinte, J. E., 1976, A Cretaceous time scale: AAPG Bulletin, v. 60, n. 4, p. 498–516.

Waage, K. M., 1953, Dakota group in northern Front Range foothills, Colorado: U. S. Geological Survey Professional Paper 274-B, p. 15–51.

Weimer, R. J., 1978, Influence of transcontinental arch on Cretaceous marine sedimentation: a preliminary report, *in* J. D. Pruit, and P. E. Coffin, eds., Energy resources of the Denver Basin: Denver, Rocky Mountain Association of Geologists Symposium, p. 211–222.

———, 1980, Recurrent movement on basement faults, a tectonic style for Colorado and adjacent areas, *in* H. C. Kent, and K. W. Porter, eds., Colorado geology: Denver, Rocky Mountain Association of Geologists Symposium, p. 23–35.

———, 1982, Tectonic influence on sedimentation, Early Cretaceous, east flank, Powder River basin, Wyoming and South Dakota: Colorado School of Mines Quarterly, v. 77, n. 4.

———, 1983, Relation of unconformities, tectonics, and sea-level changes, Cretaceous of Denver basin and adjacent area, *in* M. Reynolds, and E. Dolly, eds., Mesozoic paleogeography of west-central United States: Denver, Society of Economic Paleontologists and Mineralogists, Rocky Mountain Section Special Publication, p. 359–376.

Weimer, R. J., and C. B. Land, Jr., 1972, Field guide to Dakota Group (Cretaceous) stratigraphy Golden-Morrison area, Colorado: The Mountain Geologist, v. 9, n. 2 and 3, p. 241–267.

Weimer, R. J., and S. A. Sonnenberg, 1982, Wattenberg field, paleostructure-stratigraphic trap, Denver basin, Colorado: Oil and Gas Journal, v. 80, n. 12, p. 204–210.

Weimer, R. J., J. J. Emme, C. L. Farmer, L. U. Anna, T. L. Davis, and R. L. Kidney, 1982, Tectonic influence on sedimentation, Early Cretaceous, east flank, Powder River basin, Wyoming and South Dakota: Colorado School of Mines Quarterly, v. 77, n. 4.

Fluvial Systems of the Upper Cretaceous Mesaverde Group and Paleocene North Horn Formation, Central Utah: A Record of Transition from Thin-Skinned to Thick-Skinned Deformation in the Foreland Region

T. F. Lawton
Sohio Petroleum Company
Dallas, Texas

Synorogenic nonmarine strata of the upper part of the Mesaverde Group and North Horn Formation exposed between the Wasatch Plateau and the Green River in central Utah record a late Campanian tectonic transition in the foreland region from thrust belt deformation to basement-cored uplift. Thick Mesaverde sections in the Wasatch Plateau on the west and the Book Cliffs near the Green River on the east are separated by the San Rafael Swell, a basement uplift across which the late Campanian section is thinned by erosion. The sedimentary sequence in the Wasatch Plateau was deposited by east- and northeast-flowing braided and meandering rivers. Time-equivalent sections in the east comprise a lower sequence of mixed braided fluvial, tidal flat, and marine deposits overlain by an upward-coarsening sequence that grades upward from meandering river deposits to pebbly braided river deposits. Paleocurrent data indicate that rivers of the lower sequence flowed eastward, while those of the upper sequence flowed northeastward.

Sandstones in the upper part of the Mesaverde Group can be divided into two distinct compositional suites, a lower quartzose and an upper lithic petrofacies, which aid in lithostratigraphic correlation across the San Rafael Swell. Lithic grain populations of the upper petrofacies are dominated by sedimentary rock fragments on the west and volcanic rock fragments on the east. Sedimentary lithic grains were derived from the thrust belt, while volcanic lithic grains were derived from a more distant volcanic terrane to the southwest. Tributary streams carrying quartzose detritus from the thrust belt entered the main northeast-flowing trunk system and caused a basinward dilution of volcanic detritus. During most of Campanian time, sediment transport was eastward and northeastward away from the thrust belt. Simultaneous disappearance of volcanic grains and local changes in paleocurrent directions at the top of the section reflect a change of drainage patterns in latest Campanian time that marked initial growth of the San Rafael Swell and possibly other thick-skinned uplifts to the south. Depositional onlap across the Mesaverde Group by a largely posttectonic assemblage of fluvial and lacustrine strata (North Horn Formation) indicates a minimum age of late Paleocene for uplift of the San Rafael Swell.

INTRODUCTION

The Mesaverde Group, exposed on the east flank of the Wasatch Plateau and the Book Cliffs (Figure 1), forms a progradational clastic wedge at the top of the Upper Cretaceous section in central Utah. Deposition occurred in an eastward-thinning foreland basin that lay parallel to an active thrust belt on the west. The Mesaverde Group coarsens upward from siltstone and sandstone deposited in

delta front and delta plain settings (Flores and Marley, 1979; Balsley, 1980) into a sandstone-dominated sequence deposited in fluvial environments (Van De Graaff, 1972; Keighin and Fouch, 1981; Fouch et al., 1983). Most of the sandstone within the fluvial section represents detritus shed eastward from thrust-related uplift of the Sevier orogenic belt (Spieker, 1946; Harris, 1959; Armstrong, 1968). However, changes in dispersal direction, sandstone composition, and depositional patterns within the

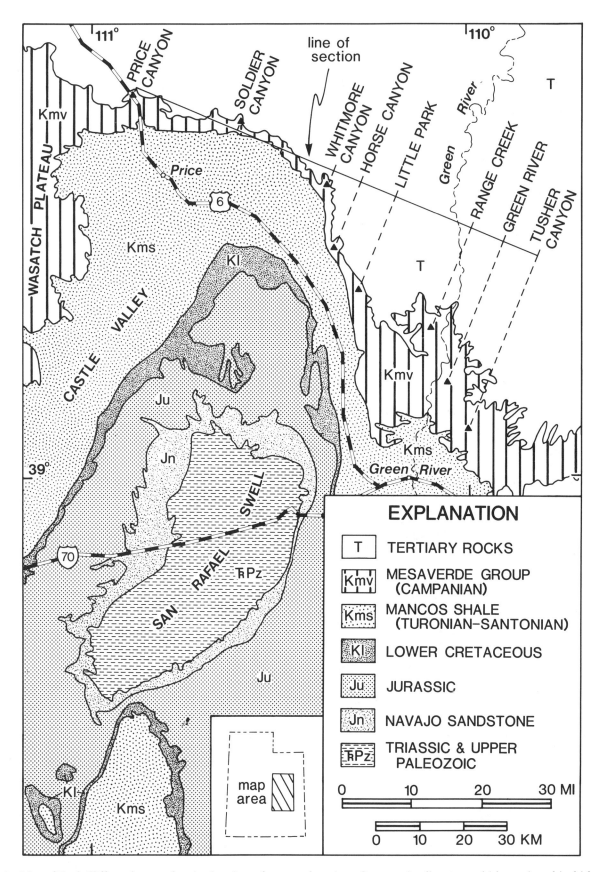

Figure 1—Map of Book Cliffs study area, showing locations of measured sections. Cross section line, onto which stratigraphic thicknesses of Figure 11 are projected, is indicated by the solid south-southeast-trending line. Kms includes the Dakota Sandstone of Late Cretaceous age. Joes Valley location lies 8 km (5 mi) west of the western map boundary where the Wasatch Plateau runs off the map.

uppermost part of the Mesaverde Group reflect an eastward shift in deformation from the thrust belt to basement-controlled uplift within the foreland basin in latest Campanian time.

The Mesaverde Group is exposed nearly continuously along cliffs that run eastward from the Wasatch Plateau and then southeastward to the Green River (Figure 1). The exposures thus provide an extensive section almost parallel to the dominant dispersal direction of the Late Cretaceous fluvial systems and permit sedimentologic analysis over a broad region. Stratigraphic and sedimentologic relationships in the upper part of the Mesaverde Group within the study area (Figure 1) define the timing of basement uplifts within the basin and shed light on latest Cretaceous paleogeography in central Utah.

Patterns of deposition recorded by the Upper Cretaceous-Paleocene section in the Wasatch Plateau suggest that the San Rafael Swell, a thick-skinned plateau uplift, existed by Maestrichtian time (Fouch et al., 1983). Stratigraphic, sedimentologic, and petrographic data presented here document that the earliest growth of the San Rafael Swell, and probably two other related uplifts, the Circle Cliffs uplift and the Monument upwarp, occurred in latest Campanian time. The data represent a refinement of previous Laramide age assignments for the block uplifts of the Colorado Plateau (Kelley, 1955; Davis, 1978).

STRATIGRAPHIC RELATIONSHIPS OF THE UPPER MESAVERDE GROUP AND NORTH HORN FORMATION

Formations of the upper Mesaverde Group described in this study are shown in Figure 2. The lowermost unit of the fluvial sections in both the western and eastern part of the study area is the Castlegate Sandstone. The Castlegate rests sharply on the Blackhawk Formation throughout the area, and the contact has been interpreted as disconformable as far east as the Green River (Spieker, 1946; Fouch et al., 1983). The hiatus (if one exists), however, is of short duration based on faunal evidence (Fouch et al., 1983). The contact may represent an abrupt change from delta plain to sandy alluvial plain sedimentation that resulted as the Castlegate fluvial system overrode the Blackhawk delta as both environments prograded eastward.

Spieker (1946) recognized that the North Horn Formation overlying the Mesaverde Group preserves a record of continuous deposition across the Cretaceous-Tertiary boundary in the central Wasatch Plateau. He inferred the contact of the North Horn and underlying Price River formations to be gradational as well. More recent work has shown, however, that the contact is unconformable (Fouch et al., in press). Where it is exposed throughout the Wasatch Plateau and eastward, the contact is marked by intense bleaching and limonitization of the underlying sandstone. Large rootlet traces are locally present in coarse-grained sandstone underlying the contact. Plagioclase grains are extensively argillized and kaolinite is often pseudomorphous after the grains in the upper few tens of meters of the Price River and Tuscher formations. Where pebbly lithologies occur high in the Cretaceous section,

particularly in exposures of the Tuscher Formation near the Green River, lags of reworked pebbles are present at the base of the North Horn Formation. Elsewhere in the study area east of the Wasatch Plateau the basal North Horn lacks extraformational pebbles. Palynomorph and faunal assemblages indicate a hiatus at the contact, as discussed in a later section.

Nomenclature

The stratigraphic terminology used here is basically that of Fisher et al. (1960) as amended by Fouch et al. (1983). The nomenclature used for the eastern facies (Figure 2) is carried westward from the Green River, through lateral facies changes and unit pinchouts within the Tavaputs Plateau, to the approximate axis of the San Rafael Swell at Soldier Creek (Figure 1). The swell is a structural boundary west of which relationships in the Castlegate-Bluecastle section are difficult to trace because of poor exposure and difficult access. The post-Bluecastle section beneath the hiatus at the base of the North Horn Formation is almost entirely absent at Soldier Creek (Fisher et al., 1960). Direct surface correlation of post-Bluecastle beds in the Price River Formation with its age equivalents farther east is thus impossible across the axis of the San Rafael Swell.

Age and Correlation

The absolute ages of zone fossils that bracket the fluvial section are taken from the chronostratigraphic correlations of Fouch et al. (1982, 1983) (Figure 3), which utilize new decay constants for radiometric age determinations. Because the units are largely of fluvial origin, direct molluscan age control is sparse. The section between the base of the Castlegate Sandstone and the top of the Bluecastle Tongue of the Castlegate was deposited during the time spanned by the zones of *Baculites asperiformis* and *Exiteloceras jennyi* (74–79 m.y.). Extensive palynomorph data (Fouch et al., 1983) indicate that the post-Bluecastle section ranges in age from middle late to latest Campanian (approximately 73–74 m.y., ending within the *B. cuneatus* zone), and provide the basis for correlations of post-Castlegate Cretaceous rocks in Price Canyon with the section east of the San Rafael Swell.

The upper part of the Mesaverde Group is thus equivalent to the upper member of the Sixmile Canyon Formation of the Indianola Group, as defined by Lawton (1982) (Figure 3). The Sixmile Canyon Formation is in part Santonian and in part Campanian in age (Fouch et al., 1982). The upper Tuscher Formation is probably equivalent to the Ohio Creek Member of the Hunter Canyon Formation of western Colorado (Johnson and May, 1980). Rocks equivalent to the Tuscher to the southwest in the Kaiparowits region probably include part of the Kaiparowits Formation and the Canaan Peak Formation (Bowers, 1972).

The North Horn Formation, which unconformably overlies the Mesaverde Group, varies widely in age and lithic character (Figure 3). A Maestrichtian age for basal North Horn strata in Price Canyon is based on freshwater molluscs, ostracodes, and pollen (Fouch et al., in press). Near the Green River, North Horn strata are indicated to be of late Paleocene age on the basis of freshwater molluscs, ostracodes, and pollen (Fouch et al., 1982; T. D. Fouch, personal communication, 1982). The base of the North

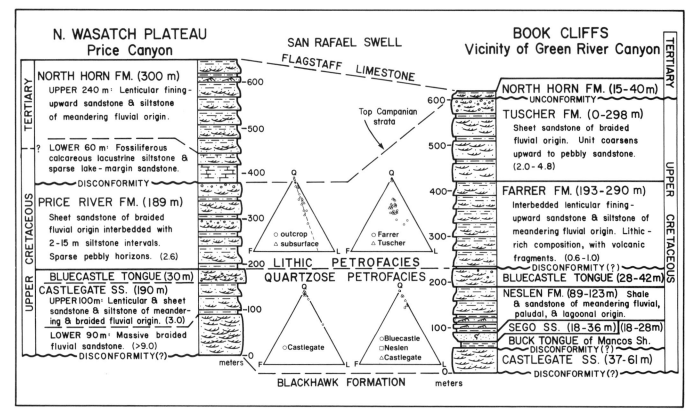

Figure 2—Stratigraphy and sandstone composition of upper Mesaverde Group strata west and east of the San Rafael Swell. Numbers in parentheses are sandstone/(sandstone + siltstone) values for the units. At all localities, the Bluecastle is a tongue of the Castlegate Sandstone (Fouch et al., 1983). Used with permission of the Society of Economic Paleontologists and Mineralogists, Rocky Mountain Section.

Horn Formation is thus time transgressive, becoming younger to the east. The hiatus separating Cretaceous and Tertiary rocks east of the San Rafael Swell represents at least 15 m.y. (Fouch et al., 1982).

Stratigraphy

Castlegate Sandstone

The Castlegate Sandstone, as used in this study, follows the usage of Spieker (1931). At its type locality in Price Canyon, the Castlegate Sandstone includes a lower massive sandstone, which forms a continuous cliff throughout the study area, and an upper sequence of interbedded sandstone and siltstone, which forms ledgy outcrops (Spieker, 1931). The entire unit is 190 m (623 ft) thick and is capped by a ledge of pebbly coarse-grained sandstone 20–30 m (66–98 ft) thick. The lower cliff-forming unit (the Castlegate Sandstone of Van De Graaff, 1972) thins eastward from 90 m (295 ft) in Price Canyon to 37 m (121 ft) in Tusher Canyon, and it has been tentatively recognized as a unit 17–20 m (56–66 ft) thick on the northeastern flank of the Uinta basin (Gill and Hail, 1975). Above the cliff-forming unit, the ledge-forming sequence grades eastward into the Neslen Formation as shown by palynomorph data (Fouch et al., 1983) and petrographic data. The uppermost pebbly unit of the Castlegate grades eastward into the Bluecastle Tongue of the Castlegate Sandstone east of the San Rafael Swell

(Figures 2 and 3) and is tentatively recognized here as the Bluecastle Tongue in Price Canyon. The Castlegate–Price River contact intersects the floor of Price Canyon at the mouth of Sulphur Canyon (SE¼, sec. 22, T. 12 S., R. 9 E., Kyune, Utah 7.5′ quadrangle).

The lower cliff-forming part of the Castlegate Sandstone consists of lenticular sandstone bodies or bedsets dominated by trough cross beds, tabular bedsets with planar cross beds and slightly inclined plane beds, and thin inclined interbeds of sandstone and siltstone. The lenticular and tabular sandstone bedsets range from 15 cm to 2.5 m (0.5–8 ft) thick. The lenticular bedsets rest on erosive bases and form stacked sequences. Large coaly plant fragments, including tree branches and trunks, and intraformational clasts of mudstone and siltstone are abundant. The tabular bedsets have sharp bases and are often found in stacked sequences of two or three beds above a complex of lenticular beds. The inclined interbeds of sandstone and siltstone range from 2 to 30 cm (1–12 in.) thick and contain ripples, climbing ripples, horizontal lamination, and contorted lamination. Siltstones are typically horizontally laminated and rippled; macerated plant debris is common. The thin interbeds of sandstone and siltstone occasionally form sequences 10 m (33 ft) or more thick and continuous for over 100 m (305 ft) laterally (Figure 4). The thin beds of sandstone and siltstone occasionally grade vertically or interfinger laterally into extensive massive to mottled dark gray siltstones with coal fragments and rare silicified bone fragments. Sometimes the inclined beds of siltstone and sandstone fill broad

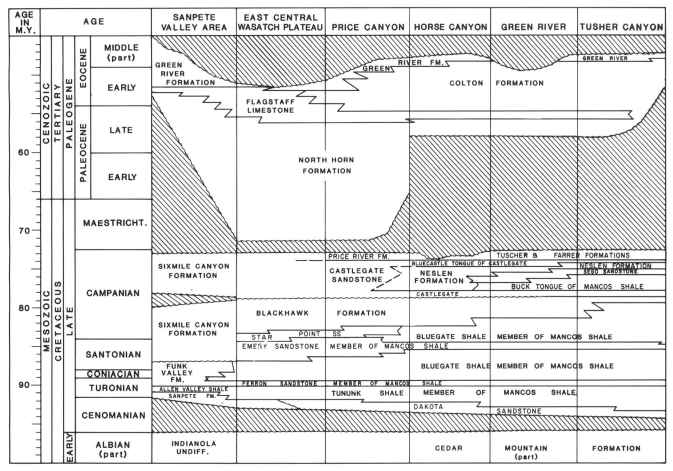

Figure 3—Time-stratigraphic chart showing nomenclature and correlation of upper Albian to middle Eocene rock units from Sanpete Valley in central Utah east to Tusher Canyon. Modified from Fouch et al. (1983). Used with permission of the Society of Economic Paleontologists and Mineralogists, Rocky Mountain Section.

channelform depressions up to 10 m (33 ft) deep and 50 m (165 ft) across. Other times they are truncated beneath a sequence of trough cross-bedded sandstone bodies. The sandstone within the Castlegate Sandstone is dominantly fine to medium grained in Price Canyon and very fine to fine grained in eastern exposures. Pebble lags and tree trunks are common in the formation at Joes Valley (see Figure 5 for location) and indicate a grain size increase southwestward along the Wasatch Plateau.

The sequences and structures in the Castlegate Sandstone are similar to the vertical sequence described by Cant and Walker (1978) for the sandy braided South Saskatchewan River. The lenticular bedsets of trough cross beds represent channel complex deposits, and the tabular beds with low-angle lamination and planar cross beds represent sand flat deposits formed by amalgamation of large transverse and longitudinal bars (T. R. Clifton, personal communication, 1984). The thinly bedded sandstone and siltstone beds probably represent splay deposits into abandoned channels, as represented by the fill of channelform depressions, and on floodplains, as represented by lateral gradation into mottled siltstone deposits. The regional geometry of the Castlegate Sandstone indicates that it was probably deposited by many different rivers on a sandy alluvial plain that paralleled the general trend of the Campanian shoreline

(Van De Graaff, 1972). Paleocurrent data indicate consistent east–southeast sediment transport within the lower cliff-forming part of the Castlegate Sandstone (Figure 5).

The upper part of the Castlegate Sandstone in Price Canyon, which forms ledges below the Bluecastle Tongue, is dominated by trough cross-bedded, fine-grained sandstone beds 2–5 m (7–16 ft) thick and rich in mudstone ripup clasts and large plant fragments. The beds grade up into interbeds of very fine grained sandstone and siltstone with depositional dips as high as 15°. The inclined beds are overlain in turn by sequences of poorly exposed gray siltstone and dark brown carbonaceous shale. Thin, very fine grained sandstone beds 1 m (3 ft) and less thick occur in the siltstone and shale sequences. The sandstone beds contain ripple lamination or are massive due to bioturbation. Rootlet traces and small plant fragments are common, and bed bases are sharp and, in places, load casted.

The upper ledge-forming part of the Castlegate Sandstone is interpreted to have been deposited in a meandering stream environment. The trough cross-bedded sandstones probably represent channel and lower point bar deposition, while overlying inclined units of finer sandstone and siltstone were deposited by lateral migration of point bars. Thin-bedded sandstone beds in the silty floodplain sequences represent crevasse splay deposits.

Figure 4—Upper 32 m (105 ft) of the Castlegate Sandstone exposed on the west wall of Green River Canyon. Thinly bedded lateral accretion or splay deposits at extreme base of photo and near top of unit alternate with irregularly bedded, dominantly trough cross-bedded (channel complex) sandstone of braided fluvial origin. Unit composed of inclined beds at top is approximately 4.5 m (15 ft) thick.

Buck Tongue of Mancos Shale and Sego Sandstone

The Buck Tongue of the Mancos Shale consists of 20–28 m (66–92 ft) of thinly bedded, medium gray siltstone and mudstone and tan, very fine grained sandstone that overlie the Castlegate Formation at the Green River and Tusher Canyon localities (Figure 1). I have placed the basal contact in this study at a horizon of nodular hematitic mudstone and chert continuous between the above localities and thought to be a paleosol horizon developed above the Castlegate. The unit thins westward and pinches out between the Green River and Horse Canyon localities (Fisher et al., 1960) (Figure 3). Time-equivalent strata lie beneath present levels of exposure at Range Creek. At the Green River and Tusher Canyon localities, the Buck Tongue grades vertically into an upward-thickening and -coarsening sequence of very fine grained sandstone and siltstone of the overlying Sego Sandstone. Thin beds low in the sequence are capped by 3–5 m (10–16 ft) of very fine to fine-grained, well-sorted sandstone. The thinly bedded sandstone contains hummocky stratification and ripple lamination, while a thick (4 m [12 ft]) sandstone capping the upward-coarsening cycle contains trough cross beds and occasional tabular cross beds. *Ophiomorpha* is present in the sandstone, which is overlain by wavy bedded sandstone and siltstone containing oyster shells. A similar but thinner upward-coarsening sequence overlies the described sequence (Lawton, 1983).

The Sego Sandstone was apparently deposited in transition zone to middle and upper shoreface environments (Harms et al., 1975; Balsley, 1980) by progradation of subaqueous bars across the Mancos shelf. Fossiliferous tidal flats that lay landward of the bars subsequently prograded across the bar deposits. The cycles, and thus the bar heights, are 10–12 m (33–39 ft) thick and contrast sharply with similar cycles in the older Blackhawk Formation, which are 25 m (82 ft) thick (Balsley, 1980; Lawton, 1983).

Neslen Formation

The Neslen Formation was defined by Fisher (1936) to include coal-bearing strata above the Sego Sandstone. He considered it to be a member of the Price River Formation.

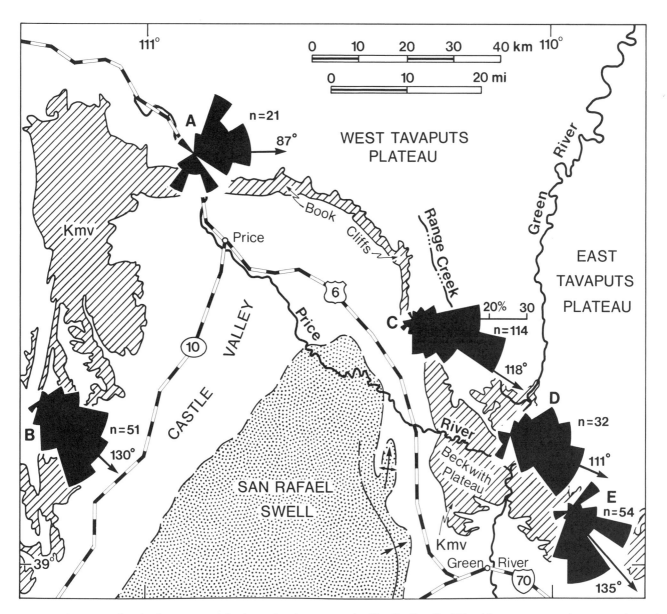

Figure 5—Paleocurrent data for lower part of Castlegate Sandstone (see also Van De Graaff, 1972). All measurements from axes of trough cross bedding. Localities: A. Price Canyon; B. Joes Valley; C. Horse Canyon; D. Green River; E. Tusher Canyon. Ruled area (Kmv) shows outcrop area of Mesaverde Group rocks of this study.

The unit was given formational status by Cobban and Reeside (1952), but was first formally raised to formational rank by Fisher et al. (1960).

The basal contact of the Neslen Formation is gradational at the Green River and Tusher Canyon localities. It is placed above the uppermost upward-coarsening sandstone cycle of the Sego Sandstone where the latter is present (Fisher et al., 1960). At Horse Canyon, where both the Buck Tongue and the Sego Sandstone are absent, the basal Neslen contact is arbitrarily placed at the top of the cliff-forming Castlegate Sandstone. Fisher et al. (1960) considered the unit above the Castlegate at Horse Canyon to be the Price River Formation, although the beds grade westward into strata of the upper part of the Castlegate Sandstone (described previously). The presence of thin coal beds, gastropod-bearing sandstone, and flaser bedded mudstone and sandstone at Horse Canyon render the interval lithologically more like the Neslen Formation of Fisher (1936) than the thick interbeds of sandstone and siltstone of the upper part of the Castlegate farther to the west. Consequently, the stratigraphic interval between the Castlegate Sandstone and the Bluecastle Tongue of the Castlegate in the region between the Green River and Horse Canyon is considered to be the Neslen Formation for purposes of this study (Figure 3). The unit is probably disconformable above the Castlegate, which is based on the evidence for paleosol development farther east.

The Neslen Formation consists of roughly 50% siltstone, which is commonly associated with wavy bedded and flaser bedded sandstone, burrowed and rooted lenticular sandstone beds as thick as 1 m (3 ft), and stringers of coal. Channelform, very fine to fine-grained sandstone beds 2–7.5 m (6.6–25 ft) thick are interbedded with the siltstone. The sandstone beds are characterized by scoured basal contacts and abundant mudstone clasts, and they grade upward into inclined beds of sandstone 30–100 cm (12–40 in.) thick separated by siltstone beds as much as 10 cm (4 in.) thick. Burrows are common in sandstones, and load casts are present locally.

The Neslen Formation was apparently deposited dominantly on a coastal plain of low relief. As interpreted here, the unit forms a regressive sequence consisting of tidal flat deposits near its base, passing up through possible distributary sandstone, into a section dominated by meander belt deposition. Meandering streams are suggested by common inclined lateral accretion deposits above the channel sandstone beds, and frequent burrowed and rooted crevasse-splay deposits in the siltstone sequences. The westward transition of coeval Neslen and upper Castlegate depositional environments from tidal flat to meander belt deposits records a lateral coarsening of lithologies toward the Sevier orogenic belt.

Bluecastle Tongue of Castlegate Sandstone

The Bluecastle Tongue gradationally overlies the Neslen Formation and has a relatively uniform thickness between 30 m (98 ft) (Price Canyon) and 42 m (138 ft) (Range Creek) within the study area, but thins farther to the east (Fisher, 1936). The Bluecastle of the study area separates the Neslen and Farrer formations. Farther east, however, beds above the Bluecastle have been included in the Neslen Formation (Fisher, 1936; Cobban and Reeside, 1952; Fisher et al., 1960).

The Bluecastle Tongue forms a persistent ledge in the eastern part of the study area. The unit weathers light brown and creates a distinctive marker horizon in the vicinity of the Green River. The Bluecastle Tongue contains abundant lenticular beds of trough cross-bedded sandstone and thin inclined beds of siltstone and sandstone similar in aspect to the bedding present in the lower Castlegate Sandstone. Consequently, the units are very similar in appearance. Inclined interbeds of sandstone and siltstone become more common eastward. West of the Green River trough cross-bedded sandstone dominates the Bluecastle, which is interrupted locally by broad lenses of siltstone and mudstone as thick as 5 m. Tabular sandstone bedsets with plane laminated beds dipping 15° or less and tabular cross beds are present but less common than in the Castlegate Sandstone. Rootlet marks are common in sandstone bed tops. The Bluecastle is coarser grained than the massive lower Castlegate and consists of poorly sorted, fine- to coarse-grained sandstone at all localities. Pebbles of chert, very fine grained quartz arenite, and punky gray claystone are common in western localities but become sparser eastward. Maximum pebble size is 15 mm at Price Canyon and 4 mm (0.2 in.) at Tusher Canyon.

Like the Castlegate Sandstone, the Bluecastle Tongue was deposited on a fluvial coastal plain. The presence of siltstone and mudstone lenses interpreted to have been deposited in abandoned channels and fewer unambiguous transverse bar deposits suggests that Bluecastle streams may have been sinuous, even meandering. Inclined sandstone and siltstone beds may represent lateral accretion of point bar deposits

rather than splays, as interpreted in the Castlegate Sandstone. Mean vectors on trough cross-bed axes in the Bluecastle Tongue are consistently northeast-directed at all localities (Figure 6).

Price River Formation

The Price River Formation originally included all Upper Cretaceous strata above the Blackhawk Formation (Spieker and Reeside, 1925; Clark, 1928). The unit is now defined as a section of ledge-forming sandstone and siltstone 189 m (620 ft) thick between the Castlegate Sandstone and the North Horn Formation in Price Canyon (Fouch et al., 1983; Lawton, 1983). Its basal contact is probably conformable, but its upper contact with the North Horn Formation is disconformable.

The lower half of the unit consists of continuous sheetlike sandstone beds interbedded with siltstone intervals. The section is approximately 75% sandstone (Lawton, 1983). Sandstone beds range from 10 to 15 m (33–48 ft) thick and consist of poorly sorted, medium-grained sandstone. Grain sizes fine upward slightly within individual sandstone sheets, which are dominated by trough cross beds. Chaotic mudchip lags are common near the bases of the sandstone beds, and ripple lamination is occasionally present near bed tops. The sheetlike sandstone beds occasionally terminate laterally in gently dipping sandstone and siltstone beds adjacent to poorly exposed lenses of siltstone and shale. The sandstone beds grade upward into poorly exposed siltstone intervals.

Sheetlike sandstones of the upper half of the Price River Formation in Price Canyon tend to be thicker and coarser than sandstone horizons of the lower half. Fining-upward cycles within the sheets range from 15 to 20 m (48–66 ft) thick and sometimes merge to form beds as thick as 30 m (98 ft). The beds commonly have rooted tops and grade rapidly into siltstone intervals 5–13 m (16–43 ft) thick. Planar cross beds are present in the sequences but are subordinate to trough cross beds. In contrast to underlying beds, the uppermost sandstone unit in Price Canyon is pebbly, contains abundant scour-and-fill structures, and lacks a distinct fining-upward trend. Measurements of trough cross bed axes throughout the Price River Formation indicate northeastward transport of sandstone deposited in Price Canyon (see Figures 7 and 8).

The dominance of trough cross beds, fining-upward trends, and inclined lateral accretion beds adjacent to abandoned channel plugs indicates that the Price River Formation was most likely deposited by sinuous to meandering rivers. The thickness of the fining-up sheet sandstones suggests that the rivers ranged from 10 to 20 m (33–66 ft) deep, using the channel depth criteria of Leeder (1973). The rivers flowed northeast on a broad alluvial plain that extended eastward into Colorado (Fouch et al., 1983). The uppermost pebbly sandstone bed may represent a shift to braided fluvial deposition in the Price Canyon area.

Farrer Formation

The Farrer Formation was defined as a member of the Price River Formation by Fisher (1936), and formally raised

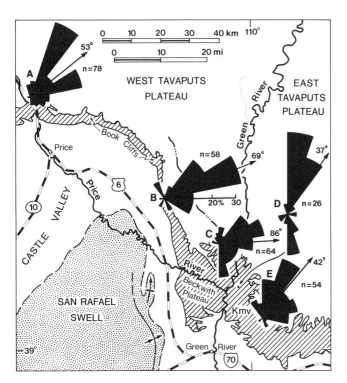

Figure 6—Paleocurrent data from Bluecastle Tongue of Castlegate Sandstone. All measurements from axes of trough cross bedding. Localities: A. Price Canyon; B. Horse Canyon; C. Range Creek; D. Green River; E. Tusher Canyon.

to formation rank by Fisher et al. (1960), although it was previously designated as a formation by Cobban and Reeside (1952). In the study area it is restricted to sections east of the San Rafael Swell, where it lies on the variably bleached and oxidized upper surface of the Bluecastle Tongue of the Castlegate Sandstone. The nature of the surface and an abrupt change to lithic sandstone (discussed later) indicate that the contact may be a disconformity, although palynomorph data suggest that little time is represented by the contact (Fouch et al., 1983). The unit is 290 m (951 ft) thick in Tusher Canyon, thins westward to 90 m (295 ft) at Horse Canyon, and is at most 40 m (131 ft) thick at Soldier Canyon, where it is directly overlain by the North Horn Formation.

The Farrer Formation as used here includes that portion of the post-Bluecastle section rich in siltstone and mudstone, although sandstone makes up 40–50% of the section. The lower part of the Price River Formation in Price Canyon, which is the western equivalent of the Farrer Formation, contains much more sandstone, up to 75%. Correlatives of the Farrer Formation thus appear to coarsen westward toward the thrust belt, but precise time equivalence is difficult to confidently establish between the eastern and western sections.

The Farrer consists of interbedded sandstone and thick siltstone and silty sandstone beds. The sandstone beds are 5–18 m (16–59 ft) thick and broadly lenticular with scoured bases. Grain size and size of primary structures diminish upward within the greenish- to brownish-gray beds of very fine to medium-grained micaceous sandstone. Mudstone chips are common in scour-and-fill structures and are

dispersed on foresets of trough cross beds low in the sandstone beds. Trough cross beds occur above the scoured bases in the lower 1–7 m (2–23 ft) of each sandstone unit and are common throughout the sandstone beds. Trough cross beds are associated with convolute lamination and planar cross beds in the middle parts of the beds and give way upward to horizontal and ripple lamination near sandstone bed tops.

Siltstone deposits gradationally overlie the sandstone beds. In general, the siltstone beds are light greenish gray and interbedded with poorly exposed dark brown shale and silty shale. Sparse interbeds of lenticular to tabular very fine to fine-grained silty sandstone 10–250 cm (4–100 in.) thick occur in the siltstone sequences. The sandstone beds both coarsen and fine upward in different examples and contain ripple lamination as well as planar and trough cross beds. Mudstone chips are common in the silty sandstone beds; bed tops are commonly burrowed and rooted and contain abundant leaf impressions.

The fining-upward sandstone sequences and thick siltstone sequences of the Farrer Formation are characteristic of meandering river deposits. The sandstones were most likely deposited in channel complexes or point bars, while the siltstone sequences represent floodplain deposits. The thin silty sandstone beds in the siltstone facies are interpreted as crevasse splay deposits. Thicknesses of the sandstone units suggest that Farrer rivers were large, with bankfull channel depths of as much as 18 m (59 ft). Paleocurrent measurements taken from many horizons indicate consistent sediment transport to the east and northeast (Figure 7).

Tuscher Formation

The sandstone-dominated sequence above the Farrer Formation and below the North Horn Formation was named the Tuscher Formation (Fisher, 1936) for exposures in a canyon east of the Green River. The canyon's name is now spelled Tusher on topographic quadrangle maps. The Tuscher Formation was assigned a Tertiary(?) age by Fisher (1936) and a Late Cretaceous age by Cobban and Reeside (1952) and Fisher et al. (1960). Keighin and Fouch (1981) discussed the confusion that exists over the actual stratigraphic limits of the Tuscher Formation and considered the lower Tuscher to be of Late Cretaceous age and a pebbly zone within the upper 50 m (164 ft) of the unit to be of Paleocene age. Pollen collected recently from two localities in the uppermost 25 m of the unit at Range Creek indicate a latest Campanian age for the top of the Tuscher Formation there (Fouch et al., 1983).

The Tuscher Formation is 280 m (919 ft) thick at Tusher Canyon, 183 m (600 ft) thick on the Green River, 109 m (358 ft) thick at Range Creek, and absent in more western localities, although the "pebbly beds" described below are believed to be equivalent to the upper pebbly part of the unit. The lower gradational contact with the Farrer Formation was picked in this study where the sandstone content of the section exceeds 50%. Some thickness variation within the study area may be due to the gradational nature of the contact.

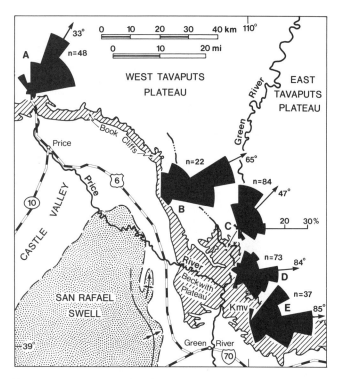

Figure 7—Paleocurrent data from Farrer Formation (east of San Rafael Swell) and lower 90 m of Price River Formation (Price Canyon). All measurements from axes of trough cross bedding. Localities: A. Price Canyon; B. Horse Canyon; C. Range Creek; D. Green River; E. Tusher Canyon. Ruled area indicates Mesaverde Group (Kmv).

The Tuscher Formation consists of tan-weathering sandstone sheets 7–15 m (23–50 ft) thick interbedded with siltstone and shale intervals 5–25 m (16–83 ft) thick. Within the lower part of the unit, the sheets typically consist of a lower medium- to fine-grained sandstone interval 5–10 m (16–33 ft) thick that fines upward to interbeds of very fine to fine-grained sandstone and siltstone that have depositional dips of as high as 20°. The inclined sandstone and siltstone beds were observed to terminate laterally in lenticular deposits of laminated shale and siltstone. The inclined sandstone beds are dominantly ripple laminated and range from 5 to 30 (2–12 in.) cm thick; interbedded siltstone is 5 to 10 cm (2–4 in.) thick. Bed sets (and laterally equivalent shale and siltstone lenses) reach 4 m (13 ft) thick. In some cases, the sandstone sheets grade upward into siltstone and shale intervals; in other cases, trough cross-bedded sandstone of the overlying sheet rests directly on any lithofacies in the lower sheet. The base of each sheet is a sharp erosional contact, and mudstone clasts are common in trough cross beds of the lower 2–5 m (7–16 ft) of each sheet. Trough cross beds are the most common primary structure. The sandstone is poorly sorted, with grain size ranging from fine to coarse.

Gray siltstone and claystone compose less than 50% of the unit. Broadly lenticular, very fine grained sandstone beds as much as 0.5 m (1.6 ft) thick occur within the siltstone intervals. The sandstone beds contain bone fragments, rootlet holes, and more rarely, the trace fossil *Palaeophycus*, possibly formed by insect burrowing (Stanley and

Fagerstrom, 1974; Ratcliff and Fagerstrom, 1980). Transitions from sandstone to siltstone are abrupt.

At Range Creek, the Green River, and Tusher Canyon, the Tuscher Formation coarsens upward into pebbly sandstone. Pebbles initially appear in the section dispersed in trough cross bedding and increase in abundance up-section. The upper 50 m (164 ft) of the section on the Green River consists of pebbly coarse sandstone with low-angle cross beds, often in lenticular scour-and-fill structures as much as 1 m (3 ft) thick. The pebbles are rounded, a maximum of 2.5 cm (1 in.) in diameter, and consist of gray and tan banded chert, black chert, very fine grained white quartz arenite, gray quartzite, pink monocrystalline quartz, pale green felsite(?), and white mudstone.

The sheetlike sandstone bodies within the lower part of the Tuscher Formation were most likely deposited by meandering rivers, as evidenced by the inclined lateral accretion beds associated with lenticular abandoned channel siltstone and shale, the fining-upward grain sizes, and the silty floodplain deposits that separate the sheets. The change upward to coarse-grained pebbly sandstones with low-angle cross beds and pebble lenses records a shift to deposition by pebbly braided streams. Paleocurrent measurements in the Tuscher Formation (Figure 8 C, D, and E) indicate northeastward transport in the lower part of the unit. Paleocurrents in the upper pebbly part of the unit are also northeasterly, with a north–northwest mean at the Range Creek locality (Figure 9). Cross bed measurements in the upper Price River Formation at Price Canyon and Joes Valley likewise indicate northeast-directed transport (Figure 8 A and B).

Pebbly Beds

West of Range Creek, the Tuscher Formation pinches out beneath Tertiary sedimentary rocks. At Little Park, only 120 m (394 ft) of Farrer rocks are preserved beneath the North Horn Formation of late Paleocene age. Local mapping of the Little Park locality shows that the North Horn Formation overlies the Farrer Formation with a few degrees of angular discordance (Figure 10). The angular relationship and the absence of Tuscher beds east of the Range Creek measured section are used to infer the regional truncation above the Mesaverde Group (Figure 11).

Broad, white-weathering lenses of coarse granule and small pebble sandstone occur between the Farrer Formation and the North Horn Formation. This white pebbly sandstone horizon occurs discontinuously along strike and is also present in the Horse Canyon section and at Whitmore Canyon. The basal contact is sharp and erosional. The sandstone ranges from 18 m (59 ft) thick at Little Park to 6 m (20 ft) thick at Horse Canyon and appears to be concordant with the underlying Farrer beds. It consists of poorly sorted, fine to coarse-grained, quartzose to sublithic sandstone in lenticular beds with discontinuous accumulations of 4–5 mm (0.16–0.2 in.) pebbles and angular granules above erosional bedding contacts. Pebble lithologies are dominated by light gray, very fine grained quartz sandstone with subordinate dark to light gray chert, monocrystalline white and pink quartz, and white mudstone. Rounded clasts, 5–10 cm (2–4 in.) in diameter, of fine-

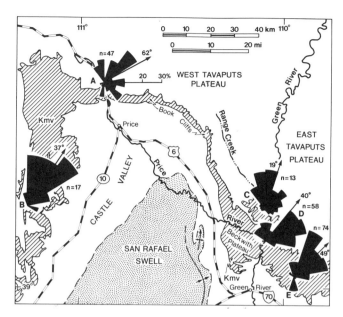

Figure 8—Paleocurrent data for upper part of Mesaverde Group. Measurements from trough cross-bedding axes of the Upper Price River Formation at Price Canyon (A) and Joes Valley (B), and the nonpebbly part of the Tuscher Formation at Range Creek (C), the Green River (D), and Tusher Canyon (E). Extent of Mesaverde Group outcrops is indicated by diagonal ruled pattern.

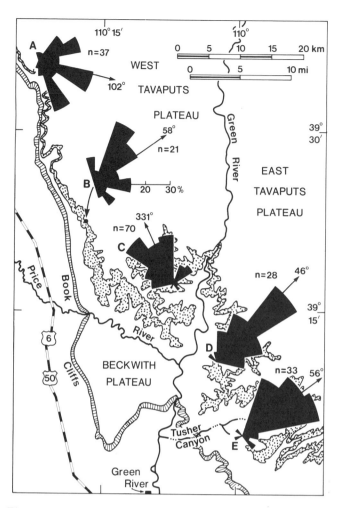

Figure 9—Paleocurrent data from "pebbly beds" at Whitmore Canyon (A) and Little Park (B), and pebbly part of Tuscher Formation at Range Creek (C), Green River (D), and Tusher Canyon (E). All measurements on axes of trough cross bedding. Distribution of Mesaverde Group is indicated by stipple. Ruled pattern indicates location of Book Cliffs.

grained quartzose sandstone are common in some trough lags. Trough cross beds and contorted lamination are the most common structures in the unit.

Measurements on trough cross bed axes in the discontinuous white sandstone indicate easterly to northeasterly paleocurrents (Figures 9 and 10). Pebble size and lithology in westernmost exposures of upper Tuscher beds are more similar to those of pebbles in the pebbly beds than to those in the upper Tuscher Formation at Tusher Canyon and elsewhere.

Because of their stratigraphic position and lithologic characteristics, the pebbly beds are believed to have been deposited at the same time as the uppermost part of the Tuscher Formation, and they thus probably represent east-flowing tributaries of the main northeast-flowing Tuscher river system. Because they are not dated, the possibility must be considered that they are actually reworked Cretaceous rocks deposited during the Tertiary prior to onlap by the North Horn Formation. Geologic relationships noted previously, however, indicate structural concordance with the Campanian section and stratigraphic equivalence with some part of the Tuscher Formation. In addition, a potential stratigraphic counterpart intermediate in age between Tuscher and North Horn beds remains unrecognized or is absent within the study area and for 20 km (12 mi) to the east, in the vicinity of Thompson Canyon, where conglomeratic sandstone beds of Paleocene age are present (T. D. Fouch, personal communication, 1983).

POST-CAMPANIAN OVERLAP STRATA (NORTH HORN FORMATION)

The North Horn Formation overlies the Price River Formation west of the San Rafael Swell and the Tuscher Formation east of the swell. The base of the unit is broadly diachronous, as discussed earlier, and varies in both lithic type and thickness. In the central Wasatch Plateau, the type North Horn Formation consists of 500 m (1640 ft) of variegated fluvial and lacustrine rocks deposited during Maestrichtian and Paleocene time, which is overlain by the Flagstaff Limestone (Spieker, 1946). At Joes Valley 10 km (6 mi) north of the type locality, tan fluvial sandstone of the North Horn overlies bleached feldspathic sandstone of the Price River Formation on a sharp contact marked by a conglomerate lag and grades upward into a section dominated by siltstone.

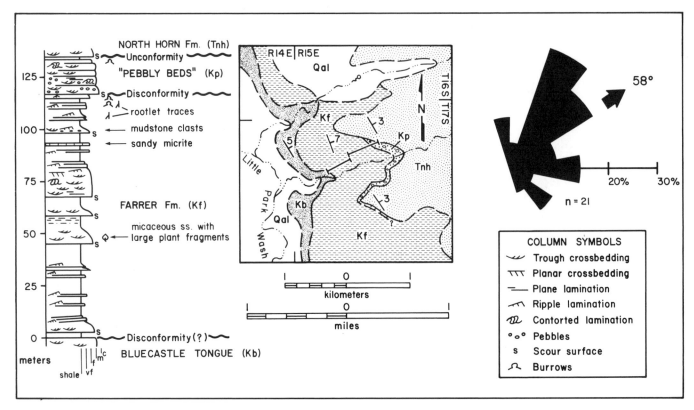

Figure 10—Stratigraphic and structural relationships of Farrer Formation, "pebbly beds," and North Horn Formation at Little Park locality, showing overlap of Tertiary strata. Paleocurrents measured in pebbly beds indicate northeastward to eastward flow on flank of San Rafael Swell, represented by dips in Cretaceous rocks, and contrast with strong northeastward to northward current directions nearby at Range Creek (Figure 9). Location of measured section is shown by line segments on map.

Thin-bedded, dark gray calcareous siltstone and mudstone of lacustrine origin form the basal 65 m (213 ft) of the North Horn Formation in Price Canyon (Lawton, 1983). The gray beds are overlain by 220 m (721 ft) of interbedded sandstone and siltstone. The sandstone and siltstone beds form upward-fining cycles that average 10 m (33 ft) in thickness. The sandstone beds have sharp bases; detrital micrite clasts and mudstone chips are common in the lower parts of the beds, which consist of fine- to medium-grained sandstone. Trough cross beds near the base decrease in amplitude upward and grade into ripple lamination. Very fine grained sandstone beds, 10–50 cm (4–20 in.) thick and interbedded with thin siltstone beds, cap the sandstone intervals and grade into thick siltstone intervals. In the upper part of the section, the fining-upward cycles are interrupted by intervals of very fine to fine-grained, rippled, and burrowed sandstone 5 m (16 ft) thick.

The gray calcareous siltstone beds in the lower part of the North Horn Formation in Price Canyon were deposited in a lacustrine environment. Lake sedimentation was succeeded by meandering stream deposition that characterized the North Horn Formation in Paleocene time (Fouch et al., 1983). Bioturbated and rippled horizons near the top of the unit represent lake margin sandstones deposited late in Paleocene time prior to deposition of the Flagstaff Limestone.

East of the San Rafael Swell, in contrast, no thick calcareous siltstone or shale is present at the base of the

North Horn section. North Horn-type clastics grade rapidly upward into and interfinger with Flagstaff carbonate strata, making the units difficult to differentiate. However, an interval of sandstone and shale from 15 to 40 m (49–131 ft) thick occurs beneath limestone-dominated lithologies in the areas studied. Sandstone beds are lenticular, as much as 10 m (33 ft) thick, and fine upward from medium- to fine-grained sandstone. The sandstone and shale were probably deposited in a meandering stream environment.

SANDSTONE COMPOSITION

A change from quartz-rich to lithic-rich sandstone compositions occurs within the Mesaverde Group above the Bluecastle Tongue of the Castlegate Sandstone. This change serves as a basis for dividing the nonmarine strata of the upper part of the Mesaverde Group into two compositional subdivisions, a quartzose petrofacies and a lithic petrofacies. The quartzose petrofacies includes a stratigraphic interval dominated by sandstones composed of greater than 80% total framework quartz. This interval includes the Castlegate Sandstone in Price Canyon and the Castlegate Sandstone, the Neslen Formation, and the Bluecastle Tongue of the Castlegate Sandstone in eastern exposures. The lithic petrofacies comprises a stratigraphic interval having sandstones generally containing less than 80% framework quartz. The lithic petrofacies includes the Price River

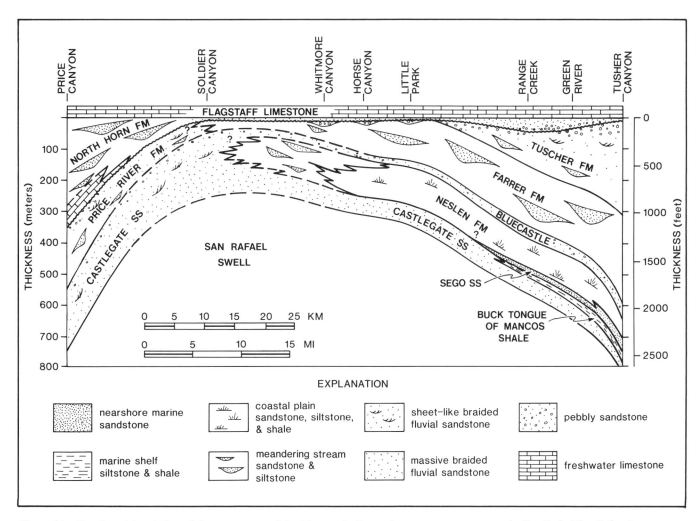

Figure 11—Stratigraphic relation of the upper part of the Mesaverde Group from west to east across the San Rafael Swell. Section thicknesses at Soldier and Whitmore canyons are from Fisher et al. (1960).

Formation in Price Canyon and the Farrer and Tuscher formations near the Green River. At the top of the Mesaverde section, a return to quartzose compositions occurs in both the Price River and Tuscher formations. A summary of mean modal compositions of Mesaverde Group sandstones is given in Table 1.

Although the petrologic shift occurs regionally at the same stratigraphic position, the lithic grain population present in the Price River Formation contrasts with that of the Farrer and Tuscher formations. Sandstone east of the San Rafael Swell is feldspathic litharenite (following the usage of McBride, 1963), with a maximum abundance of mica and volcanic lithic fragments occurring low in the section. Sedimentary lithic fragments increase in abundance up-section in the Tuscher Formation. Sandstone west of the swell is litharenite and sublitharenite with only small amounts of potassium feldspar. The contrast in grain types is shown in Figure 12 by the use of a triangular plot (Graham et al., 1976) of the relative abundances of polycrystalline quartz including chert (Qp), sedimentary lithic fragments (Ls), and volcanic lithic fragments (Lv). To ensure statistical significance of the lithic grain fraction plotted in Figure 12, only sandstone samples containing a

framework lithic population of greater than 25% are plotted. The plot indicates that the lithic grain population of sandstones in the Price River Formation is dominated by sedimentary rock fragments, whereas a variable but significant fraction of the lithic population of the Farrer and Tuscher formations consists of volcanic grain types. Lithic fragments are again uncommon in quartz-rich strata of the uppermost part of both the Price River and Tuscher formations.

Compositional trends are apparent within the petrographic subsuites. The percentage of total quartz in sandstones of the Price River Formation varies along a linear trend at a nearly constant mean L/(L + F) ratio of 0.80, where L is lithic fragments and F is feldspar (calculated from the mean modal composition in Table 1) (see Figure 2), with more quartzose samples occurring in the upper 90 m (295 ft) of the section. A similar linear trend in lithic sandstone of the Black Warrior and Ouachita basins has been explained as a result of addition of quartz grains from outside sources (multiple source dilution) and variable compositional maturity of the sandstones (Graham et al., 1976). Both processes may have operated in the Utah foreland, but the work of Pollack (1961) on compositional

Table 1—Mean modal compositions of Mesaverde Group sandstones and definition of grain parameters.[a,b]

Unit	n	Q	F	L	Qm	Lt	K	P	Qp	Ls	Lv	Detrital CO_3
Castlegate Sandstone	12	89.3 (12.7)	0.9 (1.2)	9.8 (13.0)	84.0 (11.8)	14.2 (12.3)	0.8 (1.1)	0.1 (0.2)	4.4 (3.4)	9.4 (12.8)	0.2 (0.6)	7.6 (12.9)
Neslen Formation	1	77.0	5.0	18.1	66.8	28.3	3.0	2.0	10.2	10.0	2.0	3.9
Bluecastle Tongue	8	96.8 (3.7)	0.5 (0.7)	2.6 (3.3)	92.7 (6.7)	6.7 (6.4)	0.1 (0.2)	0.3 (0.6)	2.0 (3.0)	2.6 (3.3)	0	2.0 (3.0)
Price River Formation	21	75.5 (17.2)	5.1 (3.8)	19.4 (13.7)	66.4 (18.9)	28.5 (15.4)	4.7 (3.9)	0.4 (0.6)	9.1 (4.0)	12.6 (8.2)	3.4 (5.1)	3.5 (4.3)
Farrer Formation	10	51.3 (8.0)	18.0 (4.4)	30.7 (10.4)	37.2 (7.0)	44.8 (9.1)	9.1 (2.7)	8.9 (2.8)	14.1 (4.2)	14.5 (4.3)	13.3 (9.7)	5.0 (3.7)
Tuscher Formation	13	64.1 (11.7)	15.4 (5.5)	20.5 (7.2)	51.5 (10.2)	33.1 (7.6)	10.1 (3.6)	5.3 (3.6)	12.6 (5.2)	10.5 (4.3)	7.2 (6.1)	1.2 (2.3)

[a]Numbers in parentheses are standard deviations.
[b]Q + F + L = 100%; Qm + F + Lt = 100%. Q is total framework quartz (Q = Qm + Qp), where Qm is monocrystalline quartz, including uniform extinction and undulose extinction; and Qp is polycrystalline quartz, including chert, polycrystalline quartz of sedimentary, igneous, and metamorphic origin, and aggregate quartz of indeterminate origin. F is total framework feldspar (F = K + P), where K is potassium feldspar and P is plagioclase feldspar. L is framework lithic fragments (for QFL plot; L = Ls + Lv + Lo), where Ls is sedimentary lithic fragments, including argillite (shale), very fine grained feldspathic sandstone, and detrital carbonate; Lv is volcanic (hypabyssal lithic fragments), including microlitic volcanic rock fragments, hypabyssal volcanic rock fragments, and felsite; and Lo is lithic fragments of other origin, including plutonic rock fragments, polycrystalline quartz + white mica, and polycrystalline white mica. Lt is total framework lithic fragments (Lt = L + Qp).

trends in sand of the South Canadian River suggests that transport distances significantly greater than the distance between the Wasatch Plateau and the Green River would have been required to alter modal compositions. Thus, multiple sources may have been a more important factor in determining petrographic variability.

A petrographic trend is also indicated for the lithic component of Farrer and Tuscher sandstones, in which the Lv content appears to vary along a linear trend at a nearly constant mean Qp/(Qp + Ls) ratio of 0.52 on the QpLvLs plot (calculated from mean modal compositions for the combined Farrer and Tuscher formations in Table 1) (see Figure 12). The Lv trend suggests either (1) selective disappearance of volcanic lithic fragments during transport of an originally volcanic-rich clastic suite derived from a source outside the thrust belt, or (2) dilution of an originally volcanic-rich clastic suite by addition of Ls and Qp components in a nearly constant ratio, possibly determined by the source area. In either case, it is clear that Price River fluvial systems did not contribute the observed Lv component. Moreover, paleocurrent data show that neither the Farrer nor the Tuscher Formation lies in a downcurrent direction from Price River sites studied, although more southern locations in the Wasatch Plateau may have contributed detritus.

TECTONIC AND PALEOGEOGRAPHIC EVOLUTION OF THE FORELAND REGION

The observed stratigraphic relationships and trends in depositional facies, dispersal directions, and sandstone compositions were a response to changing tectonic conditions in Late Cretaceous to early Tertiary time in the Utah foreland. Late in Mesaverde time, primary deformation and uplift shifted from the thrust belt to sites previously characterized by subsidence and clastic deposition within the foreland basin. The following discussion is an interpretation of tectonic and paleogeographic events for Campanian, Maestrichtian, and much of Paleocene time.

Middle to Late Campanian Time

During the time bracketed by the range fossils *Baculites asperiformis* and *Exiteloceras jennyi* (74–79 m.y.), the uplifted terrane of the thrust belt was the primary source of quartzose clastic detritus that was shed eastward and southeastward into the subsiding foreland basin. The lower cliff-forming part of the Castlegate Sandstone has long been interpreted as a depositional response to an episode of major uplift in the thrust belt (Spieker, 1946, 1949; Van De Graaff, 1972). Fouch et al. (1983) suggested that uplift coincided with movement on the Meade–Crawford thrust system in northern Utah and Wyoming. It may represent early movement on the Charleston–Nebo thrust and uplift of Precambrian quartzite in thrust sheets above the Pavant thrust farther south (Lawton, 1983). The Castlegate Sandstone records fluvial deposition on a sandy coastal plain that extended for at least 130 km (80 mi) (the present exposed length of the Castlegate Sandstone on the Wasatch Plateau) parallel to the thrust front. Dominantly braided rivers flowed directly away from the linear uplifted front to a shoreline located near the Colorado–Utah border (Van De Graaff, 1972).

Decreased sediment influx or more rapid subsidence in the basin following deposition of the lower part of the Castlegate resulted in a return of marine conditions to the

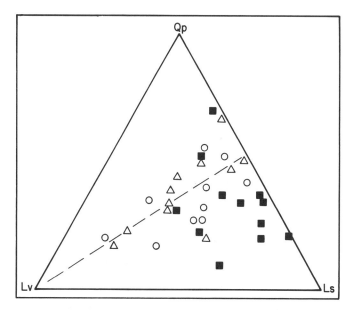

Figure 12—Triangular QpLsLv compositional plot of sandstones of Price River, Farrer, and Tuscher formations. Qp, polycrystalline quartz, including chert; Ls, sedimentary lithic grains, including detrital carbonate; Lv, volcanic lithic grains. Dashed line is Qp/(Qp + Ls) = 0.52, calculated from mean modal values in Table 1.

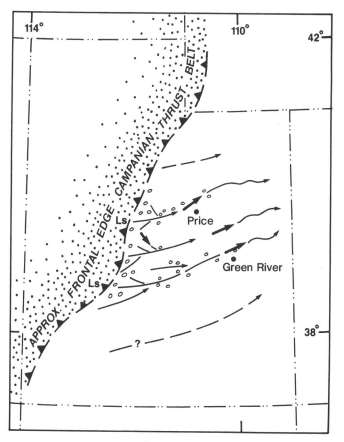

Figure 13—Inferred coastal plain fluvial system developed during deposition of the Bluecastle Tongue of Castlegate Sandstone. Bold arrows show paleocurrent locations of Figure 6, except for arrow nearest thrust front, taken from data for uppermost Sixmile Canyon Formation (Indianola Group) (Lawton, 1982). Symbol within source area indicates derivative grain type: Ls, sedimentary lithic fragments.

Green River area. Time-equivalent rocks of the middle part of the Castlegate Sandstone in Price Canyon were deposited by meandering rivers.

The Bluecastle Tongue of the Castlegate Sandstone forms a clastic blanket continuous throughout the study area, and constitutes the last major pulse of coarse sediment from the thrust belt that can be traced petrologically from the northern Wasatch Plateau into the Green River area (Figure 13). Post-Bluecastle rivers transported detritus northeastward from the vicinity of Price Canyon. The Bluecastle Tongue was deposited by braided rivers that increased in sinuosity eastward away from the source area. Like the lower part of the Castlegate Sandstone, the Bluecastle Tongue was deposited on a sandy coastal plain, but stream flow was obliquely away from the thrust belt. The Bluecastle Tongue is thinner than the lower part of the Castlegate Sandstone, but composed of coarser sandstone, factors that suggest derivation from a discrete uplift event in the thrust belt and a source more proximal than the source of the Castlegate Sandstone. This uplift is postulated by Fouch et al. (1983) to be coincident with movement on the Absaroka thrust in Wyoming and Utah north of the Uinta Mountains. Folding and ramping of the Charleston–Nebo structure at this time probably exposed quartzite cobble conglomerates of the Upper Cretaceous Indianola Group, which contributed to the coarse quartzose nature of the Bluecastle.

Late Late Campanian Time

Evidence exists in the sedimentary record for modification and eventual termination of foreland basin deposition between the zones of *Exiteloceras jennyi* and *Baculites cuneatus* (73–74 m.y.). Thrust belt deformation west of the Price Canyon area continued, probably as late uplift of the Charleston–Nebo structure and folding above more eastern

blind thrusts (Lawton, 1985). Paleocurrent data from the Price River and Farrer formations indicate that the major fluvial systems were longitudinal or subparallel to the thrust front. Northeast-flowing meandering rivers deposited the Price River Formation, which consists of sandstone rich in sedimentary lithic fragments from the thrust belt. Coeval meandering rivers that deposited the Farrer Formation also flowed northeastward, but sandstone compositions indicate that these rivers tapped volcanic sources lying far to the southwest (Figure 14). The Farrer rivers thus probably constituted a long trunk system. Tributary streams draining the thrust belt transported sedimentary lithic grains and monocrystalline quartz grains to the trunk system. The amount of detrital material derived from the thrust belt probably varied as tributary systems evolved and resulted in the variable percentage of volcanic lithic grains observed on the QpLsLv plot (Figure 12). Although the fluvial pattern apparently consisted of a trunk–tributary system oriented subparallel to the thrust belt, a north to northwest-trending shoreline lay approximately 160 km (100 mi) to the east in present-day Colorado and southwestern Wyoming (Kiteley, 1983; Molenaar, 1983); thus, the system probably retained some characteristics of a coastal plain as well. The fluvial pattern developed during deposition of the Farrer

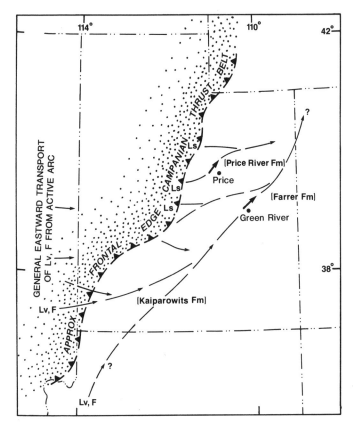

Figure 14—Inferred trunk–tributary fluvial system and source terranes during deposition of Farrer and lower Tuscher formations and their equivalents. Bold arrows show paleocurrent localities of Figures 7 and 8. Symbols within source areas indicate derivative grain types: Ls, sedimentary lithic fragments; Lv, volcanic lithic fragments; F, feldspar.

Formation persisted during deposition of most of the Tuscher Formation.

Time-equivalent units at the top of the Cretaceous section in the Kaiparowits Plateau region of southwestern Utah are richer in feldspar and volcanic fragments than the Farrer and Tuscher formations (Bowers, 1972). These units, the Kaiparowits and overlying Canaan Peak formations, are inferred to have been deposited by upstream reaches of the Farrer and Tuscher fluvial trunk system more proximal to the volcanic source rocks that probably lay beyond the thrust belt (Figure 14).

Depositional changes in the uppermost part of the Mesaverde section indicate a change in the trunk–tributary fluvial system. Braided fluvial deposition in the uppermost part of the Price River Formation of Price Canyon probably records the final stages of folding in the easternmost thrust belt. In the east, braided rivers that deposited the upper part of the Tuscher Formation succeeded the earlier meandering streams. As mentioned previously, sandstone of the uppermost part of the Tuscher Formation is quartzose. The decrease and disappearance up-section of volcanic lithic grains within the upper part of the Tuscher Formation is interpreted to signal progressive isolation of the Tuscher river systems from the volcanic terrane to the southwest. Mesozoic strata of the San Rafael Swell, Circle Cliffs uplift,

and the Monument upwarp probably began to provide a sedimentary source terrane for the uppermost Tuscher sandstone (Figure 15). With the initiation of the thick-skinned uplifts, the fluvial systems were altered, becoming intermontane trunk–tributary systems. The intraformational nature of the quartz arenite and sublitharenite clasts found in the pebbly beds indicates recycling of foreland basin sandstone from the San Rafael Swell. Moreover, a contrast in pebble lithology, size, and shape between the upper part of Tuscher Formation and the pebbly beds suggests different sources for the pebble populations. Pebbles of the pebbly beds were most likely recycled from uplifted pebbly strata of the Price River Formation.

On the basis of this study, the favored interpretation for the age of the pebbly beds, and hence, uplift of the San Rafael Swell, is latest Campanian. The interpretation requires that the synorogenic pebbly beds are equivalent to the uppermost pebbly strata of the Tuscher Formation. This inference is supported by the stratigraphic equivalence of the two units beneath the pre-North Horn unconformity. Onlap of basal North Horn strata across truncated older beds indicates that structural relief on the swell was attained prior to North Horn deposition (Figure 10).

Growth of the San Rafael Swell and development of intermontane fluvial systems clearly occurred between latest Campanian and late Paleocene time. The Campanian age interpreted here is consistent with stratigraphic and sedimentologic relationships in the study area, particularly in the absence of an intermediate Maestrichtian or lower Paleocene unit between the Tuscher and North Horn formations. Because such a unit appears to be present east of the study area, as mentioned previously, careful mapping of the unconformity below the North Horn and direct dating of the pebbly beds may be necessary to unequivocally resolve the age question.

Post-Campanian Time

Campanian rocks are succeeded by an unconformity everywhere in the study area and probably throughout the eastern Utah and western Colorado regions (Hansley and Johnson, 1980). Sedimentation resumed in the study area in the vicinity of the Wasatch Plateau in Maestrichtian time at about 70 m.y. ago (Spieker, 1946; Fouch et al., 1983). Deposition first occurred in meandering stream and lacustrine environments as drainage ponded in the structural depression between the thrust belt and the San Rafael Swell. East-directed meandering fluvial deposition resumed and continued into Paleocene time. Sandstone compositions and the stratigraphic position of the lower Tertiary fluvial strata indicate equivalence to the quartzite- and carbonate-clast conglomerates that form a basal facies of the North Horn Formation to the west and onlap deformed rocks of the thrust belt (Merrill, 1972; Pinnell, 1972; Young, 1976; Weiss, 1982). The coarse-grained strata may record final isostatic uplift of the thrust terrane as tectonically thickened crust was thinned by erosion. The deposition that began in Maestrichtian time expanded westward from the Wasatch Plateau and onlapped the frontal deformed belt of the thrust terrane by late Paleocene time (Figure 16) (Stanley and Collinson, 1979; Fouch et al., 1982). The depositional relationships of the overlap sequence thus indicate that

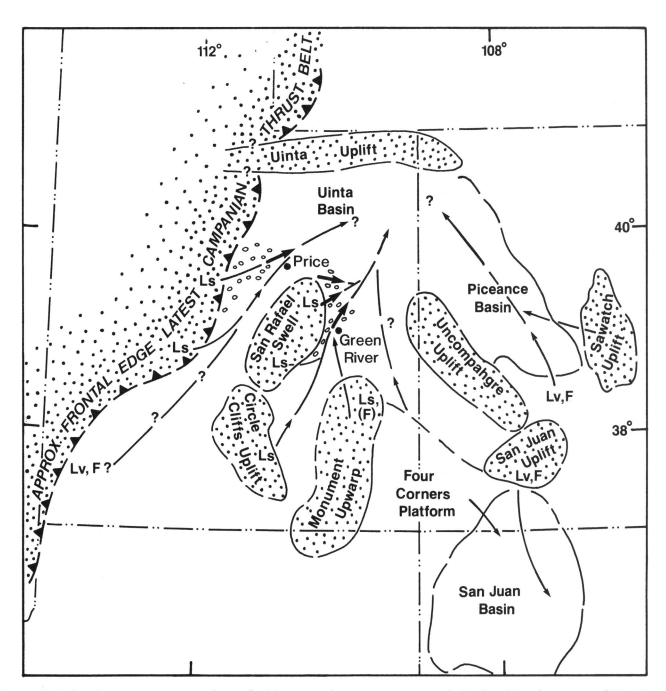

Figure 15—Inferred intermontane trunk–tributary fluvial system and extant source terranes during deposition of upper part of Price River Formation, pebbly Tuscher Formation, pebbly beds, and equivalents. Equivalents include the Canaan Peak Formation in southwest Utah (Bowers, 1972), the Ohio Creek Member of the Hunter Canyon and Mesaverde formations in the Piceance Creek basin (Johnson and May, 1980), and the Fruitland and Kirtland formations of the San Juan basin. Bold arrows show relevant paleocurrent localities of Figures 8 and 9. Symbols within source areas indicate derivative grain types: Ls, sedimentary lithic fragments; Lv, volcanic lithic fragments; F, feldspar. Inferred source areas for Piceance Creek basin are from Hansley and Johnson (1980); for the San Juan basin from M. A. Klute (personal communication, 1982). Structural elements of Colorado Plateau and Rocky Mountain regions are after Kelley (1955).

thrusting was unequivocally complete by Paleocene time, but probably terminated during the Maestrichtian. North Horn rocks also onlapped the arch of the San Rafael Swell farther east as the structural basin filled with sediment. Onlap of the swell was complete by latest Paleocene time (Fouch et al., 1982).

CONCLUSIONS

Patterns of sedimentation and sandstone compositions in central Utah record an eastward expansion of tectonism from the thrust belt into the foreland region during late Campanian time. Foreland basin sedimentation was

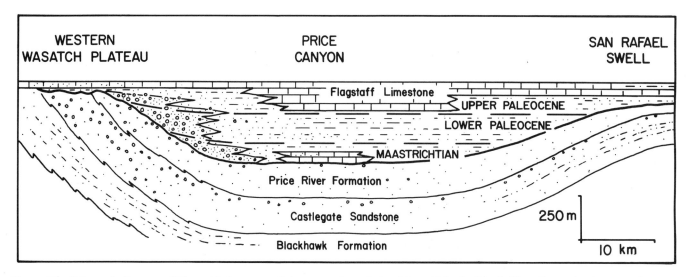

Figure 16—Schematic diagram of Maestrichtian–upper Paleocene onlap relationships within the North Horn Formation between the thrust belt and San Rafael Swell. Western half of figure is after Stanley and Collinson (1979).

terminated in latest Campanian time as deformation in the thrust belt waned and uplift occurred within the basin. Intrabasin uplift took the form of individual domes and arches over basement faults (Davis, 1978). Latest stages of folding in the thrust belt are interpreted to have been contemporaneous with uplift of the San Rafael Swell. Coeval, but spatially separate, thin-skinned and thick-skinned deformation is consistent with structural interpretations inferred from detailed sedimentologic study of the Hoback basin. This basin is situated between the Wyoming thrust belt and the Wind River uplift, which show overlapping periods of active deformation (Dorr et al., 1977). Both the thrust belt and the San Rafael Swell were onlapped by sedimentary strata of late Paleocene age.

ACKNOWLEDGMENTS

This study represents part of a dissertation project conducted at the University of Arizona. I thank W. R. Dickinson and T. D. Fouch for suggestions and comments throughout the study. C. W. Keighin generously provided thin sections of core samples he collected at a well site in Price Canyon. Reviews by R. M. Flores, C. W. Keighin, and F. R. Van De Graaff improved the manuscript. In addition to his careful review, Flores suggested terminology for the types of evolving fluvial systems described in this paper. All research was supported by Earth Sciences Division, National Science Foundation Grant EAR-7926379. Drafting by Rick Brokaw was funded in part by the Laboratory of Geotectonics, Department of Geosciences, University of Arizona. Paleocurrent data were processed on ROSE3, Programming Library, Laboratory of Geotectonics, developed by G. L. Cole.

REFERENCES CITED

Armstrong, R. L., 1968, Sevier orogenic belt in Nevada and Utah: Geological Society of America Bulletin, v. 79, p. 429–458.

Balsley, J. K., 1980, Cretaceous wave-dominated delta systems, Book Cliffs, east central Utah: Amoco Production Co. Field Guide, 163 p.

Bowers, W. E., 1972, The Canaan Peak, Pine Hollow, and Wasatch formations in the Table Cliff region, Garfield County, Utah: U.S. Geological Survey Bulletin 1331-B, 39 p.

Cant, D. J., and R. G. Walker, 1978, Fluvial processes and facies sequences in the sandy braided South Saskatchewan River, Canada: Sedimentology, v. 25, p. 625–648.

Clark, F. R., 1928, Economic geology of the Castlegate, Wellington, and Sunnyside quadrangles, Carbon County, Utah: U.S. Geological Survey Bulletin 793, 165 p.

Cobban, W. A., and J. B. Reeside, Jr., 1952, Correlation of the Cretaceous formations of the Western Interior of the United States: Geological Society of America Bulletin, v. 63, p. 1011–1044.

Davis, G. H., 1978, Monocline fold pattern of the Colorado Plateau, *in* V. Matthews III, ed., Laramide folding associated with basement block faulting in the western United States: Geological Society of America Memoir 151, p. 215–233.

Dorr, J. A., Jr., D. R. Spearing, and J. R. Steidtmann, 1977, Deformation and deposition between a foreland uplift and an impinging thrust belt: Geological Society of America Special Paper 177, 82 p.

Fisher, D. J., 1936, The Book Cliffs coal field in Emery and Grand counties, Utah: U.S. Geological Survey Professional Paper 332, 80 p.

Fisher, D. J., C. E. Erdman, and J. B. Reeside, 1960, Cretaceous and Tertiary formations of the Book Cliffs, Carbon, Emery, and Grand counties, Utah, and Garfield and Mesa counties, Colorado: U.S. Geological Survey Professional Paper 332, 80 p.

Flores, R. M., and W. E. Marley III, 1979, Physical stratigraphy and coal correlation of the Blackhawk Formation and Star Point Sandstone, sections A–A''' and B–B''', near Emery, Utah: U.S. Geological Survey Miscellaneous Field Studies Map MF-1068.

Fouch, T. D., J. H. Hanley, R. M. Forester, C. W. Keighin, J. K. Pitman, and D. J. Nichols, in press, Chart showing lithology, mineralogy, and paleontology of the nonmarine North Horn Formation, and Flagstaff Member of the Green River Formation, Price Canyon, central Utah: a principal reference section: U.S. Geological Survey Miscellaneous Investigations Chart I-1797-A, 2 sheets, 9 printed text pages.

Fouch, T. D., T. F. Lawton, D. J. Nichols, W. B. Cashion, and W. A. Cobban, 1982, Chart showing preliminary correlation of major Albian to middle Eocene rock units from the Sanpete Valley in central Utah to the Book Cliffs in eastern Utah, *in* D. L. Nielson, ed., Overthrust belt of Utah: Utah Geological Association Publication 10, p. 267–272.

————, 1983, Patterns and timing of synorogenic sedimentation in Upper Cretaceous rocks of central and northeast Utah, *in* M. W. Reynolds, and E. D. Dolly, eds., Symposium 2—Mesozoic paleogeography of west-central United States: Society of Economic Paleontologists and Mineralogists, Rocky Mountain Section, p. 305–336.

Gill, J. R., and W. J. Hail, Jr., 1975, Stratigraphic sections across Upper Cretaceous Mancos Shale–Mesaverde Group boundary, eastern Utah and western Colorado: U.S. Geological Survey Oil and Gas Investigations Chart OC-68.

Graham, S. A., R. V. Ingersoll, and W. R. Dickinson, 1976, Common provenance for lithic grains in Carboniferous sandstones from Ouachita Mountains and Black Warrior basin: Journal of Sedimentary Petrology, v. 46, p. 620–632.

Hansley, P. L., and R. C. Johnson, 1980, Mineralogy and diagenesis of low-permeability sandstones of Late Cretaceous age, Piceance Creek basin, northwestern Colorado: Mountain Geologist, v. 17, p. 88–129.

Harms, J. C., J. B. Southard, D. R. Spearing, and R. G. Walker, 1975, Depositional environments as interpreted from primary sedimentary structures and stratification sequences: Society of Economic Paleontologists and Mineralogists Short Course Notes, n. 2, 161 p.

Harris, H. D., 1959, Late Mesozoic positive area in western Utah: AAPG Bulletin, v. 43, p. 2636–2652.

Johnson, R. C., and F. May, 1980, A study of the Cretaceous–Tertiary unconformity in the Piceance Creek basin, Colorado—the underlying Ohio Creek Formation (Upper Cretaceous) redefined as a member of the Hunter Canyon or Mesaverde Formation: U.S. Geological Survey Bulletin 1482B, 27 p.

Keighin, C. W., and T. D. Fouch, 1981, Depositional environments and diagenesis of some nonmarine Upper Cretaceous reservoir rocks, Uinta Basin, Utah, *in* F. G. Ethridge, and R. M. Flores, eds. Recent and ancient nonmarine depositional environments: models for exploration: Society of Economic Paleontologists and Mineralogists Special Publication 31, p. 109–125.

Kelley, V. C., 1955, Monoclines of the Colorado Plateau: Geological Society of America Bulletin, v. 66, p. 789–804.

Kiteley, L. W., 1983, Paleogeography and eustatic–tectonic model of late Campanian (Cretaceous) sedimentation, southwestern Wyoming and northwestern Colorado, *in* M. W. Reynolds, and E. D. Dolly, eds., Symposium 2—Mesozoic paleogeography of west-central United States: Society of Economic Paleontologists and Mineralogists, Rocky Mountain Section, p. 273–303.

Lawton, T. F., 1982, Lithofacies correlations within the Upper Cretaceous Indianola Group, central Utah, *in* D. L. Nielson, ed., Overthrust belt of Utah: Utah Geological Association Publication 10, p. 199–213.

————, 1983, Tectonic and sedimentologic evolution of the Utah foreland basin: Ph.D. thesis, University of Arizona, Tucson, 217 p.

————, 1985, Style and timing of frontal structures, thrust belt, central Utah: AAPG Bulletin, v. 69, p. 1145–1159.

Leeder, M. R., 1973, Fluviatile fining-upwards cycles and the magnitude of paleochannels: Geological Magazine, v. 110, p. 265–276.

McBride, E. F., 1963, A classification of common sandstones: Journal of Sedimentary Petrology, v. 33, p. 664–669.

Merrill, R. C., 1972, Geology of the Mill Fork area, Utah: Brigham Young University Geology Studies, v. 19, p. 65–88.

Molenaar, C. M., 1983, Major depositional cycles and regional correlations of Upper Cretaceous rocks, southern Colorado Plateau and adjacent areas, *in* M. W. Reynolds, and E. D. Dolly, eds., Symposium 2—Mesozoic paleogeography of west-central United States: Society of Economic Paleontologists and Mineralogists, Rocky Mountain Section, p. 201–224.

Pinnell, M. L., 1972, Geology of the Thistle quadrangle, Utah: Brigham Young University Geology Studies, v. 19, p. 89–130.

Pollack, J. M., 1961, Significance of compositional and textural properties of South Canadian River channel sands, New Mexico, Texas, and Oklahoma: Journal of Sedimentary Petrology, v. 31, p. 15–37.

Ratcliffe, B. C., and T. A. Fagerstrom, 1980, Invertebrate lebensspuren of Holocene floodplains: their morphology, origin and paleoecological significance: Journal of Paleontology, v. 54, p. 614–630.

Spieker, E. M., 1931, The Wasatch coal field, Utah: U.S. Geological Survey Bulletin 819, 210 p.

————, 1946, Late Mesozoic and early Cenozoic history of central Utah: U.S. Geological Survey Professional Paper 205-D, p. 117–161.

————, 1949, The transition between the Colorado Plateau and the Great Basin in central Utah: Utah Geological Society Guidebook to the Geology of Utah, n. 4, 106 p.

Spieker, E. M., and J. B. Reeside, Jr., 1925, Cretaceous and Tertiary formations of the Wasatch Plateau, Utah:

Geological Society of America Bulletin, v. 36, p. 435–454.

Stanley, K. O., and J. W. Collinson, 1979, Depositional history of Paleocene–Lower Eocene Flagstaff Limestone and coeval rocks, central Utah: AAPG Bulletin, v. 63, p. 311–323.

Stanley, K. O., and J. A. Fagerstrom, 1974, Miocene invertebrate trace fossils from a braided river environment, western Nebraska, U.S.A.: Palaeogeography, Palaeoclimatology, Palaeoecology, v. 15, p. 63–82.

Van De Graaff, F. R., 1972, Fluvial-deltaic facies of the Castlegate Sandstone (Cretaceous), east-central Utah: Journal of Sedimentary Petrology, v. 42, p. 558–571.

Weiss, M. P., 1982, Structural variety on the east front of the Gunnison Plateau, *in* D. L. Nielson, ed., Overthrust belt of Utah: Utah Geological Association Publication 10, p. 49–63.

Young, G. E., 1976, Geology of Billies Mountain quadrangle, Utah County, Utah: Brigham Young University Geology Studies, v. 23, p. 205–286.

Biostratigraphic Units and Tectonism in the Mid-Cretaceous Foreland of Wyoming, Colorado, and Adjoining Areas

E. A. Merewether
W. A. Cobban
U.S. Geological Survey,
Denver, Colorado

Chronolithologic units and unconformities in mid-Cretaceous formations of the central Rocky Mountains region indicate widespread marine transgressions and regressions as well as recurrent deformation of the foreland in the Western Interior during Cenomanian, Turonian, and Coniacian time (88–96 m.y. ago). The stratigraphic record of the widely recognized Cenomanian–early Turonian transgression, middle Turonian regression, and late Turonian–Coniacian transgression was modified in several areas by episodes of slight uplift and attendant erosion. The most evident tectonism was in western Montana during middle to late Cenomanian time (93–94 m.y. ago); in western Wyoming and adjoining areas during the early Turonian–earliest middle Turonian (90–91 m.y. ago); in north-central Colorado, eastern Wyoming, and northwestern Wyoming in the early late Turonian (89.8 m.y. ago); and in northeastern Colorado, Wyoming, and southwestern Montana in the late late Turonian (89.3 m.y. ago). Crestlines of most of the swells trend generally either northwest or northeast. The tectonism of the mid-Cretaceous foreland corresponds in age to displacements of thrusts in the Sevier orogenic belt of southwestern Wyoming and southeastern Idaho. Furthermore, much of the foreland deformation probably reflects episodes of eastward thrusting in the thrust belt.

INTRODUCTION

Lateral changes in mid-Cretaceous formations and associated unconformities of the central Rocky Mountains region (Figure 1) reflect both the regional tectonism and the widespread marine transgressions and regressions of Cenomanian, Turonian, and Coniacian time (88–96 m.y. ago). Nonmarine and nearshore marine siliciclastic units in the Mancos Shale of central Utah and in the Frontier Formation of northern Utah and western Wyoming grade eastward into offshore marine siliciclastic and carbonate units in the Graneros Shale, Belle Fourche Shale, Greenhorn Limestone, Carlile Shale, and Niobrara Formation of eastern Colorado, eastern Wyoming, and adjacent areas. These sequences of mid-Cretaceous strata range in thickness from at least 2360 m (7750 ft) in northern Utah (Hale, 1962; Ryer, 1977) to about 195 m (640 ft) in south-central Colorado (Scott, 1969). Hiatuses in the sequences, which are commonly determined from collections of molluscan index fossils, have been recognized throughout large elongate areas that trend generally northwest, north, and northeast in Utah, Colorado, Wyoming, and Montana. Presumably, these hiatuses are generally largest along the crests of structural swells where the amount of truncation was greatest. Ages of the uplifts were determined at localities where the hiatuses are smallest.

The main purpose of this paper is to describe tectonic events of the mid-Cretaceous foreland in the Rocky Mountains by interpreting time-stratigraphic units and hiatuses of mid-Cretaceous age. Data were obtained from outcrops, cores, and borehole logs. The chronostratigraphic units of this study were derived largely from collections of marine molluscan fossils, which were identified by W. A. Cobban and N. F. Sohl of the U.S. Geological Survey, and from the succession of index fossils established for mid-Cretaceous time by Cobban (Figure 2). Palynomorphs from outcrops and cores in the eastern part of the thrust belt in Wyoming and Montana were identified and interpreted by D. J. Nichols of the U.S. Geological Survey. Potassium-argon ages of several molluscan fossil zones were supplied by Obradovich and Cobban (1975) and modified by Lanphere and Jones (1978).

CHRONOSTRATIGRAPHIC DIAGRAMS

Regional changes in the lithofacies and age of mid-Cretaceous rocks and in the magnitude of associated hiatuses

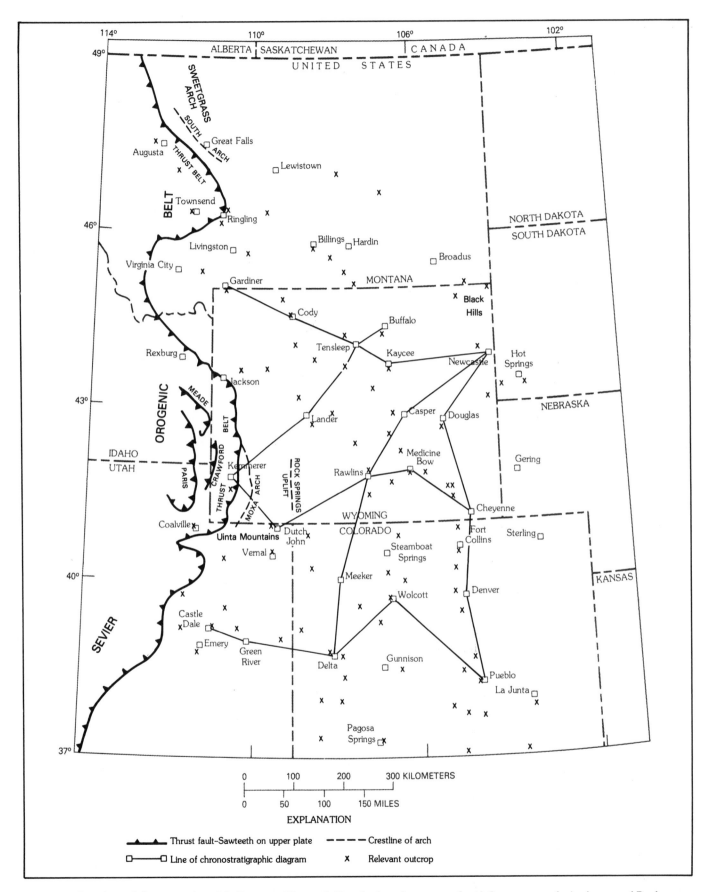

Figure 1—Locations of chronostratigraphic diagrams (Figures 3–8) and selected outcrops of mid-Cretaceous rocks in the central Rocky Mountains region. Structural features from Grose (1972), U.S. Geological Survey and AAPG (1962), Armstrong and Oriel (1965), Cobban et al. (1976), and Royse et al. (1975).

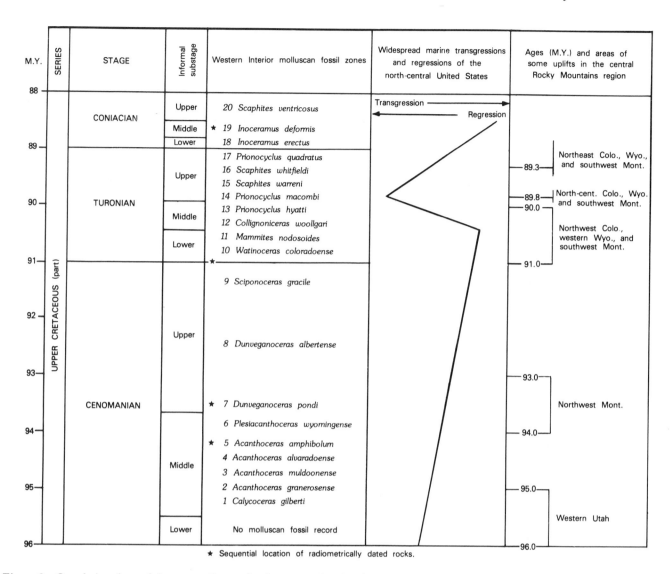

Figure 2—Correlation chart of the age, molluscan fossil zones, and major shoreline movements for mid-Cretaceous strata and the age and location of some mid-Cretaceous uplifts in the central Rocky Mountains region.

are depicted in a series of chronostratigraphic diagrams (see Figures 3–8). Outcrop sections from scattered localities represented on the diagrams were correlated largely by means of invertebrate fossils in the marine beds.

Castle Dale, Utah, to Pueblo, Colorado

The mid-Cretaceous rocks depicted on Figure 3 thin eastward from about 550 m (1800 ft) near Castle Dale to about 195 m (640 ft) near Pueblo. Nonmarine strata and nearshore marine sandstone of Cenomanian age in the Dakota Sandstone onlap eroded beds of Early Cretaceous age toward the west in east-central Utah and western Colorado (Fouch et al., 1983). The Dakota in central Utah is conformably overlain by marine shale of latest Cenomanian–middle Turonian age in the Tununk Member of the Mancos Shale. Near Castle Dale, the lower part of the Tununk was deposited during the last part of the marine transgression represented by the Dakota. This transgression is also recorded in southeastern Colorado by offshore marine strata that include, from oldest to youngest, noncalcareous

shale of Cenomanian age in the Graneros Shale and calcareous shale and limestone of Cenomanian–early Turonian age in the Greenhorn Limestone (Scott, 1969).

Near Castle Dale, marine shale of early to middle Turonian age in the upper part of the Tununk Member and nearshore marine sandstone of late middle Turonian age in the overlying Ferron Sandstone Member of the Mancos Shale were deposited during a widely recognized marine regression. Southwest of Castle Dale in the vicinity of Emery, Utah (Figure 1), the Ferron consists of marine and nonmarine rocks, including coal (Ryer, 1981). Offshore marine strata of similar age in the Carlile Shale of southeastern Colorado near Pueblo include, in ascending order, lower middle Turonian calcareous shale of the Fairport Chalky Shale Member, upper middle Turonian silty shale of the Blue Hill Shale Member, and upper middle Turonian sandstone of the Codell Sandstone Member (Scott, 1969). Hiatuses in these mid-Cretaceous sequences are of early Turonian age near Delta, Colorado, and of late Cenomanian–late Turonian age near Wolcott, Colorado (Figure 1).

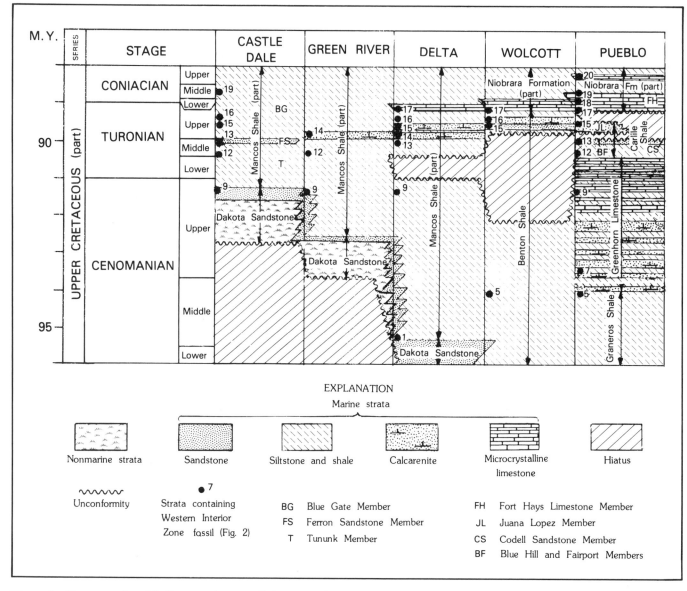

Figure 3—Chronostratigraphic diagram of some mid-Cretaceous formations in Utah and Colorado (localities shown in Figure 1).

In the Mancos Shale near Castle Dale, the Ferron Sandstone Member is conformably overlain by marine shale and siltstone of late Turonian–Coniacian age in the Blue Gate Member. These upper Turonian strata and the uppermost Turonian–middle Coniacian limestone units depicted on Figure 3 for the Colorado localities apparently record a widespread marine transgression and a local marine regression, respectively. In the Carlile Shale near Pueblo in southeastern Colorado, upper Turonian calcarenite of the Juana Lopez Member locally either disconformably overlies the Codell Sandstone Member or is absent. The hiatus between the Codell and the Juana Lopez is earliest late Turonian. The Juana Lopez is, in turn, locally either disconformably overlain by or occurs as clasts within limestone of latest Turonian age in the Fort Hays Limestone Member of the Niobrara Formation. The hiatus between the Juana Lopez and the Fort Hays is of late Turonian age. Limestone and calcareous shale of Coniacian age in the

upper part of the Fort Hays are overlain by calcareous shale, chalk, and limestone of Coniacian to Campanian age in the Smoky Hill Shale Member of the Niobrara (Scott, 1969).

Kemmerer to Cheyenne, Wyoming

The strata of mid-Cretaceous age depicted on Figure 4 are about 2200 ft (670 m) thick near Kemmerer, about 1100 ft (335 m) thick at Rawlins, and about 550 ft (170 m) thick near Cheyenne. In southwestern Wyoming in the vicinity of Kemmerer, nonmarine rocks of latest Albian–early Turonian age in the Chalk Creek Member of the Frontier Formation (Myers, 1977) overlie and interfinger with marine strata in the Lower Cretaceous Aspen Shale (M'Gonigle, 1982). The Chalk Creek is conformably overlain by nearshore marine sandstone of early Turonian age in the Coalville Member of the Frontier. The Coalville is conformably overlain by marine shale of early to middle Turonian age in the Allen Hollow Member of the Frontier. This marine and

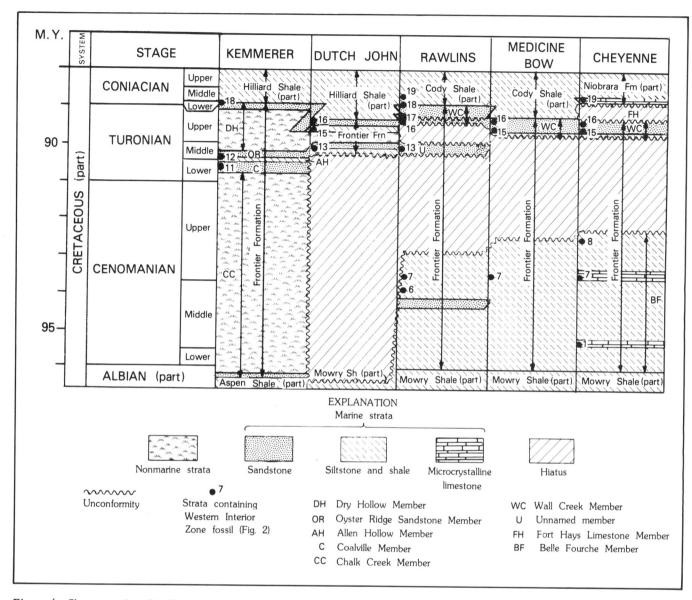

Figure 4—Chronostratigraphic diagram of some mid-Cretaceous formations in northeastern Utah and southern Wyoming (localities shown in Figure 1).

nonmarine sequence seems to indicate a prograding shoreline in early and middle Cenomanian time followed by a marine transgression in late Cenomanian and early Turonian time. Eastward of Kemmerer, the sequence is represented by marine strata and significant hiatuses. At Dutch John, Utah, on the north flank of the Uinta Mountains, a hiatus extends from late Albian to middle Turonian time. In southeastern Wyoming near Cheyenne, the Frontier includes offshore marine shale and limestone of Cenomanian age in the Belle Fourche Member, a hiatus of late Cenomanian–late Turonian age, and marine sandstone of late Turonian age in the Wall Creek Member.

The upper part of the Frontier in southwestern Wyoming contains, from oldest to youngest, marine shale of the Allen Hollow Member, nearshore marine sandstone of middle Turonian age in the Oyster Ridge Sandstone Member, nonmarine beds of Turonian–early Coniacian age in the Dry Hollow Member, and nearshore marine sandstone of early

Coniacian age at the top of the Dry Hollow. Conformably overlying the Frontier is marine shale of early Coniacian age in the lower part of the Hilliard Shale. This sequence of marine and nonmarine rocks apparently represents a marine regression in middle Turonian time and a marine transgression in late Turonian and early Coniacian time. Hale (1962) has suggested that the lower contact of the Dry Hollow Member is a disconformity. If he is correct, the disconformity would represent at least part of late middle Turonian to late late Turonian time. Near Rawlins, nearshore marine sandstone of late middle Turonian age in an unnamed member of the Frontier is disconformably overlain by marine sandstone of middle late Turonian age in the lower part of the Wall Creek Member of the Frontier. These middle upper Turonian beds are, in turn, disconformably overlain by upper upper Turonian beds of the Wall Creek. At outcrops in the vicinity of Cheyenne, marine sandstone of late Turonian age in the Wall Creek

Member of the Frontier is unconformably overlain by limestone of middle Coniacian age in the Fort Hays Limestone Member of the Niobrara Formation.

Gardiner, Montana, to Newcastle, Wyoming

The nonmarine and nearshore marine strata of mid-Cretaceous age in northwestern Wyoming, near Gardiner, Montana (Figure 5), are about 270 ft (80 m) thick. In northeastern Woming near Newcastle, the mid-Cretaceous sequence is about 1200 ft (370 m) thick. Outcrops in the vicinity of Gardiner include, in ascending order, nonmarine beds possibly of early and middle Cenomanian age in the Frontier, a hiatus for middle Cenomanian–early Turonian time, marine sandstone of early middle Turonian age in the Frontier, a hiatus for late middle Turonian–early Coniacian time, and marine shale of middle Coniacian to at least early Santonian age in the Cody Shale.

The Frontier of north-central Wyoming near Tensleep includes, from oldest to youngest, shallow marine sandstone and shale of early(?) and middle Cenomanian age, marine siltstone and shale largely of late Cenomanian age, a late Cenomanian–late Turonian hiatus, and a shallow marine sandstone of late Turonian age at the top of the formation. Conformably overlying the Frontier is marine shale of late Turonian age in the Cody Shale. The depositional environments of the rocks near Tensleep seem to reflect a local marine regression in early to middle Cenomanian time, a marine transgression in the middle to late Cenomanian, one or more episodes of marine regression and truncation during late Cenomanian to late Turonian time, and a marine transgression in the latest Turonian.

In northeastern Wyoming near Newcastle, mid-Cretaceous rocks were deposited in offshore marine environments. This sequence includes, in ascending order, shale of Cenomanian age in the Belle Fourche Shale; shale and limestone of late Cenomanian–early Turonian age in the Greenhorn Formation; shale and siltstone of middle Turonian age in the Pool Creek Member of the Carlile Shale; a middle to late Turonian hiatus; sandstone, siltstone, and shale of late Turonian age in the Turner Sandy Member of the Carlile; and shale mainly of Coniacian age in the Sage Breaks Member of the Carlile. Middle or upper Coniacian shale at the top of the Carlile apparently is disconformably overlain by chalk and calcareous shale of early Santonian age in the Niobrara Formation (Evetts, 1976). The widespread marine transgression of middle Cenomanian–early Turonian time, represented by the Belle Fourche and Greenhorn, was followed by the widely recognized marine regression of middle Turonian time, represented by the Pool Creek Member of the Carlile and by the unconformity at the top of the Pool Creek. Sandstone of late Turonian age in the Turner Sandy Member of the Carlile overlies the eroded surface and reflects the marine transgression of late Turonian to at least middle Coniacian time.

Kemmerer to Buffalo, Wyoming

The mid-Cretaceous rocks depicted on Figure 6 are about 2200 ft (670 m) thick near Kemmerer and about 900 ft (275 m) thick near Buffalo. Nonmarine strata of middle Cenomanian age in the Chalk Creek Member of the Frontier Formation at Kemmerer grade northeastward into shallow marine shale, siltstone, and sandstone of the Frontier in north-central Wyoming (Figure 6). Most of the nonmarine and marine rocks of late Cenomanian–early Coniacian age in the upper part of the Frontier at Kemmerer are represented in north-central Wyoming by disconformities. At those localities on Figure 6 where the Frontier is marine, the sandstone units in the lower part of the formation may indicate a locally shallowing sea in early and middle Cenomanian time and a deepening sea in late Cenomanian time. Lithologies, hiatuses, and fossils in the upper part of the formation apparently reflect a marine regression near Buffalo in the early Turonian, a marine transgression and following regression near Lander in the middle Turonian, and a transgression from Buffalo to Lander in late Turonian–middle Coniacian time.

Delta, Colorado, to Newcastle, Wyoming

Sequences of mid-Cretaceous strata depicted on Figure 7 are about 650 ft (200 m) thick in western Colorado in the vicinity of Delta and about 1200 ft (365 m) thick in northeastern Wyoming near Newcastle. Near Delta, the Cenomanian rocks are the nearshore marine Dakota Sandstone and the overlying marine Mancos Shale. Rocks of Cenomanian age in central Wyoming at Rawlins and Casper consist of offshore marine shale and sandstone in the Belle Fourche Member of the Frontier Formation.

The early Turonian hiatus in the Mancos Shale near Delta increases in magnitude northeastward to central Wyoming. At Rawlins, a corresponding hiatus in the Frontier Formation is late Cenomanian–middle Turonian age. No correlative hiatus was found in the sequence near Newcastle in northeastern Wyoming.

Shale, siltstone, calcarenite, and microcrystalline limestone of middle Turonian–middle Coniacian age in the Mancos near Delta and Meeker are represented at Rawlins and Casper, in ascending order, by marine sandstone and shale of middle Turonian age in the unnamed member of the Frontier, a hiatus of middle and late Turonian age, and marine sandstone and shale of late Turonian–Coniacian age in the Wall Creek Member of the Frontier and in the overlying Cody Shale.

Pueblo, Colorado, to Newcastle, Wyoming

The mid-Cretaceous strata depicted on Figure 8 are, as previously indicated, about 640 ft (195 m) thick at Pueblo and about 1200 ft (365 m) thick near Newcastle. Most of these rocks were deposited in offshore marine environments; consequently, lateral and vertical changes in lithologies commonly indicate widespread marine transgressions and regressions. The noncalcareous shale of Cenomanian age in the Graneros Shale of eastern Colorado and the Belle Fourche Shale of northeastern Wyoming, as well as the overlying calcareous rocks of Cenomanian and Turonian age in the Greenhorn Formation, record a significant marine transgression. A middle Turonian sequence of (in ascending order) shale, siltstone, and locally sandstone in the lower part of the Carlile and a disconformity at the top of the sequence near Pueblo, Denver, and Newcastle (Figure 8) indicate a marine regression. In an area near Denver, however, beds of early Turonian age in the Greenhorn are disconformably overlain

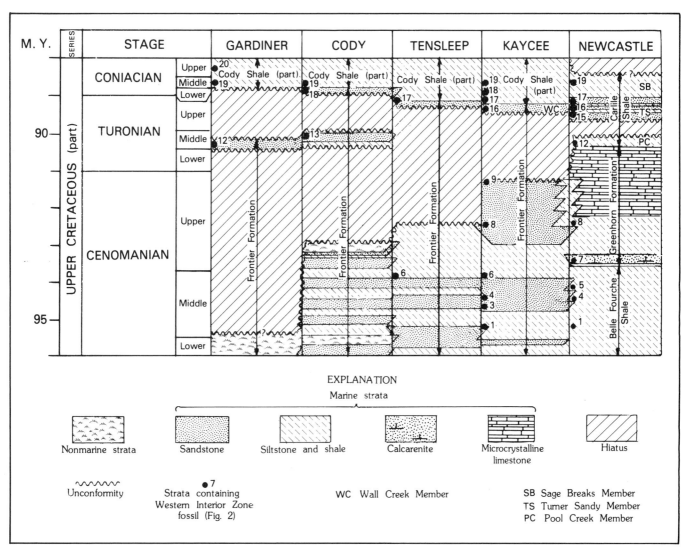

Figure 5—Chronostratigraphic diagram of some mid-Cretaceous formations in southwestern Montana and northern Wyoming (localities shown in Figure 1).

by beds of late middle Turonian age in the lower part of the Carlile. Strata of late Turonian age, which include calcarenite of the Juana Lopez Member of the Carlile in Colorado and marine sandstone of the Wall Creek Member of the Frontier and the Turner Sandy Member of the Carlile in eastern Wyoming, disconformably overlie middle Turonian and older rocks. The hiatus represented by the disconformity is for latest middle Turonian and early late Turonian time, although this hiatus, when combined with others, spans late Cenomanian–late Turonian time in southeastern Wyoming. In Colorado and southeastern Wyoming, calcarenite and sandstone of late Turonian age in the Carlile and the Frontier are unconformably overlain by limestone of latest Turonian–middle Coniacian age in the Niobrara Formation. In the Carlile of northeastern Wyoming near Newcastle, marine sandstone of late Turonian age in the Turner Sandy Member is conformably overlain by marine shale of Coniacian age in the Sage Breaks Member. These sequences of late Turonian–middle Coniacian strata and associated hiatuses apparently record a widespread marine

transgression and some brief and local marine regressions. The calcareous shale and limestone of late Coniacian age in the upper part of the Niobrara may have been deposited in a regressing sea.

CHRONOSTRATIGRAPHIC MAPS

Maps of the lithofacies for increments of mid-Cretaceous time in the central Rocky Mountains were compiled from outcrop sections, core descriptions, records of index fossils, and borehole logs. The regional distribution of nonmarine strata and marine sandstone, shale, calcarenite, and microcrystalline limestone is depicted for several units of time, which are defined mainly by the range spans of marine mollusks of Cenomanian, Turonian, and Coniacian ages (Figure 2). In an area where a time unit is represented only by a disconformity, the crest of the causative uplift is placed at the localities that have the largest hiatus.

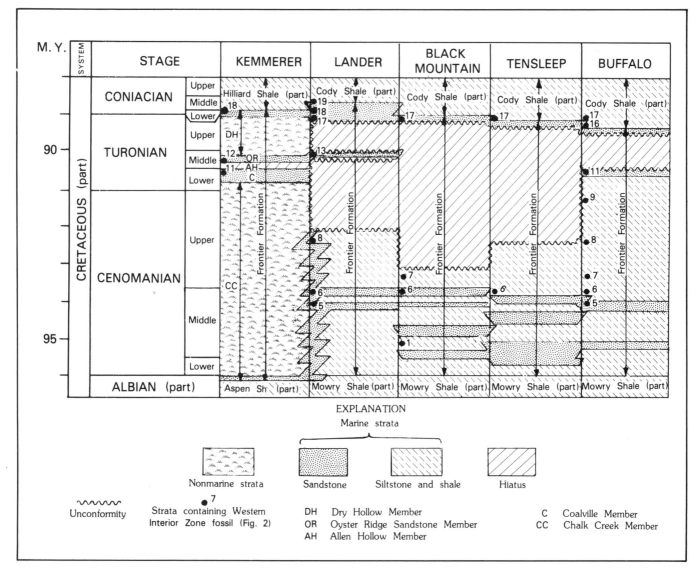

Figure 6—Chronostratigraphic diagram of some mid-Cretaceous formations in western and north-central Wyoming (localities shown in Figure 1).

Middle Cenomanian Lithofacies

The rocks of middle Cenomanian age outlined on Figure 9 were deposited in the time of fossil zones 4 and 5 (Figure 2) during a widespread marine transgression. They include nonmarine beds of the Chalk Creek Member of the Frontier Formation in central southwestern Wyoming and marine sandstone of the Frontier in central Wyoming and of the Dakota Sandstone in southwestern Colorado. Siltstone and shale of the same age are in the Frontier Formation of western Wyoming, the Belle Fourche Shale of northeastern Wyoming, the Cody Shale of south-central Montana, the Mancos Shale of western Colorado, and the Graneros Shale of eastern Colorado. Middle Cenomanian rocks, however, have not been confirmed in an area in southwestern Montana and adjacent northwestern Wyoming. Moreover, scattered fossiliferous outcrops in the area indicate that fossil zones 4 and 5 are represented by an unconformity. Strata of the age of fossil zones 4 and 5 are also missing from sequences of mid-Cretaceous rocks in southwestern

Wyoming (Merewether, 1983), northeastern Utah (C. M. Molenaar, personal communication 1980), and northwestern Colorado (Reeside, 1955; Sharp, 1963). Within this north-northwest-trending oval area (shown in Figure 9), the unconformity representing the strata has been located at outcrops and on borehole logs. In both areas, middle Cenomanian beds were truncated during mid-Cretaceous time.

Lower Turonian Lithofacies

Beds of early Turonian age, which accumulated in the span of fossil zones 10 and 11 (Figure 2) and near the end of a widely recognized marine transgression, are composed mainly of marine sandstone in the Sanpete Formation and Allen Valley Shale of central Utah and in the Frontier Formation of southwestern and north-central Wyoming (Figure 10). The lower Turonian rocks are dominantly marine shale of the Mancos Shale in eastern Utah and of the Cody Shale in south-central Montana and are largely

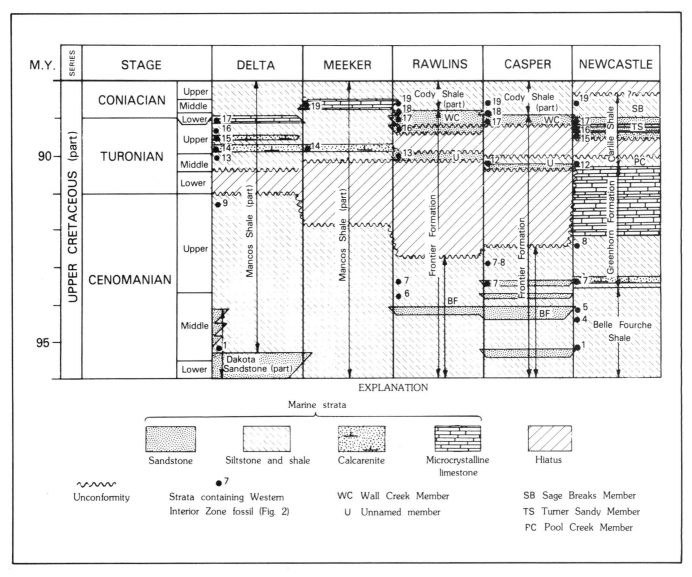

Figure 7—Chronostratigraphic diagram of some mid-Cretaceous formations in western Colorado and eastern Wyoming (localities shown in Figure 1).

limestone of the Greenhorn Limestone in much of Colorado and in western South Dakota and adjoining areas. Lower Turonian strata, however, are represented by an unconformity in a large northwest-trending area that extends from northwestern Colorado, through Wyoming, and into southwestern Montana (Figure 10). The discontinuous regional extent of the lower Turonian lithofacies indicates that early Turonian deposition was followed by at least one period of mid-Cretaceous truncation.

Lower Middle Turonian Lithofacies

The strata of early middle Turonian age depicted on Figure 11 were deposited during a major marine regression in the time of fossil zone 12 (Figure 2) and include marine shale and discontinuous units of marine sandstone. The sandstone is assigned to the Allen Valley Shale in central Utah, to part of the Mancos Shale in western Colorado, to the Oyster Ridge Sandstone Member of the Frontier Formation in western Wyoming, and to an unnamed

member of the Frontier Formation in central Wyoming. Shale of the same age is within the Mancos Shale of eastern Utah and southwestern Colorado and within the lower part of the Carlile Shale in central Colorado and northeastern Wyoming. Rocks of early middle Turonian age are represented by an unconformity within an irregularly shaped, north- to northwest-trending region that includes northwestern Colorado, most of central Wyoming, and part of southwestern Montana. The disconformity and the isolated bodies of sandstone on the margins of this region indicate that mid-Cretaceous erosion followed early middle Turonian deposition.

Upper Middle Turonian Lithofacies

Rocks of late middle Turonian age, which are depicted on Figure 12, accumulated during the span of fossil zone 13 (Figure 2) and reflect a widespread marine regression. Nonmarine facies have been assigned to the Funk Valley Formation (Fouch et al., 1983) and to the Ferron Sandstone

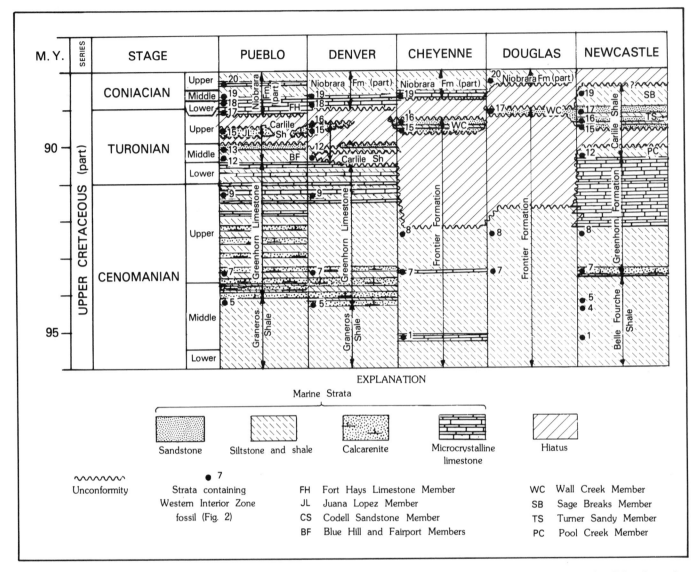

Figure 8—Chronostratigraphic diagram of some mid-Cretaceous formations in eastern Colorado and eastern Wyoming (localities shown in Figure 1).

Member (Ryer, 1981) of the Mancos Shale in central Utah and to the Dry Hollow Member (Myers, 1977) and an unnamed member (Merewether, 1983) of the Frontier Formation in western Wyoming. Nearshore marine sandstone of late middle Turonian age is included in the Ferron Sandstone Member of the Mancos in central Utah and in an unnamed member of the Frontier in south-central Wyoming. Laterally equivalent offshore marine sandstone has been assigned to the Codell Sandstone Member of the Carlile Shale in eastern Colorado (Cobban and Reeside, 1952). Marine shale of late middle Turonian age is part of the Mancos Shale in eastern Utah and western Colorado, is in the Carlile Shale of western South Dakota and adjoining areas, and is within the Cody Shale of south-central Montana. Rocks of this age are represented by disconformities in several areas in northern Colorado and southeastern and northern Wyoming, indicating erosion of those areas after late middle Turonian time.

Lower Upper Turonian Lithofacies

The lower upper Turonian strata mapped on Figure 13 were deposited in the time of fossil zone 15 (Figure 2) near the beginning of a major marine transgression. Marine sandstone of this age is included in the Frontier Sandstone Member of the Mancos Shale in northeastern Utah (Kinney, 1955), in the Wall Creek Member of the Frontier Formation in southeastern Wyoming (Merewether et al., 1979), and in the Turner Sandy Member of the Carlile Shale in northeastern Wyoming (Robinson et al., 1964; Merewether, 1980). Rice (1977) recognized the Turner Sandy Member in the subsurface of western South Dakota. Marine siltstone and shale of early late Turonian age are assigned to the Mancos Shale in eastern Utah, to the Frontier Formation in southern Wyoming, to the Carlile Shale in northeastern Wyoming, and to the Cody Shale in south-central Montana. Calcarenite of the same age is enclosed in the Mancos Shale in western Colorado, in the Benton Shale of north-central

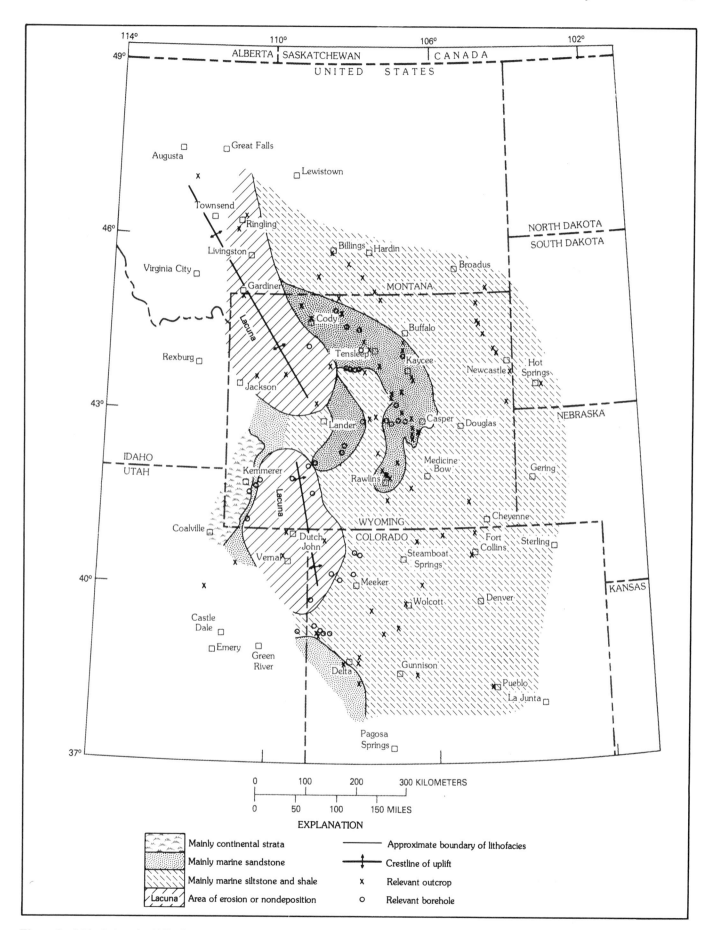

Figure 9—Lithofacies of middle Cenomanian fossil zones 4 and 5 (Figure 2) in the central Rocky Mountains region.

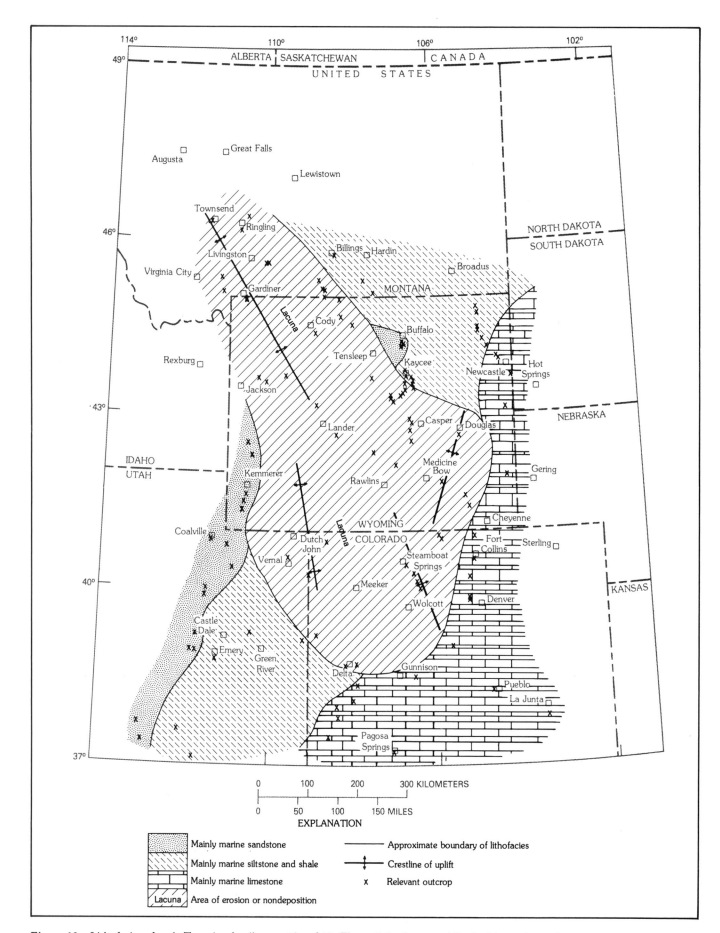

Figure 10—Lithofacies of early Turonian fossil zones 10 and 11 (Figure 2) in the central Rocky Mountains region.

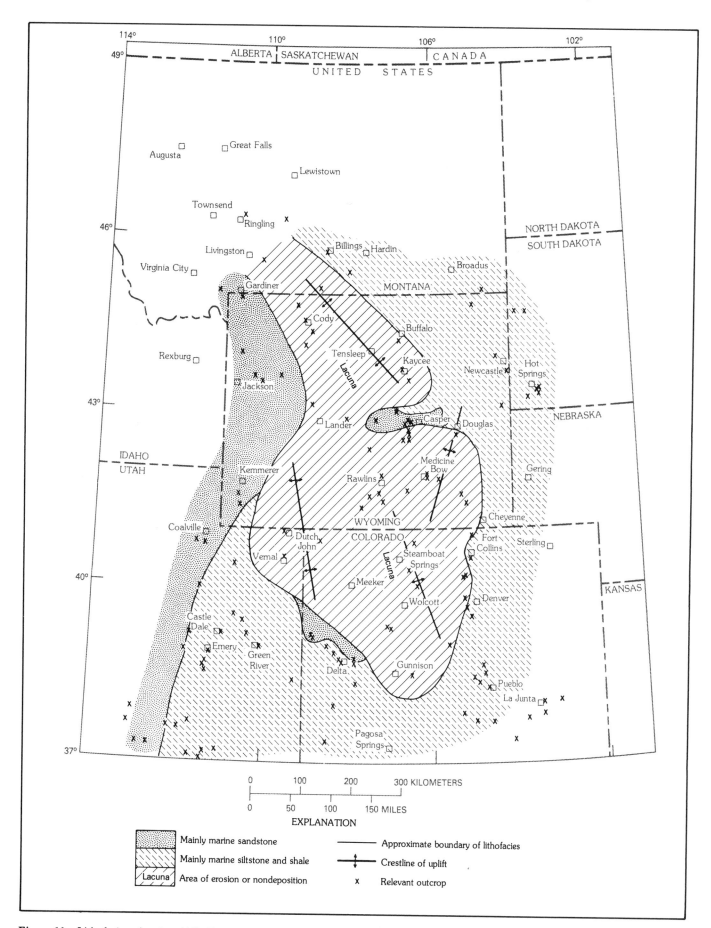

Figure 11—Lithofacies of early middle Turonian fossil zone 12 (Figure 2) in the central Rocky Mountains region.

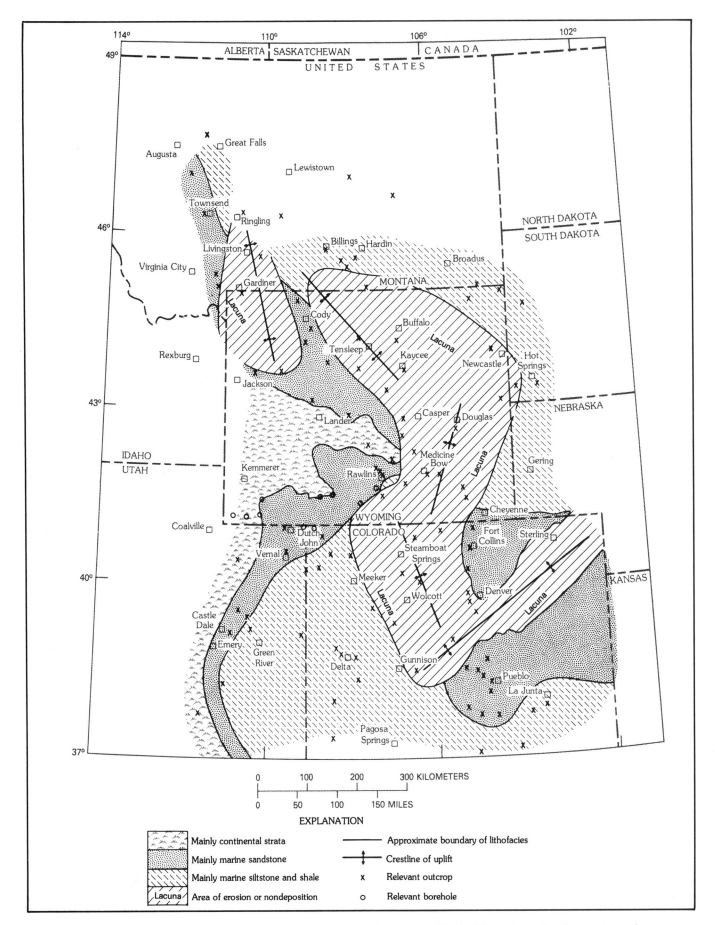

Figure 12—Lithofacies of late middle Turonian fossil zone 13 (Figure 2) in the central Rocky Mountains region. Lacuna in northeastern Colorado from Weimer and Sonnenberg (1983).

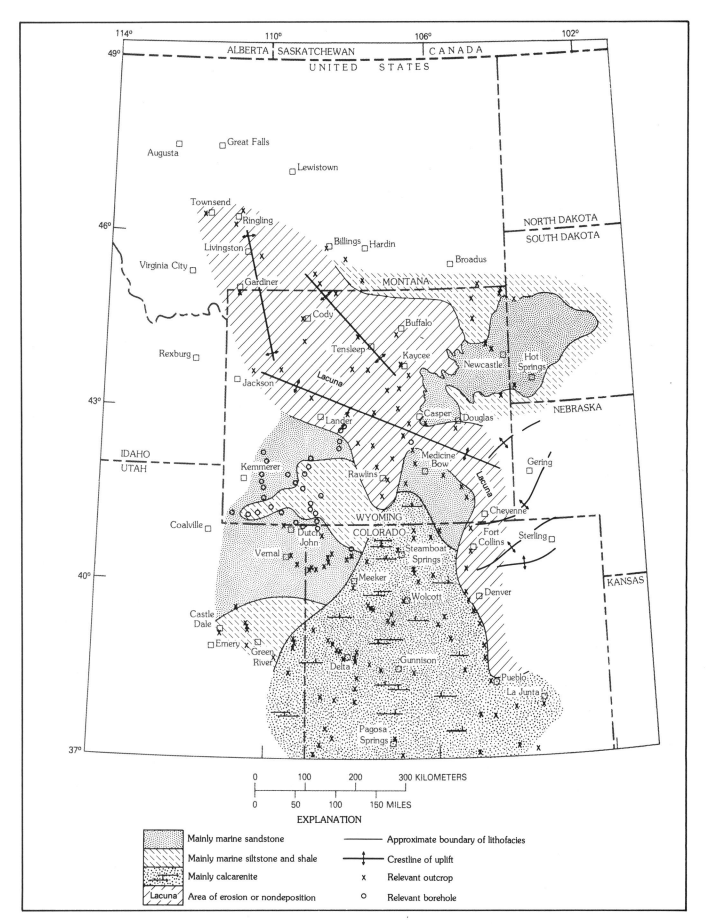

Figure 13—Lithofacies of early late Turonian fossil zone 15 (Figure 2) in the central Rocky Mountains region. Crestlines in northeastern Colorado, southeastern Wyoming, and western Nebraska from Weimer (1983).

Colorado, and in the Juana Lopez Member of the Carlile Shale in eastern Colorado. Rocks of the age of fossil zone 15 are represented by an unconformity in a very irregularly shaped area that extends northwest from northeastern Colorado, across Wyoming, to southwestern Montana. This area separates bodies of early late Turonian strata and was evidently eroded in subsequent mid-Cretaceous time.

Middle Upper Turonian Lithofacies

Marine strata of middle late Turonian age, indicated on Figure 14, accumulated in the span of fossil zone 16 (Figure 2), which is in the early part of a widespread marine transgression. Sandstone of this age is assigned to the Frontier Sandstone Member of the Mancos Shale in northeastern Utah and northwestern Colorado, to the Wall Creek Member of the Frontier Formation in central Wyoming, and to the Turner Sandy Member of the Carlile Shale in northeastern Wyoming. Shale in this unit is included in the Mancos Shale of eastern Utah and western Colorado, in the Carlile Shale of northeastern Wyoming, and in the Cody Shale of north-central Wyoming and south-central Montana. The calcarenite of middle late Turonian age is in the Benton Shale of northwestern Colorado and in the Juana Lopez Member of the Carlile Shale in eastern Colorado. Beds recording the time of fossil zone 16 are represented by a disconformity in an area that extends from eastern Colorado northwestward to southwestern Montana. This area, which resembles the area of truncation depicted for the early late Turonian (Figure 13), separates sandstone bodies of middle late Turonian age and apparently was eroded in later mid-Cretaceous time.

Uppermost Upper Turonian Lithofacies

Rocks of latest Turonian age, which are mapped on Figure 15, were deposited in the span of fossil zone 17 during a widely recognized marine transgression. Some nonmarine strata of this age are included in the Dry Hollow Member of the Frontier Formation in southwestern Wyoming and northeastern Utah (D. J. Nichols, personal communication, 1981). Laterally equivalent marine sandstone is assigned to the Frontier Sandstone Member of the Mancos Shale in northeastern Utah and to the Wall Creek Member of the Frontier Formation in central Wyoming. Marine shale of latest Turonian age is part of the Mancos Shale in eastern Utah and western Colorado, part of the Sage Breaks Member of the Cody Shale in central Wyoming, and part of the Turner Sandy Member of the Carlile Shale in northeastern Wyoming and adjoining areas. Marine limestone of the same age is commonly assigned to the Fort Hays Limestone Member of the Niobrara Formation in central Colorado. Strata of this age are represented by unconformities in large areas in eastern Colorado and southeastern Wyoming and in northwestern Wyoming and southwestern Montana. These areas were eroded in mid-Cretaceous time, evidently after Turonian time.

Lower Coniacian Lithofacies

The lithologies depicted on Figure 16 accumulated in the time of fossil zone 18 in early Coniacian time during the latter part of a major marine transgression. Marine sandstone of this age is included in the Dry Hollow Member of the Frontier Formation in northeastern Utah and southwestern Wyoming (Myers, 1977), in the Wall Creek Member of the Frontier in south-central Wyoming, and in an unnamed member at the top of the Frontier in northwestern Wyoming. Marine shale of early Coniacian age is assigned to the Mancos Shale in eastern Utah and western Colorado, to the Sage Breaks Shale and the Sage Breaks Member of the Cody Shale in central Wyoming, to the Sage Breaks Member of the Carlile Shale in northeastern Wyoming, and to the Cody Shale of south-central Montana. The laterally equivalent limestone in eastern Colorado is part of the Fort Hays Limestone Member of the Niobrara Formation. Lower Coniacian rocks are represented by unconformities in two areas, one in southeastern Wyoming and the other in northwestern Wyoming and southwestern Montana. Shale and limestone of early Coniacian age probably were deposited and presumably were uplifted and eroded in southeastern Wyoming near Douglas. Lower Coniacian sandstone may onlap the area of uplift in northwestern Wyoming and southwestern Montana.

Middle Coniacian Lithofacies

The middle Coniacian strata depicted on Figure 17 were deposited in the span of fossil zone 19 near the end of a widespread marine transgression. Marine sandstone of this age is included in the Funk Valley Formation of central Utah and in the upper part of the Frontier Formation of northern Utah and northwestern Wyoming. Marine shale of middle Coniacian age is assigned to the Mancos Shale in eastern Utah and western Colorado, to the Sage Breaks Member of the Cody Shale in central Wyoming, to the Sage Breaks Member of the Carlile Shale in northeastern Wyoming, and to the Niobrara Shale Member of the Cody Shale in south-central Montana. The laterally equivalent limestone in Colorado and southeastern Wyoming is generally part of the Fort Hays Limestone Member of the Niobrara Formation. Rocks of middle Coniacian age are represented by an unconformity in the mid-Cretaceous sequence of a small area near Douglas in southeastern Wyoming. Presumably, the lower and middle Coniacian hiatus of that area indicates local deformation and erosion in Coniacian time.

UNCONFORMITIES, HIATUSES, AND LACUNAS

The mid-Cretaceous rocks of central and eastern Utah enclose a significant unconformity, which was described in a regional stratigraphic report by Fouch et al. (1983). Rocks of late Albian age are disconformably overlain by beds of middle Cenomanian age (fossil zones 4 and 5) near the Utah–Colorado border (Figure 3) and by beds of early Turonian age (fossil zone 10) in central Utah. The magnitude of the hiatus increases westward, which reflects an early Cenomanian uplift in the north-northeast-trending Sevier orogenic belt of western Utah (Figure 1) (Armstrong, 1968) and the contemporaneous and subsequent truncation of older Cenomanian and Albian strata in central and eastern Utah.

In western Montana, the mid-Cretaceous sequence includes an important and similar unconformity, which was noted by Cobban et al. (1976) and by Mudge (1972). Near

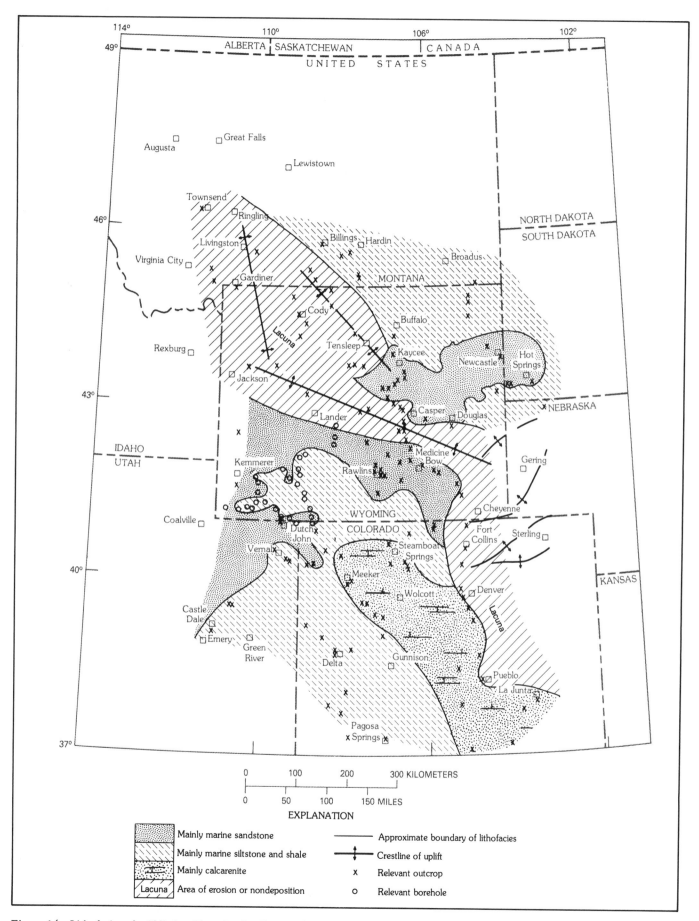

Figure 14—Lithofacies of middle late Turonian fossil zone 16 (Figure 2) in the central Rocky Mountains region. Crestlines in northeastern Colorado, southeastern Wyoming, and western Nebraska from Weimer (1983).

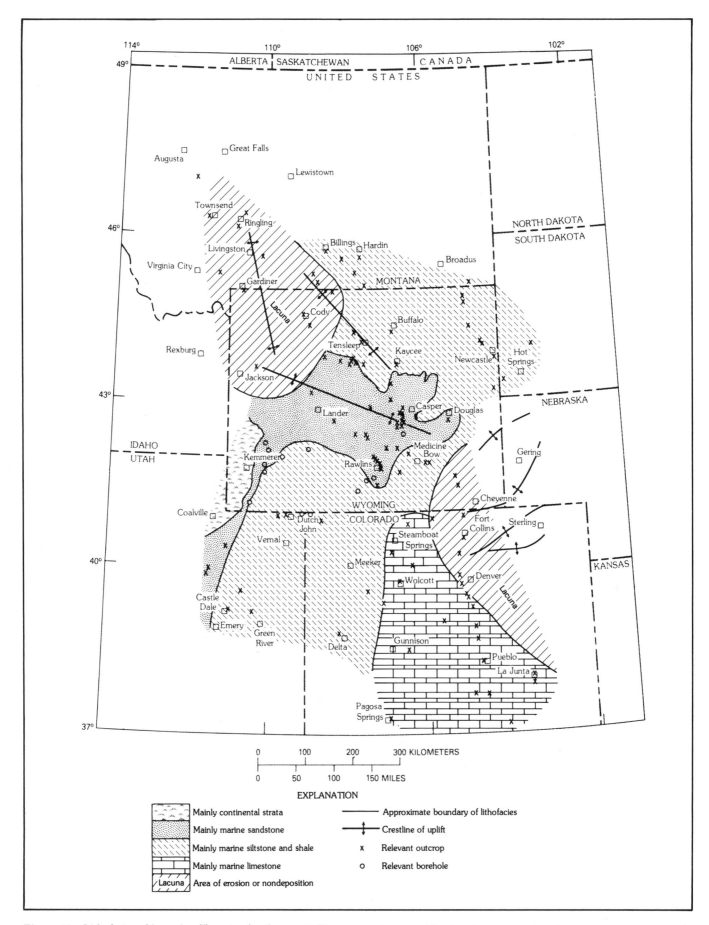

Figure 15—Lithofacies of latest late Turonian fossil zone 17 (Figure 2) in the central Rocky Mountains region. Crestlines in northeastern Colorado, southeastern Wyoming, and western Nebraska from Weimer (1983).

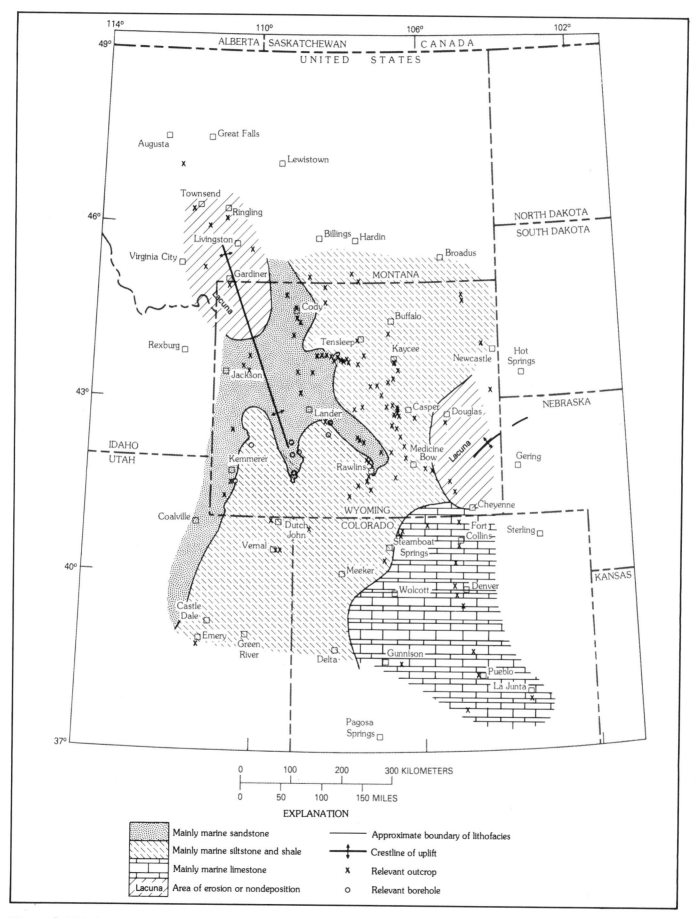

Figure 16—Lithofacies of early Coniacian fossil zone 18 (Figure 2) in the central Rocky Mountains region. Crestline in southeastern Wyoming and western Nebraska from Weimer (1983).

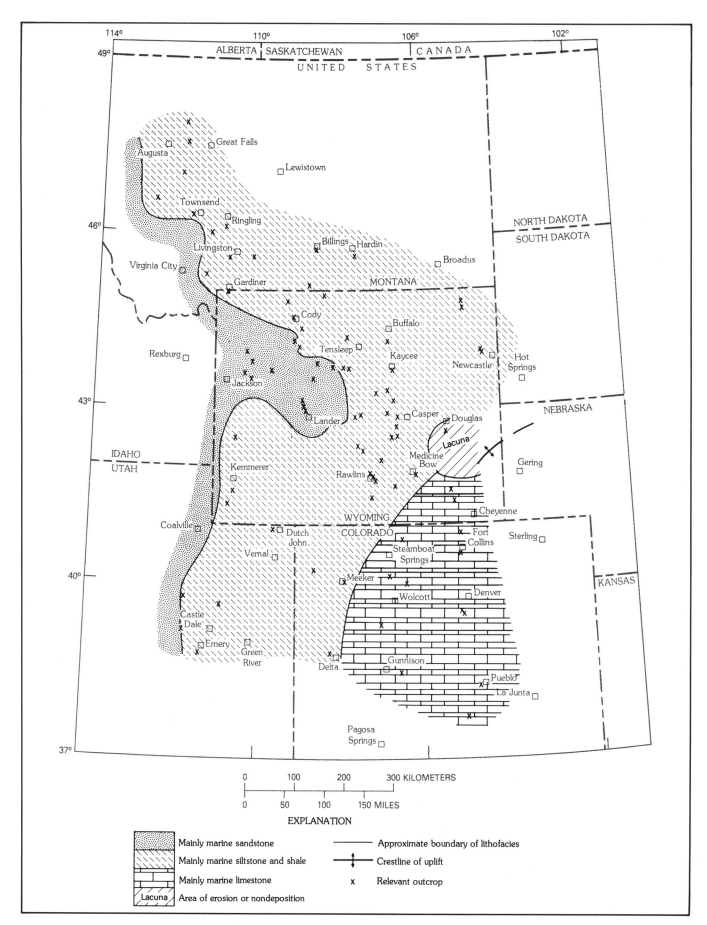

Figure 17—Lithofacies of middle Coniacian fossil zone 19 (Figure 2) in the central Rocky Mountains region. Crestline in southeastern Wyoming and western Nebraska from Weimer (1983).

Great Falls on the south arch of the Sweetgrass arch (Figure 1), marine sandstone of late Albian age in the Bootlegger Member of the Blackleaf Formation is disconformably overlain by marine shale and siltstone of late Cenomanian age (fossil zone 8) in the Floweree Member of the Marias River Shale (Cobban et al., 1976). South of Great Falls, in the vicinity of Ringling, marine siltstone and sandstone of late Albian age are disconformably overlain by marine shale and siltstone that are probably late Cenomanian. East of these localities, in central Montana, the marine sequence of mid-Cretaceous age apparently does not include this disconformity (Cobban, 1951). Rice (1981) depicts no mid-Cretaceous unconformity in north-central Montana and southwestern Saskatchewan but indicates a disconformity on the northern part of the Sweetgrass arch in Alberta. West of Great Falls and Ringling, the presence of Cenomanian beds has not been confirmed by collections of fossils, although palynomorphs of Cenomanian–Turonian age were found in nonmarine strata southwest of Virginia City in southwestern Montana (Perry et al., 1983). Rocks of Albian and middle Turonian ages, however, were recognized near Townsend (Klepper et al., 1957). Mudge (1972) reports a disconformity between the Lower Cretaceous Blackleaf Formation and the Upper Cretaceous Marias River Shale near Augusta (Figure 1), but the magnitude of the associated hiatus has not been determined. The regional extent and age of this hiatus near Great Falls is evidence of uplift and attendant erosion in middle to late Cenomanian time along the north-northwest-trending Sweetgrass arch in Montana and Alberta.

Middle Cenomanian strata are represented by disconformities in two areas in the central Rocky Mountains region (Figure 9). At the disconformity in southwestern Wyoming and adjoining parts of Utah and Colorado, beds as old as late Albian are overlain by beds as old as early middle Turonian. At outcrops near the northwest corner of Colorado, upper Albian shale of the Mowry Shale is overlain by upper middle Turonian siltstone of the Mancos Shale (Reeside, 1955). The hiatus in the area is probably largest along a north-northwest-trending line that locates the crest of an early to middle Turonian structural swell.

In northwestern Wyoming and southwestern Montana, Cenomanian strata are unconformably overlain by middle Turonian strata. Marine rocks mainly of early Cenomanian age are disconformably overlain by marine sandstone of middle Turonian age at outcrops about 22 mi (35 km) northwest of Lander, Wyoming. In the area of Cody, Wyoming (Figure 5), dominantly marine beds of Cenomanian age (Eicher, 1967) are disconformably overlain by largely marine beds of middle Turonian age (Merewether et al., 1975). Near the northwest corner of Wyoming (Figure 5), rocks probably of early Cenomanian age are overlain by marine sandstone of early middle Turonian age. In southwest Montana in the vicinity of Townsend, beds probably of Albian age (Klepper et al., 1957) are overlain by upper middle Turonian sandstone. These and other fossiliferous outcrops indicate that the hiatus is probably greatest along a line extending north-northwest from the area of Lander to the area of Townsend. The line also depicts the approximate crest of an elongate swell of early to middle Turonian age.

Marine strata of early Turonian age (Figure 10) are absent in a large region that includes northwestern Colorado, most of Wyoming, and adjoining areas of Utah, Idaho, and Montana. In western Colorado near Delta (Figure 3), calcareous shale of latest Cenomanian age is unconformably overlain by shale and siltstone of middle Turonian age. In south-central Wyoming in the vicinity of Rawlins and Casper (Figure 7), shale of late Cenomanian age is disconformably overlain by middle Turonian sandstone (Merewether et al., 1979). Similarly, at Cody in north-western Wyoming (Figure 5) shallow marine rocks of late Cenomanian age (Eicher, 1967) are unconformably overlain by marine siltstone and shale of middle Turonian age (Merewether et al., 1975). The hiatus at these scattered localities was probably caused by arching and erosion within latest Cenomanian to earliest middle Turonian time. However, some of the truncation of lower Turonian strata depicted on Figure 10 was associated with later mid-Cretaceous tectonic activity.

Strata of early middle Turonian age (fossil zone 12) are missing from a region that includes northwestern Colorado, central Wyoming, and part of southwestern Montana (Figure 11). Their absence was probably caused by both nondeposition and erosion. Areas near the northwestern corner of Colorado, near Rawlins, Wyoming, and perhaps near Lander and Cody, Wyoming, were probably above wave base and were not receiving sediments during early middle Turonian time. This local lack of sedimentation presumably was a consequence of early Turonian tectonism and of the widespread marine regression in the middle Turonian. Lower middle Turonian and upper middle Turonian beds are conformable but disconformably overlie upper Cenomanian rocks at outcrops in the areas of Delta, Colorado, and Casper, Wyoming, near the boundary of the lacuna (Figure 11). At outcrops between those areas, lower middle Turonian beds are absent and upper middle Turonian beds disconformably overlie older rocks of late Cenomanian–Albian age. Presumably, the truncated older rocks were onlapped during the middle Turonian. In other parts of the region, strata of early middle Turonian age were eroded later in mid-Cretaceous time during periods of structural deformation. At outcrops between Denver and Fort Collins, Colorado (Figure 8), lower middle Turonian rocks are either unusually thin or absent and are discon-formably overlain by beds of late middle Turonian age. The associated hiatus may indicate a slight uplift in north-central Colorado in middle Turonian time. More commonly, truncated lower middle Turonian strata are succeeded by upper Turonian beds, indicating tectonism and erosion in the early late Turonian. Rocks of early middle Turonian age are disconformably overlain by rocks of late Turonian age near Casper, Wyoming (Figure 7). The location of the crests of the early late Turonian swells is not readily determined because of earlier and later tectonism.

Rocks of latest middle Turonian age (fossil zone 13) are absent in much of northern Colorado and southeastern Wyoming and in north-central and northwestern Wyoming and adjoining areas (Figure 12). They are missing in the subsurface of northeastern Colorado within an elongate, northeast-trending area (Weimer and Sonnenberg, 1983), which was probably uplifted and eroded intermittently in late Turonian time. At Wolcott in central Colorado, Cenomanian calcareous shale is unconformably overlain by upper Turonian shale and siltstone (Figure 3). Similarly, near Medicine Bow, Douglas, and Tensleep in Wyoming, upper Cenomanian rocks are disconformably overlain by upper Turonian siltstone and sandstone. These disconformities and hiatuses represent tectonism and truncation mainly in the early late Turonian. The crestlines of the swells seem to trend north-northwestward from the Wolcott area to the vicinity of Rawlins, north-northeastward from near Medicine Bow to Douglas, and northwestward from near Kaycee, Wyoming, to the area of Cody, Wyoming. In northwestern Wyoming and adjacent parts of Montana and Idaho (Figure 12), where lower middle Turonian beds are disconformably overlain by strata of Coniacian age (Figure 5), the hiatus seems to indicate the presence of a north-northwest-trending arch in late Turonian time.

Lower upper Turonian lithofacies are absent because of nondeposition and subsequent mid-Cretaceous erosion in an irregularly shaped area that seems to extend northwest from southeastern Wyoming to southwestern Montana (Figure 13). At outcrops near Pueblo in southeastern Colorado, calcarenite of early late Turonian age in the Juana Lopez Member of the Carlile Shale disconformably overlies middle Turonian beds and is disconformably overlain by limestone of latest Turonian age in the Fort Hays Limestone Member of the Niobrara Formation (Figure 8). At other outcrops near Pueblo, clasts of Juana Lopez are enclosed in the basal bed of the Fort Hays Limestone. At outcrops in the vicinity of Cheyenne, Wyoming, lower upper Turonian shale and sandstone at the base of the Wall Creek Member of the Frontier Formation unconformably overlie upper Cenomanian strata of the Frontier (Figure 8). Locally, the shale and sandstone either are conformably overlain by younger upper Turonian beds or are disconformably overlain by the Coniacian Fort Hays Limestone Member. In the area of Rawlins, upper middle Turonian (fossil zone 13) strata are disconformably overlain by middle upper Turonian (fossil zone 16) beds and the lower upper Turonian (fossil zones 14 and 15) rocks are represented by the unconformity (Figure 7). Apparently, the lower upper Turonian strata onlap the eroded surface of the area in central Wyoming (Figure 13) that was uplifted in earliest late Turonian time. In subsequent late Turonian time, additional arching in the area that extends from southeast to northwest Wyoming (Figure 13) caused local truncation of the lower upper Turonian beds.

Middle upper Turonian rocks are missing in an area (Figure 14) that resembles the area where lower upper Turonian beds are absent (Figure 13) and that trends northwestward from southeastern Wyoming to southwestern Montana. The rocks of middle late Turonian age are missing locally because of nondeposition and are missing regionally because of later deformation and erosion. In eastern Colorado, middle upper Turonian strata are represented by disconformities. Near Pueblo, where lower upper Turonian calcarenite of the Juana Lopez is overlain by uppermost Turonian limestone of the Fort Hays (Figure 8), the disconformity indicates late Turonian tectonism. In south-central Wyoming near Rawlins (Figure 7), sandstone of middle late Turonian (fossil zone 16) age in the Wall

Creek Member of the Frontier Formation onlaps an erosional surface and seemingly is disconformably overlain by beds of latest Turonian (fossil zone 17) age. Near Casper, Wyoming, middle Turonian (fossil zone 12) shale is disconformably overlain by uppermost Turonian (fossil zone 17) sandstone (Figure 7). The sequences at these localities and others record late Turonian deformation in central Wyoming. In north-central Wyoming near Kaycee, middle upper Turonian sandstone, siltstone, and shale of the Wall Creek rest unconformably on upper Cenomanian strata and are conformably overlain by uppermost Turonian shale of the Cody Shale (Figure 5).

Rocks of latest Turonian age have not been recognized in most of the mid-Cretaceous sequences of northeastern Colorado and southeastern Wyoming and of northwestern Wyoming and southwestern Montana (Figure 15). Outcrops in the areas of Denver and Cheyenne, where middle upper Turonian strata are unconformably overlain by uppermost Turonian to middle Coniacian limestone (Figure 8), indicate arching in late Turonian time and subsequent transgressive overlap in latest Turonian–middle Coniacian time. At outcrops of the Frontier near Lander, Wyoming, uppermost Turonian beds unconformably overlie middle Turonian rocks and are conformably overlain by lower Coniacian strata (Figure 6). The Frontier of northwestern Wyoming and adjacent southwestern Montana seems to record late Turonian doming and a following latest Turonian and Coniacian marine transgression. In the areas of Jackson and Cody, Wyoming, the outcropping Frontier contains middle Turonian beds disconformably overlain by Coniacian sandstone (Figure 5).

Lower Coniacian beds are absent in parts of southeastern Wyoming (Figure 8) and northwestern Wyoming and adjoining southwestern Montana (Figure 16). The outcropping Frontier Formation near Cheyenne and Douglas contains upper Turonian sandstone and siltstone that are disconformably overlain by calcareous rocks of Coniacian age (Figure 8). At Cheyenne, the hiatus is of latest Turonian and early Coniacian age and probably reflects late Turonian arching and associated erosion. However at Douglas, the hiatus spans the early and middle Coniacian, which seems to indicate local tectonism and truncation in Coniacian time. The Frontier in northwestern Wyoming and southwestern Montana commonly includes, in ascending order, beds of middle Turonian age, a hiatus, and beds of early Coniacian or middle Coniacian age (Figure 5). Events indicated by these sequences are two episodes of uplift and erosion in the late Turonian, followed by the transgressive onlap of strata in Coniacian time.

Strata of middle Coniacian age are missing only from the mid-Cretaceous sequence near Douglas (Figure 17). As previously described, rocks of latest Turonian age at the top of the Frontier Formation are disconformably overlain by rocks of late Coniacian age at the base of the Niobrara Formation (Figure 8). The hiatus may represent local Coniacian tectonism.

SUMMARY OF REGIONAL TECTONISM

Mid-Cretaceous formations of the central Rocky Mountains region reflect a succession of structural events that apparently began in early Cenomanian time (95–96 m.y. ago) and continued at least into late Turonian time (89–90 m.y. ago). The spatial and temporal magnitude of the hiatuses in these strata indicates the approximate location and age of several swells. Fouch et al. (1983) presented stratigraphic evidence of early Cenomanian (95–96 m.y. ago) deformation in the north-northeast-trending Sevier orogenic belt of western Utah (Figures 1 and 18) (Armstrong, 1968). Cobban et al. (1976) provided evidence of structural movement in middle to late Cenomanian time (about 93–94 m.y. ago) along the northwest-tending South arch of the Sweetgrass arch in western Montana (Figures 1 and 18). Disconformities recognized in younger rocks in northwestern Colorado (Reeside, 1955; Sharp, 1963) and southwestern Wyoming (Hale, 1962; Weimer, 1962) and in northwestern Wyoming and southwestern Montana indicate north-northwest-trending swells in early to early middle Turonian time (90–91 m.y. ago) (Figures 9 and 18). The lower and middle Turonian strata of northeastern Colorado locally enclose a small hiatus (Figure 8), which was caused by middle Turonian (90–90.5 m.y. ago) erosion and possibly by middle Turonian deformation. A more significant hiatus separates middle Turonian and upper Turonian stratigraphic units in large parts of Colorado and Wyoming (Figure 12). This hiatus and some causative early late Turonian (89.8 m.y. ago) swells (Figure 18) have been recognized in north-central Colorado and south-central Wyoming where the crestline of a swell trends north-northwestward, in southeastern Wyoming where a crestline trends north-northeastward, and in north-central Wyoming where a crestline trends northwestward. A separate truncated area in northwestern Wyoming (Figure 12) indicates a north-northwest-trending swell of early late Turonian or possibly late late Turonian age. The hiatus between middle upper Turonian strata and upper upper Turonian strata in central Wyoming (cf. Figures 14 and 15) was caused by uplift of late late Turonian age (about 89.3 m.y. ago) in an area that extends diagonally across Wyoming from northeastern Colorado (Weimer and Sonnenberg, 1983) to southwestern Montana (Figure 18). Sparse evidence indicates erosion and perhaps uplift in middle Coniacian time (about 88.6 m.y.) near Douglas in southeastern Wyoming (Figure 8).

The most evident of these tectonic events (Figure 18) are (1) the early Cenomanian (95–96 m.y. ago) deformation in the Sevier orogenic belt of western Utah, (2) the middle to late Cenomanian (about 93–94 m.y. ago) arching in the foreland of western Montana, (3) the early to early middle Turonian (90–91 m.y. ago) swells of northwestern Colorado and western Wyoming, (4) the early late Turonian (89.8 m.y. ago) arching in north-central Colorado and southeastern and northwestern Wyoming, and (5) the late late Turonian (about 89.3 m.y. ago) uplift in northeastern Colorado and in southeastern, central, and northwestern Wyoming. Those four events in the mid-Cretaceous foreland span most of middle Cenomanian to latest Turonian time (about 89.3–94 m.y. ago).

The mild swells of the mid-Cretaceous foreland in western Wyoming and western Montana trend north-northwestward and northwestward (Figure 18). Several of the crestlines apparently radiate from an area in western Montana, which may indicate rotational components in the structural movements of western Montana and Wyoming

during mid-Cretaceous time. Moreover, many of the foreland swells are progressively younger toward the east, from 90–91 m.y. ago in southwestern Wyoming to perhaps 89.3 m.y. ago in southeastern Wyoming and northeastern Colorado. Most of the foreland swells in eastern Wyoming and northeastern Colorado trend north-northeastward and northeastward. Some of these conform with the Transcontinental arch and some are approximately parallel with it (Weimer, 1978, 1983).

In the thrust belt of the central Rocky Mountains region, the displacements of the Crawford thrust in southwestern Wyoming and north-central Utah and of the Meade thrust in eastern Idaho (Figure 1) have been assigned Cenomanian–early Santonian ages (Armstrong and Oriel, 1965; Jordan, 1981) and, subsequently, Coniacian ages (Wiltschko and Dorr, 1983). Movements on the Paris thrust in eastern Idaho and northern Utah in late Albian and Turonian time were indicated by Schmitt et al. (1981), Sippel et al. (1981), and Wiltschko and Dorr (1983). Tectonism of latest Albian–middle Cenomanian age in that area is represented by the marine beds and overlying nonmarine beds of the Aspen Shale and Frontier Formation, respectively, in southwest Wyoming (Figure 4) (Ryer, 1977). Later structural activity in the area in early middle Turonian time might be recorded by the prograding marine sequence of the Allen Hollow Shale and the Oyster Ridge Sandstone Member (Figure 4) (Ryer, 1977). The concurrent deformation of the eastward moving thrust plates in the thrust belt and of the migrating swells in the foreland probably indicates a genetic relationship between the structural events in the two regions. Lorenz (1982) reported that the Sweetgrass arch of western Montana in Mesozoic time was a peripheral upwarp or forebulge, which was separated from the thrust belt by a foredeep. He concluded that episodes of slight uplift along the Sweetgrass arch were contemporaneous with episodes of thrusting in the thrust belt. Consequently, the upwarping of the south arch of the Sweetgrass arch in middle to late Cenomanian time (93–94 m.y. ago) would indicate local thrusting in the thrust belt at nearly the same time. Similarly, upwarping in the vicinity of the Moxa arch and the Rock Springs uplift of southwest Wyoming (Figures 1 and 18) in early to early middle Turonian time (90–91 m.y. ago) might reflect the early middle Turonian tectonism in the thrust belt reported by Ryer (1977) or the Turonian movement on the Paris thrust indicated by Schmitt et al. (1981), Sippel et al. (1981), and Wiltschko and Dorr (1983). Presumably, some of the foreland deformation in the late Turonian also corresponds in age to structural events in the thrust belt.

POTENTIAL STRATIGRAPHIC TRAPS FOR OIL AND GAS

The flanks of the mid-Cretaceous foreland uplifts are characterized by stratigraphic pinch-outs, which were caused by differential and progressive truncation and marine onlap. Mid-Cretaceous strata on several of the structural swells include mudstone and sandstone, which probably are source rocks and reservoir beds, respectively, for hydrocarbons. Many areas could contain oil- and gas-bearing stratigraphic traps. For example, in southwestern Wyoming, largely

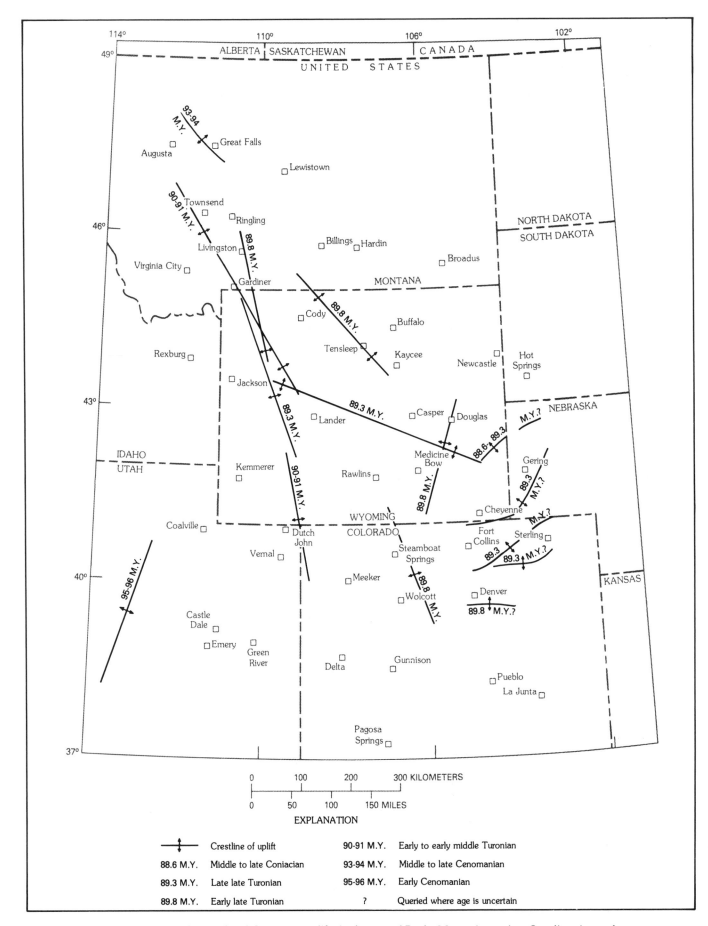

Figure 18—Inferred locations and ages of mid-Cretaceous uplifts in the central Rocky Mountains region. Crestlines in northeastern Colorado, southeastern Wyoming, and western Nebraska from Weimer (1983).

marine strata of Cenomanian age were uplifted and locally eroded in early Turonian–earliest middle Turonian time and thereafter were buried by deltaic and shallow marine sediments in middle Turonian time. The upper part of the Frontier Formation on the Moxa arch (Figure 1) includes source rocks for oil and gas and gas-bearing sandstone units (Merewether et al., 1984). Moreover, in central and northwest Wyoming, truncated marine and nonmarine strata of Cenomanian and Turonian ages are onlapped toward the northwest by shallow marine sandstone units of late Turonian and Coniacian age. Locally, carbonaceous shale and siltstone of middle Turonian age are disconformably overlain by porous sandstone of late Turonian and Coniacian ages.

ACKNOWLEDGMENTS

Financial support for this investigation was provided by the Western Gas Sands Project of the U.S. Department of Energy and was administered by C. W. Spencer of the U.S. Geological Survey (Interagency Agreement DE-AI21-83MC20422). The manuscript of this report was reviewed by R. B. Powers, C. M. Molenaar, and W. J. Perry, Jr., of the U.S. Geological Survey, and by Patricia St. Clair of Marathon Oil Co.

REFERENCES CITED

Armstrong, F. C., and S. S. Oriel, 1965, Tectonic development of Idaho–Wyoming thrust belt: AAPG Bulletin, v. 49, n. 11, p. 1847–1866.

Armstrong, R. L., 1968, Sevier orogenic belt in Nevada and Utah: Geological Society of America Bulletin, v. 79, p. 429–458.

Cobban, W. A., 1951, Colorado shale of central and northwestern Montana and equivalent rocks of Black Hills: AAPG Bulletin, v. 35, n. 10, p. 2170–2198.

Cobban, W. A., and J. B. Reeside, Jr., 1952, Correlation of the Cretaceous formations of the Western Interior of the United States: Geological Society of America Bulletin, v. 63, n. 10, p. 1011–1043.

Cobban, W. A., C. E. Erdmann, R. W. Lemke, and E. K. Maughan, 1976, Type sections and stratigraphy of the members of the Blackleaf and Marias River formations (Cretaceous) of the Sweetgrass arch, Montana: U.S. Geological Survey Professional Paper 974, 66 p.

Eicher, D. L., 1967, Foraminifera from Belle Fourche Shale and equivalents, Wyoming and Montana: Journal of Paleontology, v. 41, n. 1, p. 167–188.

Evetts, M. J., 1976, Microfossil biostratigraphy of the Sage Breaks Shale (Upper Cretaceous) in northeastern Wyoming: The Mountain Geologist, v. 13, n. 4, p. 115–134.

Fouch, T. D., T. F. Lawton, D. J. Nichols, W. B. Cashion, and W. A. Cobban, 1983, Patterns and timing of synorogenic sedimentation in Upper Cretaceous rocks of central and northeast Utah, in M. W. Reynolds, and E. D. Dolly, eds., Mesozoic paleogeography of the west-central United States: Society of Economic Paleontologists and Mineralogists Rocky Mountain Section, Rocky Mountain Paleogeography Symposium 2, p. 305–336.

Grose, L. T., 1972, Tectonics, in Geologic atlas of the Rocky Mountain region, United States of America: Denver, Rocky Mountain Association of Geologists, p. 35–44.

Hale, L. A., 1962, Frontier Formation—Coalville, Utah and nearby areas of Wyoming and Colorado: Wyoming Geological Association 17th Annual Field Conference Guidebook, p. 211–220.

Jordan, T. E., 1981, Thrust loads and foreland basin evolution, Cretaceous, western United States: AAPG Bulletin, v. 65, n. 12, p. 2506–2520.

Kinney, D. M., 1955, Geology of the Uinta River–Brush Creek area, Duchesne and Uintah counties, Utah: U.S. Geological Survey Bulletin 1007, 185 p.

Klepper, M. R., R. A. Weeks, and E. T. Ruppel, 1957, Geology of the southern Elkhorn Mountains, Jefferson and Broadwater counties, Montana: U.S. Geological Survey Professional Paper 292, 82 p.

Lanphere, M. A., and D. L. Jones, 1978, Cretaceous time scale from North America, in G. V. Cohee, M. F. Glaessner, and H. D. Hedberg, eds., Contributions to the geologic time scale: AAPG Studies in Geology, n. 6, p. 259–268.

Lorenz, J. C., 1982, Lithospheric flexure and the history of the Sweetgrass arch, northwestern Montana, in R. B. Powers, ed., Geologic studies of the Cordilleran thrust belt: Denver, Rocky Mountain Association of Geologists, p. 77–89.

Merewether, E. A., 1980, Stratigraphy of mid-Cretaceous formations at drilling sites in Weston and Johnson counties, northeastern Wyoming: U.S. Geological Survey Professional Paper 1186-A, 25 p.

———, 1983, The Frontier Formation and mid-Cretaceous orogeny in the foreland of southwestern Wyoming: The Mountain Geologist, v. 20, n. 4, p. 121–138.

Merewether, E. A., W. A. Cobban, and R. T. Ryder, 1975, Lower Upper Cretaceous strata, Bighorn Basin, Wyoming and Montana: Wyoming Geological Association 27th Annual Field Conference Guidebook, p. 73–84.

Merewether, E. A., W. A. Cobban, and E. T. Cavanaugh, 1979, Frontier Formation and equivalent rocks in eastern Wyoming: The Mountain Geologist, v. 16, n. 3, p. 67–102.

Merewether, E. A., P. D. Blackmon, and J. C. Webb, 1984, The mid-Cretaceous Frontier Formation near the Moxa arch, southwestern Wyoming: U.S. Geological Survey Professional Paper 1290, 29 p.

M'Gonigle, J. W., 1982, Interfingering of the Frontier Formation and Aspen Shale, Cumberland Gap, Wyoming: The Mountain Geologist, v. 19, n. 2, p. 59–61.

Mudge, M. R., 1972, Pre-Quaternary rocks in the Sun River Canyon area, northwestern Montana: U.S. Geological Survey Professional Paper 663-A, 142 p.

Myers, R. C., 1977, Stratigraphy of the Frontier Formation (Upper Cretaceous), Kemmerer area, Lincoln County, Wyoming: Wyoming Geological Association 29th Annual Field Conference Guidebook, p. 271–311.

Obradovich, J. D., and W. A. Cobban, 1975, A time-scale for

the Late Cretaceous of the Western Interior of North America, *in* W. G. E. Caldwell, ed., The Cretaceous system in the Western Interior of North America: Geological Association of Canada Special Paper 13, p. 31–54.

Perry, W. J., Jr., B. R. Wardlaw, N. H. Bostick, and E. K. Maughan, 1983, Structure, burial history, and petroleum potential of frontal thrust belt and adjacent foreland, southwest Montana: AAPG Bulletin, v. 67, n. 5, p. 725–743.

Reeside, J. B., Jr., 1955, Revised interpretation of the Cretaceous section on Vermilion Creek, Moffat County, Colorado: Wyoming Geological Association 10th Annual Field Conference Guidebook, p. 85–88.

Rice, D. D., 1977, Stratigraphic sections from well logs and outcrops of Cretaceous and Paleocene rocks, northern Great Plains, North Dakota and South Dakota: U.S. Geological Survey Oil and Gas Investigations Chart OC-72.

———, 1981, Subsurface cross section from southeastern Alberta, Canada, to Bowdoin dome area, north-central Montana, showing correlation of Cretaceous rocks and shallow, gas-productive zones in low-permeability reservoirs: U.S. Geological Survey Oil and Gas Investigations Chart OC-112.

Robinson, C. S., W. J. Mapel, and M. H. Bergendahl, 1964, Structure and stratigraphy of the northern and western flanks of the Black Hills uplift, Wyoming, Montana, and South Dakota: U.S. Geological Survey Professional Paper 404, 134 p.

Royse, F., Jr., M. A. Warner, and D. L. Reese, 1975, Thrust belt structural geometry and related stratigraphic problems, Wyoming–Idaho–northern Utah: Rocky Mountain Association of Geologists Guidebook, p. 41–54.

Ryer, T. A., 1977a, Age of Frontier Formation in north-central Utah: AAPG Bulletin, v. 61, n. 1 p. 112–116.

———, 1977b, Coalville and Rockport areas, Utah, *in* E. G. Kauffman, ed., Cretaceous facies, faunas, and paleoenvironments across the Western Interior basin: The Mountain Geologist, v. 14, p. 105–128.

———, 1981, Deltaic coals of Ferron Sandstone Member of Mancos Shale: predictive model for Cretaceous coal-bearing strata of Western Interior: AAPG Bulletin, v. 65, n. 11, p. 2323–2340.

Schmitt, J. G., K. N. Sippel, and D. B. Wallem, 1981, Upper Jurassic through lowermost Upper Cretaceous sedimentation in the Wyoming–Idaho–Utah thrust belt—I. Depositional environments and facies distribution (abs.), *in* Sedimentary tectonics: principles and applications: University of Wyoming Department of Geology, Wyoming Geological Survey, and Wyoming Geological Association Conference Notes, p. 26–27.

Scott, G. R., 1969, General and engineering geology of the northern part of Pueblo, Colorado: U.S. Geological Survey Bulletin 1262, 131 p.

Sharp, J. V., 1963, Unconformities within basal marine Cretaceous rocks of the Piceance basin, Colorado: Ph.D. thesis, University of Colorado, Boulder, 170 p.

Sippel, K. N., J. A. Schmitt, D. B. Wallem, and M. E. Moran, 1981, Upper Jurassic through lowermost Upper Cretaceous sedimentation in the Wyoming–Idaho–Utah thrust belt—II. Depositional environments and facies distribution (abs.), *in* Sedimentary tectonics: principles and applications: University of Wyoming Department of Geology, Wyoming Geological Survey, and Wyoming Geological Association Conference Notes, p. 26–27.

U.S. Geological Survey and AAPG, 1962, Tectonic map of the United States, scale 1:2,500,000.

Weimer, R. J., 1962, Late Jurassic and Early Cretaceous correlations, south–central Wyoming and northwestern Colorado: Wyoming Geological Association 17th Annual Field Conference Guidebook, p. 124–130.

———, 1978, Influence of Transcontinental arch on Cretaceous marine sedimentation: a preliminary report, *in* J. D. Pruitt, and P. E. Coffin, eds., Energy Resources of the Denver basin: Rocky Mountain Association of Geologists Symposium, p. 211–222.

———, 1983, Relation of unconformities, tectonics, and sea level changes, Cretaceous of the Denver basin and adjacent areas, *in* M. W. Reynolds, and E. D. Dolly, eds., Mesozoic paleogeography of the west-central United States: Society of Economic Paleontologists and Mineralogists Rocky Mountain Section, Rocky Mountain Paleogeography Symposium 2, p. 359–376.

Weimer, R. J., and S. A Sonnenberg, 1983, Codell Sandstone, new exploration play, Denver basin, *in* Mid-Cretaceous Codell Sandstone Member of Carlile Shale, eastern Colorado: Society of Economic Paleontologists and Mineralogists Rocky Mountain Section, Spring Field Trip Guidebook, p. 26–48.

Wiltschko, D. V., and J. A. Dorr, Jr., 1983, Timing of deformation in overthrust belt and foreland of Idaho, Wyoming, and Utah: AAPG Bulletin, v. 67, n. 8, p. 1304–1322.

Subtle Middle Cretaceous Paleotectonic Deformation of Frontier and Lower Cody Rocks in Wyoming

W. H. Curry, III
Consulting Geologist
Casper, Wyoming

An isopach map of the Frontier Formation and the lower part of the Cody Formation in parts of Wyoming shows (1) a north-trending area of thickening in a western paleobasin associated with the thrust belt, (2) a north-trending area of thinning on a western arch, and (3) a northwest-trending area of thickening in eastern Wyoming that includes a composite of several different thickening patterns. Paleotectonic activity during Turonian time was most intense in the area of the western basin and the western arch and caused varying thicknesses of preserved strata in the lower part of the Frontier below a regional unconformity at the base of the Wall Creek Member. Less intense paleotectonic deformation is recorded in central Wyoming by the unconformities at the base of the Wall Creek Member and by the subtle variations of thickness of a bentonite-bounded unit of marine shale in the lowermost Cody Formation. These variations in thickness in the lower Cody may have started late during the time of Wall Creek sandstone deposition, and they may be related to deposition of the uppermost sandstones of the Wall Creek, which produce hydrocarbons from stratigraphic traps in the deep part of the Powder River basin.

REGIONAL ISOPACH MAP

The isopach map of the sedimentary rocks in the Frontier Formation and in the lower part of the Cody Formation (Figure 1), from the top of the Mowry Shale to the base of a more resistive bed in the Cody shale, shows variations in thickness that are evidence of a subsiding western basin, a mildly uplifted western arch, and a moderately subsiding complex of eastern "basins."

A cross section of these broad paleotectonic features (Figure 2) indicates that most of the thickness variation is below the unconformity at the base of the Wall Creek Member. The locations of boreholes and the tops of the Mowry Shale depicted on the cross sections are listed in Table 1.

Top of the Mowry Shale

All borehole logs used in this study are referenced to the Clay Spur bentonite at the top of the Mowry Shale, because it is the most persistent lithologic and time-stratigraphic unit in the middle Cretaceous stratigraphic sequence. The top of the Mowry Shale is recognizable except in the western basin where a less resistive facies forms the upper part of the Mowry Shale.

Lower Part of the Frontier Formation

Informal members A through D are coarsening-upward sequences that start at the base with low log resistivity marine shale and grade upward into higher log resistivity sandstone and conglomerate. Using the top of the Mowry as a datum for correlation, these members are easily identified in many areas and they seem to be widespread. In the western basin, however, the members are not well developed, probably because of the presence of nonmarine strata, as at the LaBarge gas field. In eastern Wyoming, the members are also poorly defined because they are thinner and lack the siltstone and sandstone that causes higher log resistivities. Bentonite layers in this eastern shaly facies (Belle Fourche Member) can be used for correlating units and to extend some informal members into eastern Wyoming.

Wall Creek Member of the Frontier Formation

The most difficult lithologic and borehole log correlations of the entire middle Cretaceous sequence are those for the sandstone units of the Wall Creek Member and the basal Wall Creek unconformities. Regionally, these sedimentary rocks must be combined, but at outcrops containing ammonites or in areas where borehole logs are abundant, it is sometimes evident that there are two unconformities near the base of the Wall Creek Member and that they bound the informal member E (see Figure 13) or "unnamed member" of Merewether et al. (1979). The Wall Creek Member, which overlies the E member, may contain a third unconformity, as near the western end of Casper Mountain (Merewether et al., 1979). Detailed log correlations in the vicinity of known stratigraphic oil traps show considerable lateral variation in the distribution of sandstone bodies in the Wall Creek.

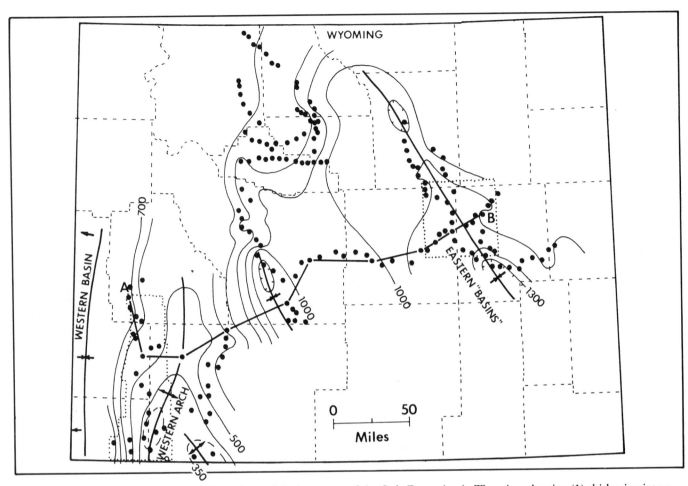

Figure 1—Isopach map of the Frontier Formation and the lower part of the Cody Formation in Wyoming, showing (1) thickening into a western basin that is associated with the thrust belt, (2) thinning across a western arch, and (3) thickening eastward off the western arch. Contour interval is 100 ft (30 m). Line A–B is location of cross section A–B on Figure 2. Black dots show location of borehole logs used to construct the map.

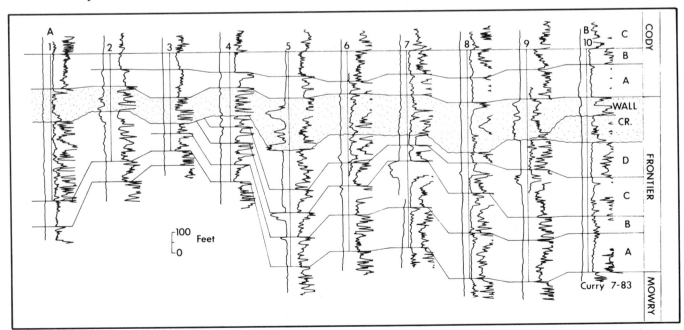

Figure 2—Cross section A–B showing isopached interval from top of Mowry Shale to top of informal member B of the Cody Formation, and the varying thickness of the lower part of the Frontier below the Wall Creek (stippled) unconformity. For all cross sections, the borehole locations and log tops of the Mowry Shale are listed in Table 1. Distance between boreholes is shown on isopach map in Figure 1.

Table 1—Locations of boreholes and log tops of the Mowry Shale depicted on cross sections.

Well Number	Location	Depth to Mowry Shale (ft)
1	SE¼SE¼, sec.8, T. 29 N., R. 113 W.	8,000
2	SE¼SE¼, sec.21, T. 23 N., R. 112 W.	10,883
3	SE¼SE¼, sec.14, T. 23 N., R. 108 W.	15,288
4	SW¼NE¼, sec.25, T. 26 N., R. 103 W.	18,028
5	NW¼SW¼, sec.25, T. 29 N., R. 96 W.	3,480
6	SW¼SW¼, sec.3, T. 33 N., R. 93 W.	5,630
7	NW¼SE¼, sec.8, T. 33 N., R. 86 W.	10,505
8	SE¼NW¼, sec.11, T. 34 N., R. 81 W.	2,620
9	NE¼SW¼, sec.18, T. 36 N., R. 77 W.	7,480
10	SE¼SE¼, sec.27, T. 38 N., R. 74 W.	13,530
11	SW¼SW¼, sec.34, T. 14 N., R. 116 W.	15,208
12	NE¼SE¼, sec.8, T. 14 N., R. 114 W.	13,262
13	SW¼NE¼, sec.20, T. 15 N., R. 113 W.	12,897
14	SE¼SW¼, sec.35, T. 15 N., R. 113 W.	12,712
15	NE¼NW¼, sec.6, T. 14 N., R. 112 W.	12,650
16	SW¼NE¼, sec.29, T. 15 N., R. 112 W.	12,540
17	SE¼NW¼, sec.20, T. 15 N., R. 111 W.	13,950
18	SW¼SE¼, sec.7, T. 15 N., R. 110 W.	15,737
19	NW¼NE¼, sec.16, T. 20 N., R. 114 W.	12,703
20	SW¼SE¼, sec.23, T. 20 N., R. 114 W.	12,275
21	SW¼SE¼, sec.25, T. 20 N., R. 113 W.	11,650
22	NE¼SW¼, sec.29, T. 20 N., R. 112 W.	11,440
23	CSW¼, sec.26, T. 20 N., R. 112 W.	11,680
24	S½SW¼, sec.25, T. 20 N., R. 112 W.	11,868
25	CSE¼, sec.20, T. 20 N., R. 111 W.	12,245
26	SE¼NW¼, sec.9, T. 27 N., R. 113 W.	7,282
27	NE¼NW¼, sec.14, T. 27 N., R. 113 W.	7,130
28	SE¼SE¼, sec.33, T. 27 N., R. 112 W.	8,192
29	NE¼SW¼, sec.4, T. 26 N., R. 111 W.	9,426
30	SW¼NE¼, sec.17, T. 28 N., R. 111 W.	11,568
31	SE¼SE¼, sec.29, T. 18 N., R. 113 W.	12,282
32	SE¼, sec.33, T. 17 N., R. 113 W.	12,290
33	NW¼NW¼, sec.21, T. 16 N., R. 113 W.	12,385
34	NE¼SE¼, sec.13, T. 16 N., R. 113 W.	12,548
35	NW¼NW¼, sec.19, T. 16 N., R. 112 W.	12,558
36	SW¼SW¼, sec.30, T. 16 N., R. 112 W.	12,536
37	SW¼SE¼, sec.16, T. 16 N., R. 112 W.	12,440
38	NW¼SW¼, sec.12, T. 38 N., R. 83 W.	1,700
39	NW¼NW¼, sec.4, T. 36 N., R. 81 W.	2,970
40	NE¼SW¼, sec.15, T. 37 N., R. 80 W.	3,883
41	NW¼SW¼, sec.7, T. 37 N., R. 79 W.	5,160
42	NW¼SW¼, sec.2, T. 37 N., R. 79 W.	5,978
43	SE¼SE¼, sec.30, T. 38 N., R. 78 W.	6,267
44	NE¼SE¼, sec.16, T. 38 N., R. 78 W.	5,210
45	NW¼NE¼, sec.17, T. 38 N., R. 77 W.	7,332
46	SE¼SW¼, sec.22, T. 38 N., R. 76 W.	13,036
47	SE¼SW¼, sec.14, T. 38 N., R. 75 W.	13,675
48	SE¼NW¼, sec.19, T. 39 N., R. 74 W.	13,422
49	SE¼SE¼, sec.16, T. 39 N., R. 73 W.	12,473
50	SE¼NW¼, sec.3, T. 39 N., R. 72 W.	11,310
51	NE¼NW¼, sec.33, T. 36 N., R. 77 W.	8,370
52	SW¼SE¼, sec.8, T. 37 N., R. 76 W.	11,865
53	SE¼SE¼, sec.18, T. 39 N., R. 75 W.	13,740

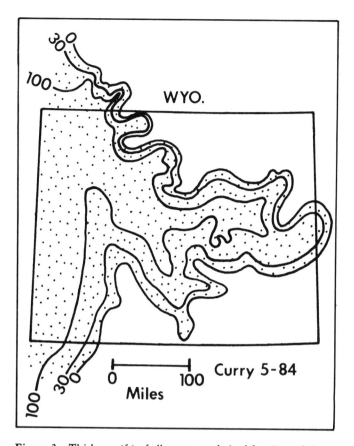

Figure 3—Thickness (ft) of all quartz sand-sized fractions of the Frontier Formation in Wyoming, compiled from borehole sample logs. The area of thicker sandstones is attributed to a series of deltas that were deposited along the western side of a shale and siltstone filled marine seaway.

Lower Part of the Cody Formation

Informal members A through C of the Cody Formation can be correlated easily where bentonites are present, as in central Wyoming, but where the A member includes sandstone, as in the Lander area, correlations are less certain.

Top of Isopach Interval

The stratigraphic top of the mapped interval (Figure 1) is the top of a relatively persistent low resistivity marine shale in the Cody B member and the base of the more resistive Cody C member. The more resistive Cody C member is a more calcareous shale and is laterally equivalent to part of the Niobrara Formation on the east side of the study area. The lithologic origin of the comparatively high log resistivity of the Cody C member in western Wyoming is uncertain. At the LaBarge gas field, the "First Frontier sandstone" is in the Cody C member or slightly higher stratigraphically.

FRONTIER SANDSTONE DISTRIBUTION

The thickness distribution of sand-sized clastic sediments, as described from borehole sample logs (Figure 3), generally

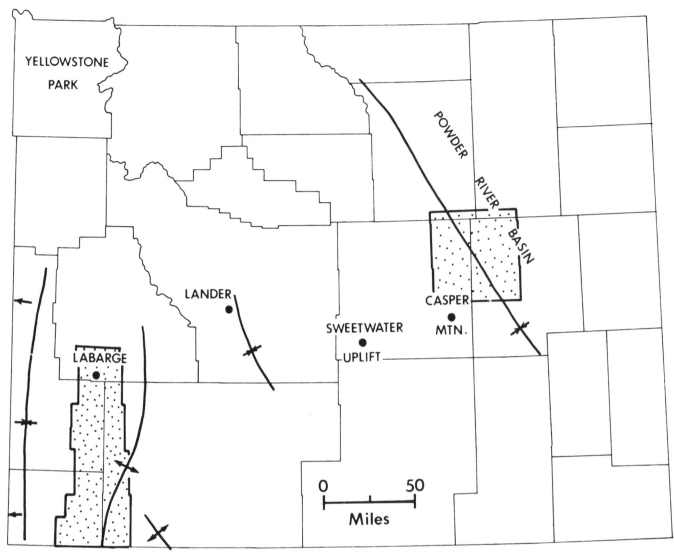

Figure 4—Index map of Wyoming showing axes of thicker and thinner Frontier and lower Cody rocks; selected localities mentioned in the text and the location (stippled) of the more detailed isopach maps are shown in Figure 5 (SW) and Figure 11 (E).

describes the distribution of the Frontier Formation in Wyoming. The western area of coarser clastics grades eastward into marine shale and siltstone (Goodell, 1962). The formation is composed mainly of gray shale and siltstone, with a much smaller content of sandstone, conglomerate, and coal. The Frontier Formation has long been considered the result of an ancient delta complex (Masters, 1952). More recent studies have concentrated on subdividing the formation into thinner and better described stratigraphic units that can be interpreted more specifically (Merewether et al., 1979). The considerable stratigraphic complexity, paucity of diagnostic fossils, thickness of the sequence, internal unconformities, mountain uplifts that have caused erosion of the Frontier Formation, and lack of borehole control in the deepest parts of most basins have all complicated efforts to combine all the evidence into one coherent stratigraphic history.

WESTERN ARCH

An isopach map of the interval between the top of the Wall Creek Member (second Frontier sandstone of the Moxa arch) and the top of the Mowry Shale (Figures 4 and 5) shows a Cretaceous hingeline between the rapid thickening in the western basin and the thinning in a broad area on the western arch. Three areas of unusual thinning along the hingeline are interpreted as paleotectonic highs. Cross sections (Figures 6, 7, and 8) of these positive areas show that most of the thinning is in the lower part of the Frontier below the Wall Creek unconformity. These relationships are thought to indicate moderate uplift and erosion before Wall Creek deposition.

Only the southern paleotectonic feature near Church Buttes (T. 16 N., R. 112 W.) shows similar thinning of the older Mowry, "Dakota," and Lakota sequence (Figures 9 and

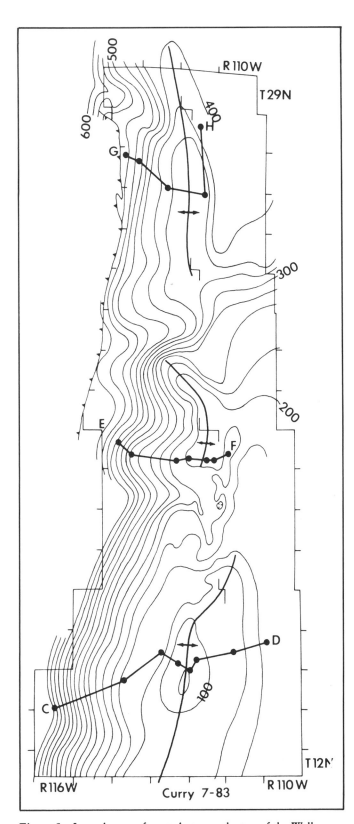

Figure 5—Isopach map of strata between the top of the Wall Creek Member and the top of the Mowry Shale showing a hingeline along the eastern edge of the western basin and three areas of unusual thinning that are interpreted as paleotectonic highs. Contour interval is 25 ft (7.5 m). Tick marks are 6 mi (10 km) apart. Lines C–D, E–F, and G–H are locations of cross sections in Figures 6, 7, and 8.

10). It should be emphasized that, except for the thickening in the western basin, all of these thickness variations are subtle.

EASTERN BASIN

An isopach map of the informal member A of the Cody shale in the vicinity of the eastern basin (Figure 11) shows several areas of subtle thickening and thinning. If the isopach interval were not bounded at the top and base by distinctive bentonites, the validity of the thickness variations could be doubted. It should be emphasized that the isopached member conformably overlies the Wall Creek Member. Furthermore, the only area of thinning on this map (Figure 11) that corresponds to an area of demonstrable thinning below the unconformities at the base of the Wall Creek is the stippled feature near the southeast corner of the map.

Although these thickness variations are slight and they are in rocks stratigraphically higher than the Wall Creek Member on the western arch, they probably reflect subtle paleotectonic activity. The paleotectonic activity indicated by the Cody member A may have started during latest Wall Creek deposition and influenced the distribution of oil- and gas-bearing sandstone reservoirs at the top of the Wall Creek Member in the deep part of the Powder River basin.

Cross section K-L (Figure 12) shows that Frontier members A through C and Cody member B thin eastward, as does the underlying Mowry Shale, in part because of differential rates of subsidence and sedimentation. In contrast, the interval from Frontier member D through the Cody A member thins westward, largely as a result of the unconformities and regional paleotectonic events described previously. The gross thickening of the Frontier and the lower part of the Cody (well #44, Figure 12) in east-central Wyoming thus does not represent a single paleotectonic basin.

Evidence of the unconformities near the base of the Wall Creek Member is based mainly on collections of fossils and on the ammonite zonations of Cobban (*in* Merewether et al., 1979). Once the unconformities were established from fossils at selected outcrops, the recognition of the unconformities at other outcrops was based on the chert pebble conglomerate at dated unconformities near the base of sandstone units. Chert pebble conglomerate at the top of sandstone units is not associated with significant unconformities, but probably represents the culmination of the coarsening-upward sequences that are mapped here as informal members of the Frontier Formation. Some conglomerates (as in SE¼SE¼, sec. 16, T. 8 N., R. 4 E.) are overlain by root zones and lignitic mudstone beds, but most are overlain by marine shale.

Unconformities located at outcrops in central Wyoming by Merewether et al. (1979), using fossils and basal chert pebble conglomerate, can be recognized on logs of nearby boreholes and then mapped in the subsurface using logs. This was done by Merewether and co-workers in the vicinity of Casper with some success. In the subsurface, the two unconformities depicted by them on the well log at

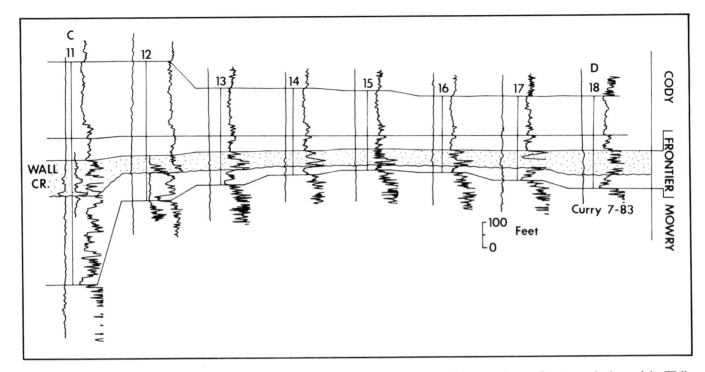

Figure 6—Cross section C–D showing Frontier Formation and thinning of rocks between interpreted unconformity at the base of the Wall Creek Member and the top of the Mowry Shale across the paleotectonic high near Church Buttes gas field (T. 16 N., R. 112 W.).

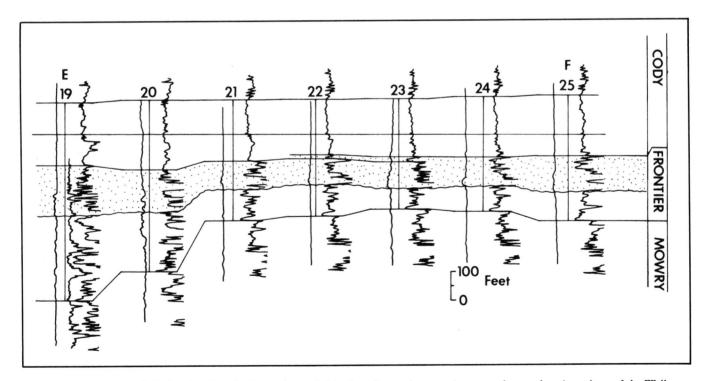

Figure 7—Cross section E–F showing Frontier Formation and thinning of strata between interpreted unconformity at base of the Wall Creek Member and the top of the Mowry Shale across the paleotectonic high near Wilson Ranch gas field (T. 19 N., R. 112 W.).

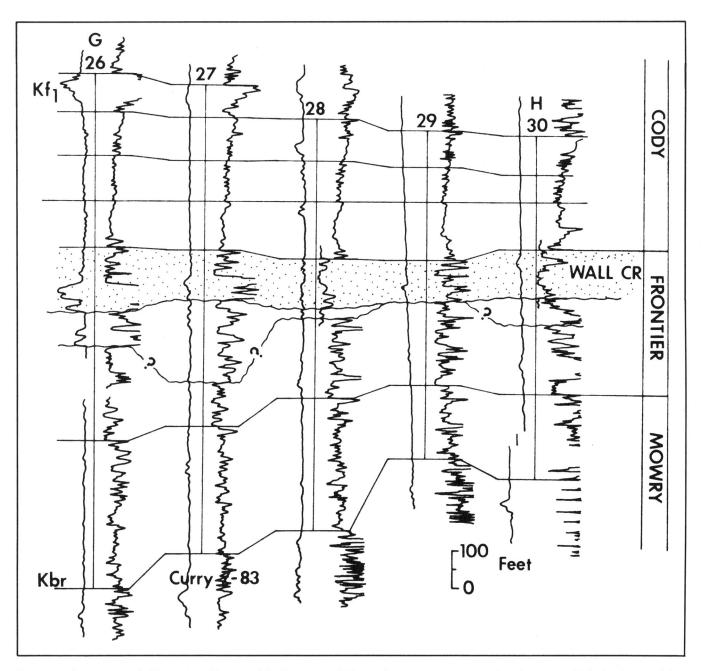

Figure 8—Cross section G–H showing thinning of the Frontier and Mowry formations across a subtle paleotectonic high that is east of the LaBarge gas field (T. 27 N., R. 113 W.). Note the location on the left side of the cross section at LaBarge of the "First Frontier sandstone" in the lower part of the Cody Formation. In general, lithologic zonation and correlation of well logs are more difficult in this area than in other parts of Wyoming.

NE¼NW¼ sec. 17, T. 34 N., R. 77 W. can be mapped in the adjacent area with considerable confidence. Sandstone beds designated "Unit I through Unit III" at the cited reference log are the capping sandstones of informal members A through C.

In areas more remote from outcrops, the location of at least one unconformity can be mapped with some confidence at the base of the Wall Creek Member where underlying units exhibit thickening and thinning that is consistent with the unconformity and paleotectonic concepts. In other areas, however, the unconformity has to be picked at the base of

the Wall Creek Member, and where the base of the Wall Creek Member is difficult to identify on borehole logs, the subsurface pick of the basal Wall Creek unconformity is best described as an educated guess.

Accordingly, in areas where the subsurface data allow the mapping of two unconformities (Figure 13), member E can be isolated from the overlying Wall Creek Member, but in the subsurface of other areas, locating the base of the Wall Creek is difficult. For example, the position of the basal contact of the Wall Creek Member in boreholes #38 and #39 of Figure 12 is uncertain and the Wall Creek, as shown,

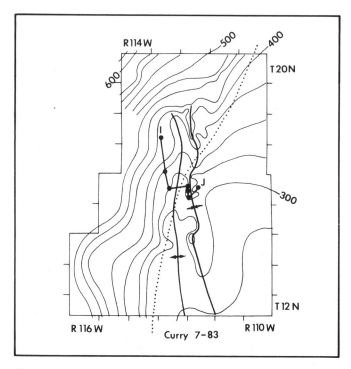

Figure 9—Isopach map of the strata between the top of the Mowry Shale and the top of the "Dakota" sandstone, showing the hingeline along the eastern edge of the western basin and the thinning over the paleotectonic high near Church Buttes. Contour interval is 25 ft (7.5 m). Line I–J is the location of cross section I–J in Figure 10.

probably includes the E member. This inconsistency of correlation and nomenclature is undesirable, but is difficult to remedy.

Cross section M–N (Figure 13) shows the persistent but subtle thinning of the shale in the Cody A member, which is bounded at its top and base by bentonite markers. The thin area (Figure 11) is interpreted as evidence of a paleohigh near Spearhead Ranch oilfield (T. 38 N., R. 74 W.). Note the lack of evidence of thinning below the Wall Creek Member over the Spearhead paleohigh. Although the relationship between the distribution of the uppermost Wall Creek sandstone beds and the paleotectonic features indicated by the Cody A member requires further documentation, the current evidence suggests that thicker sandstone units were deposited on the flanks of the paleohighs, rather than on the tops of the paleohighs. The distribution and development of Wall Creek sandstone bodies in central Wyoming appear to reflect the pattern of paleocurrents that transported sand from a northwesterly source area within a framework of subtle paleotectonic features.

POSSIBLE RELATIONSHIP OF PALEOTECTONIC ACTIVITY TO THRUSTING

The basal Wall Creek unconformity has been dated by ammonites as approximately middle Turonian in age (~90 m.y. old) (Merewether, 1983). The Paris thrust sheet apparently moved intermittently from the beginning of

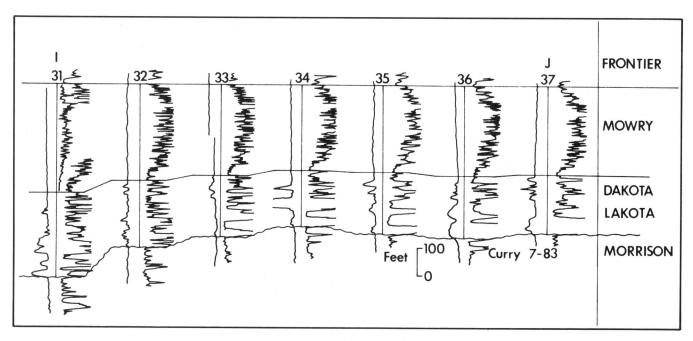

Figure 10—Cross section I–J showing subtle thinning of the Mowry, "Dakota," and Lakota formations across the paleohigh near Church Buttes gas field. This thinning indicates Early Cretaceous tectonic movement that was probably associated with early movement of the Paris thrust sheet.

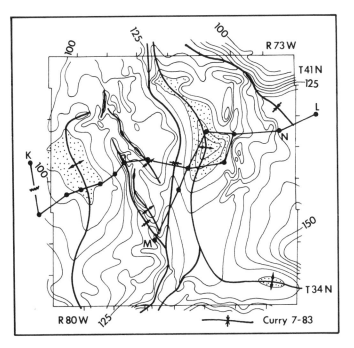

Figure 11—Isopach map of the Cody A member between two distinctive bentonite beds in the eastern basin, showing several subtle variations in thickness. Three areas (stippled) where the member is thin are interpreted as subtle paleotectonic highs that may have started early enough to have influenced deposition of the sandstones in the upper part of the Wall Creek. Some of these sandstones form hydrocarbon-bearing stratigraphic traps in the deep part of the Powder River basin. Contour interval is 5 ft (1.5 m). Tick marks are 6 mi (10 km) apart. Line K–L is location of cross section K–L in Figure 12.

Cretaceous time (141 m.y. ago) until middle Turonian time, and the Crawford and Meade thrust sheets subsequently moved in Coniacian time (~88–89 m.y. ago) (Wiltschko and Dorr, 1983; Palmer, 1983). The Targhee uplift (Love, 1973) rose west of the Yellowstone Park area and was a source of sediments in Coniacian time (~88–89 m.y. ago) (Wiltschko and Dorr, 1983).

Because the start of movement on the Crawford and Meade thrusts, the start of the Targhee uplift, and the deformation that caused the basal Wall Creek unconformity were all closely related in the time period from 87 to 90 m.y. ago, it is tempting to interpret them as a related series of paleotectonic events. However, the thinning over the Church Buttes arch in Mowry, "Dakota," and Lakota rocks, which are poorly dated as 98 to 144 m.y. old (Palmer, 1983), also suggests an even earlier period of deformation that might have been related to earlier movements on the Paris thrust.

The Turonian western basin and western arch of Wyoming (Figure 1) are apparently parts of the same paleotectonic features indicated by the total thickness of all Upper Cretaceous rocks throughout Wyoming and Utah (Figure 14) (McGookey et al., 1972, from work by Weimer and Haun, 1960 and Reeside, 1944).

Loading of the crust by thrust sheets during the Sevier and Laramide orogenies was probably a major cause of the depression of the western basin; however, the later isostatic rise (Strange and Woollard, 1964) of the western basin also suggests that a tectonic root, which was formed during Cretaceous compression, caused a postcompressional isostatic rise. Note the strong arc geometry (Figure 14) between the western basin, which is associated with the thrust belt, the western arch and the eastern arch.

The eastern arch and associated basins of McGookey et al. (1972) (Figure 14) are not obvious in the distribution of the Frontier Formation of Wyoming, either because they did not extend northward into Wyoming during Frontier deposition or because they were undetected by our investigation. There is evidence that parts of the Sweetwater uplift of Wyoming had several stages of uplift: a Late Jurassic to Early Cretaceous period of moderate uplift (Love et al., 1945), an Albian pre-Muddy sandstone period (Curry, 1959), a middle Turonian period (Merewether, 1983), and a late Campanian to early Maestrichtian stage (Reynolds, 1966). More detailed mapping of Frontier strata in southern Wyoming may well detect the eastern paleoarch of McGookey et al. (1972).

The subtle paleotectonic features indicated by the Cody A member in the eastern "basin" are of interest mainly because of their possible origin during the deposition of the upper Wall Creek sandstone and their possible influence on the distribution and character of the uppermost sandstones of the Wall Creek, which are currently being explored for stratigraphically entrapped oil and gas.

During Cretaceous time, from Albian through Maestrichtian, thousands of feet of sediments were deposited near sea level in Wyoming as the crust and depositional surface slowly subsided. Accordingly, many of the subtle thickness variations of units of the Frontier in Wyoming were probably caused by differential rates of subsidence. The basal Wall Creek unconformity and associated paleotectonic features, however, appear to represent an unusual paleotectonic event for the middle Cretaceous of this area. This unusual paleotectonic event may have been related to thrusting in the adjacent Sevier orogenic belt.

REFERENCES CITED

Curry, W. H., III, 1959, Stratigraphy and paleogeography of Upper Jurassic and Lower Cretaceous rocks of central Wyoming: Ph.D. thesis, Princeton University, Princeton, New Jersey, 216 p.

Goodell, H. G., 1962, The stratigraphy and petrology of the Frontier Formation of Wyoming: Wyoming Geological Association 17th Annual Field Conference Guidebook, p. 173–210.

Love, J. D., 1973, Harebell Formation (Upper Cretaceous) and Pinyon Conglomerate (Uppermost Cretaceous and Paleocene), northwestern Wyoming: U.S. Geological Survey Professional Paper 734-A, 54 p.

Love, J. D., R. M. Thompson, C. O. Johnson, H. H. R. Sharkey, H. A. Tourtelot, and A. D. Zapp, 1945, Stratigraphic sections and thickness maps of Lower Cretaceous and nonmarine Jurassic rocks of central Wyoming: U. S. Geological Survey Oil and Gas Investigations Preliminary Chart 13.

Masters, J. A., 1952, The Frontier Formation of Wyoming:

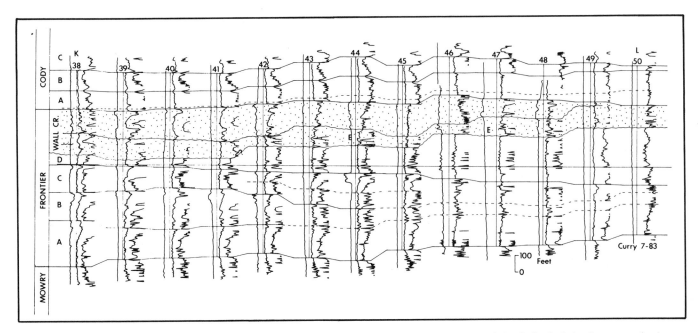

Figure 12—Cross section K–L of sedimentary rocks in the Frontier Formation and the lower part of the Cody shale in the eastern basin showing anomalous westward thinning of informal members D and E of the Frontier, the Wall Creek Member, and the Cody A member. This thinning, the basal Wall Creek unconformities, and subtle thickness variations of the Cody A member reflect the less intense Turonian paleotectonic activity in this part of Wyoming.

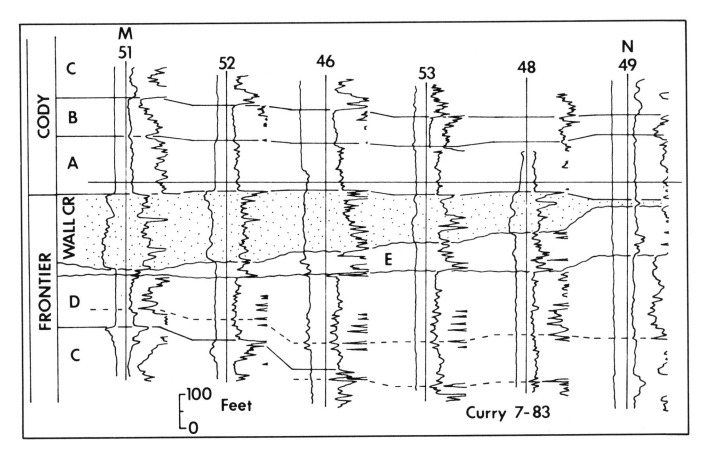

Figure 13—Cross section M–N showing the subtle thinning of the Cody A member between two bentonite beds over the Spearhead paleohigh. It is possible to map two unconformities near the base of the Wall Creek Member; however, the eastward stratigraphic rise of the upper unconformity is not accepted by all interested workers.

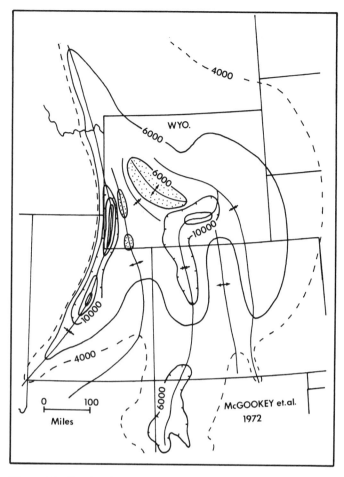

Figure 14—Regional isopach map of Upper Cretaceous strata adapted from McGookey et al. (1972) showing axes of thicker preserved rocks in the western basin associated with the thrust belt and the thinner preserved rocks on the western and eastern paleoarches. The western basin and western arch of this study appear to be parts of the same paleotectonic features interpreted from the isopach map of Upper Cretaceous strata.

Wyoming Geological Association 7th Annual Field Conference Guidebook, p. 63–66.

McGookey, D. P., J. D. Haun, L. A. Hale, H. G. Goodell, D. G. McCubbin, R. J. Weimer, and G. R. Wulf, 1972, Cretaceous System, *in* Geologic Atlas of the Rocky Mountain Region: Denver, Rocky Mountain Association of Geologists, p. 190–228.

Merewether, E. A., 1983, The Frontier Formation and mid-Cretaceous orogeny in the foreland of southwestern Wyoming: The Mountain Geologist, v. 20, n. 4, p. 121–138.

Merewether, E. A., W. A. Cobban, and E. T. Cavanaugh, 1979, Frontier Formation and equivalent rocks in eastern Wyoming: The Mountain Geologist, v. 16, n. 3, p. 67–101.

Palmer, A. R., 1983, 1983 geologic time scale: Geology, v. 11, p. 503–504.

Reeside, J. B., Jr., 1944, Maps showing thickness and general character of the Cretaceous deposits in the western interior of the United States: U.S. Geological Survey Oil and Gas Investigations Preliminary Map 10.

Reynolds, M. W., 1966, Stratigraphic relations of Upper Cretaceous rocks, Lamont–Bairoil area, south-central Wyoming: U.S. Geological Survey Professional Paper 550-B, p. 69–76.

Strange, W. E., and G. P. Woollard, 1964, The use of geologic and geophysical parameters in the evaluation, interpretation and production of gravity data: Aeronautical Chart and Information Center, U.S. Air Force, St. Louis, Missouri, Contract AF 23 (601)-3879 (Phase 1).

Weimer, R. J., and J. D. Haun, 1960, Cretaceous stratigraphy, Rocky Mountain region, U.S.A., *in* Regional paleogeography: Copenhagen, 21st International Geological Congress, Norden, pt. XII, p. 178–184.

Wiltschko, D. V., and J. A. Dorr, Jr., 1983, Timing of deformation in overthrust belt and foreland of Idaho, Wyoming and Utah: AAPG Bulletin, v. 67, n. 8, p. 1304–1322.

Laramide Tectonics and Deposition of the Ferris and Hanna Formations, South-Central Wyoming[1]

D. E. Hansen
U.S. Geological Survey
Denver, Colorado

The Upper Cretaceous–lower Tertiary Ferris and Hanna formations in south-central Wyoming constitute an interval of continental rocks that contain principal coal beds in the Hanna and Carbon intermontane basins. The interval in the Hanna foreland basin consists of plane-bedded arkosic and lithic conglomerate with sandstone and mudstone interpreted as piedmont slope deposits. They wedge southward into central basin planar and trough cross-bedded sandstone and coeval parallel bedded sandstone and siltstone, dark-colored mudstone and shale, and coals interpreted as fluvial and floodplain deposits. The adjacent Carbon basin contains the fluvial and floodplain deposits. Deposition occurred in a syncline situated between systems of uplifts that resulted from compressional deformation. Upper Cretaceous sediments were derived mostly from uplifts to the south, and Paleocene and lower Eocene sediments were derived mostly from north of the Hanna basin. Lower Eocene and older rocks were folded into the footwalls of large, east-west-trending middle Eocene uplifts that were thrust southward over the northern Hanna basin margin.

Increased structural complexity from Late Cretaceous through middle Paleocene to middle Eocene time resulted from increasing uplifts and associated counterclockwise rotation from east-west- to north-south-directed couple stresses. This rotation paralleled the movement of the North America plate as it overrode the Farallon plate. The north-south couple stresses produced thrusting of middle Eocene uplifts surrounding the basins, separation of the Hanna and Carbon basins, and thrusting that was at about right angles to Late Cretaceous thrust patterns. Crustal shortening was followed by extensional deformation and general regional uplift.

INTRODUCTION

The Upper Cretaceous–Paleocene Ferris Formation and the Paleocene–Eocene Hanna Formation contain the principal coal beds of the Hanna coal field in south-central Wyoming. Mineable coal beds of these formations occur in the Hanna and Carbon intermontane basins where the formations represent vestiges of a greater area of continental sedimentation that existed during the Laramide orogeny. The two basins have a combined area of slightly less than 1200 mi^2 (3100 km^2). These structural basins lie between the Medicine Bow–Sierra Madre mountains and the Granite–Seminoe–Shirley mountains (Figure 1).

This study was undertaken to determine the geologic history of the structural deformation and how it affected the deposition of Ferris and Hanna coal beds and their confining sediments in the Hanna and Carbon basins. In the Hanna basin, the Ferris and Hanna formations constitute a coal-bearing interval that is as much as 16,000 ft (4900 m) thick. In contrast, this coal-bearing interval is less than 4200 ft (1300 m) thick in the Carbon basin. The record of Laramide tectonism is more comprehensive in the Hanna basin.

Precambrian crystalline rocks of the Hanna basin apparently subsided to great depths during Laramide tectonism. These rocks are now more than 31,000 ft (9450 m) below mean sea level datum. Similar Precambrian rocks are now exposed more than 7300 ft (2200 m) above mean sea level datum in the uplift that has been thrust faulted over the north edge of the Hanna basin (Figure 2). The areal relationship of the Hanna basin to the surrounding major uplifts is shown on Figure 3. Though relatively small, the Hanna basin has undergone the same geologic events that formed the larger Wind River and other deep, asymmetric, thrust-fault-bounded basins in the Wyoming foreland.

The significance of Laramide compressional structures in the Wyoming foreland was not understood until Berg (1962) described the magnitude and form of the southwest thrust fault of the Wind River uplift. Gries (1983a) demonstrated that compressional deformation was common to the large Laramide uplifts in the foreland. Blackstone (1983) thoroughly described the numerous structures formed by Laramide compressional deformation in southeastern Wyoming.

[1]Publication authorized by the Director, U.S. Geological Survey.

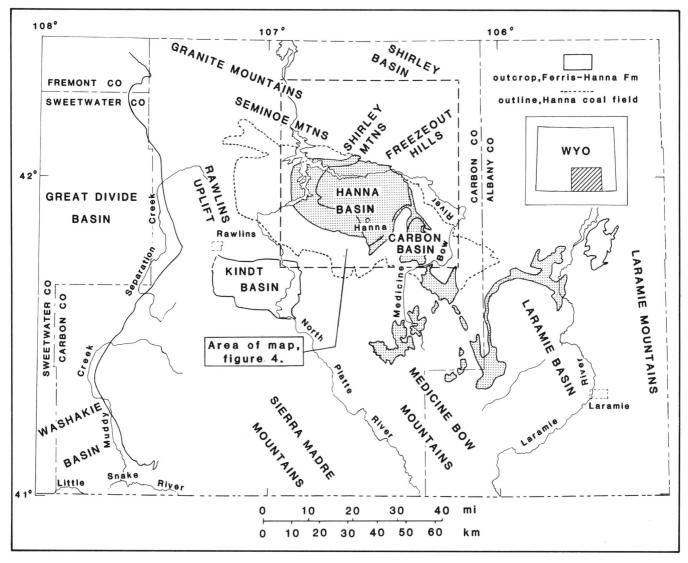

Figure 1—Location of the intermontane Hanna and Carbon basins, south-central Wyoming. In these basins the Ferris and Hanna formations contain the mineable coal beds of the Hanna coal field.

Another regional report by Gries (1983b) stressed the time and direction of the Laramide compressional deformation in the Wyoming foreland. The age of the different uplifts can be assigned by the direction they trend: (1) the Late Cretaceous structures trend north–south, (2) the Paleocene structures trend northwest–southeast, and (3) the Eocene structures trend west–east. Furthermore, the movements of the three ages of deformation have been equated by Gries (1983b) with movements of the North America plate. The movements of the plate were caused by the opening of the North Atlantic Ocean area during seafloor spreading in Late Cretaceous time and by the opening of the Arctic Ocean area during seafloor spreading in Paleocene to late Eocene time.

Rocks older than the Ferris and Hanna formations were folded into the basins and eroded off the nearby uplifts during Laramide time. Except for the Upper Cretaceous Medicine Bow Formation, the older formations are shown as combined units on the geologic map (Figure 4) and are listed in the explanation of map units. These Paleozoic and

Mesozoic sedimentary rocks are about 22,000 ft (6700 m) thick, but Upper Cretaceous rocks are the major component of the succession. The Upper Cretaceous rocks form an interval about 18,500 ft (5700 m) thick directly below the Ferris Formation. This interval of marine and nonmarine sedimentary rocks apparently lies within a Late Cretaceous depocenter in the area of the Hanna and Carbon basins (Krumbien and Nagel, 1953). At the top of this interval the upper Medicine Bow Formation consists of coarse-grained sandstone of continental origin that is conformable with the overlying Ferris Formation.

Deposited during Laramide time, the Ferris and Hanna rocks are chiefly of alluvial origin. These formations, as defined by Bowen (1918), contain basal beds of conglomeratic sandstone and thin conglomerate. Other previous investigators (Veatch, 1907; Dobbin et al., 1929; Dorf, 1938) also regarded the occurrence of beds of conglomeratic sandstone and conglomerate as requiring a division of these terrestrial rocks into stratigraphic units. The lower contact of the Ferris Formation is conformable, as

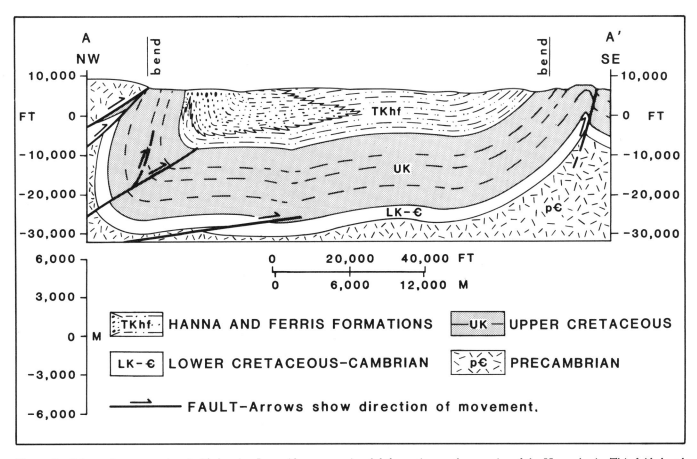

Figure 2—Schematic cross section A–A' showing Laramide compressional deformation on the margins of the Hanna basin. This folded and thrust-faulted deformation is characteristic of the Wyoming foreland. Location of cross section is shown on geologic map in Figure 4.

previously mentioned. However, the Ferris–Hanna contact in the Carbon basin and along the eastern margin of the Hanna basin is unconformable and lies below conglomeratic sandstone of the basal Hanna. This relationship led Bowen (1918) and Dobbin et al. (1929) to the perception of a general unconformable Ferris–Hanna contact.

Except in the Carbon basin and along the eastern margin of the Hanna basin, the unconformable Ferris–Hanna contact mapped by Dobbin et al. (1929) does not exist. Along the western outcrops, the coal-bearing parts of the two formations show continuity, and the Ferris–Hanna contact is found to be conformable, as reported by Blanchard and Comstock (1980). Along the northern margin of the basin, the Ferris and Hanna formations also show continuity, and both formations grade laterally into the thick sequence of conglomerates that Knight (1951) measured. In the western and northern Hanna basin, these formations, as mapped in this report (Figure 4), together form a common rock interval consisting of the two general and distinct lithologic units that are outlined below.

STRATIGRAPHY OF THE FERRIS AND HANNA ROCK INTERVAL

The two distinct lithologic units are (1) northern Hanna basin conglomerate unit that wedges southward into (2) a central basin, coal-bearing mudstone and sandstone unit that

is as much as 16,000 ft (4900 m) thick. These successions were deposited during Late Cretaceous–Eocene time, as shown in the stratigraphic columns (Figures 5 and 6). Upward increases of Precambrian rock fragments changed the aspect of the conglomerate and the coal-bearing rocks; this allowed the separation of the Ferris and Hanna formations (Bowen, 1918; Dobbin et al., 1929). But the important distinction is between the northern conglomerates and the coal-bearing mudstone and sandstone. These two distinct lithologies can be used to demonstrate how deposition of the rocks was affected by Laramide tectonism.

The general lithologies (conglomerate and coal-bearing mudstone and sandstone) can also be subdivided into units each showing a variation in aspect of the general lithology (Figure 7). The general lithologies and their subdivisions are mapped as areas having general boundaries (Figure 8), because the lateral contacts of the units are intertonguing and gradational. For the most part, the subdivisions contain intervals of rock that show variations of the principal rock types (Figure 9).

Coal-Bearing Mudstone and Sandstone

The coal-bearing rocks show a variation of color and of lithology. These rocks are subdivided into a basal dark-colored sandstone unit; units of thick coal; and units of thin coal, as previously illustrated. There are no mappable coal beds within the dark-colored sandstone unit and the thin

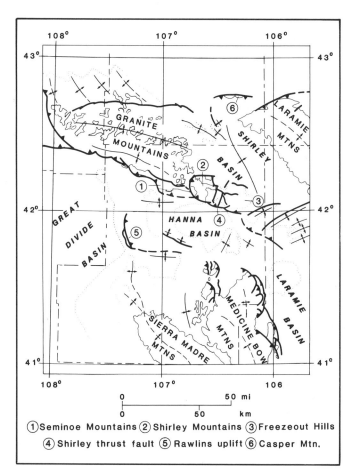

Figure 3—Location of Hanna basin and regional structures. Mountain uplifts show the regional trough aspect of the Hanna–Laramie basins. Shirley thrust fault system is on south side of Shirley Mountains–Freezeout Hills uplift. South Granite Mountains thrust fault system extends eastward along Seminoe Mountains. Dotted lines show Tertiary and Upper Cretaceous rocks that crop out and outline the uplifts and basins. General trends of synclinal and anticlinal forms are indicated by lines and arrows; barbs indicate upthrown side of thrust faults.

coal unit. The latter is mapped only in areas of thin, discontinuous, and impure coal beds.

Fossils associated with the coal-bearing rocks are invertebrates, including freshwater gastropods, bivalves, and ostracods (Glass, 1981); vertebrates, including bone fragments, turtle shell fragments, and fish remains; and plants, including common leaf imprints, branch fragments, tree trunks and stumps, and traces of roots.

Dark-Colored Sandstone

The dark-colored unit (lower Ferris Formation) consists of beds of conglomeratic sandstone; conglomerate; parallel-bedded sandstone; dark-gray and dark-brown siltstone, mudstone, and shale; carbonaceous shale; and minor coal. The unit is generally about 1000 ft (300 m) thick, but is as much as 1600 ft (490 m) thick in the northwestern Hanna basin.

The conglomeratic sandstone is dark gray and dark brown to yellowish brown and ferruginous, fine to coarse grained

with granules and small pebbles, and trough and planar cross bedded. These sandstones can be classified as lithic arenites (Ryan, 1977). Sequences of beds of these conglomeratic sandstones are as much as 80 ft (25 m) thick in the upper part of the unit. The conglomerate is brown and consists chiefly of rounded pebbles of chert and quartzite rock fragments and rarer fragments of volcanic rocks (Bowen, 1918); the pebbles are generally less than 1 in. (2 cm) in diameter. The conglomerate occurs as pockets, lenses, and thin beds within the conglomeratic sandstone.

The parallel-bedded sandstone is dark brown, tan, and gray; and generally fine- to medium-grained. They are generally less than 4 ft (~1 m) thick. In places, the sandstone beds coarsen upward to include thin, plane-bedded conglomerate. The siltstone, shale, and mudstone beds are each generally less than 5 ft (1.5 m) thick, but their composite thickness can be as much as 12 m. The carbonaceous shale beds are thin (less than 1 ft, 0.3 m), and the impure coal beds are generally only a few inches thick, although one coal lens was found to be 2 ft (~0.5) m thick. Vertebrate remains occur in the conglomerate beds, and a few have been identified as *Triceratops* (Bowen, 1918). Plant remains are common; the microfossil assemblage from samples collected by Gill et al. (1970) is of Late Cretaceous age.

Thick Coal

The thick coal unit consists of massive sandstone with local beds of conglomerate, bedded sandstone, gray to dark-gray mudstone, dark-gray shale, carbonaceous shale, and coal beds. The lower part of the unit (upper Ferris and basal Hanna) is as much as 5000 ft (1500 m) thick. The upper part of the unit (Hanna) is as much as 7000 ft (2100 m) thick.

The massive sandstone is light gray and tan to brown, fine- to coarse-grained and granular in places, trough and planar cross-bedded, commonly slump structured, partly ferruginous, locally arkosic, and concretionary. Basal parts of the massive sandstone are scour based and contain clasts that came from the associated finer grained rocks. These sandstones are chiefly lithic arenites and locally arkosic arenites (Ryan, 1977). The massive sandstones are generally 20–100 ft (6–30 m) thick. In the middle part of the thick coal interval (lower part of the Hanna Formation) beds of massive sandstone are stacked in sequence to as much as 300 ft (90 m) thick in the Hanna basin and to as much as 600 ft (180 m) in the Carbon basin.

The parallel-bedded sandstone is tan and gray, generally fine- to medium-grained, and contains thin beds of conglomerate in the northern Hanna basin. These sandstones coarsen upward, are current ripple marked and current laminated, and may occur in irregular thin beds. The beds are generally 1–3 ft (0.3–1 m) thick where associated with the mudstone, siltstone, shale, and coal beds. Where associated with the massive sandstone deposits, they are generally less than 3 m thick.

The gray to dark-gray mudstone, siltstone, shale, and claystone deposits vary greatly in thickness. In places, mudstone beds are as thick as 70 ft (20 m), but they are generally less than 40 ft (12 m) thick. The shale, claystone, and siltstone beds are each generally less than 10 ft (3 m)

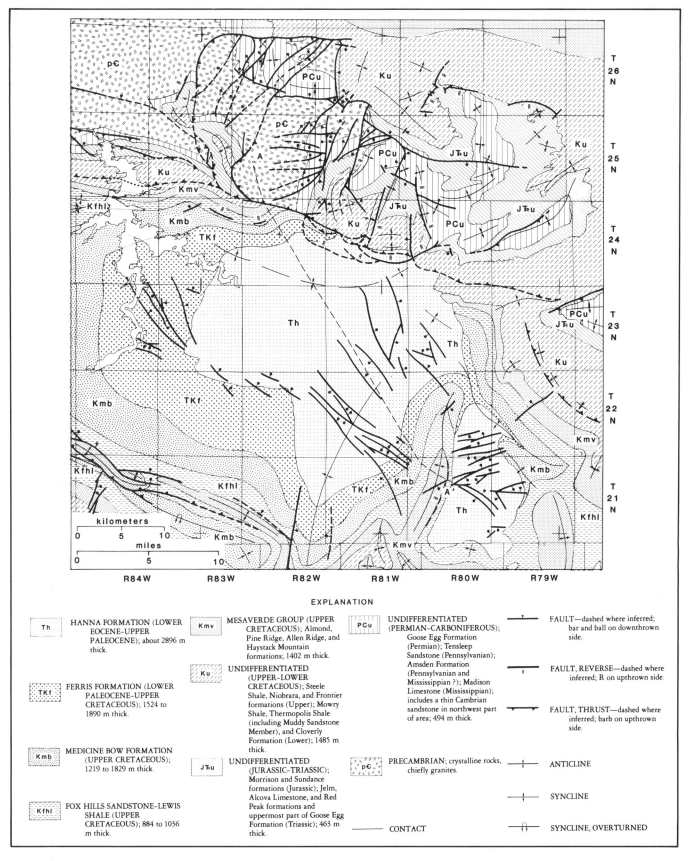

EXPLANATION

Th	HANNA FORMATION (LOWER EOCENE-UPPER PALEOCENE); about 2896 m thick.	Kmv	MESAVERDE GROUP (UPPER CRETACEOUS); Almond, Pine Ridge, Allen Ridge, and Haystack Mountain formations; 1402 m thick.	PCu	UNDIFFERENTIATED (PERMIAN-CARBONIFEROUS); Goose Egg Formation (Permian); Tensleep Sandstone (Pennsylvanian); Amsden Formation (Pennsylvanian and Mississippian ?); Madison Limestone (Mississippian); includes a thin Cambrian sandstone in northwest part of area; 494 m thick.	── FAULT—dashed where inferred; bar and ball on downthrown side.

FERRIS FORMATION (LOWER PALEOCENE-UPPER CRETACEOUS); 1524 to 1890 m thick. — TKf

UNDIFFERENTIATED (UPPER-LOWER CRETACEOUS); Steele Shale, Niobrara, and Frontier formations (Upper); Mowry Shale, Thermopolis Shale (including Muddy Sandstone Member), and Cloverly Formation (Lower); 1485 m thick. — Ku

FAULT, REVERSE—dashed where inferred; R on upthrown side.

MEDICINE BOW FORMATION (UPPER CRETACEOUS); 1219 to 1829 m thick. — Kmb

UNDIFFERENTIATED (JURASSIC-TRIASSIC); Morrison and Sundance formations (Jurassic); Jelm, Alcova Limestone, and Red Peak formations and uppermost part of Goose Egg Formation (Triassic); 463 m thick. — JŦu

PRECAMBRIAN; crystalline rocks, chiefly granites. — pЄ

FAULT, THRUST—dashed where inferred; barb on upthrown side.

FOX HILLS SANDSTONE-LEWIS SHALE (UPPER CRETACEOUS); 884 to 1036 m thick. — Kfhl

ANTICLINE

SYNCLINE

CONTACT

SYNCLINE, OVERTURNED

Figure 4—Geologic map of pre-Miocene rocks, Hanna basin area, showing areal relationships of the Shirley Mountains–Freezeout Hills uplift and the Hanna basin and other areas structures. Line A–A′ designates the cross section of Figure 2. Sources of map are Love et al. (1955), Lowry et al. (1973), and field mapping by the author.

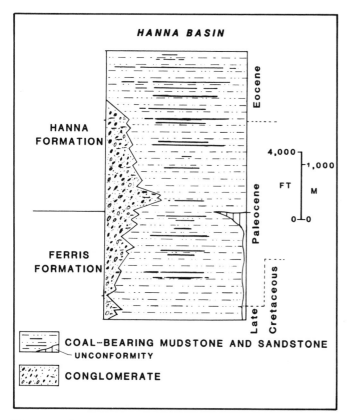

Figure 5—Generalized stratigraphic column of the coal-bearing, Ferris–Hanna interval in the Hanna basin.

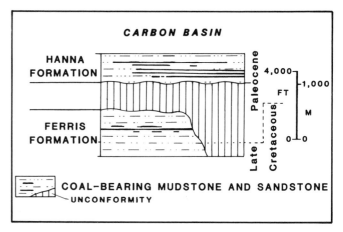

Figure 6—Generalized stratigraphic column of the coal-bearing, Ferris–Hanna interval in the Carbon basin.

thick. In the lower part of the unit (upper Ferris Formation) the carbonaceous shales are generally less than 10 ft (3 m) thick, but in the upper part of the unit (Hanna Formation) the black carbonaceous shales are generally 10–40 ft (3–12 m) thick. Coaly shale beds have been found to be as thick as 100 ft (30 m) in this upper part of the thick coal unit. The geometric relationship of typical coal, mudstone, parallel-bedded sandstone, shale, and carbonaceous shale beds to a massive sandstone complex is shown in Figure 10.

The coal beds are as much as 38 ft (12 m) thick in the Hanna basin. In the lower part of the unit (upper Ferris) about thirty coal beds are greater than 5 ft (1.5 m) thick, but only four are greater than 20 ft (6 m) thick (Glass and Roberts, 1980). Two coal beds of the lower unit occur in the Carbon basin; there the lower bed is as much as 20 ft (6 m) thick, but the upper coal bed is less than 5 ft (1.5 m) thick. In the upper part of the thick coal unit in the Hanna basin (Hanna Formation) there are twenty-four coal beds greater than 5 ft (1.5 m) thick, but no more than eight exceed 20 ft (6 m) (Glass and Roberts, 1980). In the Carbon basin, three persistent coal beds of the upper unit exceed 5 ft (1.5 m) in thickness, but only one exceeds 20 ft (6 m) in thickness. The relationship of upper unit coal zones within finer grained rock intervals and the massive sandstones of the Carbon basin is shown in Figure 11.

Microfossils from samples collected by Blanchard and Comstock (1980) and Gill et al. (1970) show that in the western Hanna basin the age of each assemblage is (1) middle Paleocene from the lower part of the unit (lower half of Ferris), (2) middle Paleocene to late Paleocene from near

the lower middle part of the unit (near top of Ferris), (3) late middle Paleocene from the upper middle part of the unit (basal Hanna), (4) late Paleocene from the middle upper part of the unit (middle Hanna), and (5) Eocene from near the top of the unit (near the top of Hanna). In the Carbon basin the age of each assemblage is (1) early middle Paleocene from the lower part of the unit (upper Ferris), (2) late Paleocene from the lower upper unit (lower Hanna), and (3) late Paleocene from the base of the upper unit (base of Hanna).

Thin Coal

The thin coal unit mapped below the base of the coal-bearing rocks (base of upper Ferris) in the western Hanna basin (Figure 8) is 800–1200 ft (~250–370 m) thick. There the rocks of the thin coal unit have a brown color that is not as evident in the overlying thick coal unit or in the other parts of the thin coal units. Similar strata, about 600 ft (180 m) thick, occur in the southeastern Hanna basin where the brown rocks form the basal part of a thin coal unit. Rocks having this brown color are massive sandstone beds that are brown to orange brown, soft, and ferruginous. The finer grained rocks of this thin coal unit are the same as those of the other thin coal and thick coal units. The coals associated with these beds of brown massive sandstone are less than 1 ft (30 cm) thick and are associated with thin 1–3 ft (30–90 cm) black and brown carbonaceous shale and claystone intervals. Microfossils from samples collected by Blanchard and Comstock (1980) show that the age of the assemblage is early to middle Paleocene.

Other parts of the thin coal unit that have been mapped occur between the thick coal units (upper Ferris and lower Hanna), lateral to the lower thick coal unit (upper Ferris), and below the upper thick coal unit (Hanna) (Figure 8). The thin coal unit is about 7000 ft (2100 m) thick where the subdivision lies below the upper thick coal unit. In the eastern Hanna basin, paleosols between the mudstone beds are indicated by white to light-gray, hackly, thin sandstone and siltstone layers. Zones of root traces were seen on the uppermost part of the massive sandstone deposits in a few places.

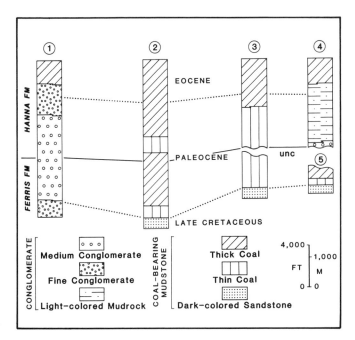

Figure 7—Stratigraphic sections of the conglomerate and coal-bearing lithologies and subdivisions of the Ferris–Hanna interval. The generalized lines of the sections that are shown on Figure 8 are in the Hanna basin, except section 5, which is on the north side of the Carbon basin.

Conglomerate

Beds of conglomerate along the northern margin of the Hanna basin are chiefly arkosic, but lithic conglomerate is also present. The components of the arkosic conglomerate are fragments of granite, black and gray chert, gray quartzite, gray sandstone, the Lower Cretaceous light-gray Mowry Shale, vein quartz, the Lower Cretaceous Cloverly Formation, and metamorphic rocks. The arkosic conglomerate beds are interbedded with conglomeratic sandstone, muddy sandstone, and thin mudstone. Diameters of clasts within this thick conglomerate succession of the northern Hanna basin (Figures 5 and 7) are seen to increase upward to a maximum and then to decrease upward. A corresponding lateral decrease in diameter is seen to occur southward from the northern periphery of the Hanna basin. Corresponding vertical and lateral changes also occur in the bedding.

The lithic conglomerate consists chiefly of clasts of gray sandstone, gray quartzite, light-gray Mowry Shale, and chert. These conglomerates are found only in small, local areas of short vertical extent relative to the arkosic conglomerates. The lithic conglomerate is interbedded with sandstone and mudstone deposits. These same mudstones are dominant lithology of the light-gray beds that overlie the lithic conglomerate in northeastern Hanna basin.

The conglomerate has been subdivided into "medium" and "fine" categories at arbitrary boundaries because of the lateral gradations and intertonguing. The original definitions were as follows: the medium conglomerates consisted of clasts having diameters roughly between 2 and 6 in. (5 and 15 cm) and the fine conglomerates had clasts with diameters less than 2 in. (~5 cm). In practice, the mapping showed that defining fine conglomerate as having clasts less

than 1.5 in. (4 cm) in diameter is more useful. Bedding characteristics of the conglomerate and of the associated lithologies change as these clast diameters change.

Arkosic Medium Conglomerate

The lower 3900 ft (1200 m) (upper Ferris) of the medium conglomerate unit consists of light-gray to white, tan and brown to yellow-brown conglomerate and conglomeratic sandstone sequences that crop out between intervals of gray, pebbly sandstone. The sequences appear as layers of small to large pebbles in the lower part of the unit. The overall grain size coarsens upward, and near the top, small cobbles (4 in., or 10 cm diameter or less) are a dominant part of the sequence. Where dominant, some of the small cobble beds are clast supported with infillings of feldspar and quartz grains. The sequences are as much as 25 ft (8 m) thick; the individual beds of the sequences are generally from 0.5 to 3 ft (15 to 90 cm) thick. In a few places, the beds showed some indistinct internal bedding and imbrication. The intervals of pebbly sandstone include thin beds and lenses of light-gray and white to gray sandy mudstone and claystone. The pebbly sandstone intervals have plane bedding in a few places. These intervals of pebbly sandstone are generally 20–40 ft (6–12 m) thick.

The upper 4000 ft (1200 m) (Hanna Formation) of the medium conglomerate unit is chiefly white to light gray and tan. The interval continues to coarsen upward to about the lower 1000 ft (300 m) of this upper part. Clast-supported conglomerates become common. The feldspar content of the upper part of the conglomerate also increases relative to the lower part. This upper part merges into the overlying arkosic fine conglomerate.

Arkosic conglomerate (Hanna) crops out above light-colored mudstone beds between Troublesome and Difficulty creeks in T. 24 N., R. 81 W. The residuum of weathered granite boulders is present along the concealed Shirley thrust fault in Sec. 22, T. 24 N., R. 81 W. The conglomerate lies within a small syncline south of the thrust fault. The medium conglomerate here is much the same as that of the larger, arkosic medium conglomerate unit already described, except the intervals of pebbly sandstone are generally less than 10 ft (3 m) thick and mudstone beds are not present. The thickness of the unit may be as much as 600 ft (180 m).

Arkosic Fine Conglomerate

The unit of arkosic fine conglomerate (Hanna) is about 3000 ft (900 m) thick. The fine conglomerate consists of a sequence of gray and tan beds of arkosic conglomerate, conglomeratic sandstone, and coarse sandstone interspersed in gray, sandy mudstone deposits. The clasts in these conglomerates are generally less than 1 in. (~2 cm) in diameter. The conglomerate units occur as plane beds and small lenses, whereas the sandstone units are chiefly trough cross-bedded and have the appearance of irregular lenses. The sequences of conglomerate and sandstone form lenticular bodies that are thin (less than 15 ft, or 4.5 m) relative to their width of several tens of meters. These bodies have a scoured contact with the underlying mudstone. The fine conglomerate grades laterally into the coal-bearing unit. Where this gradation takes place, the sandstone beds

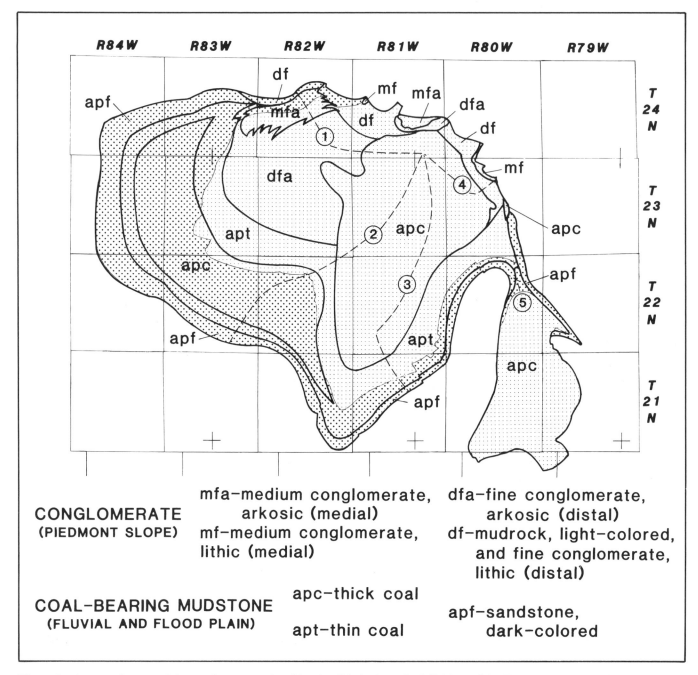

Figure 8—A composite map of the conglomerate and coal-bearing lithologies and subdivisions of the Ferris–Hanna interval, showing the environments of deposition. Note how the synorogenetic piedmont slope deposits grade into fluvial and floodplain deposits. Proximal and upper medial piedmont slope deposits are missing because of later uplift and erosion. Numbered lines indicate sections shown in Figure 7. Dot patterns outline the Ferris and Hanna formations as shown in Figure 4.

become trough and planar cross-bedded and the conglomerates occur only locally. The gray, sandy mudstone beds are indistinct units that generally occur correspondingly in regular beds about 5–10 ft (1.5–3 m) thick and locally contain thin sandstone and carbonaceous claystone and shale beds.

The medium conglomerate unit that crops out between Troublesome and Difficulty creeks in T. 24 N., R. 81 W. grades laterally into fine conglomerate that has the same bedding style as that seen in the larger unit of arkosic fine conglomerate already described. The fine conglomerate unit

may grade southward into a unit of sandstone and pebbly sandstone, but the relationship is obscured by faulting and rock dips that are in opposite directions.

Lithic Medium Conglomerate

These conglomerates occur in small areas that have been intensely deformed. Beds of gray and brown conglomerate (Hanna?) occur in the middle of a thick, light-colored mudstone unit in T. 24 N., R. 81 W. This conglomerate consists chiefly of gray sandstone and Mowry Shale pebbles

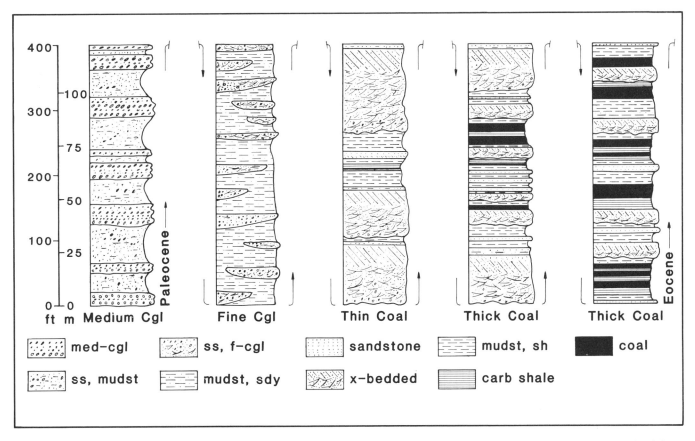

Figure 9—Selected examples of vertical sections showing subdivisions of the general lithologies. The subdivisions contain intervals of the principal rock types. All examples are from the upper part (Hanna Formation) of the Ferris–Hanna interval in the Hanna basin. The rocks in the sections are middle Paleocene and Eocene in age.

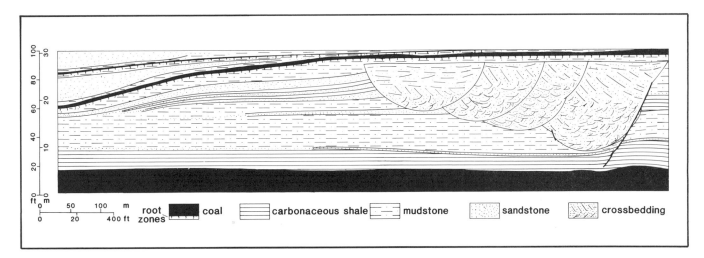

Figure 10—Schematic sketch of units of fluvial and floodplain deposits (Hanna Formation) as exposed in the highwall of cast mine in Hanna basin. Datum is at the base of the large, lower coal bed. The cross section shows the lateral and vertical relationship of a leftward migrating channel system to levee and overbank splay sandstone deposits, flood basin mudstone beds, and splay sandstone with coal beds. The Sandstone layers at upper left are later deposits. The fault to the right of the channel-fill complex developed during differential compaction as the interval was deposited, but the attitude of the bed also suggests that some structural warping of strata occurred prior to and during deposition of the interval.

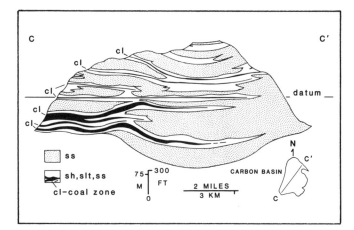

Figure 11—Schematic cross section of the Hanna Formation in Carbon basin. The interval of strata shows the predominance of massive, channel-fill sandstone sequences. The interval from the basal coal zone to the top of the formation is late Paleocene in age, but the channel sandstone below the basal coal zone probably was deposited during the middle Paleocene, a time of uplift and erosion southeast of the Hanna basin.

and small cobbles interbedded with sandstone and sandy mudstone. Particles from Niobrara marlstone were also found within the conglomerate. Clasts are as much as 6 in. (15 cm) in diameter. Beds of conglomerate are as thick as 4 ft (1.2 m) within this sequential unit. This sequential unit may be as much as 600 ft (180 m) thick.

The unit of lithic conglomerate on the northeast margin of the Hanna basin occurs in a structurally complex area. The conglomerates lie at the base of a thick, light-colored mudstone unit, where they chiefly consist of pebbles and small cobbles of gray sandstone and quartzite interbedded with layers of sandstone. The beds of conglomerate are generally less than 4 ft (1.2 m) thick, but a few beds are as thick as 10 ft (3 m). This sequential unit of conglomerate and sandstone is at least 400 ft (120 m) thick.

Lithic Fine Conglomerate

Occupying an area chiefly in T. 24 N., R. 82 W., this unit of conglomerate contains clasts that are as much as 2 in. (5 cm) in diameter at the base. This conglomerate, the lateral equivalent of the dark-colored conglomeratic sandstone, consists of about 1600 ft (500 m) of light-gray, tan, brown, and dark-gray conglomerate and sandstone sequences cropping out between intervals of gray mudstone. In a few places, sequences of conglomerate and sandstone are as much as 20 ft (6 m) thick, but for the most part, the sequences are less than 15 ft (4.5 m) thick. The individual conglomerate beds are generally planar, irregular, and 0.5 to 4 ft (15 cm to 1.2 m) thick. The internal structure is generally indistinct. The sandstone beds are plane bedded and trough cross-bedded; planar bedding was seen in one exposure. The intervals of mudstone occur in units as thick as 20 ft (6 m), but this thickness is difficult to determine because the conglomerate and sandstone sequences are interspersed within the mudstone. Carbonaceous claystone beds generally less than 1 ft (30 cm) thick occur within the mudstone. The lithic conglomerate is overlain by the thick arkosic conglomerate interval.

Light-Colored Mudrock

Lateral to and overlying the lithic medium conglomerate, the light-colored mudrock unit consists of a thick sequence of mudstone, cross-bedded sandstone (which is thick in places), parallel-bedded sandstone, and concretion beds. The unit may be as much as 6000–7000 ft (1800–2100 m) thick. The mudstone deposits are generally gray but are locally greenish gray to medium dark gray and bentonitic. The mudstone beds have been found to be as much as 200 ft (60 m) thick, but their general thickness is from 5 to 40 ft (1.5 to 12 m). The cross-bedded sandstone is light gray and tan, fine- to medium-grained and locally conglomeratic, and generally less than 15 ft (4.5 m) thick, although they are locally as thick as 30 ft (9 m). These trough cross-bedded sandstones appear as narrow channel fillings with scour bases. Parallel-bedded sandstone is tan to brown and light gray, generally fine grained, and locally concretionary and calcareous. These beds of sandstone occur as thin, widespread units generally 1–2 ft (30–60 cm) thick, but can also occur as thick as 11 ft (3.5 m). The purple and brown ironstone concretions are the most common horizontal beds within the mudstone; beds of tan, calcareous claystone concretions as much as 2 ft (60 cm) thick are locally common. Zones of root traces and thin paleosols can be found in places.

The thick sequence of mudstones includes equivalents of both the Ferris and Hanna formations in the northern Hanna basin. The Ferris part disappears under the Shirley thrust plate within a few miles east of the contact with the arkosic conglomerate. In the eastern Hanna basin, the light-colored mudstones thin to about 3000 ft (900 m), where they grade laterally southward into the coal-bearing lithologies.

SEDIMENTATION AND TECTONISM

An increase in Laramide uplift in the Wyoming foreland resulted in the areal restriction of continental sedimentation. Erosion of the uplifts resulted in drainage systems that deposited large loads of sediments in adjacent sedimentary basins where these sediments were differentiated by grain size and degree of sorting into a variety of deposits.

Deposition Environments

Various depositional environments existed in the Hanna and Carbon basins during the Laramide orogeny. The conglomerate containing sandstone and mudstone beds is interpreted as alluvial piedmont slope deposits. The coal-bearing mudstone and sandstone beds are thought to be fluvial and floodplain deposits.

Figure 12 shows a model of the restored piedmont slope deposits. Here a portion of an alluvial fan is shown with its proximal, medial, and distal parts, which are differentiated by their grain size, sorting, and bedding type. The stream systems generally have straight channels in the proximal and upper medial fan and a lateral moving system of channels in the medial fan; the distal fan has a highly braided system of channels. From the distal fan a series of tributaries leads to a large channel system that crosses the alluvial plain.

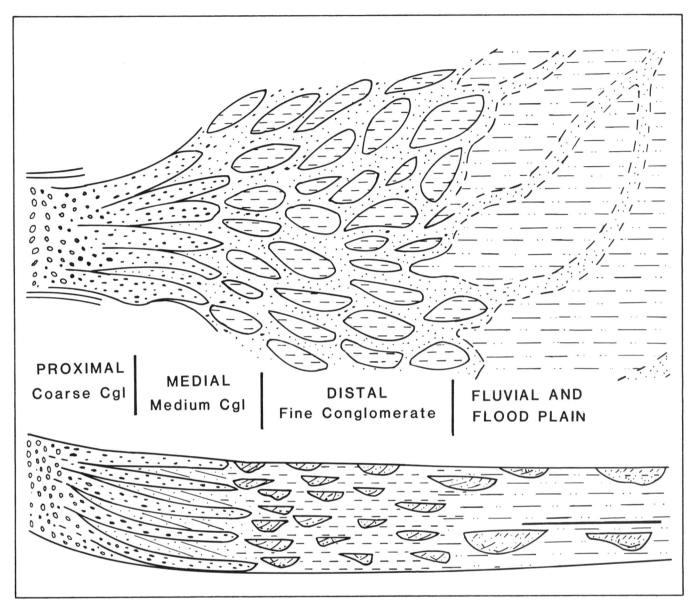

Figure 12—Idealized model of alluvial fan system and associated fluvial and floodplain system of northern and central Hanna basin. Proximal and upper medial parts of the fan are restored.

The piedmont slope deposits seen in the Hanna basin appear to be of the medial and distal parts of an alluvial fan. In the medial part, the sequences of thin-bedded, arkosic and lithic medium conglomerates and sands were deposited as bars. Within the arkosic medium conglomerates, the intervals of pebbly sand may have been wedges of sand sheets that were deposited during a different flow regime. Within the lithic medium conglomerates, the intervals of mudstone and sandstone were apparently sheet deposits. The distal deposits were short sequences of arkosic and lithic fine conglomerate and sandstone deposited in shallow channels eroded into the sandy mudstone. The sandy mudstones could have been the finer, downslope equivalents of the pebbly sandstone deposits.

The light-colored mudrock unit is included in the piedmont slope model as distal deposits lateral to the sequences of lithic medium conglomerate and the intervals of sandstone and mudstone. They were probably deposited as fine-grained, mudrich fans rather than the laterally equivalent, more normal, conglomeratic and sandy alluvial fans. Sources of the clasts in this northeastern part of the piedmont slope were from lesser uplifts that were eroded from a terrain that was mainly thick shale deposits.

The various lithologies in the coal-bearing mudstone and sandstone units are fluvial channel deposits that pass laterally into floodplain deposits, which form as intervals of coal-bearing strata. The massive sandstone beds were apparently deposited in large channels and tributaries.

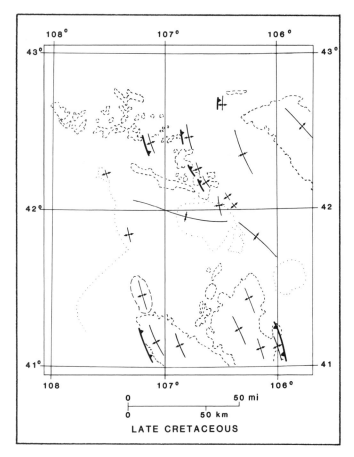

LATE CRETACEOUS

Figure 13—Regional structure pattern that formed at the end of Late Cretaceous time. Regional disconformable or unconformable relationships have not been recognized between the Upper Cretaceous and lower Paleocene strata in the Hanna basin area. Dotted lines show Tertiary and Upper Cretaceous rocks that crop out and outline the uplifts and basins. The general trends of synclinal and anticlinal forms are indicated by lines and arrows; barbs indicate the upthrown side of thrust faults.

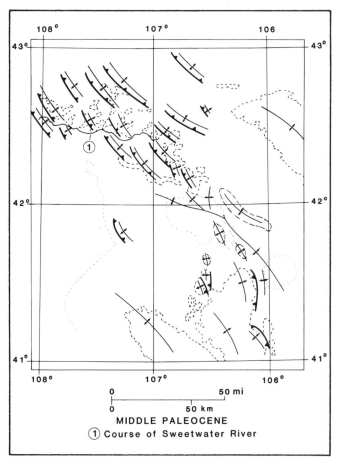

MIDDLE PALEOCENE
① Course of Sweetwater River

Figure 14—Regional structure pattern that had formed by the end of middle Paleocene time when the maximum Paleocene Laramide deformation had occurred. Sweetwater River drainage pattern indicates control by relict middle Paleocene fractures in Precambrian rocks of the Granite Mountains. See Figure 13 legend for explanation of symbols.

Associated with the channel deposits are flood levee and overbank sandstone beds and mudstone partings, and in one or two places, channel splay sandstones. The adjoining floodplain deposits include mudstone, thin lacustrine mudstone and shale, thin sandstone and siltstone channel splay deposits, and coal and carbonaceous shale beds deposited in the backswamps. Coal and carbonaceous shale deposits occur above and lateral to the other floodplain deposits.

Sedimentation and Uplifts

A regional syncline between large uplifts in south-central Wyoming formed during deposition of the dark-colored conglomeratic sandstone (lower Ferris) during Late Cretaceous time. For the greater part of the Hanna basin area, the syncline was filled with sediments that came from the southwest. This direction was determined mainly from paleocurrent data first reported by Ryan (1977). The source areas in this direction were most likely the early uplifts of the Medicine Bow and the Sierra Madre mountains. For the smaller area of the large syncline, the source of sediments

was from the early uplifts of the Granite and Shirley mountains that were in general north–south trends (Figure 13). These and later trends are based on those regional trends of Laramide uplift described by Gries (1983b). The Late Cretaceous uplifts were probably part of a northwest-trending system of regional uplift in central Wyoming (Love, 1970). The trend of uplifts north of the Hanna basin was the source of the first fan deposits of the basin. The fan systems apparently existed in that area well into Eocene time, although the trends of the structures and the magnitude of the uplifts changed through Paleocene and early Eocene time.

Granite fragments became a common constituent of the lower thick conglomerate interval (Ferris) during early Paleocene. These early uplifts of the Granite Mountain and Freezeout Hills began to increase in extent, number, and height as the uplifts formed in northwest–southeast trends characteristic of the Paleocene structures of the Wyoming foreland (Figure 14). The confluence of fans off the northern structures was developed to their maximum areal extent by late middle Paleocene time (Figure 5). These

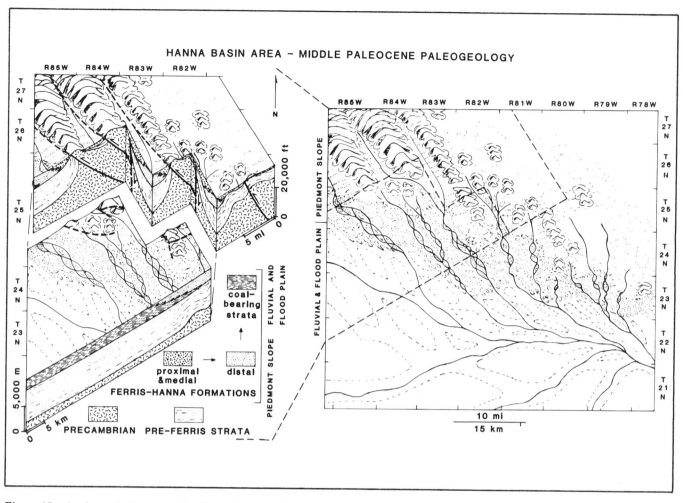

Figure 15—A schematic diagram of the Hanna basin area in late middle Paleocene time showing an example of the peripheral piedmont slope and central basin, fluvial and floodplain deposition during Laramide sedimentation in this area. Thrust-faulted, northwest-trending uplifts shed sediments into frontal synclines, which segregated the proximal part of the coalesced fans.

asymmetric uplifts north of the Hanna basin had become the predominant source area of sediments for the basin (Figure 15).

During middle Paleocene time, regional uplifts southeast of the Hanna basin resulted in complete erosion of the dark-colored sandstone and lower coal-bearing mudstone and sandstone deposits (lower and upper Ferris) in that area. The eastern margins of the Hanna and Carbon basins were affected (Figures 5 and 6). This erosion was followed by deposition during late Paleocene time of upper coal-bearing sediments (Hanna) on top of older rocks southeast of the Hanna basin. Sedimentation in the greater part of the Hanna basin, however, was continuous from early Paleocene through early Eocene time.

The last of the arkosic conglomerates was deposited between Troublesome and Difficulty creeks during late early Eocene. Deformation of this conglomerate followed when the Shirley Mountains–Freezeout Hills uplift moved south along the east–west Shirley thrust fault. This movement deformed the Ferris–Hanna rock interval and older formations (the footwall), probably during middle Eocene time (Figure 2). The time for this deformation is primarily

based on late early and middle Eocene uplift of the northern Granite Mountains, as described by Love (1970). The net slip on part of this fault was estimated by Van Ingen (1978) to be 14,600 ft (4450 m), but this is probably a minimal slip. Crustal shortening by southward movement of the Granite Mountains and Shirley Mountains–Freezeout Hills uplifts, eastward movement of the Elk Mountain (Beckwith, 1941) and Medicine Bow uplifts, and northward movement of the north Laramie Range uplift caused crowding that resulted in complex structures. For example, during these movements the Shirley Mountains–Freezeout Hills uplift overrode the east-plunging end of the Granite Mountains uplift in a short lateral movement. The middle Eocene structures were the last of the large compressional deformations in the area of the Hanna basin (Figure 16). Sets of normal faults developed in the Hanna and Carbon basins during the extensional deformation and regional uplift that followed.

CONCLUSIONS

Increased structural complexity from Late Cretaceous through middle Paleocene to middle Eocene time resulted

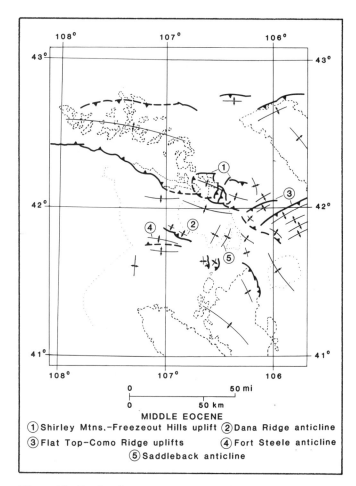

MIDDLE EOCENE
①Shirley Mtns.-Freezeout Hills uplift ②Dana Ridge anticline
③Flat Top-Como Ridge uplifts ④Fort Steele anticline
⑤Saddleback anticline

Figure 16—Regional structure pattern that formed during middle Eocene time. East-west Granite Mountains uplift and slightly overlapping Shirley Mountains–Freezeout Hills uplift moved south over thrusts. Asymmetry of present structures was established during this time. Crowding by large compressional uplifts north, south, and east of Hanna basin resulted in complicated structure when couple stresses caused counterclockwise movement around the basins. See Figure 13 legend for explanation of symbols.

from increasing uplifts and associated counterclockwise rotation from east-west- to north-south-directed couple stresses. This rotation parallels the movement of the North America plate as it overrode the Farallon plate (the movement as treated by Gries, 1983b). During this convergence of the plates, the couple stresses produced assymetric uplifts in a counterclockwise manner around the Hanna and Carbon basins (see Figure 13, 14, and 16). The earlier patterns are obscured by the last compressional deformation. This is partly because the middle Eocene structures north of the Hanna basin were the nearest and largest uplifts of the Laramide. It is also, because the north–south couple stresses produced thrusting of the middle Eocene uplifts surrounding the basin, separation of the Hanna and Carbon basins, and thrusting that was largely at right angles to Late Cretaceous thrust patterns.

During these deformations, the thick interval of the Ferris and Hanna formations was deposited as piedmont slope, and fluvial and floodplain deposits in a deeply subsiding depocenter that developed within a regional syncline.

Conformable deposition over thick Upper Cretaceous rocks within the depocenter resulted in the deep but small Hanna basin during the last episode of compressional deformation. The Ferris and Hanna formations are the regional correlative coal-bearing units earlier workers strove to identify. The coal-bearing Ferris–Hanna interval is the dominant rock unit.

REFERENCES CITED

Beckwith, R. H., 1941, Structure of the Elk Mountain district, Carbon County, Wyoming: Geological Society of America Bulletin, v. 52, p. 1445–1486.

Berg, R. R., 1962, Mountain flank thrusting in Rocky Mountain foreland, Wyoming and Colorado: AAPG Bulletin, v. 46, p. 2019–2032.

Blackstone, D. L., Jr., 1983, Laramide compressional tectonics, southeastern Wyoming, *in* Contributions to Geology: University of Wyoming, Laramie, v. 22, n. 1, p. 1–38.

Blanchard, L. F., and M. C. Comstock, 1980, Geologic map and coal sections of the Pats Bottom Quadrangle, Carbon County, Wyoming: U.S. Geological Survey Open-File Report 80-052, 2 sheets.

Bowen, C. F., 1918, Stratigraphy of the Hanna Basin, Wyoming: U.S. Geological Survey Professional Paper 108-L, p. 227–235.

Dobbin, C. E., C. F. Bowen, and H. W. Hoots, 1929, Geology and coal and oil resources of the Hanna and Carbon basins, Carbon County, Wyoming: U.S. Geological Survey Bulletin 804, 88 p.

Dorf, E., 1938, Upper Cretaceous floras of the Rocky Mountain region; 1. Stratigraphy and paleontology of the Fox Hills and Lower Medicine Bow formations of southern Wyoming and northwestern Colorado: Carnegie Institution of Washington Publication 508, p. 1–78.

Gill, J. R., E. A. Merewether, and W. A. Cobban, 1970, Stratigraphy and nomenclature of some Upper Cretaceous and Lower Tertiary rocks in south-central Wyoming: U.S. Geological Survey Professional Paper 667, 53 p.

Glass, G. B., 1981, Field guide to the coal geology of the Hanna Coal Field, *in* Geological Survey of Wyoming, Miscellaneous Publication: prepared for Annual Spring Conference, University of Wyoming, Wyoming Geological Association, and Geological Survey of Wyoming, 23 p.

Glass, G. B., and J. T. Roberts, 1980, Coals and coal-bearing rocks of the Hanna coal field, Wyoming: Geological Survey of Wyoming Report of Investigation, n. 22, 43 p.

Gries, R., 1983a, Oil and gas prospecting beneath Precambrian foreland thrust plates in Rocky Mountains: AAPG Bulletin, v. 67, n. 1, p. 1–28.

————, 1983b, North–south compression of Rocky Mountain Foreland structures, *in* Lowell, J. D., ed., Rocky Mountain Foreland basins and uplifts: Denver, Colorado, Rocky Mountain Association of Geologists, p. 9–32.

Knight, S. H., 1951, The Late Cretaceous–Tertiary history of the northern portion of the Hanna basin—Carbon County, Wyoming, *in* Wyoming Geological Association Guidebook 6th Annual Field Conference, south-central Wyoming: Casper, Wyoming, Wyoming Geological Association, p. 45–53.

Krumbein, W. C., and F. G. Nagel, 1953, Regional stratigraphic analysis of "Upper Cretaceous" rocks of Rocky Mountain Region: AAPG Bulletin, v. 37, n. 5, p. 940–960.

Love, J. D., 1970, Cenozoic geology of the Granite Mountains area, central Wyoming: U.S. Geological Survey Professional Paper 495-C, 154 p.

Love, J. D., J. L. Weitz, and R. K. Hose, 1955, Geologic map of Wyoming: U.S. Geological Survey, scale, 1:500,000.

Lowry, M. E., S. J. Rucker, and K. L. Wahl, 1973, Water resources of the Laramie, Shirley, and Hanna basins and adjacent areas, southeastern Wyoming: U.S. Geological Survey Hydrologic Investigations Atlas HA-471, 4 sheets.

Ryan, J. D., 1977, Late Cretaceous and early Tertiary provenance and sediment dispersal, Hanna and Carbon basins, Carbon County, Wyoming: Geological Survey of Wyoming Preliminary Report, n. 16, 17 p.

Van Ingen, L. B., III, 1978, Structural geology and tectonics— Troublesome Creek Valley area, Carbon County, Wyoming: Master's thesis, University of Wyoming, Laramie, Wyoming, 70 p.

Veatch, A. C., 1907, Coal fields of east-central Carbon County, Wyoming: U.S. Geological Survey Bulletin 316-D, p. 244–260.

The Upper Cretaceous Vernal Delta of Utah— Depositional or Paleotectonic Feature?

T. A. Ryer
J. R. Lovekin
RPI/Colorado
Boulder, Colorado

A conspicuous seaward bulge of the middle to late Turonian shoreline of the Cretaceous seaway in northeastern Utah and southwestern Wyoming has been identified by previous authors as the Vernal delta. Strata of the Frontier Formation and the Ferron Sandstone Member of the Mancos Shale that form the Vernal delta consist largely of fluviodeltaic facies. The delta, however, is not recognizable as a locus of Turonian sedimentation; there is no substantial isopach thickness associated with it. The Vernal delta was apparently a large feature, encompassing an area of at least 6250 mi^2 (16,250 km^2). Comparison between the depositional setting and paleogeography of northeastern Utah during the Late Cretaceous and a structurally similar present-day area on the east flank of the Andes in Colombia suggests that a feature the supposed size of the Vernal delta could not have been produced by a single river. Strata of the Vernal delta overlie the ancestral Uinta Mountain uplift, an area where Cenomanian marine shale was entirely removed by what appears to have been submarine erosion during early Turonian time. When the shoreline prograded eastward across this area during middle Turonian time, the sediment load caused the area to subside, but at a rate slower than rates of subsidence to the north and south. We hypothesize that this differential subsidence is the cause of the shoreline bulge. Although it includes deltaic facies, the Vernal delta was probably not an actual delta, but a feature produced primarily as the result of gentle tectonic movement of the ancestral Uinta Mountain uplift.

INTRODUCTION

Deltas have long been recognized as important depositional features in the stratigraphic record (Gilbert, 1885; Barrell, 1912). It has only been within the last 10–20 years, however, that a substantial body of information describing the diagnostic features of deltas has been available. This information has come primarily from studies of modern deltas, particularly that of the Mississippi River (see Wright, 1978, for a succinct description of the properties of deltas). The recognition and mapping of ancient deltas is an important matter for energy geologists inasmuch as they contain disproportionately large quantities of both oil and gas (Rainwater, 1975) and coal (Galloway and Hobday, 1983).

The substantial volume of information that exists regarding depositional processes, characteristics, and sediments of modern deltas facilitates recognizing deltaic facies in the stratigraphic record. But what are the criteria by which the outlines of ancient deltas can be distinguished and mapped? Moore and Asquith (1971) define a delta as "the subaerial and submerged contiguous sediment mass deposited in a body of water (ocean or lake) primarily by the action of a river." This is a broad definition; it avoids any statement about shape or size. Galloway and Hobday (1983)

note that several important corollaries are implicit in this broad definition: (1) deltas are progradational (see also Wright, 1978); (2) the bulk of the sediments in a delta are delivered at one or more point sources, these being the mouth or mouths of the river; (3) delta systems develop around the margins of large basins; and (4) a delta system typically defines a locus of deposition. Defining an ancient delta in the subsurface usually requires, in practice, meeting two conditions: (1) Is there a recognizable shoreline bulge? (i.e., is there evidence of more rapid rates of progradation at the site of the delta than in adjacent areas?); and (2) If present, is the bulge associated with a greater isopach thickness? (i.e., is there a recognizable locus of deposition?)

A large number of authors have recognized deltaic deposits within strata that accumulated along the western margin of the interior Cretaceous seaway of western North America. In Utah and Wyoming, deltaic facies have been recognized and described by Katich (1953), Hale and Van de Graaff (1964), Maione (1971), Hale (1972), Thomaidis (1973), Cotter (1975), De Chadenedes (1975), Peterson and Ryder (1975), Myers (1977), Uresk (1979), Balsley (1980), Winn and Smithwick (1980), Ryer (1981), Lawrence (1982), and Fouche et al. (1983) among others. Few of these authors, however, have attempted to areally define the deltas in which the strata they studied were deposited. Among these few are Hale and Van de Graaff (1964) and Hale (1972), who defined and mapped two contemporaneous deltas of middle to late Turonian age in Utah and southern Wyoming (Figure 1). The northern of these, the Vernal delta, is the subject of this paper.

THE VERNAL DELTA

The Vernal delta, as described by Hale and Van de Graaff (1964), includes predominantly sandy strata of the Ferron Sandstone Member of the Mancos Shale in central and east-central Utah and the Frontier Formation in northern Utah and southwestern Wyoming. It is worth noting that initially the Vernal delta was not fully acknowledged in the literature. Hale and Van de Graaff (1964, p. 129) initially described it as "an eastward bulging deltalike feature" that they "somewhat arbitrarily termed the 'Vernal delta'." Hale (1972) was less hesitant in his naming of the Last Chance delta farther to the south in central Utah. He made no mention of the Vernal delta, however, in his 1972 paper.

It was not until 4 years later that Cotter (1976) elevated the hypothetical Vernal delta of Hale and Van de Graaff (1964) to a fully recognized feature, seeing in the autocyclic shifting of the delta a means of explaining some of the depositional features that he had observed in the lower part of the Ferron Sandstone Member on the flanks of the San Rafael Swell. Cotter did not, himself, study deposits of the Vernal delta, having restricted his studies to outcrops of the Ferron Sandstone Member in east-central Utah.

The sandy strata of the Vernal delta are underlain by offshore marine shale of the Tununk Shale Member of the Mancos Shale and its equivalents (Figure 2). These shales were deposited at the transgressive maximum of the Greenhorn cycle of Kauffman (1977), during which the shoreline of the Cretaceous seaway encroached westward nearly to the edge of the Sevier orogenic belt (Armstrong, 1968).

The regressive phase of the Greenhorn cycle, which culminated in deposition of the Vernal delta, is recognizable throughout the Western Interior of North America. In the area of this study, it is represented by the lower part of the Funk Valley Formation (Fouch et al., 1983), the Ferron Sandstone Member (Ryer and McPhillips, 1983), and the Frontier Formation (Hale, 1960; Ryer, 1977; Myers, 1977; Merewether, 1983; Merewether et al., 1983). Hancock (1975) and Hancock and Kauffman (1979) presented evidence that this middle to late Turonian regression can be recognized worldwide and attributed it to eustatic lowering of sea level.

The Vernal delta was bounded on the south by an embayment of the shoreline that existed in east-central Utah (Ryer and McPhillips, 1983). The easternmost position of the strand, as identified by the limit of coal-bearing rocks in the Ferron Sandstone Member and the Frontier Formation, defines a northeast-trending shoreline extending from the vicinity of Price, Utah, at the western edge of the Book Cliffs, through the Uinta basin, to the area of Vernal and Dinosaur National Monument, near the Utah–Colorado state line (Figure 3). From this area, the shoreline is believed to have trended northwestward (McGookey et al., 1972), although reconstruction of the northern edge of the delta is

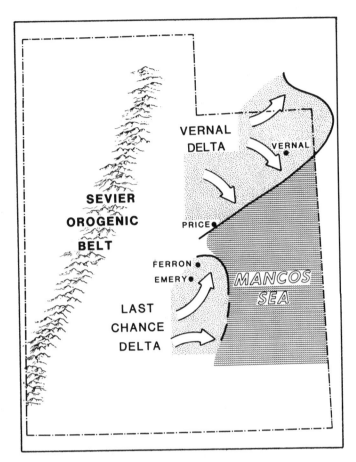

Figure 1—The Upper Cretaceous Vernal and Last Chance deltas. Arrows indicate inferred directions of sediment transport. Modified from Cotter (1976), who based his figure on paleogeographic maps by Hale (1972) and Hale and Van de Graaff (1964). Used with permission of the Department of Geology, Brigham Young University and the Utah Geological Association.

sketchy, mainly because of the lack of outcrop control on the thrusted northern flank of the Uinta Mountains. Thus defined, the Vernal delta covered an area that today includes much of the Uinta basin, the western part of the Uinta Mountains, and the southernmost part of the greater Green River basin. If the delta plain of the Vernal delta is considered to extend from the shoreline described above westward to the limits of the underlying marine shale, it covered an area of about 12,000 mi^2 (31,200 km^2).

The strata of the Vernal delta constitute a clastic wedge that thins and interfingers eastward into marine shale. The properties of these rocks are best known from outcrops of the Frontier Formation along the south flank of the Uinta Mountains. The Frontier consists primarily of nonmarine rocks in the western part of this outcrop belt. Walton (1944) identified an area of extensive interfingering between the upper part of the Frontier Formation and the Mancos Shale in the Tabiona area. The older, lower part of the wedge extends eastward to Vernal (Figure 4), where substantial quantities of coal have been mined from the Frontier Formation (Doelling and Graham, 1972). Brackish water and fluvial beds disappear eastward in Dinosaur National Park (Maione, 1971) and only platy marine sandstone is present at Blue Mountain, just east of the Utah–Colorado border (Cobban and Reeside, 1952). The facies and depositional features described by these authors are entirely compatible with a deltaic origin for the Turonian portion of the Frontier Formation.

The isopach map in Figure 5, which incorporates selected data both from published outcrop studies and analysis of electric logs, shows the combined thicknesses of pre-dominantly sandy delta front, delta plain and alluvial plain strata that compose the Vernal delta. The delta is best distinguished on the basis of the 100-ft (30-m) isopachous line. Outcrop data from east-central and northeastern Utah indicate that this line approximates the seaward limit of coal-bearing strata; uncertainty exists regarding the position of the shoreline on the northern flank of the delta. Although recognizable as a seaward bulge of the shoreline, the Vernal delta did not coincide with a major locus of sediment accumulation.

It is important to keep in mind that Figure 5 is a facies isopach map; the form of the Vernal delta would certainly be less distinct and might even disappear entirely if time-equivalent marine shale deposited to the east of the sandy deltaic facies was included in the isopached interval. Unfortunately, the detailed biostratigraphic or chrono-stratigraphic information required to construct such a map does not presently exist.

The absence of a pronounced depocenter associated with the Vernal delta is surprising, considering the supposed tremendous size of the feature. Although the existence of a shoreline bulge constitutes evidence that the Vernal delta is truly a delta, the absence of a well-defined depocenter associated with it contradicts one of the basic properties of deltaic sedimentation and casts doubt on the validity of the deltaic interpretation.

A Question of Size

As noted earlier, the Vernal delta was a very large feature that extended as much as 115 mi (185 km) in the dip direction and about 130 mi (210 km) in the strike direction and that formed a delta plain that covered an area of approximately 12,000 mi^2 (31,200 km^2). Even if the area of the Vernal delta is more conservatively measured as that part of the regressive wedge contained within the shoreline bulge itself and limited on the seaward side by the 100-ft (30-m) isopachous line (as in Figure 5), it would still have had an area of approximately 6250 mi^2 (16,250 km^2).

Is it possible that a single river draining eastward from the Sevier orogenic belt into what is now northeastern Utah and southwestern Wyoming could have produced a feature as large as the Vernal delta? A useful comparison can be made with the rivers that exist today on the eastern flank of the Andes in South America. Figure 6 pictures the approximate situation that would exist in South America if eustatic sea level were to rise about 1600 ft (500 m), as it did in the Cretaceous. Such a sea level rise would flood much of South America, producing an epeiric seaway strikingly similar to that which occupied the Western Interior of North America during Cretaceous time (compare witn paleo-geographic maps by Williams and Stelck, 1975). A Cretaceous sea level rise of about 1700 ft (510 m) was calculated by Hays and Pitman (1973), although Pitman (1978) later revised the figure downward to about 1150 ft

(345 m) above present sea level. Hancock and Kauffman (1979) speculated that Cretaceous sea level may have reached as much as 2100 ft (630 m) above its present level.

Even allowing for differences in latitude, climate, and types of vegetation, the rivers that flowed eastward from the Sevier orogenic belt during Cretaceous time could not, we believe, have been too much different than those that now occupy the eastern flank of the Andes. In Figure 7, the Vernal and Last Chance deltas are superimposed on portions of Colombia, Ecuador, and Peru. This location was chosen because the curve of the Andean mountain chain that occurs just south of the Colombia–Ecuador border is strikingly similar to the curve of the Sevier orogenic belt associated with the Grand Canyon Bight in northernmost Arizona (Stokes and Heylmun, 1963; Moir, 1974). If the spacings of Cretaceous rivers in northern Utah and southwestern Wyoming in fact resembled those of the rivers that today exist on the east flank of the Andes, it seems highly unlikely that the Vernal delta could have been formed by just one river. But if the feature was formed by more than one contemporaneous river, can it be considered a delta? More importantly, why should such a large delta exist when the paleogeography reconstructed for the area indicates that it should have included numerous small rivers whose headwaters lay no more than about 250 mi (400 km) to the west? A possible answer lies in the paleotectonic setting of the Vernal delta.

THE UNCONFORMITY AT THE BASE OF THE FRONTIER

An unconformity lies at the base of the Frontier Formation in the vicinity of the Uinta Mountains (Figure 8). On the northern flank of the Uintas at Flaming Gorge (Kinney, 1955), along the southern flank of the Uinta Mountains (Walton, 1944; Cobban and Reeside, 1952), and eastward into Colorado (Reeside, 1955; Weimer, 1962; Sharp, 1963), the lower marine shale unit of the Frontier, which contains fossils of middle Turonian age, uncon-formably overlies the upper Albian Mowry Shale. Merewether et al. (1983) have demonstrated that this unconformity extends throughout much of south-central Wyoming. There the time span of the unconformity is much less, however, and it is more difficult to recognize. The westward extent of the unconformity is unknown. It is possible, as speculated by Merewether (1983; see also Merewether et al., 1984), that the rapid eastward disappearance of the Chalk Creek and Coalville Members of the Frontier Formation between the Overthrust Belt and the southern part of the Moxa Arch may be the result of bevelling of these units against the unconformity (as indicated in Figure 2). Similarly, the southward extent of the unconformity within the Uinta basin remains unknown. The absence of the distinctive zone of *Pycnodonte newberryi* and the presence of a bed of chert pebble conglomerate at the top of the Dakota Sandstone in the vicinity of the San Rafael Swell of east-central Utah (Katich, 1954) indicate that it extends that far to the south, although the time span represented by the unconformity in this area is certainly small compared to that in the area of the Uinta Mountains.

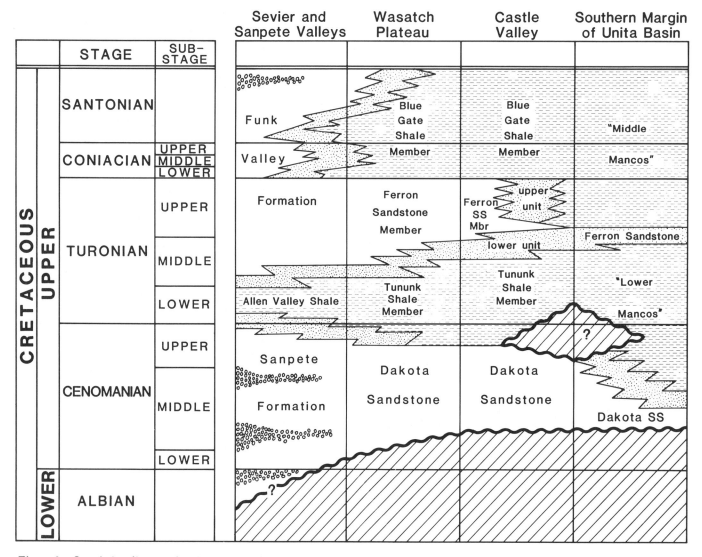

Figure 2—Correlation diagram showing stratigraphic positions and ages of units contained in and associated with the Vernal delta. Coniacian age units in the Coalville area are highly generalized.

Weimer (1962) and, more recently, Merewether (1983) utilized subsurface data from areas adjacent to the eastern end of the Uinta Mountains to demonstrate that the unconformity between the Frontier Formation and the underlying Mowry Shale is angular and that as much as 300 ft (90 m) of Cenomanian (Belle Fourche age) marine shale had been uplifted and eroded prior to Frontier deposition. Weimer (1962, p. 129) postulated that this erosion was the result of gentle upwarping of an "embryonic Uinta Mountain uplift." An isopach map (Figure 9) of the thickness of the marine shale unit that underlies the regressive wedge of the Vernal delta delineates this ancestral Uinta uplift. Such an isopach map is made possible by the fact that the contact between the highly siliceous Mowry Shale and the overlying nonsiliceous Frontier shale is relatively easy to recognize both in outcrop (Figure 4) and on electrical logs (Figure 8). Although there can be little doubt that the erosion of Cenomanian shale resulted from crustal upwarping, the agent or agents of erosion remain

unclear. Nowhere has evidence of subaerial erosion or weathering on this surface been found. Very low relief on the erosional surface and an absence of fluvial strata, soils, and transgressive lags suggest that the unconformity was of submarine origin.

A likely scenario (Figure 10) is that the upwarping of the ancestral Uinta uplift occurred during early Turonian time when the shoreline of the Cretaceous seaway lay far to the west. The upwarp produced an east-west-trending shoal area where Cenomanian marine shale was gradually stripped as it was elevated to water depths where wave-induced currents were capable of regularly working the bottom. The absence of any sandy deposits on the surface of the unconformity apparently indicates that the shoal must have been separated from the contemporaneous shoreline by an area of deeper water, precluding introduction of sand to the shoal. This area of deeper water probably coincided with the area of pronounced thickening of the marine shale section evident in Figure 9. This greater isopach thickness marks the

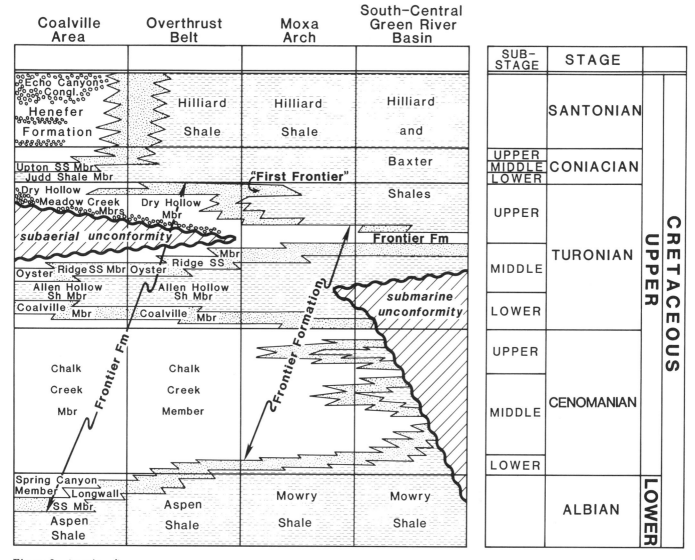

Figure 2—(continued).

eastern edge of the foreland basin that developed immediately to the east of the Sevier orogenic belt. Thinning of the shale to the west of the isopach thickness is the result of a facies change to shoreline and nonmarine facies.

PALEOTECTONIC CONTROL OF THE VERNAL DELTA

If the unconformity at the base of the Frontier developed as described above, a quite different explanation for the origin of the feature known as the Vernal delta becomes possible. A comparison of Figures 5 and 9 indicates that strata of the Vernal delta immediately overlie the ancestral Uinta uplift. The superposition of these two features strongly suggests that they share a common causal mechanism, this mechanism possibly being gentle tectonic

upwarping during Cretaceous time of the area overlying the present-day Uinta Mountains.

By way of explanation, we continue the scenario begun in the preceding section as follows. Immediately after the transgressive maximum of the Greenhorn cycle in early middle Turonian time, the shoreline of the interior Cretaceous seaway began to shift eastward. The regression was the result of two factors: eustatic lowering of sea level and an abundant supply of clastic sediment delivered by the numerous small rivers that drained the Sevier orogenic belt. Having gradually prograded eastward across the subsiding foreland basin, the shoreline encountered the area of the ancestral Uinta uplift. The arching of this structure had decreased by this time, and it was no longer an area of uplift and erosion. The change from erosion to deposition, in fact, was probably the result of crustal loading by the weight of sediments in the eastward-extending clastic wedge (Jordan, 1981). Nonetheless, the ancestral Uinta uplift remained

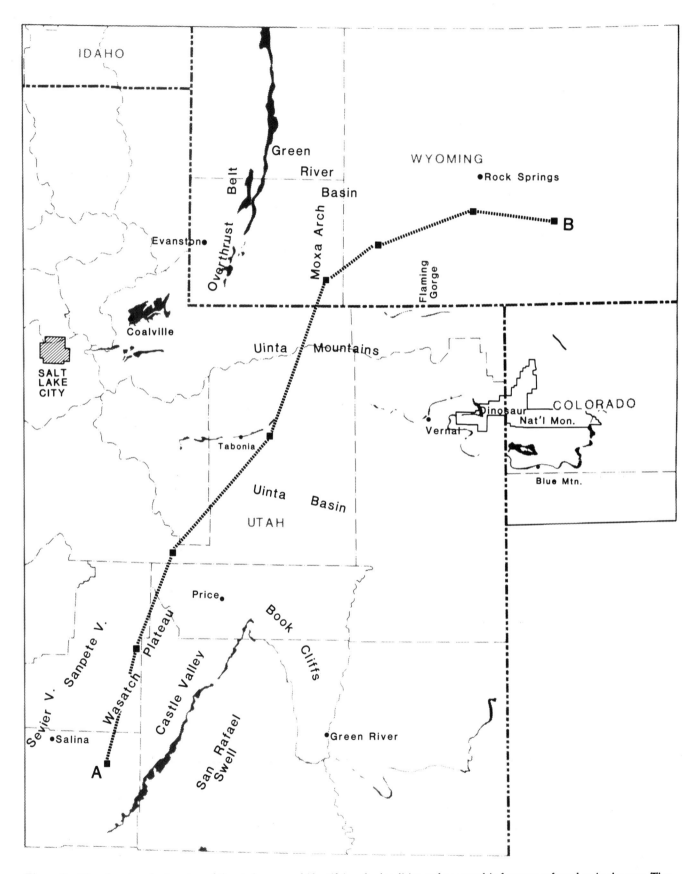

Figure 3—Map showing the location of the study area and identifying the localities and geographic features referred to in the text. The location of stratigraphic cross section A–B in Figure 8 is indicated.

Figure 4—Exposures of the Mowry Shale and the Frontier Formation near Vernal, Utah. Light-colored, siliceous shale of the Mowry (Kmry) is unconformably overlain by drab marine shale of the lower part of the Frontier (Kfl-ms). Flat-bedded delta front sandstone of the upper part of the Frontier (Kfu-df) forms the prominent ledge; it is erosionally overlain by more massive fluvial sandstone deposited by a meandering channel (Kfu-fl). The uppermost, coal-bearing strata of the Frontier do not appear in this photo.

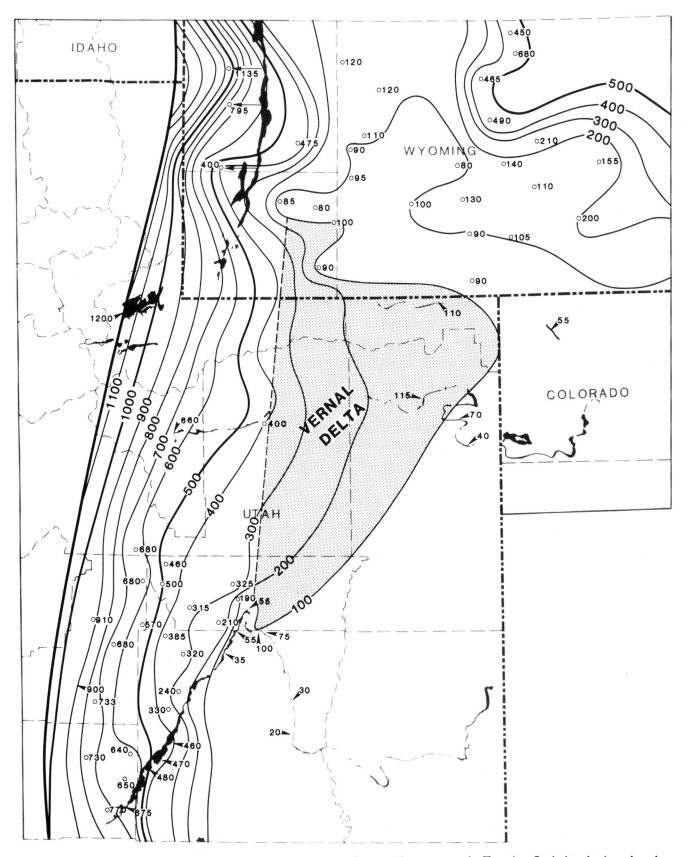

Figure 5—Isopach map (contour interval in feet) of predominantly sandy strata that compose the Turonian–Coniacian clastic wedge: the Ferron Sandstone Member; the upper part of the Frontier Formation; and in the Overthrust Belt, the Oyster Ridge, Dry Hollow, and Meadow Creek members of the Frontier. Shading indicates the area, conservatively defined, of the deltaic plain of the Vernal delta at the peak of regression. The heavy line that truncates the isopachous lines marks the western limit of the marine shale of the Niobrara cycle; the Turonian–Coniacian clastic wedge cannot be distinguished west of this line.

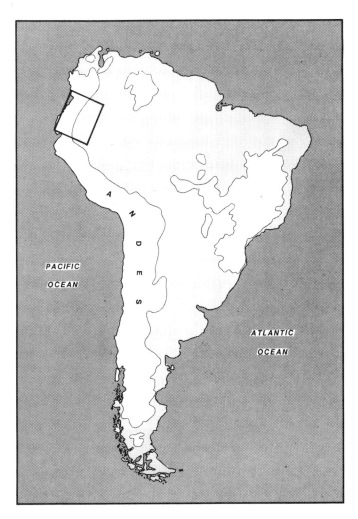

Figure 6—Shading indicates approximate area of South America that would be flooded by a 1650 ft (500 m) rise in sea level. No attempt has been made to compensate for subsidence that would result from isostatic loading by water. The hypothetical shoreline is approximately the present-day 1650 ft (500 m) contour. The shoreline position has been smoothed across valleys to simulate the effects of fluvial aggradation during sea level rise. Box indicates the location of Figure 7.

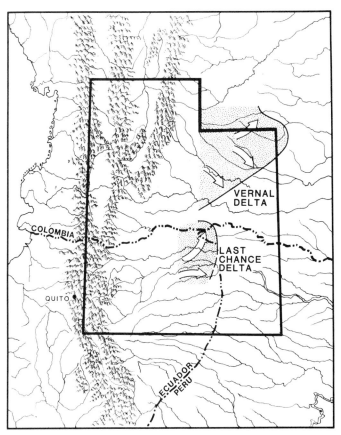

Figure 7—The Vernal and Last Chance deltas superimposed on the eastern flank of the Andes. The Cretaceous Sevier orogenic belt is made to coincide with the location of the Andes. The outline of Utah is added to aid in orientation.

slightly positive relative to adjacent areas; it subsided less rapidly than did areas to the north and south, as evidenced by thinning (possibly reflecting an onlapping relationship) of the marine shale unit of the Frontier in the vicinity of the uplift (Figures 8 and 9). As the shoreline encroached eastward onto the ancestral Uinta uplift, the shoreline began to bulge seaward. The bulge continued to grow as progradation continued, reaching the eastern edge of the ancestral Uinta uplift before being transgressed by the sea in late Turonian–early Coniacian time.

The transgression that marked the demise of the Vernal delta, and that marked the beginning of the Niobrara transgressive–regressive cycle (Kauffman, 1977), was primarily the result of rising sea level. The initial westward transgression of the sea across the Vernal delta was rapid. It was followed by a period of gradual transgression

characterized by extensive interfingering between deltaic and offshore marine deposits (Figures 2 and 8). Outcrop data from east-central (Ryer, 1981) and northeastern Utah (Walton, 1944) indicate that the position of the shoreline at the time this interfingering occurred approximately coincides with the 400-ft (120-m) isopachous line on Figure 5. The form of the Vernal delta is not recognizable on the basis of this line, indicating that the feature had ceased to exist by late Turonian time. Only strata in the lower part of the Turonian–Coniacian clastic wedge contribute to the Vernal delta.

The Vernal delta, we believe, owes its origin to the slower rates of basin subsidence that prevailed in the vicinity of the ancestral Uinta uplift during middle Turonian time. A given volume of sandy sediment will produce a blanket that is thinner but more extensive in an area of slower subsidence. In the case of the ancestral Uinta uplift, predominantly sandy sediment delivered eastward by rivers from the Sevier orogenic belt produced a thinner but more widespread clastic wedge within the area of more gradual subsidence associated with the uplift. It must be noted that this conclusion is not borne out by the isopach map of the clastic wedge presented in Figure 5. This map, however, isopachs the entire clastic wedge rather than just its lower part, and as noted above, the Vernal delta is recognizable only in this part. Until such

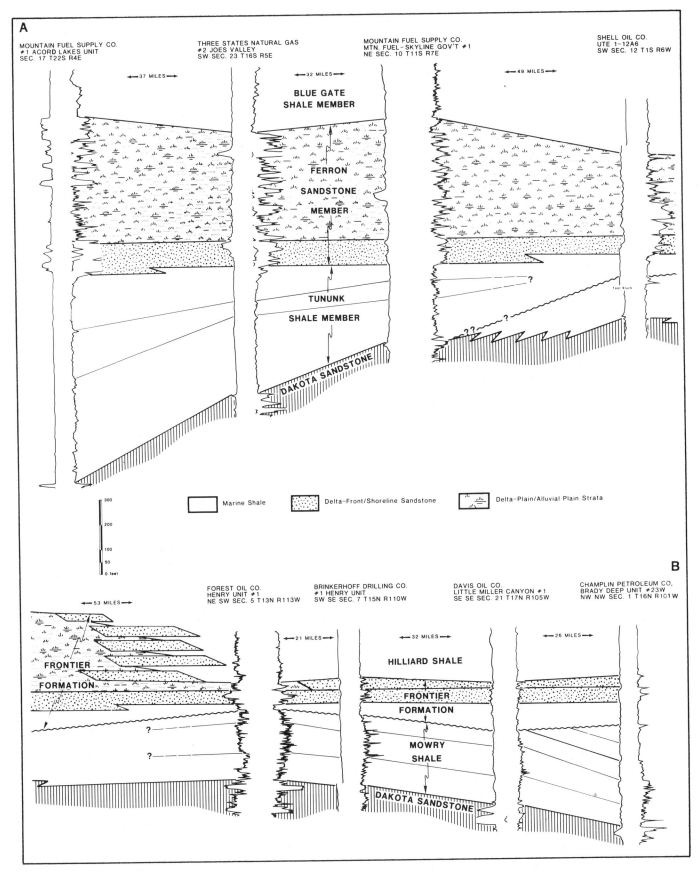

Figure 8—Electric log cross section showing the stratigraphic relationships, including interpreted relationships across the present-day Uinta Mountains, between units in and associated with the Turonian–Coniacian clastic wedge in Utah and Wyoming. The location of the cross section is shown in Figure 3.

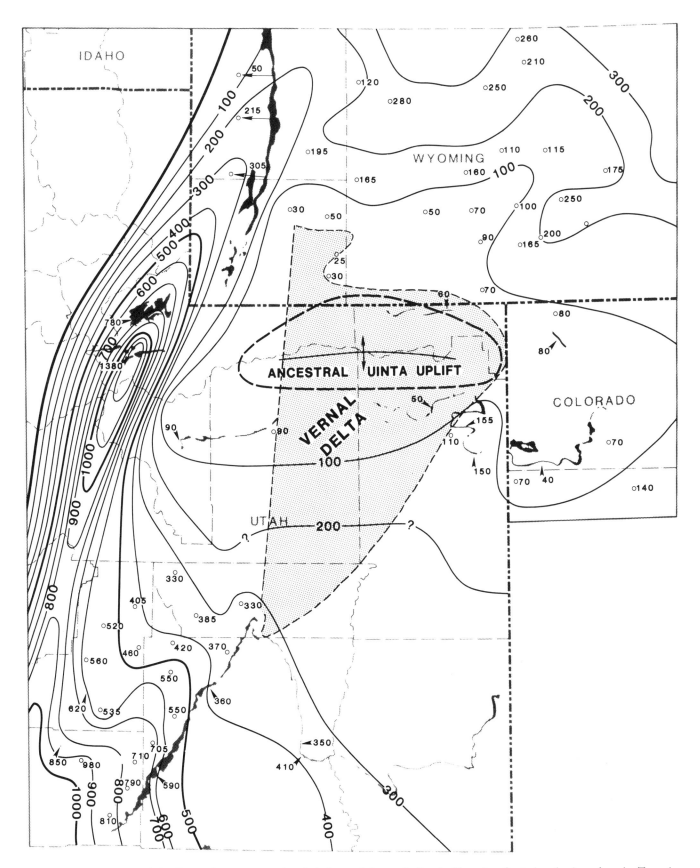

Figure 9—Isopach map of the Upper Cretaceous marine shale interval that underlies the Turonian–Coniacian clastic wedge: the Tununk Shale Member, the "lower Mancos" shale, the lower shale unit of the Frontier, and the Allen Hollow Shale Member of the Frontier. The ancestral Uinta uplift coincides with an area of pronounced thinning of this shale interval. The shaded area, as in Figure 5, is the deltaic plain of the Vernal delta.

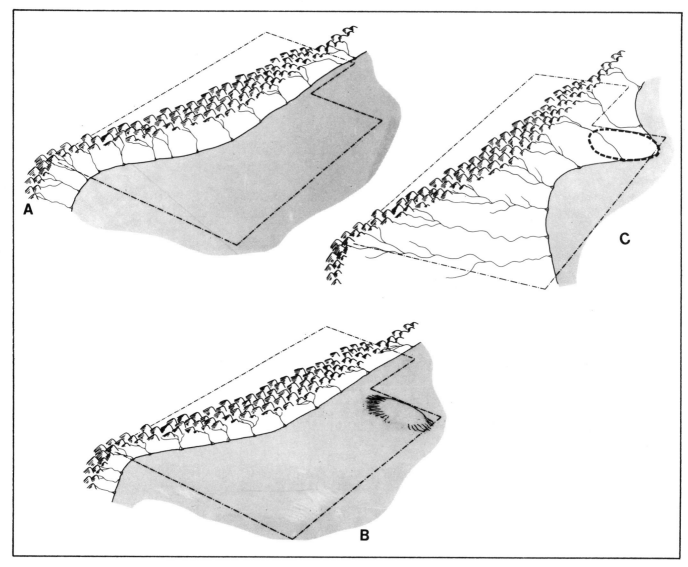

Figure 10—Diagrammatic representation of the formation of the Vernal delta. A. The Greenhorn transgression, which reached its peak in early Turonian time, caused the Cretaceous epeiric sea to encroach nearly to the edge of the Sevier orogenic belt. B. Upwarping of the ancestral Uinta uplift, possibly as a result of eastward compression associated with the Sevier orogeny, led to erosion of previously deposited marine shale. Erosion may have been entirely submarine; alternatively, the surface of the uplift may have risen above sea level. In either case, the presence of deeper water between the shoreline and the uplift apparently precluded introduction of sand to the area of the uplift. C. Eastward progradation of the shoreline during middle Turonian time and subsequent loading of the crust resulted in subsidence of the ancestral Uinta uplift, but at a somewhat slower rate than in adjacent areas. The shallower and more slowly subsiding area accumulated a thinner layer of the sediment than did areas to the north and south. The result was the eastward bulge of the shoreline identified as the Vernal delta.

time as sufficient data exist to facilitate mapping of the lower part of the Turonian–Coniacian clastic wedge throughout the study area, the proposed relationship between rates of subsidence and the thickness and lateral extent of this part of the wedge must remain hypothetical.

Although presently unprovable, the hypothesis that the Vernal delta owes its origin to the ancestral Uinta uplift satisfactorily explains the superposition of these two features. It accounts for the absence of a pronounced depocenter associated with the "delta" and is entirely compatible with more than one river supplying sediment to

the shoreline bulge. Despite the fact that the Vernal delta may consist largely or even predominantly of deltaic sediments, it was probably not an actual delta, but rather a feature that was primarily attributable to its tectonic setting.

REFERENCES CITED

Armstrong, R. L., 1968, Sevier orogenic belt in Nevada and
    Utah: Geological Society of America Bulletin, v. 79, p.
    429–458.

Balsley, J. K., 1980, Cretaceous wave-dominated delta systems—Book Cliffs, east central Utah: AAPG Field Seminar Guidebook, 163 p.

Barrell, J., 1912, Criteria for the recognition of ancient delta deposits: Geological Society of America Bulletin, v. 23, p. 377–446.

Cobban, W. A., and J. B. Reeside, Jr., 1952, Frontier Formation, Wyoming and adjacent areas: AAPG Bulletin, v. 36, p. 1913–1961.

Cotter, E., 1975, Deltaic deposits in the Upper Cretaceous Ferron Sandstone in Utah, *in* M. L. S. Broussard, ed., Deltas—models for exploration: Houston Geological Society, p. 471–484.

———, 1976, The role of deltas in the evolution of the Ferron Sandstone and its coals: Brigham Young University Geology Studies, v. 22, pt. 3, p. 15–41.

De Chadenedes, J. F., 1975, Frontier deltas of the western Green River basin, Wyoming, *in* Rocky Mountain Association of Geologists, Deep drilling frontiers in the central Rocky Mountains: Denver, p. 149–157.

Doelling, H. H., and R. L. Graham, 1972, Eastern and northern Utah coal fields—Vernal, Henry Mountains, Sego, La Sal–San Juan, Tabby Mountain, Coalville, Henrys Fork, Goose Creek, and Lost Creek: Utah Geological and Mineral Survey Monograph Series, n. 2, 409 p.

Fouch, T. D., T. F. Lawton, D. J. Nichols, W. B. Cashion, and W. A. Cobban, 1983, Patterns and timing of synorogenic sedimentation in Upper Cretaceous rocks of central and northeast Utah, *in* M. W. Reynolds and E. D. Dolly, eds., Mesozoic paleogeography of west-central United States: Society of Economic Paleontologists and Mineralogists Rocky Mountain Section, Rocky Mountain Paleogeography Symposium 2, p. 305–336.

Galloway, W. E., and D. K. Hobday, 1983, Terrigenous clastic depositional systems—applications to petroleum, coal, and uranium exploration: New York, Springer-Verlag, 423 p.

Gilbert, G. K., 1885, The topographic features of lake shores: U.S. Geological Survey Annual Report, v. 5, p. 75–123.

Hale, L. A., 1960, Frontier Formation—Coalville, Utah, and nearby areas of Wyoming and Colorado: Wyoming Geological Association 15th Annual Field Conference Guidebook, p. 137–146.

———, 1972, Depositional history of the Ferron Formation, central Utah, *in* Plateau–Basin and Range transition zone: Utah Geological Association, p. 29–40.

Hale, L. A., and F. R. Van de Graaff, 1964, Cretaceous stratigraphy and facies patterns—northeastern Utah and adjacent areas: Intermountain Association of Petroleum Geologists 13th Annual Field Conference Guidebook, p. 115–138.

Hancock, J. M., 1975, The sequence of facies in the Upper Cretaceous of Northern Europe compared with that in the Western Interior, *in* W. G. E. Caldwell, ed., Cretaceous System in the Western Interior of North America: Geological Association of Canada Special Paper 13, p. 83–118.

Hancock, J. M., and E. G. Kauffman, 1979, The great transgressions of the Late Cretaceous: Journal of the Geological Society of London, v. 136, p. 175–186.

Hays, J. D., and W. C. Pitman III, 1973, Lithospheric plate motions, sea level changes and climatic and ecological consequences: Nature, v. 246, p. 18–22.

Jordan, T. E., 1981, Thrust loads and foreland basin evolution, Cretaceous, western United States: AAPG Bulletin, v. 65, p. 2506–2520.

Katich, P. J., Jr., 1953, Source direction of Ferron Sandstone in Utah: AAPG Bulletin, v. 37, p. 858–862.

———, 1954, Cretaceous and early Tertiary stratigraphy of central and south-central Utah with emphasis on the Wasatch Plateau area: Intermountain Association of Petroleum Geologists 5th Annual Field Conference Guidebook, p. 42–54.

Kauffman, E. G., 1977, Geological and biological overview—Western Interior Cretaceous basin, *in* E. G. Kauffman, ed., Cretaceous facies, faunas, and paleoenvironments across the Western Interior basin: The Mountain Geologist, v. 6, p. 227–245.

Kinney, D. M., 1955, Geology of the Uinta River–Brush Creek area, Duchesne and Uintah counties, Utah: U.S. Geological Survey Bulletin 1007, 185 p.

Lawrence, D. T., 1982, Influence of transgressive–regressive pulses on coal-bearing strata of the Upper Cretaceous Adaville Formation, southwestern Wyoming, *in* K. D. Gurgel, ed., Proceedings of the Fifth Symposium on the Geology of Rocky Mountain Coal: Utah Geology and Mineral Survey Bulletin 118, p. 32–49.

Maione, S. J., 1971, Stratigraphy of the Frontier Sandstone Member of the Mancos Shale (Upper Cretaceous) on the south flank of the eastern Uinta Mountains, Utah and Colorado: Earth Science Bulletin, v. 4, p. 27–58.

McGookey, D. P., J. D. Haun, L. A. Hale, and H. G. Goodell, 1972, Cretaceous System, *in* D. P. McGookey, ed., Geological atlas of the Rocky Mountain region, U.S.A.: Denver, Rocky Mountain Association of Geologists, p. 190–228.

Merewether, E. A., 1983, The Frontier Formation and mid-Cretaceous orogeny in the foreland of southwestern Wyoming: The Mountain Geologist, v. 20, n. 4, p. 121–138.

Merewether, E. A., C. M. Molenaar, and W. A. Cobban, 1983, Maps and diagrams showing lithofacies and stratigraphic nomenclature for formations of Turonian and Coniacian age in the middle Rocky Mountains, *in* Society of Economic Paleontologists and Mineralogists Mid-Cretaceous Codell Sandstone Member of Carlile Shale, Eastern Colorado: Rocky Mountain Section, Field Trip Guidebook, p. 19–25.

Merewether, E. A., P. D. Blackmon, and J. C. Webb, 1984, The mid-Cretaceous Frontier Formation near the Moxa Arch, southwestern Wyoming: U.S. Geological Survey Professional Paper 1290, 29 p.

Moir, G. J., 1974, Depositional environments and stratigraphy of the Cretaceous rocks, southwestern Utah: Ph.D. thesis, University of California, Los Angeles, 427 p.

Moore, G. T., and D. O. Asquith, 1971, Delta—term and concept: Geological Society of America Bulletin, v. 82, p. 2563–2568.

Myers, R. C., 1977, Stratigraphy of the Frontier Formation

(Upper Cretaceous), Kemmerer area, Lincoln County, Wyoming: Wyoming Geological Association 29th Annual Field Conference Guidebook, p. 271–311.

Peterson, F., and R. T. Ryder, 1975, Cretaceous rocks in the Henry Mountains region, Utah, and their relation to neighboring regions: Four Corners Geological Society 8th Annual Field Conference Guidebook, p. 167–189.

Pitman, W. C., III, 1978, Relationship between eustacy and stratigraphic sequences of passive margins: Geological Society of America Bulletin, v. 89, p. 1389–1403.

Rainwater, E. H., 1975, Petroleum in deltaic sediments, *in* M. L. Broussard, ed., Deltas—models for exploration: Houston Geological Society, p. 3–12.

Reeside, J. B., Jr., 1955, Revised interpretation of the Cretaceous section on Vermillion Creek, Moffat County, Colorado, *in* Rocky Mountain thrust belt geology and resources: Wyoming Geological Association 29th Annual Field Conference Guidebook, p. 271–311.

Ryer, T. A., 1977, Patterns of Cretaceous shallow-marine sedimentation, Coalville and Rockport areas, Utah: Geological Society of America Bulletin, v. 88, p. 177–188.

———, 1981, Deltaic coals of Ferron Sandstone Member of Mancos Shale—predictive model for Cretaceous coal-bearing strata of Western Interior: AAPG Bulletin, v. 65, p. 2323–2340.

Ryer, T. A., and M. McPhillips, 1983, Early Late Cretaceous paleogeography of east-central Utah, *in* M. W. Reynolds and E. D. Dolly, eds., Mesozoic paleogeography of the west-central United States: Society of Economic Paleontologists and Mineralogists Rocky Mountain Section, Rocky Mountain Paleogeography Symposium 2, p. 253–272.

Sharp, J. V. A., 1963, Unconformities within basal marine

Cretaceous rocks of the Piceance basin, Colorado: Ph.D. thesis, University of Colorado, Boulder, 170 p.

Stokes, W. L., and E. B. Heylmun, 1963, Tectonic history of southwestern Utah, *in* Guidebook to the geology of southwestern Utah: Intermountain Association of Petroleum Geologists, p. 19–25.

Thomaidis, N. D., 1973, Church Buttes Arch, Wyoming and Utah: Wyoming Geological Association 25th Annual Field Conference Guidebook, p. 35–39.

Uresk, J., 1979, Sedimentary environments of the Cretaceous Ferron Sandstone near Caineville, Utah: Brigham Young University Geological Studies, v. 26, p. 2, p. 81–100.

Walton, P. T., 1944, Geology of the Cretaceous of the Uinta basin, Utah: Geological Society of America Bulletin, v. 55, p. 91–130.

Weimer, R. J., 1962, Late Jurassic and Early Cretaceous correlations, south-central Wyoming and northwestern Colorado: Wyoming Geological Association 17th Annual Field Conference Guidebook, p. 124–130.

Williams, N. C., and C. R. Stelck, 1975, Speculations on the Cretaceous paleogeography of North America, *in* W. G. E. Caldwell, ed., The Cretaceous System in the Western Interior of North America: Geological Association of Canada Special Paper 13, p. 1–20.

Winn, R. D., and M. E. Smithwick, 1980, Lower Frontier Formation, southwestern Wyoming—depositional controls on sandstone compositions and on diagenesis: Wyoming Geological Association, 32nd Annual Field Conference Guidebook, p. 137–153.

Wright, L. D., 1978, River deltas, *in* R. A. Davis, Jr., ed., Coastal sedimentary environments: New York, Springer-Verlag, p. 5–68.

Part IV
SOUTHERN ROCKY MOUNTAINS

The Paradox: A Pull-Apart Basin of Pennsylvanian Age

G. M. Stevenson
Consultant
Denver, Colorado

D. L. Baars
Consultant
Evergreen, Colorado

The Paradox basin of the east-central Colorado Plateau Province is an intracratonic depression developed on continental crust. The elongate, northwest-trending, rhombic-shaped salt basin of Middle Pennsylvanian age is bounded on the northeast by the Uncompahgre and San Luis segments of the Ancestral Rocky Mountains and on the southwest by the less prominent Four Corners lineament. We have previously demonstrated that the basin sagged along intersecting basement fracture zones by strong east–west extension during Middle Pennsylvanian time. The master fracture system was the northwest-trending Olympic–Wichita structural lane that lies along the eastern margin of the Paradox salt.

Oblique divergent strike-slip faulting along the Uncompahgre–San Luis segment created a tension-releasing bend where the Paradox pull-apart basin nucleated and subsequently developed throughout Middle Pennsylvanian time. Smaller subbasins developed by orthogonal spreading along intersecting northeast-trending transform faults, where the rate of basin floor subsidence was related to combinations of normal and strike-slip faulting. The greater Paradox basin was episodically deepened during Middle Pennsylvanian time by rejuvenated extensional basement faulting. Vertical displacement was greatest along the Uncompahgre front, which caused tilting of the basin and contemporaneous development of an asymmetrically thick sedimentary sequence with great thicknesses of salt and arkose.

By middle Desmoinesian time, during deposition of the Desert Creek stage, the rate of divergent strike-slip faulting slowed considerably. Folds caused by minor wrench movements provided shoaling conditions along the southwest shallow carbonate shelf where porous algal mounds developed. Meanwhile, continued tectonic movement and space reduction of the basin floor may have triggered salt flowage and diapirism in the deep eastern pull-apart trough. As wrench tectonism diminished in late Desmoinesian–Early Permian time, the eastern part of the basin continued to subside at a faster rate than the western part and was filled with marine and nonmarine sediments.

INTRODUCTION

The Paradox basin is an elongate, northwest-trending, evaporite basin that extends from northwesternmost New Mexico to east-central Utah. It covers an area of approximately 17,000 mi^2 (27,200 km^2) in southwestern Colorado and southeastern Utah (Figure 1). The highly petroliferous tectonic depression was most fully developed in Middle Pennsylvanian (Desmoinesian) time, but had a long history that spanned the interval from Proterozoic through Paleozoic time. The basin is very complex, both structurally and stratigraphically. Fortunately, there are now sufficient surface and subsurface data available to attempt a reasonable interpretation of its origin and exploration potential.

Although the presence of the basin has been known for decades, knowledge of the actual configuration of the depression has evolved very slowly. The Paradox basin was first mapped as an ovate, asymmetric depression adjacent to the northeastern border of the Uncompahgre uplift (Wengerd and Matheny, 1958). It was identified as a typical taphrogeosyncline of Kay (1951), in that it is a "sediment-filled deeply depressed rift block(s), bounded by one or more high angle faults..." (Kay, 1951, p. 107). In the following fifteen years it was gradually realized through geophysical studies and deep drilling that the depression stepped down from southwest to northeast along major northwest-trending deep-seated normal faults (Baars, 1966) and that the fringing shallow shelves of the southern and southwestern margins were tectonically controlled. Gorham (1975) termed the Paradox an *aulacogen*, and Baars (1976) followed suit. However, subsequent refinements in the definition of that term by plate tectonocists makes it inappropriate for the Paradox basin. Furthermore, it is now

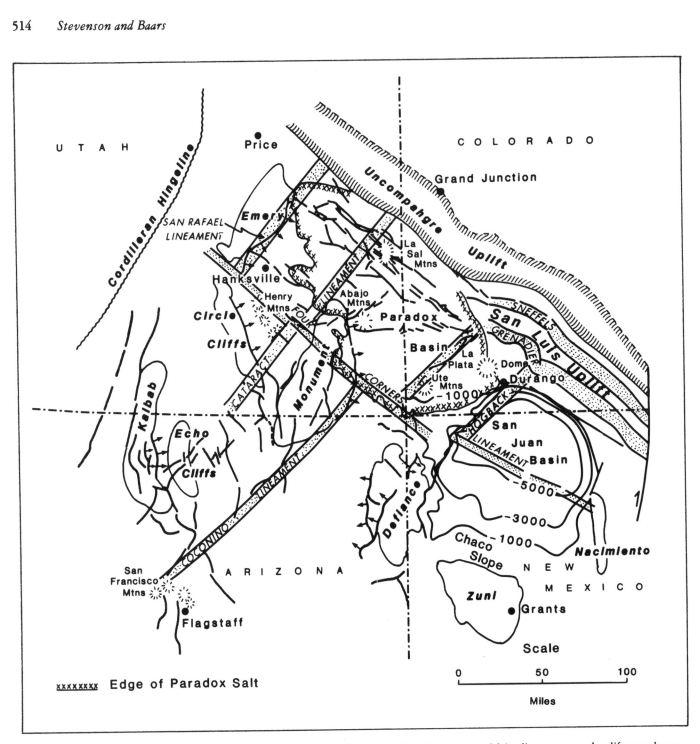

Figure 1—Location map showing major paleostructural features of the greater Four Corners area. Major lineaments and uplifts are shown in stippled patterns; Paradox basin outline is defined by the distal limit of Paradox salt. Structural contours in the San Juan basin are on a Middle Pennsylvanian marker horizon.

known to be a far more complexly faulted basin than previously realized.

The best interpretation of the presently known structural and stratigraphic relationships in the Paradox basin indicates that it is a complex pull-apart basin of large proportions. Such features are well documented in present-day and ancient depressions worldwide. These tectonic depressions form along irregularities of wrench fault zones in places where the faults first are convergent and then divergent along regional strike of the overall strike-slip

system. In the case of the Paradox basin, the depression opened along the northwest-trending Olympic–Wichita lineament of Baars (1976) at the conjugate northeast-trending Colorado lineament of Warner (1978). As originally defined, both basement lineaments are so broadly drawn as to encompass all of the basement fractures of the Paradox basin (Figure 2). However, to understand the detailed relationships present, the broader structural features must be refined and individual rift trends must be differentiated. Consequently, in this paper we will name the

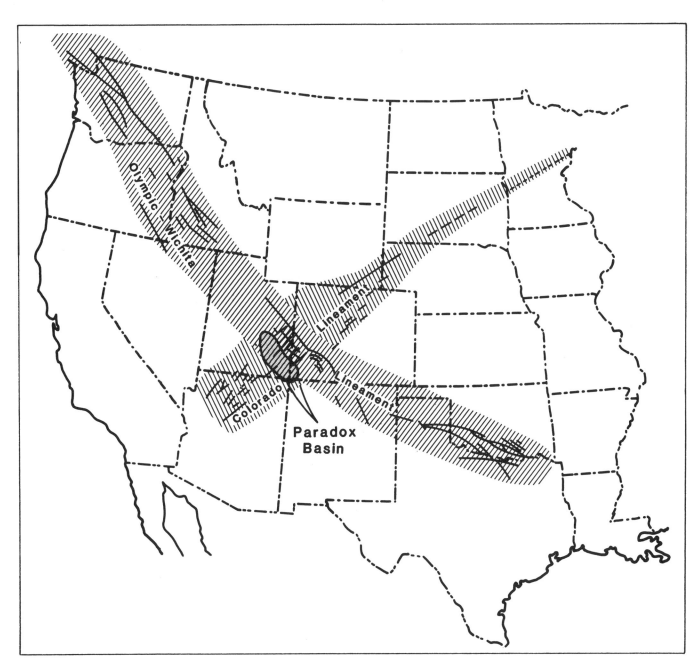

Figure 2—Map showing major continental-scale basement wrench fault sets of the western United States; northwest-trending lineaments are right-lateral; northeast-trending lineaments are left-lateral. Modified after Baars (1976). Used with permission of the International Basement Tectonics Association.

more important individual basement trends to better explain the details of basin development and sedimentation in the Paradox region through Pennsylvanian time. It is well known that the northwest-trending fractures were the dominant control of basin configuration, but basin spreading occurred along the conjugate northeasterly set. The major structural features, geographic areas, and oil fields discussed in this paper are shown in Figure 3.

Regional Setting

North–south compression caused orthogonal fracturing of the brittle crust about 1700–1600 m.y. ago. The resulting northwest- and northeast-trending rift zones dominated the

tectonic fabric of the continents, and one set of each orientation passed through what would become the Paradox basin. The northwesterly swarm of Precambrian wrench faults can now be bracketed to the time interval 1675–1460 m.y. ago in the San Juan Mountains of southwestern Colorado (Baars and Ellingson, 1984). The northeasterly fractures were dated at 1700 m.y. by Warner (1978), leading Baars (1976) to conclude that the two sets were an essentially contemporaneous conjugate pair. The northwest-trending set comprises dextral offsets, while the northeast-trending set indicates sinistral strike-slip displacement (Baars, 1976; Warner, 1978). The interpreted regional extent of these major lineaments is shown on Figure 2.

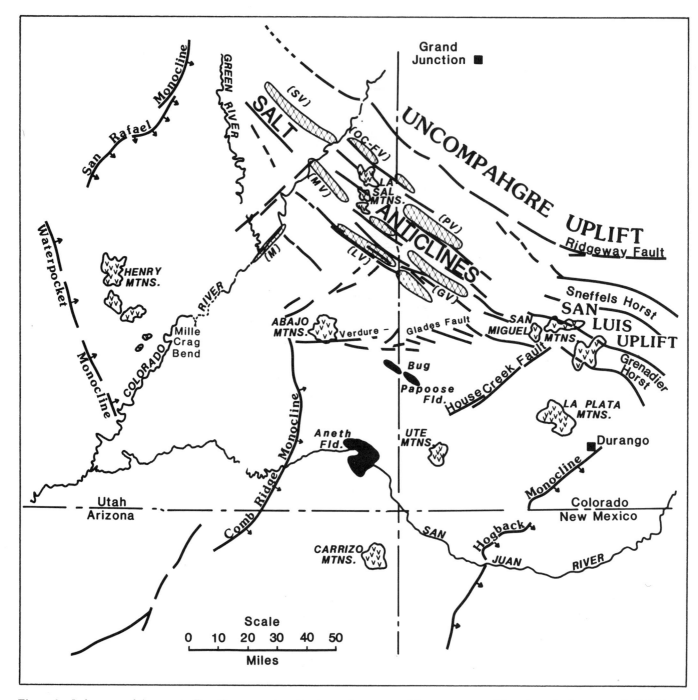

Figure 3—Index map of the greater Four Corners region showing major structural features, geographic points of reference, and oil fields discussed. Symbols for the salt anticlines are as follows: Salt Valley, SV; Onion Creek–Fisher Valley, OC-FV; Moab Valley, MV; Lisbon Valley, LV; Gypsum Valley, GV; Paradox Valley, PV; Meander anticline, M.

These major lineaments consist of swarms of fractures and wrench faults whereby the individual components can usually be distinguished (Figure 1). The Paradox basin formed by crustal extension that occurred at the intersection of the Olympic–Wichita lineament and the Colorado lineament (Baars and Stevenson, 1982). Although minor orthogonal fracture sets are ubiquitous throughout the region, several important individual trends were responsible for the development of several subprovinces or subbasins within the overall Paradox pull-apart basin.

Although the basement structural fabric was established in early Proterozoic time, it was repeatedly rejuvenated throughout Phanerozoic time (Baars, 1966; Baars and See, 1968; Spoelhof, 1976; Baars and Stevenson, 1982). Minor structural movements were sufficient to control facies distribution in Cambrian, Late Devonian, and Mississippian time (Baars, 1966; Baars and See, 1968). It was in Pennsylvanian time, however, that the most violent rejuvenation occurred, forming the Ancestral Rockies uplifts and the Paradox pull-apart basin. The extensive opening of the Paradox basin at that time provided the necessary conditions for restricted, evaporitic marine conditions to exist, highly organic petroleum source rocks to be deposited, and structurally localized stratigraphic traps to build (Baars and Stevenson, 1982). The Aneth shelf, a broad, northwest-trending restricted carbonate shelf bounded by two major lineaments, developed along the southwestern margin of the Paradox basin. This shelf is the site of a multitude of stratigraphic carbonate traps from which petroleum production is of major importance. Extensive current exploration of this shelf continues, and major oil fields are still being discovered.

But what is the basis for this pull-apart hypothesis?

NORTHWEST TRENDS

San Luis Uplift

The northwest-trending Uncompahgre and San Luis uplifts form the northeastern edge of the Paradox pull-apart basin. These two Precambrian fault blocks lie en echelon and were bounded by the master divergent fault sets during Pennsylvanian extensional tectonism (Figure 1).

Baars and Stevenson (1984) have shown that the San Luis uplift is a Precambrian structure separate and distinct from the Uncompahgre uplift. The San Luis uplift, as now recognized, extends from about Rico, Colorado, south-eastward through the San Juan Mountains (Grenadier and Sneffels horsts) into the Tusas and La Madera areas of north-central New Mexico (Figure 4).

Precambrian rocks of the San Luis uplift are pre-dominantly quartzite in lithology (Uncompahgre Quartzite). The blocks were positive topographic features through most of the Paleozoic. Baars (1966) and Baars and Stevenson (1981, 1984) suggested that the east-west-trending "kink" in the Sneffels and Grenadier blocks was caused by large-scale drag folds initially developed in late Precambrian time at the compressional intersection of the San Luis uplift and the Four Corners platform (Figures 5A and B). Tewksbury (1981) tends to corroborate this hypothesis, in that she concluded that the Uncompahgre Quartzite comprises a Late Precambrian allochthonous block that has been thrusted from north to south over the older Precambrian igneous and metamorphic complex (Figure 5B).

Coarse, quartzite pebble to boulder fanglomerates derived from the Precambrian highland and preserved in the Late Cambrian Ignacio Formation in southwestern Colorado attest to the early Paleozoic emergence of the San Luis uplift. Following a long period of early Paleozoic tectonic stability, Late Devonian tidal flat deposits lapped onto the uplift, and by Mississippian time (Ouray Formation) the area was covered by a shallow, normal marine sea (Baars, 1966). Isolated remnants of a basal quartz conglomerate in the Mississippian Espiritu Santo Formation (a shallow marine carbonate unit) in the Nacimiento Mountains in north-central New Mexico represents the easternmost advance of the Mississippian sea over the San Luis uplift (Armstrong and Mamet, 1977).

The San Luis uplift was again emergent in Atokan time (early Middle Pennsylvanian) when quartzose sandstone was deposited adjacent to its flanks (Baars and Stevenson, 1984) (Figure 6). Spoelhof (1976) demonstrated continued emergence through early Desmoinesian time by innumerable intraformational unconformities in the Hermosa Group in the San Juan Mountains. The youngest carbonates of the Hermosa Group exposed in the San Juan Mountains are chronostratigraphically equivalent to the upper Ismay Stage of the Paradox Formation in the Paradox basin. Neither Virgilian nor Missourian fossils have been described. Stratigraphic correlations and terminology of Paleozoic rock units in the Paradox basin are shown in Figure 7.

Uncompahgre Uplift

The Uncompahgre uplift is a well-known source of arkosic clastic sediments, and it makes up the westernmost uplift of the classic Permian–Pennsylvanian Ancestral Rocky Mountains. The northwest-trending uplift forms the prominent Uncompahgre Plateau physiographic feature from Ridgeway, Colorado, to Cisco dome, Utah (Figure 4). Both the southwestern and northeastern sides are bounded by faulted monoclines. The southeastern termination of the Uncompahgre front (Elston and Shoemaker, 1960) is marked by the Ridgeway fault, an east-west-trending normal fault that appears to offset the structure left-laterally to the east (Weimer, 1982). The Ridgeway fault is subparallel to the kinked part of the Sneffels and Grenadier horsts and is undoubtedly related to the Paradox pull-apart tectonism (Figures 5B and C). To the southeast, thick Tertiary volcanic rocks bury the Uncompahgre structure. Because of the distribution of similar arkosic facies in southern Colorado and north-central New Mexico to those adjacent to the exposed uplift, the Uncompahgre is believed to continue on to the southeast where it is terminated by the Picuris–Pecos fault (Figure 4).

White and Jacobson (1983) show the structural relief along the south-southwestern flank of the uplift in a series of subsurface cross sections across the northwestern part of the Uncompahgre uplift. Dip slip is seen to have been distributed along a series of subparallel reverse faults. The faults decrease in throw northward, and most of the vertical displacement is transferred to a large bounding fault near Gateway, Colorado, where nearly 26,000 ft (7880 m) of structural relief exists between the deepest part of the Paradox basin and the Uncompahgre uplift.

Several wells drilled on the Colorado portion of the Uncompahgre uplift show repeated sections of lower Paleozoic strata, documenting the presence of high-angle reverse faults of Permian–Pennsylvanian age. In Utah, Mobil Oil drilled nearly 14,000 ft (4240 m) of Precambrian crystalline basement rock before encountering overturned(?) Paleozoic rocks (Frahme and Vaughn, 1983), further substantiating the presence of high-angle reverse faults on the southwestern margin of the uplift.

Little is known about the Uncompahgre uplift northwest of Cisco dome where it plunges abruptly into the subsurface. Both gravity and magnetic intensities abruptly diminish northward, possibly because of the abrupt plunge of the uplift or termination along the northeast-trending Cataract and/or San Rafael lineaments (Figure 8). Aeromagnetic investigations by Johnson (1983) conclude that the

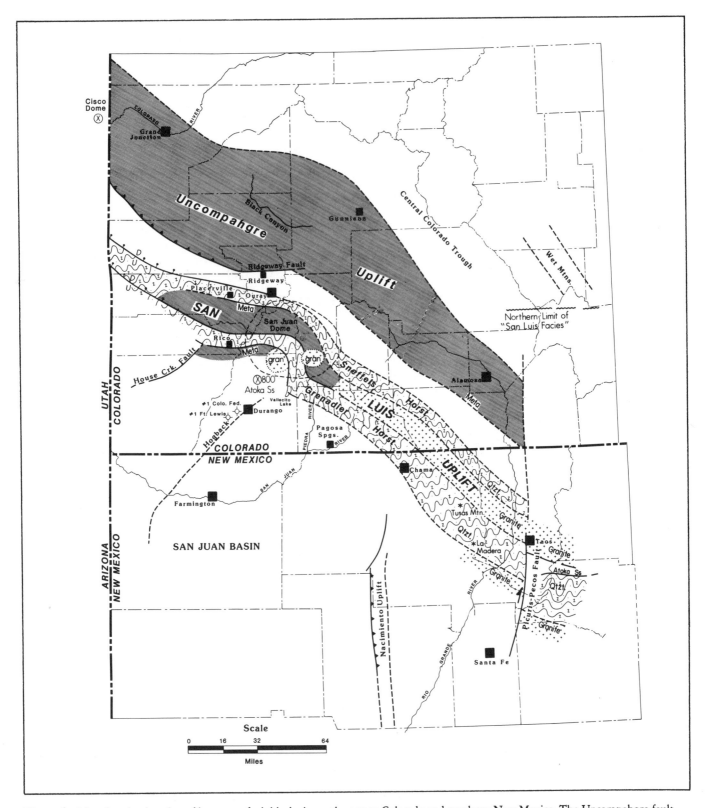

Figure 4—Map showing location of basement fault blocks in southwestern Colorado and northern New Mexico. The Uncompahgre fault block is composed of Precambrian metamorphic rocks, whereas the San Luis uplift is composed of Precambrian quartzite and granite.

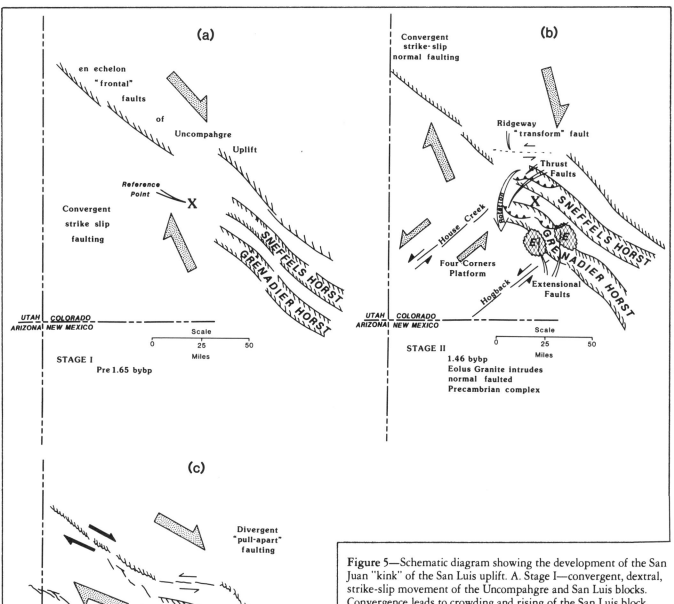

Figure 5—Schematic diagram showing the development of the San Juan "kink" of the San Luis uplift. A. Stage I—convergent, dextral, strike-slip movement of the Uncompahgre and San Luis blocks. Convergence leads to crowding and rising of the San Luis block resulting in north–south shortening. B. Stage II—convergent, dextral, wrench faulting continues leading to multiple phases of thrust faulting. Left-lateral motion along the Ridgeway fault (a possible transform) and the intersecting House Creek and Hogback faults has caused counterclockwise rotation and bending of the Sneffels and Grenadier horsts. Following thrusting, the Z shaped Grenadier block demonstrates strong east–west extension (normal faulting) along the north–south-trending kink of the bent fault block where the Eolus granite (E) intrudes the Grenadier block and cuts the bounding faults. C. Stage III—divergent, dextral, strike-slip faulting during the Pennsylvanian Ancestral Rocky Mountains epeirogeny was reactivated along this preexisting Precambrian basement fabric. Slight additional counterclockwise bending of the San Luis uplift may have occurred, but vertical tensional block faulting predominated due to overall east–west extension of the continental crust. Extensional faulting was relayed across the complex releasing bend of the San Juan kink along the intersecting House Creek and Hogback faults. The younger Precambrian intrusions (E) are faulted in a northwest and northeast-oriented conjugate set of fractures.

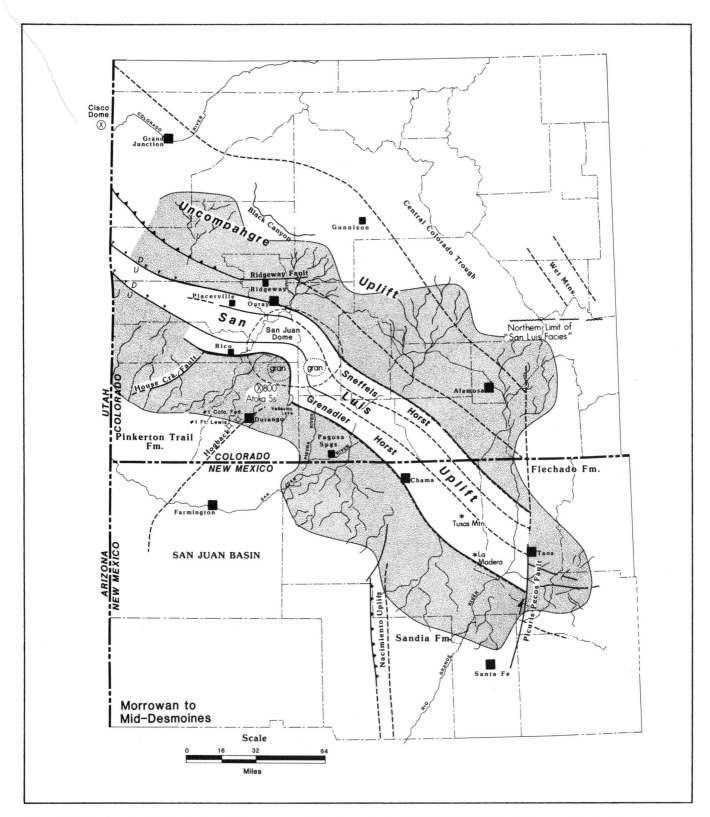

Figure 6—Map showing the distribution of Morrowan to middle Desmoinesian "San Luis facies" (stippled pattern) derived from the predominantly quartzite fault blocks of the San Luis uplift. Note that clean quartzose sandstone overlies the younger Uncompahgre uplift.

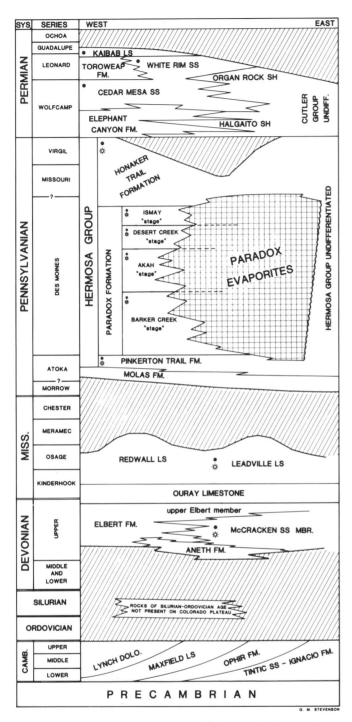

Figure 7—Paleozoic correlation chart of the Paradox basin and Four Corners region.

northwest-trending fracture sets are the oldest and are offset left-laterally by the younger northeast-trending sets. Heyman (1983) provided evidence supporting the interpretation that pre-Laramide basement weaknesses controlled the location and orientation of faults during Laramide uplift.

Four Corners Lineament

Several northwest-trending, en echelon, structural "segments" of the Four Corners lineament have acted singly

or in unison to control numerous Phanerozoic depositional patterns throughout the Paradox and San Juan basins. The southwestern edge of the Paradox pull-apart basin is defined by the Four Corners lineament (Figure 1). The "Aneth flex" of Szabo and Wengerd (1975) coincides with part of the Four Corners lineament.

The Four Corners lineament extends from the northern end of the Nacimiento uplift (in north-central New Mexico) northwestward 275 mi (438 km) through the Four Corners area and into south-central Utah where it splits the Circle Cliffs and Emery (San Rafael) uplifts (Figure 1). Numerous northwest-trending structures associated with rock units of various ages coincide in the subsurface and surface along this feature.

The Four Corners lineament is intersected by the northeast-trending Hogback, House Creek, Coconino, Cataract, and San Rafael lineaments (Figure 8), all of which demonstrate varying degrees of right-lateral offset across this northwest-trending fault system. Structurally, the Four Corners lineament separates the present-day deepest part of the San Juan basin to the northeast from the homoclinal Chaco Slope on the southwest (Figure 1). Isopach and lithofacies mapping of Jurassic strata (C. Peterson, personal communication, 1983) and Cretaceous strata (B. Black, personal communication, 1982) affirms that the reactivated basement fault zone appreciably affected continental and marine clastic sedimentation during post-Paleozoic time.

In southeastern Utah, the Four Corners lineament is coincident with the southwestern distal limit of all Paradox salt cycles (maximum restricted marine environment) (Figure 8). The lineament also coincides with the southwestern edge of upper Paradox anhydrites (semirestricted marine environment) in the Aneth area. During the deposition of the Desert Creek on the southwestern flank of the Blanding subbasin, the Four Corners lineament separated this part of the Paradox basin into two distinctive depositional settings. To the northeast the Desert Creek was deposited on a tectonically collapsing platform where shallow marine, restricted conditions prevailed. To the southwest normal marine conditions prevailed across an open marine shelf. (For additional details the reader is referred to the discussion about the Greater Aneth area later in this paper.)

Farther to the northwest, the Four Corners lineament divides the Monument upwarp into two separate structural features, which again show different depositional settings throughout much of Pennsylvanian time (Figure 9). The Early Pennsylvanian pre-Paradox interval is demonstrably thinner northeast of the Four Corners lineament than to the southwest, suggesting that the southwestern part of the upwarp sagged considerably throughout Atokan time (Figure 8). During Paradox deposition, evaporites (both salt and anhydrite) are limited to the area northeast of the Four Corners lineament, whereas normal marine conditions existed southwest of it. Local thinning of the Pennsylvanian section occurs at the Fish Creek anticline located on the Four Corners lineament at the intersection with the Comb Ridge monocline.

The Four Corners lineament extends to the northwest through the intensely fractured Mille Crag Bend of the Colorado River (Figure 8). It is offset left-laterally by the northeast-trending Cataract lineament, and conversely, the

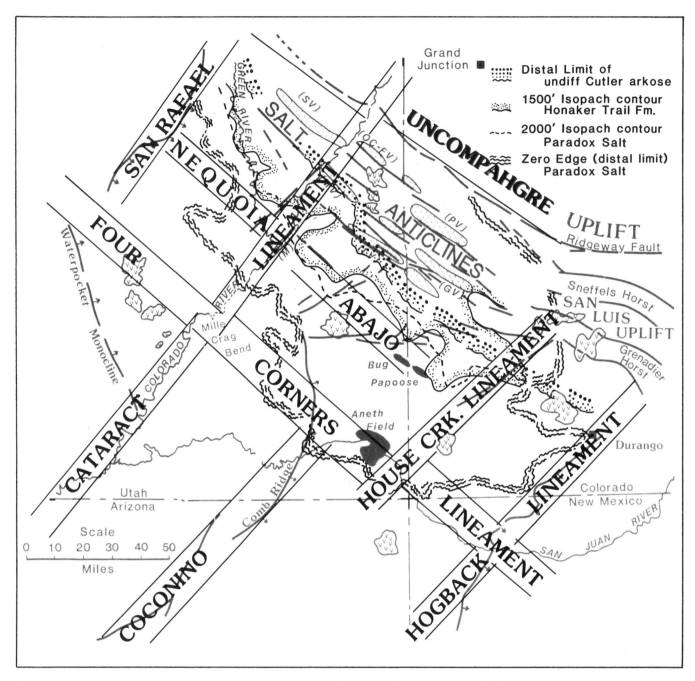

Figure 8—Map of the greater Four Corners region showing major lineaments that actively controlled Pennsylvanian sedimentation (see explanation of patterns on map). The Paradox basin is defined by the northwest-trending Uncompahgre and San Luis uplifts and the Four Corners lineament. The basin is also bounded by the northeast-trending San Rafael, House Creek, and Hogback lineaments. Faults are shown by dark lines; monoclines are shown by dark lines with arrows.

Cataract lineament is offset right-laterally by the Four Corners lineament. This relationship is made obvious by surface faulting and drainage patterns. Both Pennsylvanian and Permian stratigraphic sequences drastically change in thickness and in lithofacies across the Four Corners lineament. The Four Corners lineament continues in a northwesterly direction along the northwest-oriented Henry Mountains (Tertiary intrusive) and divides the large monoclinal folds of the Circle Cliffs and Emery (San Rafael) uplifts. Farther northwest it becomes obscured where it

intersects the northwestern edge of the Colorado Plateau Province (Cordilleran hingeline) (Figures 1 and 8).

Paradox evaporites occur only northeast of the Four Corners lineament, while normal marine carbonates prevailed to the southwest. Although beyond the scope of this paper, the Permian lithofacies (particularly the White Rim Sandstone and Elephant Canyon Formation) were dramatically affected by recurrent basement faulting coincident with this lineament. Both the Emery and Circle Cliffs uplifts reflect a late Paleozoic period of tectonism;

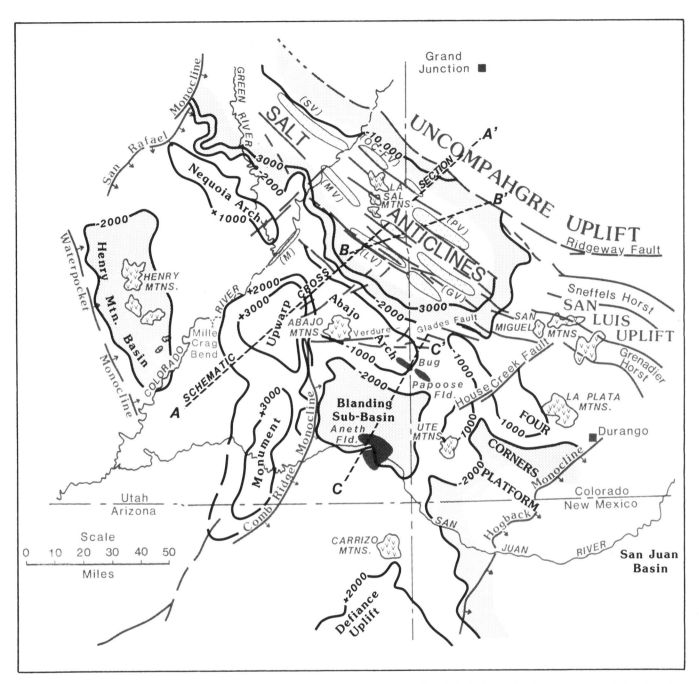

Figure 9—Map of the greater Four Corners region with structural contours on top of the Mississippian Leadville Formation. Note that the —2000-ft (—600-m) contour defines the rhomboidal-shaped Blanding subbasin, the salt anticline tectonic pull-apart trough, and the wedge-shaped Henry Mountains basin. Both the Nequoia and Abajo arches are shown as prominent structural ridges or noses. The Monument upwarp is shown as two structural closures separated by the Four Corners lineament. Compare this map to Figure 8. Location of cross sections A–A′, B–B′ and C–C′ in Figures 10, 11, and 17, respectively, are shown.

however, stratigraphic relationships in the deeper subsurface of the Emery uplift demonstrate movement as early as Late Cambrian time.

The Four Corners lineament separates the structural platform of the western Paradox basin from the wedge-shaped Henry Mountains basin (Figures 8 and 9). High levels of seismicity have been recorded along this segment of the Four Corners lineament leading Humphrey and Wong (1983) to conclude that unusual conditions exist. There is

nothing particularly unexplainable going on here; if anything, the Four Corners lineament is still active today.

Nequoia and Abajo Arches

In Middle Pennsylvanian time, the Nequoia and Abajo arches formed a prominent structural rib through the middle of the collapsing Paradox basin parallel with the Four Corners and Uncompahgre master faults (Figures 8 and 9). To the northeast of these arches, the area is characterized by large-scale complex block faults, rapidly expanded Paradox salt cycles and diapiric salt anticlines, and a complex pre-Pennsylvanian tectonic and stratigraphic history (Baars, 1966). The southwestern part of the basin is characterized by much smaller scale faulting on a predominantly carbonate platform (Figure 10).

Subsurface structural maps on any Paleozoic stratigraphic marker delineate the Nequoia segment as a prominent structural arch or "nose" extending southeast from the San Rafael lineament (southeast bounding edge of the Emery uplift) approximately 40 mi (65 km) to the intersection with the Cataract lineament (Figure 9). Present-day surface structure is obscured by wind-blown sands of the Green River desert. Surface faults showing only 50 ft (15 m) of dip slip at the surface (Entrada Sandstone juxtaposed against the overlying Carmel Formation) have nearly 4000 ft (1200 m) of vertical displacement at the top of the Mississippian Leadville Formation. Stratigraphy of the Leadville shows that the area was slightly emergent during deposition in Osagian time, as suggested by crinoidal buildups and oolitic carbonates in wells drilled in the area. A similar tectonic setting during the time of Mississippian deposition as that described by Baars (1966) for the salt anticline region is suggested for the Nequoia area.

Thickening of Early Pennsylvanian strata across the area indicates that the Nequoia arch was tilted and sagged prior to salt deposition. The arch was emergent in Desmoinesian time as indicated by Paradox salt thinning onto and wrapping around the reactivated fault block (Figure 8). The Nequoia arch remained relatively high through the end of Pennsylvanian time, followed by pre-Permian erosional beveling. In early Wolfcampian time, the Nequoia block sagged and was submerged by a restricted shallow marine sea in which carbonates and evaporites of the Elephant Canyon Formation were deposited (Baars, 1962).

The Abajo arch is a similar prominent subsurface Paleozoic structural ridge extending southeastward of the Cataract lineament approximately 35 mi (55 km) to where it is obscured by the east-west-trending Verdure–Glades fault zone (Figure 9). Kelley (1955) identified a number of en echelon surface features, such as the Gibson dome, Boulder Knoll anticline, and Dove Creek anticline, that also define the arch. Potter and McGill (1978) recognized a northwest-trending surface lineament from Chesler Park to Spanish Bottom near the confluence of the Green and Colorado rivers that they interpreted as a basement fault lying directly on trend with the Abajo arch.

Throughout Middle Pennsylvanian time, the Abajo arch actively controlled salt deposition. In fact, the 2000 ft (600-m) salt isolith contour coincides with the Nequoia–Abajo structure, which acted much like a hinge (Figure 8). To the northeast, salt thickens dramatically and is less stable and more plastic. Faulting is large scale and complex where diapiric salt anticlines developed along extensional strike-slip grabens (Figures 10 and 11). Southwest of the arch, the salt cycles thin abruptly and the youngest cycles change facies to porous carbonate rocks and thinly bedded sulfate deposits of a restricted marine platform (Figure 10). Thinning of the overlying Honaker Trail Formation (late Desmoinesian–early Missourian) further suggests that the area southwest of the Abajo arch remained high through much of Late Pennsylvanian time (Figure 8).

Paradox Fold and Fault Belt

A belt of northwest-trending elongate salt diapirs occurs in the deep trough (the "Uncompahgre trough" of some writers) between the Nequoia–Abajo hinge and the Uncompahgre uplift (Figures 8 and 9). It has been speculated by various authors since Stokes' (1948) study of the Gypsum Valley salt structure that deep-seated folds or faults must control the localization and orientation of the major diapirs. Subsequently, it has become clear, through geophysical studies and deep drilling, that major faulting predates the Paradox salt (Baars, 1966). Some fault-bounded blocks are half-grabens in the subsurface (e.g., Paradox Valley), whereas others are full grabens within the overall larger half-grabens (e.g., Onion Creek–Fisher Valley) (Figure 9). These fault blocks, where measured by deep drilling across Paradox Valley, document a vertical displacement of over 5000 ft (1500 m). The thickness of the Paradox salt ranges from zero on the southwest flank of Paradox Valley to nearly 15,000 ft (4500 m) in the core of the diapir (Figure 11). Baars (1966) demonstrated that these faults are a lateral extension of Precambrian faults of the Grenadier and Sneffels faults exposed in the San Juan Mountains. They are a splayed extension of the fault swarm associated with the San Luis uplift. The Paradox basement faults either terminate or are displaced in a left-lateral sense along the Cataract lineament (Figure 8).

Depositional thickness of the Paradox salt in the deeply faulted trough is unknown because of extensive flowage in and around the diapirs, such as described at Paradox Valley. Paradox salt definitely exceeded 3000 ft (900 m) and may have attained thicknesses of well over 5000 ft (1500 m), judging from the volume of salt in the major diapirs and the unknown volume of salt leached from the near-surface crests of the structures. The faulted troughs beneath the diapirs were in evidence in early Desmoinesian time (Figure 11) because the oldest known salt cycles in the Paradox basin occur beneath the structures (Hite, 1960). The flowage-triggering mechanism probably resulted from periodic rejuvenation of the basement structures as the basin pull-apart proceeded, combined with differential asymmetric loading by clastic sediments from the Uncompahgre uplift to the northeast. In any case, salt flowage was continuous or episodic from late Paradox time through Jurassic time, as shown by formational thinning and complex local angular unconformities adjacent to the margin of the structures (Figure 11).

NORTHEAST TRENDS

Four Corners Platform

The northeast-trending Four Corners platform of Kelley (1955) borders the southeastern edge of the Paradox basin and the northwestern edge of the San Juan basin. The Four Corners platform is bounded on the northwest by the House Creek fault zone and Comb Ridge segment of the Coconino lineament and on the southeast side by the Hogback monocline and Defiance uplift (Figures 8 and 9).

The Coconino lineament, as defined by Kelley (1955) and modified by Davis (1978), extends from the Coconino Point monocline in the eastern Grand Canyon area of northern Arizona northeastward to the Four Corners lineament in

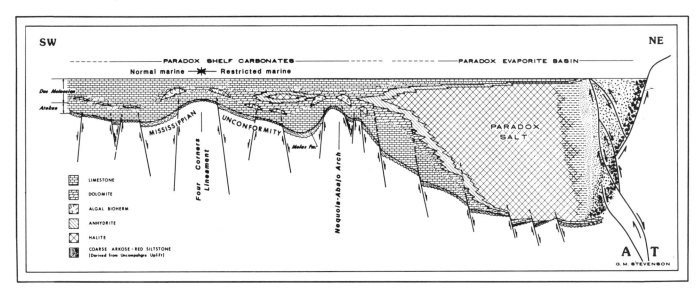

Figure 10—Schematic cross section (A–A′ in Figure 9) across Paradox basin palinspastically reconstructed at late Middle Pennsylvanian time, showing relationship of shelf carbonates to evaporite facies. Compare to Figures 8 and 9 for tectonically controlled sedimentation patterns and structural configuration. Location of faults is diagrammatic. Modified after Baars and Stevenson (1982).

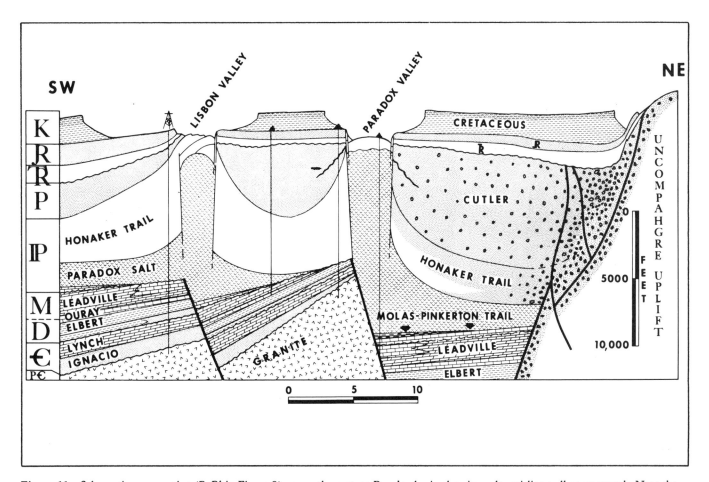

Figure 11—Schematic cross section (B–B′ in Figure 9) across the eastern Paradox basin showing salt anticline pull-apart trough. Note the relationship between rejuvenated basement faults and post–early Desmoinesian diapiric structures. Frontal faults of the Uncompahgre uplift are shown as complex braided network of wrench faults, which are locally high-angle reverse faults. Compare to Figures 8 and 9 for tectonically controlled sedimentation patterns and structural configuration. Modified after Baars (1966).

southeastern Utah (Figure 1). The Coconino lineament marks the southern terminus of the Heather, Grandview, East Kaibab, Echo Cliffs, Red Lake, and Organ Rock monoclines. It also marks the southern terminus of the Kaibab uplift, Echo Cliffs uplift, and Monument upwarp and the northwest terminus of the Black Mesa basin in northeastern Arizona (Davis, 1978).

Davis (1978) concluded that fold patterns of monoclines of the Colorado Plateau suggest that the Coconino lineament served as a major partitioning element in the basement separating different structural domains to the northwest and southeast.

Case and Joesting (1972) provided substantial geophysical support for a basement fracture zone coincident with the Coconino lineament, as well as many other fracture zones across the Colorado Plateau. They noted nearly straight zones of steepened magnetic and/or gravity gradients and lines of discontinuities of anomalies that persist for many kilometers and interpreted them to indicate a fundamental fracture pattern of Precambrian age. The trends of these fracture zones are dominantly northeast and northwest.

The Comb Ridge monocline, an element of the Coconino lineament, defines the southeastern edge of the Monument upwarp and the southwestern margin of the Blanding subbasin (Figures 8 and 9). The deep-seated fault underlying the Comb Ridge monocline lies en echelon with the House Creek fault zone, which is offset right-laterally by the Four Corners lineament (Figure 8).

The House Creek fault zone consists of several northeast-trending en echelon faults that suggest left-lateral movement. This fault zone also exhibits classic wrench fault characteristics in Paleozoic rocks at depths where the apparent upthrown block changes abruptly along strike from one side of the fault to the other. For example, the present-day downthrown block is north of the fault, yet the Paradox salt section is thicker immediately to the south of the fault, indicating at least one episode of reversal of relative dip-slip motion between Pennsylvanian and post-Laramide time. Farther to the northeast, the House Creek fault zone increases in complexity where it converges with the east-west-trending Glades fault zone (Figure 8). The two fault zones converge where the Grenadier horst is terminated in the San Juan Mountains at their complex intersection. The northeasternmost terminus of the Hogback fault occurs at the south-southeastern end of the Grenadier highland (Figures 8 and 9).

The peculiar kink in the San Juan Mountains of the Grenadier and Sneffels horsts was shown by Baars (1966) and Baars and Stevenson (1981, 1982) to be rather large-scale, left-lateral, drag folds developed at the intersection of major sets of conjugate basement fractures. The resulting compressional corner apparently buckled in late Precambrian time where allochthonous Uncompahgre Quartzite blocks were thrust (northeast to southwest) over older Precambrian metamorphic rocks (Figure 5) (Tewksbury, 1985).

The relatively large number of wells drilled to the Precambrian on the Four Corners platform in northeastern Arizona, northwestern New Mexico, and southwestern Colorado provide valuable information concerning the tectogenesis of the area. Precambrian, Cambrian, and

Devonian terranes are juxtaposed along a myriad of northwest- and northeast-trending basement fault blocks throughout the Four Corners platform (Figure 12). The structural style and geologic history are comparable to those observed in Precambrian and early Paleozoic outcrops in the San Juan Mountains in southwestern Colorado (Baars, 1966; Stevenson and Baars, 1977; Baars and Stevenson, 1982).

The Hogback monocline is underlain by northeast-trending faults showing high-angle reverse faulting in Paleozoic and Precambrian rocks. The Mesozoic and younger clastic beds are draped over the Hogback fault by west-to-east Laramide compression, thereby forming the southeast facing monocline as shown by seismic surveys (Figure 9). Present-day structural relief of the Pennsylvanian Paradox strata exceeds 4000 ft (1200 m) across the Four Corners platform into the central part of the San Juan basin.

The Four Corners platform was a slightly emergent landmass in post–Late Cambrian time, and Cambrian strata were preserved only in isolated down-dropped fault blocks (Figure 12). Throughout Cambrian–Devonian time the basement floor of the Four Corners platform was repeatedly stretched or compressed in response to minor tectonic adjustments between it and the adjacent northwest-trending San Luis uplift and the Defiance uplift southwest of the Four Corners lineament.

The Four Corners platform was relatively stable during Mississippian deposition of the Leadville Limestone. In early Middle Pennsylvanian (Atokan) time, however, the platform responded as a left-lateral transform fault zone that divided the tectonically collapsing San Juan and Paradox basins.

The nucleation site of the ensuing late Middle Pennsylvanian (Desmoinesian) pull-apart tectonism was the compressional intersection of the Four Corners platform buttressed against the kinked San Juan Mountain portion of the San Luis uplift (Figure 5C). The platform was depressed in Early Pennsylvanian time and remained low through middle Desmoinesian time, as indicated by dark, argillaceous, deeper water carbonates in the upper Paradox Formation. The limit of all Paradox salt cycles in the southeastern area of the Pennsylvanian evaporite basin was tectonically controlled at the margin of the Four Corners platform (Figure 8). From Late Pennsylvanian (Missourian) through Early Permian (Wolfcampian) time, the Four Corners platform was again relatively stable.

The Four Corners platform merges on the southwest with the north-trending Defiance uplift (Figure 12). The Defiance uplift is also segmented by northwest- and northeast-trending basement faults; the northeast-trending set appear to dominate. The uplift is strongly asymmetric to the east, bounded by the north-northeast-trending East Defiance monocline (Kelley, 1955). The northeast-trending Tsaile fault dissects the uplift and is en echelon between the Hogback fault and East Defiance monocline, terminating at the intersection with the northwest-trending Tocito horst (Figure 12).

Subsurface information indicates that the northern two-thirds of the uplift had a Precambrian and early Paleozoic history of juxtaposed fault blocks similar to those of the four Corners platform (Figure 12) (Stevenson and Baars, 1977). The southern end of the uplift remained high throughout most of Paleozoic time in an area where Permian strata

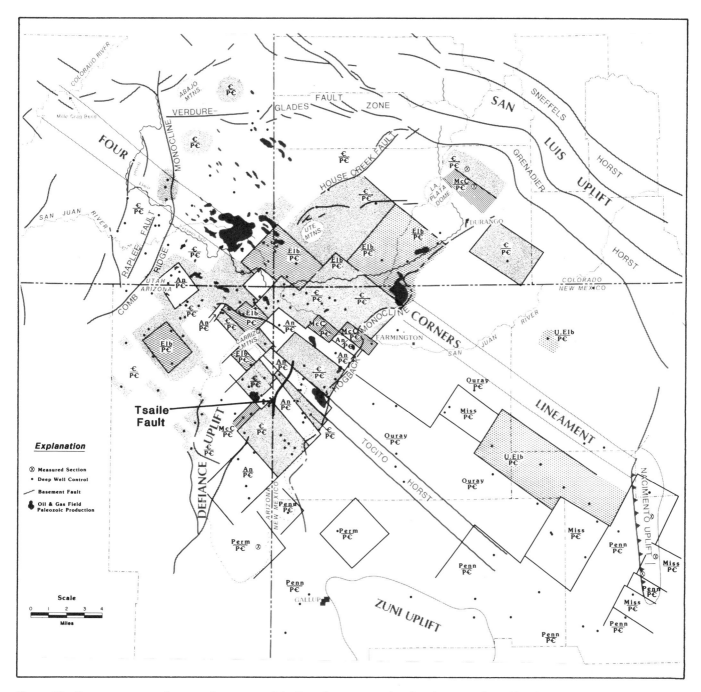

Figure 12—Basement penetration worm's eye map of the Four Corners area showing the nonconformable relationship of subsurface Paleozoic rock units with the Precambrian basement. The myriad of juxtaposed terranes, particularly on the Four Corners platform, demonstrate a subtle, yet complex, tectonic history of repeated stretching and attenuation of the southeastern pull-apart basin margin. Oil and gas fields producing from Paleozoic zones are shown in black. Note that most major fields occur near the intersection of these inferred juxtaposed blocks. Diagram updated and modified after Stevenson and Baars (1977).

nonconformably overlay the Precambrian (Figure 12). Thin Mississippian and Pennsylvanian carbonate deposits lapped onto the flanks of the Defiance positive element before burial by Permian clastic sediments.

Colorado Lineament

The Colorado lineament, as described by Warner (1978), is a belt of northeast-trending Precambrian faults that

traverse the Colorado Plateau. These basement faults cross the Rocky Mountains of Colorado, follow the Mullen Creek–Nash Fork shear zone in southeastern Wyoming, and extend northeastward across the northern Great Plains region into the southern Canadian Shield area (Figure 2). The lineament is defined by (1) gravity and aeromagnetic anomalies, (2) grain changes and compositional changes in Precambrian terranes, (3) radiometric age dates, and (4) mineralization zones such as the Colorado Mineral Belt, the Hartville iron deposits of Wyoming, and the Cuyuma Iron Range of Minnesota.

Similar observations in the Colorado Plateau and Colorado Rocky Mountains of northeast-trending basement faults have been noted by Kelley (1955), Kelley and Clinton (1960), Tweto and Sims (1960, 1963), Case et al. (1963), Anderson (1967), Mutschler and Hite (1969), Case and Joesting (1972), Hite (1975), Davis (1978), Shoemaker et al. (1978), Brill and Nuttli (1983), and Maarouf (1983).

Evidence includes northeast-trending faults with extensive breccia development, rectilinear drainage patterns with preferred northeasterly trends such as the course of the Colorado River through the salt anticline country, and the location of igneous intrusives (Hite, 1975). In the Grand Canyon, Shoemaker et al. (1978) postulated as many as nine episodes of Precambrian and Paleozoic faulting for the northeast-trending Bright Angel fault, including normal, reverse, and transcurrent displacements.

Copper and silver veins and uranium ores in the salt anticline region all follow northeast-trending fractures and probably share a similar tectonic history as that of the northeast-trending Colorado Mineral Belt (Hite, 1975). Shear zones of the Colorado Mineral Belt are Precambrian in age, but show documented tectonic episodes of reactivation through Paleozoic and Mesozoic time (Tweto and Sims, 1960, 1963).

Brill and Nuttli (1983) recognized that many of the larger earthquakes in the past 100 years in the west-central United States have occurred along the Colorado lineament. Earthquake epicenters maximum modified Mercalli intensity VI and greater are accounted for by the presence of seismic zones associated with major geologic features (e.g., the Nemaha uplift in eastern Kansas, the Rio Grande rift in central New Mexico, the Wichita Mountain uplift in the Texas panhandle and southwestern Oklahoma, the Overthrust Belt in southwestern Utah and southwestern Montana, and the Colorado lineament as shown in Figure 2). Wong and Simon (1981) concluded that an active basement wrench fault zone (the Roberts rift zone of Hite, 1975) existed along (at least) a 35-km stretch of the Colorado River southwest of Moab, Utah.

Cataract Lineament

The segment of the Colorado lineament (Warner, 1978) that is the most significant in the Paradox basin is a relatively narrow belt of northeast-trending basement faults that mainly follow the course of the Colorado River in the eastern part of the Colorado Plateau province, herein called the Cataract lineament (Figure 8). The Cataract lineament is coincident with a definite basement weakness defined by gravity and aeromagnetic data along the Colorado River (Mutschler and Hite, 1969; Case and Joesting, 1972). Strong aeromagnetic and gravity anomalies that define the Uncompahgre uplift decrease gradually between the northeast-trending Cataract and the San Rafael lineaments (Figure 8). Hite (1975) defined at least eight separate northeast-trending lineaments in the salt anticline portion of the eastern Paradox basin. He referred to the segment along the Colorado River as the Colorado lineament (Hite,

1975). The faults were shown by him to indicate left-lateral offset through the salt anticline region by the abrupt splaying of the northwest-trending faults into the northeast-trending Cataract lineament. Notably, both the Moab Valley and Castle Valley salt anticlines abruptly terminate along their intersection with the Cataract lineament (Figure 8).

Drag along the northwest-trending Salt Valley anticline further denotes the relative sense of left-lateral strike-slip movement along the Cataract lineament. A high gravity maximum, which lies immediately northeast of the Salt Valley diapir, is aligned with the Onion Creek–Fisher Valley salt structure across the Cataract lineament to the southeast. The salt structures in the Paradox basin are exemplified by high gravity minima. Left-lateral displacement of the Salt Valley and Onion Creek–Fisher Valley salt structures is on the order of 4 mi (6–7 km).

During Middle Pennsylvanian divergent wrench faulting, the northeast-trending, intrabasinal slippage along the Cataract lineament left-laterally offset the northwest-trending Four Corners lineament and Nequoia–Abajo lineaments (Figure 8). The Cataract lineament was relatively passive through much of Paleozoic time and mainly responded as a transform fault between larger northwest-trending fault zones. By late Desmoinesian time, the extreme northeastern quadrant of the greater Paradox basin, which was bounded by the northeast-trending Cataract and San Rafael lineaments and the northwest-trending Uncompahgre and Nequoia lineaments, subsided markedly. Here in this rhomboidal subbasin, the youngest salt cycles were deposited (chronostratigraphic equivalents of the upper Ismay stage carbonates of the Paradox Formation).

In late Paleozoic time, the Cataract lineament is best defined by the shoreline or zero edge of the Lower Permian (Leonardian) White Rim Sandstone (Baars, 1962). The northeast-trending Cataract lineament was sufficiently active to form a knife-edge facies change parallel with the Colorado River.

San Rafael Lineament

The San Rafael monocline trends in a northeasterly direction and defines the northwestern edge of the Paradox basin (Figures 8 and 9).

The San Rafael monocline borders the steeply dipping southeastern margin of the San Rafael Swell (the present-day counterpart of the older Paleozoic Emery uplift). Pennsylvanian strata thin abruptly onto the southeast flank of the Emery uplift. Both post-Pennsylvanian truncation and syndepositional intraformational thinning in Pennsylvanian rocks indicate that the Emery was a tectonically active, positive structural feature in late Paleozoic time. Permian clastic and carbonate strata on the crest of the Emery uplift nonconformably overlie Mississippian and Devonian rocks. Subsurface well data also indicates a post-Cambrian to pre–Late Devonian episode of uplift by marked thinning of Cambrian strata over the ancient highland.

The northern limit of Middle Pennsylvanian Paradox salt is coincident with the San Rafael lineament (Figure 8). Little is known about the Precambrian or Paleozoic tectonic or depositional history at this margin of the Paradox basin because of a near total lack of surface exposures and subsurface information. Both gravity and magnetic anomalies associated with the central segment of the Uncompahgre uplift dissipate gradually between the San Rafael and Cataract lineaments.

The youngest salt cycles in the Paradox basin are confined to a rhomboidal subbasin bounded by the San Rafael, Cataract, Uncompahgre, and Nequoia lineaments, which indicate late Desmoinesian subsidence of the area (Figure 8). Numerous boreholes into the older salt cycles of the Paradox Formation and in the pre-Pennsylvanian strata, however, commonly show high-angle reverse faults. These faults occur in that part of the Paradox basin where episodic convergence along the right-laterally displaced Uncompahgre master fault block should be expected. The San Rafael lineament is about 24 mi (38 km) shorter than the Four Corners platform, which causes compressional convergence of the rhomboidal-shaped basin along the Uncompahgre master fault and other northwest-trending faults in this quadrant (Figure 8).

THE PARADOX PULL-APART

In the past the Paradox basin has been referred to (in a generic sense) as a tectonic depression without consideration for the structural mechanics or kinematics that caused such a depression.

Szabo and Wengerd (1975) recognized several northwest-trending hinges or flex points in an overall extensional basin. Gorham (1975) and Baars (1976) referred to the Paradox basin as an aulacogen. Baars and Stevenson (1981, 1982) demonstrated that the Uncompahgre uplift was part of a continental-scale wrench fault system that was episodically active in Precambrian and Phanerozoic time. It has been shown that the northwest-trending basement fault system is intersected by an equally significant northeast-trending fracture system, which also has a Precambrian history (Kelley, 1955; Case and Joesting, 1972; Hite, 1975; Davis, 1978; Warner, 1978; Baars and Stevenson, 1981, 1982).

The tectonics and structural style of the Paradox basin strongly suggest that the basin has had a pull-apart history. Depositional sequences also support this hypothesis. The Paradox, however, is more than a pull-apart basin. It has had a polyphase tectonic history in which the basin evolved from a sag through an interior fracture basin to a wrench or shear basin (Kingston et al., 1983). Of these evolutionary stages, the wrench basin dominated in late Paleozoic time.

The tectonic fabric of the Paradox basin and Ancestral Rocky Mountain uplifts was shown to have been established in middle to late Precambrian time (about 1700 m.y. ago) (Baars, 1976). Recent mapping and age dating of the Precambrian in the San Juan Mountains in southwestern Colorado indicate that the date can be adjusted to 1600 m.y. (Baars and Ellingson, 1984). The recently adjusted date fits into a worldwide Precambrian orogenic episode named by Salop (1983) as the Vyborgian diastrophism of the second order of the lower Baikalian era (Neoproterozoic); it is the middle Proterozoic of James (1972). A three-stage sequence showing the development of the San Juan kink is schematically depicted in Figure 5. Most strike-slip movement occurred during Precambrian tectonism where

convergence of blocks caused thrusting at this restraining bend. By Pennsylvanian time, divergent strike-slip movement allowed extensional faulting of the bend as it released.

Brittle fracturing of basement rocks resulted in an orthogonal pattern of northwest- and northeast-trending fractures. Both fracture sets demonstrate strike-slip movement. The sense of displacement on the first-order northwest-trending fracture sets is right-lateral, and they are usually the dominant or master faults. The intersecting northeast-trending sets of first-order fractures are left-lateral and usually antithetic to the northwest-trending set.

From Late Cambrian through Osagian (Tournaisian) time, the eastern Colorado Plateau varied little, evolving through several cycles from an interior sag to early stages of an interior fracture basin. Faulting was minor, occurring along preexisting Precambrian fractures. By Middle Pennsylvanian time, however, the eastern Colorado Plateau had rapidly evolved into a strike-slip or pull-apart basin (Figure 13).

The Paradox differs from other basins in many respects and for that reason is aptly named. Of 97 Paleozoic evaporite basins known in the world today, the Paradox and Eagle basins are the *only* salt basins of Middle Pennsylvanian age (Zharkov, 1981). Few examples are known of preserved wrench basins older than Tertiary. Typically, most pre-Tertiary wrench basins have evolved to wrenched fold belts (Kingston et al., 1983). Wrenching occurred episodically along the Olympic–Wichita lineament, and the Middle Pennsylvanian Paradox pull-apart basin has been beautifully preserved because of arrested movement along the wrench system. Perhaps the most significant observation concerning the tectonic origin of the Paradox basin is that it developed along intersecting active basement fractures, not just along a solitary master fault system as has been indicated for most other pull-apart basins. The conjugate fracture system played an integral part in the tectogenesis of the Paradox basin. To our knowledge, only Mann and Burke (1982) have addressed the significance of intersections of conjugate strike-slip faults to the formation and development of pull-apart basins. The sequence for Pennsylvanian tectonic development of the Paradox basin is discussed below.

Early Middle Pennsylvanian (Atokan)

Divergent right-lateral strike-slip motion along the Uncompahgre–San Luis master fault system formed a releasing bend at the intersection with the Coconino–House Creek lineament (Figures 5C and 13). Much of the releasing strain was transferred along the left-lateral strike-slip motion of the intersecting House Creek fault southwestward to the Four Corners lineament. Extensional faulting progressed rapidly through Atokan time where, by early Desmoinesian time, the pull-apart geometry of the basin was well established (Figure 13).

The oldest and thickest salt cycles of the Paradox Formation were confined to the narrow, elongate rhomb-graben developed adjacent to the front of the Uncompahgre uplift (Figure 13). Southwest of the Nequoia–Abajo arch, shallow restricted marine carbonates and sulfates were deposited on a subsiding platform. Normal marine conditions prevailed southwest of the Four Corners

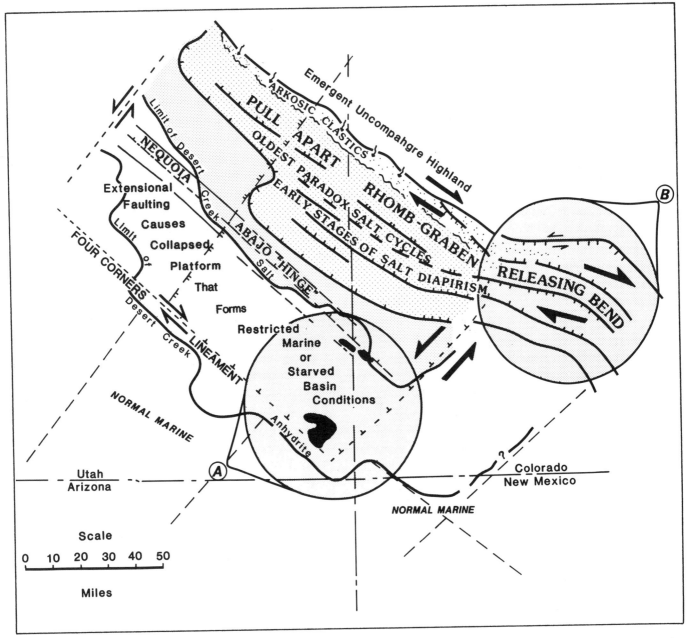

Figure 13—Paleogeologic map of the Paradox basin at the close of Desert Creek (Cherokean–middle Desmoinesian) time. Distribution of lateral time-equivalent salt cycles of the Desert Creek Stage are tectonically controlled at the hinge along the Nequoia–Abajo lineament. The Four Corners lineament controls the lateral southwestern limits of all Desert Creek anhydrite facies. Normal marine conditions prevailed outside the perimeter of the Paradox pull-apart basin. Location of Figures 14 and 17 is shown by the area encircled and labeled "A" and Figure 5 by the area labeled "B."

lineament and southeast of the Four Corners platform (Figure 13). Relatively nonfeldspathic quartzose clastic sediments accumulated along the southern border of the San Luis uplift (Baars and Stevenson, 1984) (Figure 6).

Early to Middle Desmoinesian Time

Smaller basement pull-aparts developed due to orthogonal spreading along intersecting northwest- and northeast-trending transform faults. As extensional faulting progressed across the basin floor, a complex network of coalescing subbasins developed. The narrow trough along the Uncompahgre master fault deepened, and extremely

thick salt cycles continued to accumulate. Arkosic clastic sediments and talus breccias or fanglomerates began to be shed from the emergent Uncompahgre highland. Continued collapse of the narrow trough caused grabens to develop within grabens (Figures 10 and 13). Early diapirism of the Paradox salt was initiated by synorogenic space reduction and by asymmetric differential loading of the accumulating Cutler arkose transported from the Uncompahgre uplift (Figure 11).

Southwestward across the intrabasinal Nequoia–Abajo hinge the Paradox salt cycles are markedly thinner. The Four Corners lineament actively controlled the distal limits of lower to middle Desmoinesian evaporites (Figure 8). Normal marine conditions prevailed outside the bounding faults of the Paradox pull-apart basin (Figure 13).

Middle to Late Desmoinesian Time

The degree of extensional divergence waned during middle and late Desmoinesian time and the rate of basinal collapse slowed considerably. In the deep rhomb-graben along the Uncompahgre front, salt diapirism accelerated because of continued basin collapse and continued differential loading of arkosic clastic sediments being deposited in the trough. Salt continued to be deposited northeast of the Nequoia–Abajo hinge, and restricted shallow marine carbonates, including many bioherms and sulfate rocks, were deposited on the southwest platform (Figure 13). Mild orthogonal faulting of the subsiding platform controlled the distribution of hydrocarbon reservoirs in this restricted marine setting (Baars and Stevenson, 1982). On the southwestern shelf, over 400 million bbl of oil and 500 billion cu ft of gas have been produced from carbonate stratigraphic traps (mostly in phylloid algal bioherms) (Figure 14). Normal marine conditions prevailed outside of the Paradox basin.

During latest Desmoinesian time, the northeastern quadrant of the Paradox basin, where the youngest salt beds were accumulating, continued to collapse. Restricted marine conditions prevailed elsewhere within the Paradox basin, and arkosic clastic sediments continued to amass adjacent to the Uncompahgre uplift. By the end of Desmoinesian time, normal marine conditions returned to the region, as divergent strike-slip movement waned.

Late Pennsylvanian–Early Permian

Although the eastern part of the basin continued to subside at a faster rate than the western half, normal marine conditions prevailed. The Honaker Trail Formation is considerably thicker northeast of the active Nequoia–Abajo hinge. Marine carbonate and clastic rocks intertongue with the prograding Cutler arkosic redbeds (Figure 8). Rocks of Virgilian age are known only in the eastern part of the basin.

Throughout Wolfcampian time, arkosic redbeds prograded westward across the Paradox basin and intertongued with marine sandstone and carbonate rocks (Baars, 1962). Strandlines between fluvial and marine facies were influenced by the orthogonal basement framework. The Cataract lineament and the conjugate intersecting Four Corners and Nequoia–Abajo lineaments controlled several depositional boundaries between facies of Wolfcampian and early Leonardian sediments.

GEOLOGICAL CRITERIA FOR PULL-APART BASIN DEVELOPMENT AND MECHANICS OF ORIGIN

Pull-apart basins have been recognized along major strike-slip faults worldwide since the term was first introduced by Burchfiel and Stewart in 1966 for their interpretation of Death Valley, California. Crowell (1974) interpreted the Salton Trough to be a classic pull-apart basin. In their

discussion of the evolution of pull-apart basins, Aydin and Nur (1982) cited evidence from scores of Quaternary basins along active strike-slip faults. The reader is referred to Kingma (1958), Quennell (1958), Clayton (1966), Freund (1971), Crowell (1974), Garfunkel and Freund (1981), and Hempton et al. (1983) for overviews of ancient pull-apart basins. Kinematics of strike-slip faults have been studied, most notably, by Tchalenko (1970), Rodgers (1980), and Seagall and Pollard (1980). Moody and Hill (1956), Moody (1973), Wilcox et al. (1973), Harding (1974), and Reading (1980) provide the interested reader with a complete review of strike-slip structural features and related sedimentation. Mann et al. (1983) made a comparative study of active and ancient pull-apart basins and have developed qualitative models for their continuous development.

Theoretical pull-apart models have been compared to numerous active and ancient strike-slip basins around the world. The following conclusions and observations are a synthesis of geologic criteria for pull-apart basin development.

Pull-apart basins nucleate at releasing bends of oblique fault segments (Figure 15A) along active strike-slip master faults (Freund, 1971; Crowell, 1974; Mann et al., 1983). Pull-apart basins commonly display straight margins where they are parallel to the major wrench faults and irregular borders along the pull-apart margins. The master faults are complicated in detail, although generally straight on a regional scale. The zone consists of braided fault slices, thrust blocks and detachment faults. Extensional and compressional tectonism results in complicated uncon-formities and sedimentation patterns, especially around their margins (Crowell, 1974) (Figure 16).

In rigid intracontinental settings where master fault sets are significantly nonparallel, wider pull-aparts are formed whereby several pull-aparts may nucleate simultaneously as part of a series on a segment of the principal displacement fault zone. Mann et al. (1983) determined that divergence of 10° or more from the theoretical strike-slip plane is necessary for multiple pull-aparts to develop. Pull-aparts developed along oblique-divergent wrench faults tend to be closely spaced and may coalesce to form a wider basin. Aydin and Nur (1982) determined that pull-aparts do not steadily widen as they lengthen. Their widths remain fixed by the width of the releasing bend fault segment. However, they did not consider orthogonal fracture sets and their effect on kinematics or basin geometry.

Non-parallelism of master strike-slip faults produce local basin edge deformation. At restraining bends, crowding and uplift result within convexities of the bounding lithospheric blocks. Severe deformation leads to folds, upthrusts, or bulges (Figures 15A and B). At releasing bends, gaps develop in the form of one-sided grabens or rifts (Figures 15A and C).

Floors of present-day pull-aparts show circular-shaped depressions or "deeps" suggesting that extension between the overlapped master faults is not uniform. Alternatively, in basins with large separation along master faults, a diagonal array of smaller pull-apart subbasins may develop (Aydin and Nur, 1982). Mann et al. (1983) and Mann and Burke (1982) suggested that smaller basement pull-aparts may result from orthogonal spreading along transform faults that relay extension across the floor of the basin. In

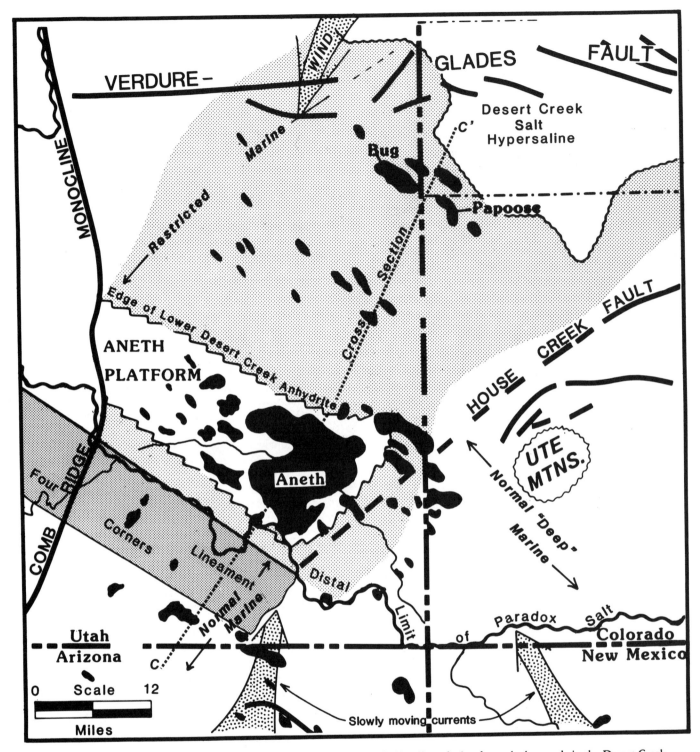

Figure 14—Map showing the location of Aneth, Bug, and Papoose Canyon fields, all producing from algal mounds in the Desert Creek Stage. Note how the lower Desert Creek anhydrite wraps around the Aneth platform. Other fields producing from Paradox carbonates are shown in black.

other words, the center of rhomboidal pull-apart basins are not necessarily the site of greatest extension and subsidence. Basin floor subsidence is related to combinations of normal and strike-slip faulting, and multiple smaller pull-apart subbasins remain difficult to distinguish without detailed subsurface structural and stratigraphic information. To this

end, the elastic dislocation theory of Rodgers (1980) demonstrates the close correspondence between predicted and theoretical fault patterns.

Increased extension of rhomboidal pull-aparts produce deep asymmetric topographic depressions bounded by prominent faults that define the edge of the basin where

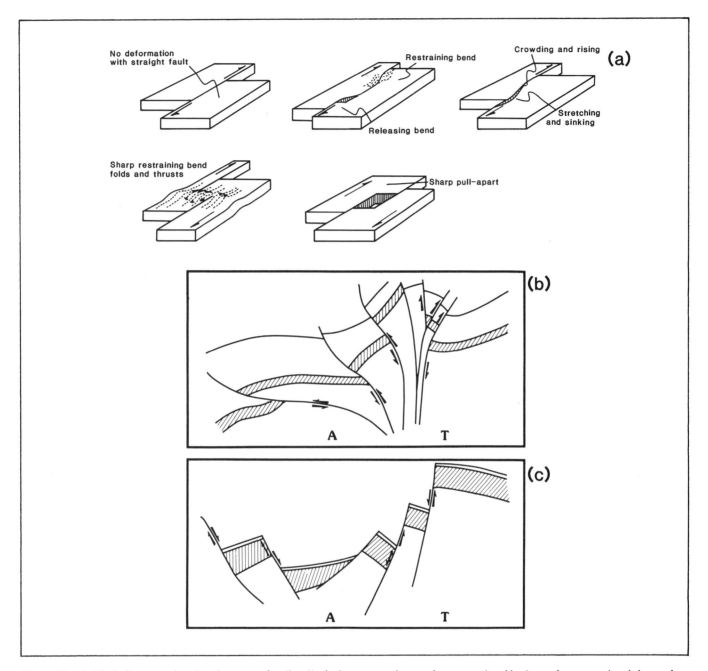

Figure 15—A. Block diagrams showing the types of strike-slip fault patterns that produce extensional basins and compressional thrusted or folded terrains. After Crowell (1974). B. Schematic cross section of a restraining bend where upthrusts or "flower structures" would be expected. C. Schematic cross section of a releasing bend where tensional downfaulted or rifted structures would be expected.

steep gradients are associated with complex zones of oblique slip faults. The rate of subsidence commonly exceeds the rate of sedimentation, resulting in fault-bounded deep marine, marine, or lacustrine depositional environments. Vertical displacement is usually greater on one of the master faults, causing tilting of the basin and development of asymmetric facies. The sedimentary sequence commonly exhibits thick, coarse-grained facies (fanglomerates) along the lateral margin of the steep master fault and finer grained facies along the gentler gradient of the complimentary master fault. Conversely, faults defining the ends of rhomboidal pull-apart basins normally show little or

no surface expression and cause only subtle changes in the basinal facies (Figure 16). Lacustrine or flood basin deposits commonly develop toward the center of the tectonically subsiding depocenter where sediments may be cut off or restricted from the prevailing depositional process (either clastic or carbonate sedimentation).

Pull-apart basins are recognized as important petroleum provinces. Strike-slip fault systems provide excellent conditions for the generation of hydrocarbons (source), reservoir rocks, seals, and a variety of structural and stratigraphic traps (Harding, 1974). During the early extensional stage of intracontinental basins, starved

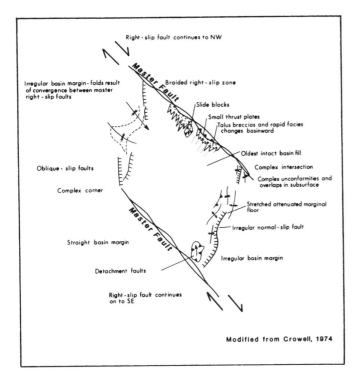

Right-slip fault continues to NW

Master Fault

Irregular basin margin - folds result
of convergence between master
right - slip faults

Braided right - slip zone

Slide blocks

Small thrust plates

Talus breccias and rapid facies
changes basinward

Oldest intact basin fill

Complex intersection

Complex unconformities and
overlaps in subsurface

Oblique - slip faults

Complex corner

Master Fault

Stretched attenuated marginal
floor

Irregular normal - slip fault

Straight basin margin

Detachment faults

Irregular basin margin

Right-slip fault continues
on to SE

Modified from Crowell, 1974

Figure 16—Idealized pull-apart basin. The model is oriented for direct comparison to the Paradox basin; see Figures 8, 9, and 13. All labeled features apply directly to the features in the Paradox basin. Modified from Crowell (1974).

conditions may allow organic-rich sediments to be deposited under reducing conditions. Rapid burial and subsidence promotes source rock maturation where higher ambient temperatures and slightly higher geothermal gradients can be expected (Wilde et al., 1978). As the basin fills, prograding continental facies tend to crowd or displace marine sediments. Lacustrine sediments, evaporite sequences, oil shale, lignite, and coal are other facies commonly found in strike-slip zones. Also, mineralization occurs along fracture zones and ore bodies are commonly located along major lineaments (Reading, 1980).

ECONOMIC SIGNIFICANCE

Introduction

The greater Aneth field is located in extreme southeastern Utah in the Blanding subbasin (Figure 9) near the intersection of the Four Corners lineament and the Coconino–House Creek lineament (Figure 14). The field produces primarily from calcareous phylloid algal buildups developed in the Desert Creek Stage of the Paradox Formation (Figure 17A). The Aneth complex is best defined by the 150-ft (45-m) isopach of the Desert Creek interval. Maximum thickness exceeds 64 m. Calculated estimates of over 1 billion bbl of oil *in place* truly qualifies Aneth as a world class giant oil field. To date, some 500 producing wells have cumulatively recovered over 320 million bbl of oil and 310 billion cu ft of associated gas from a field of about 75 mi² (195 km²).

Nearly 30 years after its discovery, however, the greater Aneth oil field complex remains somewhat of a paradox. Peterson and Ohlen (1963), McComas (1963), Peterson

(1966), and Peterson and Hite (1969) have discussed the overall geometry and internal depositional characteristics of Aneth field. Most previous workers have related the cyclic episodes of carbonate and evaporite sedimentation that characterize the upper Carboniferous to global eustatic sea level changes brought on by glaciation in the southern hemisphere. They did not realize the significance of syndepositional tectonism.

For decades following the discovery of Aneth, it was believed that deeper water semirestricted conditions extended continuously northeastward from Aneth into thick halite equivalents of the deep Paradox basin. Many unsuccessful wildcat wells were drilled northeast of Aneth, adding credence to the deeper water, restricted marine hypothesis.

The discovery of Papoose Canyon Field in 1970 and Bug Field in 1980 as well as other wells drilled in the Blanding subbasin, now provide data for an explanation as to how the Aneth complex developed. As in the case of Aneth, both the Papoose Canyon and Bug fields are productive from mounds developed in the lower Desert Creek zone (Figures 14 and 17A). However, the mounds differ from those at Aneth in that they developed in *very shallow water* in the upper reaches of the intertidal zone and may have undergone periodic subaerial exposure. The rocks in the two areas definitely exhibit distinct different diagenetic histories (Roylance, 1984). What was postulated for years as deeper water deposition of nonporous sediments was found to be a much shallower depositional facies than previously realized.

Stratigraphy of Aneth Field

At Aneth the lower part of the Desert Creek Stage is composed of a complex of stacked phylloid algal mounds. Where these algal mounds have been leached, vugular intercrystalline and interparticle porosity is well developed. High permeabilities occur in well-interconnected vugular porosity (McComas, 1963). The upper part of the Desert Creek is composed of oolite sand bars and pelletal bioclastic buildups that generally record a shoaling, high-energy environment (Figure 17B). Oomoldic porosity is commonly well developed, yet poorly interconnected, yielding low permeabilities (McComas, 1963). The oolitic pelletal facies tend to rim the inner bioclastic and lime mudstone facies (Peterson and Ohlen, 1963). The paleotopographic relief is steep on the northeast, east, and southeast flanks of the field where the limestone biohermal complex changes abruptly to dark, argillaceous dolomitic mudstone and bedded anhydrite. The dark carbonate mud and anhydrite indicate deeper water, restricted marine conditions. Maximum restriction of marine water circulation occurred during deposition of the two bedded anhydrite zones in the Desert Creek stage. The distal limits of these anhydrite beds coincide with paleostructural features that controlled their areal distribution and delineate the Aneth platform (Figure 14).

Facies changes on the west–northwest flank of the Aneth bank complex are much more gradual in comparison to the abrupt changes on the other flanks. The limestone grades to calcareous bioclastic platformal wackestone and isolated algal mounds. Peterson and Ohlen (1963) speculated that asymmetry of the facies distribution was related to paleoclimatologic effects. They considered that prevailing wind from the north–northeast influenced a prevalent open

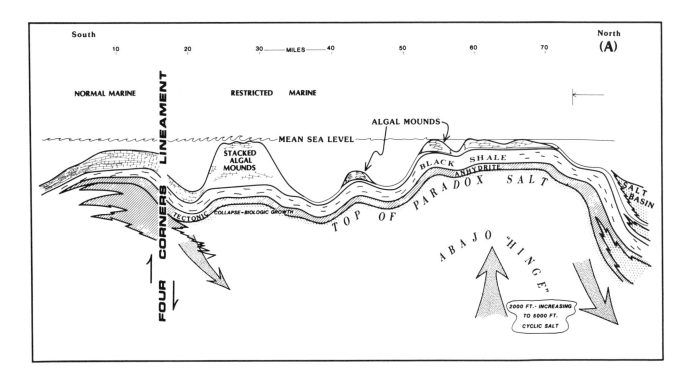

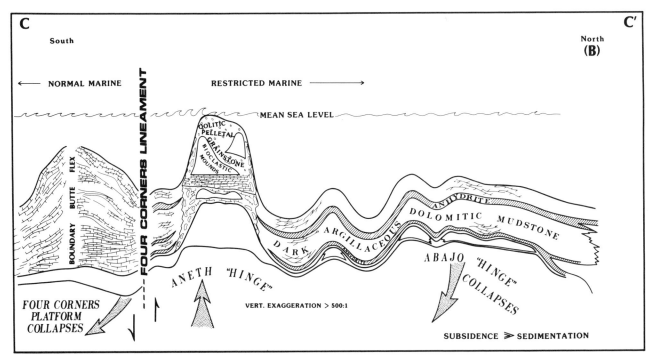

Figure 17—Stratigraphic cross section (C–C′ in Figures 9, 13, and 14) across the Blanding subbasin. A. During lower Desert Creek deposition, algal mounds stacked vertically on the tectonically collapsing Aneth platform while thin algal mounds (Bug–Papoose) coalesced on the emergent Abajo arch. Restricted marine conditions prevailed north and east of the southern margin of the Paradox pull-apart basin (Figure 13). Stagnant evaporative lagoonal conditions existed northeast of the Abajo arch where thick salt deposits accumulated in the rapidly subsiding trough. B. By the close of Desert Creek deposition, the shallow carbonate shelf rotated whereby the Abajo hinge collapsed or sagged at a faster rate than the Aneth platform. Shoaling conditions prevailed at Aneth, where oolite sand bars and bioclastic mounds accumulated atop the thick algal mound complex. Deeper water semirestricted conditions prevailed northeast of Aneth where dark, argillaceous dolomitic mud and subaqueous anhydrite beds were deposited. Anhydrite beds surround the Aneth platform and are tectonically controlled by the Four Corners lineament and House Creek fault.

marine current on the southwest or seaward side. It is unlikely that a restricted marine facies could have developed on both seaward and windward sides as postulated by Peterson and Ohlen (1963).

Tectonic Synthesis

Both the Four Corners lineament and the Coconino–House Creek lineaments have been shown to be active structural components through Pennsylvanian time and to define the southwest corner of the Paradox basin (Figure 8).

The tectonic style of the Blanding subbasin is similar to the tectonic style of the greater Paradox basin (Figure 9). Extensional faulting apparently continued during Desert Creek time causing the southwestern shelf of the Paradox to subside at varying rates, creating a seesaw motion of the platform.

In Desert Creek time, the Abajo arch appears to have acted as a fulcrum or tectonic hinge whereby the rate of basinal subsidence was greatest to the northeast toward the evaporite trough. The shallow marine shelf that extended southwestward toward Aneth subsided at a slower differential rate. Faults along the Four Corners lineament acted complimentary to the Abajo hinge (Figure 17A). Extensional subsidence also occurred along the northwest flank of the bounding orthogonal House Creek fault (Figure 13). Aneth is situated at the collapsed intersection of the Four Corners and House Creek fracture sets (Figures 8 and 14). The rate of algal growth accelerated apparently because of space availability as the Aneth platform collapsed. Paleoecologic conditions remained ideal for algal mounds to coalesce on the Abajo arch and stack vertically at Aneth as tectonically controlled paleobathymetry maintained the necessary ecologic parameters, as reviewed by Baars and Stevenson (1982) (Figure 17A). The growth of the large algal complex at Aneth further restricted waters that reached the Abajo arch. Mounds that developed on the flank of the Abajo arch aided in restricting circulation of marine waters, and hypersaline evaporative conditions prevailed northeast of these barriers (Figures 14 and 17A).

Toward the end of Desert Creek time, the Blanding block rotated and the Abajo hinge collapsed at a faster rate than did the Aneth hinge (Figure 17B). Evaporites and dark carbonate muds covered the Blanding shelf except that the Aneth platform remained a slightly positive area where bioclastic mounds and thick oolitic sand bars developed. Evaporite deposits and dark mudstones now surround the Aneth platform except that they do not occur across the Four Corners lineament to the southwest of the platform (Figure 17B).

Thus, Aneth is the result of being in the right place at the right time. The rapid growth of phylloid algae kept pace with the subsidence of the platform. As subsidence slowed, thick grainstone bars accumulated around the platform margin, while restricted marine water prevailed on three sides of the Aneth platform and further isolated the Blanding and greater Paradox evaporite basin. The steepened northeast, east, and southeast flanks of the Aneth complex are due to the proximity of intersecting basement fractures that actively affected depositional rates. The thick carbonate complex developed because of structural instability

of this part of the pull-apart basin, probably independently of prevailing winds or paleocurrent directions.

CONCLUSIONS

The Paradox basin is a pull-apart basin of major proportions that developed along the northwest-trending Olympic–Wichita wrench fault system. The basic tectonic fabric originated in Proterozoic time (about 1600–1700 m.y. ago) and was rejuvenated during Cambrian, Devonian, and Mississippian times. A major east–west extensional pulse in Middle Pennsylvanian time caused the Ancestral Rocky Mountains to emerge and the Paradox basin to subside along preexisting basement structures. The master faults were the northwest-trending San Luis and Uncompahgre fault systems along the northeast margin of the basin and the Four Corners lineament along the southwest margin. Oblique strike-slip faulting along the Uncompahgre–San Luis segment throughout Middle Pennsylvanian time allowed the structural extension of the Paradox basin. Divergence occurred along the northeast-trending Four Corners platform bounded by the House Creek and Hogback transform faults at the south end of the basin. Convergence closed the basin on the north coincident with the Cataract and San Rafael lineaments. The tectonically induced pull-apart basin caused restricted marine evaporative environments in the basin, and thick halite and anhydrite sequences filled the deeper faulted eastern trough and western subbasins. Distribution of specific evaporite cycles was controlled by structural lineaments.

A relatively stable shelf developed along the southwestern flank of the basin. The shelf was bounded on the south by the Four Corners lineament and on the north by the parallel Nequoia–Abajo arch. Significant petroleum reserves occur in carbonate bioherms along the margins of the platform. The Aneth algal biohermal complex occurs at the southern margin of the shelf, while the intertidal algal complex of the Bug Field trend lies along the Nequoia–Abajo arch to the north. Specific reservoir trends and evaporite occurrences are the result of paleotectonic influences.

ACKNOWLEDGMENTS

We thank L. A. Warner, W. R. Muehlberger, Edwin K. Maughan, W. A. Thomas, R. J. Hite, and J. A. Peterson for reviewing the manuscript and providing constructive criticism.

Precision Graphics and Printing drafted the illustrations and Sharon Kimball typed the manuscript.

REFERENCES CITED

Anderson, C. A., 1967, Precambrian wrench fault in central Arizona: U.S. Geological Survey Professional Paper 575-C, p. 60–65.

Armstrong, A. K., and B. L. Mamet, 1977, Biostratigraphy and paleogeography of the Mississippian System in northern New Mexico and adjacent San Juan Mountains of southwestern Colorado: New Mexico Geological Society 28th Field Conference Guidebook, p. 111–127.

Aydin, A., and A. Nur, 1982, Evolution of pull-apart basins and their scale independence: Tectonics, v. 1, n. 1, p. 91–105.

Baars, D. L., 1962, Permian System of the Colorado Plateau: AAPG Bulletin, v. 46, p. 149–218.

———, 1966, Pre-Pennsylvanian Paleotectonics—key to basin evolution and petroleum occurrences in Paradox basin: AAPG Bulletin, v. 50, p. 2082–2111.

———, 1976, The Colorado Plateau aulacogen: key to continental-scale basement rifting: Proceedings of the Second International Conference on Basement Tectonics, p. 157–164.

Baars, D. L., and P. D. See, 1968, Pre-Pennsylvanian stratigraphy and paleotectonics of the San Juan Mountains, southwestern Colorado: Geological Society of America Bulletin, v. 79, p. 333–350.

Baars, D. L., and G. M. Stevenson, 1981, Tectonic evolution of the Paradox basin, Utah and Colorado, *in* D. L. Wiegand, ed., Geology of the Paradox basin: Rocky Mountain Association of Geologists Guidebook, p. 23–31.

———, 1982, Subtle stratigraphic traps in Paleozoic rocks of Paradox basin, *in* M. Halbouty, ed., Deliberate search for the subtle trap: AAPG Memoir 32, p. 131–158.

———, 1984, The San Luis uplift, Colorado and New Mexico—an enigma of the Ancestral Rockies: The Mountain Geologist, v. 21, n. 2, p. 57–67.

Baars, D. L., and J. A. Ellingson, 1984, Geology of the western San Juan Mountains, *in* D. C. Brew, ed., Four Corners Geological Society Guidebook: 37th Annual Meeting Rocky Mountain Section, Geological Society of America, p. 1–32.

Brill, K. G., Jr., and O. W. Nuttli, 1983, Seismicity of the Colorado lineament: Geology, v. 11, p. 20–24.

Burchfiel, B. C., and J. H. Stewart, 1966, "Pull-apart" origin of the central segment of Death Valley, California: Geological Society of America Bulletin, v. 77, p. 439–442.

Case, J. E., and H. R. Joesting, 1972, Regional geophysical investigations in the central Colorado Plateau: U.S. Geological Survey Professional Paper 736, 31 p.

Case, J. E., H. R. Joesting, and P. E. Byerly, 1963, Regional geophysical investigations in the LaSal Mountain area, Utah and Colorado: U.S. Geological Survey Professional Paper 316-F, p. 91–116.

Clayton, L., 1966, Tectonic depressions along the Hope Fault, a transcurrent fault in North Canterbury, New Zealand: New Zealand Journal of Geology and Geophysics v. 9, p. 95–104.

Crowell, J. C., 1974, Origin of late Cenozoic basins in southern California, *in* R. H. Dott and R. H. Shaver, eds., Modern and ancient geosynclinal sedimentation: Society of Economic Paleontologists and Mineralogists Special Publication 19, p. 292–303.

Davis, G. H., 1978, Monocline fold pattern of the Colorado Plateau, *in* V. Matthews, ed., Laramide folding associated with basement block faulting in the western U.S.: Geological Society of America Memoir 151, p. 215–233.

Elston, D. P., and E. M. Shoemaker, 1960, Late Paleozoic and early Mesozoic structual history of the Uncompahgre front, *in* K. G. Smith, ed., Geology of the Paradox basin fold and fault belt: Four Corners Geological Society 3rd Field Conference Guidebook, p. 47–55.

Frahme, C. W., and E. B. Vaughn, 1983, Paleozoic geology and seismic stratigraphy of the northern Uncompahgre front, Grand County, Utah, *in* J. D. Lowell, ed., Rocky Mountain foreland basins and uplifts: Rocky Mountain Association of Geologists Guidebook, p. 201–211.

Freund, R., 1971, The Hope Fault: a strike-slip fault in New Zealand: New Zealand Geological Survey Bulletin, v. 86, p. 1–49.

Garfunkel, Z. I., and R. Freund, 1981, Active faulting in the Dead Sea Rift: Tectonophysics, v. 80, p. 1–26.

Gorham, F. D., Jr., 1975, Tectogenesis of the central Colorado Plateau aulacogen, *in* J. Fassett, ed., Canyonlands country: Four Corners Geological Society 8th Field Conference Guidebook, p. 211–216.

Harding, T. D., 1974, Petroleum traps associated with wrench faults: AAPG Bulletin, v. 58, p. 1290–1304.

Hempton, M. R., L. A. Dunne, and J. F. Dewey, 1983, Sedimentation in a modern strike-slip basin, southeastern Turkey: Journal of Geology, v. 91, p. 401–412.

Heyman, O. G., 1983, Distribution and structural geometry of faults and folds along the northwestern Uncompahgre uplift, western Colorado and eastern Utah, *in* W. R. Averett, ed., Northern Paradox basin–Uncompahgre uplift: Grand Junction Geological Society Guidebook, p. 45–57.

Hite, R. J., 1960, Stratigraphy of the saline facies of the Paradox member of the Hermosa Formation of southeastern Utah and southwestern Colorado, *in* K. G. Smith, ed., Geology of the Paradox basin fold and fault belt: Four Corners Geological Society 3rd Field Conference Guidebook, p. 86–89.

———, 1975, An unusual northeast-trending fracture zone and its relations to basement wrench faulting in northern Paradox basin, Utah and Colorado, *in* J. Fassett, ed., Canyonlands country: Four Corners Geological Society 8th Field Conference Guidebook, p. 217–223.

Humphrey, J. R., and I. G. Wong, 1983, Recent seismicity near Capitol Reef National Park, Utah, and its tectonic implications: Geology, v. 11, n. 8, p. 447–451.

James, H. L., 1972, American Committee on Stratigraphic Nomenclature, Note 40—Subdivision of Precambrian: an interim scheme to be used by U.S. Geological Survey: AAPG Bulletin, v. 56, p. 1128–1133.

Johnson, V. C., 1983, Preliminary aeromagnetic interpretation of the Uncompahgre uplift and Paradox basin, west-central Colorado and east-central Utah, *in* W. R. Averett, ed., Northern Paradox basin–Uncompahgre uplift, Grand Junction Geological Society Guidebook, p. 67–70.

Kay, M., 1951, North American geosynclines: Geological Society of America Memoir 48, 132 p.

Kelley, V. C., 1955, Regional tectonics of the Colorado Plateau and relationship to the origin and distribution of uranium: University of New Mexico Publications in

Geology, n. 5, 120 p.

Kelley, V. C., and J. N. Clinton, 1960, Fracture systems and tectonic elements of the Colorado Plateau: University of New Mexico Publications in Geology, n. 6, 104 p.

Kingma, J. T., 1958, Possible origin of piercement structures, local unconformities and secondary basins in the eastern geosyncline, New Zealand: New Zealand Journal of Geology and Geophysics, v. 1, p. 269–274.

Kingston, D. R., C. P. Dishroon, and P. A. Williams, 1983, Global basin classification system: AAPG Bulletin, v. 67, n. 12, p. 2175–2193.

Maarouf, A., 1983, Relationship between basement faults and Colorado Plateau drainage, *in* W. R. Averett, ed., Northern Paradox basin–Uncompahgre uplift, Grand Junction Geological Society Guidebook, p. 59–62.

Mann, P., and K. Burke, 1982, Basin formation at intersections of conjugate strike-slip faults: examples from southern Haiti (abs.): Geological Society of America Abstracts with Programs, v. 14, p. 555.

Mann, P., M. R. Hempton, D. C. Bradley, and K. Burke, 1983, Development of pull-apart basins: Journal of Geology, v. 91, p. 529–554.

McComas, M. R., 1963, Productive core analysis characteristics of carbonate rocks in the Four Corners area, *in* R. O. Bass and S. L. Sharps, eds., Shelf carbonates of the Paradox basin: Four Corners Geological Society 4th Field Conference Guidebook, p. 149–156.

Moody, J. D., 1973, Petroleum exploration aspects of wrench-fault tectonics: AAPG Bulletin, v. 57, p. 449–476.

Moody, J. D., and M. J. Hill, 1956, Wrench fault tectonics: Geological Society of America Bulletin, v. 67, p. 1207–1246.

Mutschler, F. E., and R. J. Hite, 1969, Origin of the Meander anticline, Cataract Canyon, Utah, and basement fault control of Colorado River drainage (abs.): Geological Society of America Abstracts with Programs, p. 57–58.

Peterson, J. A., 1966, Stratigraphic vs. structural controls on carbonate-mound hydrocarbon accumulation, Aneth area, Paradox basin: AAPG Bulletin, v. 50, p. 2068–2081.

Peterson, J. A., and H. R. Ohlen, 1963, Pennsylvanian shelf carbonates, Paradox basin, *in* R. O. Bass and S. L. Sharps, eds., Shelf carbonates of the Paradox basin: Four Corners Geological Society 4th Field Conference Guidebook, p. 65–79.

Peterson, J. A., and R. J. Hite, 1969, Pennsylvanian evaporite–carbonate cycles and their relation to petroleum occurrence, southern Rocky Mountains: AAPG Bulletin, v. 53, p. 884–908.

Potter, D. B., Jr., and G. E. McGill, 1978, Field analysis of a pronounced topographic lineament, Canyonlands National Park, Utah: Proceedings of the Third International Conference on Basement Tectonics, p. 169–176.

Quennell, A. M., 1958, The structural and geomorphic evolution of the Dead Sea Rift: Quarterly Journal of the Geological Society of London, v. 114, p. 2–24.

Reading, H. G., 1980, Characteristics and recognition of strike-slip fault systems, *in* P. F. Ballance and H. G.

Reading, eds., Sedimentation in oblique-slip mobile zones: International Association of Sedimentologists Special Publication 4, p. 7–26.

Rodgers, D. A., 1980, Analysis of pull-apart basin development produced *en echelon* strike-slip faults, *in* P. F. Ballance and H. G. Reading, eds., Sedimentation in oblique-slip mobile zones: International Association of Sedimentologists Special Publication 4, p. 27–41.

Roylance, M. H., 1984, Significance of botryoidal aragonite in early diagenetic history of phylloid algal mounds in Bug and Papoose Canyon fields, southeastern Utah and southwestern Colorado (abs.): AAPG Bulletin, v. 68, n. 4, p. 523.

Salop, L. J., 1983, Geological evolution of the Earth during the Precambrian: New York, Springer-Verlag, 459 p.

Seagall, P., and D. D. Pollard, 1980, Mechanics of discontinuous faults: Journal of Geophysical Research, v. 85, n. B8, p. 4337–4350.

Shoemaker, E. M., R. L. Squires, and M. J. Abrams, 1978, Bright Angel and Mesa Butte fault systems of northern Arizona, *in* R. B. Smith and G. P. Eaton, eds., Cenozoic tectonics and regional geophysics of the Western Cordillera: Geological Society of America Memoir 152, p. 341–367.

Spoelhof, R. W., 1976, Pennsylvanian stratigraphy and paleotectonics of the western San Juan Mountains, southwestern Colorado, *in* R. C. Epis and R. W. Weimer, eds., Studies in Colorado field geology: Professional Contributions of Colorado School of Mines, p. 159–179.

Stevenson, G. M., and D. L. Baars, 1977, Pre-Carboniferous paleotectonics of the San Juan basin, *in* W. E. Fassett and H. L. James, eds., San Juan basin, III: New Mexico Geological Society 28th Field Conference Guidebook, p. 99–110.

Stokes, W. L., 1948, Geology of the Utah–Colorado salt dome region, with emphasis on Gypsum Valley, *in* Guidebook to the Geology of Utah, 2: Utah Geological Society, 50 p.

Szabo, E., and S. A. Wengerd, 1975, Stratigraphy and tectogenesis of the Paradox basin, *in* J. Fassett, ed., Canyonland country: Four Corners Geological Society 8th Field Conference Guidebook, p. 193–210.

Tchalenko, J. S., 1970, Similarities between shear zones of different magnitudes: Geological Society of America Bulletin, v. 81, p. 1625–1640.

Tewksbury, B. J., 1981, Polyphase deformation and contact relationships of the Precambrian Uncompahgre Formation, Needle Mountains, southwestern Colorado: Ph.D. thesis, University of Colorado, Boulder, 294 p.

———, 1985, Revised interpretation of the allochthonous rocks of the Uncompahgre Formation, Needle Mountains, Colorado: Geological Society of America Bulletin, v. 96, p. 224–232.

Tweto, O., and P. K. Sims, 1960, Relation of the Colorado mineral belt to Precambrian structure, *in* Short papers in the geological sciences: U.S. Geological Survey Professional Paper 400-B, p. B8–B10.

———, 1963, Precambrian ancestry of the Colorado mineral belt: Geological Society of America Bulletin, v. 74, n. 8, p. 991–1014.

Warner, L. A., 1978, The Colorado lineament—a middle Precambrian wrench fault system: Geological Society of America Bulletin, v. 89, p. 161–171.

Weimer, P. C., 1982, Upper Cretaceous stratigraphy and tectonic history of the Ridgeway area, northwestern San Juan Mountains, Colorado: The Mountain Geologist, v. 19, p. 91–104.

Wengerd, S. A., and M. L. Matheny, 1958, Pennsylvanian System of Four Corners region: AAPG Bulletin, v. 42, p. 2048–2106.

White, M. A., and M. I. Jacobson, 1983, Structures associated with the southwest margin of the Ancestral Uncompahgre uplift, *in* W. R. Averett, ed., Northern Paradox basin–Uncompahgre uplift: Grand Junction Geological Society Guidebook, p. 33–39.

Wilcox, R. E., T. P. Harding, and D. R. Seely, 1973, Basic wrench tectonics: AAPG Bulletin, v. 57, p. 74–96.

Wilde, P., W. R. Normark, and T. E. Chase, 1978, Channel sands and petroleum potential of Monterey deep-sea fan, California: AAPG Bulletin, v. 62, p. 967–983.

Wong, I. G., and R. S. Simon, 1981, Low level historical and contemporary seismicity in the Paradox basin, Utah, and its tectonic implications, *in* D. L. Wiegand, ed., Geology of the Paradox basin: Rocky Mountain Association of Geologists Guidebook, p. 169–185.

Zharkov, M. A., 1981, History of Paleozoic salt accumulation: New York, Springer-Verlag, 308 p.

Minturn and Sangre de Cristo Formations of Southern Colorado: A Prograding Fan Delta and Alluvial Fan Sequence Shed from the Ancestral Rocky Mountains

D. A. Lindsey
U.S. Geological Survey
Denver, Colorado

R. F. Clark[1]
Union Oil Company of California
Oklahoma City, Oklahoma

S. J. Soulliere
U.S. Geological Survey
Denver, Colorado

The Middle Pennsylvanian Minturn Formation and the Pennsylvanian–Permian Sangre de Cristo Formation of the northern Sangre de Cristo Range form a thick progradational sequence of coarse clastic sediments. These sediments were deposited along the western margin of the central Colorado trough during uplift of the late Paleozoic Uncompahgre highland, a major structural and topographic feature of the Ancestral Rocky Mountains.

The Minturn is composed mostly of sandstone and shale deposited by fan deltas that prograded into the central Colorado trough. The Minturn of the northern Sangre de Cristo Range is divisible into a turbidite-bearing facies, a limestone-bearing facies, and a redbed facies. The turbidite-bearing facies is interpreted as deposits of fan deltas that prograded onto the sea bottom below wave base. The limestone-bearing facies is interpreted as deposits of fan deltas that prograded onto a shallow sea bottom above wave base. In this facies, sandstones containing deltaic foresets overlie thin shallow marine limestones and are considered diagnostic of that facies. Both facies contain thick intervals of sandstone and shale interpreted as deltaic and alluvial deposits. Where it onlaps the Uncompahgre highland, the lower part of the Minturn Formation contains quartzose and arkosic redbeds of probable alluvial origin.

The continental Sangre de Cristo Formation conformably overlies the Minturn Formation basinward, but it unconformably overlies the Minturn and Precambrian basement near the Uncompahgre highland. The Sangre de Cristo contains two facies. A sandstone facies consists of fining-upward cycles of red conglomeratic sandstone and siltstone interpreted as braided stream deposits on distal parts of alluvial fans. A conglomerate facies (Crestone Conglomerate Member) consists of poorly sorted conglomerates interpreted as debris flow and mudflow deposits and sorted conglomerate and sandstone interpreted as streamflow and sheetflow deposits on proximal parts of alluvial fans.

The fan deltas and alluvial fans that deposited the Minturn and Sangre de Cristo formations probably developed in response to faulting and uplift of the Uncompahgre highland. The highland rose at least three times from Middle Pennsylvanian to Early Permian time. Later, during the Laramide orogeny, the faulted boundary between the highland and the central Colorado trough was destroyed by thrusting. The boundary faults were not strike-slip, as shown by the distribution of clasts in conglomerates directly downstream from identical Precambrian rocks that represent part of the highland.

---

[1]Formerly U.S. Geological Survey, Denver, Colorado, and
Colorado School of Mines, Golden, CO.

INTRODUCTION

Deposits that fill intermontane basins of the late Paleozoic Ancestral Rocky Mountains contain important evidence for interpreting the history and origin of the Ancestral Rockies (Figure 1). The Middle Pennsylvanian Minturn Formation and the Pennsylvanian–Permian Sangre de Cristo Formation of the northern Sangre de Cristo Range were deposited in the central Colorado trough along the east side of the Uncompahgre highland, a major element of the Ancestral Rocky Mountains. The Minturn and Sangre de Cristo formations record uplift and erosion of the highland and subsidence and filling of the central Colorado trough from Middle Pennsylvanian to Early Permian time. Prograding sequences of sand and gravel were deposited on fan deltas in the trough during Minturn time. Sand and coarse gravel were deposited on alluvial fans that probably filled the trough during Sangre de Cristo time. Abrupt progradation of coarse detritus in the Minturn and Sangre de Cristo formations records three major episodes of uplift in the highland, probably in response to faulting along the eastern boundary.

The late Paleozoic structure of the Ancestral Rocky Mountains in Colorado was obscured during the Laramide orogeny when highlands of the Ancestral Rockies were thrust over adjacent basins. The late Paleozoic structural boundary of the east side of the Uncompahgre highland was destroyed when the highland was thrust over the central Colorado trough in Laramide time. Nevertheless, some characteristics of the boundary can be interpreted from the depositional record. In the study area, detritus was dispersed toward the northeast into the central Colorado trough, directly downstream from identical Precambrian rocks in remnants of the highland, which indicates the absence of strike-slip movement. The east side of the Uncompahgre highland was probably a zone of normal or reverse faulting in late Paleozoic time.

GEOLOGIC SETTING

The Sangre de Cristo Range from Blanca Peak northward is composed of Precambrian crystalline rocks; a thin (100 m [300 ft]) sequence of Ordovician–Mississippian sandstone, shale, and carbonate rocks; and a thick (3000–4000 m [9800–13,100 ft]) sequence of upper Paleozoic clastic sedimentary rocks that is the subject of this report (Figures 2 and 3). Rocks of Precambrian age are exposed in the northern and southern parts of the range and in a narrow strip along the west side. They include 1.7–1.8-b.y.-old gneiss; 1.7-b.y.-old quartz monzonite and related plutonic rocks correlated with the Boulder Creek Granodiorite; and 1.4-b.y.-old quartz monzonite correlated with the Silver Plume Quartz Monzonite (Johnson et al., in press). On the flanks of the range are sedimentary rocks of Mesozoic age, sedimentary and volcanic rocks of Tertiary age, alluvium of late Tertiary and Quaternary age, and glacial deposits of Quaternary age.

Precambrian and Paleozoic rocks of the Sangre de Cristo Range (and Mesozoic and lower Tertiary rocks in Huerfano Park) have been folded and thrust in Laramide (Late Cretaceous–Eocene) time. The central part of the northern Sangre de Cristo Range is traversed by northwest-trending thrusts (Figure 2) (Burbank and Goddard, 1937; Lindsey et al., 1983). These thrusts break the upper Paleozoic rocks of the study area into an autochthonous terrane and three major thrust plates: the Spread Eagle Peak plate, the Marble Mountain plate, and the Huckleberry Mountain plate (Figure 3). The Marble Mountain plate is not much displaced from the Spread Eagle Peak plate, but the Huckleberry Mountain plate is part of a large upper-plate terrane that has been thrusted over the Spread Eagle Peak plate and the autochthonous terrane.

Beginning in late Oligocene time, the opening of the Rio Grande rift disrupted the Laramide thrust zone (Tweto, 1979b). Rocks of the Sangre de Cristo Range were intruded by stocks, dikes, and sills and were metamorphosed regionally during late Oligocene and early Miocene time. The horst of the Sangre de Cristo Range and the adjacent grabens of the San Luis and Wet Mountain valleys were formed by normal faulting and erosion during Miocene and later time (Lindsey et al., 1983).

STRATIGRAPHY

The stratigraphy of the Minturn and Sangre de Cristo formations is presented here in terms of regional facies defined by lithologic features and depositional sequences (Table 1). Each facies has been interpreted in terms of depositional models. Stratigraphic relationships of the Minturn and Sangre de Cristo formations of the northern Sangre de Cristo Range were determined by identifying and mapping numerous informal stratigraphic units and by preparing palinspastic restorations of the formations (Figure 4). The restorations are based on a new geologic map of the area (Johnson et al., in press) and on analysis of structural cross sections of the area. Individual units were studied in detail to determine local stratigraphic relationships, depositional sequences, and environments of deposition (Clark, 1982; Lindsey and Schaefer, 1984; Soulliere et al., 1984; Lindsey et al., 1985). The resulting panels show the restored facies at a low angle to the paleostrike of about N 50° W (Figure 4, A–A'), and approximately parallel to the paleoslope of about N 40° E (Figure 4, B–B'), from the flank of the Uncompahgre highland into the central Colorado trough.

Minturn Formation

The Minturn Formation of the northern Sangre de Cristo Range consists of three regionally extensive facies (Table 1): (1) a turbidite-bearing facies, (2) a limestone-bearing facies, and (3) a redbed facies. Although lenses of turbidite sandstone and thin limestone beds containing shallow marine fossils are mutually exclusive and diagnostic features of the first two facies, they make up only a small part of each facies. Cross-bedded sandstone, siltstone, and shale are the most abundant components. The redbed facies is distinguished by red sandstone and siltstone; much of the facies contains abundant quartz sand and pebbles. The reference section of the Minturn Formation (Lindsey et al., 1985) represents the most basinal sequence observed; it consists of about 1500 m (4900 ft) of turbidite-bearing facies

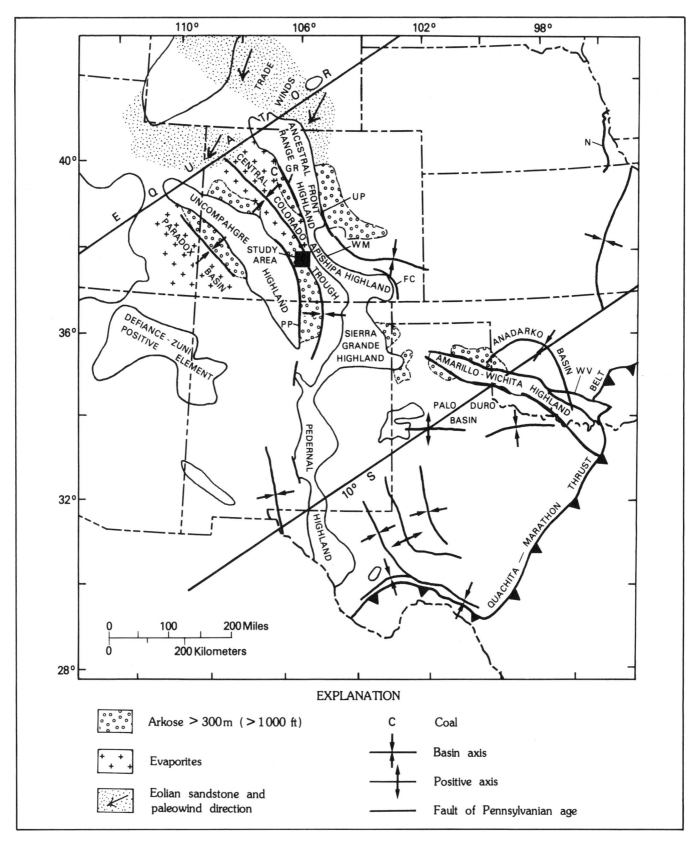

EXPLANATION

Arkose > 300m (>1000 ft)	C Coal
Evaporites	Basin axis
Eolian sandstone and paleowind direction	Positive axis
	Fault of Pennsylvanian age

Figure 1—Map showing location of study area and paleogeographic setting of the Ancestral Rocky Mountains for Middle and Late Pennsylvanian time. Modified from McKee et al. (1975). Paleoequator located approximately for Middle Pennsylvanian time from Smith et al. (1981); distribution of evaporites and coal from Mallory (1972a) and Mack et al. (1979); distribution of thick arkoses from Mallory (1972b) and Dutton (1982); and paleowind directions from Knight (1929), Opdyke and Runcorn (1960), and Poole (1962). Pennsylvanian faults: GR, Gore Range fault; UP, Ute Pass fault; WM, Wet Mountain fault; PP, Picuris–Pecos fault; FC, Freezeout Creek fault; WV, Washita Valley fault; N, Nemaha fault.

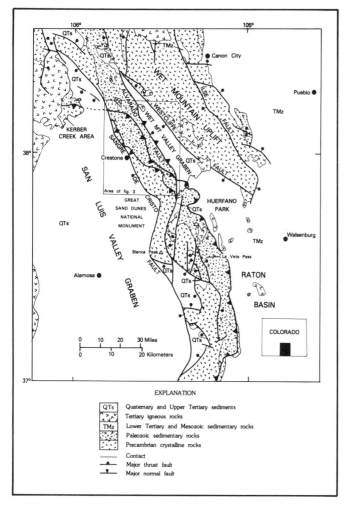

Figure 2—Map showing geologic setting of south-central Colorado and location of study area (in box). Modified from Tweto (1979a).

overlain by about 500 m (1600 ft) of limestone-bearing facies (Figure 4, right side of cross section A–A' and center of B–B'). The redbed facies was deposited to the southwest, nearest the Uncompahgre highland (Figure 4, left side of B–B'). The base of the Minturn Formation is not exposed because of faulting in the study area. The Minturn conformably overlies the Early–Middle Pennsylvanian Belden Formation in the Arkansas Valley (Taylor et al., 1975). The top of the Minturn is placed at or near the highest laterally continuous marine limestone bed.

The turbidite-bearing facies is well developed in the lower 1500 m (4900 ft) of the Minturn Formation on the east side of the Spread Eagle Peak thrust plate. Southward along the line of cross section A–A' (Figure 4), the entire 1500 m (4900 ft) of turbidite-bearing facies passes transitionally into the limestone-bearing facies. The Minturn of the Marble Mountain thrust plate contains no turbidites (Figure 4, middle of A–A') and that of the autochthon is not known to contain them, although the autochthon has not been studied closely enough to rule out the presence of turbidites there. The facies has not been recognized on the west side of the Spread Eagle Peak plate. Reconstruction of sections through the Spread Eagle Peak plate suggests that the turbidite-

bearing facies passes westward into the redbed facies (Figure 4, B–B'), but these relationships are not exposed. The lower part of the Minturn is cut by the Spread Eagle Peak thrust, so the nature of facies in the lowest Minturn is not known.

The limestone-bearing facies is widespread in the Minturn Formation. The facies occupies the upper 300–500 m (1000–1600 ft) of the Minturn in the Spread Eagle Peak thrust plate and the entire 1500 m (4900 ft) of Minturn in the Marble Mountain plate (Figure 4). The thickness of the limestone-bearing facies in autochthonous terrane to the north and east is not known, but its presence there is indicated by limestone beds in the upper part of the Minturn.

A redbed facies composed of fining-upward cycles of quartzose and arkosic sandstone and siltstone has been assigned previously to other formations, but we consider it here to be part of the Minturn Formation. Red sandstone contains abundant quartz sand, granules, and pebbles of vein quartz. Fining-upward cycles in the redbeds are tentatively interpreted as alluvial, but they have not been studied in detail. About 800 m (2600 ft) of redbeds northeast of the village of Crestone were assigned to the Pennsylvanian Kerber and Sharpsdale formations (Karig, 1964), but correlation with these formations is uncertain. Quartzose and arkosic redbeds also occur in thrust-bounded slices high in the Sangre de Cristo Range north of Blanca Peak. In Huerfano Park, southeast of the area studied, 340 m (1100 ft) of quartzose and arkosic redbeds were described under the name "Deer Creek Formation" (Bolyard, 1959), which was later renamed the Sharpsdale Formation (Chronic, 1958).

The age of the Minturn of the northern Sangre de Cristo Range is Atokan–Desmoinesian (Middle Pennsylvanian) (Lindsey et al., 1985), which is in agreement with the age of the type Minturn (Tweto and Lovering, 1977). Fusulinids of early(?) late Desmoinesian age occur near the top of biohermal limestone units (Berg, 1967) in the lower Minturn of the Marble Mountain thrust plate, and fusulinids of late Desmoinesian age occur in limestone beds in the upper part of the Minturn of the Spread Eagle Peak plate and the autochthonous terrane (R. C. Douglass, personal communication, 1981, 1982, 1983). Conodonts from the bioherms of the Marble Mountain plate and from limestone in the upper part of the Minturn suggest an Atokan age (B. R. Wardlaw, personal communication, 1980, 1982, 1983). Although the paleontologic evidence is not in perfect accord, it does indicate that all of the strata assigned to the Minturn are Middle Pennsylvanian in age.

Sangre de Cristo Formation

The Pennsylvanian–Permian Sangre de Cristo Formation contains a sandstone facies and a conglomerate facies (Crestone Conglomerate Member) (Table 1). The two facies correspond to the unnamed lower member (about 600 m [2000 ft] thick) and the Crestone Conglomerate Member (about 1100 m [3600 ft] thick), respectively, at the principal reference section in the Spread Eagle Peak thrust plate (Lindsey and Schaefer, 1984). Northeast (basinward) of the principal reference section, the Sangre de Cristo Formation consists of more than 1500 m (4900 ft) of sandstone facies; southwest (toward the highland), it consists of more than

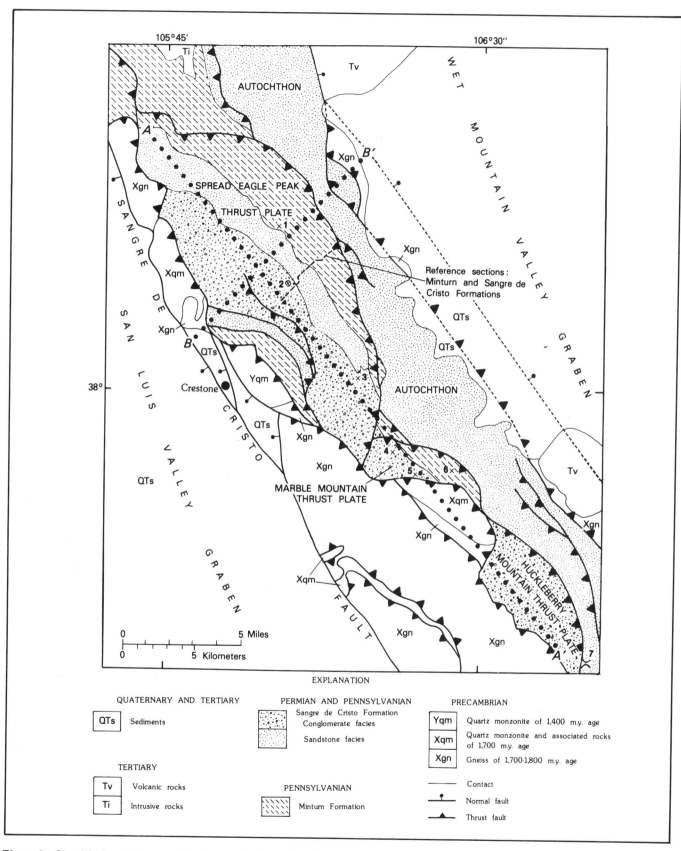

Figure 3—Simplified geologic map of study area in the northern Sangre de Cristo Range, Colorado. Small areas of lower Paleozoic and Mesozoic rocks not shown. Numbered localities: 1. Rito Alto Peak; 2. Groundhog basin; 3. Mount Adams; 4. Crestone Needle; 5. Milwaukee Peak; 6. Marble Mountain; and 7. Medano Pass. A–A' and B–B' are lines of restored stratigraphic cross sections shown in Figure 4. Modified from Lindsey et al. (1983) and Johnson et al. (in press).

Table 1—Stratigraphic units, facies, sequences, and depositional models for the Minturn and Sangre de Cristo formations, northern Sangre de Cristo Range.

Stratigraphic Unit or Interval	Facies	Characteristic Sequence	Depositional Model
Sangre de Cristo Formation Crestone Conglomerate Member	Conglomerate	Randomly interbedded, sorted and unsorted conglomerate and sandstone	Proximal alluvial fan
Lower member, Spread	Sandstone	Sandstone-dominated	Distal alluvial fan
Undivided, autochthon	Sandstone	Same as above	Same as above
Minturn Formation Upper 300–500 m, Spread Eagle Peak plate	Limestone bearing	Weakly developed coarsening-upward clastic cycles; fining-upward clastic subcycles in upper part; separated by thin marine limestones	Fan delta prograding onto sea bottom above wave base
All, Marble Mountain	Limestone	Same as above	Same as above
Lower 1500 m, east side Spread Eagle Peak plate	Turbidite bearing	Coarsening-upward clastic cycles with intervals of turbidites; fining-upward clastic subcycles in upper part	Fan delta prograding onto sea bottom below wave base
Lower 800 m, west side Spread Eagle Peak plate	Redbed	Fining-upward clastic cycles	Alluvial(?)

2000 m (6600 ft) of conglomerate facies (Figure 4).

The sandstone facies is the most widespread facies of the Sangre de Cristo Formation. It makes up the entire Sangre de Cristo Formation in the autochthonous terrane of the study area (Figure 3) and also in the Arkansas River Valley (Brill, 1952) and Huerfano Park (Rhodes, 1964). The sandstone facies conformably overlies the Minturn Formation at basinward localities, such as the east side of the Spread Eagle Peak plate and the autochthon. In the Spread Eagle Peak thrust plate, the sandstone facies interfingers southwestward into the overlying Crestone Conglomerate Member (Figure 4, right side of A–A′ and middle of B–B′). Southward, in the Marble Mountain and Huckleberry Mountain plates, the sandstone facies is absent.

The conglomerate facies (Crestone Conglomerate Member) is restricted to the west-central part of the northern Sangre de Cristo Range (Figure 3). The facies is recognized easily by the dominance of coarse red conglomerate, although locally as much as half is composed of conglomeratic sandstone. The base of the Crestone Conglomerate Member is arbitrarily placed at the lowest level of numerous thick conglomerate beds. Near the late Paleozoic Uncompahgre highland, in the Marble Mountain and Huckleberry Mountain plates, about 2000 m (6600 ft) of

Crestone Conglomerate Member makes up the entire Sangre de Cristo Formation. The Crestone Conglomerate Member unconformably overlies the Minturn Formation in the Marble Mountain plate, and unconformably overlies Precambrian rocks along the southwest side of the Huckleberry Mountain plate (Figure 4, A–A′). All of the conglomerate contains clasts derived from gneiss that is 1.7–1.8 b.y. in age, as well as distinctive clasts of other Precambrian rocks. Three conglomerate-clast subfacies are distinguished by: (1) red syenite clasts of unknown source, accompanied by gneiss clasts; (2) clasts of pink quartz monzonite porphyry and related rocks, all with distinctive large potassium feldspar phenocrysts, and derived from identical rocks correlated with the 1.7-b.y.-old Boulder Creek Granodiorite; and (3) mostly gneiss clasts derived from 1.7–1.8-b.y. old gneiss. These subfacies are referred to as (1) the syenite clast facies, (2) the quartz-monzonite clast facies, and (3) the gneiss clast facies, respectively (Figure 4).

Fossil and stratigraphic evidence from the northern Sangre de Cristo Range indicates that the lower part of the Sangre de Cristo Formation there is Middle Pennsylvanian in age. The upper part of the formation has not yielded fossils, and it may range in age from Middle Pennsylvanian to Early Permian (Lindsey and Schaefer, 1984). Conodonts

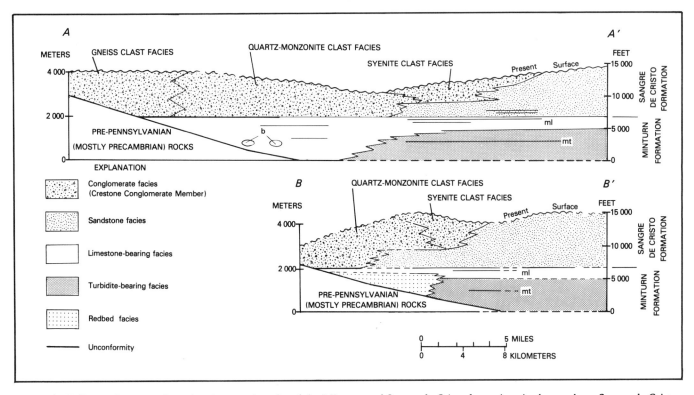

Figure 4—Palinspastic restorations showing stratigraphy of the Minturn and Sangre de Cristo formations in the northern Sangre de Cristo Range. Location of cross sections are shown on Figure 3: A–A′ is at a low angle to the paleostrike and B–B′ is approximately parallel to the paleoslope. Restoration of left side of A–A′ does not take into account unknown displacement of Huckleberry Mountain thrust plate. b, limestone bioherms; ml, marker limestone; mt, main turbidite member; unlabeled lines, other limestone markers; dashed contacts (--) indicate projections across thrusts or through unexposed strata.

from marine limestones in the lower 300 m (100 ft) of the sandstone facies of the Spread Eagle Peak plate indicate an Atokan or Desmoinesian (Middle Pennsylvanian) age (B. R. Wardlaw, personal communication, 1982). Intertonguing of the entire thickness of sandstone and conglomerate facies of the Sangre de Cristo Formation indicates a continuous succession of deposits that could have accumulated entirely during Middle Pennsylvanian time. An age younger than Middle Pennsylvanian is possible, however, because the unfossiliferous upper part of the Sangre de Cristo Formation has not been dated, and because erosion of the top of the formation may have removed strata of Late Pennsylvanian or Early Permian age. On the basis of fossil vertebrates, a Missourian (Late Pennsylvanian) age has been assigned to the middle part of the Sangre de Cristo Formation in the Arkansas River Valley of Colorado (Vaughn, 1972). In the southern Sangre de Cristo Range of New Mexico, an Early Permian age has been assigned to the lower part of the Sangre de Cristo Formation, because it overlies strata containing fusulinids of Wolfcampian age (Baltz and O'Neill, 1980). Thus, the age of the Sangre de Cristo Formation throughout its extent from the Arkansas River valley to the southern Sangre de Cristo Range in New Mexico is both Pennsylvanian and Permian.

DESCRIPTION AND INTERPRETATION OF FACIES

Turbidite-Bearing Facies of Minturn Formation

The turbidite-bearing facies of the Minturn Formation is interpreted as deposits of fluvial-dominated fan deltas that prograded onto a sea bottom below wave base (Figure 5A). The turbidite-bearing facies contains cycles 30–300 m (100–1000 ft) thick that consist mostly of (1) prodelta marine(?) shale and siltstone, (2) prodelta sandstone deposited by turbidity flows, (3) delta front conglomeratic sandstone, and (4) deltaic and alluvial plain sandstone, siltstone, and shale arranged in fining-upward subcycles (Figure 6A) (Lindsey et al., 1985). We consider the prodelta environment, represented by shale and siltstone, as a marine shelf extending seaward from the delta front (Figure 5A). The Minturn turbidites were probably deposited in the transition zone between the prodelta and the delta front and can be assigned to either environment. The delta front environment, represented mostly by conglomeratic sandstone, is the zone where alluvial sediment entered the basin and was dispersed. As defined, Minturn delta front deposits probably include those of both distributary channels and distributary-mouth bars. Deltaic and alluvial plain

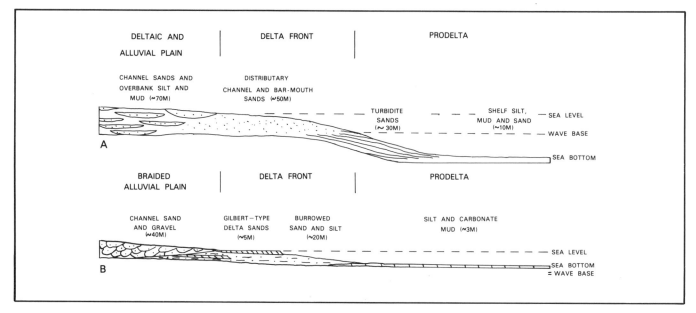

Figure 5—Diagrams of proposed depositional environments and deposits of fan deltas, Minturn Formation. A. Turbidite-bearing facies. B. Limestone-bearing facies. Approximate thicknesses are typical of each deposit (no scale).

deposits were laid down landward from the delta front. This simplified classification of deltaic depositional environments, adapted from Wright (1978) and Elliott (1978), will be followed throughout this discussion. The term "fan delta," meaning an alluvial fan prograding into a body of standing water (Holmes, 1965), is used synonymously here for "delta," because the Minturn deltas are interpreted as seaward extensions of alluvial fans.

Assignment of shale and siltstone to a prodelta marine(?) environment is uncertain because reliable criteria for interpretation of this depositional environment have not been recognized in the Minturn Formation. Fossils are sparse and confined to impressions of plant fragments and pelecypods of uncertain environmental affinity. Thin beds of sandstone in shale and siltstone intervals contain abundant current ripple marks and ripple cross lamination; some sandstone beds contain planar and trough cross bedding. Shales are dark gray to black, thinly laminated, and contain abundant silty laminae and isolated ripple marks. Intervals of shale and siltstone grade upward into turbidite sandstone.

Prodelta turbidite sandstone beds are interpreted as deposits of density currents that flowed from the delta front onto a sea bottom below wave base (Figure 5A). Most of the features described here (Figure 7) were observed in the main turbidite member of the Minturn Formation (Figure 4) (Soulliere et al., 1984), supplemented by observations of other intervals of turbidites. Turbidite sandstones form large lenses that attain a maximum thickness of 150 m (450 ft) and maximum extent of 13 km (8 mi) along the depositional strike of the Minturn Formation. The turbidite sandstones are arkosic and coarse grained, even conglomeratic. Individual turbidite beds average 50–60 cm (20–24 in.) thick but commonly range to 1 m (3 ft) thick. The sand:shale ratio is about 30:1 to 50:1; shale divisions at the tops of individual turbidites are generally only 1–2 cm (0.4–0.8 in.) thick. The turbidites closely resemble those designated as "proximal"

by R. G. Walker (1967) (facies C of Ricci-Lucchi, 1975) and indicate deposition near the source of density currents.

Intervals of these turbidites contain many examples of truncated Bouma sequences. Bouma sequences (Bouma, 1962) composed of graded and massive sandstone beds overlain by thin shale beds account for 52% of the turbidites in the main turbidite member. Bouma sequences that begin with graded or massive sandstone beds followed by parallel-laminated or ripple cross-laminated sandstone and thin shale beds account for 34% of the turbidites. *Zoophycus*, a trace fossil found in both shelf and slope facies (Seilacher, 1978), is common in turbidite sandstones of the Minturn Formation. The base of each turbidite is sharp; erosional scours having 10–20 cm (4–8 in.) of relief occur locally. Sole marks are uncommon, but small flute and groove casts are seen on the bottoms of a few beds. Load casts and flame structures, resulting from settling of sand into underlying unstable silt and mud, are common. Turbidites rarely fill channels and do not form thinning-upward aggradational sequences that would indicate deposition in channels (Ricci–Lucchi, 1975). Intervals of turbidite sandstones are transitional downward into sparsely fossiliferous, thin-bedded sandstone, siltstone, and shale, and upward into lenses of conglomeratic sandstone that attain a size comparable to the turbidite lenses.

Lens-shaped packets of turbidites in the Minturn Formation must have been deposited by density currents flowing from the delta front, as shown by the direct transition from the interval of turbidites to the overlying interval of delta front sandstone and by parallel paleocurrent directions in both intervals. Density currents may have been initiated by periods of high alluvial discharge or by slumping on the delta front. The turbidites must have been deposited below wave base because they do not show evidence of extensive reworking by currents after deposition. The absence of channels and the absence of prodelta siltstone

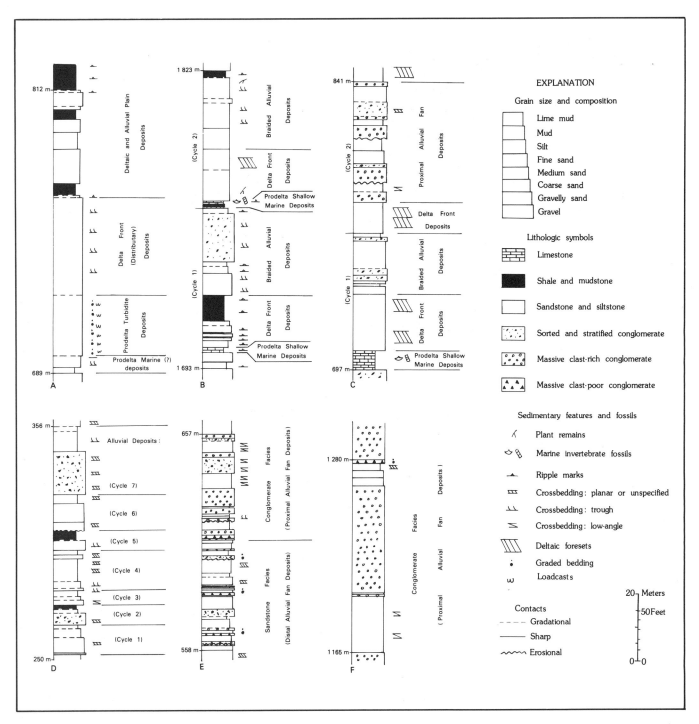

Figure 6—Facies sequences from measured sections of Minturn and Sangre de Cristo formations, northern Sangre de Cristo Range (Lindsey and Schaefer, 1984; Lindsey et al., in press). A. Cyclic sequence from turbidite-bearing facies, Minturn Formation, Spread Eagle Peak plate. B. Two cyclic sequences from limestone-bearing facies, Minturn Formation, Spread Eagle Peak plate. C. Two cyclic sequences from limestone-bearing facies, Minturn Formation, Marble Mountain plate; cycle one resembles those of (B) and cycle two contains beds of coarse alluvial conglomerate. D. Fining-upward cycles in sandstone facies, Sangre de Cristo Formation, Spread Eagle Peak plate. E. Interval of interfingering between sandstone facies (lower member, Sangre de Cristo Formation) and conglomerate facies (Crestone Conglomerate Member, Sangre de Cristo Formation), Spread Eagle Peak plate. F. Apparently random sequence from conglomerate facies, Crestone Conglomerate Member, Sangre de Cristo Formation, Spread Eagle Peak plate.

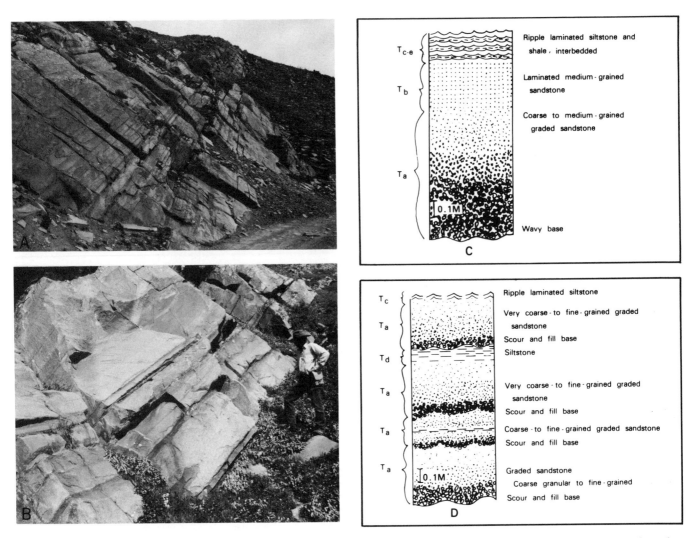

Figure 7—Photographs and drawings of turbidites deposited in front of fan deltas, turbidite-bearing facies of Minturn Formation, Spread Eagle Peak plate. A. Outcrop showing continuity of bedding, characteristic of intervals of turbidite sandstones; along Hermit Road, east of Rito Alto Peak. B. Typical turbidite sandstone beds along Hermit Road, showing sharp base, normal grading, dominance of sandy over shaley divisions, and shale clast in lowest bed. C. Nearly complete Bouma sequence in turbidite sandstone. D. Compound turbidite composed of truncated Bouma sequences and lacking shaley divisions.

and shale between the turbidites and delta front sandstones indicate that the Minturn turbidites were not deposited as submarine fans at the mouths of feeder channels in the manner described by Walker (1978), but instead were deposited marginally to the delta front. In this regard, they closely resemble the thin sequences of turbidites reported by Balsley (1982) from distributary-mouth bar facies of deltaic deposits in the Cretaceous Blackhawk Formation of Utah. Turbidites have been reported from other deltaic sequences (McBride et al., 1975) and from lateral equivalents of fan delta facies in Pennsylvanian strata of the southern Sangre de Cristo Range (Casey, 1980).

Conglomeratic and arkosic sandstone, interpreted as delta front deposits, forms lenses of approximately the same size as the turbidite lenses. Beds of conglomeratic sandstone are separated by thin partings and beds of fine-grained sandstone, siltstone, and shale. Sandstone beds have scoured bases, fine upward, and are massive or contain abundant unidirectional trough cross bedding of medium scale (20–50

cm [8–20 in.]); tops of some beds have ripple marks. Lenses (<10 cm [4 in.] thick) rich in magnetite occur locally. All of the sandstones are composed of abundant angular to subrounded grains of quartz and feldspar. Some of the sandstones are locally very coarse and contain lenses of massive and crudely stratified cobbles and boulders of quartz monzonite and gneiss. Intervals of conglomeratic sandstone overlie turbidites along a transitional contact, but evidence of scouring is common a few meters above the contact. Sandstone intervals are generally overlain abruptly by sandstone and siltstone interpreted as deltaic and alluvial plain deposits. Some thick conglomeratic sandstones are tentatively interpreted as coalescing distributary channels and distributary-mouth bars formed by braided distributaries of rivers with high bed load in the manner suggested by Wright (1978).

Intervals of alternating arkosic sandstone and shale, arranged in multiple fining-upward subcycles, are assigned mostly to undifferentiated deltaic and alluvial plain environments. The thickness of siltstone and shale generally equals or exceeds sandstone; sandstone beds are lenticular, fill small channels, contain trough cross bedding, and have rippled tops. Individual lenses and channels of sandstone interfinger with siltstone and shale interpreted as overbank deposits. A few horizons of plant trunks (generally casts of *Calamites*) in growth position and intervals of calcareous nodules of possible pedogenic origin indicate times of emergence.

Limestone-Bearing Facies of Minturn Formation

The limestone-bearing facies of the Minturn Formation is interpreted as deposits of fan deltas that prograded onto a shallow sea bottom above wave base (Figure 5B). The limestone-bearing facies contains cycles 65–160 m (210–525 ft) thick that consist of the following (from bottom to top): (1) thin prodelta and interdeltaic shallow marine limestone, shale, siltstone, and arkosic sandstone; (2) delta front conglomerate and conglomeratic sandstone; and (3) fining-upward subcycles of arkosic sandstone, siltstone and shale deposited by braided streams (Figure 6B) (Clark, 1982; Lindsey et al., 1985). Similar cyclic sequences in fan delta deposits have been described from strata of Desmoinesian age in the southern Sangre de Cristo Range of New Mexico (Casey and Scott, 1979; Casey, 1980). Cycles composed of shallow marine limestone and arkose also occur in the lower Minturn at La Veta Pass (Becker, 1978).

Fossiliferous shallow marine limestone and associated burrowed sandstone and siltstone are interpreted as marine shelf deposits formed between and seaward from fan deltas. Most limestone beds are 1–3 m (3–10 ft) thick; some that extend less than 1 km (0.6 mi) evidently represent marine sedimentation between small delta lobes. Laterally extensive beds, such as the distinctive marker limestone (Figure 4) that extends more than 13 km (8 mi) along depositional strike near the top of the Minturn in the Spread Eagle Peak thrust plate, may represent larger areas of interdeltaic seas or, possibly, local marine transgressions.

Limestone beds contain a wide variety of textures, sedimentary structures and fossils that indicate a shallow marine environment ranging from low energy to strongly agitated by waves and currents and extending up into the intertidal zone (Clark, 1982). Limestone includes packstone and wackestone rich in skeletal debris and lime mud, intraclast-bearing and stromatolitic lime mudstone, and skeletal and oolitic grainstone. Lime mudstone and grainstone occur mostly in the autochthonous Minturn; they are interpreted as offshore shelf and shoal deposits, respectively. Structures in limestone beds include planar and wavy lamination, ripple cross lamination, and trough cross bedding. Orientations of structures are diverse and include many directed parallel to the general southeast trend of the hypothetical shoreline. Fossils in the limestone include crinoid columnals, brachiopods, bryozoans, phylloid algae, gastropods, pelecypods, rugose corals, fusulinids, and conodonts.

The most distinctive features in sandstone units interpreted as delta front deposits are thick intervals of sandy foresets, commonly 5 m (16 ft) but as much as 15 m (50 ft) thick (Figure 8A). Individual foresets are 0.5 m (1.5 ft) or less thick. They dip as much as 30°, and where they

Figure 8—Photographs of fan-delta deposits in the limestone-bearing facies, upper part of Minturn Formation, Spread Eagle Peak plate. A. Large deltaic foresets (Gilbert-type deltas) in conglomeratic sandstone on the east side of Rito Alto Peak; interval of foresets is about 15 m (50 ft) thick and interfingers with calcareous siltstone containing marine fossils. B. Sequence of stromatolitic silty limestone (s) of intertidal zone, overlain by carbonaceous shale of possible coastal swamp environment and sandstone deposited by braided streams; upright *Calamites* plant (above man) is rooted in shale and engulfed by sandstone; east side of Mount Adams.

approach underlying beds, they become tangential to them. The foresets were formed by small deltas prograding into shallow standing water in the manner of the small lacustrine deltas described by Gilbert (1883) and the deltaic sedimentation units of Collinson (1968). Deltaic foresets in strata of Minturn age have also been described from the southern Sangre de Cristo Range (Casey and Scott, 1979). The deltaic foresets are distinguished from eolian cross bedding by their development in immature conglomeratic sandstone; they are distinguished from cross bedding formed by migrating point bars by interfingering with fossiliferous marine beds. Delta front sandstone beds that lack large foresets are difficult to recognize in the limestone-bearing facies; their presence is suggested by abundant burrows (Clark, 1982).

Deposits interpreted as having formed in coastal bays and swamps are sparse in the Minturn Formation. Calcareous siltstone and shale containing marine invertebrate fossils and carbonaceous plant fragments, interpreted as interdistributary bay deposits, interfinger with other delta front deposits along strike. Deposits of swamp origin are sparse; no coal beds are present, but some beds of finely laminated carbonaceous shale may have been deposited in swampy coastal areas (Figure 8B).

Conglomerate and sandstone deposited by braided streams overlie delta front deposits and, in some cases, other coastal deposits. Braided stream deposits are identified by fining-upward cycles of lenticular-bedded, channeled, and cross-bedded conglomerate and sandstone capped by thin intervals of siltstone and shale. Locally, shales contain intervals of limestone nodules interpreted as paleocaliche. Trunks of *Calamites* in growth position were found in braided stream deposits at a few localities; the 2-m-high (6-ft-high) specimen shown in Figure 8B is still rooted in shale interpreted as a coastal swamp deposit and engulfed by pebbly sand deposited by braided streams.

The limestone-bearing facies of the Minturn Formation in the Marble Mountain thrust plate resembles that of the autochthon and the Spread Eagle Peak thrust plate, except that it contains numerous beds as much as 40 m (130 ft) thick of coarse conglomerate interpreted as proximal alluvial fan deposits (Figure 6C). The conglomerate beds are massive to crudely stratified and are composed of abundant angular to subrounded cobbles and boulders in a matrix of sand and finer detritus. Massive and stratified conglomerate interfingers with sandstone containing large deltaic foresets, interpreted as delta front deposits. Locally, coarse conglomerate lies directly on marine limestone. Lozenge-shaped limestone bioherms (Berg, 1967) as much as 1.5 km (1 mi) across and 300 m (1000 ft) thick occur in the lower part of the Minturn Formation of the Marble Mountain thrust plate.

Sandstone Facies of Sangre de Cristo Formation

The sandstone facies of the Sangre de Cristo Formation is interpreted as distal alluvial fan deposits laid down by braided streams. The sandstone facies occurs throughout both the lower member and the undivided Sangre de Cristo Formation. The facies is characterized by fining-upward cycles (Figure 6D) 2–37 m (6–120 ft) thick of red conglomerate, sandstone, siltstone, and shale (Lindsey and Schaefer, 1984) called "piedmont cyclothems" by Brill (1952) and Bolyard (1959). The fining-upward cycles of the sandstone facies closely resemble those of distal braided rivers (Rust, 1978) and the Donjek braided stream model (Miall, 1978), named for middle reaches of the Donjek River in Yukon Territory, Canada. Lack of deep channels and a low proportion of overbank siltstone and shale to channel

sandstone indicate that the fining-upward cycles in the sandstone facies of the Sangre de Cristo Formation were not formed by meandering streams.

The basal contacts of cyclic sequences are generally scoured as much as 1–2 m (3–6 ft) into the underlying cycle, but deep channels were not observed. The lower parts of most cycles are composed of 1 m (3 ft) or more of trough cross-bedded conglomeratic sandstone and lenticular pebble and cobble conglomerate. Locally, the basal deposits contain red shale clasts as well as pebbles and cobbles of Precambrian igneous and metamorphic rocks. The main part of each cycle is composed of red arkosic sandstone exhibiting bedding that ranges from cross-bedded (trough and tabular) to horizontally laminated. Low-angle (<5°) cross bedding is common near the transition with the Crestone Conglomerate Member (Figure 6E). The upper parts of most cycles consist of medium- to thin-bedded sandstone and siltstone containing small-scale trough and planar cross bedding, current ripple marks, and locally, mud cracks and burrows. The most common burrows are tubular, upright to inclined structures, 0.5–1 cm (0.2–4 in.) in diameter and 2–4 cm (1–2 in.) long. Burrows are concentrated near the tops of fine-grained sandstone and siltstone beds. Thin (<1 m [3 ft]) dark red shales of overbank origin occur at the tops of a few cycles. Plant remains are scarce and limited to a few impressions of fossil stems.

The uppermost siltstone and shale beds of some cycles contain unfossiliferous calcareous nodules and beds. Some calcareous intervals are as much as 1 m (3 ft) thick; they exhibit nodular structure in their lower parts and grade into massive fine-grained limestone and calcite-cemented siltstone and shale in their upper parts. These calcareous intervals resemble the paleocaliche deposits of other redbeds (Steel, 1974; Turner, 1980) in every respect and are thus interpreted as such.

Shallow marine limestone beds, commonly silty and stromatolitic, occur at a few stratigraphic levels as high as 300 m (1000 ft) above the base of the sandstone facies of the Sangre de Cristo Formation. Limestones as much as 3 m (10 ft) thick overlie fining-upward sandstone cycles. Some beds are composed entirely of planar to crinkly algal stromatolites; columnar algal structures are sparse. The limestones contain fossils of crinoid columnals, brachiopods, bryozoans, conodonts, and locally, fusulinids.

Conglomerate Facies of Sangre de Cristo Formation

The conglomerate facies (Crestone Conglomerate Member) of the Sangre de Cristo Formation consists entirely of proximal alluvial fan deposits comparable to those described from other ancient and modern alluvial fans by Bull (1972) and Nilsen (1982). The conglomerate facies interfingers with the sandstone facies over nearly the entire 2000-m (6600-ft) thickness of the Sangre de Cristo Formation (Figure 4). It is composed of massive clast-rich and matrix-rich conglomerate, sorted and stratified conglomerate and sandstone, and minor siltstone and shale (Figure 6F); all these rocks are interbedded. The conglomerate facies lacks well-defined cyclic sequences, but instead appears to be weakly cyclic to random. At the principal reference section, fining-upward alluvial cycles resembling those of the sandstone facies occur only in a single interval within the conglomerate facies (Lindsey and

Schaefer, 1984).

Massive beds of clast-rich conglomerate (Figures 9A and B) are interpreted as debris flow deposits according to the criteria of Middleton and Hampton (1976). Beds of clast-rich conglomerate are massive to weakly stratified, lack visible clast fabric, and contain abundant matrix of sand and finer detritus (Figure 9B). They differ in all these respects from sorted and stratified conglomerates deposited by flowing water, but some grade upward into the latter. Such gradation may be due to transition from debris flow to streamflow deposition or to reworking of the upper parts of debris flow deposits by streams. Most beds of massive clast-rich conglomerate are matrix supported, but they contain abundant pebbles, cobbles, and boulders, in contrast to beds of mudflow conglomerate that contain only sparse clasts. The base of most beds is sharp; in a few places, beds fill channels cut as much as 1 m (3 ft) into underlying beds. Beds of massive clast-rich conglomerate average 1–3 m (3–10 ft) thick but are as much as 55 m (180 ft) thick; most of the thicker beds are compound in detail, composed of more than one texturally distinct division. Thick compound beds extend for more than 1 km (0.6 mi) along strike. Individual beds exhibit diverse patterns of grain size variation, including normal and reverse grading of the coarse size fraction, various combinations of normal and reverse grading, and alternation of coarse and fine intervals (Flores, 1985). Angular cobbles and boulders (as much as 2 m [6 ft] across) of Precambrian igneous and metamorphic rocks are common. Uncommonly large boulders are concentrated at the tops of some beds and project above the top, indicating a bouyancy effect during transport. Debris flow conglomerates comparable to some in the Sangre de Cristo Formation have been described from the lower part of the Pennsylvanian–Permian Cutler Formation at Gateway, Colorado (Mack and Rasmussen, 1984).

Massive beds of matrix-rich conglomerate interpreted as mudflow deposits are less abundant than clast-rich conglomerate in the conglomerate facies. Most beds of matrix-rich conglomerate are a few meters or less thick (Figure 9A), but they can be as thick as the 40-m (130-ft) thick compound bed that crops out on the east side of Crestone Needle and Milwaukee Peak in the Marble Mountain thrust plate (Figure 9C). Matrix-rich conglomerates contain clasts of Precambrian igneous and metamorphic rocks as much as 1–2 m (3–6 ft) across dispersed in a matrix of sand and finer detritus colored dark red by hematitic pigment. The distribution of the coarse size fraction shows the same diverse patterns as those seen in beds of massive clast-rich conglomerate (Flores, 1985). Faint stratification within beds is defined by variations in matrix content and clast size. Compound beds are composed of texturally distinct intervals separated by sharp contacts and, more rarely, by intercalated lenses of stratified sandstone. Concentrations of boulders occur locally at the tops of beds.

Thin beds of poorly sorted pebbly sandstone having a massive and graded structure represent a sandy variant of mudflow deposits. Pebbly sandstone contains granules and pebbles dispersed in a matrix of sand and finer detritus colored dark red by hematitic pigment (Figure 9D). Pebbly sandstone beds are interleaved with beds of massive conglomerate, sorted conglomerate, and sandstone in the conglomerate facies, and they also occur in the upper part of the sandstone facies of the Sangre de Cristo Formation. Most beds are 3–10 cm (1–4 in.) in thickness and have sharp contacts; some compound beds range up to 2 m (6 ft) in thickness. Calcareous concretions are common in some. A thin (1–5 mm [0.04–0.2 in.]) drape of red mudstone colored by hematitic pigment overlies some beds.

Intervals as much as 50 m (160 ft) thick of sorted and stratified conglomerate, conglomeratic sandstone, sandstone, and minor siltstone (Figure 9E) are interpreted mostly as streamflow and sheetflow deposits. Sorted and stratified conglomerate and sandstone are more abundant in the conglomerate facies of the Spread Eagle Peak thrust plate than in the Marble Mountain and Huckleberry Mountain thrust plates, located nearer the source. Lenses of sorted and stratified conglomerate and sandstone locally fill shallow (1–2 m [3–6 ft]) channels (Figure 9C), but many lack channels. Sorted and stratified conglomerate and sandstone contain abundant horizontal lamination and low-angle (<5°) cross bedding, lenses and laminae of black iron-oxide minerals, and large cobbles and boulders of igneous and metamorphic rock (Figure 9E). Siltstone and shale beds that might represent overbank deposits are thin to absent. All of these features are in agreement with a streamflow and sheetflow origin, possibly in broad channels that traversed the proximal and medial surfaces of alluvial fans. Horizontally stratified sheetflow deposits are common in alluvial fans (Bull, 1972; Rachocki, 1981), and low-angle cross bedding has been reported from braided alluvial deposits (Miall, 1978) that form during unconfined streamflow. Lenses and laminae of black sand (heavy minerals) form on bars in shallow channels during flash floods (Luchitta and Suneson, 1981). Intervals of sorted and stratified conglomerate and sandstone in the conglomerate facies of the Sangre de Cristo Formation resemble proximal fan deposits of the Pennsylvanian–Permian Fountain Formation near Manitou Springs and Canon City, Colorado (Langford and Fishbaugh, 1984; Shultz, 1984).

PALEOCURRENTS AND SEDIMENT DISPERSAL

Paleocurrent directions for the Minturn and Sangre de Cristo formations were determined mostly by measurement of trough cross bedding in sandstone beds. Trough axes were measured where possible, but many measurements were made on dipping cross laminae where inspection revealed the probable orientation of the trough axis; planar cross bedding was measured locally. This procedure enables collection of large numbers of measurements, but probably increases the error of measurement and hence the total variance. Rose diagrams and vector means of the measurements are useful for estimating paleocurrent directions, but estimates of variation are probably not useful for environmental interpretations. Vector strength and variance indicate dispersion of vectors about the mean, and the level of significance determined by the Rayleigh test indicates the likelihood that the vectors measured were drawn from a population distributed uniformly in all directions (Table 2).

Paleocurrent directions of the Minturn and Sangre de Cristo formations show a consistent northeast-flowing drainage system that originated on the east side of the

Figure 9—Photographs of proximal alluvial fan deposits in the conglomerate facies (Crestone Conglomerate Member), Sangre de Cristo Formation. A. Massive beds of clast-rich conglomerate (c) interpreted as debris flow deposits, interbedded with matrix-rich conglomerates (m) interpreted as mudflow deposits; man at lower right (in circle) shows scale; northwest of Medano Pass. B. Massive clast-rich conglomerate northwest of Medano Pass. C. Compound bed of matrix-rich conglomerate, 40 m (130 ft) thick, showing divisions defined by weak stratification and changes in texture, interpreted as a stack of mudflow deposits; overlain by sorted and stratified conglomerate and sandstone containing channels, interpreted as streamflow deposits; east side of Milwaukee Peak. D. Beds of pebbly sandstone, interpreted as sandy mudflow deposits; north side of Groundhog basin. E. Conglomeratic sandstone containing horizontal lamination, low-angle cross bedding, and large clasts, interpreted as streamflow or sheetflow deposit; south side of Groundhog basin.

Table 2—Summary statistics for paleocurrent directions determined from trough (and minor planar) cross bedding in the Minturn and Sangre de Cristo formations, northern Sangre de Cristo Range.

Tectonic or Stratigraphic Unit	Number of Measurements	Vector Mean (°)	Vector Magnitude (%)	Standard Deviation (°)	Level of Significance (Rayleigh test)[a]
Autochthon	372	51	21	90	$<10^{-5}$
Sangre de Cristo Formation	313	50	26	86	$<10^{-5}$
Minturn Formation	59	205	6	100	>0.80
Spread Eagle Peak plate	709	42	48	69	$<10^{-20}$
Sangre de Cristo Formation	288	45	59	59	$<10^{-20}$
Conglomerate facies	20	60	62	56	$<10^{-3}$
Sandstone facies	268	44	59	59	$<10^{-20}$
Minturn Formation	421	38	41	75	$<10^{-20}$
Limestone-bearing facies	175	55	22	88	$<10^{-3}$
Turbidite-bearing facies	246	34	56	62	$<10^{-20}$
Marble Mountain plate					
Minturn Formation	103	43	35	81	$<10^{-5}$
Summary					
Sangre de Cristo Formation	601	47	42	74	$<10^{-20}$
Minturn Formation	583	40	35	80	$<10^{-20}$
Both formations	1184	43	38	77	$<10^{-20}$

[a]From Curray (1956, figure 4).

Uncompahgre highland (Figure 10). Regionally consistent paleocurrent directions are recorded by the vector means of cross-bedding orientations from the combined formations of the autochthonous terrane (N 51° E), the Spread Eagle Peak thrust plate (N 42° E), and the Marble Mountain thrust plate (N 43° E) (Table 2). Sediment dispersal was approximately perpendicular from the margin of the highland as reconstructed by other workers (Mallory, 1972b; McKee et al., 1975); dispersal parallel to the axis of the central Colorado trough is not evident except in some marine deposits of the Minturn Formation. The consistency of paleocurrents through time is recorded by the vector means of cross-bedding orientations for the Minturn Formation (N 40° E) and the Sangre de Cristo Formation (N 47° E), and it is again demonstrated by the small range of vector means (N 34–60° E) for four facies in the Spread Eagle Peak plate (Table 2).

Paleocurrent directions indicate that the late Paleozoic Ancestral Front Range and Apishipa highlands, located east of the central Colorado trough, did not contribute sediment to the Minturn and Sangre de Cristo formations in the study area. In the northern Sangre de Cristo Range, both formations were derived from the Uncompahgre highland, west of the central Colorado trough. The sedimentary fill of the east side of the central Colorado trough was evidently

eroded after Sangre de Cristo time or was overridden by the Wet Mountain uplift during Laramide time. The east side of the trough is preserved only along the southwest flank of the Wet Mountains, where orientation of trough cross bedding in the upper part of the Sangre de Cristo Formation indicates southwesterly flow from the Apishipa highland.

Although the late Paleozoic fill of the central Colorado trough was deformed and partly overriden by the east side of the Uncompahgre highland during the Laramide orogeny, the trough fill has apparently not been moved laterally by strike-slip faulting along the highland–trough boundary. The distribution of distinctive clasts in the conglomerate facies and in conglomeratic lenses of the sandstone facies of the Sangre de Cristo Formation matches closely with identical Precambrian rocks exposed today in the northern Sangre de Cristo Range (Figure 11). Southwest of Laramide thrust plates that contain most of the Sangre de Cristo Formation, Precambrian rocks were apparently at or near the surface during late Paleozoic time, as indicated by conglomerate facies of the Sangre de Cristo Formation resting directly on Precambrian rocks and on strata of lower Paleozoic age. Thus, Precambrian rocks exposed today in some Laramide thrust plates of the range represent pieces of the east side of the Uncompahgre highland (Figure 11). Gneiss is abundant in conglomerate and conglomeratic sandstone throughout

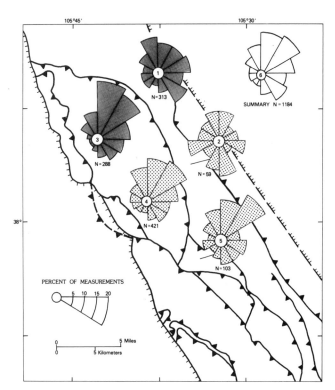

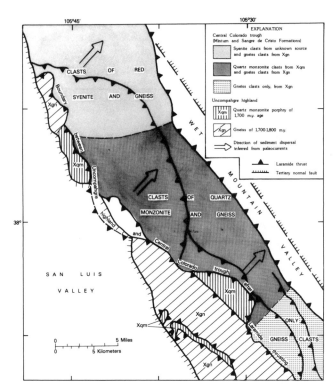

Figure 10—Map showing paleocurrents determined by measurement of cross bedding in Minturn and Sangre de Cristo formations, northern Sangre de Cristo Range. Rose diagrams: 1. Sangre de Cristo Formation, autochthon; 2. Minturn Formation, autochthon; 3. Sangre de Cristo Formation, Spread Eagle Peak plate; 4. Minturn Formation, Spread Eagle Peak plate; 5. Minturn Formation, Marble Mountain plate; and 6. summary of all measurements. Shaded rose diagrams, Minturn Formation; dotted, Sangre de Cristo Formation. *N*, number of measurements used in each rose diagram.

Figure 11—Map showing post-Laramide remnants of Uncompahgre highland and central Colorado trough in the study area. Clasts of distinctive pink quartz monzonite porphyry (Xqm) of 1.7 b.y. age occur downstream from identical rock exposed in thrust plates that represent pieces of the Uncompahgre highland. Unlabeled areas, no information on type of terrane.

the Sangre de Cristo Formation; it also crops out discontinuously throughout the Precambrian terrane that represents the highland. Clasts of pink quartz monzonite porphyry and related rocks occur only downstream from plutons of identical quartz monzonite and related rocks correlated with the Boulder Creek Granodiorite of 1.7 b.y. age (Figure 11). A stock of quartz monzonite near Crestone, correlated with the Silver Plume Quartz Monzonite of 1.4 b.y. age, probably was not exposed during Pennsylvanian time. Although clasts of the younger quartz monzonite occur in the Sangre de Cristo Formation, they probably were derived from other stocks now buried beneath the downfaulted San Luis Valley. The source of red syenite clasts also is not exposed; the probable source of the syenite has been downfaulted and concealed in the northern San Luis Valley.

IMPLICATIONS FOR ANCESTRAL ROCKY MOUNTAINS

Tectonics

The Ancestral Rocky Mountains began to form in Early Pennsylvanian time and were the dominant topographic

features of the Rocky Mountain region for the rest of Paleozoic time (Mallory, 1972b; McKee et al., 1975) (Figure 1). They rose rapidly in Middle Pennsylvanian time during the culmination of deformation in the Ouachita–Marathon thrust belt. The Ancestral Rockies may have formed in response to complex stresses transferred to the southwestern salient of Laurentia when it collided with Gondwana and produced the Ouachita–Marathon belt (Kluth and Coney, 1981a, b; 1983). A commonly assumed model of Ancestral Rockies tectonics involves mostly vertical movement along high-angle faults, giving rise to a system of tilted ranges and basins (DeVoto, 1980; Kluth and Coney, 1983; Tweto, 1983).

The structure of the Ancestral Rocky Mountains has also been interpreted as a system of wrench faults (Baars and Stevenson, 1983) or as large reverse faults and thrusts resembling those of the Laramide foreland province of the Rocky Mountains (Warner, 1983). Evidence for alternative interpretations of the structure of the Ancestral Rockies is perhaps best observed outside the area affected by the Laramide orogeny, in Oklahoma. There, a wrench interpretation of Pennsylvanian faulting is supported by evidence for left-lateral movement of the Washita Valley fault in the Arbuckle Mountains (Tanner, 1967), located southeast of the Amarillo–Wichita highland. A compressional origin by thrusting is supported by evidence from seismic reflection profiles that reveal moderately dipping thrusts separating the Amarillo–Wichita highland from the Anadarko basin of Oklahoma (Brewer et al., 1983). Although the data from Oklahoma provide the best evidence for a compressional fault system, a component of strike-slip movement is indicated by the strike-slip Washita Valley fault.

The Ancestral Rocky Mountains have been delineated by faults interpreted to have been active during Pennsylvanian time, by the absence of pre-Triassic strata on highlands, and by the presence of thick deposits of coarse detritus of Pennsylvanian and Permian age in basins adjacent to faults (Tweto, 1983). Most faults of Pennsylvanian age were also active in Laramide or later time, but on the basis of thickness changes and stratigraphic relationships, they are interpreted to have been active in Pennsylvanian time also. The Ancestral Front Range and Apishipa highlands were probably bounded by the Gore Range fault on the west side (Tweto and Lovering, 1977), the Ute Pass and Wet Mountain faults on the east side (Maher, 1953), and the subsurface Freezeout Creek fault east of the Apishipa highland (McKee, 1975) (Figure 1). Faults active in Pennsylvanian time along the margins of the Uncompahgre highland include the Picuris–Pecos fault on the southeast side (Miller et al., 1963) and a major fault system along the southwest side of the highland (Mallory, 1972b; Spoelhof, 1974; Millberry, 1983). The Picuris–Pecos, Gore Range, and perhaps other faults were active during Precambrian, late Paleozoic, and Late Cretaceous–early Tertiary times (Miller et al., 1963; Tweto and Lovering, 1977; Budnick, 1983).

Parts of two structural elements of the Ancestral Rockies, the central Colorado trough and the Uncompahgre highland, are preserved within the present Sangre de Cristo Range (Figures 1 and 11). Faults of Pennsylvanian age that may have separated the two elements are obscured by Laramide thrusts (Figure 3). Laramide thrust plates containing thick sections of Minturn and Sangre de Cristo formations represent pieces of the trough. Thrust plates composed principally of Precambrian rocks, with conglomerate facies of the Sangre de Cristo Formation resting unconformably on Precambrian and lower Paleozoic rocks, represent fragments of the highland. Faults of Pennsylvanian age may have occupied the positions of Laramide thrusts now separating the plates, but the overprint of Laramide thrusting makes recognition of earlier faults impossible. As inferred from the distribution of distinctive clasts of pink quartz monzonite porphyry in the Sangre de Cristo Formation directly downstream from their probable source (Figure 11), the boundary between the central Colorado trough and the Uncompahgre highland has not been offset by major strike-slip faulting in the study area.

The distribution of facies and unconformities in the Minturn and Sangre de Cristo formations (Figure 4) is interpreted as the record of uplift for the eastern part of the Uncompahgre highland. At least three episodes of uplift along the boundary fault are recorded by the following: (1) the onset of deposition of coarse clastic sediment at the beginning of Minturn time, (2) the abrupt transition from shallow marine limestone-bearing facies to alluvial fan

sandstone and conglomerate facies at the Minturn–Sangre de Cristo contact and the unconformable nature of the contact in the Marble Mountain thrust plate, and (3) the abrupt progradation of conglomerate over sandstone facies of the Sangre de Cristo Formation in the Spread Eagle Peak thrust plate. These episodes of uplift must have affected most of the east-central part of the Uncompahgre highland because their stratigraphic record extends throughout most of the northern Sangre de Cristo Range. Proposed correlation of similar stratigraphic records in other parts of the central Colorado trough (Brill, 1952) is possible but remains unconfirmed.

Paleoclimate

The Ancestral Rocky Mountains lay within 10–20° of the equator in Middle Pennsylvanian to Early Permian time and within the zone of easterly trade winds (Heckel, 1977; Smith et al., 1981; Parrish et al., 1983) (Figure 1). Paleowind directions determined from cross bedding in eolian sandstone (Knight, 1929; Opdyke and Runcorn, 1960; Poole, 1962) support the predicted direction of trade winds (Figure 1). Based on the distribution of evaporite minerals, coal, and plant fossils and on textural and mineralogic studies of arkose, the eastern side of the Ancestral Rocky Mountains of Colorado was probably more humid than the western side in Early Pennsylvanian time (Mack et al., 1979). Thin coals, plant fossils, and paleosols with root casts indicate humid conditions during deposition of Lower Pennsylvanian strata on the east side of the Ancestral Rockies; their absence, along with vertical changes in arkose petrography, suggest more arid conditions in later Pennsylvanian time (Suttner, 1984). Deposition of evaporite minerals had spread throughout the region by Permian time, suggesting a more arid climate than that of Early Pennsylvanian time (Mack et al., 1979). Evidence for increasingly dry conditions during Pennsylvanian time may reflect the development of monsoonal climate over extensive equatorial mountains, raised by the collision of Gondwana and Laurentia (Parrish et al., 1983).

Depositional features of the Minturn and the lower Sangre de Cristo formations reflect both wet and dry conditions on the east side of the Uncompahgre highland during Middle Pennsylvanian time; features of the Sangre de Cristo Formation may indicate dry conditions as late as Early Permian time. Abundant fragmentary remains of coal swamp plants, such as *Calamites*, *Cordaites*, and *Neuropteris* (identified by R. M. Kosanke, personal communication, 1982) in fan delta deposits of the Minturn Formation, indicate that coal swamp plants grew along streams and coastal areas during Middle Pennsylvanian time. The Pennsylvanian coal swamp flora of the eastern United States is generally regarded as good evidence for a warm, humid climate (Schopf, 1975). Evidence for dry conditions in the Minturn is sparse, but horizons of calcareous nodules interpreted as paleocaliche occur at scattered stratigraphic levels. Distal alluvial fan deposits of the Sangre de Cristo Formation, including some dated as Middle Pennsylvanian, contain numerous intervals of paleocaliche, and these deposits interfinger with proximal fan deposits that include abundant deposits of debris flows. Debris flow deposits characterize modern alluvial fans in the arid southwestern

United States (Bull, 1972), and calcareous paleosols are consistent with semiarid to arid climates (Turner, 1980), although neither is limited to arid regions.

Evidence for wet conditions in Minturn time and drier conditions during Sangre de Cristo time is consistent with the model of increasing aridity proposed by Mack et al. (1979), but humid and dry environments may have coexisted on the east side of the Uncompahgre highland during Middle Pennsylvanian time. In the study area, evidence for wet conditions is from deposits that were apparently in low-lying streams and coastal areas (Minturn Formation), and evidence for dry conditions is mostly from alluvial fan deposits (Sangre de Cristo Formation). Thus, low-lying stream courses and deltas on the east side of the Uncompahgre highland had sufficient water to support growth of coal swamp plants, but upland areas and fringing alluvial fans of the east side of the highland were probably dry.

Redbeds

Great thicknesses of arkosic detritus accumulated around the margins of the Ancestral Rocky Mountains (Maher and Collins, 1953; Mallory, 1972b; McKee et al., 1975). Most arkosic detritus, composed of unweathered quartz, feldspar, and rock fragments, was gray to pink in color at the time of deposition. Subsequently, much of the arkose has been altered red by diagenetic leaching of iron from mafic minerals (T. R. Walker, 1967). Alteration took place around the margins of basins where oxidizing ground water flowed through arkosic detritus in alluvium. Groundwater leached iron from ferromagnesian minerals such as hornblende and biotite and deposited amorphous iron oxide or hydroxide in pore spaces. Ultimately, amorphous iron compounds were converted to finely crystallized hematite that colored the rock red (T. R. Walker, 1967; Turner, 1980).

Except for quartzose sandstone in the lower part of the Minturn Formation, sandstones of the Minturn and Sangre de Cristo are arkosic. Total feldspar equals quartz, and feldspar is about evenly divided between potassium feldspar and plagioclase; these ratios are about what would be expected if unweathered detritus was eroded from Precambrian gneiss, quartz monzonite, and related plutonic rocks now exposed in the Sangre de Cristo Range. Flakes of muscovite and biotite make up a few percent of the arkose. Mica in gray and pink sandstone of the Minturn Formation is partly altered and surrounded by halos of red hematite pigment. Relict flakes of mica in the red Sangre de Cristo Formation are replaced completely by hematite pigment, and pores throughout the sandstone are filled with pigment.

Most redbeds in the Minturn and Sangre de Cristo formations are basin margin alluvial deposits. Redbeds include all of the conglomerate and sandstone facies of the Sangre de Cristo Formation, the alluvial redbed facies of the Minturn Formation, and a few intervals of alluvial and delta plain deposits in the Minturn Formation. Commonly, red and gray intervals alternate near the boundary between the Minturn and Sangre de Cristo formations.

CONCLUSIONS

The Minturn and Sangre de Cristo formations were apparently deposited on fan deltas and alluvial fans along the west side of the central Colorado trough. In the Minturn Formation, fan delta deposits that prograded onto the sea bottom below wave base are distinguished from those that prograded above wave base by the presence of turbidites in the former and shallow marine limestones and deltaic foreset beds in the latter. The fan delta deposits of the Minturn were covered by alluvial fans in Sangre de Cristo time. The Sangre de Cristo Formation contains coarse sediments deposited on the distal surfaces of alluvial fans by braided streams and on proximal fan surfaces by debris flows and streams. The Crestone Conglomerate Member of the Sangre de Cristo Formation is composed entirely of proximal alluvial fan deposits.

Faulting along the boundary between the Uncompahgre highland and the central Colorado trough initiated uplift and erosion of the highland and deposition of coarse detritus along the west side of the trough. Sediment dispersal was almost perpendicular to the eastern margin of the Uncompahgre highland. The late Paleozoic faulted boundary of the highland was destroyed by Laramide thrusting, but the boundary was probably a normal or reverse fault system. Alignment of the dispersal pattern of clasts in conglomerate with identical rocks in the source terrane indicates absence of strike-slip movement. At least three periods of uplift caused abrupt progradation of coarse detritus into the central Colorado trough. During or after deposition, alluvial deposits around the margins of the trough were altered to redbeds.

ACKNOWLEDGMENTS

We thank C. A. Brannon, B. L. DeAngelis, K. Hafner, R. A. Schaefer, and D. M. Walz for their assistance in the field; R. J. Flores for observations on the Crestone Conglomerate Member of the Sangre de Cristo Formation; and R. E. Phillips for study of thin sections; all were employed by the U.S. Geological Survey. D. L. Baars, C. Campbell, R. H. DeVoto, R. M. Flores, L. J. Suttner, O. Tweto, and T. R. Walker reviewed the manuscript; we thank them for their comments.

REFERENCES CITED

Baars, D. L., and G. M. Stevenson, 1983, Paleotectonic control of Pennsylvanian sedimentation in Paradox basin (abs.): AAPG Bulletin, v. 67, n. 8, p. 1329.

Balsley, J. K., 1982, Cretaceous wave-dominated delta systems: Book Cliffs, east central Utah: AAPG Guidebook, 219 p.

Baltz, E. H., and M. O'Neill, 1980, Preliminary geologic map of the Mora River area, Sangre de Cristo Mountains, New Mexico: U.S. Geological Survey Open-File Map 80-374.

Becker, P. J., 1978, Middle Pennsylvanian cyclical sedimentation in the Minturn Formation of south-central Colorado: Master's thesis, University of Iowa, Iowa City, 79 p.

Berg, T. M., 1967, Pennsylvanian biohermal limestones of Marble Mountain, south-central Colorado: Master's thesis, University of Colorado, Boulder, 196 p.

Bolyard, D. W., 1959, Pennsylvanian and Permian stratigraphy in Sangre de Cristo Mountains between La

Veta Pass and Westcliffe, Colorado: AAPG Bulletin, v. 43, n. 8, p. 1896-1939.

Bouma, A. H., 1962, Sedimentology of Some Flysch Deposits: a Graphic Approach to Facies Interpretation: Elsevier, Amsterdam, 168 p.

Brewer, J. A., R. Good, J. E. Oliver, L. D. Brown, and S. Kaufman, 1983, COCORP profiling across the southern Oklahoma aulacogen: overthrusting of the Wichita Mountains and compression within the Anadarko basin: Geology, v. 11, n. 2, p. 109-114.

Brill, K. G., Jr., 1952, Stratigraphy in the Permo-Pennsylvanian zeugogeosyncline of Colorado and northern New Mexico: Geological Society of America Bulletin, v. 63, n. 8, p. 809-880.

Budnick, R. T., 1983, Recurrent intraplate deformation on the Ancestral Rocky Mountain orogenic belt (abs.): Geological Society of America Abstracts with Programs, 96th Annual Meeting, p. 535.

Bull, W. B., 1972, Recognition of alluvial-fan deposits in the stratigraphic record, *in* J. K. Rigby, and W. D. Hamblin, eds., Recognition of ancient sedimentary environments: Society of Economic Paleontologists and Mineralogists Special Publication 16, p. 63-83.

Burbank, W. S., and E. N. Goddard, 1937, Thrusting in Huerfano Park, Colorado, and related problems of orogeny in the Sangre de Cristo Mountains: Geological Society of America Bulletin, v. 48, n. 7, p. 931-976.

Casey, J. M., 1980, Depositional systems and paleogeographic evolution of the late Paleozoic Taos trough, northern New Mexico, *in* T. D. Fouch, and E. R. Magathan, eds., Paleozoic paleogeography of west-central United States: Rocky Mountain Section, Society of Economic Paleontologists and Mineralogists, p. 181-196.

Casey, J. M., and A. J. Scott, 1979, Pennsylvanian coarse-grained fan deltas associated with the Uncompahgre uplift, Talpa, New Mexico, *in* R. V. Ingersoll, ed., Guidebook of Santa Fe country: New Mexico Geological Society 13th Annual Field Conference Guidebook, p. 211-218.

Chronic, B. J., Jr., 1958, Pennsylvanian rocks in central Colorado, *in* Symposium on Pennsylvanian rocks of Colorado and adjacent areas: Denver, Rocky Mountain Association of Geologists, p. 59-63.

Clark, R. F., 1982, Stratigraphy, sedimentology and copper-uranium occurrences of the upper part of the Middle Pennsylvanian Minturn Formation, Sangre de Cristo Mountains, Colorado: Master's thesis, Colorado School of Mines, Golden, 151 p.

Collinson, J. D., 1968, Deltaic sedimentation units in the upper Carboniferous of northern England: Sedimentology, v. 10, p. 233-254.

Curray, J. R., 1956, The analysis of two-dimensional orientation data: Journal of Geology, v. 64, n. 2, p. 117-131.

DeVoto, R. H., 1980, Pennsylvanian stratigraphy and history of Colorado, *in* H. C. Kent, and K. W. Porter, eds., Colorado geology: Denver, Rocky Mountain Association of Geologists, p. 71-101.

Dutton, S. P., 1982, Pennsylvanian fan-delta and carbonate deposition, Mobeetie Field, Texas panhandle: AAPG Bulletin, v. 66, n. 4, p. 389-407.

Elliott, T., 1978, Deltas, *in* H. G. Reading, ed., Sedimentary Environments and Facies: Blackwell Scientific Publications, Oxford, p. 97-142.

Flore, R. J., 1985, Sedimentation model for the Crestone Conglomerate Member of the Sangre de Cristo Formation (Pennsylvanian–Permian), south-central Colorado: Master's thesis, Indiana University, Bloomington, Indiana, 146 p.

Gilbert, G. K., 1883, The topographic features of lake shores: U.S. Geological Survey 5th Annual Report, p. 75-123.

Heckel, P. H., 1977, Origin of phosphatic black shale facies in Pennsylvanian cyclothems of mid-continent North America: AAPG Bulletin, v. 61, n. 7, p. 1054-1068.

Holmes, A., 1965, Principles of Physical Geology (2nd ed.): Ronald Press, New York, 1288 p.

Johnson, B. R., D. A. Lindsey, R. M. Bruce, and S. J. Soulliere, in press, Reconnaissance geologic map of the Sangre de Cristo wilderness study area, south-central Colorado: U.S. Geological Survey Miscellaneous Field Investigations Map MF-1635-B.

Karig, D. E., 1964, Structural analysis of the Sangre de Cristo Range, Venable Peak to Crestone Peak, Custer and Saguache counties, Colorado: Master's thesis, Colorado School of Mines, Golden, 143 p.

Kluth, C. F., and P. J. Coney, 1981a, Plate tectonics of the Ancestral Rocky Mountains: Geology, v. 9, n. 1, p. 10-15.

———, 1981b, Reply to comment on "Plate tectonics of the Ancestral Rocky Mountains": Geology, v. 9, n. 9, p. 388-389.

———, 1983, Reply to comment on "Plate tectonics of the Ancestral Rocky Mountains": Geology, v. 11, no. 2, p. 121-122.

Knight, S. H., 1929, The Fountain and the Casper formations of the Laramie basin: University of Wyoming Publications in Science: Geology, v. 1, n. 1, 82 p.

Langford, R. P., and D. A. Fishbaugh, 1984, Sedimentology of the Fountain fan-delta complex near Manitou Springs, Colorado, *in* L. J. Suttner, R. P. Langford, and A. W. Shultz, eds., Sedimentology of the Fountain fan-delta complex near Manitou Springs and Canon City, Colorado: Rocky Mountain Section, Society of Economic Paleontologists and Mineralogists 1984 Spring Field Conference Guidebook, p. 1-30.

Lindsey, D. A., and R. A. Schaefer, 1984, Principal reference section for the Sangre de Cristo Formation (Pennsylvanian and Permian), northern Sangre de Cristo Range, Saguache County, Colorado: U.S. Geological Survey Miscellaneous Field Investigations Map MF-1622-A.

Lindsey, D. A., B. J. Johnson, and P. A. M. Andriessen, 1983, Laramide and Neogene structure of the northern Sangre de Cristo Range, south-central Colorado, *in* J. D. Lowell, ed., Rocky Mountain foreland basins and uplifts: Denver, Rocky Mountain Association of Geologists, p. 219-228.

Lindsey, D. A., R. F. Clark, and S. J. Soulliere, 1985, Reference section for the Minturn Formation (Middle Pennsylvanian), northern Sangre de Cristo Range,

Custer County, Colorado: U.S. Geological Survey Miscellaneous Field Investigations Map MF-1622-C.

Luchitta, I., and N. Suneson, 1981, Flash flood in Arizona—observations and their application to the identification of flash-flood deposits in the geologic record: Geology, v. 9, n. 9, p. 414–418.

Mack, G. H., and K. A. Rasmussen, 1984, Alluvial-fan sedimentation of the Cutler Formation (Permo-Pennsylvanian) near Gateway, Colorado: Geological Society of America Bulletin, v. 95, n. 1, p. 109–116.

Mack, G. H., L. J. Suttner, and J. R. Jennings, 1979, Permo-Pennsylvanian climatic trends in the Ancestral Rocky Mountains, *in* D. L. Baars, ed., Permianland, Four Corners Geological Society, 9th Field Conference Guidebook, p. 7–12.

Maher, J. C., 1953, Permian and Pennsylvanian rocks of southeastern Colorado: AAPG Bulletin, v. 37, n. 5, p. 913–939.

Maher, J. C., and J. B. Collins, 1953, Permian and Pennsylvanian rocks of southeastern Colorado and adjacent areas: U.S. Geological Survey Oil and Gas Investigations Map OM-135.

Mallory, W. W., 1972a, Regional synthesis of the Pennsylvanian system, *in* W. W. Mallory, ed., Geologic atlas of the Rocky Mountain region: Denver, Rocky Mountain Association of Geologists, p. 111–127.

———, 1972b, Pennsylvanian arkose and the Ancestral Rocky Mountains, *in* W. W. Mallory, ed., Geologic atlas of the Rocky Mountain region: Denver, Rocky Mountain Association of Geologists, p. 131–132.

McBride, E. F., A. E. Weide, and J. A. Wolleben, 1975, Deltaic and associated deposits of Difunta Group (Late Cretaceous to Paleocene), Parras and La Popa basins, northeastern Mexico, *in* M. L. Broussard, ed., Deltas, models for exploration: Houston Geological Society, p. 485–522.

McKee, E. D., 1975, Interpretation of Pennsylvanian history, *in* E. D. McKee, E. J. Crosby, et al., compilers, Paleotectonic investigations of the Pennsylvanian system in the United States: U.S. Geological Survey Professional Paper 853, pt. 2, p. 1–21.

McKee, E. D., E. J. Crosby, et al., compilers, 1975, Paleotectonic investigations of the Pennsylvanian System in the United States: U.S. Geological Survey Professional Paper 853, 3 parts, 17 plates.

Miall, A. D., 1978, Lithofacies types and vertical profile models in braided river deposits: a summary, *in* A. D. Miall, ed., Fluvial sedimentology: Canadian Society of Petroleum Geologists Memoir 5, p. 597–604.

Middleton, G. V., and M. A. Hampton, 1976, Subaqueous sediment transport and deposition by sediment gravity flows, *in* D. J. Stanley, and D. J. P. Swift, eds., Marine Sediment Transport and Environmental Management: John Wiley, New York, p. 197–218.

Millberry, K. W., 1983, Tectonic control of Pennsylvanian fan delta deposition, southwest Colorado (abs.): AAPG Bulletin, v. 67, n. 3, p. 514.

Miller, J. P., A. Montgomery, and P. K. Sutherland, 1963, Geology of part of the southern Sangre de Cristo Mountains, New Mexico: New Mexico Bureau of Mines and Mineral Resources Memoir 11, 106 p.

Nilsen, T. H., 1982, Alluvial fan deposits, *in* P. A. Scholle, and D. Spearing, eds., Sandstone depositional environments: AAPG Memoir 31, p. 49–86.

Opdyke, N. D., and S. K. Runcorn, 1960, Wind direction in the western United States in late Paleozoic: Geological Society of America Bulletin, v. 71, n. 7, p. 959–972.

Parrish, J. T., A. M. Ziegler, and C. R. Scotese, 1983, Global paleogeography and paleoclimate in the late Carboniferous (abs.): Geological Society of America Abstracts with Programs, 96th Annual Meeting, p. 658.

Poole, F. G., 1962, Wind directions in late Paleozoic to middle Mesozoic time on the Colorado Plateau, *in* Geological Survey Research 1962: U.S. Geological Survey Professional Paper 450-D, p. D147–D151.

Rachocki, A. H., 1981, Alluvial Fans: John Wiley, New York 161 p.

Rhodes, J. A., 1964, Stratigraphy and origin of the Pennsylvanian–Permian rocks of the Huerfano Park quadrangle, Colorado: Ph.D. thesis, University of Michigan, Ann Arbor, 176 p.

Ricci-Lucchi, R., 1975, Depositional cycles in two turbidite formations of northern Appennines (Italy): Society of Economic Paleontologists and Mineralogists Journal, v. 45, n. 1, p. 3–43.

Rust, B. R., 1978, Depositional models for braided alluvium, *in* A. D. Miall, ed., Fluvial sedimentology: Canadian Society of Petroleum Geologists Memoir 5, p. 605–625.

Schopf, J. M., 1975, Pennsylvanian climate in the United States, *in* E. D. McKee, E. J. Crosby, et al., compilers, Paleotectonic investigations of the Pennsylvanian system in the United States: U.S. Geological Survey Professional Paper 853, pt. 2, p. 23–31.

Seilacher, A., 1978, Use of trace fossils for recognizing depositional environments, *in* P. B. Basan, ed., Trace fossil concepts: Society of Economic Paleontologists and Mineralogists Short Course No. 5, p. 167–181.

Shultz, A. W., 1984, Provenance and sedimentology of the Fountain Formation near Canon City, Colorado, *in* L. J. Suttner, R. P. Langford, and A. W. Shultz, eds., Sedimentology of the Fountain fan-delta complex near Manitou Springs and Canon City, Colorado: Rocky Mountain Section, Society of Economic Paleontologists and Mineralogists Guidebook 1984 Spring Field Conference Guidebook, p. 62–85.

Smith, A. G., A. M. Hurley, and J. C. Briden, 1981, Phanerozoic paleocontinental world maps: Cambridge University Press, Cambridge, Massachusetts, 104 p.

Soulliere, S. J., B. L. DeAngelis, and D. A. Lindsey, 1984, Measured sections and discussion of the main turbidite member, Middle Pennsylvanian Minturn Formation, northern Sangre de Cristo Range, Custer and Saguache counties, Colorado: U.S. Geological Survey Miscellaneous Field Investigations Map MF-1622-B.

Spoelhof, R. W., 1974, Pennsylvanian stratigraphy and tectonics in the Lime Creek–Molas Lake area, San Juan County, Colorado: Master's thesis, Colorado School of Mines, Golden, 193 p.

Steel, R. J., 1974, Cornstone (fossil caliche)—its origin, stratigraphic and sedimentological importance in the New Red Sandstone, western Scotland: Journal of Geology, v. 82, n. 3, p. 351–369.

Suttner, L. J., 1984, Sedimentologic indicators of Fountain paleoclimate, *in* L. J. Suttner, R. P. Langford, and A. W. Shultz, eds., Sedimentology of the Fountain fan-delta complex near Manitou Springs and Canon City, Colorado: Rocky Mountain Section, Society of Economic Paleontologists and Mineralogists for 1984 Spring Field Conference Guidebook, p. 87–96.

Tanner, J. H., 1967, Wrench fault movement along the Washita Valley fault, Arbuckle Mountains area, Oklahoma: AAPG Bulletin, v. 51, n. 1, p. 126–134.

Taylor, R. B., G. R. Scott, and R. A. Wobus, 1975, Reconnaissance geologic map of the Howard quadrangle, central Colorado: U.S. Geological Survey Miscellaneous Investigations Map I-892.

Turner, P., 1980, Continental red beds: Developments in Sedimentology 29: Elsevier, New York, 562 p.

Tweto, O., compiler, 1979a, Geologic map of Colorado: U.S. Geological Survey, scale 1:500,000.

———, 1979b, The Rio Grande rift system in Colorado, *in* R. E. Riecker, ed., Rio Grande rift: Tectonics and magmatism: Washington, D.C., American Geophysical Union, p. 33–56.

———, 1983, Geologic sections across Colorado: U.S. Geological Survey Miscellaneous Investigations Series Map I-1416.

Tweto, O., and T. S. Lovering, 1977, Geology of the Minturn 15-minute quadrangle, Eagle and Summit counties, Colorado: U.S. Geological Survey Professional Paper 956, 96 p.

Vaughn, P. P., 1972, More vertebrates, including a new microsaur, from the upper Pennsylvanian of central Colorado: Los Angeles County Museum Contributions to Science, n. 223, p. 1–29.

Walker, R. G., 1967, Turbidite sedimentary structures and their relationship to proximal and distal depositional environments: Society of Economic Paleontologists and Mineralogists Journal, v. 37, n. 1, p. 25–43.

———, 1978, Deep-water sandstone facies and ancient submarine fans: models for exploration for stratigraphic traps: AAPG Bulletin, v. 62, n. 6, p. 932–966.

Walker, T. R., 1967, Formation of red beds in modern and ancient deserts: Geological Society of America Bulletin, v. 78, n. 3, p. 353–368.

Warner, L. A., 1983, Comment on "Plate tectonics of the Ancestral Rocky Mountains": Geology, v. 11, n. 2, p. 120–121.

Wright, L. D., 1978, River deltas, *in* R. A. Davis, Jr., ed., Coastal Sedimentary Environments: Springer-Verlag, New York, p. 5–68.

Jurassic Paleotectonics in the West-Central Part of the Colorado Plateau, Utah and Arizona

F. Peterson
U.S. Geological Survey
Denver, Colorado

Updated isopach maps and new information on the sedimentology and distribution of facies indicate that the Jurassic was a time of slight but noteworthy crustal deformation on the Colorado Plateau that was of sufficient magnitude to exert an influence on sedimentary processes.

The tectonic history of Lower Jurassic rocks is poorly understood. The west-central part of the Colorado Plateau was a broad lowland region covered with thick eolian sand of the Wingate (Lukachukai Member) and Navajo sandstones, interrupted briefly by deposition of the fluvial Kayenta and Moenave formations. Drainage patterns obtained from cross-bedding in the fluvial beds trend generally westward, which might be a reflection of westward regional tilting.

Considerable westward regional tilting occurred in earliest Middle Jurassic time during formation of the J-2 unconformity, and westward regional tilting continued through middle Middle Jurassic time. The Monument region was an elongate north-south-trending structural bench and remained as an important structural element during the rest of the Jurassic. Downwarps formed in the Circle Cliffs and Black Mesa regions. A significant change in structural style occurred in late Middle Jurassic time coinciding with formation of the J-3 unconformity. Westward regional tilting ceased, and the Circle Cliffs and Black Mesa regions became structurally positive areas. The Kaiparowits and Henry Mountains regions subsided as basins, and uplift in the Echo Cliffs–Kaibab region defined the southwestern margin of the Kaiparowits downwarp. The Monument region continued as a structural bench.

The same general tectonic framework continued into the early Late Jurassic. The Emery uplift and Black Mesa area were small positive elements that rose above base level and contributed minor quantities of sediment to adjacent areas. Other than these two uplifts, the other active positive structures, including the Monument uplift, were covered by Jurassic sedimentary rocks and subsided at a slower rate than adjacent downwarps. Latest Jurassic tectonism is poorly known because most of these rocks were deeply eroded or removed during Early Cretaceous time prior to deposition of the Dakota Formation. The Monument uplift and Henry trough persisted, and uplift probably continued on the Circle Cliffs and Black Mesa uplifts.

Only vertical movements were identified, although the ultimate cause of the deformation was probably compressional stresses accompanying continental and oceanic plate interactions along the western continental margin of North America.

INTRODUCTION

This paper stresses the tectonic activity that occurred on the Colorado Plateau during Jurassic time, a period commonly regarded as a time of tectonic quiescence. Stokes (1954a, b) probably was the first to document it, and he and several subsequent workers (Wright and Dickey, 1963; Peterson, 1980, 1984; Santos and Turner-Peterson, in press) also noticed the relationship of sedimentologic features in some of the Jurassic rocks of the Colorado Plateau to crustal movements. Although structural activity certainly was subdued, sufficient evidence is now available to document that a small but noteworthy amount of deformation occurred in a large part of the region during the Jurassic Period.

Although the emphasis here is on the Jurassic System in a major part of the Colorado Plateau, a number of workers, including Walcott (1890), Stokes (1954a, b), Heylmun (1958), Wengerd and Matheny (1958), Fetzner (1960), Kirkland (1963), McKee (1963), Wright and Dickey (1963), Lessentine (1965), Baars (1966), Baars and See (1968), Baars and Stevenson (1977, 1981), Stevenson and Baars (1977), Peterson (1980, 1984), Doelling (1981), and Santos and Turner-Peterson (in press) have shown that many segments of the Colorado Plateau cratonic block were sites of slight but repeated tectonic movements throughout much of Precambrian, Paleozoic, Triassic, and Jurassic time. Present-day structural attitudes generally have less than 2° of dip throughout most of the region, and this small amount of deformation reflects the combined effects of Laramide tectonism as well as that which occurred in earlier times. Thus, the amount of structural deformation can be so slight that it will appear insignificant, but, nevertheless, it can produce anomalous thickness variations and can govern the distribution of sedimentary facies.

The purpose of this paper is to document the tectonic activity that occurred in the west-central part of the Colorado Plateau during Jurassic time (Figure 1). The movements probably involved the entire crust and are not related to shallow crustal processes, with the exception of the northeastern part of the study area where flowage in thick beds of Pennsylvanian salt has caused considerable local deformation (Figure 1). Because it is difficult to envision the study area as a tectonic entity isolated from the rest of the Colorado Plateau, it is highly likely that similar types of deep seated deformation occurred throughout the plateau during the Jurassic.

METHODS

For purposes of paleotectonic analysis, Jurassic formations in the west-central part of the Colorado Plateau are here grouped into six divisions corresponding roughly to informal time-stratigraphic units (Figure 2). Most of these divisions are bounded by unconformities or well-defined depositional surfaces interpreted to approximate time lines.

For clarity, major structural elements are referred to as regions (such as the Circle Cliffs region). Terms for structural style are used for the division under discussion; that style may have changed from time to time and may have been different from the structural style presently seen in the study area. For example, the Circle Cliffs region (now a positive structure) was a trough in part of Middle Jurassic time and an uplift in part of Late Jurassic time. The terms *positive area* or *uplift* are used in a sense relative to adjacent areas of subsidence. For conformable strata within divisions to have accumulated over structurally positive areas, the positive areas must have subsided below base level, but at a slower rate than that which occurred in surrounding areas.

The most helpful concepts for use in paleotectonic analysis of the region are thickness variations and facies distributions. Thicknesses of the various stratigraphic units were measured on outcrops or obtained from oil and gas or uranium drill hole logs. Additional thickness data were obtained from Gilluly and Reeside (1928), Baker (1933), Dane (1935), McKnight (1940), Hunt et al. (1953), McFall (1955), Sears (1956), Harshbarger et al. (1957), Craig et al. (1959), Phoenix (1963), Smith et al. (1963), Beaumont and Dixon (1965), Ekren and Houser (1965), O'Sullivan (1965, 1978, 1980a, b, 1981a, b, 1983, 1984), Davidson (1967), Cater (1970), Wright and Dickey (1978a, b, c), Doelling (1981), and O'Sullivan and Pipiringos (1983). In many places, thickness variations of stratigraphic units are a reflection of syndepositional folding. This concept is particularly valid for strata with inferred primary sheet geometry (e.g., coastal sabkha or tidal flat deposits). Also, earlier studies (Peterson, 1984) indicate that thickness

variations are most likely the result of crustal folding rather than differential compaction. Isopach maps of stratigraphic units can therefore provide useful criteria for interpreting the presence and shape of paleostructures. The identification of any given anomaly as a paleosyncline or paleoanticline can be strengthened considerably where confirmed by sedimentologic evidence, especially facies distributions. The use of sedimentary facies distribution data in paleostructural analysis is based on the concept that patterns of sedimentation will vary across structurally controlled topographic high and low areas. For example, lacustrine deposits are likely to be concentrated in topographically and structurally low areas such as synclines.

Sedimentologic studies of stratification ratios in fluvial strata of the Morrison Formation aid in documenting the timing of structural movements during the Late Jurassic. For these studies, the vertical thickness of sets of horizontal laminations and planar cross-beds (including tabular-planar and wedge-planar types of McKee and Weir, 1953) was determined for each fluvial sandstone bed (Peterson, 1984). From these measurements, the stratification ratio, defined as $[P/(P + H)]100$ was calculated, where P is the total thickness of planar cross-bedded sets and H is the total thickness of horizontally laminated sets (modified slightly from Smith, 1970). Ratios were determined for all fluvial sandstone beds in sections of the Salt Wash Member of the Morrison Formation. When used in conjunction with cross-bedding studies, the ratios proved especially helpful in dividing the Salt Wash alluvial complex into three parts or sequences because a fluvial sandstone marker bed approximately in the middle of the Salt Wash had consistently lower values than overlying or underlying beds. In addition, cross-bedding dip vector resultants in this bed tend to dip toward the east or southeast, whereas vector resultants in overlying and underlying beds tend to dip toward the northeast (Peterson, 1984). Stratification ratios appear to be most useful in braided stream beds; they may not be useful in studies of meandering or anastomosing streams that contain scarce planar cross-bedding sets.

A study of paleocurrent flow was also made in the Salt Wash Member. The direction of dip of cross-bedding in trough-shaped sets was determined for at least 15 sets and generally for 25 sets in each fluvial sandstone bed at each locality. These were summarized by the methods of Reiche (1938) and Peterson (1984) to give the resultant (also known as the vector mean) and the consistency factor (also known as the vector mean strength). The resultant indicates the general direction of stream flow, whereas the consistency factor is interpreted to indicate the degree of local variation in the stream flow or, in the case of braided streams, the degree of variation in pathways taken by the individual braid channels that make up the larger braided stream (Peterson, 1984).

LOWER JURASSIC SERIES

Glen Canyon Division

Stratigraphy

The Glen Canyon Group, consisting of four formations, is the oldest assemblage of rocks here included in the Jurassic

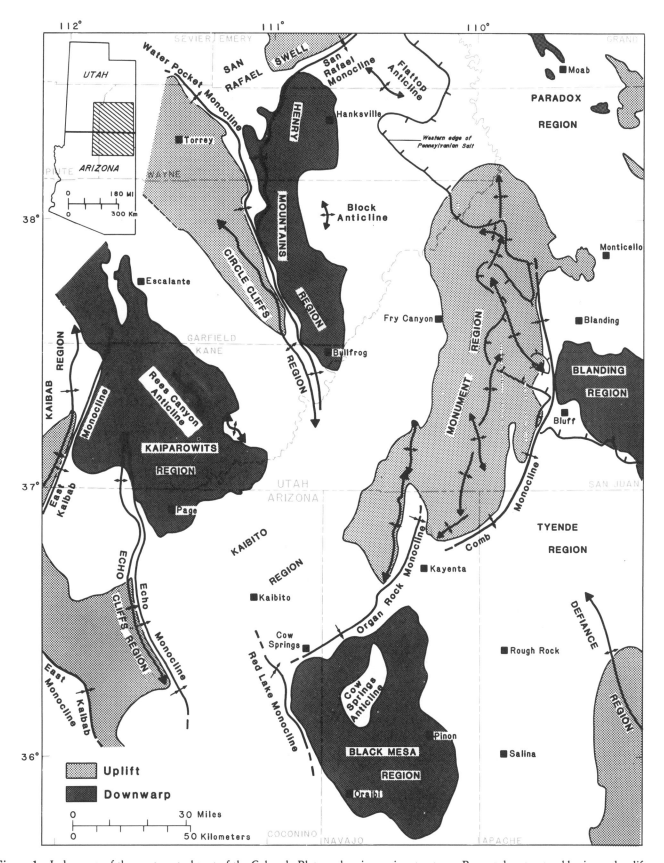

Figure 1—Index map of the west-central part of the Colorado Plateau showing major structures. Present-day structural basins and uplifts are indicated by shading of areas below the 5500 ft (1680 m) and above the 9000 ft (2750 m) structural contours drawn on the base of the Dakota Formation. Axes of downwarps are omitted for clarity. After Kelly (1955), O'Sullivan and Beikman (1963), Williams (1964), Williams and Hackman (1971), Haynes et al. (1972), Hackman and Wyant (1973), Hackman and Olson (1977), and Haynes and Hackman (1978). The western edge of Pennsylvanian salt is from G. M. Stevenson (unpublished map, 1980).

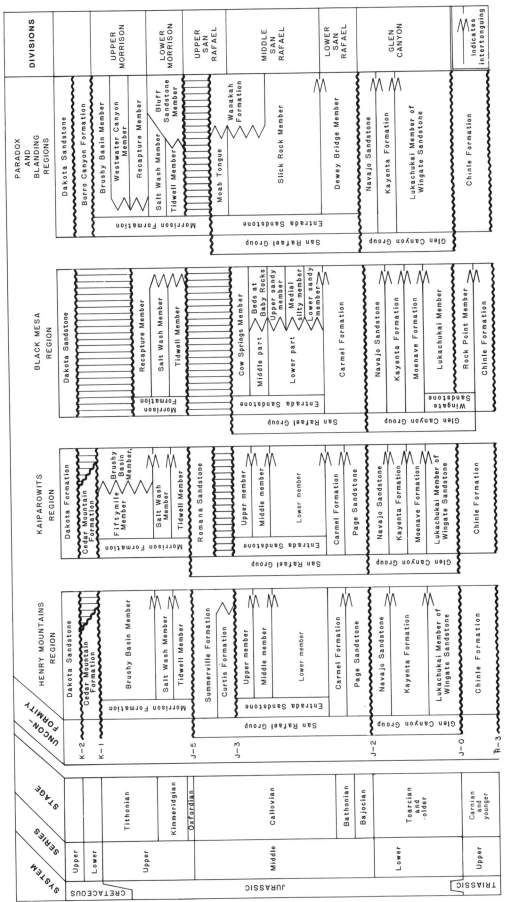

Figure 2—Stratigraphic units in the west-central part of the Colorado Plateau showing major unconformities, stratigraphic divisions, and age assignments. Not to scale.

System. From oldest to youngest, the formations and their current age assignments by the U.S. Geological Survey (Lewis et al., 1961) are the Wingate Sandstone (Late Triassic), the Moenave and Kayenta formations (both Late Triassic?), and the Navajo Sandstone (Late Triassic? and Early Jurassic). Recent paleontologic and stratigraphic studies suggest that the systemic boundary falls within the Wingate Sandstone (Figure 2) (Olsen and Galton, 1977; Peterson et al., 1977; Galton, 1978; Peterson and Pipiringos, 1979). Based on these studies, Peterson and Pipiringos (1979) have expressed the opinion that the part of the group lying above the Rock Point Member of the Wingate Sandstone is Early Jurassic (Sinemurian, Pleinsbachian, and Toarcian) in age. Although there is some disagreement concerning the age of these beds, for purposes of this report that part of the Glen Canyon Group lying above the Rock Point Member of the Wingate Sandstone is here included in the Lower Jurassic.

Jurassic rocks of the Glen Canyon division range in thickness from zero to slightly greater than 2400 ft (730 m) in the study area (Figure 3). The J-0 and J-2 unconformities of Pipiringos and O'Sullivan (1978) and Peterson and Pipiringos (1979) separate these rocks from underlying Upper Triassic and overlying Middle Jurassic strata (Figure 4). (The J-1 unconformity lies at the top of the Navajo outside the study area in southwestern Utah; see Figure 4.) Interfingering relationships and/or gradational contacts are present between the units of the division, making it difficult to interpret the structural history of the region from isopach maps of the individual units.

The Navajo Sandstone and the Lukachukai Member of the Wingate Sandstone consist largely of eolian sandstone and minor lenticular limestone deposited in small playa lakes. These two units are separated by fluvial sandstone of the Kayenta and Moenave formations that contains minor overbank floodplain sandstone and mudstone and lacustrine mudstone. Paleocurrent studies indicate that the Moenave and Kayenta were derived from highland sources to the east in the ancestral Rocky Mountain uplift (Poole, 1961) and to the southeast in the ancestral Mogollon highlands (D. P. Edwards, oral communication, 1985).

Paleotectonics

The Colorado Plateau was a broad, fairly stable platform bordered on the west by the Utah–Idaho trough, which was teconically active during Early Jurassic time and most of Middle Jurassic time. Subsidence of the trough began at least as early as Early Jurassic time, judging from thickness variations of the Glen Canyon division and the source area of some of the beds. The greatest known thickness of the division, approximately 5500 ft (1680 m), is in the Valley of Fire just east of Las Vegas, Nevada (Bohannon, 1983). Just west of Las Vegas, a western source for the basal conglomerate of the group (Marzolf, 1983) suggests deposition on the western flank of the trough, whereas farther east on the Colorado Plateau, an eastern source for fluvial sandstone of the Kayenta and Moenave formations (Poole, 1961) suggests deposition on the eastern flank of the trough.

The tectonic history of the west-central Colorado Plateau during Early Jurassic time is difficult to unravel because of the scarcity of sedimentologic studies that are of sufficient detail to show a relationship between contemporaneous sedimentation and tectonism. Westward thickening (Figure 4) and westward drainage patterns during deposition of the fluvial Kayenta and Moenave formations (Poole, 1961) may be interpreted to indicate westward regional tilting toward the Utah–Idaho trough. The Glen Canyon division is truncated eastward beneath the early Middle Jurassic J-2 unconformity (Figure 4), making it difficult to assess how much of the eastward thinning occurred during rather than after deposition.

On the basis of the isopach map (Figure 3), several structural elements may have been active during Early Jurassic time. Alternatively, these features may not have been activated until earliest Middle Jurassic time when the J-2 unconformity formed. Wide separation of isopach lines just east of the Henry Mountains and Kaiparowits regions indicates a structural bench that was an early manifestation of the Monument uplift. The north end of this feature swings northwestward across the site of the present-day Flattop anticline (or Nequoia arch) and toward the San Rafael Swell. These two structures lie on the Emery uplift and Emery spur (or Schick spur of Wengerd and Matheny, 1958) that were tectonically active in late Paleozoic time (Fetzner, 1960). The relationships suggest that the San Rafael Swell and Flattop anticline are remnants of late Paleozoic structures that were activated during the Early Jurassic and possibly the early Middle Jurassic. The small Rees Canyon anticline formed in the Kaiparowits region on another structural bench defined by widely spaced isopach lines.

MIDDLE JURASSIC SERIES

The Middle Jurassic Series consists entirely of the San Rafael Group and is here divided into three parts (Figure 2). The lower and middle San Rafael divisions have somewhat similar sedimentologic and tectonic histories, whereas the upper San Rafael is sedimentologically similar to underlying divisions of the San Rafael, but has a tectonic history more closely related to that of the overlying lower division of the Morrison Formation.

Lower Division of the San Rafael Group

Stratigraphy

The lower division of the San Rafael Group contains the widespread Carmel Formation and the Page Sandstone of more restricted distribution (Figure 5). The Dewey Bridge Member of the Entrada Sandstone is a correlative of the Carmel and Page that is present in the northeastern part of the study area (Figure 2). The division ranges in thickness from zero to greater than 1200 ft (370 m) in the study area (Figure 6) and thickens to about 3975 ft (1210 m) in the deepest part of the Utah–Idaho trough (Standlee, 1982). West of the Monument region the upper boundary of the

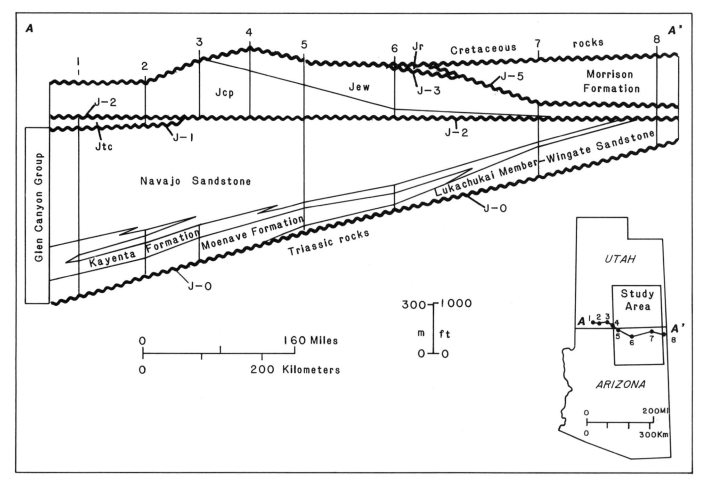

Figure 3—Stratigraphic section showing eastward truncation of the Glen Canyon Group and Temple Cap Sandstone by the J-2 unconformity. J-0, J-1, etc. indicate unconformities of Pipiringos and O'Sullivan (1978). Jtc, Temple Cap Sandstone; Jcp, Carmel and Page formations, undivided; Jew, Entrada and Wanakah formations, undivided; Jr, Romana Sandstone.

division forms a nearly isochronous depositional surface, although the contact is marked by minor interfingering involving about 6 ft (2 m) of strata in the southeastern part of the Kaiparowits region. O'Sullivan (1981a) noted interfingering along this contact east of the Monument region, but the nature of the contact in the Black Mesa region is unknown because it is rarely exposed there. Imlay (1980) assigned an early Middle Jurassic age (Bajocian, Bathonian, and early Callovian) to these rocks.

The lower San Rafael division consists largely of mudstone, sandstone, limestone, and gypsum deposited during two stages of advance and retreat of the sea into the study area (Blakey et al., 1983). Limestone and mudstone were deposited in marine environments in the westernmost part of the Henry, Circle Cliffs, and Kaiparowits regions bordering the Utah–Idaho trough (Figure 6). The marine beds are flanked by coastal sabkha deposits of mudstone and gypsum deposited on a platform farther east and southeast. During early San Rafael time, the Page Sandstone was deposited in an eolian dune field east of the sabkha deposits. At times, the dune field covered most of the Henry and Kaiparowits regions (Figure 7), and it probably extended southwestward into the area west of Black Mesa. Paleocurrent studies indicate that eolian sand in the Page

was transported by north-northwesterly winds and was derived from the shoreline of the Carmel seaway (Blakey et al., 1983).

Paleotectonics

During earliest Middle Jurassic time, the Colorado Plateau was tilted westward toward the Utah–Idaho trough, judging from eastward truncation of the Glen Canyon Group beneath the J-2 unconformity (Figures 3 and 4). The entire Glen Canyon Group and the earliest Middle Jurassic Temple Cap Sandstone, which lies above the Glen Canyon in southwestern Utah, were beveled under the J-2 unconformity, which truncates down-section to the east (Figure 4). This relationship indicates that westward regional tilting continued into early Middle Jurassic (middle Bajocian) time, although uplift or reduced base level resulted in the regional unconformity. In the southeastern part of the study area, the division was removed from the Defiance uplift during the J-2 erosion interval. Farther east in northwestern New Mexico, the informally named beds at Lupton may correlate with part of the Glen Canyon division (O'Sullivan and Green, 1973); presumably this would be with the Moenave

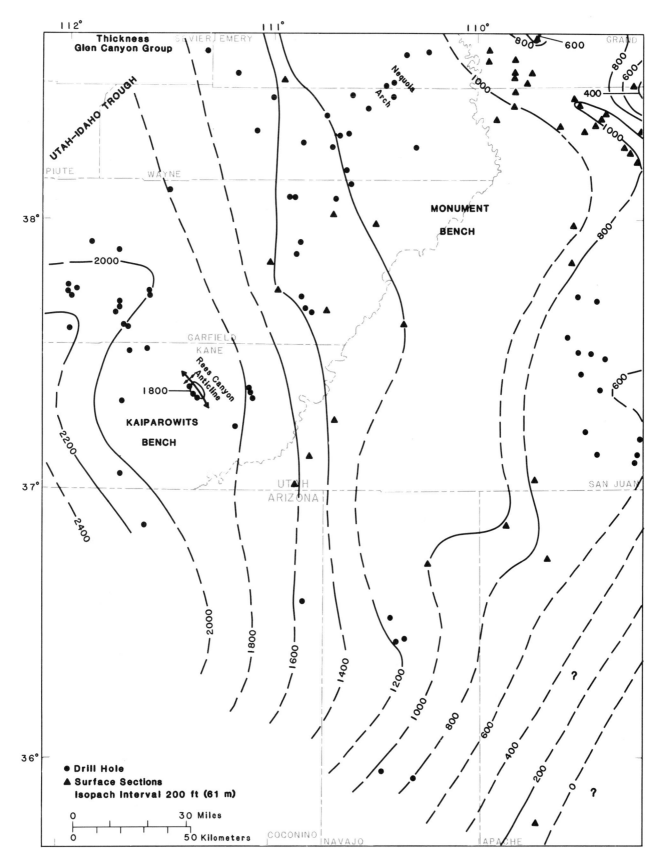

Figure 4—Isopach map of the Glen Canyon Group (excluding the Rock Point Member of the Wingate Sandstone) showing inferred structures.

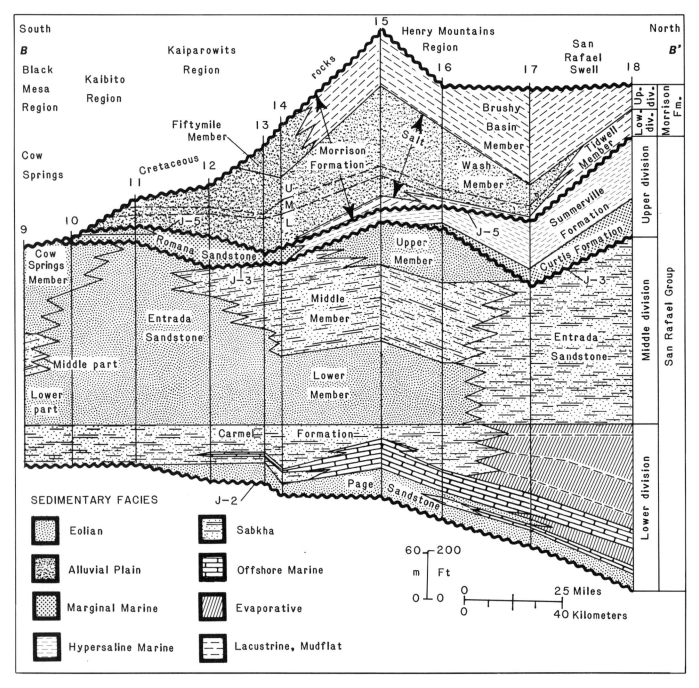

Figure 5—Stratigraphic section of the San Rafael Group and Morrison Formation from Black Mesa to San Rafael Swell showing sedimentary facies. Line of section B–B' is shown in Figure 6. Letters L, M, and U indicate the lower, middle, and upper sequences in the Morrison Formation, respectively. J-2, J-3, etc. indicate unconformities of Pipiringos and O'Sullivan (1978).

Formation. If so, it would suggest that structural movement in the Defiance region during earliest Middle Jurassic time occurred as an uplift rather than as a structural bench or platform.

Slightly later during deposition of the upper part of the lower San Rafael division in late Bajocian and Bathonian time, the west-central Colorado Plateau was dominated structurally by the Utah–Idaho trough and the Monument bench. Extensive coastal sabkha and tidal flat deposits of this division thicken to the west within and west of the Kaiparowits and Henry Mountains regions (Figure 6),

reflecting westward regional tilting that continued from earlier times. Wide separation of isopach lines in the Monument region indicates that it continued as a structural bench. The conclusion is supported by the eastward pinchout of both the Page Sandstone and limestone in the Carmel Formation on the west flank of this structural element (Figures 6 and 7).

On a smaller scale, several downwarps are apparent. Increased thicknesses of lower San Rafael strata in the Circle Cliffs and southern Henry Mountains regions suggest that this area subsided as a northwest-plunging trough. This

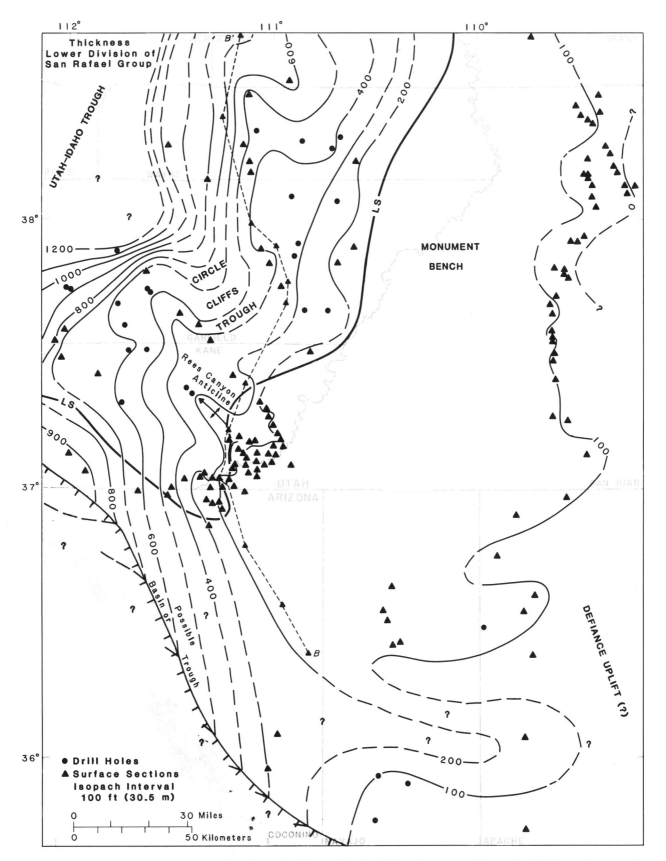

Figure 6—Isopach map of the lower division of the San Rafael Group showing inferred structures. Line marked LS shows the approximate southeastern limit of limestone in the upper part of the Carmel Formation. Shading shows where the division is unconformably overlain by the Upper Cretaceous Dakota Formation and where a complete thickness is not present. Dashed line B–B' is line of section shown in Figure 5.

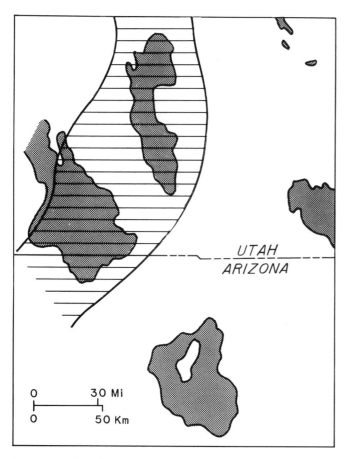

UTAH
ARIZONA

0   30 Mi

0   50 Km

Figure 7—Map of the study area showing where the Page Sandstone (lined pattern) lies directly on the basal unconformity of the lower San Rafael division.

Middle Division of the San Rafael Group

Stratigraphy

The middle division of the San Rafael Group consists of part of the Entrada Sandstone and part of the Wanakah Formation. West and south of the Monument region the division includes the entire Entrada. East of the Monument region, however, the division only includes the Slick Rock Member of the Entrada, and it also includes the lower part of the Wanakah Formation (Figures 2 and 5). In the Kaiparowits and Henry Mountains regions, the Entrada has been divided informally into lower, middle, and upper members. Strata that may correlate with the upper member are called the Cow Springs Member at Black Mesa (Peterson, in press). The division ranges in thickness from zero to approximately 1100 ft (335 m) in the western part of the study area, and it is about 1715 ft (523 m) thick farther west in the Utah–Idaho trough (Standlee, 1982). The thickness is unknown east of the Monument region because it has not yet been possible to separate it from the overlying upper San Rafael division in that area. The J-3 unconformity of Pipiringos and O'Sullivan (1978) forms the upper boundary of the division in the western part of the study area, but the unconformity fades out eastward and is not recognized east of the Monument region (O'Sullivan, 1980a). Imlay (1980) assigned a Middle Jurassic age (lower and middle Callovian) to the part of the Entrada that is here included in the middle San Rafael division.

West of the Monument region, the lower member of the Entrada contains a sabkha facies consisting of flat-bedded siltstone and sandstone and an eolian facies consisting of large-scale, well-sorted, cross-bedded sandstone (Figure 8). The sabkha facies was deposited primarily in and adjacent to the Utah–Idaho trough in a broad sabkha bordering marine environments that probably occupied the northern part of the trough. Similar rocks also occur in northeastern Arizona and southeasternmost Utah in what was probably an isolated inland sabkha (Figure 8). The eolian facies was deposited in a large dune field that stretched northeastward across the study region (Figure 8) and extended farther eastward across much of Colorado (Kocurek and Dott, 1983). Paleocurrent analyses indicate that much of the eolian sediment in the Entrada was derived from the north, which is apparently where a broad inland seaway existed (Poole, 1962; Kocurek and Dott, 1983; Peterson, in press). It is here suggested that the sand was transported along the southern edge of the seaway by longshore currents and was subsequently carried inland by generally northerly winds.

The middle member of the Entrada was deposited when the broad coastal sabkha in the Utah–Idaho trough expanded southeastward across the Henry and Kaiparowits regions and into the eastern part of the Black Mesa region (Figure 8). Cross-bedded eolian sandstone interfingers with local sabkha deposits in the Kaibito and southwestern Kaiparowits regions, suggesting that a dune field lay in the Kaibito region and probably farther southwest.

The upper member and the Cow Springs Member consist largely of cross-bedded sandstone of eolian origin and minor flat-bedded siltstone and sandstone of sabkha origin. The two members could not be correlated across the Kaibito and

conclusion is also supported by the distribution of limestone in the upper part of the Carmel Formation, which thickens and extends farther inland in the Circle Cliffs and southern Henry Mountains areas (Figure 6). The lower San Rafael division thickens west of Black Mesa toward another downwarp. The location of the axis and the areal extent of this downwarp could not be determined because the division was removed by erosion west of Black Mesa. A northwest-trending thin area in the eastern Kaiparowits region reflects uplift on the ancestral Rees Canyon anticline that continued from earlier times.

Several small folds are suggested by thickness variations in several parts of the study region. Several west-trending folds might be present in the eastern Black Mesa region, but because of the scarcity of data points, the isopach lines could be drawn differently. Some of the thickness variations could also reflect interfingering relationships between the Carmel Formation and the overlying Entrada Sandstone and therefore would not necessarily have any bearing on the structural development of the region. Lower and middle San Rafael division strata interfinger over a maximum thickness of 50–80 ft (15–24 m) east of the Monument region (O'Sullivan, 1981a). The interfingering does not involve enough strata to change the basic structural interpretations made from isopach studies of these two divisions.

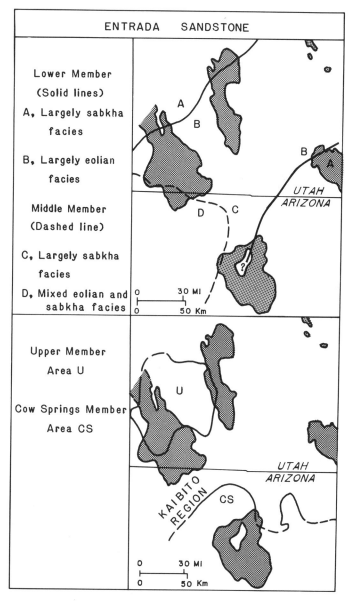

ENTRADA   SANDSTONE

Lower Member
(Solid lines)

A. Largely sabkha
facies

B. Largely eolian
facies

Middle Member
(Dashed line)

C. Largely sabkha
facies

D. Mixed eolian and
sabkha facies

Upper Member
Area U

Cow Springs Member
Area CS

Figure 8—Maps of the study area showing the distribution of members or facies in the Entrada Sandstone.

southeastern Kaiparowits regions (Figure 8). The upper member and the Cow Springs Member apparently grade into mixed eolian and lenticular sabkha strata in and near the Kaibito region, where details of the stratigraphic relationships are unclear. (For illustrative purposes, the mixed facies is shown as an eolian facies of the undivided Entrada Sandstone in Figure 5.) A further complication arises from removal of the upper part of the Entrada in the Kaibito and southeastern Kaiparowits regions during formation of the J-3 unconformity (Figures 5 and 9).

The lower part of the Wanakah Formation consists of laminated to very thin-bedded mudstone and siltstone that comprises an eastern facies of the upper part of the Entrada Sandstone in the Paradox and Blanding basins.

Paleotectonics

Most of the deformation that occurred during deposition of the lower division of the San Rafael Group continued during deposition of the middle division of the group. Westward regional thickening west of the Henry and Kaiparowits regions (where not truncated by the Dakota Formation) reflects additional subsidence in the Utah–Idaho trough (Figure 9). Based on the configuration of isopach lines, the Circle Cliffs region probably was a shallow trough that plunged gently northwestward. It is conceivable that the thickening resulted from eolian deposition of the upper member of the Entrada on a topographic high, and if so, the thickening would not reflect increased subsidence. However, the interpretation of the Circle Cliffs region as a structural trough seems most likely because: (1) the division tends to thicken toward the Circle Cliffs region in nearby areas where the upper member is absent; (2) eolian sands of the upper member are poorly lithified and most likely would have been easily beveled to a nearly horizontal surface during the advance of the late Middle Jurassic Summerville–Curtis seaway across the region; and (3) the thick area of the underlying lower San Rafael division in the same general region suggests that subsidence of the Circle Cliffs trough continued from earlier times.

In the northern part of the study area, O'Sullivan (1980a, 1981a) has shown that the J-3 unconformity, which separates the middle and upper San Rafael divisions west of the Monument region, fades out eastward, roughly over the Monument region. Thus, the boundary between these two divisions could not be identified east of the Monument region, and the two divisions were isopached together to gain an understanding of the tectonic deformation in that area. Thinning of the combined middle and upper San Rafael divisions (Figure 10) suggests that the Monument region continued as a structural bench throughout the time interval represented by both of these divisions.

Movement on several other structures is also suggested by the isopach maps. Thick areas of the middle division indicate small troughlike precursors to the Henry and Kaiparowits basins that trend toward the larger Circle Cliffs trough (Figure 9). A semienclosed thick area in part of the Black Mesa region suggests relatively rapid subsidence during or immediately after deposition of the middle San Rafael division, possibly in a basin or trough. Eastward thinning of the middle San Rafael division in the area east of the Black Mesa region suggests movement of the Defiance region as an uplift or bench. Movement on Rees Canyon anticline in the eastern Kaiparowits region is indicated by local thinning of the middle division there (Figure 9).

Upper Division of the San Rafael Group

Stratigraphy

The upper division of the San Rafael Group includes the Curtis and overlying Summerville formations in the Henry Mountains and San Rafael Swell regions, the Romana

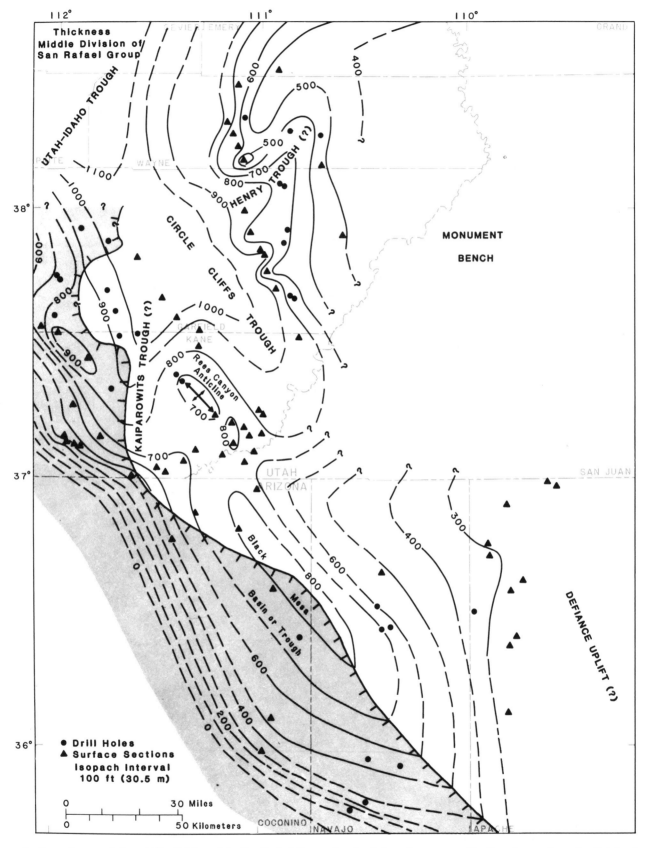

Figure 9—Isopach map of the middle division of the San Rafael Group showing inferred structures. Shading shows where the division is unconformably overlain by the Upper Cretaceous Dakota Formation and thus where a complete thickness is not present.

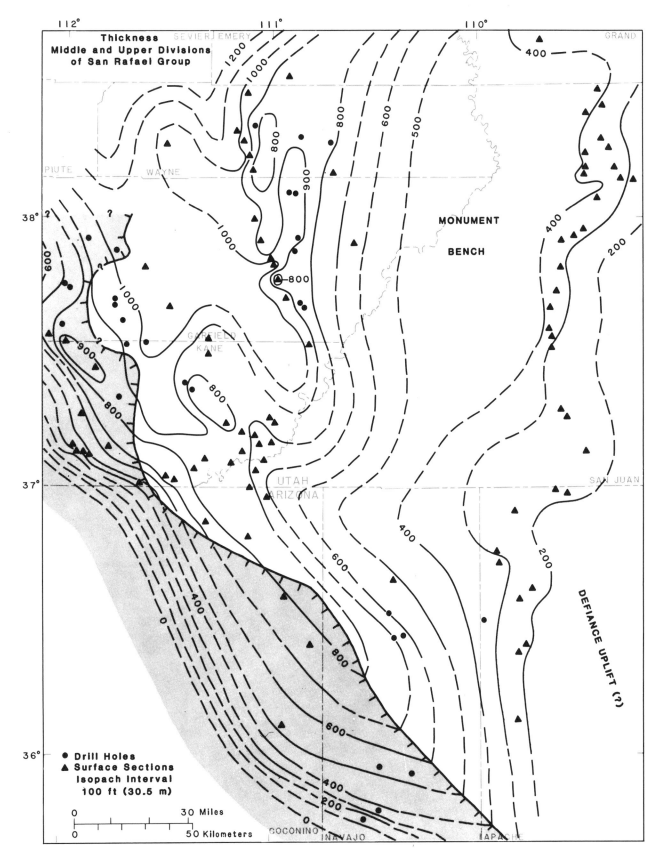

Figure 10—Isopach map of the combined middle and upper divisions of the San Rafael Group showing inferred structures in the eastern part of the study area. Shading shows where the divisions are unconformably overlain by the Upper Cretaceous Dakota Formation and thus where a complete thickness is not present.

Sandstone in the Kaiparowits and Kaibito regions, and the upper part of the Wanakah Formation and the Moab Tongue of the Entrada Sandstone in the Paradox and Blanding regions (Figures 2 and 5) (Peterson, in press). The Curtis consists largely of greenish-gray, glauconitic, flat-bedded sandstone, whereas the Summerville consists largely of dark, reddish-brown mudstone and scarce thin beds of white, pink, and green gypsum, reddish-brown mudstone, and reddish-brown sandstone. The Romana is largely composed of light gray to light greenish-gray, flat-bedded to cross-bedded sandstone that also includes, at the base, a thin marker bed of red mudstone or siltstone in the structurally deepest part of the Kaiparowits basin. The upper part of the Wanakah consists largely of red, laminated to very thin-bedded mudstone and siltstone, whereas the Moab Tongue of the Entrada is composed of light gray, well-sorted, fine-grained sandstone. The upper San Rafael division ranges in thickness from zero to about 600 ft (180 m) west of the Monument region (Figure 11); the division is about 460–530 ft (140–160 m) thick farther northwest in the area of the Utah–Idaho trough (Standlee, 1982). In the area east of the Monument region where the J-3 unconformity could not be recognized, the middle and upper San Rafael divisions could not be separated. The J-5 unconformity marks the top of the upper division throughout the study area. (Studies by Pipiringos and O'Sullivan [1978] suggest that the J-4 unconformity and the lower to middle Oxfordian strata between it and the J-5 unconformity are not present in the study area.) The upper San Rafael division is considered late Middle Jurassic (middle to late Callovian) in age by Pipiringos and Imlay (1979) and Imlay (1980).

The upper division of the San Rafael Group was apparently deposited during a southward advance and northward retreat of the Summerville–Curtis seaway (Peterson, in press). Glauconitic sandstone, scarce limestone, and rare gypsum of the Curtis Formation form the lower part of this division in the Henry Mountains region and was mostly deposited in marine environments during a transgression. Mudstone and minor gypsum of the overlying Summerville Formation were deposited in shallow, hypersaline, restricted marine conditions, generally during a regression.

The Romana Sandstone is a landward facies of the Summerville Formation that is only present in the Kaiparowits and Kaibito regions (Figure 5). It consists largely of light gray to light greenish-gray sandstone interpreted as marginal marine and eolian deposits. A red mudstone or siltstone marker bed as much as 9 ft (3 m) thick lies at the base of the Romana in much of the Kaiparowits region (Figure 11) and is considered a tongue of the Summerville based on lithologic similarity to that formation (Peterson, in press). The red marker bed was probably deposited during the greatest southward advance of the Curtis–Summerville seaway.

Paleocurrent studies indicate that eolian transportation was toward the east during deposition of the Romana Sandstone (Peterson, in press). Some of the eolian sand in the Romana may well have been derived from the Entrada, but chert pebbles as much as ½ in. (1 cm) long have been found in the Romana in the southern part of the

Kaiparowits region and must have had a different source because the Entrada only rarely contains chert pebbles of this size. Most likely the pebbles came from the Carmel Formation or Triassic and older strata west and southwest of the Kaiparowits region.

Paleotectonics

A dramatic change in the structural style of the west-central Colorado Plateau occurred at the beginning of this interval. Isopach patterns indicate that the Henry Mountains region was a north-northwest-plunging structural trough at this time, and the tendency of the Curtis Formation to occupy the structurally deepest part of the trough also supports this interpretation (Figure 11). The isopach patterns indicate a closed basin in the Kaiparowits region, but the basinal form probably developed during the J-5 erosion interval. The red mudstone marker bed in the lower part of the Romana Sandstone does not thin or grade laterally northwestward into another facies, but instead is truncated in the northwestern part of the basin. This suggests that the Kaiparowits region was a northwest-plunging trough during deposition of the upper San Rafael division and that uplift and closure of the northwestern part of the basin did not occur until the following erosion interval. Downwarping of the Kaiparowits region as a basin or trough also implies some sort of structurally positive area to the southwest that would have been the ancestral Echo Cliffs–Kaibab uplift.

Several other positive structural elements were also present in the region. The upper San Rafael division thins over the Circle Cliffs region, indicating that the former trough in this area changed to a positive structure. Rocks of the upper division are truncated by the J-5 unconformity southeastward from the Kaiparowits region and are not present at Black Mesa. The southeastward truncation and southeastward thinning of the thin red mudstone marker unit at the base of the Romana (Figure 11) also suggest that the Black Mesa region may have been a positive area at this time. Positive movement in the Monument and Circle Cliffs regions is indicated by thinning of the division and onlapping pinchout of the Curtis toward these areas. Isopach patterns of the combined middle and upper divisions of the San Rafael Group suggest that the Monument region continued to act as a structural bench (Figure 10). Eastward thinning in northeastern Arizona might also indicate movement of the Defiance region as an uplift or bench, but this could not be confirmed by facies relationships or other sedimentologic features.

An area of thin upper San Rafael in the eastern part of the Henry Mountains region is suggestive of growth on a small anticline, here named Block anticline, that lies in approximately the same area as a small, unnamed anticline now exposed at the surface. Data are insufficient to determine if the eastward pinchout of the Curtis was affected by growth on this fold. The late Middle Jurassic is one of the few times during the Jurassic Period that no movement occurred on the Rees Canyon anticline.

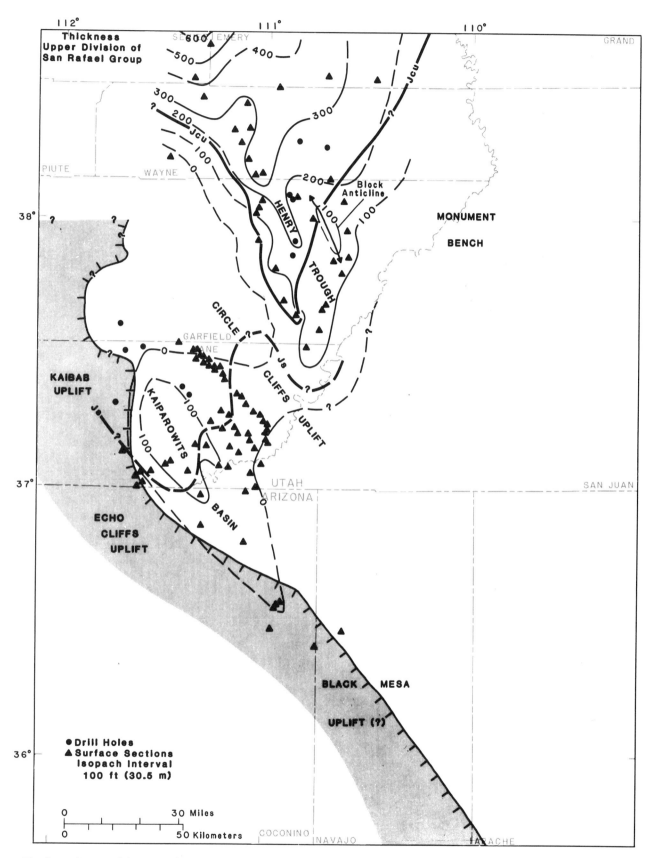

Figure 11—Isopach map of the upper division of the San Rafael Group showing inferred structures. Line marked Jcu shows the southern extent of the Curtis Formation; line marked Js indicates the southern extent of the Summerville Formation and equivalent redbeds in the lower part of the Romana Sandstone. Shading shows where the division is unconformably overlain by the Upper Cretaceous Dakota Formation and thus where a complete thickness is not present.

UPPER JURASSIC SERIES

Lower Division of the Morrison Formation

Stratigraphy

The lower division of the Morrison Formation consists of the Tidwell and Salt Wash members in the western and southern parts of the study area, but it also includes the Bluff Sandstone Member in the Paradox and Blanding regions and lowermost beds of the Recapture Member in the southeasternmost part of the study area. The lower Morrison division is bounded by a basal unconformity and a conformable and locally interfingering upper contact. Interfingering at the top probably involves less than about 30 ft (9 m) of strata, and it therefore may not be important to structural interpretations based on isopach patterns. The Tidwell and Salt Wash members are separated by a gradational and interfingering contact (Figure 5). The Tidwell consists largely of mudstone and sandstone, although it also contains thin limestone lenses and, in the northern part of the Henry Mountains region, thick beds of gypsum (Peterson, 1984). It underlies the Salt Wash throughout most of the study area, but it is missing in the southeastern Kaiparowits region where it is replaced by fluvial sandstone at the base of the Salt Wash. The Salt Wash consists largely of sandstone, but it grades westward into conglomerate in the westernmost part of the Henry Mountains region; both of these lithologies are interbedded with thin beds of mudstone. The Tidwell and Salt Wash pinch out by onlap toward two paleotopographic highs, one on the southwest side of the San Rafael Swell and the other in the Black Mesa region (Figures 12 through 18). In the southeastern part of the study area, Salt Wash and Tidwell beds grade southward into the lowermost part of the Recapture Member of the Morrison Formation. In southeastern Utah the Tidwell grades into the Bluff Sandstone Member of the Morrison, which is predominantly of eolian origin. The Bluff thickens to about 300 ft (90 m) in this area, whereas the Salt Wash thins and pinches out over the thickest part of the Bluff; presumably this occurs largely by onlap because the Bluff and Salt Wash rarely intertongue (Craig et al., 1955; O'Sullivan, 1965, 1983). The thickness of the lower Morrison division ranges from zero to about 600 ft (180 m) (Figure 12). Imlay (1980) assigned a Late Jurassic age (late Oxfordian and Kimmeridgian) to strata that comprise the lower division of the Morrison.

The Tidwell Member was deposited in broad mudflat, lacustrine, and subaqueous evaporative environments and locally in eolian environments. Scarce fluvial channel sandstone deposits in the Tidwell were deposited mainly by meandering streams. The Salt Wash Member is an alluvial complex deposited largely by braided streams west of the Monument region and largely by meandering streams east of the Monument region and in the northwestern Henry Mountains region.

Although some of the Tidwell and Salt Wash sediment probably was derived from older Mesozoic formations in the southwestern and western parts of the Colorado Plateau region, the greater part of the clastic material in these members was most likely derived from highlands 100 mi (160 km) or more southwest and 60 mi (100 km) or more

west of the study region. Chert pebbles in the Salt Wash and Tidwell contain middle to late Paleozoic fossils (Craig et al., 1955), and the streams that deposited these members flowed generally eastward to northeastward (Figures 14, 16, and 17). These pebbles most likely came from the southwest beyond the present margin of the Colorado Plateau because the entire Paleozoic section is still well preserved in the southwestern part of the Colorado Plateau where Late Jurassic streams would have flowed. Also, regional stratigraphic reconstructions suggest that the Triassic formations of the southwestern Colorado Plateau were not exposed until Tertiary time. Because older Jurassic strata on the plateau southwest and west of the study area do not contain a great abundance of small chert pebbles, it is unlikely that they were the source of the abundant pebbles in the Salt Wash.

Sand in the Bluff Sandstone Member most likely was derived from contemporaneous fluvial deposits in the lowermost part of the Salt Wash alluvial complex that lay west of the Monument region. In addition to the Bluff, other small eolian sandstone lenses have been found in the Tidwell of the Henry Mountains and Kaiparowits regions (Peterson, 1980). Paleocurrent studies of Tidwell and Bluff eolian deposits indicate that transportation was toward the east or northeast across the study area (Poole, 1962; Peterson, 1980, in press). A change in general wind direction from northerly during deposition of the Entrada and older formations to westerly or southwesterly during deposition of the Romana and Morrison formations may reflect drift of the North American continent northward across latitudinal belts possessing different wind directions. However, southeast-directed cross-bedding found in some of the eolian beds of the Morrison in southwestern Colorado and northwestern New Mexico (F. Peterson and S. M. Condon, unpublished data) makes that interpretation tenuous.

For purposes of paleotectonic analysis, the Tidwell and Salt Wash members in the Henry Mountains, Kaiparowits, and Black Mesa regions and in northeasternmost Arizona are divided into three depositional sequences that are distinguished by differences in sedimentologic features of the fluvial sandstone beds (Peterson, 1980, 1984). The lower sequence has relatively high stratification ratios and northeastward paleocurrent indicators (Figure 14), the middle sequence has relatively low stratification ratios and eastward to southeastward paleocurrent indicators (Figure 16), and the upper sequence has intermediate stratification ratios (not shown in this report) and northeastward or locally southeastward paleocurrent indicators (Figure 17; Peterson, 1984). The contact between the lower and middle sequences is relatively sharp but the contact between the middle and upper sequences has about 15 ft (5 m) of local interfingering in the southeastern part of the Henry Mountains region. Because of the interfingering and local deep scouring at the base of the upper sequence, the middle and upper sequences were grouped together to make a single isopach map (Figure 15). The three sequences could not be distinguished in the Blanding and Paradox regions, where the middle sequence could not be identified, nor in the southeasternmost part of the study area, where the Tidwell and Salt Wash are replaced by lowermost beds of the Recapture Member.

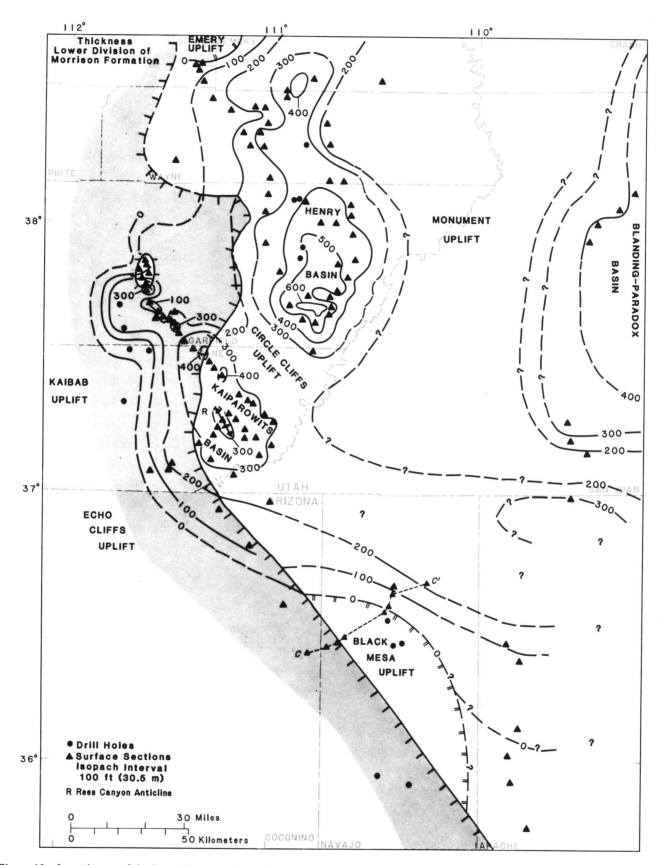

Figure 12—Isopach map of the lower division of the Morrison showing inferred structures. Shading shows where the division is unconformably overlain by the Upper Cretaceous Dakota Formation and where a complete thickness is not present. Doubly hachured lines indicate uplifts that rose above base level. Dashed line C–C′ is the line of section shown in Figure 18.

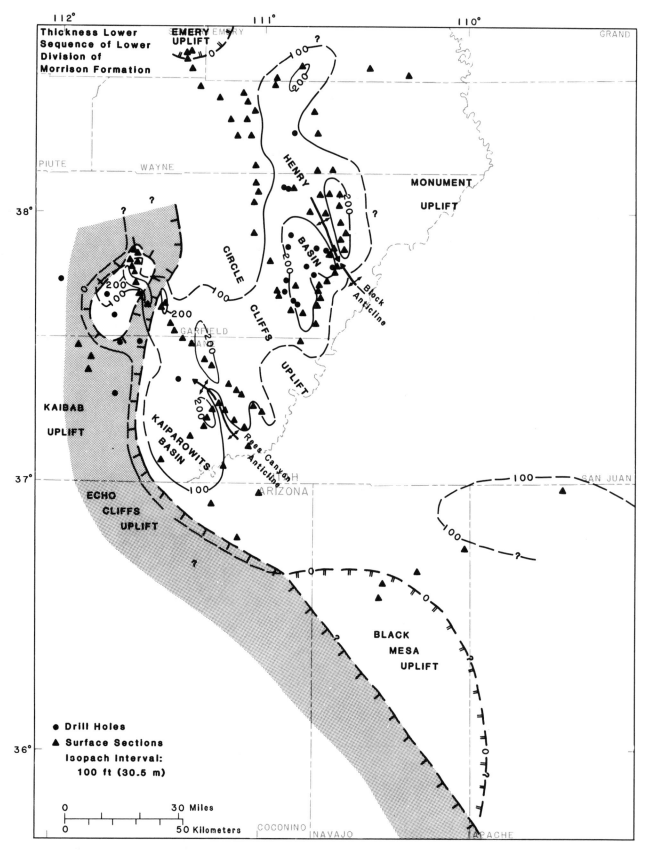

Figure 13—Isopach map of the lower sequence of the lower division of the Morrison showing inferred structures. Shading shows where the sequence is unconformably overlain by the Upper Cretaceous Dakota Formation and thus where a complete thickness is not present. Doubly hachured lines indicate uplifts that rose above base level.

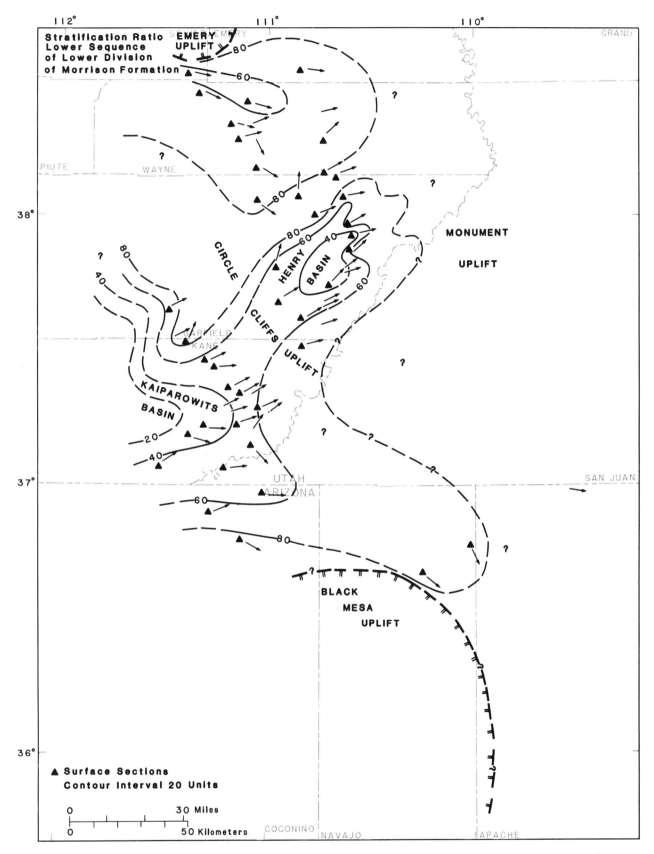

Figure 14—Isopleth map showing stratification ratios in the lower sequence of the Morrison and their relationship to inferred structures. Doubly hachured lines indicate uplifts that rose above base level. Arrows show vector resultants of paleocurrent indicators in fluvial sandstone beds of the lower sequence.

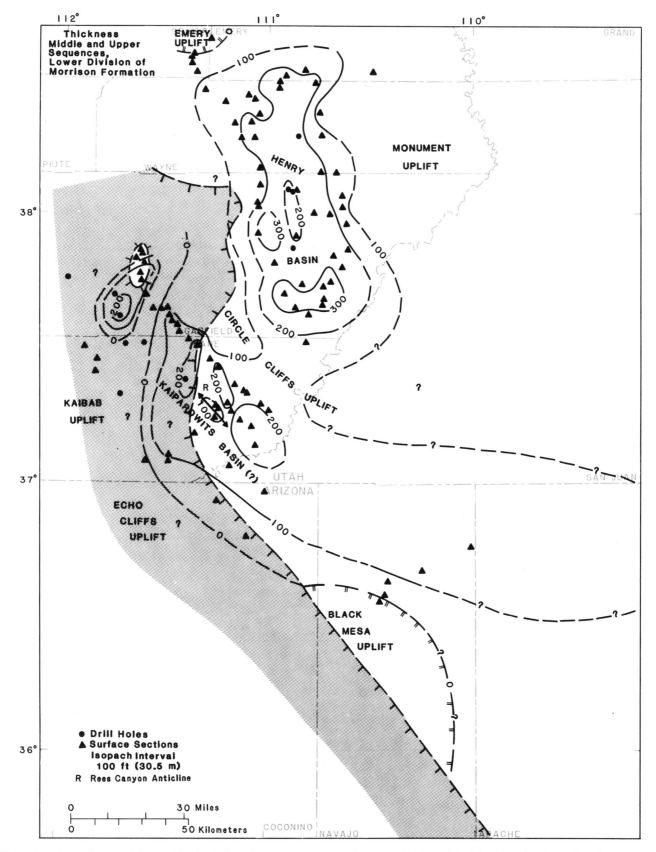

Figure 15—Isopach map of the combined middle and upper sequences of the lower division of the Morrison showing inferred structures. Shading shows where the sequences are unconformably overlain by the Upper Cretaceous Dakota Formation and thus where a complete thickness is not present. Doubly hachured lines indicate uplifts that rose above base level.

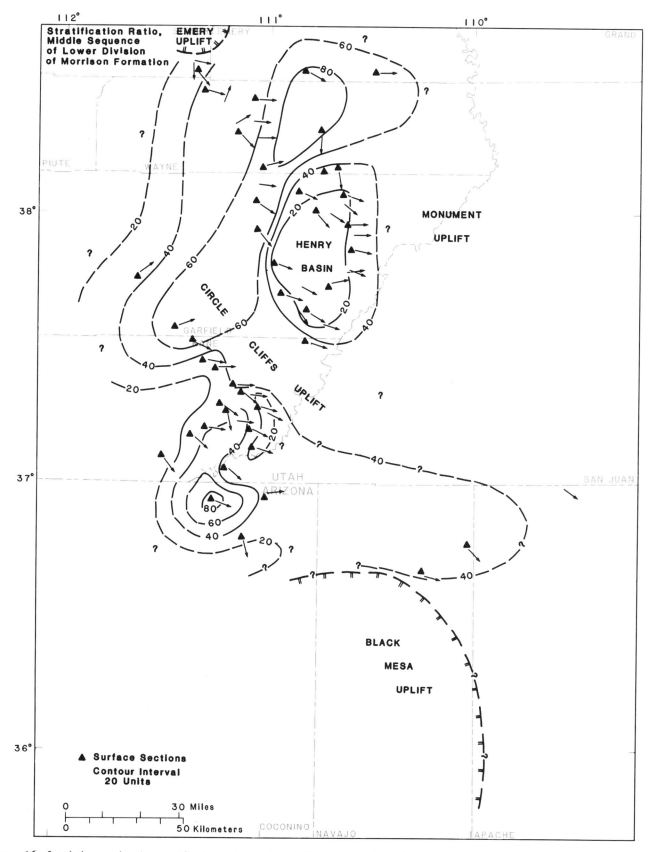

Figure 16—Isopleth map showing stratification ratios in the middle sequence of the Morrison and their relationship to inferred structures. Doubly hachured lines indicate uplifts that rose above base level. Arrows show vector resultants of paleocurrent indicators in fluvial sandstone beds of the middle sequence.

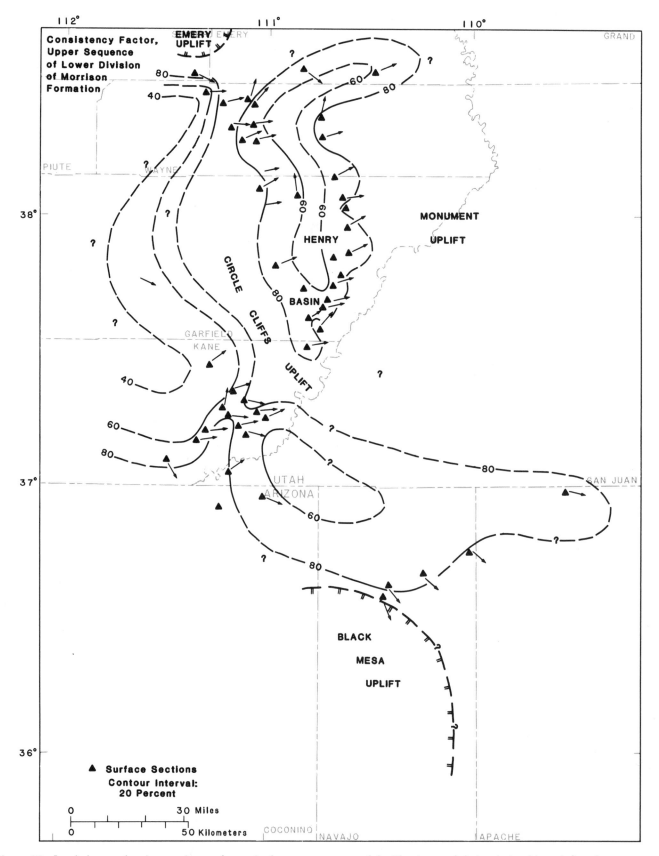

Figure 17—Isopleth map showing consistency factors in the upper sequence of the Morrison and their relationship to inferred structures. Doubly hachured lines indicate uplifts that rose above base level. Arrows show vector resultants of paleocurrent indicators in fluvial sandstone beds of the upper sequence.

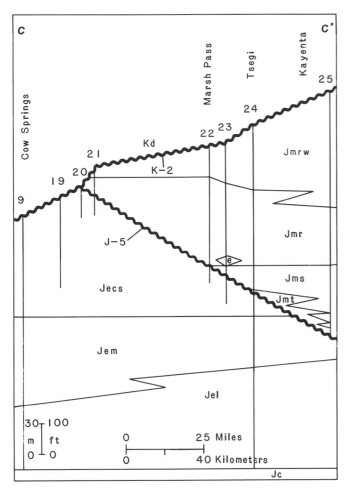

Figure 18—Stratigraphic section showing onlap of strata in the Morrison on the north side of the Black Mesa region. Entrada Sandstone: Jel, lower part; Jem, middle part; Jecs, Cow Springs Member. Morrison Formation: Jmt, Tidwell Member; Jms, Salt Wash Member; Jmr, main body of Recapture Member (e indicates a local eolian sandstone bed); Jmrw, alluvial complex in upper recapture that may include some Westwater Canyon Member at top. Line of section C–C′ is shown in Figure 12.

Paleotectonics

There is more available evidence for structural activity in the west-central Colorado Plateau during early Late Jurassic time than for any other part of the Jurassic. Detailed sedimentologic studies are particularly important in demonstrating that sedimentation and deformation were synchronous and continuous during this time interval.

The Emery and Black Mesa uplifts (Figure 12) were positive emergent areas that served as minor sediment sources. In the southwestern part of the San Rafael Swell, the Tidwell and Salt Wash thin and pinch out across a rejuvenated segment of the Emery uplift, which was an active positive structure in late Paleozoic time (Wengerd and Matheny, 1958; Fetzner, 1960). Here, mudstone of the younger Brushy Basin Member lies directly on the Summerville Formation over the crest of the upwarp between the NE¼ sec. 28, T. 25 S., R. 5 E. and the NE¼ sec. 13, T. 24 S., R. 6 E., Sevier and Emery counties, Utah, respectively.

Pinch-out of the lower Morrison division indicates that Jurassic vertical movement on the Emery uplift probably began after deposition of the Summerville, during the time interval represented by the J-5 unconformity. Movement on the Emery uplift may have ceased by the end of deposition of the lower division of the Morrison because the overlying Brushy Basin Member does not appear to thin over this fold (Figure 19). However, more data are needed to confirm this hypothesis. The lower Morrison division also thins and pinches out by onlap against the Black Mesa uplift (Figures 12 and 18), indicating that it too was a positive element during deposition of the division. Cow Springs anticline is a small fold that presently lies approximately in the middle of the Black Mesa region (Figure 1) and probably is a remnant of the considerably larger Late Jurassic positive structure.

The area of the former Utah–Idaho trough may have been uplifted at this time as evidenced by westernmost exposures of the Salt Wash locally containing large blocks of material that appear to have been derived from uplifted and eroded beds of the Tidwell Member a short distance farther west. Salt Wash conglomeratic sandstone beds containing scarce blocks of poorly lithified limestone about 3 ft (1 m) long that appear to have come from the Tidwell were found on the west side of Jones Bench in the NW¼ NW¼ sec. 21, T. 26 S., R. 5 E., Sevier County, Utah. Large rounded cobbles and blocks as much as 3 ft (1 m) long of medium gray chert that also could have come from the Tidwell were found in the drainage of Solomon Creek in the NE¼ SW¼ sec. 32, T. 25 S., R. 5 E., Sevier County, Utah. The similarity of the chert to medium gray authigenic chert masses in the Tidwell, the large size of these clasts, and the lack of similar lithologies in the underlying Summerville Formation all suggest that the blocks may have been derived from Tidwell strata that were originally deposited farther northwest in or near the Utah–Idaho trough.

It seems unlikely that such large clasts could have been derived from the eroded edges of thrust sheets farther west because: (1) thus far, only large limestone clasts have been found and a normally expected population with decreasing limestone clast sizes has not been found in these beds, and (2) most of the chert clasts that presumably were derived from highlands or thrust sheets farther west are dark gray and considerably smaller (average size clast is about 2–3 in. [5–8 cm] and length of largest cobble about 9 in. [23 cm]). Thus, the evidence suggests that Tidwell strata originally deposited farther northwest were uplifted, eroded, and incorporated into slightly younger beds of the Salt Wash when the trough was uplifted. Moreover, large slumps about 15 ft (4.5 m) high and overturned to the southeast are locally present in the Tidwell Member northwest of the Henry Mountains region in the SE¼ sec. 7, T. 27 S., R. 7 E., Wayne County, Utah, and indicate southeastward slumping away from the region of the Utah–Idaho trough.

Isopach patterns suggest that the Henry Mountains and Kaiparowits regions were folded into basins during deposition of the lower Morrison division and that the Echo Cliffs–Kaibab, Circle Cliffs, and Monument regions were positive structural elements between or adjacent to those basins (Figures 12, 13, and 15). The lower division thickens eastward from the Monument region into the Blanding–Paradox basin, indicating that the Monument region was an

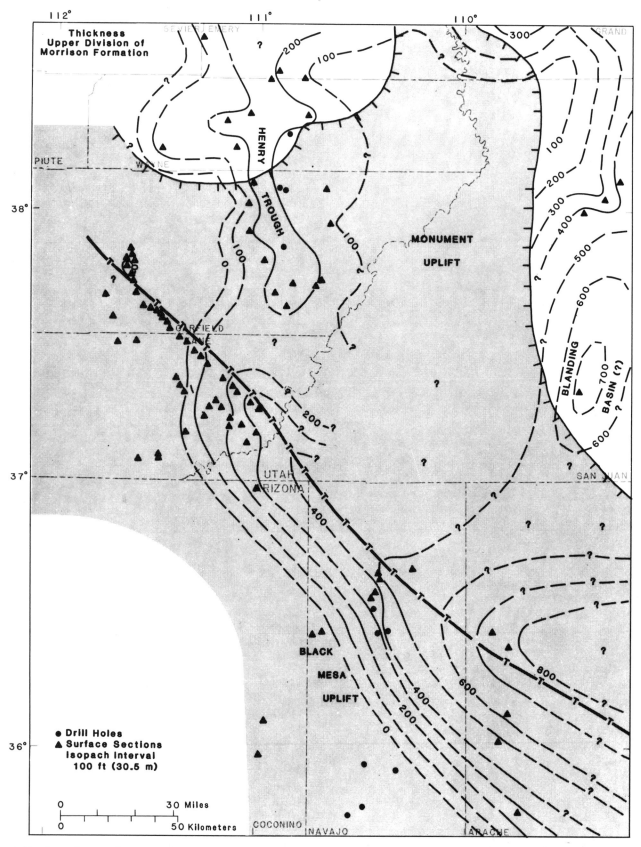

Figure 19—Isopach map of the upper division of the Morrison showing inferred structures. Shading shows where the division is unconformably overlain by the Upper Cretaceous Dakota Formation and thus where a complete thickness is not present. In the unshaded area, the upper division is overlain by the Cedar Mountain Formation west of the Monument uplift and by the Burro Canyon Formation east of the Monument uplift. Jurassic and older rocks are abruptly truncated beneath the sub-Dakota unconformity southwest of the line marked —T—.

uplift flanked by downwarps rather than the structural bench that it had been earlier in Middle Jurassic time. Detailed sedimentologic studies of the Salt Wash near Blanding, Utah, by Stokes (1954a, b) also suggest uplift of the Monument region. Stokes (1954b) also postulated that the Defiance region may have been an active positive structure, but the scarcity of thickness information in that area makes it difficult to interpret this from the isopach map (Figure 12).

Detailed sedimentologic studies of individual sequences within the lower division of the Morrison yield a better understanding on the timing of movement on the major structures and also on some of the small folds within the larger structures. Stratification ratios obtained from fluvial sandstone beds in the lower sequence support the hypothesis that the Henry basin subsided. In a tectonically stable region of sediment aggradation, these ratios tend to increase downstream in response to changes in the bed forms (Smith, 1970; Gilbert and Asquith, 1976). On the unstable Colorado Plateau, the overall downstream trend was disrupted by folds that grew during deposition of the fluvial beds. Thus, superimposed on the overall downstream trend are locally high ratios that tend to occur over structurally positive areas and locally low ratios that tend to occur over structurally negative areas. This probably reflects local changes in sinuosity and possibly gradient of the streams as they passed over the various active structural elements (Peterson, 1984).

In fluvial strata, local increases in the amount of horizontal laminations generally are attributed to two primary processes, although other processes are theoretically possible. Horizontal laminations tend to increase in areas of high velocity if water depths remain constant, such as where the gradients locally increase, or they tend to increase in abundance in areas where water depth decreases even though the velocity may not change. Anomalously low stratification ratios in the Henry basin are interpreted to be related to basinal downwarping that may have lowered the gradients of the streams that deposited the lower sequence, but more importantly, that may have allowed the streams to increase in width and become shallower (Peterson, 1984). Although the two processes may tend to produce opposite values for stratification ratios, it is thought that locally shallow depths in downwarped areas are dominant. This may account for the locally large amount of horizontal laminations in the fluvial sandstone beds in the study area. Also, lower stream gradients in the Henry basin may have allowed greater flooding there, resulting in increased quantities of horizontal laminations in the stream sediments and therefore reduced stratification ratios (Peterson, 1984). The low ratios are interpreted to indicate continued subsidence of the Henry basin during deposition of the lower sequence (Figure 14). Because ratios in the lower sequence are averages from several beds, it is possible that the subsidence may have been episodic and did not occur during deposition of every fluvial bed in this sequence.

In addition to the larger structural features, isopach patterns (Figures 12, 13, and 15) and the distribution of lake beds in space and time (Peterson, 1980, 1984) indicate that Rees Canyon anticline in the Kaiparowits basin and Block anticline in the Henry basin also were positive structures.

Thickness variations also show that vertical movement on most of these structures continued during deposition of the middle and upper sequences. An isopach map of both of these sequences shows several thick areas suggestive of an irregular basin in the Henry Mountains region and a thin area suggestive of a positive area in the Circle Cliffs region (Figure 15). The middle and upper sequences were so deeply eroded beneath the Dakota Formation in the western two-thirds of the Kaiparowits region that it is difficult to evaluate from isopach patterns whether or not that region subsided as a basin. However, an area of thin deposits in the eastern Kaiparowits region suggests that the Rees Canyon anticline was active.

The sedimentology of fluvial strata in the middle sequence also supports the hypothesis of movement on many of these structures. As with the lower sequence, stratification ratios in the middle sequence are anomalously low in the Henry Mountains region (Figure 16), suggesting continued downwarping to form a broad lowland area of shallow braided streams. Paleocurrent studies indicate that middle sequence streams tended to flow eastward or southeastward, as contrasted with the general northeastward flow of lower and upper sequence streams. It is possible that slight southward tilting of the Colorado Plateau province occurred at the beginning of middle sequence deposition and caused the brief change in stream directions.

Movement on the structures during deposition of the upper sequence of the Morrison is also supported by sedimentologic data. Stratification ratios obtained from fluvial sandstone beds in this sequence (not included here) neither support nor refute the hypothesis of movement on these folds, but paleocurrent studies do support this hypothesis. The consistency factor obtained from studies of trough cross-bedding is related to the degree of departure of individual cross-bed dip vector measurements from the resultant (Figure 17). Low consistency values indicating high angular variation in the compass direction of the individual cross-bed sets probably reflect a high degree of sinuosity of the individual braid channels that make up the braided streams that deposited the sequence. In contrast, high consistency values indicating low angular variation in the measurements probably reflect low sinuosity of the individual braid channels. In other words, areas with low consistency values probably had small braid channels that deviated more from the overall downstream trend of the stream than areas with high values (Peterson, 1984). As with previously noted conclusions obtained from stratification ratios, it is thought that low consistency values in downwarped areas reflect wide and shallow braided streams, whereas high values over positive areas reflect comparatively narrower and deeper streams.

Low consistency values in the Henry Mountains region suggest that crustal downwarping produced an area marked by relatively low stream gradients. Decreased gradients would have led to greater sinuosity of the individual small channels that make up the larger braided stream (Figure 17). Conversely, high consistency values in the Circle Cliffs region suggest a structurally positive area with relatively less sinuosity of the individual channels.

Paleocurrent studies indicate that the streams flowed directly across positive structures such as the Circle Cliffs uplift. It would not be expected that streams would flow around positive structures in a region of net sediment aggradation where the axes of the positive structures lie approximately perpendicular to the stream trend, as is the common type of setting throughout most of the study area.

A trough that trends from the Kaiparowits region southeastward across northeastern Arizona is suggested by maps of part or all of the lower division (Figures 12 through 17). For each of these maps, however, there is a paucity of data points because the beds have been eroded from much of that region; thus, the isopach or isopleth lines could possibly be contoured differently.

Upper Division of the Morrison Formation

Stratigraphy

The upper division of the Morrison Formation consists of four members that are recognized in different parts of the region. The division consists solely of the Brushy Basin Member in the Henry Mountains and northern Kaiparowits regions (Figures 2 and 5). This member consists largely of smectitic mudstone, but it also contains some sandstone and minor conglomerate west of the Monument region. The Fiftymile Member of the Morrison is laterally equivalent to the Brushy Basin and is only recognized in the southeastern Kaiparowits region. The lower part of the Fiftymile consists largely of mudstone and the upper part of sandstone (Peterson, in press). The upper part of the Recapture Member and fluvial beds, some of which may correlate with the Westwater Canyon Member, make up the upper Morrison division farther south and east in the Black Mesa region and northeasternmost Arizona. The Recapture consists mainly of interbedded mudstone and sandstone, and, at the top, a thick, fluvial sandstone unit. If the Westwater Canyon is indeed present it would be included in the fluvial sandstone complex at the top of the Recapture. The Fiftymile Member may be a northwestern correlative of the Recapture and Westwater Canyon members. The Recapture Member at Black Mesa thins southwestward by onlapping onto a paleotopographic high (Figure 18). The Salt Wash Member grades southward into lowermost strata of the Recapture in the southeastern part of the Black Mesa region and farther east in the southeasternmost part of the study area (Condon and Peterson, in press). Thus, in the southeastern part of the study area, the lower part of the Recapture includes beds equivalent in age to the lower division of the Morrison.

The upper division of the Morrison ranges in thickness from zero to slightly greater than 800 ft (245 m) (Figure 19). The division is considered latest Jurassic (Tithonian) in age by Imlay (1980). However, recent fission track dates on zircon from tuff beds suggest that the upper part of the Brushy Basin Member may be Early Cretaceous in age (Kowallis and Heaton, 1984).

The lower part of the upper division of the Morrison contains scarce fluvial deposits and was deposited mostly in lakes and on large mud flats in a broad lowland region. These beds lapped onto a slight topographic high in the vicinity of Black Mesa that probably shed minor amounts of sediment to surrounding areas (Peterson, 1984). During deposition of the upper part of the division, sand was transported into the Black Mesa and southeastern Kaiparowits regions by upper Recapture, Westwater Canyon, and upper Fiftymile streams. The fluvial sedimentary deposits grade northward and northeastward

into mud flat and possibly lacustrine deposits that form the upper part of the Brushy Basin Member. Well-preserved shards in thin, authigenic zeolite beds of the Brushy Basin indicate that its abundant smectitic clay layers are largely alteration products of volcanic ash (Bell, 1984, in press). The volcanic ash was probably transported into the region from the west and southwest by winds from volcanos or calderas in a magmatic arc that lay along the western edge of the North American continent.

Paleotectonics

The tectonic evolution of the west-central part of the Colorado Plateau during deposition of the upper Morrison division is poorly understood. Uppermost beds of the division were removed from most of the region during the erosional interval that preceded deposition of the Upper Cretaceous Dakota Formation (Figure 19) and detailed facies studies suitable for paleotectonic interpretations have not yet been made. Isopach patterns indicate a north-trending trough or possible basin in the Henry Mountains region. Because the form of the trough or basin does not change farther north where it plunges beneath Lower Cretaceous rocks, the thickness trend probably reflects downwarping during Late Jurassic rather than during Early Cretaceous time. The stratigraphy and sedimentology of the Brushy Basin Member is too poorly known to determine if the downfolding occurred during and/or immediately after deposition.

Black Mesa probably continued as a structurally positive area, at least during deposition of the lower part of the upper division, because the Recapture Member thins and pinches out southwestward by onlap onto a paleotopographic high interpreted to reflect a paleostructural high (Figure 18). Areas with thick upper division strata are present in southeastern Utah and northeastern Arizona (Figure 19). Although there is some additional control from sections farther east just outside the study area, there are too few data points to confidently determine isopach configurations in these areas. Where it is still preserved beneath Lower Cretaceous rocks, thinning of the upper division of the Morrison toward the Monument uplift indicates that it was a positive feature at this time. Similarly, thinning west of the Henry trough indicates the presence of a positive element to the west, possibly in the area of the former Utah–Idaho trough. Upper Jurassic strata were once thought to occur in the area of the trough, but recent studies there by I. J. Witkind (personal communication, 1984) suggest that these beds have been misidentified and that Upper Jurassic strata are not present there.

LOWER CRETACEOUS SERIES

Stratigraphy

The Cedar Mountain Formation (Stokes, 1952) lies above the Morrison Formation in the northwestern part of the Henry Mountains region, and in a small area in the northern part of the Kaiparowits region (Figure 2). It consists of the fluvial Buckhorn Conglomerate Member and an unnamed upper member consisting of interbedded fluvial sandstone

and overbank mudstone. The distribution of Lower Cretaceous rocks shown in Figure 19 is after Craig (1981) and
Tschudy et al. (1984), but is modified to reflect a presumed
absence of these beds from the Monument uplift. This
modification is based on absence of the Cedar Mountain
Formation in the southeastern two-thirds of the Henry
Mountains region and throughout most of the Kaiparowits
region. Cedar Mountain sediments were shed eastward
(Craig, 1981) from the Sevier highlands of western Utah
and southern Nevada. The Burro Canyon Formation is
partly correlative with the Cedar Mountain Formation and
lies east of the Monument region, roughly along the
Utah–Colorado state line. The Burro Canyon was deposited
in a separate depositional basin, largely by northward-
flowing streams (Craig, 1981). Recent palynologic studies by
Tschudy et al. (1984) indicate that the Cedar Mountain and
Burro Canyon formations are Early Cretaceous in age.

Paleotectonics

Positive movement on the Monument uplift probably was
responsible for separation of the two depositional basins in
which the Cedar Mountain Formation (to the west) and the
Burro Canyon Formation (to the east) were deposited
(Figure 19).

Following deposition of Lower Cretaceous rocks, the
southwestern part of the Colorado Plateau was raised and
tilted toward the northeast in latest Early Cretaceous time,
possibly extending into the earliest Late Cretaceous. Tilting
of the plateau accompanied broad regional upwarping of the
Mogollon highlands farther south in central Arizona. During
this event, formations older than the Dakota were beveled
southwestward, causing the Dakota to be deposited on
successively older rocks to the southwest. The part of the
modern Colorado Plateau that was involved most in the
broad northeastward regional tilting lies southwest of a line
that passes through the northeastern parts of the
Kaiparowits and Black Mesa regions (Figure 19). Jurassic
strata in the area southwest of this line thin abruptly by
truncation beneath the Dakota, which is reflected by the
sharp bends in contours on most of the isopach maps of this
area. The abruptness of this truncation of beds and the
strikingly different style of tectonism that occurred just
before deposition of the Dakota compared with earlier times
make it difficult to reconstruct the tectonic history of the
various divisions of the Jurassic in the southwestern part of
the Colorado Plateau wherever the Dakota lies directly on
these divisions.

UPPER CRETACEOUS SERIES

The lowermost Upper Cretaceous Dakota Formation
consists of sandstone, mudstone, minor coal, and conglomerate deposited in fluvial, overbank, swamp, and marginal
marine environments (Peterson et al., 1980). Clastic
materials in the formation were derived from highland
source areas in eastern Nevada, western Utah, and west-
central Arizona and were transported to the Colorado
Plateau by streams whose gradients decreased toward the
plateau. Hence, truncation of Jurassic and older strata

beneath the Dakota along the southwestern margin of the
Colorado Plateau must have been by uplift and northeastward tilting of Lower Cretaceous and older strata and
not by southwestward down-cutting beneath the Dakota.
The formation appears to represent deposits associated with
the first transgression of the Western Interior seaway onto
the Colorado Plateau.

MONOCLINES

Monoclines are the most impressive and characteristic
structural features of the Colorado Plateau (Figure 1), and an
effort was made to determine if and how these flexures
moved during the Jurassic. The effort was disappointing
because of a pressing need for more data and because the
rocks have been eroded from many critical areas. Monoclines
of the Colorado Plateau are known or thought to lie above
basement faults (Stokes, 1954a, b; Kelly, 1955; Lohman,
1965; Davis, 1978; Reches, 1978). Although there is insufficient evidence available to demonstrate whether Jurassic
movement on these flexures occurred as monoclines,
homoclines, or limbs between symmetric folds, the studies
presented in this paper indicate that the general areas
marked by the monoclines were sites of repeated differential
movements across which the sense of structural tilt may
have been reversed from time to time. In the following
paragraphs, the Jurassic structural tilt across the general area
of these structures could well have been monoclinal, but the
presently available data are too sparse or too widely spaced
to permit such a specific determination.

Evidence of crustal movement on or near the San Rafael
monocline is scant. Thickening of the entire San Rafael
Group generally toward the west suggests westward structural tilt in this region during Middle Jurassic time (Figures
6, 9, 10, and 11). Eastward thickening of the lower division
of the Morrison across the monocline suggests eastward
structural tilting during early Late Jurassic time (Figure 12).
The Brushy Basin Member of the Morrison generally
thickens westward in the vicinity of the San Rafael
monocline, suggesting that the area may have tilted back to
the west at the end of Jurassic time (Figure 19).

Waterpocket monocline was a zone of weakness that had a
complex structural history in Middle and Late Jurassic time,
but there are insufficient data to determine whether the
flexure moved as a monocline or merely as a limb between
approximately symmetric folds. The Glen Canyon division
and the lower and middle divisions of the San Rafael Group
thicken west of the Waterpocket monocline (Figures 4, 6,
and 9). Although the specific structural type of movement
could not be determined, relative movement was opposite to
that which occurred subsequently in late Middle and Late
Jurassic time (Figures 11, 12, and 19).

The East Kaibab and Echo monoclines have complex
geologic histories, if not as monoclinal flexures, at least as
areas of differential crustal movements. The Butte fault
underlies the East Kaibab monocline in the eastern Grand
Canyon and had down-to-the-west movement in late
Precambrian time, which contrasts with down-to-the-east
movement that occurred during the Laramide (Walcott,
1890). On the basis of isopach trends, it is most likely that
any movement that might have occurred on this flexure

during the Jurassic would have been with the west side down in early and middle Middle Jurassic time and with the east side down in late Middle and early Late Jurassic time (Figures 6, 9, 11, and 12). Similarly, the area west of Echo monocline subsided during early and middle Middle Jurassic time and then was uplifted relative to the area farther east during late Middle and Late Jurassic time (Figures 6, 9, 11, and 12). The available evidence is not adequate to determine if movement in the vicinity of the East Kaibab monocline was of a monoclinal type, although detailed studies of the Echo monocline in the southern part of the Kaiparowits region suggest that it flexed as a monocline during Early Cretaceous and possibly latest Jurassic time (Peterson, 1969).

The available data suggest that the area at or near Red Lake monocline was part of a much larger structure that was tilted homoclinally to the west and was not differentiated as a distinct structural entity during the Early Jurassic (Figure 4). On the basis of isopach studies, it appears that the area at or near the flexure may have been tilted westward during deposition of the lower division of the San Rafael Group and eastward during deposition of the middle division (Figures 6 and 9). However, lack of adequate data and truncation of middle San Rafael strata beneath the Dakota Formation make these interpretations tenuous.

There is a lack of even speculative evidence for Jurassic movement across the general area of the Organ Rock monocline. Southeastward thinning of the Romana Sandstone northwest of this flexure might reflect uplift in the vicinity of Black Mesa and regional structural tilt to the northwest across the area of the monocline.

The area of the Comb monocline probably was an active structural zone throughout the Jurassic. During Early and Middle Jurassic time, it acted as a hinge separating a westward-tilted eastern block that included the Blanding and Paradox regions from a more stable western block containing the Monument region (Figures 4, 6, and 10). Positive movement on the Monument uplift is indicated for Late Jurassic time (Figures 12 and 19) and might have been accompanied by eastward monoclinal flexing along Comb monocline.

REGIONAL TECTONIC RELATIONSHIPS

Although a notable amount of tectonism occurred in the Colorado Plateau province during the Jurassic, the region also was strongly influenced by tectonic processes in surrounding regions. The province appears to have acted as a mildly deformed cratonic block surrounded by regions of moderately to significantly greater structural activity during various parts of the Jurassic. Thus, the Colorado Plateau province appears to have acted as a separate crustal block during the Jurassic.

During part of Early Jurassic time, the Ancestral Rocky Mountain uplift in central Colorado and north-central New Mexico was an uplifted source area for the Kayenta and part of the Moenave fluvial systems. The Mogollon highlands in southern Arizona were another source for some of the Moenave.

In Middle Jurassic time the Colorado Plateau was a broad lowland region bordered to the east, south, and west by the Rocky Mountain, Mogollon, and Sevier uplifts, respectively (Figure 20A). None of these areas contributed significant quantities of coarse clastic debris to the study region, suggesting that they were probably lowlands undergoing only moderate erosion.

During Late Jurassic time, large quantities of coarse clastic debris were carried into the region by streams originating in the Sevier and western Mogollon uplifts to the west and southwest of the study area (Figure 20B). The Uinta Mountains region along the northern boundary of the Colorado Plateau in northeastern Utah and northwestern Colorado is a structural block that was tectonically active throughout much of Phanerozoic time (Hansen, 1965; Mallory, 1972). Isopach studies by Hansen (1965) suggest that the Uinta region may have been a westward-plunging trough during deposition of the Upper Jurassic Morrison Formation.

During Jurassic time, a discontinuous Andean type magmatic arc lay along the west coast of the conterminous United States. In Early and Middle Jurassic time, this arc was approximately along the present-day California–Nevada border (Figure 20A). In Late Jurassic time (Kimmeridgian, ~155 m.y. ago), an island arc terrane accreted to the western continental margin (Figure 20B) (Moores and Day, 1984; Schweickert et al., 1984). The accretionary event coincides reasonably well with the onset of thrusting and uplift in the Sevier highlands, dated at approximately 150 m.y. (Carr, 1980). This also coincides with late Oxfordian–Kimmeridgian deposits (~152–158 m.y. old, according to the time scale of Palmer, 1983) of the lower division of the Morrison Formation (Imlay, 1980), which were derived from the Sevier highlands. The event also marks a major change in sedimentation patterns on the Colorado Plateau and a significant change in source areas of Colorado Plateau sediments. A major structural reorientation of the Colorado Plateau occurred slightly earlier in Middle Jurassic time between the times of deposition of the middle and upper divisions of the San Rafael Group. Whether this reflects the earliest stage of accretion of the island arc terrane to the western continental margin or some other process is, at present, unknown.

Structural movements in the study area during the Jurassic were vertical, but the ultimate cause was most likely compressive stresses applied to the crust by underthrusting and collision of Pacific basin plates with western North America. Horizontal compression may be inferred from back arc thrusting in the foreland region in southern Nevada (Carr, 1980, 1983). Reversal of vertical movements in some parts of the study area may reflect changes in relative plate motions and the accompanying reorientation of stress regimes. The large quantity of tuff in the Brushy Basin Member of the Morrison reflects an outpouring of voluminous quantities of volcanic ash from eruptive centers farther west. Presumably the eruptive centers were in or near the magmatic arc along the west coast, although they could also have been behind the magmatic arc in the belt of scattered igneous centers that lay in central Nevada (Figure 20B).

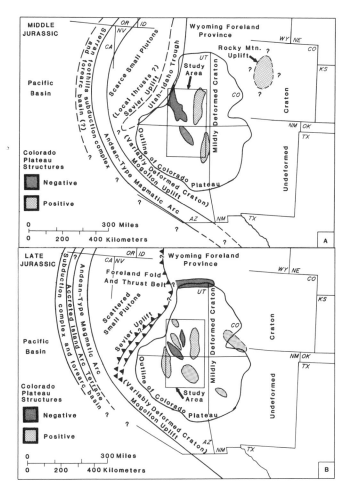

Figure 20—Maps showing the major tectonic features in the southwestern United States during Middle and Late Jurassic time. Structures shown in or near the Colorado Plateau were active during deposition of the Entrada Sandstone (Map A, Middle Jurassic) and during deposition of the lower division of the Morrison Formation (Map B, Late Jurassic). State lines not shown in many places because of palinspastic restoration. Compiled after Schweickert and Cowan (1975), Dickinson (1976, 1981), Schweickert (1976), Hamilton (1978, 1981, and personal communication, 1982), Carr (1980, 1983), Stewart (1980), Moores and Day (1984), S. Y. Johnson (personal communication, 1984), S. S. Oriel (personal communication, 1984), and Schweickert et al. (1984). Sawteeth are on upper plate of thrust sheets.

ECONOMIC RESOURCES

Jurassic strata on the Colorado Plateau contain appreciable quantities of uranium. The distribution of tabular uranium ore deposits in Upper Jurassic (Salt Wash) sandstone beds in the eastern part of the Henry basin are closely related to the distribution of nearby thin organic carbon-bearing lacustrine mudstone beds (Peterson, 1980; Peterson and Turner-Peterson, 1980). The lakes in which these mudstones were deposited formed in actively growing synclines in the Henry basin during periods of low clastic influx. The distribution of the uranium ore deposits was therefore governed, at least partly, by Late Jurassic tectonics. Although the thin organic carbon-bearing lacustrine mudstone beds have been found in

all of the other Salt Wash mining districts on the Colorado Plateau, the necessary detailed sedimentologic studies have not yet been done to determine if their distribution was controlled by contemporaneous tectonism as it was in the Henry Mountains regions.

Jurassic strata in the region are not known to contain hydrocarbons, but their tectonic history may have some bearing on the hydrocarbon potential of Paleozoic rocks, which contain significant quantities of oil and gas in parts of the Colorado Plateau region. Downwarping of the Circle Cliffs and Black Mesa regions in early and middle Middle Jurassic time makes it likely that Paleozoic hydrocarbons may have migrated updip and away from these regions if the hydrocarbons were not involved in stratigraphic traps. Roughly 6000–7000 ft (1800–2100 m) of Paleozoic and Triassic strata lie beneath the Jurassic in these regions. Unless the thermal gradient was higher than normal before or during Jurassic time, most likely only lower Paleozoic strata would have been buried to sufficient depths to generate hydrocarbons. In addition, structures that were active during the Jurassic, especially before the major structural reorientation in Middle Jurassic time, may well have been reactivated structures that previously governed the distribution of Paleozoic source and reservoir rocks. Here, a study of the Triassic tectonic history of the region may be especially helpful.

Considerable quantities of gypsum are present in the Middle Jurassic Carmel Formation in the northwestern part of the study area. These beds were deposited in and adjacent to the Utah–Idaho trough when the northern part of the seaway that occupied the trough was highly restricted or blocked, possibly by vertical crustal movements in the Uinta Mountains region. The distribution of the gypsum conforms closely to the trough and presumably was controlled by subsidence of it.

SUMMARY AND CONCLUSIONS

Although it is commonly thought that many of the structures on the Colorado Plateau originated during the Laramide orogeny, data presented in this report and summarized in Figure 21 show that significant tectonic movements occurred throughout the west-central part of the region during most of the Jurassic. That many of the structures were strongly reactivated during the Laramide is not disputed. It can now be added, however, that many of these structures had a far more complex pre-Laramide tectonic history than had previously been envisioned.

The Early Jurassic tectonic history of the region is not clear, but westward thickening and westward fluvial drainage suggest that the region may have been tilted westward during deposition of the Glen Canyon division. In addition, the large Monument bench and a smaller bench in the Kaiparowits region may have been structurally active during this time.

The Middle Jurassic was the time of greatest diversity of structural movements in the study area (Figure 21). The Colorado Plateau was tilted westward toward the Utah–Idaho trough during early and middle Middle Jurassic time. Several significant positive and negative structures

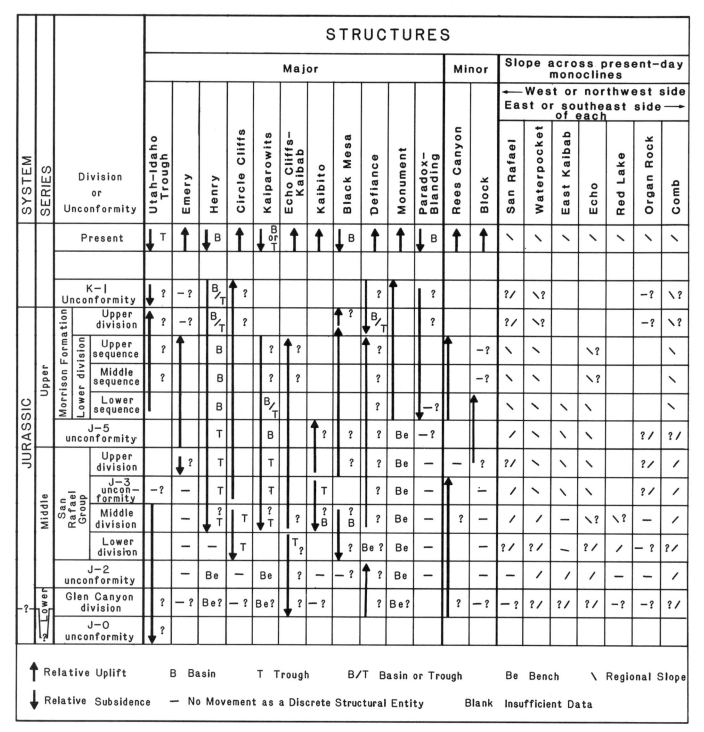

Figure 21—Summary of tectonic movements in the west-central part of the Colorado Plateau during Jurassic time showing their relationship to existing structures. The structural slope across areas of present-day monoclines is indicated, but it could not be determined if these flexures moved as monoclines in the Jurassic.

were also active during deposition of the lower and middle divisions of the San Rafael Group, the most notable of which are the Circle Cliffs trough and the Monument bench. The Monument region was by far the most persistent tectonic element in the region during the Jurassic. By the time of formation of the J-2 unconformity in early Middle Jurassic time, it was a large structural bench and it remained as such throughout Middle Jurassic time (Figure 21).

The most significant change in structural movements occurred at or near the beginning of late Middle Jurassic time, which approximately coincided with the erosion interval that formed the J-3 unconformity. Westward regional tilting apparently ceased, and several of the folds that had been downwarps earlier became positive structures (Figure 21). The study area came closest to resembling its present structural configuration at this time. The Black Mesa region, which was a positive structure at the time but is now a doughnut-shaped downwarp, is a notable exception.

A major change in source areas and sedimentologic processes, but not in the tectonic framework of the region, began in Late Jurassic time following formation of the J-5 unconformity. Streams flowed from highland source areas that lay west and southwest of the Colorado Plateau, and fluvial processes thus became dominant. The Monument region changed from a bench to a positive structure, but most of the other structures persisted from late Middle Jurassic time. An exception is the Emery uplift in the southwestern part of the San Rafael Swell, which had not been structurally active earlier in the Jurassic.

The structural activity in and adjacent to the Colorado Plateau during deposition of the lower division of the Morrison Formation marks the time of the Late Jurassic Nevadan orogeny. Studies by Schweickert et al. (1984) in California indicate that the orogeny was a brief event in late Oxfordian–Kimmeridgian time, which correlates well with the late Oxfordian–Kimmeridgian age of the lower division of the Morrison. In California, the orogeny is attributed to collision of an island arc terrane with the North American continent (Schweickert et al., 1984). This suggests that forces involved in accretionary processes may be translated several hundred miles (or kilometers) inland to produce slight, but nevertheless noticeable, structural deformation in the craton.

Curiously, the Nevadan orogeny in California as presently understood does not include the time of major change in the structural configuration of the Colorado Plateau that occurred somewhat earlier in late Callovian time during or near formation of the J-3 unconformity. The cause of this event may lie in reorientation of stress regimes in the southwestern United States accompanying changes in plate motions, and not in accretion of exotic terranes along the western continental margin. Alternatively, it may lie in accretionary events that have not yet been identified on the west coast.

Reconstruction of the tectonic history of a relatively mildly deformed tectonic region such as the Colorado Plateau demands considerably detailed regional sedimentologic studies that are closely integrated with paleotectonic concepts. Results of this investigation demonstrate a complex tectonic history for the west-central part of the region during the Jurassic. More work is needed, however, especially on the distribution of sedimentary facies, before the tectonic evolution of the Jurassic on the Colorado Plateau can be well understood.

ACKNOWLEDGMENTS

The writer thanks R. B. O'Sullivan, C. E. Turner-Peterson, and the late L. C. Craig for sharing their knowledge and thoughts on many facets of the Jurassic System on the Colorado Plateau. The report was greatly improved through constructive criticism and helpful comments by D. L. Baars, S. Y. Johnson, C. M. Molenaar, and an anonymous reviewer.

REFERENCES CITED

Baars, D. L., 1966, Pre-Pennsylvanian paleotectonics—key to basin evolution and petroleum occurrences in Paradox basin, Utah and Colorado: AAPG Bulletin, v. 50, p. 2082–2111.

Baars, D. L., and P. D. See, 1968, Pre-Pennsylvanian stratigraphy and paleotectonics of the San Juan Mountains, southwestern Colorado: Geological Society of America Bulletin, v. 79, n. 3, p. 333–350.

Baars, D. L., and G. M. Stevenson, 1977, Permian rocks of the San Juan basin, *in* J. E. Fassett, ed., Guidebook of San Juan basin III, northwestern New Mexico: New Mexico Geological Society 28th Field Conference, p. 133–138.

———, 1981, Tectonic evolution of the Paradox basin, Utah and Colorado, *in* D. L. Wiegand, ed., Geology of the Paradox basin: Rocky Mountain Association of Geologists Field Conference Guidebook, p. 23–31.

Baker, A. A., 1933, Geology and oil possibilities of the Moab district, Grand and San Juan counties, Utah: U.S. Geological Survey Bulletin 841, 95 p.

Beaumont, E. C., and G. H. Dixon, 1965, Geology of the Kayenta and Chilchinbito quadrangles, Navajo county, Arizona: U.S. Geological Survey Bulletin 1202-A, p. A1–A28.

Bell, T. E., 1984, Deposition and diagenesis of the Brushy Basin Member of the Upper Jurassic Morrison Formation, San Juan basin, New Mexico (abs.): Geological Society of America Abstracts with Programs, v. 16, n. 4, p. 214.

———, in press, Depositional and diagenesis trends in the Brushy Basin Member and upper part of the Westwater Canyon Member of the Morrison Formation, San Juan Basin, New Mexico, *in* A basin analysis case study—the Morrison Formation, Grants uranium region, New Mexico: AAPG Studies in Geology.

Blakey, R. C., F. Peterson, M. V. Caputo, R. C. Geesaman, and B. J. Voorhees, 1983, Paleogeography of Middle Jurassic continental, shoreline, and shallow marine sedimentation, southern Utah, *in* M. W. Reynolds and E. D. Dolly, eds., Mesozoic paleogeography of the west-central United States: Rocky Mountain Section, Society of Economic Paleontologists and Mineralogists Symposium, v. 2, p. 77–100.

Bohannon, R. G., 1983, Mesozoic and Cenozoic tectonic development of the Muddy, North Muddy, and western Black Mountains, Clark County, Nevada, *in* D. M. Miller, V. R. Todd, and K. A. Howard, eds., Tectonic and stratigraphic studies in the eastern Great basin: Geological Society of America Memoir 157, p. 125–148.

Carr, M. D., 1980, Upper Jurassic to Lower Cretaceous(?) synorogenic sedimentary rocks in the southern Spring Mountains, Nevada: Geology, v. 8, n. 8, p. 385–389.

———, 1983, Geometry and structural history of the Mesozoic thrust belt in the Goodsprings district, southern Spring Mountains, Nevada: Geological Society of America Bulletin, v. 94, n. 10, p. 1185–1198.

Cater, F. W., 1970, Geology of the salt anticline region in southwestern Colorado: U.S. Geological Survey

Professional Paper 637, 80 p.

Condon, S. M., and F. Peterson, in press, Stratigraphy of Middle and Upper Jurassic rocks in the San Juan basin—historical perspective, current ideas, and remaining problems: *in* A basin analysis case study—the Morrison Formation, Grants uranium region, New Mexico: AAPG Studies in Geology.

Craig, L. C., 1981, Lower Cretaceous rocks, southwestern Colorado and southeastern Utah, *in* D. L. Wiegand, ed., Geology of the Paradox basin: Rocky Mountain Association of Geologists Field Conference Guidebook, p. 195–200.

Craig, L. C., C. N. Holmes, R. A. Cadigan, V. L. Freeman, T. E. Mullens, and G. W. Weir, 1955, Stratigraphy of the Morrison and related formations, Colorado Plateau region, a preliminary report: U.S. Geological Survey Bulletin 1009-E, p. 125–168.

Craig, L. C., C. N. Holmes, V. L. Freeman, T. E. Mullens, et al., 1959, Measured sections of the Morrison and adjacent formations: U.S. Geological Survey Open-File Report 485.

Dane, C. H., 1935, Geology of the Salt Valley anticline and adjacent areas, Grand County, Utah: U.S. Geological Survey Bulletin 863, 184 p.

Davidson, E. S., 1967, Geology of the Circle Cliffs area, Garfield and Kane counties, Utah: U.S. Geological Survey Bulletin 1229, 140 p.

Davis, G. H., 1978, The monocline fold pattern of the Colorado Plateau, *in* V. Matthews III, ed., Laramide folding associated with basement block faulting in the western United States: Geological Society of America Memoir 151, p. 215–233.

Dickinson, W. R., 1976, Sedimentary basins developed during evolution of Mesozoic–Cenozoic arc–trench system in western North America: Canadian Journal of Earth Sciences, v. 13, n. 9, p. 1161–1185.

———, 1981, Plate tectonics and the continental margin of California, *in* W. G. Ernst, ed., The geotectonic development of California: Englewood Cliffs, New Jersey, Prentice-Hall, p. 1–28.

Doelling, H. H., 1981, Stratigraphic investigations of Paradox Basin structure as a means of determining the rates and geologic age of salt-induced deformation: Utah Geological and Mineral Survey Open-File Report 29, 88 p.

Ekren, E. B., and F. N. Houser, 1965, Geology and petrology of the Ute Mountains area, Colorado: U.S. Geological Survey Professional Paper 481, 74 p.

Fetzner, R. W., 1960, Pennsylvanian paleotectonics of Colorado Plateau: AAPG Bulletin, v. 44, n. 8, p. 1371–1413.

Galton, P. M., 1978, Fabrosauridae, the basal family of ornithischian dinosaurs (Reptilia: Ornithopoda): Paleontologische Zeitschrift, v. 52, p. 138–159.

Gilbert, J. L., and G. B. Asquith, 1976, Sedimentology of braided alluvial interval of Dakota Sandstone, northeast New Mexico: New Mexico Bureau of Mines and Mineral Resources Circular 150, 16 p.

Gilluly, J. L., and J. B. Reeside, Jr., 1928, Sedimentary rocks of the San Rafael Swell and some adjacent areas in eastern Utah: U.S. Geological Survey Professional Paper 150-D, p. 61–110.

Hackman, R. J., and D. G. Wyant, 1973, Geology, structure, and uranium deposits of the Escalante quadrangle, Utah and Arizona: U.S. Geological Survey Miscellaneous Investigations Series Map I-744.

Hackman, R. J., and A. B. Olson, 1977, Geology, structure, and uranium deposits of the Gallup 1° × 2° quadrangle, New Mexico and Arizona: U.S. Geological Survey Miscellaneous Investigations Series Map I-981.

Hamilton, W., 1978, Mesozoic tectonics of the western United States, *in* D. G. Howell and K. A. McDougal, eds., Mesozoic paleogeography of the western United States: Pacific Section, Society of Economic Paleontologists and Mineralogists Symposium, p. 33–70.

———, 1981, Plate-tectonic mechanism of Laramide deformation: University of Wyoming Contributions to Geology, v. 19, n. 2, p. 87–92.

Hansen, W. R., 1965, Geology of the Flaming Gorge area, Utah–Colorado–Wyoming: U.S. Geological Survey Professional Paper 490, 196 p.

Harshbarger, J. W., C. A. Repenning, and J. H. Irwin, 1957, Stratigraphy of the uppermost Triassic and Jurassic rocks of the Navajo Country: U.S. Geological Survey Professional Paper 291, 74 p.

Haynes, D. D., and R. J. Hackman, 1978, Geology, structure, and uranium deposits of the Marble Canyon 1° × 2° quadrangle, Arizona: U.S. Geological Survey Miscellaneous Investigations Series Map I-1003.

Haynes, D. D., J. D. Vogel, and D. G. Wyant, 1972, Geology, structure, and uranium deposits of the Cortez quadrangle, Colorado and Utah: U.S. Geological Survey Miscellaneous Investigations Series Map I-629.

Heylmun, E. B., 1958, Paleozoic stratigraphy and oil possibilities of Kaiparowits region, Utah: AAPG Bulletin, v. 42, n. 8, p. 1781–1811.

Hunt, C. B., P. Averitt, and R. L. Miller, 1953, Geology and geography of the Henry Mountains region, Utah: U.S. Geological Survey Professional Paper 228, 234 p.

Imlay, R. W., 1980, Jurassic paleobiogeography of the conterminous United States in its continental setting: U.S. Geological Survey Professional Paper 1062, 134 p.

Kelly, V. C., 1955, Regional tectonics of the Colorado Plateau and relationship to the origin and distribution of uranium: University of New Mexico Publications in Geology, n. 5, 120 p.

Kirkland, P. L., 1963, Permian stratigraphy and stratigraphic paleontology of a part of the Colorado Plateau, *in* R. O. Bass, ed., Shelf carbonates of the Paradox basin: Four Corners Geological Society, Fourth Field Conference, p. 80–100.

Kocurek, G., and R. H. Dott, Jr., 1983, Jurassic paleogeography and paleoclimate of the central and southern Rocky Mountain region, *in* M. W. Reynolds and E. D. Dolly, eds., Mesozoic paleogeography of the west-central United States: Rocky Mountain Section, Society of Economic Paleontologists and Mineralogists Symposium, v. 2, p. 101–116.

Kowallis, B. J., and J. Heaton, 1984, Fission track stratigraphy of Upper Jurassic and Lower Cretaceous bentonites from the Morrison and Cedar Mountain Formations of central Utah: Geological Society of America Abstracts with Programs, v. 16, n. 6, p. 565.

Lessentine, R. H., 1965, Kaiparowits and Black Mesa

basins—stratigraphic synthesis: AAPG Bulletin, v. 49, n. 11, p. 1997–2019.

Lewis, G. E., J. H. Irwin, and R. F. Wilson, 1961, Age of the Glen Canyon Group (Triassic and Jurassic) on the Colorado Plateau: Geological Society of America Bulletin, v. 72, n. 9, p. 1437–1440.

Lohman, S. W., 1965, Geology and artesian water supply, Grand Junction area, Colorado: U.S. Geological Survey Professional Paper 451, 149 p.

Mallory, W. W., 1972, ed., Geologic atlas of the Rocky Mountain region: Denver, Rocky Mountain Association of Geologists, 331 p.

Marzolf, J. E., 1983, Early Mesozoic eolian transition from cratonal margin to orogenic-volcanic arc, *in* K. D. Gurgel, ed., Geologic excursions in stratigraphy and tectonics—from southeastern Idaho to the southern Inyo Mountains, California, via Canyonlands and Arches National Parks, Utah: Utah Geological and Mineralogical Survey, Special Studies, n. 60, p. 39–46.

McFall, C. C., 1955, Geology of the Escalante–Boulder area, Garfield County, Utah: Ph.D. thesis, Yale University, New Haven, Connecticut, 180 p.

McKee, E. D., 1963, Triassic uplift along the west flank of the Defiance positive element, Arizona: U.S. Geological Survey Professional Paper 475-C, p. C28–C29.

McKee, E. D., and G. W. Weir, 1953, Terminology for stratification and cross-stratification in sedimentary rocks: Geological Society of America Bulletin, v. 64, n. 4, p. 381–389.

McKnight, E. T., 1940, Geology of area between Green and Colorado rivers, Grand and San Juan counties, Utah: U.S. Geological Survey Bulletin 908, 147 p.

Moores, E. M., and H. W. Day, 1984, Overthrust model for the Sierra Nevada: Geology, v. 12, n. 7, p. 416–419.

Olsen, P. E., and P. M. Galton, 1977, Triassic–Jurassic tetrapod extinctions—are they real?: Science, v. 197, n. 4307, p. 983–985.

O'Sullivan, R. B., 1965, Geology of the Cedar Mesa–Boundary Butte area, San Juan County, Utah: U.S. Geological Survey Bulletin 1186, 128 p.

———, 1978, Stratigraphic sections of Middle Jurassic San Rafael Group from Lohali Point, Arizona, to Bluff, Utah: U.S. Geological Survey Oil and Gas Investigations Chart OC-77.

———, 1980a, Stratigraphic sections of Middle Jurassic San Rafael Group and related rocks from the Green River to the Moab area in east-central Utah: U.S. Geological Survey Miscellaneous Field Studies Map MF-1247.

———, 1980b, Stratigraphic sections of Middle Jurassic San Rafael Group from Wilson Arch to Bluff in southeastern Utah: U.S. Geological Survey Oil and Gas Investigations Chart OC-102.

———, 1981a, Stratigraphic sections of some Jurassic rocks from near Moab, Utah, to Slick Rock, Colorado: U.S. Geological Survey Oil and Gas Investigations Chart OC-107.

———, 1981b, Stratigraphic sections of Middle Jurassic Entrada Sandstone and related rocks from Salt Valley to Dewey Bridge in east-central Utah: U.S. Geological Survey Oil and Gas Investigations Chart OC-113.

———, 1983, Stratigraphic sections of Middle Jurassic Entrada Sandstone and related rocks from Dewey Bridge, Utah, to Bridgeport, Colorado: U.S. Geological Survey Oil and Gas Investigations Chart OC-122.

———, 1984, Stratigraphic sections of Middle Jurassic San Rafael Group and related rocks from Dewey Bridge, Utah, to Uravan, Colorado: U.S. Geological Survey Oil and Gas Investigations Chart OC-124.

O'Sullivan, R. B., and H. M. Beikman, 1963, Geology, structure, and uranium deposits of the Shiprock quadrangle, New Mexico and Arizona: U.S. Geological Survey Miscellaneous Geologic Investigations Map I-345.

O'Sullivan, R. B., and M. W. Green, 1973, Triassic rocks of northeast Arizona and adjacent areas, *in* H. L. James, ed., Guidebook of Monument Valley and vicinity, Arizona and Utah: New Mexico Geological Society 24th Field Conference Guidebook, p. 72–78.

O'Sullivan, R. B., and G. N. Pipiringos, 1983, Stratigraphic sections of Middle Jurassic Entrada Sandstone and related rocks from Dewey Bridge, Utah, to Bridgeport, Colorado: U.S. Geological Survey Oil and Gas Investigations Chart OC-122.

Palmer, A. R., 1983, The Decade of North American Geology 1983 geologic time scale: Geology, v. 11, n. 9, p. 503–504.

Peterson, F., 1969, Cretaceous sedimentation and tectonism in the southeastern Kaiparowits region, Utah: U.S. Geological Survey Open-File Report, 259 p.

———, 1980, Sedimentology as a strategy for uranium exploration—concepts gained from analysis of a uranium-bearing depositional sequence in the Morrison Formation of south-central Utah, *in* C. E. Turner-Peterson, ed., Uranium in sedimentary rocks—application of the facies concept to exploration: Rocky Mountain Section, Society of Economic Paleontologists and Mineralogists, p. 65–126.

———, 1984, Fluvial sedimentation on a quivering craton—influence of slight crustal movements on fluvial processes, Upper Jurassic Morrison Formation, western Colorado Plateau: Sedimentary Geology, v. 38, n. 1/4, p. 21–49.

———, in press, Stratigraphy and nomenclature of Middle and Upper Jurassic rocks, western Colorado Plateau, Utah and Arizona: U.S. Geological Survey Bulletin 1633-B.

Peterson, F., and G. N. Pipiringos, 1979, Stratigraphic relations of the Navajo Sandstone to Middle Jurassic formations, southern Utah and northern Arizona: U.S. Geological Survey Professional Paper 1035-B, p. B1–B43.

Peterson, F., and C. E. Turner-Peterson, 1980, Lacustrine-humate model—sedimentologic model for tabular sandstone uranium deposits in the Morrison Formation, Utah, and application to uranium exploration: U.S. Geological Survey Open-File Report 80-319, 48 p.

Peterson, F., B. Cornet, and C. E. Turner-Peterson, 1977, New data bearing on the stratigraphy and age of the Glen Canyon Group (Triassic and Jurassic) in southern Utah and northern Arizona: Geological Society of America Abstracts with Programs, v. 9, n. 6, p. 755.

Peterson, F., R. T. Ryder, and B. E. Law, 1980, Stratigraphy, sedimentology, and regional relationships of the Cretaceous System in the Henry Mountains region,

Utah, *in* M. D. Picard, ed., Henry Mountains Symposium: Utah Geological Association Publication 8, p. 151–170.

Phoenix, D. A., 1963, Geology of the Lees Ferry area, Coconino County, Arizona: U.S. Geological Survey Bulletin 1137, 86 p.

Pipiringos, G. N., and R. B. O'Sullivan, 1978, Principal unconformities in Triassic and Jurassic rocks, western Interior United States—a preliminary report: U.S. Geological Survey Professional Paper 1035-A, p. A1–A29.

Pipiringos, G. N., and R. W. Imlay, 1979, Lithology and subdivisions of the Jurassic Stump Formation in southeastern Idaho and adjoining areas: U.S. Geological Survey Professional Paper 1035-C, p. C1–C25.

Poole, F. G., 1961, Stream directions in Triassic rocks of the Colorado Plateau: U.S. Geological Survey Professional Paper 424-C, p. C139–C141.

———, 1962, Wind directions in late Paleozoic to middle Mesozoic time on the Colorado Plateau: U.S. Geological Survey Professional Paper 450-D, p. D147–D151.

Reches, Z., 1978, Development of monoclines—Part I, structure of the Palisades Creek branch of the East Kaibab monocline, Grand Canyon, Arizona, *in* V. Matthews III, ed., Laramide folding associated with basement block faulting in the western United States: Geological Society of America Memoir 151, p. 235–271.

Reiche, P., 1938, An analysis of cross-lamination—the Coconino Sandstone: Journal of Geology, v. 46, p. 905–932.

Santos, E. S., and C. E. Turner-Peterson, in press, Tectonic setting of the San Juan basin in Jurassic time, *in* A basin analysis case study—the Morrison Formation, Grants uranium region, New Mexico: AAPG Studies in Geology.

Schweickert, R. A., 1976, Shallow-level plutonic complexes in the eastern Sierra Nevada, California, and their tectonic implications: Geological Society of America Special Paper 176, 58 p.

Schweickert, R. A., and D. S. Cowan, 1975, Early Mesozoic tectonic evolution of the western Sierra Nevada, California: Geological Society of America Bulletin, v. 86, n. 10, p. 1329–1336.

Schweickert, R. A., N. L. Bogen, G. H. Girty, R. E. Hanson, and C. Merguerian, 1984, Timing and structural expression of the Nevadan orogeny, Sierra Nevada, California: Geological Society of America Bulletin, v. 95, n. 8, p. 967–979.

Sears, J. D., 1956, Geology of Comb Ridge and vicinity north of San Juan River, San Juan County, Utah: U.S. Geological Survey Bulletin 1021-E, p. 167–207.

Smith, J. F., Jr., L. C. Huff, E. N. Hinrichs, and R. G. Luedke, 1963, Geology of the Capitol Reef area, Wayne and Garfield counties, Utah: U.S. Geological Survey Professional Paper 363, 102 p.

Smith, N. D., 1970, The braided stream depositional environment—comparison of the Platte River with some Silurian clastic rocks, north-central Appalachians:

Geological Society of America Bulletin, v. 81, n. 10, p. 2993–3014.

Standlee, L. A., 1982, Structure and stratigraphy of Jurassic rocks in central Utah—their influence on tectonic development of the Cordilleran foreland thrust belt, *in* R. B. Powers, ed., Geologic studies of the Cordilleran thrust belt: Rocky Mountain Association of Geologists, v. 1, p. 357–382.

Stevenson, G. M., and D. L. Baars, 1977, Pre-Carboniferous paleotectonics of the San Juan basin, *in* J. E. Fassett, ed., Guidebook of San Juan basin III, northwestern New Mexico: New Mexico Geological Society 28th Field Conference Guidebook, p. 99–110.

Stewart, J. H., 1980, Geology of Nevada: Nevada Bureau of Mines and Geology Special Publication 4, 136 p.

Stokes, W. L., 1952, Lower Cretaceous in Colorado Plateau: AAPG Bulletin, v. 36, n. 9, p. 1766–1776.

———, 1954a, Relation of sedimentary trends, tectonic features, and ore deposits in the Blanding district, San Juan County, Utah: U.S. Atomic Energy Commission, RME-3093, pt. 1, 33 p.

———, 1954b, Some stratigraphic, sedimentary, and structural relations of uranium deposits in the Salt Wash Sandstone: U.S. Atomic Energy Commission, RME-3102, 51 p.

Tschudy, R. H., B. D. Tschudy, and L. C. Craig, 1984, Palynological evaluation of Cedar Mountain and Burro Canyon formations, Colorado Plateau: U.S. Geological Survey Professional Paper 1281, 24 p.

Walcott, C. D., 1890, Study of a line of displacement in the Grand Canyon of the Colorado, in northern Arizona: Geological Society of America Bulletin, v. 1, p. 1–86.

Wengerd, S. A., and M. L. Matheny, 1958, Pennsylvanian System of Four Corners region: AAPG Bulletin, v. 42, n. 9, p. 2048–2106.

Williams, P. L., 1964, Geology, structure, and uranium deposits of the Moab quadrangle, Colorado and Utah: U.S. Geological Survey Miscellaneous Investigations Series Map I-360.

Williams, P. L., and R. J. Hackman, 1971, Geology, structure, and uranium deposits of the Salina quadrangle, Utah: U.S. Geological Survey Miscellaneous Geologic Investigations Map I-591.

Wright, J. C., and D. D. Dickey, 1963, Relations of the Navajo and Carmel formations in southwest Utah and adjoining Arizona: U.S. Geological Survey Professional Paper 450-E, p. E63–E67.

———, 1978a, East–west cross sections of the Jurassic age San Rafael Group rocks from western Colorado to central and western Utah: U.S. Geological Survey Open-File Report 78-784.

———, 1978b, North–south cross sections of the Jurassic San Rafael Group in Utah and western Colorado: U.S. Geological Survey Open-File Report 78-965.

———, 1978c, Miscellaneous cross sections of the Jurassic San Rafael Group in southern Utah: U.S. Geological Survey Open-File Report 58-966.

Structurally Controlled Sediment Distribution Patterns in the Jurassic Morrison Formation of Northwestern New Mexico and Their Relationship to Uranium Deposits

A. R. Kirk
S. M. Condon
U.S. Geological Survey
Denver, Colorado

Active Jurassic structures, inferred from structural contour and isopleth maps, significantly affected depositional patterns in the Westwater Canyon and Brushy Basin members of the Morrison Formation in the southern San Juan basin, New Mexico. Isopleth maps illustrate the geometry of the major depositional units, the distribution of sandstone depocenters, and large-scale lithofacies variations within the units. A reconstruction of topography at the base of the Westwater Canyon Member shows a series of subparallel paleotopographic highs and lows that trend east–southeast. The Westwater Canyon is thick and sandy along paleotopographic lows but is thin and has a lower sand content over the structurally controlled paleotopographic highs. These relationships suggest active structural control of facies distribution by differential subsidence along east–southeast-oriented folds or faults during deposition of the unit. Locally, east–southeast-oriented basement faults that were episodically reactivated through time have been detected by detailed seismic reflection studies. Depositional patterns and lithofacies distribution, in turn, appear to have controlled the location of uranium deposits. Primary uranium ore in the Westwater Canyon Member is restricted to depocenters defined by anomalously thick and sandy facies that trend east–southeast. Redistributed ore deposits are also localized in the vicinity of anomalously thick zones of the Westwater Canyon but in rocks with relatively low sandstone:mudstone ratios. The location of redistributed deposits, however, is much more closely controlled by the position of a regional oxidation–reduction (redox) interface whose three-dimensional configuration was apparently influenced both regionally and locally by Laramide and younger structures. Remnant ore deposits are relict primary deposits that occur in locally preserved, chemically reduced rocks. They occur updip of the redox interface and display sedimentologic controls similar to those of primary ore bodies. These remnant deposits appear to have been shielded from oxidation by an unusual stratigraphic or structural setting.

INTRODUCTION

The San Juan basin, a large structural depression formed by Late Cretaceous to middle Tertiary tectonism, is located on the southeastern side of the Colorado Plateau in northwestern New Mexico and southwestern Colorado (Figure 1). Within the San Juan basin, Pennsylvanian(?)–Tertiary sedimentary units are exposed at the surface, and rocks of Precambrian, Cambrian, Devonian, and Mississippian age are locally present in the subsurface (Hilpert, 1969). In this study we examine the Westwater Canyon and Brushy Basin members of the Upper Jurassic Morrison Formation. Stratigraphic units within and adjacent to the Morrison Formation are depicted schematically and shown on a typical geophysical log in Figure 2.

Many large uranium deposits are concentrated in the Grants uranium region or Grants mineral belt (Kelley, 1963b), which lies along the southern margin of the San Juan basin (Figure 1). The Grants uranium region includes the entire study area but extends eastward across the Mount Taylor volcanic field to the Rio Grande rift (Figure 1). Most of the ore deposits occur on a homoclinal structural feature called the Chaco slope, which dips gently (2–10°) northward away from the Zuni uplift between the Acoma and Gallup sags (Figure 1). The Grants uranium region is the largest uranium producing area in the United States; it has accounted for more than 40% of the total U.S. uranium production (Chenoweth, 1976) and contains about 53% of domestic reserves (Chenoweth and Holen, 1980). Most of the production has come from deposits hosted in sandstones

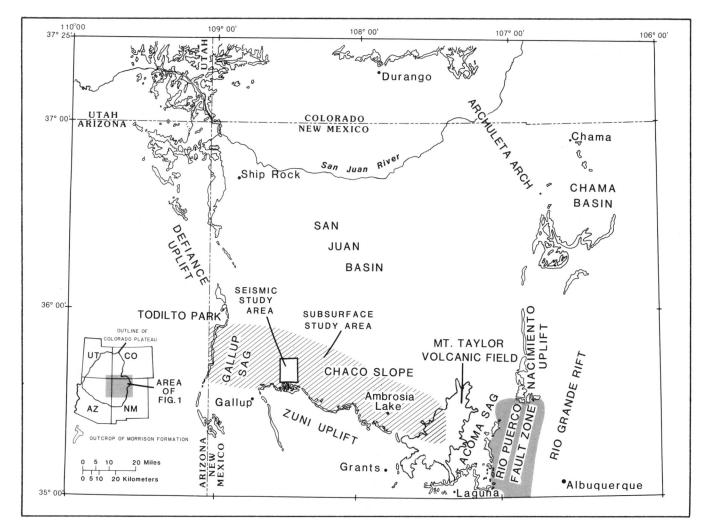

Figure 1—Map of the San Juan basin and adjacent areas, showing location of the subsurface (hachured pattern) and seismic study areas, structural elements, and outcrops of the Morrison Formation.

of the Westwater Canyon Member of the Morrison Formation (Adams and Saucier, 1981).

In this paper we present the results of a subsurface study in which we examined the distribution of lithofacies in the Westwater Canyon and Brushy Basin members of the Morrison Formation by contouring stratigraphic parameters obtained from drill hole data. Changes in the values of these parameters reflect the distribution of sedimentary facies that are structurally controlled; the locations of uranium deposits are directly related to these stratigraphic and sedimentologic variations.

GENERAL GEOLOGIC SETTING

Regional Structure and Tectonics

Evidence suggests that sedimentary rocks of the Morrison Formation were deposited on a relatively stable continental foreland slope behind an actively developing magmatic arc located to the west during Late Jurassic time (Dickinson, 1981). Numerous beds of altered volcanic ash occur throughout the upper part of the Morrison Formation of northwestern New Mexico and document coeval volcanic

activity. In the ancestral San Juan basin area, deposition of sediment appears to have been confined to a shallow, broad, open northwest–southeast oriented trough (Saucier, 1976; Turner-Peterson et al., 1980; Santos and Turner-Peterson, 1986) that was bounded on the north by a subdued ancestral Uncompahgre–San Luis highland and on the south by the actively rising Mogollon highland (Saucier, 1976). Both of these highlands were important structural features during Pennsylvanian and Permian time and appear to have been reactivated during Late Jurassic time. The Mogollon highland and the adjoining magmatic arc probably were the primary sources for the mixed detritus composing the Morrison Formation in this region (Craig et al., 1955; Adams and Saucier, 1981; Turner-Peterson, 1986; Hansley, 1986).

The dominant northwest–southeast orientation of these major structural features is believed to have been controlled by faults in the Precambrian basement having a similar orientation (Kelley, 1955; Saucier, 1976). Episodic movement on these structures occurred throughout the Phanerozoic, and sedimentary rocks of the Chaco slope were locally folded and faulted during several events related to movement of the ancestral Zuni uplift. Detailed seismic

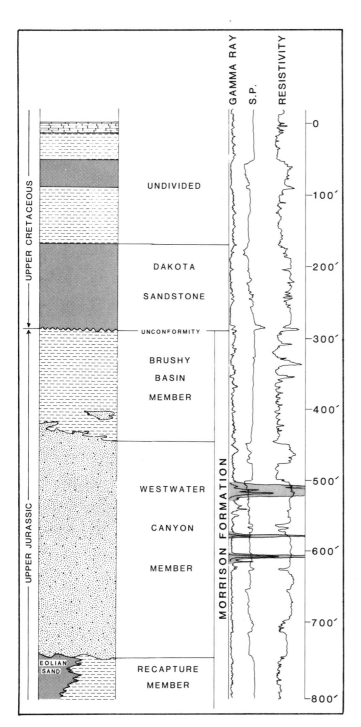

Figure 2—Schematic stratigraphic section and typical geophysical log of the Dakota Sandstone, part of the Morrison Formation, and adjacent units, showing uranium mineralization (shaded) in the Westwater Canyon Member.

studies on a limited area of the Chaco slope (Phelps et al., 1986) clearly document the continued reactivation of Precambrian basement faults in the region.

Significant deformation occurred during Late Jurassic–Early Cretaceous time and is indicated by folds in the Morrison Formation that are truncated by the overlying Upper Cretaceous Dakota Sandstone east of the study area, near Laguna (Figure 1) (Hilpert and Corey, 1957; Hilpert

and Moench, 1960; Moench and Schlee, 1967; Hilpert, 1969). These folds apparently developed contemporaneously with deposition of Upper Jurassic rocks, and they are thought to have affected depositional patterns and processes (Moench and Schlee, 1967; Hilpert, 1969).

Uplift during Late Jurassic time is also indicated by depositional thinning and local pinchouts of members of the Morrison Formation onto the northern and eastern flanks of the ancestral Zuni uplift. This was followed by tilting and erosional truncation of the Morrison and older units prior to the deposition of the Dakota Sandstone.

The San Juan basin itself did not form until the Laramide orogeny, which began in latest Cretaceous time and continued into the Eocene (80–40 m.y. ago) (Coney, 1972; Chapin and Cather, 1981; Dickinson, 1981). The severe deformation that accompanied this event in North America is similar in many respects to faulting and basement uplifts occurring in parts of present-day South America that are characterized by very low subduction angles (Jordan et al., 1983). Movements associated with the Laramide orogeny produced the uplifts, sags, and arches that defined the margins of the present San Juan basin, deepened the central part of the basin, and produced the major folds, faults, and regional dip on the Chaco slope (see Figures 1 and 3). The difference in the style of deformation between the east and west halves of the Chaco slope is striking. The east half is intensely faulted, whereas the west half has only a few faults, none of which has any significant offset.

Middle–late Tertiary uplift and faulting began in the Miocene as broad continental arching across the Colorado Plateau and as crustal extension throughout the southwestern United States (Kelley, 1955; Hilpert, 1969; Christiansen and Lipman, 1972). These events resulted in downfaulting along the Rio Puerco fault zone and rapid deposition of sediments of the Tertiary Santa Fe Group into the widening Rio Grande rift (Figure 1) and may have formed or reactivated some of the faults observed on the Chaco slope (Hilpert, 1969; Santos, 1970). Rifting was accompanied by sporadic Pliocene–Holocene dominantly basaltic volcanic activity, such as that of the Mount Taylor field (Bassett et al., 1963; Lipman and Mehnert, 1975, 1980).

Stratigraphy

Three members of the Morrison Formation, the Recapture, Westwater Canyon, and Brushy Basin, were named for exposures in southeastern Utah by Gregory (1938); this nomenclature was extended into the southern San Juan basin by Rapaport et al. (1952). The Salt Wash Member of the Morrison Formation (Lupton, 1914; Gilluly and Reeside, 1928) is not present in outcrop or in the subsurface of our study area. Two informal units are recognized in the Morrison Formation on the Chaco slope and are of interest because they contain large uranium deposits. The "Jackpile sandstone" (of economic usage) is at the top of the Brushy Basin Member, and the "Poison Canyon sandstone" is at the top of the Westwater Canyon Member.

The Recapture Member (Figure 2) is the lowest formal member of the Morrison Formation exposed in outcrop and encountered in the subsurface in our study area. The Recapture was apparently deposited in a predominantly fluvial setting as channel sandstones and overbank and lacustrine mudstones. Locally, it is interbedded with what appear to be eolian deposits of both dune and interdune playa origin and with minor, thin, lacustrine or playa limestones (Saucier, 1976; Green, 1980; Turner-Peterson et al., 1980). Between Thoreau and Gallup, New Mexico (Figure 1), the Recapture fluvial sediments are replaced laterally by eolian sandstone, and the overlying sandstones of the Westwater Canyon are deposited directly on a scoured surface in the eolian sandstone. The Recapture pinches out as a result of either depositional thinning or onlap a few miles south of Grants, New Mexico (Figure 1) (Thaden et al., 1967; Hilpert, 1969; Santos, 1970; Maxwell, 1976). Fluvial sediment transport directions have been reported as northeasterly in this unit (Craig et al., 1955; Saucier, 1976). Most data used in this subsurface study were obtained from drill holes that penetrated at least the upper part of the Recapture Member.

The contact between the Recapture and the overlying Westwater Canyon Member is usually sharp; fluvial sandstones of the Westwater Canyon lie on scour surfaces cut into the Recapture Member, but locally the lithologies are interbedded and have been mapped as intertonguing at the outcrop (Thaden et al., 1966, 1967; Robertson, 1973, 1974; Green, 1976). These contact relationships show that, at least locally, deposition of the two members appears to have occurred simultaneously and that no major time break or erosional event is represented by the scoured contacts.

The Westwater Canyon Member (Figure 2) is composed of vertically stacked and laterally coalesced sandstone beds that are interbedded with minor, thin, laterally discontinuous mudstones, ash beds, and clay clast conglomerates. The Westwater Canyon Member was apparently deposited by composite systems of moderate- to high-energy braided streams (Campbell, 1976; Galloway, 1980; Turner-Peterson et al., 1980). Historical interpretation of sediment transport directions within the Westwater Canyon Member in the southern part of the basin suggests multiple directions of transport, with northeasterly directions being the most prominent (Craig et al., 1955; Campbell, 1976). Other work suggests that northeasterly directions are most prominent in the lower and uppermost parts of the Westwater Canyon, while southeasterly components are more common in the middle part (Turner-Peterson et al., 1980).

The Westwater Canyon Member is present throughout most of the San Juan basin (Craig et al., 1955), with the exception of the area south and southwest of Gallup in the Gallup sag (Figure 1) where it was removed during Late Jurassic–Early Cretaceous time by erosion (Saucier, 1967). The Westwater Canyon has also been observed to thin and pinch out depositionally about 7 mi (11 km) south of Grants (Santos, 1970; Maxwell, 1976), and it ranges from 0 to 50 ft (0 to 15 m) near Laguna (Figure 1) (Moench and Schlee, 1967). The maximum known thickness of the Westwater Canyon at the outcrop occurs near Todilto Park (Figure 1) where it is roughly 330 ft (100 m) thick and conglomeratic (Craig et al., 1955; A. R. Kirk and A. C. Huffman, unpublished data, 1980). It attains a thickness of more than 440 ft (134 m) in the subsurface about 17 mi (27 km) northeast of Gallup.

The overlying Brushy Basin Member (Figure 2) consists predominantly of bentonitic and zeolitic mudstones, claystones, and siltstones, with minor interbedded tabular and lenticular sandstone beds that are interpreted to have been deposited in both fluvial and lacustrine playa settings (Bell, 1983). The Westwater Canyon Member intertongues with the Brushy Basin Member on both a local and regional scale. In this zone of extensive intertonguing, mudstones of the Brushy Basin become thicker and have greater lateral continuity, which makes member contacts sometimes difficult to pick on geophysical logs. The Poison Canyon sandstone is the informal name given to the uppermost sandstone bed of the Westwater Canyon Member within this zone of intertonguing at the Poison Canyon Mine (sec. 19, T. 13 N., R. 9 W.) in the Ambrosia Lake District, north of Grants (Hilpert and Freeman, 1956). For this study, we combined this and similar Westwater Canyon–like sandstones found in the zone of intertonguing with the underlying Westwater Canyon Member.

The Brushy Basin is present throughout the San Juan basin in New Mexico except south of Gallup and in another area south of Grants (Santos, 1970; Maxwell, 1976), where it was apparently removed by Late Jurassic–Early Cretaceous erosion. Throughout the study area, the Brushy Basin is overlain by the Upper Cretaceous Dakota Sandstone along a major regional unconformity, which locally has a considerable amount of relief (as much as 120 ft [36 m]). This unconformity was caused by Late Jurassic–Early Cretaceous deformation that tilted the region to the northeast and by erosion that removed successively older pre-Dakota stratigraphic units southwestward across the San Juan basin (Craig et al., 1955; Harshbarger et al., 1957; Maxwell, 1976). Thus, the total original thickness and nature of the eroded upper part of the Brushy Basin Member are unknown in the study area. East of the study area, near Laguna (Figure 1), the Jackpile sandstone is preserved at the top of the Brushy Basin (Schlee and Moench, 1961; Moench and Schlee, 1967; Adams et al., 1978) in the axis of a northeast-trending pre-Dakota fold; it contains several large uranium deposits. The Jackpile sandstone does not, however, extend into the area studied.

ACTIVE JURASSIC STRUCTURES AND RELATED SEDIMENTATION PATTERNS

Methods

Data for this subsurface study came from approximately 1800 geophysical logs of uranium and oil and gas exploration drill holes and more than 100 measured sections. About 20% of the drill holes studied also had lithologic logs, and 12 cores were examined in detail both for lithology and for correlation with the geophysical logs. A drill hole density of approximately two drill holes per 1 mi^2 (2.5 km) was maintained throughout most of the study area.

Parameters measured or computed from the measured values for members of the Morrison Formation were thickness, sandstone:mudstone ratio, net sandstone, percentage of sandstone, number of alternations between sandstone and mudstone per 100 ft (30 m) of section, and total combined thickness of the Brushy Basin and Westwater

Canyon members. Only the isopach and sandstone:mudstone ratio maps are presented here, but a summary statement of important relationships observed on other maps is included below. A detailed description and the maps themselves can be found in a paper by Kirk and Condon (1986).

A structural contour map drawn on the base of the Dakota Sandstone was also compiled (Figure 3), and this map was used to delineate and remove post-Jurassic structures. Isopach maps of the Westwater Canyon (see Figure 4) and Brushy Basin members were then used to identify potential sandstone depocenter trends, to identify potential structures that may have affected sediment distribution patterns, and to note local thick and thin parts of the two units. Because the Brushy Basin thins systematically to the south and west beneath the pre-Dakota erosion surface and because the Westwater Canyon Member thins depositionally eastward, relatively thick and thin parts of the members are more important than the absolute thickness at any point on the map. Sandstone:mudstone ratios are extremely variable, ranging from 1.6 to 300 in the Westwater Canyon Member (see Figure 5). Values larger than 30 (a large sandstone:mudstone ratio) were arbitrarily assigned a value of 30 for purposes of contouring. The ratios were used in conjunction with isopach maps to determine the character (sandiness) of thick and thin sedimentary facies.

The line of truncation of the Brushy Basin formed by pre-Dakota erosion (see Figures 3, 4, 5, 6, and 10) is important because southwest of this line, pre-Dakota erosion has removed part of the underlying Westwater Canyon Member, and the Dakota Sandstone rests unconformably on the eroded Westwater Canyon Member. Therefore, the values of all stratigraphic parameters of the Westwater Canyon contoured in this area should be considered as only approximate.

The paleotopographic map (see Figure 6) was constructed in several stages. An isopach map of the interval comprising the combined thickness of the Westwater Canyon and Brushy Basin members was first constructed. Then a correction was made to restore the material removed from the top of the Brushy Basin Member by pre-Dakota erosion (using a rate of truncation, derived earlier, of 10 ft/mi [1.9 m/km] as a correction factor). Adjustments were made to remove irregularities on the upper surface of the isopach interval that were due to erosional processes such as deep fluvial scours or paleovalleys at the base of the Dakota Sandstone. An additional correction was also made to remove the effects of a regional tilt of about 1/8 of a degree to the northeast and east–northeast. This tilt was caused either by pre–Westwater Canyon structural tilt or by depositional slopes or gradients into the basin. The contours were then renumbered so that we could refer them to an arbitrary datum rather than to the isopach interval.

A correction for compaction was considered but deemed unnecessary. An estimated loss of original thickness of siltstone and mudstone by compaction of about 20% has been documented by Peterson (1984) for similar rocks elsewhere on the Colorado Plateau. When compared with the large amount of sandstone present in the Westwater Canyon Member, an increase of 20% of the mudstone thickness does not significantly change either the overall unit thickness or the sandstone:mudstone ratio.

Sedimentation

We here examine sedimentation in the Morrison Formation through a discussion of various isopleth maps (see Figures 4–7) and a summary statement of important relationships observed on other maps. The isopach map of the Westwater Canyon Member (Figure 4) is the most useful map for understanding the overall depositional system. On this map, the Westwater Canyon Member can be seen to range in thickness from 440 ft (134 m) about 17 mi (27 km) northeast of Gallup, New Mexico, to less than 80 ft (24 m) over the McCarty arch (Figure 3, no. 13) northwest of Mount Taylor. An overall thinning occurs from west to east across the map as the Westwater Canyon fluvial system changes from extensive, laterally coalesced and vertically stacked channel sandstones in the west to isolated channel sandstones interbedded with more abundant mudstones to the east. These relationships have been observed in outcrop studies by Campbell (1976), Galloway (1980), and Turner-Peterson et al. (1980). Overall depositional thinning of the Westwater Canyon can also be observed from north to south across the map area (Figure 4). This southward depositional thinning of the Westwater Canyon strongly supports the existence of an actively rising ancestral Zuni uplift (Figure 1) against which the depositional pinchout or onlap of both the Recapture and Westwater Canyon members occurred about 7 mi (11 km) southeast of Grants. This also lends support to the idea of a Late Jurassic deformational event, which is indicated by the pre-Dakota folds elsewhere on the Chaco slope (Moench and Schlee, 1967; Hilpert, 1969). Episodic upwarping of the ancestral Zuni uplift may also explain the observed shifts between northeast and southeast fluvial transport directions of sandstone deposition during Westwater Canyon and Brushy Basin time (Turner-Peterson et al., 1980).

Over the entire map of Figure 4 alternating thick and thin zones of the Westwater Canyon trend to the east–southeast. This general orientation has also been observed in other more regional subsurface studies by Galloway (1980) and R. D. Lupe (personal communication, 1984). Because of the regional eastward and southward thinning of the Westwater Canyon, trends are defined by areas of relative thickness compared to nearby adjacent areas. For example, thick trends in the northwest part of the map area are considered to be greater than 320 ft (98 m) in thickness; but in the southeast part of the map, thick trends are 200 ft (61 m) or more in thickness. This is why we shaded different intervals from west to east on Figure 4.

We consider the thick sandstone trends to be depocenter axes along which Westwater Canyon sediments accumulated. Although geophysical log responses suggest vertical stacking of channel sandstones, the term *depocenter axis* is preferred, because without specific knowledge of sediment transport directions for this unit in the subsurface, it cannot be assumed that these axes are channel sequences indicating flow to the southeast. There are a variety of other explanations for orientation of depocenters that need not be related to sediment transport directions, such as: preexisting paleovalleys that localized channel sandstone deposition; scour at the base of the lowest sandstone of the Westwater Canyon, which is observed in outcrop and mine studies; or structural downwarping of depocenter axes. A few areas in Figure 4 that show a coincidence of thick accumulations and

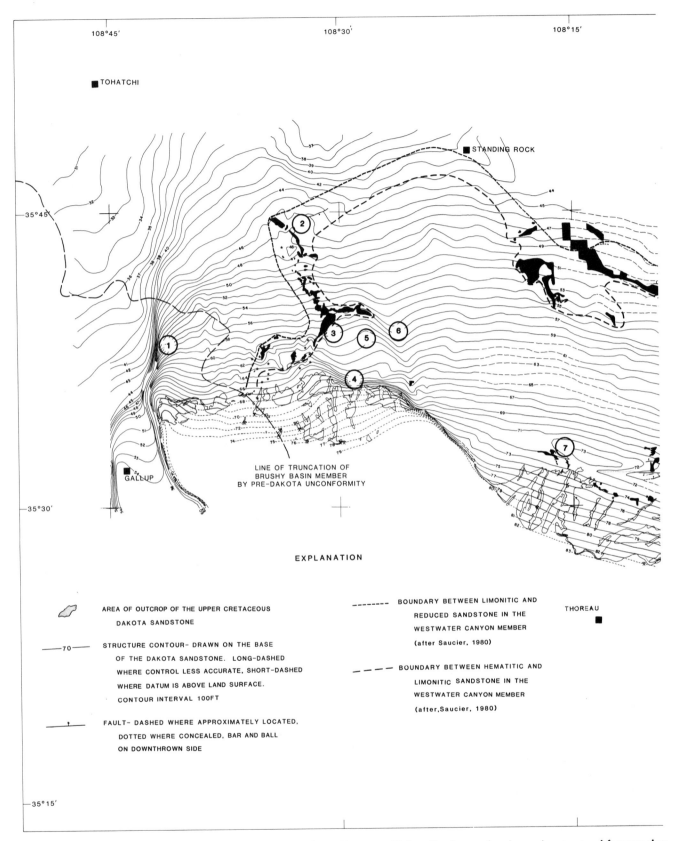

Figure 3—Structure contour map drawn on the base of the Upper Cretaceous Dakota Sandstone showing major structural features that include the following: (1) Nutria monocline, (2) Coyote Canyon anticline, (3) Pipeline fracture zone, (4) Pinedale monocline, (5) anticline, (6) syncline, (7) Mariano Lake–Ruby Wells anticline, (8) Bluewater fault zone, (9) Big Draw fault zone, (10) Ambrosia fault zone, (11)

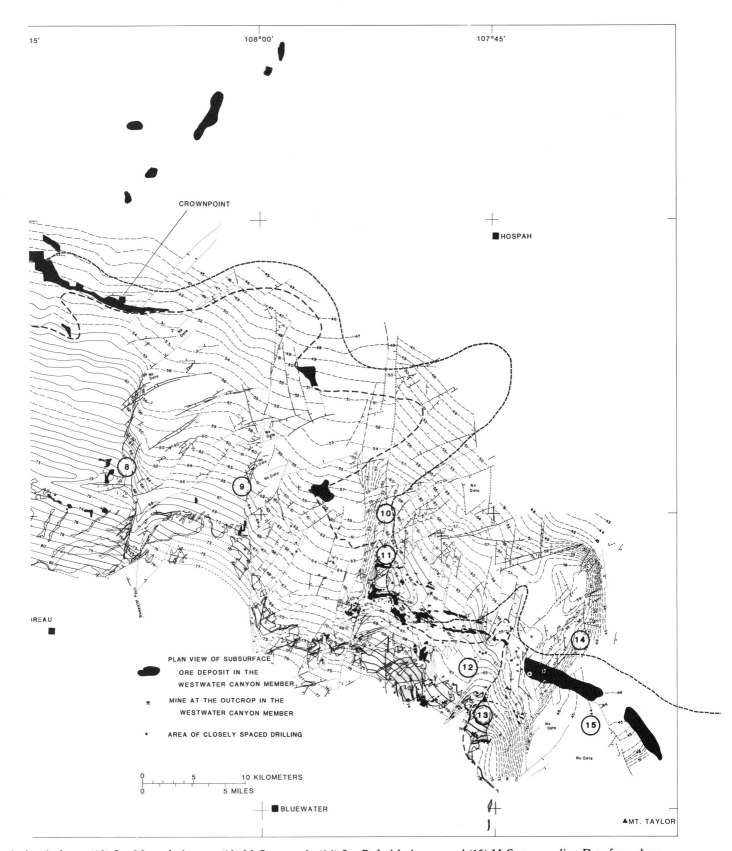

15'   108°00'   107°45'

CROWNPOINT

■HOSPAH

⑧ ⑨ ⑩ ⑪ ⑫ ⑬ ⑭ ⑮

ⓇEAU ■

PLAN VIEW OF SUBSURFACE
ORE DEPOSIT IN THE
WESTWATER CANYON MEMBER

MINE AT THE OUTCROP IN THE
WESTWATER CANYON MEMBER

AREA OF CLOSELY SPACED DRILLING

0   5   10 KILOMETERS

0   5 MILES

■BLUEWATER

▲MT. TAYLOR

Ambrosia dome, (12) San Mateo fault zone, (13) McCartys arch, (14) San Rafael fault zone, and (15) McCartys syncline. Data from about 4000 geophysical logs and about 100 measured sections were used to construct this map. Much of the eastern part of the map area is only slightly modified from earlier mapping by numerous U.S. Geological Survey personnel over a number of years.

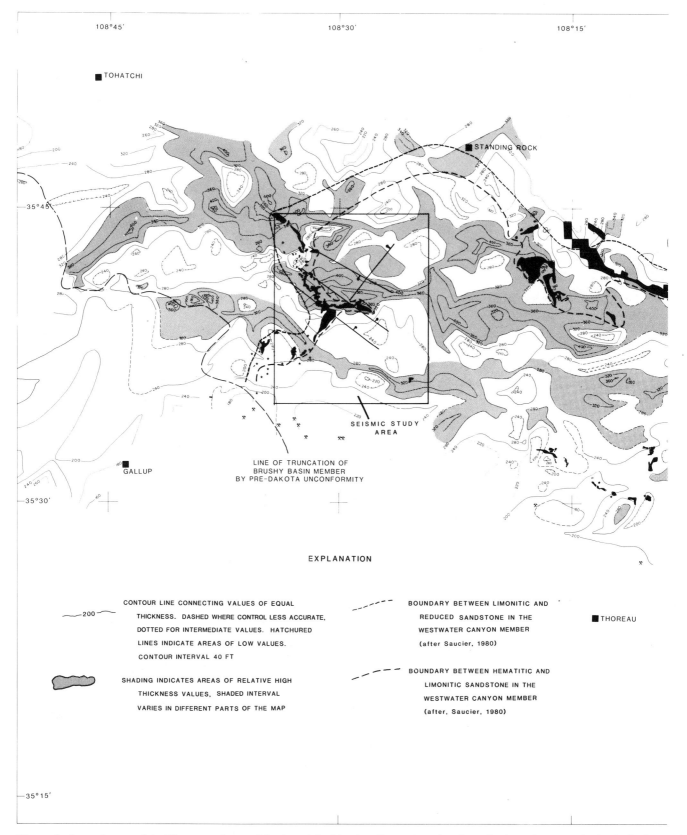

Figure 4—Isopach map of the Westwater Canyon Member of the Morrison Formation, showing seismic study area and seismically detected faults.

'15'   108°00'   107°45'

CROWNPOINT

■HOSPAH

■THOREAU

PLAN VIEW OF SUBSURFACE
ORE DEPOSIT IN WESTWATER
CANYON MEMBER

MINE AT THE OUTCROP IN
THE WESTWATER CANYON
MEMBER

AREA OF CLOSELY SPACED
DRILLING

Insufficient
Data

AMBROSIA DOME

McCARTY'S ARCH

0   5   10 KILOMETERS

0   5 MILES

■BLUEWATER

▲MT. TAYLOR

Figure 4 continued.

sand-rich sediments in east-southeast-trending depocenters also show abrupt facies changes to thinner, less sandy, laterally equivalent units. This suggests a local stacking of channel sequences along the depocenter axes. However, these thick, sandy sediment accumulations having abrupt lateral facies transitions do not occur over the entire map area, which is to be expected in the complex braided stream systems of the Westwater Canyon.

The location of depocenter axes, combined with observations from outcrop studies, allowed us to identify two prominent areas of voluminous sediment input into the Westwater Canyon depositional system. One is located near Todilto Park (Figure 1) where the Westwater Canyon is thicker (330 ft [100 m]) than anywhere else along the outcrop belt. At Todilto Park, the Westwater Canyon is also extremely conglomeratic with abundant cobble-sized clasts (Craig et al., 1955; A. R. Kirk and A. C. Huffman, unpublished data, 1980). This major Westwater Canyon depocenter axis probably passes through Todilto Park and continues southeast to join the thick sediments southeast of Tohatchi shown in Figure 4, or alternatively, it may extend northeastward into the central part of the San Juan basin. A second area of substantial sediment input is suggested by two depocenter axes north of Gallup (Figure 4) near the north end of the Gallup sag (Figure 1), one of which merges to the east with the depocenter axis southeast of Tohatchi. This area of proposed sediment input was recognized by earlier workers from both surface (Craig et al., 1955; Saucier, 1976) and subsurface studies (Galloway, 1980).

On the sandstone:mudstone ratio map (Figure 5) there is also an overall east-southeast trend to the contoured values. Many of the areas having large sandstone:mudstone ratios coincide with areas of thick sediment; this occurs, for example, in the proposed areas of significant sediment input along the west side of the study area southeast of Tohatchi and in the vicinity of the Gallup sag.

A zone of relatively low sandstone:mudstone values crosses the central part of the western half of the Figure 5 map. This zone coincides with a rather thick zone of the Westwater Canyon (Figure 4). Both the thick sediment zone and the relatively low sandstone:mudstone ratios result from the presence of sandstones interbedded with mudstones at the top of the Westwater Canyon Member in the zone of intertonguing with the Brushy Basin.

The percentage of sandstone and the net sandstone content of the Westwater Canyon (Kirk and Condon, 1986) show lithofacies patterns that lend additional support to the existence of east-southeast-oriented depocenter axes and to the existence of the two proposed areas of sediment input on the west side of the study area. The percentage of sandstone parameter is also useful because it is a measure of sandstone content that is independent of total thickness. If we use percentage of sandstone instead of thickness, the effect of overall depositional thinning of the Westwater Canyon from west to east becomes less apparent and the depocenter axes with high sandstone content become more apparent. The net sandstone content of the Westwater Canyon also clearly shows a decrease in sandstone content in the Westwater Canyon from west to east. Values for net sandstone content in the west tend to be greater than 280 ft (85 m) and decrease to values of less than 160 ft (48 m) at the east end of the study area.

The number of alternations between sandstone and mudstone beds per 100 feet of stratigraphic section in the Westwater Canyon Member (Kirk and Condon, 1986) is a measure of the number of mudstone interbeds in the section. If an area has a low sandstone:mudstone ratio, the number of mudstone interbeds enables one to determine whether this low value results from many thin interbeds or a few thick interbeds. This parameter reflects the transmissivity of the interval. In general, there are fewer mudstone interbeds in the northwestern part of the study area and more in the central and eastern parts. The strongly developed east-southeast trends, evident on other maps, are not obvious on the number of alternations map; however, the areas of voluminous sediment input along the west side of the study area contain very few mudstone interbeds.

Thickness variations in the Brushy Basin Member (Kirk and Condon, 1986) show the effect of regional truncation of the member beneath the pre-Dakota erosion surface. We computed the approximate rate of truncation (mentioned earlier) for the Brushy Basin beneath this unconformity to be about 10 ft/mi (1.9 m/km) with truncation proceeding from the east-northeast toward the west-southwest.

The net sandstone content of the Brushy Basin Member (Kirk and Condon, 1986) shows only two distinct trends. One is an east-west-trending sandstone thickness just north of the outcrop in the central part of the map area west of Thoreau, and the other trends almost due north and is located north of the town of Thoreau.

The combined thickness of the Westwater Canyon and Brushy Basin members (Kirk and Condon, 1986) shows the regional depositional thinning toward the south across the study area and thinning due to pre-Dakota truncation in the southwest. East-southeast oriented depocenter axes are also clearly suggested.

STRUCTURAL CONTROLS ON SEDIMENTATION

Overview

Once the striking continuity of east-southeast depositional patterns and other regional changes in the distribution of the sedimentary facies were recognized, we looked at the isopleth maps to determine if these patterns might have been produced by structural control of sedimentation. We began by analyzing the distribution of facies with respect to Laramide structures and found that, at least in the areas around Ambrosia dome and McCartys arch (Figure 4), these structural features seemed to control patterns of sediment distribution in the Morrison Formation. Both Hilpert (1969) and Santos (1970) constructed geologic cross sections across McCartys arch and Ambrosia dome. Both of these geologic sections and our isopach map (Figure 4) show dramatic thinning of the Westwater Canyon over these structures. Santos (1970) suggested that the thinning over the Ambrosia dome was due to intertonguing of a red mudstone unit with the base of the Westwater Canyon Member. He also showed, using the base of the Dakota Sandstone as a datum for a geologic section, that individual mudstone and sandstone beds were not folded over the dome prior to Dakota deposition and, thus, that this structure probably was not a pre-Dakota fold. Two lines of evidence, however, lead us to believe that the Ambrosia dome (see Figures 3 and 7) was an actively growing structure during deposition of the Morrison Formation. Not only does the Brushy basin thin over the dome (thinned depositionally or by pre-Dakota erosion, or both), but Westwater Canyon mudstones deposited over the dome are contemporaneous with thick sandstones deposited adjacent to it. This structure probably deflected major channel sandstone systems of the Westwater Canyon around the dome along downwarping or subsiding depocenter axes while fine-grained overbank sediments accumulated over the topographically elevated crest of the dome (see Figures 4, 5, and 7).

Noting that some of the structures that were active during Morrison deposition were reactivated during the Laramide deformation, we searched for other less obvious pre-Dakota folds or faults that might have affected the Morrison depositional system. To do this we first constructed a corrected paleotopographic map drawn on the base of the Westwater Canyon Member (discussed above) and then examined this surface for paleotopographic highs and lows. We then compared these highs and lows with the isopleth maps and looked for variations in facies distribution that were consistent with structural or paleotopographic control of sedimentation. For example, if a paleotopographic high at the base of the Westwater Canyon caused consistent thinning and predominant shale deposition within overlying depositional units or members, it was hypothesized that this was a feature that was active and had affected deposition of the overlying sediments. Similarly, areas adjacent to inferred paleotopographic highs were frequently anomalously low paleotopographic areas and the sites of deposition of a thick sequence of sandstone. In this manner, many paleo-topographic highs and lows from our reconstruction appear to have affected sedimentation patterns and are proposed to have been active structures during deposition of the Morrison Formation.

We use the term "actively rising structure" to denote a sense of relative movement. We know that the depositional area as a whole was subsiding, as is apparent from its ability to continue to accommodate sediments. However, some areas (positively or actively rising) were subsiding at a rate slower than in adjacent areas.

Paleotopographic Reconstruction

The paleotopographic map (Figure 6) depicts the approximate location of interpreted topographic highs and lows at the beginning of deposition of the Westwater Canyon. The fundamental assumption used was that at least some topographic highs and lows were coincident with structural highs and lows. This assumption is apparently valid in light of the facies distribution described above, at least for the examples from the Ambrosia dome and McCartys arch areas. Local broad and open folding that may have occurred during latest Jurassic through Early Cretaceous time could not be differentiated, but it is thought to have been minimal (as none has been identified west of Laguna [Santos, 1970]) and not to have produced significant changes in thickness.

In the paleotopographic reconstruction in Figure 6, one can see a distinct east–southeast alignment of paleo-topographic trends, including a large east–southeast paleotopographic low area running through the center of the map. The two proposed areas of high sediment input

along the western side of the study area (one southeast of Tohatchi and the other in the vicinity of the Gallup sag) are clearly associated with paleotopographic lows, or depocenters (Figure 6), whereas the Ambrosia dome and the McCartys arch show up as positive paleotopographic features.

We reasoned that hills and valleys could be produced by several processes. On the one hand, hills could be structurally controlled by actively rising structures, and thus, continued activity during sedimentation would affect subsequent sedimentation patterns. On the other hand, they could simply be paleotopographic erosional remnants or hills that would become buried with continued deposition. In the latter case, the effect of the paleotopography should decrease dramatically during late Westwater Canyon deposition and certainly should not affect overlying Brushy Basin lacustrine playa deposition. In contrast, valleys or paleotopographic lows could be actively subsiding regions, pre–Westwater Canyon paleotopographic valleys produced by erosion, or scours at the base of the Westwater Canyon Member that resulted from progradation of the high-energy fluvial system of the Westwater Canyon over the Recapture.

To discriminate these features, a synoptic map (Figure 7) was constructed by overlaying each of the isopleth maps (Figures 4 and 5; plus additional maps published elsewhere [Kirk and Condon, 1986]) on the paleotopographic reconstruction (Figure 6). We noted the cumulative effects of the paleotopographic highs and lows on facies distribution, and we delineated on Figure 7 the actively rising structures (a subset of the larger set of paleotopographic highs) separated by sandstone depocenter axes.

Several actively rising structures (Figure 7) affected the Brushy Basin and Westwater Canyon members in the following ways (Table 1): Westwater Canyon thickness values are low over these structures relative to adjacent areas (Figure 4); sandstone:mudstone ratios (Figure 5), net sandstone, and percentage of sandstone values are low relative to adjacent areas; and the number of mudstone beds (alternations) per 100 ft tend to be higher over the structures. In addition, the total thickness of the combined Westwater Canyon and Brushy Basin members (Kirk and Condon, 1986) is lower over all of the proposed actively rising structures than over the surrounding depocenter axes. Actively rising structures also appear to have localized Westwater Canyon sandstone deposition along depocenter axes around the structures (Figure 7).

Comparison of the paleostructure map (Figure 7) with the Westwater Canyon isopach map (Figure 4) shows that all areas of proposed actively rising structures are coincident with areas of anomalously thin deposits of the Westwater Canyon. Thick accumulations of sediment are everywhere coincident with depocenter axes shown in Figure 7. There are, however, more depocenter axes than those defined by thickness alone.

The Brushy Basin is thinner only over some of the rising structures that affected the Westwater Canyon deposition, such as the Ambrosia dome and the Church Rock structures (Table 1), but it has a wide range of thickness over other structures. This lack of correlation may be due to pre-Dakota erosion, which resulted in incomplete preservation of the Brushy Basin, or more likely, it suggests that some of the structures were no longer active during the time of Brushy

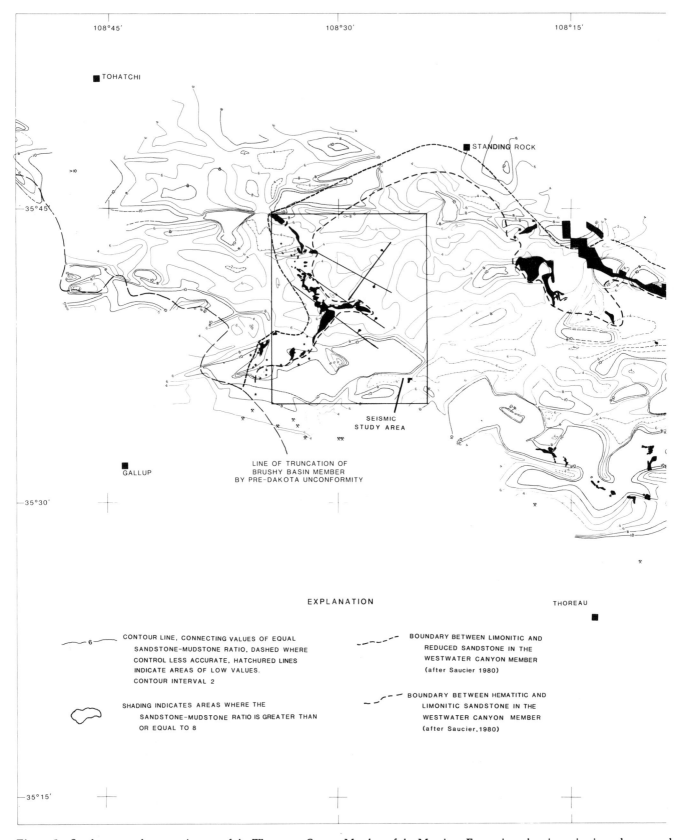

108°45'   108°30'   108°15'

■ TOHATCHI

■ STANDING ROCK

35°45'

■ GALLUP

35°30'

SEISMIC
STUDY AREA

LINE OF TRUNCATION OF
BRUSHY BASIN MEMBER
BY PRE-DAKOTA UNCONFORMITY

EXPLANATION

THOREAU

——6——   CONTOUR LINE, CONNECTING VALUES OF EQUAL
SANDSTONE-MUDSTONE RATIO, DASHED WHERE
CONTROL LESS ACCURATE, HATCHURED LINES
INDICATE AREAS OF LOW VALUES.
CONTOUR INTERVAL 2

SHADING INDICATES AREAS WHERE THE
SANDSTONE-MUDSTONE RATIO IS GREATER THAN
OR EQUAL TO 8

– – – –   BOUNDARY BETWEEN LIMONITIC AND
REDUCED SANDSTONE IN THE
WESTWATER CANYON MEMBER
(after Saucier 1980)

– – –   BOUNDARY BETWEEN HEMATITIC AND
LIMONITIC SANDSTONE IN THE
WESTWATER CANYON MEMBER
(after Saucier, 1980)

35°15'

Figure 5—Sandstone:mudstone ratio map of the Westwater Canyon Member of the Morrison Formation, showing seismic study area and seismically detected faults.

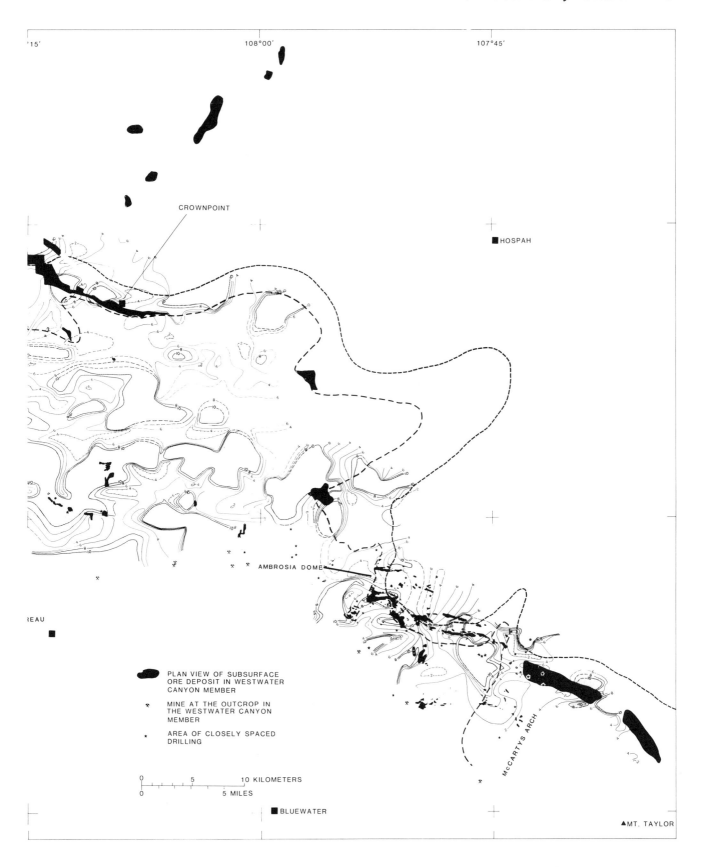

'15'   108°00'   107°45'

CROWNPOINT

■ HOSPAH

REAU

■

AMBROSIA DOME

■

McCARTYS ARCH

🝙 PLAN VIEW OF SUBSURFACE
   ORE DEPOSIT IN WESTWATER
   CANYON MEMBER

✳ MINE AT THE OUTCROP IN
   THE WESTWATER CANYON
   MEMBER

✶ AREA OF CLOSELY SPACED
   DRILLING

0   5   10 KILOMETERS

0   5 MILES

■ BLUEWATER

▲ MT. TAYLOR

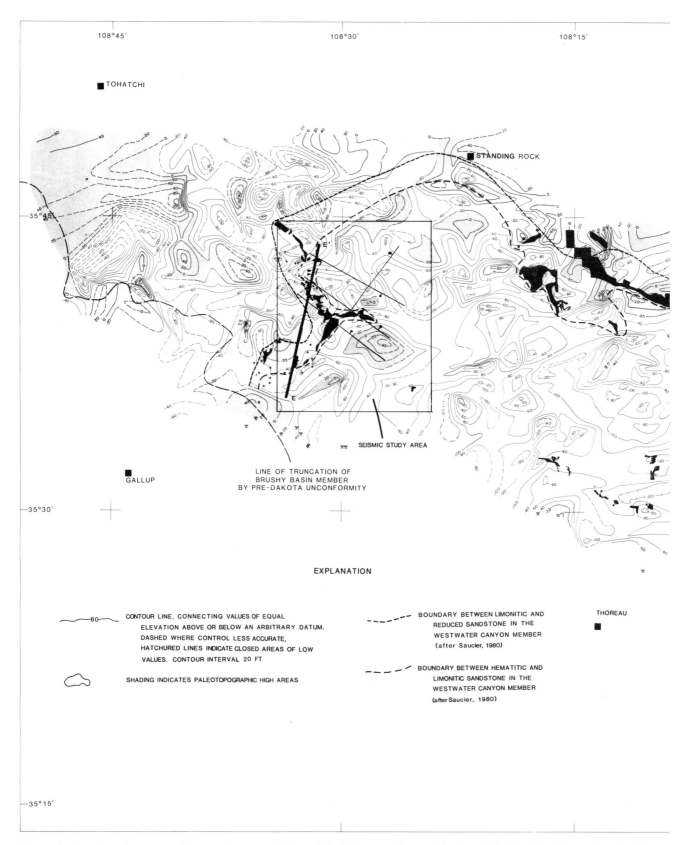

Figure 6—Jurassic paleotopographic map drawn on the base of the Westwater Canyon Member of the Morrison Formation, showing seismic study area and seismically detected faults.

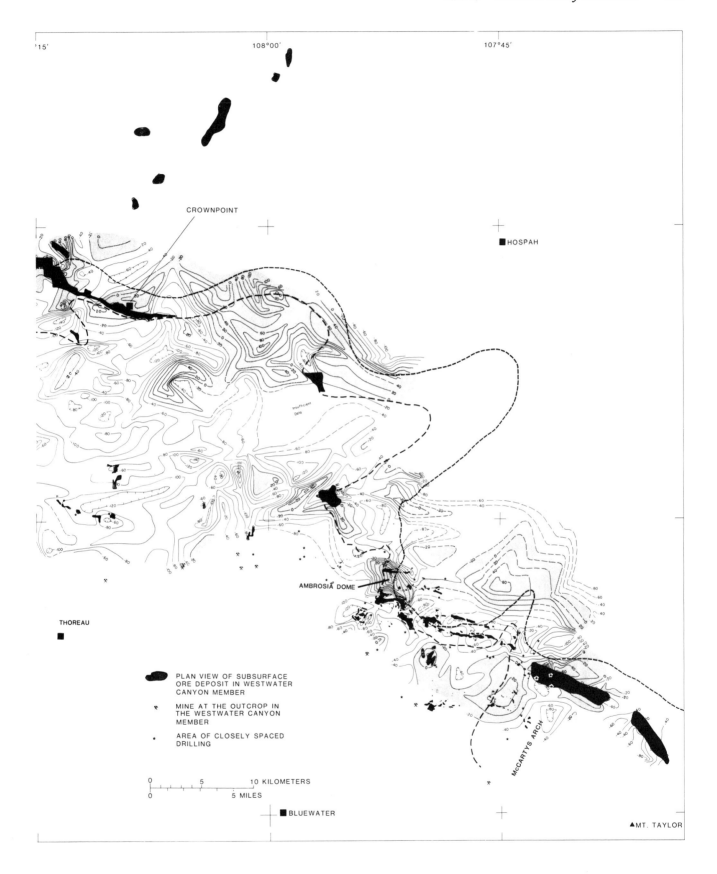

'15'   108°00'   107°45'

CROWNPOINT

■HOSPAH

Insufficient
Data

AMBROSIA DOME

THOREAU
■

PLAN VIEW OF SUBSURFACE
ORE DEPOSIT IN WESTWATER
CANYON MEMBER

MINE AT THE OUTCROP IN
THE WESTWATER CANYON
MEMBER

AREA OF CLOSELY SPACED
DRILLING

McCARTYS ARCH

0   5   10 KILOMETERS

0   5 MILES

■BLUEWATER

▲MT. TAYLOR

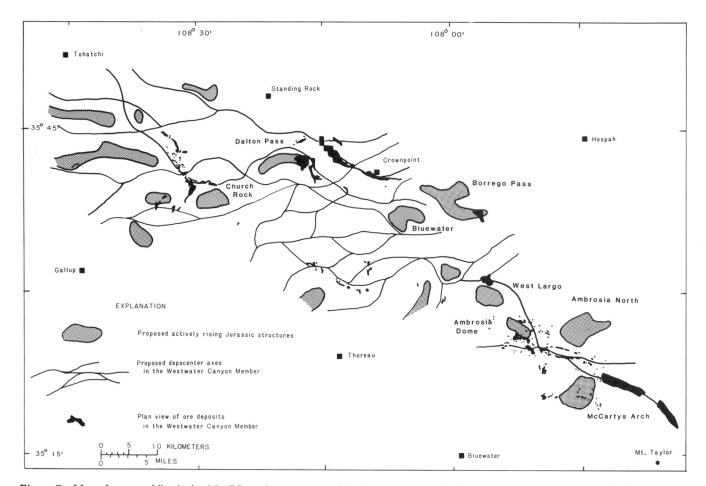

Figure 7—Map of proposed "actively rising" Jurassic structures and depocenter axes in the Westwater Canyon Member of the Morrison Formation.

Table 1—Relationship of proposed actively rising structures to subsurface parameters.

Positive Jurassic Structures[a]	Westwater Canyon Member (Jmw)						Brushy Basin Member (Jmb)			
	Structure Contour	Isopach	Sandstone:Mudstone Ratio	Net Sandstone	Alternations	Percentage of Sandstone	Isopach	Net Sandstone	Total Thickness Jmb and Jmw	Post-Dakota Structure
Church Rock	High	Thin	Mostly low, mixed	Low	3–5	Mostly low	Thin	—	Thin	—
Dalton Pass	High	Thin	—	Low	3	Moderate	Moderate	Low	Thin	—
Blue Water	High	Thin	—	Low	2–3	Low	—	Low	Thin	Flat area
Borrego Pass	High	Thin	Low	Low	3–4	Low	Variable to thick	Low	Thin	Big Draw fault
West Largo	High	Thin	High	Moderate	3–4	Moderate	Variable to thick	Low	Thin	—
Ambrosia Dome	High	Thin	Low	Low	4–5	Low	Thin	—	Thin	Dome
Ambrosia North	High	Thin	Low	Low	4–5	Low to High	Variable	Low	Thin	—
McCarty Arch	High	Thin	Low	Low	6–7	—	Thick	High	Thin	Anticline

[a]See Figure 7 for locations.

Basin deposition.

Negative paleotopographic features at the base of the Westwater Canyon Member and localization of depocenter axes were more difficult to interpret than the positive structures because they could result from several mechanisms: true paleotopography, the effect of which should become muted as the landscape is eroded and buried; scour at the base of the Westwater Canyon Member as the high-energy fluvial system prograded over the unconsolidated Recapture Member; or the result of actively subsiding structures. A seismic reflection study (Phelps et al., 1986) allows us to evaluate these different possibilities in the Church Rock mine area (Figure 1). Phelps and co-workers suggest that paleotopography in this area was controlled by a graben bounded by northwest–southeast-trending faults. These faults offset the Precambrian basement and Paleozoic and lower Mesozoic sedimentary rocks as much as 325 ft (100 m), but offset the Middle Jurassic Todilto Limestone only about 80 ft (24 m). Only one of the faults cuts the lower part of the Westwater Canyon Member with about 40 ft (12 m) of displacement occurring at the lower contact. This fault also apparently caused normal displacement along listric surfaces in the overlying Upper Cretaceous rocks. Although only one of these faults actually shows displacement at the Westwater Canyon–Recapture contact, the dip of the Westwater Canyon changes across all of the faults. These changes in dip are the result of syndepositional draping of sediments across a sag that formed in response to movement along faults deeper in the section (Phelps et al., 1986). Clearly, these northwest–southeast-trending fault zones originate in the Precambrian basement and are probably related to the ancestral Zuni uplift. Evidence suggests that they were reactivated at least once in the late Paleozoic or early Mesozoic and again during deposition of the Morrison Formation, and that at least one fault was reactivated again during the Late Cretaceous. This graben locally coincides with an east–southeast-trending depocenter axis (Figure 7) in the Westwater Canyon Member. A good correlation also exists between the position of thick sediments (Figure 4), low values of sandstone:mudstone ratios (Figure 5), anomalously high net sandstone values, and the location of the graben (Figure 8).

The effects of the graben on formation thickness, the sandstone:mudstone ratio, the net sandstone values of the combined members, and the configuration of the paleotopographic profile can be seen in cross section in Figure 9. These relationships indicate that in the Church Rock mine area, depocenter axes are related to an actively subsiding graben formed by movement of Precambrian basement blocks along preexisting faults that were reactivated through time. This graben controlled both the orientation of the depocenter axis and the facies deposited in and adjacent to the axis. Because depositional patterns, sediment character, and paleotopography are similar along other depocenter axes, we conclude that at least some, and perhaps all, of the other depocenter axes were also structurally controlled.

RELATIONSHIP OF ORE DEPOSITS TO SEDIMENTATION AND STRUCTURE

Introduction

Uranium deposits in sandstones of the Morrison Formation can be divided into three types: primary, redistributed, and remnant (Figure 10). This division is made on the basis of the geometry of the ore deposits, the postulated ore-forming process, the organic content, and the location of the deposits with respect to the regional oxidation–reduction (redox) interface (Saucier, 1980).

The regional redox interface within the Westwater Canyon is important to the understanding of uranium deposits and consists of the following: a zone of

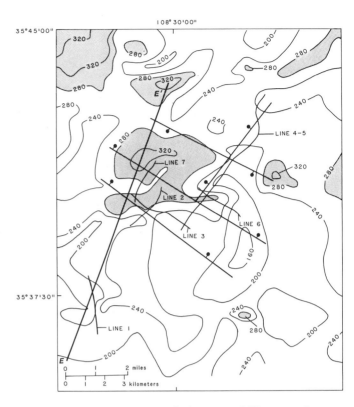

Figure 8—Seismically detected fault traces and Westwater Canyon net sandstone values from the vicinity of the Church Rock ore deposits. Shading indicates areas with large values of net sandstone (>280). Downthrown sides of faults are indicated by dots.

hematitically altered sandstone (south of the southernmost dashed line on Figure 10), a zone of limonitically altered sandstone (between the two dashed lines), and a zone of reduced sandstone (north of the northernmost line). In the Grants uranium region we consider the location of the redox interface to lie at the contact between limonitically altered and reduced rock. The redox interface, however, is difficult to locate precisely because of extensive intertonguing of hematitically and limonitically altered and reduced rock, and we consider it to be a redox zone rather than a sharp interface.

The formation of redistributed ore deposits and the preservation of remnant deposits are related to the alteration and modification of primary ores updip of this redox interface, and therefore, many ore deposits exhibit characteristics of more than one deposit type. To simplify our discussion, however, we have separated the deposit types into three fundamental classes: (1) remnant ore deposits that occur updip (south) of the interface, (2) redistributed deposits that occur along the regional redox interface (Figure 10), and (3) primary deposits predominantly downdip (north) of the line designating hematitically altered rocks. Primary and remnant ore deposits are also intimately associated with an organic-rich kerogen or humate(?) cement (Granger et al., 1961; Squyres 1967, 1980; Leventhal, 1980; Natcher et al., 1986).

The empirically observed relationships among the contoured stratigraphic parameters and the distribution of specific types of ore deposits are summarized in Table 2.

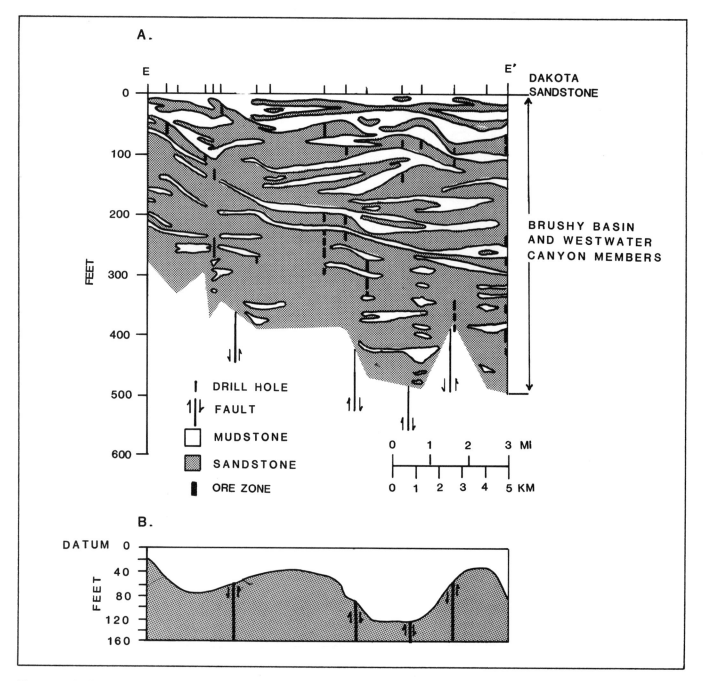

Figure 9—A. Cross section constructed from drill hole data along line E–E′ (Figure 8) in the Church Rock mine area, using the base of the Dakota Sandstone as a datum. Faults located in the seismic study and lithologies in the Morrison Formation are shown. Cross section is oriented transverse to depocenter axis.   B.   A paleotopographic profile (shaded area) drawn on the base of the Westwater Canyon, along line E–E′ (Figure 8) and constructed from data on Figure 6.

Primary Ore Deposits

Primary ore deposits have been called prefault, tabular, or trend-type deposits (Granger et al., 1961; Santos, 1970) and are considered to represent the original form of ore deposition in the Grants uranium region. They are elongate, tabular, organic-rich uranium deposits that occur suspended in reduced sandstone and are aligned along a series of subparallel trends oriented S 70–80° E (Santos, 1970). Mineralization trends, such as those of the Ambrosia Lake area north of Grants (Figure 10), range in width from 1000

to 3000 ft (300 to 920 m) and extend for distances as great as 14 km (Adams and Saucier, 1981). East–southeast-oriented thick accumulations of sediments are parallel to and locally coincident with these observed ore trends. Primary ore in the main Ambrosia Lake area and ore in the Crownpoint trend (Figure 10) are developed in or adjacent to the thickest parts of the Westwater Canyon Member along depocenter axes that were probably structurally controlled (Figures 4 and 7). These ore deposits are spatially associated with organic-rich kerogen (Leventhal, 1980) or humate bodies (Granger et al., 1961; Squyres, 1967, 1980; Natcher et al., 1986) that are commonly stacked vertically as elongate, locally sinuous deposits (Santos, 1970; Saucier, 1976; Adams and Saucier, 1981) within depocenter axes. Humate is only present locally at Crownpoint, and these deposits are considered to be extensively altered and remobilized primary ore (Wentworth et al., 1980; Hansley, 1986). Other geologists have noted the relationship between the thickness of the Westwater Canyon Member and the ore, both at individual mines (Fitch, 1980; Wentworth et al., 1980) and on a more regional scale (Santos, 1970; Galloway, 1980). More specifically, the Ambrosia Lake ore trend (Figure 10) is associated with a depocenter axis defined by a paleotopographic low and a thick accumulation of sediment. The general occurrence of ore deposits along depocenter axes is striking.

Primary ore deposits are also commonly associated with zones having large sandstone:mudstone ratios (generally >10). The occurrence of primary ore in zones having thick sediments and large sandstone:mudstone ratios suggests a relationship between the distribution of primary ore and zones of high transmissivity, which were probably important in controlling the deposition and localization of the organic-rich humate material that is intimately associated with uranium deposits. These observations suggest that the presence of thick build ups of highly transmissive, clean sandstone with very little interbedded mudstone may be conducive to, and possibly a requirement for, ore deposition or preservation. In the Ambrosia Lake area, sandstone:mudstone ratios are locally variable (Figure 5 and Table 2), which suggests that although the groundwater flow was generally concentrated along zones of good transmissivity, the presence of some mudstone may be necessary to allow precipitation of humate. Isotopic dating of deposits (Brookins, 1980; Ludwig et al., 1984) and the field observation that primary ore in the Jackpile sandstone is truncated by the overlying Dakota Sandstone (Nash and Kerr, 1966) both suggest that primary deposits formed shortly after deposition of the host sediments. This would be the optimum time for mobilizing humic acids derived from the degradation of organic material and for leaching of uranium from interbedded volcanic material. The most efficient transport of these materials would probably occur by groundwater migrating along depocenter axes that were the transmissive conduits within sandstone bodies—conduits that were already established along structurally controlled low areas (Figure 6). Precipitation and flocculation of the humic acids may have occurred as a result of changes in salinity down the hydrologic flow paths (Galloway, 1980) at solution interfaces lower in the hydrologic system (Granger et al., 1961; Adams and Saucier, 1981) or by disruption of the transmissivity by mudstone interbeds. Uranium may have been transported and deposited with the humic material and the deposits enriched by further precipitation of uranium from solution during the subsequent passage of uranium-bearing groundwater.

Redistributed Ore Deposits

Redistributed uranium deposits are formed by the partial to complete destruction of primary ore by oxidizing groundwater that enters the host rock, dissolves the organic

material, and remobilizes the uranium downdip along an oxidation–reduction interface (Granger et al., 1961). Most of these deposits are located along the limonite–hematite interface (Figure 10) (Saucier, 1980). They have an irregular C-shaped or roll-type geometry similar to that of Wyoming roll-type deposit; this geometry is caused by hydrologic flow between confining aquicludes (Granger and Warren, 1969; Harshman, 1972; Adler, 1974). However, redistributed deposits in the Grants uranium region are much more irregular in shape than the Wyoming deposits because individual deposits are composed of multiple C-shaped rolls that occur in isolated, vertically stacked sandstone beds. The wide lateral distribution of redistributed ore that can be seen on a plan view results from differences in transmissivity between isolated sandstone beds, along which the redox interface has been able to migrate varying distances downdip. The shapes can be further modified and complicated by the regional groundwater flowing perpendicular to, rather than along, sandstone thickness trends. The regional redox interface is thus more diffuse than that of typical Wyoming roll-type deposits, and the ore deposits are spread over greater distances updip of the redox interface. Saucier (1976, 1980) considers redistributed deposits to be the result of destructive modification of primary deposits; in other words, the deposits were not formed everywhere along the regional redox interface, but occur only downdip or down flow of preexisting primary deposits. Thus, uranium was probably redistributed only over short distances of about 1000–3000 ft (300–900 m) (Adams and Saucier, 1981). Redistributed deposits may also be stacked vertically along preexisting fault zones to form the stacked or postfault ore of Granger et al. (1961).

Redistributed ore seems to be associated with thick accumulations of sediments (Figure 4 and Table 2) and is generally associated with zones of lower sandstone:mudstone ratios (<6) (Figure 5), but these ratios are variable and usually the result of only a few thick mudstone interbeds. The Crownpoint ore trend (Figure 10) appears to be made up of primary ore deposits that have been extensively altered by Tertiary oxidation but not remobilized over great distances (Wentworth et al., 1980; Hansley). This may explain the occurrence of ore in thick sediments but in a zone of mixed but generally large sandstone:mudstone ratios, which are more similar to those associated with primary deposits. The lower sandstone:mudstone ratios associated with redistributed deposits may indicate that reduced transmissivity, resulting from a greater number of impermeable mudstones, retarded the advancing tongues of oxidizing groundwater and caused the reprecipitation of uranium along the margins of the depocenter axis.

An important control on the localization and form of the regional redox interface is the position of the ore bodies themselves. Saucier (1980) noted that the downdip migration of the redox interface appears to be slowed or retarded in areas along which it must oxidize organic matter and/or uranium deposits before migrating farther downdip. The result is that redistributed uranium deposits occur in upflow salients or reentrants along the redox interface.

Laramide structures have been important in controlling the Tertiary oxidation pattern within the Westwater Canyon

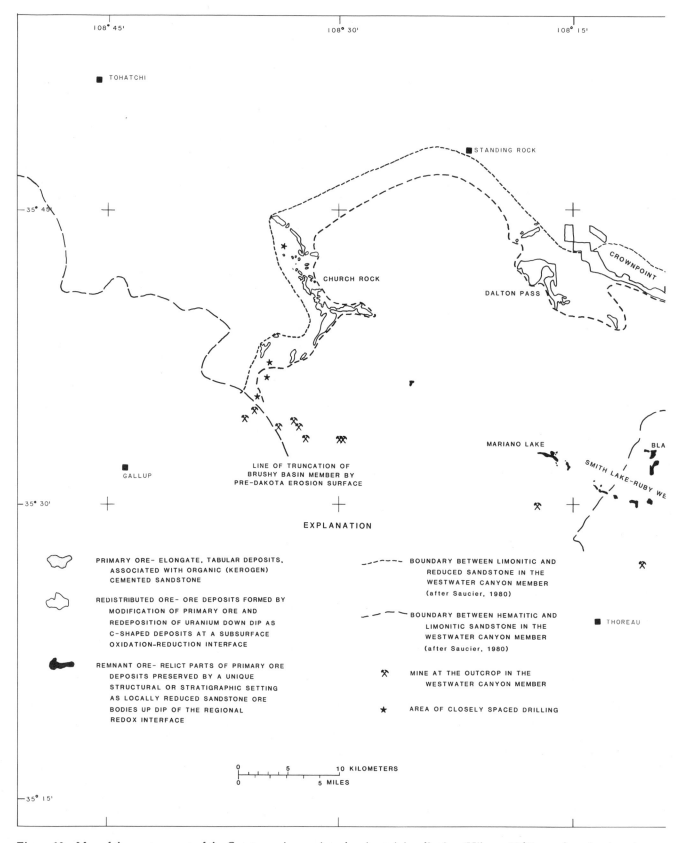

Figure 10—Map of the western part of the Grants uranium region, showing mining districts (Hilpert, 1969), ore deposits, deposit types, names of groups of deposits, and location of the regional redox interface of Saucier (1980). Plan views of individual ore deposits are of varying quality and accuracy; sources for these data include Kelley (1963a), Hilpert (1969), Santos (1970), Rautman (1980), Adams and Saucier (1981), and various company mining and milling plans. Specific property names for deposit-type designations can be found in papers by Hilpert (1969), Santos (1970), and McCammon et al. (1986).

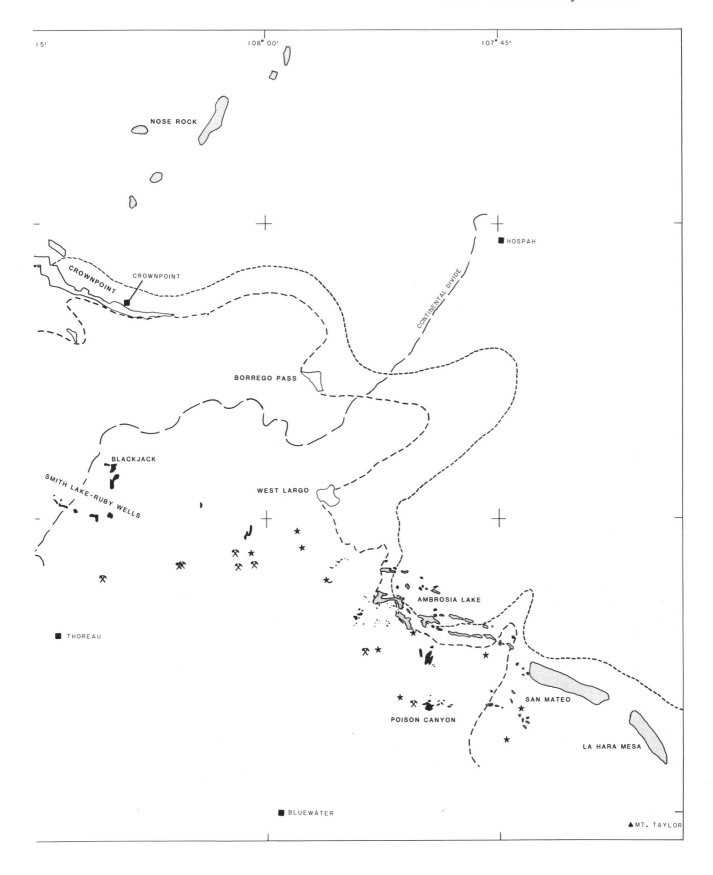

Table 2—Relationship of types of uranium deposits to contoured subsurface parameters of the Westwater Canyon Member.

Ore Type	Isopach	Sandstone:Mudstone Ratio	Depocenter Axes	Number of Alternations	Percentage of Sandstone	Net Sandstone
Primary	Thick	Mostly >10	On thick axes	<3	>80	High
Remnant	—	Mostly >10	On axes	<3	>85	Variable
Redistributed	Thick	Low but variable	On thick axes	Mostly <3	Low but variable	Variable

Member and thus the localization of redistributed ore deposits as well. Hematitically and limonitically altered sandstone extends in a broad arcuate lobe northward downdip from the outcrop of the Morrison Formation and is approximately symmetric about the point where the outcrop crosses the Continental Divide, northwest of Thoreau (Saucier, 1980) (Figures 3 and 10). This alteration extends downdip from the outcrop 17 mi (27 km) to depths of about 2600 ft (790 m) in the vicinity of Standing Rock, New Mexico. In the subsurface, the redox interface is not sharp, but is rather a zone of lateral and vertical intertonguing between hematitic, limonitic, and locally reduced rock.

The hematitically altered sandstone zone is thought to be related to middle Tertiary uplift and erosion (Saucier, 1980). During this event, the Westwater Canyon Member north of the Zuni Mountains was exposed, and recharge of this aquifer permitted oxidizing waters to migrate downdip (northward) toward a discharge area along the San Juan River in northwestern New Mexico (Figure 1). Radiometric dating based on uranium–lead isotopes gives ages ranging from Miocene to Holocene, which indicates that this was also the period during which primary deposits were remobilized (Ludwig et al., 1984).

In a number of places the location of this redox interface appears to be structurally controlled. For example, many of the large north–northeast trending fault zones were zones of higher transmissivity that permitted oxidizing groundwater to move farther downdip to produce extended, lobate tongues of oxidation such as those associated with the Bluewater, Big Draw, and San Mateo fault zones (Figure 3, nos. 8, 9, and 12). The slight easterly skew to the lobate oxidation fronts associated with these fault zones may be the result of very transmissive and thick sandstone depocenters trending east–southeast in the Westwater Canyon that also acted as conduits for oxidizing groundwater.

The Ambrosia fault zone (Figure 3, no. 10) also has a similar, but smaller, lobate tongue at its north end, which probably formed under circumstances slightly different from those mentioned above. The main Ambrosia Lake District is a horst block bounded on the east and west by the San Mateo and Ambrosia fault zones, respectively (Figure 3) (Santos, 1970). If recharge was largely centered at the outcrop along the Continental Divide, then groundwater movement to the north and east may have encountered the Ambrosia fault zone where its flow was retarded and largely diverted toward the north. Moreover, the diversion of oxidizing groundwater coupled with limited recharge along the Morrison outcrop at the southern edge of this horst block may be largely responsible for preservation of relatively unmodified primary deposits in the main Ambrosia Lake District (Santos, 1970; Saucier, 1980; Granger and Santos,

1981). Saucier (1980) suggested that groundwater flow along east–southeast-oriented transmissive zones may have destroyed uranium deposits between the Continental Divide and the Ambrosia Lake District (Figure 10) and transported uranium to enrich the primary ore deposits of the main Ambrosia Lake District.

Another example of structural control of the position of the redox interface occurs in the vicinity of the Church Rock deposits (Figure 10). Here, Saucier (1980) suggests that the Pipeline fracture zone (Figure 3, no. 3), which trends northeast along the southeast side of the Church Rock deposits, may have retarded the northwesterly flow of oxidizing groundwater. We think that it is more likely that groundwater recharge south of the Church Rock area was restricted by limited outcrop area and was locally diverted to the northeast by the Pinedale monocline (Figure 3, no. 4).

Remnant Ore Deposits

Remnant ore deposits are relict parts of primary ore deposits bypassed by the regional redox interface. They occur in zones of locally reduced sandstone that are enclosed in hematitically altered rock updip of the redox interface (Figure 10). Remnant deposits were apparently preserved because of special local properties of the host rock or the geologic and hydraulic setting that enabled them to remain chemically reduced in an otherwise oxidized environment. Special aspects of the host rock or setting include the following: (1) relatively impermeable and insoluble organic-rich ores associated with adjacent zones of highly transmissive sandstone that permitted the oxidizing groundwater to bypass the deposit (Smith and Peterson, 1980); (2) the stratigraphic pinchout of the host sandstone into mudstone in an updip direction before the host sandstone reached the present surface, thereby limiting the access of oxidizing groundwater into the sandstone; (3) the stratigraphic pinchout of the host sandstone into mudstone in a downdip direction, which lowered the transmissivity of the unit; (4) the blocking or diverting of oxidizing groundwater access by fault-bounded structures; and (5) the presence of cementation that occurred after ore formation, which locally reduced the transmissivity of the host sandstone and effectively isolated the ore from oxidizing groundwater.

Remnant deposits generally do not seem to be related to thick accumulations of sediments (Figure 4 and Table 2). However, remnant deposits in the Mariano Lake–Ruby Wells trend (Figure 10) occur in an isolated upper sandstone tongue of the Westwater Canyon Member. Thus, it may be necessary to contour this sandstone tongue thickness independently to determine the local relationship between ore distribution and thickness.

Preservation of remnant deposits in the Mariano Lake–Ruby Wells trend probably results from the complete enclosure of the host sandstone by mudstone in the upper part of the Westwater Canyon and from the occurrence of the ore bodies in a tongue of the Westwater Canyon that does not extend southward far enough to reach the surface. This sandstone is therefore connected with the main body of the Westwater Canyon only downdip to the north, where it scours out the intervening mudstone, and it has no direct connection to the surface for oxygenated groundwater recharge.

CONCLUSIONS

Regional depositional trends during Middle and Late Jurassic time were controlled by two northwest-trending positive features: the Mogollon highland in central Arizona (Craig et al., 1955) and the Uncompahgre–San Luis highland in southwestern Colorado (Saucier, 1976). The Morrison Formation was deposited in a northwest–southeast-elongated, broad, open trough between these highlands (Saucier, 1976; Santos and Turner-Peterson, 1986). Sediments were transported into the trough by fluvial systems flowing toward the northeast and southeast (Craig et al., 1955; Campbell, 1976; Turner-Peterson et al., 1980) down the paleotopographic slope. The ancestral Zuni uplift appears to have been an actively rising, positive topographic area during deposition of the Morrison Formation. This is indicated by the depositional thinning of the Recapture and Westwater Canyon members beneath the Brushy Basin Member onto the north flank of the Zuni uplift and the thinning to a depositional pinchout of the Recapture and Westwater Canyon members on the northeast flank of the uplift about 7 mi (11 km) south of Grants, New Mexico (Hilpert, 1969; Santos, 1970; Maxwell, 1976).

We have shown in this report that syndepositional deformation in Late Jurassic time apparently controlled, to a great extent, the distribution of sedimentary facies in the Westwater Canyon and Brushy Basin members of the Morrison Formation. This syndepositional deformation resulted in subsidence along localized east–southeast-oriented trends that were adjacent to "actively rising" structures. Actively rising structures identified within the depositional area of the Morrison Formation appear to be east–southeast-elongated domes or broad, low-amplitude, doubly plunging anticlines that may have resulted from movement transmitted vertically through the overlying sedimentary sequence from fault blocks that remained high in comparison to adjacent downdropped grabens. These east–southeast-oriented zones of subsidence or grabens apparently became the loci of deposition of Westwater Canyon sandstones and formed depocenter axes, while the actively rising structures acted as paleotopographic highs. Evidence suggests that these highs deflected Westwater Canyon streams into the depocenter axes. Sediment distribution in the Morrison Formation over these actively rising structures generally shows thinning of sediments, low sandstone:mudstone ratios, low net sandstone values, and a low percentage of sandstone for the Westwater Canyon Member, and low net sandstone and local thinning of the overlying Brushy Basin Member of the Morrison Formation. Thus, we suggest that these structures affected the distribution of facies in the Westwater Canyon Member, although most of them have no expression in post-Jurassic rocks. The Ambrosia dome and McCartys arch, however, are exceptions; they were reactivated during Laramide deformation and occur as anticlinal or domal structures on structure contour maps drawn on the base of the Dakota Sandstone.

In turn, the distribution of these facies appears to have largely controlled the localization of primary uranium ore deposits that lie along the depocenter axes in thick accumulations of sandstone-rich sedimentary rocks. The location of redistributed ore, although locally controlled by sedimentary facies, is largely controlled by Laramide structures and subsequent erosion that initiated from a middle–late Tertiary oxidation event. At that time oxidized groundwater entered the Westwater Canyon and redistributed ore from preexisting primary deposits along a regional oxidation–reduction interface. Remnant deposits exhibit many of the stratigraphic controls of primary ore deposits and have probably been preserved by locally unusual stratigraphic or structural settings that have prevented them from being oxidized by passage of the regional redox interface.

ACKNOWLEDGMENTS

We would like to express our appreciation to L. A. Indelicato and R. S. Zech for assistance in preparing the structure contour map. J. O. Kork and N. J. Bridges offered many hours of their time to aid in data entry, storage, and retrieval. H. C. Day did the final line work on the maps. We would like to thank M. W. Green, E. S. Santos, Ken Hon and D. Frishmen for their thoughtful reviews of the manuscript. We would like to express particular thanks to E. S. Santos for providing us with a set of geophysical logs from the Ambrosia Lake area, for his valuable discussions of problems that arose during the project, and to A. E. Saucier for sharing with us not only specific data but his knowledge of the deposits and the effects of Tertiary oxidation. We also sincerely appreciate the cooperation and assistance of approximately 25 mining and exploration companies, which provided us with drilling data to help resolve both a research and exploration problem. Publication authorized by the Director, U.S. Geological Survey, March 7, 1984.

REFERENCES CITED

Adams, S. S., H. S. Curtis, P. L. Hafen, and H. Salek-Nejad, 1978, Interpretation of postdepositional processes related to the formation and destruction of the Jackpile–Paguate uranium deposit, northwest New Mexico: Economic Geology, v. 73, n. 8, p. 1635–1654.

Adams, S. S., and A. E. Saucier, 1981, Geology and recognition criteria for uraniferous humate deposits, Grants uranium region, New Mexico: U.S. Department of Energy Open-File Report GJBX-2(81), 226 p.

Adler, H. H., 1974, Concepts of uranium ore formation in

reducing environments in sandstones and other sediments, *in* Formation of uranium ore deposits: Vienna, International Atomic Energy Agency, p. 141–168.

Bassett, W. A., P. F. Kerr, O. A. Schaeffer, and R. W. Stoenner, 1963, Potassium–argon ages of volcanic rocks near Grants, New Mexico: Geological Society of America Bulletin, v. 74, n. 2, p. 221–226.

Bell, T. E., 1983, Depositional and diagenetic trends in the Brushy Basin and upper Westwater Canyon Members of the Morrison Formation in northwest New Mexico and its relationship to uranium deposits: Ph.D. thesis, University of California, Berkley, 102 p.

Brookins, D. G., 1980, Periods of mineralization in the Grants mineral belt: geochronological evidence, *in* C. A. Rautman, compiler, Geology and mineral technology of the Grants uranium region, 1979: New Mexico Bureau of Mines and Mineral Resources Memoir 38, p. 52–58.

Campbell, C. V., 1976, Reservoir geometry of a fluvial sheet sandstone: AAPG Bulletin, v. 60, n. 7, p. 1009–1020.

Chapin, C. E., and S. M. Cather, 1981, Eocene tectonics and sedimentation in the Colorado Plateau–Rocky Mountain area, *in* W. R. Dickinson and W. D. Pagne, eds., Relations of tectonics to ore deposits in the Southern Cordillera: Arizona Geological Society Digest, v. 14, p. 173–198.

Chenoweth, W. L., 1976, Uranium resources of New Mexico, *in* L. A. Woodward and S. A. Northrop, eds., Tectonics and mineral resources of southwestern North America: New Mexico Geological Society Special Publication 6, p. 138–143.

Chenoweth, W. L., and H. K. Holen, 1980, Exploration in the Grants uranium region since Memoir 15, *in* C. A. Rautman, compiler, Geology and mineral technology of the Grants uranium region, 1979: New Mexico Bureau of Mines and Mineral Resources Memoir 38, p. 17–21.

Christiansen, R. L., and P. W. Lipman, 1972, Cenozoic volcanism and plate tectonic evolution of the western United States, Part 2, late Cenozoic: Philosophical Transactions of the Royal Society of London, v. 271, p. 249–284.

Coney, P. J., 1972, Cordilleran tectonics and North American plate motion: American Journal of Science, v. 272, p. 603–628.

Craig, L. C., C. N. Holmes, R. A. Cadigan, et al., 1955, Stratigraphy of the Morrison and related formations, Colorado Plateau Region: a preliminary report: U.S. Geological Survey Bulletin 1009-E, p. 125–168.

Dickinson, W. R., 1981, Plate tectonic evolution of the southern Cordillera, *in* W. R. Dickinson and W. D. Payne, eds., Relations of tectonics to ore deposits in the southern Cordillera: Arizona Geological Society Digest, v. 14, p. 173–198.

Fitch, D. C., 1980, Exploration for uranium deposits in the Grants Mineral Belt, New Mexico, *in* C. A. Rautman, compiler, Geology and mineral technology of the Grants uranium region, 1979: New Mexico Bureau of Mines and Mineral Resources Memoir 38, p. 40–51.

Galloway, W. E., 1980, Deposition and early hydrologic evolution of the Westwater Canyon wet alluvial fan system, *in* C. A. Rautman, compiler, Geology and

mineral technology of the Grants uranium region, 1979: New Mexico Bureau of Mines and Mineral Resources Memoir 38, p. 59–69.

Gilluly, J., and J. B. Reeside, Jr., 1928, Sedimentary rocks of the San Rafael Swell and some adjacent areas in eastern Utah: U.S. Geological Survey Professional Paper 150-D, 49 p.

Granger, H. C., and C. G. Warren, 1969, Unstable sulfur compounds and the origin of roll-type uranium deposits: Economic Geology, v. 64, p. 160–171.

Granger, H. C., and E. S. Santos, 1981, Geology and ore deposits of the section 23 mine, Ambrosia Lake district, New Mexico: U.S. Geological Survey Open-File Report 81-702, 70 p.

Granger, H. C., E. S. Santos, B. G. Dean, and F. B. Moore, 1961, Sandstone-type uranium deposits at Ambrosia Lake, New Mexico—an interim report: Economic Geology, v. 56, p. 1179–1210.

Green, M. W., 1976, Geologic map of the Continental Divide quadrangle, McKinley County, New Mexico: U.S. Geological Survey Geologic Quadrangle Map GQ-1338, scale 1:24,000.

————, 1980, Disconformities in the Grants mineral belt and their relationship to uranium deposits, *in* C. A. Rautman, compiler, Geology and mineral technology of the Grants uranium region, 1979: New Mexico Bureau of Mines and Mineral Resources Memoir 38, p. 70–74.

Gregory, H. E., 1938, The San Juan Country: U.S. Geological Survey Professional Paper 188, 123 p.

Hansley, P. L., 1986, Regional diagenetic trends and uranium mineralization in the Morrison Formation across the Grants mineral belt, New Mexico *in* C. E. Turner-Peterson and E. S. Santos, eds., A basin analysis case study—the Morrison Formation, Grants uranium region, New Mexico: AAPG Studies in Geology No. 22.

Harshbarger, J. W., C. A. Repenning, and J. H. Irwin, 1957, Stratigraphy of the uppermost Triassic and the Jurassic rocks of the Navajo Country: U.S. Geological Survey Professional Paper 291, 74 p.

Harshman, E. N., 1972, Geology and uranium deposits, Shirley Basin area, Wyoming: U.S. Geological Survey Professional Paper 745, 82 p.

Hatcher, P. G., E. C. Spiker, W. H. Orem, et al., 1986, Organic geochemical studies of uranium associated organic matter from the San Juan Basin—A new approach using solid state 13C nuclear magnetic resonance *in* C. E. Turner-Peterson and E. S. Santos, eds., A basin analysis case study—the Morrison Formation, Grants uranium region, New Mexico: AAPG Studies in Geology No. 22.

Hilpert, L. S., 1969, Uranium resources of northwestern New Mexico: U.S. Geological Survey Professional Paper 603, 166 p.

Hilpert, L. S., and V. L. Freeman, 1956, Guides to uranium deposits in the Morrison Formation, Gallup–Laguna area, New Mexico, *in* L. R. Page et al., U.S. Geological Survey Professional Paper 300, p. 299–302.

Hilpert, L. S., and A. F. Corey, 1957, Northwest New Mexico, *in* Geological investigations of radioactive deposits, semi-annual progress report, December 1, 1956, to May 31, 1957: U.S. Atomic Energy Commission

TEI-690, p. 366–381.

Hilpert, L. S., and R. H. Moench, 1960, Uranium deposits of the southern part of the San Juan Basin, New Mexico: Economic Geology, v. 55, n. 3, p. 429–464.

Jordan, T. E., B. L. Isacks, R. W. Allmendinger, et al., 1983, Andean tectonics related to geometry of subducted Nazca plate: Geological Society of America Bulletin, v. 94, p. 341–361.

Kelley, V. C., 1955, Regional tectonics of the Colorado Plateau, and relationship to the origin and distribution of uranium: University of New Mexico Publications in Geology, n. 5, 120 p.

———, ed., 1963a, Geology and technology of the Grants uranium region: New Mexico Bureau of Mines and Mineral Resources Memoir 15, 277 p.

———, 1963b, Tectonic setting, *in* V. C. Kelley, ed., Geology and technology of the Grants uranium region: New Mexico Bureau of Mines and Mineral Resources Memoir 15, p. 19–20.

Kirk, A. R., and S. M. Condon, 1986, Structural control of sedimentation patterns and the distribution of uranium deposits in the Westwater Canyon Member of the Morrison Formation, northwestern New Mexico *in* C. E. Turner-Peterson and E. S. Santos, eds., A basin analysis case study—the Morrison Formation, Grants uranium region, New Mexico: AAPG Studies in Geology No. 22.

Leventhal, J. S., 1980, Organic geochemistry and uranium of the Grants mineral belt, *in* C. A. Rautman, compiler, Geology and mineral technology of the Grants uranium region, 1979: New Mexico Bureau of Mines and Mineral Resources Memoir 38, p. 75–84.

Lipman, P. W., and H. H. Mehnert, 1975, Late Cenozoic basaltic volcanism and development of the Rio Grande depression in the southern Rocky Mountains, *in* B. F. Curtis, ed., Cenozoic history of the southern Rocky Mountains: Geological Society of America Memoir 144, p. 119–154.

———, 1980, Potassium–argon ages from the Mt. Taylor volcanic field, New Mexico: U.S. Geological Survey Professional Paper 1124-B, 8 p.

Ludwig, K. R., K. R. Simmons, and J. A. Webster, 1984, U–Pb isotope systematics and apparent ages of uranium ores, Ambrosia Lake and Smith Lake districts, Grants mineral belt, New Mexico: Economic Geology, v. 79, n. 2, p. 322–337.

Lupton, C. T., 1914, Oil and gas near Green River, Grand County, Utah: U. S. Geological Survey Bulletin 541, p. 115–134.

McCammon, R. B., W. I. Finch, J. O. Kork, and N. J. Bridges, 1986, Estimation of uranium endowment in the Westwater Canyon Member, Morrison Formation, San Juan Basin, New Mexico, using a data-directed numerical method, *in* C. E. Turner-Peterson and E. S. Santos, eds., A basin analysis case study—the Morrison Formation, Grants uranium region, New Mexico: AAPG Studies in Geology No. 22.

Maxwell, C. H., 1976, Stratigraphy and structure of the Acoma region, New Mexico, *in* L. A. Woodward, and S. A. Northrup, eds., Tectonics and mineral resources of southwestern North America: New Mexico Geological Society Special Publication, n. 6, p. 95–101.

Moench, R. H., and J. S. Schlee, 1967, Geology and uranium deposits of the Laguna District, New Mexico: U.S. Geological Survey Professional Paper 519, 117 p.

Nash, J. T., and P. F. Kerr, 1966, Geologic limitations on the age of uranium deposits in the Jackpile sandstone, New Mexico: Economic Geology, v. 61, n. 7, p. 1283–1287.

Peterson F., 1984, Fluvial sedimentation on a quivering craton: influence of slight crustal movements on fluvial processes, Upper Jurassic Morrison Formation, western Colorado Plateau: Sedimentary Geology, v. 38, p. 21–49.

Phelps, W. T., R. S. Zech, and A. C. Huffman, Jr., 1986, Seismic studies in the Church Rock uranium district, southwest San Juan Basin *in* C. E. Turner-Peterson and E. S. Santos, eds., A basin analysis case study—the Morrison Formation, Grants uranium region, New Mexico: AAPG Studies in Geology No. 22.

Rapaport I., J. P. Hadfield, and R. H. Olson, 1952, Jurassic rocks of the Zuni uplift, New Mexico: U.S. Atomic Energy Commission RMO-642, 60 p.

Rautman, C. A., compiler, 1980, Geology and mineral technology of the Grants uranium region, 1979: New Mexico Bureau of Mines and Mineral Resources Memoir 38, 400 p.

Robertson, J. F., 1973, Geologic map of the Thoreau quadrangle, McKinley County, New Mexico: U.S. Geological Survey Open-File Report 1973, scale 1:24,000.

———, 1974, Preliminary geologic map of the Pinedale quadrangle, McKinley County, New Mexico: U.S. Geological Survey Open-File Report 74-224, scale 1:24,000.

Santos, E. S., 1970, Stratigraphy of the Morrison Formation and structure of the Ambrosia Lake district, New Mexico: U.S. Geological Survey Bulletin 1272-E, 30 p.

Santos, E. S., and C. E. Turner-Peterson, 1986, Tectonic setting of the San Juan Basin in Jurassic time *in* C. E. Turner-Peterson and E. S. Santos, eds., A basin analysis case study—the Morrison Formation, Grants uranium region, New Mexico: AAPG Studies in Geology No. 22.

Saucier, A. E., 1967, The Morrison and related formations in the Gallup region: Master's thesis, University of New Mexico, Albuquerque, 106 p.

———, 1976, Tectonic influence on uraniferous trends in the Late Jurassic Morrison Formation, *in* L. A. Woodward and S. A. Northrup, eds., Tectonics and mineral resources of southwestern North America: New Mexico Geological Society Special Publication 6, p. 151–157.

———, 1980, Tertiary oxidation in the Westwater Canyon Member of the Morrison Formation, *in* C. A. Rautman, compiler, Geology and mineral technology of the Grants uranium region, 1979: New Mexico Bureau of Mines and Mineral Resources Memoir 38, p. 116–121.

Schlee, J. S., and R. H. Moench, 1961, Properties and genesis of the "Jackpile" sandstone, Laguna, New Mexico, *in* J. A. Peterson, and J. C. Osmond, eds., Geometry of sandstone bodies: AAPG Research Committee, p. 134–150.

Smith, D. A., and R. J. Peterson, 1980, Geology and recognition of a relict uranium deposit in Section 28, T

14 N, R 10 W, southwest Ambrosia lake area, McKinley County, New Mexico, *in* C. A. Rautman, compiler, Geology and mineral technology of the Grants uranium region, 1979: New Mexico Bureau of Mines and Mineral Resources Memoir 38, p. 215–225.

Squyers, J. B., 1967, Origin and depositional environment of uranium deposits of the Grants region, New Mexico: Ph.D. thesis, Stanford University, Stanford, California, 228 p.

———, 1980, Origin and significance of organic matter in uranium deposits of the Morrison Formation, San Juan Basin, New Mexico, *in* C. A. Rautman, compiler, Geology and mineral technology of the Grants uranium region, 1979: New Mexico Bureau of Mines and Mineral Resources Memoir 38, p. 86–97.

Thaden, R. E., E. S. Santos, and E. J. Ostling, 1966, Geologic map of the Goat Mountain quadrangle, McKinley County, New Mexico: U.S. Geological Survey Geologic Quadrangle Map GQ-682, scale 1:24,000.

Turner-Peterson, C. E., 1986, Fluvial sedimentology of a major uranium host sandstone—a study of the Westwater Canyon Member of the Morrison Formation, San Juan Basin, New Mexico, *in* C. E. Turner-Peterson and E. S. Santos, eds., A basin analysis case study—the Morrison Formation, Grants uranium region, New Mexico: AAPG Studies in Geology No. 22.

Turner-Peterson, C. E., L. C. Gundersen, D. S. Francis, and W. M. Aubrey, 1980, Fluvio-lacustrine sequences in the Upper Jurassic Morrison Formation and the relationship of facies to tabular uranium ore deposits in the Poison Canyon area, Grants mineral belt, New Mexico, *in* C. E. Turner-Peterson, ed., Uranium in sedimentary rocks—application of the facies concept to exploration: Rocky Mountain Section, SEPM Short Course Notes, p. 177–211.

Wentworth, D. W., D. A. Porter, and H. N. Jensen, 1980, Geology of the Crownpoint Section 29 uranium deposit, McKinley County, New Mexico, *in* C. A. Rautman, compiler, Geology and mineral technology of the Grants uranium region, 1979: New Mexico Bureau of Mines and Mineral Resources Memoir 38, p. 139–144.

Eocene Depositional Systems and Tectonic Framework of West-Central New Mexico and Eastern Arizona

S. M. Cather[1]
B. D. Johnson[2]
University of Texas at Austin
Austin, Texas

The deposits of the Baca and Carthage–La Joya basins record late Laramide (Eocene) sedimentation in west-central New Mexico and eastern Arizona. Sedimentation in these basins was influenced by tectonism (both intrabasinal and in surrounding uplifts) and by the prevailing semiarid climate.

A well-exposed facies tract through the central part of the Baca basin reveals a broad spectrum of depositional paleoenvironments. The braided alluvial-plain system, consisting of the deposits of coalesced humid alluvial fans, is widespread throughout the basin. Sedimentary structures present in these deposits indicate that discharge was increasingly flashy toward the basin center. Meander belt sedimentation was predominantly restricted to an actively subsiding, fault-bounded block in the eastern part of the basin; examples of both fine- and coarse-grained point-bar deposits are present in this area. Lacustrine sedimentation occurred in small flood-plain lakes and in a large, shallow, closed lake near the eastern end of the basin. Cyclic progradation, abandonment, and subsidence of fine-grained deltas and fan deltas well out into the lake basins prevented development of extensive lacustrine basin-center deposits.

Only rocks of the braided alluvial-plain system are present in the poorly preserved deposits of the Carthage–La Joya basin. Updip retreat of facies tracts in both the Carthage–La Joya and Baca basins reflects waning Laramide deformation in adjacent uplifts during late Eocene time.

INTRODUCTION

The Eocene Baca Formation of New Mexico and correlative Eagar Formation and Mogollon Rim gravels of Arizona make up a sequence of conglomerate, sandstone, mudstone, and claystone that crops out in a discontinuous west-trending belt from near Socorro, New Mexico, to the Mogollon Rim of Arizona (Figure 1). These rocks were deposited in two separate, yet structurally related, basins that formed during late Laramide time. The largest and westernmost of these basins, here termed the Baca basin (Figure 2), is represented by sedimentary rocks of the Mogollon Rim gravels, the Eagar Formation, and that part of the Baca Formation that crops out west of the Rio Grande. Maximum exposed thickness of Eocene deposits in the Baca basin is about 580 m, although as much as 760 m

has been reported in the subsurface (Snyder, 1971). Deposits of the Carthage–La Joya basin, the smaller of the two basins, are represented by a series of small exposures of the Baca Formation that crop out to the east of the Rio Grande and attain a maximum exposed thickness of about 315 m. In this paper, we delineate the depositional setting within these two basins and briefly describe the Eocene structural elements that influenced sedimentation within the basins. To facilitate discussion of various parts of the Eocene outcrop belt, we have divided the exposure areas into geographic segments (Figure 1).

The deposits of the Baca basin unconformably overlie strata ranging in age from Late Cretaceous to Pennsylvanian. Rocks of Late Cretaceous age underlie the majority of the Baca basin, but basin-fill deposits onlap progressively older rocks along the basin margins (such as in the Mogollon Rim and Lemitar Mountains areas). Subsurface relationships within the Carthage–La Joya basin appear to be similar to those in the Baca basin; however, interpretation of many geologic features of the Carthage–La Joya basin is greatly hindered by erosional stripping and extensive structural disruption due to subsequent development of the Rio Grande rift. Throughout most of west-central New Mexico

[1]Present address: New Mexico Bureau of Mines and Mineral Resources, Campus Station, Socorro, NM 87081.

[2]Present address: Florida Exploration, P. O. Box 5025, Denver, CO 80217.

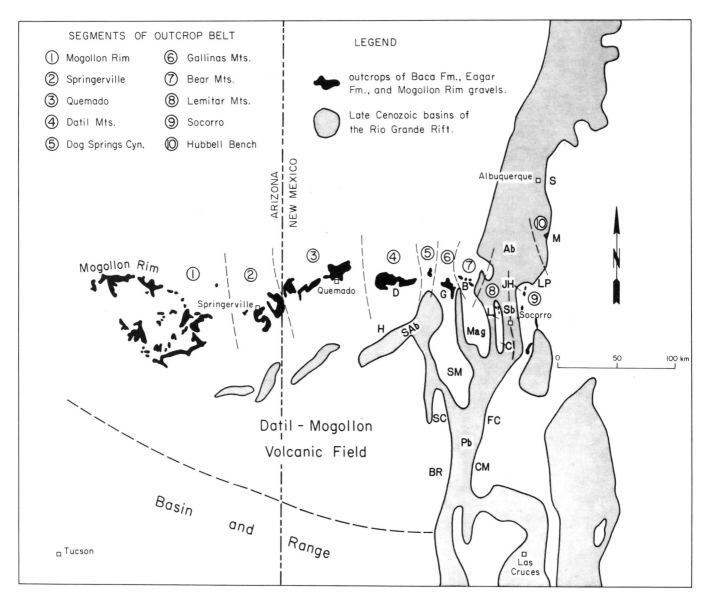

Figure 1—Map of western New Mexico and eastern Arizona showing basins of Rio Grande rift, segments of Eocene outcrop belt, and localities mentioned in text. S = Sandia Mountains, M = Manzano Mountains, Ab = Albuquerque Basin, LP = Los Pinos Mountains, JH = Joyita Hills, D = Datil Mountains, G = Gallinas Mountains, B = Bear Mountains, L = Lemitar Mountains, Sb = Socorro Basin, Mag = Magdalena Mountains, C = Chupadera Mountains, H = Horse Mountain, SAb, San Agustin Basin, SM = San Mateo Mountains, SC = Sierra Cuchillo, FC = Fra Cristobal Mountains, Pb = Palomas Basin, BR = Black Range, CM = Caballos Mountains. Base map after Chapin and others (1978). Used with permission of the New Mexico Geological Society.

and eastern Arizona, the Baca Formation and its equivalent units are conformably and transitionally overlain by andesitic volcaniclastic deposits of the latest Eocene Spears Formation (Cather, 1986). In eastern Arizona, however, an unknown thickness of the upper part of the Mogollon Rim gravels was removed by erosion beginning in Oligocene (?) time (Peirce et al., 1979). In this area, the rim gravels are locally disconformably overlain by volcanic rocks of late Oligocene and younger age.

The Baca and Eagar Formations and the Mogollon Rim gravels have been assigned ages ranging from Late Cretaceous to early Oligocene. Most determinations, however, indicate an Eocene age for these units (see

summaries in Johnson, 1978; Cather, 1980; Lucas et al., 1981). Volcanic clasts in the lower part of the overlying Spears Formation have been radiometrically dated at 38.6 ±1.5 m.y. and 39.6 ±1.5 m.y. in the western Gallinas Mountains (C. E. Chapin, unpublished data, 1980), 38.6 ±1.5 m.y. in the Datil Mountains (Bornhorst et al., 1982), and 37.1 ±1.5 in the Joyita Hills (Burke et al., 1963). Thus, in these areas the Baca Formation can be no younger than late Eocene. In their summary of the Eocene biostratigraphy of New Mexico, Lucas et al. (1981, p. 962) state ". . .the fossiliferous outcrop of the Baca Formation east of the Rio Grande is of Bridgerian age, whereas the fossiliferous outcrops of the Baca Formation west of the Rio Grande are

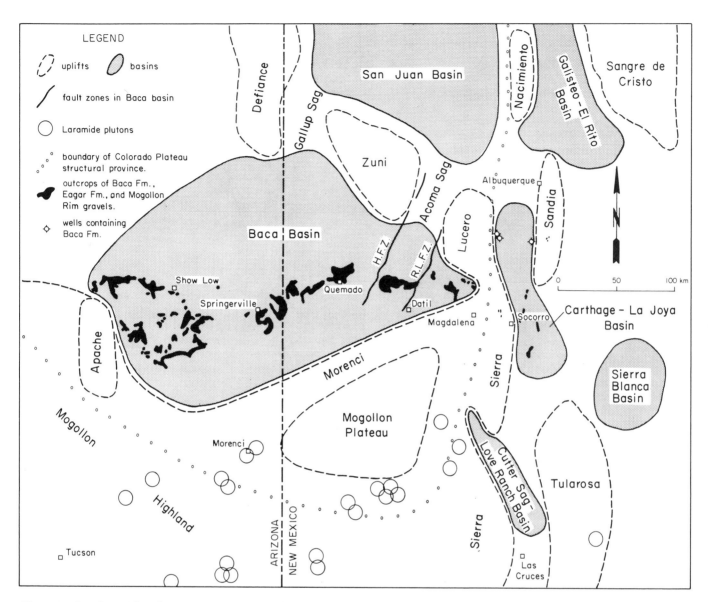

Figure 2—Late Laramide paleotectonic map showing distribution of Eocene uplifts and basins in western New Mexico and eastern Arizona (map coverage same as Figure 1). Base map and outcrop data modified from Dane and Bachman (1965) and Wilson et al. (1969), Laramide pluton data from Chapin et al. (1978), Red Lake fault zone (RLFZ) and Hickman fault zone (HFZ) simplified from Wengerd (1959) and Chamberlin (1981).

significantly younger, being of Duchesnean age." Paleontologic data for the Baca Formation, however, are sparse, and temporal relationships between the Baca and Carthage–La Joya basins cannot be firmly established at this point. In fact, deposits in parts of both the Carthage–La Joya and Baca basins were derived from the same uplift (the Sierra uplift; see Figure 2), and the deposits of both these basins are transitionally overlain by the latest Eocene Formation. This implies that sedimentation in these two basins may have been largely synchronous.

The quality of exposure of the sedimentary rocks in the Carthage–La Joya basin and the sedimentologic perspective it allows contrast sharply with those of the Baca basin. The majority of the exposures of Carthage–La Joya basin deposits occur in a north-trending belt of small, discontinuous

outcrops that approximately parallels the depositional strike of the western flank of the basin (Figure 2). Because of the effects of subsequent rift-related tectonism, large amounts of sediment in the Carthage–La Joya basin (especially deposits in distal environments) have been erosionally removed or buried beneath Neogene basinal sediments of the Rio Grande rift.

In contrast, deposits of the Baca basin are well-preserved, and exposures representative of the majority of the facies tract are present along the outcrop belt. The Baca outcrop belt transects the axis of the basin and trends approximately parallel to the depositional dip of the basin-fill sediments, thus affording us an informative, perhaps unique, perspective of the downdip succession of paleoenvironments within a Laramide foreland basin.

EOCENE TECTONIC FRAMEWORK

Tectonic Framework

The Laramide orogeny occurred between about 80 and 40 m.y. ago (Coney, 1971) and resulted in the creation of numerous uplifts and basins in the western North American foreland. Three general structural provinces and associated styles of deformation can be delineated. These are (1) the "thin-skinned," low-angle thrusts and folds of the Cordilleran thrust belt; (2) the asymmetric, basement-cored, anticlinal uplifts of the classic Laramide Rocky Mountains; and (3) the monocline-bounded uplifts of the Colorado Plateau (Figure 3).

Structural depressions between and adjacent to these uplifts were the sites of accumulation of large volumes of early Tertiary continental deposits. During Eocene time, the Laramide foreland basins consisted of three types (Chapin and Cather, 1981): (1) Green River type—large, elliptical to equidimensional basins that exhibit a quasi-concentric zonation of sedimentary facies and that commonly contain lacustrine deposits; (2) Denver type—large, asymmetric, open basins bounded along one margin by an uplift and exhibiting a unidirectional, proximal-to-distal piedmont facies tract; and (3) Echo Park type—small, highly elongate basins of strike-slip origin with throughgoing drainages. Some basins, such as the Hanna and Laramie basins, may have a hybrid character and may exhibit features of more than one of the above end-member types.

Maximum strain rates and the structural culmination of Laramide deformation began in latest Paleocene–early Eocene time (Chapin and Cather, 1981); development of the Baca and Carthage–La Joya basins is related to this late phase of Laramide tectonism. Unlike many of the Laramide basins in the western United States, the Baca and Carthage–La Joya basins do not contain any known Paleocene deposits. Sedimentation within these basins was influenced by intrabasin structures and by adjacent uplifts. The following is a synopsis of these structural features.

Lucero Uplift

The northern boundary of the Baca basin is defined by the Lucero, Zuni, and Defiance uplifts (Figure 2), which are monocline-bounded flexures typical of Laramide deformation on the Colorado Plateau (Kelley, 1955). The Lucero uplift lies on the eastern margin of the Colorado Plateau and consists of broadly arched, west-dipping strata, except along its eastern flank where it is marked by a complexly faulted and folded section of steeply east-tilted strata (Callender and Zilinski, 1976). Pennsylvanian–Triassic sedimentary rocks compose the majority of the exposures on the uplift. The structurally low area between the Lucero and Zuni uplifts is called the Acoma sag (Kelley, 1955).

Zuni Uplift

The Zuni uplift trends northwest and is markedly asymmetric; its northeastern flank dips gently into the San Juan basin, while its southwestern flank (the Nutria monocline) dips steeply toward the Baca basin and exhibits as much as 1400 m of structural relief (Edmunds, 1961). The

majority of the exposures on the Zuni uplift consists of sedimentary rocks of Permian–Triassic age. However, Precambrian granite, metarhyolite, gneiss, schist, and metaquartzite (Goddard, 1966; Fitzsimmons, 1967) crop out along the crest of the uplift. The structurally low area between the Zuni and Defiance uplifts is called the Gallup sag. More than 2450 m of structural relief exists between the highest part of the uplift and the Gallup sag. Structural relief between the uplift and the deepest part of the San Juan basin is more than 4000 m (Kelley, 1955).

Defiance Uplift

The north-trending Defiance uplift is asymmetric, and its steep eastern flank (the Defiance monocline) faces the Gallup sag and the San Juan basin. The Defiance monocline is highly sinuous because of the presence of a series of en echelon, southeast-plunging anticlines and synclines in the eastern part of the uplift (Kelley and Clinton, 1960). These en echelon folds are suggestive of right slip along the Defiance monocline, and Kelley (1967) has documented approximately 13 km of dextral offset of Jurassic facies in this area. Although the precise timing of Laramide deformation along the Defiance monocline has not been established, it may be related to dextral wrench faulting in the southern Rocky Mountains that culminated in early Eocene time (Chapin and Cather, 1981).

With the exception of a few small outcrops of Precambrian quartzite, the majority of the exposures on the Defiance uplift are of Permian–Triassic sedimentary rocks. Structural relief between the highest part of the uplift and the Gallup sag is at least 2150 m (Kelley, 1955).

Mogollon Highland and Apache Uplift

The Mogollon highland of southern Arizona and southwestern New Mexico forms the southwestern boundary of the Baca basin and was the dominant contributor of detritus to the basin. The early Tertiary structural style of the Mogollon highland is the subject of much ongoing controversy. Various interpretations of the nature of Laramide tectonism in this area include large-scale overthrusting (Drewes, 1976, 1978), crustal extension and differential uplift (Rehrig and Heidrick, 1972, 1976; Lowell, 1974), and vertical uplift of large, basement-cored blocks during strong regional compression (Keith and Barrett, 1976; Davis, 1979). Keith (1982) has compiled evidence for large-scale Eocene underthrusting of the Colorado Plateau in this area.

The magnitude of Laramide structural relief in the Mogollon highland is difficult to estimate, because of the superposition of numerous episodes of orogeny in this area. However, the dominance of sediments derived from the Mogollon highland in the Baca basin suggests that topographic relief, and hence structural relief, was considerable. Lithologies exposed in southern Arizona during Baca time were highly diverse and ranged in age from early Tertiary to Precambrian.

The Apache uplift is a monocline-bounded uplift having at least 1800 m of Laramide structural relief (Davis et al., 1982) which appears to be contiguous with the region of

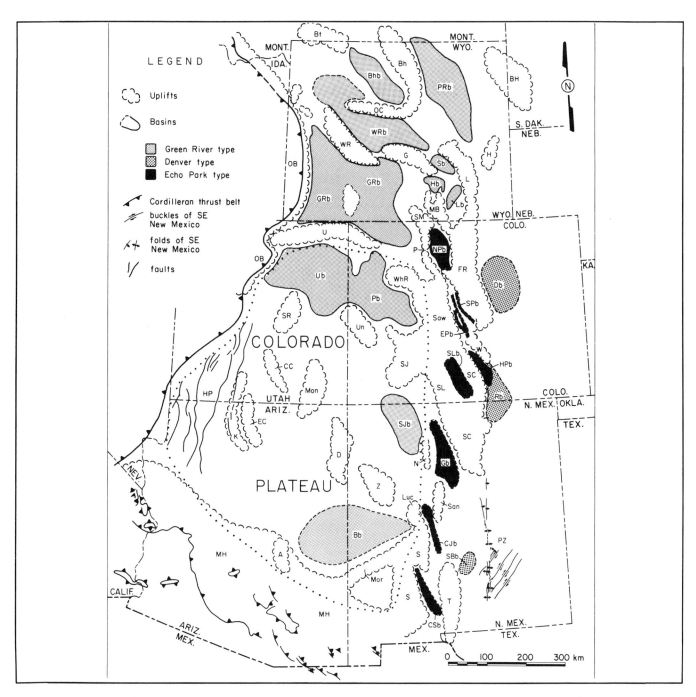

Figure 3—Map of the Colorado Plateau and Rocky Mountain area showing Eocene uplifts, basins, and selected structural features. The dotted line indicates the approximate boundary of the Colorado Plateau in Eocene time. The overthrust belt (OB) and the Mogollon Highland (MH) are two thrust-related uplifts that border the area to the west and south, respectively. Monoclinal uplifts of the Colorado Plateau are the White River (WhR), San Rafael (SR), Uncompahgre (Un), San Juan (SJ), Circle Cliffs (CC), Monument (Mon), Kaibab (K), Echo Cliffs (EC), Defiance (D), Zuni (Z), Lucero (Luc), Apache (A), and Morenci (Mor) uplifts. Basement-cored uplifts of the Laramide Rocky Mountains are the Beartooth (Bt), Bighorn (Bh), Black Hills (BH), Owl Creek (OC), Wind River (WR), Granite (G), Laramie (L), Hartville (H), Uinta (U), Sierra Madre (SM), Park (P), Medicine Bow (MB), Front Range (FR), Sawatch (Saw), San Luis (SL), Sangre de Cristo (SC), Wet (W), Nacimiento (N), Sandia (San), Sierra (S), and Tularosa (T) uplifts. Green River-type basins are the Bighorn (Bhb), Powder River (PRb), greater Green River (GRb), Shirley (Sb), Hanna (Hb), Laramie (Lb), Uinta (Ub), Piceance (Pb), San Juan (SJb), and Baca (Bb) basins. Denver-type basins are the Denver (Db), Raton (Rb), and Sierra Blanca (SBb) basins. Echo Park type basins are the North Park–Middle Park (NPb), South Park (SPb), Echo Park (EPb), Huerfano Park (HPb), San Luis (SLb), Galisteo-El Rito (Gb), Carthage-La Joya (CJb), and Cutter Sag-Love Ranch (CSb) basins. Pecos zone (PZ) data from Kelley (1971, 1972). Data for High Plateaus faults (HP) from Lovejoy (1976). Distribution of Eocene southwest-vergent thrust faults in southern Arizona and adjacent areas from Keith (1982). Apache uplift data from Davis et al. (1982). Base map modified from the tectonic map of North America by King (1969). From Chapin and Cather (1981).

general Laramide uplift to the south (the Mogollon highland). The Apache uplift may have influenced sedimentation in the westernmost part of the Baca basin, but such effects cannot be documented with available data.

Morenci Uplift

The probable existence of a Laramide drainage divide along the southeastern margin of the Baca basin, which was subsequently obscured by thick accumulations of middle Tertiary volcanic rocks of the Datil–Mogollon volcanic field, was pointed out to us by C. E. Chapin (1980, personal communication). The presence of large exposures of Laramide volcanic and plutonic rocks in southwestern New Mexico and adjacent Arizona (Figure 2) and the nearly complete lack of volcaniclastic detritus in the eastern part of the Baca outcrop belt (Cather, 1980) imply the existence of an intervening uplift that prohibited input of volcanic detritus into the eastern part of the Baca basin. The Morenci lineament (Chapin et al., 1978) provides a likely site along which such a drainage divide may have developed.

A series of en echelon, northeast-trending fault zones and grabens, the largest of which is the San Agustin basin (Figure 1), presumably represent portions of the Morenci uplift that collapsed during middle and late Tertiary extension. Stratigraphic relationships also support the existence of a Laramide positive area along the Morenci lineament. At Horse Mountain, along the northwestern margin of the San Agustin basin, volcaniclastic rocks of the uppermost Eocene Spears Formation overlie Triassic–Permian strata with apparent unconformity (Stearns, 1962). In the Magdalena Mountains at the northeastern end of the Morenci uplift, upper Paleozoic strata are unconformably overlain by the Spears Formation (Krewedl, 1974; Blakestad, 1978). Thus, in these areas a period of Laramide uplift and erosion preceded the onset of middle Tertiary volcanism.

The Morenci uplift separated two regions of relative crustal stability: the Mogollon Plateau and the Colorado Plateau. In this paper, the Mogollon Plateau is considered to be a subregion of the Laramide Colorado Plateau on the basis of the relatively undeformed nature of its central portion (Coney, 1976) and its dissimilarity to adjacent, more structurally complex areas to the south and east. The structural style of the Morenci uplift is poorly known because of the extensive cover of middle Tertiary volcanic rocks and partial collapse during middle–late Tertiary extension. Because of its association with relatively stable Colorado Plateau-type crust, however, the Morenci uplift may have been monoclinal. Davis (1978) stated that Colorado Plateau monoclines were produced by reactivation along basement fracture zones (lineaments) during early Tertiary tangential compression, and he recognized two dominant trend directions (N 20° W and N 55° E) of lineaments that controlled monocline development. The Morenci lineament of Chapin et al. (1978), which appears to control the Morenci uplift, closely parallels the northeasterly trend direction of Davis (1978). In detail, however, the Morenci uplift may have consisted of several en echelon segments, as suggested by the en echelon arrangement of the San Agustin basin and its associated extensional features.

If the en echelon pattern of late Tertiary extensional structures in this area were inherited from a Laramide compressional precursor, right shift may have taken place along the Morenci lineament during Laramide time. Magnitude of Laramide structural relief along the Morenci uplift is impossible to estimate at present, although it must have been appreciable, because significant amounts of Precambrian detritus are present in Baca Formation deposits derived from this uplift (Cather, 1980).

The existence of a Morenci uplift would also explain the increased abundance of volcanic clasts in the western part of the Baca basin (Johnson, 1978; Peirce et al., 1979). Laramide volcanic rocks in the vicinity of Morenci, Arizona, would not have been isolated by the postulated drainage divide and may have supplied detritus to the western part of the Baca basin.

Sierra Uplift

Eardley (1962) proposed the existence of a Laramide positive area adjacent to the eastern margin of the Colorado Plateau in southern New Mexico and termed it the Sierra uplift. Sedimentologic data from the Baca Formation indicate that the uplift extended northward into the present-day Socorro vicinity approximately 70 km beyond the northern boundary of the uplift as depicted by Eardley. Large areas of this uplift apparently subsided to form basins of the Rio Grande rift during middle and late Tertiary time; thus, documentation of a Laramide positive area in this region must depend on evidence derived from synorogenic sediments, regional stratigraphic relationships, and the structural style of the remaining unsubsided areas of the uplift.

The only detailed sedimentologic studies and facies analyses of sediments derived from the Sierra uplift are those done for the Baca Formation by Snyder (1971), Johnson (1978), and Cather (1980, 1982). Upper Cretaceous–Eocene deposits of the Cutter Sag–Love Ranch basin (McRae and Love Ranch formations) (Seager, 1975, 1981) are present adjacent to the central and southern parts of the uplift, but to date, no systematic study of these rocks has been attempted. Paleocurrent and facies distribution data from the Baca Formation discussed in this paper clearly demonstrate that the northern part of the Sierra uplift encompassed much of what is now the Socorro basin of the Rio Grande rift.

Stratigraphic relationships suggest that much of the rift south of present-day Socorro was a Laramide positive area (central and southern Sierra uplift). In the ranges bounding part of the west side of the rift (Magdalena Mountains, San Mateo Mountains, and Sierra Cuchillo), uppermost Eocene volcanic rocks typically overlie upper Paleozoic strata (Farkas, 1969; Krewedl, 1974; Blakestad, 1978; Jahns et al., 1978). In the Black Range, similar relationships exist, but they are complicated by the local presence of a suite of older volcanic and hypabyssal rocks, some of which are as old as Late Cretaceous (Hedlund, 1974). Along the eastern shoulder of the Palomas basin, upper Eocene–lower Oligocene volcanic rocks unconformably overlie Paleozoic strata in the southern Caballos Mountains (Kelley and Silver, 1952; Kottlowski et al., 1969; Seager, 1975), and deposits possibly correlative to the McRae Formation overlie Precambrian gneiss in the northern Fra Cristobal Mountains (Kelley and McCleary, 1960). These stratigraphic relationships differ greatly from those in adjacent Laramide basins, where thick deposits of lower Tertiary sediments typically overlie Upper Cretaceous strata.

The Sierra uplift implies a Laramide compressional phase prior to the onset of middle and late Tertiary extension. This early phase of compressive tectonism has been noted by Kelly and Clinton (1960, p. 55). They state:

> ...there is much evidence along the Rio Grande trough of an early tectonic compressional history. The structures of the Caballos Mountains (Kelley and Silver, 1952, p. 136–146) are clearly early Tertiary and compressional. Low-angle (15°) faults in which gneiss rests on Cambrian limestone are clearly displayed within a few hundred yards of the Rio Grande trough in the Fra Cristobal Range of Sierra County, New Mexico (Jacobs, 1957, p. 257). Thrusts and overturns have been mapped in many places in the Sandia and Manzano uplifts...

Thrust faults of Laramide age have also been reported along the eastern margin of the rift in the vicinity of Socorro (Wilpolt and Wanek, 1951), and Seager (1975, 1981) documents Laramide compressive structures and uplifts in the Las Cruces area.

Baca sediments derived from the northern part of the Sierra uplift are arkosic and contain locally abundant clasts of Precambrian granite, schist, and metaquartzite. Only a few small exposures of Precambrian rocks are present today along the Rio Grande rift near Socorro. Thus, Precambrian detritus in the Baca Formation in this area must have been derived largely from source terranes that were subsequently downfaulted and buried in the rift. In the Chupadera Mountains, which are the southern part of an intrarift horst adjacent to the Socorro basin, Precambrian rocks are locally overlain by lower Oligocene volcanic and volcaniclastic deposits (Eggleston, 1982), indicating at least 1.5 km of Laramide uplift and erosional stripping in that area.

In the Laramide Cutter Sag–Love Ranch basin to the south, Precambrian-derived detritus is present in the McRae Formation (Kelley and Silver, 1952; Bushnell, 1955) and in an unnamed lower Tertiary(?) sandstone overlying the Love Ranch Formation (Seager, 1981). Clasts of Precambrian lithologies also occur in deposits possibly equivalent to the McRae Formation in the northernmost part of the Fra Cristobal Mountains (Kelley and McCleary, 1960). Although these deposits may have been derived largely from the Sierra uplift, determination of the provenance and precise dating of these units awaits further study.

Detailed analysis of the structural style and geometry of the Sierra uplift is not possible with the data available at present. However, on the basis of trend, geographic location, presence of thrust faults and associated compressional structures along its east flank, and evidence for significant exposures of Precambrian rocks along its crest, we hypothesize that the Sierra uplift represents a basement-cored anticlinal uplift (or series of uplifts) similar in structural style to the classic Laramide Front Range uplifts of Colorado and southern Wyoming. Chapin and Cather (1981) have proposed that north–northeastward translation of the Colorado Plateau, culminating during early Eocene time, resulted in the creation of an en echelon series of

uplifts and basins along its eastern margin (including the Sierra uplift and Carthage–La Joya and Cutter Sag–Love Ranch basins) and caused severe crustal shortening in the Wyoming province to the north.

Sandia Uplift

The existence of a Laramide positive area, here called the Sandia uplift, in the vicinity of the present-day Sandia, Manzano, and Los Pinos Mountains has been proposed by several workers (Eardley, 1962; Kelley and Northrop, 1975; Kelley, 1977; Chapin and Cather, 1981) on the basis of structural data and stratigraphic and regional tectonic relationships. The Sandia uplift is separated from the Lucero uplift by the northern part of the Carthage–La Joya basin, which contains as much as 1146 m of Baca Formation red beds in the subsurface beneath the southern Albuquerque basin (Foster, 1978).

Sedimentologic evidence for the Sandia uplift is restricted to a few isolated outcrops of the Baca Formation on the Hubbell bench (Kelley, 1977, 1982). These exposures occur near the center of T. 5 N., R. 4 E. and consist of conglomerate and sandstone composed of detritus derived solely from upper Paleozoic units (dominantly Abo Formation sandstone and siltstone). The highly conglomeratic nature of the Baca deposits on the Hubbell bench and the presence of well-developed pebble imbrication indicative of westward paleoflow suggest a source nearby to the east. Kelley (1982) postulated that the axis of Laramide uplift passed somewhat east of the present crest of the Manzano Mountains, in the vicinity of the Montosa and Paloma reverse faults. The lack of Precambrian detritus in Baca exposures on the Hubbell bench implies that structural relief was only moderate. The Sandia uplift may have been monoclinal in nature, with its steep, east-facing limb locally broken by reverse faults (Kelley, 1977). The deposits of the Galisteo–El Rito basin to the north (Gorham and Ingersoll, 1979) and of the Baca Formation in the Socorro area show no evidence of derivation from the Sandia uplift.

Intrabasin Structures

Structural features within the Baca basin were locally important determinants of the synorogenic sedimentation style and the thickness of basin-fill deposits. Throughout most of the Baca basin, the Laramide style of deformation is characterized by low-amplitude, gently plunging, broad folds and associated minor faulting. These features are best developed in Cretaceous and older rocks that underlie the basin, but are also present to a lesser extent in the Eocene basin-fill deposits. Although these low-amplitude folds and faults appear to have had only minor effect on Eocene sedimentation, Chamberlin (1981) suggests that they may have provided important controls on uranium mineralization in a lateritic weathering profile that underlies the Baca Formation in the Datil Mountains area.

Two fault zones in the eastern part of the Baca basin, the Hickman and Red Lake fault zones (Figure 2) (Wengerd, 1959; Chamberlin, 1981), were major intrabasin structures active during Eocene time. The Hickman and Red Lake fault zones are two north-northeast-trending systems of normal faults and folds that form the western and eastern boundaries, respectively, of a large, late Cenozoic "synclinal horst" (Wengerd, 1959) in the Datil Mountains area. Thickness variations in the Baca Formation, however, demonstrate that the area between these fault zones was downthrown during Eocene time. At Mariano Mesa, approximately 25 km west of the Hickman fault zone, the Baca Formation is about 185 m thick (Guilinger, 1982). East of the Red Lake fault zone, in the Dog Springs Canyon, Gallinas Mountains, and Bear Mountains area, the Baca attains thicknesses of about 335, 285, and 230 m, respectively. The greatest exposed thickness of the Baca basin-fill deposits (about 580 m; Chamberlin, 1981), however, occurs *between* the Hickman and Red Lake fault zones, in the Datil Mountains area. The greater thickness of the Baca Formation between these fault zones and the nearly complete restriction of low-gradient meanderbelt deposits to this same area (see below) strongly suggest that the block bounded by the Hickman and Red Lake fault zones was actively subsiding during the time of Baca sedimentation.

The Hickman and Red Lake fault zones apparently represent two north-northeast-trending systems of Laramide reverse faults that have been reactivated as late Cenozoic normal faults. The south-plunging syncline present in the presently uplifted block between these fault zones may be a relict Laramide feature, and it appears to merge northward with the Acoma sag (Wengerd, 1959; Chamberlin, 1981). An en echelon series of northwest-trending folds present between the Hickman and Red Lake fault zones (Wengerd, 1959, fig. 4) suggests that right slip may have accompanied Laramide reverse faulting, although this cannot be confirmed with available data.

The Puertecito fault system, a north-trending structural zone between the Bear Mountains and the Gallinas Mountains (Figure 1), may also represent a Laramide intrabasin wrench fault (Johansen, 1983). Deformation along this zone, however, appears to have had little effect on Baca sedimentation.

Documentation of Laramide intrabasin structures in the Carthage–La Joya basin is difficult because of poor exposure quality and the discontinuous nature of Baca Formation outcrops in this area. It is probable, however, that much of the complex faulting and folding present along the eastern shoulder of the Rio Grande rift in the Socorro area are relict or reactivated Laramide structures (e.g., Smith, 1983).

DEPOSITIONAL SYSTEMS

Criteria used in the discrimination of paleoenvironments include lithofacies geometry, lateral and vertical variation in grain size and sedimentary structures, nature of contacts between lithofacies, petrographic data, and fossils. Since subsurface data in the study area are extremely limited, our method of study consisted of the following steps: (1) measurement, description, and interpretation of numerous partial and complete stratigraphic sections throughout the outcrop belt; (2) local detailed mapping, especially in the Gallinas Mountains area (Cather, 1980); (3) petrographic and x-ray study of sandstone and mudstone; (4) measurement and analysis of paleocurrent indicators; (5)

conglomerate lithology and clast-size studies (Johnson, 1978; Cather and Johnson, 1984); and (6) regional reconnaissance.

Braided Alluvial-Plain System

The braided alluvial-plain system constitutes the most widespread depositional system within the Baca basin, and it is the only system present in those parts of the Carthage–La Joya basin that have been preserved. We have arbitrarily divided this system into proximal fan facies (predominantly conglomerate), mid-fan facies (subequal sandstone and conglomerate), and distal fan facies (predominantly sandstone). This classification has been applied to an ancient alluvial fan sequence by McGowen and Groat (1971), but differs from that used in studies of modern proglacial outwash fans (Boothroyd and Ashley, 1975; Boothroyd and Nummedal, 1978), which delineate proximal, mid-, and distal fan environments on the basis of clast size. We also discuss a variant of the mid-fan facies, which we call the arroyo-fill facies.

Six segments of the braided alluvial-plain system are exposed along the Eocene outcrop belt of west-central New Mexico and eastern Arizona:

1. Widespread exposures of proximal, mid-, and distal fan deposits occur in the Mogollon Rim, Springerville, Quemado, and Datil Mountains areas (Figure 1). These deposits represent an extensive, east-facing, braided alluvial plain that drained part of the Mogollon highland in Arizona.
2. Mid-fan and distal fan deposits occur in the basal part of the Datil Mountains section and were derived from the Zuni uplift by a southeast-flowing tributary system.
3. Lenticular, canyon-fill conglomerates in the Lemitar Mountains (Chamberlin, 1982) and fan delta and braided stream deposits in the Bear Mountains area represent the extreme proximal and distal ends of a poorly preserved, coalesced fan system that drained the western part of the Sierra uplift and the southwestern flank of the Lucero uplift.
4. Distal fan deposits of northeast-flowing braided streams are present in the Gallinas Mountains area and consist of detritus derived from the Morenci uplift.
5. Discontinuous exposures of proximal fan and mid-fan deposits are present in the Socorro area and are indicative of eastward drainage off the Sierra uplift into the Carthage–La Joya basin.
6. Small, isolated deposits of westward-flowing proximal braided streams in the Hubbell bench area represent drainage from the Sandia uplift.

Proximal Fan Facies

Characteristics—The proximal fan facies is exposed in the Mogollon Rim and Socorro areas. The conglomerates exposed in the Lemitar Mountains and the Hubbell bench areas are also included in this facies, although they apparently represent canyon-filling braided stream deposits updip of fan systems (Cather and Johnson, 1984). On the basis of clast composition, paleocurrent indicators, and apparent continuity of depositional characteristics, we include the gravels near Blue House Mountain, Arizona, in

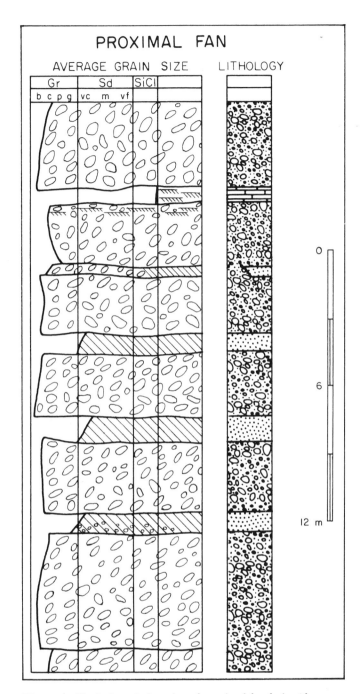

Figure 4—Typical vertical section of proximal fan facies (data from roadcut on U.S. Highway 60 near Blue House Mountain, Mogollon Rim area). See Figure 5 for explanation of symbols.

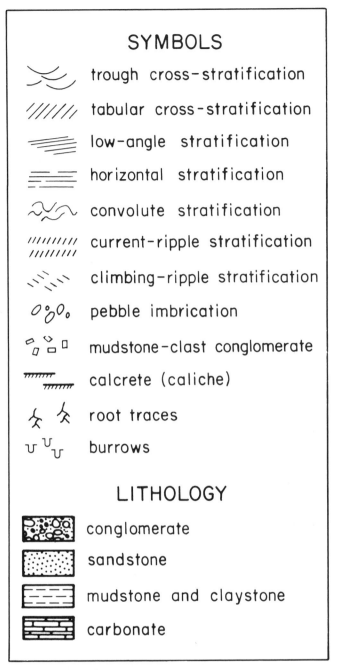

Figure 5—Explanation of symbols used in vertical sections in this paper. Average grain size: Gr = gravel; Sd = sand; SiCl = silt and clay; b = boulders; c = cobbles; p = pebbles; g = granules; vc = very coarse grained; m = medium grained; vf = very fine grained.

the proximal fan facies, although these sediments are interpreted as younger deposits by Peirce et al. (1979).

A typical vertical section of the proximal fan facies (Figure 4) shows a dominance of massive to horizontally stratified conglomerate with interbedded lenses of very coarse-grained sandstone. Figure 5 is an explanation of symbols used in this and subsequent vertical sections. Conglomerate clasts are generally pebble to boulder size, clast supported, crudely imbricated, and lie within a sand matrix. Conglomerate units are laterally extensive and are as much as 6 m thick in the Mogollon Rim area (Figure 6).

Their thickness in the Socorro, Lemitar, and Hubbell bench areas, however, is generally less than 1.5 m.

Sandstone occurs in lens-shaped beds as much as 1 m thick and dominantly exhibits horizontal lamination and medium to large-scale planar foreset bedding. Contacts between units are generally sharp and do not show high-relief scour surfaces. Local dense layers of microcrystalline calcium carbonate occur in proximal fan deposits in the Mogollon Rim area; however, we did not observe these in such deposits in other areas.

Figure 6—Horizontally stratified conglomerate and sandstone lenses of the proximal fan facies, Mogollon Rim area. Increments on scale are 0.3 m.

Figure 7—Channel-fill conglomerate with large-scale foresets, midfan facies, Datil Mountains area. Increments on scale are 0.3 m.

Depositional Processes—The conglomerates of the proximal fan facies are equivalent in structure, texture, and depositional geometry to perennial braided stream deposits in the upper reaches of modern proglacial outwash fans (Boothroyd and Ashley, 1975; Nummedal and Boothroyd, 1976; Boothroyd and Nummedal, 1978). Using these modern fans as a depositional analog, we interpret the proximal fan conglomerates as deposits of laterally extensive longitudinal bars that were active during infrequent high floods.

Deposition of sandstone beds occurred by: (1) construction of foresets on longitudinal bar margins (Rust,

1972; Boothroyd and Ashley, 1975); (2) infilling of shallow scours and channels; and (3) downstream migration of linguoid bars (Collinson, 1970; Miall, 1977, 1978). Sand deposition took place during waning flood stages.

Matrix-supported debris flow deposits, typical of alluvial fans in arid regions (Hooke, 1967; Bull, 1972), are very rare in the proximal fan facies. Local, dense calcium carbonate layers exposed in the Mogollon Rim area are interpreted as paleocaliche horizons. These zones formed pedogenically, presumably resulting from a shift in the axis of sedimentation and long-term exposure of the fan surface.

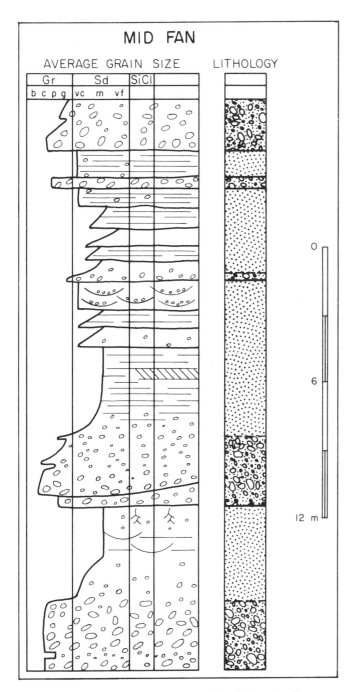

Figure 8—Typical vertical section of mid-fan facies (data from NW ¼ sec. 28, T. 10 N., R. 30 E., Springerville area). See Figure 5 for explanation of symbols.

Mid-fan Facies

Characteristics—This facies is present in the Springerville, Datil Mountains, Bear Mountains, and Socorro areas. In contrast to the proximal fan facies, mid-fan conglomerates occur as lens- or channel-shaped units as much as 30 m wide and 5 m thick. Conglomerates typically exhibit crude horizontal stratification and imbrication, but large-scale cross-bedding is locally present (Figure 7).

Sandstones are very coarse- to fine-grained and occur in laterally extensive sheetlike deposits and in channel-shaped units as wide as 50 m. Sedimentary structures present within sandstone beds include horizontal stratification, trough cross-stratification, low-angle stratification, and rare planar cross-stratification. Graded laminae and graded beds as much as 2 m thick were commonly observed. A typical vertical section of the mid-fan facies is presented in Figure 8.

Depositional Processes—The horizontally stratified conglomerates of the mid-fan facies are longitudinal bar and channel-lag deposits and are analogous to such deposits in modern braided streams described by Williams and Rust (1969), Rust (1972), and Church and Gilbert (1973). Large-scale tabular cross-bedding in conglomerate units formed by migration of gravelly transverse bars (Hein and Walker, 1977; Miall, 1977), by point-bar accretion of gravels (Martini, 1977), and more rarely, by deposition on slip faces of longitudinal bars (Boothroyd and Nummedal, 1978; Rust, 1978). Channel-margin accretionary bedding, such as that shown in Figure 7, was probably formed by deposition on a gravel point bar.

The mid-fan facies contains more channel-shaped conglomerate units than does the proximal fan facies. This is consistent with the downdip bifurcation and increase in number of channels noted on proglacial outwash fans by Boothroyd and Nummedal (1978).

Sand was deposited dominantly by migrating bedforms (primarily dunes) in shallow braided channels and by plane bed aggradation in overbank areas. The sandstone units of the mid-fan facies are very similar to those of the distal fan facies, and are discussed in detail in that section.

Arroyo-Fill Facies

Characteristics—The arroyo-fill facies is exposed in the southern part of the Socorro area near the Carthage coal field. Sandstone and conglomerate occur in subequal proportions in this facies, thus we consider it to be a variant of the mid-fan facies. A typical vertical section (Figure 9) shows fine-grained sandstone and sandy mudstone interbedded with very coarse- to medium-grained sandstone and conglomerate. The common occurrence of steep-walled, rectangular, channel-shaped units of conglomerate and coarse sandstone as much as 100 m wide (Figure 10) distinguish this facies from the mid-fan facies. Channel-fill deposits are poorly sorted and exhibit abrupt lateral changes in texture and sedimentary structures. Sedimentary structures within channel units include crudely stratified conglomerate, horizontally laminated sandstone, and more rarely, trough cross-stratified sandstone and medium-scale, tabular, cross-stratified sandstone. Convolute stratification in sandstone beds was also noted at several localities.

Channel-shaped units are incised into laterally persistent sandy mudstone beds and thin sandstone beds. Burrows and root traces are common in the fine-grained deposits and may be largely responsible for the general lack of stratification in these beds. Faint horizontal stratification, however, is locally present in some of the nonchannelized deposits.

Depositional Processes—We interpret the arroyo-fill facies to represent deposition by unconfined vertical aggradation of the fan surface alternating with episodes of arroyo incision and infilling. The vertical-walled aspect of

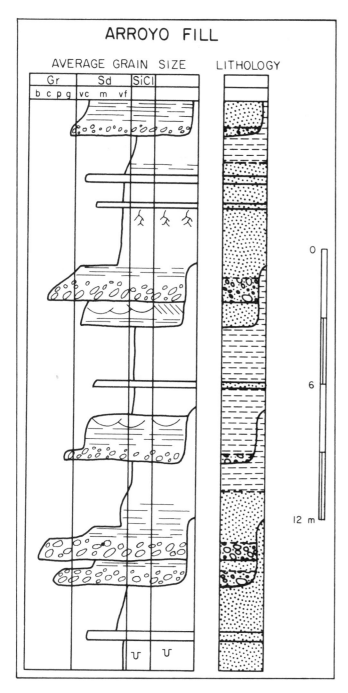

Figure 9—Typical vertical section of arroyo-fill facies (composite data from localities near Carthage coal field, Socorro area). See Figure 5 for explanation of symbols.

the channels in this facies and the coarse-grained nature of the channel fill is similar to modern arroyos throughout the study area and to the arroyos described by Leopold and Miller (1956), Schumm and Hadley (1957), Leopold et al. (1964), and Love (1977). Love (1977) described alternating arroyo cutting and unconfined flooding in Chaco Canyon, New Mexico, which may be a modern analogy to the depositional processes of the arroyo-fill facies.

On the basis of preserved sedimentary structures, gravel-bar formation and plane-bed aggradation during conditions of upper flow regime were the dominant sedimentation

processes in the arroyo channels. Dune migration and progradation of transverse bars occurred locally. The poor sorting of the channel-fill deposits suggests high rates of deposition, possibly due to rapid infiltration of flood waters. Lack of large-scale lateral accretion surfaces, such as those described by Shepard (1978) for the degradational deposits of Tapia Canyon, New Mexico, imply that the incised channels of the arroyo-fill facies were not highly sinuous.

The fine-grained, laterally continuous units of the arroyo-fill facies are evidence of deposition of overbank flood deposits and sheet wash during periods of unconfined flow. The abundance of root traces in some of these beds indicates at least seasonally wet conditions.

Distal Fan Facies

Characteristics—The distal fan facies is exposed in the Springerville, Quemado, Datil Mountains, Gallinas Mountains, and Bear Mountains areas. A typical vertical section (Figure 11) shows a dominance of very coarse- to fine-grained sandstone with subordinate conglomerate, mudstone, and claystone. The prevalent mode of occurrence of sandstone in this facies is sheetlike bodies that contain abundant horizontal laminations (Figure 12). Individual laminae are generally not graded, but graded laminae and larger scale upward-fining units as much as 2 m thick were sometimes observed. Low-angle stratification and parting lineation are commonly associated with horizontally stratified sandstone beds.

Trough cross-stratification is common in the distal fan facies and generally occurs as cosets of mutually cross-cutting, medium- to large-scale, festoon cross-bedding in broad, shallow, channel-shaped sandstone units as wide as 50 m. More rarely, solitary trough-shaped scour fill and minor planar cross-stratification occur within the horizontally stratified sandstone bodies. Conglomerates form very thin channel and sheetlike deposits as much as 15 m wide and also occur as small lenses at the base of trough-shaped scours. Conglomerates locally contain abundant clasts of intraformational mudstone and claystone, commonly exhibit crude horizontal stratification and imbrication, and may be clast supported or occur as scattered pebbles in a stratified sand matrix.

In contrast to the proximal fan and mid-fan facies, the distal fan facies locally contains appreciable amounts of mudstone and claystone that occur as thin, laterally continuous beds. Rare horizons of nodular caliches and locally abundant root traces also occur.

Depositional Processes—Vertical textural trends and stratification types in this facies are comparable to those described by McKee et al. (1967) for recent flood deposits along ephemeral Bijou Creek, Colorado. Large-scale, upward-fining units in horizontally stratified sandstone beds record deposition by plane-bed aggradation during waning flood stages. Low-angle stratification represents filling of shallow scours during high-velocity flow (Rust, 1978). Graded laminae, common in the horizontally stratified sandstones of the distal fan facies, may be the result of minor velocity fluctuations during plane-bed aggradation or the migration of very low-amplitude sandwaves in shallow water (Smith, 1971).

Figure 10—Coarse-grained, arroyo-fill deposits incised into laterally continuous fine sandstones and mudstones, Socorro area. Paleoflow was toward viewer. Socorro Peak on skyline at center.

The dominance of medium- to large-scale trough cross-bedding in channel-fill deposits indicates that dune migration was the primary depositional process in the distal fan channels. Dune development in channels occurred dominantly during waning flood stages (McKee et al., 1967) as did minor dune and transverse bar activity on the adjacent floodplain. Present-day analogs of dune-dominated channel systems include the shallow, ephemeral braided streams of Colorado (McKee et al., 1967), central Australia (Williams, 1971), and Israel (Karcz, 1972). The scarcity of planar cross-bedding in channel deposits of the distal fan facies differs from perennial braided stream models (Smith, 1970; Costello and Walker, 1972; Rust, 1972; Cant and Walker, 1976).

Thin mudstones and claystones present in the distal fan facies represent floodplain and ephemeral pond deposits. Reworking of these fine-grained sediments by subsequent floods produced the clay clasts commonly observed in conglomerates that were deposited as channel bottom lags and in scours leeward of advancing dunes.

Braided Alluvial-Plain Model

The braided alluvial-plain system described here consists of the deposits of a series of coalesced humid alluvial fans. Although discrete fans probably existed directly adjacent to surrounding uplifts, the lack of radial paleocurrent indicators within basin-fill deposits suggests basinward integration of fans into broad, braided alluvial plains or piedmont slopes.

The most commonly cited present-day examples of humid alluvial fans are the proglacial braided-outwash fans of Alaska and Iceland (Gustavson, 1974; Boothroyd and Ashley, 1975; Boothroyd and Nummedal, 1978) and the Kosi tropical fan of India (Gole and Chitale, 1966). Humid alluvial fans differ from arid fans in several respects:

1. Flow in braided streams on humid fans is perennial or seasonal, in contrast to the ephemeral discharge characteristics typical of arid fans.
2. Humid alluvial fans exhibit smooth, gentle downstream gradients, ranging from about 6 m/km (proximal) to 3 m/km (distal) in proglacial outwash fans (Boothroyd and Nummedal, 1978) and 1–0.2 m/km in the Kosi fan (Gole and Chitale, 1966).
3. Humid fans are much more areally extensive than their arid counterparts. The relatively uniform flow characteristics of humid fans result in steady, long-term deposition, which favors development of large, low-gradient surfaces on these fans.
4. Unlike arid alluvial fans, humid fans consist almost exclusively of braided stream sediments and characteristically lack significant debris flow and sieve deposits. The abundance of debris flow and sieve deposits

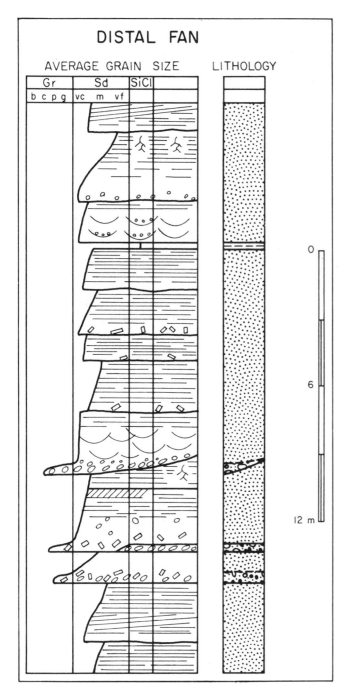

Figure 11—Typical vertical section of distal fan facies (data from NW ¼ sec. 19, T. 1 N., R. 18 W., Quemado area). See Figure 5 for explanation of symbols.

in arid fans is due to rapid erosion and runoff resulting from sporadic, intense rainfall events typical of arid climates.

5. Humid fans exhibit a systematic downstream decrease in grain size. In arid fans, the sporadic supply of coarse, unsorted sediments to all parts of the fan by debris flows hinders development of such downstream textural trends (J. C. Boothroyd, 1979, personal communication).

Certain aspects of the proglacial outwash fans of Alaska and Iceland and the Kosi fan of India provide useful modern analogs for parts of the braided alluvial-plain system discussed in this paper. Several important differences, however, exist between these modern fans and the ancient braided alluvial-plain systems present in the Baca and Carthage–La Joya basins. First, although the downdip extent of the large alluvial plain present in the western part of the Baca basin is similar in size (~150 km long) to that of the Kosi fan, it is much larger than typical proglacial braided-outwash fans (10–35 km long). The smaller size of the outwash fans can probably be explained by the relatively short duration during which they have been active (late Pleistocene–Holocene), in contrast to the Baca fan system which was presumably active for several million years during the Eocene. The size of modern outwash fans is roughly similar to those of the poorly exposed fan systems present in the Carthage–La Joya basin and the eastern part of the Baca basin.

Second, important sedimentologic differences exist between the Baca fan systems and the recent examples described in the literature. Although the proximal part of the Kosi River fan contains pebble to boulder gravels similar to those in the proximal fan facies, the lower reaches of the Kosi fan contain much more fine-grained sediment than do analogous parts of the Baca braided alluvial-plain system. Similarly, textures and stratification types present in the upper reaches of modern proglacial outwash fans are very much like the proximal fan facies of the Baca and Carthage–La Joya basins. However, the mid-fan and distal fan facies show evidence of non-uniform, perhaps ephemeral, flow and thus differ markedly from proglacial examples.

Sedimentary structures present in braided alluvial-plain deposits of the Baca and Carthage–La Joya basins indicate discharge was increasingly flashy basinward. The proximal fan conglomerates are similar to the perennial "Scott-type" of Miall (1977, 1978), whereas the distal fan facies is much like Miall's ephemeral "Bijou Creek-type," although containing more trough cross-bedding and conglomerate. The mid-fan facies seems to be transitional between these two end-members. The downdip change in paleohydrologic regime implied by this facies succession may be due to: (1) the presence of a wetter microclimate in the highlands surrounding the Baca and Carthage–La Joya basins, giving rise to seasonal or perennial flow near the mountain fronts; and (2) increased infiltration and evapotranspiration in the basins that would tend to increase the discharge peakedness in the distal fan environment. Evidence for downfan increase in discharge "flashiness" is best developed in the large fan system present in the western part of the Baca basin, although it also occurs to a lesser extent in other areas of the Baca and Carthage–La Joya basins.

Evidence suggests that the climate in west-central New Mexico during Baca time was semiarid and savannalike (Cather and Johnson, 1984). Climatic effects seem to best explain the differences between the braided alluvial-plain system discussed in this paper and the proglacial outwash model and seem to provide controls on other modes of deposition within the Baca basin.

Figure 12—Horizontally stratified flood deposits, distal fan facies, Quemado area. Several graded intervals are present.

Fluvial Meanderbelt System

Fine-Grained Meanderbelt Facies

Characteristics—The fine-grained meanderbelt facies is exposed in the Datil Mountains and Dog Springs Canyon segments of the outcrop belt. This facies consists of tabular sandstone units as much as 30 m thick and 1 km wide interbedded with mudstone and siltstone beds; fine-grained rocks generally make up >50% of this facies. A typical vertical section (Figure 13) shows a pronounced stacking of genetic cycles within the sandstone lithofacies (multistory sandstone units). Individual cycles average about 4–6 m in thickness and display upward-fining textural trends and relatively consistent vertical arrays of sedimentary structures. Thin interbeds of bioturbated, fine-grained deposits locally occur between superimposed sandstone cycles. The base of a genetic cycle consists of an erosional, high-relief scour surface overlain by massive to horizontally stratified conglomerate (commonly intraformational) and very coarse-grained sandstone. This is overlain by upward-fining sandstone, which displays (in order of ascending stratigraphic position) large-scale trough cross-stratification, medium-scale trough cross-stratification and horizontal laminations, and current-ripple and climbing-ripple laminations. Where present, the uppermost part of a typical cycle consists of fine- to very fine-grained sandstone and siltstone that exhibit abundant root traces and common burrows. Because of the stacked, multistory nature of genetic units in the fine-grained meanderbelt facies, the upper portions of individual cycles are commonly erosionally truncated by the base of superjacent cycles. Where interfingering of sandstone and mudstone lithofacies occurs, individual sandstone units representing complete genetic cycles are commonly preserved. Large-scale epsilon cross-bedding (Allen, 1965) is locally well-developed (Figure 14).

The mudstone lithofacies consists mainly of thick, tabular mudstone units with minor siltstone and thin interbeds of very fine- to medium-grained sandstone, although local channel-shaped mudstone units are also present. Mudstone and siltstone beds are typically structureless, but locally may show current-ripple and climbing-ripple lamination, burrows, and root traces. Sandstone beds range from 5 cm to 0.6 m in thickness and exhibit horizontal lamination, current-ripple lamination, and rare root traces and burrows.

Depositional Processes—Numerous criteria for the recognition of point-bar accretion in a meandering river environment have been proposed in the literature. The fine-grained meanderbelt facies exhibits all of the major characteristics, including upward-fining textural sequences (Allen, 1964; Masters, 1967; Visher, 1972), vertical decrease in scale of sedimentary structures (Allen, 1964; Belt, 1968; Bernard et al., 1970; Visher, 1972;), and the presence of large-scale epsilon cross-bedding (Allen, 1965; Moody-Stuart, 1966; Cotter, 1971;). Bernard et al. (1970) described the distribution of grain sizes and sedimentary structures in a modern meander of the Brazos River, Texas. The vertical sequence of structures and textures resulting from point-bar accretion in the Brazos River is similar to that exposed in sandstones of the fine-grained meanderbelt facies.

The fine-grained rocks locally interbedded between point-bar sandstones are top-stratum deposits (Allen, 1965). The highly bioturbated, horizontally laminated and ripple-laminated deposits locally preserved at the top of point-bar cycles probably represent levee deposits. They commonly show abrupt textural variation between fine sandstone, siltstone, and mudstone that characterize modern levees (Fisk, 1944).

Truncation of the upper parts of the genetic sandstone cycles results from multistory stacking of point-bar deposits during downdip migration of individual meander loops within the meanderbelt system. Comparable ancient stacked

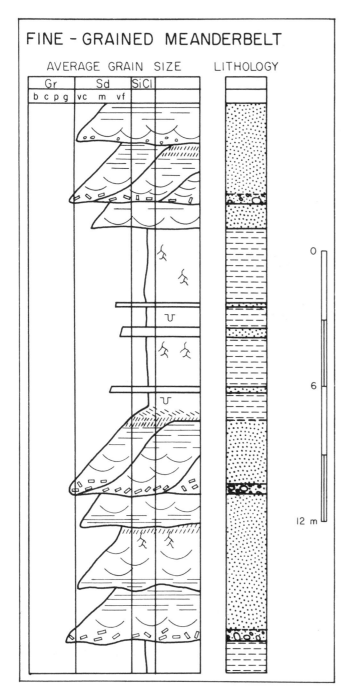

Figure 13—Typical vertical section of fine-grained meanderbelt facies (data from Remuda Canyon, Datil Mountains area). See Figure 5 for explanation of symbols.

grained sandstone interbeds are interpreted as crevasse-splay deposits that entered the flood basin through breaches in levees during floods. The thickness of these sandstone beds and the sedimentary structures and root traces within them are similar in recent crevasse splays. Channel-shaped mudstone units, commonly associated with point-bar sandstone sequences, represent abandoned-channel fill. Interbedding of lacustrine delta deposits (see discussion below) and point-bar sequences in the Datil Mountains area suggests that the floodbasins were periodically occupied by lakes.

The size of epsilon cross-bedding can be used to estimate paleochannel width and depth. The width of a cross-bed set is approximately two-thirds of the bankfull channel width (Allen, 1965; Moody-Stuart, 1966; Cotter, 1971). Bankfull depth in straight reaches of the paleochannel can be estimated by the following relationship (Ethridge and Schumm, 1978):

$$\text{bankfull depth} = \text{epsilon cross-bed height} \times 0.585/0.9$$

The measured width of multistory sandstone units provides a rough estimate of average meanderbelt width.

Using these criteria in conjunction with paleocurrent data, we delineate two distinct fine-grained meanderbelt paleoenvironments in the Datil Mountains area. Their paleohydrologic characteristics are summarized in Table 1. We interpret the smaller, stratigraphically lower, point-bar sequences as deposits of southeast-flowing tributary streams that drained the Zuni uplift. The larger paleomeanderbelt system represents an east-flowing, integrated river system that occupied the axial part of the Baca basin. In addition to these fluvial trends, Chamberlin (1981) notes the presence of a northeast-flowing tributary system in the southwestern part of the Datil Mountains area.

Coarse-Grained Meanderbelt Facies

Characteristics—This facies is restricted to the upper–middle part of the Baca Formation in the Datil Mountains area and is overlain and underlain by the fine-grained meanderbelt facies. A typical vertical section of the coarse-grained meanderbelt facies is presented in Figure 15. The sandstone lithofacies shows poorly developed stacking of genetic cycles (multilateral sandstone units), scarcity of ripple-laminated fine-grained sandstone and siltstone, poorly developed upward-fining textural trends, abundance of horizontally laminated sandstones, and local large-scale foreset bedding as much as 2 m thick. These features distinguish the coarse-grained meanderbelt facies from its fine-grained counterpart. The presence of significant amounts of mudstone and the local occurrence of epsilon cross-bedding in the coarse-grained meanderbelt facies differs from the braided stream deposits found elsewhere in the Baca basin. The mudstone lithofacies of the coarse-grained meanderbelt facies is similar to, but less abundant than, that of the fine-grained meanderbelt facies.

Depositional Processes—Sedimentologic characteristics of sandstones in this facies indicate deposition by chute-modified point bars (McGowen and Garner, 1970); these are commonly termed coarse-grained point bars, because in these the fine-grained, upper part of the point bars are

point-bar sequences have been described by Allen (1964), Belt (1968), and Steel (1974). Puigdefabregas and van Vliet (1978) suggest that the occurrence of multistory sandstone sequences, as opposed to isolated point-bar deposits encased within floodplain mudstone, indicate that the meanderbelt occupied a relatively stable position within the floodbasin and that vertical aggradation rates were not high relative to the rate of meander migration.

Laterally persistent mudstone and siltstone beds represent vertically accreted flood-basin deposits. The thin, fine-

Figure 14—Point-bar sandstone showing large-scale epsilon cross beds. Fine-grained meanderbelt facies, Datil Mountains area. The sandstone unit shown is about 4 m thick.

Table 1—Paleohydrologic characteristics of the fine-grained meanderbelt facies.

Type of Paleostream	Bankfull Paleochannel Size		Ancient Meanderbelt Width
	Width	Depth	
Axial (east-flowing)	65 m[a]	4.2 m[a]	525 m[c]
Tributary (southeast-flowing)	35 m[b]	2.2 m[b]	185 m[c]

[a]Based on 3 examples.
[b]Based on 4 examples.
[c]Based on 16 examples.

replaced by coarse-grained chute-fill and chute-bar deposits. The large-scale foreset cross-bedding in this facies records avalanche-face accretion on chute and transverse bars that formed in response to sediment transport across the point bars during floods. Similar foresets can form in large bedforms in a fine-grained meanderbelt environment, such as scroll bars (Jackson, 1976) and spillover lobes (Bernard et al., 1970). However, the large foresets in the coarse-grained meanderbelt facies probably cannot be attributed to these types of bedforms. The overall change in point-bar characteristics and the decrease in the sandstone:mudstone ratio in this facies, as well as the restriction of this facies to a specific stratigraphic interval, support a change to a coarse-grained meanderbelt regime.

The abundant horizontally laminated sandstones of this facies were deposited during upper-flow regime plane-bed aggradation on point bars and floodplains. McGowen and Garner (1970) described similar horizontally laminated flood deposits on modern coarse-grained point bars of the Amite River, Louisiana. Isolated scour troughs associated with the horizontally laminated sandstones may have been generated by turbulence around vegetation on point-bar surfaces and on the floodplain. This interpretation is supported by the abundance of root traces in the horizontally laminated sandstones.

Coarse-grained point bars occur in modern rivers characterized by unstable, rapidly shifting channels (McGowen and Garner, 1970). Lack of channel stabilization may explain the poorly developed stacking of chute-bar deposits over lower point-bar sandstones in the coarse-grained meanderbelt facies. Instead, the sandstone lithofacies commonly consists of multilateral sequences of (1) trough cross-stratified lower point-bar sandstones, (2) large-scale foreset deposits of chute bars, and (3) horizontally laminated upper point-bar and floodplain sandstones.

Conditions favoring chute and chute-bar development in the modern Amite and Colorado rivers are high bedload: suspended load ratio, moderate gradients, short peak flood duration, and sandy banks that are stabilized by vegetation (McGowen and Garner, 1970). Sediment load and discharge characteristics would support a braided stream regime, but a meandering pattern is preserved by vegetative stabilization of point bars. In contrast, fine-grained meandering rivers tend to have relatively uniform flow, low gradients, low bedload:suspended load ratio, fine-grained, cohesive bank materials, and channels well-stabilized by vegetation (Leopold and Wolman, 1957; Schumm and Kahn, 1972)

The coarse-grained meanderbelt facies in the Baca basin is both overlain and underlain by fine-grained meanderbelt deposits. Possible reasons for the temporary shift to a

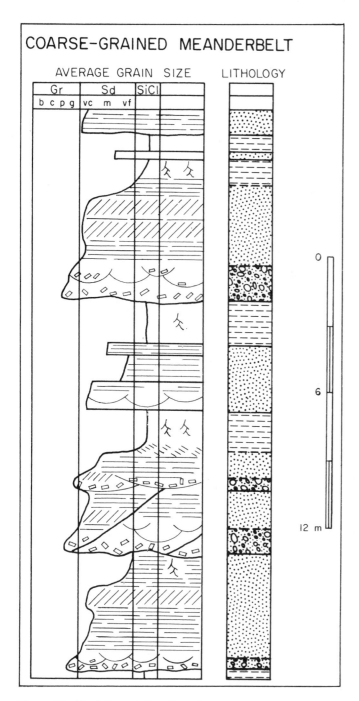

Figure 15—Typical vertical section of coarse-grained meanderbelt facies (composite data from localities in Red Canyon, Datil Mountains area). See Figure 5 for explanation of symbols.

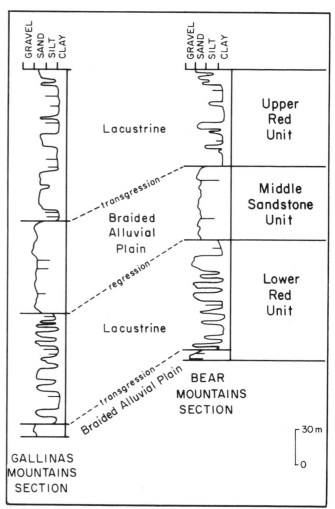

Figure 16—Generalized stratigraphic columns of Baca Formation in Gallinas and Bear mountains areas, showing relationships between informal units, shoreline migrations, and depositional systems.

coarse-grained meanderbelt regime are (1) a pulse of tectonism in source regions adjacent to the Baca basin or in intrabasin areas, giving rise to increased stream gradients and greater supply of bedload sediments; and (2) a change to a drier climate causing less uniform discharge characteristics and decreased stabilization of channels by vegetation. Although neither of these possible mechanisms can be ruled out, climatic change appears to be the probable cause of a major regressive phase of Eocene lacustrine sedimentation in the eastern part of the Baca basin. These low-stand,

lacustrine deposits occupy a stratigraphic position similar to that of the coarse-grained meanderbelt facies in the Datil Mountains area. Evaluation of their time-stratigraphic relations, however, is not yet possible due to sparse paleontologic control.

Lacustrine System

The lacustrine system is widespread throughout the Baca basin. Lacustrine sedimentation took place in two general settings: (1) small, inpermanent lakes situated on fluvial-system floodplains in the Springerville, Quemado, and Datil mountains areas; and (2) a large, shallow, persistent lake located in the Dog Springs Canyon, Gallinas Mountains, and Bear Mountains areas.

In the Bear Mountains area, Potter (1970) divided the Baca Formation into three informal members, which he called (in ascending stratigraphic order) the lower red, middle sandstone, and upper red units. Cather (1980, 1982) extended this terminology to the Gallinas Mountains area, and emphasized that these informal units are genetically related to lake-level fluctuations in a large Eocene lacustrine

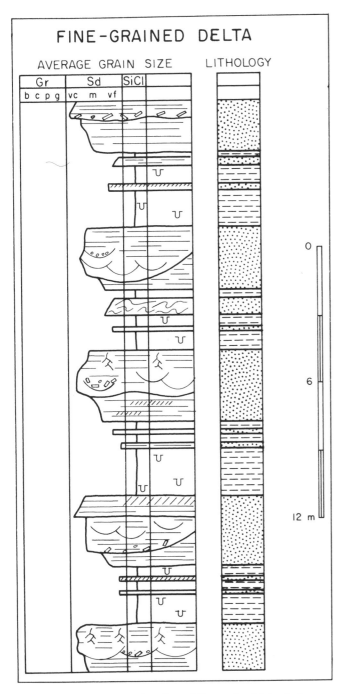

Figure 17—Typical vertical section of fine-grained delta facies (data from NW ¼ sec. 13, T. 1 N., R. 6 W., Gallinas Mountains area). See Figure 5 for explanation of symbols.

The lacustrine deposits of the Baca basin consist of both delta and basin deposits. We have divided the delta deposits into two types (fine-grained delta and fan delta), although a continuous spectrum of delta characteristics exists between these two end-members.

Fine-Grained Delta Facies

Characteristics—Exposures of the fine-grained delta facies are present in the Datil Mountains, Dog Springs Canyon, Gallinas Mountains, and Bear Mountains areas. Cyclic upward-coarsening sequences characterize this facies (Figure 17). Individual cycles range in thickness from about 6 to 35 m and average about 9 m.

The basal portion of a typical cycle consists of a laterally persistent calcareous mudstone or claystone layer intercalated with thinly bedded (generally <25 cm thick), very fine- to medium-grained sandstones. Carbonate content of the mudstone (determined by weighing samples before and after acidization with cold, dilute hydrochloric acid) generally ranges between 10 and 15% by weight. Mudstone units rarely exhibit laminations and are usually structureless and homogeneous, with the exception of burrowing, which is pervasive. In contrast to the sandstones in the lower parts of deltaic cycle, mudstones do not exhibit well-defined burrows, but rather have a churned, curdled texture both megascopically and microscopically that gives rise to the homogeneous nature of the mudstones. Rare horizons of mudcracks and pedogenic calcite nodules were also observed.

Structures present within the thinly bedded sandstones intercalated with the burrowed mudstones include horizontal laminations, current-ripple laminations, parting-step lineation, occasional normal-graded beds, and burrows. Burrows are vertical, horizontal, and oblique, range in diameter from 1 to 5 cm, and commonly exhibit knobby surface ornamentation and scoop-shaped backfill laminae. These burrows are similar to *Scoyenia* sp., which is common in nonmarine red beds (Hantzschel, 1975) and is believed to have been formed by polychaete worms. The thinly bedded sandstones in the lower part of cycles are often inclined, forming large-scale foresets with dip angles as high as 15° in rare instances. Dip angles are more commonly only a few degrees (Figure 18) and are often so gently inclined that the angularity is not readily apparent in a solitary exposure. Soft-sediment deformation is locally well developed in the lower part of deltaic cycles.

The units described above are transitionally overlain by a horizontal- and current-ripple laminated, laterally continuous, fine- to coarse-grained sandstone that averages about 1.5 m in thickness. Orientation of current-ripple cross-laminations generally indicates a direction of flow at high angles to that shown by other paleocurrent indicators within the same deltaic cycle. The well-sorted, nearly homogeneous nature of these sandstones gives rise to a quasi-spheroidal weathering habit (Figure 18). Coloration of this and superjacent sandstones within a given cycle may be red or yellowish gray, whereas the previously described intercalated mudstone and sandstone is almost always red.

Up-section, the next part of an ideal cycle will commonly exhibit large, symmetric, channel-shaped sandstone units that have erosional bases. Channels may be as much as 15 m wide and 5 m deep, but are generally much smaller.

system in the eastern part of the Baca basin. In the Bear and Gallinas mountains areas, the lower and upper red units record intervals of predominantly lacustrine sedimentation, whereas the middle sandstone unit represents an interval of braided stream sedimentation during a low stand of the lake (Figure 16). Although complicated by later faulting and sedimentary cover, a similar tripartate division of the Baca Formation in the Dog Springs Canyon area is possible. In this area, however, the middle sandstone unit is made up of the stacked sandstone units and intervening mudstones of fine-grained meanderbelt facies (Robinson, 1981).

Figure 18—Gently inclined prodelta foreset beds overlain by quasi-spheroidally weathered, delta-front deposits. Fine-grained delta facies, Gallinas Mountains area. Hammer handle points in direction of delta progradation.

Intrachannel sedimentary structures include medium-scale trough cross-bedding and plane beds. Where symmetric channel-shaped units are not present, the base of the upper part of each cycle is represented by an irregular, low-relief, erosional surface. The remainder of the upper portion of individual deltaic cycles comprise the deposits of either the distal fan facies or the fine-grained meanderbelt facies.

Depositional Processes—We interpret the cyclical deposits of the fine-grained delta facies to record alternate deltaic progradation and abandonment in a shallow lake. Geometry of lithofacies and vertical and lateral sequences of textures and sedimentary structures are similar to those found in lobate high-constructive marine deltas (Fisher et al., 1969). Depositional processes are inferred to be the same as in these marine deltas.

The lower intercalated mudstones and thinly bedded sandstones were deposited in a prodelta and distal delta-front environment. Silt and clay were deposited by settling from suspension. The thin sandstone beds appear to be frontal splay deposits, probably representing prodelta turbidites. Normal-graded beds, a typical feature of turbidites, are occasionally seen in these sandstones. Mudcracks and caliches in the lower part of deltaic cycles are interpreted to represent low stands of lake level.

The laterally continuous, quasi-spheroidally weathered, horizontally laminated and rippled sandstones are delta front deposits. These sandstones were deposited in channel mouth bars or in longshore-current redistributed bars. Primary processes on the delta front were plane-bed aggradation and ripple migration. The large divergence between paleocurrent directions shown by the delta-front ripples and that of other indicators within a given deltaic cycle suggests that ripple migration direction was predominantly controlled by longshore currents.

The large, symmetric, channel-shaped sandstones commonly present in the upper part of deltaic cycles are distributary channel deposits. Sedimentary structures indicate that plane beds and subaqueous dunes were the dominant bedforms within the distributary channels.

Vegetative stabilization of stream channels on the delta platform near the lake may have allowed for the development of large, symmetric distributary channels with a relatively low width-to-depth ratio. Deltaic deposits that lack distributary channels apparently indicate nonstabilization of delta platform stream channels, as is common in classic "Gilbert-type" delta or fan-delta deposits (Gilbert, 1885; Theakstone, 1976). In the Gallinas and Bear mountains areas, the remainder of the upper part of the fine-grained delta deposits is composed of delta-platform sandstones and minor conglomerates deposited by braided stream processes identical to those of the distal fan facies. In the Datil Mountains and Dog Springs Canyon areas, however, delta-platform deposits include sediments of the meanderbelt system.

Fan Delta Facies

Characteristics—The lacustrine fan-delta facies is exposed in the Springerville, Quemado, and Bear mountains sections of the outcrop belt. Cyclical upward-coarsening deposits characterize this facies (Figure 19). Sediment caliber and sandstone:mudstone ratio are generally greater than those of the fine-grained delta facies.

The interbedded red sandstones and mudstones in the lower part of the upward-coarsening fan-delta cycles range from subhorizontal beds to steeply inclined foresets (20–30°) as much as 4 m in thickness. These deposits exhibit bedding geometries and sedimentary structures similar to those of analogous parts of fine-grained deltas. In the Quemado area, steeply inclined lower delta deposits commonly appear to be filling depressions scoured into older strata. Up-section, these sandstones and mudstones are transitionally overlain by horizontal- to current-ripple laminated sandstones and thinly bedded conglomerates as much as 2 m thick and form massive, rounded outcrops. These are, in turn, overlain by sandstones and conglomerates of the mid-fan or distal fan facies (Figure 20).

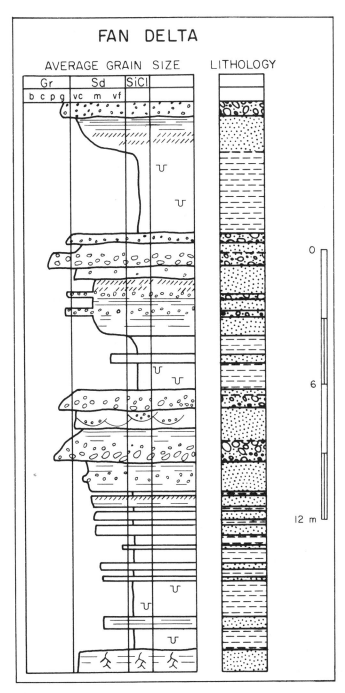

Figure 19—Typical vertical section of fan-delta facies (composite data from Baca Canyon and vicinity, Bear Mountains area). See Figure 5 for explanation of symbols.

Depositional Processes—The upward-coarsening genetic units of the fan-delta facies record deltaic progradation of bedload (braided) streams into lakes, forming small fan deltas. Fan-delta sequences with large-scale, steeply inclined bedding are classic Gilbert-type deltas (Gilbert, 1885). These deltas typically display well-defined bottomset, foreset, and topset beds. Steep foreset bedding, however, is present in only a few localities in the Baca lacustrine rocks; possible controls on deltaic foreset development are discussed below.

Depositional unit geometry and the vertical sequence of sedimentary structures and textures in the lower part of fan-

delta deposits are similar to, although somewhat coarser grained than, those of the prodelta and delta front part of the fine-grained delta facies. Depositional processes are thus inferred to be similar. The style of fluvial sedimentation on the fan deltas, however, appears to have differed significantly from that of the fine-grained delta-end member. Well-developed distributary channel deposits, which are common in the fine-grained deltas, were rarely observed in the fan-delta deposits. This may be a result of decreased vegetative stabilization of braided stream channels on the fan deltas or of the coarser grained, less cohesive nature of the fan delta deposits.

Basin Facies

Characteristics—Basin facies deposits occur in the Bear and Gallinas mountains areas and are particularly well developed in the lower part of the Dog Springs Canyon section. Where present, the lacustrine basin facies occurs directly beneath the above-described delta facies and consists of calcareous, generally structureless mudstone and claystone with sparse, thin interbeds of very fine sandstone and siltstone. Evidence of burrowing is abundant. The basin facies is identical to the lower prodeltaic portion of the fine-grained delta deposits, with the exception that the sandstone and siltstone interbeds of the basin facies are never inclined and are generally thinner and less abundant than those of the delta facies. The boundary between the two facies is arbitrary.

Depositional Processes—Basin depositional processes are exactly the same as those of the lower prodeltaic portion of the fine-grained delta facies, and include settling of silt and clay from suspension, deposition of silt and sand by turbidity currents, and homogenization of mud and clay by burrowing. The thin, sparse nature of the sand and silt interbeds indicates deposition far from nearshore sources of coarse sediments.

Characteristics of the Lacustrine System

Delta and basin sediments in the large lake that was present in the Dog Springs Canyon, Gallinas Mountains, and Bear Mountains areas show evidence of deposition in a closed lacustrine basin. High concentrations of early authigenic (precompaction) calcite and dolomite in lower delta and basin sandstones and mudstones suggest evaporative concentration of solutes in a closed lake environment. The restricted megafaunal assemblage in the Baca lacustrine rocks also favors a closed lake system. Only a few scattered ostracods were observed. Langbein (1961) notes that closed lakes usually exhibit widely fluctuating water levels and are found exclusively in arid and semiarid regions. The climate during Baca time in west-central New Mexico was probably semiarid and savannalike (Cather and Johnson, 1984), and rare caliches and mud-cracked horizons in Baca prodelta and basin deposits are indicative of fluctuating lake levels. The presence of the fluvially dominated middle sandstone unit between the predominantly lacustrine lower and upper red units indicates a large-scale regression that was probably caused by a drastic drop in lake level due to climatic change (see discussion below).

Figure 20—Upward-coarsening prodelta deposits (foreground) overlain by delta-front and braided-stream deposits (across arroyo). Fan-delta facies, Bear Mountains area.

Steeply inclined Gilbert-type foresets are rare in the fine-grained delta facies and only locally well developed in the fan-delta facies. Three factors contribute to the development of foresets: (1) homopycnal flow (inflow density approximately equal to lake water density), which causes three-dimensional mixing and an abrupt decrease in current velocity, resulting in rapid deposition of sediment (Bates, 1953); (2) deltaic progradation into deep water (Axelsson, 1967; McGowen, 1970); and (3) transport of coarse bed-load sediments, of which Gilbert-type foresets are composed, to the distributary mouth (Axelsson, 1967; Smith, 1975). These conditions were uncommon during deposition of fine-grained deltas in the large lake present in the eastern part of the Baca basin. Sediment caliber was rarely coarser than coarse sand, and the water depth, as indicated by the thickness of prodelta deposits, was generally less than 6 m. The probable closed nature of the lake implies that the lake waters were more dense (due to salinity) than inflowing river water, producing hypopycnal flow and plane jet formation (Bates, 1953). Lack of three-dimensional mixing caused the plane jet to maintain its velocity over a relatively long distance basinward, resulting in deposition of suspended sediments over a considerable distance from the distributary mouth. This led to the development of a gently sloping prodelta surface, which in turn produced the typical shallowly inclined prodelta foresets seen in the fine-grained delta facies.

The Gilbert-type deltas in the Bear Mountains area probably formed in response to possible deeper water conditions and the input of coarse, conglomeratic sediments derived from the nearby Sierra uplift. Many of the fan-delta deposits in the Quemado area that display well-developed bottomset, foreset, and topset beds appear to have filled deep scours along the margins of impermanent, interfluvial lakes. These deposits probably record local drainage incision resulting from a drop in lake level followed by infilling of the scour by prograding fan-delta sediments during subsequent high stands. Foreset development in these deposits thus appears to have been related to the presence of local deep water conditions and perhaps also to the probable occurrence of homopycnal flow and three-dimensional mixing in these small, freshwater lakes.

The laterally persistent nature of the delta-front and delta-platform sandstones and the general lack of destructional phase features indicate that Baca fine-grained deltas were mainly high-constructive lobate (Fisher et al., 1969), which suggests progradation into relatively shallow water. Thin, destructional phase shoreface sequences produced by reworking of upper delta sediments by waves and longshore currents following deltaic abandonment and subsidence are rarely seen. The paucity of destructional phase deposits suggests relatively low-energy conditions within the lake basin. Attenuation of wave energy resulting from shallow water depths may explain the rarity of destructional phase features.

The large, shallow lake system that was present during Eocene time in the eastern part of the Baca basin area would have tended to be polymictic (i.e., frequent overturn) since surface mixing by eddy diffusion would be expected to penetrate at least several tens of feet. Evidence of widespread burrowing in basin and prodelta deposits indicates that lake-bottom sediments were oxygenated, thus supporting a polymictic regime. Preservation of laminations in lake-bottom sediments is usually restricted to oligomictic and meromictic (permanently stratified) lakes, in which lack of oxygen inhibits the activities of burrowing organisms.

Most of the red coloration of the Baca Formation (in both lacustrine and fluvial deposits) appears to be due to intrastratal solution of iron-bearing minerals, precipitation of hydrated iron oxides, and subsequent dehydration of these oxides resulting in the development of hematite pigment (e.g., Walker, 1967). The diagenetic reddening of Baca lacustrine sediments was apparently initiated by dissolution of unstable, iron-bearing minerals in a positive Eh setting. This may have taken place in the oxygenated, lake-bottom environment or via subaerial exposure during periods of lake-level low stands.

In contrast to many lacustrine systems, the Baca deltas were not restricted to the periphery of the lake but appear to have prograded well out into the shallow lake basin. The occurrence of intermittent deltaic deposition throughout the lake basin precluded the development of widespread, chemically precipitated and/or strongly reduced, fine-grained clastic deposits typical of lacustrine basin-center environments.

Three possible explanations exist for the large-scale transgressions and regressions represented by the lower red, middle sandstone, and upper red units in the eastern part of the Baca basin: (1) shifting of the locus of lacustrine sedimentation resulting from tectonism within the basin; (2) increased erosion caused by tectonic activity in the source area, with resultant large-scale regression due to progradation of alluvial aprons basinward; and (3) climatic fluctuations with resultant large-scale transgressive and regressive phases. The relative importance of each of these hypotheses is difficult to evaluate. One line of evidence, however, suggests that climatic changes were a major cause of the lake level fluctuations. The lacustrine mudstone units in the Gallinas Mountains area (with the exception of the prodelta mudstone of the basal deltaic cycle of the upper unit) are highly calcareous and are probably indicative of saline, closed lake conditions. The essentially noncalcareous nature of the basal mudstone of the upper red unit (2% calcite by weight as compared with 10–15% in other mudstones) implies a temporary change to more freshwater conditions during the beginning of the second transgressive phase. If the transgression were due to a change to a wetter climate, the low-saline characteristics could be easily explained by the introduction of large amounts of fresh water to the lake. Neither of the tectonic alternatives can explain the noncalcareous nature of the basal mudstone of the upper red unit. Large-scale lake level fluctuations in contemporaneous Eocene Lake Gosiute, Wyoming, have also been attributed to climatic changes by Surdam and Wolfbauer (1975).

Several other lines of evidence support a lacustrine interpretation for large parts of the Baca Formation in the Bear and Gallinas mountains area. These include (1) the presence of rare oolites and ostracods and common intraclasts that were observed in thin sections from the Gallinas Mountains area (Cather, 1980); (2) the occurrence of limestone beds in the lower red unit of the Baca Formation in the Bear Mountains area, which are interpreted by Massingill (1979) to be of lacustrine origin; and (3) palynologic data. In his work on the palynology of the Baca Formation in the Bear Mountains vicinity, Chaiffetz (1979) states that the predominantly pink-gray Baca deposits (i.e., the lower and upper red units, which we interpret to be dominantly lacustrine) yielded only a few poorly preserved pollen grains and the freshwater alga *Pediastrum*. However, a greenish-gray shale sample from a thin mudstone in the fluvially dominated middle sandstone unit produced spores and pollen from a wide variety of upland flora, including conifera. Chaiffet (1979, p. 268) further states that the presence of *Pediastrum* "... perhaps requires some standing fresh-water bodies in the region at that time."

The existence of deltaic deposits in the basal Spears Formation indicates that the lake persisted for a short time during the beginning of Oligocene volcanism. The cause of the final demise of the lake is not known. Likely possibilities include climatic change and rapid infilling of the lake with volcaniclastic sediments.

Lacustrine Model

The limited exposures of the Baca lacustrine system make comparison to modern and ancient analogs difficult. However, any model of the large lake present in the eastern part of the Baca basin must account for the following characteristics: (1) shallow water depth; (2) fluctuating lake levels; (3) rarity of lacustrine megafauna; and (4) high concentrations of early authigenic carbonate cements in lacustrine deposits.

Eugster and Surdam (1973), Eugster and Hardie (1975), and Surdam and Wolfbauer (1975) have proposed a closed-basin, playa-lake model for Eocene Lake Gosiute in Wyoming. Many aspects of the marginal, clastic-dominated areas of this lake adequately fit the characteristics of the Baca lacustrine system. Lake Gosiute was a large, shallow, closed lake that exhibited widely fluctuating lake levels. With the exception of periods of lake-level low stands during which large volumes of trona were deposited in basin-center areas, chemical sedimentation within Lake Gosiute was dominated by precipitation of calcite and dolomite. Megafaunal diversity in Lake Gosiute was greater than that in the Baca lacustrine system and included ostracods, mollusks, algal reefs, and fish. The more restricted assemblage of the Baca lacustrine system may be due to lack of potentially more fossiliferous basin center deposits and/or higher salinity resulting from higher evaporation rates in the more southerly Baca lacustrine system. Surdam and Wolfbauer (1975) suggest that modern Deep Springs Lake in Inyo County, California, may be a modern (although much smaller) analog of Lake Gosiute.

Certain aspects of Lake Chad, Africa (Mothersill, 1975), are similar to those inferred for the Baca lacustrine system,

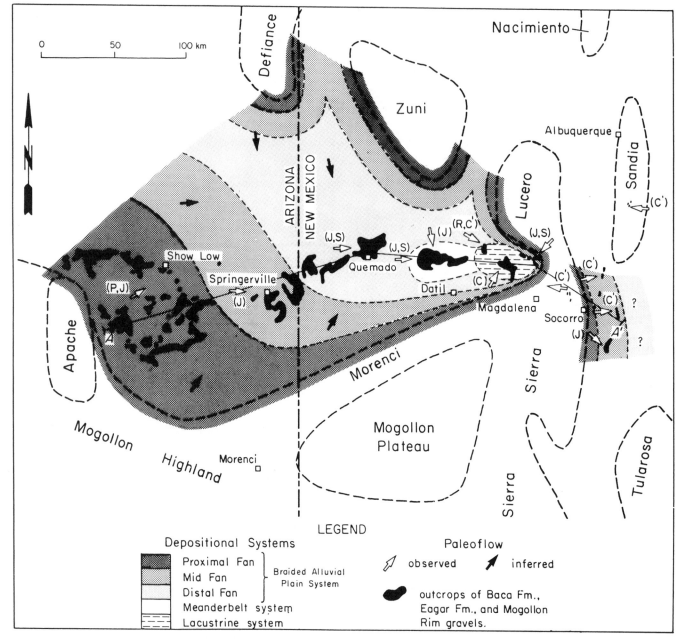

Figure 21—Model for distribution of depositional systems and paleocurrents in the Baca and Carthage–La Joya basins during early Baca time (high lake stand). Paleocurrent data sources (in parentheses): J = Johnson (1978), P = Peirce et al. (1979), S = Snyder (1971), R = Robinson (1981), C = Cather (1980), C′ = Cather (ongoing research). A–A′ is section line for Figure 22.

including a polymictic regime, the closed nature of the lake and shallow water depth (less than 5 m).

SUMMARY

Figure 21 is a model for the regional distribution of paleoflow and depositional environments for both the Baca and Carthage–La Joya basins; it is based on paleocurrent and facies distribution data obtained from outcrops of Eocene rocks in west-central New Mexico and eastern Arizona. An east–west cross-section (Figure 22) depicts facies geometries and subsurface relationships in these basins.

The interplay of two factors, climate and tectonism, can account for most of the sedimentologic attributes of the Baca and Carthage–La Joya basins. Tectonism (both intrabasinal and extrabasinal) was the fundamental determinant of basin geometries, regional topography (and therefore paleoflow), subsidence rates and thickness variations of basin-fill deposits, and many aspects of the gross distribution of paleoenvironments within the basins. The prevailing semiarid climate provided strong controls on styles of fluvial and lacustrine sedimentation by affecting rates of erosion and sediment supply, volume and peakedness of discharge, evaporation rates, and amount of vegetative cover.

Sedimentation in the Baca and Carthage–La Joya basins

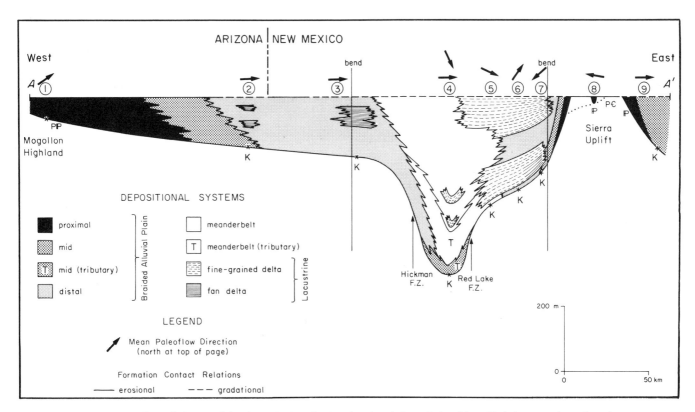

Figure 22—Cross-section through Baca and Carthage–La Joya basins showing facies relationships. Circled numerals are keyed to outcrop belt segments in Figure 21. Line of section A–A' is shown in Figure 21. Crosses at lower contacts of basins show thickness control points; symbols beneath basins indicate age of subcrop. FZ = fault zone; P = Permian, ℙ = Pennsylvanian, K = Cretaceous, P∈ = Precambrian.

began in Eocene time on a low-relief erosion surface developed on Upper Cretaceous rocks. Initiation of sedimentation in the Carthage–La Joya basin may have occurred somewhat earlier than in the Baca basin, but establishment of precise temporal relationships between the two basins is not possible with available data. Although the uplifts adjacent to these basins probably began to rise prior to Baca time, petrographic data (Cather, 1980) indicate that maximum structural development and widespread exposure of Precambrian rocks in at least two of the uplifts (the Morenci and Sierra uplifts) did not occur until early Baca time. Many aspects of the timing and style of Eocene deformation and sedimentation in west-central New Mexico are consistent with the late Laramide wrench fault model of Chapin and Cather (1981).

Braided alluvial-plain, meanderbelt, and lacustrine environments of deposition are represented in a well-exposed facies tract through the axial portion of the Baca basin. These deposits record lower stream gradients, decreased sediment caliber, and decreased sand:mud ratio toward the basin center. Only rocks of the braided alluvial-plain system are present in the poorly preserved remnants of the Carthage–La Joya basin.

Deposits of the braided alluvial-plain system are widespread in both the Baca and Carthage–La Joya basins. These deposits demonstrate that perennial or seasonal flow in the proximal fan facies gave way to more intermittent, perhaps ephemeral, flow in the distal environs. An extensive braided alluvial plain dominated the western two-

thirds of the Baca basin and represents detritus shed from the Mogollon highland.

Examples of both coarse- and fine-grained point-bar deposits are present in the Baca basin. Meanderbelt deposits are dominantly restricted to a structurally low area between the Red Lake fault zone and the Hickman fault zone that was actively subsiding during the time of Baca sedimentation (Figure 22). Paleohydrologic reconstruction in this area indicates the presence of one or more large, east-flowing, integrated meandering river systems in the axial part of the basin, as well as smaller, tributary meandering streams that drained the Zuni uplift.

Lacustrine sedimentation took place in small, temporary lakes on fluvial-system floodplains and in a large, shallow, closed lake in the eastern part of the Baca basin. Sedimentation in these lakes was characterized by cyclical progradation, abandonment, and subsidence of both fine-grained deltas and fan deltas. Deltas prograded well out into the lake basins and thus precluded the development of widespread, lacustrine, basin-center deposits. Major fluctuations in lake level in the large, closed lake present in the eastern part of the Baca basin may have resulted from climatic change. Interestingly, the deposits of this lake did not occupy the structurally lowest part of the basin. This is apparently a reflection of the general easterly paleoslope in the central and western parts of the Baca basin that was largely due to the dominance in those areas of sediments derived from the Mogollon highland.

The deposits of the Baca and Carthage La Joya basins

record the waning of Laramide deformation in adjacent uplifts. A cross-section through both basins (Figure 22) shows an onlapping relationship of component facies of the basin fill. This probably resulted from the updip retreat of the facies tract in response to gradual erosion of source areas. Decreased source-area relief during late Baca time is further supported by the general up-section decrease in grain size and in the sandstone:mudstone ratio in the deposits of both basins. Laramide positive areas persisted during the onset of middle Tertiary volcanism (Cather, 1986), and the distribution of sedimentary facies in the basal volcaniclastic rocks of the Spears Formation is very much like that of the underlying Laramide deposits in most areas.

ACKNOWLEDGMENTS

We benefited greatly from discussions with V. R. Baker, R. M. Chamberlin, C. E. Chapin, R. L. Folk, W. E. Galloway, S. J. Johansen, S. G. Lucas, G. R. Osburn, and A. J. Scott. The manuscript was improved by reviews from E. H. Baltz, R. M. Chamberlin, C. E. Chapin, R. L. Folk, R. W. Foster, W. E. Galloway, and D. O. Snyder. Financial support was provided by the New Mexico Bureau of Mines and Mineral Resources (F. E. Kottlowski, Director) and the Geology Foundation of the University of Texas at Austin.

REFERENCES CITED

Allen, J. R. L., 1964, Studies in fluviatile sedimentation: Six cyclothems from the Lower Old Red Sandstone, Anglo–Welsh Basin: Sedimentology, v. 3, p. 163–198.

———, 1965, The sedimentation and palaeogeography of the Old Red Sandstone of Anglesey, North Wales: Yorkshire Geological Society Proceedings, v. 35, p. 139–185.

Axelsson, V., 1967, The Laiture delta—a study in deltaic morphology and processes: Geografiska Annaler, v. 49, p. 1–127.

Bates, C. C., 1953, Rational theory of delta formation: AAPG Bulletin, v. 37, p. 2119–2162.

Belt, E. S., 1968, Carboniferous continental sedimentation, Atlantic Provinces, Canada, *in* G. deV. Klein, ed., Late Paleozoic and Mesozoic continental sedimentation, northeastern North America: Geological Society of America Special Paper 106, p. 127–176.

Bernard, H. A., C. F. Major, B. S. Parrott, and R. J. Leblanc, 1970, Recent sediments of southeast Texas: University of Texas at Austin, Bureau of Economic Geology Guidebook 11, 120 p.

Blakestad, R. B., 1978, Geology of the Kelly mining district, Socorro County, New Mexico: Master's thesis, University of Colorado, Boulder, 127 p.

Boothroyd, J. C., and G. M., Ashley, 1975, Processes, bar morphology, and sedimentary structures on braided outwash fans, northeastern Gulf of Alaska, *in* A. V. Jopling and B. C. McDonald, eds., Glaciofluvial and glaciolacustrine sedimentation: SEPM Special Publication, n. 23, p. 193–222.

Boothroyd, J. C., and D. Nummedal, 1978, Proglacial braided outwash: a model for humid alluvial-fan deposits, *in*

A. D. Miall, ed., Fluvial sedimentology: Canadian Society of Petroleum Geologists, Memoir 5, p. 641–668.

Bornhorst, T. J., D. P. Jones, W. E. Elston, et al., 1982, New radiometric ages on volcanic rocks from the Mogollon-Datil volcanic field, southwestern New Mexico: Isochron/West, n. 35, p. 13–15.

Bull, W. B., 1972, Recognition of alluvial-fan deposits in the stratigraphic record, *in* J. K. Rigby and W. K. Hamblin, eds., Recognition of ancient sedimentary environments: SEPM Special Publication, n. 16, p. 63–83.

Burke, W. H., G. S. Kenny, J. B. Otto, and R. D. Walker, 1963, Potassium-argon dates, Socorro and Sierra Counties, New Mexico: New Mexico Geological Society 14th Field Conference Guidebook, Socorro Region, p. 224.

Bushnell, H. P., 1955, Mesozoic stratigraphy of south-central New Mexico: New Mexico Geological Society 6th Field Conference Guidebook, p. 81–87.

Callender, J. F., and R. E. Zilinski, Jr., 1976, Kinematics of Tertiary and Quaternary deformation along the eastern edge of the Lucero uplift, central New Mexico, *in* L. A. Woodward and S. A. Northrop, eds., Tectonics and mineral resources of southwestern North America: New Mexico Geological Society Special Publication, n. 6, p. 53–61.

Cant, D. J., and R. G. Walker, 1976, Development of a braided-fluvial facies model for the Devonian Battery Point Sandstone, Quebec: Canadian Journal of Earth Sciences, v. 13, p. 102–119.

Cather, S. M., 1980, Petrology, diagenesis, and genetic stratigraphy of the Eocene Baca Formation, Alamo Navajo Reservation, Socorro County, New Mexico: Master's thesis, University of Texas, Austin, 243 p.

———, 1982, Lacustrine sediments of the Baca Formation (Eocene), western Socorro County, New Mexico: New Mexico Geology, v. 4, p. 1–6.

———, 1986, Volcano-sedimentary evolution and tectonic implications of the upper Eocene–lower Oligocene Datil Group, west-central New Mexico: Ph.D. thesis, University of Texas, Austin.

Cather, S. M., and B. D. Johnson, 1984, Eocene tectonics and depositional setting of west-central New Mexico and eastern Arizona: New Mexico Bureau of Mines and Mineral Resources Circular 192, 33 p.

Chaiffetz, M. S., 1979, Palynological age and paleoecology of the Baca Formation, northwestern Socorro County, central-western New Mexico (abs.): Geological Society of American Abstracts with Programs, v. 11, p. 268.

Chamberlin, R. M., 1981, Uranium potential of the Datil Mountains–Pie Town area, Catron County, New Mexico: New Mexico Bureau of Mines and Mineral Resources Open-File Report 138, 51 p.

———, 1982, Geologic map, cross sections, and map units of the Lemitar Mountains: New Mexico Bureau of Mines and Mineral Resources Open-File Report 169, 3 sheets.

Chapin, C. E., and S. M. Cather, 1981, Eocene tectonics and sedimentation in the Colorado Plateau–Rocky Mountain area, *in* W. R. Dickinson and W. D. Payne, eds., Relations of tectonics to ore deposits in the southern Cordillera: Arizona Geological Society Digest, v. 14, p. 173–198. (Reprinted in *Rocky Mountain Foreland*

Basins and Uplifts, J. D. Lowell, ed., 1983, Rocky Mountain Association of Geologists, p. 33–56.)

Chapin, C. E., R. M. Chamberlin, G. R. Osburn, et al., 1978, Exploration framework of the Socorro geothermal area, New Mexico, *in* C. E. Chapin and W. E. Elston, eds., Field guide to selected cauldrons and mining districts of the Datil–Mogollon volcanic field, New Mexico: New Mexico Geological Society Special Publication 7, p. 115–130.

Church, M., and R. Gilbert, 1973, Proglacial fluvial and lacustrine environments, *in* A. V. Jopling and B. C. McDonald, eds., Glaciofluvial and glaciolacustrine sedimentation: SEPM Special Publication, n. 23, p. 22–100.

Collinson, J. D., 1970, Bedforms in the Tana River, Norway: Geografiska Annaler, v. 52A, p. 31–55.

Coney, P. J., 1971, Cordilleran tectonic transitions and motion of the North American plate: Nature, v. 233, p. 462–465.

———, 1976, Structure, volcanic stratigraphy, and gravity across the Mogollon Plateau, New Mexico, *in* W. E. Elston and S. A. Northrop, eds., Cenozoic volcanism in southwestern New Mexico: New Mexico Geological Society Special Publication, n. 5, p. 29–41.

Costello, W. R., and R. G. Walker, 1972, Pleistocene sedimentology, Credit River, southern Ontario: a new component of the braided river model: Journal of Sedimentary Petrology, v. 42, p. 389–400.

Cotter, E., 1971, Paleoflow characteristics of a Late Cretaceous river in Utah from analysis of sedimentary structures in the Ferron Sandstone: Journal of Sedimentary Petrology, v. 41, p. 129–138.

Dane, C. H., and G. O. Bachman, 1965, Geologic map of New Mexico: U. S. Geological Survey, 2 sheets, scale 1:500,000.

Davis, G. A., 1978, Monocline fold pattern of the Colorado Plateau, *in* V. Matthews III, ed., Laramide folding associated with basement block faulting in the western United States: Geological Society of America Memoir 151, p. 215–233.

———, 1979, Laramide folding and faulting in southeastern Arizona: American Journal of Science, v. 279, p. 543–569.

Davis, G. A., S. R. Schowalter, G. S. Benson, et al., 1982, The Apache uplift, heretofore unrecognized Colorado Plateau uplift (abs.): Geological Society of America Abstracts with Programs, v. 14, p. 472.

Drewes, H. A., 1976, Laramide tectonics from Paradise to Hells Gate, southeastern Arizona: Arizona Geological Society Digest, v. 10, p. 151–167.

———, 1978, The Cordilleran orogenic belt between Nevada and Chihuahua: Geological Society of America Bulletin, v. 89, p. 641–657.

Eardley, A. J., 1962, Structural geology of North America (2nd ed.): New York, Harper and Row, 743 p.

Edmunds, R. J., 1961, Geology of the Nutria monocline, McKinley County, New Mexico: Master's thesis, University of New Mexico, Albuquerque, 100 p.

Eggleston, T. L., 1982, Geology of the central Chupadera Mountains, Socorro County, New Mexico: Master's thesis, New Mexico Institute of Mining and Technology, Socorro, 161 p.

Ethridge, F. G., and S. A. Schumm, 1978, Reconstructing paleochannel morphologic and flow characteristics: Methodology, limitations, and assessment, *in* A. D. Miall, ed., Fluvial sedimentology: Canadian Society of Petroleum Geologists Memoir 5, p. 703–722.

Eugster, H. P., and L. A. Hardie, 1975, Sedimentation in an ancient playa-lake complex: The Wilkins Peak Member of the Green River Formation of Wyoming: Geological Society of America Bulletin, v. 86, p. 319–334.

Eugster, H. P., and R. C. Surdam, 1973, Depositional environment of the Green River Formation of Wyoming: a preliminary report: Geological Society of America Bulletin, v. 84, p. 1115–1120.

Farkas, S. E., 1969, Geology of the southern San Mateo Mountains, Socorro and Sierra Countries, New Mexico: Ph.D. thesis, University of New Mexico, Albuquerque, 137 p.

Fisher, W. L., L. F. Brown, Jr., A. J. Scott, and J. H. McGowen, 1969, Delta systems in the exploration for oil and gas: University of Texas at Austin Bureau of Economic Geology, Research Colloquium, 212 p.

Fisk, H. N., 1944, Geological investigation of the alluvial valley of the lower Mississippi River: Vicksburg, Mississippi River Commission, 78 p.

Fitzsimmons, J. P., 1967, Precambrian rocks of the Zuni Mountains: New Mexico Geological Society 18th Field Conference Guidebook, p. 119–121.

Foster, R. W., 1978, Selected data for deep drill holes along the Rio Grande rift in New Mexico: New Mexico Bureau of Mines and Mineral Resources Circular 163, p. 236–237.

Gilbert, G. K., 1885, The topographic features of lake shores: U.S. Geological Survey Annual Report, n. 5, p. 69–123.

Goddard, E. N., 1966, Geologic map and sections of the Zuni Mountains Fluorspar district, Valencia County, New Mexico: U.S. Geological Survey Miscellaneous Investigations Map I-454, scale 1:31,680.

Gole, C. V., and S. V. Chitale, 1966, Inland delta building activity of the Kosi River: Proceedings of the American Society of Civil Engineers, Journal of Hydraulics Division, HY-2, p. 111–126.

Gorham, T. W., and R. V. Ingersoll, 1979, Evolution of the Eocene Galisteo basin, north-central New Mexico: New Mexico Geological Society 30th Field Conference Guidebook, p. 219–224.

Guilinger, D. R., 1982, Geology and uranium potential of the Tejana Mesa–Hubbell Draw area, Catron County, New Mexico: Master's thesis, New Mexico Institute of Mining and Technology, Socorro, 129 p.

Gustavson, T. C., 1974, Sedimentation on gravel outwash fans, Malaspina Glacier Foreland, Alaska: Journal of Sedimentary Petrology, v. 44, p. 374–389.

Hantzchel, W., 1975, Treatise on invertebrate paleontology, Part W, Trace fossils and problematica (2nd ed.): University of Kansas, Lawrence, 269 p.

Hedlund, D. L., 1974, Age and structural setting of base-metal mineralization in the Hillsboro–San Lorenzo area, southwestern New Mexico (abs.): New Mexico Geological Society 25th Field Conference Guidebook, p. 378–379.

Hein, F. J., and R. G. Walker, 1977, Bar evolution and

development of stratification in the gravelly, braided, Kicking Horse River: Canadian Journal of Earth Sciences, v. 14, p. 562–570.

Hooke, R. L., 1967, Processes on arid region alluvial fans: Journal of Geology, v. 75, p. 453–456.

Jackson, R. G., II, 1976, Depositional model of point bars in the lower Wabash River: Journal of Sedimentary Petrology, v. 46, p. 579–594.

Jacobs, R. C., 1957, Geology of the central front of the Fra Cristobal Mountains, New Mexico: New Mexico Geological Society 8th Field Conference Guidebook, p. 256–257.

Jahns, R. H., D. K. McMillan, J. D. O'Brient, and D. L. Fisher, 1978, Geologic section in the Sierra Cuchillo and flanking areas, Sierra and Socorro Counties, New Mexico: *in* C. E. Chapin and W. E. Elston, eds., Field guide to selected cauldrons and mining districts of the Datil-Mogollon volcanic field, New Mexico Geological Society Special Publication, n. 7, p. 130–138.

Johansen, S., 1983, Abrupt changes in the Gallup Sandstone (Cretaceous) across the Puertecito fault system in northwest Socorro County: New Mexico Geological Society 34th Field Conference Guidebook, Socorro Region II, p. 45–46.

Johnson, B. D., 1978, Genetic stratigraphy and provenance of the Baca Formation New Mexico, and the Eagar Formation, Arizona: Master's thesis, University of Texas, Austin, 150 p.

Karcz, I., 1972, Sedimentary structures formed in flash floods in southern Israel: Sedimentary Geology, v. 7, p. 161–182.

Keith, S. B., 1982, Evidence for late Laramide southwest-vergent underthrusting in southeast California, southern Arizona, and northeast Sonora (abs.): Geological Society of America Abstracts with Programs, v. 14, n. 4, p. 177.

Keith, S. B., and L. F. Barrett, 1976, Tectonics of the central Dragoon Mountains: a new look: Arizona Geological Society Digest, v. 10, p. 169–204.

Kelley, V. C., 1955, Regional tectonics of the Colorado Plateau and relationships to the origin and distribution of uranium: University of New Mexico Publications in Geology, n. 5, 120 p.

———, 1967, Tectonics of the Zuni–Defiance region, New Mexico and Arizona: New Mexico Geological Society 18th Field Conference Guidebook, p. 28–31.

———, 1971, Geology of the Pecos country, southeastern New Mexico: New Mexico Bureau of Mines and Mineral Resources Memoir, n. 24, 75 p.

———, 1972, New Mexico lineament of the Colorado Rockies front: Geological Society of America Bulletin, v. 83, p. 1849–1852.

———, 1977, Geology of the Albuquerque basin, New Mexico: New Mexico Bureau of Mines and Mineral Resources Memoir 33, 60 p.

———, 1982, Diverse geology of the Hubbell bench, Albuquerque basin, New Mexico: New Mexico Geological Society 33rd Field Conference Guidebook, p. 159–160.

Kelley, V. C., and N. J. Clinton, 1960, Fracture systems and tectonic elements of the Colorado Plateau: University of New Mexico Publications in Geology, n. 4, 104 p.

Kelley, V. C., and J. T. McCleary, 1960, Laramide orogeny in south-central New Mexico, AAPG Bulletin, v. 44, p. 1419–1420.

Kelley, V. C., and S. A. Northrop, 1975, Geology of the Sandia Mountains and vicinity: New Mexico Bureau of Mines and Mineral Resources Memoir, n. 29, 136 p.

Kelley, V. C., and C. Silver, 1952, Geology of the Caballos Mountains: University of New Mexico Publications in Geology, n. 5, 286 p.

King, P. B., 1969, Tectonic map of North America: U.S. Geological Survey, scale 1:5,000,000, 2 sheets.

Kottlowski, F. E., R. H. Weber, and M. E. Willard, 1969, Tertiary intrusive-volcanic episodes in the New Mexico region (abs.): Geological Society of America Abstracts with Programs, part 7, p. 278–280.

Krewedl, D. A., 1974, Geology of the central Magdalena Mountains, Socorro County, New Mexico: Ph.D. thesis, University of Arizona, Tucson, 128 p.

Langbein, W. B., 1961, Salinity and hydrology of closed lakes: U.S. Geological Survey Professional Paper 412, 50 p.

Leopold, L. B., and J. P. Miller, 1956, Ephemeral streams— hydraulic factors and their relation to the drainage net: U.S. Geological Survey Professional Paper 282-A, 37 p.

Leopold, L. B., and M. G. Wolman, 1957, River channel patterns: braided, meandering, and straight: U.S. Geological Survey Professional Paper 282-B, 85 p.

Leopold, L. B., M. G. Wolman, and J. P. Miller, 1964, Fluvial processes in geomorphology: San Francisco, W. H. Freeman and Co., 522 p.

Love, D. W., 1977, Dynamics of sedimentation and geomorphic history of Chaco Canyon National Monument, New Mexico: New Mexico Geological Society 28th Field Conference Guidebook, San Juan Basin III, p. 291–298.

Lovejoy, E. M. P., 1976, Cedar Pocket Canyon (Grand Wash)–Shebit–Garlock fault complex, northwestern Arizona and southwestern Utah: reinterpretation of the time of faulting: Arizona Geological Society Digest, v. 10, p. 133–150.

Lowell, J. D., 1974, Regional characteristics of porphyry copper deposits of the southwest: Economic Geology, v. 69, p. 601–617.

Lucas, S. G., R. M. Schoch, E. Manning, and C. Tsentas, 1981, The Eocene biostratigraphy of New Mexico: Geological Society of America Bulletin, v. 92, p. 851–867.

Martini, I. P., 1977, Gravelly flood deposits of Irvine Creek, Ontario, Canada: Sedimentology, v. 24, p. 603–622.

Massingill, G. L., 1979, Geology of the Riley–Puertecito area, southeastern margin of the Colorado Plateau, Socorro County, New Mexico: D. G. S. thesis, University of Texas, El Paso, 301 p.

Master, C. D., 1967, Use of sedimentary structures in determination of depositional environments, Mesaverde Formation, Williams Fork Mountains, Colorado: AAPG Bulletin, v. 51, p. 2033–2043.

McGowen, J. H., 1970, Gum Hollow fan delta, Nueces Bay, Texas: University of Texas at Austin Bureau of Economic Geology Report of Investigations 69, 91 p.

McGowen, J. H., and L. E. Garner, 1970, Physiographic features and stratigraphic types of coarse-grained point bars: Sedimentology, v. 14, p. 77–111.

McGowen, J. H., and C. G. Groat, 1971, Van Horn Sandstone, west Texas: an alluvial fan model for mineral exploration: University of Texas at Austin Bureau of Economic Geology Report of Investigations 72, 57 p.

McKee, E. D., E. J. Crosby, and H. L. Berryhill Jr., 1967, Flood deposits, Bijou Creek, Colorado, June 1965: Journal of Sedimentary Petrology, v. 37, p. 829–851.

Miall, A. D., 1977, A review of the braided-river depositional environment: Earth Sciences Review, v. 13, p. 1–62.

———, 1978, Lithofacies types and vertical profile models in braided river deposits: a summary, *in* A. D. Miall, ed., Fluvial sedimentology: Canadian Society of Petroleum Geologists Memoir, n. 5, p. 597–604.

Moody-Stuart, M., 1966, High- and low-sinuosity stream deposits, with examples from the Devonian of Spitsbergen: Journal of Sedimentary Petrology, v. 36, p. 1102–1117.

Mothersill, J. S., 1975, Lake Chad: geochemistry and sedimentary aspects of a shallow polymictic lake: Journal of Sedimentary Petrology, v. 45, p. 295–309.

Nummedal, D., and J. C. Boothroyd, 1976, Morphology and hydrodynamic characteristics of terrestrial fan environments: Coastal Research Division, Department of Geology, University of South Carolina Technical Report, n. 10-CRD, 61 p.

Peirce, H. W., P. E. Damon, and M. Shafiqullah, 1979, An Oligocene(?) Colorado Plateau edge in Arizona: Tectonophysics, v. 61, p. 1–24.

Potter, S. C., 1970, Geology of Baca Canyon, Socorro County, New Mexico: Master's thesis, University of Arizona, Tucson, 54 p.

Puigdefabregas, C., and A. van Vliet, 1978, Meandering stream deposits from the Tertiary of the southern Pyrenees, *in* A. D. Miall, ed., Fluvial sedimentology: Canadian Society of Petroleum Geologists Memoir, n. 5, p. 469–486.

Rehrig, W. A., and T. L. Heidrick, 1972, Regional fracturing of Laramide stocks of Arizona and its relationship to porphyry copper mineralization: Economic Geology, v. 67, p. 198–213.

———, 1976, Regional tectonic stress during the Laramide and late Tertiary intrusive periods, Basin and Range Province, Arizona: Arizona Geological Society Digest, v. 10, p. 205–228.

Robinson, B. R., 1981, Geology of the D-Cross Mountain Quadrangle, Socorro and Catron Counties, New Mexico: Ph.D. thesis, University of Texas, El Paso, 213 p.

Rust, B. R., 1972, Structure and process in a braided river: Sedimentology, v. 18, p. 221–245.

———, 1978, Depositional models for braided alluvium, *in* A. D. Miall, ed., Fluvial sedimentology: Canadian Society of Petroleum Geologists Memoir, n. 5, p. 605–625.

Schumm, S. A., and R. F. Hadley, 1957, Arroyos and the semiarid cycle of erosion: American Journal of Science, v. 255, p. 161–174.

Schumm, S. A., and H. R. Kahn, 1972, Experimental study of channel patterns: Geological Society of America

Bulletin, v. 83, p. 1755–1770.

Seager, W. R., 1975, Cenozoic tectonic evolution of the Las Cruces area: New Mexico Geological Society 26th Field Conference Guidebook, p. 241–250.

———, 1981, Geology of the Organ Mountains and southern San Andres Mountains, New Mexico: New Mexico Bureau of Mines and Mineral Resources Memoir, n. 36, 97 p.

Shepard, R. G., 1978, Distinction of aggradational and degradational fluvial regimes in valley-fill alluvium, Tapia Canyon, New Mexico, *in* A. D. Miall, ed., Fluvial sedimentology: Canadian Society of Petroleum Geologists Memoir, n. 5, p. 277–286.

Smith, C. T., 1983, Structural problems along the east side of the Socorro constriction, Rio Grande rift: New Mexico Geological Society 34th Field Conference Guidebook, Socorro Region II, p. 103–109.

Smith, N. D., 1970, The braided stream depositional environment: Comparison of the Platte River and some Silurian clastic rocks, north-central Appalachians: Geological Society of America Bulletin, v. 81, p. 2993–3014.

———, 1971, Pseudo-planar stratification produced by very low amplitude sand waves: Journal of Sedimentary Petrology, v. 41, p. 69–73.

———, 1975, Sedimentary environments and late Quaternary history of a low-energy mountain delta: Canadian Journal of Earth Sciences, v. 12, p. 2004–2013.

Snyder, D. O., 1971, Stratigraphic analysis of the Baca Formation, west-central New Mexico: Ph.D. thesis, University of New Mexico, Albuquerque, 158 p.

Stearns, C. E., 1962, Geology of the north half of the Pelona quadrangle, Catron County, New Mexico: New Mexico Bureau of Mines and Mineral Resources Bulletin, v. 78, 46 p.

Steel, R. J., 1974, New Red Sandstone floodplain and piedmont sedimentation in the Hebridean Province, Scotland: Journal of Sedimentary Petrology, v. 44, p. 336–357.

Surdam, R. C., and C. A. Wolfbauer, 1975, The Green River Formation: a playa-lake complex: Geological Society of America Bulletin, v. 86, p. 335–345.

Theakstone, W. H., 1976, Glacial lake sedimentation, Austerdalsisen, Norway: Sedimentology, v. 23, p. 671–688.

Visher, G. S., 1972, Physical characteristics of fluvial deposits, *in* J. K. Rigby and W. K. Hamblin, eds., Recognition of ancient sedimentary environments: SEPM Special Publication, n. 16, p. 84–97.

Walker, T. R., 1967, Formation of red beds in modern and ancient deserts: Geological Society of America Bulletin, v. 78, p. 353–368.

Wengerd, S. A., 1959, Regional geology of the Lucero region, west-central New Mexico: New Mexico Geological Society 10th Field Conference Guidebook, West-Central New Mexico, p. 121–134.

Williams, G. E., 1971, Flood deposits of the sand-bed ephemeral streams of central Australia: Sedimentology, v. 17, p. 1–40.

Williams, P. F., and B. R. Rust, 1969, The sedimentology of a braided river: Journal of Sedimentary Petrology, v. 39,

p. 649–679.

Wilpolt, R. H., and A. A. Wanek, 1951, Geology of the region from Socorro to San Antonio east to Chupadera Mesa, Socorro County, New Mexico: U.S. Geological Survey Oil and Gas Investigation Map OM-121, scale 1:62,500.

Wilson, E. D., R. T. Moore, and J. R. Cooper, 1969, Geologic map of Arizona: U.S. Geological Survey, scale 1:500,000.

Paleozoic Paleotectonics and Sedimentation in Arizona and New Mexico

C. A. Ross[1]
Gulf Oil Exploration and Production Company
Houston, Texas

J. R. P. Ross
Western Washington University
Bellingham, Washington

During Paleozoic time, the paleotectonic and sedimentation patterns of the southern Ancestral Rocky Mountains extended southward through New Mexico and Arizona into adjacent parts of Chihuahua and Sonora. During Cambrian–Middle Devonian time, this was a particularly stable part of the North American craton. Slow deposition of shelf clastics and dolomitic carbonates was interrupted by several long erosional hiatuses. Major tectonic features first appeared in Late Devonian time, and at least one depositional basin formed west of the Defiance–Zuni uplift. Thin Lower Mississippian shelf carbonates and evaporites covered nearly the entire region.

The most significant tectonic activities started in late Chesterian time and extended with increasing magnitude until the end of Wolfcampian time. Local basins and uplifts date from this interval and occurred in two belts. One belt was about 80 mi (130 km) wide along the western sides of the Diablo and Pedernal uplifts and along both sides of the Uncompahgre uplift. Another belt extended northwest from the Pedregosa basin into southeastern Arizona. Major tectonic events initiated the Morrowan, Atokan, and Missourian epochs and occurred twice in the Wolfcampian. Leonardian, Guadalupian, and Ochoan epochs were times of tectonic stability. During the Leonardian, sediments from the Uncompahgre uplift gradually covered all the other uplifted blocks. The timing of these Paleozoic tectonic events suggests a cause and effect relationship with plate tectonic histories that brought North America and northern Europe together in Late Devonian time (Acadian orogeny) and Euramerica and northwestern Gondwana together in Late Mississippian through Early Permian time (Appalachian orogeny).

INTRODUCTION

During late Paleozoic time, the major tectonic patterns of the Ancestral Rocky Mountains extended southward from the Sierra Grande and Uncompahgre uplifts through central and western New Mexico, eastern Arizona, and into northern Chihuahua and Sonora in Mexico (Figure 1). In sharp contrast, during early Paleozoic time this region was a stable low-lying shelf having relatively little relief. It was underlain by a complex Precambrian basement that formed a southwestward extension of the North American cratonic shield. Beginning in Late Devonian time, a few high-angle faults began to break across parts of this stable shelf to form uplifts and basins. By Late Mississippian (Chesterian) time many additional uplifts and basins appeared, and these became the features that were further displaced during the Pennsylvanian and early parts of the Early Permian (Wolfcampian). Leonardian and Late Permian were again times of tectonic stability.

Paleozoic tectonic and sedimentary patterns have been obscured by post-Paleozoic structures in many parts of this region mainly as a result of Laramide and younger Cenozoic tectonic events. These later events reactivated movement on many Paleozoic structures, but they did not affect many others. These variations reflect differences between the directions of forces and stress of the post-Paleozoic structural events and those of the Paleozoic. In this paper, we include in the region (Figure 1) the area south of the Paradox basin, west of the Sierra Grande–Pedernal–Hueco uplifts, and those parts of the southern structurally distributed belt south of the Colorado Plateau for which we have information. We do not include adjacent and related Paleozoic uplifts and basins in the panhandle of Texas and in Oklahoma, such as the Amarillo, Wichita, and Arbuckle mountains or the western Texas basins and platforms, although these are important parts of the larger tectonic

---

[1]Present Address: Chevron U.S.A., Inc., Houston, TX.

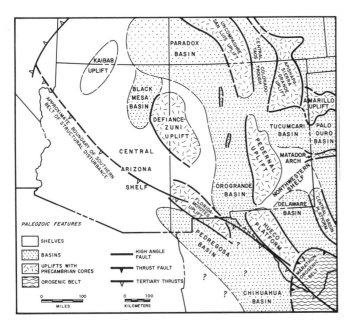

Figure 1—Major late Paleozoic structural features in the southern part of the Ancestral Rocky Mountain region. Compiled from Wilson (1958), Kottlowski (1960, 1962, 1963b), Baars (1962, 1973), Lopez-Ramos (1969), Greenwood et al. (1970), Foster et al. (1972), Mallory et al. (1972), Ross (1973), Woodward (1973), Tovar-Rodriquez and Valencia (1974), Chronic (1979), Blakey (1980), Casey (1980), Peterson (1980), Rawson and Turner-Peterson (1980), Kluth and Coney (1981), Kashfi (1983), and Baars and Stevenson (1984).

picture of the late Paleozoic.

Our understanding of the Paleozoic geology of northern Mexico is incomplete. Paleozoic rocks are exposed in scattered mountain ranges in northern Chihuahua and Sonora where they have been caught up in late Mesozoic–early Cenozoic tectonic features (Diaz and Navarro, 1964; Lopez-Ramos, 1969). These rocks suggest that Paleozoic cratonic shelf facies extended as far south as the latitude of the city of Chihuahua (Bridges et al., 1964; Wilson et al., 1969; Tovar-Rodriquez and Valencia, 1974; Thompson et al., 1978). Most of the southwestern end of the North American Paleozoic craton is obfuscated by Mesozoic geosynclinal sediments, Cenozoic volcanics, and post-Paleozoic complex, transform, and transcurrent faulting (Anderson and Schmidt, 1983).

GENERAL STRATIGRAPHY

The stratigraphic succession in the region is outlined in Figure 2. The region formed part of the tectonically stable North American craton throughout early Paleozoic time. Those lower Paleozoic sediments that are preserved are generally thin and are mainly eroded remnants of what appears to have been more extensive, thin, blanket-like deposits. They were deposited on an erosional surface that generally had low relief of only a few tens of feet (2–4 m), although relief of several hundred feet (30–60 m) is known locally (Hayes, 1975). The Precambrian rocks are diverse

and schists and gneisses are the most widespread. In central Arizona, southern New Mexico, and western Texas, slightly younger Precambrian sedimentary rocks are also preserved beneath this unconformity. As pointed out by Hayes (1975), only the relatively large area of Precambrian Apache Group and Troy Quartzite sedimentary rocks in central Arizona had much effect on Cambrian sedimentary patterns. There the Cambrian Bolsa Quartzite was in part derived from these older rocks. Our present limited knowledge of the distribution, types of rocks, and structural relationships of these Precambrian rocks does not permit us to recognize any direct relationship between features in this old cratonic basement and those in late Paleozoic and younger tectonic patterns.

Cambrian and Lower Ordovician

Cambrian and Lower Ordovician rocks (Figure 2) formed a major depositional succession across southern Arizona, New Mexico, and western Texas (Figure 3) (Hayes, 1975). The lithologies are time transgressive from west to east. The basal sandstone is possibly as old as late Early Cambrian in central southern Arizona and as young as latest Late Cambrian or earliest Early Ordovician in the Franklin Mountains near El Paso. Similar Lower Cambrian basal sandstone units are known from southern California and northwestern Sonora (Caborca).

Two facies make up this succession. At the base, a trangressive quartzose sandstone ranges from 0 to 750 ft (0 to 230 m) thick. This is generally overlain by a dolomitic carbonate that can reach 1000 ft (300 m) in thickness and within which a number of members and formations have been recognized (Hayes, 1975). In central southern Arizona the succession is divided into the Bolsa Quartzite and the Abrigo Formation. In southeastern Arizona and part of southwestern New Mexico the names Coronado Sandstone and El Paso Limestone are applied to a younger part of the transgressive succession. In southern New Mexico and western Texas, the names Bliss Sandstone and El Paso Group are applied to a yet younger part of this trangressive succession.

Minor unconformities occur within the Cambrian–Lower Ordovician transgressive succession. The top of the El Paso, however, is marked by a major regional unconformity that represents nearly all of Middle Ordovician time (about 15 m.y.). Erosional and solution features are common at this unconformity.

Upper Ordovician

The Upper Ordovician Montoya Formation overlies a Middle Ordovician unconformity in most of the southern part of the region (Figure 4). It is as much as 500 ft (165 m) thick locally, but is usually about 300 ft (100 m) thick or less. The Montoya has a thin basal sandstone member, the Cable Canyon Sandstone, and three additional carbonate members (Kottlowski, 1963b; Hayes, 1975). The Montoya represents all of Late Ordovician time in nearly continuous deposition. The top of the Montoya is usually shown as an unconformity of short duration, and deposition may have been locally continuous into the overlying Fusselman Dolomite.

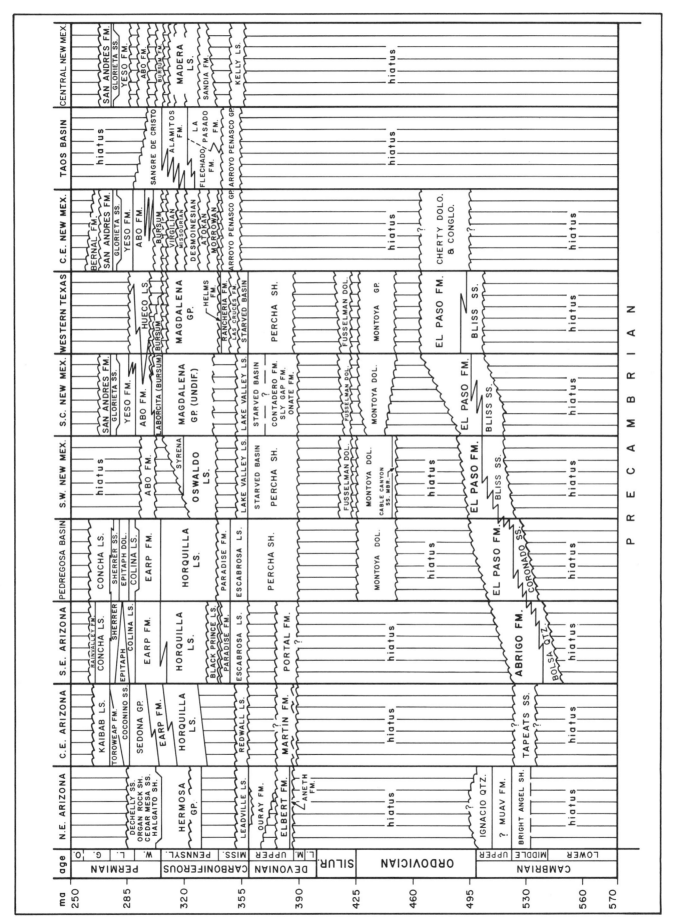

Figure 2—Stratigraphic successions in Arizona and New Mexico. Compiled from many sources; see references cited for Figures 3 to 10.

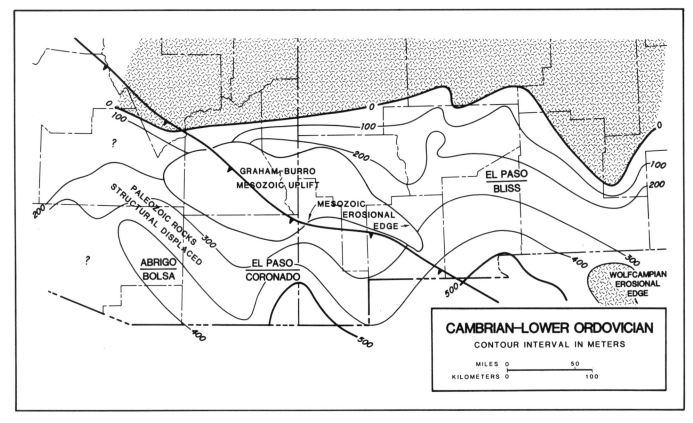

Figure 3—Distribution of Cambrian–Lower Ordovician strata in southern Arizona and southern New Mexico. After Hayes (1975).

Silurian

The Silurian Fusselman Dolomite (Figure 5) is a wedge of strata that extends into the southern part of New Mexico. Just east of the region in the Carlsbad area of New Mexico, it thickens to as much as 1000 ft (300 m). The Fusselman is more than 1000 ft (300 m) thick near the Mexico–New Mexico border and thins to an erosional edge about 80–100 mi (130–160 km) to the north. It also is removed by pre–Late Devonian erosion in westernmost New Mexico. This post–Fusselman unconformity is one of the longest hiatuses (about 30 m.y.) in this region of the cratonic shelf.

Upper Devonian

The Upper Devonian Percha Shale (Figure 6) forms a transgressive succession that originally covered much of the southern Ancestral Rocky Mountain region (Kottlowski, 1963b). Toward the margins of this cratonic shelf, such as in the Sacramento Mountains, slightly older, late Middle Devonian (Givetian) strata are known above the post-Fusselman unconformity. This marine transgression moved across a very broad, low-lying, nearly planar, regional erosional surface with only local relief. The resulting inland sea had the same general features of restricted (poorly oxygenated) water as several other time equivalent units, such as the Exshaw Shale in Alberta, the New Albany Shale in the Illinois Basin, and the Chattanooga Shale in the southern part of the Appalachian Basin.

In central Arizona (Huddle and Dobrovolny, 1952; Beus, 1980), these dark shales pass westward and northwestward into the Martin Limestone and then into the Temple Butte Dolomite in the Grand Canyon area. In the Four-Corners area, this succession is divided into the Aneth, McCracken, Elbert, and Ouray formations. These mainly carbonate units were apparently deposited on shelf areas in normal marine water. Local relief included the initial appearance of the Defiance uplift in Late Devonian time, which was a local source of clastics for the McCracken Sandstone (Kashfi, 1983) which spread as a sheet over the dark Aneth Shale. The timing of this uplift suggests a relationship to the beginnings of the Roberts Mountains orogeny farther west.

Mississippian

Mississippian carbonates (Figure 7) overlie Upper Devonian beds with little apparent disconformity throughout southern Arizona and southern New Mexico (Armstrong, 1963; Armstrong et al., 1979, 1980). Lower and Middle Mississippian (Kinderhookian, Osagean, and lower Meramecian = Tournaisian and lower Visean) strata form some of the most widespread beds in the region. To the south these include the Escabrosa Limestone Group; Lake Valley Limestone (undifferentiated); Caballero Formation, Lake Valley Limestone (restricted), and Rancheria and Las Cruces formations (the latter two being basinal facies) (Kottlowski, 1963b). To the north (Peirce, 1979; Kent and Rawson, 1980), the Redwall Limestone Group includes the

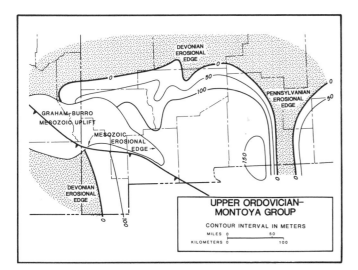

Figure 4—Distribution of Upper Ordovician strata (Montoya Group) in southern New Mexico. After Kottlowski (1963a) and Hayes (1975).

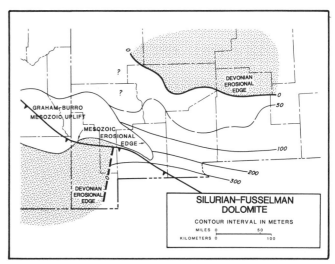

Figure 5—Distribution of Silurian strata (Fusselman Dolomite) in southern New Mexico. After Kottlowski (1963a). Used with permission of the New Mexico Geological Society.

Espiritu Santo Formation, Tererro Formation, Caloso Formation, Kelly Limestone, and Leadville Limestone. These units include a number of widespread unconformities that appear to be related to fluctuations in sea level during Mississippian time. Within these Mississippian beds, a widespread collapsed breccia of limestones marks an early Visean evaporite interval that was dissolved during Chesterian (Late Mississippian) or Early Pennsylvanian time. Along the southern margin of this extensive carbonate shelf, deeper water dark shales formed a thin sequence of starved basin deposits (Lane and DeKeyser, 1980).

Upper Meramecian and Chesterian rocks are less widely distributed (Armstrong et al., 1979) and were deposited mainly in two areas (Figures 2 and 7): a southeastern extension of the Paradox basin that reached into northwestern New Mexico and northeastern Arizona and an east–west belt across southern New Mexico and Arizona. A connection between these depositional areas may have existed between the Pedernal and Zuni highlands and around the western side of the Defiance–Zuni highlands. Three lithologic facies are common in these beds. In the northern part of the region, an upper Meramecian sandy and shaly, pelletoid limewackestone (Tererro Formation, Hachita Formation, and lower part of the Paradise Formation) is overlain by the Chesterian Log Springs Formation, an argillaceous and arenaceous wackestone and packstone that contains some red beds. In the southern part of the region, Chesterian facies include oolitic limestone and sandstone (Helms and Paradise formations).

In general, these higher Mississippian stratigraphic units offlap the Defiance–Zuni, Sierra Grande, and Pedernal highlands. They also form a broad belt across central and northern New Mexico and east-central Arizona (Figure 7). On the exposed shelves, Tournaisian and early Visean carbonates and evaporites were removed by late Mississippian erosion prior to late Atokan and early Desmoinesian marine transgressions that lapped on to these areas again. The Mississippian has fewer depositional

hiatuses and is more completely represented in the Pedregosa basin in southeastern Arizona and southwestern New Mexico than elsewhere in the region.

Pennsylvanian

Pennsylvanian strata (Figure 8) are widely distributed in the region (Kottlowski, 1960, 1962; Lopez-Ramos, 1969; Foster et al., 1972; Ross, 1973; Thompson et al., 1978; Armstrong et al., 1979; Pierce, 1979; Casey, 1980). Unlike the Upper Devonian and Lower Mississippian deposits, which were fairly uniform with gradual changes in facies and thickness, the Pennsylvanian deposits are characterized by abrupt facies and thickness changes (Figure 7). In central New Mexico, a series of four small basins and two small uplifts formed within a nearly north–south belt that extended from the southeastern end of the Paradox basin to the Orogrande basin. These geographically small basins and uplifts apparently had several hundred meters of topographic relief causing the carbonate shelf edge deposits to pass abruptly into dark, deep-water shale facies. During much of Pennsylvanian time, the central parts of many of these basins were starved of sediment and most had sills that prevented the circulation of oxygenated water in their deeper parts. Movement along these structural features was irregular and not uniform; for example, the Joyita Hills uplift appears to have formed near the end of Desmoinesian time and was reactivated in the middle part of Wolfcampian time (Kottlowski and Stewart, 1970). Several of these basins accumulated more than 800 m of Pennsylvanian sediments and still maintained significant topographic relief, as indicated by their abrupt facies changes.

The eastern boundary of this central belt of deformation was marked by the locally irregular, western fault escarpments of the Uncompahgre and Pedernal uplifts. This line of fractures extended farther south into western Texas along the western margin of the Diablo platform where it is now largely obscured by the southern structurally disturbed

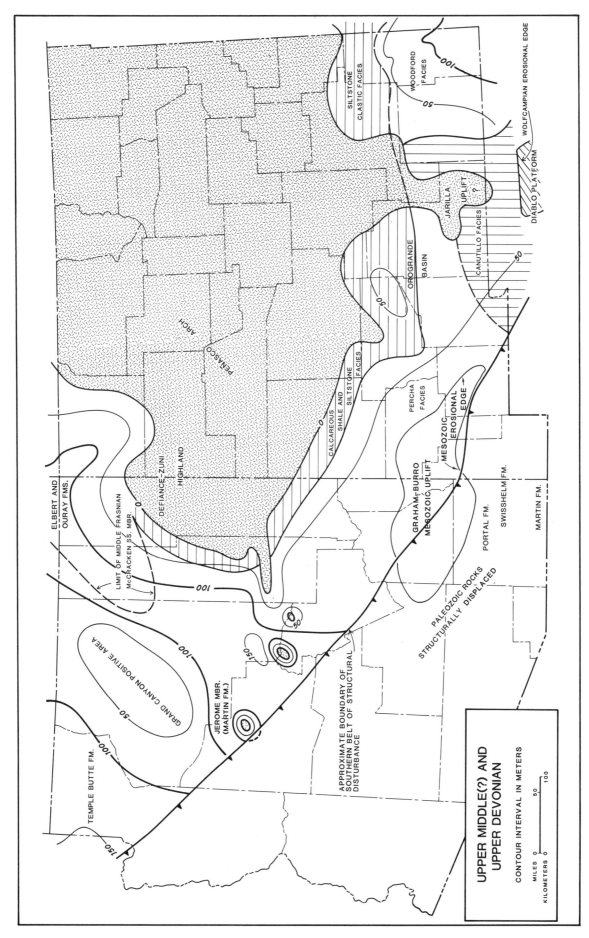

Figure 6—Distribution of Upper Middle(?) and Upper Devonian strata in Arizona and New Mexico. Thicknesses for this interval are not shown southwest of the disturbed belt. Compiled from Huddle and Dobrovolny (1952), Sabins (1957a, b), Kottlowski (1963b), Zeller (1965, 1970), Foster et al. (1972), Beus (1980), and Kashfi (1983).

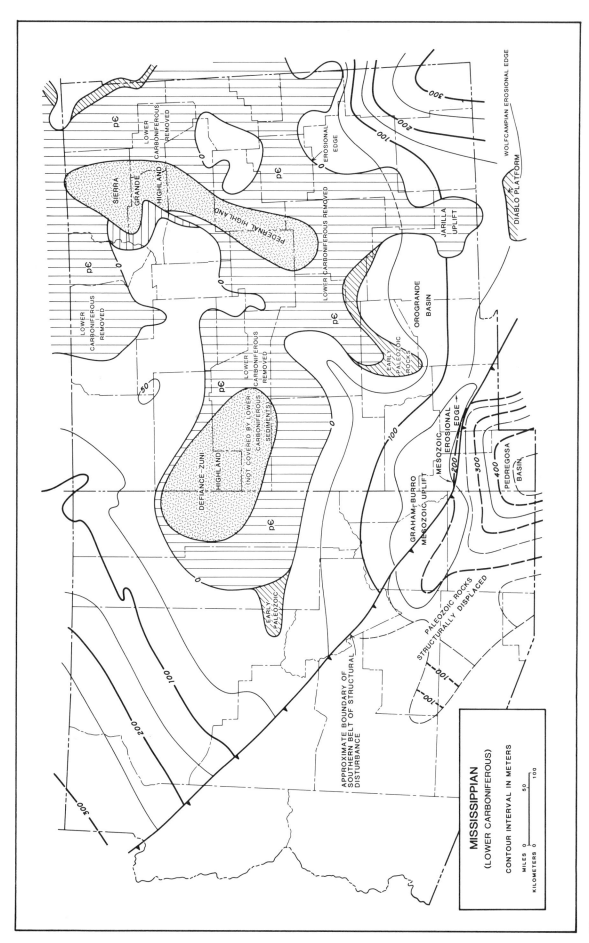

Figure 7—Distribution of Mississippian strata in Arizona and New Mexico. Compiled from Huddle and Dobrovolny (1952), Sabins (1957a, b), Armstrong (1963), Kottlowski (1963b), Zeller (1965, 1970), Foster et al. (1972), Armstrong et al. (1979, 1980), Chronic (1979), Peirce (1979), Kent and Rawson (1980), and Lane and DeKeyser (1980).

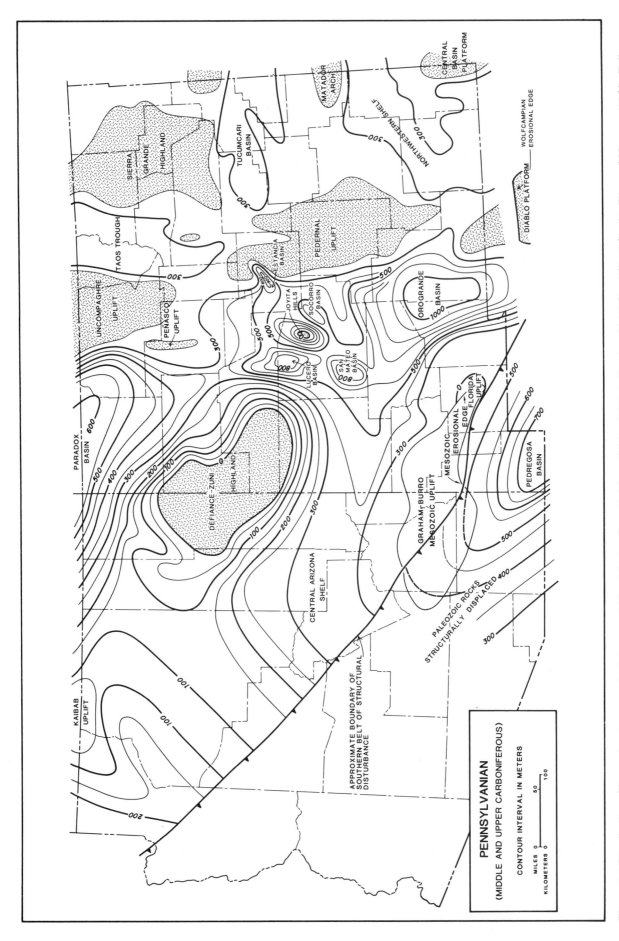

Figure 8—Distribution of Pennsylvanian strata in Arizona and New Mexico. Compiled from Sabins (1957a), Kottlowski (1960, 1962, 1963a, b), Diaz and Navarro (1964), Zeller (1965, 1970), Lopez-Ramos (1969), Wilson et al. (1969), Greenwood et al. (1970, 1977), Foster et al. (1972), Mallory (1972), Ross (1973), Woodward (1973), McKee and Crosby (1975), Thompson et al. (1978), Armstrong et al. (1979), Chronic (1979), Peirce (1979), Blakey (1980), Casey (1980), and Kues (1984).

belt and Cenozoic volcanics. Farther east, the Sierra Grande highland, Amarillo Mountains, Central Basin platform, and Matador arch were active features at a number of times during Pennsylvanian time. Although the amount of displacement and even the direction of movement on the faults that bounded these uplifts varied considerably, the actual timing of the movements on many of these features is commonly synchronous.

To the west in southeastern Arizona, generally similar relationships existed (Ross, 1973). In the area north of the southern structurally disturbed belt, Pennsylvanian marine deposits gradually lapped on to the nearly stable flanks of the Defiance–Zuni highland in a series of transgressions. Late in Pennsylvanian time, the supply of red clastic debris from the Defiance–Zuni highlands (and from the Uncompahgre uplift farther north) increased markedly and gradually displaced the shoreline southward. Within the southern disturbed belt of southeastern Arizona, local changes in thickness and facies in Pennsylvanian beds suggest that small basins and uplifts existed, each of which also had local histories of irregular vertical movement. Although the amount of relief at this time was less than in the central New Mexico belt of deformation, the timing on many of these uplifts and depressions is similar. During late Pennsylvanian time, the broad outlines of the Pedregosa basin were well established, although deposition was not particularly thick. A reefal carbonate shelf formed a prominent facies on the northeastern side of the basin and perhaps stood as much as several hundred feet (50–100 m) above a semistarved depositional basin. Thompson (1980) used the term *Pedregosa basin* for the basin of subsidence that includes these shelf margins and the term *Alamo Hueco basin* for the marine, deeper water basin between the shelves.

Lower Permian

In the Wolfcampian part of Early Permian time (Figure 9), marine deposition was progressively eliminated from most of the region (Baars, 1962; Kottlowski, 1963b; Ross, 1973; Blakey, 1980; Casey, 1980; Peterson, 1980). In the north, the Paradox basin was filled with red beds from the reactivated Uncompahgre uplift. To the east, between the Uncompahgre uplift and the Sierra Grande highland, the Taos trough received more than 1000 m of red clastic deposits that spilled into the Tucumcari basin filling it with fluvial red beds. The Pedernal uplift was also reactivated, but it soon became almost completely covered by nonmarine clastics. The Defiance–Zuni positive area was buried by eolian and beach sandstone deposits and fluvial red beds, and only a few earliest Wolfcampian marine beds extended northward into east central Arizona.

After middle Wolfcampian time, normal marine deposition in the region was largely confined to the Pedregosa basin and its margins, to the margins of the Diablo platform, to the southern part of the Orogrande basin, and to a complex of basins parallel to the western edge of the Diablo platform including the Marfa basin in western Texas. The Alamo Hueco part of the Pedregosa basin during Wolfcampian time (Thompson et al., 1978) became a deeper, much narrower, and more elongate feature than its Pennsylvanian predecessor because of growth of shelf areas into the basin. The eastern flank of this basin

developed thick reefal carbonate deposits along the bounding structure, and the deeper parts of the basin received black shale and sandstone sediments in an euxinic closed basin. Where they have been dated, the major movements on these structural features along the southern margin of the southern Rocky Mountain region are middle Wolfcampian (i.e., post-Bursum and post-Neal Ranch), which suggests a close relationship in timing to the last major thrusting events (middle Wolfcampian) in the Marathon orogenic belt.

Leonardian (upper Lower Permian) and higher Permian strata in most of the region are more difficult to correlate, because they are generally nonmarine (mainly eolian and fluvial) clastics that lack abundant diagnostic fossils or laterally traceable marker beds. Only in and around the Delaware basin, in southeasternmost Arizona, and in northwestern Arizona do normal marine beds intertongue into the edges of these extensive clastic facies. These facies changes are commonly associated with evaporites and/or beach sandstones (Baars, 1962). Facies and thickness changes in rocks assigned to the Leonardian (Figure 10) are gradual and lack the abrupt transitions seen in Pennsylvanian and Wolfcampian strata. These features suggest a lack of major tectonic activity in the region, but with continued erosion of existing uplifts during Leonardian time. Clastic sediments mainly from the Uncompahgre uplift continued to blanket nearly the entire region and finally buried the Pedernal uplift. Rocks of this age have been eroded adjacent to the southern disturbed belt. Within the disturbed belt, however, marginal marine and nonmarine deposits and evaporites are distributed around the shelf margins of the Pedregosa basin. Also, within the Pedregosa basin, shallow marine limestones of early Leonardian age suggest a continuation of a persistent marine embayment that lay just west of the Diablo platform.

Upper Permian

Upper Permian strata (Figure 11) are even more difficult to correlate than Leonardian strata. Three limestones of latest Leonardian to earliest Guadalupian age that have restricted marine faunas encroached into the region. From the east, the San Andres Limestone reached into south central New Mexico (Kottlowski, 1963b). Its age is variously considered latest Leonardian and/or earliest Guadalupian, and, if its deposition is a result of a rapid transgression and gradual regression, then it may be diachronous across the Leonardian–Guadalupian boundary.

In northwestern and north-central Arizona, the Kaibab Limestone has somewhat similar age relationships to those of the San Andres Limestone. In northwestern Arizona, the Kaibab is considered to be latest Leonardian, and in areas to the northwest in Utah and Nevada, it is considered earliest Guadalupian. Kaibab deposition may have also resulted from a rapid transgression followed by a gradual regression that crosses the Leonardian–Guadalupian boundary. In the Glass Mountains of western Texas, the Road Canyon and Word limestones have similar relationships.

In the disturbed belt of southeastern Arizona, the lower part of the Concha Limestone contains latest Leonardian–early Guadalupian faunas. The Concha is overlain by limestones of the Rainvalley Formation. These

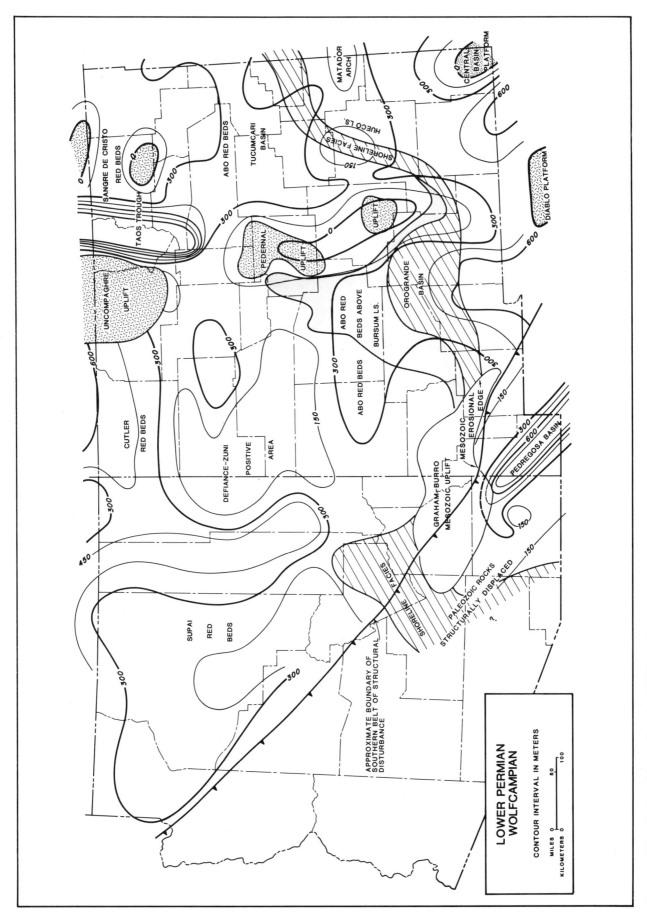

Figure 9—Distribution of Lower Permian (Wolfcampian) strata in Arizona and New Mexico. Compiled from Sabins (1957a), Baars (1962, 1973), Kottlowski (1963b), Diaz and Navarro (1964), Wilson et al. (1969), Greenwood et al. (1970, 1977), Foster et al. (1972), Rascoe and Baars (1972), Ross (1973), Woodward (1973), Thompson et al. (1978), Blakey (1980), and Peterson (1980).

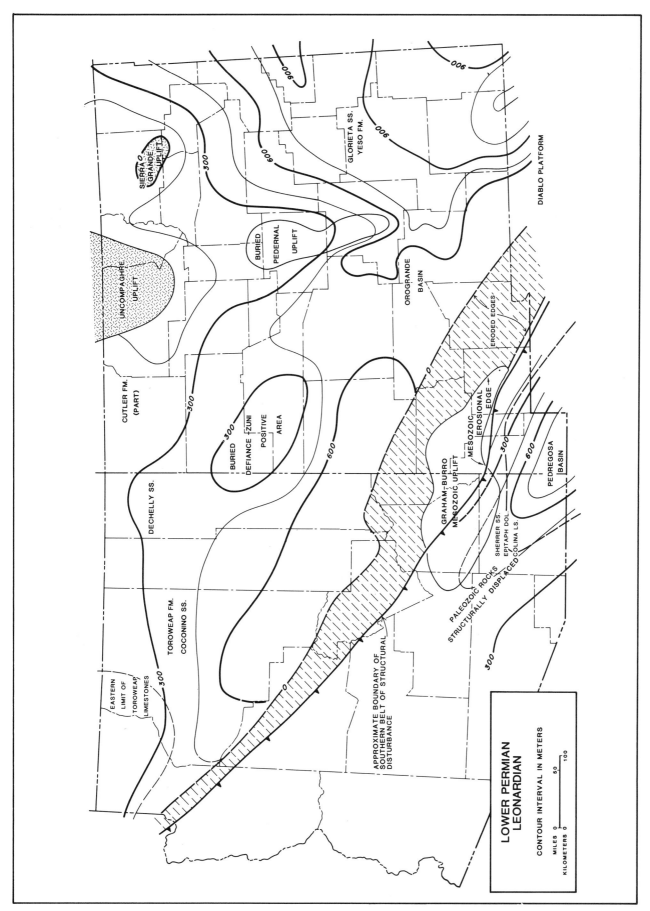

Figure 10—Distribution of upper Lower Permian (Leonardian) strata in Arizona and New Mexico. Compiled from Sabins (1957a), Kottlowski (1963b), Zeller (1965, 1970), Greenwood et al. (1970, 1977), Foster et al. (1972), Woodward (1973), Thompson et al. (1978), Blakey (1980), Peterson (1980), and Rawson and Turner-Peterson (1980).

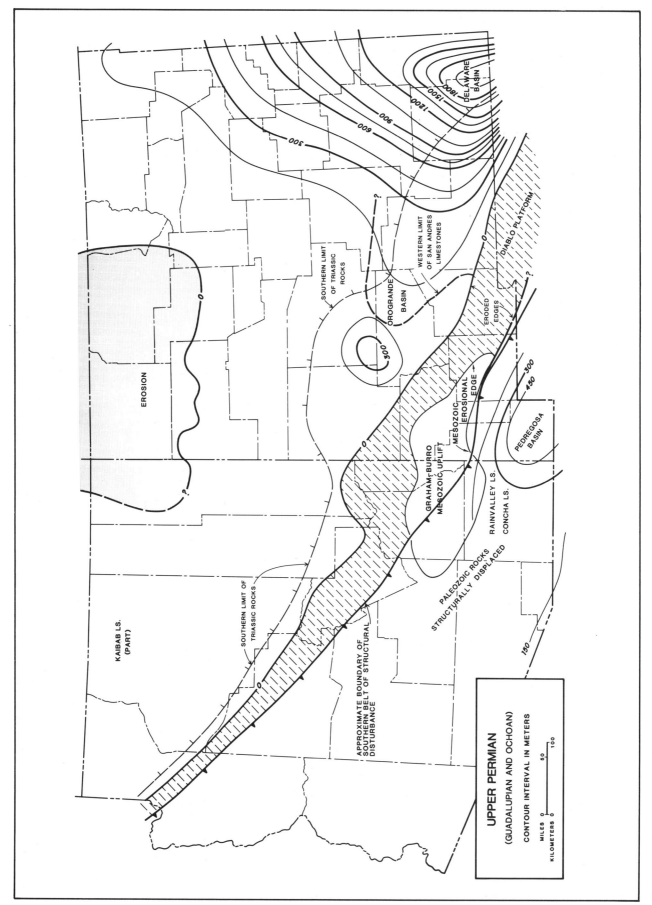

Figure 11—Distribution of Upper Permian (Guadalupian and Ochoan) strata in Arizona and New Mexico. Compiled from Kottlowski (1963b), McKee et al. (1967), Foster et al. (1972), and Peterson (1980).

limestones also suggest a rapid latest Leonardian–earliest Guadalupian marine trangression followed by a slower regression. Although Mesozoic and Cenozoic erosion has removed the Guadalupian and Ochoan strata in a broad belt just north of the southern disturbed belt (Figure 11), the thickest Upper Permian deposition occurs only in the depocenters associated with these three marginal marine embayments.

PALEOZOIC TECTONIC EVENTS

During Paleozoic time, the southern Ancestral Rocky Mountain region experienced at least five general changes in patterns of basin sedimentation, which can be broadly related to major patterns of Paleozoic tectonic activity. Circumstantial evidence strongly suggests that these changes in patterns are directly related to the position, movement, and relationship of the Paleozoic North American craton and its crustal plate boundaries with adjacent plates and spreading centers at different times during the Paleozoic.

The earliest pattern dates from the latest Precambrian and early Paleozoic and represents a time during which the western margin of the North America craton was passive and the eastern oceanic edge of the plate was an active margin closing with northern Europe to eliminate the ancient Iapetus oceanic crust by subduction. Closure of the Iapetus ocean basin was essentially completed by the end of Silurian or early Devonian time.

A second pattern emerges with the collision of the northern European craton against part of the North American craton during Devonian time. This brought about major changes in paleotectonic relationships. Instead of a cratonic plate edge overriding an oceanic plate, the two cratonic masses were being actively driven together. Initially, this impact was absorbed by deformation of the thin and irregular edges of the cratons. By the later part of the Devonian, however, the two cratons had begun to act as a single unit and moved as a unit westward. The western edge of the North American craton was changed from a passive margin to an active margin that was pushed westward over the edge of the Panthalassa ocean basin. This is represented by the initial stages of the Antler orogeny in eastern Nevada.

In the northern part of the southern Ancestral Rocky Mountain region, Late Devonian sedimentary features were strongly influenced by north-south- to northwest-southeast-oriented, fault-bounded uplifts and basins along the western side of the craton. These were apparently aligned roughly at a high angle to the western edge of the Paleozoic cratonic margin, but were nearly perpendicular to the relative motion of North America at that time.

During Early and Middle Mississippian time, the southern Ancestral Rocky Mountain region was a fairly stable cratonic region. In this third pattern of depositional history, the region was alternately slightly submerged or slightly exposed as eustatic changes of sea level and cratonic tectonics caused seas either to wash against or over this southern extension of the North American Transcontinental arch. Shallow shelf carbonate and evaporite deposits and local clastic sediments are characteristic of this part of the

stratigraphic record. Although the clastic basins were relatively small during this time interval, many of them were precursors of more extensive clastic wedges and basinal filling in Pennsylvanian and Early Permian time.

The relative stability of the southern Ancestral Rocky Mountains region during Mississippian time was disturbed during latest Chesterian or early Morrowan (Pennsylvanian) through Wolfcampian (early Early Permian) time. Deep structural basins and adjacent fault-bounded highlands were progressively and irregularly developed (Kluth and Coney, 1981). In this fourth pattern of deposition, tectonic activities in the Ouachita deformation belt (Thomas, 1983) are associated with Early Pennsylvanian tectonic basins and the initial development on the craton of the major outlines of the Palo Duro and Tucumcari basins and the Matador arch. These were clearly formed by the early part of Middle Pennsylvanian (Atokan) time. The major structural features to the south, such as the Midland basin, were progressively defined so that by the end of Wolfcampian time, all of the major late Paleozoic basins and platforms of the western Texas, New Mexico, and southeastern Arizona regions had been delineated and formed (Ross, 1973). Most of the major uplifts are bounded by high-angle faults that have had various directions of motion. At different times in their history, they have been normal faults, reverse faults, and strike-slip faults. Many of the west-northwest-aligned cratonic faults along the Amarillo–Wichita–Arbuckle uplift (Figure 12) show major vertical movements that date from late Chesterian to early Morrowan time. These faults were reactivated in late Desmoinesian, late Missourian, and late Virgilian time and again in middle Wolfcampian time. In central New Mexico, Arizona, and western Texas nearly all of the north to north–northwest faults became very active only in late Desmoinesian through middle Wolfcampian time. The greatest amount of movement in south-central New Mexico, southeastern Arizona, and the Diablo platform part of Texas was late Virgilian to middle Wolfcampian in age.

The relationship of this fourth phase of deposition of tectonics is closely tied to the Hercynian–Ouachita–Marathon orogenic belt (Thomas, 1983). During this time the northwestern corner of Gondwana collided with the southeastern and southern margins of Euramerica. Contact first was against the southeastern (Appalachian) edge. As this collision progressed, the Euramerican and Gondwana cratons either rotated against one another by several degrees to close first the Ouachita geosyncline and then the Marathon geosyncline as a series of jawlike sutures or the cratons themselves fractured along northwest-trending zones of structural weakness. The forces of rotation or the direction of forces on the series of fault-bounded cratonic blocks slightly changed as collision progressed. In addition to horizontal forces, the collision of Euramerica and Gondwana resulted in the southern margin of western Euramerica being overridden by allochthonous oceanic and clastic wedge sediments, as shown by the progressive displacement of foredeep sediments onto cratonic shelf margin facies. The Dugout and Marathon allochthons lie on top of Diablo platform shelf facies and show this relationship well. Therefore, not only were there major horizontal forces acting on the fractured cratonic margin, but also

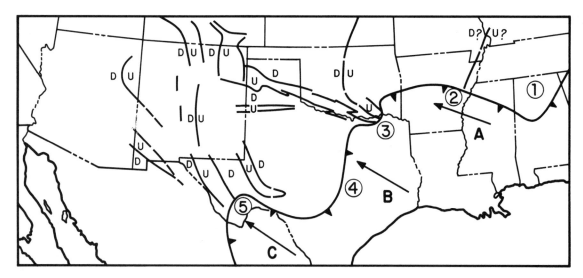

Figure 12—The southern part of the North American (Euramerican) craton during the collision with northwestern South America (Gondwana) during late Paleozoic time. 1. Alabama recess; 2. Mississippi Valley reach; 3. Ouachita salient; 4. Texas recess; 5. Marathon salient. Modified after Thomas (1983). Geometry of the fault patterns that bound the cratonic uplifts and basins is modified after Kluth and Coney (1981) and others.

considerable vertical forces as well, which resulted in an "ice breaker" effect. By Early Permian time these changes in direction of forces had resulted in a loosening of compressive forces in many northwest-southeast-aligned, graben–horst fault systems. The result was a great and abrupt deepening of many preexisting basins and a corresponding uplift of adjacent highlands.

By the end of Wolfcampian time, the faults that formed a grid pattern along the margins of these various cratonic blocks had become grid-locked. After the collision of the Euramerican and Gondwanan continents, there was apparently no longer any major differential motion along the various arcs that formed their sutured margins. In addition, the combined continents may have had less total motion after this time.

By the beginning of Leonardian time (late Early Permian), the collision of Gondwana and Euramerica was completed and the fifth depositional pattern became established. The southern margin of the Ancestral Rocky Mountain region became a positive area that received mostly nonmarine deposits for the remainder of Permian time. Some structural adjustments continued to occur in the region during Leonardian and the remainder of Permian time, but these were minor and related mainly to the filling of existing basins and the continued erosion of uplifted areas.

ACKNOWLEDGMENTS

We thank F. E. Kottlowski and S. Thompson III for reviewing this manuscript and offering a number of helpful suggestions. C. A. Ross thanks Gulf Oil Exploration and Production Company for permission to publish this paper.

REFERENCES CITED

Anderson, T. H., and V. A. Schmidt, 1983, The evolution of Middle America and the Gulf of Mexico–Caribbean Sea region during Mesozoic time: Geological Society of America Bulletin, v. 94, p. 941–966.

Armstrong, A. K., 1963, Biostratigraphy and paleoecology of the Mississippian System, west-central New Mexico: New Mexico Geological Society Fourteenth Field Conference Guidebook, p. 112–122.

Armstrong, A. K., F. E. Kottlowski, W. J. Stewart, B. L. Mamet, E. H. Baltz, Jr., W. T. Siemers, and S. Thompson III, 1979, The Mississippian and Pennsylvanian (Carboniferous) Systems in the United States—New Mexico: Geological Survey Professional Paper 1110-W, p. W1–W27.

Armstrong, A. K., B. L. Mamet, and J. E. Repetski, 1980, The Mississippian System of New Mexico and southern Arizona, *in* T. D. Fouch, and E. R. Magathan, eds., Paleozoic paleogeography of the west-central United States: Rocky Mountain Section, Society of Economic Paleontologists and Mineralogists, Rocky Mountain Paleogeography Symposium 1, p. 82–100.

Baars, D. L., 1962, Permian System of Colorado Plateau: AAPG Bulletin, v. 46, n. 2, p. 149–218.

———, 1973, Permianland: the rocks of Monument Valley: New Mexico Geological Society Twenty-fourth Field Conference Guidebook, p. 68–71.

Baars, D. L., and G. M. Stevenson, 1984, The San Luis uplift Colorado and New Mexico—an enigma of the Ancestral Rockies: Mountain Geologist, v. 21, p. 57–61.

Beus, S. S., 1980, Late Devonian (Frasnian) paleogeography and paleoenvironments in northern Arizona, *in* T. D. Fouch, and E. R. Magathan, eds., Paleozoic

paleogeography of the west-central United States:
Rocky Mountain Section, Society of Economic
Paleontologists and Mineralogists, Rocky Mountain
Paleogeography Symposium 1, p. 55–70.

Blakey, R. C., 1980, Pennsylvanian and Early Permian
paleogeography, southern Colorado Plateau and
vicinity, *in* T. D. Fouch, and E. R. Magathan, eds.,
Paleozoic paleogeography of the west-central United
States: Rocky Mountain Section, Society of Economic
Paleontologists and Mineralogists, Rocky Mountain
Paleogeography Symposium 1, p. 239–258.

Bridges, L. W., 1964, Stratigraphy of Mina Plomosas–Placer
de Guadalupe Area, *in* Geology of Mina
Plomosas–Placer de Guadalupe area, Chihuahua,
Mexico: West Texas Geological Society Publication 64-
50, p. 50–59.

Casey, J. M., 1980, Depositional systems and paleogeographic
evolution of the late Paleozoic Taos trough, northern
New Mexico, *in* T. D. Fouch, and E. R. Magathan, eds.,
Paleozoic paleogeography of the west-central United
States: Rocky Mountain Section, Society of Economic
Paleontologists and Mineralogists, Rocky Mountain
Paleogeography Symposium 1, p. 181–196.

Chronic, J., 1979, The Mississippian and Pennsylvanian
(Carboniferous) Systems in the United States—
Colorado: Geological Survey Professional Paper 1110-
M-DD, p. V1–V25.

Diaz, G. T., and G. A. Navarro, 1964, Lithology and
stratigraphic correlation of the Upper Paleozoic in the
region of Palomas, Chihuahua, *in* Geology of Mina
Plomosas–Placer de Guadalupe area, Chihuahua,
Mexico: West Texas Geological Society Publication 64-
50, p. 65–84.

Foster, R. W., R. M. Frentress, and W. C. Riese, 1972,
Subsurface geology of east-central New Mexico: New
Mexico Geological Society Special Publication, n. 4, p.
1–22.

Greenwood, E., F. E. Kottlowski, and A. K. Armstrong,
1970, Upper Paleozoic and Cretaceous stratigraphy of
the Hidalgo County area, New Mexico: New Mexico
Geological Society Twenty-first Field Conference
Guidebook, p. 33–44.

Greenwood, E., F. E. Kottlowski, and S. Thompson III, 1977,
Petroleum potential and stratigraphy of Pedregosa
basin: comparison with Permian and Orogrande basins:
AAPG Bulletin, v. 61, p. 1448–1469.

Hayes, P. T., 1975, Cambrian and Ordovician rocks of
southern Arizona and New Mexico and westernmost
Texas: U.S. Geological Survey Professional Paper 873,
98 p.

Huddle, J. W., and E. Dobrovolny, 1952, Devonian and
Mississippian rocks of Central Arizona: U.S. Geological
Survey Professional Paper 233-D, p. 67–112.

Kashfi, M. S., 1983, Upper Devonian source and reservoir
rocks in the Black Mesa basin, northeastern Arizona:
Oil and Gas Journal, December 12, p. 151–158.

Kent, W. N., and R. R. Rawson, 1980, Depositional
environments of the Mississippian Redwall Limestone
in northeastern Arizona, *in* T. D. Fouch, and E. R.
Magathan, eds., Paleozoic paleogeography of the west-
central United States: Rocky Mountain Section, Society

of Economic Paleontologists and Mineralogists, Rocky
Mountain Paleogeography Symposium 1, p. 101–110.

Kluth, C. F., and P. J. Coney, 1981, Plate tectonics of the
Ancestral Rocky Mountains: Geology, v. 9, p. 10–15.

Kottlowski, F. E., 1960, Summary of Pennsylvanian sections
in southwestern New Mexico and southeastern Arizona:
New Mexico Bureau of Mines and Mineral Resources
Bulletin 66, 187 p.

———, 1962, Pennsylvanian rocks of southwestern New
Mexico and southeastern Arizona, *in* C. C. Branson, ed.,
Pennsylvanian System in the United States: Tulsa,
Oklahoma, AAPG Symposium Volume, p. 331–371.

———, 1963a, Pennsylvanian rocks of Socorro County, New
Mexico: New Mexico Geological Society Fourteenth
Field Conference, p. 102–111.

———, 1963b, Paleozoic and Mesozoic strata of
southwestern and south-central New Mexico: New
Mexico Bureau of Mines and Resources Bulletin 79,
100 p.

Kottlowski, F. E., and W. J. Stewart, 1970, The Wolfcampian
Joyita uplift in central New Mexico: New Mexico
Bureau of Mines and Mineral Resources Memoir 23,
82 p.

Kues, B. S., 1984, Pennsylvanian stratigraphy and
paleontology of Taos area, north-central New Mexico,
in W. S. Baldridge, P. W. Dickerson, R. E. Riecker, and
J. Zidek, eds., Rio Grande rift, northern New Mexico:
New Mexico Geological Society Guidebook 35, p.
107–114.

Lane, H. R., and T. L. DeKeyser, 1980, Paleogeography of
the Late Early Mississippian (Tournaisian) in the
central and southwestern United States, *in* T. D. Fouch,
and E. R. Magathan, eds., Paleozoic paleogeography of
the west-central United States: Rocky Mountain
Section, Society of Economic Paleontologists and
Mineralogists, Rocky Mountain Paleogeography, p.
149–162.

Lopez-Ramos, E., 1969, Marine Paleozoic rocks of Mexico:
AAPG Bulletin, v. 53, p. 2399–2417.

Mallory, W. W., 1972, Pennsylvanian System, *in* W. W.
Mallory, et al., eds., Geological atlas of the Rocky
Mountain region: Denver, Rocky Mountain Association
of Geologists, p. 111–127.

Mallory, W. W., et al., eds., 1972, Geological atlas of the
Rocky Mountain region: Denver, Rocky Mountain
Association of Geologists, 331 p.

McKee, E. D., and E. J. Crosby, coordinators, 1975,
Paleotectonic investigations of the Pennsylvanian
System in the United States: U.S. Geological Survey
Professional Paper 853, pt. I, 349 p.; pt. II, 192 p.

McKee, E. D., S. S. Oriel, et al., 1967, Paleotectonic maps of
the Permian System: U.S. Geological Survey
Miscellaneous Geological Investigations Map I-450.

Peirce, H. W., 1979, The Mississippian and Pennsylvanian
(Carboniferous) Systems in the United States—
Arizona: U.S. Geological Survey Professional Paper
1110-Z, p. Z1–Z20.

Peterson, J. A., 1980, Permian paleogeography and
sedimentary provinces, west Central United States, *in*
T. D. Fouch, and E. R. Magathan, eds., Paleozoic
paleogeography of the west-central United States:

Rocky Mountain Section, Society of Economic Paleontologists and Mineralogists, Rocky Mountain Paleogeography Symposium 1, p. 271–292.

Rascoe, B., Jr., and D. L. Baars, 1972, Permian System, *in* W. W. Mallory, et al., eds., Geologic Atlas of the Rocky Mountain Region: Denver, Rocky Mountain Association of Geologists, p. 143–165.

Rawson, R. R., and C. E. Turner-Peterson, 1980, Paleogeography of northern Arizona during the deposition of the Permian Toroweap Formation, *in* T. D. Fouch, and E. R. Magathan, eds., Paleozoic paleogeography of the west-central United States: Rocky Mountain Section, Society of Economic Paleontologists and Mineralogists, Rocky Mountain Paleogeography Symposium 1, p. 341–352.

Ross, C. A., 1973, Pennsylvanian and Early Permian depositional history, southeastern Arizona: AAPG Bulletin, v. 57, p. 887–912.

Sabins, F. F., Jr., 1957a, Stratigraphic relations in Chiricahua and Dos Cabezas mountains, Arizona: AAPG Bulletin, v. 41, n. 3, p. 466–510.

Thomas, W. A., 1983, Continental margins, orogenic belts, and intracratonic structures: Geology, v. 11, p. 270–272.

Thompson, S., III, J. C. Tovar-Rodriquez, and J. N. Conley, 1978, Oil and gas exploration wells in the Pedregosa basin: New Mexico Geological Society Twenty-ninth Field Conference Guidebook, p. 331–342.

Tovar-Rodriquez, J. C., and R. J. Valencia, 1974, Road log, Ojinaga to Chihuahua City, *in* Geologic field trip through the States of Chihuahua and Sinaloa, Mexico: West Texas Geological Society Publication 74-63, p. 7–43.

Wilson, J. L., A. Madrid-Solis, and R. Malpica-Cruz, 1969, Microfacies of Pennsylvanian and Wolfcampian strata in southwestern U.S.A. and Chihuahua, Mexico: New Mexico Geological Society Twentieth Field Conference Guidebook, p. 80–90.

Wilson, J. M., 1958, Stratigraphy and geological history of the Pennsylvanian sediments of southeastern Colorado: Denver, Rocky Mountain Association of Geologists Symposium on Pennsylvanian Rocks of Colorado and Adjacent Areas, p. 69–79.

Woodward, L. A., 1973, Structural framework and tectonic evolution of the Four Corners region of the Colorado Plateau: New Mexico Geological Society Twenty-fourth Field Conference Guidebook, p. 94–98.

Zeller, R. A., Jr., 1965, Stratigraphy of the Big Hatchet Mountains area, New Mexico: New Mexico Bureau of Mines and Mineral Resources Memoir 16, 128 p.

———, 1970, Stratigraphy of the Big Hatchet Mountains Area, New Mexico: New Mexico Geological Society Twenty-first Field Conference Guidebook, p. 45–58.

Laramide Paleotectonics of Southern New Mexico

W. R. Seager
G. H. Mack
New Mexico State University
Las Cruces, New Mexico

The chief mode of Laramide deformation of the foreland area of south-central and adjacent parts of southwestern New Mexico was uplift of relatively simple basement blocks and subsidence of complementary basins that were similar in style to but smaller in size than some of those of the central Rocky Mountains. Uplifts trend generally northwestward to west–northwestward and are asymmetric with southwest-dipping, thrust- or reverse-faulted northeastern margins. Broad, less deformed southwestern flanks plunge into complementary basins filled with lower Tertiary clastic rocks about 600–2100 m (2000–6800 ft) thick. In south-central New Mexico the reconstructed uplifts and basins resemble Wind River-type uplifts and basins; they display evidence of strong horizontal compression and significant crustal shortening.

The general style of deformation extends into the northern margin of terrane previously regarded as part of the Cordilleran "overthrust" belt. In this region, however, strike-slip as well as reverse movement distinguishes uplift marginal fault zones. Thus, uplifts in this region may be a product partly of convergent wrenching and partly of northeast–southwest compression, and adjacent basins may be a mix of Wind River (compressional) and Echo Park (transpressional) types. Thrust faulting adjacent to steep, uplift marginal faults may also be viewed as a consequence of convergent wrenching rather than regional overthrusting.

Breakup of the southern New Mexico craton into thrust-bounded blocks may be a result of the attempt by the North American craton, driving actively southwestward, to underthrust a resisting magmatic arc. Strike-slip on faults in southwestern New Mexico may be a result of oblique subduction of the Farallon plate or a product or reactivation of suitably oriented pre-Laramide faults.

INTRODUCTION

Although Laramide deformation in southern New Mexico has been recognized for many years (e.g., Lasky, 1947; Kelley and Silver, 1952; Zeller, 1970; Corbitt and Woodward, 1973; Seager, 1975; Brown and Clemons, 1983), the age, location, trends, and structural style of Laramide uplifts and basins have proven difficult to define with any confidence. The problem is that Laramide structures were buried by volcanic and sedimentary rocks and dissected by Basin and Range normal faults during the Cenozoic. Today, Laramide structures and sedimentary rocks are exposed only in Basin and Range fault blocks and then only where the volcanic or sedimentary cover has been removed by erosion (Figure 1).

In recent years, the search for petroleum in southern New Mexico has fostered a renewed interest in the Laramide structural style. Interpretations are in dispute largely because the data are fragmentary and the locally exposed structural relationships are often complex. In the northeastern part of the study area (Figure 1), mapping favors basement-cored

block uplifts (e.g., Seager, 1981). In the southwestern part of the region, however, both overthrust and block uplift models have been interpreted (Corbitt and Woodward, 1973; Drewes, 1978; Woodward and DuChene, 1981; Brown and Clemons, 1983). Furthermore, strike-slip movement on some Laramide faults has been recognized (Drewes and Thorman, 1980a, b; Thorman and Drewes, 1980; Seager, 1983). In this paper we offer our interpretation of the Laramide structural style and paleotectonics in southern New Mexico.

Our account of Laramide deformation is based both on structural data and on data from synorogenic and postorogenic sedimentary rocks. Much of our information was acquired from field studies. "Laramide style" folds and faults exposed in several Basin and Range horsts are key reference points for reconstruction of Laramide uplifts and interpretation of the deformational style. Sedimentologic data were also used to define the locations of boundary faults and intermontane basins, as well as to provide a record of the growth and erosional history of the uplift. Coarse-grained, proximal alluvial fan facies and sediment dispersal

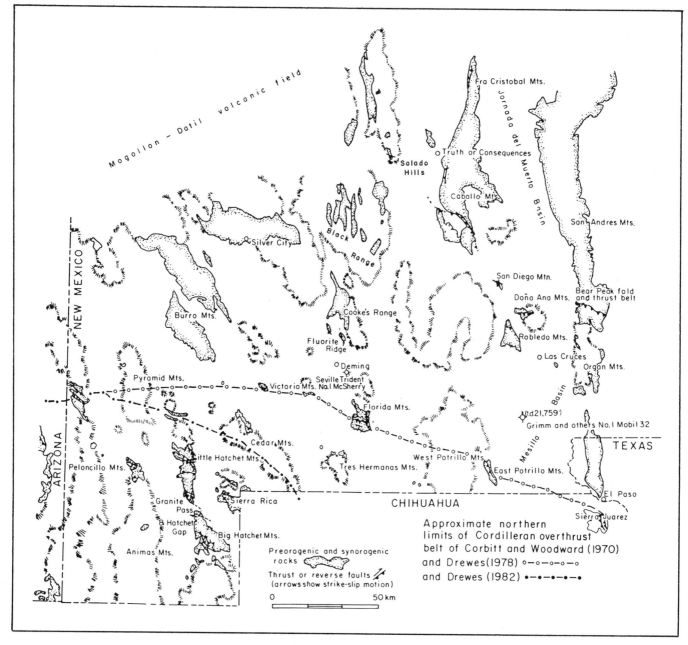

Figure 1—Location map of study area, southwestern New Mexico.

patterns constrain the location of boundary faults. Thickness distribution of Laramide clastic wedges may also provide information on the location and trend of intermontane basins. The sedimentologic data serve to "flesh out" the paleogeography between structural reference points. Although field studies comprise our largest body of data, a few widely spaced deep oil tests provide additional information, especially about Laramide basins.

An important conclusion of this report is that Laramide compression in southern New Mexico produced a series of northwest- to west-northwest-trending, asymmetric basement uplifts and complementary basins. In southwestern New Mexico, high-angle faults, which we consider to be boundary faults of block uplifts, were previously considered to be examples of regional "sled

runner" type overthrusts. At least some tight folds and low-angle thrusts, also used to support previous models of regional overthrusting, may also be interpreted as transpressional structures associated with important strike-slip movement on some uplift boundary faults.

PALEOTECTONIC SETTING

Precambrian granite and metamorphic rocks form the basement of all of south-central and southwest New Mexico and adjacent parts of Arizona. During Paleozoic time, the region was buried by a southward and southwestward thickening wedge of marine rocks, mostly carbonates. The Paleozoic section is more than 4000 m (13,500 ft) thick in the southwest corner of New Mexico and about 2500 m (8200 ft) thick near El Paso, and it thins northward to less than 1000 m (3300 ft) thick along a line between Silver City and Truth or Consequences (Kottlowski, 1963; Greenwood et al., 1977; Thompson, 1982). Epeirogenic uplift in Middle Ordovician, Early Silurian, Late Silurian–Middle Devonian and late Mississippian time is represented by unconformities in the sedimentary record; these unconformities are partly responsible for the overall northward thinning. Thickness variations also reflect Paleozoic subsidence of the Pedregosa basin in the southwest corner of New Mexico and adjacent parts of Arizona, as well as subsidence of the Orogrande basin in the south-central part of New Mexico (Kottlowski, 1965; Greenwood et al., 1977). Approximately 3000 m (10,000 ft) of Pennsylvanian and Permian marine strata accumulated in the Pedregosa basin and half that amount in the Orogrande basin. Between the two basins, Pennsylvanian and Permian rocks thin across a positive area near the Florida Mountains (Kottlowski, 1958, 1960).

Triassic and Jurassic rocks generally are absent over most of southern New Mexico, although marine Jurassic rocks are known from a deep oil test southwest of Las Cruces (Thompson and Bieberman, 1975; Uphoff, 1978; Thompson, 1982). These Jurassic carbonates probably thicken southward into the Chihuahua trough where Jurassic evaporites are diapiric and may be responsible for Laramide décollement and thin-skinned folding in the Chihuahua tectonic belt, including the Sierra Juarez (Haenggi and Gries, 1970; Gries, 1980).

In Early Cretaceous time the Burro uplift–Deming axis (Elston, 1958; Turner, 1962) extended northwestward from the West Potrillo Mountains through the Florida Mountains into Arizona. This uplift was the major source of thick (4500 m [15,000 ft]) Lower Cretaceous marine and nonmarine rocks that were deposited south of the uplift (Mack et al., in press; Zeller, 1965, 1970; Hayes, 1970). North of the Burro uplift, Lower Cretaceous rocks are either thin or missing altogether (Kottlowski, 1963).

In Late Cretaceous and early Tertiary time during the Laramide orogeny, a magmatic arc extended northwestward from southwesternmost New Mexico into southeastern Arizona (Coney, 1978a; Dickinson, 1981), but most of southern New Mexico was a cratonic foreland area behind the arc. As much as 980 m (3200 ft) of Upper Cretaceous marine and nonmarine rocks have been preserved in Laramide intermontane basins. Still greater thicknesses of uppermost Cretaceous and lower Tertiary fanglomerates, redbeds, and sandstones are present in the same basins (Doyle, 1951; Kelley and Silver, 1952; Bushnell, 1953; Zeller, 1965, 1970; Seager, 1981). Arc rocks in the southwestern part of the state are generally intermediate composition volcanic and intrusive rocks ranging in age from Late Cretaceous to Paleocene (Marvin et al., 1978).

AGE OF LARAMIDE DEFORMATION IN SOUTHERN NEW MEXICO

The age of Laramide deformation in southern New Mexico is fixed primarily by the age of syntectonic and posttectonic terrigenous sediment deposited along the flanks of uplifts and in intermontane basins. These deposits set the age of Laramide orogenesis between Late Cretaceous and late Eocene, although Paleocene to early Eocene appears to be the time of most active deformation. We have data from the Little Hatchet Mountains area of southwestern New Mexico, the Florida Mountains region, and the Rio Grande area near Las Cruces (Figure 1).

In the Little Hatchet Mountains, the Ringbone Formation and overlying Hidalgo volcanic rocks appear to be synorogenic Laramide deposits (Zeller, 1970). A Late Cretaceous age for the Ringbone has been suggested based on the fossil palm *Sabal* (Zeller, 1970). However, the range of *Sabal* extends into the Tertiary, so the Late Cretaceous age for the Ringbone is equivocal. Radiometric ages of the Hidalgo volcanics range from 70 to about 55 m.y. (Marvin et al., 1978), although dates from the Little Hatchet Mountains are about 62 m.y. (Mobil Oil Corp., J. M. Cys, personal communication, 1970, cited in Chapin et al., 1975). Loring and Loring (1980) dated a thrust fault cutting the Hidalgo volcanics as Paleocene or possibly latest Cretaceous in age.

Beneath the Ringbone Formation, clasts in thick sandstones and fanglomerates of the Hell-to-Finish and Mojado formations record pre-Laramide (Early Cretaceous) uplift and erosion of the Burro uplift (Mack et al., in press; Elston, 1958) and/or deformation to the southwest or west of New Mexico (Zeller, 1965, 1970; H. Drewes, personal communication, 1983). Movement phases during the Laramide are indicated by angular unconformities between the Mojado (Lower Cretaceous) and Ringbone Formation, between the Ringbone and Hidalgo volcanics, and between the Hidalgo and comparatively undeformed middle Tertiary volcanic rocks. Thus, in southwestern New Mexico, semicontinuous orogenesis between Early Cretaceous and early Tertiary time seems clear, although deformation apparently culminated in latest Cretaceous–early Tertiary time.

Laramide orogenic deposits in the Florida Mountains and Cooke's Range area are the Lobo Formation (Darton, 1928; Lemley, 1982) and the Starvation Draw Member, which is the basal unit of the Rubio Peak Formation (Clemons, 1982). The Lobo has been dated paleomagnetically as late Paleocene in age or younger (Lemley, 1982); but the age of the Starvation Draw Member has been dated only as older than 44–38 m.y., the age of lava flows in the Rubio Peak Formation (Clemons, 1982). Although the Starvation Draw Member and Lobo Formation appear to be largely postorogenic, the Lobo Formation is synorogenic locally. For example, at the southern end of the Florida Mountains, the Lobo accumulated as fan debris adjacent to the Florida Mountains "upthrust"; subsequent movement on the fault displaced the Lobo fan (Lemley, 1982; Clemons and Brown, 1983). Petrofacies studies of the Lobo and Starvation Draw deposits and of overlying volcaniclastic rocks, described in the next section of this paper, suggest that Laramide deformation had ceased by 44–38 m.y. ago as Eocene volcanism commenced.

In the Rio Grande region of south-central New Mexico, orogenic deposits associated with Laramide deformation include the McRae and Love Ranch formations (Bushnell, 1953; Kottlowski et al., 1956). The McRae Formation, located in the Elephant Butte area north of the Caballo

Mountains, has been divided into the lower Jose Creek and upper Hall Lake members (Bushnell, 1953). The Jose Creek Member, which contains a fanglomeratic facies derived from volcanic and hypabyssal porphyritic rocks, has been assigned a Late Cretaceous age on the basis of four plant species. Latest Cretaceous *Triceratops* remains have been uncovered from lower parts of the Hall Lake member, a unit consisting of sandstone and shale. Upper parts of the Hall Lake may be Tertiary.

The Love Ranch Formation appears to be more widespread and to contain a larger volume of fanglomerate than the McRae Formation; it probably documents the main phase of Laramide uplift in south-central New Mexico. Unfortunately, it is not closely dated. Lower parts of the Love Ranch appear to interfinger with McRae beds along the eastern slopes of the Caballo Mountains, indicating either a Late Cretaceous or early Tertiary age for lower parts of the Love Ranch. Upper parts of the Love Ranch interfinger with volcanic rocks (Palm Park Formation) dated 43–40 m.y. old (Seager and Clemons, 1975). The Love Ranch may therefore range in age from latest Cretaceous to late Eocene.

Lower beds of the Love Ranch Formation are synorogenic, while upper parts are largely postorogenic. For example, along the Bear Peak reverse fault in the southern San Andres Mountains, oldest Love Ranch fanglomerate records initial uplift and erosion of the eastern part of the Rio Grande uplift. Subsequently, these fanglomerates were folded and faulted and later buried by younger parts of the Love Ranch clastic wedge (Figure 2). In fact, upper parts of the Love Ranch are comparatively undeformed everywhere in the region; they overlap the margins of deeply eroded Laramide uplifts and locally overlie Precambrian rocks. In central areas of Laramide basins, however, drilling suggests that upper and lower parts of Love Ranch strata—mostly fine-grained clastic sediments and coal in this setting—are conformable and probably represent complete Laramide sections ranging in age from Late Cretaceous to late Eocene (Seager, 1983).

Regionally, Laramide deformation appears to have swept eastward from southeastern Arizona into central New Mexico and western Texas. Drewes (1981) documents an early (Pimian) phase about 75–69 m.y. ago and a late (Helvetian) phase about 65–52 m.y. ago in southeastern Arizona, which collectively comprise his Cordilleran orogeny (Drewes, 1978). Similar long-ranging tectonic movements seem well documented in southwestern New Mexico (Zeller, 1965, 1970). In south-central New Mexico, western Texas, and northeastern Chihuahua, Mexico, strong tectonism has been reported only for latest Cretaceous to early Tertiary time (Wilson, 1965; Haenggi and Gries, 1970), although several movement phases can be demonstrated locally (Seager, 1981). This eastward decrease in age is accompanied by a general decrease in structural complexity across the region.

TEMPORAL RELATIONSHIP BETWEEN LARAMIDE DEFORMATION AND EOCENE VOLCANISM

Laramide terrigenous sedimentary rocks interfinger in most places with overlying Eocene volcanic rocks,

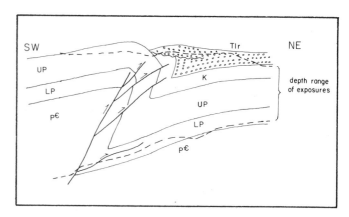

Figure 2—Cross section showing stratigraphic and structural relationships of Love Ranch fanglomerate to the Bear Peak fold and fault zone in the southern San Andres Mountains. The fold and fault zone is the margin of the Rio Grande uplift in this area. PЄ, Precambrian rocks; LP, lower Paleozoic rocks; UP, upper Paleozoic rocks; K, Cretaceous rocks; Tlr, Love Ranch fanglomerate. After Seager (1981) and Seager (1983). Used with permission of the New Mexico Bureau of Mines and Mineral Resources.

demonstrating a close relationship in space and time between Laramide deformation and late Eocene volcanism. Were Laramide blocks still actively being uplifted when volcanism began? To evaluate this question, we collected data from the Lobo Formation of the Florida Mountains and from the Starvation Draw Member in the southern Cooke's Range (Mack et al., 1983).

A transition from arkose to volcanic arenite petrofacies in the upper part of the Lobo and Starvation Draw sections reflects a change in dominant source rock from Precambrian crystalline rocks in the cores of Laramide uplifts (arkose petrofacies) to andesitic volcanic rocks (volcanic arenite petrofacies) (Figure 3). The volcanic arenite petrofacies contains as much as 30% andesitic volcanic rock fragments and 50% plagioclase, but also has minor perthite and quartzofeldspathic rock fragments, indicating that the Laramide uplifts supplied sediment concomitant with andesitic volcanic source rocks.

To determine whether or not Laramide deformation had ceased by the time late Eocene volcanism had commenced, we compared vertical grain size change in the clastic wedge with the grain size changes at the boundary of the arkose and volcanic arenite petrofacies. Active uplift along range marginal faults result in coarsening-upward sequences on the scale of 20–100 m (65–330 ft) thick in the adjacent clastic wedge (Steel and Wilson, 1975). Conversely, tectonic quiescence and weathering back of the mountain front produces a fining-upward sequence in the clastic wedge. The transition from arkose to volcanic arenite petrofacies in the Lobo and Starvation Draw sections is associated with a fining-upward sequence. This suggests that, although the Laramide uplifts were still exposed and acting as sediment sources, active uplift within these blocks had essentially ceased at the onset of Eocene volcanism. Similar studies on other Laramide clastic wedges are needed before this interpretation can be applied on a regional scale.

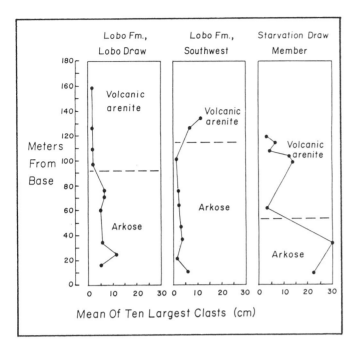

Figure 3—Vertical changes in grain size of the Lobo Formation, Florida Mountains, and the Starvation Draw Member of the Rubio Peak Formation, Cooke's Range. The boundaries of sandstone petrofacies (arkose and volcanic arenite) are also shown (Mack et al., 1983).

LOCATION AND TREND OF LARAMIDE UPLIFTS

Using a combination of structural and sedimentologic data, we recognize at least three major northwest-trending uplifts bordered by intermontane basins (Figure 4). All appear to be asymmetric, northeast-facing structures. Uplifts have thrust- or reverse-faulted northeastern margins and broad, less deformed southwestern flanks that dip into Wind River-type basins or into basins that may have characteristics of both Wind River and Echo Park types (Chapin and Cather, 1981). The trends of uplifts and basins are based on (1) trends of uplift boundary fault zones exposed in modern fault block ranges; (2) depth of pre–late Eocene erosion, which on Laramide uplifts extends into rocks of Precambrian to Permian age; and (3) distribution of Cretaceous rocks and lower Tertiary clastic wedges preserved in Laramide basins.

Rio Grande Uplift and Potrillo Basin

Fragments of the Rio Grande uplift are exposed in ranges adjacent to the Rio Grande in south-central New Mexico: the Robledo, Organ, San Andres, and Caballo mountains; the southern Black Range; the Salado Hills; and San Diego Mountain. In all of these areas, upper Eocene volcanic rocks or Love Ranch fanglomerate overlie rocks ranging in age from Precambrian to Permian. The northern margin of the uplift, defined by southwest-dipping, reverse or thrust faults, is exposed in the southern San Andres Mountains (Bear Peak thrust; Seager, 1981), San Diego Mountain (Seager et

al., 1971), Caballo Mountains (Kelley and Silver, 1952), and Salado Hills (unpublished map, W. R. Seager). Love Ranch fanglomerate overlies Precambrian rocks at several places along this uplift margin (Figure 4), demonstrating as much as 2500 m (8200 ft) of Laramide uplift and pre–late Eocene erosion.

Although shown in Figure 4 as a single simple uplift, the Rio Grande uplift may in fact be a composite of several smaller uplifts. Furthermore, the segments of boundary faults exposed in various ranges may not be continuous. They may also be interpreted as fault zones at the margins of individual uplifts that collectively comprise the Rio Grande uplift. In any case, our mapping suggests that the uplifts are strikingly asymmetric in north–south cross sections.

Structures and unconformities between San Diego Mountain and the Robledo Mountains, along with deep drill hole information in southern Dona Ana County, provide clues to the cross-sectional geometry of the Rio Grande uplift, at least along long. 106° 50′ to 107° 00′ W (Figure 5). The northern margin of the uplift at San Diego Mountain is a west-northwest-striking thrust fault dipping southward 30°. Precambrian granitic rocks form the hanging wall, and a small exposure of silicified Paleozoic carbonate rock forms the footwall. Near this marginal thrust, Love Ranch fanglomerate depositionally overlies Precambrian rocks, indicating enough uplift on the thrust to allow erosion of about 2500 m (8200 ft) of Paleozoic and Mesozoic rocks. Farther south over an area 35 km wide, including the modern Robledo Mountains, thin postorogenic Love Ranch deposits overlap successively younger Paleozoic strata up to the Lower Permian Hueco Limestone. This overlap indicates erosional truncation of Paleozoic rocks, probably caused by a southerly tilt of the Rio Grande uplift. An additional 30 km (18 mi) farther south, in the Grimm et al. American Arctic Limited No. 1 Mobil 32 oil test, Cretaceous and Jurassic rocks overlie Permian rocks, and basinal clastic rocks (sandstone, fine-grained red beds, and coal) correlative with the Love Ranch Formation are more than 2000 m (6500 ft) thick (Thompson and Bieberman, 1975; Uphoff, 1978; Thompson, 1982). Thus, the Rio Grande uplift at this longitude is strikingly asymmetric, with a broad flank that dips perhaps 10–12° southward into a deep complementary basin—the Potrillo basin (Figures 4 and 5). Overall, the structure is similar to the Wind River uplift and complementary Wind River basin of Wyoming.

Love Ranch Basin

North of the exposed Rio Grande uplift marginal faults, Cretaceous McRae and Love Ranch strata are as much as 1800 m (6000 ft) thick where they are preserved in the Laramide Love Ranch basin (Figure 4). Love Ranch fanglomerate in the basin is widely exposed along the northeastern flank of the Caballo Mountains where clasts of Precambrian and lower Paleozoic rocks comprise a large percentage of the fanglomerate. Clearly these clasts were eroded from Rio Grande uplift terrane now exposed in the southern Caballo Mountains or farther southwest. In fact, Love Ranch fanglomerate was locally deposited on Precambrian granite (Figure 4). Love Ranch fanglomerate about 600 m thick also accumulated along the northern

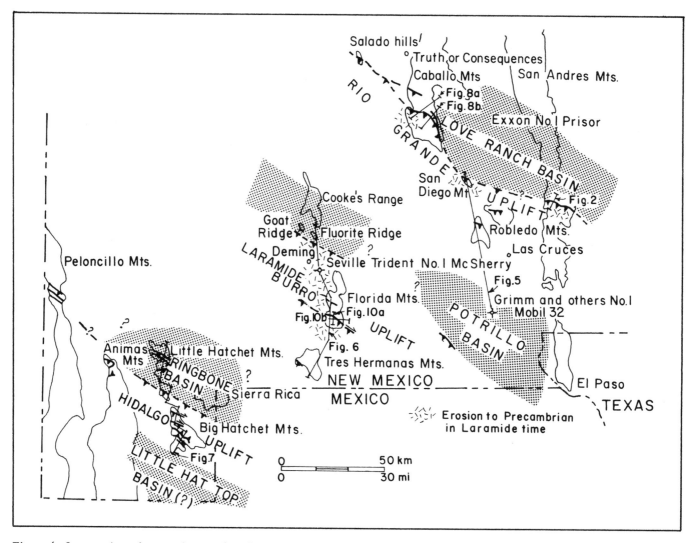

Figure 4—Interpretive paleotectonic map of southwestern New Mexico in Laramide time showing inferred west-northwest-trending block uplifts and basins described in text. Note location of cross sections shown in Figures 2, 5, 6, 7, 8, and 10.

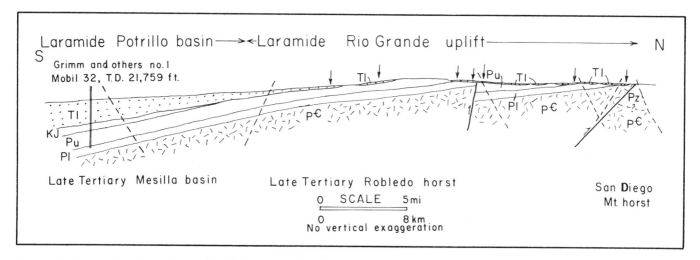

Figure 5—Cross section through Laramide Rio Grande uplift and adjacent Potrillo basin along long. 106° 51′ to 107° 00′ W. Note overlap and thinning of Tl unit (Love Ranch and correlative rocks) onto uplift, and truncation of upper and lower Paleozoic units and Precambrian rocks below the Tl unconformity. Arrows show location of outcrop control for the unconformity beneath Tl. Effects of late Tertiary faults (dashed) have been removed. Tl, lower Tertiary Love Ranch fanglomerate, sandstone, and coal; KJ, Cretaceous–Jurassic rocks; Pu, upper Paleozoic rocks; Pl, lower Paleozoic rocks; pε, Precambrian rocks; Pz, Paleozoic rocks, undifferentiated. See Figure 4 for location.

margin of the Bear Peak reverse fault zone in the southern San Andres Mountains (Figure 2). Most clasts at this locality are Paleozoic limestone, but the upper 10 m record the erosional unroofing of Precambrian crystalline basement (Seager, 1981). Near the center of the Love Ranch basin (Figure 4), about 600 m (1200 ft) of fine-grained redbeds containing minor conglomerate and sandstone were penetrated by the Exxon No. 1 Prisor well. These basinal rocks overlie Cretaceous sedimentary rocks and underlie middle Tertiary volcanic rocks; we consider them to be a basinal facies of the Love Ranch Formation.

Reactivated Laramide Burro Uplift

We have based our reconstruction of the Laramide Burro uplift (Figure 4) on structures and fanglomerates exposed in the Florida Mountains, in the southern Cooke's Range, and in the Tres Hermanas Mountains. However, these exposures provide only a transverse cross-sectional view of the uplift; its extent northwest or southeast along strike is unknown. In the transverse section the uplift consists of at least three major basement blocks bordered by steep faults stair-stepped down to the north (Figure 6). On these fault blocks, thin Lobo fanglomerate overlies Precambrian or lower Paleozoic rocks. The southern flank of the broad uplift is seemingly marked by the full Paleozoic section exposed in the Tres Hermanas Mountains and the northern margin by thick proximal fan deposits of the Starvation Draw Member of the Rubio Peak Formation in the Cooke's Range–Fluorite Ridge area.

Along the northern margin of the Laramide Burro uplift, the Starvation Draw Member of the Rubio Peak Formation (Clemons, 1982) overlies Cretaceous and Paleozoic rocks in the Fluorite Ridge–Cooke's Range area and consists of about 400 m of fanglomerate. A debris flow facies exposed on Fluorite and Goat Ridge consists of boulders of Precambrian granite and gneiss as much as 3 m in diameter. These proximal fan deposits indicate the existence of a Laramide boundary fault involving Precambrian rocks in the area immediately south of Fluorite and Goat Ridge, which is an area now covered by late Cenozoic bolson gravels. Thrust faults involving Paleozoic and Cretaceous strata on Goat Ridge may be associated with this uplift boundary fault (Figure 4). Northward, away from the inferred uplift boundary, the Starvation Draw Member becomes finer grained. Thus, we interpret the unit as proximal to midfan deposits at the margin of a major intermontane basin, which adjoins the reactivated Burro uplift on the north and which aligns with the Potrillo basin to the southeast.

Across the central part of the Laramide Burro uplift structural relationships are revealed by outcrops in the Florida Mountains and by a deep drill hole near Deming. Two high-angle, northwest-trending Laramide faults have been mapped in the Florida Mountains, one in the northern part and one in the southern part of the range (Brown, 1982; Clemons, 1985; Clemons and Brown, 1983). Both faults bound major Burro uplift blocks and both are downthrown to the north. The southern fault—the Florida Mountains "upthrust" (Brown and Clemons, 1983)—is clearly more significant and is associated with a lower Tertiary clastic wedge, the Lobo Formation (Lemley, 1982; Mack et al., 1983). Near the Florida Mountains upthrust the

Lobo consists of coarse-grained proximal and midfan facies, and one fanglomerate section was involved in recurrent movement on the fault zone. Farther north in the Florida Mountains, exposures of the Lobo are finer grained. Here they apparently constitute midfan or distal fan facies, suggesting northward sediment dispersal away from the Florida Mountains upthrust. This finer grained part of the Lobo fan overlapped Precambrian and lower Paleozoic rocks previously juxtaposed by movement on the more northerly fault zone. A gradual change upsection in the composition of the Lobo Formation from a basal sedarenite petrofacies to an arkose petrofacies records the successive removal of Paleozoic and Mesozoic rocks and the unroofing of the Precambrian crystalline core of the uplift (Mack et al., 1983).

Compared to the Starvation Draw Member and to the rocks filling the Love Ranch, Potrillo, and Ringbone basins, the Lobo clastic wedge is thin (200 m [660 ft]) across the Laramide Burro uplift. Locally, it is absent altogether. For example, the Seville Trident No. 1 McSherry well, located near Deming north of the Lobo outcrops (Figure 4), penetrated Precambrian rocks at 3535 m (11,590 ft) directly beneath Tertiary–Quaternary basin fill and volcanic rocks. No lower Tertiary rocks were found above the Precambrian (Figure 6) (S. Thompson, personal communication, 1983; R. Clemons, personal communication, 1984). The thinness and local absence of the Lobo fanglomerate indicate that it was deposited in an intra-uplift basin within the Laramide Burro uplift, rather than in a major intermontane basin such as the Love Ranch or Potrillo basin.

Thus, the broad Laramide Burro uplift spans the region from just south of Fluorite Ridge to the Tres Hermanas Mountains, an area where Precambrian and lower Paleozoic rocks were deeply eroded and widely exposed in early Tertiary time. Individual fault blocks stair-stepped down to the north or northeast, and their intramontane clastic wedges (Lobo Formation) accumulated on and overlapped downdropped blocks. Thicker orogenic deposits (Starvation Draw Member) accumulated along an inferred northeastern margin south of Fluorite Ridge. Although we presume the uplift had the same northwesterly trend as uplift faults, the actual extent of the uplift to the northwest and southeast of the Deming area is not constrained. The location of the Laramide Burro uplift along strike with the Lower Cretaceous Burro uplift (Elston, 1958; Turner, 1962) suggests that the Laramide uplift may have been reactivated along Burro uplift structures.

Hidalgo Uplift and Ringbone Basin

Exposures in the Big and Little Hatchet Mountains and Sierra Rica (Figures 1 and 4) give evidence for the third major uplift and complementary basin: the Hidalgo uplift and Ringbone basin (Figures 4 and 7). The Hidalgo uplift includes all of the modern Big Hatchet Mountains whose structurally high Paleozoic and Mesozoic strata dip generally southwestward, although there are both broad and tight folds (Zeller, 1975). We presume the attitude of these rocks reflects Laramide rather than late Tertiary deformation inasmuch as the rocks strike northwesterly, obliquely across the modern fault block, parallel to major Laramide faults in the Granite Pass–Hatchet Gap area. These faults in the Granite Pass–Hatchet Gap area border the Hidalgo uplift on

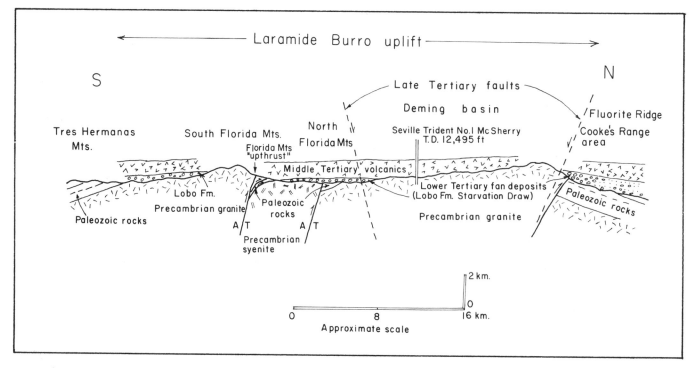

Figure 6—Cross section of reactivated Laramide Burro uplift from southern Cooke's Range through the Florida Mountains to the Tres Hermanas Mountains; effects of late Tertiary faulting and middle Tertiary intrusive activity have been omitted to emphasize the Laramide structure. Major feature is a broad, block-faulted uplift, stepped down by steep faults toward the north and exposing in Laramide time a broad area of Precambrian and lower Paleozoic rocks. Evidence for strike slip along faults is discussed in text. A, strike slip away from viewer; T, strike slip toward viewer. See Figure 4 for location.

the northeast; they are steep to moderately southwest-dipping reverse faults that juxtapose Precambrian granite of the uplift with Lower Cretaceous sedimentary rocks of the adjacent Ringbone basin. A Laramide age for the fault zone is interpreted because of (1) its reverse fault geometry; (2) its northwest trend, which is truncated by late Tertiary, north-trending, range boundary faults; and (3) its intrusion by middle Tertiary granite (Zeller, 1970). This zone of complex faulting that marks the northeastern edge of the Hidalgo uplift may also be exposed farther northwest in the Animas (Soule, 1972) and Peloncillo mountains (Drewes and Thorman, 1980a, b; Armstrong et al., 1978) and to the southeast at the southern edge of Sierra Rica (Zeller, 1975).

North of the Hidalgo uplift boundary fault zone, thick Cretaceous and Tertiary sedimentary rocks and Hidalgo volcanic rocks of the Little Hatchet Mountains were preserved in the Ringbone basin, viewed here as a complementary basin to the Hidalgo uplift (Figure 4). The Upper Cretaceous–lower Tertiary Ringbone Formation of the Little Hatchet Mountains may be interpreted as a syntectonic and posttectonic clastic wedge (about 2000 m [6000 ft] thick) derived from the Hidalgo uplift. Thick Cretaceous rocks of the Sierra Rica may also have been preserved in the Ringbone basin. In contrast to the thick syntectonic to posttectonic fanglomerates of the Ringbone basin, lower Tertiary fanglomerates along the southern edge of the Hatchet uplift (Little Hat Top fanglomerate of Zeller, 1965) appear to be much thinner. These we interpret to be deposits that overlapped the southwestern flank of the Hidalgo uplift from the next intermontane Laramide basin to the south

STRUCTURAL FEATURES ALONG UPLIFT BASIN MARGINS

In the previous section, we interpreted the three Laramide uplifts as asymmetric structures with narrow, faulted and folded northeastern margins and broad, less deformed southwest-dipping flanks. Adjacent basins exhibit asymmetry complementary to the uplifts. In this section we examine more closely the nature of the faulting and folding along the uplift basin margins.

Rio Grande Uplift

The northeastern margin of the Rio Grande uplift is best exposed in the southern San Andres Mountains (Bear Peak fault zone; Seager, 1981), in the central to southern Caballo Mountains (Kelley and Silver, 1952), and at San Diego Mountain (Seager et al., 1971). In these areas the uplift margin is distinguished by narrow zones of low-angle to moderately dipping (30–60°) reverse and thrust faults that dip southwest and that transect Precambrian basement and modify overturned flexures in sedimentary cover rocks. Uplift along these structures ranges from 1000 to 5000 m (35,000–17,000 ft), and the folds are seemingly compressive structures at the edge of or adjacent to basement fault blocks. Clearly, substantial crustal shortening and vertical uplift are indicated by the faults and associated folds. The Bear Peak reverse fault zone is a fairly simple example, and it also illustrates some of the subsidiary thrusts that can develop in the compressed beds on the downthrown side of the fault zone (Figure 2).

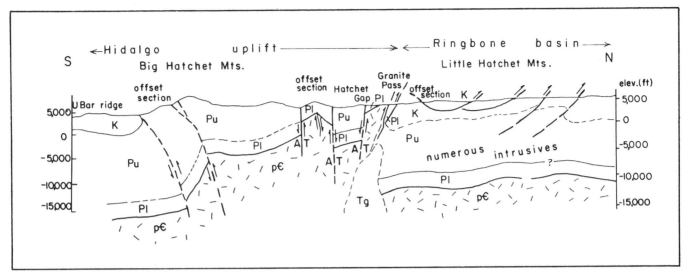

Figure 7—Diagrammatic section through the Hidalgo uplift and adjacent Ringbone basin as seen today in the Big and Little Hatchet Mountains. Tg, Tertiary granite; K, Cretaceous rocks; Pu, upper Paleozoic rocks; Pl, lower Paleozoic rocks; p∈, Precambrian rocks; A, strike slip away from viewer; T, strike slip toward viewer. See Figure 4 for location.

In the central and southern Caballo Mountains, the nature of the uplift margin changes along strike from simple, like the Bear Peak fault zone, to much more complex (Figure 8). Locally, the boundary fault zone divides into two or more strands, each associated with thrust-faulted folds overturned to the northeast (locally almost recumbent). At higher levels, thrusts and associated overturned beds dipping 30–40° or less are typically developed. Erosion is deep enough to show that some faults along uplift margins transgress downward into Precambrian rocks, transecting them at an angle of about 60° (30° at San Diego Mountain; see Figure 5). The flatter, high-level thrusts must either steepen downward, merging into the 60° dipping fault zone or, more likely, intersecting the steeper fault zone at an angle (Figure 8). In either case, comparatively low-angle dips of uplift boundary faults (<60°) and the strong to moderate overturning of sedimentary cover rocks indicates important crustal shortening as well as substantial vertical uplift. The folds are not merely drapes across the edges of uplifted basement blocks. We have found no evidence for strike-slip movement on any of the faults.

Laramide Burro Uplift

Structures in the reactivated Laramide Burro uplift are especially well exposed in the central and southern Florida Mountains. The primary structure is the steep (85°), southwest-dipping, northwest-trending Florida Mountains upthrust (Figure 9). Although this structure was previously interpreted to be the "sled runner" front of a regional overthrust (Corbitt and Woodward, 1973; Woodward and DuChene, 1981), mapping by Brown (1982), Brown and Clemons (1983), and Clemons (1985) has suggested the structure is the upthrust margin of a basement block. Brown and Clemons (1983) argued for predominantly vertical movement along the fault. However, compelling evidence for substantial right slip is also indicated by large-scale drag of Paleozoic rocks in the footwall adjacent to the fault

(Figure 9). The southern limb of the drag fold seemingly has been segmented into complex thin slices, which are bordered by low-angle faults and are stacked and sheared in the footwall along the length of the fault to the northwest (Figure 10). Most of the faulting resulted in younger rocks displaced over older rocks (low-angle normal faults), but a few produced typical thrust relationships of older over younger rocks. Both styles of faulting "root" downward into the major reverse fault so that the overall structure resembles half of a "flower" structure (Figure 10) (Lowell, 1972; Harding and Lowell, 1979). We consider the system of low-angle faults to be a product of transpression in the Florida Mountains upthrust zone. Another indication of strike-slip movement along the Florida Mountains upthrust is the dissimilar Precambrian rocks on either side of the fault (Figure 9), which also yield different K-Ar ages (Clemons and Brown, 1983). A component of northeast motion, however, is also indicated by thrust-faulted, northeast-vergent folds (Figure 10b) located in footwall rocks.

Hidalgo Uplift and Ringbone Basin

Evidence for right slip along the boundary fault zone of the Hidalgo uplift also exists in the Granite Pass–Hatchet Gap area between the Big Hatchet and Little Hatchet mountains. Zeller (1970, 1975) mapped drag and second-order folds south of Hatchet Gap that are apparently associated with a steep northwest-trending fault (Figure 11). A few miles farther south in Chaney Canyon, Zeller (1975) interpreted oblique right slip on another major fault with the same trend. These faults and the major southwest-dipping reverse fault that juxtaposes Precambrian granite and Lower Cretaceous sedimentary rocks at Granite Pass are major elements of the fault zone that elevated the Hidalgo uplift as a basement block and tilted it southwestward (Figure 7). A significant component of right-slip wrenching is clearly indicated in the fault zone.

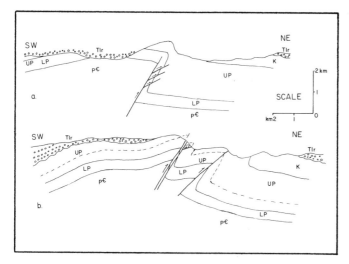

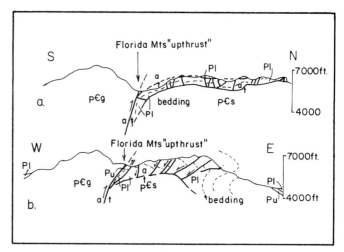

Figure 8—Cross sections through northeastern faulted margin of Rio Grande uplift in southern Caballo Mountains; Effects of late Tertiary faulting have been removed. Tlr, Tertiary Love Ranch Formation; K, Cretaceous rocks; Pu, upper Paleozoic rocks; Pl, lower Paleozoic rocks; p∊, Precambrian rocks. The top cross section (a) is located northwest of the bottom one (b); see Figure 4 for locations.

Figure 10—South–north (a) and west–east (b) cross sections across Florida Mountains "upthrust" (oblique-slip reverse fault). Thrusting in footwall rocks seems to be result of transpression. Tl, Lobo Formation; Pu, upper Paleozoic rocks; Pl, lower Paleozoic rocks; p∊s, Precambrian syenite; p∊g, Precambrian granite; A, strike slip away from viewer; T, strike slip toward viewer. See Figure 4 for locations. From Seager (1983). Used with permission of the New Mexico Bureau of Mines and Mineral Resources.

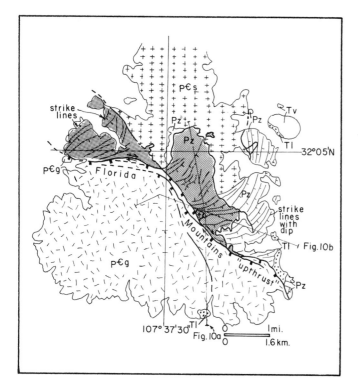

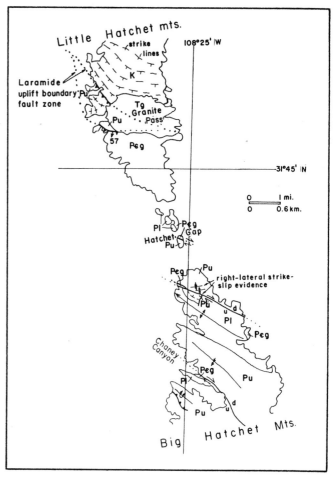

Figure 9—Geologic map of the southern part of the Florida Mountains; geology is after Brown (1982), Clemons and Brown (1983), and Clemons (1985). Lines with tick marks are strike lines showing dip direction. Note large drag in Paleozoic strata at southeastern end of range and different Precambrian rocks on either side of Florida Mountains "upthrust," both of which are taken as evidence of significant right slip on the upthrust. The shaded area is complexly broken by closely spaced normal and thrust faults; it is bounded by low-angle faults that steepen downward to merge with the Florida Mountains upthrust (see Figure 10). Tv, middle Tertiary volcanic rocks; Tl, Lobo Formation; Pz, Paleozoic sedimentary rocks; p∊s, Precambrian syenite; p∊g, Precambrian granite.

Figure 11—Geologic map of Granite Pass–Hatchet Gap area between Little and Big Hatchet Mountains. Tg, Tertiary granite; K, Cretaceous rocks; Pu, upper Paleozoic rocks; Pl, lower Paleozoic rocks; p∊g, Precambrian granite. After Zeller (1970, 1975).

Thrust faulting within the Hidalgo uplift and Ringbone basin may be related to the uplift of the Hidalgo block or to transpression along the boundary fault zone rather than to regional overthrusting. For example, in the Paleozoic strata of the Hidalgo uplift, fold axes and northeast-dipping thrusts indicate tectonic transport to the southwest over a fairly broad area, a relationship that seems more consistent with either horstlike block uplift (Figure 7) or convergent wrenching along the uplift boundary fault zone. These folds and thrusts are exposed in the central and southern Big Hatchet Mountains (Zeller, 1975). Thrust faults in the Little Hatchet Mountains support previous interpretations of regional overthrusting (Corbitt and Woodward, 1973), but they may also be interpreted as products of transpression adjacent to the major oblique-slip reverse faults at Granite Pass and Hatchet Gap. The situation here may be analogous to the low-angle faults adjacent to the south Florida Mountains fault, with the exception that the Florida Mountain fault has undergone a deeper level of erosion and indicates some transtension as well as major transpression caused by strike slip on the nearby oblique-slip reverse fault. Alternatively, the thrusts in the Little Hatchet Mountains may have resulted from southwestward underthrusting of the Ringbone basin relative to the Hidalgo uplift. In any case, large-scale regional overthrusting may not be required.

SUMMARY AND DISCUSSION

Our main objective in this paper is to show that available evidence from southern New Mexico favors interpretation of Laramide deformation in terms of basement block uplift models. We have reconstructed parts of three west-northwest- to northwest-trending uplifts and their complementary basins (Figure 4). The uplifts are seemingly asymmetric, northeast-facing structures bordered by southwest-dipping reverse or thrust fault zones. Less deformed southwestern flanks plunge into complementary basins filled with uppermost Cretaceous and lower Tertiary clastic rocks about 600 m to more than 2000 m (2000–6800 ft) thick. We interpret uplifts and basins in the foreland region of south-central New Mexico to be similar in style (but smaller in structural relief) to the Wind River uplift and Wind River basin. The dip of boundary thrusts (which cut basement) and the geometry of associated overturned folds point to uplift of the New Mexico ranges as rigid basement blocks accompanied by substantial crustal shortening. Unlike the Wind River uplift, however, the south-central New Mexico uplifts face the craton and are in a Laramide backarc setting, differences that may be important in any dynamic interpretation. In southwestern New Mexico, uplift boundary faults have a significant component of strike slip as well as vertical slip, which indicates that a tangential component of stress (parallel to Laramide structural grain) was somehow generated in that intraarc region. In south-central New Mexico evidence favors vertical slip only under a regional compressional stress.

Several mechanisms that might produce, by compression, basement block uplifts far inland of a trench have been suggested in recent years. Those possibly applicable to southern New Mexico include the following: coupling of (or

drag between) upper and lower plates during low-angle subduction (e.g., Dickinson and Snyder, 1978; Brewer et al., 1980); transpression in zones of regional wrench faults (Chapin and Cather, 1981; Chapin, 1983); and underthrusting of an arc by an actively driving upper cratonic plate (e.g., Coney, 1972; Burchfiel and Davis, 1975; Scholten, 1982; Suarez et al., 1983).

We do not view low-angle subduction and transpression as the most important factors in Laramide deformation of southern New Mexico. Although there is evidence for Laramide flattening of the Benioff zone across the American southwest and Mexico (Coney and Reynolds, 1977; Damon et al., 1981), it apparently never flattened sufficiently to shut off arc magmatism. Consequently, the effectiveness of upper and lower plate coupling or drag to produce tangential compression in southern New Mexico is in doubt. Similarly, although we recognize the effectiveness of transpression in great wrench fault zones in producing uplifted basement blocks, we doubt that the mechanism applies in a significant way to the block uplifts of southern New Mexico. Transpression effects are present in the form of folds and thrusts along the margins of oblique-slip faults in southwestern New Mexico, and uplift of basement blocks in this region may even be partly the result of this convergent wrenching. In general, however, uplifts and boundary faults are approximately orthogonal to relative Laramide plate convergence directions, and therefore the vertical-slip movements prominently displayed by these faults are likely the result of tangential compression, not strong transpression. It seems to us more likely that basement thrusts of southern New Mexico represent zones of localized yielding of the craton as the craton attempted to underthrust (imperfectly subduct) the magmatic arc.

Three lines of evidence favor the underthrust model. First, boundary faults in south-central New Mexico dip southwestward toward the arc so that cratonward (footwall) blocks are relatively underthrusted. By our reckoning, the basinal (footwall) blocks may have been the active underthrusting blocks. A second line of evidence comes from an analogy of Laramide southern New Mexico with backarc parts of the central and northern Andes where Benioff zones dip eastward about 15–30°. Andean upper plate earthquakes with thrust mechanisms occur extensively east of the magmatic arc, but not across the arc or in forearc regions (Stauder, 1975; Barazangi and Isaacs, 1979; Suarez et al., 1983). Apparently, compressive stresses are not effectively transmitted from the downgoing oceanic plate across thermally weakened arc terrane into the backarc region (Coney, 1972). In view of this, compressive earthquake activity in a backarc area, such as the sub-Andean zone of Peru, Bolivia, and northern Argentina, likely results from active underthrusting of the Brazilian shield against or beneath the resisting arc (Jordan et al., 1983; Suarez et al., 1983). Similarly, the southern New Mexico craton may have been the active tectonic element during the Laramide, deriving compressive stresses by driving southwestward against the magmatic arc and attempting to underthrust it.

Rapid Laramide absolute and relative plate motions may also favor craton underthrusting. Reconstruction of plate motions indicate rapid northeast-southwest relative convergence of the Farallon and North American plates during the Laramide (Coney, 1978b; Engebretson et al.,

1984; Jurdy, 1984). More important for the craton underthrust model may be the rapid absolute movement of the North American plate southwestward toward the arc and trench (Coney, 1971, 1972, 1976). The actively driving North American craton then might yield by breakup into rigid splinters as it attempted to underthrust the arc (Figure 12). The lack of a thick sedimentary prism in southern New Mexico favored breakup of the basement into thrust-bounded rigid blocks rather than thin-skinned regional overthrusting.

The oblique-slip reverse faults that border uplifts in southwestern New Mexico indicate that a tangential shear component (parallel to the northwest-oriented structural grain) as well as northeast–southwest compression (normal to the structural grain) affected the rocks of that region during the Laramide. The oblique-slip faults are part of a broad belt of similar faults that trend northwestward into southeastern Arizona (Sabins, 1957; Drewes, 1972, 1980, 1981, 1982), and that, for the most part, are intraarc structures (e.g., Coney, 1978a; Dickinson, 1981). A few early workers in Arizona considered the northwest-trending zone of faults as part of the Texas lineament, most often interpreted as a right-lateral fault system (Hill, 1902; Ransome, 1915; Mayo, 1958). Left-lateral movement during the Laramide has been inferred by more recent work in Arizona (e.g., Drewes, 1978, 1980, 1981), in contrast to the right slip we propose for faults of the Laramide Burro and Hidalgo uplifts of New Mexico. Although emphasis in the literature has been on the strike-slip nature of these faults, Davis (1979), Drewes (1981), and Seager (1983) point out that vertical movement during the Laramide is at least as important as strike slip and is probably predominant along many segments of the faults. Vertical slip is the natural consequence of the nearly orthogonal relationship (especially for the "late" Laramide of Chapin and Cather, 1981) between the Laramide structural trend and the greatest principal stress direction (σ_1) as determined by Rehrig and Heidrick (1976) and Heidrick and Titley (1982). Strike-slip movement is more difficult to account for.

Although we do not have enough information to evaluate thoroughly the role of strike-slip faulting in Laramide tectonics, the following points seem worth making:

- Most strike-slip faults in southeastern Arizona and southwestern New Mexico are probably reactivated older faults (e.g., Titley, 1976, 1981; Drewes, 1981; Bilodeau, 1984). Locally, these faults may have been suitably oriented in the Laramide stress field for right- or left-oblique slip.
- Early Laramide greatest principal stress was approximately ENE–WSW ($\pm 20°$) according to data of Rehrig and Heidrick (1976) and Heidrick and Titley (1982). This direction would yield left slip on northwest-trending faults (Heidrick and Titley, 1982). "Late" Laramide compression was apparently northeast-southwest, orthogonal to Laramide structural trends, as indicated by data from Morenci, Santa Rita, Safford, and the Victorio Mountains (Chapin and Cather, 1981; Heidrick and Titley, 1982).
- Right slip could only be developed where pre-Laramide faults trended northward or where Laramide compressive

stress was more NNE–SSW than indicated by Rehrig and Heidrick's (1976) or Heidrick and Titley's (1982) data. In this regard, Gries (1982) describes geologic evidence from the Rocky Mountains for north–south compression in Early to Late Cretaceous time, and maps by Smith et al. (1981) indicate south–southwest movement of North America during the Laramide. This kind of plate motion and compression might promote right-slip northwest-trending on pre-Laramide fractures of southern Arizona and New Mexico.

- Oblique subduction may result in strike-slip faulting in the arc region as shown by Fitch (1972) and Beck (1983). Carlson (1982) and Jurdy (1984) present evidence for oblique subduction of the Farallon plate beneath North America during the Laramide. Carlson (1982) indicates a right component of oblique subduction during Late Cretaceous and early Tertiary time, whereas Jurdy (1984) in her preferred model suggests that right slip was important in Late Cretaceous time, changing to left slip at the beginning of Tertiary time. The point here is that a right- or left-tangential component of oblique subduction may have affected the arc terrane of the North American southwest during the Laramide, resulting in strike-slip movement on new or reactivated faults.

The points listed above emphasize our incomplete understanding of Farallon–North America plate motions during the Laramide. Precise directions of relative and absolute convergence are in doubt, as is the amount and sense of oblique convergence and the degree to which plate motion changed through Laramide time. A full appreciation of the role of northwest-trending, strike-slip faulting in Laramide tectonics of southwestern North America may depend on resolution of Laramide plate kinematics. On the other hand, the central New Mexico wrench fault zone of Chapin and Cather (1981) seems better understood, although its extension into southern New Mexico poses problems.

We find little evidence in southern New Mexico for a zone of right-lateral shearing following the Rio Grande, which Chapin and Cather (1981) suggest may have accommodated northward crowding of the Colorado Plateau relative to the Great Plains (Figure 13). Farther north in New Mexico, Chapin and Cather (1981) and Chapin (1983) make a convincing case for its existence. Laramide basins south of the Caballo Mountains are viewed as essentially compressional Wind River types, and not as Echo Park types resulting from a north–south couple as suggested by Chapin and Cather (1981) for their Cutter basin. We do recognize, however, that basins in southwestern New Mexico may be partly transpressional in origin, but these clearly are not related to a central New Mexico wrench fault zone.

An unresolved question is how Chapin and Cather's (1981) wrench fault zone affected the basement block uplift terrane of southwestern New Mexico, if the fault zone does indeed extend south of Socorro and is the magnitude they indicate (100 km [62 mi] offset). There seem to be three possibilities (C. E. Chapin, personal communication, 1984). First, the central New Mexico wrench fault zone may continue southward through the basement uplifts of south-central New Mexico. Second, the zone may veer

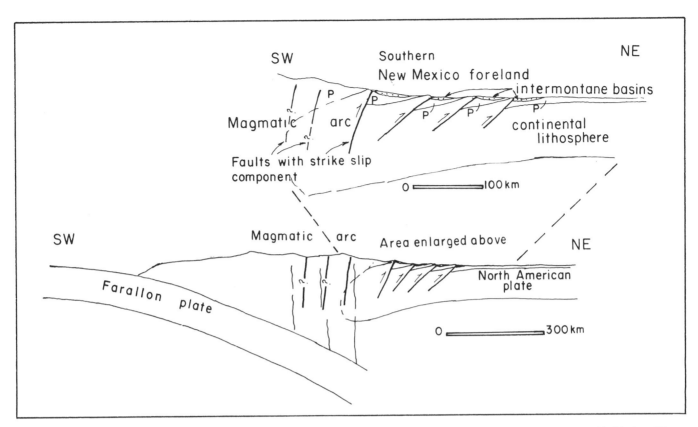

Figure 12—Diagrammatic northeast–southwest cross sections illustrating possible origin of southern New Mexico Laramide block uplifts as a product of compression caused by an attempt of the craton to underthrust the magmatic arc.

southwestward from Socorro and be hidden beneath the Mogollon–Datil volcanic field. Third, in southern New Mexico, the fault zone may be distributed across numerous faults from the Pecos "buckles" (Kelley, 1971) on the east to beneath the Mogollon–Datil field on the west. None of these alternatives have support from field relationships, although it should be remembered that the crucial evidence may lie hidden beneath the vast areas of post-Laramide cover in southern New Mexico.

By contrast with the largely hypothetical extension of the central New Mexico wrench fault zone into southern New Mexico, the right-lateral buckles on the Pecos slope are well known (Kelley, 1971). These fractures may be interpreted as tears separating uncompressed crust of southeastern New Mexico from compressed and shortened crust of southwestern New Mexico (Figure 13) (C. E. Chapin, personal communication, 1983). In this respect, the Pecos buckles accommodated the differential movements between southwestern and southeastern New Mexico in exactly the same manner as Chapin and Cather's (1981) central New Mexico wrench fault zone accommodated differential movement between the Colorado Plateau and Great Plains (Figure 13).

ACKNOWLEDGMENTS

Many of the ideas expressed in this paper matured through discussion with Glen Brown, C. E. Chapin, R. E. Clemons, and S. Thompson III. We thank Exxon for permission to use drill hole data from their No. 1 Prisor well and S. Cather and F. E. Kottlowski for their review of an earlier version of this manuscript. V. C. Kelley, John Gries, Eugene Greenwood, and C. E. Chapin substantially improved the manuscript by their thoughtful reviews. We especially want to thank the New Mexico Bureau of Mines and Mineral Resources and F. E. Kottlowski, Director, for their continuing support of our geologic research in southern New Mexico.

REFERENCES CITED

Armstrong, A. K., M. L. Silberman, V. R. Todd, W. C. Hoggatt, and R. B. Carten, 1978, Geology of central Peloncillo Mountains, Hidalgo County, New Mexico: New Mexico Bureau of Mines and Mineral Resources Circular 158, p. 1–19.

Barazangi, M., and B. L. Isacks, 1979, Subduction of the Nazca plate beneath Peru: evidence from spatial distribution of earthquakes: Geophysical Journal, v. 57, p. 537–555.

Beck, M. E., Jr., 1983, On the mechanism of tectonic transport in zones of oblique subduction: Tectonophysics, v. 93, p. 1–11.

Bilodeau, W. L., 1984, Laramide sedimentation and tectonics in southeastern Arizona and southwestern New Mexico: reactivation of pre-Laramide basement structures (abs.):

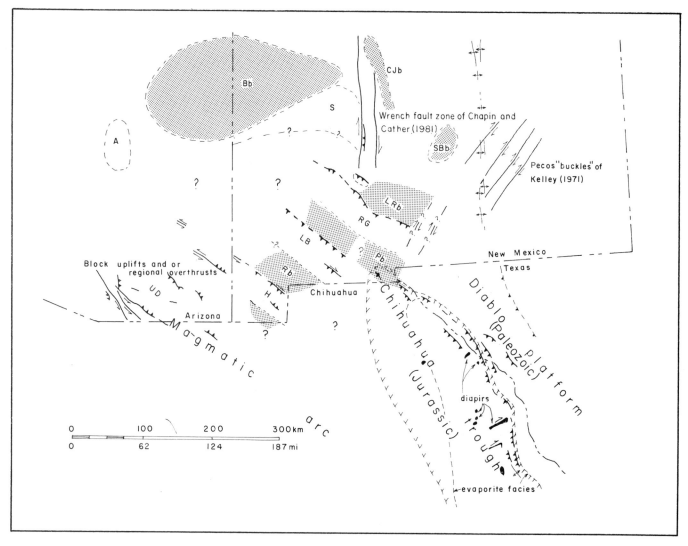

Figure 13—Map showing major Laramide structural features in western Texas, northern Chihuahua, southern New Mexico, and southeastern Arizona. Uplifts: A, Apache uplift; S, Sierra uplift; RG, Rio Grande uplift; LB, Laramide Burro uplift; H, Hidalgo uplift, UD, uplift of Davis (1979). Basins: Bb, Baca basin; CJb, Carthage–La Joya basin; SBb, Sierra Blanca basin; LRb, Love Ranch basin; Pb, Potrillo basin; Rb, Ringbone basin.

Geological Society of America Abstracts with Programs, v. 16, p. 445.

Brewer, J. A., S. B. Smithson, J. E. Oliver, S. Kaufman, and L. D. Brown, 1980, The Laramide orogeny: evidence from COCORP deep crustal seismic profiles in the Wind River Mountains, Wyoming: Tectonophysics, v. 62, p. 165–189.

Brown, G. A., 1982, Geology of the Mahoney mine–Gym Peak area, Florida Mountains, Luna County, New Mexico: Master's thesis, New Mexico State University, Las Cruces, 82 p.

Brown, G. A., and R. E. Clemons, 1983, Florida Mountains section of southwest New Mexico overthrust belt—a reevaluation: New Mexico Geology, v. 5, p. 26–29.

Burchfiel, B. C., and G. A. Davis, 1975, Nature and controls of Cordilleran orogenesis, western United States: extension of an earlier hypothesis: American Journal of Science, v. 275-A, p. 363–395.

Bushnell, H. P., 1953, Geology of the McRae Canyon area, Sierra County, New Mexico: Master's thesis, University of New Mexico, Albuquerque, 106 p.

Carlson, R. L., 1982, Cenozoic convergence along the California coast: a qualitative test of the hot-spot approximation: Geology, v. 10, p. 191–196.

Chapin, C. E., 1983, An overview of Laramide wrench faulting in the southern Rocky Mountain with emphasis on petroleum exploration, *in* J. D. Lowell, ed., Foreland basins and uplifts: Rocky Mountain Association of Geologists, p. 169–179.

Chapin, C. E., and S. M. Cather, 1981, Eocene tectonics and sedimentation in the Colorado Plateau–Rocky Mountain area, *in* W. R. Dickinson and W. D. Payne, eds., Relations of tectonics to ore in the southern Cordillera: Arizona Geological Society Digest, v. 14, p. 173–198.

Chapin, C. E., W. T. Siemers, and G. T. Osburn, 1975, Summary of radiometric ages of New Mexico rocks: New Mexico Bureau of Mines and Mineral Resources Open-File Report 60.

Clemons, R. E., 1982, Geology of Massacre Peak quadrangle, Luna County, New Mexico: New Mexico Bureau of

Mines and Mineral Resources Geologic Map 51.

_____, 1985, Geology of the South Peak quadrangle, Luna County, New Mexico: New Mexico Bureau of Mines and Mineral Resources Geologic Map 59.

Clemons, R. E., and G. A. Brown, 1983, Geology of Gym Peak quadrangle, Luna County, New Mexico: New Mexico Bureau of Mines and Mineral Resources Geologic Map 58.

Coney, P. J., 1971, Cordilleran tectonic transitions and motion of the North American plate: Nature, v. 233, p. 462–465.

_____, 1972, Cordilleran tectonics and North American plate motion: American Journal of Science, v. 272, p. 603–628.

_____, 1976, Plate tectonics and the Laramide orogeny, *in* L. A. Woodward and S. A. Northrop, eds., Tectonics and mineral resources of southwestern North America: New Mexico Geological Society Special Publication 6, p. 5–10.

_____, 1978a, The plate tectonic setting of southeastern Arizona: New Mexico Geological Society 29th Annual Fieldtrip Guidebook, p. 285–290.

_____, 1978b, Mesozoic–Cenozoic Cordilleran plate tectonics, *in* R. B. Smith and G. P. Eaton, eds., Cenozoic tectonics and regional geophysics of western Cordillera: Geological Society of America Memoir 152, p. 33–50.

Coney, P. J., and S. J. Reynolds, 1977, Cordilleran Benioff zones: Nature, v. 270, p. 403–406.

Corbitt, L. L., and L. A. Woodward, 1973, Tectonic framework of Cordilleran foldbelt in southwestern New Mexico: AAPG Bulletin, v. 57, p. 2207–2216.

Damon, P. E., M. Shafiqullah, and K. F. Clark, 1981, Age trends of igneous activity in relation to metallogenesis in the southern Cordillera, *in* W. R. Dickinson and W. D. Payne, eds., Relations of tectonics to ore deposits in the southern Cordillera: Arizona Geological Society Digest, v. 14, p. 137–154.

Darton, G. H., 1928, "Red beds" and associated formations in New Mexico, with an outline of the geology of the state: U.S. Geological Survey Bulletin, v. 794, p. 1–356.

Davis, G. H., 1979, Laramide folding and faulting in southeastern Arizona: American Journal of Science, v. 279, p. 543–569.

Dickinson, W. R., 1981, Plate tectonic evolution of the southern Cordillera, *in* W. R. Dickinson and W. D. Payne, eds., Relations of tectonics to ore deposits in the southern Cordillera: Arizona Geological Society Digest, v. 14, p. 113–135.

Dickinson, W. R., and W. S. Snyder, 1978, Plate tectonics of the Laramide orogeny, *in* V. Matthews III, ed., Laramide folding associated with basement block faulting in the western United States: Geological Society of America Memoir 151, p. 355–365.

Doyle, J. C., 1951, Geology of the northern Caballo Mountains, Sierra County, New Mexico: Master's thesis, New Mexico Institute of Mining and Technology, Socorro, 51 p.

Drewes, H., 1972, Structural geology of the Santa Rita Mountains, southwest of Tucson, Arizona: U.S. Geological Survey Professional Paper 748, p. 1–35.

_____, 1976, Laramide tectonics from Paradise to Hells Gate, southeastern Arizona: Arizona Geological Society Digest, v. 10, p. 151–167.

_____, 1978, The Cordilleran orogenic belt between Nevada and Chihuahua: Geological Society of America Bulletin, v. 89, p. 641–657.

_____, 1980, Tectonic map of southeastern Arizona: U.S. Geological Survey Miscellaneous Geological Investigations Map I-1109.

_____, 1981, Tectonics of southeastern Arizona: U.S. Geological Survey Professional Paper 1144, p. 1–96.

_____, 1982, Some general features of the El Paso–Wickenburg transect of the Cordilleran orogenic belt, Texas to Arizona, *in* H. Drewes, ed., Cordilleran overthrust belt, Texas to Arizona: Rocky Mountain Association of Geologists, p. 87–96.

Drewes, H., and C. H. Thorman, 1980a, Geologic map of the Steins quadrangle and the adjacent part of the Vanar quadrangle, Hidalgo County, New Mexico: U.S. Geological Survey Miscellaneous Geological Investigations Map I-1220.

_____, 1980b, Geologic map of the Cotton City quadrangle and the adjacent part of the Vanar quadrangle, Hidalgo County, New Mexico: U.S. Geological Survey Miscellaneous Geological Investigations Map I-1221.

Elston, W. E., 1958, Burro uplift, northeastern limit of sedimentary basins of southwestern New Mexico and southeastern Arizona: AAPG Bulletin, v. 42, p. 2513–2517.

Engebretson, D. C., A. Cox, and G. A. Thompson, 1984, Correlation of plate motions with continental tectonics: Laramide to Basin-Range: Tectonics, v. 3, p. 115–119.

Fitch, T. J., 1972, Plate convergence, transcurrent faults, and internal deformation adjacent to southeast Asia and the western Pacific: Journal of Geophysical Research, v. 77, p. 4432–4461.

Greenwood, E., F. E. Kottlowski, and S. Thompson III, 1977, Petroleum potential and stratigraphy of Pedregosa basin—comparison with Permian and Orogrande basins: AAPG Bulletin, v. 61, p. 1448–1469.

Gries, J. C., 1980, Laramide evaporite tectonics along the Texas–northern Chihuahua border: New Mexico Geological Society 31st Annual Fieldtrip Guidebook, p. 93–100.

Gries, R., 1983, North–south compression of Rocky Mountain foreland structures, *in* J. D. Lowell, ed., Rocky Mountain foreland basins and uplifts: Rocky Mountain Association of Geologists, p. 9–32.

Haenggi, W. T., and J. C. Gries, 1970, Structural evolution of northeastern Chihuahua tectonic belt: Permian Basin section, Society of Economic Paleontologists and Mineralogists Publication 70-12, p. 55–69.

Harding, T. P., and J. D. Lowell, 1979, Structural styles, their plate-tectonic habitats and hydrocarbon traps in petroleum provinces: AAPG Bulletin, v. 63, p. 1016–1058.

Hayes, P. T., 1970, Cretaceous paleogeography of southeastern Arizona and adjacent areas: U.S. Geological Survey Professional Paper 658-B, p. 1–42.

Heidrick, T. L., and S. R. Titley, 1982, Fracture and dike patterns in Laramide plutons and their structural and tectonic implications, *in* S. R. Titley, ed., Advances in geology of the porphyry copper deposits, southwestern North America: Tucson, University of Arizona Press, p. 73–92.

Hill, R. T., 1902, The geographic and geologic features, and their relation to the mineral products of Mexico: American Institution of Mining Engineers Transactions, v. 32, p. 163–178.

Jordan, T. E., B. L. Isacks, R. W. Allmendinger, J. A. Brewer, V. A. Ramos, and C. V. Ando, 1983, Andean tectonics related to geometry of subducted Nazca plate: Geological Society of America Bulletin, v. 94, p. 341–361.

Jurdy, D., 1984, The subduction of the Farallon plate beneath North America as derived from relative plate motions: Tectonics, v. 3, p. 107–113.

Kelley, V. C., 1971, Geology of the Pecos country, southeastern New Mexico: New Mexico Bureau of Mines and Mineral Resources Memoir 24, p. 1–75.

Kelley, V. C., and C. Silver, 1952, Geology of the Caballo Mountains: University of New Mexico Publications in Geology Series, n. 4, p. 1–286.

Kottlowski, F. E., 1958, Pennsylvanian and Permian rocks near the late Paleozoic Florida islands: Roswell Geological Society 11th Annual Guidebook, p. 79–87.

———, 1960, Summary of Pennsylvanian sections in southwestern New Mexico and southeastern Arizona: New Mexico Bureau of Mines and Mineral Resources Bulletin, v. 66, p. 1–187.

———, 1963, Paleozoic and Mesozoic strata of southwest and south-central New Mexico: New Mexico Bureau of Mines and Mineral Resources Bulletin, v. 79, p. 1–100.

———, 1965, Sedimentary basins of south-central and southwestern New Mexico: AAPG Bulletin, v. 49, p. 2120–2139.

Kottlowski, F. E., R. H. Flower, M. L. Thompson, and R. W. Foster, 1956, Stratigraphic studies of the San Andres Mountains, New Mexico: New Mexico Bureau of Mines and Mineral Resources Memoir 1, p. 1–132.

Lasky, S. G., 1947, Geology and ore deposits of the Little Hatchet Mountains, Hidalgo and Grant Counties, New Mexico: U.S. Geological Survey Professional Paper 208, p. 1–101.

Lemley, I. S., 1982, The Lobo Formation and lithologically similar units in Luna and southwestern Dona Ana counties, New Mexico: Master's thesis, New Mexico State University, Las Cruces, 95 p.

Loring, A. K., and R. B. Loring, 1980, Age of thrust faulting, Little Hatchet Mountains, southwestern New Mexico: Isochron/West, n. 27, p. 29–30.

Lowell, G. D., 1972, Spitsbergen Tertiary orogenic belt and the Spitsbergen fracture zone: Geological Society of America Bulletin, v. 83, p. 3091–3102.

Mack, G. H., R. E. Clemons, and W. R. Seager, 1983, Paleogene sandstone petrofacies in south-central New Mexico: evidence for change from Laramide deformation to volcanism (abs.): Geological Society of America Abstracts with Programs, v. 15, p. 633.

Mack, G. H., W. B. Kolins, and J. A. Galemore, in press, Lower Cretaceous stratigraphy, depositional environment, and sediment dispersal in southwestern New Mexico: American Journal of Science.

Marvin, R. F., C. W. Naeser, and H. H. Mehnert, 1978,

Tabulation of radiometric ages—including unpublished K-Ar and fission-track ages—for rocks in southeastern Arizona and southwestern New Mexico: New Mexico Geological Society 29th Annual Fieldtrip Guidebook, p. 243–252.

Mayo, E. B., 1958, Lineament tectonics of some ore districts in the southwest: Mining Engineering, v. 10, p. 1169–1175.

Ransome, F. L., 1915, The Tertiary orogeny of the North American Cordillera and its problems, *in* W. N. Rice et al., eds., Problems of American geology: New Haven, Yale University Press, p. 287–376.

Rehrig, W. A., and T. L. Heidrick, 1976, Regional tectonic stress during the Laramide and late Tertiary intrusive periods, Basin and Range province: Arizona Geological Society Digest, v. 10, p. 205–228.

Sabins, F. F., 1957, Geology of the Cochise Head and western part of the Vanar quadrangles, Arizona: Geological Society of America Bulletin, v. 68, p. 1315–1342.

Scholten, R., 1982, Continental subduction in the northern United States Rockies—a model for back-arc thrusting in the western Cordillera: Rocky Mountain Association of Geologists, p. 123–136.

Seager, W. R., 1975, Cenozoic tectonic evolution of the Las Cruces area: New Mexico Geological Society 26th Annual Fieldtrip Guidebook, p. 297–321.

———, 1981, Geology of the Organ Mountains and southern San Andres Mountains, New Mexico: New Mexico Bureau of Mines and Mineral Resources Memoir 36, p. 1–97.

———, 1983, Laramide wrench faults, basement-cored uplifts, and complementary basins in southern New Mexico: New Mexico Geology, v. 5, p. 69–76.

Seager, W. R., and R. E. Clemons, 1975, Middle to Late Tertiary geology of the Cedar Hills-Selden Hills area, Dona Ana County, New Mexico: New Mexico Bureau of Mines and Mineral Resources Circular 133, p. 1–23.

Seager, W. R., J. W. Hawley, and R. E. Clemons, 1971, Geology of San Diego Mountain area, Dona Ana County, New Mexico: New Mexico Bureau of Mines and Mineral Resources Bulletin, v. 97, p. 1–38.

Smith, A. G., A. M. Hurley, and J. C. Briden, 1981, Phanerozoic paleocontinental world maps: London, Cambridge University Press, p. 1–102.

Soule, J. M., 1972, Structural geology of northern part of Animas Mountains, Hidalgo County, New Mexico: New Mexico Bureau of Mines and Mineral Resources Circular 125, p. 1–15.

Stauder, W., 1975, Subduction of the Nazca plate beneath Peru as evidenced by focal mechanisms and by seismicity: Journal of Geophysical Research, v. 80, p. 1053–1063.

Steel, R. J., and A. C. Wilson, 1975, Sedimentation and tectonism (?Permo-Triassic) on the margin of the North Minch basin, Lewis: Geological Society of London Quarterly Journal, v. 131, p. 183–202.

Suarez, G., P. Molnar, and B. C. Burchfiel, 1983, Seismicity, fault plane solutions, depth of faulting, and active tectonics of the Andes of Peru, Ecuador, and southern Columbia: Journal of Geophysical Research, v. 88, p. 10403–10428.

Thompson, S., III, 1982, Oil and gas exploration wells in

southwestern New Mexico, *in* H. Drewes, ed., Cordilleran overthrust belt, Texas to Arizona: Rocky Mountain Association of Geologists, p. 136–153.

Thompson, S., III, and R. A. Bieberman, 1975, Oil and gas exploration wells in Dona Ana County, New Mexico: New Mexico Geological Society 26th Annual Fieldtrip Guidebook, p. 171–174.

Thorman, C. H., and H. Drewes, 1980, Geologic map of the Victorio Mountains, Luna County, southwestern New Mexico: U.S. Geological Survey Miscellaneous Field Studies Map MF-1175.

Titley, S. R., 1976, Evidence for a Mesozoic linear tectonic pattern in southeastern Arizona: Arizona Geological Society Digest, v. 10, p. 71–101.

———, 1981, Geologic and geotectonic setting of porphyry copper deposits in the southern Cordillera, *in* W. R. Dickinson and W. D. Payne, eds., Relations of tectonics to ore deposits in the southern Cordillera: Arizona Geological Society Digest, v. 14, p. 79–97.

Turner, G. L., 1962, The Deming axis, southeastern Arizona, New Mexico, and Trans-Pecos Texas: New Mexico Geological Society 13th Annual Fieldtrip Guidebook, p. 59–71.

Uphoff, T. L., 1978, Subsurface stratigraphy and structure of the Mesilla and Hueco bolsons, El Paso region, Texas and New Mexico: Master's thesis, Universiy of Texas, El Paso, 55 p.

Wilson, J. W., 1965, Cenozoic history of the Big Bend area, west Texas: West Texas Geological Society Field Conference Guidebook, n. 65-51, p. 34–36.

Woodward, L. A., and H. R. DuChene, 1981, Overthrust belt of southwestern New Mexico—comparison with Wyoming–Utah overthrust belt: AAPG Bulletin, v. 65, p. 722–729.

Zeller, R. A., Jr., 1965, Stratigraphy of the Big Hatchet Mountains area, New Mexico: New Mexico Bureau of Mines and Mineral Resources Memoir 16, p. 1–128.

———, 1970, Geology of the Little Hatchet Mountains, Hidalgo and Grant counties, New Mexico: New Mexico Bureau of Mines and Mineral Resources Bulletin, v. 96, p. 1–22.

———, 1975, Structural geology of Big Hatchet Peak quadrangle, Hidalgo County, New Mexico: New Mexico Bureau of Mines and Mineral Resources Circular 146, p. 1–23.

Index

A reference is indexed according to its important, or "key" words.

Three columns are to the left of a keyword entry. The first column, a letter entry, represents the AAPG book series from which the reference originated. In this case, ME stands for AAPG Memoir Series. Every five years, AAPG will merge all its indexes together, and the letters ME will differentiate this reference from those of the Studies in Geology Series (ST) or from the AAPG Bulletin (B).

The following number is the series number. In this case, 41 represents a reference from AAPG Memoir 41. The third column lists the page number of this volume on which the reference can be found.